ISBN 978-0-282-47240-5
PIBN 10852893

THE NORWEGIAN AURORA POLARIS EXPEDITION 1902–1903

VOLUME I

ON THE CAUSE OF MAGNETIC STORMS AND THE ORIGIN OF TERRESTRIAL MAGNETISM

BY

KR. BIRKELAND

FIRST SECTION

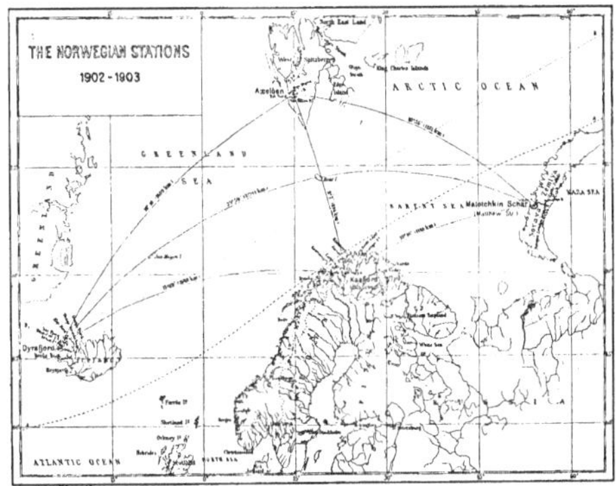

CHRISTIANIA
H. ASCHEHOUG & CO.

LEIPZIG
JOHANN AMBROSIUS BARTH

LONDON, NEW YORK
LONGMANS, GREEN & CO.

PARIS
C. KLINCKSIECK

CHRISTIANIA A. W. BRØGGERS PRINTING OFFICE 1908

PREFACE.

The knowledge gained, since 1896, in radio-activity has favoured the view to which I gave expression in that year, namely, that magnetic disturbances on the earth, and aurora borealis, are due to corpuscular rays emitted by the sun.

During the period from 1896 to 1903 I carried out, in all, three expeditions to the polar regions for the purpose of procuring material that might further confirm this opinion. I have moreover, during the last ten years, by the aid of numerous experimental investigations, endeavoured to form a theory that should explain the origin of these phenomena. It is the results of these investigations that are recorded in this work, the first volume of which treats of terrestrial magnetic phenomena and earth-currents, this section forming the first two thirds of the volume. The second volume will treat of aurora and some results of meteorological observations made at our stations.

The leading principle that I have followed in this work has been to endeavour always to interpret the results of the worked-up terrestrial-magnetic observations, and the observations of aurora, upon the basis of my above-mentioned theory.

Thus the magnetic storms, for instance, have been studied in such a manner, that on the one hand we have formed from our observation-material a field of force which gives as complete a representation as possible of the perturbing forces existing on the earth at the times under consideration. On the other hand, by experimental investigations with a little magnetic terrella in a large discharge-tube, and by mathematical analysis, we have endeavoured to prove that a current of electric corpuscles from the sun would give rise to precipitation upon the earth, the magnetic effect of which agrees well with the magnetic field of force that was found by the observations on the earth.

Although our observation-material of magnetic storms was, I may safely say, the largest that has ever been dealt with at one time, it was deficient in certain points, as might well be expected.

We generally had at our disposal in 1902—1903, magnetic registerings from 25 observatories scattered all over the world, among them being our 4 Norwegian stations on Iceland, Spitsbergen, Novaja Semlja, and in Finmark.

We have moreover treated separately certain well-marked magnetic storms in 1882—1883, from the observations in the reports of the international polar expeditions.

In addition to the deficiencies in our observation-material, there are also defects in the experimental and mathematical investigations; but notwithstanding all this, the results are so satisfactory that I can hardly be mistaken in my belief that we are on the right road.

Besides making clear the origin of important terrestrial phenomena, the investigations give promise of the possibility of drawing, from the energy of the corpuscular precipitation on the earth, well-founded conclusions regarding the conditions on the sun.

The disintegration theory, which has proved of the greatest value in the explanation of the radio-active phenomena, may possibly also afford sufficient explanation as to the origin of the sun's heat. The energy of the corpuscular precipitation that takes place in the polar regions during magnetic storms seems indeed to indicate a disintegration process in the sun of such magnitude, that it may possibly clear up this most important question in solar physics.

Future researches in the paths here entered upon, which I believe will lead to the solution of some of the most attractive scientific problems of our age, e. g. the origin of terrestrial magnetism, and the origin of the sun's heat, may be carried out upon a far wider basis than I have been able to employ, without making the expenses connected therewith too great a deterrent.

In 1902—1903 I had the great good fortune of having twenty-five observatories with me; but on a future occasion it will be necessary to have double the number.

We should then have to send out small expeditions with, say, ten stations suitably distributed about each of the magnetic poles, and make sure of getting magnetic registerings for the same period from all the observatories in the world.

As the position of the stations, within certain limits, may be chosen with tolerable freedom, the end would be best attained by accompanying whalers, or, as I once had to do, equipping such vessels one's self for certain places.

The mathematical investigations, which, together with my experiments, are intended to make clear the movement of electric corpuscles from the sun to the earth, have been carried out, with a perseverance and ingenuity worthy of all admiration, by my friend, Professor STØRMER, who will publish the complete results of his investigations in a special part of the present work. These results, however, will be known to some extent from the papers he has already published.

In concluding this first section, I have to thank those persons who have so greatly assisted me in my work. In Mr. L. VEGARD I have had an invaluable collaborator, whom I have to thank for many excellent suggestions. Great merit is also due to Mr. DIETRICHSON and Mr. KROGNESS for their share in this work; and I would further thank Messrs. RUSSELTVEDT, NORBY and IRGENS, for their energetic labour.

The translation, which I consider very successful, has been performed by Miss JESSIE MUIR.

Christiania; October, 1908

Kr. Birkeland.

CONTENTS.

INTRODUCTION.

Page

Art. 1. The first Expedition, 1897 . . . 1
" 2. The second Aurora Expedition, 1899—1900 . . . 5

THE EXPEDITION OF 1902—1903 . . . 9
" 4, 5. The Auroral Station in Kaafjord . . . 10
" 6, 7. The Auroral Station in Dyrafjord, Iceland . . . 18
" 8, 9. The Auroral Station in Spitsbergen . . . 24
" 10, 11. The Auroral Station in Novaja Semlja . . . 31
" 12. The Working-up of the Material . . . 37

PART I.

MAGNETIC STORMS, 1902—1903.

INVESTIGATIONS BY MEANS OF DIURNAL REGISTERINGS FROM 25 OBSERVATORIES.

CHAPTER I.

PRELIMINARY REMARKS CONCERNING OUR MAGNETIC RESEARCHES.

" 13. Our Aim and our Method of Working . . . 41
" 14. On the Calculation of the Perturbing Force . . . 44
" 15. On the Separation of Simultaneous Perturbations . . . 47

CALCULATION OF THE SCALE-VALUES FOR THE REGISTERINGS AT THE NORWEGIAN STATIONS.

" 16. Determination of the Scale-Values for the Declinometer . . . 48
" 17. Determination of the Sensibility of the Variometers for the Horizontal and Vertical Intensity 48
" 18. Determinations of Sensibility for Kaafjord and Bossekop . . . 50
" 19. Determinations of Sensibility for Dyrafjord . . . 51
" 20. Determinations of Sensibility for Axeløen . . . 53
" 21. Determinations of Sensibility for Matotchkin Schar . . . 54
" 22. Temperature Coefficients for the Registerings . . . 55
" 23. Explanation of the Charts . . . 56
" 24. The Copies of the Magnetic Registerings, Explanation and General Remarks . . . 58

CHAPTER II.

ELEMENTARY PERTURBATIONS.

" 25. General Remarks . . . 61
" 26. **The Equatorial Perturbations** . . . 62
" 27. The Positive Equatorial Perturbation. The Perturbation of the 26th January 1903 . . 63
" 28, 29. The Perturbations of the 9th December, 1902 . . . 70
" 30. The Perturbation of the 23rd October, 1902 . . . 76

		Page
Art. 31.	Concerning the Cause of the Positive Equatorial Perturbation	78
„ 32.	The Negative Equatorial Storms	83
„ 33.	**The Polar Elementary Storms**	84
„ 34.	The Typical Field for the Polar Elementary Storms	85
„ 35.	The Perturbation of the 15th December, 1902	87
„ 36.	Concerning the Cause of the Perturbation	95
„ 37, 38.	The Perturbation of the 10th February, 1903	106
„ 39.	Concerning the Cause of the Perturbation	113
„ 40—43.	The Perturbations of the 30th and 31st March, 1903	115
„ 44—47.	The Perturbations of the 22nd March, 1903	127
„ 48.	The Perturbations of the 26th December, 1902	137
„ 49.	**Cyclo-Median Storms**	144
„ 50, 51.	The Perturbation of the 6th October, 1902	145
„ 52.	Concerning the Cause of the Perturbation	149
„ 53.	Further Comparison with Størmer's Mathematical Theory	158

CHAPTER III.

COMPOUND PERTURBATIONS.

„ 54.	The Perturbation of the 29th and 30th October, 1902	161
„ 5	The Perturbation of the 25th December, 1902	164
„ 56:	The Perturbation of the 28th December, 1902	169
„ 57, 58.	The Perturbations of the 15th February, 1903	172
„ 59, 60.	The Perturbations of the 7th and 8th February, 1902	187
„ 61, 62.	The Perturbations of the 27th and 28th October, 1902	209
„ 63, 64.	The Perturbations of the 28th and 29th October, 1902	222
„ 65, 66.	The Perturbations of the 31st October and 1st November, 1902	230
„ 67.	How these Perturbations may be explained	234
„ 68.	The Perturbations of the 11th and 12th October, 1902	251
„ 69.	Concerning the Cause of the Perturbations. **Positive and negative Polar Storms**	268
„ 70, 71.	The Perturbations of the 23rd and 24th November, 1902	272
„ 72, 73.	The Perturbations of the 26th and 27th January, 1903	286
„ 74.	Further Comparison with the Terrella-Experiments	297

CHAPTER IV.

CONCERNING THE INTENSITY OF THE CORPUSCULAR PRECIPITATION IN THE ARCTIC REGIONS OF THE EARTH.

„ 75.	Development of General Formulæ	303
„ 76—79.	Numerical Values for Height and Strength of Current	306
„ 80.	The Energy of the Corpuscular Precipitation. The Source of the Sun's Heat	311

PLATES.

Pl. I. The Perturbation of the 6th October, 1902

Pl. II. The Perturbations of the 11th and 12th October, 1902.

Pl. III. The Perturbation of the 23rd October, 1902.

Pl. IV. The Perturbations of the 27th and 28th October, 1902.

Pl. V. The Perturbations of the 28th and 29th October, 1902.

Pl. VI. The Perturbations of the 29th and 30th October, 1902.

Pl. VII. The Perturbations of the 31st October and 1st November, 1902.

Pl. VIII. The Perturbations of the 23rd and 24th November, 1902.

Pl. IX. The Perturbations of the 9th December, 1902.

Pl. X. The Perturbation of the 15th December, 1902.

Pl. XI. The Perturbation of the 25th December, 1902.

Pl. XII. The Perturbation of the 26th December, 1902.

Pl. XIII. The Perturbation of the 28th December, 1902.

Pl. XIV. The Perturbation of the 26th January, 1903.

Pl. XV. The Perturbations of the 26th and 27th January, 1903.

Pl. XVI. The Perturbation of the 8th February, 1903.

Pl. XVII. The Perturbations of the 8th February, 1903.

Pl. XVIII. The Perturbation of the 10th February, 1903.

Pl. XIX. The Perturbation of the 15th February, 1903.

Pl. XX. The Perturbations of the 22nd March, 1903,

Pl. XXI. The Perturbations of the 31st March, 1903.

ERRATA.

Page 44, line 14 from above: For "in front of the special treatment of the separate perturbations", read "at the end of this volume".

" 59: As the table shows, ε_v is not determined for Wilhelmshaven. By comparing the vertical curves with those from Potsdam, we found by deduction that $\varepsilon_v = 10\ \gamma$ per mm. might not be so far from the right value. This value has been used in the calculations. On a later inquiry at the observatory, we obtained the value $\varepsilon_v = 20\ \gamma$ per mm., but it was rather uncertain. This value, however, we have not made use of, for in what we had to consider it was the direction of the vertical component and its variation that were of the most importance, and not the actual amount of P_v.

" 67, line 1 from below: For "Chap. III", read "Part II, Chap. I".

" 68—208, On the Charts, for "V_p", read "P_v".

" 70, line 12 from below: For "negative", read "positive".

" 71, lines 12 & 13 from above: For "must be of a somewhat local character", read "must belong to another system".

" 96, " 7 & 6 " below: After "positive vortices" add "of the negative rays".

" 96, " 6 & 5 " " For "divergence", read "convergence", and vice versa.

" 121, Table XVIII, Zi-ka-wei, P_v line 14: For "$5.83 \times 10\ \gamma$", read "$5.8\ \gamma$".

" 128, line 3 from below: For "Chapter III", read "Part II, Chapter I".

" 198, Table XXX, Christchurch, P_v line 1: For "$-1.5\ \gamma$", read "$+1.5\ \gamma$".

INTRODUCTION.

THE EXPEDITION of which the results are here given, is the third of a series which the author, with the aid of the Norwegian State, the University and the Scientific Society in Christiania, and private persons, got together and led, with the object of investigating the aurora borealis and magnetic disturbances in the polar regions.

1. The first expedition, in February and March, 1897, was a failure, partly owing to unfortunate circumstances, but chiefly to a lack of experience. The idea was to make it a reconnoitring expedition, in order that we might gather knowledge and prepare for a larger expedition; but it was also our special aim to find out whether the northern lights could, as frequently asserted, come right down to the tops of the mountains in the district between Bossekop and Kautokeino on the Finland border of Norway, and to make atmospheric-electric and magnetic measurements high up on the mountains during the occurrence of aurora.

The expedition has not been described before, because it was such a sad adventure; but now that time has drawn a veil of melancholy oblivion over the misfortune that befell us, I will briefly relate some of our experiences. An acquaintance with these may be of some interest to those who may think at some future time of making investigations in the winter on the mountains in the far north.

Besides myself, there were two excellent students, B. HELLAND-HANSEN and K. LOWS, who shared in the investigations. They had offered themselves as assistants solely out of interest in the matters to be dealt with.

We set off from Christiania on the 2nd February, and by the 8th were ready to ascend the mountain from the well-known polar station Bossekop in Finmark. We had procured reindeer to take us and our traps, and a first-class guide in the old Finn "postvappus" (postman), CLEMET ISAKSEN HÆTTA.

After a quick run in brilliant moonlight, we arrived at the mountain hut of Gargia, 25 kilometres south of Bossekop.

The reindeer, each with its pulk, were fastened together in a line one behind another, called a "raide", and the pace, especially down-hill, was something tremendous.

The next morning, the 9th February, there was a little wind, but we all got ready for the start, both those who were going to Kautokeino, those who were returning to Bossekop, and we who were going up to Lodikken Hut on Beskades, 16 kilometres from Gargia. The temperature that day was — 25° C.

When we got up on to Beskades, the snow was drifting a little, but not at first in any alarming degree; and we went on up the comparatively gently sloping mountain, passing cairn after cairn on the Kautokeino road, up which we went at a walking pace for a distance of about 10 kilometres. The wind howled a good deal in the old, weather-beaten guide-posts with their outstretched arms, that showed that day both where the wind came from and where the road went to, as we passed them one by one; but we did not interpret it as a warning. The storm increased, however, and we asked the *vappus* several times if it were safe to proceed, and whether he was sure of the way, to which he answered "Yes".

We then left the road with the cairns, to go up towards Lodikken on the wild mountain, having then 5 kilometres to reach the hut. But the storm increased with frightful rapidity. The guide had to lead the reindeer, or they would not face the wind; and it was impossible to sit in the pulk, as at that height from the ground we were pelted with bits of ice and even small stones, which did not reach our face when we were on our feet.

We worked our way on; but while the storm increased, our strength diminished.

At last the *vappus* cried that we should have to turn back, but the next moment said, "No, we must go on. We can't have more than 2 kilometres to go, and perhaps it will be more difficult to go down than up."

So on we went. Progress was very slow, and I felt that I was approaching a critical state of weariness. Immediately after, Helland-Hansen's nose and chin were frost-bitten, but nothing could be done. Fortunately the affected parts were soon covered with a protecting mask of ice, beneath which they gradually thawed, whereupon the ice was removed.

Later on we were all more or less frost-bitten in the exposed parts of our face, the *vappus* in particular, a large part of his face being white with frost-bite.

It was not long before some of the reindeer lay flat down, and the *vappus* thereupon threw himself upon a pulk, declaring that he could go no farther, and could not find the way. "You must go on by yourselves, and keep the wind in your face," he said.

Fig. 1.
Postvappus C. I. Hætta.

Under these circumstances there was no question of continuing our way; the only thing to be done was to make what arrangements we could, and get into our sleeping-bags as quickly as possible. We agreed, however, to try and build a barricade with the pulks and our baggage, and behind it to put up a little low tent upon a piece of hard snow.

While thus engaged, Helland-Hansen got his hands frost-bitten. It was done in a few minutes. We then got into our sleeping-bags with all possible speed, Lows being the last, as he had been the toughest, and was the least exhausted.

Fig. 2. A "Raide" of Reindeer.

The twenty hours we lay thus were a dismal time for us. We passed it partly in lying and thinking our own thoughts, partly in struggle, first Helland-Hansen's desperate and vain attempts to bring life into his fingers, and then our endeavours to prevent our being buried in the snow; for wherever there was a little shelter from the wind, the snow would heap itself up into a thick, compact drift, in which you sat as in a vice if you let it grow.

After the long night, it at last began to grow light; but the wind was almost as strong. The *vappus* had lain all the time in his Finn furs under a pulk. I shouted to him from my bag until at last he heard and crawled up to me. I said we must try to get down to Gargia again, and asked him to take all the baggage and instruments off the pulks. His only answer was that he was so fearfully cold; and nothing was done until Lows crept out of his bag, and set things going. Lows was the one who had kept up best, but then before he lay down he had had the good sense to rip up a bag of bread with his knife, and take out a loaf. He divided it into two, and threw one half over to me; but I did not hear him shout when he did this, and thus had none. He had gnawed at his half during the night, and of course it had strengthened him; and he was the only one of us who had tasted food since we left Gargia.

At last we started, each in our pulk, after the guide had solemnly asked us if it were really our intention to try to get back to Gargia in this weather. We could not see more than a few yards in front of us, but we were quite determined to try.

The couple of hours spent in the descent were the most exciting I have ever gone through. It was now that our guide showed himself to be the adept that I had been told he was. It was wonderful to see the way he ran to the right or to the left, to find tracks or take a course, and how he drilled

Fig. 3. Lodikken Hut on Beskades.

the reindeer when they became unmanageable and suddenly set off up in the face of the wind again. The energy he developed when once he had thawed was incredible. At last we had the good fortune to run almost up against a cairn with a sign-post on the Kautokeino road, and then we knew we were alright.

We got back to Gargia at 4 p. m., 31 hours after we had left it. Here Helland-Hansen's hands, which were white and stiff to the wrists, were immediately put into ice-cold water, and kept there until they thawed; and by this means the circulation returned to his hands, except the end joints of eight fingers. We then at last got something to eat, not having tasted food all through the terrible journey; and then we once more turned our attention to Helland-Hansen's hands, which were in a terrible state, and dressed them as well as in the mean time we were able. And in spite of everything, our spirits now rose high, in our intense delight at having at any rate not lost our lives.

Next morning I went to Bossekop for a doctor, who came and bandaged Helland-Hansen's hands properly; but he could not of course give any opinion as to how it would end. Under his ægis, Helland-Hansen was taken to Bossekop, whence he went on as soon as possible with Lows, who took charge of him, to Hammerfest, and went to the hospital (1). I remained at Gargia to await an opportunity of going with the guide to look for our things, the instruments in particular. The first time we set out on the search, the wind was so high that we had to come down again.

The journey back down the Beskades hills with fresh reindeer, was the wildest piece of driving one can imagine. The animals flew like the wind, and galloped along in places where a horse would have gone carefully step by step. We had five reindeer fastened together in a *raide*, and I sat in the last pulk, firmly lashed to it. Occasionally the pulk was thrown over the edge of the slope, notwithstanding that I put on all the brake that I possibly could with my elbows, which were well protected with fur. Once indeed my reindeer itself fell, wonderfully sure-footed though it was; but after being dragged along by the others for a few moments, it managed to struggle to its feet without assistance.

The day after this unsuccessful attempt, we once more went up. There was a little wind in the morning, very much as it had been on the 9th; but this time, instead of increasing, it gradually dropped as we ascended; and when we began to beat up and down in the neighbourhood of the place in which our things might be supposed to be, the sun shone out brightly, and there was no more wind than that the Finn could light his pipe.

We found the things at last, nearly all of them buried in the snow, scarcely more than one kilometre from Lodikken hut, where we had thought of staying.

We dug out nearly all our things, and got safely back to Gargia with them.

That evening there was bright aurora, and I therefore unpacked some instruments, and had the good fortune to make an interesting observation, which I have described in the report of my 2nd aurora expedition (2).

We had previously, also on our first expedition, made a very interesting observation of a rare, but very significant, auroral phenomenon, which I will here briefly describe. To myself it is of special interest from the fact of its being my first auroral observation of any importance. Moreover it immediately appeared to me that the observation was a confirmation of the hypothesis put forward by me in 1896 regarding the origin of the aurora, namely that the northern lights are due to cathode rays or similar rays emitted by the sun, these rays being drawn in from space towards the earth by the terrestrial-magnetic forces.

It was ten minutes to six on the evening of the 5th February, when we were some miles from Hammerfest, the weather clear and the moon shining, when there appeared a sharply-defined arc of light from east to west through the zenith. From the very first, the arc was very intense, but very narrow, right above our heads. Notwithstanding the bright moonlight, the aurora, which soon began to pass through various phases of development with draperies and sheaves of rays, was visible up to half past seven, when it disappeared.

At Hammerfest the next day, the weather was just as clear; and at five minutes past six, the same arc suddenly appeared again, though considerably fainter. Its manner of development and its disappearance were so similar to those of the arc of the preceding day, that the phenomena left a decided impression that the position of the sun or the moon in relation to the earth must play a direct part in them.

It may, as we know, not infrequently be seen in the registering of magnetic disturbances, not only that well-defined perturbations reappear on two or more consecutive days, which in other respects may be fairly calm magnetically, but that these well-defined perturbations can be so wonderfully uniform in

(1) HELLAND-HANSEN is now Director of the Biological Station at Bergen.

(2) Expédition Norvégienne de 1899—1900 pour l'étude des aurores boréales, par KR. BIRKELAND, p. 76. Videnskabs-Selskabets Skrifter 1901, No. 1.

character, that the impression they leave is similar to that of the above-mentioned auroral observation. We shall return to this parallelism between aurora and magnetic disturbances later.

2. The second aurora expedition, from September, 1899, to April, 1900, had stations upon the top of two mountains about 3000 feet in height, Sukkertop and Talviktop, situated in the mountain district of Haldde, on the west side of the Alten Fjord, between Kaafjord and Talvik.

Fig. 4. Sukkertop and Talviktop.

As long before as the autumn of 1897, after my unsuccessful first expedition, I had again been up in Finmark to find a mountain that would do for my auroral investigations. After ascending and examining six of the highest mountains about Kaafjord, and the Lang Fjord, I decided on Sukkertop and Talviktop — the latter situated at a distance of 3·4 kilometres to the north of the first-named mountain — as most suitable for my purpose.

I then obtained a grant from the State in order to build two small mountain observatories on these summits. They were built of stone and cement, and were finished in September, 1899; so upon those Haldde mountains, right in the southern margin of the auroral zone, there now stand two of the best auroral observatories in the world. In clear weather everything that takes place in the sky can be observed, from the point where it begins to that where it leaves off. The view is uninterrupted, and from both observatories, but especially the highest and northernmost, there is a panorama stretching from the sharp, blue peaks of the Kvænang mountains in the west, to the softer outlines of the Porsanger mountains in the east, and from the precipitous cliffs of Lang Fjord and Stjerne Island in the north to the mountain plateau in the south, stretching inland in undulating lines as far as the eye can see, in towards the winter home of the mountain Lapps. And far below lies the fjord like a dark channel that is continued in the Alten valley itself and its numerous branches.

Fig. 5. On the way to Sukkertop.

The expedition of 1899—1900 was furnished, *inter alia,* with self-registering barometers, thermometers, and hygrometers, and also with apparatus for the photographic registration of the three components of terrestrial magnetism, and of the electric condition of the atmosphere. On Sukkertop we had kites, with self-registering instruments for investigations high up in the atmosphere; but the wind was almost always very strong up on the mountain, and we very soon lost them. The members of the expedition were myself, SEM SÆLAND, amanuensis at the University Physical Institute, E. BOYE, a student, K. KNUDSEN, telegraphic engineer, and a cook.

Fig. 6. The Observatory on Sukkertop.

The results of the expedition's magnetic investigations and of the auroral observations have been already published in the above-mentioned work, whereas the meteorological observations have unfortunately not yet been worked up.

Many of our experiences during our stay upon these mountain-tops were such as others have probably not passed through; for as far as is known no one has ever before passed a winter upon the highest mountain-summits in Finmark.

It is my intention, however, not to relate here much more about our life and our difficulties in the second and third expeditions than may serve to show the development in these undertakings, but to tell enough to give those who may make future expeditions in the same regions, the benefit of our experience to build upon.

The natural force with which we especially had to battle with up in Haldde was the wind; for it sometimes blew fearfully. We were unable to measure the highest velocities, but once we measured one of 46 metres per second. For this we used two good little hand anemometers of Richard Frères; but they were certainly not intended for such great wind-velocities, and what the error may have been in these extreme measurements, I cannot say.

Fig. 7. The observatory on Talviktop.

We often had much greater hurricanes, however, than the one mentioned which we measured. The wind sometimes roared so against the houses, that you would have thought you were sitting at the foot of a waterfall; and the floors trembled and everything shook. We soon got to be able to gauge relatively the storm outside by the noise within. Our measuring apparatus, as I have said, did not allow of our determining the greatest wind-velocities, and often we could not get out of the house ourselves for several days. One strong anemograph we had put up was blown to pieces in the course of a few days, and we found pieces of it from 50 to 100 metres from the place where it had been put up. The reason of this was probably that at the same time as the wind, the air was at times so saturated or supersaturated with moisture, that ice formed upon everything. In nine or ten hours, ice-formations the length of one's finger would be formed, always pointing towards the wind. Suspended telephone wires would become as thick as a man's arm with ice. It was probably a heavy coating of ice such as this that destroyed our very strongly built anemometer in a hurricane. In high winds it was impossible to go out, and more than once, on Sukkertop, it took three men with a great effort to close our little door.

After storms such as this, there were of course many changes to be seen. We have seen a layer of snow a metre thick, and so hard that you could jump on it without sinking in, practically disappear from the summit in the course of nine or ten hours. It may be imagined then what a whirling and drifting there was in a wind, when the snow was comparatively fresh, and not pressed into such a compact mass.

Fig. 8. Going to measure the Wind-velocity.

For the sake of comparison it may be mentioned that the greatest wind-velocity observed by the Nansen Expedition in three years was only 18 metres. This is an interesting circumstance, for it shows that on the ice-fields of the polar regions in a more restricted sense a comparative stillness prevails in the atmosphere.

As a rule the wind on the Haldde mountains was not especially cold, but it *could* be sometimes. On the 20th February, 1900, when the temperature was —33·5° C., the wind-velocity was about 20 metres. The greatest wind-velocities observed upon the Haldde mountains are given below.

Temperatures of — 20° accompanied by winds with a velocity of from 20 to 30 metres were pretty frequent both in January and February, 1900.

	Wind-velocity in metres per second	Direction of Wind	Temperature C.
Nov. 17, 1899	37	NW	— 7·2°
Dec. 30, 1899	38	SSE	—13°
Jan. 20, 1900	38	S	— 16°
Feb. 28, 1900	35	NNE	— 10°
March 3, 1900	41	NW	— 5°
March 4, 1900	46	WNW	— 4°

No one who has not tried it can imagine what it is to be out in such weather. Knudsen, for instance, once had one hand frost-bitten in the few minutes he was out to take a reading, although he had on thick woollen gloves. He had neglected the precaution of having fur gloves over them. Frost-bite such as this, however, is not serious when you can go at once into a warm house, and get ice-water for your hands.

The wind would sometimes come like a rushing river at the one station, while it was fairly calm at the other. On the 19th January, for instance, on Sukkertop, the velocity of a wind from the SSW was found to be 36 metres, while on Talviktop at the same time there was no wind, this being ascertained by telephone. The wind was heard on Talviktop, however, as a tremendous rushing from the south; and an hour and a half later the wind blew with tremendous force over both mountains.

In extreme cold and a high wind, it was uncomfortable on Talviktop. Water once froze there a couple of yards from a glowing stove; and the lamp was blown out on the table in the middle of the room, although in a general sense the house was well enough built.

The worst trouble was the repeated breaking of our telephone-wires, occasioned by the snow-storms. At first the telephone wires between the two summits were hung upon poles in the usual manner; but this proved to be useless. Either the wires themselves were blown to pieces, or the insulators

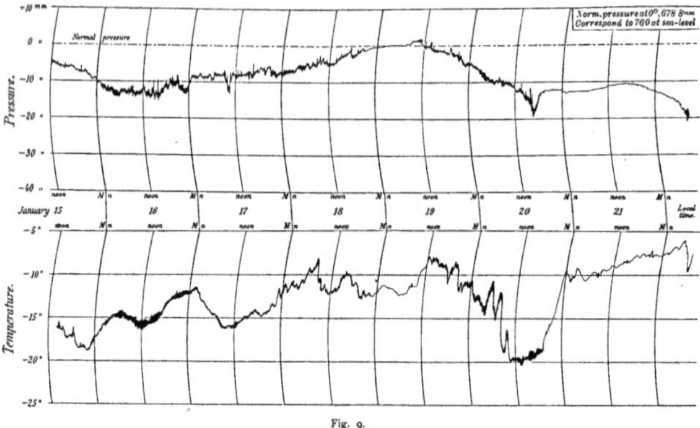

Fig. 9.

torn down, and the line in either case destroyed. On the other hand, the wires, when laid upon the ground, keep fairly well, except on hills, where great snow-drifts are heaped up upon them. In such places they often came to grief; and our first work after a fall of snow and storms used to be to get them repaired.

In the same way we at first had a double line from Sukkertop down to Kaafjord; but here too the wires were often broken, and we had great difficulty in repairing them.

A couple of hours before violent winds came over Haldde, great changes were generally observed in the barometer, which sometimes went up and down at intervals of a few seconds; and when this occurred, we knew that it would not be safe to start from one observatory to go to the other.

During the storms this vibration of the barometer, owing to dynamical causes, was very considerable, as will be seen from the barograms, and could serve as a relative gauge for the violence of the storm.

Figure 9 shows a couple of correlated barograph and thermograph curves drawn on Sukkertop. They show the conditions during these very January storms mentioned, which moreover were the cause of many casualties on the coast of Norway that year.

In spite of our barograph predictions of storms, our postman, a sturdy little Finmark man, now and again happened to come in for dangerous weather when he came with the post from Kaafjord once or twice a week. We were often afraid for him, but he was always alright, though sometimes so covered with ice when he arrived, that he was quite unrecognisable. I once asked him if he were never frightened when the weather was so bad. At first he did not answer, but sat quietly down to thaw; but a little while after he said: "I'm too stupid to be frightened".

Sad to say, our second aurora expedition was also destined not to terminate without a great misfortune, which occurred just a week before we thought of packing up.

The very road that our postman traversed every week as long as the expedition lasted, was to be the scene of the death of two clever men, an avalanche having overwhelmed in Sivertdalen five persons who were on their way to visit the observatory in Haldde on the 16th March.

The two who perished were our good comrade, E. Boye, and Captain Lange, master of the Kaafjord Mines' steamer; the other three escaped without injury. There had been an unusually heavy snowstorm the night before, preceded by frost.

THE EXPEDITION OF 1902—1903.

3. The treatment of the observations that were collected during the 2nd aurora expedition, the results of which have been published in the previously-mentioned work, showed with perfect clearness that in order to solve the problem of the cause of the aurora and magnetic perturbations, it was necessary to have at our command simultaneous magnetograms and observations from several suitable polar stations at distances of about 1000 kilometres from one another, and also corresponding material from as many other stations all over the world as it was possible to obtain.

I demonstrated namely, that certain well-defined magnetic perturbations that occurred over large portions of the earth might be naturally explained as the effect of electric currents, which, it might be supposed, in the polar regions flowed approximately parallel with the surface of the earth at heights of several hundred kilometres, and strengths of up to a million ampères, if they could be measured by their effect as galvanic currents. These currents in the polar regions were well defined and greatly concentrated, and often passed for the most part *between* two neighbouring stations, as, for instance, Bossekop and Jan Mayen (see "Expédition", etc., l. c., p. 27), in such a way that Bossekop lay quite on the one side of the current, and Jan Mayen on the other; and the magnetic effect of the currents in the polar regions was not infrequently as much as 20 times stronger than in Central Europe. The investigation of these phenomena would necessarily, of course, require simultaneous registrations of the magnetic elements at several uniformly equipped polar stations.

By such registrations, other important, unexplained phenomena that are very characteristically developed in the polar regions, might be excellently studied, e. g. the tremendous changes in the magnetic components, which often occur at short intervals, especially during an aurora. A rapid registering of the magnetic elements and of the earth-currents appearing simultaneously, would greatly assist the study of these conditions.

It was with these things in my mind that from the beginning of 1901 I began to work for the sending out of a new aurora expedition, with stations in Finmark, Iceland, Spitsbergen and Novaja Semlja, so as to obtain observations simultaneously from both sides of the auroral zone.

On this occasion also, the Norwegian Government looked upon my plans with favour, a grant of 20,000 krones being made by the Storthing towards a new expedition. The president of the Storthing,

GUNNAR KNUDSEN, J. FABRICIUS, a landed proprietor, and A. SCHIBSTED, editor of the "Aftenposten" then contributed 6000 krones each; and the remainder, which amounted to about 30,000 krones, I have furnished myself.

It may safely be said that economy is one of the virtues of Norwegians as a nation, perhaps one may say a virtue of necessity; but the nation's idealism often turns the balance in delightful non-conformity with economy. The grants to my aurora expeditions are an instance of this. I will take this opportunity of offering my respectful thanks to the government authorities, the scientific institutions, and the private men who have given their support to these undertakings.

The preparations for the expedition were pushed on with the greatest energy for a year, and in this I was ably assisted by my assistant of the 2nd expedition, Hr. S. Sæland. After a search in the four lands mentioned above, for the purpose of finding suitable dwelling-houses with as easy access from Christiania as possible, I fixed upon the following as my four stations: Kaafjord in Finmark, Dyrafjord in Iceland, Axeløen in Spitsbergen, and Matotchkin Schar in Novaja Semlja.

The expedition was ready to start about the Ist July, 1902.

THE AURORAL STATION IN KAAFJORD.

4. This station was in the province of Finmark, close to the Kaafjord Copper Mines, in 69° 56′ N. Lat. and 22° 58′ E. Long.

The members of the expedition were RICH. KREKLING, a science graduate, and O. EGENÆS, an engineer. The station was under my special supervision; during my absence it was managed by Krekling.

Sæland, Krekling and Egenæs set out for Kaafjord with their equipment on the 10th July, 1902, and arrived at their destination on the 17th.

The first investigations that were made here during this expedition were simultaneous registerings of the terrestrial-magnetic components, with two exactly similar sets of registering apparatuses. The one set was placed in the mountain observatory on Talviktop, the other in a mine, 100 metres in under the mountain. Sæland registered in the mine, while the other two men worked at the summit from the 26th July to the 15th August.

Fig. 10. The Kaafjord Station.

The second series of investigations comprised magnetic and earth-current observations, and in the next place meteorological and atmospheric-electric measures. These were made in Kaafjord during the period from the 18th August, 1902, to the 13th March, 1903.

The third series of investigations, magnetic and earth-current registering, was made, for reasons given below, at Bossekop, during the period from the 15th March to the 1st April, near the locality of the polar station in 1882 and 1883.

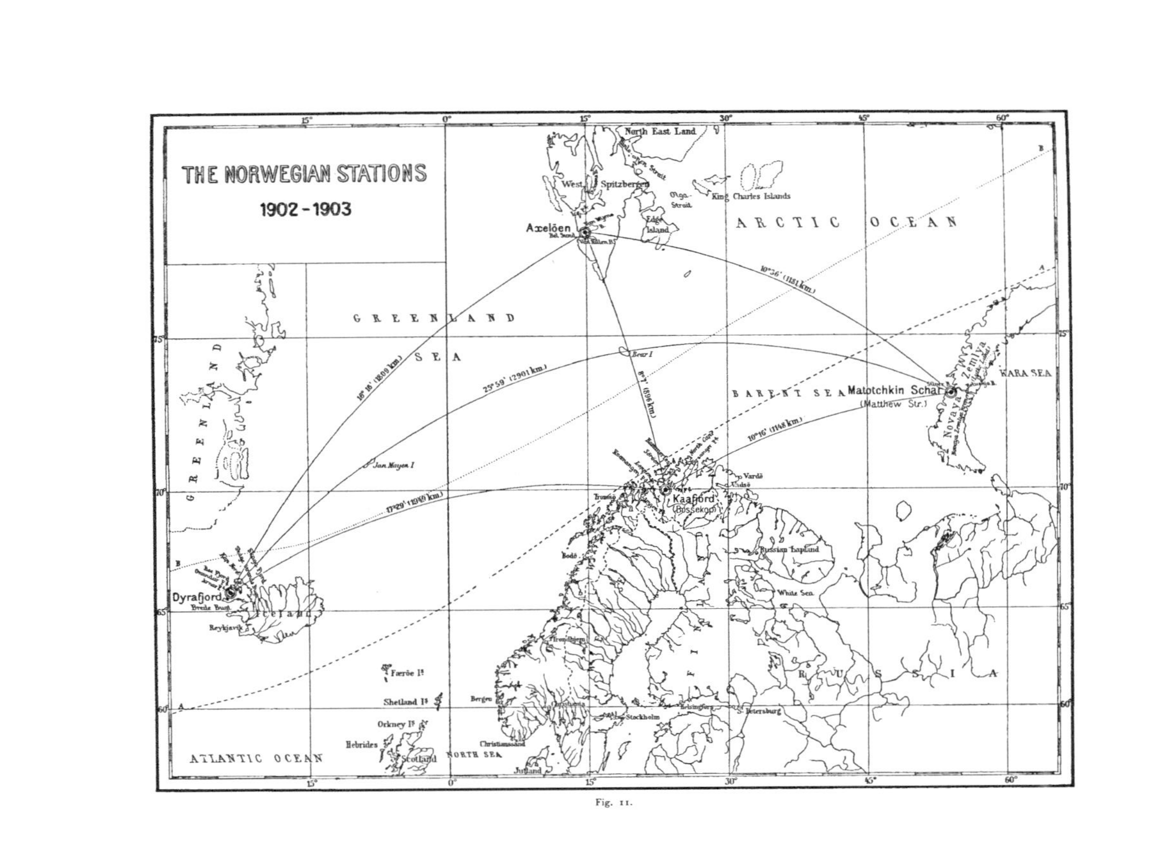
THE NORWEGIAN STATIONS
1902 - 1903
North East Land
West Spitzbergen
King Charles Islands
Olga Strait
Edge Island
Axelöen
ARCTIC OCEAN
10°56' (1181 km.)
GREENLAND
SEA
GREENLAND
16°16' (1809 km.)
25°59' (2901 km.)
Bear I
8°7' (896 km.)
BARENT SEA
Matotchkin Schar
(Matthew Str.)
Novaya Zemlya
KARA SEA
10°16' (1148 km.)
Jan Mayen I
17°29' (1949 km.)
Vardö
Vadsö
Tromsö
Kaafjord
(Bossekop)
Russian Lapland
White Sea
Bodö
Dyrafjord
Iceland
Reykjavik
Trondhjem
Færöe Is
Shetland Is
Bergen
Orkney Is
Hebrides
Scotland
NORTH SEA
Christiansand
Jutland
Stockholm
Helsingfors
St. Petersburg
RUSSIA
ATLANTIC OCEAN

Fig. 11.

EQUIPMENT.

Magnetic Instruments.

A set of terrestrial-magnetic variation instruments with photographic registering apparatus and lamp reflector of the Eschenhagen pattern from Otto Toepfer's, Potsdam.

An Eliott Brothers' unifilar magnetometer, belonging to the observatory in Christiania.

An inclinatorium, lent by Professor Rydberg, and previously used on the "Vega" expedition.

An earth-inductor, from G. Schulze in Potsdam, with galvanometer made by O. Pluth, Potsdam.

Earth-current Apparatuses.

Two Deprez-d'Arsonval galvanometers from Keiser & Schmidt, Berlin. As these instruments proved to be bad, one of them afterwards had to be exchanged for one from Hartmann & Braun, Frankfort-on-the-Main.

A registering apparatus with accessories, resistance-boxes, cables with rubber insulation, etc.

Meteorological Apparatuses.

A mercurial barometer.

A thermometer-screen with its thermometers, and spare thermometers.

A large barograph.

A large thermograph with forms, from the Meteorological Institute in Christiania.

A cloud-measuring apparatus, an anemometer from Richard Frères, Paris, etc.

Electrical Apparatuses.

An Elster & Geitel's electroscope with accessories for observations of dissipation of electricity in the air.

A Zamboni battery, with wires, insulators and tightly-closing drum, from Günther & Tegetmeyer, Brunswick.

Astronomical Instruments.

The station had no permanent theodolite, as it was in telegraphic communication with the astronomical observatory in Christiania. The azimuth of the mark (the spire of Kaafjord Church) was found by Sæland with a large theodolite in the autumn of 1902, before he left for Iceland.

The expedition had borrowed from the Military College in Christiania a box-chronometer, Kessel 1390.

They also took with them books, papers, etc., rifles, ammunition and provisions, as some time was to be spent at the Haldde observatory. In Kaafjord, the members of the expedition put up at the Kaafjord Copper Mines.

Fig. 12. At the Astronomical Pillar.

BUILDINGS.

Upon the arrival of the expedition, the following buildings and contrivances were put up:

The terrestrial magnetic register-observatory.
The observatory for absolute determinations.
Hut for the registering of earth-currents.
Thermometer-hut.
Pillar for astronomical measurements, etc.

The Terrestrial Magnetic Register-Observatory.

This was a stone cellar, divided into two rooms, the outer of which served as an entry, the north, inner room being the real observatory. (Plan Fig. 13). Here there were 4 stone pillars, the same as were used in the polar year 1882—83, for the instruments and the registering apparatus. P, P^I and P^{II}, are the pillars for the three instruments, P^{III} the one for the registering apparatus. L is the lamp reflector, R the registering apparatus, V, D, and H the variometers for respectively the vertical intensity, declination, and horizontal intensity. δ^I and δ^{II} are the two doors.

The drawing beside the plan, on the magnetic meridian arrow, represents the position of the magnets in the instruments in relation to the meridian. The magnets in the drawing are about one fifth of their actual size.

The Observatory for the Absolute Determinations.

This observatory was a house of the same kind as that in Spitsbergen, the drawing of which will therefore serve to illustrate this one. There was only one difference, namely that the stone pillar upon which the various magnetic instruments and the earth-inductor were set when in use, was placed in the middle of the house. The azimuth of the pillar was determined by triangulation, the pillar forming one vertex of a triangle of which the two other vertices were the *astronomical pillar* (marked on the map (1), and mentioned above under the heading 'Buildings'), and the *spire of Kaafjord Church.*

Hut for the Registering of Earth-Currents.

This hut was built of wood, and stood beside the magnetic register-cellar, as shown on the map. The purpose of these earth-current investigations was to obtain photographic curves showing the variations in the earth-currents, especially during magnetic storms.

Four insulated cables of a length of 200 metres were laid down in the directions north, east, south, and west. Their ends were connected with the earth by filling deep holes with coal-dust, which was pressed firmly down round a bright copper wire.

In the register-house the two cables, north and south, were connected, with a suitable shunt, with one galvanometer Deprez-d'Arsonval, and the east and west cables similarly connected with another exactly similar galvanometer. The oscillations of the galvanometer were registered photographically.

Unfortunately these galvanometers, supplied by Keiser & Schmidt, Berlin, were very bad, so that at last, after prolonged trial, we had to reject one and replace it with one from Hartmann & Braun, of Frankfort. When subsequently we succeeded in obtaining good photograph curves, an electromagnetic contrivance for the time-marks was arranged for all magnetic and earth-current registerings, in order to facilitate comparison with the magnetic curves. Down in the dwelling-house, by the side of the chronometer, the time could be marked on all the photograms by pressing an electric button. This, especially during the rapid march of the registering apparatuses, was of very great importance.

As it appeared that the earth-currents in Kaafjord had a predominant direction which seemed to indicate that local conditions such as the proximity of the coast-line, etc., had something to do with it,

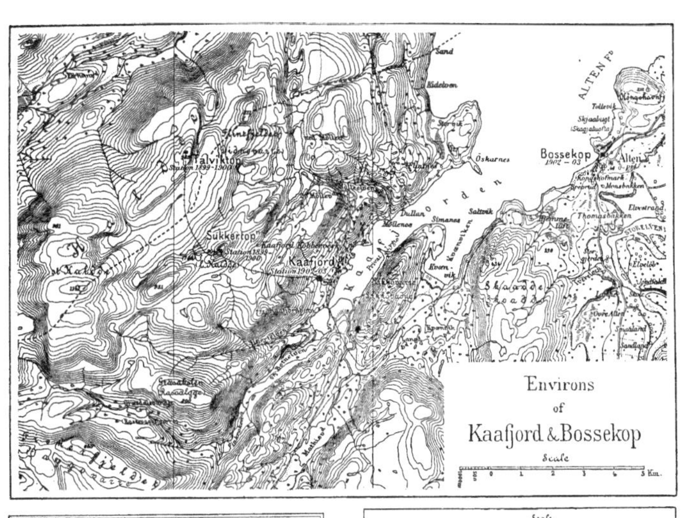
Environs
of
Kaafjord & Bossekop
Scale
Km.
Bossekop
1902 – 03
Alten
Kaafjord
Sukkertop
Talviktop
ALTEN Fd
Sand
Öskarnes
Saltvik
Thomasbakken
Elvestrand
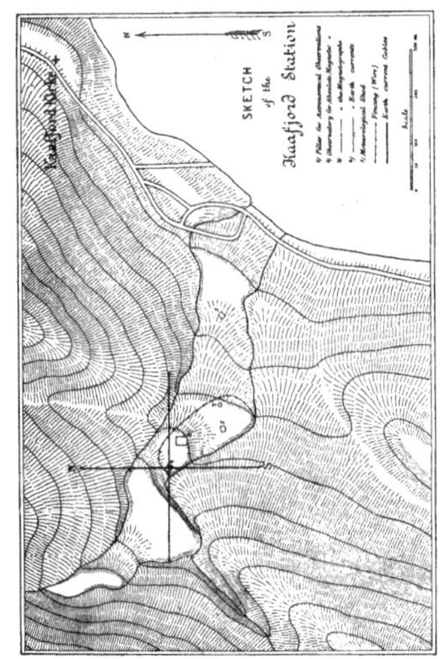
SKETCH
of the
Kaafjord Station
Scale
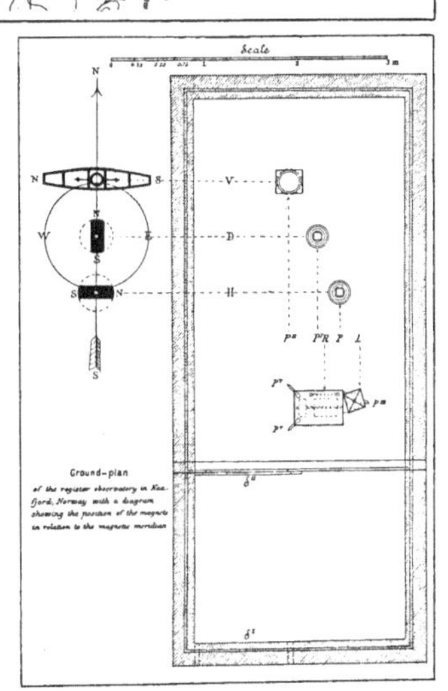
Scale
N
S
W
E
V
D
H
Ground-plan
of the register observatory in Kaafjord, Norway with a diagram showing the position of the magnets in relation to the magnetic meridian

Fig. 13.

the whole auroral station, as already stated, was moved to Bossekop, in the vicinity of the polar station of 1882—83, on the 13th March, 1903. Before many days had passed, all the instruments were again in operation.

The Thermometer-Hut.

This was built like an English hut of wood, and large enough to contain the thermometer-screen and the thermograph. The arrangement was the ordinary one. By the thermometer-hut was placed a weather-vane, with which measurements were taken 3 times daily of the velocity of the wind, with the aid of an anemometer Richard.

The barograph was placed in an unused room in the dwelling-house. Near it stood the cloud-measuring apparatus, especially for use in determinations respecting polar bands and cirrus clouds.

The electric measurements with Elster and Geitel's apparatus, were also made in the vicinity of the dwelling-house, in order that wind and weather should not have too disturbing an influence.

5. During our stay at the stations Haldde and Kaafjord, a journal was kept of the meteorological elements, and of the aurora and cirrus-bands observed. These observations cover a period extending from the 28th August to the end of February. For the last month, March, there are no records of this description, as the entire day was taken up with registering, especially rapid registering with changing of the photographic paper on the instruments every two hours.

The meteorological observations were made regularly 3 times a day — at 8 a. m., 2 p. m., and 8 p. m.

These observations show that the weather, as is usual in these regions at this time, has been very variable. The sky has very seldom been quite clear, but was as a rule covered with clouds, a circumstance which has to some extent hindered us in our observations of aurora.

Some idea of the weather-conditions at this time may be obtained by looking at the table below, in which the highest and lowest temperatures and barometer-readings, and the highest wind-velocity observed at the above-mentioned hours are given for each month.

Month	Temperature		Barometer-reading		Wind-velocity
	Max.	Min.	Max.	Min.	Max.
	C°	C°			Metres per sec.
August	11·6	6·8	766·7	763·0	3·5
September	14·0	− 1·0	767·8	731·6	6·2
October	5·2	−12·1	766·9	733·4	9·8
November	6·6	−16·4	771·7	726·1	13·3
December	6·7	−16·8	766·7	731·6	15·0
January	6·6	−20·3	768·7	721·0	19·0
February	4·6	−13·9	758·7	711·0	12·4

In August and the first half of September, the atmospheric pressure was fairly low, but with little precipitation to speak of. The temperature remained, on an average, at about 3° C. In the latter half of September, there was high pressure with rain. On the 27th September, the first snow fell, the temperature at the time being about 2·2°.

During the first half of October there was low atmospheric pressure with frequent falls of snow, often accompanied by high wind. Throughout the latter half of the month the pressure was higher, with sleet and snow, the latter sometimes very thick.

In November the weather was variable, without much precipitation, but sometimes with high winds. The temperature was not very low, having kept at about 0°.

During the first half of December, the sky was alternately clear and overcast, but there was little precipitation. Towards the end of the month, the pressure was lower. High winds were frequent, though they did not attain a higher velocity than 15 metres per second.

During the first week of January, the weather was cold and calm, the lowest temperature being −20·3°. Later on a lower pressure supervened, with mild weather and high wind.

From the 8th to the 15th February, we had the lowest pressures that were observed. It went right down to 711.0 mm. and remained at about that height for several days. With the exception of a couple of days in the middle and end of the month, the atmospheric pressure throughout February was unusually low, with a cloudy sky and some snow.

In the course of the autumn and winter, 27 auroral phenomena, some of them very well developed and of long duration, were observed and described. It appears that almost without exception, they make their appearance in the afternoon and during the evening, generally disappearing soon after midnight.

They usually develope from the northern sky, but not infrequently, especially during a bright manifestation, they appear on the southern sky. This was observed in the cases of the bright, exceedingly beautiful and long-lasting auroras of the 11th, 24th and 31st October, and 24th November, which took place simultaneously with some of the very greatest magnetic storms that were observed during that period.

The aurora of the 24th November in particular was one of extreme beauty. It developed into an auroral corona, which lasted some minutes, and then dissolved into a great number of intensely brilliant, red streamers. These moved backwards and forwards across the heavens for some time, making the sky glow with red.

Considering that there was so much cloudy weather in October, it must be admitted that we were exceptionally fortunate in being able to observe these beautiful auroral phenomena. On the other hand, it is not improbable that the overcast sky from the 8th to the 15th February may have caused some auroral phenomena to escape our attention, as at that time, owing to magnetic conditions, bright aurora might have been expected.

The weather on the whole must be said to have been not unfavorable. The violent storms experienced on former occasions up at the mountain observatories, we that winter escaped by keeping down in the valley at Kaafjord. The greatest wind-velocity measured was not more than 19 metres per second.

AURORAL STATION IN DYRAFJORD, ICELAND.

6. The station was situated upon a promontory, Höfdaodden, on the north side of Dyra Fjord (see Fig. 17). Its latitude was 66° 15′ N., and longitude 22° 30′ W., equivalent to 1 hour and 30 minutes before Greenwich time.

The members of the expedition were Sem Sæland (leader), amanuensis to the University Physical Institute, and Larus Bjørnsson (assistant). Sæland left Christiania with his equipment on the 10th October, and arrived in Iceland on the 10th November, 1902. The voyage was satisfactorily accomplished, but the vessel was delayed a fortnight by snow-storms.

EQUIPMENT.

Magnetic Instruments.

A set of terrestrial-magnetic variation instruments with photographic registering apparatus of the Eschenhagen pattern, supplied by Otto Toepfer, Potsdam.

A universal magnetometer (travelling instrument), capable of being used for the absolute determination of intensity, declination and inclination; supplied by L. Tesdorpf, Stuttgart.

Meteorological Apparatuses.

An aneroid barometer from the Norwegian Meteorological Institute.

A thermometer-screen with its thermometers, and spare thermometers, from the Meteorological Institute.

A meteorograph (baro-thermo-hygrograph) from the Physical Institute.

A cloud-measuring apparatus, recently procured.

Electrical Apparatuses.

An Elster & Geitel's electroscope with accessories, for measuring the conductivity of the air.

A Zamboni battery (high-tension battery) with wires, insulator, and tightly-closing drum, for investigating the radio-activity of the atmosphere; supplied by Günther & Tegetmeyer.

An Elster & Geitel's high-tension electroscope.

Astronomical Instruments.

A large theodolite with broken axis, borrowed from the Astronomical Observatory in Christiania.

A box-chronometer, Hohwii No. 639, and a pocket-chronometer Michelet, also from the Astronomical Observatory.

Books were also taken, paper, forms, etc., some tools, besides rifles and ammunition. As regards food, only some delicacies were taken, as the members of the expedition lodged at Berg's whaling-station, which lay at the extreme end of the promontory, as shown in the sketch.

BUILDINGS.

After Sæland's arrival, the following were erected:

The magnetic variation observatory.
The observatory for absolute determinations.
Thermometer-hut.
Pillar for cloud-measuring apparatus.
The mark.

The Magnetic Register-Observatory.

The observatory was erected farthest from the other buildings, a little way from the shore (see Fig. 17). It was built of wood (framework), and was completely sunk in the loose, brown sand of which the ground consisted. The house was divided into 3 rooms, in order to obtain as even a temperature in the north, innermost room as possible. The first room (entry) was provided with a descending flight of stairs, and was separated from the inner room by a sliding door, δ^{I}, that room being separated from the register-room by a similar door, δ^{II}. In the middle room, various requisites were kept.

In the innermost room, six pillars were imbedded in the earth, two large ones for the three variation instruments, and three smaller for the three legs of the registering apparatus. The pillars were cut from a mast-tree, and set deep down under the floor in a large hole, which was afterwards filled up with stones.

Fig. 14. Observation-Huts at Dyrafjord Station.

Wooden pillars of this kind, buried in this manner, and exposed to fairly constant humidity, and, as in this case, beyond the reach of the frost, have proved quite satisfactory. The instruments must be placed directly upon the end-grain.

P, *PI*, and *PII* are the pillars for the registering apparatus, *PIII*, *PIV*, and *PV*, those for the magnetometers. *H*, *D*, and *V* are the variometers for respectively horizontal intensity, declination, and vertical intensity, *R* is the registering apparatus and *L* the lamp reflector.

The drawing on the right of the plan shows how the magnets were placed in the instruments, in what direction the north pole of the magnets pointed, and the size and shape of the magnets. The scale is about one/fifth of the actual size.

The Observatory for Absolute Determinations.

This was very well and practically made. The drawing gives a plan and elevation, and shows how the whole was arranged. It will be seen that the house was partially buried in the sand. The part above the ground was almost entirely of glass. A square hole was dug in the ground, and into the corners and sides of this were driven 12 posts, upon which rested a frame, a similar frame connecting their lower ends upon the earth beneath the floor. The floor rested upon the latter frame, and from it, and up to the surface of the ground, were nailed boards, which thus formed the walls of the underground portion. Above the ground, grooves were cut up the sides of the posts, into which were fitted glazed window-frames. The windows were kept in their place by bolts. In the drawing, one of these is marked *K*. The roof was formed of three window-frames, which were wedged into the beams of the roof in the same way as the side windows. The roof windows were kept in their place by two overlapping clamped beams, one end of which was attached by hinges, h, h^1, the other end being held fast by the clamps l, l^1, which could be unhooked, and thereby allow the beams to be raised, and one or all of the windows to be removed. The side windows could be removed in a similar manner. Thus the great advantage of this observatory was that

Fig. 15. View from Dyrafjord Station; by moonlight.

Fig. 16. View from Dyrafjord Station; by moonlight.

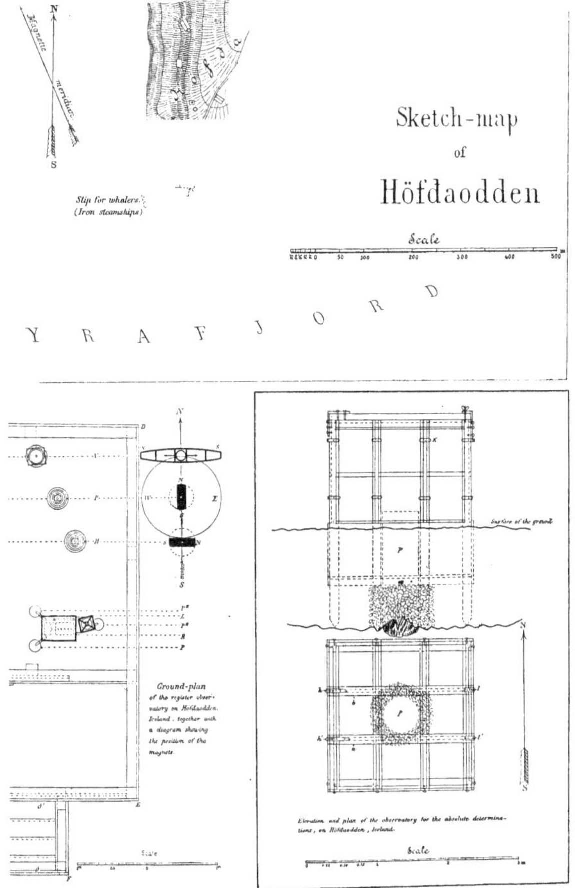
N
S
Magnetic meridian
Slip for whalers.
(Iron steamships)
Sketch-map
of
Höfdaodden
Scale
0
50
100
200
300
400
500
Y R A F J O R D
Ground-plan
of the register observatory on Höfdaodden, Iceland, together with a diagram showing the position of the magnets.
Surface of the ground
Elevation and plan of the observatory for the absolute determinations, on Höfdaodden, Iceland.
Scale

Fig. 17.

there was abundant light, and that the telescope could be pointed in any direction desired, as any window could be removed.

In the middle of the room was a solid wooden pillar, fixed in the same manner as those in the register-observatory. The pillar is marked *P*.

The Thermometer-Hut (see the sketch).

A perfectly plain hut was erected between the observatory for absolute determinations and the pillar for the cloud-measuring apparatus.

The Pillar for the Cloud-Measuring Apparatus was a wooden pillar sunk in the earth, with stones round it.

The Mark was a wooden pole.

There was also here, as at the other stations, a mark at a greater distance from the station. For this Sæland had chosen a prominent point on the other (western) side of Dyra Fjord.

No accidents occurred during the winter, either to instruments or buildings. It appeared that Sæland in his completely closed and underground register-observatory, was no more inconvenienced by the condensation of moisture on the instruments than was Russeltvedt in Spitsbergen, where a slow, practical ventilation was contrived.

7. The expedition to Dyra Fjord was carried out much later than had been planned, as Sæland had to make a journey of inspection to Novaja Semlja in September, instead of Professor Birkeland, who had the misfortune to be bitten by a dog at Archangel under such suspicious circumstances, that he was advised by the doctors to go to Moscow to be treated at the Pasteur Institute there. Further delay was caused by the very stormy weather experienced on the voyage to Iceland in the latter part of October and beginning of November.

Both in the erection of the observation-houses and in other ways, our expedition received valuable assistance from Captain Berg's whaling-station.

The general impression of the weather during the winter was that it was much more uncertain than it usually is in Dyra Fjord. The sky was almost constantly overcast from the beginning of November to the end of January. Snow-storms from the NW alternated unceasingly with a south wind and deluges of rain; and if, between whiles, the wind dropped for a day or so, we always had to be prepared for a fresh gale. In February, however, we did get a little clearer, frosty weather, and when in March the drift-ice came in-shore, we had clear, cold winter weather for about a fortnight.

At times the wind was exceedingly strong. On the night of the 13th November, for instance, a large portion of the roof of the whaling-station was blown off, and a number of houses in the surrounding district suffered more or less damage. The barometer readings were throughout extraordinarily low. On the 19th February, a reading of 693 mm. was noted on the aneroid barometer of the expedition. The day before, according to Icelandic papers, a correspondingly low reading had been noted in Vestmaneyarne.

It is obvious that with such weather there were comparatively few opportunities of observing aurora. We kept regular watch in the evening; but as a rule only very small patches of sky were visible, and what auroras were observed, were therefore usually observed piecemeal.

Opportunities of observing the typical development of auroral arcs at right angles to the magnetic meridian, with a slow ascent from the northern horizon up towards the zenith, were rare. This may to some extent be due to the above-mentioned conditions; but on the other hand, it was far more usual here than, for instance, at Haldde in Alten, to see aurora in the south, and also it was our impression that among the various forms of aurora, the *corona* is far more general in Iceland than at Haldde.

On the whole, however, the aurora in Dyra Fjord also, is seen far more frequently in the north than in the south. In this particular, it does not quite seem to carry out the current theory as to the position of the auroral zone being to the south of Iceland.

THE AURORAL STATION IN SPITSBERGEN.

8. The station was situated on the Axel Islands in Belsund, West Spitsbergen. The expedition was stationed, as the map shows, (Fig. 18), at the southern end of the largest, most northerly island (Hovedøen). The astronomical pillar near the dwelling-house has a latitude of 77° 41′ 21,5″ N., and a longitude of 14°50′ E., equivalent to 0 hrs. 59 min. 20 sec. by Greenwich mean time.

The head of the station was Nils Russeltvedt, assistant at the Meteorological Institute in Christiania; and there was only one permanent assistant, namely, H. Hagerup, an electrotechnicist. They went, however, with a hunting expedition, under the command of Captain Hagerup from Tromsø; and the members of the latter expedition were bound to render ours whatever assistance they required.

EQUIPMEMT.

Magnetic Instruments.

For a continuous record of the terrestrial-magnetic elements, 2 registering apparatuses were taken, and 2 unifilar magnetometers of the Eschenhagen pattern by Otto Toepfer, Potsdam, and a Lloyd's balance from Charpentier, Paris.

For the absolute determination of the terrestrial-magnetic elements there was a Fox's circle, and a Dover's inclinatorium, and also some requisites and spare parts. During his stay at the station, the leader of the expedition made a special instrument for the determination of the declination.

Meteorological Apparatuses.

For meteorological uses there were 2 thermographs, 1 barograph, 1 mercurial barometer, 1 aneroid barometer, 6 thermometers 1/5° C., 2 sling-thermometers, 1 large thermometer-screen, 4 minimum thermometers, an anemometer Richard, and a cloud-measuring apparatus, besides books, forms, etc., some of them placed at our disposal by the Norwegian Meteorological Institute. A thermometer and thermograph hut was made at the place, and a weather-vane.

Electrical Apparatuses.

For measurements of the dissipation of the electricity in the air, there was an Elster & Geitel's electroscope, with accessories.

Astronomical Instruments.

For astronomical uses we had a theodolite and a large sextant belonging to the Astronomical Observatory in Christiania. There were also 2 chronometers, a Lacklan & Son No. 512 and an Arnold No. 152.

Some instrument-maker's tools were also taken, as also guns and ammunition. To the vessel's equipment belonged a camp forge and smith's tools, some carpenter's tools, etc.

Russeltvedt left Christiania on the 3rd July — taking with him the instruments and the tinned provisions that were required — to join the other members of the expedition at Tromsø, and to attend to the equipment of the ship. The ship, which was to winter in Spitsbergen, was a large coaster called "Jasai".

When everything was arranged, the expedition started from Tromsø on the 24th July, and arrived in Spitsbergen on the 7th August.

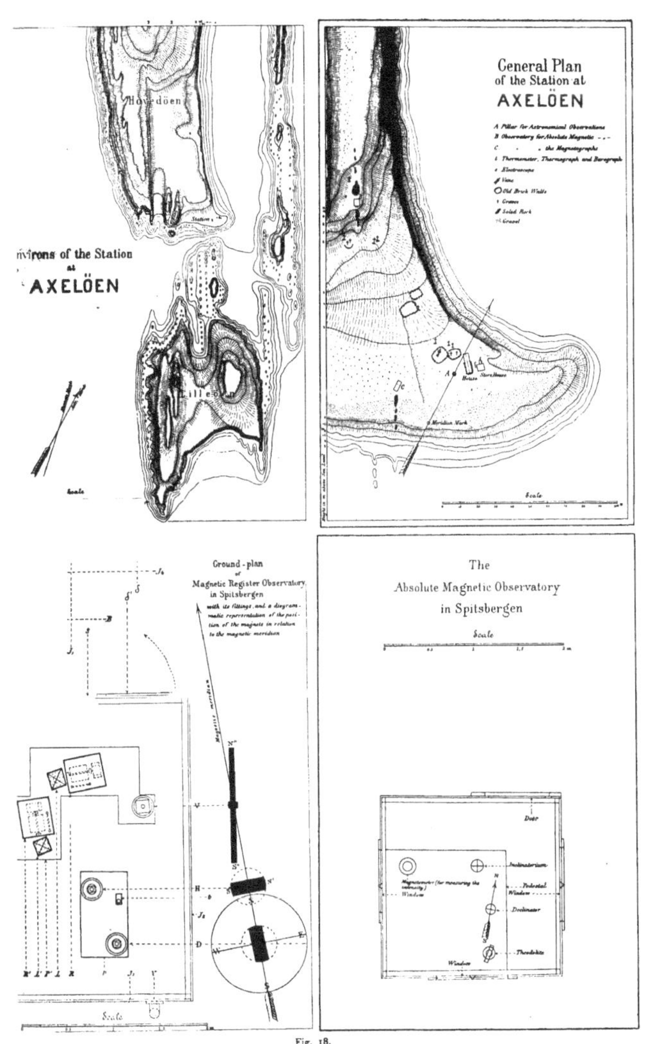
Hovedöen
Station
nvirons of the Station
at
AXELÖEN
Scale
General Plan
of the Station at
AXELÖEN
Meridian Mark
Store House
Scale
Ground-plan
of
Magnetic Register Observatory
in Spitsbergen
Scale
The
Absolute Magnetic Observatory
in Spitsbergen
Scale
Door
Inclinatorium
Podestal
Window
Declinator
Theodolite
Window

Fig. 18.

The following buildings were repaired and erected at the station:

The magnetic register-observatory

The observatory for absolute magnetic determinations

A dwelling-house

A storehouse, to which were attached a thermometer-screen (t) and the electroscope-hut (e).

Fig. 19. Dwelling-house of the Expedition.

The Magnetic Register-Observatory.

The building was quite a plain wooden house (frame-house). It was sunk down into the earth as far as the underlying rock would permit. (See sketch Fig. 18). Some earth was thrown up against the walls; but owing to the lack of loose, light earth, it could not be covered entirely over with earth. Stones were laid upon the roof to prevent its being torn off by the wind. The observatory was divided into two rooms. The first, more northerly, was fitted up for developing; the inner, more southerly, was the register-room.

On the ground-plan are the following:

J_1, J_2, J_3, J_4, indicate respectively the north, east, south and west walls. The door, δ, opens into the front room, where B is the bench upon the west wall. Upon this were kept various chemicals, and implements for the keeping in order of the instruments. The arrow, V^1, shows the direction of the ventilating air. In the north outside wall some holes were bored, through which the air was admitted under the bench in the front room, where the snow, etc., that accompanied it was separated from it, and the air could pass through the holes in the partition-wall, S, in a pure condition, free from snow. The snow that blew in could easily be taken away from under the bench.

Fig. 20. Observation-huts at Axeløen Station.

The door, δ^1, led into the register-room. Here were built two solid cement pillars upon the firm rock. They were of the form shown in the ground-plan at P and P^1. Upon them were placed the registering apparatuses, which consisted of 2 photographic registering apparatuses, R and R^1, with their benzine reflectors, L and L^1. The 3 magnetometers (variometers), D, H, and V, are respectively the declination variometer, the horizontal intensity variometer, and the vertical intensity variometer.

At the side of the plan, upon a line which indicates the magnetic meridian, three magnets are drawn one third of their actual size. Their position is here shown in relation to the meridian.

t is a thermograph which registers the temperature in the observatory.

The ventilating air, which, as already said, entered the register-room through the partition-wall, *S*, passed out through the draught-pipe, *V*, into the open air. As this ventilation was only for the purpose of eliminating the moisture produced by the benzine lamps, and to provide fresh air for the latter, it did not need to be particularly strong. Too strong ventilation is injurious, as with a change in the weather it may occasion a deposit of hoar frost.

The Observatory for Absolute Magnetic Determinations.

The house was a frame-house, and, like the register-observatory, was roofed with tarred paper. The foundations were dug down to the solid rock, and the walls shored up with earth and stones. As will be seen from the sketch, there is, a door to the north, and a window in each of the other three walls. There is only a single large pillar cemented on to the rock; but this was so large that the instruments kept their place unchanged all the year through. Their places can best be seen in the sketch (Fig. 18). When one of the magnetic instruments was being used for observation, the magnets were removed from the others, and were then kept in their cases in an empty barrel a little to the north of the observatory. The theodolite was also removed, if it was not down at the dwelling-house at the time. The south window was so arranged that one or more of the four panes could be taken out when observations were being made with the theodolite or the declinator.

Fig. 21. Hut for Absolute Magnetic Measures, and the Coaster "Jasai".

The Dwelling-house.

This consisted of two rooms. Of these, the south one served as a living-room and office. It had a door leading to the north room, which not only did duty as an anteroom, but also as a workshop and storehouse for various things. The north room had two exits, one to the east and one to the west. The house was built of stone, with wood pannelling inside (frame-work). Between the frame and the stone wall there was a close internal layer of birch-bark, and externally a 6-inch layer of moss. On the roof also there was first a layer of birch-bark, then moss, and on the top of that a layer of gravel; and finally, the whole was roofed with slates. In this way, the house was both substantial and warm.

The Storehouse.

This was a little square house with door on the north side. It served as the storeroom for the most necessary of our things, such as food, ammunition, etc., so that, in case of fire, we should not be left without the necessaries of life. Outside the north wall stood *the thermometer-screen, (t)* It was divided into two compartments, one for the thermometers and one for the registering apparatuses. It was also arranged so that the draught of air could be reduced to a minimum. The air was admitted through holes in the bottom. The draught was reduced when it was snowing, in order to hinder the snow from blowing in and filling the screen.

The electroscope-hut, (*e*) (Fig. 18) was a kind of cupboard on the west wall of the storehouse. In this cupboard, which was ventilated while the observations were being made, observations could be made in almost all kinds of weather. Observations of the electric conductivity of the air were taken three times a day, together with the meteorological observations. If time permitted, observations were moreover made every quarter of an hour during rapid registering.

As the observations were made, in the hut *(e)* and were thus not exposed to the full force of the wind, it should be remarked that the observations cannot be directly compared with observations made in other places in the open air. In this case, however, this was of minor importance, as the main object was to obtain the variations in the local electric conditions. Had the observations been made in the open air, only a small number would have been successful. As it was, it was only in the worst weather that the observations had to be suspended.

The arrangements of the other things is best seen in the detail-map. The only remark to be made in conclusion is that the auroral observations were made from a board that was nailed to the bottom of an empty barrel, which was placed between the dwelling-house and the register-observatory.

9. A few adventures and occurrences of the expedition are related here.

Captain Hagerup, accompanied by the members of our expedition, left Tromsø on the 24th July; but as the wind was unfavorable, they did not get to sea until the 27th.

On the 2nd August, Bell Sound was sighted, but also, at the same time, the ice, which appeared to form an impenetrable barrier. On the 7th, it looked as if the ice had become slacker, and at last there was room for the ship to advance a little, though not sufficiently to allow of her getting in to the Axel Islands. She was therefore compelled to seek a haven on the west side of the main island about 800 metres from the winter haven.

Here they remained, passing the time in hunting. On the night of the 12th, an open line was seen in the ice between the islands. A whaling-boat was immediately lowered, and filled with building materials. Two boat-loads were taken ashore. On the way back at 4 in the morning, they only just managed to get the boat back. All hands, except two, were then on shore and worked the two following days. The observatory for the registering apparatuses was set up on a rocky knoll, small enough for the house to surround it, and thus have a splendid foundation.

This house was soon put to the proof, for on the 17th there blew such a hurricane, that it was impossible to stand on deck. No attempt to go ashore could be made. The magnetic register-observatory was then finished except for the stones and earth along the walls. It was blown down and broken to splinters. The heavy boarding of which the house was built was torn from the framework, and some of it flung to a distance of more than 100 metres.

On the 18th the wind had gone down, and it was possible to venture ashore. The work of restoring the ruined house was started, and at 11 p. m. it was quite completed and literally loaded with stones, both on the roof and along the walls. The sleepers, moreover, were cemented to the rock.

The ice had now drifted away, so that the ship could be taken into a safe harbour. On the 19th, the instruments were brought ashore, and on the 20th the installation of the magnetic apparatuses was begun, and was completed without any accident.

The instruments were considerably out of order, but everything was capable of being put right. The balance for the determination of variations in the vertical intensity occasioned some trouble, but that too was set right. On the 29th, the registering was begun regularly, slight changes being made subsequently; and the work at this comparatively poorly equipped station was executed to my entire satisfaction.

It may serve to give some idea of the peculiar difficulties with which the expedition to Spitsbergen had to contend, if we begin by describing a stormy period such as there were a score of during the time the expedition lasted, most of them in the winter.

It must be in a great measure to the tremendously varying conditions of weather, that the immense loss of life on West Spitsbergen is due. It is no exaggeration to say that all round about our station was one great graveyard.

It is for this reason that of late no one has ventured to winter in Spitsbergen; it is only during the last three or four years that it has been done once more, for the polar bear hunting.

It was fine during the first few days of January. The sky was clear, and the temperature was more or less steady at — 30° C. But then the temperature began to rise, and the weather became unsettled, with short stormy periods, all the rest of the month. On the night of the 13th January, the temperature was — 34° C. with a hazy atmosphere. On the morning of the 14th it had risen, however, to —°19, and on the evening of the 15th, 0·6° was recorded. The wind was fresh but tolerably steady from the SSW. The precipitation was in the form of a rapidly varying mixture of snow, rain and soft hail. In the night, however, the temperature fell to — 7° again, and snow was continuous. It may be mentioned that on the Axel Islands it can quite well pour with rain with a temperature 5° or 6° below zero. The wind changed, however, in the course of the 16th, through the west to north, while the temperature slowly sank, and at midday on the 17th, we had quite a soft east-north-east wind with a temperature of — 15·4°. Good weather had been expected again; but the black, threatening atmosphere that rolls in from the sea (the Gulf Stream) in the west, when a storm is brewing, hung over us, heavy and unchanged.

Fig. 22. Celebrating a National Festival.

The temperature began to rise again, and we had five or six hours' storm from the east on the night of the 17th. In the morning — 9·5° C was recorded, and by midday the temperature was about 0° again, together with a south-west wind with rain, snow and sleet.

During the 18th, 19th and 20th, the temperature sank again slowly, while the wind kept in the south. The sky was an inky black, and it snowed and rained now and again. In the evening of the 19th, it rained with a temperature of — 4·8° C. By the evening of the 20th, it had sunk to — 14·5°, and the atmosphere was a little lighter than it had been for a long time, so that the hope of fine weather this time was well-founded, as the wind also had gone over to NNE again. On the morning of the 21st, however, the temperature was up to — 9°, and later in the morning the wind was due south with a very variable temperature with an average of 0·4°. That night there began a regular Spitsbergen storm in all its wildness and greatness. We were awakened by the roar and noise occasioned by wind, ice and rain. In the morning the storm reached its height. There was an average temperature of 2° C. The wind was from the south, but its velocity varied incessantly; at one moment there was none, or a slight breeze, the next it was blowing the wildest hurricane. It was these fearful gusts of wind, which often occur in the stormy periods, that were dangerous to any one going out, for it is impossible to keep one's balance in such a wind. During a storm of this kind, every condition varies by fits and starts — wind, temperature and precipitation. You hear boom after boom, now in the distance and now so close that you are in the very middle of it, and hear a roar as of a torrent around you; and gravel, stones and snow are whirled about. The gusts often last only a few seconds. You can hear them coming and then dying away in the distance. This may sometimes be followed by a heavy deluge of rain, but the rain may also come during a lull. The sky is no longer an even black, but dark clouds of every possible form are being driven along.

On the 5th February there began the most violent snowstorm that we had during our stay there, and it lasted almost uninterruptedly until the 9th. While it was going on, it was exceedingly difficult to carry out the meteorological observations. The thermometer-screen stood only four or five metres from the door, but on one occasion five vain attempts were made to get a reading of the thermometers.

It was especially during the dark season, which lasted about four months, that the storms raged worst; but October too was a bad month. The calmest and most beautiful time was July, August, and part of September.

It will be easily understood that weather such as this placed enormous difficulties in the way of the observations. It was, for instance, impossible, with the few means at our disposal, to prevent even great changes in temperature and humidity occurring in the magnetic register-room. The warm, damp air found its way into the observatory through the ventilators, and precipitated its moisture upon the instruments, dimmed the glasses, etc. Even the bases for the instruments, which were built into the rock, were not altogether beyond the possibility of change.

THE AURORAL STATION IN NOVAJA SEMLJA.

10. The station was situated on Matotchkin Schar, on the western side of the island, in a bay in the strait. The latitude of the place is 73° 16′ 38″ N, and its longitude 53° 57·1′ E. No map was made, but the accompanying sketch will make the conditions intelligible.

The members of the expedition were H. Riddervold, science graduate (chief), and H. Schaanning and J. Koren as assistants.

EQUIPMENT.

Magnetic Instruments.

For magnetic measurements we had a set of terrestrial-magnetic registering apparatuses of the Eschenhagen pattern, made by Otto Toepfer, Potsdam. For the absolute measurements of the magnetic elements, a unifilar magnetometer of the Kew pattern, made by Eliott Brothers, and a Dover's inclinatorium.

Meteorological Apparatuses.

For meteorological uses there were a mercurial barometer, 6 thermometers 1/5° C., 2 sling-thermometers, a thermometer-screen, 4 minimum thermometers, a cloud-measuring apparatus, and an anemometer Richard, besides forms, etc.

Electrical Apparatuses.

For electric measurements (atmospheric electricity) we had an Elster & Geitel's electroscope with a Zamboni battery and other accessories.

Astronomical Instruments.

For astronomical uses we had a theodolite and two box-chronometers, a Poulsen No. 5 and a Kessel No. 1280.

There were also some tools, guns and ammunition, and the necessary provisions.

On the 14th August, our instruments, baggage, coal and wood were landed and brought to the station. The instruments had suffered little on the whole, and could be set up without much difficulty.

BUILDINGS.

Two observation-houses we had brought with us were erected, namely:

The magnetic register-observatory, and

The observatory for absolute measures.

The Magnetic Register-Observatory.

The observatory, as the sketch and accompanying plan shows (Fig. 24), was erected to the SSW of the dwelling-house. There was no rock foundation there, so the house could be sunk some way into the earth. As the plan shows, the observatory is quite a plain wooden house, divided into two rooms, both dark rooms. The front, more southerly one merely forms the necessary anteroom to the inner, north room which is the register-room.

The following is an explanation of the plan:

δ^{I} and δ^{II} are the two doors by which the register-room is entered. To the right of the entrance is the vertical intensity variometer, V, then the declination variometer, D, and finally the variometer for horizontal intensity, H. These instruments are placed upon a wooden board, T, which rests upon two solid wooden posts, P and P^{1}, which are sunk far down into the earth and surrounded with stones. Farthest in is the register, R, with the reflector, L. It stood on the ground, upon the long legs belonging to the instrument.

The drawing beside the above is a diagrammatic representation — scale two fifths — of the position of the magnets during the registering. The arrow through it gives the magnetic meridian. The letters on the magnets give the direction in which the poles pointed. A wind-rose is drawn round the declination variometer.

Fig. 23. Our Station at Matotchkin Schar.

The Observatory for Absolute Measures.

This was a house exactly similar to that erected in Spitsbergen. Instead of the cement pillar, however, there was a solid wooden post about 35 centimetres in diameter in the middle of the house, properly sunk into the earth and surrounded with stones.

11. The other buildings shown in the sketch were already there, and were placed at our disposal with great willingness by the Russian government. The dwelling-house, which had been built for the Russian painter, Borisoff, was a good, substantial house, fully furnished and in good condition. The Russian authorities were most kind in the assistance they gave to our expedition. The Governor, Rimski Korsakoff, showed us his good-will in many ways. We were even carried free of charge from Archangel to Matotchkin Schar and back, with all our baggage; and the steamer "Wladimir" had instructions to land all our cases at Borisoff's house. We further received permission to make use, if necessary, of the depot that is intended for shipwrecked sailors who may come ashore there. There was also a thermometer-hut and a weather-vane there already; all we had to do was to put in the thermometer-screen, and to put the whole thing into a state of efficiency.

The electroscope was not observed regularly, and when it was, it was done in the open without protection. The Zamboni battery got out of order during the time of observation.

In August and part of September, it was summer in Matotchkin Schar; but it was cold and inclement, and there was rarely more than 10 degrees of heat. It was almost always cloudy and damp, and the sun was seldom visible.

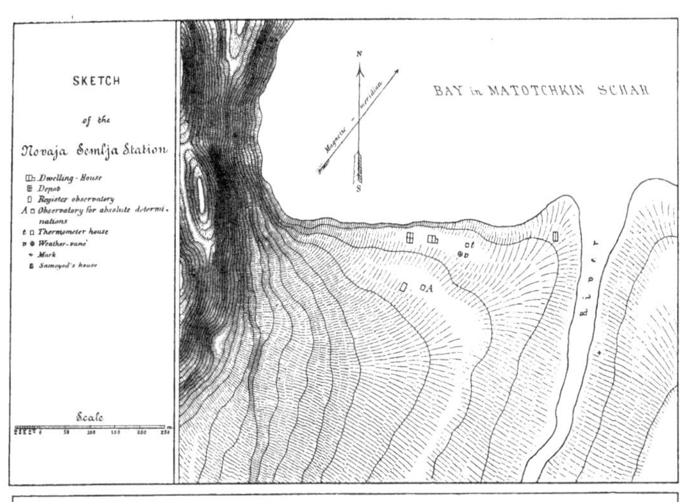
SKETCH
of the
Novaja Semlja Station
Dwelling-House
Depot
Register observatory
A Observatory for absolute determinations
t Thermometer house
v Weather-vane
+ Mark
Samoyed's house
Scale
BAY in MATOTCHKIN SCHAR
N
S
Magnetic meridian
River

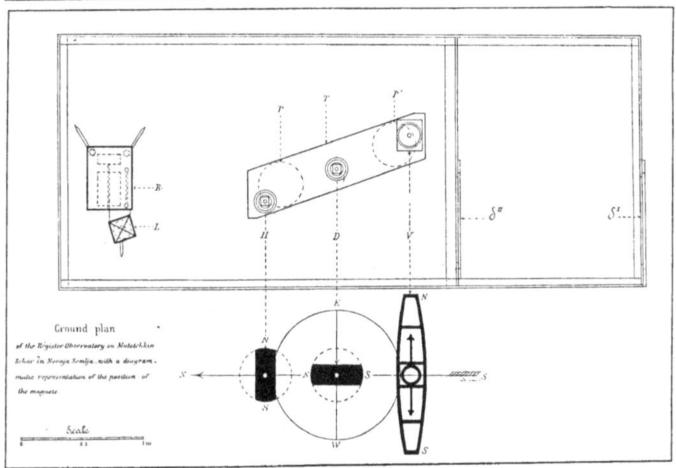
Ground plan
of the Register Observatory on Matotchkin Schar in Novaja Semlja with a diagrammatic representation of the position of the magnets
Scale
H
D
V
N
S
E
W

Fig. 24.

On the 28th September, the "Wladimir" came again, bringing Sæland to inspect the station. The vessel remained for three days, and it soon appeared that she had been none too early in getting away, as the winter came unusually early. About a week after her departure, ice covered the sea after a snow-storm and a week of cold weather had cooled the water.

The first part of the winter was severe. As early as November, the thermometer showed as a rule between 20 and 30 degrees of frost. There was, however, comparatively more clear weather than at other times of the year. But it was the same here as in other places; calm weather and from 30 to 40 degrees of cold gave no inconvenience. It was worse, however, when there were about —20 degrees C. and a snow-storm, which might continue for a week or two at a time.

We had a great deal of aurora during the first part of the winter. It would begin with an arc low down in the north, which gradually moved upwards and increased in brightness, and at last often stood almost magnetic east and west through the zenith. There then sometimes developed several large arcs, with a flaming rosette in the zenith; now and then the entire northern heavens seemed like a sea of fire. Sometimes the reflection would be so bright, that every object upon the ground could be distinctly seen.

Fig. 25. Hut for Magnetic Observations.

As the winter advanced, the days became quickly shorter. From November, the sun was always below the horizon, and in the latter half of November, in December and January, we had to burn lamps all day long. At first there was no difficulty in doing without daylight, but as it continues, the constant darkness has a depressing effect.

The severest part of the winter was the month of January. We then had for long periods at a time from 30 to 40 degrees of frost. It is strange that even in this severe part of the winter, a wind from the south could send the thermometer up above freezing-point. The lowest temperature observed was —42° C.

Fig. 26. Samoyed and Team.

On the 22nd February, a very remarkable thing occurred. The barometer suddenly fell to the lowest level of the year. In the morning, when we looked out of the window, the whole mass of ice in the strait, which had been fast since November, and was very thick, was drifting westwards. Soon after we had open water everywhere. The wind, which otherwise is the most important cause of changed ice-conditions, had nothing to do with this freezing of the ice. At the beginning of March, the weather again became cold, the strait froze over once more, and the ice became fast as before.

In the latter half of February, the polar bear appeared. This animal, while at other seasons of the year remaining in the north of the Kara Sea, wanders farther afield in the latter half of the winter, Matotchkin Strait being one of its favorite haunts.

The first two bears were seen on the 18th February; they were jogging quietly along the west coast of the island. In a great, deep snow-drift they had dug themselves a big lair, which looked very nice. Large bear-paths led from it in several directions, which showed that the bears must have been living there for some time. The hunt now began, and two days later the two bears, a she-bear and a year-old young one, were brought down.

It was not long before there was a continuation of the bear-hunting. The very next day, three bears passed our door; we seized our rifles, and in another instant the three bears lay stretched upon the ground, each with a well-aimed bullet in its body. (See figure 23).

The bear-hunting also brought a welcome addition to our larder. Our supply of meat, which besides tinned things, for which one soon gets a distaste, consisted only of gulls and other sea-birds preserved from the autumn shooting, had now become very small.

The weather as regards February, March and part of April, may be most correctly described as one long storm, now and then broken by calm intervals. Now and then, too, the wind increased to a hurricane.

The first harbinger of spring came on the 12th May. On that day the first bird of passage arrived, the snow bunting; and after it came gradually the others — larks, swans, geese, etc.

Winter still held on obstinately for some time, and the snow in most places did not disappear until June or July. Through the greater part of June we had frost, with calm, foggy or cloudy weather. Not until July was there any summer warmth.

Fig. 27. The Observer as Hunter.

In the middle of July, after the conclusion of the observations, the members of the expedition met with a disagreeable adventure.

They had gone out with a rowing-boat several miles from home, and had landed on the farther side of a little river, which at that time could be waded without much difficulty. The boat was moored to the bank.

When they had been there a few days, quite unsuspecting of danger, a fearful storm broke; the lightning flashed and the thunder roared — a very rare occurrence in those regions. At the same time the east wind broke loose in earnest, with oppressive heat. The consequences were not long in being noted. When the storm had abated, evidences were visible of the effect of the heat and the wind in the melting of snow, for the river was changed into a foaming torrent. The entire tongue of land upon which the boat had lain, was washed away; and the boat was nowhere to be seen; it had drifted out to sea with the east wind.

The question now was, what was to be done? With no boat, and the river, which was many miles long and very broad, now impossible to wade. Of provisions there were none, and no matches. Fortunately the members had brought their guns farther inland, so they set out on a hunting-expedition and shot some birds, which were immediately skinned and eaten raw. The following day they attempted to go along the river, in the hope that its upper part might be more easily crossed; but after wading for 20 or 30 miles, the attempt was abandoned. They then went back to the sea, and tried for several days in every possible way to get across, but all in vain.

It was clear, however, that they must at all costs manage to get home. The fare was not first-class; it still consisted of the one dish — raw bird. With some old rope and some drift-wood they made a kind of raft, and also found some boards that could be used as oars. It was an exceedingly poor vessel; even when all three men rowed with all their might, it made only the slowest progress. They nevertheless put out from the shore; but when they got into the river-current, they were carried rapidly out to sea, and were soon several kilometres from the shore. They rowed with all their might in order to cross the current and get into the counter-current that was formed on the border between the current and the still water. The worst of it was that the raft began to go to pieces, so that one man had to hold it together with his hands and feet while the others rowed.

After a hard struggle they at length reached a little iceberg that was grounded, where they at any rate did not drift away from the shore. Once more they took to the oars, and were fortunate enough to get into the counter-current, which carried them shorewards, while at the same time a gentle sea-breeze also helped a little. The row in was therefore easier than they had ventured to hope, and at last they all reached land safe and sound.

But when they were safe on terra firma they saw how great the danger had really been; for a fog as dense as a wall came pouring down from the north. If this had come a little sooner while they were rowing, it is highly probable that they would have gone on rowing in a circle all the time while the stream would have driven them farther and farther out; and the result would then have been very doubtful. But now they were on familiar ground; they had only a few miles to go, and six hours after landing, they were all at home.

A week later, on the 21st July, at 2 in the morning, the "Wladimir" steamed into the haven, and the expedition broke up hastily, and on the 3rd August reached Archangel.

12. The Working-up of the Material. From the four Norwegian polar stations here described, a quantity of material was gathered in 1902 and 1903, which has been in process of working up for a long time; but, principally for financial reasons, the publication of the results has not been practicable until now.

For the gain to science which our auroral expedition has brought, we owe a debt of gratitude not only to those who guaranteed the undertaking financially, but also to others, especially the directing heads of a large number of magnetic and meteorological observatories all over the world.

Experience from earlier work in this field had clearly shown me that if light was to be thrown upon the phenomena that we had set ourselves to study, it would be of the greatest importance — necessity, I may say — to obtain simultaneous observations from most parts of the earth. This applies to a certain extent both to cloud-observations and to observations of aurora; but it is of special importance in the study of the magnetic storms, for they, as is generally known, are usually of a universal character.

With the object of getting, if possible, several observatories to co-operate in these researches, I sent out a circular, dated May, 1902, from Christiania, before the departure of the expedition, to a number of observatories all over the world.

I will here confine myself to giving a brief extract from this circular[1].

"As leader of the expedition started by the Norwegian Government for the study of Earth-Magnetism, Polar Aurora and Cirrus clouds, I beg to inform you that during the time from August 1st 1902 until June 30th 1903, four Norwegian Stations will be erected, viz. at Bossekop (Finmarken), at Dyrafjord (Iceland), at Axel Island (Spitzbergen) and Matotchkin-Schar (Novaja Zemlja)."

"The above-mentioned expedition has assumed the task of determining the connection existing between earth-magnetical perturbations, boreal lights and cirrus-clouds."

"To obtain a happy solution of this task, it is absolutely necessary to get the requisite facts from the largest number of points of observation distributed as widely as possible over the whole earth.»"

---

(1) Terr. Magn. and Atm. Electr. June, 1902, pp. 81.

"At these four Norwegian stations the three magnetic elements will be registered photographically. To this effect registering instruments will be employed like those used by the contemporary Antarctic Expeditions. The three elements will also be determined absolutely. The term observations stipulated by the Antarctic Expedition will also be carried through at these stations.

The special subject prosecuted by our expedition, and for the fulfillment of which we solicit your kind support are:

The determination of the cause and progress of different magnetic perturbations, as discussed by me in my report: *"Expédition Norvégienne de 1899—1900 pour l'étude des aurores boreales. Résultats des recherches magnétiques"*.

Some information was added as to the best way of making the observations for the purpose desired; and the times for the rapid registerings were fixed.

As it would have been impossible to have material sent from all the observatories for the whole of this period, it was necessary to confine ourselves to a few fixed days. As soon as we had observations from two of our stations, I sent out a new circular from Christiania, to the same observatories, dated June, 1903(¹). Part of this was as follows

"After comparing photograms from Bossecop with corrresponding ones from Potsdam, I selected thirty days, on which general magnetic disturbance was great, as those which most suited my purpose and I have, consequently, determined to adopt these as the basis of my investigations. I now take the liberty of asking all those who are in the position to do so, to give or lend me copies — photographic preferred — of photograms of magnetic disturbances that may have occurred on those thirty days, and urge them, in the interest of science, not to mind facing the considerable amount of trouble which must be undertaken in order to fill such a request; and, if required, I am willing to refund any expence necessarily incurred in connection with it. In the work that I intend to publisch, I shall reproduce so far I can by photography, a very large number of such photograms after they have been reduced to a uniform scale as regards time, so that any one may be able to check the results arrived at, by me, from my manipulation of the materials to hand. The variations of most value for my work, are those of the two horizontal elements. In respect to the thirty days in question, when the vertical intensity shows marked variations, it will be, likewise, very important to me to obtain copies of photograms relating to vertical intensity."

We have in this way, in response to our request, received numerous photographic reproductions of magnetograms and tables of magnetic observations for comparison with simultaneous observations from our 4 stations, from each of the following 23 observatories: Honolulu, Sitka, Baldwin, Toronto, Cheltenham, San Fernando, Stonyhurst, Kew, Val Joyeux, Uccle, Wilhelmshaven, Munich, Potsdam, Pola, Pawlowsk, Tiflis, Jekaterinburg, Bombay, Dehra Dun, Irkutsk, Batavia, Zi-ka-wei, Christchurch.

We have further received observations of occurrences of cirrus bands — these being made, while the expedition lasted according to a common plan — from the meteorological observatories at Valencia (Ireland), Falmouth, Aberdeen, Kew, Aix-la-Chapelle, Von der Heydt-Grube (b. Saarbrücken), Bremen, Uslad, Celle, Brocken, Christiania, Potsdam, Grünberg, Schneekappe, Neustettin, Budapest, Königsberg.

For this extreme readiness on the part of my honoured confrères to give their assistance, I would here offer them my warmest thanks.

It is my hope that the importance of this material to our work will be fully apparent from the subsequent treatment of the subject.

To one man more particularly, if he had lived, this expression of gratitude would have been addressed, namely the late Geheimrath von Bezold. It was especially through his valuable aid that I succeeded in obtaining such ready response from observatories all over the world as I finally did.

(¹) Terr. Magn. and Atm. Electr. June, 1903, pp. 74.

PART I.

MAGNETIC STORMS, 1902—1903.

INVESTIGATIONS BY MEANS OF DIURNAL REGISTERINGS
FROM 25 OBSERVATORIES.

CHAPTER I.

PRELIMINARY REMARKS CONCERNING OUR MAGNETIC RESEARCHES.

13. Our Aim and our Method of Working. It has, as is generally known, been ascertained that there exists a close connection between sunspots and the magnetic conditions upon the earth. As early as 1852, SABINE discovered, almost simultaneously with GAUTIER and WOLF, that in years when sun-spots were numerous, the magnetic storms were more frequent and more violent than in years when there were few sun-spots. By comparison with the period of magnetic oscillations pointed out by LAMONT in 1850, it was discovered that maxima and minima in the magnetic period coincided with maxima and minima in the sun-spot period.

These and kindred circumstances have since been carefully investigated. It has been found that the magnetic constants have secular variations, which, with convincing exactitude, follow the simultaneous variations in the occurrence of sun-spots; and further, that there are periods for the frequency of magnetic storms and for aurora, which correspond with the so-called undecennial period of the sun-spots.

From the very first, when these relations were discovered, attempts were naturally made to find out the connecting mechanism between these phenomena, so that the physical cause might become clear; but these have not as yet been entirely successful.

It has gradually come to be acknowledged that aurora and magnetic perturbations should be regarded as rather moderate manifestations — at present the only ones there are for us to observe — of an unknown cosmic agent of solar origin, and quite different from light, heat or gravitation. It has long been supposed that this unknown agent was in some way or other of an electrical nature. The elder BECQUEREL even, gave expression to some very interesting ideas on this subject.

With regard to the magnetic storms in particular, it is clear that the observed changes in force can be formally explained by an infinity of assumptions with distribution of fitting agents that generate magnetic forces; but nevertheless it may safely be said that up to the present not one definitely formulated hypothesis has been put forward, which explains all the phenomena so simply and naturally, that the hypothesis becomes satisfactory.

In the following pages it will be shown how far I have succeeded in explaining the above-mentioned and several kindred relations, starting with the assumption which, viewed from the present standpoint of natural philosophy, is a legitimate one, namely, that the sun, and especially the spots on the sun, send out into space cathode or kindred rays.

In order to gain definite conceptions of the effect of such rays in the vicinity of the earth, I have again and again had recourse to analogisms from my previously-described experiment in which a magnetic terrella is suspended in a large discharge-tube[1], and exposed to cathode rays.

The experiment, which was originally made for the purpose of finding points of support for a hypothesis for the formation of aurora, has proved a veritable mine of wealth, in which I have constantly made valuable discoveries.

---

[1] Expédition Norvégienne de 1889—1900, etc., l. c., pp. 39 et seq.

The experiment in various forms has been repeated a great many times in the course of the last few years, and I have succeeded in photographing all the light-phenomena appearing. The results of these experiments will be fully described in a later part, and the light-phenomena illustrated by numerous photographs.

There are light-phenomena produced by the rays that beat directly down upon the terrella, and which, in my opinion, answer most nearly to the light-phenomena and certain magnetic storms in the auroral zone on the earth.

There are light-phenomena produced by rays made to fall upon a movable screen, for the purpose of ascertaining how those rays behave that do not fall directly upon the terrella, but move about in its immediate neighbourhood. I think that such rays can give a natural explanation of the cause of certain universal magnetic disturbances and sometimes to aurora polaris, if the ray-stream comes near enough to the atmosphere of the earth.

Finally, there is a flat, detached bright ring round the magnetic equator of the terrella, which immediately recalls Saturn's ring.

It seems as if this bright ring might bring us almost to the solution of other most important terrestrial magnetic problems.

In a lecture *"On the Cause of Magnetic Storms, and the Origin of Terrestrial Magnetism"*, given before the Scientific Society in Christiania, on the 25th January, 1907, I gave a sketch of the results of the terrestrial-magnetic investigations which will be produced in the present work.

The conformity discovered by Sabine and others between sun-spots and magnetic perturbations, as also aurora, has become apparent through observation and the summing up of a large number of single phenomena. It must necessarily be supposed from this conformity, that also in single cases it must be possible to prove a connection between these phenomena. This has often, especially in more recent times, been observed in particularly marked cases.

It will therefore be an important task to endeavour to discover the course of the process which at times takes place in the neighbourhood of the sun-spots, and gives rise subsequently to aurora and magnetic perturbations, and thus show that these terrestrial and solar phenomena are only different phases in a continuous process.

In order to solve this problem, one is naturally led to take one of two ways. The most rational, if the necessary material were forthcoming, would be to start from the sun, where the process begins. This is the way I have formerly taken. Starting with the hypothesis that the sun-spots are the source for the emission of cathode rays, I have endeavoured to follow the process from the sun to the earth, and by analogy with the above-mentioned experiment see how some of the rays strike the earth, and some glance past it under the influence of terrestrial magnetism. This is moreover the way my friend, Professor STORMER, has taken in his mathematical investigations of the path of such rays from the sun to the earth. He has published the complete results of his investigations in a special part of the present work; but these results will already be to some extent known from his earlier papers. Here, for the first time, a detailed mathematical treatment of the aurora problem and kindred problems will be found.

The other way is to start with the conditions upon the earth, study a single perturbation, seek for the terrestrial processes that might be able to influence them directly, and follow these up until, if possible, we are stopped at the point when the cause can no longer be sought upon the earth, but in the arrival of something from without; and here the two ways may meet.

It is by going both ways, employing both methods, that we have thought we might have the best prospect of solving our problem.

That which, at a certain spot on the earth, and at a given moment, characterises a magnetic perturbation, is the strength and direction of the perturbing force.

In order, therefore, to obtain a clear conception of the perturbation, such as it actually appears on the earth, there are in particular two important points upon which enlightenment is to be sought, namely,

(1) How is the force distributed upon the earth at a definite point of time during the perturbation?

(2) How does the distribution of force change with time?

The investigation of these two points has formed one of our principal tasks.

Our investigations were thus in the first place directed towards finding out how an individual perturbation developes, and what course it takes. We find that for the solution of this problem it has been particularly important to study with special exactitude the *simplest* phenomena, those in which the course is simple and with no great, sudden changes, as at the outset it seems probable that we are here face to face with elementary phenomena, which together may form the multiplicity of magnetic storms.

As, however, there will, as a rule — notwithstanding the many great similarities — always be many individual peculiarities in each perturbation, which should be specially mentioned, we have decided to treat each perturbation separately, each accompanied by a description. We have, however, tried to arrange them together in groups according to their special character, in such a way that the various elementary types come first, after which the more compound perturbations will be treated.

There may also be a question of finding average characteristics of a large number of perturbations at one particular place on the earth. It appears, however, that there are several kinds of perturbations, and in order to pick out the average characteristics, it is necessary to keep to one particular kind. Moreover, the course of the perturbations in one place will be greatly dependent upon the time of day. It will thus also be necessary, starting from this point of view, first to proceed to a close investigation of the distribution and course of the perturbations.

In the treatment of the separate perturbations, we have, in accordance with the above remarks, employed the following mode of procedure.

The horizontal and vertical components of the perturbing force are calculated for all the observatories for a series of points of time within the period in which the perturbation appears, and the result is given in tables.

In order to obtain a clear idea of the distribution of force, we have employed a synoptic representation on charts. The direction of the horizontal component of the perturbing force, which was originally determined in relation to the magnetic meridian, is fixed in relation to the astronomical, by the aid of declination.

Now it might seem reasonable to pick out the perturbing forces themselves, and place them, with their particular direction and magnitude, on the charts. We have, however, instead of the perturbing forces themselves, to mark so-called "current-arrows". These would give the direction of the horizontal current that would produce, above the place, a magnetic force in the direction of the perturbing force. The size of the current-arrows is proportional in every case to the perturbing force, and gives the force in magnetic units.

This mode of representation is specially chosen out of regard to the Norwegian stations; for there, during a whole series of the greatest polar perturbations, the force will undoubtedly be produced by currents that flow almost horizontally; and the current-arrow then nearly gives the direction of the horizontal current. We have, moreover, other groups of perturbations, e. g. those which we have called equatorial perturbations and cyclo-median perturbations, which are also best represented by current arrows.

This mode of marking also presents advantages with regard to the geometrical representation of the vertical component of the perturbing force.

It must not, however, be assumed that the current-arrow indicates that a current is actually flowing in the direction stated, all over the place. The perturbing force may, in the first place, be generated by several simultaneously operating current-systems; it may moreover be the effect of far distant systems that are not even always horizontal. The current-arrow is simply and solely a geometrical representation of the perturbing force.

With regard to the number of charts that should be worked out for each perturbation, it will be a matter of opinion how many should be taken in each case. We have, however, throughout made it a rule that for perturbations in which the perturbing force undergoes slow changes, the time between each chart shall be longer than for those in which the perturbing force oscillates.

By a comparison of the charts, a clear idea of the development of the perturbation will be obtained.

In this way we can, however, represent the perturbation only for certain separate points. In order to obtain a representation of the connected course of the perturbation, a plate will be drawn of each perturbation, reproducing on a somewhat reduced scale the actual registered curves for variations in H, D, and V.

These copies of curves from all the observatories will be found all together, arranged according to date, in front of the special treatment of the separate perturbations.

To ensure the best possible result being obtained from this method, material should be collected from a large number of stations distributed over all parts of the world. The best material for the purpose would include registerings of all three magnetic elements from a ring of stations round both poles of the earth, and a number of other stations more or less evenly distributed over the rest of the world — as many as possible.

We have no such material at our disposal. Our simultaneous observations of 1902 and 1903 are all, with the exception of the registerings from Batavia and Christchurch, New Zealand, confined to the northern hemisphere. In the arctic regions, moreover, we have observations only from our own four stations; and although we think that these four stations were admirably situated for their object, yet the material has not proved quite sufficient for a comprehension and elucidation of the perturbation-conditions in the regions around the so-called auroral zone.

In order to throw more light upon these conditions in the auroral zone itself, we have made a special investigation of the conditions in these regions, and for this purpose have made use of the material from the polar year, 1882—83.

Our study of the universal character of the magnetic perturbations thus divides into two sections.

The first section comprises the working-up of the material from 1902 and 1903. In the course of this, an attempt is made, by the employment of the previously-cited method, to throw light both upon the conditions in lower latitudes, and upon the possible connection of these conditions with the storms occurring at the same time at our four stations near the auroral zone.

The second section comprises an investigation by the same method, which is more especially directed to the conditions in the arctic regions in and about the auroral zone. We have moreover, for the sake of completeness, and in order to be better able to compare the results of these two sections, also included in our investigations of the polar observations from 1882—83, observations from a few stations that have a more southerly situation, namely, Christiania, Göttingen and Pawlowsk.

14. On the Calculation of the Perturbing Force. For the calculation of the perturbing force, there are registerings of the variations in horizontal intensity and declination, and for some stations in vertical intensity also. When there are only the first two, only the horizontal component of the perturbing force can be determined.

When no perturbations occur, the curves will have an even course, having only a slight bend owing to the daily variation. If the curve has a marked divergence from this line, which must be ascribed to the alteration in the magnetic constants, we then have a perturbation.

It need hardly be said that instances will necessarily occur in which it will be difficult to decide whether the curve is normal or not. No exact definition of a perturbation can therefore be given; but we shall always try to keep to cases in which there is no doubt about the matter.

We will call the magnetic force that is actually found at a given moment, F_t, and the force we should have had at the time, without perturbation, F_n.

The perturbing force P is the force which, together with F_n, has F_t as its resultant.

We resolve all the forces along 3 axes at right angles to one another — one vertical, one along the magnetic meridian, and one perpendicular to these, and we designate

the components of F_t as F_{th}, F_{td}, F_{tv}
» — » F_n » F_{nh}, F_{nd}, F_{nv}
» — » P » P_h, P_d, P_v.

We thus obtain

$$\left.\begin{aligned} P_h &= F_{th} - F_{nh} = H_t - H_n \\ P_d &= F_{td} - F_{nd} = F_{td} \\ P_v &= F_{tv} - F_{nv} = V_t - V_n, \end{aligned}\right\} \quad (1)$$

introducing the customary denotations for the horizontal and vertical components of terrestrial magnetism.

We will call the horizontal component of the perturbing force P_1, and we have

$$P_1 = \sqrt{P_h^2 + P_d^2} \text{ and}$$
$$P = \sqrt{P_1^2 + P_v^2}.$$

It appears from equations (1), that it is only necessary to know the difference between the components of F_t and F_n, and not their absolute value; and this difference is found by the curves, a "normal line" being drawn upon the magnetogram, which gives the course of the curve, if no perturbation has taken place.

If we denote the ordinate from the base-line to the curve and to the normal line at a given moment, as O_t, and O_n, and if a deviation of one length-unit on the magnetogram answers to a magnetic force ε, then

$$\begin{aligned} P_h &= \varepsilon_h\ (O_{th} - O_{nh}) = \varepsilon_h\, l_h \\ P_d &= \varepsilon_d\ (O_{td} - O_{nd}) = \varepsilon_d\, l_d \\ P_v &= \varepsilon_v\ (O_{tv} - O_{nv}) = \varepsilon_v\, l_v, \end{aligned}$$

the differences of the ordinate being denoted by l_h, l_d and l_v,

According to our definition-equations (1), we shall have P_h and P_v becoming positive in the same direction as the corresponding total forces. H is positive towards the north, and V is assumed to be positive downwards. We hereby obtain the following rule for the sign of ε_h and ε_v.

ε_h *is positive* when increasing ordinate corresponds to increasing horizontal intensity.

For ε_v we obtain

(1) In the northern hemisphere,

ε_v *positive*, when increasing ordinate corresponds to *increasing* numerical value of V.

(2) In the southern hemisphere.

ε_v *positive*, when increasing ordinate corresponds to *decreasing* numerical value of V.

With regard to ε_d it should be noted that in general it is not directly given. On the other hand, the number of minutes of arc, β, that the declination is altered by oscillations of one length-unit is given.

A simple mechanical reflection then shows that

$$\varepsilon_d = \frac{\pi}{180 \cdot 60} \beta H_t = \omega_d H_t.$$

If we resolve to reckon P_d positive towards the west, we obtain the following rule for the sign of ε_d. ε_d is positive when increasing ordinate corresponds to *increasing* westerly, or *decreasing* easterly, declination.

In taking out the ordinate-differences, a purely graphic method has been adopted, the normal line being drawn upon the magnetogram itself, and the ordinate-differences taken out directly by measurement.

One thing which here often causes some difficulty, is the placing of the normal line. It may sometimes happen, especially when the perturbation is of long duration, that doubt may arise with regard to its situation, and in this way a corresponding fault may arise in the determination of the perturbing force.

In a series of perturbations, however, this doubtful territory is small, so that the position of the normal line is decided almost without question.

It will immediately be seen that the strong, brief perturbations, which appear somewhat suddenly on an otherwise calm day, will be particularly favorable in this respect. Here the normal line will be a line that connects the calm districts before and after, in such a manner that its further course is ruled by the curve on the nearest calm days. Perturbations such as these, in which the situation of the normal line can be easily fixed, will be indicated as well-defined perturbations. The study of these short, well-defined perturbations will also, as already remarked, be advantageous for the reason that we are here possibly face to face with elementary phenomena, which together may form the multiplicity of the perturbations.

If the perturbation is of long duration, if it extends over the whole magnetogram, which generally represents 24 hours, there will very likely be some uncertainty. If, for instance, there is a part of the curve that is normal, part of the normal line will thereby also be determined. Its absolute distance from the base-line will then be ascertained, and its further course over the perturbed region must be determined by the form of the curve on the nearest calm days. We must here notice whether, if the temperature has varied during the period under consideration, it has approximately varied in the same manner throughout the day; should this not be the case, we should have to find, by the aid of the temperature coefficient, the form for the neighbouring curves, that corresponds to the temperature on the day under consideration.

If there is no part of the magnetogram calm, the normal line must be determined, both as to its form and to its absolute distance from the base-line, by the aid of the curves on the nearest calm days. And here regard must be paid to differences in temperature. If we are to avoid corrections for temperature, it will not be sufficient that the temperature-curve has the same course during the two days; the temperature must also have the same absolute value at the same hour. As a rule, the temperature in the observatory will be fairly constant, so that in most cases by this method there will be no need of correction for temperature, unless it were actually to affect the sensibility.

As the curves from day to day in other respects — presupposing the same circumstances — do not repeat themselves altogether congruently, there is liable to be some arbitrariness in their situation. If therefore we are to be able to count upon obtaining values for the perturbing force with a reasonable error-percentage, these protracted perturbations must also be strong, if the calculation is to yield any return; and it will frequently happen that in such cases the direction and strength of the perturbing force cannot be greatly relied upon, when the magnitude of the force is small.

This is, in the main, what can in general be said with regard to the placing of the normal line. In certain cases special circumstances may arise which may make it necessary to take other things into consideration, our material being somewhat imperfect for these determinations, as we have only magnetograms for separate days from the foreign observatories, and these separate days are just some of the perturbed ones. Fortunately, in the case of several places, there are several curves upon one magnetogram, so that in this way the neighbouring curves accompany them, a circumstance which has been of great importance to us.

On the Plates in which the magnetograms are reproduced, the normal line that has been employed in the calculation is generally drawn.

13. On the Separation of Simultaneous Perturbations. The perturbing force calculated according to the above-mentioned method, will give us the resultant of all the perturbing forces that are present at the moment. Now it will often happen that we at any rate have one system of perturbations which is predominant, so that the total perturbing force gives us directly the effect of this system. But it may also frequently happen that at the same time we have to do with several perturbations, that, in other words, we have in the actual field the superposition of fields from several current-systems. It may then be important to find the effect of each separate one — in other words to decompose the total perturbing force into several partial forces, each of which is the effect of an independent current-system, or is at any rate due to relatively independent causes.

A decided rule for the permissibility of such a decomposition can in general scarcely be given. The reasons that favour the interpretation of the total perturbation as the resultant effect of several simultaneously acting systems, must be apparent from the single case in question.

We will here, however, draw particular attention to two circumstances, which will be of some importance.

(1) When the perturbing force during a protracted calm perturbation suddenly changes its direction and strength, only to return once more, after some time, to its original value, it will be natural to conclude that a change such as this is due to an independent system appearing at the same time.

If this sudden change in P for all places on the earth is only a change in strength, there will, on the other hand, be little reason for assuming the presence of an independent system.

(2) Another thing which may lead to the settlement of this question is the examination of those places on the earth in which the perturbing force is greatest.

If, during a perturbation that is strongest at one particular place on the earth, a sudden change takes place that is greatest at a spot situated at a great distance from the first-named place, this must of necessity be regarded as two separate phenomena that work into one another.

It will thus often happen that during a perturbation that is highly developed at the equator, there appears a change, which increases towards the north pole. Here then, we have undoubtedly to do with two different current-systems, one with its point of departure in the polar regions, and one equatorial current-system.

Frequently, however, the existence of independent systems may be recognised, although, with the material at our disposal, we may not have the means wherewith to discriminate their magnetic effect.

It will often be a matter of judgement, whether to undertake a decomposition of the total perturbing force or not.

It is very fortunate when a protracted perturbation is very quiet and uniform in direction, and the intermediate one is relatively strong and not of very long duration. In such a case, it would be natural to take out the effect of the intermediate storm by drawing a normal line that harmoniously connects the curves before and after it.

It should at once be remarked that it is the total force that can be calculated almost after an objective method. The components, or partial forces, as we will call them, will as a rule be less exact. A decomposition will nevertheless be of value in throwing light upon the development of the perturbations.

CALCULATION OF THE SCALE-VALUES FOR THE REGISTERINGS AT THE NORWEGIAN STATIONS.

16. A. Determination of the Scale-Values for the Declinometer. The declinometer consists principally of a magnet suspended by a quartz thread. Fixed to the magnet is a mirror. Light from a fixed source is reflected in the mirror, and is focussed by a lens into a spot of light upon the photographic paper. If the fibre had no torsion, the turning of the mirror would give directly the change of declination.

But the fibre has torsion, and its effect must be determined.

The effect of the torsion is found by twisting a certain angle α minutes of arc, and measuring the corresponding deviation on the paper.

The scale-value, or the angle in absolute measurement, which answers to a length-unit in deviation, is determined by the following formula:

$$\omega_d = \frac{1}{2\,r_d}(1 + \varkappa)$$

where $\varkappa$ is given by the equation

$$\varkappa = \frac{x}{2\,k\alpha r_d - x}$$

r_d is the distance from the mirror of the declinometer to the cylinder with the photographic paper.

$k\alpha$ is the angle in radians about which the twisting is done.

x is the deviation on paper, answering to the torsion.

When the angle in the equation for $\varkappa$ is measured in minutes, and x in millimetres, we can put for our apparatuses for the numerical calculation, approximately,

$$2\,kr_d = 1, \qquad \text{and}$$

$$\varkappa = \frac{|x|}{|\alpha| - |x|}.$$

For the calculation of the perturbing force perpendicular to the meridian, we obtain the following scale-value:

$$\varepsilon_d = \frac{1}{2\,r_d}(1 + \varkappa)\,H_t\,.$$

H_t is the horizontal component actually existing at the moment.

17. B. Determination of the Sensibility of the Variometers for the horizontal and vertical Intensity. The main principle here consists in seeking the deflection corresponding to a known magnetic force f.

If a deflection of n length-units on the photogram answers to f, then the scale-value is

$$\varepsilon_h = \frac{f}{n_h}$$

$$\varepsilon_v = \frac{f}{n_v}$$

f is to act as the deflecting force for the horizontal variometer along the line of direction of the horizontal component, for the variometer for vertical intensity, in a vertical direction.

f is determined in relation to H_θ, or the horizontal component of the magnetic force during the determination of sensibility. This is done by letting the deflecting magnet, as before, deflect the declination-needle. During the determination, care must be taken that the deflecting magnet in all three cases is at the same distance from the observation-magnet.

If the declination-needle undergoes a deflection answering to n_d length-units, we obtain

$$f = \frac{n_d}{2\, r_d}(1 + \varkappa)\, H_\theta$$

If this is inserted, we obtain, employing the equation for ω_d,

$$\varepsilon_h = \frac{n_d}{n_h} \cdot \omega_d\, H_\theta$$

$$\varepsilon_v = \frac{n_d}{n_v} \cdot \omega_d\, H_\theta$$

If we do not here demand greater exactness from ε_h and ε_v than 1 per cent. of the amount, we can in general, as long as the declinometer has the same thread and the same distance, consider ω_d as constant, $\varkappa$ being small in proportion to the unit. We can then generally, instead of H_θ, choose a mean value, H_0, of the horizontal component. This we can safely do here, as a determination of sensibility made during a great perturbation ought not to be employed.

We then obtain

$$\varepsilon_d = \omega_d\, H_t$$

$$\varepsilon_h = \frac{n_d}{n_h} \cdot \omega_d\, H_0$$

$$\varepsilon_v = \frac{n_d}{n_v} \cdot \omega_d\, H_0$$

For slighter perturbations, we can put, with the same accuracy as before,

$$H_t = H_0.$$

This assumption, which we can probably always make with more southerly stations, is not always permissible for our Norwegian stations in the treatment of perturbations; for at the latter the horizontal component of the magnetic force is very small, while at the same time the variations in it on account of the perturbation may go up to 500 γ or even more. We can now put

$$H_t = H_0 + \Delta H_t.$$

In general we have

$$\Delta H_t = R_h + P_h,$$

where R_h is the reduction from the mean value to the normal value for the point of time under consideration. P_h is the perturbing force in the direction of the magnetic meridian.

In the cases in which the equation will come to be employed, P_h is preponderant in relation to R_h, and if we put

$$\frac{P_h}{H_0} = \delta,$$

we obtain

$$\varepsilon_d = \omega_d \, H_0 \, (1 + \delta)$$

$$\varepsilon_h = \frac{n_d}{n_h} \, \omega_d \, H_0$$

$$\varepsilon_v = \frac{n_d}{n_v} \, \omega_d \, H_0$$

Our calculations have been made according to these formulæ.

In the determination of sensibility, the following mode of operation has been used in the main for all the four stations:

TABLE I.

1	2	3	4	5
Torsion	Declinometer	H. I. Variometer	V. I. Variometer	Declinometer
Equilibrium	Equilibrium	Equilibrium	Equilibrium	Equilibrium
$+\alpha^\circ$	Magn. W, North pole W	Magn. N, North pole N	Magn. Up, N. pole Up	Magn. E, N. pole E
$-\alpha^\circ$	„ „ „ „ E	„ „ „ „ S	„ „ „ „ Down	„ „ „ „ W
$+\alpha^\circ$	„ E, „ „ W	„ S, „ „ N	„ Down, „ „ Up	„ W, „ „ E
Equilibrium	„ „ „ „ E	„ „ „ „ S	„ „ „ „ Down	„ „ „ „ W
	Equilibrium	Equilibrium	Equilibrium	Equilibrium

DETERMINATIONS OF SENSIBILITY FOR KAAFJORD AND BOSSEKOP.

18. The apparatuses were in position at Kaafjord from the 19th August, 1902, up to, and including, the 13th March, 1903. During this time they underwent no changes of any importance.

From the 15th March, 1903, to the 2nd April following, the apparatuses were set up at Bossekop. During this time, considerable changes were made in them, a new thread having been put in on the 25th March, in the H. I. variometer, and 6 astatising magnets placed beneath the declinometer.

On the 29th March, another new thread was put in the H. I. variometer, and the astatising magnets were moved higher up. These alterations were made for the purpose of increasing the sensibility.

In the table below will be found the quantities that come into the formulæ, and the calculated scale-values, for the determinations of sensibility that were made, as also the date of the determinations, and the temperature at the beginning of each measurement.

As a unit for scale values we use $1\gamma = 10^{-5}$ abs. magn. units, referred to 1 mm. deviation on the magnetogram. See art. 14.

TABLE II.

Scale-values for Kaafjord.

$H_0 = 0.1248 \quad r_d = 1708$ mm. $\quad \varkappa = 0.00465 \quad \omega_d H_0 = 3.67$

Date	n_d	n_h	n_v	ε_h	ε_v	Temp.
Sept. 9, 1902 . . .	37.1 mm.	22.9 mm.	28.6 mm.	5.95	4.76	+ 9.5°
„ 26, „ . . .	36.8 „	23.0 „	28.9 „	5.87	4.68	+ 8.3°
Dec. 19, „ . . .	36.4 „	22.5 „	18.9 „	5.95	7.07	− 4.3°
Jan. 22, 1903 . . .	36.1 „	21.6 „	17.0 „	6.13	7.83	− 1.2°
March 13, „ . . .	36.2 „	21.7 „	(4.9?) „	6.12	(27.1 ?)	− 5.0°

The table shows that the scale-values for H and V are not constant; in the case of ε_v in particular there is a considerable increase, and in the determination of the 3rd March, 1903, the balance was almost immovable. This abnormal circumstance seems, however, to have been only of a temporary nature, as will be seen from the curves before and after. We have not employed any smoothing formula here for ε_h, but have found the scale-values by interpolating between two successive observations.

We have employed the following formula for ε_v:

$$\varepsilon_v = 4.76 + 0.0285\, t,$$

t indicating the number of days reckoned from the 1st October.

TABLE III.

Scale-values for Bossekop.

$r_d = 1740$ mm.

H_0' indicates the magnetic force, to which the declinometer-needle is actually subjected. During the first determination of sensibility it is only terrestrial magnetism that is acting.

The force acting on the declinometer-needle, during the 2nd observation, may be determined in two ways. We can either use the deflection by the torsion, having the same thread and the same position in both cases; or we can employ the deflection with the deflecting magnet, the magnet having been placed at the same distance on the deflection-rod in both cases. The two methods give about the same result. The value given is the mean value. For the period from the 25th to the 29th March, we have no scale-value, a determination of sensibility that was made on the 27th having been unsuccessful.

DETERMINATIONS OF SENSIBILITY FOR DYRAFJORD.

19. The registering at Dyrafjord was begun on the 25th November, 1902, and was continued almost without interruption until the 15th April, 1903.

During that time, neither the declinometer nor the variometer for horizontal intensity underwent any change, except that the torsion-head of the variometer for the horizontal intensity was a little twisted on the 1st December, 1902.

As we shall presently notice more fully, the variometer for the vertical intensity, in the course of the above-mentioned period, underwent a few small changes, which, however, have had no perceptible influence upon the scale-value.

As the torsion in the thread of the declinometer is slight, z will be small. The torsion has therefore only been determined 3 times, namely on the 28th November, and 8th December, 1902, and the 16th January, 1903. As the mean of these three, it is found that

$$\gamma = 0.00164.$$

TABLE IV.

Scale-values for Dyrafjord.

$H_0 = 0.120$ (1) $r_d = 1734$ mm. $\omega_d H_0 = \varepsilon_d = 3.47$.

Date		n_d	n_h	n_v	ε_h	ε_v	Temperature.
1902							
Nov.	29	34.0	21.4	21.4	5.50	5.50	4.93 °
Dec.	2	39.4	24.7	28.2	5.54	4.85	4.3 °
"	8	45.7	28.9	33.2	5.48	4.78	5.1 °
"	11	{ 55.6	35.4		5.46		6.7 °
		{ 46.4	29.2	35.4	5.51	4.56	
"	16	{ 47.3	30.0		5.47		4.5 °
		{ 33.6		28.0		4.16	
1903							
Jan.	2	40.4	25.1	28.7	5 63	4.93	0.9 °
"	16	39.2	25.3	27.5	5.40	4.97	2.8 °
"	19	40.3	25.6	29.7	5.48	4.74	2.0 °
"	24	40.4	25.5	29.0	5.53	4.86	2.2 °
Feb.	5	40.2	25.6	30.3	5.47	4.63	1.4 °
"	21	39.8	24.8	29.2	5.57	4.85	1.5 °
"	27	40.2	25.4	28.1	5.52	5.00	1.4 °
March	11	40.5	25.8	28.2	5.51	5.03	−0.5 °
"	24	40.3	25.8	27.7	5.44	5.07	−0.3 °
"	31	40.8	25.7	27.6	5.54	5.16	0.5 °
April	11	40.7	25.5	26.6	5.58	5.33	1.1 °

Note. December 1, 1902, the sensibility of the variometer for V. I. made a little greater.
December 15, 1902, compensation for the variometer for V. I. altered.
February 23, 1903, the curve longer from the base-line for V. I.; otherwise unaltered.
January 27, 1903, fixed new mirror for V. I.; otherwise unaltered.

It appears from the above table that the sensibility for H has remained nearly constant all the time. No decided variation in the temperature is noticeable, nor yet any decided variation with time. It is therefore most natural to let ε_h be constant all the time. The mean of the scale-value is

$$\varepsilon_h = 5.51.$$

ε_v also remains fairly constant all the time. For ε_v we obtain the following:

Nov. 25, 1902, to Dec. 1, 1902, $\varepsilon_v = 5.50$
Dec. 1, 1902, to Dec. 15, 1902, $\varepsilon_v = 4.73$
Dec. 15, 1902, to Jan. 27, 1903, $\varepsilon_v = 4.92$
Jan. 27, 1903, to Apr. 15, 1903, $\varepsilon_v = 4.61 + 0.0094, t.$

For this last period from the 27th January to the 15th April, we have a fairly regular increase of ε_v with time. The formula set up is calculated by the method of least squares.

t here stands for the number of days reckoned from the 27th January.

(1) It must be remarked that this value is somewhat uncertain; for owing to the illness of Sæland, whose knee became stiff while at Dyrafjord, no complete absolute determination was made. A deflection experiment was made, and this, combined with a knowledge of the magnetic moment of the magnet employed, gave the value here given, which moreover is in accordance with the terrestrial-magnetic charts.

DETERMINATIONS OF SENSIBILITY FOR AXELØEN.

20. The registering apparatuses on Axeløen were in operation from the 30th August, 1902, without interruption until the 7th June, 1903.

Neither the variometer for the horizontal intensity, the declinometer, nor the balance were changed during that time.

No determinations of sensibility were made on Axeløen for the variometer for the vertical intensity, as this apparatus was without deflection rods. The position of the movable parts of the magnet, and the arrangement of the balance, were however accurately noted. Determinations of sensibility were made at the Physical Institute, Christiania, after the return of the Expedition, the conditions that had prevailed on Axeløen being reproduced as exactly as possible. No alteration in the magnetic moment is to be apprehended, as the magnet was several years old. The balance-magnet was of the form shown in the figure. The movable parts consist of a small weight B, which can be screwed backwards and forwards along a small, horizontal, brass rod, and a weight A, capable of being moved in a vertical direction. By moving B, the magnet can be adjusted horizontally. It is easy to see that a small change in B will have no great influence upon the sensitiveness, as the centre of gravity of the system is neither raised nor lowered thereby to any noticeable extent. By screwing A, on the other hand, the sensitiveness is altered, as the height of the centre of gravity is thereby altered.

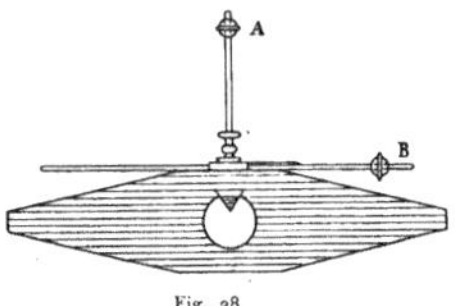

Fig. 28.

As the position of A is not so easy to find again accurately, two determinations were made, the weight A being placed in the highest and lowest positions possible in the case in question.

The determinations gave the following result:

A in lower position	Distance of deflecting magnet	56.4 cm.	$\varepsilon_v = 25.6$
	» » » »	47.05 »	$\varepsilon_v = 24.9$
		Mean	$= 25.25$
A in upper position	Distance of deflecting magnet	47.05 cm.	$\varepsilon_v = 23.85$
		Mean	$= 24.6\ \gamma$

As we use the mean value, the error should not exceed 4 per cent.

TABLE V.

Scale-values for Axeløen.

$H_0 = 0.0941 \quad r_d = 1733$ mm. $\quad \varkappa = 0.0079 \quad \omega_d H_0 = 2.736$

Date	n_d	n_h	ε_h	T
Sept. 12, 1902	43.9	26.2	4.59	3.0°
Nov. 16, »	43.75	25.9	4.63	0.5°
Dec. 12, »	44.1	26.2	4.61	−10.0°
March 1, 1903	43.6	25.8	4.62	− 8.0°

The table shows that the scale-value for the variometer for horizontal intensity has remained constant all the time, and has not altered perceptibly with the temperature.

We therefore employ the mean value, viz.

$$\varepsilon_h = 4.613\,\gamma \text{ per mm. deviation.}$$

DETERMINATIONS OF SENSIBILITY FOR MATOTCHKIN SCHAR.

21. The registering apparatuses here were in operation from the 30th August, 1902, to the 11th March, 1903. The first month, from the 30th August to the 30th September, was spent in trial registering: for it proved to be very difficult to get the balance compensated for variations in temperature. Compensation of the balance was effected on September 9th, 10th, 11th, 12th and 27th, and October 6th. The sensitiveness of the balance was altered on the 23rd September and the 9th October, being increased both times.

The H. I. variometer acted almost without change; it only now and then underwent small corrections with regard to the position of the base-line. These cannot, however, be supposed to have had any special influence upon the sensibility. The declinometer acted without alteration all the time. Astatising magnets were not employed.

It will be seen from the diagram below, that the thread in the declinometer was very stiff, the effect of this being that $\varkappa$ is very large, and therefore has to be determined very exactly. At the same time the H. I. variometer has a sensibility, which, especially considering the violent storms that occur here, must be characterised as disproportionately great. It seems as though the threads for the two variometers have been interchanged.

TABLE VI.

Scale-values for Matotchkin Schar.

$H_0 = 0.1113 \quad \omega_d H_0 = 3.23\,(1 + \varkappa).$

Date	α	x	$\varkappa$	n_d	n_h	n_v	ε_h	ε_v	$\omega_d H_0$	T
Sept. 20, 1902 . .	4°	67.0	0.387	28.2	77.0		1.64		4.48	2.2°
Oct. 17, „ . .	4°	66.1	0.380	29.9	90.9	19.0	1.47	7.02	4.46	− 4.6°
Nov. 16, „ . .	4°	62.8	0.354	29.5	74.4	21.1	1.73	6.06	4.37	− 2.4°
Dec. 22, „ . .	4°	65.5	0.375	27.7	76.6	16.8	1.61	7.33	4.44	− 5.8°
Feb. 12, „ . .	4°	65.3	0.373	28.2	76.5	14.7	1.63	8.52	4.44	−13.8°

It will be seen that ε_d and ε_h keep fairly constant, and exhibit no decided variation with time and temperature. We make use of the mean, putting

$$\varepsilon_h = 1.62$$

$$\omega_d H_0 = 4.44.$$

The value of ε_v are found from a curve, which together with the observed values, is shown in the following figure.

Fig. 29.

TEMPERATURE COEFFICIENTS FOR THE REGISTERINGS.

22. The temperature at our four arctic stations was registered all the time, simultaneously with the magnetic elements. At the stations at Dyrafjord, Kaafjord, and Matotchkin Schar, the temperature was registered upon the magnetogram itself. At Axeløen, it was registered by an ordinary thermograph. The temperature moreover is generally given at the beginning and end of each magnetogram.

Lloyd's balance, on all stations, except at Axeløen, were compensated for changes in temperature by means of magnets which were placed at a suitable distance under the balance. The compensation was tried by artificial warming of the rooms by means of hot bricks.

In order to be able to correct the curve for changes in temperature, we must be acquainted with the following particulars:

ε_t, or the number of degrees centigrade that answer to a deflection in the temperature-curve of 1 mm.
θ_h = the number of mm. the H. I. curve is displaced in relation to the base-line per degree centigrade.
θ_d = the number of mm. the D curve is displaced in relation to the base-line per degree centigrade.
θ_v = the number of mm. the V. I. curve is displaced in relation to the base-line per degree centigrade.

We call these quantities positive, when the curve, by an increase in temperature, is sent upwards on the magnetogram.

The values found for our four arctic stations are given in the table below.

TABLE VII.

	Kaafjord	Dyrafjord	Axeløen	Matotchkin Schar
ε_t	0.088	0.055		0.062
θ_h	−0.57	−1.38	−1.56	−0.54
θ_d	1.94	1.56	−0.67	0.00
θ_v	2.71	5.53	1.34	0.15

ε_t is found by comparing the temperatures read with the ordinates to the temperature-curve.

In the case of the three other quantities, we have employed a somewhat different mode of procedure in the calculation. For Dyrafjord they are found by the aid of the change in temperature that will always take place during a determination of sensibility, and which can be determined by the temperature-curves. In order to be sure that the displacement of the curve is due to the temperature, it must be calm before and after. The diurnal variation must, moreover, be taken into consideration. The values given are means of 10 determinations distributed over the various months.

At the other three stations, a method has been employed by which we escape having to consider the diurnal variation. Under normal conditions, the ordinates to the curve in points lying 24 hours from one another — provided the temperature is the same — should be of the same length. The majority of our magnetograms cover a period of 24 hours.

We have now selected a series of registerings for the very calmest days with a great difference in temperature between the beginning and the end. The required temperature coefficients are then found from the differences between the ordinates to the terminal points and the difference in temperature. This method is very suitable, as the temperature for Axeløen is read directly, at the beginning and end of each magnetogram. At Matotchkin Schar, it is only at the beginning. At Kaafjord, on the other hand, the temperature in the register-house was read only a few times in the course of the winter, and there we have had to keep to the registered temperature-curve only.

The values given in the table are in each case the mean of 12 such determinations. In the calculation of the mean, we have assigned different weights to the determinations, according to the amount of difference in temperature, and the calmness of the twenty-four hours.

The temperature-coefficients of our registerings — with the exception of those for Matotchkin Schar — are not inconsiderable; and as the temperature at these temporary stations undergoes great changes, it has been necessary for us, in our calculations, in each case to direct our attention to its effect.

EXPLANATION OF THE CHARTS.

23. Our investigations of the distribution and course of the magnetic perturbations, divide, as already mentioned, into two sections, the one embracing the whole earth, the other more especially confined to the regions round the North Pole.

We have here found it most practical to employ two different charts in the synoptic representation.

For the universal part we have employed a map of the world on Mercator's projection. The advantage of this projection is that it is orthomorphic, so that angles upon the earth can be marked directly upon the chart.

For the second section we have used a polar map in the equidistant zenithal-projection. This projection is not orthomorphic; but the angular deformation in the polar regions is very slight. For all stations except that of Cape Thordsen we have, however, taken this deformation into account. As for Cape Thordsen the deformation is less than the accuracy with which the angles can be determined.

The previously explained current-arrows are marked on the maps, representing geometrically the perturbing forces calculated for a particular point of time. The time is stated at the top of the map. The length of the arrows is proportional in each chart to the perturbing forces. At the foot of the chart a scale is marked, by means of which the magnitude of the perturbing force can easily be taken directly from the chart. As the unit of magnetic force we have employed $1\ \gamma = 10^{-5}$ absolute units.

It has proved inexpedient to make all the arrows on one chart to the same scale, as the perturbing forces at the northern stations are often more than 10 times as great as over the other parts of the earth during the same period.

We have therefore in general employed different scales for the arctic regions and for the rest of the earth. On the Mercator chart, the scale given is the one employed for the more southerly stations. The scale for the four Norwegian stations is only a fraction — generally $^1/_5$ — of that given on the chart.

In order to indicate this, we have written beside the arrow the fraction by which the scale marked on the chart must be multiplied in order to find out the scale employed for the place. When, for instance, the fraction $^1/_5$ is found on the chart, this signifies that each length-unit of the arrow is equivalent to a force 5 times as great as that which would be directly indicated by the scale given on the chart.

On the polar chart, on the other hand, the conditions are reversed. There we have given the scale that is employed for the polar stations, that is to say for the places where the perturbation is strongest; and the scale for the more southerly stations is given in the same manner by a multiplier.

In order to make the charts easy of comprehension and give a direct idea of the course of the perturbation, the same scale has as far as possible been kept for the whole of a perturbation. On the other hand, the scale will not be the same for all perturbations, as it must be chosen so as to give the arrow on an average a suitable length.

As the vertical intensities are of the greatest importance for a complete determination of the character of the perturbation, they are also placed upon the charts, in order that both their magnitude and direction may be taken thence. They are represented by lines drawn out from the place at right angles to the current-arrows, and are marked on the same scale as the latter. Their direction is determined in the following manner. If we imagine ourselves to be standing on the place in question, and looking out in the direction of the current-arrows, the vertical arrow is placed on the left if P_v is turned downwards, on the right if it is turned upwards. Or we might express it as follows: Let P_v be turned 90° with the hands of a clock, the observer facing the direction of the current-arrow.

It appears from Ampère's law, that when the perturbation at a place is due to a horizontal current-system above the earth, the vertical arrow will point out towards the places where the current has its greatest density.

This law has a special application to the arctic stations.

As the current-arrow, however, very often does not give the direction for a horizontal current, but is only a representative of the perturbing force, the vertical arrow loses this significance; but it gives, at any rate, P_v in magnitude and direction.

For the purpose of distinguishing the vertical arrow from the current-arrow, the latter is made a little thicker and with an arrow-point.

It is only from a very few stations, however, that there are registerings of variations in vertical intensity. As a rule, arrows will be marked for the following:

The Norwegian stations Kaafjord, Dyrafjord, Axeløen, and Matotchkin Schar;

and also

Christchurch,	Tiflis,
Munich,	Val Joyeux,
Pawlowsk,	Wilhelmshaven,
Pola,	Zi-ka-wei,
Potsdam,	

and sometimes for Irkutsk and Jekaterinburg. In general, no oscillation will be noticed in the V. I. curve for Zi-ka-wei, partly on account of the small sensibility. Upon the whole, moreover, P_v will be small, often imperceptible, in southern latitudes.

The following signs will also occur on the charts:

(?) indicates that the perturbing force cannot be determined, owing to lack of material.

(*) See note in the text. The perturbation is then often ill-defined, and so small that the perturbing force cannot be calculated with any advantage.

(0) The perturbing force is imperceptible.

(☉ & ☾) Indicate respectively the sun and the moon, and these signs are placed where the sun and moon respectively, in the epoch under consideration, stand in the zenith.

(⊕) Indicates the point in which the magnetic axis of the earth intersects the earth's surface, i. e. the axis of the elementary magnet to which the earth approaches for infinitely great distances. At the new year, 1903, this point was determined thus: North latitude 78° 20′, West longitude 71° 11′. On the Mercator's chart, the equator line for this pole-point will also often be marked.

(◎) The magnetic north pole.

To show the position of the so-called auroral zone, two curves, from FRITZ's aurora chart, are drawn on all the polar charts and on a few of the Mercator's charts. The most southerly gives the places of the greatest frequency of observed aurora. The most northerly connects points where aurora is seen as frequently in the south as in the north.

It sometimes happens, in the case of the northern stations, especially Matotchkin Schar, 1902 and 1903, that the patch of light, owing to the strength of the perturbations and the great sensitiveness of the apparatus, passes out of the paper, returning again in a little while. We know then that the deflection is at least as great as to the edge of the paper. This minimum value of the perturbing force, obtained by measuring to the edge of the paper, is then placed upon the chart as a dotted arrow; and at its point is placed an arrow, to give the direction in which the current-arrow really has its point.

In cases in which the total perturbing force is resolved into two partial forces, the corresponding current-directions will be given with dotted arrows, while their resultant is drawn in full.

THE COPIES OF THE MAGNETIC REGISTERINGS.

EXPLANATION AND GENERAL REMARKS.

24. As already mentioned, there will be a plate belonging to each perturbation, containing copies of the magnetograms obtained.

As it is important, when reading the descriptions, to have the curves themselves before one, it might have been better if the latter could have been in the same place as the descriptions. The fact that, notwithstanding this, we have considered it advisable to keep all the curves together, is mainly due to circumstances of a purely technical nature.

The curves will follow one another in chronological order.

Upon the district in which the perturbation is found, the normal line will be drawn, according to the previously given rules, as a dotted line.

With a knowledge of the scale-value, it will thus be possible, if desired, to find out the perturbing force at any point of time.

The scale-value is given graphically by lines placed at the end of each curve, and giving the length of oscillation of a particular force. At the head of the column are the signs L^n_h, L^n_d, and L^n_v, which indicate the length of a deflection in H, D and V respectively, corresponding to magnetic force, $n.\gamma$, operating in the respective directions. In the middle of the line is an arrow-head, which gives the direction of increasing H. I. increasing westerly declination, and increasing vertical intensity.

The scale in relation to the original magnetograms is so arranged that all the magnetograms shall have the same time-length. The scale-value is thus increased in the same proportion as the time-length is diminished.

In the table below, the scale-values appear as they were given us direct, as also the length of one hour upon the original magnetograms.

TABLE VIII.

Observatory	ε_h	$\omega_d H_0$	ε_v	Length of 1 hour	Remarks
				mm.	
Axeloen	+ 4.61	+ 2.74	− 24.6	20.06	
Baldwin	+ 3.595 − 0.0125 t. (*)	+ 6.36		19.97	(*) t. = Temp. in degrees centigrade.
	+ 3.56	− 12.0		15.58	
Bombay	+ 5.12	+ 12.3	+ 16.1	15.36	
Cheltenham . . .	+ 1.959 − 0.03 (t. − 21) (*)	+ 5.94		19.92	(*) t. = Temp. in degrees centigrade.
Christchurch *) . .	4.6	7.43	3.12	15.36	(*) Sign changes. Given on the plates.
Dehra Dun	+ 3.94	− 9.85		14.74	
	+ 5.51	+ 3.47	+ (*)	19.94	(*) See table of scale-values.
Honolulu	(1) From Nov. 25, 02 + 2.24 + 0.0058 h.	+ 8.3		19.95	(*) { h = ordinate in mm. Average ε_h = 2.56.
Kaafjord	+ (*)	+ 3.67	+ (*)	19.89	(*) See table of scale-values.
Kew . .	+ 5.1	+ 4.68		15.01	
Matotchkin Schar	+ 1.62	+ 4.44	+ (*)	19.94	(*) See table of scale-values.
	− 5.0	− 7.61	− 3.78	20.35	
Pawlowsk	+ 5.03	− 4.6	− 7.48	14.99	
Pola	− 4.48	+ 6.94	+ 2.07 (*)	20.00	(*) Exactly { Oct. 10 − 23, 02 = 2.09; Oct. 23, 02 − Feb. 20, 03 = 2.12; Feb. 20 − Mar. 30, 03 = 2.00
Potsdam	− 3.165	− 5.08	− 3.00	20.48	
San Fernando . .	{ 1902 = + 6.4; 1903 = + 7.4 }	− 8.2		15.4	
	+ (*)	+ 4.51		19.98	(*) { Oct. 10 − Dec. 19, 02 = 1.937 − 0.143 t. Dec. 21, 02 − Mar. 31, 03 = 1.76.
Stonyhurst	+ 5.1	+ 5.71		15.24	
Tiflis	{ To Dec. 23, 02 = − 2.14; From " 24, 02 = − 2.21 }	− 3.71	− 2.55	15.58	
Toronto	+ 4.5	+ 6.02		18.22	
Val Joyeux	− 8.0	− 8.37	− (*)	9.94	(*) Oct. 8.0 | Nov. 9.0 | Dec. 10.0 | Jan. 11.0 | Feb. 9.0 | Mar. 9.0
Wilhelmshaven .	+ 4.67	+ 6.11	(*)	15.40	(*) ε_v not determined.
Zi-ka-wei	− 6.00	− 5.00	− (*)	15.50	(*) ε_v varies greatly. The values will be given for each curve.

(1) On Sept. 30th the value was 4.38 and increased 0.01 per diem up to Oct. 31st, after which it was constant up to Nov. 25th.

For convenience in the Plates, the sign is here fixed as follows:

ε_h, $\omega_d H_0$ and ε_v are indicated by +, when a deflection upwards answers respectively to increasing H. I., increasing westerly declination, or increasing numerical value of V. I.

The reduction of the magnetograms has been effected by a pantograph belonging to the Geographical Survey of Norway. The reduction to equal hour-length, and also the drawings, have been executed by a very skillful cartographer, Mr. J. Natrud of the Geographical Survey.

As mentioned in my circular of June, 1903, it was my original intention to publish some of the magnetic records by means of photographic reproduction. This mode of procedure, however, has proved to be very unsuitable for the arrangement of curves from so large a number of observatories; but I think that the method of reproduction chosen by us will be of equal value to science.

A list of the perturbations that will be treated in the following chapter is given below in Table IX. The great generosity and interest shown by the heads of all the previously-mentioned observatories, without which the exceptionally valuable material relating to magnetic storms contained in these twenty-one plates could not have been collected, would be worthy of emulation in all branches of science.

TABLE IX.

No. of Pert.	Date	No. of Plate	Class of Perturbation
1	Jan. 26, 1903	XIV	Equatorial
2	Dec. 9, 02	IX	—»—
3	Oct. 23, 02	III	—»—
4	Dec. 15, 02	X	Elementary polar
5	Feb. 10, 03	XVIII	—»—
6	Mar. 31, 03	XXI	—»—
7	Mar. 22, 03	XX	—»—
8	Dec. 26, 02	XII	—»—
9	Oct. 6, 02	I	Cyclo-median
10	Oct. 30, 02	VI	Compound
11	Dec. 25, 02	XI	—»—
12	Dec. 28, 02	XIII	—»—
13	Feb. 15, 03	XIX	—»—
14	Feb. 8, 03	XVI & XVII	—»—
15	Oct. 27, 02	IV	—»—
16	Oct. 28, 02	V	—»—
17	Oct. 31, 02	VII	—»—
18	Oct. 11, 02	II	—»—
19	Nov. $\frac{22}{23}$, 02	VIII	—»—
20	Jan. $\frac{26}{27}$, 03	XV	—»—

CHAPTER II.

ELEMENTARY PERTURBATIONS.

25. It will be our endeavour, as stated in the introduction to this section, while studying the perturbations, to find out their extent and course in each case. We consider it to be of the greatest importance for the attainment of this object that what has taken place should be viewed as directly as possible, at moments during the perturbation as numerous and close together as is practicable. This then has guided us in our calculation of the perturbing force, and we considered that we arrived most easily at the truth by placing the normal line actually on the magnetogram, in accordance with the previously mentioned rules.

In connection with this, it should be mentioned that it would be expedient, when reading the description, to have the curves before one, as there the conditions appear as directly as it is possible to have them.

With this object in view, our purpose is best served by dividing the perturbations into groups, which seem to have comparatively well-defined properties.

After the experience we have gained through the treatment of this material, it is our hope that also other natural philosophers will feel convinced that we have taken the right road, a road that leads to a clear comprehension of the laws of perturbations.

It must not be imagined, however, that these groups stand as altogether separate phenomena.

A complete acquaintance with the nature of the perturbations will assuredly lead to the assumption that there is a certain genetic connection between the various groups. It is moreover our opinion that this is the case, at any rate as regards the majority of the most important groups, as the physical agents that consitute the currents are supposed to have in the sun their common source.

The following treatment of perturbations will include the most important of those that occur in the registerings of the thirty days [1] for which we have received material from a number of observatories — mentioned previously — all over the world. This choice of days is based upon observations from Kaafjord and Potsdam. The qualities that have guided the selection have principally been strength and distinctness; but on the other hand, the selection was made without regard to the character of the perturbation in other respects. As, however, the choice was based upon observations from one particular region of the earth, this circumstance could not but cause the perturbations that appear especially strong about the Norwegian stations, to receive a prominent place; but this, far from being a drawback, must, in our opinion, be considered an advantage, as the material collected by us in our arctic expedition will thereby be turned to best account. This one-sidedness, moreover, in the material is considerably reduced by the circumstance that for each of the hours mentioned in the circular, we have always received registerings for at least one day, and in the case of several of the observatories even for several days. We

---

[1] Circular of June, 1903.

have thus had an opportunity of studying a number of perturbations that do not belong to those specially mentioned in the circular.

It would be impossible, if we are to treat the perturbations upon the lines we have laid down, to take notice of all the deviations that might indeed be worthy of mention. We have had to confine ourselves to the study of the greatest and longest, or at any rate to perturbations of a universal character.

We will here mention a circumstance that confirms us in our opinion that we have succeeded in treating a number of the most important of the perturbations that have taken place during this period.

Being aware of the one-sidedness there might possibly be in our material, we wrote on the 9th March, 1907, to the Director of the Coast and Geodetic Survey of the United States with a request that he would send us magnetograms of some of the greatest perturbations that had occurred at Sitka and in North America during the period from the autumn of 1902 to the spring of 1903. The Superintendent, Mr. O. H. Tittmann, and the Director, Mr. L. A. Bauer, were good enough to comply with our request. The perturbations, however, which had been selected with regard to Sitka for ten days in which "the magnetic perturbations were remarkably distinct, powerful and simple", proved to be of no very different kind or magnitude from those we had already studied. It was principally a series of perturbations in January that were comparatively great in those regions. We shall go more fully into these conditions farther on, as, with the aid of the material from the polar stations of 1882—83, we may draw important conclusions regarding the position of the storm-centres about the auroral zone at various times of the day.

A similar request was sent to the Director of the Observatory at Christchurch (New Zealand), whence we also once more received magnetograms for 19 days of the period observed, in which the perturbations at that place were remarkably distinct, powerful and simple. In 16 cases, however, the perturbations were coincident with some we had previously received and discussed.

THE EQUATORIAL PERTURBATIONS.

26. It appears that magnetic storms of any considerable strength, are most frequently of a kind in which the force increases towards the poles. It also appears, however, that it is not unusual to find perturbations that are best developed and most powerful at the equator. It has even been found that these perturbations in the regions about the equator, act principally upon the horizontal intensity, in such a manner that the current-arrows point along the magnetic parallels.

As regards the lower latitudes, the circumstances of the perturbation often exhibit symmetry both with respect to the magnetic axis and to the equator. Such perturbations we have chosen to call equatorial perturbations.

Of these there are again two kinds possible, namely, such as produce an increase in the horizontal intensity, and such as produce a diminution. Both of these occur.

The first of these we have called *positive equatorial perturbations;* the second kind we have called *negative equatorial perturbations.*

The reason for this separation is not merely the more formal one that the force is in opposite directions; but it goes deeper, the two perturbations having quite a different character and course. The positive equatorial perturbation in particular is strongly characterised, so much so that if attention has once been drawn to it, it will always be recognised with the first glance at the registered curves. Its more detailed characterisation will come out best in the treatment of the separate typical cases.

THE POSITIVE EQUATORIAL PERTURBATION.
THE PERTURBATION OF THE 26th JANUARY, 1903.

Pl. XIV.

27. For the study of this perturbation, there are magnetograms from all the stations. As the curves show, only the latter half of the perturbation has been obtained at most of the European stations.

The perturbation appears quite suddenly upon a quiet day. It begins at 8^h 52^m, and lasts until 14^h 20^m. (The time, when not otherwise stated, is Gr. M. T., 0^h = midnight).

It is particularly well developed and well defined in the equatorial regions; its effect is not confined to any single district, but it appears all round the equator. If, for instance, we look at the curves for Dehra Dun, Batavia and Honolulu, we see that at these three places the perturbation agrees down to the smallest details. We further notice immediately that it appears only in the horizontal intensity, and in such a way that all the time the perturbing force is directed northwards, i. e. in the direction of the magnetic meridian.

If we pass from the equator towards the poles, we see that the character of the perturbation is maintained, the only difference being that the deflections become a little smaller. As far south as Christchurch, which is our most southerly station, and as far north as Toronto in America, and Stonyhurst and Pawlowsk in Europe, the perturbation preserves in the main its character of appearing only in the horizontal intensity. When we come, however, to our most northerly stations, we find that it also appears in the declination, which means that here in the north the direction of the perturbing force is no longer along the magnetic meridian. At the same time, the average deflection becomes considerably less for these stations. This, together with the more disturbed course of the curve, makes it difficult to measure the perturbing force. The perturbation here acquires to some extent the character of marked oscillations about the mean line.

In glancing at the curves, we also notice at once their jagged character during the perturbation, answering to a great variability in the strength of the perturbing force. If we compare the serrations in the curves for the various stations, we find them to a great extent repeated from place to place. We further notice that as we approach the poles, the serrations become more acute and larger, and of a somewhat local character. A sudden change in the curve answers to a great change in the perturbing force, which again must be produced by a great change in the perturbing impulses.

It might now be asked whether these perturbing impulses reach the various parts of the earth simultaneously, or whether they require an appreciable time to be transmitted from one station to another.

The very fact that the serrations can be distinctly identified in the different curves, makes it natural to expect that they appear simultaneously, as it would be difficult to imagine that an impulse of this kind during a comparatively slow motion, could preserve its character unchanged.

In order to throw light upon this circumstance, we have reckoned the times at all the stations, for a series of points that allow of easy identification. The result is given in the Table below, where the points are indicated by the numbers 1, 2, etc., and will be found marked on the curve for Dehra Dun.

The following table shows that the time varies so little with the geographical position, that it would be premature to draw conclusions from it.. The slight differences that appear rather irregularly, may be ascribed to inaccuracies in the determinations of time on the magnetograms; for we see that if a difference in time for a certain point appears between two places, this difference is maintained for all the points, a circumstance which seems best to be explained by an inaccuracy in the statement of the time. We may conclude from this that the serrations appear simultaneously, or rather, the differences in time are less than the amount that can be detected by these registerings. Characteristic serrations such as these may therefore often be of great use in controlling the time of the magnetograms.

TABLE X.

Observatory	1	2	3	4	5	6
Honolulu	8^h 50.'8	10^h 50.'1	11^h 29.'8	12^h 29.'7	13^h 49'	14^h 13.'0
Baldwin (1)	8^h 52.'4	10^h 53.'5				
Toronto (1)	8^h 52.'6					
Cheltenham (1)	8^h 54.'9	10^h 54'				
Kaafjord	8^h 52.'6	10^h 52.'9	11^h 30.'4	12^h 32.'3	13^h 54.'9	14^h 18.'8
Pawlowsk	Wanting			12^h 29.'8	13^h 51.'7	14^h 15.'7
Stonyhurst	»	»	11^h 27.'1	12^h 32.'5		14^h 17.'5
Kew	»	»	11^h 32.'9	12^h 32.'7	13^h 55.'8	14^h 19.'5
Wilhelmshaven	»	»	11^h 30'	12^h 29.'9	13^h 53.'6	14^h 17.'3
Potsdam	»	»		12^h 33.'2	13^h 53.'8	14^h 18.'9
San Fernando	8^h 54.'3	»	11^h 32.'5	12^h 33.'6	13^h 54.'5	14^h 19.'1
Tiflis	»	»	11^h 31.'1	12^h 31.'6	13^h 51.'7	14^h 16'
Dehra Dun	8^h 52.'7	10^h 51.'6	11^h 31.'4	12^h 32'	13^h 54.'2	14^h 18.'4
Bombay	8^h 53.'9	10^h 53'	11^h 31.'5	12^h 33.'3	13^h 52.'6	14^h 19.'7
Zi-ka-wei		10^h 54.'1	11^h 33.'5	12^h 34.'9	13^h 56'	14^h 20.'8
Batavia	8^h 54.'9	10^h 54.'9	11^h 33.'9	12^h 35.'5	13^h 56'	14^h 19.'5
Christchurch	8^h 54.'8	10^h 53.'2	11^h 33.'6	12^h 33.'2	13^h 55.'4	14^h 19.'9

The above question, which is of great importance, cannot be definitely decided until we are in possession of rapid registerings, as usual of 12 times the rapidity of the daily registerings. By this means we should see if the apparent difference in time, as shown in Table X, between, for instance, Honolulu and Batavia, is a real one.

The perturbing force is calculated for a number of hours, the results being given in the annexed Table. It should be remarked that as the perturbation is of rather long duration, the perturbing force will be somewhat uncertain for the middle part of the perturbation. It will be seen from the Table that the horizontal component of the perturbing force is directed, as already mentioned, along the magnetic meridian, except as regards the most northerly situated stations. Further, the force decreases somewhat in strength from the equator to the poles, as the charts very distinctly show.

If we compare the force on the two sides of the equator, we see that the course is similar, but that the force has a smaller value at Honolulu than at Dehra Dun, Bombay and Batavia.

The curve for the magnetic equator, or rather the line of intersection of the plane perpendicular to the magnetic axis, with the earth, is also drawn on the charts. We see that the direction of the arrows is on the whole parallel with this line.

As compared with the horizontal component, the vertical component of the perturbing force is exceedingly small; and this proportion continues as far as Pawlowsk, as far, indeed, as the Norwegian stations about the auroral zone. There is, however, in the south, namely, at Christchurch, an undoubted deflection in the vertical-intensity curve, answering to a force-component directed downwards, and

(1) The curves for Baldwin, Toronto and Cheltenham are so finely serrated as to make identification difficult

not exceeding the value 2.5 γ in magnitude. In the north, it is almost imperceptible at Pawlowsk. Even at Tiflis, where the sensibility is very great (ε_v = 2.55 γ), the deflections in the vertical curve may best be characterised as small vibrations about the mean line; while at the same time, the horizontal component has values going up to 24 γ. The directions of the vertical components are indicated on the charts by dotted lines, as they are too small to allow of their size being marked.

It would appear from the above that we here have a perturbation of a very characteristic and peculiar kind, a species of perturbation with which we shall very often meet. As a rule, however, it will appear together with other phenomena, which disturb its regular development; but here we seem to have the perturbation almost alone, and on a quiet day.

It will often happen that during a perturbation that is powerful at the equator, great storms will occur in the north, of which the effect makes its way southwards, but is weakened towards the equator. Here too, there is an indication of conditions such as these, of which we shall later on have several examples. At Sitka, for instance, a sudden change in the curves occurs between 11 and 12.30. It is another phenomenon altogether that here makes its appearance, and which has its focus in the polar regions, its effect being almost imperceptible in the vicinity of the equator. It is fairly distinct at the Norwegian stations, and its effect may also be traced in Central Europe. On the chart for 12 o'clock, this current direction represents the total force resolved into one that should answer to the equatorial current; the other component, which answers to the polar current, will then be directed towards the south-west, answering to a current towards the north-west.

While we allow this perturbation to serve as a typical example of these perturbations, the positive equatorial perturbations may be more fully characterised as follows.

The perturbation appears with greatest strength in the regions round the equator. It is true that for a short time the deflections may be greater at the poles than at the equator; but the force does not remain constant for so long a time. The conditions at the poles are frequently characterised as an oscillation about the mean line, of a somewhat local character.

The perturbing force in southern latitudes, and more especially in the neighbourhood of the equator, is directed northwards in the direction of the magnetic meridian.

The perturbations appear simultaneously all round the equator, and with a similar course, but not always with the same strength.

The curves for the horizontal intensity, where the perturbation mainly shows itself, present a characteristically serrated appearance. The serrations may very frequently be recognised all over the earth, and in such case occur simultaneously.

TABLE XI.

The Perturbing Forces on the 26th January, 1903.

	Honolulu		Sitka[1]		Baldwin		Toronto		Cheltenham	
Gr. M. T.	P_h	P_d	P_h	P_d	P_h	P_d	P_h	P_d	P_h	P_d
h m										
9 0	+ 6.4 γ	0	?	?	+ 5.3 γ	0	+ 4.5 γ	0	+ 5.3 γ	0
10 0	+ 5.9 »	0	?	?	+ 4.6 »	0	+ 5.4 »	0	+ 5.0 »	0
11 0	+ 4.1 »	0	— 4.1 γ	0	+ 4.3 »	0	+ 7.2 »	E 1.2 γ	+ 5.3 »	0
12 0	+ 6.2 »	0	— 6.7 »	W 9.8 γ	+ 5.3 »	W 3.2 γ	+ 8.1 »	W 4.2 »	+ 5.9 »	W 4.1 γ
30	+ 16.7 »	W 3.3 γ	+ 1.1 »	» 13.4 »	+ 13.5 »	» 8.2 »	+ 11.3 »	» 9.4 »	+ 10.6 »	» 7.7 »
13 0	+ 16.7 »	0	+ 8.9 »	E 4.5 »	+ 15.6 »	» 5.7 »	+ 17.1 »	» 8.5 »	+ 13.8 »	» 4.1 »
30	+ 13.9 »	0	+ 8.3 »	W 3.1 »	?	?	+ 18.0 »	» 6.7 »	+ 11.7 »	0
14 0	+ 12.9 »	0	+ 5.1 »	E 8.0 »	+ 13.5 »	0	+ 24.3 »	0	+ 16.1 »	E 2.4 »

(1) As we have only the close of the perturbation, the choice of normal lines is somewhat difficult.

TABLE XI (continued).

	Matotchkin Schar			Kaafjord			Pawlowsk			Stonyhurst	
Gr. M. T.	P_h	P_d	P_v	P_h	P_d	P_v	P_h	P_d	P_v	P_h	P_d
h m											
9 0	+ 10.3 γ	0	+ 7.3 γ	+ 3.7 γ	W 2.6 γ	— 5.5 γ	?	?	No perturbation	+ 3.6 γ	0
10 0	+ 4.9 »	W 4.4 γ	0	?	?	?	?	?		+ 4.6 »	0
11 0	+ 0.8 »	» 1.3 »	— 5.1 »	+ 3.0 »	» 2.6 »	— 3.9 »	?	?		+ 8.2 »	W 2.9 γ
12 0	+ 7.1 »	» 3.1 »	+ 10.2 »	0	» 4.0 »	0	+ 5.5 »	0		+ 7.6 »	0
30	0	E 11.1 »	+ 13.9 »	— 3.7 »	» 7.0 »	+ 6.3 »	+ 8.0 »	W 2.3 γ		+ 8.5 »	» 2.3 »
13 0	+ 11.3 »	W 4.9 »	+ 5.1 »	+ 6.1 »	» 9.2 »	— 3.1 »	+ 17.0 »	» 4.6 »		+ 17.8 »	» 1.7 »
30	+ 14.5 »	» 8.9 »	0	+ 9.8 »	» 6.2 »	— 3.9 »	+ 19.1 »	» 3.2 »		+ 17.8 »	E 0.6 »
14 0	+ 22.5 »	» 9.3 »	0	+ 5.5 »	» 17.2 »	— 6.3 »	+ 22.5 »	» 2.3 »		+ 15.4 »	W 5.1 »

TABLE XI (continued).

	Kew		Val Joyeux			Wilhelmshaven		Potsdam		San Fernando	
Gr. M. T.	P_h	P_d	P_h	P_d	P_v	P_h	P_d	P_h	F_d	P_h	P_d
h m											
9 0	?	?	+ 6.4 γ	0	No perceptible perturbation; no curve after 12h.	?	0	?	?	+ 9.6 γ	Oscillations of a duration of about 4 minutes, but too small to allow of being measured.
10 0	?	?	+ 5.6 »	0		?	0	?	?	+ 4.4 »	
11 0	+ 3.8 γ	E 1.9 γ	+ 6.3 »	E 5.0 γ		+ 4.2 γ	0	?	?	+ 3.7 »	
12 0	+ 5.1 »	» 2.8 »	+ 9.6 »	0		+ 4.9 »	E 3.1 γ	+ 4.7 γ	0	+ 11.1 »	
30	+ 6.6 »	W 1.9 »	?	?		+ 4.7 »	» 4.9 »	+ 7.5 »	W 2.0 γ	+ 17.0 »	
13 0	+ 16.3 »	» 1.4 »	+ 16.8 »	?		+ 16.3 »	» 0.6 »	+ 17.1 »	» 6.0 »	+ 23.0 »	
30	+ 15.1 »	» 0.5 »	+ 16.8 »	0		+ 16.8 »	0	+ 16.1 »	» 3.0 »	+ 22.9 »	
14 0	+ 16.3 »	» 3.3 »	+ 18.0 »	» 4.2 »		+ 17.5 »	» 3.7 »	+ 14.2 »	» 5.0 »	+ 22.0 »	

TABLE XI (continued).

	Munich			Pola		Tiflis			Dehra Dun	
Gr. M. T.	P_h	P_d	P_v	P_h	P_d	P_h	P_d	P_v	P_h	P_d
h m										
9 0	+ 7.0 γ	W 2.3 γ	V. decreases slightly between 12h 45m and 13h 45m	+ 8.5 γ	W 3.4 γ	?	?	?	+ 11.8 γ	No measurable deflection.
10 0	+ 3.0 »	» 0.4 »		+ 2.5 »	» 4.8 »	?	?	?	+ 8.3 »	
11 0	+ 6.0 »	0		+ 3.6 »	» 4.8 »	+ 6.0 γ	?	?	+ 7.9 »	
12 0	+ 9.5 »	0		+ 9.4 »	0	+ 8.8 »	0	?	+ 9.8 »	
30	+ 7.8 »	0		+ 9.0 »	0	+ 11.0 »	W 0.4 γ	?	+ 12.6 »	
13 0	+ 16.8 »	0	— 0.8 γ	+ 19.0 »	» 1.4 »	+ 22.5 »	» 5.2 »	?	+ 23.6 »	
30	+ 15.5 »	0	— 0.8 »	+ 15.0 »	0	+ 21.6 »	» 3.3 »	0	+ 21.7 »	
14 0	+ 15.0 »	0	0	+ 13.4 »	» 3.4 »	+ 19.4 »	» 5.9 »	+ 1.8 γ	+ 20.1 »	

TABLE XI (continued).

	Bombay			Zi-ka-wei			Batavia		Christchurch	
Gr. M. T.	P_h	P_d	P_v	P_h	P_d	P_v	P_h	P_d	P_h	P_d
h m										
9 0	+ 9.7 γ			+ 10.0 γ	0		+ 9.6 γ	0	+ 10.6 γ	0
10 0	+ 8.2 »	No mea-	No visible	+ 7.8 »	0		+ 8.2 »	0	+ 10.1 »	0
11 0	+ 6.7 »	surable	distur-	+ 8.8 »	W 1.3 γ	No per-	+ 7.6 »	0	+ 8.3 »	0
12 0	+ 9.2 »	perturba-	bance.	+ 14.5 »	E 2.0 »	turbation.	+ 10.0 »	E 2.4 γ	+ 15.6 »	W 0.7 γ
30	+ 11.8 »	tion.	Sensibility	+ 18.0 »	» 3.0 »		+ 20.0 »	0	+ 28.1 »	0
13 0	+ 22.5 »		small.	+ 24.8 »	» 2.0 »		+ 23.1 »	0	+ 23.0 »	» 0.7 »
30	+ 21.0 »			+ 22.8 »	» 1.3 »		+ 20.8 »	0	+ 18.4 »	» 0.7 »
14 0	+ 19.5 »			+ 21.2 »	» 1.3 »		+ 15.3 »	0	+ 15.2 »	» 3.0 »

Axeløen.

Only small oscillations about the normal line, without any distinct deflection.

Dyrafjord.

The declination-curve not drawn here. The horizontal intensity oscillates about the normal line.

For Wilhelmshaven and Pola P_v directed upwards, for Christchurch directed downwards. In all cases too small to allow of being measured.

Figures 30 and 31 give the position of the current-arrows corresponding to the perturbation on the 26th January, 1903. The current-arrows are constructed in the manner explained in Art. 23, by the aid of the values for P_h, P_d and P_v given in Table XI.

With regard to the times employed, it should be said that the first is chosen immediately after the commencement of the perturbation, and thus represents the magnitude of the perturbing forces that at that hour suddenly make their appearance upon the earth. After this hour the oscillations diminish somewhat — as Table XI and Plate XIV show — until at about $11^h\ 20^m$ in many places they have already become 0. Between 9^h and 12^h, the conditions at the various stations are on the whole only slightly changed, and remain fairly constant, with small perturbing forces. For this intermediate period therefore, no charts have been constructed. After 12^h, however, the oscillations begin to increase, attain their highest value a little before 14^h, and then rapidly decrease to zero. These conditions will be found represented on the last three charts. On Chart IV the length of the arrows in certain tracts is a little abnormal, as the way in which the force increases towards the equator is not very clearly distinguishable. This is partly accounted for by the fact that the force at this time varies so greatly, that a slight displacement in time may cause considerable changes. Even the small polar precipitations, moreover, will exert an influence. They will possibly assert themselves most in North America — Toronto and Sitka (cf. the perturbation of the 15th Dec., 1882; chap. III).

Current-Arrows for 26th January, 1903; Chart I at 9h and Chart II at 12h.

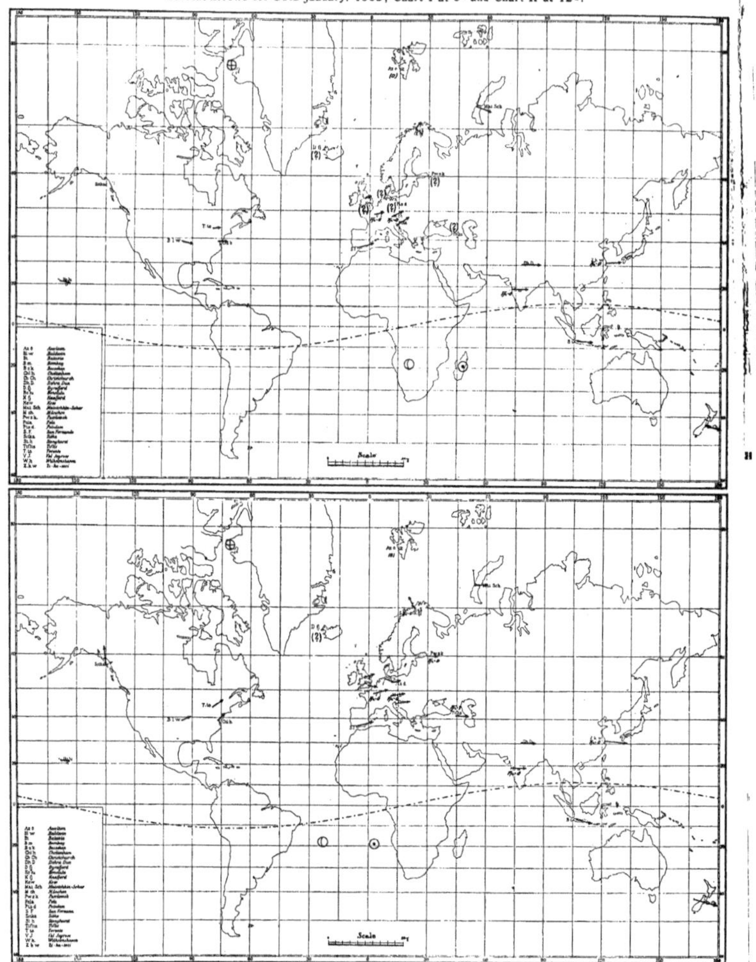

Fig. 30.

Current-Arrows for 26th January, 1903; Chart III at 13h and Chart IV at 14h.

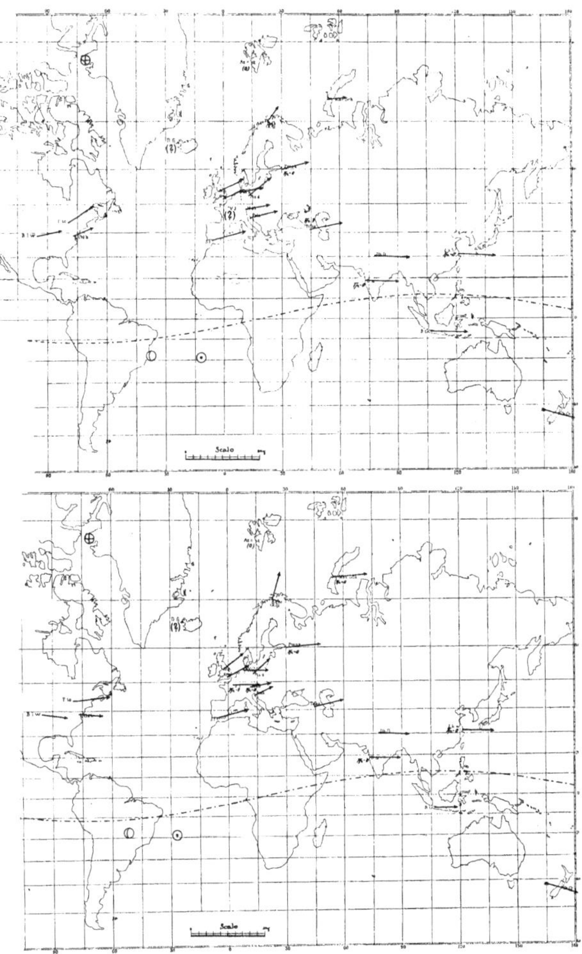

Fig. 31.

THE PERTURBATIONS OF THE 9th DECEMBER, 1902.

(Pl. IX).

28. These perturbations may be briefly characterised as follows.

They begin with a lengthy perturbation, which is relatively weak, but is especially developed at the equator, where it appears only in *H*, and on the whole exhibits all the properties that characterise the positive equatorial perturbations.

It commences quite suddenly, simultaneously all over the earth, at $5^h\ 40{\cdot}6^m$ Greenwich mean time. At the equator it appears only in *H*, and the deflection answers to an increase in *H*. In the vicinity of the poles, this condition is altered, while at the same time the mean deflection becomes smaller. From $5^h\ 40^m$ on to 13^h, the deflection in *H* is continued in the direction mentioned, but with varying strength. The character of the curve is somewhat quieter than usual. At the Norwegian stations there is a particularly strong and characteristic impulse at the commencement. At Matotchkin Schar, for instance, it is partly of an undulating form, answering to a rapid turning round of the perturbing force. Subsequently the perturbation at the three westernmost of the Norwegian stations is chiefly characterised by small oscillations about the normal line, interrupted by smaller, sometimes brief, impulses of a more local polar nature. Between 15^h and 18^h, the character of the perturbation-conditions is essentially changed. It is this feature that we continually find repeated, namely, that when the equatorial storm has lasted for some hours, polar systems appear.

It is early apparent from the curves at our Norwegian stations, that we have to do chiefly with polar storms during this period. The system, however, is of the very simplest kind. At Dyrafjord and Kaafjord the deflections in *D* and *H* are in a direction opposite to that usual during storms that commence on the midnight side. When we come to Matotchkin Schar, we get the deflection that characterises the nocturnal perturbations.

As this perturbation during several hours is of a typical equatorial character, we have preferred to class it among such. Even the polar storm with which it concludes, is a phenomenon that often seems allied to this equatorial type.

THE FIELD OF FORCE.

(1). The Equatorial Part.

29. The field during the period is given on two charts, Chart I for $6^h\ 0^m$, and Chart II for $9^h\ 0^m$.

This field is of the typical form for negative equatorial perturbations. It is most powerful on the sun-side, and becomes weaker towards the poles. On Chart II, the arrows have a direction that indicates that they are circling round the magnetic pole.

Chart III represents the conditions at $12^h\ 15^m$, and at 15^h. At the first-named hour, the perturbation is still mainly equatorial in character. At Axeløen and Sitka, only small polar disturbances are observable. At the second hour named, we are just at the transition to the polar field.

(2). The Field during the Polar Storm.

Charts IV, V and VI show the field as it appears, in the main, during the polar storm.

Chart IV shows the field at two hours, namely, $16^h\ 0^m$ and $16^h\ 45^m$. At the first of these, the perturbation was especially noticeable in Europe and Asia, where it forms a considerable area of divergence. At Dyrafjord, Kaafjord and Matotchkin Schar, the force is now very small. It appears, from the form of the field in southern latitudes(1), that the storm-centre is situated to the east of our Norwegian

(1) See "Polar Elementary Storms".

stations. At $16^h 45^m$, the perturbation on the whole has greatly increased in strength. We now have very powerful perturbations at our Norwegian stations. We recognise the form of field as the typical one for the polar elementary storms. The current-arrow in the storm-centre is now directed eastwards along the auroral zone; and in the district of Europe and North America, the field forms an area of divergence.

Chart V; $17^h 0^m$.

The field in southern latitudes has mainly the same character as at $16^h 45^m$; but at the Norwegian stations the conditions have changed.

At Dyrafjord and Kaafjord we still have a current-arrow directed eastwards along the auroral zone; but as regards Kaafjord, the force is considerably less. At Axeløen, where we now have registerings, the conditions are of a character altogether different from those of the two first-named stations. The current-arrow at Axeløen points almost due west. This indicates that the perturbations here must be of a somewhat local character. At Matotchkin Schar the direction of the arrow is reversed, and is now almost exactly opposite to that at Kaafjord. This indicates the existence of a new storm-centre, which is advancing from the east. These districts to the east of Matotchkin Schar are now upon the night side, and we find moreover that the current-arrow about the storm-centre af this system is directed westwards along the auroral zone.

Chart VI.

In lower latitudes the field is almost unchanged, except at Sitka, where a remarkable difference occurs. The conditions at Dyrafjord and Kaafjord are much the same as before; while at Axeløen and Matotchkin Schar the force has turned.

We see that this storm has a tendency to form a field similar to that described for the polar elementary storms. The circumstances are not, however, of the simplest. There is no doubt that we have to do with several simultaneous polar precipitations of electric corpuscles.

TABLE XII.

The Perturbing Forces on the 9th December, 1902.

Gr. M. T.	Honolulu		Sitka		Baldwin		Toronto		Cheltenham	
	P_h	P_d	P_h	P_d	P_h	P_d	P_h	P_d	P_h	P_d
h m										
6 0	+ 7.1 γ	0	+ 3.5 γ	W 2.2 γ	+ 5.3 γ	0	+ 5.4 γ	0	+ 3.4 γ	0
7 45	+ 6.3 »	0	+ 2.8 »	0	+ 2.5 »	0	+ 3.6 »	0	+ 1.3 »	0
9 0	+ 7.4 »	0	+ 4.1 »	» 1.8 »	+ 4.6 »	0	+ 6.3 »	0	+ 2.6 »	0
12 15	+ 6.6 »	0	+ 3.0 »	E 15.8 »	+ 9.2 »	W 3.8 γ	+ 8.1 »	W 4.8 γ	+ 3.9 »	W 4.1 γ
15 0	+ 3.8 »	0	− 11.2 »	W 9.5 »	+ 4.9 »	» 5.7 »	+ 6.3 »	» 3.6 »	+ 2.5 »	» 4.1 »
16 0	− 3.1 »	0	− 3.0 »	» 1.8 »	− 3.5 »	0	− 1.8 »	» 4.8 »	− 1.5 »	0
45	+ 3.1 »	W 5.0 γ	− 12.3 »	» 29.4 »	− 9.6 »	» 12.1 »	− 9.9 »	» 15.0 »	− 8.5 »	» 11.3 »
17 0	+ 5.6 »	» 3.3 »	+ 6.5 »	» 5.4 »	− 0.7 »	» 18.4 »	− 6.3 »	» 21.6 »	− 4.2 »	» 16.0 »
15	+ 3.3 »	» 2.5 »	+ 3.7 »	E 9.5 »	0	» 15.3 »	0	» 19.2 »	0	» 16.0 »

TABLE XII (continued).

Gr. M. T.	Dyrafjord			Axeløen			Matotchkin Schar		
	P_h	P_d	P_v	P_h	P_d	P_v	P_h	P_d	P_v
h m									
6 0	+ 2.7 γ	E 3.1 γ	+ 3.3 γ	0	E 3.8 γ	+ 9.8 γ	+ 5.6 γ	0	+ 5.8 γ
7 45	+ 2.7 »	» 6.2 »	− 3.7 »	0	W 12.3 »	− 17.2 »	+ 4.5 »	W 7.5 γ	− 10.2 »
9 0	+ 3.8 »	» 1.3 »	− 3.3 »	0	» 7.4 »	0	+ 7.1 »	» 10.6 »	− 2.9 »
12 15	?	?	?	+ 20.7 γ	0	+ 2.4 »	?	?	?
15 0	?	?	?	+ 16.6 »	» 7.4 »	0	+ 1.4 »	» 1.3 »	+ 5.1 »
16 0	+ 13.0 »	0	+ 9.8 »	?	?	?	+ 20.0 »	» 7.5 »	+ 14.6 »
45	+ 136.0 »	E 17.0 »	− 37.5 »	?	?	?	+ 27.6 »	» 60.0 »	− 175.0 »
17 0	+ 127.0 »	» 31.0 »	− 14.2 »	− 154.0 »	W 42.2 »	− 177.0 »	− 41.2 »	E 29.6 »	− 46.8 »
15	+ 73.0 »	» 8.3 »	− 17.8 »	− 5.1 »	» 27.3 »	− 160.0 »	+ 41.7 »	» 51.5 »	+ 41.0 »

TABLE XII (continued).

Gr. M. T.	Kaafjord			Pawlowsk			Stonyhurst		Wilhelmshaven		
	P_h	P_d	P_v	P_h	P_d	P_v	P_h	P_d	P_h	P_d	P_v
h m											
6 0	+ 4.1 γ	E 2.2 γ	0	?	?	?	+ 6.1 γ	0	+ 5.6 γ	E 2.4 γ	0
7 45	+ 3.5 »	W 4.8 »	0	?	?	?	+ 5.6 »	0	+ 7.4 »	» 2.4 »	0
9 0	+ 1.2 »	» 4.8 »	0	?	?	?	+ 14.3 »	0	+ 9.8 »	» 1.2 »	0
12 15	0	» 6.2 »	0	+ 6.0 γ	W 6.0 γ	0	+ 13 2 »	0	+ 7.9 »	0	0
15 0	0	» 1.4 »	+ 7.7 γ	0	» 0.9 »	0	+ 4.0 »	E 1.2 γ	+ 3.7 »	» 1.8 »	0
16 0	0	» 8.0 »	+ 15.5 »	− 7.0 »	» 1.8 „	0	− 2.5 »	W 2.8 »	− 6.1 »	W 3.7 »	0
45	+ 155.0 »	» 64.0 »	+ 11.3 »	− 13.1 »	E 7.3 »	+ 3.0 γ	− 17.8 »	» 4.0 »	− 20.5 »	0	+ 4.0 γ
17 0	+ 57.0 »	» 19.8 »	+ 19.0 »	+ 36.2 »	» 5.5 »	+ 5.2 »	− 14.3 »	» 15.4 »	− 27.5 »	» 26.3 »	+ 4.0 »
15	+ 32.0 »	0	+ 20.4 »	+ 22.6 »	» 5.5 »	+ 6.0 »	− 14.3 »	» 9.7 »	− 24.2 »	» 10.4 »	0

TABLE XII (continued).

Gr. M. T.	Kew		Potsdam		Val Joyeux			Munich		San Fernando	
	F_h	P_d	P_h	P_d	P_h	P_d	P_v	P_h	P_d	P_h	P_d
h m											
6 0	?	?	?	?	?	?	?	?	?	+ 8.3 γ	E 4.1 γ
7 45	?	?	?	?	?	?	?	?	?	+ 7.0 »	0
9 0	?	?	?	?	+ 9.6 γ	0	0	+ 8.5 γ	W 2.3 γ	+ 14.7 »	0
12 15	+ 9.7 γ	0	+ 6.6 γ	?	+ 9.6 »	0	0	+ 7.5 »	» 1.5 »	+ 5.7 »	0
15 0	+ 2.5 »	E 1.8 γ	+ 3.1 »	E 1.0 γ	+ 8.0 »	E 2.5 γ	+ 3.0 γ	+ 3.5 »	E 1.5 »	+ 5.1 »	» 4.9 »
16 0	− 2.0 »	W 1.4 »	− 4.7 »	W 3.0 »	− 1.6 »	0	+ 4.0 »	− 4.0 »	W 1.5 »	− 5.7 »	0
45	− 17.8 »	» 2.8 »	− 17.0 »	0	− 16.8 »	W 0.8 »	+ 5.0 »	− 16.0 »	0	− 14.1 »	W 4.9 »
17 0	− 16.3 »	» 12.2 »	− 26.5 »	» 13.0 »	− 17.6 »	» 12.5 »	+ 4.0 »	− 19.5 »	» 11.4 »	− 14.1 »	» 7.3 »
15	− 13.3 »	» 10.3 »	− 17.0 »	» 4.5 »	− 13.6 »	» 8.3 »	+ 3.0 »	− 14.5 »	» 8.4 »	− 8.9 »	» 5.7 »

TABLE XII (continued).

I. T.	Tiflis			Dehra Dun		Zi-ka-wei		Batavia		Christchurch	
	P_h	P_d	P_v	P_h	P_d	P_h	P_d	P_h	P_d	P_h	P_d
m											
0	?	?	?	+ 15.4 γ	W 4.9 γ	+ 13.2 γ	E 4.0 γ	+ 14.2 γ	0	+ 11.5 γ	0
45	?	?	?	+ 11.0 »	» 9.8 »	+ 7.3 »	» 2.0 »	+ 7.5 »	E 6.0 γ	+ 4.1 »	0
0	?	?	?	+ 15.8 »	» 6.9 »	+ 14.4 »	» 1.0 »	+ 11.0 »	» 2.4 »	+ 14.2 »	0
15	+ 6.9 γ	0	+ 0.5 γ	+ 10.6 »	0	+ 8.4 »	0	+ 7.8 »	0	+ 9.6 »	W 5.2 γ
0	+ 3.0 »	E 2.9 γ	0	+ 2.7 »	0	+ 1.2 »	0	+ 3.9 »	0	0	0
0	− 9.2 »	» 3.7 »	+ 2.3 »	− 10.2 »	0	− 7.2 »	0	− 7.8 »	0	− 8 7 »	E 1.5 »
45	− 15.4 »	» 11.9 »	+ 3.0 »	− 10.2 »	E 7.9 »	0	» 7.0 »	− 4.3 »	W 2.4 »	+ 10.6 »	» 3.0 »
0	− 23.0 »	» 12.2 »	+ 5.3 »	− 18.1 »	» 11.8 »	− 6.0 »	» 5.0 »	− 8.9 »	» 3.6 »	+ 11.0 »	W 4.5 »
15	− 17 8 »	» 8.5 »	+ 3.0 »	− 13.4 »	» 6.9 »	− 7.2 »	» 4.0 »	− 7.8 »	» 1.2 »	+ 6.0 »	» 3.0 »

Current-Arrows for the 9th December, 1902; Chart I at 6h.

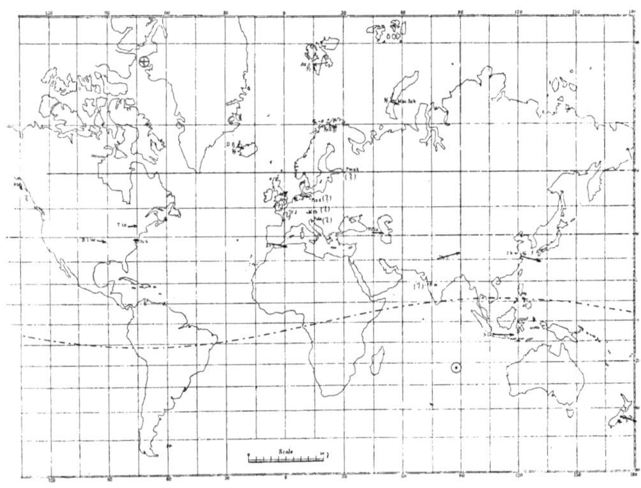

Fig. 32.

Current-Arrows for the 9th December, 1902; Chart II at 9h, and Chart III at 12h 15m, and 15h.

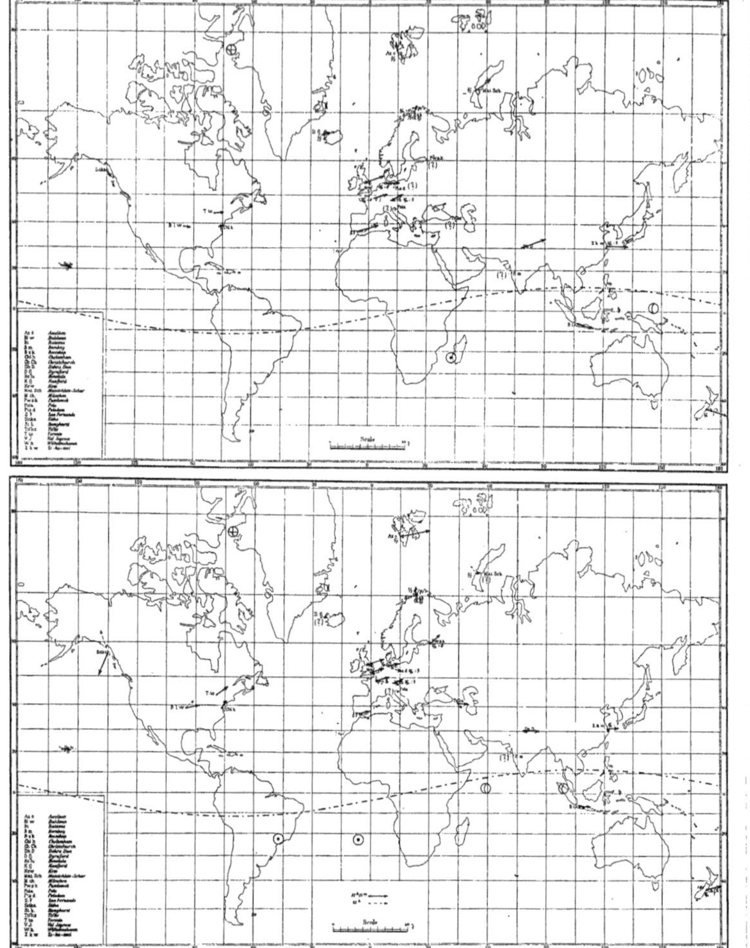

Fig. 33.

nt-Arrows for the 9th December, 1902; Chart IV at 16^h and 16^h 45^m, and Chart V at 17^h.

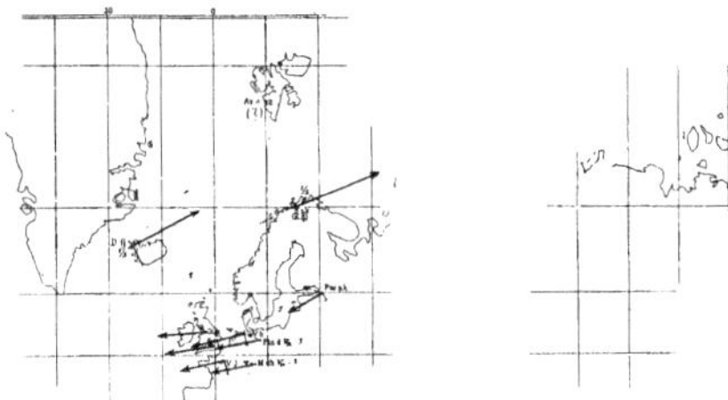

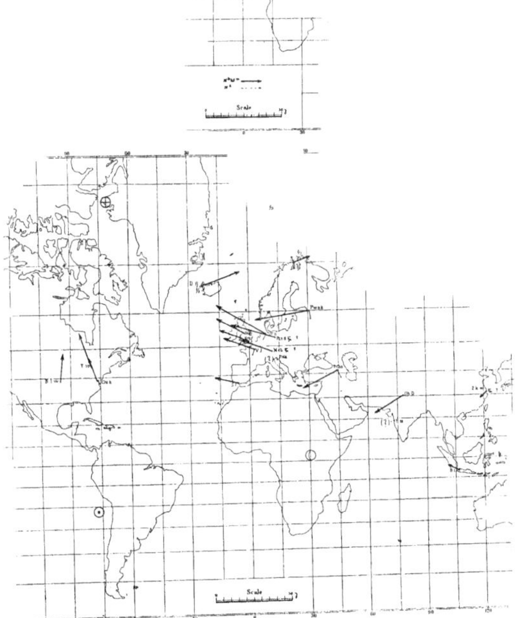

Fig. 34.

Current-Arrows for the 9th December, 1902; Chart VI at 17^h 15^m.

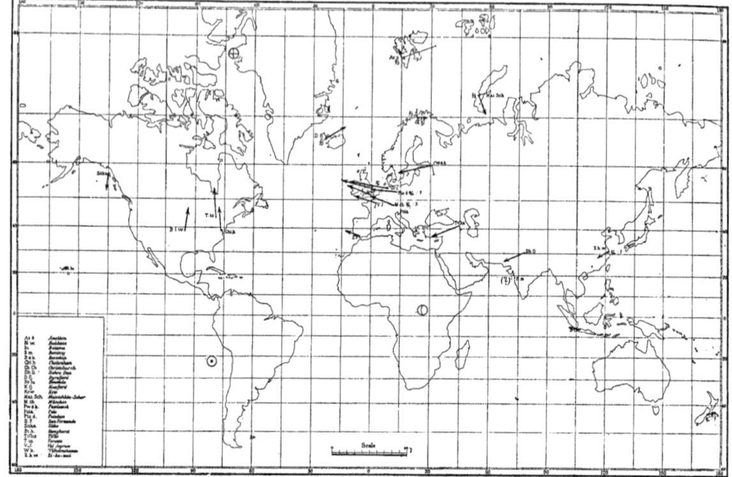

Fig. 35.

THE PERTURBATION OF THE 23rd OCTOBER 1902.

(Pl. III).

30. This perturbation does not belong to those mentioned in the circular, and is therefore from only a small number of stations.

It is especially developed about the equator, and is there characterised as a positive equatorial perturbation. It commences suddenly at 19^h 11^m, simultaneously all over the earth. The curve is serrated in character, and appears only in *H*, in which it occasions an increase.

About $1^1/_2$ hours later, a polar storm, not, indeed, violent, but characteristic, simple and well-defined, appears around the Norwegian stations. It is especially distinct at Matotchkin Schar. This storm has at the same time the properties that characterise the polar elementary storms. The current arrow points westward along the auroral zone, indicating that the storm-centre, which is situated in the region about Matotchkin Schar, now lies on the midnight side.

The field of force is shown on a chart, which represents the conditions at 19^h 16^m and 22^h 22.5^m.

At the first-named hour, the field exhibits a typical equatorial character; at the last-named, it is the effect of the polar system, which, at any rate in somewhat more northerly latitudes, is most conspicuous (see the polar elementary storms).

TABLE XIII.

The Perturbing Forces on the 23rd October, 1902.

Gr. M. T.	Toronto		Axeløen			Matotchkin Schar.		
	P_h	P_d	P_h	P_d	P_v	P_h	P_d	P_v
h m								
19 16	+ 20.0 γ	W 6.0 γ	0	W ca. 12 γ	0	+ 10.3 γ	W 14.2 γ	0
22 22.5	+ 8.5 »	0	+ 5.5 γ	» 21.5 »	+ 242.0 γ	− 110.0 »	E 27.6 »	− 37.0 γ

TABLE XIII (continued).

Gr. M. T.	Kaafjord			Munich		Pola		
	P_h	P_d	P_v	P_h	P_d	P_h	P_d	P_v
h m								
19 16	+ 9.6 γ	W 6.7 γ	ca. − 10 γ	+ 17.0 γ	W 3.8 γ	+ 14.8 γ	W 3.5 γ	+ 0.8 γ
22 22.5	− 57.0 »	E 33.0 »	− 138.0 »	+ 13.5 »	E 8.3 »	+ 13.9 »	E 8.3 »	0

TABLE XIII (continued).

Gr. M. T.	San Fernando		Dehra Dun		Bombay		Christchurch	
	P_h	P_d	P_h	P_d	P_h	P_d	P_h	P_d
h m								
19 16	+ 15.4 γ	E 3.0 γ	+ 17.0 γ	0	+ 14 γ	?	?	?
22 22.5	+ 8.3 »	» 6.4 »	+ 16.2 »	0	+ 13.3 »	?	+ 5.5 γ	0

Current-Arrows for the 23rd October, 1902, at 19h 16m and 22h 22.5m.

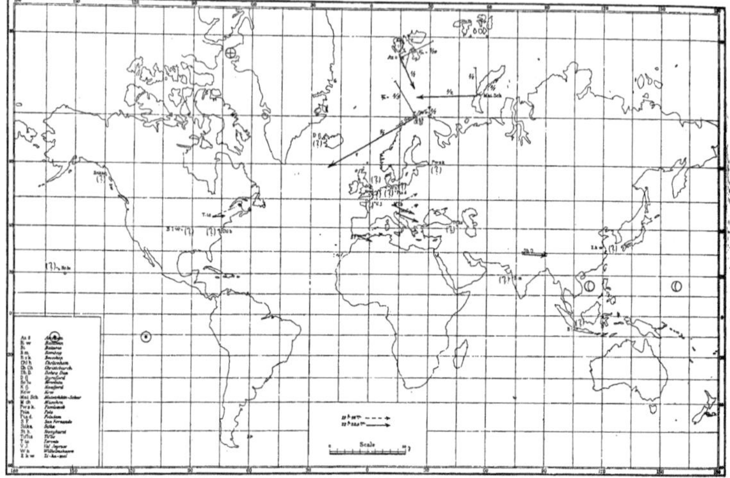

Fig. 36.

CONCERNING THE CAUSE OF THE POSITIVE EQUATORIAL PERTURBATION.

31. The fact that this type of perturbation exhibits such great simplicity with regard to the distribution of the force, and also that it shows such a tendency to repeat itself from time to time, indicates that these perturbations might have a simple explanation.

As already remarked in the introduction, it will always be possible, in a purely formal manner, to satisfy the properties of the field in several ways. It is our intention here to mention some of the possibilities that might perhaps explain these perturbations, and we will in the first place find out what magnetic systems might be assumed to have produced the field.

(1) We cannot assume a variation in the terrestrial-magnetic field itself, which would explain the field about the equator; for as we go north, the perturbing force is no longer directed along the total intensity. P is directed horizontally almost everywhere; in the south its direction is somewhat downwards, in the north often upwards. In the far north, moreover, P_1 (see p. 45) is no longer directed along the magnetic meridian.

(2) As we shall subsequently see, current systems will undoubtedly appear in the polar regions during a series of polar perturbations. It might then be reasonable to try whether this equatorial

perturbation might not also be explained by a polar current-system. Considering that the perturbation may be due to currents of a cosmic nature that approach the earth under the influence of terrestrial magnetism, there would be a possibility of the existence of current-systems that consisted of current-spirals, which stretched down at the poles, and in this way acted as though magnet poles were put down. Poles such as these, however, though they might explain the principal features in the form of the field, would not be reconcilable with the fact that the force increases towards the equator.

We are therefore of necessity led to seek the explanation in currents that have their greatest density in low latitudes near the magnetic equator. We thus naturally come to consider the two possibilities — the perturbation either has its direct cause in currents at the surface of the earth, or in currents above the earth.

It seems hardly likely that the phenomenon is due to earth-currents. These currents, it is true, would explain the small vertical intensity as regards magnitude, as it might be assumed that the current was distributed over a large portion of the earth's surface; but a wide-spread system of earth-currents such as this would hardly explain the other properties of the perturbation. The direction of the earth-currents must, in such a case, be from east to west, the reverse of the direction of the current-arrows marked; and it would then be difficult to explain how the force P has a component directed upwards north of the equator, and downwards south of the equator. Such earth-currents, if, as independent phenomena, they are to be able to explain the perturbations, cannot be induced currents, but must depend upon conditions in the earth itself. As, however, the direct cause must be sought in processes in the earth itself, it is incomprehensible how these currents can have so universal a character, and maintain so constant a direction with so singular a form. It seems especially impossible to explain the simultaneous serrations; for the perturbing force would then at each place principally be determined by that part of the current that passed beneath the place. From a physical point of view there are greater difficulties in assuming that different parts of a wide-spread current-system such as this, which should have its direct cause in the earth itself, should act rhythmically, and that the alteration of current-density with the latitude at each point of time should take place so regularly and connectedly. The question might, indeed, be settled, if they were surface-currents, by looking at the registerings for the earth-current. If the perturbation were conditioned by surface-currents on the earth, the curve of the earth-current should exhibit a course similar to that of the curve on the magnetograms. If, on the other hand, the perturbation is due to currents lying outside the earth, the curve for the earth-current will look like vibrations about the normal line, as the rapid changing in the perturbing force would produce corresponding induced alternating impulses.

We have no complete set of earth-current registerings, however, for any station except Kaafjord. Here, indeed, we do find that the earth-current curves are of the character described. They are undoubtedly for the most part induced currents, but their direction is mainly determined by the local conditions. as for instance the conductivity of the soil in the various directions.

When the great perturbations show maximal deviation, the earth-currents usually pass a 0 value.

As we shall see later on, it is easy to reconcile the existence of such conditions in the polar regions with the fact that certain magnetic disturbances in southern latitudes, far away from the storm-centre, may often in great part be caused by earth-currents.

The earth-currents will be treated in a subsequent part of this work.

We have already mentioned that this equatorial perturbation often comes as a precursor of polar storms; and indeed, we have really never met with an entire perturbation of this kind with which there have not, within the same period, been polar storms. The necessary consequence of this must be that these two kinds of perturbations should be closely connected with one another. Now there is no doubt, as will be shown later on, that the polar storms are due to currents above the earth; if so, this should also be the case with the equatorial perturbations now under consideration.

According to this, we must necessarily seek the cause of the perturbations in currents above the surface of the earth. If the current is to be sought at a distance from the surface of the earth that is small in comparison to the earth's dimensions, we must, in order to explain the field, have a wide-spread plane current circulating round the earth. Our being obliged to have a wide-spread plane system is a consequence of the fact that otherwise the fields would be limited more rapidly. If, as is the case, the effect is extended to all parts of the earth, there must also be currents in those regions. A system of this kind, however, if it is to satisfy the actual conditions, is inadmissible; for we meet here with difficulties similar to some of those in the way of the acceptance of the earth-current theory.

The first of these is that the *relative* strength of the current in the various districts of the earth should remain fairly constant throughout long periods, notwithstanding that the field, as already mentioned, is remarkable for great variableness in strength: the variations take place in all districts in about the same proportion. It seems, moreover, impossible, if we are not to have recourse to the mysterious, but keep to the well-known possibilities of physics for the production of cosmic currents, to have the stability of the current explained; for the current, as we know, if composed of free portions, is deflected by terrestrial magnetism, the separate bearers of the electric charge — whatever the physical nature of the latter — moving in spirals about the magnetic lines of force, or being carried out into space, if the corpuscular current-rays are stiffer.

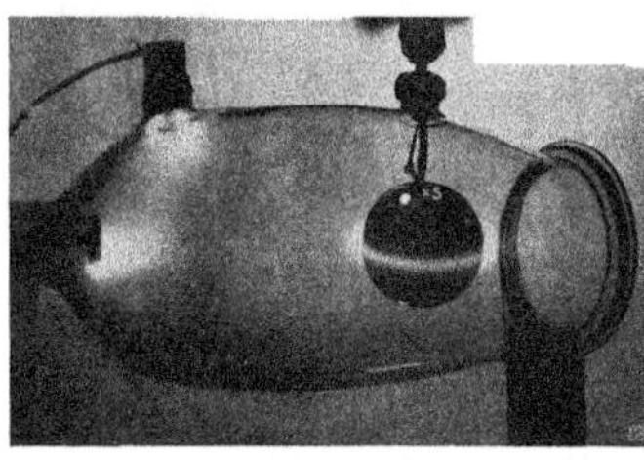

Fig. 37.

The only possibility then left is that the positive equatorial perturbations are due to the effect of a current-system, whose distance from the earth is of the same order as the dimensions of the earth. Owing to the distribution of force in the field, and the symmetry that is found, as a rule, with regard to the equator, this current, as already mentioned, must have its greatest effect about the plane of the equator; and on account of the direction of the perturbing force, the current-lines, at any rate in the region nearest the earth, must lie in planes that are approximately parallel with the plane of the magnetic equator.

There are still two essentially different cases possible here,

(1) that the current passes round the earth, and

(2) that the earth is quite outside the system that in the main conditions the perturbation.

When, on account of the field, the currents must be sought at so considerable a distance from the earth, we are compelled, with the knowledge we at present have of the physical qualities, to assume that these currents are corpuscular in constitution. The systems that may then be formed must be such as may arise when a magnet is subjected to corpuscular electric radiation of some kind or other.

In order to become better acquainted with the systems that may arise under these conditions, a little attention should be given to the experiments I have made, in which a magnetic terrella is exposed to cathode rays. These will be fully treated in Volume II, and illustrated by numerous photographs; but even here we will draw attention to a few important circumstances.

In addition to the polar precipitations there are still in particular two characteristic phenomena.

(1) Under certain circumstances there is formed round the terrella a very steady, luminous ring. As the system itself is confined within the form of a flat torus, the trajectories of the corpuscles in consequence form approximately entire circles (see fig. 37). Owing to terrestrial magnetism, such negative corpuscles in space, coming from the sun, must then move from west to east round the earth.

(2) At some height above the terrella, and on the side turned towards the cathode, we shall be able to get very well characterised systems. The existence of these systems may be shown by a phosphorescent screen, as illustrated in fig. 38 *a*, *b*, *c*, where the terrella is placed in three different positions in relation to the screen, as indicated by the diagram below the images.

The precipitations appear only on one side of this screen, and their inner border is sharply defined. The system is of considerable breadth. It does not remain in the neighbourhood of the equator, but extends on both sides, and fades away towards the poles, or unites with the polar system.

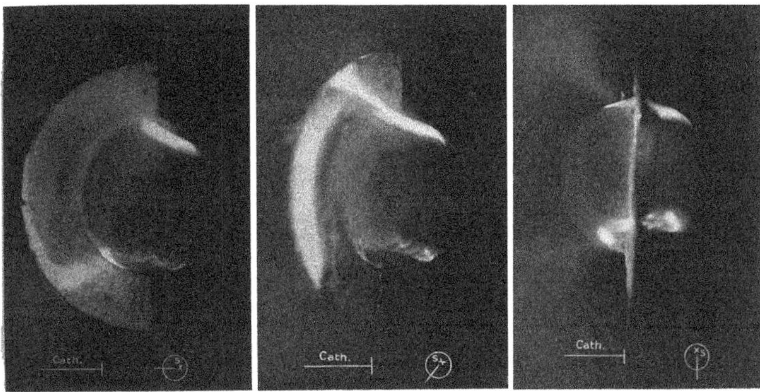

a. *b.* *c.*

Fig. 38.

The three figures, 38 *a*, *b* and *c*, show how cathode rays are drawn in towards a highly magnetic terrella. Both terrella and fixed screen are covered with phosphorescent substances. In position *a*, the screen points straight towards the cathode, that is to say, the plane of the screen is perpendicular to that of the cathode. In position *b*, the planes make an angle of 45° with one another; and in position *c*, the screen is parallel with the cathode-surface.

We can see how the rays are drawn in in rings or zones round the magnetic poles on the terrella itself; but the phenomenon to which we shall here pay special attention, is the strong light that is found only on the east side of the screen, and which is due to cathode rays that turned back before reaching the terrella. They are caught by the screen, however, and rendered visible. It will be seen that the mass of the rays turn back and come into contact with the screen in position *b*, answering to the afternoon side of the terrella. Professor Stormer has calculated the trajectories of electrically charged corpuscles sent by the sun towards the earth, and has, amongst other things, studied the course of the trajectories at the earth's magnetic equator.

Fig. 39 is taken from his paper, "Sur les trajectoires des corpuscules électrisés dans l'espace sous l'action du magnétisme terrestre"(¹). It will here be seen that rays answering to $\gamma = -0.5$ and -0.7 fall in towards the earth very much as do the greater number of the rays in the experiment with the terrella in position *b*. There will also certainly be rays coming in towards the terrella, that answer to the other mathematically possible paths; but it is not so easy to demonstrate them with this experimental arrangement with a screen.

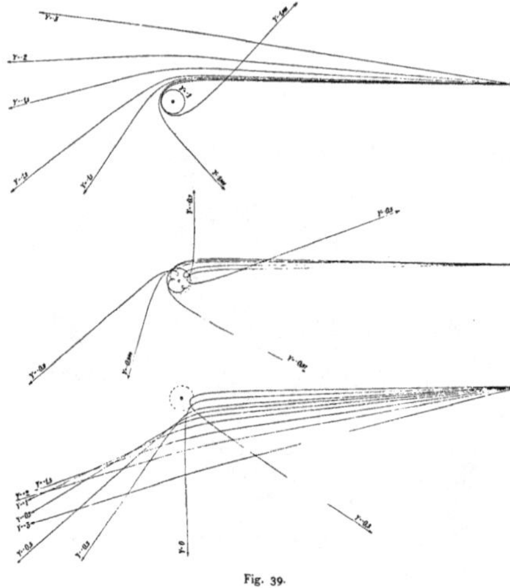

Fig. 39.

The phenomenon represented in fig. 37, of the ring of light round the equator, should answer to paths where $\gamma =$ about -1. The stronger the magnetism, the larger will the ring be. In the experiments shown in fig. 38, the magnetism is so strong that the equator-ring is not formed, owing to the glass walls of the discharge-tube. By the experiment with the terrella, it is also easy to show a phenomenon that is most easily explained by the presence of rays answering to the calculated paths for $\gamma =$ between -0.5 and -1. I have mentioned in a former work(²) that, just within the equator-ring, the terrella sometimes has a clearly phosphorescent line along the equator. I had formerly to have recourse to the assumption of secondary rays in order to explain this phenomenon; but it is now explained most naturally by rays answering to Størmer's calculated trajectories.

What we have to notice, however, is that the bulk of the rays in the experiment turn round in front of the terrella on the afternoon side. The mathematical treatment has hitherto given only the mathematically possible trajectories, but has not stated where the bulk of the rays pass the earth, partly because the nature of the rays emitted by the sun is not sufficiently known.

As the current-arrows during our perturbations are directed towards the east, the perturbation cannot be explained by a ring such as this round the earth. If, on the other hand, we assumed the permanent existence of such a ring, we might imagine the perturbation to be explained by a diminution in the strength of this current. This explanation is very improbable and unnecessary. It seems necessary, owing to the connection of these perturbations with the polar storms, to suppose that the equatorial

(¹) Archives des Sciences Physiques et Naturelles. Geneva. Vol. XXIV, 1907, chap. IV.
(²) Expédition Norvégienne de 1899—1900, l. c., p. 46.

perturbations under consideration are also due to the rising of new, independent systems, and do not merely indicate a weakening of that which may already exist.

On the other hand, it is our opinion that the positive equatorial perturbations find their natural explanation in the second of the two systems mentioned. At the place in which the earth is found, the system will have a force directed towards the north. If the system is far off in proportion to the earth's dimensions, the force round the equator can be almost constant. If the system is nearer, there will be a stronger effect upon the evening side. This is also what we find in reality, as the effect about Dehra Dun is somewhat stronger than at Honolulu. It must be remembered, however, that the observed force is also dependent upon the magnetic induction in the earth.

It would be useless to attempt here a more detailed description of these current-systems. It seems probable that at times they may have a somewhat different character, being at one time fairly symmetrical about the equator, and at another pushed out more towards the north or the south.

The experiment shows that the system may extend considerably in directions north and south. This, together with the effect of the magnetic induction of the earth, will account for the smallness of the vertical components.

We have observed certain impulses in the north that appear to be of a local character, as the force about the auroral zone might diverge greatly in direction at two adjacent stations, and receive a marked, opposite twist. The equatorial perturbation of the 22nd March, 1903, is an instance. This agrees very well with our view, as at times radial impulses may come right down to the earth about the poles. In the experiment, moreover, we see that the equatorial system finally unites with the polar; and we shall often have great polar precipitations of corpuscles. For this reason, a number of these perturbations will be found described under the polar storms.

THE NEGATIVE EQUATORIAL STORMS.

32. On several occasions in the course of our investigations of the composite magnetic storms, we shall meet with conditions in the field of force, which naturally lead to the assumption that the perturbing force in the polar regions, on account of its independence of the polar systems, must be due to systems that have their greatest strength in the equatorial regions. They differ, however, distinctly from the previously-described equatorial perturbations in two very important respects, namely:

(1) The perturbing force is directed southwards, answering to a current-arrow towards the west, and

(2) The curve has not the characteristic, serrated appearance that marks the positive equatorial perturbations. The latter generally appear very suddenly, whereas those now under consideration appear more gradually.

We have not succeeded, however, in finding in our material of this kind of perturbation, sufficiently distinct types to enable us to class them under any elementary form. In the treatment of the composite perturbations, we shall repeatedly have opportunities of examining more closely the reasons that determine the assumption of such perturbations. We may here mention as instances the perturbations of the 31st October, 1902, and the 8th February, 1903.

These, like the positive equatorial perturbations, have a very wide distribution, as the conditions of perturbation alter slowly from place to place. This, together with the quiet character of the curve, shows that the systems that are to condition the perturbation, must be sought at a considerable height above the earth. While we are thus led to suppose them to be corpuscular currents, we shall naturally be obliged to connect this perturbation with the circular systems, which, according to the theoretical investigations of the trajectories of electric corpuscles, can exist, and the possibility of which we have also proved experimentally by the previously-mentioned ring (see fig. 37).

THE POLAR ELEMENTARY STORMS.

33. One cannot look long at the curves for the registered magnetic elements without observing a regularity in a number of details, especially in the behaviour of the great storms. This, strange to say, is not least apparent at the stations round about the auroral zone, and especially in the storms that have occurred at our Norwegian stations during the period in which the magnetic conditions have been observed by us. In the first place, it appears that the great majority of storms of short duration are at their height at our stations at about midnight by local time; and when they make their appearance at that time, it is found that they nearly always cause oscillations in the same direction for the horizontal intensity and declination. We further find that the direction of the oscillation in the vertical curve, especially in the case of Axel Island and of Kaafjord, is also repeated time after time. We get a direct impression that, notwithstanding little accidental circumstances, the magnetic storms, in their formation and course, are controlled by very limited conditions, and that these conditions are pre-eminently fulfilled in very limited areas in the polar regions. This impression is opposed to the theory upheld by Ad. SCHMIDT(1) and other terrestrial-magnetists — that the magnetic storms are produced by free cyclonic electric current-systems.

In the well-known paper mentioned below, Professor SCHMIDT says:

"Electric currents have hitherto principally been accepted as the cause of perturbations, either currents in the ground or in the air, especially in the upper, probably better conducting strata of the atmosphere. Although no great clearness prevails as to the physical conditions under which such currents may occur, yet we shall venture to maintain this hypothesis, notwithstanding the objections raised against it by BIGELOW, the rather that no doubt can any longer exist as to the reference of the diurnal variation to such currents. Regarded from this point of view, these centres of action can hardly be anything else but current-phenomena that stand out with a certain distinctness from the current-system of the whole earth, on account of their intensity and individual limitation, in fact wandering current-vortices that, in the simplicity of the elementary perturbation, we may also expect as the normal, like the cyclones and anti-cyclones of the atmosphere".

The violent storms in the north are always accompanied by simultaneous perturbations, that are observable right to the equator; and as a rule we shall find, by direct study of the curves, that in general the effect becomes slighter towards the equator.

The important question now presents itself: In what way are the perturbations in southern latitudes connected with the perturbations in the north? Is there any simple connection at all?

In order to throw light upon these questions, we have made a careful investigation of a number of very simple storms. At the outset it is only natural to suppose that when we have a perturbation that runs the simplest possible course, this phenomenon will be particularly well adapted for throwing light upon the laws of the perturbation.

The next section will deal with a number of simple polar storms such as this, which we have picked out and called polar elementary storms. These, independently of any hypothesis, can be characterised as follows:

(1) They are comparatively strong at the poles. The simultaneously perturbing forces, even as far north as the 60th parallel, have already sunk to about a tenth of their strength in the auroral zone.

(2) They are of short duration, frequently lasting not more than two or three hours.

---

(1) Ueber die Ursache der magnetischen Störme. Meteorologische Zeitschrift, Sept., 1899.

(3) The conditions before and after are comparatively quiet.

(4) The oscillations at the polar stations, especially the more southern ones, run a simple course. At the poles, they are often characterised by a simple increase to a maximum, and decrease to zero. We may sometimes, even at the northern stations, have to some extent an undulating form, answering to a slow turning of the perturbing force.

It follows from this, that these perturbations must be well-defined, and thus afford an opportunity for an exact determination of the perturbing force.

THE TYPICAL FIELD FOR THE POLAR ELEMENTARY STORMS.

34. It proves — as the aggregate treatment of these elementary types of perturbations shows — that the same field of force is repeated almost exactly from perturbation to perturbation. It will therefore be most convenient for its description, to note, even at this point, its typical form, in order thereby to avoid too many repetitions. We shall then keep principally to the horizontal perturbing force, and the field that it forms upon the earth's surface.

In the auroral zone we have very great perturbing force, and we will call the regions about those places where the perturbation is strongest, the perturbation-centre or storm-centre. If we imagine ourselves moving along the surface of the earth, so as always to follow the direction of the horizontal component of the perturbing force, we should be moving along some curve or other upon the earth, which we will call a line of force.

Supposing we were to move in such a way as always to advance in the direction of the current-arrows, we should get another set of curves, which we will call current-lines. The one set of curves will intersect the other at right angles.

We will now suppose that we project these two sets of curves upon the earth's surface, upon a plane by some kind of zenithal projection, which at the same time is conform, and in such a way that the plane of projection is tangent to the earth in the storm-centre. The two sets of curves will thus be projected orthogonally.

If we imagine this done for the field of the various polar elementary storms, we shall obtain a system of lines, which, in the main, is of the form represented in figure 40 (p. 86). The continuous lines are the lines of force, the broken lines are the current-lines. C is the projection of the storm-centre, and the figure is symmetrical round it, as also on both sides of two axes, A and B, at right angles to one another. The former we will call the principal axis of the system, the latter the transverse axis. On the transverse axis, and symmetrical as regards the principal axis, are two points, from one of which the lines of force issue, while in the other they terminate. We will call the point from which they issue the point of divergence, and that to which they converge the point of convergence. The immediate surroundings of these points we will call respectively the field of divergence and the field of convergence. We find that the current-lines in these two fields form respectively positive and negative vortices. The field of force has some formal resemblance to the field induced by two opposite poles; but this resemblance disappears when we consider the strength of the force. At the two points in which the lines of force here meet, the horizontal force equals 0. In the neighbourhood of these points we have a neutral area. The perturbing force, then, should stand, at these points, perpendicular to the surface of the earth.

With regard to the vertical component, it may generally be said that except in the regions, nearest to the centre, it is exceedingly small in proportion to the horizontal. It is only in the points of divergence and convergence that P_v will predominate, athough it is generally comparatively small.

In order to obtain an idea of the conditions for P_v, we will consider the values along the transverse axis B. In the centre, C, P_v will equal 0. Starting from this point, P_v will rapidly rise to a maximum. On the side on which the point of convergence lies, the direction of P_v will be upwards, and on the other side downwards. After reaching the maximum, P_v again drops quite rapidly to a trifling value (see lower diagram, fig. 40).

With regard to the position of the point of convergence, we may note the following.

If we imagine an observer swimming out from the centre in the direction of the current-arrows, and with face turned towards the earth, the point of convergence will be to his left.

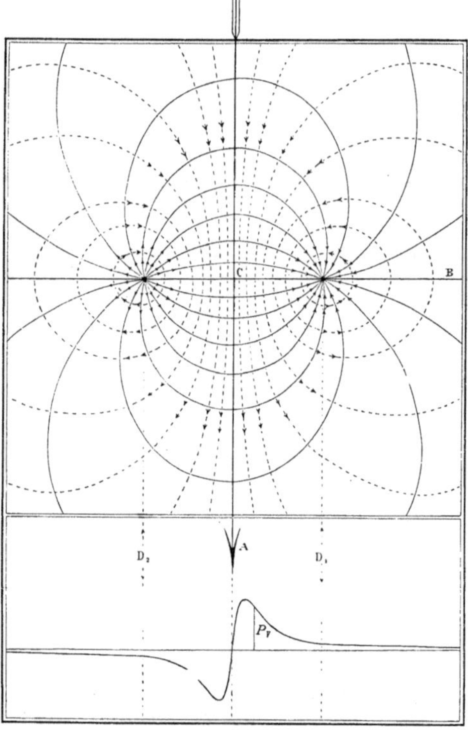

Fig. 40.

This then, in an idealised form, is the appearance of the field which has a tendency to develope during the polar elementary storms. It is not founded upon any sort of hypothesis, but is merely a collocation of what almost invariably takes place, and of which demonstration will be given in the treatment of the separate storms, when we shall also have an opportunity of going into the question of the forms of current that may be assumed to have produced a field such as this.

In comparing the above with the charts, we must remember that we there employ current-arrows. We must then compare these with the current-lines in fig. 40.

THE PERTURBATION OF THE 15th DECEMBER 1902.

35. This magnetic disturbance makes its appearance upon an otherwise very calm day. It begins, as the copies of the curves show, without any preceding equatorial perturbation, with a great storm in the north, about Dyrafjord and Axeløen, and is accompanied by a perturbation, small indeed, but well-defined, which is observed in northern America and Europe. The effect increases as we approach the above-named Norwegian stations. It is only just perceptible at Dehra Dun, and not at all at Zi-ka-wei, Batavia and Honolulu. There are unfortunately no magnetograms for that day from Christchurch.

The perturbation is of rather short duration. It is first observed at Dyrafjord about $0^h\ 10^m$, and reaches its maximum at $1^h\ 8^m$ with a perturbing force of 386 γ. At about $3^h\ 15^m$ the storm is over; but for a little while there are still slight oscillations to the opposite quarter.

On Axeløen the storm does not make its appearance until about 35 minutes later than at Dyrafjord, reaches its maximum at $1^h\ 46^m$ with a perturbing force of 193 γ, and is over at about $3^h\ 45^m$.

The strange thing is that the oscillations at Kaafjord and Matotchkin Schar are comparatively so small. At the first-named station, the perturbation begins at about the same time as on Axeløen, and reaches its maximum at $1^h\ 45^m$ with a perturbing force of only 39.6 γ. At Matotchkin Schar it begins at about $0^h\ 51^m$. The perturbing force reaches its maximum at about $1^h\ 9^m$, with 27 γ.

At the stations Toronto, Baldwin and Cheltenham, a peculiarity is apparent, in that the perturbation is not of equal duration in the horizontal intensity and the declination. In the horizontal intensity it takes place between $0^h\ 40^m$ and $3^h\ 3^m$, a period which coincides almost exactly with the time of the storm in the north. In the declination, on the other hand, the oscillation is of shorter duration, as it begins at $0^h\ 55.5^m$, but is well-defined and by no means inconsiderable. The oscillation in declination thus takes place at the time when the storm in the north is at its height.

In Europe, on the other hand, it begins rather suddenly at $0^h\ 45^m$, and simultaneously in the horizontal intensity and the declination. It lasts about 3 hours, but the time of its termination cannot be exactly fixed, as the oscillations decrease little by little.

This perturbation, as will appear from the above, has its origin in the northern regions. Its sphere of action, which is rather limited, is concentrated about the neighbourhood of Dyrafjord and Axeløen. The shortness of its duration, as also the comparatively calm character of the curves even during the perturbation, seems to indicate that this is a polar-elementary storm of the most typical nature; it appears to be produced by a coherent impulse, which increases to a certain size, and then again decreases to 0 during the course of the perturbation. At the same time, as the perturbation does not make its appearance at all places simultaneously, the perturbing cause must be supposed to move with a somewhat continuous motion.

The perturbing forces for the various places are calculated for a series of times (see the table), and there is a series of charts representing current-arrows answering to simultaneous perturbing forces. In studying the charts, the significance of the multiplier beside the current-arrows must always be kept in mind (see Art. 23).

TABLE XIV.

The Perturbing Forces on the 15th December, 1902.

Gr. M. T.	Sitka		Baldwin		Toronto		Cheltenham	
	P_h	P_d	P_h	P_d	P_h	P_d	P_h	P_d
h m								
1 0	− 7.0 γ	E 4.0 γ	− 5.7 γ	E 7.6 γ	− 9.9 γ	E 10.6 γ	− 7.1 γ	E 8.3 γ
15	− 9.0 »	» 0.4 »	− 7.1 »	» 7.6 »	− 9.9 »	» 16.4 »	− 6.8 »	» 14.2 »
30	− 10.6 »	0	− 7.4 »	» 10.1 »	− 9.0 »	» 18.4 »	− 6.2 »	» 15.3 »
45	− 9.7 »	W 2.7 »	− 6.3 »	» 4.4 »	− 7.0 »	» 10.9 »	− 6.2 »	» 9.4 »
2 0	− 7.9 »	» 3.6 »	− 6.3 »	» 0.6 »	− 8.1 »	» 0.9 »	− 7.4 »	» 2.4 »
15	− 5.8 »	0	− 4.6 »	0	− 4.7 »	0	− 5.6 »	» 1.2 »
30	− 5.3 »	0	− 5.7 »	» 1.9 »	− 5.9 »	0	− 5.0 »	0
45	− 3.9 »	0	− 4.6 »	0	− 6.3 »	0	− 5.3 »	0

TABLE XIV (continued).

Gr. M. T.	Dyrafjord			Axeløen			Matotchkin Schar		
	P_h	P_d	P_v	P_h	P_d	P_v	P_h	P_d	P_v
h m									
12 30	− 56.9 γ	0	+ 19.1 γ	− 7.8 γ	E 8.7 γ	+ 46.7 γ	− 4.9 γ	E 6.6 γ	0
45	−141.5 »	W 50.3 γ	+ 35.8 »	− 7.8 »	» 13.6 »	+ 66.3 »	− 2.1 »	» 3.1 »	0
1 0	−345.7 »	» 19.1 »	+ 12.5 »	− 24.8 »	» 23.1 »	+ 103.0 »	− 16.8 »	0	− 10.7 γ
15	−273.5 »	» 6.9 »	− 1.7 »	− 27.6 »	» 42.3 »	+ 135.0 »	− 18.7 »	» 4.4 »	− 4.3 »
30	−206.2 »	» 8.7 »	− 22.5 »	− 70.4 »	» 69.9 »	+ 184.0 »	− 20.4 »	» 2.7 »	− 2.8 »
45	−237.4 »	E 33.0 »	− 61.6 »	−158.2 »	» 109.1 »	+ 159.5 »	− 13.4 »	» 3.1 »	− 9.9 »
2 0	−171.2 »	» 17.4 »	− 75.7 »	−158.7 »	» 79.4 »	+ 137.5 »	− 6.8 »	» 8.9 »	− 17.8 »
15	−114.9 »	» 17.4 »	− 63.6 »	−101.2 »	» 68.3 »	+ 132.5 »	− 4.7 »	» 12.0 »	− 24.1 »
30	− 70.0 »	» 7.9 »	− 34.9 »	− 78.2 »	» 49.0 »	+ 122.6 »	0	» 7.1 »	− 21.3 »
45	− 58.0 »	» 9.7 »	− 21.2 »	− 59.3 »	» 32.1 »	+ 98.5 »	+ 2.1 »	» 3.5 »	− 24.1 »
3 0	− 35.0 »	» 7.9 »	− 20.8 »	− 39.6 »	» 28.8 »	+ 63.0 »	+ 7.3 »	» 2.7 »	− 22.0 »

TABLE XIV (continued).

Gr. M. T.	Kaafjord			Pawlowsk			Stonyhurst	
	P_h	P_d	P_v	P_h	P_d	P_v	P_h	P_d
h m								
12 30	− 4.2 γ	E 2.6 γ	The balance has probably stuck, or has been out of order in some other way, as there is only a very slight perturbation in V.					
45	− 7.1 »	» 1.8 »						
1 0	− 23.8 »	W 15.4 »		− 5.0 γ	W 15.6 γ	0	+ 7.7 γ	W 16.8 γ
15	− 33.3 »	» 15.0 »		− 2.5 »	» 15.2 »	0	+ 10.2 »	» 9.4 »
30	− 38.1 »	» 9.5 »		0	» 14.3 »	− 0.7 γ	+ 10.7 »	» 6.3 »
45	− 29.8 »	E 9.5 »		+ 5.0 »	» 6.4 »	− 1.5 »	+ 8.2 »	E 1.7 »
2 0	− 21.4 »	» 18.4 »		+ 6.0 »	0	− 4.1 »	+ 4.1 »	» 6.3 »
15	− 11.9 »	» 19.1 »		+ 4.0 »	E 3.7 »	− 4.1 »	0	» 4.3 »
30	− 7.7 »	» 17.6 »		+ 1.5 »	» 3.7 »	− 3.4 »	− 3.1 »	» 2.9 »
45	0	» 11.4 »		0	» 3.2 »	− 1.9 »	− 5.1 »	0
3 0	+ 2.4 »	» 11.0 »						

TABLE XIV (continued).

Gr. M. T.	Kew		Val Joyeux			Wilhelmshaven		
	P_h	P_d	F_h	P_d	P_v	P_h	P_d	P_v
h m								
1 0	+ 8.2 γ	W 13.6 γ	+ 11.2 γ	W 12.2 γ	− 4.0 γ	+ 4.2 γ	W 20.2 γ	Small ne-
15	+ 8.9 »	» 11.0 »	+ 11.6 »	» 10.5 »	− 5.0 »	+ 8.9 »	» 15.3 »	gative de-
30	+ 9.2 »	» 6.5 »	+ 12.0 »	» 5.8 »	− 4.5 »	+ 12.1 »	» 10.4 »	flection.
45	+ 6.6 »	0	+ 8.8 »	0	− 4.0 »	+ 13.6 »	0	
2 0	+ 4.1 »	E 6.1 »	+ 5.6 »	E 4.6 »	− 3.5 »	+ 11.7 »	E 7.0 »	
15	− 0.5 »	» 5.6 »	0	» 3.4 »	− 2.0 »	+ 5.6 »	» 5.5 »	
30	− 3.8 »	» 3.0 »	− 1.6 »	» 1.7 »	− 1.5 »	0	» 3.4 »	
45	− 3.1 »	0	− 1.6 »	0	− 1.0 »	− 0.9 »	0	

TABLE XIV (continued).

Gr. M. T.	Potsdam		San Fernando		Munich		Dehra Dun	
	P_h	P_d	P_h	P_d	P_h	P_d	P_h	P_d
h m								
1 0	+ 3.2 γ	W 16.8 γ	+ 6.4 γ	W 2.0 γ	+ 6.0 γ	W 7.6 γ	− 2.8 γ	W 4.9 γ
15	+ 6.6 »	» 12.4 »	+ 13.4 »	» 3.3 »	+ 8.5 »	» 13.0 »	− 2.0 »	» 3.9 »
30	+ 9.1 »	» 8.6 »	+ 12.1 »	» 2.5 »	+ 9.0 »	» 9.9 »	− 0.8 »	» 3.9 »
45	+ 9.1 »	» 1.0 »	+ 11.5 »	0	+ 9.0 »	» 4.9 »	+ 2.0 »	3.4 »
2 0	+ 7.9 »	E 4.1 »	+ 8.3 »	E 5.7 »	+ 7.5 »	E 1.9 »	+ 3.5 »	» 3.0 »
15	+ 2.2 »	» 3.6 »	+ 4.5 »	» 3.3 »	+ 3.5 »	» 3.0 »	+ 3.2 »	» 3.0 »
30	− 1.3 »	» 2.0 »	0	» 3.3 »	0	» 1.5 »	+ 1.6 »	» 3.0 »
45	− 2.2 »	0	− 1.9 »	0	− 2.3 »	» 0.8 »	0	» 3.0 »

TABLE XIV (continued).

Gr. M. T.	Ekaterinburg		
	P_h	P_d	P_v
h m			
1 0	− 4.0 γ	W 9.5 γ	0
15	− 2.4 »	» 9.5 »	0
30	+ 2.0 »	» 7.0 »	− 1.0 γ
45	+ 4.5 »	» 4.2 »	− 1.8 »
2 0	+ 5.0 »	» 1.1 »	− 2.0 »
15	+ 5.0 »	0	− 1.8 »
30	+ 5.0 »	0	− 1.0 »

From Pola and Christchurch no magnetograms were received.

At Batavia, Zi-ka-wei, and Honolulu, the perturbation was so slight that the perturbing force cannot letermined. On the charts it is marked 0.

For Bombay and Tiflis there is no declination-curve. In the case of Tiflis there is a noticeable urbation in the horizontal intensity.

Current-Arrows for the 15th December, 1902; Chart I at 1h, Chart II at 1h 15m.

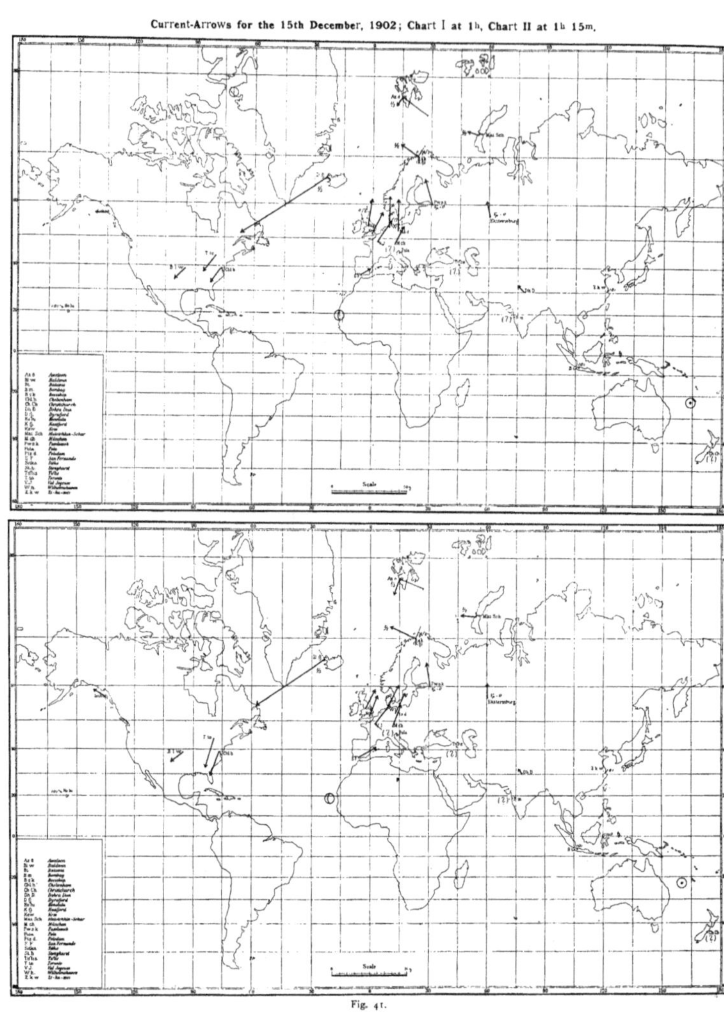

Fig. 41.

Current Arrows for the 15th December, 1902; Chart III at 1h 30m, Chart IV at 1h 45m.

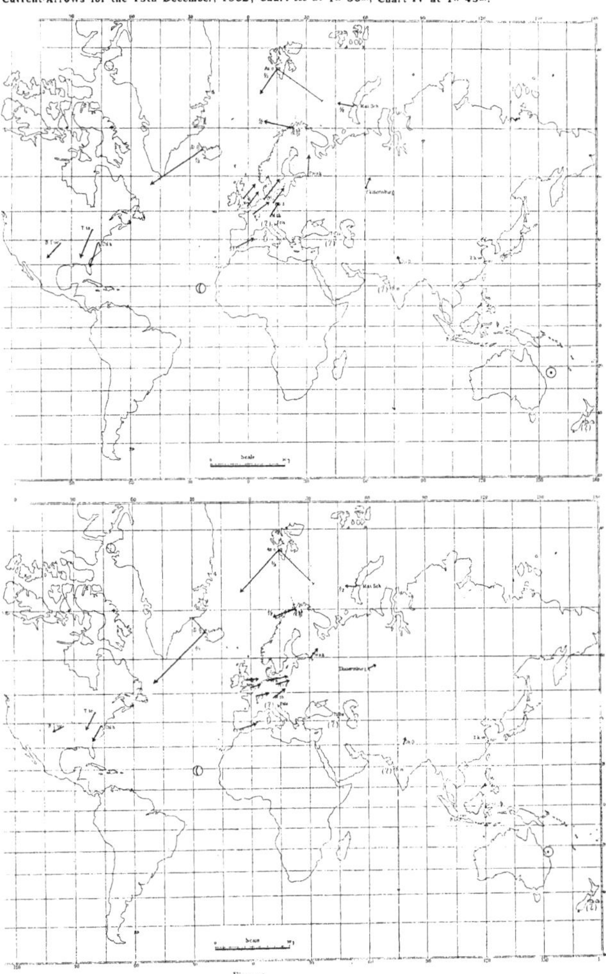

Fig. 42.

Current-Arrows for the 15th December, 1902; Chart V at 2h, Chart VI at 2h 15m.

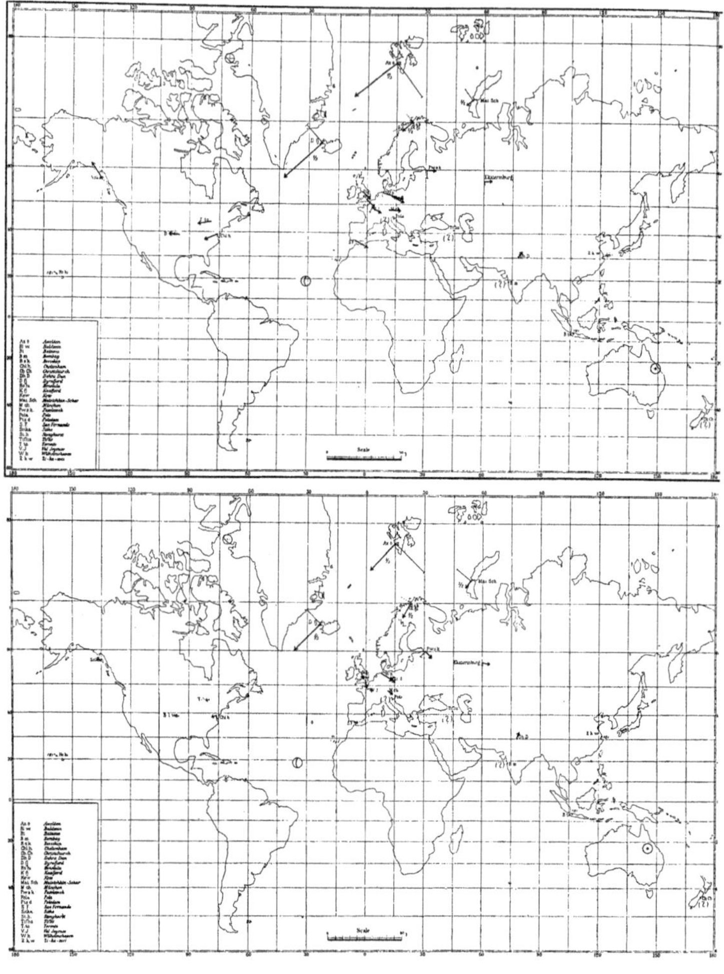

Current-Arrows for the 15th December, 1902; Chart VII at 2^h 45^m.

Fig. 44.

Chart I shows the conditions at 1^h, or about the time when the perturbing force for Dyrafjord has its maximum; and we see that it has a direction characteristic of this place, namely south of west. At the other Norwegian stations, the perturbing force is small at the same time, notwithstanding that these stations are situated about the line of direction of the current-arrow at Dyrafjord. We notice further that the current-arrows at these three stations converge towards one point.

Taking the European stations, the current-arrows show that the perturbing force for San Fernando at this hour has a north-westerly direction, while farther north it goes almost due west. As far north as Pawlowsk, its direction is WSW, and at Bossekop SW.

The perturbing force at Toronto, Cheltenham and Baldwin, is directed towards the SE, as is usual during those polar storms which are especially violent at the Norwegian stations. At Sitka, its direction is S. We notice that the arrows for these four places appear to issue from the same spot at the south point of Greenland.

Chart II. Time 1^h 15^m.

The conditions as a whole are the same as in Chart I. The force has increased in strength at Axeløen, and decreased at Dyrafjord, while the directions are the same. In the mean time P_e at Dyrafjord has changed its direction.

The arrows for Sitka and Baldwin, and still more for the European stations, have turned a little in direction from the left towards the right. This direction we will designate as the positive direction.

Chart III. Time 1^h 30^m.

The arrow at Axeløen has increased and assumed a direction more in accordance with Dyrafjord, where the force has decreased in strength, but is unchanged in direction. P_2 for Dyrafjord is directed upwards, for Axeløen downwards.

The conditions in America are very much like those at 1^h 15^m. In Europe, the arrows have turned farther in the same direction.

Chart IV. Time 1^h 45^m.

The perturbing force at Axeløen is now of about the same magnitude as at Dyrafjord. The condition of the vertical components is the same. The arrow for Kaafjord has turned a little in direction, so that it is more in accordance with Dyrafjord and Axeløen; but the force is still small.

The conditions in America are almost unchanged, except that the forces have diminished in strength. In Europe, the turning is continued in a positive direction. At Dehra Dun, where the horizontal component of the perturbing force has been directed towards SW, the force has now also taken part in the turning. The direction is now WNW.

Chart V. Time 2^h.

The force at Axeløen is now greater than at Dyrafjord. The condition of the vertical components is the same as before. At Kaafjord and Matotchkin Schar, the direction of P_1 is now in accordance with the two first-named stations, and P_2 for both is directed upwards.

In the rest of Europe, the turning of P_1 is continued in the same direction. In America also, the horizontal forces are turned a little in the positive direction.

Chart VI. Time 2^h 15^m.

The distribution of force is the same, but the intensity is less. The turning in Europe is continued a little.

Chart VII. Time 2^h 45^m.

The force on the whole weaker, except in America, where it seems to be somewhat greater than it was at 2^h 15^m. Otherwise the distribution of force the same.

We see, on the whole, that at each separate point of time, the field presents in its main features the typical form mentioned in the introduction to this chapter. The position of this field is determined in the following manner.

The principal axis is tangent to the auroral zone, and the current-arrow is directed towards WSW. As we have seen, the spot of the greatest effect moves in the direction from Dyrafjord and Axeløen, or, in other words, the centre moves eastwards along the auroral zone, but in such a manner that the principal axis always keeps its direction. While this strong impulse in the north is moving, *the field in lower latitudes moves with it.*

The district of Central Europe here comes in the area of convergence, and outside the point of convergence. The regular turning of the force, both in this district and at Kaafjord, has its simple explanation in the actual circumstance that the field in its entirety is moving forwards.

CONCERNING THE CAUSE OF THE PERTURBATION.

36. The cause of the great magnetic disturbance at Dyrafjord, and subsequently at Axeløen also, must mainly be sought in the effect of a horizontal current. This follows from the fact that the places of the greatest effect are found for a long distance in the direction of the current-arrow, while in the direction perpendicular to it, the effect very quickly diminishes. At $1^h\ 45^m$, for instance, the perturbing force at Dyrafjord is 240 γ, at Axeløen 193 γ, and the direction about the same, reckoned from the meridian of the place. At the same time, the strength at Kaafjord and Matotchkin Schar is respectively 39.6 γ and 26.6 γ, and the distance between Dyrafjord and Axeløen is 1809 kilometres, while between Axeløen and Kaafjord it is only 896 kilometres (see fig. 11).

In the district between Dyrafjord and Axeløen we must assume a horizontal current, which ought to flow fairly close to the earth for a long distance; for, owing to the rapid diminution in the effect out towards the sides, the current must flow rather low in relation to the earth's dimensions. We shall return to this later on.

We may conclude from the vertical intensities that it must be a current above the earth's surface. This is proved in the case of similar storms (see February 10th and March 31st, 1903), also by a consideration of the earth-current curve; but this is unfortunately wanting for the day under discussion.

With regard to the further course of the current, there are two possibilities that may be considered.

(1) The entire current-system belongs to the earth. The current-lines are really lines where the current flows upon the earth's surface, or rather at some height above it.

(2) The current is maintained by a constant supply of electricity from without. The current will consist principally of vertical portions. At some distance from the earth's surface, the current from above will turn off and continue for some time in an almost horizontal direction, and then either once more leave the earth, or become partially absorbed by its atmosphere.

According to the first assumption, the total current-volume at Dyrafjord and Axeløen should be squeezed together so that the greater part of it must pass through a comparatively small section, while the electricity, both before and after, should be spread over a wider section. In this case the current-lines drawn on fig. 40 would possess a physical reality, as there should actually be currents above the earth, somewhat in the direction of the current-arrow, answering to these current-lines.

It is true that systems of plane currents can always be arranged for a given field, which, from a purely mathematical point of view, would be able to explain the field; but when we consider the physical conditions for the formation of such a system, we meet with great difficulties, for it is not easy to comprehend what terrestrial processes would be able to maintain a current with this peculiar form, which moreover remains constant for several hours.

In my report of the 2nd Aurora Expedition — "Expédition Norvégienne de 1899—1900", etc. — I assumed such a system of horizontal currents in order to explain the magnetic perturbations. But the currents there are imagined as having come into existence mainly as a secondary effect of the electric corpuscles from the sun drawn in out of space, and thus far come under the second of the possibilities mentioned above. With observations from Pawlowsk, Copenhagen, Potsdam, Paris, Greenwich and Toronto as a foundation, I have drawn up a chart of the ordinary current-directions at midnight, Greenwich mean time, which is reproduced in fig. 45. It will be seen how well these current-directions fit into the current-lines in the idealised diagram, fig. 40.

There does not appear, however, to be any special reason why a current-system upon the earth should maintain such fixed directions and such a motion. If this were only a single case, one might perhaps regard it as a freak of nature. Among all the phenomena that occur from time to time, some will

Current-Lines at Midnight.

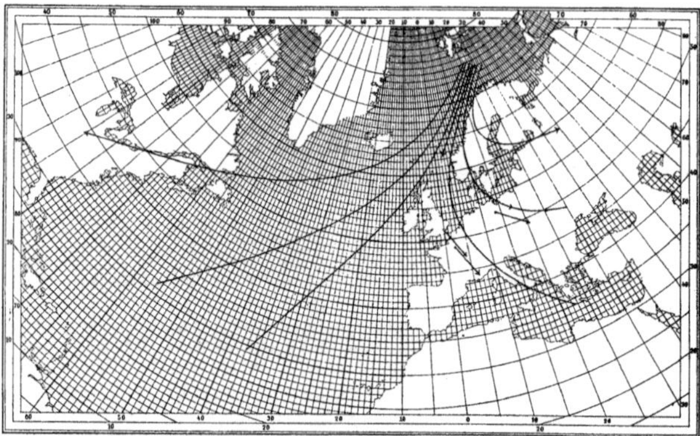

Fig. 45.

assume strange forms. But this is not an isolated case; as the entire treatment of these great polar storms will show, we shall always, in them, find again the same direction for the current about the Norwegian stations. We know, however, no circumstances connected with the earth itself and its immediate surroundings, that are sufficient to explain why one direction should so persistently predominate. A current such as this, moreover, which is a surface-current, would have to keep in the higher strata of the earth's atmosphere. It would have to be a corpuscular current in a medium in which these corpuscles can freely move out to the sides. The direction of the current would thereby be compelled to conform to the laws for the deflection of such currents in the terrestrial-magnetic field. But with an acquaintance with the laws for these movements, it is immediately evident that quite different forms would then be produced.

If such plane currents were possible at all, one would have to assume that the corpuscles, on account of some properties belonging to the upper strata of the atmosphere, would be obliged to move within a spherical shell situated at some distance above the earth's surface; for if the electric rays are at all pliable, they will in the main follow the lines of force, and from the polar regions these issue quite vertically. The rays might either go out into space, or back to the south pole of the earth. If the rays were very stiff, they would certainly for a time be able to keep approximately horizontal, but would at last have to run out into space, so that no entire circle of the above-mentioned kind would be formed.

Those rays, moreover, that move approximately horizontally at the poles, would have to turn off to the same side; or, in other words, on the northern hemisphere there would only be positive vortices, or areas of divergence for the perturbing force. But, as we see, we also have areas of convergence of a very simple form.

This brings us to the necessity of considering more closely the second possibility, namely, that the current is fed by a fairly constant supply from without, lasting for several hours. The supply must then, in the first place, be given in the regions in which the perturbation is strongest; and the strong perturbations in the north ought to be a direct effect of the descending current, which acts as

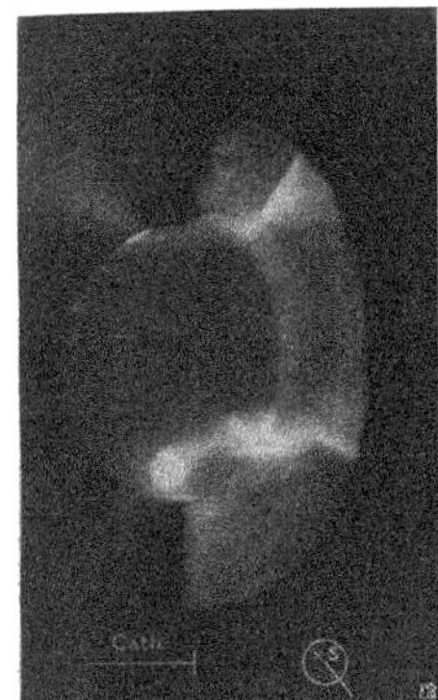

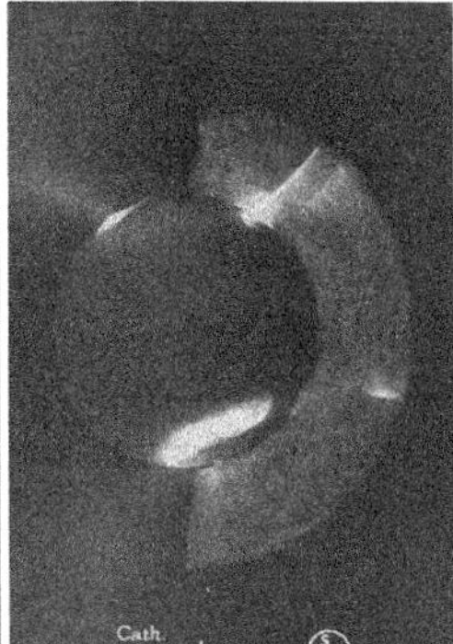

a. *b.* *c.*

Fig. 46.

a horizontal current for a long distance between Dyrafjord and Axeløen. This would satisfactorily explain the constant direction that the perturbation in this and other similar cases shows.

In order to obtain a clear conception of the conditions, we will once more have recourse to my experiments with the terrella. The experiments shown in fig. 46, *a*, *b* and *c*, follow directly on to those in fig. 38, *a*, *b* and *c*. In fig. 46 *a*, the terrella is so turned that the screen forms an angle of 135° with its first position (fig. 38 *a*). In the next experiment (fig. 46 *b*), the angle is 180°. The angles are here measured from west to east. Fig. 46 *c* shows how the cathode rays strike the terrella; when the latter is not magnetic, but is in the same position as in the experiment given in fig. 46 *b*, only the half that is turned towards the cathode becomes luminous with phosphorescence.

It will be seen from figs. 46 *a* & *b* how the cathode rays behave when the terrella is very powerfully magnetised.

We will here especially direct our attention to the luminous wedge that is thrown upon the screen at about the 70th parallel of latitude north.

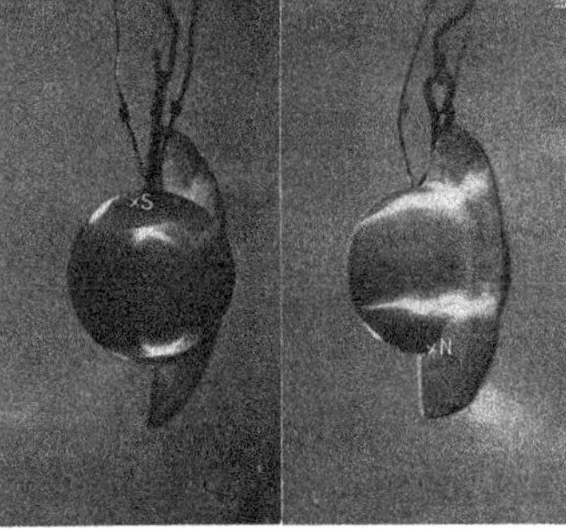

a. *b.*

Fig. 47.

In figs. 47 *a* & *b*, we have a confirmation of the way in which the rays whirl round the terrella in the above-mentioned wedge-shaped spaces about the poles. The screen here forms in both cases an angle of 270° with its original position (fig. 38 *a*), and the photographs are now taken from directions that form angles of respectively 120° and 240° with the plane of the screen in its original position, and not, as all the previous ones, from a direction making an angle of 90° with the screen in its original position.

The way in which the photographs were generally taken was to first expose the plate for about five seconds during the cathode-light experiment, and then, in order to obtain a picture of the terrella itself, to expose the latter for several minutes, illuminated by lamplight.

These experiments clearly show by analogy how, for instance, cathode rays from the sun will force their way towards the earth in the auroral zone, in such a manner, however, that the bulk of the rays are inclined to slip past it on the night side. The magnetic effect of the rays upon the earth would then be comparable to an ordinary electric current above the earth, whose direction is the reverse of that of the rays, thus approximately from east to west.

In order to find out whether currents of rays such as these are actually capable of explaining the multiplicity of magnetic perturbations, we must first try to obtain an idea of the exact course of the rays in the vicinity of the earth, and of the relative strength of the bundles of rays.

Owing to its deflection by terrestrial magnetism, the current from without can, as we have seen, only enter very limited districts, which will alter according as the magnetic axis assumes various positions in relation to the point on the sun that is the source of the rays.

We must therefore expect to find constant conditions for the current, which, when circumstances are favorable, can force its way down to the earth; at any rate, it will be easy to understand that distinct directions may thereby occur, as the electric rays, in order to come in, must follow paths whose initial direction lies within narrow limits.

Further, if the rays come from bodies lying outside the earth, the variation in the position of the points of radiation in relation to the magnetic axis, which is occasioned by the rotation of the earth, could give an explanation of the entire movement of the system, as the initial conditions are thereby continually varied.

If we assume, as, from a physical point of view, we might legitimately do, that the current is of a cosmic nature, and consists of negatively or positively charged corpuscles, the trajectories of the separate corpuscles must, as already stated, more or less approximately follow the magnetic lines of force, moving in spirals about them.

This will at any rate be the case with the hitherto known rays of this kind, such as ordinary cathode rays, β rays and α rays, and within a distance from the earth a few times greater than the diameter of the earth.

We should then, in this perturbation of the 15th December, have to consider the effect of a long vertical current, which, in the case of negative corpuscles, must come near to the earth at about Dyrafjord, or somewhat west of it, answering to an ascending galvanic current. A little above the surface of the earth it turns eastwards, or rather the aggregate effect of the cosmic current relative to the earth is as that of a galvanic current that is directed westwards, or more accurately towards the south-west.

In this descent of electric corpuscles, some will occasionally come so near the earth that they will be partially absorbed by its atmosphere, and will then eventually give rise to aurora. If the earth were able to retain an electric charge, we should have approximately horizontal currents, which would be necessary for the production of electrical equilibrium. But secondary electric radiation ought also to begin, and then, as it is still influenced by terrestrial magnetism, give rise to vertical ray-currents. The bulk of the corpuscles, however, must be imagined, as shown by experimental and theoretical investigations, as able to return, owing solely to this very influence of terrestrial magnetism, and give rise to reversed electric currents. Starting from physical considerations, we are thus naturally led to seek to explain the field by a system, which, in its average effects, has the character of two vertical currents in opposite directions, connected by a horizontal part.

In their main features, the conditions for such ray-currents can approximately be settled, as there is a long series of experimental and theoretical investigations on the course of cathode rays in a magnetic field. It will be sufficient for our purpose to refer to papers by POINCARÉ (1), myself(2), STØRMER (3) and VILLARD (4).

In accordance with the facts learned from the above-mentioned papers, I have here put forward a hypothesis regarding the course of the rays in the vicinity of the earth, by which, as it will be seen, the magnetic fields of force observed during magnetic storms are explained in a simple manner.

Figure 48 illustrates by diagram this hypothesis, which is to the effect that the rays — which are drawn in towards the earth in the sharply wedge-shaped space in the polar regions, always whirling around the magnetic lines of force, (fig. 48 *a*) — either, as generally happens, pass the earth with an average curvature such as is shown by the curve *b*, or, less frequently, with a loop such as curve *c* shows.

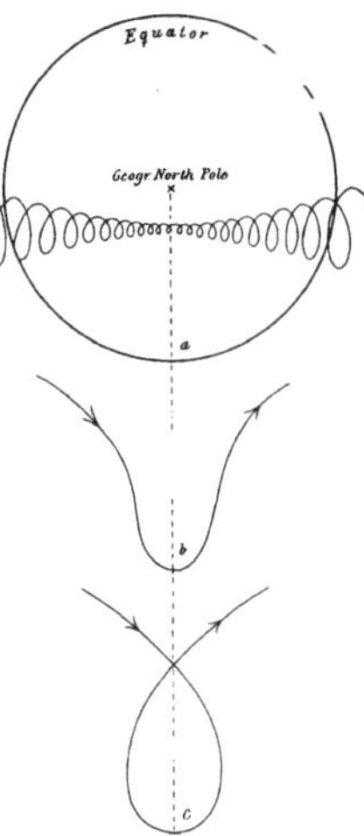

Fig. 48.

In those regions of the earth in the auroral zone, that lie close beneath the rays, the rays in the lowest bend of the curves *b* and *c* will mainly condition the magnetic disturbances; and the perturbing forces produced will be in reverse directions in the two cases. This will mean that the current-arrows for this area will generally point from east to west along the auroral zone (answering to the form of curve *b*), while less frequently the reverse direction may occur (corresponding to the form of curve *c*).

In the equatorial perturbation of the 9th December, 1902, it is mentioned that the direction of the polar storm that finally supervenes, is the reverse of our ordinary polar night storms. We thus have before us a field that can be explained by a current-system, the effect of which is the same as that produced by a linear current of about the same form as the loop in fig. 48 *c*.

We shall farther on meet again and again with these reversed polar storms. Fields similar to that of the 9th December will often be formed, principally on the noon and afternoon side, frequently breaking suddenly in upon an ordinary polar storm, only to disappear again as suddenly, when the first storm once more resumes its course.

In reality, the violent deflections that are found in nearly all magnetograms from the polar regions during a storm, are probably due to "loops" appearing locally, and repeatedly coming and going nearly over the place of observation.

---

(1) POINCARÉ. Remarque sur une expérience de M. BIRKELAND. Comptes Rendus 123, p. 930, 1896.
(2) KR. BIRKELAND. Archives des Sciences Phys. et Nat. Geneva (4) p. 497, 1896; and September, 1898.
(3) C. STØRMER, Sur le Mouvement d'un Point, etc. Videnskabsselsk. skrifter i Mathem.. Naturvidensk. Cl. No. 3. 1904.
(4) Comptes Rendus, June 11 & July 9, 1906.

At great distances from the polar regions, e. g. in the south of Europe, only the mean magnetic effect of the precipitation in those regions will make itself felt.

The question that now presents itself for closer consideration is, Will a galvanic current such as this give rise to a field such as we have found for the storm now under discussion?

By the aid of the elementary law for the effects of electric currents, it will be easy to see that such will be the case.

At great distances it will be mainly the two long vertical parts of the current that will be of decisive effect. In the vicinity of the storm-centre, the effect on P_1 of the vertical parts will be opposite to that of the horizontal part; but as the latter lies nearest the earth, it will predominate in these regions. If, however, we come out along the transverse axis of the system, we shall reach a point at which the horizontal component will equal 0, and farther out its direction will be reversed.

As approximately the long vertical portions of the current are a necessity for the appearance of these polar storms in the auroral zone, and as it is they which should especially give rise to the universal part of the perturbation, this explains in a simple manner the fact that the polar storms are always accompanied by perturbations in lower latitudes. It also gives an explanation of a circumstance which is especially distinct in this perturbation, namely, that the variations in the field with time are called forth by the motion of a field with a constant form.

This current-system further explains the following typical properties of the polar storms:

(1) That during the storm the curves for the arctic stations undergo great and sudden changes with time and place, in accordance with our supposition that the current in these regions really comes near the earth.

(2) That the curves in lower latitudes, during the great polar elementary storms, exhibit a very even course, that the form of the curve may be preserved over comparatively large regions of the earth, and that the transitions take place very gradually. The explanation of this is simple, namely that the magnetic disturbances are the effect of a comparatively distant system. The variations that will appear in certain parts of the current-system, and which give to the curves their very jagged character around the storm-centre, are not observable at great distances, as we then only get the average effect outwards of that which takes place within the current-space.

(3) It explains the peculiarity which these elementary polar storms exhibit, in appearing with such comparatively great strength around the auroral zone, while we find, as a rule, that southwards the strength suddenly drops to a small fraction of what it is at the centre.

(4) It explains an exceedingly characteristic quality of the magnetic storms, namely, that it is only around the storm-centre that the vertical component of the perturbing force has a magnitude of the same order as the horizontal component; while in lower latitudes, it will, as a rule, even during the greatest storms, be only just perceptible with the apparatuses generally employed. Its value in Central Europe seldom exceeds 8γ. The only place where P_v may have a greater value in relation to P_1 (see Art. 14) is near the points of convergence and divergence, where P_1 equals 0.

It is easy to see that our current-system must give rise to a condition such as this. In the neighbourhood of the storm-centre, the effect will be mainly determined by the horizontal part. If we consider the effect of this portion of the current out, for instance, along the transverse axis, the direction of the magnetic force, which was horizontal immediately beneath the current, gradually becomes more vertical. At the two points, one on each side of the principal axis, in which the tangential plane through the horizontal current touches the surface of the earth, the force will be exactly perpendicular to that surface, and thus the horizontal component = 0.

Farther along the transverse axis, the effect in the horizontal plane will be the reverse of those previously found, and P_1, as those points are passed, turns round to the opposite direction.

If we assume the two other portions of the current to be perfectly vertical, they will only give rise to a magnetic force that is perpendicular to them, and thus everywhere horizontal, if the earth is considered as a homogeneous sphere.

In the storm-centre and its immediate surroundings, these vertical currents will counteract the horizontal portion of the current. Farther out along the transverse axis, we shall reach two points situated symmetrically in relation to the principal axis, at which the effect of the horizontal portion in a horizontal direction will be neutralised by those of the vertical currents. These two points then, answer to those that we have previously designated as the points of convergence and divergence. Still farther away from the storm-centre, from the moment of passing the points of tangency already mentioned, the horizontal and the resultant of the two vertical portions will act in the same direction, and thus strengthen one another.

From the points of convergence and divergence then, P_1 will increase rapidly; at a certain distance it will attain a maximum, and then once more decrease.

With regard to P_v, we find that it is only the horizontal portion that can produce a force such as this. One would expect, moreover, to find the vertical components strongest along the transverse axis, at two points situated one on each side of the principal axis, and not far from the storm-centre. At the point of convergence, P_v should be directed upwards, at the point of divergence downwards.

Along the principal axis, it will be chiefly the horizontal current that acts, at any rate in the district that comes between the two vertical currents. In this district, the vertical currents will act contrary to the horizontal. As we pass the points in which the vertical currents produced will meet the principal axis, the nearest vertical portion will act in the same direction as the horizontal.

In the quadrants enclosed between these axes, the effect of the nearest vertical portion at rather greater distances will predominate; and the distribution of force will be as shown in fig. 40.

We have thus seen that the chief features of the form of the field in such a system, answer completely to those that are typical of an elementary polar storm.

We cannot, however, without more ado, draw any conclusion as to the distribution of intensity; it is possible that these fields corresponded only qualitatively, not quantitatively. I have therefore made a calculation of the effect along the transverse axis of some systems such as this. This is sufficient, as the form of the field is thereby given accurately enough. The actual current-conditions do not answer so exactly to these assumed linear currents with two vertical portions and one horizontal, as to make it worth while going into details.

If we consider, in the first place, the magnetic effect of an infinitely narrow rectilinear piece of current on a magnetic mass 1 $cm^{\frac{3}{2}}\ g^{\frac{1}{2}}\ sec^{-1}$, we find that

$$K = \int_a^b \frac{i}{10} \cdot \frac{ds}{r^2} \cdot \sin\alpha = \frac{i}{10}\int_a^b \frac{y\,ds}{(y^2 + s^2)^{\frac{3}{2}}}$$

$$= \frac{i}{10y} \left. \frac{s}{\sqrt{y^2 + s^2}} \right|_a^b ,$$

y being the distance from the point under consideration to the current, and r and α respectively the distance of the point from the current-element under consideration, and the angle made by the element with the direction to the pole. The direction of the force is found by Ampère's rule, and as limits, must be inserted the distances of the terminal points from the perpendicular that can be dropped from the point under consideration to the current-line.

Here i is assumed to be expressed in amperes, therefore K in dynes. This we will apply to a current-system of the form mentioned above, assuming that the horizontal portion of the current lies at a height h above the storm-centre, and has a length of $2l$.

The distance from the storm-centre in degrees along the transverse axis, we will designate ψ, the horizontal magnetic force-component, produced by the portions of current I, II and III along the transverse axis, respectively $P_{I\psi}$, $P_{II\psi}$ and $P_{III\psi}$, and the other magnitudes as given in fig. 49. We will call the force positive when it is directed towards the storm-centre, if we are on the same side as the point of convergence, and negative if we are on the opposite side.

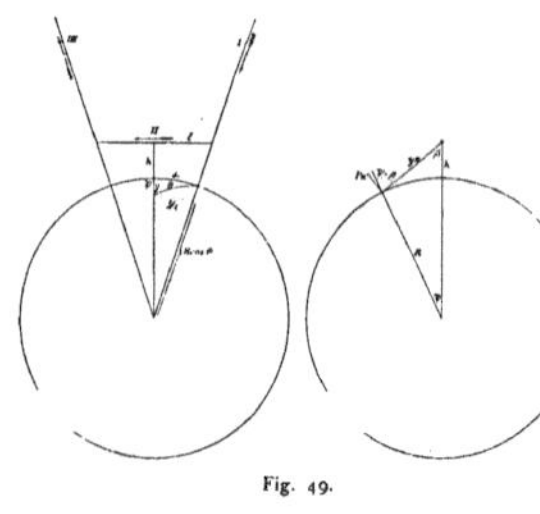

Fig. 49.

We then obtain

$$P_{II\psi} = -\frac{i}{10y}\left|\frac{s}{\sqrt{y^2+s^2}}\right|_{-l}^{+l} \cdot \cos(\psi+\beta) =$$

$$= -\frac{i}{5y} \cdot \frac{l}{\sqrt{y^2+l^2}} \cos(\psi+\beta)$$

Here $y = R\dfrac{\sin\psi}{\sin\beta}$,

where R is the radius of the earth. β is determined by the equation

$$\tan\left(\frac{\psi}{2}+\beta\right) = \frac{2R+h}{h} \cdot \tan\frac{\psi}{2}.$$

The equation can thus be written in the form

$$P_{II\psi} = -\frac{i}{5R} \cdot \frac{\sin\beta}{\sin\psi} \cdot \frac{l}{\sqrt{y^2+l^2}} \cos(\psi+\beta).$$

In the storm-centre itself we have

$$P_{II_0} = -\frac{i}{5h}\frac{l}{\sqrt{h^2+l^2}}$$

We further obtain

$$P_{I\psi} + P_{III\psi} = 2\,\frac{i}{10y}\left|\frac{s}{\sqrt{y^2+s^2}}\right|_{a}^{\infty} \cdot \sin\gamma$$

where

$$a = \frac{l}{\sin\alpha} - R\cos\theta,$$

$$y = R\sin\theta,$$

$$\sin\gamma = \frac{\sin\alpha}{\sin\theta},$$

and

$$\cos\theta = \cos\alpha \cdot \cos\psi.$$

If ζ is determined by the equation

$$\tan\zeta = \frac{R\sin\theta}{\dfrac{l}{\sin\alpha} - R\cos\theta} = \frac{R\sin\theta\sin\alpha}{l - R\cos\theta\sin\alpha},$$

we obtain

$$P_{I\psi} + P_{III\psi} = \frac{i}{5R} \cdot \frac{\sin\alpha}{\sin^2\theta}\left[1 - \cos\zeta\right] = \frac{i}{5R}\,\frac{2\sin\alpha \cdot \sin^2\frac{\zeta}{2}}{\sin^2\theta}.$$

n has been made in three cases, and the result is given in the tables below.

$$R = 6366 \text{ km.}$$

mean radius of the earth.

values, given in the table, correspond to a current-strength of 10^6 amperes, and the are expressed in γ.

TABLE XV

$h = 200$ km.; $2\,l = 1600$ km.

ψ	$P_{II\,\psi}$	$P_{I\,\psi} + P_{III\,\psi}$	P_{ψ}
0°	− 970.13	+ 166.78	− 803.35
10°	− 7.08	+ 61.69	+ 54.61
30°	+ 2.81	+ 9.74	+ 12.55
45°	+ 2.24	+ 4.29	+ 6.53

$h = 300$ km.; $2\,l = 2500$ km.

ψ	$P_{II\,\psi}$	$P_{I\,\psi} + P_{III\,\psi}$	P_{ψ}
0°	− 648.26	+ 100.77	− 547.49
10°	− 21.00	+ 58.57	+ 37.57
30°	+ 3.59	+ 13.08	+ 16.67
45°	+ 3.22	+ 6.07	+ 9.29

$h = 300$ km.; $2\,l = 5000$ km.

ψ	$P_{II\,\psi}$	$P_{I\,\psi} + P_{III\,\psi}$	P_{ψ}
0°	− 661.91	+ 48.30	− 613.61
10°	− 26.08	+ 40.36	+ 14.28
30°	+ 6.15	+ 17.23	+ 23.38
45°	+ 5.93	+ 9.43	+ 15.36

$h = 200$ km.; $l = \infty$

ψ	0°	10°	30°	45°
P_{ψ}	− 1000.00	− 12.32	+ 12.09	+ 14.05

$\psi = 0°$

$2\,l$	h	$P_{II\,\psi}$	$P_{I\,\psi} + P_{III\,\psi}$	P_{ψ}
1 000	300	− 571.66	+ 177.55	− 394.11
400	200	− 706.94	+ 353.99	− 352.95
200	200	− 417.21	+ 204.21	− 212.97

It will be seen from the above that there is also a quantitative correspondence between the actual field and that which is produced by the calculated systems.

The first answers to a system in which the horizontal portion of the current lies at a height of 200 kilometres, its length being a little less than double the distance between Kaafjord and Axeløen, or than the distance from these two stations to Dyrafjord; it is thus a comparatively low, compressed system. It appears that the force here diminishes a little more quickly than it is found to do during our most typical elementary storms.

In the second system, on the other hand — in which the horizontal portion of the current is at a height of 300 kilometres, its length being 2500 kilometres — the distribution of force shows a great resemblance to that found during the polar elementary storms. The length of the horizontal portion is here a little less than the distance between Dyrafjord and Matotchkin Schar, which is roughly 3000 kilometres.

For the value $\psi = 10°$, we have passed, as the table shows, the point of convergence or divergence, and the perturbing force is about $\frac{1}{15}$ of what we find at the storm-centre. At greater distances from this, the force varies in a manner corresponding fairly well with that found during the polar elementary storms.

In the third system the horizontal portion of the current is 5000 kilometres in length, and at the same height above the storm-centre as in the preceding case. The points of convergence and divergence are now situated at a rather greater distance from the storm-centre; and for greater values of ψ, the forces are now of a comparatively greater strength than before.

On the whole, the fields produced by the last two current-systems correspond fairly exactly with those found during the polar elementary storms.

In order, in the next place, to investigate the effect of the horizontal part, if that part became very long, we have calculated the effect for $l = \infty$. We then see distinctly how the directions change at the above-mentioned points of tangency.

On a closer examination, it will be easily seen that $P_{II\varphi}$ at the storm-centre, and its immediate surroundings, will always be greater than $P_{I\varphi} + P_{III\varphi}$. In order to inquire into the manner in which the latter change in relation to one another, we have, in the next place, calculated the effect at the storm-centre of some systems of various forms, where the horizontal portion of the current is made comparatively short.

We see, that for the small values of l, i.e. $2\,l = 400$ and 200 km., with the horizontal part lying at a height of 200 km. above the storm-centre, the proportion between $P_{I\varphi} + P_{III\varphi}$ and $P_{II\varphi}$ is about 1 : 2. For the third system, $2\,l = 1000$ and $h = 300$ km., the proportion is somewhat less.

Finally, we have calculated some forces along the principal axis, in order to obtain a general idea of the way in which the forces change here. The formulæ that will be employed are developed in a manner exactly similar to the previous ones; all that has to be done is to insert in the general formula some other values for distance and limits.

There is no need for a more careful investigation here, and we have therefore contented ourselves with calculating a few values for the system $2\,l = 1600$, $h = 200$. We have chosen this especially, in order that the changes might be more noticeable.

For this system we have found $\alpha = 6°\ 56'.8$.

In the storm-centre, and at the distances 2° 30′ and 5° from it, we have found the respective values

$$-803.35, \quad -756.13 \quad \text{and} \quad -603.06.$$

Here too, then, the change is not so great when we keep between the two vertical currents. If we withdraw farther to the other side of one vertical current, however, the force will diminish more rapidly.

If we look specially at the perturbation under discussion, we see, true enough, that the vertical components at the Norwegian stations have about the same magnitude as the horizontal component.

The conditions at these stations at 1 a.m. have already been mentioned. From these it appears that the total perturbing force at Kaafjord, Axeløen and Matotchkin Schar may be explained as the effect of a galvanic current, which drops at a certain angle towards the earth in a direction from Axeløen towards Dyrafjord. The current here is so near the stations, that the nearest part will be the important part. We make use of the law that when we approach an infinitely thin conductor, in which a stationary current is flowing, the effect will be approximately that which would be obtained if the system were replaced by an infinitely long current of the same strength, which passed through the nearest point on the conductor.

The conditions which we have educed from our current-system for the vertical components in more southerly latitudes, are corroborated in a striking manner by comparing the conditions at the few other stations from which we have received the vertical curves for this perturbation. In accordance with our hypothesis, the vertical components in these latitudes are very small in comparison with the horizontal. For instance, at 1^h and $1^h\ 15^m$, P_v for Pawlowsk and Ekaterinburg is imperceptible, whereas at the same time, in the case of Val Joyeux, which is situated nearer the point of convergence, the oscillation in the vertical curve is distinct, although faint, and answers to a perturbing force directed upwards.

Now when the current-system moves towards ENE, we should expect that the vertical intensity would also become noticeable at the two first-named stations, since, by the movement, they would be brought into the area in which vertical components might be expected. This is confirmed by the actual circumstances.

In the following charts, we find a noticeable vertical component for Pawlowsk and Ekaterinburg, while at the same time it diminishes in the case of Val Joyeux, but is directed upwards in all three.

As the effect is so limited, and the vertical components so great, the width of the current must be small in proportion to, for instance, 1000 kilometres. I have supposed a maximal width of 500 km. in my report, "Expédition Norvégienne", etc., l. c., p. 26, although it is probable that the boundary is not sharply defined. It must therefore be understood that it is the main body of the current that has this narrow width.

It follows from the cosmic constitution of the whole current, that the form we have assumed for the current-system that shall be able to explain the field, is only an ideal form, which in its main features characterises the system; and further it is to be understood that it is the total effect outwards that in its principal features is explained by a system such as this. It does not follow from this, of course, that the trajectories of the separate corpuscles must coincide with the direction of the assumed system.

The field at each place is in reality the sum of the magnetic effect of the separate corpuscles at each moment.

It is evident, both from my experiments and Størmer's calculations, that a drawing-in of rays generally takes place over areas of greater or less extent; and we will here only suggest that the effect of a bundle of rays in which the course of the rays is, as shown in fig. 50 *a* & *b*, very near, will be the same as that of a linear current consisting of two vertical currents connected by a horizontal one.

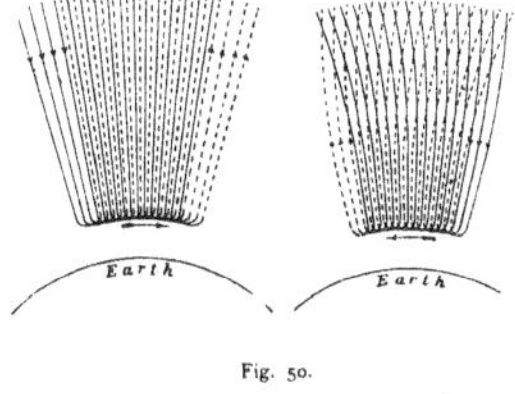

Fig. 50.

In the more central parts it is evident that the downward and upward-going rays destroy each the others' effect, so that only the effect of the outer parts is left. In the figure, we have made the direction of the arrow indicate the direction in which the negatively-charged corpuscles should move; and the galvanic currents must be imagined flowing in the opposite direction.

The paths of the separate corpuscles do not, indeed, coincide with those here indicated; but on the whole a system of rays such as this might not be so far removed from those that actually produce the magnetic storms.

We have hereby only wished to prove that these two systems of rays fully explain the principal features in the two typical fields found in the polar elementary storms. Fig. 50 *a* represents those in which the current-directions at the storm-centre are directed westwards, and 50 *b* those in which the currents move eastwards.

Such cosmic current-systems in the polar regions as are here assumed, will of course induce a very complicated system of currents all over the earth itself, this being a conducting sphere composed of sea and land.

In a later part of this work we shall deal with this question, and see how such earth-currents would affect the magnetic instruments in different places.

THE PERTURBATION OF THE 10th FEBRUARY, 1903.
(Pl. XVIII.)

37. This magnetic disturbance is brief, and commences without any previous equatorial perturbation on an otherwise very quiet day. First a small disturbance appears rather suddenly at about $21^h\ 6^m$. This precursor of the real storm partakes on the whole of the latter's character. It is most powerful at the northern stations, especially at Matotchkin Schar, but is also perceptible in Europe and North America. After about 30 minutes, the conditions are once more almost normal; but disquiet still prevails at the northern stations, and at the other European stations a slight deflection is noticeable, especially in the declination.

The powerful perturbation, with which we are especially concerned,and which we shall now follow, does not commence until 23^h.

As the copies of the curves show, it is very powerful, and especially so at the four arctic stations; while southwards, in Europe and America, there are simultaneous relatively powerful, violent perturbations.

After about an hour and three quarters, the storm is over. At most of the stations, the conditions have now become quite normal, the arctic ones only being still somewhat disturbed. At $2^h\ 30^m$ on the 11th February, another short, slight perturbation appears, which is especially remarkable for the sharply-defined northern limits of its sphere of action (cf. perturbation of 15th Dec., 1902). Thus while fairly powerful at Axeløen and Dyrafjord, it is almost imperceptible at Matotchkin Schar and Kaafjord; while it is tolerably distinct in America, and less powerful on the continent of Europe.

This storm belongs to the class of perturbations that we have called elementary storms, and has a peculiar resemblance to the perturbations of the 15th December, 1902, and the 31st March, 1903; but the curves for the northern stations in this perturbation are of a more disturbed character than those of the perturbation of the 15th December.

It is difficult to say exactly when the powerful perturbation begins; but we shall see from the curves that in the case of most of the stations, the time when the perturbation begins to be *very* powerful can be approximately given. The time when the perturbation is at its height can also be determined with tolerable accuracy; but, as in so many other cases, that of its cessation is difficult to decide.

In the table below is given the hour at which the perturbation commences, as also the time at which the horizontal component of the perturbing force has its highest value, and the magnitude of its maximal strength, and further the time at which the perturbing force has sunk to about five per cent. of its maximal amount, this hour being given as the time when the perturbation ceases. This determination cannot lay claim to any great accuracy, and is therefore found by an estimate.

TABLE XVI.

Observatory	Begins	Reaches Max.	$P_{1(max)}$	Ends
	h m	h m		h m
Matotchkin Schar . .	23 0	23 45	373 γ	0 36
Dyrafjord	» 8	» 50	372 »	
Axeløen	» 16	24 3	354 »	1 46
Kaafjord	» 8	23 48	238 »	0 40
Wilhelmshaven . . .	» 0.5	» 17	47.4 »	1 0
Stonyhurst	» 2	» 16	41.5 »	1 0
Potsdam	» 0	» 17	37.4 »	1 0
Kew	» 1	» 16	35.8 »	1 0
Val Joyeux	» 1.5	» 18	34.4 »	1 0
Pawlowsk	» 0	» 15	29.1 »	1 0
Pola	» 1	» 18	28.0 »	1 0
San Fernando . . .	» 0	» 18	27.5 »	1 0
Munich	» 1	» 20	26.0 »	1 0
Toronto	» 3	» 21	19.4 »	0 8
Cheltenham	» 4	» 22	18.0 »	0 6
Baldwin	» 0	» 19	17.4 »	0 16
Tiflis	» 0	» 25	17.3 »	0 48
Sitka	» 0	» 18	16.3 »	0 20
Dehra Dun	» 2	» 38	13.5 »	0 20—0 50
Christchurch	» 2	» 20	12.0 »	
Honolulu	» 0	» 19	7.6 »	
Zi-ka-wei	» 0	» 15	7.5 »	
Batavia			< 5.0 »	

It appears from the Table, as also directly from the curves, that at the northern stations the perturbation occupies a peculiar position in relation to the other stations.

The times of the commencement and of the maximum of the perturbation, it will be seen, are very different at our four Norwegian stations. At Axeløen the perturbation commences about a quarter of an hour, and at Dyrafjord and Kaafjord about eight minutes, later than at Matotchkin Schar, although the distance between the stations is only from 900 to 1800 kilometres. It should also be mentioned in this connection, that at the arctic stations the curves exhibit great variableness from place to place.

In marked contrast to this, we find that at all the other stations scattered over the northern hemisphere, the perturbation commences simultaneously. The slight differences in time, which do not exceed three minutes, need not imply an actual difference in time, but may be ascribed to inaccuracy in determining the time on the magnetograms. The hour for the maximum is also the same for wide districts of the earth; and the form of the curve is repeated almost without change from station to station, the variation in form being gradual. All the stations of Central and Southern Europe have the same characteristic form of curve. The *H*-curve at Tiflis forms the transition to that at Dehra Dun. The comparatively high value at Wilhelmshaven seems to have been due to local conditions, as this station always shows a greater force than the surrounding stations.

The conditions at Pawlowsk do not appear to allow of a similar explanation, the comparatively small force there being accounted for by the peculiar nature of the perturbation in question, a circumstance to which we shall return later on.

We must here mention one more peculiarity. Although at Batavia the perturbation is almost imperceptible, we find, on coming as far south as Christchurch, that there is a distinct perturbation in the horizontal intensity, appearing simultaneously with that in the northern hemisphere, and resembling in its course the perturbations at the American stations.

It is usual for Christchurch to occupy a peculiar position such as this, and frequently the forms appearing in these southern districts are quite different. This may be explained by the fact that the perturbation in the arctic regions is often accompanied by simultaneous perturbations in the antarctic regions, and it is the effect of these latter that is noticed in Christchurch. Our material does not, however, allow of certain conclusions being drawn in this matter.

THE PERTURBING FORCES.

38. This perturbation, as we have said, has a great resemblance to the previously-described perturbation of the 15th December, 1902. This resemblance is also apparent in the perturbing forces. If we compare the charts of the two perturbations, we find a great similarity, as for instance in the direction of the horizontal and vertical components of the perturbing force. The chief difference is that the force at Kaafjord and Axeløen on the 15th December was very small in proportion to that at the other places.

The perturbing force elsewhere in Europe moreover exhibits a similar though smaller turn clockwise. The smaller extent of the turn seems undoubtedly to be connected with the circumstance that at the commencement of this perturbation, the direction of the perturbing forces coincides with that at 2 a. m. on the 15th December, at which time, on that occasion, the perturbation was far past its maximum.

As the force in these perturbations does not seem to continue to turn after the current-arrows in Europe have become almost uniform in direction with those at the arctic stations, it is evident that the perturbing force in this perturbation of the 10th February must have a smaller area to turn in.

For this perturbation four charts have been drawn, at intervals of a quarter of an hour. They give a clear idea of the distribution of the force, and its changes during the progress of the perturbation.

TABLE XVII.
The Perturbing Forces on the 10th February, 1903.

Gr. M. T.	Honolulu		Sitka		Baldwin		Toronto		Cheltenham	
	P_h	P_d	P_h	P_d	P_h	P_d	P_h	P_d	P_h	P_d
h m										
23 0	− 0.5 γ	0	0	0	0	0	0	0	0	0
15	− 7.6 »	0	− 9.4 γ	0	− 10.5 γ	0	− 11.3 γ	0	− 14.7 γ	0
30	− 6.6 »	W 2.9 γ	− 13.7 »	W 5.9 γ	− 15.6 »	0	− 17.5 »	E 1.2 γ	− 15.4 »	E 1.8 γ
45	− 6.0 »	» 2.9 »	− 15.0 »	» 5.9 »	− 13.6 »	0	− 13.5 »	» 1.8 »	− 11.3 »	» 2.4 »
24 0	− 3.9 »	» 0.8 »	− 6.9 »	» 1.8 »	− 7.0 »	E 2.5 γ	− 6.8 »	» 6.7 »	− 1.7 »	» 4.5 »
0 15	− 1.3 »	0	+ 3.2 »	0	− 1.4 »	0	0	0	0	0
30	0	0	+ 3.4 »	0	0	0	0	0	0	0

TABLE XVII (continued).

	Dyrafjord		Axeløen			Matotchkin Schar		
	P_d	P_v	P_h	P_d	P_v	P_h	P_d	P_v
− 24.7 γ	W 8.7 γ	− 30.8 γ	− 16.1 γ	E 15.2 γ	− 24.6 γ	− 96.2 γ	E 49.7 γ	− 59.5 γ
− 172.0 »	» 39.8 »	− 16.1 »	− 18.4 »	» 74.8 »	− 9.8 »	− 244.0 »	» 100.0 »	− 161.0 »
− 273.0 »	» 101.0 »	− 73.5 »	− 69.0 »	» 77.3 »	+ 81.0 »	− 321.0 »	» 113.0 »	− 244.0 »
− 370.0 »	» 68.0 »	− 96.5 »	− 186.0 »	» 74.0 »	+ 150.0 »	− 359.0 »	» 100.0 »	− 191.0 »
− 304.0 »	» 32.6 »	− 28.0 »	− 202.0 »	» 76.7 »	+ 196.0 »	− 311.0 »	» 106.0 »	− 156.0 »
− 106.0 »	» 2.4 »	− 78.0 »	− 298.0 »	» 76.4 »	+ 164.0 »	− 292.0 »	86.5 »	− 131.0 »
− 122.0 »	E 13.2 »	− 149.0 »	− 345.0 »	» 76.4 »	+ 158.0 »	− 196.0 »	» 29.3 »	− 85.0 »
− 128.0 »	» 23.2 »	− 93.5 »	− 232.0 »	» 78.9 »	+ 208.0 »	− 141.0 »	» 48.4 »	− 47.6 »
− 115.0 »	» 33.2 »	− 32.8 »	− 138.0 »	» 46.2 »	+ 172.0 »	− 65.0 »	28.8 »	− 21.6 »
− 88.0 »	0	− 9.2 »	− 85.0 »	» 76.2 »	+ 49.2 »	− 18.0 »	» 17.8 »	− 14.5 »
− 141.0 »	W 11.8 »	+ 89.2 »		P_d somewhat uncertain between 23^h 22.5^m and 23^h 45^m, owing to the indistinctness of the curve.				
− 240.0 »	» 81.2 »	− 164.0 »						

TABLE XVII (continued).

Gr. M. T.	Kaafjord		
	P_h	P_d	P_v
h m			
23 0	− 58.0 γ	E 28.6 γ	− 49.9 γ
7.5	− 73.2 »	0	− 101.0 »
15	− 154.0 »	W 5.5 »	− 195.0 »
22.5	− 233.0 »	E 29.4 »	− 229.0 »
30	− 232.0 »	» 48.4 »	− 230.0 »
37.5	− 189.0 »	» 66.1 »	− 253.0 »
45	− 176.0 »	» 64.2 »	− 258.0 »
52.5	− 134.0 »	» 62.4 »	− 217.0 »
24 0	− 94.6 »	» 73.4 »	− 168.0 »
0 15	− 15.3 »	» 57.2 »	− 113.0 »

TABLE XVII (continued).

			Stonyhurst		Kew		Val Joyeux		
		P_v	P_h	P_d	P_h	P_d	P_h	P_d	P_v
+ 1.0 γ	E 5.5 γ	− 1.5 γ	− 3.6 γ	E 10.8 γ	− 2.0 γ	E 6.8 γ	− 3.2 γ	E 5.9 γ	The deviation very small, about + 5.5 γ at 11^h 15^m.
+ 22.6 »	W 18.4 »	− 3.7 »	+ 27.5 »	» 30.8 »	+ 25.4 »	» 25.0 »	+ 20.0 »	» 27.7 »	
+ 21.2 »	» 2.8 »	− 11.6 »	+ 19.9 »	» 34.8 »	+ 16.8 »	» 28.2 »	+ 21.6 »	» 26.8 »	
+ 12.6 »	E 3.7 »	− 15.7 »	+ 10.2 »	» 36.5 »	+ 7.1 »	» 30.8 »	+ 10.0 »	» 31.9 »	
+ 2.5 »	» 20.2 »	− 16.4 »	− 5.6 »	» 28.5 »	− 10.7 »	» 28.2 »	− 6.4 »	» 24.3 »	
− 5.0 »	» 21.6 »	− 12.0 »	− 17.9 »	» 17.2 »	− 14.3 »	» 16.4 »	− 12.8 »	» 13.4 »	
− 4.5 »	» 10.6 »	− 6.7 »	− 7.7 »	» 8.0 »	− 7.1 »	» 8.0 »	− 7.2 »	» 5.5 »	

TABLE XVII (continued).

Gr. M. T.	Wilhelmshaven			Potsdam			San Fernando	
	P_h	P_d	P_v	P_h	P_d	P_v	P_h	P_d
h m								
23 0	− 2.3 γ	E 6.1 γ	V. I. variometer showing little sensitiveness. There is, however, a slight deflection in the positive direction, with maximum at 11h 20m.	− 1.9 γ	E 6.1 γ	+ 0.5 γ	− 2.9 γ	0
15	+ 42.0 »	» 22.0 »		+ 34.8 »	» 13.2 »	− 4.7 »	+ 11.8 »	E 16.6 γ
30	+ 28.0 »	» 23.8 »		+ 25.9 »	» 18.8 »	− 4.4 »	+ 17.0 »	» 21.6 »
45	+ 16.8 »	» 29.3 »		+ 14.5 »	» 20.8 »	− 4.1 »	+ 7.4 »	» 19.1 »
24 0	− 8.2 »	» 22.8 »		− 6.3 »	» 19.3 »	− 1.4 »	− 5.9 »	» 11.6 »
0 15	− 16.3 »	» 13.4 »		− 13.3 »	» 13.7 »	− 0.2 »	− 12.6 »	» 4.2 »
30	− 11.2 »	» 1.2 »		− 8.5 »	» 4.1 »	− 0.5 »	− 5.9 »	» 3.3 »

TABLE XVII (continued).

Gr. M. T.	Munich			Pola			Tiflis		
	P_h	P_d	P_v	P_h	P_d	P_v	P_h	P_d	P_v
h m									
23 0	− 3.0 γ	E 5.6 γ	Very small, almost imperceptible. The force is directed *upwards*.	− 2.2 γ	E 7.6 γ	+ 0.4 γ	1.8 γ	E 1.9 γ	+ 0.5 γ
15	+ 22.0 »	» 13.9 »		+ 21.1 »	» 18.2 »	0	+ 12.4 »	W 7.4 »	− 2.6 »
30	+ 21.0 »	» 15.0 »		+ 20.2 »	» 18.8 »	0	+ 16.8 »	» 4.8 »	− 3.1 »
45	+ 14.0 »	» 22.5 »		+ 13.4 »	» 22.2 »	− 2.1 »	+ 14.1 »	» 1.1 »	− 2.8 »
24 0	− 2.5 »	» 18.0 »		0	» 18.8 »	− 1.7 »	+ 4.9 »	E 4.6 »	− 1.0 »
0 15	− 10.0 »	» 15.4 »		− 6.3 »	» 15.6 »	− 2.5 »	− 1.8 »	» 7.4 »	0
30	− 7.0 »	» 6.4 »		− 3.1 »	» 6.9 »	− 2.3 »	− 3.3 »	» 2.8 »	+ 0.5 »

TABLE XVII (continued).

Gr. M. T.	Dehra Dun		Zi-ka-wei		Christchurch	
	P_h	P_d	P_h	P_d	P_h	P_d
h m						
23 0	− 2.9 γ	0	The deflection in *H* is not measurable here. There are however small irregularities.	0	0	0
15	+ 5.0 »	W 7.8 γ		W 7.0 γ	− 12.0 γ	E 1.9 γ
30	+ 8.3 »	» 7.8 »		» 7.5 »	− 11.0 »	» 1.9 »
45	+ 7.0 »	» 5.9 »		» 4.0 »	− 8.3 »	» 1.8 »
24 0	+ 2.8 »	» 2.9 »		» 1.5 »	− 6.4 »	» 1.6 »
0 15	− 1.5 »	0		0	− 3.2 »	» 1.5 »
30	− 2.7 »	0		0	0	» 1.1 »

The only magnetograms from Bombay are for *H*. The conditions of the perturbation are similar to those at Dehra Dun.

At Batavia the perturbation is noticeable, but very faint and ill-defined, so that no perturbing force can be determined.

Current-Arrows for the 10th February, 1903; Chart I at 23h 15m, and Chart II at 23h 30m.

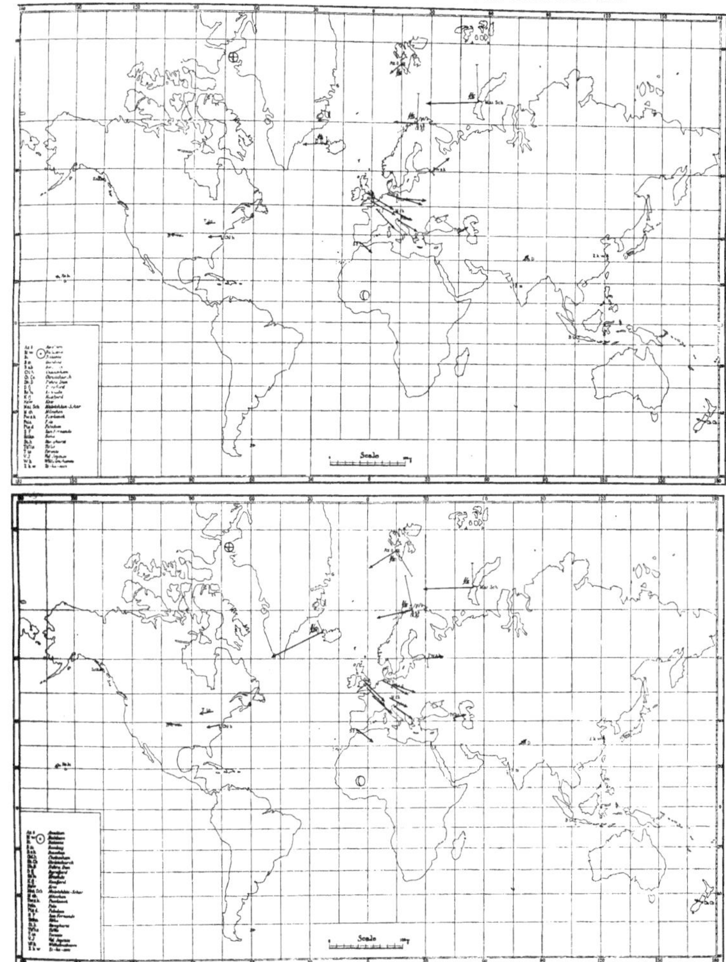

Fig. 51.

Current-Arrows for the 10th February, 1903; Chart III at 23h 45m, and Chart IV at 24h.

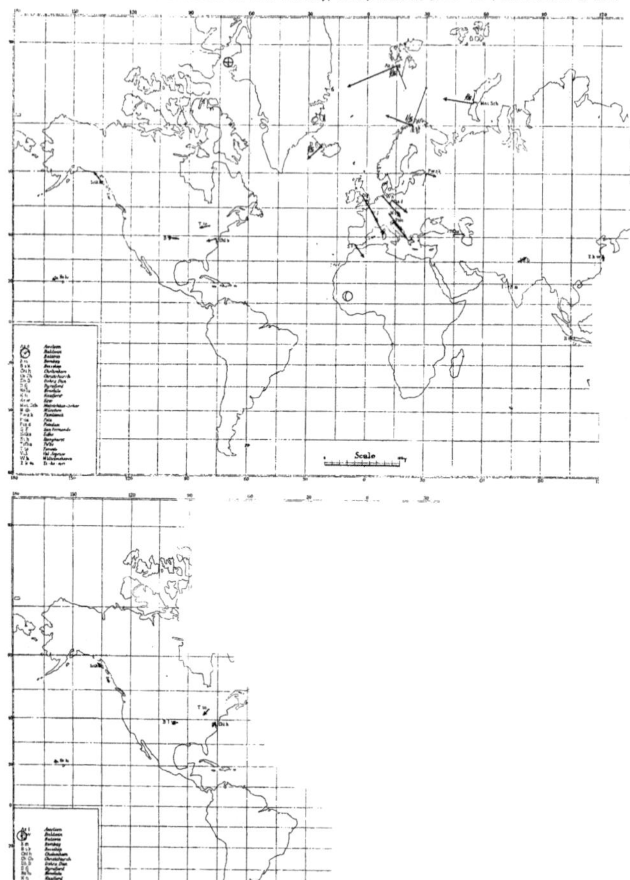

Fig. 52.

Chart. I. Time 23^h 15^m.

The field of perturbation here shows itself to be of the typical form that is always to be found during the polar elementary storms. The principal axis of the system falls, as shown by the chart, along the auroral zone; and the storm-centre seems to lie a little nearer to Matotchkin Schar than to the other Norwegian stations, though its position cannot be given more exactly. The rest of Europe is in the vicinity of the system's area of convergence. Judging from the force at Pawlowsk, the point of convergence itself should be situated a little to the north of that place. In America we again find the usual directions for the current-arrows, namely, west at the three more easterly stations, and north-west at Sitka.

Chart. II. Time 23^h 30^m.

The conditions are not essentially different from those of the preceding chart. The principal axis of the system is more conspicuous in the forces at the Norwegian stations, where they are now more or less of the same strength. It still lies along the auroral zone between Kaafjord and Axeløen, and a little to the north of Dyrafjord and Matotchkin Schar, judging from the vertical intensities. In the southern European stations, the forces are more or less uniform in direction with those to be found on Chart I, except that at Pawlowsk there is a slight turn clockwise. The point of convergence still lies a little to the north of the last-named station.

Chart III. Time 23^h 45^m.

The storm-centre seems to have moved eastwards, the force at Dyrafjord being considerably smaller than before. At the same time the forces at the southern stations in Europe have turned considerably, clockwise.

Chart IV. Time 24^h 0^m.

The forces have diminished considerably everywhere, as the close of the perturbation is now approaching. At the southern European stations, the turning is continued in the same direction as before, so that the current-arrow is now directed distinctly southwards. In other respects, the form of the field is in all essentials the same as before.

CONCERNING THE CAUSE OF THE PERTURBATION.

39. By reasoning as in the case of the perturbation of the 15th December, 1902, we here too arrive at the conclusion that the perturbation at the four arctic stations is mainly due to the effect of a horizontal current-system, which keeps fairly close to the surface of the earth in the area over which the storm is most violent. In this case therefore, it should be mainly a horizontal current from Matotchkin Schar to Dyrafjord. As it is more or less horizontal in this district, the direction of the current must in a large measure coincide with that of the current-arrows drawn on the chart. It follows from the vertical components, that the main volume of the current must flow north of Matotchkin Schar, passing in a WSW direction between Kaafjord and Axeløen, and on to the north of Dyrafjord. This is in the main the same course as that taken by the current on the 15th December.

We should mention, in this connection, that the earth-currents during this perturbation have been very beautifully registered (see Part III, Earth-Currents). This is most fortunate, as this perturbation is so simple in its course, increasing to a maximum and decreasing to zero. The earth-current, on the other hand, as the curve shows, takes the following course. While the magnetic storm is *increasing* to its maximum, the current flows in the same direction, increasing to a maximum and decreasing to zero; during the second part of the perturbation, while *decreasing*, the direction of the earth-current is reversed, its volume increasing to a maximum and decreasing to zero. This furnishes a direct proof that the primary cause of the perturbation is to be found in currents above the earth, since the current in the earth is evidently an induced current produced by the magnetic storm. The latter must therefore have its cause in a current-system above the surface of the earth, if, as may be considered certain in the case of these perturbations, it is conditioned by electric currents at all.

Owing to the rapid weakening of the effect southwards, these horizontal currents must lie at a comparatively little height above the earth. The perturbations must be of a local character in the north, a fact that is immediately apparent from the already-mentioned great variation in the nature of the perturbation from place to place. The perturbations in the southern districts are in strong contrast to this, as they there show a slow, continuous change in their character.

The perturbations in the southern districts are not of such a character that they can be regarded as the effect of adjacent systems; their cause must necessarily be sought in the average effect of that which takes place in the more distant systems, a circumstance which explains the quiet, regular character of the curve.

In discussing the perturbation of the 15th December from to some extent other points of view, we arrived at the same result, as the explanation of the effect of the force outwards at great distances from the arctic regions, must be sought in that of vertical currents in an opposite direction, connected with the low-lying, horizontal portion of the current, which gave rise to the powerful perturbations in the north.

By a generally continuous movement of this system, the turning of the perturbing force is precisely explained. On that day, the sphere of action in the north being more than ordinarily local, this movement may be clearly proved by the fact that the perturbation made its appearance much later at Axeløen than at Dyrafjord.

This perturbation is greater, and its influence is almost equally strong at all the four stations. It commences quite as early in the regions about Matotchkin Schar as at Dyrafjord.

Thus, although we cannot prove, from the times at which the perturbation began in the north, that there was any movement eastwards along the auroral zone, the current-arrows on Charts II and III at Dyrafjord indicate that such a movement really took place there, as already mentioned in the description of the charts. Outwards there is also the same distribution of force and turning of the perturbing force, as in the perturbation of the 15th December.

As we have said, the distribution of force at 2^h on that day answers to that at 23^h 15^m on this 10th February. If we now imagine the system to be moving on eastwards, it will be easily seen that the European stations would be passed by the magnetic field in a district in which the direction of the perturbing force alters only slightly, and the turning would be with the hands of a clock.

In this perturbation the current-system may be assumed on the whole to have a more easterly position than in that of the previous 15th December, in accordance with the fact that the latter appeared later in the day.

The field of force on the surface of the earth indicates that our current-system should generally have two symmetrically-situated points, the points of convergence and divergence, one on each side of the horizontal portion of the current, for the horizontal component, two neutral districts in which the horizontal component was very small.

We have not yet seen both these points during the same perturbation; for when one of them is in Europe, e. g. in the neighbourhood of Pawlowsk, as we shall generally find it during the polar elementary storms that have their storm-centre near our Norwegian stations, the other should be situated symmetrically on the other side of the auroral zone, or in the most northern parts of Greenland.

During the perturbations of the 15th December and the 10th February, we have found the area of convergence, and during that of the 9th December we have found the area of divergence. In our researches on the storms of 1882—83 in the polar regions (Part II), we shall also sometimes find a field on the other side of the auroral zone, that appears to indicate an area of divergence, at the same time as the forces in the southern parts of Europe form an area of convergence.

This fully explains a circumstance mentioned in the description of the first part of the perturbation, namely that Pawlowsk has a very small horizontal component considering the northerly situation of the place. During the beginning of the perturbation, the direction of the current-arrow is almost the reverse of that of the horizontal portion of the current. During that time therefore, the station ought to lie nearer to the neutral district than later, when, owing to the movement of the system, the perturbing force is turned more in accordance with the conditions in Central Europe.

In this perturbation also, the vertical components are very small in the regions outside the arctic district, a circumstance that accords perfectly, as we have already said, with our explanation of the perturbation, as those components should mainly be conditioned by the horizontal portion of the current. In the vicinity of the neutral district, P_2 only should be of considerable size in proportion to P_1. At Pawlowsk there is actually a considerable vertical component directed upwards all the time. In the cases of Potsdam and Pola, it is much smaller, but directed upwards; and at Val Joyeux it is almost imperceptible.

THE PERTURBATIONS OF THE 30th & 31st MARCH, 1903.

(Pl. XXI).

40. For the study of these perturbations, we have magnetograms for the horizontal intensity and declination from all the stations marked on the chart with the exception of Matotchkin Schar, where, on that day, the registering apparatuses were not acting. The declination-curve for Bombay is also wanting. The observations from Ekaterinburg and Irkutsk are only for every hour; and as the perturbation is short, there will here be little use in taking out intermediate values.

At Bossekop, the needle in the variometer for horizontal intensity during the perturbation was deflected out of the field, and did not return. The perturbing force here can only be taken for the first part of the perturbation.

In addition to the horizontal intensity and declination curves, there are also vertical intensity curves for the Norwegian and some other stations.

The time during which this violent perturbation is acting at the Norwegian stations is very short. The deflections, moreover, are uniform in direction. The character of the curve in the north is as usual very disturbed, and varies greatly from place to place, indicating that the current-systems that condition the phenomenon here, must come comparatively near the earth.

Simultaneously with this exceedingly powerful, brief storm round the Norwegian stations, distinct perturbations are noticed at all the observatories from which observations have been received. The curves immediately show that the perturbations outside the arctic district are of a universal character, as the form of the curve remains very nearly constant over large districts, and the transition takes place gradually — conditions with which we meet in most of the polar storms.

It will be seen from the magnetograms from the districts visited by this perturbation, that in advance of this elementary storm in the north, there is a long perturbation that is especially powerful and distinct at the stations near the equator, and occurs chiefly in H. We also note the jagged character of the curve, and that the serrations occur simultaneously all over the earth. That the perturbation between 24^h and 2^h is connected with that in the north is probable from the fact that it then rather suddenly becomes comparatively powerful in D, and also that the horizontal intensity curve oscillates greatly at this time. The perturbation moreover becomes more powerful with an approach to the northern stations.

powerful southwards towards the equator.

We may therefore safely assume that we here have two phenomena to be dealt with, one connected with the storm in the north, and before it an equatorial perturbation of a kind similar to that of the 26th January, 1903.

The placing of the normal line on the magnetograms has occasioned no special difficulty. The storms are fairly powerful and well-defined at all the stations with the exception of Christchurch and Honolulu; the perturbing force can therefore be taken out with very satisfactory accuracy. The following circumstances are taken into consideration in the drawing of the normal line. In declination the conditions are simple, as there the perturbation is of short duration. The quiet parts before and after the perturbation are connected in such a manner that the form of the curve corresponds with that at the same hour on the nearest calm days. The conditions in the horizontal intensity are somewhat more difficult, as there, as we have said, there is a long perturbation in front of the one under consideration. In this, judging from things in general, the curve for most of the stations is normal at about 3^h, and for an hour afterwards. The absolute distance of the normal line from the base-line on the magnetogram will thereby be determined; and its further course is regulated by the nearest calm days.

THE EQUATORIAL PERTURBATION.

41. As early as 19^h, those little, sudden, very variable perturbations are noticed, which occur simultaneously all over the earth, and symmetrically as regards the magnetic axis. It will be seen from the copies of the *H*-curve that the conditions at Dehra Dun, Batavia and Honolulu entirely correspond with one another. The force is mainly directed northwards. The perturbation appears to be over at about $23^h\ 12^m$. From $21^h\ 28^m$ to 23^h, the force remains almost constant both in magnitude and direction. The perturbing forces are calculated for 22^h, and the corresponding current-arrows are marked upon the chart.

Current-Arrows for the 30th March, 1903, at 22^h.

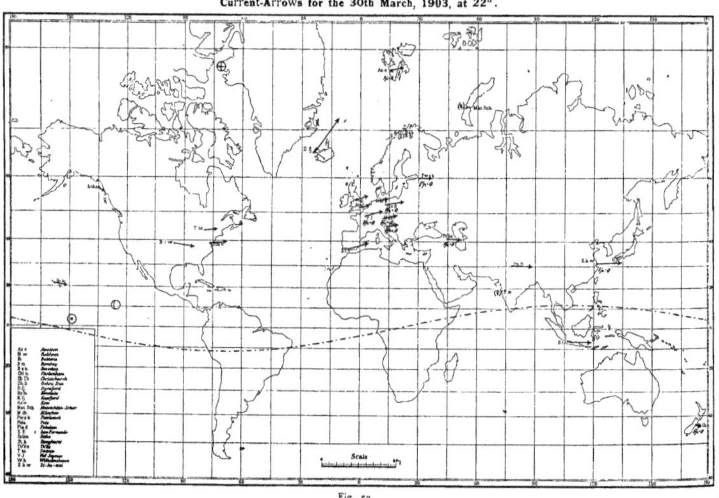

Fig. 53.

We here distinctly see that except as regards the arctic stations, one circumstance is very conspicuons, namely, that the perturbing force is strongest in the equatorial regions, and decreases towards the poles. Honolulu is an exception to this; but, as mentioned under the perturbation of the 26th January, this may be ascribed to local conditions. The arrows point along the magnetic parallels from west to east.

In the arctic regions, especially at Dyrafjord, the conditions are different, owing to polar disturbances. In these regions, indeed, there is hardly ever calm. The distribution of force, and the perturbation as a whole, are of exactly the same character as that of the 26th January; we therefore refer the reader to the description of the latter, for its most probable explanation.

At about $23^h\ 12^m$, after this equatorial perturbation has ceased, comparatively normal conditions appear to supervene, at any rate in latitudes lower than 60°; and these are maintained for three quarters of an hour. At the stations nearest to the equator, however, there is now a distinct deflection in H to the opposite side. There is thus now for a time a slight equatorial perturbation, corresponding to a current-system resembling the previous one, but in the opposite direction.

THE POLAR PERTURBATION.

42. The storm about the auroral zone is very powerful and well-defined, especially at Dyrafjord, where it appears very suddenly at $0^h\ 24^m$, and concludes almost as suddenly at $2^h\ 16^m$.

At Axeløen the perturbation is observed a little earlier, but the really powerful storm nevertheless commences later here than at the other arctic stations.

At Kaafjord the perturbation begins very much earlier than at the two previously-mentioned stations, especially in H. As early as 23^h, the deflections in H begin to increase continuously. At $0^h\ 24^m$, that is to say at the same time as the storm at Dyrafjord begins, the point of light swings out of the field, to return no more. The reason of this great deflection must partly be that at this hour the sensibility was made very great, the magnet being suspended by a thread with small moment of torsion. But if otherwise, on this occasion, the apparatus acted properly, it would at any rate appear that the perturbation began with a low-lying current about Kaafjord, which then developed further into a more extended system, at the same time moving northwards. That the system really moves in a northerly direction seems also to be shown by the very interesting vertical intensity curve at Kaafjord; for at about $0^h\ 36^m$, P_v, from being directed downwards, turns upwards, corresponding to the flowing of the horizontal portion of the current past the place from south to north. The curve exactly resembles that in the lower diagram in fig. 40.

The fact that the point of light does not return — i. e. that the magnet goes round to another position of equilibrium — prevents our concluding very much from this circumstance; for it is not impossible that the enormous deflection is partly due to the almost neutral equilibrium of the apparatus over a large area.

At about $23^h\ 50^m$, the effect of the polar perturbation is noticed at all the southern stations throughout the world. At $2^h\ 10^m$, the normal conditions have reappeared in these latitudes.

In this, as in so many other instances, Christchurch occupies a peculiar position, inasmuch as conditions appear there, which have no parallel in the northern hemisphere. A distinct perturbation is also observable there, however, which to some extent coincides with the perturbation in the northern hemisphere, which it also resembles in its course. At the three American stations, Toronto, Cheltenham and Baldwin, a peculiarity is observable, namely, that the perturbation apparently lasts longer in H than in D.

In declination there is a brief, well-defined, powerful perturbation, which takes place at the time when the storm about the auroral zone is at its height. In this case it lasts from $0^h\ 12^m$ to $1^h\ 16^m$. In reality this only means that the perturbing force has turned. A similar condition was observed on the 15th December. These two perturbations on the whole resemble one another in a striking degree, a circumstance that is undoubtedly connected with the fact that they both occur at about the same time of day.

Observatory	Time of Max.	P_1 (max.)	Observatory	Time of Max.	P_1 (max.)
	h m			h m	
Dyrafjord	0 58	546 γ	Pawlowsk . . .	0 30	41.1 γ
Axeløen . . .	0 50	280 »	Val Joyeux . .	0 30	38.4 »
Toronto . . .	0 39	65.5 »	Munich	0 37	33.4 »
Cheltenham . .	0 39	50.0 »	Pola	0 30	30.7 »
Baldwin . . .	0 39	39.7 »	San Fernando .	0 37.5	26.6 »
Sitka	0 45	22.0 »	Tiflis	0 23	16.5 »
Honolulu . . .	0 58	12.1 »		0 45	16.3 »
Wilhelmshaven .	0 30	63.0 »		1 0	
Stonyhurst . .	0 30	47.8 »	Dehra Dun . .	0 30	16.5 »
Potsdam . . .	0 30	45.8 »		1 15	16.3 »
Kew	0 30	41.5 »	Zi-ka-wei . . .	1 0	15.1 »
			Batavia	0 30	10.5 »

THE DISTRIBUTION OF FORCE.

43. In the above table the time of the maximum of the horizontal perturbing force is given as the value of P_1 (max.) at that time.

The maximum occurs, strangely enough, earliest at the European mainland stations, where it is very distinct and well defined. At Tiflis and the Asiatic stations, the force remains for some time almost constant in magnitude. At $0^h\ 39^m$ the maximum occurs at the three American stations; and last of all it occurs at the northern stations, together with Honolulu and the Asiatic stations.

The earlier occurrence of the maximum on the continent of Europe and in North America than at the source itself round the auroral zone, is a peculiar circumstance that, regarded superficially, might lead to the belief that the phenomena in the arctic regions were separate from those in more south-lying districts. We shall return to this subject later.

The maximal force, as we see, is strongest at Dyrafjord, where it attains the rather unusually large value of 546 γ. The table clearly shows that the force increases with proximity to the district about this station, independently of the direction of its approach.

After the arctic district, the force is greatest at Toronto, where it attains the comparatively large value, 65.5 γ. On the whole, this perturbation is stronger at Toronto and the two stations in the United States than at the European stations, as will best be seen from the charts.

Next to Toronto comes Wilhelmshaven, which thus, on this occasion also, occupies a comparatively prominent place, a circumstance to be accounted for by local conditions (see the 10th February, 1903).

The perturbing forces are calculated for a series of times, given in the table, and are synoptically represented by a number of charts. As the reasons which led us, on the 15th December, to our assumption of the current-system, are also present in this case, we will, in describing each separate chart, compare the field of force with our current-system.

TABLE XVIII.

The Perturbing Forces on the 30th & 31st March, 1903.

Gr. M. T.	Honolulu		Sitka		Baldwin		Toronto		Cheltenham	
	P_h	P_d	P_h	P_d	P_h	P_d	P_h	P_d	P_h	P_d
h m										
22 0	+ 9.2 γ		+ 5.0 γ	0	+ 13.9 γ	0	+ 9.0 γ	0	+ 12.0 γ	0
0 0	− 2.42 »		+ 5.3 »	W 9.00 γ	− 7.3 »	0	− 8.5 »	W 4.8 γ	− 6.5 »	W 1.77 γ
7.5	− 5.32 »		− 6.5 »	» 1.80 »	− 12.2 »	0	− 10.8 »	» 6.7 »	− 9.9 »	» 4.13 »
15	− 7.50 »		+ 3.5 »	» 7.20 »	− 7.8 »	E 3.78 γ	− 6.7 »	E 1.2 »	− 6.4 »	E 4.13 »
22.5	− 6.29 »	No perturbation observable in declinatio.	+ 8.1 »	» 9.90 »	− 9.0 »	» 17.01 »	− 12.6 »	» 30.3 »	− 1.9 »	» 24.19 »
30	− 6.05 »		− 5.3 »	» 5.85 »	− 16.4 »	» 30.24 »	− 17.1 »	» 55.8 »	− 2.8 »	» 46.61 »
37.5	− 8.95 »		− 19.4 »	E 0.90 »	− 19.8 »	» 34.65 »	− 18.4 »	» 63.0 »	− 7.5 »	» 49.56 »
45	− 10.16 »		− 21.0 »	W 1.35 »	− 24.0 »	» 28.98 »	− 22.9 »	» 51.5 »	− 12.2 »	» 42.48 »
52.5	− 10.89 »		− 18.8 »	» 4.50 »	− 21.9 »	» 30.87 »	− 18.0 »	» 46.3 »	− 14.2 »	» 37.76 »
1 0	− 12.10 »		− 22.1 »	» 1.35 »	− 19.2 »	» 17.01 »	− 18.4 »	» 31.5 »	− 9.9 »	» 23.60 »
15	− 6.78 »		From 1h to 2h, the curves have not been drawn.		− 19.2 »	W 1.89 »	− 18.9 »	W 3.0 »	− 19.9 »	0
30	− 2.66 »				− 17.1 »	» 5.04 »	− 19.8 »	» 6.1 »	− 19.6 »	W 2.36 »
45	0				− 12.2 »	» 5.04 »	− 15.3 »	» 5.5 »	− 14.6 »	» 3.54 »
2 0	0				− 8.7 »	» 2.52 »	− 11.1 »	0	− 11.2 »	» 0.59 »

r. M. T.	Dyrafjord		
	P_h	P_d	P_v
h m			
22 0	− 25.0 γ	W 8.5 γ	− 7.7 γ
0 0	0	0	+ 38.2 »
15	+ 8.3 »	0	+ 45.4 »
30	− 99.7 »	E 45.1 »	+ 46.4 »
45	− 443.2 »	» 208.2 »	+ 155.1 »
52.5	− 482.0 »	» 65.9 »	+ 258.0 »
1 0	− 565.1 »	» 121.5 »	+ 237.4 »
7.5	− 515.2 »	» 31,2 »	+ 263.2 »
15	− 398.9 »	» 27.8 »	+ 160.0 »
22.5	− 382.3 »	W 12.1 »	+ 72.2 »
30	− 243.8 »	0	+ 23.2 »
45	− 138.5 »	» 15.6 »	+ 5.2 »
2 0	− 72.0 »	0	+ 41.3 »
0 59	− 637.1 »	E 104.1 »	+ 242.5 »
0 42.7	− 382.3 »	» 255.0 »	+ 211.6 »

Gr. M. T.	Axeløen			Bossekop		
	F_h	P_d	P_v	P_k	P_d	P_v
h m						
22 0	+ 9.4 γ	0	0	?	?	?
0 0	+ 20.7 »	W 9.0 γ	0	− 352 γ	0	− 34.9 γ
7.5	+ 11.0 »	0	0	− 343 »	0	− 40.5 »
15	− 4.6 »	E 5.4 »	0	− 477 »	E 7.5 γ	− 52.4 »
22.5	− 16.1 »	» 28.3 »	+ 12.3 γ	?	» 28.1 »	− 24.4 »
27				?	» 45.9 »	+ 19.9 »
30	− 21.1 »	» 55.8 »	+ 36.8 »	?	» 18.8 »	0
37.5	− 31.3 »	» 108.8 »	+ 110.5 »	?	W 23.4 »	− 209.4 »
39				?	» 48.8 »	− 202.4 »
45	− 75.0 »	» >163.2 »	+ 187.0 »	?	E 41.3 »	− 172.8 »
52.5	− 186.0 »	» >163.2 »		?	» 76.9 »	− 150.1 »
54.7	− 223.0 »	» 163.2 »	+ 162.0 »	?		
1 0	− 163.0 »	» 121.3 »	+ 157.0 »	?	» 93.8 »	− 134.3 »
7.5	− 189.0 »	» 84.3 »	+ 127.5 »	?	» 91.3 »	− 115.2 »
15	− 169.0 »	» 87.0 »	+ 157.0 »	?	» 66.6 »	− 136.1 »
22.5	− 150.0 »	» 125.1 »	+ 127.5 »	?	» 51.0 »	− 147.5 »
30	− 115.0 »	» 68.0 »	+ 110.5 »	?	» 43.1 »	− 129.1 »
45	− 80.5 »	» 42.2 »	+ 73.5 »	?	» 34.7 »	− 89.0 »
2 0	− 47.9 »	» 42.4 »	+ 73.5 »	?	» 28.1 »	− 75.0 »
15	− 30.8 »	» 35.4 »	+ 49.2 »			
30	− 11.5 »	» 28.8 »	+ 39.3 »			

There are no observations from Matotchkin Schar for this date.

TABLE XVIII (continued).

Gr. M. T.	Pawlowsk		Stonyhurst		Kew		Val Joyeux		
	P_h	P_d	P_h	P_d	P_h	P_d	P_h	P_d	P_v
h m									
22 0	+ 7.8 γ	0	+ 8.7 γ	0	− 6.6 γ	0	+12.0 γ	0	0
0 0	− 5.5 »	E 2.3 γ	0	E 4.3 γ	− 3.1 »	E 4.5 γ	− 3.2 »	0	+ 1.8 γ
7.5	− 7.0 »	» 6.9 »	− 7.7 »	» 8.8 »	−10.2 »	» 8.7 »	− 8.0 »	E 5.9 γ	+ 2.3 »
15	−10.1 »	» 4.6 »	− 8.4 »	» 8.6 »	−10.2 »	» 2.3 »	− 9.7 »	» 8.4 »	+ 1.8 »
22.5	−25.2 »	W 11.5 »	− 2.6 »	W 19.7 »	− 3.1 »	W 23.7 »	−10.5 »	W 11.8 »	+ 1.4 »
30	−32.2 »	» 25.3 »	+ 6.6 »	» 47.9 »	+ 7.4 »	» 41.0 »	0	» 38.6 »	0
37.5	− 9.0 »	» 33.1 »	+14.3 »	» 35.4 »	+13.8 »	» 30.9 »	+ 9.7 »	» 35.7 »	− 1.8 »
45	+ 7.5 »	» 23.0 »	+13.8 »	» 11.4 »	+15.3 »	» 8.2 »	+16.2 »	» 17.1 »	− 4.5 »
52.5	+12.6 »	» 11.0 »	+ 9.2 »	E 6.3 »	+12.7 »	E 8.2 »	+20.2 »	0	− 5.0 »
1 0	+ 7.5 »	E 1.0 »	0	» 12.5 »	+ 4.6 »	» 15.9 »	+12.2 »	E 8.6 »	− 3.6 »
15	+ 5.0 »	» 21.2 »	− 5.6 »	» 17.7 »	− 4.1 »	» 16.8 »	+ 3.2 »	» 16.7 »	− 2.7 »
30	+ 1.5 »	» 16.1 »	− 5.6 »	» 12.0 »	− 5.1 »	» 12.3 »	− 4.1 »	» 12.2 »	− 2.7 »
45	− 1.5 »	» 12.9 »	− 5.1 »	» 6.6 »	− 6.4 »	» 7.3 »	− 3.2 »	» 8.4 »	− 1.8 »
2 0	+ 1.0 »	» 9.2 »	− 3.6 »	» 6.3 »	− 5.1 »	» 6.4 »	− 1.6 »	» 6.7 »	− 1.4 »

TABLE XVIII (continued).

Gr. M. T.	Wilhelmshaven			Potsdam			San Fernando	
	P_h	P_d	P_v	P_h	P_d	P_v	P_h	P_d
h m								
22 0	+ 7.9 γ	0	0	+12.6 γ	0	0	+13.7 γ	0
0 0	− 3.3 »	E 1.2 γ	− 2.5 γ	− 3.48 »	E 2.54 γ	+ 2.7 »	−10.4 »	0
7.5	− 9.3 »	» 7.2 »	− 3.0 »	− 9.0 »	» 5.6 »	+ 2.7 »	−14.8 »	0
15	− 7.9 »	» 9.05 »	− 2.0 »	− 8.22 »	» 3.56 »	+ 2.7 »	−16.2 »	0
22.5	−19.6 »	W 25.9 »	− 2.0 »	−16.12 »	W 20.3 »	+ 4.5 »	− 6.6 »	W 13.3 γ
30	−22.5 »	» 59.1 »	− 3.0 »	−15.48 »	» 43.2 »	+ 3.2 »	+16.2 »	» 15 »
37.5	0	» 44.6 »	− 5.0 »	+ 5.69 »	» 34.0 »	− 1.5 »	+23.7 »	» 14.9 »
45	+16.4 »	» 13.9 »	− 2.0 »	+21.5 »	» 13.7 »	− 5.9 »	+25.5 »	» 7.5 »
52.5	+21.5 »	E 6.03 »	− 4.0 »	+24.96 »	E 4.5 »	− 6.0 »	+22.2 »	E 5.0 »
1 0	+15.4 »	» 18.7 »	− 3.0 »	+18.64 »	» 11.2 »	− 3.4 »	+14.1 »	» 5.8 »
15	+ 1.4 »	» 24.1 »	− 4.0 »	+ 6.32 »	» 17.8 »	− 3.3 »	+ 3.7 »	» 6.3 »
30	− 5.1 »	» 15.7 »	− 4.0 »	0	» 10.2 »	− 2.1 »	− 3.0 »	?
45	− 6.5 »	» 7.85 »	− 4.5 »	0	» 6.4 »	− 1.7 »	− 3.7 »	?
2 0	− 5.1 »	» 6.6 »	− 4.5 »	0	» 5.6 »	− 1.5 »	− 2.2 »	?

TABLE XVIII (continued).

T.	Munich			Pola			Tiflis		
	P_h	P_d	P_v	P_h	P_d	P_v	P_h	P_d	P_v
m									
0	+ 9.8 γ	0	0	+ 7.6 γ	0	0	+ 9.5 γ	0	0
0	− 5.0 »	E 3.8 γ	+ 1.1 γ	− 4.5 »	E 3.4 γ	+ 1.7 γ	− 8.6 »	E 1.8 γ	+ 3.8 γ
7.5	− 8.5 »	» 4.5 »	+ 1.5 »	− 9.0 »	» 5.5 »	+ 2.5 »	− 13.3 »	» 5.6 »	+ 3.8 »
15	− 8.5 »	» 6.8 »	+ 1.7 »	− 8.5 »	0	+ 1.05 »	− 15.7 »	» 7.4 »	+ 3.6 »
2.5	− 7.5 »	W 3.8 »	+ 2.3 »	− 7.6 »	W 20.8 »	− 4.2 »	− 15.9 »	» 4.4 »	+ 3.6 »
0	− 3.0 »	» 27.8 »	+ 2.8 »	0	» 30.5 »	− 4.0 »	− 11.3 »	W 4.1 »	+ 3.3 »
37.5	+ 6.0 »	» 33.0 »	+ 2.3 »	+ 5.8 »	» 22.2 »	− 0.8 »	− 7.1 »	» 12.2 »	+ 2.3 »
45	+ 16.5 »	» 23.2 »	+ 0.8 »	+ 14.8 »	» 7.6 »	+ 1.9 »	0	» 16.3 »	− 0.5 »
52.5	+ 18.5 »	» 5.3 »	0	+ 14.8 »	0	+ 3.6 »	+ 6.6 »	» 13.9 »	− 1.9 »
0	+ 14.5 »	E 4.5 »	0	+ 11.2 »	E 11.1 »	+ 3.1 »	+ 11.1 »	» 10.4 »	− 2.8 »
15	+ 6.0 »	» 14.2 »	0	+ 4.0 »	» 13.9 »	+ 1.9 »	+ 10.6 »	E 1.1 »	− 2.8 »
30	+ 1.7 »	» 11.3 »	0	0	» 9.0 »	0	+ 5.3 »	» 4.8 »	− 1.3 »
45	0	» 7.9 »	0	0	» 4.1 »	0	+ 2.2 »	» 4.1 »	0
0	0	» 5.2 »	0	0	» 3.4 »	0	+ 0.9 »	» 3.3 »	0

TABLE XVIII (continued).

M. T.	Dehra Dun		Bombay	Zi-ka-wei			Batavia		Christchurch	
	P_h	P_d		P_h	P_d	P_v	P_h	P_d	P_h	P_d
m										
0	+ 13.4 γ	0		+ 17.3 γ	0		+ 16.0 γ	0	+ 4.1 γ	
0	− 12.6 »	0	There is only P_h for this date, but in that the perturbation is distinct, with a course similar to that at Dehra Dun.	− 5.4 »	W 2.0 γ	From 10 a. m. to 2 p.m., there is little increase in vertical intensity. Its maximum at noon amounted to $5.83 \times 10\,\gamma$	− 8.9 »	E 4.8 γ	− 8.7 »	
7.5	− 15.4 »	0		− 6.6 »	» 5.5 »		− 12.8 »	» 3.6 »	− 10.6 »	
15	− 15.7 »	0		− 9.0 »	» 5.0 »		− 11.0 »	» 1.2 »	− 11.5 »	
22.5	− 16.5 »	W 1.5 γ		− 9.6 »	» 4.5 »		− 10.7 »	0	− 11.5 »	No perceptible perturbation.
30	− 15.7 »	» 4.9 »		− 7.8 »	» 4.5 »		− 10.7 »	(0?)	− 11.5 »	
37.5	− 9.0 »	» 9.3 »		− 7.2 »	» 7.0 »		?	?	− 16.1 »	
45	− 3.9 »	» 12.3 »		− 3.6 »	» 7.5 »		− 8.9 »	» 2.4 »	− 17.5 »	
52.5	+ 2.0 »	» 14.7 »		0	» 10.5 »		− 3.6 »	» 4.8 »	− 17.2 »	
0	+ 5.1 »	» 12.8 »		+ 10.2 »	» 11.0 »		− 2.5 »	» 2.4 »	− 16.6 »	
15	+ 11.4 »	» 11.8 »		+ 10.8 »	» 7.5 »		+ 5.7 »	» 2.4 »	− 7.4 »	
30	+ 7.9 »	» 8.9 »		+ 10.8 »	» 6.0 »		+ 4.6 »	» 1.8 »	− 1.4 »	
45	+ 4.3 »	» 6.9 »		+ 10.2 »	» 1.5 »		+ 2.8 »	» 1.2 »	+ 3.2 »	
0	+ 2.8 »	» 4.9 »		+ 8.4 »	0		+ 1.8 »	0	+ 7.8 »	

Current-Arrows for the 31st March, 1903; Chart I at $0^h\ 15^m$, and Chart II at $0^h\ 30^m$.

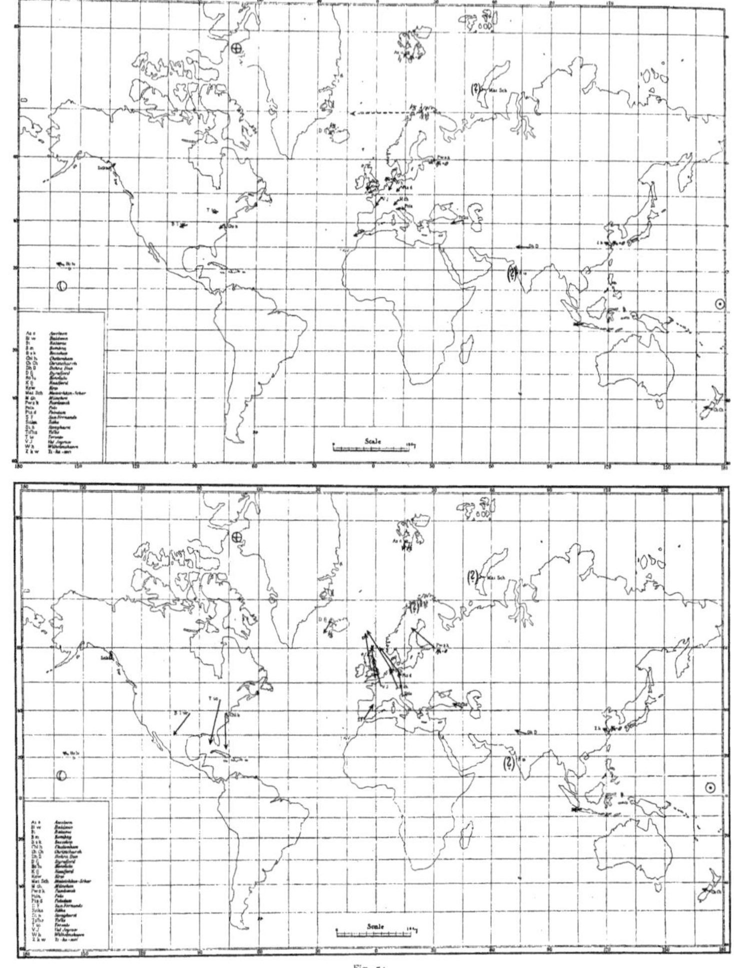

Fig. 54.

Current-Arrows for the 31st March, 1903; Chart III at $0^h\ 45^m$, and Chart IV at $1^h\ 0^m$.

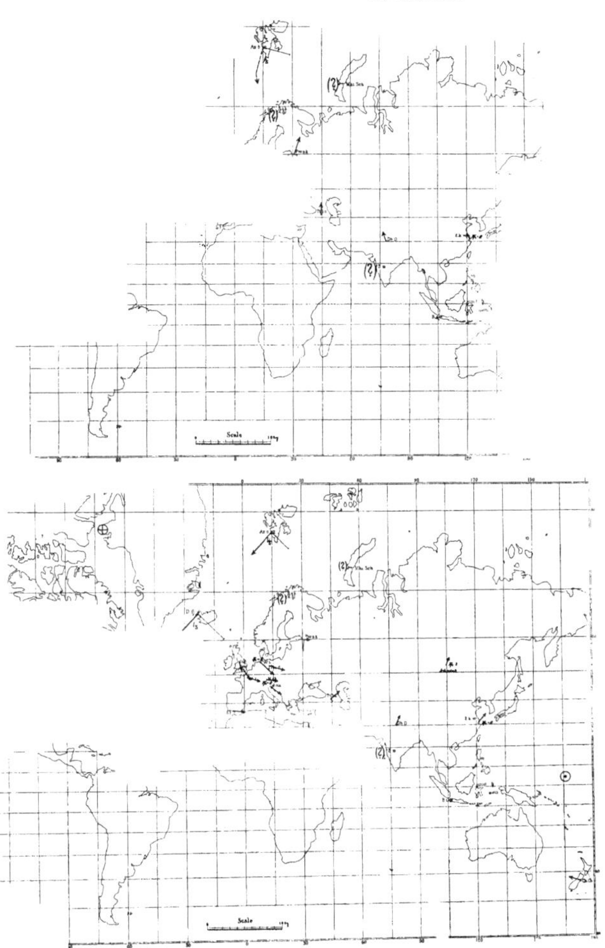

Current-Arrows for the 31st March, 1903; Chart V at $1^h\ 15^m$, and Chart VI at $1^h\ 30^m$.

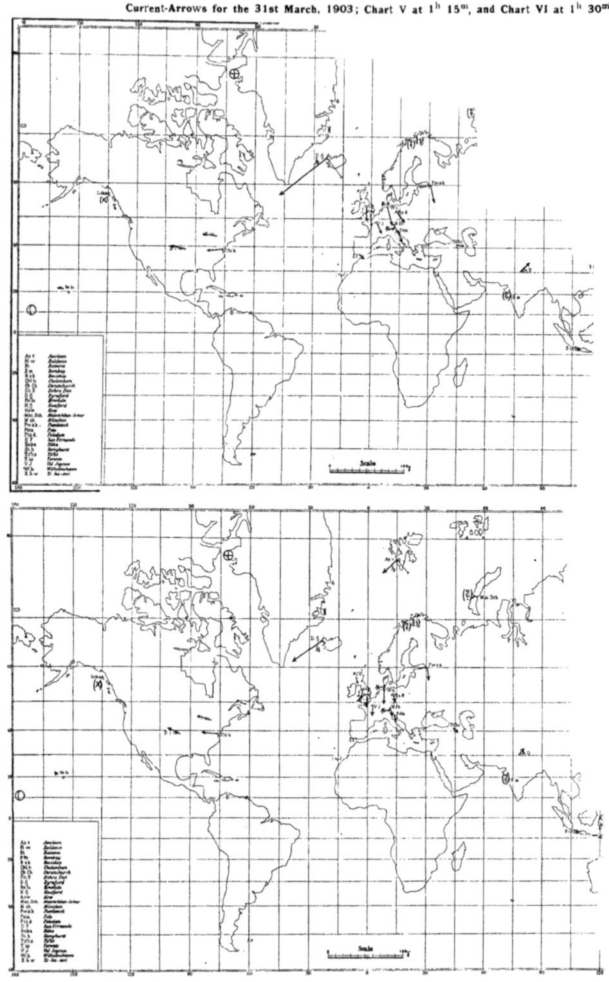

Chart I. Time 0^h 15^m.

In the regions nearest to the equator, the current-arrow points from E to W, while in Central and Southern Europe it has a more southerly direction.

The northernmost stations differ greatly in this respect, the conditions at Kaafjord, in particular, being quite peculiar. If the perturbation really has attained to such magnitude by this time, it must be the result of purely local occurrences, or rather, the effect must be so strong on account of the proximity to the currents that bring about the phenomenon.

Leaving the most northerly stations out of consideration, the force is strongest at the equator. From this we may conclude that we still, at this hour, have to a great extent the effect of the previously-mentioned equatorial perturbation, which commenced at 23^h 12^m, and which had a current-system the reverse of that shown on the chart for 22^h 0^m.

In the period under consideration, what we are concerned with is thus a slight equatorial perturbation together with the incipient polar storm.

Chart II. Time 0^h 30^m.

The effect of the polar storm is now altogether predominant. In Europe the current-arrows have already reached their maximum by this time. The directions of the arrows in Europe and the United States show distinctly that the field of force for the horizontal component has a point of convergence that is situated somewhere in the North Atlantic, probably a little south of the point of Greenland. There, according to our assumption with regard to the cause of these perturbations, the horizontal force should equal 0. We notice also the direction of the current-arrows at Toronto and in the United States, converging as they do to a point in the north of Labrador.

On the whole we may say that outwards the field is explained by the assumption that the current with negative particles descends towards the earth in the direction of the north of Labrador. It then turns off almost along the auroral zone, and leaves the earth in the district between the southern point of Greenland and Iceland. Judging from the form of the outer field of force, the current-system should have its centre at the southern point of Greenland, or a little to the west of it.

If we look at the conditions in the vertical intensity, we should expect, if this were the only system, to find P_v negative at all the stations in Europe and Asia, or possibly zero at certain places. On the contrary, however, we find that at several places there are positive values of P_v, e. g. at Potsdam, Val Joyeux and Tiflis; while at Pola and Wilhelmshaven they are in the opposite direction. The conditions at Bossekop, moreover, at these hours, are rather peculiar in the two components that we have; for just before, these two turn round in the opposite direction, and P_d remains for a time in a westerly direction, and P_v for a shorter time positive. This opposite deflection takes place slightly earlier in the vertical intensity than in the declination. The forces otherwise are so strong that they can hardly be explained by this system alone. Other perturbing causes seem to assert themselves, but of what kind it is impossible to determine, as nothing can be concluded as to the conditions in the horizontal intensity. It is possible that these two circumstances are connected with one another; but as we have said, the data necessary for the determination of this question are wanting.

Chart III. Time 0^h 45^m.

The storm has now also become powerful at Axeløen, in fact it is at its height. The arrows in the western hemisphere are about the same in direction and size as in the preceding chart. The arrows in Europe, on the other hand, have made quite a considerable turn clockwise. The perturbing forces at Dyrafjord and at the stations in England, Germany and France, have the reverse direction, and point downwards towards the same point. The central point of the system must thus be situated somewhere to the south-east of Dyrafjord, almost in the south-east of Iceland. The point of convergence must lie somewhere in the regions between Iceland and Stonyhurst, probably nearer the latter station, as the force there is so small.

Taking for granted the point of convergence, the horizontal force should first increase along the transverse axis of the system from 0, and then slowly decrease; and we do indeed find that the force increases from Stonyhurst towards Munich, Val Joyeux and San Fernando, and then becomes smaller towards Tiflis and Dehra Dun. The change, moreover, corresponds fairly well with that which we find in the two calculated systems in Table XV, where the horizontal part of the current lies at a height of 300 km.

The vertical components at Val Joyeux, Wilhelmshaven, Potsdam and Pawlowsk are all directed upwards, just as we should expect. At Pola there was earlier a fairly considerable vertical component directed upwards; but it now about equals 0.

Chart IV. Time $1^h\ 0^m$.

The arrows in Europe and Asia have continued to turn. In the United States also, the arrows have now turned a little. The alteration in the field is fully explained by the assumption that our current-system has moved a little farther in the direction from Dyrafjord to Axeløen.

The directions of the arrows in Europe show that the point of convergence of the horizontal components ought now to be found a little to the north of Pawlowsk. At Pawlowsk, as we should expect from its lying near the point of convergence, P_h is exceedingly small, only 7.5 γ; but on the other hand P_v = 14 γ, and is directed upwards.

At Tiflis and Potsdam, P_v is directed upwards, but is rather small. At Wilhelmshaven, $P_v = 0$. At Pola, a small force is directed downwards.

It is in harmony with our assumption that we also find a larger horizontal force south of Pawlowsk than at that station itself. It is greater even at Dehra Dun, Irkutsk and Zi-ka-wei. At the last-named station, $P_v = 0$. It appears from the vertical forces at our stations, that the principal axis of our system should lie to the south-east of Dyrafjord and Axeløen, as P_v is there directed downwards. At Kaafjord, however, we find P_v directed upwards, which also indicates that the axis lies between the two first-named stations and the latter.

Chart V. Time $1^h\ 15^m$.

The current-arrows in the United States are turned so that their direction is now about west, answering to a southward direction of P_h.

In Europe, P_h is turned farther in the same direction, and is now directed eastwards. The field during this period resembles that at the conclusion of the perturbation of the 15th December, or those of the 22nd March and 10th February.

At Pawlowsk there is still a considerable vertical component directed upwards. The point of convergence should now have moved farther east.

Chart VI. Time $1^h\ 30^m$.

The distribution of force is as in the preceding chart, but the forces are much smaller. In the case of the European stations, the turning is continued a little.

During this great but gradual alteration in the outer field, the conditions at Dyrafjord and Axeløen, notwithstanding small local irregularities, have remained very constant. At both stations the current-arrows have been directed all the time south-west; and the vertical component all the time has been directed downwards. At Kaafjord, on the contrary, the vertical force has been directed upwards all the time, with the exception of a short time at about $0^h\ 28^m$, and attains a magnitude of 209 γ.

This circumstance, together with the fact that the effect at the side, at right angles to the current. arrows, ceases before very long, can only be explained by the assumption of a comparatively low-lying horizontal part of the current, which passed between Axeløen and Kaafjord, and a little to the south of Dyrafjord. This horizontal part of the current forms the connection between the upward and downward flowing vertical currents. Perhaps at about 0ʰ 28ᵐ, the current has passed south of Kaafjord, but has then turned off over this place to take up the above-named position. The curve for P_v at Kaafjord seems to indicate the transverse passage of the current over this place at the beginning of the polar perturba. tion. We have seen, moreover, that the field may always be assumed to have been produced by a system such as this, which, in order to explain the variation of the field with time, must be supposed to be moving eastwards along the auroral zone (see the perturbation of the 15th December).

We have mentioned the remarkable fact of the maximum occurring earlier in Central and Western Europe and the United States than at the arctic stations. This is a necessary consequence of our assumption. At 0ʰ 30ᵐ, when the perturbation is at its height on the continent of Europe, these stations lie considerably to the east of the point of convergence, which, on account of the direction of the forces, must be looked for in the region of the North Atlantic. Owing, however, to the movement of the system, the stations on the mainland of Europe, at the time the perturbation in the north is at its height, will be situated in the neighbourhood of the neutral area. This same movement of the system will also cause it to withdraw farther and farther from the American stations. This again will cause the maximum of the perturbing force at these stations to occur before the time at which the current-strength of the system has reached its maximum. This displacement must be greatest at those stations which lie nearest to the current-system; and this we also find to be the case. The displacement, as will be seen from the table, is less at Sitka than at Toronto; and at Honolulu it is imperceptible, as the time of the maximum coincides with that at the northern stations.

While this perturbation was going on, remarkable aurora was observed, and earth-currents were registered at Kaafjord. These will be discussed under the special treatment of these phenomena.

THE PERTURBATIONS OF THE 22nd MARCH, 1903.

(Pl. XX.)

44. The perturbation of the 22nd March is in reality, like that of the 31st March, composed of two principal phenomena, an equatorial perturbation and a short, well-defined, comparatively powerful elementary polar storm. As the equatorial one is rather slight, it will not have a greatly disturbing effect upon the polar storm, of which the properties can therefore be fairly accurately determined. As it is the polar storm to which, on account of its simple course, we have especially turned our attention, we have thought it best to class it among the polar elementary storms.

THE EQUATORIAL PERTURBATION.

45. This perturbation begins quite suddenly at 12ʰ 58ᵐ, with an oscillation that is noticed simultaneously all over the world. In the equatorial regions, this sharp deflection is uniform in direction, and appears principally in *H*. About the auroral zone the curve oscillates, and the perturbation is noticeable both in *D* and *H*. This first oscillation is shown on

Chart I, at 13ʰ 4ᵐ,

which is the time when it reaches its maximum. About the equator the arrows are comparatively large, and run about parallel with the magnetic equator. In the south and centre of Europe, the current-arrow points considerably towards the north, as compared with what is generally the case during these equatorial perturbations. At Kaafjord also, the direction of the arrow is in accordance with the rest of Europe, but the force is somewhat greater. At Dyrafjord and Axeløen we find a peculiar circumstance, namely, that in the course of a few minutes the force oscillates very violently. A number of arrows are placed upon the chart, answering to various hours, the scale being the same for the southern stations as for the northern. At Dyrafjord, the current-arrow makes a negative turn of about 180° from SW to NE. At $13^h\ 4^m$, the direction of the force is uniform with that of the arrows in the south of Europe. At the same time, the arrows on Axeløen turn from S in a positive direction, until at $13^h\ 7^m$ they point NE as at the other European stations.

We will not here attempt to give an explanation of this peculiar circumstance, but will only say that this turning in different directions at two places so near to one another, must necessarily lead to the conclusion that in the north at any rate, the perturbation is to some extent of a local character.

We thus see that while there is a current that acts powerfully and almost symmetrically on both sides of the equator, there will be exactly simultaneous perturbations of a local character in the north. These currents in the north, which are very slight, cannot, on account of the extent of the perturbation, be the cause of the perturbation as a whole; for, as we see, the force diminishes from the poles southwards as far as Tiflis and San Fernando, whereupon it increases, and even at Christchurch is great. In the vertical intensity this first oscillation is noticed, in southern latitudes, only at Tiflis, where it indicates a force directed upwards. The reason why it is not felt at Zi-ka-wei can only be that the sensibility there is so small; but on the other hand, it seems stranger that nothing is noticed at Pawlowsk and Pola, where the sensibility is fairly great.

After the first deflection, the equatorial perturbation continues with a small deflection in H, answering to a perturbing force directed northwards along the magnetic meridian. Judging from the characteristically serrated appearance of the H-curve in low latitudes, the perturbation seems to last until the polar storm is over, or from about $12^h\ 57^m$ until midnight.

The distribution of force, as it is on the whole maintained on account of this equatorial perturbation, is shown on

Charts II and III, for $15^h\ 0^m$ and $19^h\ 50^m$.

The current-arrows in somewhat more southern latitudes lie, as we see, almost parallel with the magnetic parallels, and the force there is comparatively great. We notice that the force at the Central European stations varies greatly in magnitude. We must not, however, immediately draw conclusions from this circumstance; for it may be accounted for partly by the difficulty there is in determining the normal line for so long an interval, and partly by the fact that, owing to the rapid changes in the deflections, a mistake in the time will easily occasion a mistake in the determination of the perturbing force, of which the percentage becomes all the larger, when the perturbing force is small.

At the arctic stations the force is comparatively great, and we see that the current-arrow bends northwards, and indicates a circle round the magnetic pole, showing that it is not the axis of the earth, but the magnetic axis, that determines the phenomenon. At Sitka too, the current-arrows are somewhat abnormal, as we also found them to be in previous equatorial perturbations. This must be due to the polar precipitation that is always present during these storms. If we look at fig. 37, we see that the light parts in the terrella's auroral zone, come more or less in the region answering to the north of N. America. It is possible that this drawing-in of rays may also to some extent be the cause of the abnormal smallness of the perturbing force at Baldwin on Chart I. We shall find this confirmed in the conditions during the equatorial storm of the 15th December, 1882, described in Chapter III.

In the vertical intensity, the perturbation is almost imperceptible, being only slight at Tiflis, where it is directed upwards at the moment of observation. At Pawlowsk it is not noticeable, and at Dyrafjord very slight. We should notice this circumstance with regard to the vertical components. On the whole, this perturbation is in accordance with the usual equatorial perturbations, and to these we may refer for the explanation of its cause.

THE POLAR STORM.

46. The polar storm, as the curves show, is very well defined and brief. It is especially worthy of notice that the deflections, which, in the Central European field, are particularly powerful in the declination, keep to one direction all the time. Even at the arctic stations, the deflections, both in H and in D, are nearly uniform in direction, Dyrafjord alone having an oscillation in declination. In the Table XIX will be found the times of the commencement and termination of the polar storm, as also the time of the maximum of the horizontal component, and the value of the latter at the moment. Since, as we have said, an equatorial perturbation appears in advance of, and presumably simultaneously with, the polar storm, it would seem difficult to decide when the polar storm commences and terminates. In the northern regions, however, the polar storm will make its appearance with such strength, that the effect of the equatorial perturbation will be comparatively minimal. At the arctic stations, we have therefore taken the times when the great storm commences and ceases. As regards the southern districts of Europe, we are aided by the circumstance that the polar storm appears mainly in the declination, while the previous storm has kept principally to the horizontal intensity. In the United States and Honolulu, on the other hand, they both appear in H, but there the effect of the polar storm is marked as a decided undulation.

The position of the normal line for Sitka was somewhat difficult to determine, and there is therefore also some difficulty in accurately determining the commencement of the perturbation. At the Asiatic stations, both the perturbations appear in H, so that neither beginning nor end can be determined to any advantage.

It will be seen that the perturbing force on the whole diminishes with increasing distance from the region of the Norwegian stations. Wilhelmshaven, as usual, comes out of its order in the series, being before Stonyhurst, and with a very much greater maximal force. At most of the stations, the storm lasted, as we see, for about $2^1/_2$ hours.

We find, as usual, that the perturbation appears first at Bossekop, then at Dyrafjord, and then at Axeløen. In the central and southern districts of Europe, the maximum occurs at about $22^h\ 10^m$; in the United States and at Honolulu it is later—about $22^h\ 40^m$. The maximum on the whole is not well defined, but the force remains for a fairly long time almost constant. This even applies to the arctic stations, and we have therefore set no definite point of time here.

It appears from the Table, as also from direct observation of the copies of the curves, that the perturbations at all the places are connected with one another, as they appear simultaneously, and their course is somewhat similar. We find again, moreover, a very characteristic feature of these polar storms, namely, that whereas the perturbation in the arctic districts changes very much from one time to another, and from one place to another, the conditions in lower latitudes vary more slowly with time and place. This must necessarily lead to the assumption that the perturbation in lower latitudes must be due to the same cause as that in the arctic districts. The perturbation in southern latitudes can, moreover, only be the distant effect of the same current-systems that come nearer to the earth about the auroral zone.

The circumstances are represented on Charts IV—VII, for the hours $22^h\ 0^m$, $22^h\ 15^m$, $22^h\ 30^m$, and $23^h\ 0^m$.

On the whole, the distribution of force remains constant all the time. There is the same system of lines of force, the intensity alone varying.

This time also, however, the force in Central and Southern Europe makes a distinct, though very slight, turn clockwise. The field is of the same typical form as that of the polar elementary storms already described.

THE PERTURBING FORCES.

47. The total of the perturbing forces is calculated for a number of hours, and marked upon charts.

As the polar perturbation is so much greater than the equatorial, the field of force shown, perhaps with the exception of the equatorial regions, is mainly conditioned by the polar system.

TABLE XIX.

Observatory	Begins	Reaches Max.	$P_{t\,max}$	Ends
	h m	h m h m		h m
Axeløen	21 12	22 4	ca. 370 γ	23 45
Dyrafjord	21 0	22 0-22 48	ca. 330 »	23 44
Bossekop	20 30	21 54 22 34	ca. 220 »	23 45
Wilhelmshaven	21 9	22 6	52 »	23 47
Stonyhurst	21 10	22 8	47 »	23 44
Val Joyeux	21 14	22 14	45.4 »	23 48
Potsdam	21 10	22 8	44.7 »	ca. 23 42
Kew	21 9	22 8	44 »	23 46
Pawlowsk	21 10	22 9	42 »	24 0
Sitka	20 39	22 12	41.3 »	ca. 24
Munich	21 12	22 10	39 »	ca. 23 48
Pola	21 9	22 9	34.7 »	ca. 23 45
Baldwin	ca. 21	22 39	30 »	ca. 24
Toronto	ca. 21	22 40	25.2 » [1]	ca. 23 50
Cheltenham	ca. 21 15	22 36	24 »	ca. 23 50
Tiflis	21 11	22 10	22.5 »	ca. 23 50
Dehra Dun	indeterminable	21 45	17.5 »	ca. 23
Honolulu	21 30	22 36	15.5 »	24 0
Bombay	indeterminable	ca. 22	14.3 »	ca. 23
Zi-ka-wei	»	ca. 22 20	12.4 »	ca. 23
Batavia	»	21 49	12.4 »	ca. 23
Christchurch	»	indeterminable	indeterminable	indeterminable

TABLE XX.

The Perturbing Forces on the 22nd March, 1903.

Gr. M. T.	Honolulu		Sitka		Baldwin		Toronto		Cheltenham	
	P_h	P_d	P_h	P_d	P_h	P_d	P_h	P_d	P_h	P_d
h m										
13 4	+ 6.6 γ	0	+ 9.0 γ	W 10.0 γ	+ 4.2 γ	? [2]	?	?	? [2]	? [2]
15 0	+ 7.1 »	0	+ 3.5 »	0	? [2]	?	?	?	?	?
18 10	+ 3.8 »	0	+ 6.2 »	E 4.5 »	?	?	+ 11.2 γ	0	?	?
19 50	+ 3.2 »	0	0	W 10.0 »	?	?	+ 14.0 »	0	?	?
21 45	− 0.9 »	0	− 20 2 »	» 9.5 »	− 10.5 »	W 3.2 γ	− 11.3 »	?	− 16.2 γ	W 5.9 γ
22 0	− 7.0 »	W 3.3 γ	− 32.2 »	» 14.4 »	− 20.0 »	» 5.7 »	− 20.7 »	?	− 21.2 »	» 6.5 »
15	− 11.9 »	» 3.3 »	− 38.3 »	» 16.7 »	− 21.1 »	» 7.0 »	− 24.3 »	?	− 25.2 »	0
30	− 14.4 »	» 4.2 »	− 31.7 »	» 24.8 »	− 21.4 »	» 2.5 »	− 20.3 »	?	− 24.4 »	E 2.4 »
45	− 14.9 »	» 3.3 »	− 34.2 »	» 31.2 »	?	?	− 24.8 »	?	− 25.2 »	W 1.8 »
23 0	− 14.4 »	0	− 23.7 »	» 32.5 »	− 19.0 »	» 4.5 »	− 20.7 »	?	− 22.7 »	» 2.4 »
15	− 13.0 »	0	− 19.7 »	» 23.8 »	− 14.0 »	» 1.9 »	− 18.9 »	?	− 14.7 »	0

(1) The value of P_h, there being no declination curve.
(2) The normal line somewhat uncertain.

TABLE XX (continued).

	Gr. M. T.	Axeløen			Kaafjord		
		P_h	P_d	P_v	P_h	P_d	P_v
	h m						
0	12 57	0	E 12.5 γ				
+ 2.0 γ	59	− 37.5 γ	W 4.2 »	Slightly			
− 2.6 »	13 3	+ 12.0 »	» 18.0 »	negative			
− 1.5 »	4				+ 20.5 γ	W 15.5 γ	?
− 4.6 »	7	+ 31.7 »	» 16.7 »				
− 6.1 »	15	0	» 12.5 »				
− 6.1 »	15 0	+ 38.0 »	» 26.0 »	Slightly neg.	+ 23.5 »	0(?)	?
− 4.6 »	18 10	+ 19.0 »	» 17.0 »	0	+ 3.8 »	0	?
− 2.6 »	19 50	+ 12.7 »	» 16.5 »	+ 24.6 γ	+ 10.0 »	E 25.0 »	?
Possibly	21 45	− 83.0 »	E 30.0 »	+ 334.0 »	− 153.0 »	» 75.0 »	− 149.0 γ
slightly	22 0	− 265.0 »	» 171.0 »	+ 408.0 »	− 182.0 »	» 99.0 »	− 205.0 »
negative	15	− 327.0 »	» 177.0 »	+ 492.0 »	− 185.0 »	» 101.0 »	− >205.0 »
+ 31.0 γ	30	− 133.0 »	» 84.0 »	+ 484.0 »	− 200.0 »	» 92.5 »	− >205.0 »
− 115.0 »	45	− 140.0 »	» 49.0 »	+ 396.0 »	− 125.0 »	» 47.0 »	− >205.0 »
− 184.0 »	23 0	− 136.0 »	» 101.0 »	+ 266.0 »	− 91.0 »	» 23.0 »	− >205.0 »
0	15	− 78.0 »	» 74.0 »	+ 202.0 »	− 38.0 »	» 16.0 »	− <222 >205 γ
+ 51.0 »							
+ 77.0 »							
+ 38.0 »							

TABLE XX (continued).

owsk		Stonyhurst		Kew		Val Joyeux		
d	P_v	P_h	P_d	P_h	P_d	P_h	P_d	P_v
9.2 γ	0	+ 15.8 γ	W 9.7 γ	+ 12.5 γ	W 7.6 γ	+ 14.4 γ	W 8.4 γ	0 (?)
4.0 »	0	+ 20.4 »	0	+ 20.3 »	0	+ 19.5 »	0	0
9.2 »	0	+ 17.8 »	0	+ 12.0 »	0	+ 20.0 »	0	0
4.6 »	0	+ 12.2 »	0	+ 13.0 »	0	+ 23.2 »	0	0
0.8 »	− 7.5 γ	− 9.7 »	E 38.3 »	− 8.7 »	E 34.6 »	− 3.2 »	E 30.2 »	
6.8 »	− 11.2 »	− 5.6 »	» 46.2 »	− 10.7 »	» 42.8 »	+ 4.0 »	» 37.8 »	V
6.8 »	− 14.2 »	− 12.8 »	» 45.0 »	− 11.2 »	» 42.8 »	− 2.4 »	» 45.4 »	increases
7.6 »	− 15.0 »	− 10.2 »	» 35.5 »	− 11.2 »	» 33.7 »	0	» 35.3 »	a little.
9.8 »	− 15.0 »	− 10.2 »	» 34.9 »	− 10.7 »	» 35.0 »	0	» 32.8 »	
5.2 »	− 16.5 »	− 11.7 »	» 27.0 »	− 11.2 »	» 34.1 »	− 5.6 »	» 28.6 »	
1.5 »	− 10.5 »	− 12.2 »	» 16.9 »	− 10.2 »	» 19.6 »	0	» 21.0 »	

TABLE XX (continued).

Gr. M. T.	Wilhelmshaven			Potsdam		San Fernando		Munich		
	P_h	P_d	P_v	P_h	P_d	P_h	P_d	P_h	P_d	P_v
	+ 16.3 γ	W 9.0 γ		+ 12.6 γ	W 6.6 γ	+ 11.1 γ		12.5 γ	9.5 γ	0
15 0	+ 19.6 »	0		+ 16.5 »	0	11		18.0 »	0 (?)	0
18 10	+ 16.0 »	0		+ 15.2 »	0	9.		15.0 »		0
19 50	+ 18.6 »	0		+ 19.0 »	0	+ 13.3 »		13.0 »		0
21 45	+ 13.1 »	E 47.6 »	A slight positive deflection.	+ 12.6 »	E 35.6 »			9.0 »	31.5 »	A slight positive deflection, with maximum answering to $P_v = +1.9\gamma$
22 0	+ 12.6 »	» 49.4 »		+ 12.3 »	» 41.7 »	The curve for D coincides with the base-line, the deflection in both curves being so slight that nothing is taken out.		10.0 »	35.3 »	
15	+ 2.8 »	» 48.2 »		+ 5.1 »	» 41.7 »			5.0 »	39.0 »	
30	— 2.3 »	» 35.6 »		+ 4.7 »	» 30.5 »				30.0 »	
45	— 4.7 »	» 32.6 »		0	» 27.9 »			4.0 »	27.0 »	
23 0	— 11.2 »	» 23.5 »		— 4.7 »	» 18.3 »				18.6 »	
15	— 13.5 »	» 15.1 »		— 4.7 »	» 14.2 »				15.0 »	

TABLE XX (continued).

Gr. M. T.	Pola			Tiflis			Dehra Dun		Bombay	
	P_h	P_d	P_v	P_h	P_d	P_v	P_h	P_d	P_h	P_d
h m										
13 4	+ 12.0 γ	W 7.0 γ	0	+ 8.9 γ	W 3.7 γ	— 1.3 γ	+ 13.4 γ	0	+ 10.0 γ	No curve.
15 0	+ 12.0 »	0 (?)	0	+ 9.5 »	0 (W 1.5 γ?)	0	+ 13.0 »	0	+ 11.8 »	
18 10	+ 11.0 »	0	0	+ 12.2 »	0	0	+ 12.4 »	0	+ 11.5 »	
19 50	+ 11.0 »	0	0	+ 18.8 »	0	— 2.7 »	+ 17.3 »	0	+ 16.0 »	
21 45	0	E 25.4 »	+ 5.0 γ	+ 18.8 »	E 9.3 »	— 3.8 »	+ 15.7 »	0		
22 0	0	» 33.7 »	+ 5.0 »	+ 18.3 »	» 13.4 »	— 3.8 »	+ 15.0 »	0	Nothing taken out as P_d is wanting.	
15	— 2.2 »	» 33.7 »	+ 2.8 »	+ 14.1 »	» 13.0 »	— 2.6 »	+ 9.8 »	0		
30	0	» 26.8 »	+ 1.2 »	+ 8.8 »	» 11.1 »	— 1.3 »	+ 4.7 »	0		
45	— 2.2 »	» 24.0 »	+ 1.0 »	+ 3.3 »	» 10.8 »	0	+ 1.6 »	0		
23 0	— 4.5 »	» 19.2 »	0	0	» 9.3 »	0	— 0.8 »	0		
15	— 4.5 »	» 13.7 »	— 1.0 »	0	» 5.9 »	0	0	0		

TABLE XX (continued).

Gr. M. T.	Zi-ka-wei		Batavia		Christchurch		Irkutsk		
	P_h	P_d	P_h	P_d	P_h	P_d	P_h	P_d	P_v
h m									
13 4	+ 10.0 γ	0	+ 11.4 γ	W 3.6 γ	+ 9.2 γ	0			
15 0	+ 11.0 »	0	+ 10.0 »	0	+ 9.2 »	0			
18 10	+ 6.5 »	0	+ 6.5 »	0	+ 10.4 »	0			
19 50	+ 8.5 »	0	+ 13.5 »	0	+ 13.8 »	0			
21 45	+ 10.2 »	W 7.0 γ	+ 13.2 »	0					
22 0	+ 9.2 »	» 7.0 »	+ 11.0 »	0	Owing to the difficulty in determining the normal line, nothing is taken out.		+ 12 γ	W 17 γ	— 3 γ
15	+ 2.4 »	» 5.0 »	+ 9.5 »	0			+ 10 »	» 15 »	— 4 »
30	0	» 5.0 »	+ 3.2 »	0			+ 5 »	» 8 »	— 4.6 »
45	0	» 2.5 »	+ 2.5 »	0					
23 0	0	0	0	0					
15	0	0	0	0					

for the 22nd March, 1903; Chart I at $13^h\ 4^m$, and Chart II at 15^h.

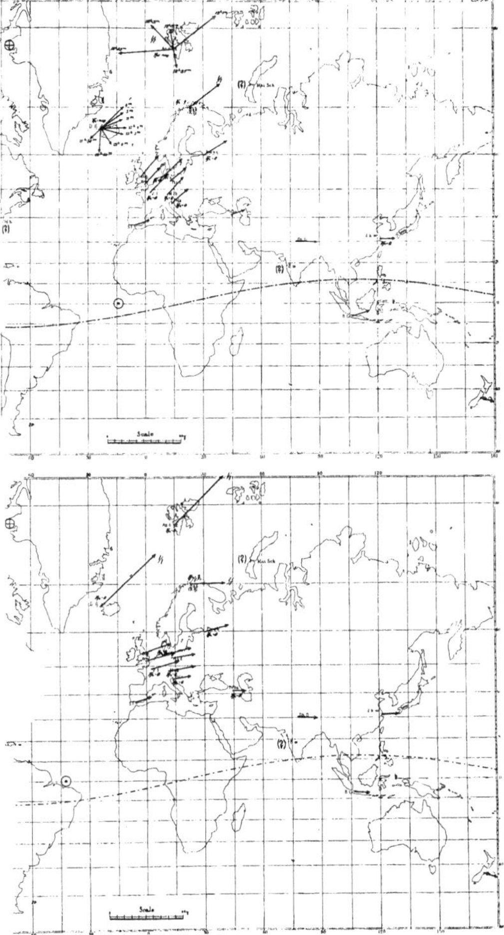

Fig. 57.

Current-Arrows for the 22nd March, 1903; Chart III at $19^h 50^m$, and Chart IV at 22^h.

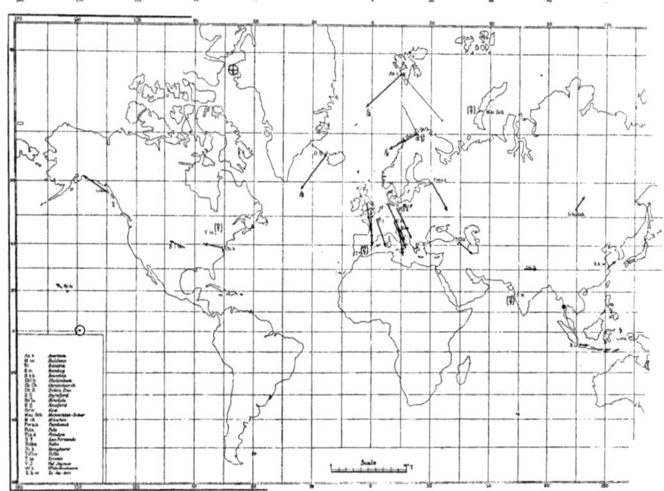

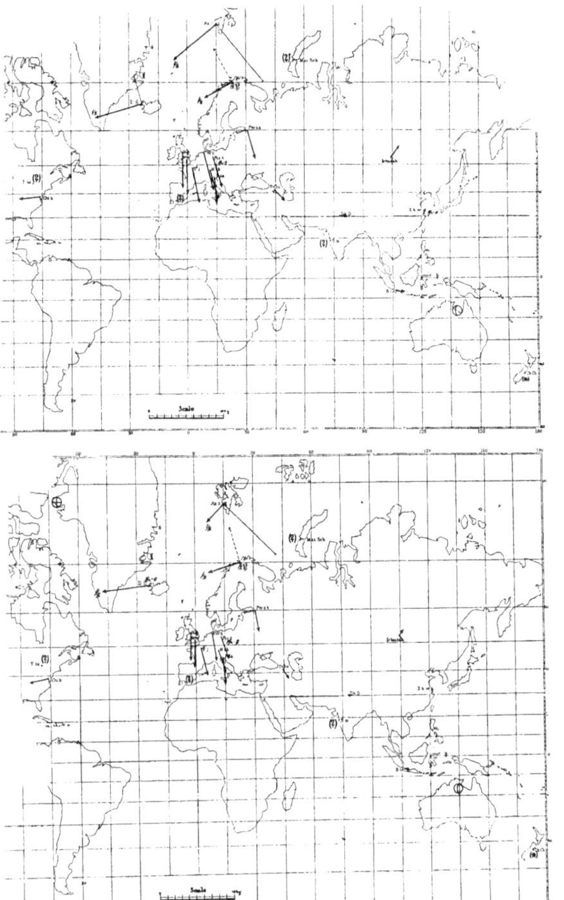

Fig. 59.

Current-Arrows for the 22nd March, 1903; Chart VII at 23h.

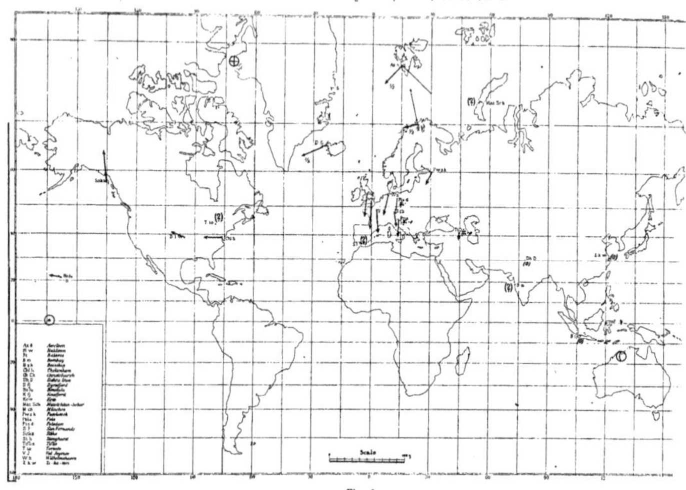

Fig. 60.

The current-arrows indicate very decidedly two current-vortices, a positive vortex in the north of North America, and a negative one about the river Obi in Siberia, answering respectively to areas of divergence and convergence of the perturbing forces.

As we have no observations from places near the points of convergence, we cannot here recognise the characteristic perpendicular position of the total force in relation to the earth's surface.

As regards the cause, we may confine ourselves to referring to the previously-described elementary storms. Here too it is difficult to understand how the perturbation in lower latitudes can be mainly due to plane currents, as in that case this peculiarly formed current-system should retain its form and position for nearly 2½ hours.

The slight oscillation of the force in Europe on this date, is in accordance with the fact that the point of convergence is now far to the east, and this is certainly connected with the circumstance that the perturbation appears so early in the night, reckoning by Greenwich time. At the Norwegian stations about the auroral zone, the current-arrows point in the characteristic direction westwards along the zone. On this date the vertical components at Bossekop and Axeløen are exceedingly powerful and in opposite directions, answering to a current passing between the two stations. At Dyrafjord, P_v is comparatively smaller, indicating that the current should pass north of this station.

We have no observations for this date from Matotchkin Schar. From Potsdam no curves for V were received. For Ekaterinburg nothing can be taken out.

THE PERTURBATIONS OF THE 26th DECEMBER, 1902.

(Pl. XII).

48. The perturbations to which we have especially turned our attention are two successive, brief, well-defined storms, that are particularly powerful at our Norwegian stations, more especially Dyrafjord and Matotchkin Schar.

The first of these two well-characterised polar storms is especially powerful at Matotchkin Schar, where P_1 attains a value of $248\,\gamma$. At Axeløen there is a perturbation that is quite distinct in all three components. At Kaafjord there is simultaneously a very distinct perturbation, but one that is very small both in D and H, whereas in V it is considerably stronger. At Dyrafjord, the curve shows clearly that this brief polar storm occurs simultaneously with a more lengthy perturbation. Its effect, on the whole, at Dyrafjord, is contrary to that of the longer storm. A decomposition of the perturbing force may here be effected.

The same conditions, although less marked, are found on the continent of Europe, where the H-curve shows a faint, but long perturbation. There too, the course of the intermediate perturbation is the reverse of that of the longer storm; but as the former is much more powerful, it will predominate during the time in which it occurs.

The second storm is especially powerful between $22^h\ 30^m$ and 24^h. It also occurs in the north as a characteristic polar elementary storm, which is particularly powerful at Dyrafjord. This is in accordance with the fact that it appears later.

At the stations in lower latitudes, we notice in the case of both storms simultaneous but comparatively slight perturbations; and the effect becomes weaker with an approach to the equator. At Sitka, the perturbation is only of the same magnitude, and has the same course, as in the rest of America.

According to this, it is natural to consider these two storms as two successive polar elementary storms, in which the storm-centre is situated somewhat differently. This will be still more apparent on a closer examination of the field of force.

The *field of force* during the first storm is shown on *Charts I and II*, for the hours $20^h\ 45^m$, and 21_h respectively.

The form of the field is in the main the same in both cases, as also the relative strength. This clearly indicates that the system in question is one that on the whole preserves its form and its position, and only varies in strength. The arrows at Axeløen and Matotchkin Schar form exceptions in this respect, the force at these stations being almost as great at 21^h as at $20^h\ 45^m$. This does not necessarily, however, alter our view of the conditions; for, owing to the local character of the perturbations in these regions, very slight movements of the system may here have a great effect, and thus the force at one place may very well have its greatest value at a time other than that at which the system as a whole is strongest.

The form of the field is that typical of the polar elementary storms. The storm-centre is situated in the region north-east of Matotchkin Schar, and the area of convergence in north-eastern Russia. The current-arrow about the centre is as usual directed WSW. There is an area of divergence in America, which seems to belong to another storm-system, this being also confirmed by the arrows at Dyrafjord.

As regards the vertical intensities, we find at Pawlowsk a perturbing force directed upwards, just as we should have expected. At Wilhelmshaven, Pola and Tiflis, on the contrary, we find positive values for P_z. The deflections, it is true, are only slight, but still are sufficiently distinct. They cannot be due to the system that we have assumed to be at our easterly stations, as that system can produce only negative values of P_z in the area of convergence.

It is difficult to decide what forces here play a part. The system that produced the area of divergence in America, may indeed possibly be supposed to exert an influence here too; and this would also of course produce positive values of P_1. But it seems difficult to imagine that its effect may be traced as far off as at Tiflis.

It seems more natural to explain the conditions by rays that come rather near to the earth in lower latitudes, as in the cyclo-median perturbations. The considerable strength of the current-arrows in Europe, as shown on Chart I, seems to point in this direction, although the increased strength may possibly be chiefly due to the fact that the two polar systems are here acting in about the same direction.

On account of the quiet character of the deflections, and the small perturbing forces, these currents must nevertheless lie fairly high; and it is possible that they are connected with one of the polar systems, probably that in America.

The *field of force* during the second storm is shown on *Charts III, IV, V and VI,* for the hours $23^h\ 0^m$, $23^h\ 15^m$, $23^h\ 30^m$, and $23^h\ 52.5^m$, respectively.

Chart III shows the conditions at the beginning of the second storm. It is only at Dyrafjord that the perturbing force has reached any magnitude. The arrows for the European stations represent a very curious field of force; but as they are small, the determination is somewhat uncertain, owing to the inaccuracy in the determination of the normal line.

The field in Charts IV and V shows very distinctly the form that is typical for the polar elementary storms.

At Dyrafjord, the force is exceedingly great, and is directed westwards along the auroral zone. The storm-centre, which is presumably situated very near Dyrafjord, is now about 145° east of the sun. The field to the south exhibits a well-marked area of convergence. There is probably, however, not only precipitation round Dyrafjord; but it also seems as if there were local currents round the other Norwegian stations, as the force there is also comparatively strong.

At $23^h\ 52.5^m$, Chart VI, the strength of the field is considerably less. At Dyrafjord the direction of the arrow is different, being now south.

We notice a peculiar circumstance, namely, that with the turning of the arrow at Dyrafjord, the whole field turns.

The arrow at Kaafjord, and at the more southerly European stations from Kew to Tiflis, indicates an area of convergence. Judging from the shape of the field, the centre of this area should be about Pawlowsk; and in fact we find that at this moment the force there equals 0.

We thus see that the conditions in more southern latitudes are in very close connection with those round the auroral zone. This circumstance, as we have said, may be explained in a very simple way, the perturbations in low latitudes, in these cases, being assumed to be produced by the action, at a great distance, of the systems that are necessary to the production of the perturbations about the auroral zone.

In all the elementary polar storms described, it will generally have been remarked (1) that all the current-arrows in lower latitudes turn clockwise during the perturbation, and (2) that in the same latitudes, the simultaneous current-arrows turn clockwise, if one moves from eastern to western stations. These assertions I have already made in my earlier work, 'Expédition Norvégienne de 1899—1900', etc., pp. 32 & 33.

In this earlier work, I assumed that these assertions were explained by a current-system like that in fig. 45, and by the fact that this current-system, starting in the polar regions, was there deflected westwards during the perturbation. We have here maintained a somewhat different view, as, instead of the horizontal current-system, we have supposed a system that, idealised, consists of two vertical branches connected by a horizontal portion, and that this current-system has a district of precipitation in the polar regions, with its principal axis along the auroral zone. The current-line system (see Art. 34 and fig. 40) is however even now similar to the formerly assumed real current-sytem. The turning s in lower latitudes is then occasioned by the eastward movement of the storm-centre along zone, with the principal axis always keeping its direction (see p. 94). When it is desired to verify on all the charts this movement of the storm-centre during the course of the perturbation, it is necessary, as we have said several times, to remember that the size of the current-arrows at the four Norwegian stations, is not always a certain guide to the position of the storm-centre (see p. 137). This travelling of the storm-centres is possibly caused by the effect of terrestrial magnetism upon the current. , and by the alteration in the earth's magnetic axis during the perturbation. We shall return to subject later on.

TABLE XXI.

The Perturbing Forces on the 26th December, 1902.

Gr. M. T.	Honolulu		Sitka		Baldwin		Cheltenham	
	P_h	P_d	P_h	P_d	P_h	P_d	P_h	P_d
h m								
20 45	− 1.8 γ	E 5.8 γ	− 6.5 γ	Disturbed vibrations, but nothing can be determined.	− 5.7 γ	E 2.5 γ	− 6.2 γ	0
21 0	0	» 7.5 »	− 6.5 »		− 4.7 »	» 2.5 »	− 4.4 »	0
23 0	− 8.5 »	» 11.6 »	− 1.6 »		− 1.8 »	» 2.5 »	− 4.4 »	0
15	− 8.5 »	» 10.0 »	− 6.5 »		− 4.7 »	» 3.2 »	− 6.5 »	E 3.5 γ
30	− 6.2 »	» 10.0 »	− 6.2 »		− 6.5 »	» 3.2 »	− 7.9 »	» 5.9 »
52.5	− 3.4 »	» 6.6 »	0		− 1.4 »	» 1.3 »	− 0.8 »	0

TABLE XXI (continued).

Gr. M. T.	Dyrafjord			Axeløen			Matotchkin Schar		
	P_h	P_d	P_v	F_h	P_d	P_v	P_h	P_d	P_v
h m									
20 45	+ 12.1 γ	E 13.8 γ	− 43.0 γ	− 164.0 γ	0	+ 150.0 γ	− 202.0 γ	E 146.2 γ	− 202.0 γ
21 0	+ 12.1 »	» 25.2 »	− 44.0 »	− 191.0 »	E 12.0 γ	+ 290.0 »	− 153.0 »	» 158.0 »	− 128.0 »
23 0	− 154.0 »	W 53.3 »	+ 33.3 »	+ 2.3 »	W 12.0 »	+ 34.3 »	− 27.3 »	0	− 31.5 »
15	− 247.0 »	» 159.0 »	+ 21.5 »	− 11.5 »	» 30.7 »	+ 34.3 »	− 56.3 »	» 1.8 »	− 63.5 »
30	− 225.0 »	» 74.2 »	− 35.2 »	+ 7.3 »	» 50.2 »	+ 86.0 »	− 54.6 »	W 2.7 »	− 41.0 »
52.5	+ 48.8 »	E 18.7 »	− 17.1 »	0	» 10.7 »	+ 41.0 »	− 20.8 »	» 6.2 »	− 5.1 »

TABLE XXI (continued).

Gr. M. T.	Kaafjord			Pawlowsk			Wilhelmshaven		
	P_h	P_d	P_v	P_h	P_d	P_v	P_h	P_d	P_v
h m									
20 45	− 13.2 γ	E 24.5 γ	− 80.2 γ	+ 24.2 γ	E 9.2 γ	− 2.2 γ	+ 10.2 γ	E 38.5 γ	+ 4.0 γ
21 0	− 15.4 »	» 11.7 »	− 88.0 »	+ 7.5 »	» 3.7 »	− 3.7 »	+ 7.0 »	» 15.9 »	+ 4.0 »
23 0	− 28.9 »	0	− 38.0 »	− 4.0 »	W 5.5 »	0	+ 1.8 »	W 3.6 »	Possibly a slight negative tendency.
15	− 61.3 »	W 10.2 »	− 45.0 »	− 3.0 »	» 15.6 »	0	+ 9.3 »	» 14.6 »	
30	− 61.3 »	» 8.4 »	− 86.0 »	+ 1.5 »	» 15.6 »	− 2.2 »	+ 15.4 »	» 3.6 »	
52.5	− 25.3 »	E 4.2 »	− 6.2 »	0	0	− 3.7 »	+ 4.2 »	E 3.0 »	

TABLE XXI (continued).

Gr. M. T.	Kew		Potsdam		Val Joyeux		Munich	
	P_h	P_d	P_h	P_d	P_h	P_d	P_h	P_d
h m								
20 45	− 12.7 γ	E 22.0 γ	+ 8.5 γ	E 27.5 γ	− 9.6 γ	E 22.5 γ	+ 6.0 γ	E 19.0 γ
21 0	− 5.1 »	» 17.3 »	+ 7.2 »	» 15.2 »	− 3.2 »	» 17.6 »	+ 2.5 »	» 10.6 »
23 0	0	0	+ 1.9 »	W 1.0 »	0	» 4.2 »	+ 4.0 »	0
15	+ 10.7 »	W 4.6 »	+ 7.2 »	» 9.6 »	+ 9.6 »	W 3.3 »	+ 12.5 »	W 6.8 »
30	+ 12.2 »	» 1.4 »	+ 13.9 »	» 6.1 »	+ 12.2 »	0	+ 10.5 »	» 3.0 »
52.5	0	E 6.1 »	+ 5.0 »	E 3.0 »	0	E 2.5 »	+ 5.5 »	E 2.3 »

TABLE XXI (continued).

Gr. M. T.	Pola			San Fernando		Tiflis		
	P_h	P_d	P_v	P_h	P_d	P_h	P_d	P_v
h m								
20 45	+ 2.2 γ	E 17.3 γ	+ 3.1 γ	− 8.3 γ	E 11.4 γ	+ 7.5 γ	E 1.8 γ	+ 3.0 γ
21 0	+ 3.1 »	» 9.7 »	+ 1.0 »	− 5.1 »	» 11.4 »	+ 4.9 »	» 1.1 »	+ 1.2 »
23 0	+ 1.8 »	» 2.0 »	0	− 2.5 »	» 2.4 »	− 2.9 »	W 3.7 »	0
15	+ 8.5 »	W 6.2 »	− 1.0 »	+ 5.1 »	» 4.1 »	0	» 9.3 »	− 0.7 »
30	+ 14.0 »	» 3.5 »	− 0.8 »	+ 12.1 »	» 5.0 »	+ 3.3 »	» 10.4 »	− 1.2 »
52.5	+ 5.8 »	E 2.0 »	0	+ 3.2 »	» 2.4 »	+ 1.1 »	» 3.7 »	− 0.7 »

TABLE XXI (continued).

Gr. M. T.	Dehra Dun		Zi-ka-wei		Batavia		Christchurch		
	P_h	P_d	P_h	P_d	P_h	P_d	P_h	P_d	P_v
h m									
20 45	+ 5.9 γ	W 6.9 γ	+ 2.4 γ (1)	W 5.0 γ	+ 4.2 γ	0	− 4.6 γ	0	0
21 0	+ 3.5 »	» 3.9 »	+ 2.4 » (1)	» 3.0 »	+ 4.2 »	0	− 2.8 »	0	0
23 0	− 1.6 »	» 2.9 »	− 2.4 » (1)	» 3.0 »	0	W 12.0 »	− 5.5 »	0	+ 2.2 γ
15	0	» 4.9 »	0	» 4.0 »	0	» 15.6 »	− 5.5 »	0	+ 2.2 »
30	+ 1.6 »	» 4.9 »	0	» 5.0 »	− 1.8 »	» 15.6 »	− 7.8 »	0	+ 1.9 »
52.5	+ 2.7 »	0	+ 2.4 » (1)	» 2.0 »	− 3.5 »	» 18.0 »	− 2.8 »	0	+ 3.1 »

(1) Uncertain value.

Current-Arrows for the 26th December, 1902; Chart I at $20^h\ 45^m$, and Chart II at 21^h.

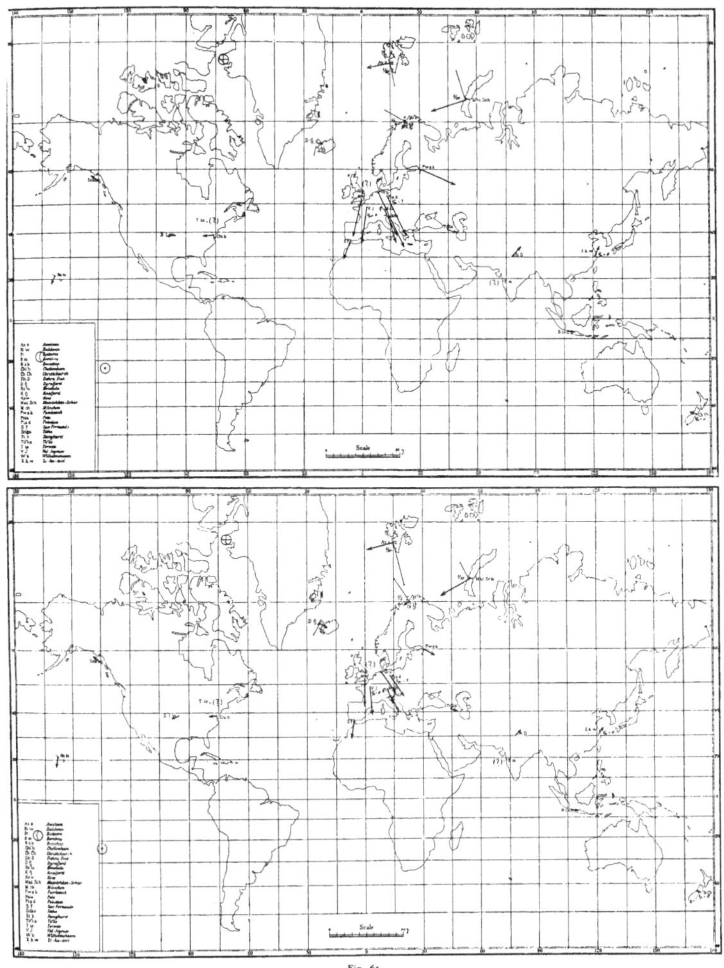

Fig. 61.

Current-Arrows for the 26th December, 1902; Chart III at 23^h, and Chart IV at 23^h 15^m.

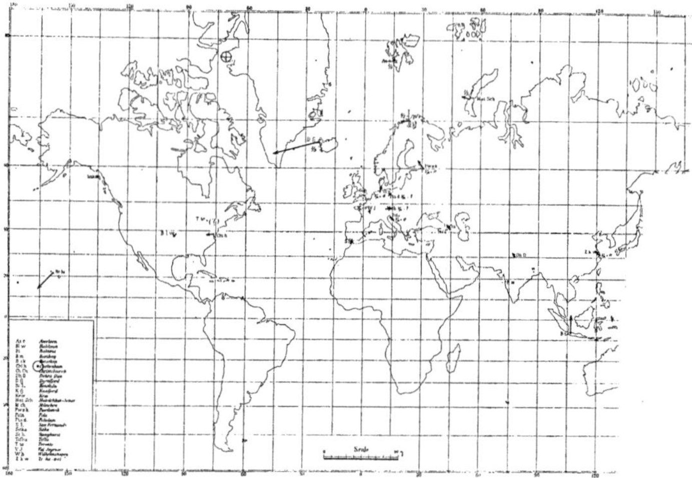

Current-Arrows for the 26th December, 1902; Chart V at $23^h\ 30^m$, and Chart VI at $23^h\ 52.5^m$.

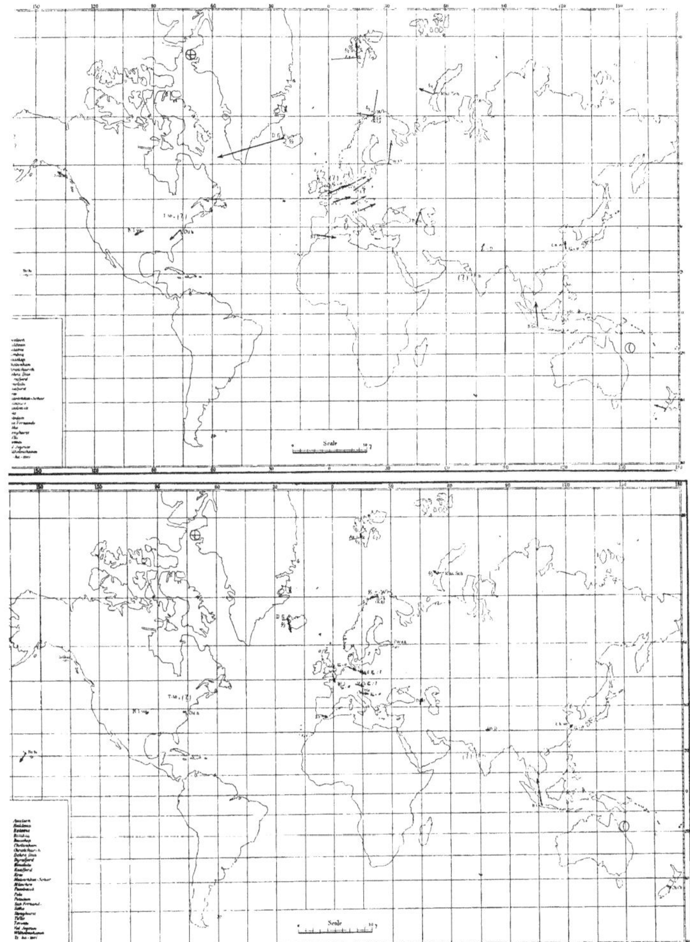

Fig. 63.

THE CYCLO-MEDIAN STORMS.

49. The idea that seems to have gained most adherents regarding the nature of the currents that should produce the great magnetic perturbations is that the magnetic storms should be conditioned, so to speak, by electric cyclones, wandering over the earth's surface.

This view is upheld very positively by Ad. Schmidt. We will here give a brief extract from his previously-mentioned well-known paper, "Ueber die Ursache der magnetischen Stürme" (1).

"The most characteristic thing of all, however, is the continual change that prevails in all these respects. Surprising similarity is followed in the course of a few minutes by a complete difference or a decided contrast; a great deflection in one curve answers to a scarcely perceptible jag or bend in the other, while soon after in the one calm ensues, and in the other the liveliest motion.

"These well-known properties of magnetic storms, as especially the longer and more intense disturbances have aptly been called, point unmistakably to prevailing local occurrences as the likeliest cause of these phenomena — occurrences of varying strength and extent, which, appearing now here, now there, perhaps also simultaneously at different places, probably exert a magnetic influence over the whole earth at the same moment, and attain an intense influence, but for the most part only over a more or less limited area".

This characterisation of the perturbation-conditions during great magnetic storms will do sufficiently well as far as the arctic regions are concerned. As regards lower latitudes, on the other hand, our impression of the conditions is very often as nearly as possible the contrary. There, at any rate during the *great* storms, the circumstance that attracts most attention is the similarity that the perturbation presents at the various places. As a rule, for instance, the curve for the entire district, Stonyhurst to Pola and Wilhelmshaven to San Fernando, exhibits in the main the same form. The conditions at Tiflis also, often constitute a transition form to those at Dehra Dun. The difference in the forms of curve often only depends upon a gradual turning of the field.

In conformity with this, our view of the great magnetic storms will be quite a different one, since we assume that the storm is often only of a local nature in the regions around the auroral zone, while the simultaneous perturbations in lower latitudes are probably, as we have seen in the treatment of the polar elementary storms, due to the effect of distant systems. It appears, however, that there is a class of perturbations that are due to current-systems which appear in lower latitudes at a height above the earth that is small in proportion to the earth's dimensions. These systems, however, seldom seem to appear with any great strength, at any rate not in 1902—03. Whether, by following up the perturbations in their smallest details, we should often find a component that must be due to current-systems of a local character, is a question that we cannot here go into; but it seems probable that when we come to the very small perturbations, we shall find much to be of a local character. This follows indeed from the fact that there are almost always more or less alternating earth-currents, and also, on account of the current-systems during the great storms, and simultaneously with them, currents must be induced in the earth, and this will give the perturbations in lower latitudes a local component.

In the whole of our material, we have not found more than one considerable perturbation that in its entirety must be due to systems that come near to the earth in lower latitudes. This was on the 6th October, 1902.

It appears, however, so clearly and distinctly on an otherwise calm day, that its properties can be all the more carefully studied; and it can also be traced over a considerable area. There is always a possibility that such systems may also to some extent co-operate with the polar storms.

(1) Meteorologische Zeitschrift, September, 1899.

THE PERTURBATION OF THE 6th OCTOBER, 1902.

(Pl. I).

50. This perturbation appears quite suddenly upon an otherwise very calm day. As far as one can decide from the magnetograms, it makes its appearance simultaneously in all parts of the area over which it is observable. Only at Axeløen, and to some extent at the other Norwegian stations, has the perturbation a somewhat peculiar character. At the other stations at which it is noticeable, its course is as follows.

It makes its appearance at $14^h\ 13.5^m$ simultaneously in both D and H. The deflection increases suddenly, and about 5 minutes later reaches its maximum, this also occurring simultaneously in the two curves. The deflection thereupon decreases in both, first rather suddenly, afterwards more slowly, until about $14^h\ 48^m$, when no deflection is observable.

It will be seen from the copies of the magnetograms, that the geographical distribution of the perturbation is within fairly sharply-defined limits. The effect is greatest in Europe, especially at the more westerly stations up to and including Wilhelmshaven and Pola; but even at Pawlowsk, where the perturbation is distinctly perceptible, it is only slight. If we compare simultaneous perturbing forces in Pawlowsk and Wilhelmshaven, we see that at the latter station they are about four times as great as at the former. At Tiflis the perturbation is only just perceptible.

At Dehra Dun, Zi-ka-wei, Batavia and Christchurch, the H-curve, as the perturbation makes its appearance, gives a little leap, which means that H receives a small, and as it appears, permanent increase. These stations are marked (o) on the chart, as no definite perturbing force can be taken out.

At the three American stations, Toronto, Baldwin and Cheltenham, the perturbation runs nearly the same course as in Europe, except that the deviation in declination is to the opposite side. From these stations the effect diminishes greatly westwards. At Sitka it is almost, and at Honolulu quite, imperceptible.

At our Norwegian stations it appears as follows. At Kaafjord it is distinctly noticed, but its course is somewhat different, especially as regards the latter half. At Matotchkin Schar a disturbance is noticeable, but no measurable deflection. On Axeløen there is simultaneously a perturbation of about the same duration and strength as on the continent of Europe; it takes place on the whole within the same period, but its course is different. On the other hand it is of about the same magnitude as the perturbation in the south-west of Europe, or perhaps a little smaller.

From Dyrafjord we unfortunately have no observations; but it seems likely, judging from the course of the current-lines as shown by the charts, that this station would have been the most important.

THE FIELD OF FORCE.

51. During the perturbation the form of the field is maintained unaltered, the strength alone varying. We have therefore found it sufficient to work out two charts, namely, for the hours $14^h\ 22.5^m$ and $14^h\ 30^m$.

We have made the calculation, however, for several hours, and these will be found in Table XXII. With a view to increased accuracy, we have had all the curves enlarged photographically to five times their original size.

Fig. 64 shows these enlarged copies of curves from Wilhelmshaven.

In the area from which we have observations, the greatest effect is in the south-west of Europe, and the east of North America. It occurs, as we see, upon the day side. The current-arrow indicates very distinctly a negative vortex, which should go round the North Atlantic Ocean; in reality we have an area of convergence for the perturbing force. Whether the vortex is closed, whether—in

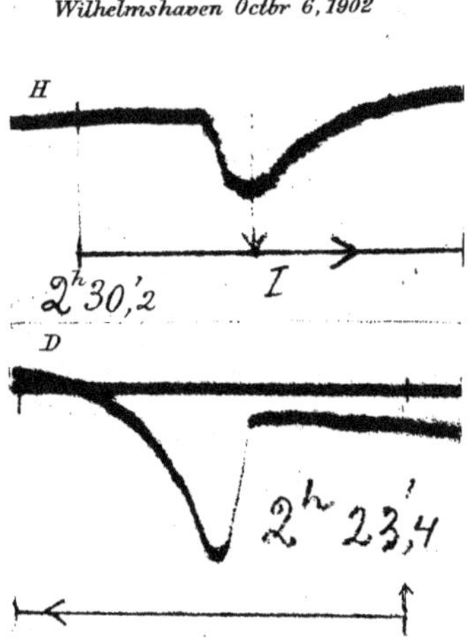

Fig. 64.

other words—the forces converge from *all* directions towards some region or other in the Atlantic, we are unable to say with certainty, as we lack material from the more southerly regions. A knowledge of the conditions in West Africa and the east of South America would be of special importance.

We note the fact that the effect of the force seems to keep rather constant in the same direction as the West European and American current-arrows, while the strength of the field decreases very rapidly perpendicular to the direction of the arrows outwards from the vortex-centre. Thus in Europe the effect decreases very rapidly eastwards, the force being very small both at Pawlowsk and Tiflis. At Kaafjord and Sitka also, the force is small.

Among the other Norwegian stations, only Axeløen can show a perturbing force that is at all great; but there its direction is almost due north, and it thus does not appear to join the field of force in more southerly latitudes.

TABLE XXII.

The Perturbing Forces on the 6th October, 1902.

Gr. M. T.	Baldwin		Cheltenham		Toronto		Axeløen		
	P_h	P_d	P_h	P_d	P_h	P_d	P_h	P_d	P_v
h m									
14 15	− 1.6 γ	E 3.9 γ	0	W 0.6 γ	− 2.2 γ	E 0.6 γ	+ 8.1 γ	W 0.3 γ	+ 2.4 γ
18.8	− 3.5 »	» 10.2 »	− 6.0 γ	E 6.5 »	− 6.4 »	» 12.3 »	+ 16.5 »	E 0.6 »	+ 19.7 »
22.5	− 4.4 »	» 8.7 »	− 11.1 »	» 12.7 »	− 7.1 »	» 14.4 »	+ 22.5 »	» 2.0 »	+ 17.2 »
26.3	− 3.4 »	» 6.5 »	− 9.7 »	» 8.6 »	− 2.6 »	» 8.5 »	+ 24.5 »	» 0.1 »	0
30	− 2.0 »	» 5.0 »	− 7.5 »	» 5.6 »	− 4.7 »	» 4.8 »	+ 23.5 »	W 1.1 »	− 14.8 »
33.8	− 1.1 »	» 3.0 »	− 5.9 »	» 5.2 »	− 3.2 »	» 2.6 »	+ 16.4 »	» 4.6 »	− 19.7 »
37.5	− 0.6 »	» 2.5 »	− 4.1 »	» 2.4 »	− 1.9 »	» 1.7 »	+ 11.2 »	» 6.4 »	− 14.8 »
41.3	− 0.1 »	» 1.3 »	− 2.7 »	» 0.8 »	− 0.7 »	0	+ 7.1 »	» 4.9 »	− 12.3 »
45	0	» 0.4 »	− 1.3 »	» 0.2 »	0	0	+ 4.5 »	» 2.6 »	− 9.8 »

TABLE XXII (continued).

Gr. M. T.	Kaafjord			Pawlowsk			Stonyhurst		Wilhelmshaven		
	P_h	P_d	P_v				P_h	P_d	P_h	P_d	P_v
h m											
14 15	− 2.9 γ	W 3.3 γ	0	− 0.5 γ	W 5.5 γ	A slight perturbation; P_v max. = + 3.7 γ at about $14^h\ 24^m$.	− 5.6 γ	W 9.4 γ	0	W 1.9 γ	0
18.8	− 3.1 »	» 5.5 »	+ 1.1 γ	− 5.5 »	» 5.5 »		− 13.0 »	» 23.5 »	− 6.2 γ	» 31.3 »	0
22.5	− 1.9 »	» 0.5 »	+ 5.0 »	− 5.0 »	» 2.8 »		− 11.8 »	» 17.2 »	− 13.1 »	» 28.3 »	− 3.0 γ
26.3			+ 7.2 »	− 2.0 »	» 0.9 »		− 9.8 »	» 11.8 »	− 12.7 »	» 16.8 »	− 5.5 »
30			+ 6.2 »	− 1.0 »	» 0.5 »		− 8.2 »	» 7.8 »	− 8.2 »	» 9.6 »	− 4.0 »
33.8	1	» 30 »	+ 6.2 »	− 1.0 »	0		− 6.7 »	» 4.7 »	− 5.2 »	» 4.9 »	0
37.5	1	» 28 »	+ 6.2 »	− 0.5 »	0		− 5.4 »	» 2.7 »	− 2.9 »	» 2.2 »	0
41.3	0	» 27 »	+ 5.5 »	0	0		− 4.5 »	» 1.5 »	− 1.1 »	» 0.4 »	0
45	0	» 15 »	+ 1.0 »	0	0		− 3.9 »	» 0.8 »	− 0.6 »	0	0

TABLE XXII (continued).

Gr. M. T.	Kew		Potsdam			Val Joyeux		
	P_h	P_d	P_h	P_d	P_v	P_h	P_d	P_v
h m								
14 15	0	W 16.1 γ	− 0.6 γ	W 10.4 γ	− 0.6 γ	− 2.1 γ	W 8.2 γ	Perhaps a slight neg. deflection. The curve somewhat indistinct.
18.8	− 4.4 γ	» 24.3 »	− 4.4 »	» 25.5 »	0	− 3.2 »	» 16.2 »	
22.5	− 7.1 »	» 17.3 »	− 5.8 »	» 16.2 »	+ 0.6 »	− 3.5 »	» 14.4 »	
26.3	− 7.5 »	» 11.4 »	− 5.1 »	» 9.6 »	+ 0.6 »	− 3.5 »	» 7.9 »	
30	− 6.2 »	» 8.2 »	− 3.6 »	» 4.9 »	0	− 3.3 »	» 5.3 »	
33.8	− 4.8 »	» 5.4 »	− 2.3 »	» 2.6 »	0	− 3.1 »	» 3.4 »	
37.5	− 4.4 »	» 3.8 »	− 1.7 »	» 1.0 »	0	− 2.5 »	» 2.1 »	
41.3	− 4.1 »	» 2.8 »	− 1.1 »	0	0	− 2.3 »	» 1.4 »	
45	− 3.5 »	» 2.1 »	− 0.7 »	0	0	− 1.8 »	» 0.8 »	

TABLE XXII (continued).

Gr. M. T.	Munich		Pola			San Fernando	
	P_h	P_d	P_h	P_d	P_v	P_h	P_d
h m							
14 15	+ 2.0 γ	W 15.0 γ	+ 0.4 γ	0	From $14^h\ 16^m$ to $14^h\ 28^m$ a slight perturbation in V. At $14^h\ 20^m$ P_v max. = − 2.1.	0	0
18.8	+ 2.5 »	» 30.0 »	+ 1.8 »	W 13.1 γ		+ 10.4 γ	W 27.2 γ
22.5	0 (1)	» 19.5 »	+ 3.9 »	» 22.7 »		+ 10.3 »	» 25.9 »
26.3	0 (1)	» 11.3 »	+ 3.0 »	» 15.1 »		+ 4.5 »	» 14.9 »
30	+ 1.10 » (1)	» 7.5 »	+ 0.4 »	» 9.3 »		+ 1.2 »	» 9.9 »
33.8	+ 3.0 » (1)	» 3.8 »	+ 0.2 »	» 5.2 »		+ 0.5 »	» 6.2 »
37.5	0	» 2.2 »	0	» 2.6 »		0	» 3.2 »
41.3	0	0	0	» 1.0 »			» 1.5 »
45	0	0	0	0		0	» 0.7 »

(1) The curious form of the H-curve is due to work that was going on in the observatory at the time. The corresponding values of P_h are therefore rather uncertain.

Current-Arrows for the 6th October, 1902; Chart I at $14^h\ 22.5^m$, and Chart II at $14^h\ 30^m$.

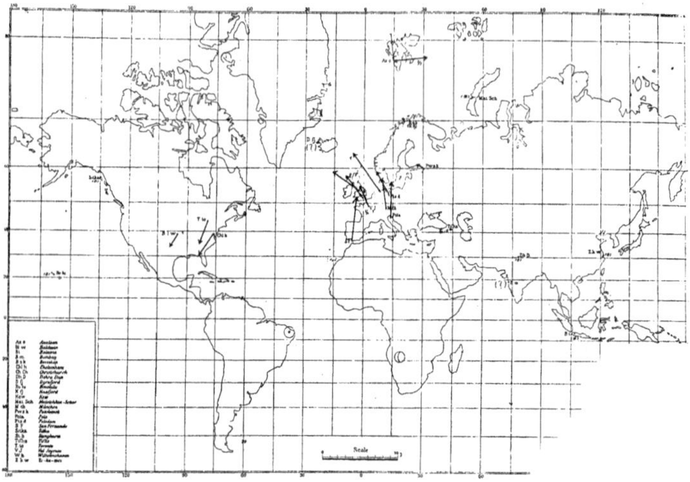

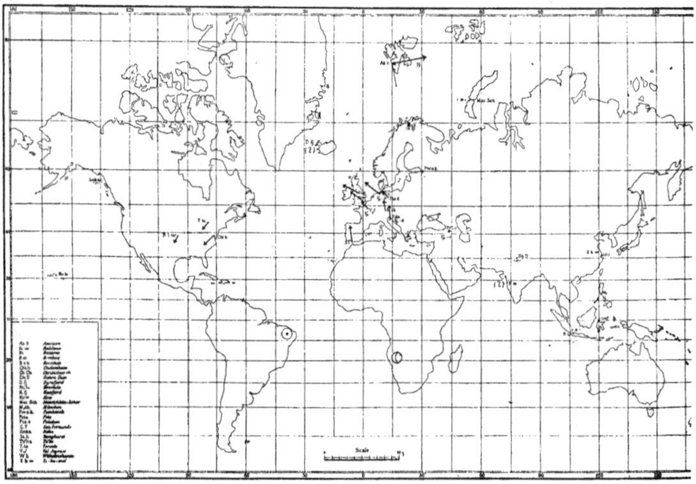

Fig. 65.

CONCERNING THE CAUSE OF THE PERTURBATION.

52. Notwithstanding its simplicity, this perturbation possesses rather peculiar properties, which make it difficult to refer it to any of the other types. In the first place, the perturbation at Axeløen, owing to the difference in its course, and to the direction of the force, must be ascribed to the effect of a relatively independent system. In more southerly latitudes, the field forms, as we have seen, an area of convergence. This immediately brings to mind the polar elementary storms. There are, however, strong reasons against such a view.

On account of the form of the field, we should expect to have the storm-centre somewhere about the south of Greenland, and the current-arrow might here be expected to be directed westwards along the auroral zone. In the ordinary polar elementary storms, we shall then find the strongest force-effect in this current-arrow's line of direction, around the main axis of the system, while the effect should become less inwards towards the area of convergence. This time we come upon a peculiarity, namely that the effect at Kaafjord and Pawlowsk is very small in proportion to that, for instance, at Val Joyeux and San Fernando, which should lie almost at the same distance from the storm-centre, but much nearer the area of convergence. A knowledge of the conditions at Dyrafjord would have enabled us to settle the question; for if the perturbation should be referred to the same type as the polar elementary storms, we should have found the effect very strong at Dyrafjord.

It might be thought, as an explanation of the smallness of the force at Kaafjord and Pawlowsk, that the system that brought about the perturbation on Axeløen, counteracted the southern system. This has, indeed, to some extent been the case, especially at Kaafjord. It does not, however, explain it entirely; for then the counter effect of the northern system would be as great at Pawlowsk as at Axeløen. But everything seems to indicate that the perturbation at Axeløen is of a very local character. The vertical component, for instance, changes its direction. And at Matotchkin Schar, nothing at all is noticeable.

It does not thus seem possible to refer this perturbation to the polar elementary storms. In favour of this conclusion, there is also the fact that if it were so referable, it would have its storm-centre in the sun's meridian, while the storms that have the current-arrow directed westwards along the auroral zone, generally appear about midnight. But this is not all. From the calm conditions at the stations round the auroral zone, it does not even seem to be of a polar nature. The cause of the perturbation in lower latitudes must also be sought in occurrences in those lower latitudes.

The cause of the magnetic storms must however be sought in electric currents, of whose form and kind we shall endeavour to obtain a clear idea by the aid of the experiments with the terrella.

The system with two vertical current-portions connected by a horizontal part, cannot satisfy the field of force. It is then most natural to seek an explanation of the phenomenon in currents moving for long distances along the surface of the earth, either on it, or at some height above it. It here seems natural to suppose, after glancing at the chart, that we have had a current that, at any rate in the North Atlantic region, has assumed the character of a real current-vortex.

The perturbing force in the south-west of Europe, as we see, converges greatly. If we were to produce all the forces until they intersected one another, the district of the greatest density — the point of intersection — would lie only a little to the north-west of Spain. The force in North America, on the other hand, has not such a great convergence. If we imagined ourselves moving over the earth's surface in such a manner that we always advanced in the direction of the current-arrow, we should describe some sort of curve, which we might call a current-line. What these current-lines are like in our case, our material does not allow us to judge with certainty. There can be no doubt that those from North America turn east, and unite with the conditions in the south-west of Europe, always, as they do so, curving to the right, and always, the nearer they approach towards Europe, with a greater curvature. Two things might now be possible; either the curvature might continue to increase, when we should obtain a spiral, or it might decrease, and the lines pass westwards through the South Atlantic, and thus form elliptical paths. We may conclude from the rapid decrease of the perturbation out towards the sides, both eastwards in Europe and westwards in America, that the current-system must appear both in the neighbourhood of the American stations and in that of the stations in the west of Europe; or to speak more precisely, the bulk of the system ought to lie at a distance from the West European stations that is small in proportion to the distance between Pola and Tiflis, or between Wilhelmshaven and Pawlowsk, as the perturbation at Pawlowsk is only a fourth part of that at Wilhelmshaven, and the perturbation at Tiflis is almost imperceptible.

It will be seen that the effect over the district Wilhelmshaven, San Fernando, Stonyhurst, Pola, is of about the same magnitude. As this constitutes an area that has a section almost equal to the distance between Pola and Tiflis, we should be able to conclude that the current-system itself has its greatest density in this district.

In order to draw conclusions from the vertical intensity at Pawlowsk, which is directed downwards, they must be electric currents *above* the earth's surface, with which we have to do.

These currents would then have to be sought at a height that was small in proportion to the earth's dimensions, small indeed in proportion to the distance between Pola and Tiflis.

We can draw similar conclusions for the stations in the western hemisphere.

On account of the convergence of the forces, it might perhaps be natural to seek an explanation of the system in the effect of a south pole situated in their point of convergence. But the effect from this point would not be able to account for the properties of the field. While this pole should be acting strongly, both in America and in Europe, we see that the force from Pola to Tiflis passes from a value that lies near the maximum of the values observed, to an almost imperceptible amount. The bulk of the current itself must thus pass over the place in about the direction given by the current-arrows.

If we assume the current to be of a cosmic nature, and consisting of electrically charged particles in motion, we see that it is deflected in just such a manner as would result from the movement of the current in the magnetic field, as in the northern hemisphere we must get vortices with a movement contrary to that of the hands of a clock.

The simple course of this perturbation enables it to be very carefully studied. The form of field also exhibits conditions of a simple nature. The perturbation cannot be referred either to the equatorial or to the polar storms, but is of a special type. Its chief characteristics are that it is as great in medium as in high latitudes, and that the current-lines are vortical in form. For this reason, we have called these perturbations *cyclo-median*.

The perturbation of the type now under discussion, does not, however, appear as a free current-vortex.

However the system may be constituted, it is almost stationary all through the time of its appearance, the relative strength of the perturbation remaining constant all the time.

With the material at our disposal, it is impossible to draw any certain conclusions as to the composition of the current.

From the stability and immobility of the system, it must necessarily follow that it is ruled by higher laws.

It is difficult to suppose that such a system might arise and be maintained only by means of processes on the earth, as in that case other more variable and compound forms would be brought into action. It is probable, on the contrary, that the current-systems in question are produced by the emission from the sun of very stiff rays of electric corpuscles; for then all the corpuscles that reach the earth will have travelled nearly the same way, and in a short space of time the relative positions of the sun and the earth, which should be decisive for the form of the system, would undergo only slight alteration.

With reference to this cyclo-median perturbation, I have made a number of experiments with my magnetic terrella, and will here give some of the results of these.

With a suitable proportion between the stiffness of the cathode rays and the intensity of the magnetisation, the rays strike the terrella in lower latitudes, and form a well-defined luminous area.

Fig. 66 shows an area such as this. In making the experiments, an influence-machine was used as the source of electricity, and a discharge-tube similar to that shown in fig. 37. The four positions of the terrella, shown in the four photographs in fig. 66, were such that in No. 1, the magnetic south pole (answering to the terrestrial-magnetic north pole) was in such a position that, considering the cathode as representing the sun, there was noon there. In the positions 2, 3, and 4, the terrella is so turned that at the same south pole it is respectively 6 p. m., midnight, and 6 a. m.

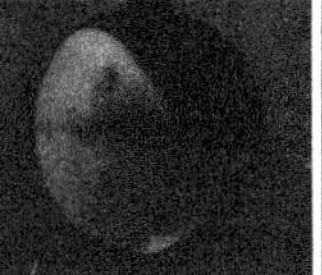
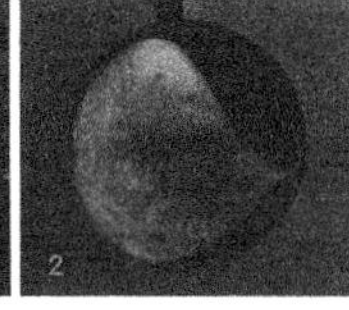

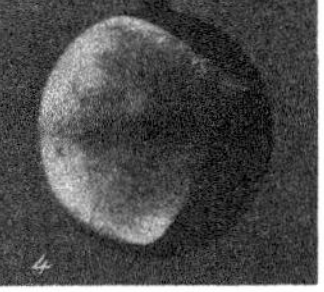

Fig. 66.

The tension employed between the anode and the cathode was about 10,000 volts. The terrella was magnetised with a current of 3.2 amperes, and the gas-pressure in the tube was 0.0011 mm.

The photographs were taken from the same position in all four cases, i. e. so that the line from the centre of the terrella to the camera made an angle of 45° with the line from the centre to the central point of the cathode. The characteristic changes undergone by the luminous area during the turning of the terrella, are distinctly seen. It is especially noticeable that the strength of the light is greatest in the polar regions, and that the luminous point towards the east near the equator moves from southern to northern latitudes during the turning of the terrella.

By studying this phenomenon more closely, I have found out that under certain circumstances, several such characteristic luminous areas may be obtained on the terrella.

By employing an inductorium as the source of electricity, and a very strong current for the magnetisation of the terrella, I have found three distinct, and possibly more, such areas, arranged one after the other round the terrella from west to east. In order to make sure that these different luminous areas were not due to the almost simultaneous appearance of cathode rays of various degrees of stiffness, during each discharge from the inductorium (1), I have repeated all the experiments, employing as the source of the current a high-tension direct-current machine, system *Thury*, Geneva. This machine, when in regular work, can supply $^1/_8$ ampere at 15,000 volts, but with lower current strength can go up to 20,000 volts.

It now turned out that I obtained exactly the same kind of light-figures on the terrella as I did when employing the inductorium as the source of the current.

(1) See Birkeland, "Sur un spectre des rayons catodiques". Comptes Rendus. 28 Sept., 1896.

a. Discharge-tube with terrella. *b.b.* 20,000-volt generator with motor. *c.* Static Kelvin volt-meter. *d, d, d.* Photographic apparatuses. *e, e.* Oil-pump with motor. *f.* Mercurial pump. *g.* Gaede pump with motor.

Fig. 67.

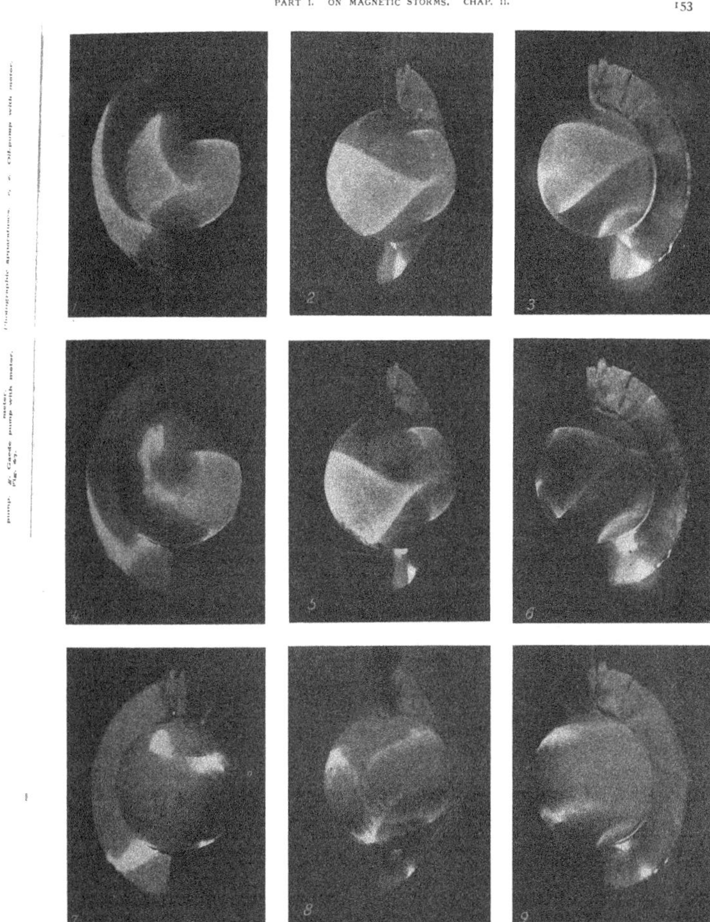

Fig. 68.

These figures will throw considerable light upon the questions we are endeavouring to solve.

The arrangements for the experiments are made plain by fig. 67, in which *a* is the discharge-tube with terrella, *b-b* a 20,000-volt generator with motor, *c* a static Kelvin volt-meter up to 20,000 volts, *d*, *d*, *d* are photographic apparatuses, *e-e* is an oil-pump with motor, from Siemens-Schuckert, *f* a mercurial pump worked by hydraulic pressure, for measuring the gas-pressure in the discharge-tube, and *g* a Gaede pump with motor from Leybold's Nachfolger, the best mercurial pump that I know of for obtaining a high degree of exhaustion in large tubes.

The nine photographs in fig. 68 are taken with the terrella always in the same position, but under three different electric and magnetic experimental conditions. The photographs are taken, as fig. 69 shows, simultaneously from three sides of the terrella. The photographs 1, 2, 3, fig. 68, belong to one experiment, 4, 5, 6 to another, and 7, 8, 9, to a third. In all the experiments, it is noon at the magnetic south pole, the cathode representing the sun.

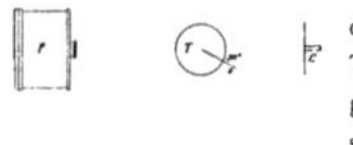

Fig. 69.

The intention of the three experiments is to show how the descent of rays upon the terrella alters when the stiffness is continually decreasing. The first experiment shows the result when the stiffness of the rays is very great in proportion to the magnetisation employed upon the terrella. The stiffness of the rays is altered most simply by altering the pressure of the gas in the discharge-tube. With an exceedingly low pressure, however, the disadvantage is that so much gas is evolved from the cathode during the experiment, that it is not easy to photograph the phenomena, as they change.

In the first experiment (1, 2, 3) therefore, I have been obliged, for the sake of the photographing, to keep a comparatively high pressure in the discharge-tube, but on the other hand I have employed a lower magnetising current upon the terrella than in the next two experiments (4, 5, 6 and 7, 8, 9). It has, however, been proved with certainty that the light-figures will be the same if, in the first experiment, the same high degree of magnetisation be employed as in the second and third experiments, when the discharge-tube is exhausted sufficiently.

In the first experiment, the magnetising current was 15 amperes, answering to a magnetic moment M, of the terrella, of 6200 C.G.S. The pressure in the discharge-tube was 0.018 mm., the discharge current was 8.9 milliamperes, and the difference of potential between the electrodes was 4200 volts.

In the second experiment the magnetising current was 33 amperes, answering to about $M = 10{,}000$. The pressure was about 0.006 mm., the current 9.5 milliamperes, and the tension 5 500 volts.

In the third experiment $M = 10{,}000$, as in the second. The pressure was 0.03 mm., the strength of the current 8 milliamperes, and the tension 3300 volts.

As most of the experiments described in this volume were made with the same terrella,—marked No. 5—there may be some interest in seeing the curve for its magnetic moment at about 20° C. for various intensities of the magnetising current. Fig. 70 shows this moment-curve.

The values for high current-intensities are not very exact, owing to the great changes of temperature during the measurements.

There are various circumstances that appear in the experiments represented in fig. 68, to which we will pay special attention.

It should first be remarked that if the rays become still more pliant than in experiment 3, the conditions in the fundamental experiment represented in fig. 47 can be exactly obtained. In that experiment, three regions for the descent of the cathode rays were distinctly seen in a zone round each of the magnetic poles.

It is easy to follow the development from experiment 1 up to the last-mentioned, represented in fig. 47. In experiment 1 we see distinctly three characteristic light-areas round the terrella. In the succeeding experiments these light-areas undergo several important changes. First the strength of the light diminishes in the middle of the areas, so that the edges come out more distinctly. Then the edges also partly disappear, except in the polar regions, where the light increases in intensity.

The first figure in the three rows (1, 4 & 7) shows the light reduced to two patches, the lower of which, however, has coincided with a descent of rays upon the screen, indicating rays that have been deflected and have turned back before they reached the terrella (see fig. 39, third example).

The second figure in the three rows illustrates clearly the development mentioned above.

The third figure, as photograph 9 shows, changes into polar bands that have possibly been produced by the covering over of more light-areas than the three mentioned. These zones of light are best seen in fig. 47. Other light-phenomena are also seen in photographs 3, 6 & 9, fig. 68, about the magnetic north pole and on the screen.

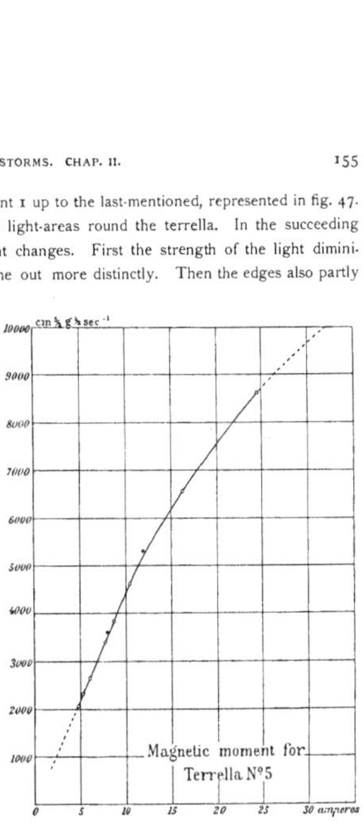

Fig. 70.

These consecutive light-areas round the terrella have some resemblance to other light-phenomena observed by me during the study of the trajectories of cathode rays under the influence of *one* magnetic pole (¹). With one magnetic pole, the consecutive figures became constantly smaller and smaller, while here they are all nearly of the same size.

From these experiments we shall draw comparisons both now, while discussing the cyclo-median perturbations, and subsequently in the treatment of the observations from 1882—83, Vol. I, Part II, where the question of districts of precipitation in the polar regions for magnetic storm-centres is discussed, and lastly in the treatment of the observations of aurora and of cirrus clouds (Vol. II).

The experiments described in connection with figs. 47 and 68, are of fundamental importance to our theory of magnetic disturbances. Concluding by analogy from these, we should never expect to have purely elementary magnetic perturbations upon the earth, as, among other things, the experiments show that there are several districts of precipitation at the same time upon the earth for the electric rays from the sun. In the preceding pages also, it has frequently been indicated that the magnetic currents are never purely elementary, like, for instance, the idealised polar form represented in fig. 40.

As regards polar storms, we have only been able to study those with the district of precipitation in the neighbourhood of the four Norwegian stations.

In order to obtain a clear understanding of the circumstances, we ought to have simultaneous observations from stations right round the auroral zone, and if possible also from the antarctic regions. A year's simultaneous observations from all the acting magnetic observatories in the world, and from, for instance, 10 stations in a zone round the terrestrial-magnetic north pole, and from as many as far south

(¹) Archives des Sciences Physiques et Naturelles. Quatrième période, t. VI. Geneva, Sept., 1898.

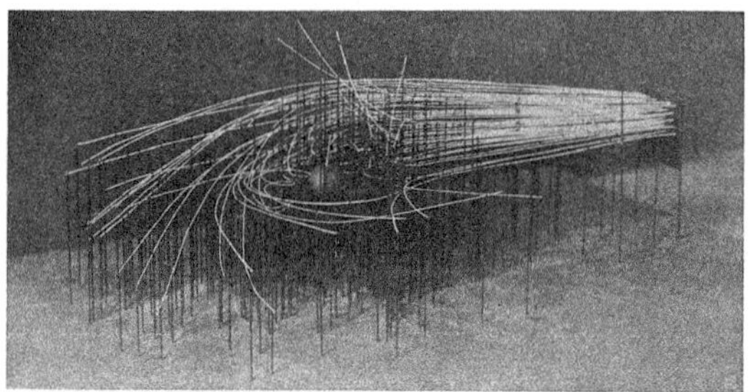

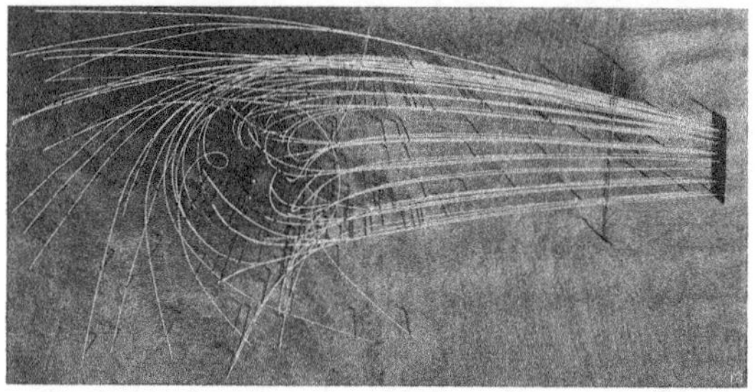

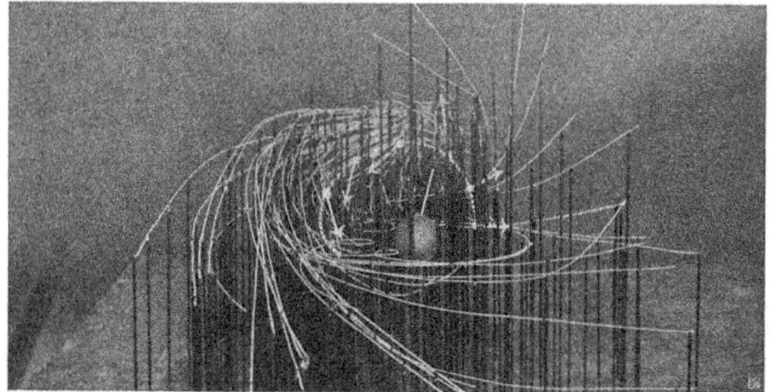

Fig. 71.

in the southern hemisphere as could practically be reached without too great expense, by accompanying hunting-expeditions, would without doubt raise the veil that obscures the great question of the origin of terrestrial magnetism, which has hitherto been one of the greatest mysteries of Nature.

In order to illustrate clearly the course of the rays in the case illustrated in fig. 66, Størmer has calculated the trajectories of cathode corpuscles answering to those in this experiment, and has shown the result in a wire model, which is photographed in three positions in fig. 71.

Størmer has added some remarks upon this model, which he kindly allows me to quote.

"This wire model (fig. 71) represents a number of trajectories of negatively-charged corpuscles, moving under the influence of an elementary magnet.

"The trajectories are constructed on a graphic method of integration, worked out for the occasion, which will be more fully described in the second part of this work [1].

"The model was specially made for Birkeland's experiments, and the sphere therefore represents the terrella, and the plate on the right the cathode. The elementary magnet is placed in the centre of the sphere, with its axis parallel to the black rods, and the south pole upwards, the latter being marked with a cross. The sphere is fitted with a rod representing the earth's axis of rotation.

"The lowest layer of rays consists of plane curves lying in the magnetic equatorial plane; they are calculated exactly, and are a good check upon the others, which are constructed graphically. Above this lowest layer of trajectories lie four other layers, so that the model shows more than 50 different paths. To each path in the model, there is also a corresponding one that is symmetrical with the first with reference to the magnetic equatorial plane; but all the trajectories thus produced are omitted so as not to make the model too intricate.

"The ring is clearly seen that answers to the luminous ring round the terrella in Birkeland's experiment. If we call the moment of the elementary magnet M, and express the characteristic constant [2] of the corpuscles by $H_0\varrho_0$, then the radius of the ring equals $\sqrt{\frac{M}{H_0\varrho_0}}$ cm.

"On the third photograph are marked the points in which the trajectories intersect a sphere concentric with that in the model, and with a radius rather less than that of the ring. At the points of intersection, the tangents to the trajectories have also been drawn. It will be seen how the directions of the tangents form a vortex; and symmetrical with this, there is a vortex on the other hemisphere, below the magnetic equatorial plane. If arrows are marked all over the sphere in directions the reverse of those of the above-mentioned tangents, we obtain the accompanying figure 72 in which the sphere is seen from without. The figure is only diagrammatic. We see that the part upon which the corpuscles impinge has the same form as that visible in the experiment; and above this there are two contrary cyclonic current-vortices in the direction of the arrows, situated symmetrically with reference to the magnetic equatorial plane, and answering to the positive currents that might produce cyclo-median perturbations.

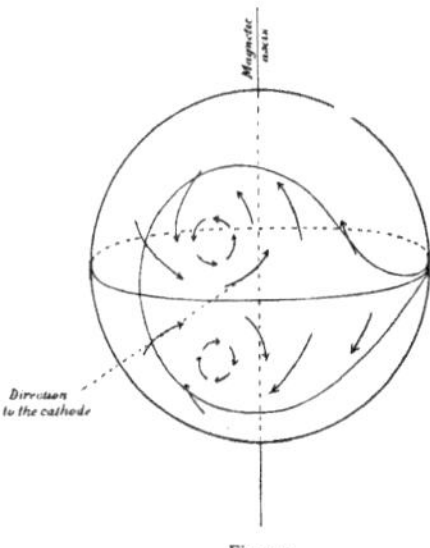

Fig. 72.

"The trajectories that have been chosen in the wire model are especially those that approach the elementary magnet, and then once more recede to an infinite distance, and not such as

(1) Cf. "On the Graphic Solution of Dynamical Problems", by Carl Størmer. Videnskabsselskabets Skrifter; Christiania, 1908.
(2) Cf. Carl Størmer's "Sur les Trajectoires des Corpuscules Électrisés dans l'Espace, etc." Archives de Genève, July—October, 1907.

come very near, or go right up to the elementary magnet. These paths will receive a special demonstration in other models, which will be described in the detailed treatment of the experiments with the terrella (see also fig. 73).

"From the form of the cyclo-median perturbations, and comparison with experiment and theory, we find that the radius of the ring is here about 1.5 that of the earth. Now since the magnetic moment of the earth, M, is 8.52×10^{25}, this gives, for the corpuscles that cause the cyclo-median perturbations, $H_0 \varrho_0 = \frac{8.52 \times 10^{25}}{[1.5 \times 6.37 \times 10^8]^2} = 93$ millions, approximately.

"In other words, the rays in these perturbations must be excessively stiff."

It thus appears from Størmer's calculations that two cyclonic vortices, symmetrical with reference to the equator, are produced, such that if we reckon with positive current-directions, the vortex north of the equator is counter-clockwise, that south of the equator, clockwise. This is in accordance with our observations in as far as the cyclo-median perturbation formed a counter-clockwise vortex. Judging from the light effects produced by the experiments with the terrella, a cyclo-median perturbation should also have a somewhat stronger effect in the polar regions. This assumption, unfortunately, cannot be verified, as we have no observations from Dyrafjord for the 6th October.

It should be remarked that the length of the arrows in fig. 72 has nothing to do with the intensity of the current or of the magnetic effect.

I have not yet proved the existence of electric current-vortices such as these experimentally, but shall try to do so later on. This is a case in which mathematical analysis has shown a superiority to experimental investigations. It is generally, as we know, only after the experimental discoveries have been made that analysis steps in to explain and enlarge the comprehension of the results obtained; and this has hitherto also been the case here.

The discovery of the various districts of precipitation in the polar regions is experimental, and from the results of the observations from the expeditions in 1882—83, we have found such simultaneous districts of precipitation for the magnetic storms. This subject will be discussed in Part II of this volume. Later on, in Vol. II, a corresponding investigation of the distribution of simultaneous aurora will be made, in which both our own collected material will be employed, and also that from the expeditions of 1882—83.

FURTHER COMPARISON WITH STØRMER'S MATHEMATICAL THEORY.

53. It seems as if Størmer's investigations would be of great importance in the problem of finding theoretically also, the various districts of precipitation in the polar regions. This is apparent from the following remarks, which Størmer allows me to quote:

"All these remarkable light-phenomena, shown in figs. 47 and 68, can doubtless be explained theoretically by my mathematical investigations of the paths of electrically charged corpuscles in the field of an elementary magnet. We shall return to this subject in a subsequent section of this work. At present I will only point out that the patches of light about the poles, obtained by sufficiently strong magnetism, are probably due to cathode corpuscles flung out into paths in the immediate proximity of those which, theoretically, would strike the elementary magnet in the centre of the terrella, and whose field, at great distances, represents the magnetic field of the terrella.

"As I have previously calculated a series of the simplest of such paths, all that is now necessary for the re-finding of the districts of precipitation visible on the terrella is to employ these calculations. Fig. 73 shows a wire model constructed for the case occurring in the experiments shown in fig. 47.

Several bundles of rays are here seen issuing from two points, one of which is in the magnetic equatorial plane, and the other a little above it, the rays being directed towards the terrella.

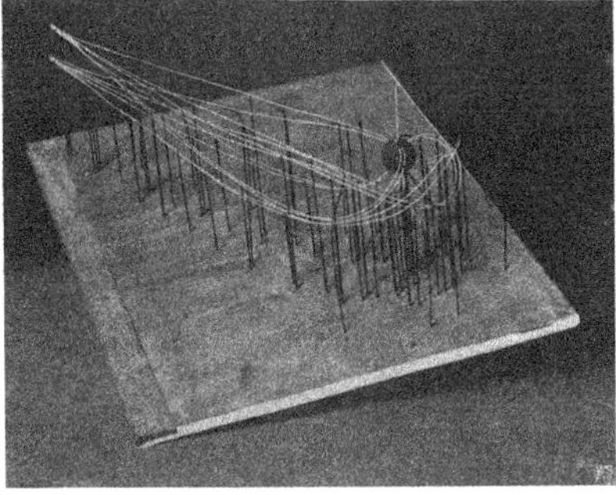

Fig. 73.

"Fig. 74 shows a comparison between the observed and the theoretical districts of precipitation[1]. It will be seen that the similarity is striking.

"In this connection I will mention that the same calculations may be employed as regards the earth, for the purpose of finding the districts of the precipitation of electric (negative) corpuscles coming from the sun. All the data necessary for such a calculation will be found in my Geneva paper (l. c., chap. IV).

"Let O (fig. 75) be the centre of the earth, P the north pole, OM the earth's magnetic axis, OAB the magnetic equatorial plane, and OS the direction to the centre from which the corpuscles emanate (the sun). OS is calculated from the time of the phenomenon, by well-known formulæ from spherical astronomy. The angle ψ is thereby found, i. e. the altitude of the sun above the magnetic equatorial plane, or in other words, the sun's altitude above the horizon at the point M.

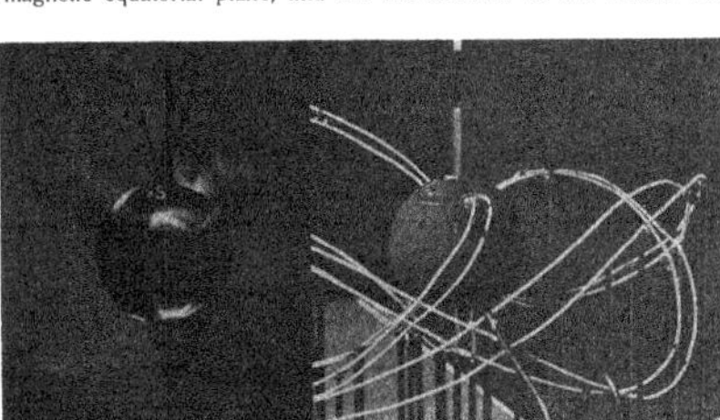

Fig. 74.

"The angle of deflection Φ (calculated positive[2] westwards) answering to ψ, is now obtained, as regards the simplest trajectories, with sufficient accuracy by the tables given in §§ 14 & 15 of my paper. They give the following curves, in which Φ is the abscissa and ψ the ordinate (fig. 76).

"The continuous line is the curve for the northern hemisphere, the broken line, symmetrical with the first, that for the southern.

"For each value of ψ, we generally find that there are several values of Φ answering to various trajectories from the same point of emanation; and this gives correspondingly various districts of precipitation[3].

(1) See "Sur les Trajectoires des Corpuscles Électrisés", etc., by Carl Størmer, § 16, Archives de Genève, July—October, 1907; and a lecture on the same subject given at the International Mathematical Congress at Rome, April, 1908

(2) For positive rays, Φ must be calculated positive eastwards. — (3) l. c. §§ 14, 15 & 18.

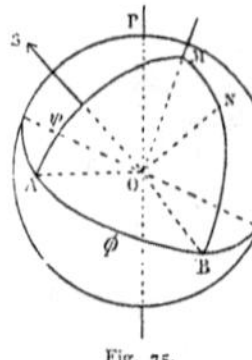

Fig. 75.

"As regards the angle MON, much will depend upon the stiffness of the rays (see my paper, § 17); with constant stiffness, however, point N will approach M when ψ increases. Before further data can be obtained for the stiffness of the rays that cause aurora and magnetic perturbations, we may assume, in accordance with the observations, that N is situated in the auroral zones.

"The appearance and disappearance of the various districts of precipitation, and their movements along the north and south auroral zones, according as the altitude ψ of the sun above the magnetic equatorial plane changes with time, can then be calculated by the above. We shall return to this subject in a later section of this work."

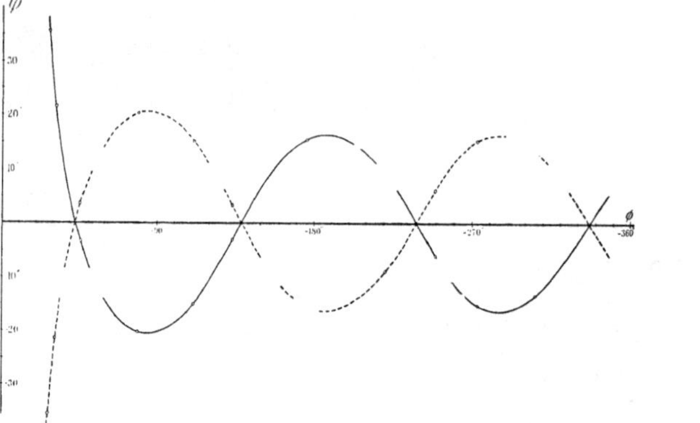

Fig. 76.

CHAPTER III.

COMPOUND PERTURBATIONS.

THE PERTURBATIONS OF THE 29th & 30th OCTOBER, 1902.

(Pl. VI).

54. These storms consist of two principal phenomena, first appearing at the equator mainly as a positive equatorial perturbation, which commences suddenly at 16^h 52^m. At what hour it ceases it is difficult to say, as perturbations of another kind soon begin. The perturbation at the equator is especially powerful at about 1^h 30^m on the 30th October. It seems to be directly apparent from the curves that this is really an equatorial perturbation. Unfortunately there are no observations for this date from Honolulu and several other stations, as the time was not given in my Circular (p. 38). Simultaneously with this perturbation, there are powerful storms round the Norwegian stations, that at Matotchkin Schar being particularly so, and of long duration. The positive equatorial perturbations observed by us are *always* accompanied by polar storms. As a rule, the polar storms do not begin until a little while after the equatorial; but on this occasion they begin almost simultaneously, that at Matotchkin Schar lasting from 16^h 40^m to about midnight.

The almost simultaneous appearance of the polar storm and the positive equatorial perturbation has been already mentioned as of frequent occurrence. The explanation of the positive equatorial perturbation given in Art. 31, also at once suggests the connection. Fig. 38 *b* shows the descent upon the screen of those rays that would turn back before reaching the terrella. It was these rays which we assumed to be the cause of the positive equatorial perturbation. The figure also distinctly shows, however, that this descent of rays upon the screen occurs simultaneously, and is connected, with the descent in the polar regions on the terrella.

The field of force for the perturbation in question is shown in Table XXIII and in the two charts following.

TABLE XXIII.

The Perturbing Forces on the 29th & 30th October, 1902.

Gr. M. T.	Toronto		Axeloen			Matotchkin Schar		
	P_h	P_d	F_h	P_d	P_v	P_h	P_d	P_v
h m								
17 30	+ 3.1 γ	0	− 224.0 γ	E 28.5 γ	− 300 γ	− 27.7 γ	E 3.1 γ	− 31.5 γ
18 52.5	+ 3.6 »	0	− 131.0 »	W 36.7 »	+ 148 »	− 217.0 »	» 136.0 »	?
20 30	− 2.2 »	0	− 74.5 »	E 26.1 »	+ 205 »	− 237.0 »	» 91.0 »	?
21 30	+ 1.8 »	0	+ 6.4 »	W 7.6 »	+ 143 »	− 113.0 »	» 69.5 »	− 18.7 »
23 15	+ 9.5 »	0	− 3.7 »	E 7.6 »	+ 94 »	− 41.5 »	» 46.5 »	− 18.7 »
1 0	+ 19.8 »	E 2.4 γ	?	?	+ 37 »	+ 9.3 »	W 5.7 »	0
1 30	+ 10.8 »	» 2.4 »	?	?	+ 118 »	+ 9.4 »	E 19.0 »	0

TABLE XXIII (continued).

Gr. M. T.	Kaafjord			Stonyhurst		Wilhelmshaven		
	P_h	P_d	P_v	P_h	P_d	P_h	P_d	P_v
h m								
17 30	?	?	?	+ 6.1 γ	E 5.1 γ	+ 11.7 γ	E 5.5 γ	0
18 52.5	?	?	?	− 10.7 »	» 8.6 »	− 4.2 »	» 19.5 »	0
20 30	?	?	?	− 5.6 »	» 18.2 »	+ 4.6 »	» 21.3 »	0
21 30	?	?	?	+ 3.0 »	» 4.5 »	+ 2.8 »	» 4.3 »	− 5.0 γ
23 15	− 46.0 γ	E 24.8 γ	− 89.3 γ	+ 3.5 »	» 8.6 »	+ 10.7 »	» 10.4 »	− 5.0 »
1 0	− 1.2 »	W 8.4 »	− 56.3 »	+ 11.7 »	W 4.5 »	+ 17.2 »	W 5.5 »	0
1 30	− 3.0 »	E 20.2 »	− 40.0 »	+ 4.6 »	E 10.8 »	+ 17.2 »	E 16.5 »	0

TABLE XXIII (continued).

Gr. M. T.	Kew		Munich		San Fernando		Dehra Dun	
	P_h	P_d	P_h	P_d	P_h	P_d	P_h	P_d
h m								
17 30	+ 3.0 γ	E 6.5 γ	+ 5.5 γ	E 3.8 γ	+ 7.6 γ	E 6.5 γ	+ 5.1 γ	0
18 52.5	− 9.7 »	» 8.4 »	− 6.5 »	» 13.7 »	− 3.2 »	» 6.5 »	+ 6.7 »	E 2.9 γ
20 30	− 3.5 »	» 14.1 »	+ 2.0 »	» 16.8 »	+ 3.2 »	» 9.8 »	+ 6.7 »	0
21 30	+ 3.0 »	» 3.3 »	+ 2.5 »	» 6.1 »	+ 7.6 »	» 4.1 »	+ 4.7 »	0
23 15	+ 7.1 »	» 8.4 »	+ 4.5 »	» 8.4 »	+ 9.6 »	» 6.5 »	+ 13.0 »	W 2.9 »
1 0	+ 13.7 »	W 3.7 »	+ 11.5 »	W 2.3 »	+ 17.9 »	W 2.5 »	+ 26.0 »	» 9.8 »
1 30	+ 7.6 »	E 12.2 »	− 8.0 »	E 12.2 »	+ 11.5 »	E 9.0 »	+ 33.5 »	» 13.8 »

TABLE XXIII (continued).

Gr. M. T.	Zi-ka-wei			Batavia		Christchurch		
	P_h	P_d	P_v	P_h	P_d	P_h	P_d	P_v
h m								
17 30	+ 4.8 γ	0	No noticeable deflection	?	?	?	?	?
18 52.5	+ 6.2 »	0		?	?	?	?	?
20 30	+ 2.4 »	0		?	?	?	?	?
21 30	+ 4.8 »	0		?	?	?	?	?
23 15	+ 16.8 »	E 9.0 γ		?	?	+ 27.2 γ	E 8.9 γ	− 2.5 γ
1 0	+ 24.0 »	» 9.0 »		+ 29.2 γ	W 19.2 γ	+ 37.2 »	» 8.9 »	0
1 30	+ 28.8 »	» 4.0 »		+ 36.0 »	» 24.0 »	+ 38.6 »	» 9.7 »	0

and 30th October, 1902; Chart I at 18^{h} 52.5^{m} and 20^{h} 30^{m} on the 29th, and Chart II at 1^{h} on the 30th.

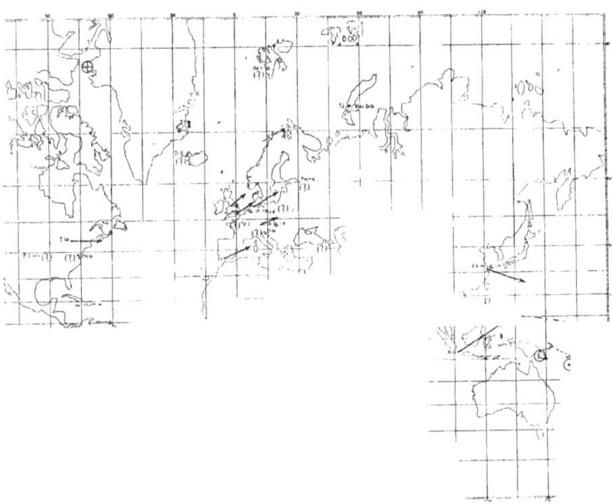

Fig. 77.

Chart I shows the conditions at 18^h 52.5^m and 20^h 30^m on the 29th October.

At these hours, it is the polar systems that give the field its character. There is a polar system in its centre presumably in the neighbourhood of Matotchkin Schar. The direction of the current-arrow is westward along the auroral zone, indicating that the storm-centre is on the midnight side. In lower latitudes there is an area of convergence. On the mainland of Europe, the field is turning counter-clockwise as in the polar regions.

Chart II shows the conditions at 1^h on the 30th October.

The field is now mainly conditioned by the equatorial perturbation, which at this hour is very powerful.

This is an example of a composite perturbation of the very simplest kind, in which there is the simultaneous occurrence of a very simple equatorial perturbation, and a polar storm that also exhibits very simple forms.

THE PERTURBATION OF THE 25th DECEMBER, 1902.

(Pl. XI).

55. It is a brief, but powerful and well-defined perturbation, particularly marked at the observatories in North America, that has here attracted attention. It commences there at 3^h 14^m, increases rapidly, and reaches a maximum at 3^h 21^m, after which it decreases more slowly, and at 3^h 57^m the conditions are once more almost normal.

We notice especially that the perturbation appears with much greater strength at Toronto than at Baldwin and Cheltenham. At Toronto, the horizontal component of the perturbing force attains a value of 45.3 γ, and at Baldwin and Cheltenham values of 23 and 25.4 γ respectively. At Sitka the perturbation is noticed distinctly, but it is very faint. The perturbation that, on account of its course, should be connected with the above, there attains a strength of 7.5 γ.

During the time under consideration, perturbations occur all over the world. At our Norwegian stations, there are storms of considerable magnitude, and elsewhere in Europe slight, but distinct perturbations.

These perturbations, however, run an altogether different course from those in America. At Dyrafjord, there is a perturbation of medium strength, but of much longer duration than those in America; it has considerable strength as early as about 1^h, and lasts almost until 5^h. There is moreover a fairly powerful storm at about midnight.

At Axeløen, the conditions resemble those at Dyrafjord, except that the course of the perturbation differs still more in its conditions from those in America. At Dyrafjord, during the time in which the short perturbation in America is taking place, we can notice a distinct variation in the form of the curve, especially that for *H*, which almost coincides with that for the perturbation in America. At Axeløen, on the contrary, nothing special is noticed. There the perturbation has at that time already passed its maximum, which occurs at 2^h 32^m. At Axeløen also, there is a comparatively powerful perturbation at about midnight, commencing later, namely at 23^h 45^m on the 24th, and continuing fairly powerful right on to 5^h on the 25th.

The conditions at Kaafjord on this date are particularly interesting, in that during the time in which powerful storms are occurring in the north, there are only very faint perturbations here, such as might best be characterised as slightly disturbed conditions. We notice, however, a perturbation that appears simultaneously with, and runs the same course as, the perturbation in America. Its strength is also about the same, if anything a little less.

In Europe as a whole, the conditions are slightly disturbed from 23^h on the 24th to 5^h on the 25th. There are especially distinct perturbations about midnight, and from $2^h\ 30^m$ to 4^h. We thus see that the conditions there are in the main connected with the polar storms at the Norwegian stations. If we look at the curve for the declination, we see, moreover, that exactly at the time when the brief, powerful perturbation is occurring in America, there is a perturbation in Europe with very much the same course; it commences exactly at $3^h\ 15^m$, increases to a maximum, which occurs at $3^h\ 21^m$, and then slowly decreases until about 4^h, when it is at an end.

At Tiflis, Dehra Dun, Batavia and Zi-ka-wei, this perturbation in the main occurs simultaneously and runs a course similar to, the perturbation in America. It occurs in H only.

The field of force is shown in two charts for four different hours.

TABLE XXIV.

The Perturbing Forces on the 25th December, 1902.

Gr. M. T.	Sitka		Baldwin		Toronto		Cheltenham	
	P_h	P_d		P_d	P_h	P_d	P_h	P_d
h m								
3 0	− 0.8 γ	W 2.2 γ	− 4.6 γ	0	− 6.7 γ	0	− 3.5 γ	0
15	− 3.6 »	E 3.1 »	− 4.6 »	E 8.9 γ	− 5.8 »	E 15.0 γ	+ 4.1 »	E 12.5 γ
20	− 7.5 »	» 4.0 »	− 4.3 »	» 22.3 »	+ 1.3 »	» 44.0 »	+ 5.8 »	» 24.4 »
30	− 7.3 »	» 7.6 »	+ 2.8 »	» 20.3 »	+ 5.0 »	» 31.3 »	+ 3.2 »	» 15.4 »
40	− 2.1 »	» 5.4 »	+ 1.4 »	» 8.9 »	0	» 12.6 »	+ 0.9 »	» 8.3 »
4 0	− 0.7 »	» 2.7 »	− 3.2 »	0	− 9.0 »	» 1.8 »	− 3.0 »	» 2.3 »

TABLE XXIV (continued).

Gr. M. T.	Dyrafjord			Axeloen			Kaafjord		
	P_h	P_d	P_v		P_d	P_v	P_h	P_d	P_v
h m									
3 0	− 110 γ	E 2.4 γ	+ 23.5 γ	− 105.0 γ	E 62.5 γ	+ 76.0 γ	0	E 12.8 γ	− 21.2 γ
15	− 218 »	» 5.9 »	+ 53.3 »	− 96.3 »	» 64.0 »	+ 81.0 »	− 6.7 γ	» 5.1 »	− 21.2 »
20	− 206 »	» 12.1 »	+ 59.9 »	− 93.0 »	» 58.0 »	+ 83.5 »	− 14.7 »	0	− 17.3 »
30	− 106 »	W 13.1 »	− 6.4 »	− 59.8 »	» 48.6 »	+ 88.5 »	− 16.0 »	» 2.5 »	− 18.0 »
40	− 105 »	E 23.8 »	− 7.8 »	− 44.2 »	» 30.4 »	+ 71.2 »	− 9.8 »	0	− 15.7 »
4 0	− 44 »	W 9.7 »	− 3.4 »	− 34.5 »	» 27.2 »	+ 14.7 »	− 4.3 »	W 2.2 »	− 14.1 »

TABLE XXIV (continued).

Gr. M. T.	Pawlowsk			Stonyhurst		Wilhelmshaven			Kew	
	P_h	P_d	P_v	P_h	P_d	P_h	P_d	P_v	P_h	P_d
h m										
3 0	0	E 3.2 γ	No noticeable perturbing forces.	− 3.0 γ	E 4.0 γ	− 1.4 γ	E 3.0 γ	A slight neg. deflection.	− 3.0 γ	E 4.6 »
15	− 6.5 γ	0		− 5.6 »	» 2.3 »	− 2.8 »	» 2.4 »		− 3.5 »	» 1.9 »
20	− 10.0 »	W 3.2 »		− 6.6 »	W 8.0 »	− 10.8 »	W 15.2 »		− 4.0 »	W 6.5 »
30	− 9.0 »	0		− 5.6 »	» 11.4 »	− 10.3 »	» 13.4 »		− 6.1 »	» 9.8 »
40	− 7.0 »	» 2.3 »		0	» 8.5 »	− 5.6 »	» 11.0 »		− 2.0 »	» 7.0 »
4 0	+ 1.0 »	» 1.4 »		+ 3.5 »	» 2.3 »	+ 3.7 »	» 3.0 »		+ 3.0 »	» 2.3 »

TABLE XXIV (continued).

Gr. M. T.	Potsdam		Val Joyeux			Munich		San Fernando	
	P_h	P_d	P_h	P_d	P_v	P_h	P_d	P_h	P_d
h m									
3 0	− 0.6 γ	E 3.0 γ	− 1.6 γ	E 5.0 γ	A very slight neg. deflection about $3^h 30^m$.	− 1.5 γ	E 2.3 γ	− 3.2 γ	0
15	− 2.2 »	W 2.5 »	− 2.4 »	0		− 5.0 »	W 5.3 »	− 3.8 »	W 4.8 γ
20	− 8.2 »	» 9.6 »	− 3.2 »	W 10.0 »		− 4.5 »	» 7.6 »	− 2.5 »	» 6.5 »
30	− 6.9 »	» 8.6 »	− 4.8 »	» 7.5 »		− 6.0 »	» 6.1 »	0	» 4.8 »
40	− 4.4 »	» 7.1 »	− 2.4 »	» 4.2 »		− 1.0 »	» 5.3 »	0	0
4 0	+ 3.2 »	» 1.5 »	+ 3.2 »	0		+ 20 »	» 1.5 »	+ 1.2 »	0

TABLE XXIV (continued).

Gr. M. T.	Tiflis			Dehra Dun		Zi-ka-wei		Batavia		Christchurch	
	P_h	P_d	P_v	P_h	P_d	P_h	P_d	P_h	P_d	P_h	P_d
h m											
3 0	−13.2 γ	E 1.4 γ	Slight deflections to small to allow of being measured.	− 1.6 γ	No noticeable deflections.	− 3.6 γ	No noticeable deflections.	− 3.2 γ	No noticeable deflections.	− 1.3 γ	No noticeable deflections.
15	− 1.5 »	0		− 3.9 »		− 9.6 »		− 6.4 »		− 4.6 »	
20	− 3.9 »	0		− 4.7 »		− 9.6 »		− 6.0 »		− 7.8 »	
30	− 4.4 »	W 1.4 »		− 4.7 »		− 7.2 »		− 1.3 »		− 5.5 »	
40	− 4.4 »	» 0.7 »		− 1.9 »		− 2.4 »		0		− 2.3 »	
4 0	− 0.8 »	0		0		− 2.4 »		− 1.8 »		0	

Current-Arrows for the 25th December, 1902; Chart I at 3^h 15^m and 3^h 20^m.

Fig 78.

Current-Arrows for the 25th December, 1902; Chart II at 3^h 30^m and 3^h 40^m.

Fig. 79.

Chart I; Time 3^h 15^m and 3^h 20^m.

At the first hour named, the conditions are similar to those prevailing at the time when the power-perturbation in America commences. In the United States the current-arrows are directed southwards, with some divergence; at Sitka, westwards. In Europe the direction of the arrows varies greatly from place to place. This may certainly in a great measure be accounted for, partly by the fact that when the arrows are small, their direction is rather liable to error, as the normal line cannot be so positively determined, and partly that an inaccuracy in the time-determination, owing to the great variableness of the conditions at this point of time, will result in a large error in the force.

Turning to the Norwegian stations, we find the force to be especially strong at Dyrafjord and Axeløen, and at both these places the current-arrow, as is usual in such circumstances, is directed WSW along the auroral zone.

At 3^h 20^m the perturbation in America is at its height, and the field of force in southern latitudes is now in the main determined by this brief perturbation.

The field of force in Europe and North America now shows a strong resemblance to that during the cyclo-median storm of the previous 6th October, the chief difference being that the latter was more restricted in area, its remarkable field of action being principally confined to North America and Europe. At Dyrafjord and Axeløen the conditions are almost as at 3^h 15^m.

Chart II; Time 3^h 30^m and 3^h 40^m.

The form of the field is on the whole unaltered, except that the strength is less.

Although it may appear, from a glance at the curves, as if the perturbation were fairly simple, it is in reality of a rather composite character. In the district from Axeløen to Dyrafjord, there is polar precipitation. There is, on the whole, a current-system acting as a horizontal current flowing almost in the direction from Axeløen to Dyrafjord. The system should have its greatest density to the south of these two stations. On account of the comparatively quiet conditions at Kaafjord, the powerful effect at Dyrafjord and Axeløen must be due to the fact that the currents causing the perturbation must come comparatively close to these stations. These currents remain in the north rather a long time with varying strength, but in about the same position from about midnight until 5^h.

While these currents are acting in the north, and directly or indirectly producing very faint perturbations southwards in Europe, a peculiar perturbation occurs, well-defined and powerful, but of short duration, and remarkable for its universal distribution. It is the more remarkable that there is no place at which it seems to be accompanied by storms of great violence, but appears to be as powerful in lower as in higher latitudes.

We have said that the field of this perturbation resembles in its main features that of the previous 6th October. In addition to this, its course is on the whole the same. The two perturbations are about equal in duration, increase suddenly to a maximum, and then more slowly decrease to 0; and their strength is about equal. The only difference is that this perturbation is most powerful in North America, while that of the 6th October was most powerful in Western Europe.

This brief storm must thus, it seems, be classed with those perturbations which we have called cyclo-median.

We might suppose that the field of force in this short perturbation was produced by a descent of rays towards the earth, similar to that towards the terrella, which occasioned the appearance of one of the areas of light that we find in fig. 68. We will examine a little more closely into the resemblance of the field of force observed, to that which was to be expected according to the experiments and Størmer's calculations. We will however draw attention to the fact that we have not yet any experiments that are exactly suited to this perturbation as regards date and hour.

At Zi-ka-wei, Dehra Dun and Tiflis, the arrows are directed westwards, answering to the conditions near the point at the eastern end of the patch of light. Fig. 79 distinctly shows the direction of the current to be as one would expect. The north-westerly direction of the arrows in Central and Northern Europe, the south-westerly at Dyrafjord, and southerly in eastern America, correspond again to the rest of the path; but there is nothing answering to Axeløen.

It is natural to look upon the whole field of force as a composite field, imagining it to be partly formed by polar precipitation round Axeløen and Dyrafjord, but also by precipitation in lower latitudes of stiffer rays, and probably chiefly conditioned by the latter.

We may also mention the fact that some of the polar elementary storms already described, and described only as elementary, sometimes have fields that may be regarded as the production of cyclo-median storms. The best example of this will be found on Chart II for the 31st March, 1903 (p. 122), where it is of exactly the same shape as that now under discussion.

By assuming a composite field such as this, we also find an explanation of the positive values of P_v, which occurred in the system's area of convergence, and which thus seem to be at variance with the assumption of a single polar elementary system in the auroral zone.

We have also subsequently met with a similar disagreement as regards P_v, e. g. on the 26th December, where we have indicated the probability that there the rays came comparatively near to the earth in lower latitudes. This had special reference to the rays that occur in cyclo-median storms.

THE PERTURBATION OF THE 28th DECEMBER, 1902.

(Pl. XIII.)

56. This perturbation is not one of those that it was originally intended to describe, and the time is therefore not given in my circular dated June 1903. There are thus only a few more or less chance observations besides those from the Norwegian stations. What has determined us nevertheless to describe it is the peculiarity we find on comparing the curves for Dyrafjord with those for the American stations. The perturbation occurrs chiefly between 4^h 40^m and 6^h, that is to say about midnight, local time, at the three easterly North American stations.

The well defined deflection in the curves for Dyrafjord indicates that the storm could be a polar elementary one, of which the district of precipitation perhaps is in the vicinity of that station. The time of the perturbation, however, differs from that generally found in the best examples of polar elementary storms at the Norwegian stations. The conditions at Kaafjord and Matotchkin Schar also show with sufficient distinctness that there is no field of precipitation at those stations, the perturbing forces there being quite inconsiderable. At Axeløen, on the other hand, there are more powerful perturbing forces, and the perturbation there is of somewhat longer duration than at Dyrafjord, as it begins earlier and concludes at about the same time. The character of the curve too, is so different that it is difficult to decide whether the perturbing forces at these two stations arise from two separate systems or not; but this question is of no great actuality in our study of this storm. The main thing is to prove the connection between the perturbations at Dyrafjord and the American stations. The form of the curves has a very great resemblance to those found in Europe during the polar elementary storms occurring at about midnight on, for instance, the 15th December. We should therefore imagine that in this instance, the field on the midnight side was similar to that previously found at the Norwegian stations; and a closer investigation seems to verify that so is the case.

On *Chart I, for* 4^h 45^m *and* 5^h, there appears to be an area of convergence in the east of North America, and adjoining part of the Atlantic, and in the west of Europe. This should indicate that in the neighbourhood of Dyrafjord, possibly a little to the west of it, there should be a stormcentre with current-arrows directed westwards. It is impossible to determine the size and position of the field of precipitation more precisely with the comparatively few data that we have to go upon; but the conditions at Sitka indicate that it must extend comparatively far westwards in North America. Judging from the curves for Sitka, we may suppose that the same system is at work there as at Toronto and Cheltenham. The similarity between the curves at these places is great enough to allow such an assumption. The centre of gravity, so to speak, of the field of precipitation may be assumed to be about the south of Greenland. Sitka should be situated almost on the main axis.

The rest of the course of the perturbation may now be very simply explained by a westward movement of this storm-centre. On surveying the curves closer, we see that at Toronto P_h turns from positive to negative a little earlier than P_d from E to W. In consequence of this the arrows will turn with the hands of a clock, the current-arrow from S by W to N. Their size at this time, about 5^h 20^m, is very small. In Cheltenham P_h and P_d change the sign at nearly exactly the same time, so that here one does not get a rotation, but more a sudden change of direction from S to N of the current-arrow. Thus in Toronto the conditions are such, as if the point of convergence passes just a trifle south of the place, while the conditions in Cheltenham indicate that the point just passes the same. At any rate we may conclude from this that the point of convergence will pass near these stations. But to determine its course more exactly is difficult, as precision in the fixing of time here plays an important part.

At Sitka the directions of the arrows are at first rather constant, but then turn with a counterclockwise movement, showing that as the system moves westwards, the place comes into the area of convergenee. If we suppose that the principal axis of the system is always almost tangent to the auroral zone, it corresponds exceedingly well with what one would expect.

Chart II for 5^h 15^m, 5^h 30^m, and 5^h 45^m, shows the conditions as they subsequently develope. We can here distinctly follow the movement described above.

The storm-centre now is entering North America, and at the last two hours named it is perhaps situated a little to the west of Hudson Bay; for it may be concluded from the arrows that the transverse axis must pass between Sitka on the one side and the eastern stations on the other.

The force at Dyrafjord in the mean while has decreased considerably, showing that the storm-centre has moved away. At Axeløen, too, the forces are considerably less than before.

We thus have in this perturbation an instance of a polar elementary storm that occurs at a different time of day, and has a somewhat different course, from those described previously. There may, moreover, possibly be other perturbing forces in Europe. The declination-curve for Stonyhurst, wich we have, points indeed in this direction; but we have not the material to enable us to study this more closely. We have therefore not included this among the elementary storms.

The movement of the system in America that we have here met with, will be also investigated more throughly in the material from 1882—83; and we shall there find similar conditions during nearly all the perturbations that occur in this region at about this time of day.

TABLE XXV

The Perturbing Forces on the 28th December, 1902.

Gr. M. T.	Sitka		Toronto		Cheltenham		Dyrafjord		
	P_h	P_d	P_h	P_d	P_h	P_d	P_h	P_d	P_v
h m									
4 45	− 13.9 γ	E 16.2 γ	+ 21.6 γ	E 15.1 γ	+ 17.4 γ	E 5.9 γ	− 154.2 γ	E 43.7 »	− 14.7 γ
5 0	− 15.4 »	» 24.8 »	+ 11.3 »	» 18.1 »	+ 13.2 »	» 8.9 »	− 124.5 »	» 36.8 »	− 26.5 »
15	− 10.2 »	» 8.1 »	− 4.5 »	» 12.0 »	0	» 7.1 »	− 71.6 »	» 31.2 »	− 36.9 »
30	+ 25.2 »	» 26.6 »	− 6.8 »	W 33.1 »	− 1.8 »	W 19.0 »	− 63.3 »	» 19.8 »	− 26.5 »
45	+ 10.6 »	» 14.4 »	− 3.6 »	» 28.3 »	− 1.5 »	» 19.0 »	− 57.8 »	» 9.0 »	− 36.9 »
6 0	− 2.2 »	» 7.7 »	+ 5.9 »	» 6.0 »	+ 5.3 »	» 4.8 »	0	» 4.5 »	− 47.2 »
15	− 7.5 »	» 8.6 »	+ 4.1 »	0	+ 4.1 »	» 1.2 »	− 6.6 »	» 10.4 »	− 47.2 »
30	− 6.9 »	» 6.3 »	0	0	0	» 1.2 »	0	» 3.5 »	− 24.1 »

TABLE XXV (continued).

Gr. M. T.	Axeløen			Matotchkin Schar			Kaafjord		
	P_h	P_d	P_v	P_h	P_d	P_v	P_h	P_d	P_v
h m									
4 45	− 86.0 γ	E 68.5 γ	+ 19.7 γ	− 8.9 γ	0	− 12.5 γ	− 8.4 γ	E 5.5 γ	0
5 0	− 65.7 »	» 52.1 »	+ 4.9 »	− 10.1 »	E 2.6 γ	− 6.6 »	− 8.4 »	» 4.7 »	− 4.2 γ
15	− 68.0 »	» 39.2 »	− 22.1 »	0	0	+ 8.8 »	− 4.8 »	0	− 2.1 »
30	− 10.1 »	» 19.7 »	− 49.2 »	+ 14.4 »	0	+ 19.8 »	+ 3.0 »	» 8.4 »	+ 6.4 »
45	− 40.8 »	» 17.0 »	− 49.1 »	− 5.8 »	W 3.5 »	0	− 1.8 »	» 2.2 »	0
6 0	0	» 7.8 »	− 49.2 »	+ 7.4 »	0	+ 25.7 »	0	W 1.8 »	+ 4.2 »
15	+ 11.5 »	» 44.8 »	− 51.5 »	− 5.6 »	» 3.1 »	0	0	E 3.6 »	+ 9.6 »
30	− 8.3 »	» 9.2 »	− 24.6 »	− 5.2 »	E 1.3 »	+ 18.3 »	»	» 9.5 »	+ 5.7 »

Current-Arrows for the 28th December, 1902; Chart I at $4^h\ 45^m$ and 5^h, and Chart II at $5^h\ 15^m$, $5^h\ 30^m$ and $5^h\ 45^m$.

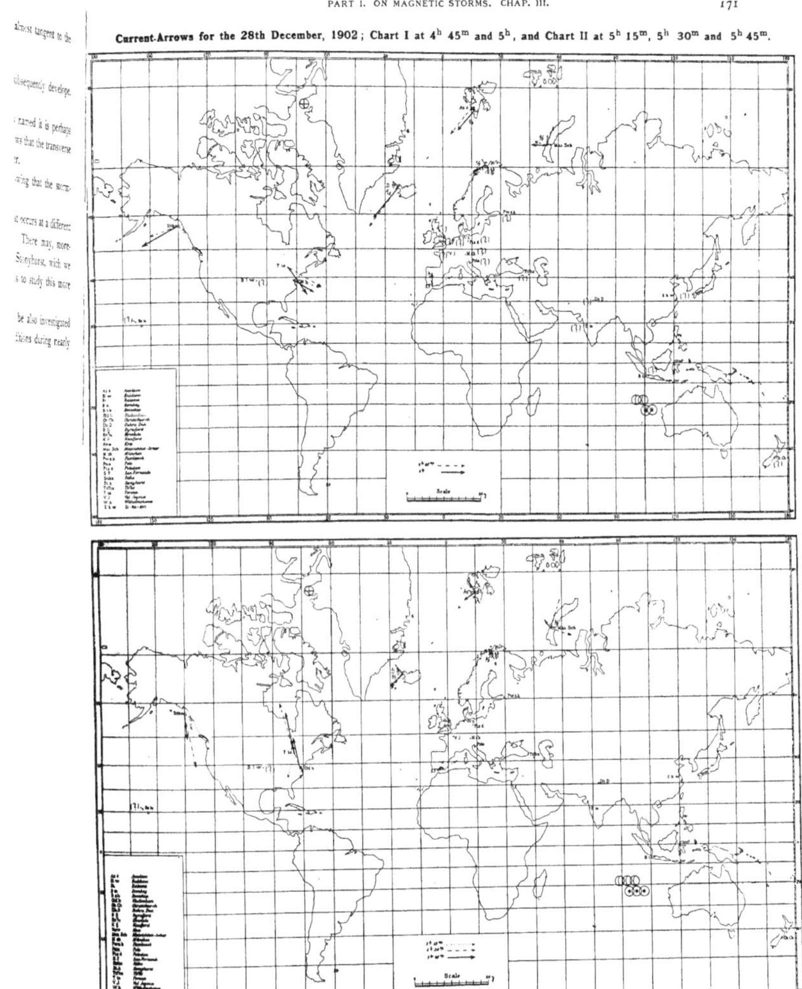

Fig. 80.

Table. XXV (continued).

Gr. M. T.	Stonyhurst		San Fernando	
	P_h	P_d	P_h	P_d
h m				
4 45	No copy recieved.	W 6.3 γ	0	W 9.0 γ
5 0		» 6.9 »	+ 1.3 γ	» 7.4 »
15		» 4.0 »	+ 5.1 »	» 4.1 »
30		» 1.1 »	+ 3.8 »	0
45		» 1.7 »	+ 3.2 »	» 0.8 »
6 0		» 8.5 »	+ 4.5 »	» 5.7 »
15		» 5.7 »	+ 7.7 »	0
30		» 2.3 »	+ 9.0 »	E 0.8 »

THE PERTURBATION OF THE 15th FEBRUARY, 1903.

(Pl. XIX).

57. This perturbation appears on an otherwise very quiet day. It is of fairly long duration, commencing at about 2 p. m. Greenwich mean time, and lasting about $4^1/_2$ hours. It is nevertheless very well defined, and in most cases the normal line can be easily determined, as the conditions before and after are rather normal. In this respect, however, the conditions in North America present some difficulty, as the normal line commences at the moment when the curve shows a marked curvature owing to the diurnal variation; and it appears that, even assuming that conditions are quiet, the form of the curve is not repeated exactly from day to day.

We have drawn up a table for this perturbation, giving the times of its commencement and termination and of the P_1 maximum, as also the value of the last-named. It appears, as regards the European stations in particular, that the perturbation does not begin and cease simultaneously in D and H; and we have therefore determined these times separately.

We see that in Central and Southern Europe the perturbation begins almost two hours sooner in H than in D, and ends about half an hour earlier in D than in H; but as a set-off, it is on the whole very strong in D as long as it lasts. We further see from the table that on the whole the maximum occurs almost simultaneously everywhere, somewhere about $16^h\ 40^m$. It should be remarked, however, that the time of the maximum cannot be exactly determined, as the maximal point is not sharply defined.

Axeløen, Sitka and Tiflis form exceptions in this respect. Axeløen, as the curve shows, has no well-defined maximum; but the force is maintained, with occasional violent oscillations, in great strenght from $16^h\ 15^m$ until $17^h\ 30^m$. Before the great storm, however, there is a fairly well defined, but much slighter perturbation. Its course is almost similar to that of the first perturbation appearing at Sitka; it occurs at about 14^h, and has its maximum at about $14^h\ 40^m$.

At Sitka, the impression given by the curve is that of two almost separate perturbations, each with its well-defined maximum. The first last from $14^h\ 10^m$ to $16^h\ 10^m$, and the second from $16^h\ 10^m$ to about 18^h, the peculiarity here being that, in contrast to the other parts of the world, the first part is the more powerful of the two.

At Kaafjord the conditions on the whole are similar to those farther south in Europe, with the exception that the conditions in D and H are interchanged, the perturbation in H at Kaafjord almost corresponding with that in D farther south. During the first part, from $13^h\ 45^m$ to $15^h\ 35^m$, it is a

TABLE XXVI.

Observatory	Beg. in H	Beg. in D	Time of max.	P_1 (max.)	End in H	End in D
Axeløen	14^h 6^m	16^h 14^m[1]	16^h 27^m	392 γ	18^h 19^m	17^h 36^m[1]
Matotchkin Schar .	15 50[1]	15 48[1]	16 45	280 »	17 30[1]	18 15
Kaafjord	15 40	14 15	16 28	141 »	17 40	18 15
Dyrafjord	ca. 14 40	16 15[1]	16 33	140 »	ca. 18 45	ca. 17 50
Pawlowsk	14 15	13 45	16 30	65 »	18 20	18 30
Wilhelmshaven . .	14 15	16 5	16 39	58.5 »	18 19	17 44
Stonyhurst	14 16	16 10	16 42	50 »	18 16	17 48
Potsdam	14 15	16 5	16 38	43.5 »	18 18	17 45
Munich	14 15	16 15	16 45	39 »	18 18	ca. 18
Val Joyeux	14 14	16 7.5	16 37.5	39 »	18 18	17 45
Kew	14 14	16 7.5	16 45	38 »	18 16	17 45
Sitka	14 9[1]	14 45[1]	15^h & 17^h 15^m	35 & 31 »	17 45	18 0
Pola	14 15	16 12	16 38	29 »	18 21	17 45
Baldwin	ca. 13 15	indeterm.	17 0	25.2 »	18 18	indeterm.
Cheltenham	ca. 13	»	16 36	25.2 »	18 8	»
San Fernando . . .	14 15	16 7.5	16 45	25 »	18 15	18 0
Toronto	ca. 13 27	ca. 15	ca. 16 40	21.5 »	18 6	17 50
Tiflis	ca. 13	ca. 14	15 45	20.5 »	ca. 18 30	ca. 18 30
Dehra Dun	ca. 15	ca. 16	16 37.5	18 »	17 20	ca. 17 30
Batavia	14 20	16 12	16 37.5	16.4 »	17 25	ca. 17 30
Bombay	ca. 15 7.5	?	16 37.5	12.4 »	17 15	?
Honolulu	ca. 15 30	16 30	16 33	10.6 »	18 15	17 54

[1] Respectively the beginning and end of the actual storm.

well-defined perturbation, occurring almost exclusively in D and V, and having a course similar to that of the already-mentioned perturbation which occurs at Sitka during this period.

Neither at Dyrafjord nor Matotchkin Schar is any perturbation with a course such as this to be observed between 13^h 45^m and 15^h 35^m.

At Tiflis a peculiarity appears, in that the maximum occurs much earlier than in Central Europe; and when the maximum is reached there, there is nothing of that kind at Tiflis, or at any rate only a small secondary. At the time that the powerful perturbation in D commences in Central Europe, the declination conditions at Tiflis are undergoing no particular change. The H-curve, on the other hand, forms a bend similar to that appearing in D farther north; but this deflection is in the opposite direction to that before and after it, its only effect being to cause the perturbing force to become smaller and make an oscillation.

At Dehra Dun, Bombay and Batavia also, the H-curve is about of the same form, the only difference being that this deflection in an opposite direction is so prominent that the total force P_1, becomes greater than that with the previously reverse direction, and the maximum comes at the given place after all.

The conditions are probably most likely to be understood as follows. While the perturbation in Central Europe is great in D, we are concerned with the effect of at least two simultaneously acting, principal systems. One of the perturbations is of long duration, and in low latitudes the form of its field remains fairly constant. While it is going on, a comparatively powerful storm commences, with a somewhat different distribution of force.

There are fairly powerful perturbations all this time at the Norwegian stations. We also receive a distinct impression that a perturbation commences during the time in which the great deviation takes place in more southern latitudes. The conditions before and after the intermediate storms, however, are somewhat different. Before it, both at Axeløen and Kaafjord, there is apparently a comparatively independent system occuring simultaneously, with a course similar to that of the first powerful perturbation at Sitka, which has its maximum at 15^h.

It must thus be assumed that these are in the main polar perturbations; but the conditions are not simple, indicating, as they do, both in the arctic regions and in lower latitudes, that there are a number of systems acting to some extent simultaneously. This then is not an elementary storm, but must be classed among the simplest compound storms.

According to the above, we may consider it beyond a doubt that during the time from 16^h to $17^h\ 30^m$, we have the effect of an intermediate perturbation with a field of force of its own, the latter differing considerably, especially in Europe and Asia, from the field before and after.

We have worked out a plate for this perturbation from 14^h to 18^h, showing the perturbing forces at one place at various times (fig. 81).

On considering the conditions in Europe and Asia, we get a direct impression that in the above-mentioned period the effect apparent is that of an independent system.

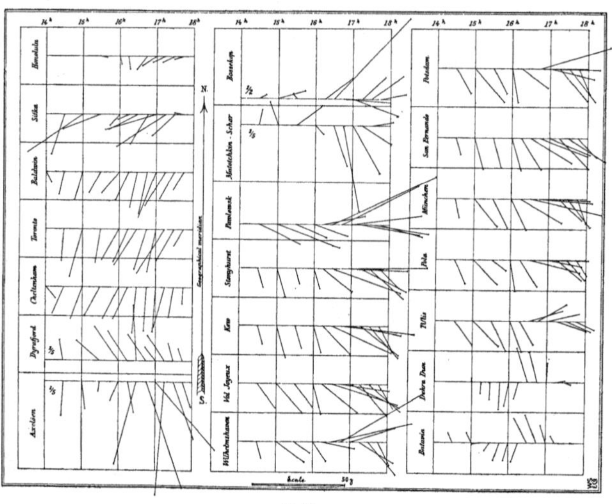

Fig. 81.

As regards Europe and Asia, the circumstances on the whole justify the decomposition of the perturbing force. In America the forces act the whole time almost in one direction, so that decomposition there cannot be effected.

THE PERTURBING FORCES.

58. In giving a detailed deseription of the field of force, we will divide the subject into three separate sections, viz.

(1) from the commencement of the perturbation up to $16^h\ 15^m$,

(2) „ $16^h\ 15^m$ to $17^h\ 15^m$, that is, during the powerful intermediate storm, and

(3) „ $17^h\ 15^m$ to its termination.

The conditions during the *first section* are shown on the *Charts I, II, III, and IV* for the hours $14^h\ 30^m$, $15^h\ 0^m$, $16^h\ 0^m$, and $16^h\ 15^m$.

During this period the field of force in southern latitudes, and also at Dyrafjord and Kaafjord, remains fairly constant. At Dyrafjord the current-arrow points along the auroral zone, but in an easterly direction. At Kaafjord its direction is SE and E, and at Pawlowsk SSW. At the stations in Central and Western Europe their direction is WSW, and in the United States WNW.

We thus see that the current-arrows in these districts during this period maintain the form of a positive vortex, which means that there is here an area of divergence for the perturbing force.

It will be seen that the arrows at Dyrafjord and Stonyhurst are in opposite directions, indicating that the point of divergence must lie between these stations, that is to say somewhat to the north-west of Scotland. In the vicinity of the point of divergence, $P_1 = 0$. We find moreover that the arrows in the district between Pola and Stonyhurst decrease throughout, and even at Wilhelmshaven are comparatively small. In accordance with our theory, the vertical arrows at Kaafjord have downward direction. The arrows at Ekaterinburg and Irkutsk indicate further that there is also an area of convergence for the perturbing force with a storm-centre lying in the north-east of Siberia. During the first part, hardly any perturbation is noticeable at the equatorial stations, the force on the charts at $14^h\ 30^m$ and 15^h being either zero or very small.

In the district about Dehra Dun, distinct perturbations do not begin until about 15^h, and at Honolulu half an hour later, indicating the existence of a perturbing force directed almost due south, along the magnetic meridian.

It appears from the curves, as also from Charts III & IV, that the perturbations at Dehra Dun and at Batavia are very similar both in magnitude and course.

The current-arrows moreover are very different in direction from what one would expect if the direction were to harmonise with the field farther north, that is to say if it were a direct effect of polar systems. For this reason it seems probable that it is not exclusively polar systems that we have to do with here. On looking at the charts (III & IV), we receive a very decided impression that in addition to the polar system, which undoubtly exists, there is an equatorial system, or more correctly speaking a system of which the greatest effect is to be looked for in low latitudes. The fact that the system in north has lasted for a appreciable time before anything is noticed at the equator also goes to prove that the perturbation in the south is due to something relatively independent.

The conditions at the Norwegian stations, Dyrafjord and Kaafjord, have been already mentioned. The perturbation there is rather slight, and the curve quiet in character. The conditions moreover are closely connected with those farther south.

As regards Matotchkin Schar, the current-arrow is at first eastward in direction, along the auroral zone, that is to say in direction similar to that at Dyrafjord. At $16^h\ 0^m$ the force has already changed.

The curve on the whole is much more disturbed; and at $16^h\ 15^m$ the instruments oscillate so violently that we were unable to determine any perturbing force. These great disturbances shows that we are now in the vicinity of the current-systems; indeed there are indications of precipitation close to the station.

At Axeløen the arrow during this period is on the whole westward in direction. It oscillates backwards and forwards about this mean direction.

The form of the field, as we have seen, remains unchanged during this period in medium latitudes; in other words, the course of the lines of force is retained. The conditions, however, are not such as can be explained by the assumption of the existence of a simple, stationary system with constant form, that has only altered in strength in the course of that time. Were this the case, the relative distribution of strenght would remain constant all the time. This is not so, however. Sitka, for instance, shows a very marked maximum in the perturbing force during this period, a maximum that we have already found at Axeløen, Kaafjord, and Pawlowsk, and of which there is an indication in North America, but which is not found in the south of Europe.

The polar storm thus seems to be somewhat variable in character; but there appear on the whole to be fields with the characteristic properties of the polar elementary storms. We find especially two areas that are characteristic of the polar elementary storms, the area of divergence in Europe and America, and the area of convergence in Asia.

If we imagine these two to belong to the same system, and the transverse axis to be drawn in that system, this axis would pass from a point in the vicinity of Iceland, right across the Pole, to the district of east Siberia. If we imagine a plane passing through the sun and the magnetic axis of the earth, the above-mentioned line will almost coincide with the line of intersection of this plane with the earth. The point of divergence lies nearest to the sun, the point of convergence far from it; and the field of force shows that as the negatively charged particles sweep down to the earth, they turn off to the left, as viewed from the sun.

It is difficult to imagine, however, that these are only the effects of a single field of precipitation. It seems far more probable that the precipitation is concentrated about various areas, and that each of these produces its characteristic field of precipitation in the north of Asia, which should produce the area of convergence that we find. The direction of the current-arrows in this storm-centre must be westerly. The current-arrow at Axeløen indicates, too, a continuation of this system, and thus seems to confirm our assumption. But in addition to this system, we must assume a weaker one that should produce the area of divergence in Europe and America, where the direction of the current-arrows in the storm-centre is easterly, the centre being situated somewhat north of Dyrafjord. Whether we have further to assume perturbing forces that act principally in lower latitudes, it is impossible to decide; and we will therefore content ourselves with establishing the fact that these two fields of precipitation account, in the main, for the fields before us. That we are justified in assuming two such systems is perhaps not shown with sufficient clearness by the observations we here can bring forward; but in the chapter on the perturbations in 1882—83, we shall find that this is the view to be taken of the conditions. It should be possible to account for the direction of the current-arrows in the centre of the weaker system north of Dyrafjord by rays out of space that are drawn in the manner, shown in fig. 38 b. To make the matter still more clear, we may refer the reader also to the second case in fig. 39, with values of γ about—0.7 and further to fig. 50 b.

The *second section*, from $16^h\ 15^m$ to $17^h\ 15^m$.

At most of the stations from which we have observations, the storm is at its height during this period, and its pronounced polar character is now very marked. We here at least have the effect of two systems, as the field in low latitudes, as described under the first section, is supposed to continue through this period also.

In the intermediate storm, the form of the field in America will be very much as before, the effect of the force there being rather slight as compared with that in Europe. The perturbing forces also, which appear during the intermediate storm, and are conditioned by it, form an area of divergence in this district. An endeavour has been made to separate the field of force of the intermediate storm in the district of Europe and Asia from the total field. The result of the decomposition is given in Charts V, VI, & VII. This has not been done in Chart VIII, but the effect of the intermediate storm is still distinct. This field has the following course. The current-arrow passes through Europe in a SSE direction, and turns eastwards through India. We here have a distinctly-marked area of convergence, lying much farther west than in the previous field. The neutral field should be in the region about the river Obi or perhaps somewhat farther to the east.

This accords well with the conditions at the Norwegian stations. At the north-easterly stations, Axeløen and Matotchkin Schar, the storm is very violent; and this fact, together with the rapid alternation with time and place, in the curves, shows that the current system must have approached those stations. Even at Kaafjord we find conditions quite different to those at the two stations named, the force at the former being much smaller, and its direction very different.

The current-arrows at Axeløen and Matotchkin Schar on Chart V, for 16^h 30^m, are somewhat different in direction, that at Axeløen being WNW, and that at Matotchkin Schar WSW. On the following charts, they have become almost parallel, a fact which points decidedly to a westward movement of the current-system along the auroral zone. This condition is rather unusual, for the ordinary polar elementary storms that we have treated up to the present, and which have had their centre between Dyrafjord and Axeløen, move eastwards (see 15th December, 1902). This storm, however, occurs earlier than the above mentioned; and we shall find from the material from 1882—83 that this is to be regarded as the normal condition at this time of day. In southern latitudes the corresponding movement in perturbations such as that of the 15th December, is a turning of the force clockwise. This time we should have expected a turning in the oposite direction, and on looking at three charts in succession, we do find a slight counter-clockwise turning in Central and Southern Europe.

At Matotchkin Schar, during the intermediate storm, the balance makes a distinct deflection in one direction, such as would imply a vertical component directed upwards. The centre of the current-system should therefore lie almost to the north of this station. At Axeløen the balance oscillates up and down about its mean position. The force is at first directed upwards, then downwards. If this effect is mainly due to the system under consideration, it would mean that the greater part of the current-system at first lay somewhat to the north, and afterwards somewhat to the south, of this station. In accordance with this, P_1 is generally more powerful at Axeløen than at Matotchkin Schar. The total force at the latter station, however, is somewhat smaller than the force that is due to the intermediate storm, as the two systems probably counteract one another.

At Kaafjord and Dyrafjord the perturbation is much weaker, P_1 attaining at both places at the most about 140 γ. The direction of P_1 is particularly worthy of notice. At Dyrafjord the direction of the current-arrow all the time is ENE along the auroral zone, that is to say exactly the reverse of the arrow at the two north-eastern stations. At Kaafjord it has an intermediate direction. At first it is southeast in direction, and thus has a tendency to be regulated by the conditions at Dyrafjord. It changes afterwards to SSW, more in accordance with the conditions at Matotchkin Schar. But on the whole the conditions at Kaafjord form the transition to the conditions farther south.

If we seek a simple explanation of the fields formed during this second section of the storm, we find that it is only necessary to assume a further development of the systems that were supposed to have produced the fields during the first section. We saw, that the system on the midnight-side had a westward motion, and the conditions at Dyrafjord may be considered as produced by a system similar to that assumed in the first section of the storm, that is to say by rays that descended upon the day-side and were deflected, perhaps in a manner resembling that shown in fig. 50 b on p. 105.

Here, too, the same difficulties present themselves as on several previous occasions. At Tiflis, for instance, we find positive values of P_h, at any rate at first in Charts V and VI; and we are therefore compelled to assume that, as already mentioned, perturbing forces also appear in lower latitudes, possibly produced by systems similar to those producing the cyclo-median storms. We cannot, however, go into this subject, as the fields do not furnish us with any reliable information concerning these systems. In any case, the perturbation clearly shows the great variableness of the storm in the region about the auroral zone, a condition which plainly proves that during this storm the current must come comparatively near the earth.

The *third section.*

The field is given in two charts, IX and X, for the hours 17^h 30^m and 17^h 45^m respectively. The form of the field is the same, on the whole, as during the first period. The chief difference is that hardly any disturbance is now noticeable at Dehra Dun and Batavia. The conditions at the Norwegian stations also are the same. At Matotchkin Schar the current-arrow is in the act of swinging round to the opposite quarter counter-clockwise; and at 17^h 30^m its direction is SSE. There is no current-arrow for this station on Chart X, the magnetogram-paper having been changed at that hour. The curves show, however, that the force ends by being directed northwards along the magnetic meridian. It thus seems reasonable to assume that all through the intermediate storm; the effect of this system, which we find before and after, has been perceptible.

Upon the whole we recognise in the current the characteristic feature of these perturbations, namely, greatly varying local conditions in the arctic regions, while in lower latitudes they vary less rapidly with time and place. We conclude from this that the perturbation there must be due to a distant system.

There is another circumstance connected with this perturbation, that may be worth noticing. If we look at the H-curve in the district from Stonyhurst to Pola during the intermediate storm, we notice three types of curves. The first of these is formed at the stations Stonyhurst, Kew, and Val Joyeux, the second at Wilhelmshaven and Potsdam, and the third at Munich and Pola. The curves of the first and second types both have a marked undulating form; while in the 3rd type there is a single, uniformly-directed deflection. This last condition is also found at Asiatic stations.

In accordance with the undulating form in the first two types, there is a more pronounced turning of the current-arrow. In this there is possibly a resemblance to the previously-described polar elementary storms. There, too, the turning of the current-arrow was most pronounced at the stations whose curves were classed under the first two types, and less pronounced in southern latitudes; and the cause would then be sought for in a movement of the current-system that produced the effect. I have already drawn attention to this circumstance in my report "Expédition Norvégienne 1899—1900," pp. 32 & 33.

TABLE XXVII.

The Perturbing Forces on the 15th February, 1903.

Gr. M. T.	Honolulu		Sitka		**Baldwin**		Toronto		Cheltenham	
	P_h	P_d	P_h	P_d	P_h	P_d	P_h	P_d	P_h	P_d
h m										
13 30	0	0	− 4.1 γ	E 10.6 γ	− 5.0 γ	E 4.4 γ	− 3.1 γ	E 4.5 γ	− 5.5 γ	E 6.3 γ
14 0	0	0	0	0	− 5.0 »	» 3.8 »	− 6.5 »	» 5.0 »	− 6.4 »	» 4.9 »
30	0	0	−21.9 »	» 4.5 »	−12.6 »	W 1.2 »	−17.1 »	W 3.0 »	−13.8 »	W 7.1 »
15 0	0	0	−31.5 »	W 15.8 »	−15.5 »	» 2.5 »	−24.7 »	» 4.5 »	−16.6 »	» 7.7 »
30	0	W 3.3 γ	−17.0 »	» 14.4 »	−12.0 »	» 5.1 »	−17.6 »	» 4.6 »	−14.2 »	» 9.9 »
16 0	− 4.0 γ	E 1.7 »	− 7.8 »	» 1.8 »	−12.5 »	» 8.2 »	−15.3 »	» 11.4 »	−16.2 »	» 10.2 »
15	− 6.6 »	» 1.7 »	− 5.3 »	» 4.5 »	−15.8 »	» 5.1 »	−16.7 »	» 10.2 »	−16.1 »	» 9.9 »
30	−10.1 »	0	−14.9 »	E 9.9 »	−19.2 »	» 1.9 »	−17.6 »	» 14.4 »	−24.8 »	» 4.9 »
45	− 8.0 »	W 4.15 »	−18.0 »	W 1.3 »	−22.4 »	» 1.9 »	−20.3 »	» 8.7 »	−23.3 »	» 3.3 »
17 0	− 6.1 »	» 6.6 »	−20.9 »	» 20.3 »	−25.2 »	» 6.36 »	−21.6 »	» 6.6 »	−23.7 »	» 3.8 »
15	− 4.5 »	» 6.6 »	−27.3 »	» 14.9 »	−19.0 »	» 8.9 »	−21.6 »	» 17.5 »	−21.0 »	» 8.2 »
30	− 4.5 »	» 6.2 »	−15.9 »	» 5.0 »	−15.3 »	» 6.3 »	−15.3 »	» 11.4 »	−15.3 »	» 4.4 »
45	− 2.65 »	» 5.0 »	− 1.8 »	» 5.0 »	−10.0 »	» 4.4 »	− 8.1 »	» 6.0 »	− 8.6 »	» 1.1 »

TABLE XXVII (continued).

Gr. M. T.	Dyrafjord			**Axeløen**			**Matotchkin Schar**			Kaafjord		
	P_h	P_d	P_v	P_h	P_d	P_v	P_h	P_d	P_v	P_h	P_d	P_v
h m												
14 30	+ 47.6 γ	W 19.8 γ	+ 28.7 γ	− 87.4 γ	W 23.1 γ	0	+ 49.5 γ	E 66.6 γ	− 12.0 γ	+ 4.3 γ	E 8.0 γ	+56.5 γ
15 0	?	?	?	− 73.7 »	» 5.6 »	0	+ 79.5 »	» 38.2 »	+ 56.0 »	+ 8.6 »	» 31.8 »	+81.0 »
30	+ 78.0 »	» 21.0 »	+ 28.7 »	− 25.3 »	» 9.5 »	0	+ 70.0 »	» 2.2 »	+ 40.0 »	+ 5.5 »	W 6.6 »	+44.8 »
16 0	+109.5 »	» 3.1 »	+ 34.2 »	− 46.0 »	» 33.0 »	+ 17.2 γ	− 16.0 »	» 29.0 »	−176.0 »	+93.2 »	» 7.3 »	+57.2 »
15	+ 93.5 »	0	+ 10.8 »	− 83.0 »	0	+ 17.2 »	Violent oscillations.		−440.0 »	+21.5 »	E 29.3 »	+33.7 »
30	+126.0 »	E 74.0 »	+ 8.0 »	−202.0 »	» 103.0 »	−135.0 »	− 92.0 »	E 67.0 »	−500,0 »	+86.0 »	» 80.7 »	+44.8 »
45	+ 84.0 »	W 1.4 »	− 34.7 »	−294.0 »	E 43.5 »	−123.0 »	−201.0 »	» 109.0 »	−517.0 »	− 3.6 »	» 50.3 »	−27.4 »
17 0	+ 45.5 »	E 2.7 »	0	−205.0 »	» 129.0 »	+172.0 »	− 96.0 »	» 107.0 »	−296.0 »	−17.2 »	» 48.8 »	−30.6 »
15	+ 70.5 »	0	+ 12.2 »	−290.0 »	W 81.5 »	−135.0 »	− 23.0 »	» 89.0 »	−216.0 »	+23.3 »	» 25.0 »	−10.2 »
30	+ 46.6 »	» 15.2 »	+ 32.0 »	−159.0 »	E 27.2 »	+ 17.2 »	+ 28.0 »	» 53.0 »	−148.0 »	+24.5 »	» 34.2 »	+19.6 »
45	+ 35.0 »	» 9.0 »	+ 41.2 »	− 69.0 »	» 8.1 »	+ 22.2 »	?	?	?	+ 5.5 »	» 26.4 »	+25.8 »

TABLE XXVII (continued).

Gr. M. T.	Pawlowsk			Stonyhurst		Kew		Val Joyeux		
	P_h	P_d	P_v	P_h	P_d	P_h	P_d	P_h	P_d	P_v
h m										
14 30	− 10.1 γ	E 23.0 γ	+ 1.5 γ	− 10.7 γ	0	− 10.2 γ	0	− 15.2 γ	E 5.8 γ	
15 0	− 12.1 »	» 34.1 »	+ 4.5 »	− 15.8 »	0	− 17.8 »	0	− 18.8 »	» 10.0 »	
30	− 7.0 »	» 17.2 »	+ 6.6 »	− 8.7 »	0	− 12.7 »	0	− 14.4 »	» 6.3 »	
16 0	− 6.0 »	» 18.4 »	+ 6.0 »	− 11.2 »	0	− 14.8 »	0	− 16.4 »	0	No noticeable deflection.
15	+ 4.5 »	» 17.5 »	+ 4.5 »	− 16.3 »	E 11.4 γ	− 18.9 »	E 10.3 γ	− 19.2 »	» 14.7 »	
30	+ 25.7 »	» 55.2 »	+ 1.5 »	− 23.0 »	» 31.4 »	− 23.0 »	» 26.9 »	− 19.6 »	» 31.8 »	
45	+ 21.1 »	» 41.4 »	+ 1.5 »	− 13.3 »	» 44.2 »	− 15.3 »	» 36.5 »	− 10.4 »	» 24.3 »	
17 0	+ 5.0 »	» 37.7 »	− 1.5 »	− 10.2 »	» 29.4 »	− 12.8 »	» 25.7 »	− 18.8 »	» 22.6 »	
15	− 5.0 »	» 31.3 »	0	− 18.9 »	» 20.0 »	− 20.4 »	» 15.2 »	− 20.8 »	» 10.0 »	
30	− 4.5 »	» 81.3 »	0	− 20.4 »	» 12.8 »	− 20.4 »	» 10.1 »	− 18.4 »	» 12.6 »	
45	− 6.0 »	» 23.0 »	0	− 12.8 »	» 7.4 »	− 15.3 »	» 3.7 »	− 14.4 »	» 4.2 »	

TABLE XXVIII (continued).

Gr. M. T.	Uccle			Wilhelmshaven			Potsdam	
	P_h	P_d	P_v	P_h	P_d	P_v	P_h	P_d
h m								
14 30	−10.0 γ	W 0.9 γ	0	− 9.3 γ	0	A slight pos. deflection with max. 16h 30m	−14.2 γ	E 5.6 γ
15	−16.0 »	» 0.5 »	+ 4.5 γ	−15.0 »	E 11.6 γ		−19.9 »	» 12.7 »
30	−11.0 »	» 0.3 »	+ 7.0 »	− 8.4 »	» 6.8 »		−14.2 »	» 5.1 »
16	−20.0 »	0	+10.0 »	−12.1 »	» 4.2 »		−18.3 »	0
15	−17.7 »	E 27.1 »	+ 4.0 »	− 8.4 »	» 16.5 »		−12.6 »	» 12.7 »
30	−24.7 »	» 52.6 »	+ 1.1 »	− 3.3 »	» 45.0 »		− 6.6 »	» 36.6 »
45	−11.6 »	» 56.2 »	+ 4.7 »	+11.7 »	» 50.7 »		+ 4.7 »	» 38.6 »
17	−10.6 »	» 41.4 »	+14.6 »	0	» 25.7 »		− 9.2 »	» 22.3 »
15	−23.9 »	» 24.7 »	+15.0 »	−16.3 »	» 14.7 »		−19.0 »	» 11.7 »
30	−23.9 »	» 14.4 »	+ 9.8 »	−21.0 »	» 9.2 »		−19.6 »	» 11.2 »
45	−17.7 »	» 1.3 »	+ 9.2 »	−14.0 »	» 2.4 »		−14.2 »	» 3.6 »

TABLE XXVII (continued).

Gr. M. T.	San Fernando		Munich		
	P_h	P_d	P_h	P_d	P_v
h m					
14 30	− 9.3 γ	0	−10.0 γ	0	A slight deflection only just perceptible
15 0	−17.7 »	0	−16.5 »	E 9.9 γ	
30	−16.3 »	0	−16.5 »	» 5.3 »	
16 0	−17.8 »	0	−17.5 »	» 5.3 »	
15	−19.2 »	E 6.1 γ	−16.2 »	» 19.8 »	
30	−18.5 »	» 18.0 »	−15.5 »	» 36.5 »	
45	−16.3 »	» 22.1 »	− 7.5 »	» 27.4 »	
17 0	−14.8 »	» 18.5 »	− 5.0 »	» 21.7 »	
15	−20.0 »	» 11.1 »	−13.5 »	» 10.6 »	
30	−17.8 »	» 10.7 »	−17.0 »	» 9.1 »	
45	−12.6 »	» 6.6 »	−13.5 »	» 2.3 »	

TABLE XXVII (continued).

Gr. M. T.	Pola			Dehra Dun		Tiflis		
	P_h	P_d	P_v	P_h	P_d	P_h	P_d	P_v
h m								
14 30	−13.5 γ	E 3.5 γ	+ 3.2 γ	0	0	−12.4 γ	E 10.4 γ	0
15 0	−14.8 »	» 10.4 »	+ 1.2 »	+ 1.5 γ	0	−13.3 »	» 14.8 »	− 2.8 γ
30	− 7.2 »	» 6.9 »	0	− 9.9 »	0	−15.9 »	» 8.5 »	0
16 0	−16.6 »	0	0	−12.6 »	W 2.0 γ	−15.9 »	» 7.4 »	− 2.5 »
15	−15.2 »	» 6.9 »	+ 2.7 »	− 9.1 »	» 2.0 »	−11.9 »	» 8.3 »	− 1.8 »
30	−11.2 »	» 25.7 »	+ 4.7 »	+17.3 »	» 6.9 »	+ 6.2 »	» 18.5 »	+ 3.1 »
45	− 4.5 »	» 26.4 »	+ 2.1 »	+15.8 »	» 4.9 »	+ 9.4 »	» 15.6 »	+ 1.8 »
17 0	− 6.3 »	» 17.4 »	+ 1.2 »	+ 9.9 »	0	+ 4.4 »	» 17.4 »	0
15	−12.5 »	» 11.1 »	0	+ 1.1 »	» 4.9 »	− 4.9 »	» 16.0 »	− 2.3 »
30	−11.6 »	» 7.6 »	0	− 1.1 »	» 3.0 »	− 5.2 »	» 13.4 »	− 1.3 »
45	− 9.0 »	» 4.2 »	0	0	» 1.0 »	− 4.0 »	» 10.0 »	0

TABLE XXVII (continued).

Gr. M. T.	Bombay		Batavia		Ekaterinburg			Irkutsk		
	P_h	P_d	P_h	P_d	P_h	P_d	P_v	P_h	P_d	P_v
h m										
14 30	0		+ 4.9 γ	W 2.4 γ	0	E 17.5 γ		+ 11.3 γ	E 9.5 γ	
15 0	0		+ 4.9 »	» 2.4 »	0	» 28.5 »		+ 15.0 »	» 11.8 »	
30	− 7.2 γ		− 4.3 »	» 3.0 »	0	» 20.0 »		+ 16.0 »	» 11.0 »	
16 0	−10.75 »		− 9.8 »	» 3.0 »	0	» 5.6 »		+ 16.0 »	» 9.4 »	
15	−11.2 »	Wanting.	− 10.3 »	» 3.6 »	+ 1.5 γ	» 8.0 »	No noticeable deflection.	+ 16.3 »	» 8.7 »	Indeterminable.
30	+ 10.2 »		+ 12.8 »	» 7.2 »	+ 5.0 »	» 14.0 »		+ 17.0 »	» 5.5 »	
45	+ 12.0 »		+ 14.6 »	» 6.0 »	+ 10.0 »	» 20.0 »		+ 17.5 »	» 5.3 »	
17 0	+ 7.4 »		+ 9.3 »	» 3.0 »	+ 13.0 »	» 22.5 »		+ 18.0 »	» 3.5 »	
15	+ 1.0 »		+ 3.9 »	» 2.4 »	+ 12.5 »	» 21.4 »		+ 16.3 »	» 3.0 »	
30	− 1.0 »		0	» 3.0 »	+ 10.0 »	» 17.4 »		+ 12.5 »	» 2.8 »	
45	0		0	» 1.8 »	+ 1.5 »	» 12.6 »		+ 7.5 »	» 2.8 »	

TABLE XXVIII.

Partial Perturbing Forces on the 15th February, 1903.

Gr. M. T.	Pawlowsk		Stonyhurst		Kew		Wilhelmshaven		Potsdam		Val Joyeux	
	P'_h	P'_d	P'_h	P'_d	P'_h	P'_d	P'_h	P'_d	P'_h	P'_d	P'_h	P'_d
h m												
16 0	0	0	0	0	0	0	0	E 4.2 γ	0	0	0	0
15	+ 12.1 γ	W 2.3 γ	0	E 11.4 γ	− 1.5 γ	E 10.3 γ	+ 3.3 γ	» 16.5 »	+ 6.0 γ	E 12.7 γ	0	E 14.7 γ
30	+ 32.2 »	E 34.5 »	− 5.1 γ	» 31.4 »	− 4.1 »	» 26.9 »	+ 8.0 »	» 45.0 »	+ 12.0 »	» 36.6 »	0	» 31.8 »
45	+ 27.3 »	» 15.6 »	+ 2.5 »	» 44.2 »	− 3.0 »	» 36.5 »	+ 24.0 »	» 50.7 »	+ 21.8 »	» 38.6 »	+ 7.2 γ	» 24.3 »
17 0	+ 11.6 »	» 12.9 »	+ 9.7 »	» 29.4 »	+ 8.6 »	» 25.7 »	+ 13.0 »	» 25.7 »	+ 7.9 »	» 22.3 »	− 2.4 »	» 22.6 »
15	0	» 6.0 »	+ 1.5 »	» 20.0 »	+ 2.0 »	» 15.2 »	− 1.8 »	» 14.7 »	− 2.2 »	» 11.7 »	0	» 10.0 »
30	+ 2.0 »	» 6.0 »	0	» 12.8 »	0	» 10.1 »	− 7.0 »	» 9.2 »	− 3.1 »	» 11.2 »	0	» 12.6 »

TABLE XXVIII (continued).

Gr. M. T.	Munich		Pola		San Fernando		Tiflis		Dehra Dun		Batavia	
	P'_h	P'_d	P'_h	P'_d	P'_h	P'_d	P'_h	P'_d	P'_h	P'_d	P'_h	P'_d
h m												
16 0	0	E 5.2 γ	0	0	0	0	0	0	0	W 2.0 γ	0	W 3.0 γ
15	+ 2.0 γ	» 19.8 »	+ 3.1 γ	E 6.9 γ	0	E 6.1 γ	+ 18.1 γ	0	0	» 2.0 »	0	» 3.6 »
30	+ 4.5 »	» 36.5 »	+ 6.7 »	» 25.7 »	0	» 18.0 »	+ 20.1 »	E 9.2 γ	+ 24.3 γ	» 6.9 »	+ 22.0 γ	» 7.2 »
45	+ 11.5 »	» 27.4 »	+ 12.6 »	» 26.4 »	+ 3.7 γ	» 22.1 »	+ 12.2 »	» 4.8 »	+ 24.3 »	» 4.9 »	+ 21.4 »	» 6.0 »
17 0	+ 13.0 »	» 21.7 »	+ 10.3 »	» 17.4 »	+ 5.2 »	» 18.5 »	+ 2.6 »	» 3.3 »	+ 14.6 »	0	+ 12.8 »	» 3.0 »
15	+ 4.5 »	» 10.6 »	+ 2.7 »	» 11.1 »	0	» 11.1 »	+ 0.8 »	» 1.1 »	+ 4.3 »	» 4.9 »	+ 5.3 »	» 2.4 »
30	0	» 9.1 »	0	» 7.6 »	0	» 10.7 »	0	0	0	» 3.0 »	0	» 3.0 »

Current-Arrows for the 15th February, 1903, Chart I at 14^h 30^m; and Chart II at 15^h.

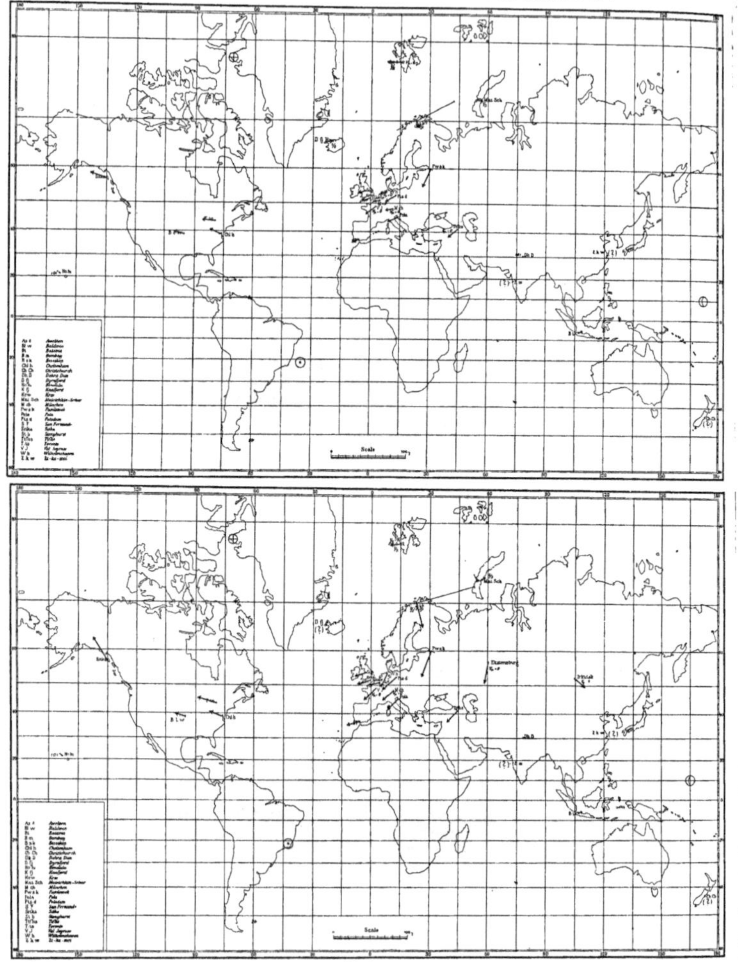

Fig. 82.

Current-Arrows for the 15th February, 1903; Chart III at 16^h, and Chart IV at 16^h 15^m.

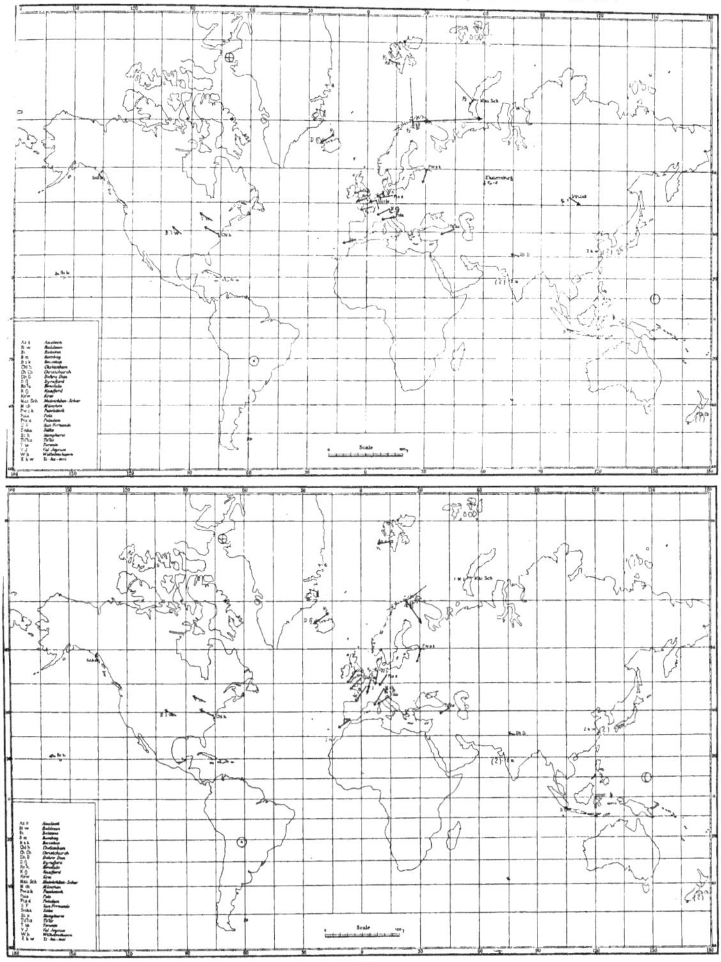

Fig. 83.

at 16^{h} 45^{m}.

Current-Arrows for the 15th February, 1903; Chart VII at 17^h, and Chart VIII at $17^h\ 15^m$.

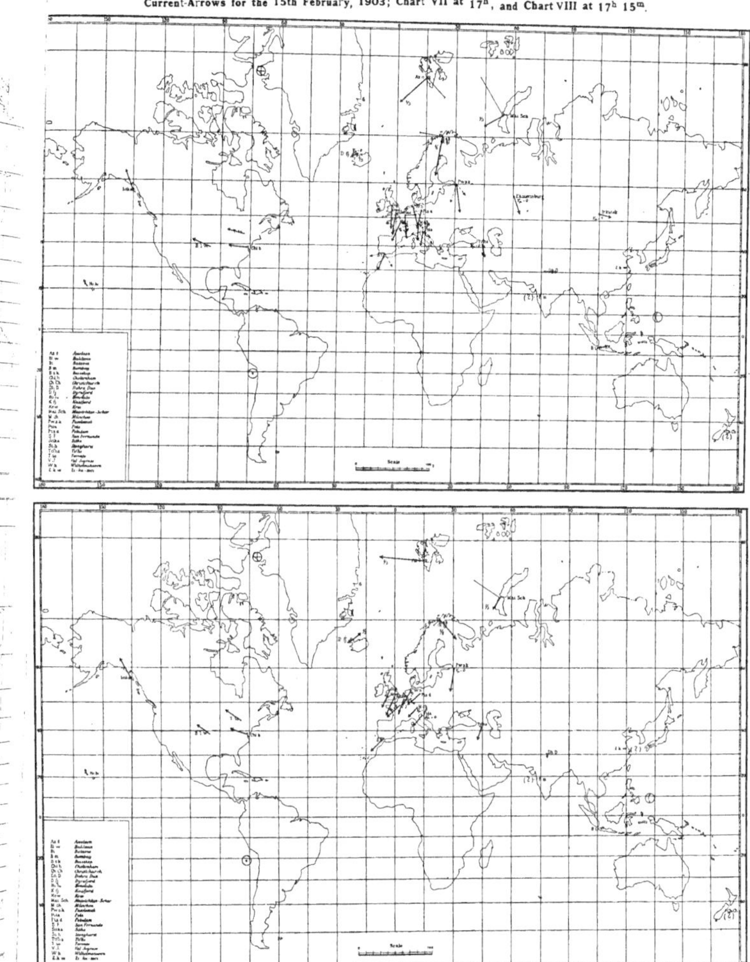

Fig. 85.

Current-Arrows for the 15th February, 1903; Chart IX at $17^h\ 30^m$, and Chart X at $17^h\ 45^m$.

Fig. 86.

THE PERTURBATIONS OF THE 7th & 8th FEBRUARY, 1903.

(Pl. XVI & XVII).

59. The storms now to be described, some of them powerful ones, break in upon a very long period of calm, which may be said to have lasted with single exceptions since the cessation of the storms at the end of November, 1902.

This interruption of the quiet conditions occurs suddenly at the Norwegian stations with a fairly powerful storm, commencing at $21^h 5^m$, on the 7th February, and lasting, at Kaafjord, until about 1 a. m. on the 8th Februay.

The first perturbation on the 7th does not belong to the series of perturbations mentioned in the circular, and our material is therefore not sufficiently complete to allow of our investigating it more fully in southern latitudes. As it happens, however, registerings for this date have also been received from a few stations in addition to the Norwegian stations, namely from Kew, Wilhelmshaven, Munich, Toronto and Christchurch. Judging from the conditions at these places, we here have a typical polar elementary storm, with its centre near the Norwegian stations.

This storm is not succeeded by calm, however. Towards morning on the following day, there are varying precipitations about the auroral zone. Between 2^h and 5^h for instance, there are powerful storms round Axeløen; and they are also very powerful in Toronto. In southern latitudes too, there is constant disturbance as time passes.

From 9^h to 11^h on the 8th there is a perturbation that is especially powerful at Sitka and the American station, and is accompanied by simultaneous perturbations all over the northern hemisphere and over the southern right down to Christchurch.

Commencing with this perturbation, we will study the conditions more carefully, although in the first place it is the powerful polar storm, with a maximum at about $19^h 25^m$ on the same day, to which we have especially turned our attention, and which is given in the circular.

As we must confine ourselves to a study of the chief features of the perturbations, we shall here mainly give our attention to three periods of time, in which the perturbations are particulary powerful. It will easily be seen from the conditions at Sitka that a division such as this is the natural one, the three sections being:

(1) the above-mentioned perturbation from 9^h to 11^h,

(2) a perturbation between 14^h and 18^h, and

(3) the period from 18^h to 23^h.

The curves for the second and third periods are shown on the same plate, those for the first being separate.

THE PERTURBING FORCES.

60. The *first section* (Pl. XVI).

The perturbation is particularly powerful at Sitka, and is especially violent from 9^h to $9^h 35^m$. Simultaneously at the other stations in the New World, there are fairly powerful perturbations; and we see directly from the curves that the conditions vary greatly from place to place. We shall find, for instance, a considerable difference if we compare the H-curves for the three stations, Toronto, Cheltenham and Baldwin. At Toronto there is a long, rather powerful perturbation, as also at Cheltenham, both showing a diminution in H. At Baldwin, on the other hand, H remains almost normal, if anything a little too great during the perturbation. At Honolulu there is a faint but distinct perturbation in declination, coinciding with the perturbation farther north. In H too, there is some resemblance in the form of the curve to that of the declination-curves for the three eastern stations in North America, as a comparison with the declination-curve for Cheltenham will at once show. A peculiarity is now apparent, however, inasmuch as the normal line lies in such a position that while the perturbation is at its height, H is almost normal. At one of the Norwegian stations, Kaafjord, the perturbation is only just perceptible, the reason of this probably being that only at Kaafjord are the conditions so quiet that the comparatively slight effect is observable. At Axeløen and Dyrafjord. the conditions are very disturbed before and after. This disturbed condition is also observable in southern latitudes, and is instrumental in making this perturbation less clearly defined.

It is the conditions in southern latitudes in Europe and Asia that contribute to make those of this period especially worthy of remark. A very well-defined perturbation makes its appearance there in H, with a simple course. The force gradually increases to a maximum, after which it once more diminishes to zero. Throughout this district, the deflection represents a diminution in H.

The table below shows the hour at which the perturbation commences and terminates, and that at which the maximum is reached, as also the value of P_1 at the last named hour.

TABLE XXIX.

Observatory	Commences	Reaches max.	P_1 max.	Terminates
	h m	h m		h m
Sitka	8 45	9 16,5	123 γ	11 0
Toronto	9 0([1])	$9^h 15^m - 10^h 0^m$	41 γ—39 γ	indeterminable
Baldwin	9 0([1])	9 15 — 10 0	37 »—30.5 »	indeterminable
Cheltenham	9 0([1])	9 15 — 10 0	27 »—29 »	indeterminable
Wilhelmshaven	8 33	10 8	35 »	
Kew	8 34	10 0	30.7 »	10 49
Val Joyeux	8 36	10 0	30.0 »	10 48
San Fernando	8 35	10 2	29.0 »	10 50
Munich	8 38	10 7	27.0 »	10 51
Pola	8 35	10 0	24.4 »	10 47
Dehra Dun	8 33	10 6	22.8 »	10 48
Batavia	8 38	10 10	22.5 »	10 52
Zi-ka-wei	8 38	10 5	22.5 »	10 52

([1]) The time of the commencement is here taken from the D-curve.

It will be seen, that the conditions at Sitka are rather peculiar as regards the course of the perturbation. The three stations in the east of North America come nearest to Sitka. The simple conditions found between San Fernando in the west and Zi-ka-wei in the east, and between Kew in the north and Batavia in the south, form a strong contrast to these variable conditions. In the latter district, the perturbation is throughout chiefly in H. It is well defined, and as far as we can determine, commences everywhere simultaneously at about $8^h\ 35^m$. The maximum is not very distinct, but the time of its occurence nevertheless does not vary greatly. It terminates simultaneously at about $10^h\ 50^m$. As the force is practically constant for several minutes about the maximum, P_1 max. will represent simultaneous perturbing forces. The strength, it is true, is throughout somewhat greater in Europe than in the Asiatic district; but nevertheless, between Kew and Zi-ka-wei and Batavia it does not vary more than from about 30.7 γ to 22.5 γ. This time the force is comparatively great at Wilhelmshaven too, a circumstance that may be due to local conditions.

The conditions are represented in three charts for the hours $9^h\ 15^m$, $9^h\ 36^m$ and 10^h.

From 9^h to $9^h\ 30^m$ at Sitka, there is a great current-arrow directed almost due south, as shown on Chart I. Subsequently the current-arrow becomes smaller and is directed westwards along the auroral zone. This condition continues from $9^h\ 30^m$ to the conclusion of the perturbation.

In the United States, the conditions are fairly uniform all the time. The current-arrows show a great convergence of the perturbing force.

Owing to the above-mentioned similarity between the form of the curve at Honolulu and that at the three eastern American stations, we may conclude that this polar storm must have an effect in Honolulu. It is impossible to take out any decided values; but a glance at the curve will show that the effect consists in a perturbing force directed towards the north-east. The current-arrow, inasmuch as it is dependent upon the polar system, thus comes to be directed towards the south-east. In this way the force at Honolulu completes the area of convergence.

In the above-mentioned equatorial district on the eastern hemisphere, the forces are directed along the magnetic parallels.

With regard to the wiew to be taken of this perturbation, it may in the first place be considered probable that the conditions in the north of America are mainly determined by a polar elementary storm at first not very far north-east of Sitka. The centre afterwards travels westwards. It may be remarked that during the perturbation this district passes midnight. The current-arrow about the centre is probably directed westwards along the auroral zone. The storm is in the main of a character similar to those that usually occur a little before midnight, with their centre near our Norwegian stations, and almost always travelling eastwards.

As regards the simultaneous perturbation over the district between Kew and Batavia, it seems impossible, both on account of the form of the field and of the magnitude of the force, that this storm can be a direct effect of the polar system. On the other hand, the field must immediately suggest the thought of the current round the earth as the cause of the perturbation. Some doubt may be felt on this hand owing to the disturbing influence occasioned by the polar storm in the western hemisphere. We have previously mentioned conditions, however, especially as regards Honolulu, which indicate that there two systems appear simultaneous in H, counteracting at one another. The polar system, from the form of the curve, must be assumed to act in a northerly direction, when the other must act i a southerly direction in order to compensate the former, in which case the conditions in Honolulu should be in accordance with those in the eastern hemisphere.

According to this, it is not improbable that simultaneously with the polar storm there is a perturbation answering to a current round the earth from east to west, a perturbation of the type we have called negative equatorial storms. Owing to the slight variation of the force from place to place, and to the uniform course of the perturbation, this current may be assumed to lie at a distance from the earth of at least a magnitude equal to the radius of the earth; and symmetry would point to the regions round the plane of the magnetic equator as its situation.

The main features in the form of the field may thus be explained, as we have seen, fairly simply in the above manner. If we look at the charts, however, we see, that the field bears an unmistakable resemblance to those that we should expect to find during the cyclo-median storms. Under such an assumption, the perturbing forces that appear at Sitka at about $9^h\ 15^m$ also receive quite a simple explanation. It is only necessary to refer to the photographs of the terrella, when, if we compare the light-area in fig. 68, 1 with our field, we find the resemblance is striking, if we imagine Sitka as being near the uppermost angle. If we then imagine the field moved westwards with the sun, we have more or less the conditions of Charts II and III. The arrow at Christchurch on Chart II is worthy of notice. It answers to that part of the light-area that falls upon the southern hemisphere; and the direction of the arrow is also in accordance with what we should expect to find if the system on Chart I were moved westwards. There may well be some doubt as to the view to be taken of the conditions. Perhaps the most probable is that at first the perturbation partakes most of the nature of a cyclo-median storm, and subsequently changes into a more purely polar one.

The *second section*, from $14^h\ 0^m$ to 18^h (Pl. XVII).

(a) *The conditions in northern latitudes.*

At Dyrafjord, beginning at $13^h\ 40^m$, there is a rather long, not violent, but still considerable perturbation, which acts principally upon *H*, tending to increase it. This condition lasts until the commencement of the violent storm about $18^h\ 35^m$, and is continued for some time after the conclusion of the latter at $22^h\ 15^m$.

At Kaafjord the conditions are more variable, giving almost the impression of two separate storms, the first with maximum at $14^h\ 45^m$, the second lasting from $15^h\ 30^m$ until the commencement of the great storm. All three elements are here about equally disturbed, *H* however most.

At Axeloen the conditions assume the nature of a fairly long perturbation, which maintains more or less the same character from $14^h\ 0^m$ until the commencement of the great storm. The perturbation is strongest between 14^h and 15^h, and at about $18^h\ 0^m$.

At Matotchkin Schar the conditions between 14^h and the commencement of the great storm, are very variable. They very much resemble those at Kaafjord. There is first a very well defined storm between $13^h\ 45^m$ and $15^h\ 15^m$, with maximum at $14^h\ 35^m$, after which, in the course of a few minutes, comparative calmness, and then once more the storm leaps up with oscillations principally in the same direction as during the first part of the perturbation.

In connection with these conditions at the Norwegian stations, we will examine those at Sitka. Here the perturbation is particularly powerful from $14^h\ 24^m$ to $15^h\ 45^m$, the maximum being at $14^h\ 45^m$. Thus this storm commences during the same period of time, and has its maximum at the same hour as the first powerful impulse, which was especially well defined at Kaafjord and Matotchkin Schar. We find, however, that on the whole it apears somewhat later at Sitka. After this first powerful storm there is comparative quiet, and then once more a slight perturbation appears, principally affecting *H*, and lasting from $16^h\ 30^m$ to 18^h.

(b) *The conditions in lower latitudes.*

In Europe the conditions assume the character of a lengthy perturbation, which begins to be particularly perceptible at about $13^h\ 45^m$. In declination the conditions vary a good deal, the curve being now above, now below, the normal line. In the horizontal intensity the conditions remain more constant. All the time, until the powerful storm commences, there is an oscillation in *H*, answering to a diminution there, this condition being also continued after the cessation of the powerful storm, and lasting until past midnight. Here too we notice a particularly powerful perturbation with maximum at $14^h\ 42^m$. This augmentation occurs at the same time as the previously-mentioned, particularly powerful storm at the northern stations. This characterisation of the conditions is also applicable to Tiflis, and indeed, especially as regards *H*, also to the district from Dehra Dun to Batavia.

At Dehra Dun there is quite a powerful perturbation in *H*. Here too *H* remains on the whole below the normal, right up to the commencement of the great storm; and this condition continues after the latter has ceased. In declination, especially as regards Dehra Dun, there are small oscillations towards the east.

At Christchurch also, perturbations occur throughout the period under consideration. In *H* the conditions here are nearly the reverse of those at Dehra Dun, as *H* throughout has too large a value. The already-mentioned perturbation with maximum at $14^h\ 45^m$ is very marked here too, and is quite powerful both in *H* and in *D*, and quite distinct even in *V*. Here too its maximum is at $14^h\ 45^m$; but it is of shorter duration than in the northern hemisphere.

There is some disturbance in the United States, but strange to say no particularly well defined oscillations such as at Sitka and the European stations.

At Honolulu the conditions are very quiet, with the exception of the period about $14^h\ 45^m$. If we look at the *H*-curve about the time mentioned, we shall find some similarity between its course here and at Christchurch, a similarity which may lead to the bringing of the perturbations here and at Christchurch into connection with one another.

The field during this second section is given on three charts (IV, V and VI).
Chart IV represents the conditions at $14^h\ 45^m$,
" *V* at three hours, viz. $16^h\ 10^m$, 17^h, and $17^h\ 30^m$, and
VI at $18^h\ 0^m$.

As we see from the curves, the perturbations within this period cannot be regarded as consisting mainly of a single perturbation, but as a series of short, principally polar impulses with somewhat changing centre.

Axeløen occupies a peculiar position, the perturbing force there remaining throughout fairly constant both in magnitude and direction. The conditions here do not in any way resemble those at the other Norwegian stations, the force at Axeløen being almost equally strong, but opposite in direction, and the current-arrow principally directed towards the west. The conditions at Axeløen, moreover, show an entirely independent course, in which there is nothing answering to the successive maxima and minima that we notice, for instance, at Kaafjord.

On Chart IV, for $14^h\ 45^m$, we find at the three southernmost Norwegian stations, current-arrows of considerable strength directed eastwards along the auroral zone. In Europe and the west of Asia, there is now a corresponding area of divergence. At Sitka there is a fairly strong current-arrow directed towards the north-west; and at the same time, the other American stations indicate that there is an area of convergence. It would appear from the form of this area that we had before us the effect of polar precipitation with the storm-centre a little to the west of Sitka, that is to say in a district situated on the night-side. The direction of the current-arrows round this district must then be westerly.

The field as it appears on this chart thus seems to be somewhat complicated, but the form is not an unknown one. If we compare these conditions with those, for instance, shown on Charts IV and V for $16^h\ 45^m$ and 17^h on the 9th December, 1902 (p. 75), we find that the resemblance is striking. The time, moreover, should be noted at which these two storms commence. The conditions remain more or less constant throughout this period, the changes consisting principally only in a certain amount of variation in the strength of the forces, but little in their direction, so that the form of the field is not essentially changed, at any rate in higher latitudes. The changes that do occur can all be accounted for by the translocation of the systems. The period extends, as we have said, from 14^h to 18^h and we thus here too find a resemblance to the 9th December.

In the preceding perturbation on the 15th February, we also found exactly analogous conditions at these stations during the first two sections. There does not seem to be any essential difference between the fields on these two days, the only ones being that on the present occasion the stormcentre, with its eastward-pointing arrows at the more southerly Norwegian stations, stands out more distinctly, and that the system extends farther east than in the preceding storm. The current-arrows are also stronger, and the area of divergence is more distinct.

The resemblance between the fields is so great that it is impossible to regard it as chance; and we involuntarily receive the impression that the field before us is possibly typical of the polar storms that appear at this time of day, just as we have previously found the typical form of the field that forms about midnight, Greenwich time. In what way, in my opinion, the field is to be understood has been already indicated in the description of the preceding storm, and I will therefore only refer the reader to it.

In the perturbations that follow, we shall moreover have an opportunity of studying the fields that form at this time of day; and we shall see that conditions similar to those that we have here pointed out will be continually repeated.

At the three hours shown on Chart V—16^h 10^m, 17^h, and 17^h 30^m—we also find on the whole the same conditions as at 14^h 45^m, the only difference, besides a diminution in the strength of the forces, being a change in the direction of the arrows in the eastern hemisphere, as if the precipitation on the day-side were moving westwards with the sun. The change, however, also may be due only to the diminution in the strength of this system upon the night-side.

We have previously mentioned that the curve for the field now under discussion gives the impression of several relatively independent systems succeeding one another. In this case therefore, it would perhaps be natural to consider the one system as vanishing, and new systems being formed, in such a manner that they advance towards the west. The curves for Dyrafjord seem perhaps to make such an assumption of *new* systems doubtful, as the conditions there remain fairly constant. The movement may also be explained by the assumption that the night-system moves westwards, and little by little destroys the effect of the eastern part of the day-system.

The conditions at Sitka and Honolulu indicate, though only faintly, an area of convergence answering to a precipitation on the night-side. At Baldwin, Cheltenham and Toronto, there is a very small force. It appears, from investigations of the material from 1882—83, that systems on the night-side have a west-ward motion. The reason why the forces in eastern America are so small in the present instance, may therefore possibly be that the storm-centre has now moved too far away. This, moreover, is in accordance with the fact that its effect in Europe becomes more noticeable.

On Chart IV, for 18^h 0^m, the same conditions continue at Axeløen. Matotchkin Schar also seems now to be mainly influenced by this precipitation.

At Dyrafjord the force is now particulary strong, and the current-arrow is still directed towards the east. It seems to be this precipitation on the day-side, which now lies farther west that especially gives to the field in lower latitudes its character, as there is here an area of divergence. At Kaafjord the force is smaller, but seems mainly to be determined by the precipitation at Dyrafjord.

The *third section*, from 18^h to 23^h.

We have already, in the preceding section, had an opportunity of observing that the powerful storm breaks in upon one of long duration. This we found to be the case both at the Norwegian stations and, on the whole, at stations in the eastern hemisphere. This is a well-known circumstance, and we will only refer to the perturbation of the 15th February. With the same reason as on that day, we can, by drawing a normal line that forms a harmonious connection between the conditions before and after, obtain a more exact determination of the perturbation, in so far as it is dependent upon the powerful polar storm. It will be in the main for the horizontal component as the perturbations in D at most places seem to be chiefly connected with the polar system.

(a) *The conditions at the Norwegian stations.*

The violent storm is powerful at all the four Norwegian stations simultaneously, most powerful at Axeløen and Matotchkin Schar. It is very varied in its details, but the oscillations retain in the main an uniformity of direction.

At Dyrafjord the powerful storm commences at 18^h 33^m, and is over at 22^h 17^m. After this time, perturbations still appear for a time; but they are principally in accordance with the conditions beforehand. The perturbation is at its height between 19^h 8^m and 20^h 14^m. At about 20^h 37^m the oscillations are relatively very small, both in declination and in horizontal intensity, while they remain very powerful in *V*. The oscillations in *H* and *D*, however, immediately become stronger again.

At Kaafjord the storm becomes powerful at $19^h\ 5^m$, with a deflection that is particularly marked in declination. It does not become really great in *H* until $19^h\ 22^m$. At about $21^h\ 41^m$ the conditions are quiet for a time, after which there is only a very slight perturbation; and at $22^h\ 40^m$ comparative calm has supervened. In all the three curves the deflections are uniform in direction all the time, and towards the side that is typical for these powerful polar storms. The deflection in the *V*-curve is particularly marked.

At Axeløen we also get an impression that the storm makes its appearance while other disturbances are taking place. The actuel storm begins here very decidedly at $19^h\ 7^m$. It suddenly increases, and ten minutes later it is at its height. Right on to $21^h\ 0^m$, it continues very violent; but from that time until its close at $22^h\ 33^m$ there is only a small perturbation.

At Matotchkin Schar the powerful storm is of longer duration than at the other stations. In *H* it sets in with considerable strength as early as $18^h\ 37^m$, and in the *D*-curve at $18^h\ 58^m$. The perturbation principally affects the *H*-curve, where it lasts until $22^h\ 21^m$. Considering the violence of the storm, the oscillations in the *D*-curve are very small and variable. What is especially remarkable is that the perturbation throughout has so little effect upon *V*. It does, it is true, generally decrease *V*; but the oscillations are not great and sometimes to the opposite side of the mean line.

The oscillations at the Norwegian stations, with the exception of those in declination at Dyrafjord, which are deflected towards the west, have the directions characteristic of those storms, which occur before midnight at the Norwegian stations, and are powerful and of short duration.

(b) *The conditions in southern latitudes.*

Simultaneously with the storm in the north, a powerful perturbation is noticed on the continent of Europe. It is especially powerful after $19^h\ 5^m$, and increases in the course of a few minutes to a maximum, which occurs at $19^h\ 18^m$. At $20^h\ 34^m$ it is once more comparatively slight, and at $22^h\ 48^m$ it ceases in declination, although it still continues for a long time in *H*.

At Potsdam, and still more at Pawlowsk, there is a well-defined perturbation in *V*. The deflection is always in one direction, and answers to a diminution of *V*.

At Munich a small deviation from the normal is just perceptible. Here, too, *V* becomes less.

At Pola there is a greater effect in *V*, and principally on the opposite side.

The conditions at Tiflis form the transition to those at Dehra Dun and the Asiatic district. On the one side they very much resemble those farther north in Europe; but on the other hand, the variation in the *H*-curve at Tiflis exhibits a close correspondance to the variations in the district between Dehra Dun, Zi-ka-wei, and Batavia, which exactly correspond with these in the storm in the auroral zone. We notice, for instance, the sudden great change that took place in *H* about $19^h\ 5^m$, indicating that the polar storm at the Norwegian stations makes its appearance at this hour. We here find conditions that justify a decomposition of the perturbing force. We will in the first place remark that there are variations in *H*, which in the main closely correspond with simultaneous variations in the perturbation-conditions at the Norwegian stations. We find, for instance, at $19^h\ 6^m$, a sudden change in the *H*-curve, *H* having risen, in the course of twenty minutes, from a value that is $14\,\gamma$ below the normal, to its highest value, which is $28\,\gamma$ above the normal. The oscillation then decreases a little in strength, and then once more increases, attaining a new maximum at $20^h\ 10^m$. The perturbation then gradually decreases, and about $20^h\ 40^m$, the *H*-curve coincides with the normal line. In the course of an hour, the horizontal intensity has become almost normal, and continues to decrease, remaining below the normal until far into the night. There is, as we see, an oscillation which actually accompanies more or less simultaneously the storm in the north; and in order to bring out the conditions that belong to these storms, we must, if possible, consider as an effect of the polar storm the deviations from the conditions before and after the period in which the polar storm occurs. In this way the conditions are certainly elucidated, as will best be seen when we come to consider the field of force. If we look at the total force as belonging to the polar storms, we here find a change in the direction of the force that has no parallel farther north, where, as we shall see, it remains almost constant in direction throughout the perturbation.

At Christchurch too, there is a very considerable and well-defined perturbation, which is particularly well developed in *H*, and exhibits a course that in the main resembles that at Dehra Dun, but has perhaps a still greater resemblance to those in North America.

In the western hemisphere we also find simultaneous considerable perturbations, which are especially powerful at Sitka, but also of no little strength in the United States; while even at Honolulu there is a very considerable effect on that day.

We will first consider the four northernmost stations.

In the *H*-curve, in particular, the course of the perturbation exactly corresponds with that at our Norwegian stations. It commences with some strength at about $19^h\ 0^m$, increases rather rapidly to a maximum, and remains fairly powerful for about an hour, after which it diminishes, but then once more increases somewhat, and forms a new, secondary maximum at $21^h\ 30^m$. We have then first a powerful maximum and then a weaker one—a condition we observed at all the Norwegian stations. In declination, on the other hand, the conditions here are somewhat peculiar. A perturbation appears at the three stations in the east of North America, at $17^h\ 56^m$, answering to a deflection westwards, and remains, excepting for a short interval when the polar storm is at its height, almost constant for several hours, only ceasing at about $23^h\ 0^m$. Whatever this deflection may be due to, we must assume that it cannot be the effect of the system we are now considering, as this does not begin to act until more than an hour later.

At Honolulu a distinct variation is noticed especially in the *H*-curve, coinciding with the polar storm; but on drawing the mean line, it appears that there are perturbations both before and after. Before, *H* is greater than the normal, while after, it has a value that, is considerably below the normal.

The field during the powerful storm is shown on ten charts. The first represents the conditions at 19^h, the last at $22^h\ 30^m$. In southern latitudes a decomposition of forces has been effected on the charts from $19^h\ 15^m$ to $21^h\ 30^m$, but at the Norwegian stations and Sitka this has not been done. At the latter places the powerful storm is so dominant that the total forces are principally conditioned by the powerful polar storm. The field at these northermost stations remains, as we see, fairly constant in its form throughout. At the Norwegian stations the current-arrows on the whole are directed westwards along the auroral zone.

At Dyrafjord the current-arrows at first have the very usual direction, WSW (see the chart for $19^h\ 15^m$), but afterwards turn northwards, and remain almost the whole time pointing towards the west, or even farther towards the north. The vertical component of the perturbing force is directed upwards all the time.

At Axeløen and Kaafjord we have the field that is typical of these storms. The current-arrows are almost parallel—except at about $19^h\ 15^m$—, and WSW in direction. The horizontal component of the perturbing force is greatest at Axeløen; but on the other hand, the vertical component at Kaafjord is greater throughout, and is directed upwards at this station, and downwards at the former. At about $19^h\ 15^m$ a peculiarity makes its appearance at Kaafjord, namely, that the horizontal component becomes about 0, while at the same time the vertical is very powerful. To explain this, it is natural to conclude that there is a local perturbation at Kaafjord of contrary effect. Sharp local deflections such as these are very frequent in these regions. This impression is also confirmed by a study of the copies of the curves.

At Matotchkin Schar the current-arrow maintains the characteristic direction, making oscillations about the main direction.

Up to the chart for 20^h, the force is almost as strong at Dyrafjord as at Matotchkin Schar; but on the next chart, that for 20^h 15^m, the field in the north shows that the storm-centre has moved eastwards. The force at Matotchkin Schar has increased, while that at Dyrafjord has diminished. At the same time the current-arrows for Axeløen and Kaafjord have acquired a distinct divergence.

In southern latitudes the field is decomposed. The dotted arrows represent the field as it is before and after the polar storm. As regards this field, we will only state that it has on the whole the same character as that in the previously-mentioned perturbation from 9^h to 11^h. The current-arrow in the eastern hemisphere is directed westwards, and that in the United states towards NNW.

That which here especially interests us, however, is the field in so far as it is connected with the storm in the north. The current-arrows to represent this force are drawn with broken lines. The field, as we see, may be characterised in a few words by referring to the previously-described polar elementary storms e. g. of the 15th December, 1902, and the 10th February and the 22nd March, 1903. This holds good, at any rate during the time when the storm is at its height, and the perturbing forces can be most accurately determined. There is a distinctly-marked area of convergence in the eastern hemisphere, and a distinct area of divergence in the western. In Europe the direction of the current-arrows is at first south-west; but between 19^h 15^m and 19^h 46^m, they turn a little counter-clockwise. They then, however, turn back, a turning that is in accordance with the eastward movement of the field, which we deduced from the conditions at the Norwegian stations. Simultaneously with this, there is also a clockwise motion of the arrow at Sitka.

Although the conditions in the main are similar to those found during the usual polar elementary storms that appear at this time of day, there are also certain deviations from the typical conditions. The force in Europe, for instance, at about 20^h and 20^h 30^m, seems to be comparatively small, while at Sitka at the same time it is comparatively great, and turns, as we have said, in a positive direction. The distribution of force cannot here be explained by the assumption of a single elementary system. The comparatively great force at Sitka indicates that there is a simultaneous precipitation on the day-side; and it seems as if in Europe at this time—20^h 0^m—there are possibly two systems counteracting one another.

We will look more closely into this peculiar variableness of the conditions in Central Europe.

While the direction varies greatly from place to place, the force is small. There is no doubt that the direction of the force, especially during lengthy perturbations, becomes uncertain, when the absolute value of the force is small, as the unavoidable error in the placing of the mean line with small forces will have a great influence; but nevertheless when we look at the curves, there is a very noticeable change from place to place. This difference is especially evident in the *H*-curve. Here there are three types of curves; the first is found at the stations Stonyhurst, Kew and Val Joyeux, the second at Wilhelmshaven and Potsdam, and the third at Munich and Pola. Within each type the form of the curve is very similar.

It has been already said that this storm exhibits many points of resemblance to the storm of the 15th February, 1903, and in this respect also, there is now a complete accordance between the two days. On that day also the *H*-curve showed exactly similar differences in the European field; and the stations were separated into exactly the same three groups, a circumstance which strongly confirms our opinion that this is not a chance resemblance.

TABLE XXX.

The Perturbing Forces on the 8th February, 1903.

Gr. M. T.	Honolulu		Sitka		Baldwin		Toronto		Cheltenham	
	P_h	P_d	P_h	P_d	P_h	P_d	P_h	P_d	P_h	P_d
h m										
9 15	+ 3.5 γ	E 7.5 γ	+74.3 γ	E 97.8 γ	+ 5.1 γ	W 36.9 γ	−28.4 γ	W 29.5 γ	− 6.3 γ	W 26.1 γ
36	+ 1.5 „	„ 4.1 „	−48.4 „	„ 18.0 „	+ 1.4 „	„ 22.9 „	−28.4 „	„ 22.9 „	− 3.7 „	„ 25.5 „
10 0	+ 2.3 „	„ 5.0 „	−39.5 „	„ 22.6 „	− 1.1 „	„ 30.5 „	−24.7 „	„ 30.2 „	− 7.3 „	„ 28.5 „
14 45	+ 7.1 „	W 7.4 „	−54.8 „	W 34.7 „	− 7.9 „	„ 12.7 „	−14.4 „	„ 16.3 „	− 6.3 „	„ 8.9 „
16 10	0	0	− 1.0 „	E 14.0 „	+ 9.7 „	E 3.7 „	+ 1.8 „	0	+ 4.0 „	„ 2.3 „
17 0	+ 2.0 „	„ 8.3 „	−13.1 „	W 5.4 „	+ 1.8 „	„ 5.7 „	− 0.9 „	E 4.8 „	− 2.3 „	E 2.3 „
30	+ 4.1 „	„ 12.4 „	−11.7 „	„ 10.4 „	− 1.8 „	0	− 3.6 „	0	− 7.9 „	„ 3.0 „
18 0	+ 7.9 „	„ 12.4 „	+ 1.8 „	„ 6.3 „	+ 1.4 „	W 4.4 „	− 4.0 „	W 10.2 „	− 5.2 „	W 5.3 „
19 0	+ 1.8 „	„ 10.0 „	−14.2 „	„ 13.5 „	−10.7 „	„ 12.7 „	− 9.0 „	„ 16.3 „	− 9.2 „	„ 11.3 „
15	− 5.4 „	„ 10.0 „	−23.5 „	„ 6.3 „	−21.5 „	„ 11.4 „	−15.3 „	„ 15.6 „	−22.4 „	„ 13.1 „
30	−13.0 „	„ 10.8 „	−40.0 „	„ 19.0 „	−38.4 „	„ 14.0 „	−38.6 „	„ 17.5 „	−42.6 „	„ 14.3 „
45	−15.8 „	„ 11.6 „	−41.5 „	„ 13.5 „	−37.0 „	„ 13.3 „	−29.7 „	„ 11.4 „	−40.4 „	„ 7.1 „
20 0	−16.9 „	„ 10.0 „	−45.1 „	„ 22.5 „	−42.0 „	„ 1.9 „	−34.2 „	E 15.1 „	−43.0 „	E 14.8 „
15	−15.1 „	„ 10.8 „	−42.5 „	„ 47.4 „	−37.0 „	„ 5.7 „	−25.2 „	„ 4.8 „	−37.5 „	„ 1.8 „
30	−10.1 „	„ 8.3 „	−26.2 „	„ 60.0 „	−27.2 „	„ 11.4 „	−17.1 „	W 6.0 „	−28.3 „	W 7.1 „
21 0	− 8.2 „	„ 1.6 „	−15.1 „	„ 32.5 „	−15.1 „	„ 14.6 „	− 4.0 „	„ 9.6 „	−16.0 „	„ 11.9 „
30	−10.0 „	E 2.5 „	−18.1 „	„ 34.3 „	−21.2 „	„ 18.4 „	−10.3 „	„ 18.6 „	−21.0 „	„ 17.8 „
22 0	− 9.2 „	„ 5.8 „	−11.7 „	„ 23.4 „	−19.4 „	„ 16.5 „	− 8.1 „	„ 13.2 „	−16.0 „	„ 11.9 „
30	−11.5 „	„ 5.8 „	− 3.9 „	„ 2.7 „	−12.5 „	„ 14.0 „	− 1.3 „	„ 10.8 „	− 8.7 „	„ 10.1 „

TABLE XXX (continued).

Gr. M. T.	Dyrafjord			Axeløen			Matotchkin-Schar			Kaafjord		
	P_h	P_d	P_v	P_h	P_d	P_v	P_h	P_d	P_v	P_h	P_d	P_v
h m												
9 15	− 13.7 γ	W 20.5 γ	− 58.6 γ	+ 55.1 γ	E 15.8 γ	+ 44.2 γ	?	?	?	− 13.5 γ	E 8.4 γ	− 3.9 γ
36	0	„ 11.4 „	− 21.6 „	+ 42.3 „	„ 28.8 „	+ 22.0 „	?	?	?	− 7.3 „	„ 9.5 „	0
10 0	+ 23.1 „	„ 28.5 „	− 21.6 „	0	„ 29.4 „	− 36.8 „	?	?	?	− 9.2 „	„ 13.2 „	+ 13.8 „
14 45	+ 77.0 „	E 12.1 „	+ 15.5 „	− 88.0 „	W 50.8 „	− 19.6 „	+134.0 γ	W 58.0 γ	− 47.7 γ	+145.0 „	W 43.8 „	+ 60.3 „
16 10	+118.0 „	„ 2.7 „	− 8.0 „	− 57.0 „	„ 31.8 „	+ 14.7 „	+ 87.0 „	„ 69.2 „	+183.0 „	+ 78.0 „	„ 15.0 „	+ 50.4 „
17 0	+107.0 „	„ 2.7 „	− 1.9 „	− 61.1 „	„ 21.2 „	0	+133.0 „	E 7.1 „	+165.0 „	+ 70.5 „	0	+ 62.0 „
30	+ 47.8 „	„ 6.2 „	+ 6.6 „	−108.0 „	„ 24.7 „	+ 37.0 „	+ 75.5 „	„ 5.3 „	− 5.9 „	+ 30.0 „	E 18.7 „	+ 64.0 „
18 0	+ 97.0 „	„ 5.2 „	+ 13.6 „	−129.0 „	„ 12.8 „	+118.0 „	− 5.4 „	„ 26.6 „	− 89.3 „	+ 30.0 „	„ 11.0 „	+ 28.2 „
19 0	− 35.7 „	W 26.1 „	−134.0 „	− 47.5 „	„ 54.8 „	+ 22.1 „	−223.0 „	„ 144.0 „	− 57.0 „	+ 23.9 „	W 44.7 „	− 7.0 „
15	−262.0 „	E 21.5 „	−213.0 „	−509.0 „	E 59.0 „	+211.0 „	−294.0 „	„ 136.0 „	−282.0 „	+ 48.5 „	E 52.5 „	−190.0 „
30	−226.0 „	W 96.1 „	−188.0 „	−428.0 „	„ 121.0 „	+211.0 „	−282.0 „	„ 30.6 „	−106.0 „	−226.0 „	0	−242.0 „
45	−175.0 „	„ 163.0 „	−258.0 „	−480.0 „	„ 116.0 „	+492.0 „	−292.0 „	„ 81.0 „	− 71.5 „	−282.0 „	„ 37.0 „	−196.0 „
20 0	−199.0 „	„ 195.0 „	− 96.0 „	−297.0 „	„ 38.9 „	+334.0 „	−330.0 „	„ 207.0 „	− 85.2 „	−346.0 „	„ 86.5 „	−282.0 „
15	− 59.3 „	„ 102.0 „	−108.0 „	−299.0 „	0	+354.0 „	−292.0 „	„ 172.0 „	−134.0 „	−288.0 „	„ 141.0 „	−355.0 „
30	− 7.7 „	„ 64.9 „	−148.0 „	−230.0 „	„ 104.0 „	+302.0 „	−221.0 „	„ 44.3 „	−136.0 „	−182.0 „	„ 87.5 „	−253.0 „
21 0	− 66.5 „	„ 50.7 „	−199.0 „	−110.0 „	0	+343.0 „	−252.0 „	„ 190.0 „	− 87.0 „	− 77.2 „	„ 29.0 „	−232.0 „
30	− 94.0 „	„ 50.7 „	− 71.0 „	− 46.5 „	W 38.9 „	+208.0 „	−142.0 „	„ 160.0 „	+ 22.2 „	−128.0 „	„ 104.0 „	− 48.5 „
22 0	+ 36.3 „	„ 42.3 „	−159.0 „	− 71.5 „	E 9.8 „	+187.0 „	− 18.7 „	„ 67.0 „	− 51.1 „	− 54.0 „	„ 30.5 „	− 69.5 „
30	+ 72.0 „	E 12.8 „	− 92.0 „	− 21.2 „	0	+145.0 „	+ 22.0 „	„ 20.3 „	− 11.9 „	− 2.4 „	„ 19.4 „	− 54.0 „

TABLE XXX (continued).

Gr. M. T.	Pawlowsk			Stonyhurst		Kew		Val Joyeux		
	P_h	P_d	P_v	P_h	P_d	P_h	P_d	P_h	P_d	P_v
h m										Two small deviations about 19h 20m and 20h, but nothing can be taken out.
9 15	?	?	?	?	?	−20.9 γ	E 1.9 γ	−16.0 γ	0	
36	?	?	?	?	?	−26.4 „	0	−27.2 „	0	
10 0	?	?	?	?	?	−32.5 „	W 4.7 „	−30.4 „	0	
14 45	−34.7 γ	W 6.4 γ	+ 6.7 γ	−17.3 γ	W 24.0 γ	−15.8 „	„ 10.7 „	−26.4 „	W 8.4 γ	
16 10	− 7.0 „	„ 18.0 „	+ 1.0 „	0	0	− 3.0 „	„ 4.2 „	−10.4 „	„ 7.5 „	
17 0	−11.5 „	0	+ 1.0 „	−10.7 „	0	−11.7 „	0	−15.2 „	0	
30	− 1.5 „	E 9.6 „	0	−13.3 „	E 8.6 „	−10.2 „	E 13.1 „	−11.2 „	E 11.7 „	
18 0	− 7.5 „	0	0	− 4.1 „	„ 1.1 „	− 8.7 „	„ 8.4 „	− 4.8 „	„ 6.7 „	
19 0	−26.6 „	W 2.3 „	+ 3.0 „	− 9.7 „	W 10.9 „	−12.8 „	W 6.1 „	−14.4 „	W 4.2 „	
15	+20.2 „	E 39.1 „	+ 1.5 „	−11.7 „	E 45.1 „	−24.0 „	E 30.9 „	−10.4 „	E 20.8 „	
30	+30.6 „	„ 10.6 „	− 9.0 „	+25.4 „	„ 49.7 „	+19.4 „	„ 52.9 „	+12.8 „	„ 62.0 „	
45	+25.1 „	0	−16.4 „	+19.4 „	„ 39.4 „	+16.8 „	„ 41.1 „	+23.2 „	„ 43.5 „	
20 0	− 7.0 „	„ 16.6 „	−20.2 „	+ 5.1 „	„ 8.6 „	+ 4.1 „	„ 15.9 „	+20.0 „	„ 20.8 „	
15	+10.6 „	„ 32.6 „	−26.2 „	− 8.1 „	„ 47.5 „	−11.7 „	„ 44.9 „	− 2.4 „	„ 50.2 „	
30	− 8.0 „	„ 30.8 „	−23.9 „	−15.3 „	„ 31.4 „	−18.8 „	„ 33.6 „	−12.0 „	„ 35.2 „	
21 0	−21.6 „	„ 15.2 „	−15.7 „	−11.7 „	„ 9.1 „	−16.8 „	„ 14.5 „	−13.6 „	„ 19.2 „	
30	−21.6 „	„ 34.5 „	−11.2 „	−20.5 „	„ 11.4 „	−25.5 „	„ 14.5 „	−25.6 „	„ 20.9 „	
22 0	−18.6 „	„ 19.8 „	− 3.7 „	−13.7 „	„ 8.6 „	−18.3 „	„ 9.3 „	−18.4 „	„ 15.1 „	
30	−12.6 „	„ 10.5 „	− 1.5 „	−12.2 „	„ 2.8 „	−14.3 „	„ 5.1 „	−18.4 „	„ 10.9 „	

TABLE XXX (continued).

Gr. M. T.	Wilhelmshaven			Potsdam			San Fernando	
	P_h	P_d	P_v	P_h	P_d	P_v	P_h	P_d
h m								
9 15	−23.7 γ	0	0	?	?	?	−17.0 γ	W 1.6 γ
36	−19.6 „	E 1.2 γ	0	?	?	?	−17.0 „	„ 8.2 „
10 0	−36.0 „	0	0	?	?	?	−26.4 „	„ 12.3 „
14 45	−41.1 „	W 30.6 „	0	−36.0 γ	W 16.3 γ	+ 3.6 γ	−26.4 „	„ 21.3 „
16 10	−13.1 „	„ 14.1 „	+ 1.0 γ	−10.7 „	„ 11.2 „	0	− 8 9 „	„ 9.8 „
17 0	−18.2 „	„ 1.2 „	+ 1.0 „	−15.8 „	E 2.5 „	0	−14.8 „	0
30	−12.1 „	E 16.5 „	+ 3.0 „	− 7.6 „	„ 14.2 „	− 0.6 „	−14.8 „	E 1.6 „
18 0	− 8.9 „	„ 3.6 „	+ 2.0 „	−12.3 „	„ 4.0 „	+ 0.6 „	−17.0 „	W 3.3 „
19 0	−25.2 „	W 9.2 „	+ 2.0 „	−23.3 „	W 4.0 „	+ 2.7 „	−25.2 „	„ 8.2 „
15	− 3.7 „	E 62.3 „	+ 6.0 „	− 2.5 „	E 45.7 „	+ 0.6 „	−26.6 „	E 13.1 „
30	+57.0 „	„ 66.0 „	+12.0 „	+35.1 „	„ 39.6 „	− 4.5 „	0	„ 27.8 „
45	+24.7 „	„ 39.1 „	+ 4.0 „	+24.3 „	„ 29.5 „	− 3.6 „	+11.8 „	„ 24.6 „
20 0	−13.5 „	„ 8.6 „	− 2.0 „	− 8.5 „	„ 13.2 „	0	− 1.5 „	„ 13.9 „
15	+ 3.7 „	„ 57.5 „	+ 3.0 „	+ 3.5 „	„ 46.6 „	− 5.4 „	−13.3 „	„ 18.8 „
30	−17.2 „	„ 37.3 „	0	−14.8 „	„ 30.5 „	− 2.7 „	−22.2 „	„ 4.1 „
21 0	−24.7 „	„ 10.4 „	0	−20.5 „	„ 11.7 „	− 2.1 „	−25.2 „	0
30	−32.2 „	„ 19.6 „	0	−32.2 „	„ 20.3 „	0	−32.5 „	0
22 0	−22.9 „	„ 12.2 „	0	−24.6 „	„ 12.2 „	0	−25.8 „	0
30	−19.6 „	„ 8.6 „	0	−22.7 „	„ 8.1 „	0	−22.3 „	0

TABLE XXX (continued).

Gr. M. T.	Munich			Pola			Tiflis			Dehra Dun	
	P_h	P_d	P_v	P_h	P_d	P_v	P_h	P_d	P_v	P_h	P_d
h m											
9 15	− 11.5 γ	0	0	− 11.5 γ	0	0	?	?	No copy received.	− 15.7 γ	W 11.8 γ
36	− 11.5 „	0	0	− 18.8 „	W 0.7 γ	0	?	?		− 17.3 „	„ 5.9 „
10 0	− 22.5 „	0	0	− 24.2 „	0	0	?	?		− 21.2 „	„ 3.0 „
14 45	− 23.5 „	„ 12.9 γ	0	− 32.2 „	„ 16.0 „	− 5.1 γ	− 33.2 γ	E 5.9 γ		− 14.5 „	E 15.8 „
16 10	− 10.5 „	„ 7.6 „	0	− 12.6 „	„ 9.9 „	− 1.9 „	− 17.4 „	W 7.8 „		− 12.2 „	„ 3.9 „
17 0	− 13.0 „	0	0	− 15.3 „	0	+ 1.7 „	− 14.8 „	E 4.8 „		+ 1.9 „	„ 7.9 „
30	− 9.5 „	E 9.1 „	0	− 9.4 „	E 6.9 „	+ 1.9 „	− 7.7 „	„ 7.8 „		+ 6.3 „	„ 3.0 „
18 0	− 5.0 „	„ 4.6 „	0	− 9.8 „	„ 3.5 „	0	− 7.7 „	„ 3.3 „		0	„ 4.9 „
19 0	− 19.0 „	W 8.4 „	0	− 21.1 „	W 2.8 „	+ 2.9 „	− 25.0 „	„ 10.8 „		− 13.4 „	„ 13.8 „
15	− 9.5 „	E 16.8 „	0	− 13.9 „	E 29.2 „	+ 11.9 „	− 14.6 „	„ 10.8 „		+ 9.0 „	„ 6.9 „
30	+ 22.0 „	„ 41.2 „	0	+ 16.6 „	„ 29.9 „	+ 3.8 „	+ 24.3 „	„ 10.8 „		+ 26.7 „	„ 1.0 „
45	+ 23.0 „	„ 28.2 „	+ 0.7 γ	+ 19.7 „	„ 22.9 „	− 0.8 „	+ 21.7 „	„ 5.2 „		+ 18.5 „	„ 3.0 „
20 0	+ 11.0 „	„ 7.6 „	+ 1.5 „	+ 6.7 „	„ 22.9 „	+ 2.7 „	+ 11.3 „	„ 6.3 „		+ 16.9 „	„ 1.0 „
15	+ 4.0 „	„ 37.4 „	+ 1.5 „	+ 2.2 „	„ 35.4 „	+ 10.6 „	+ 13.9 „	„ 17.8 „		+ 17.7 „	„ 4.9 „
30	− 6.0 „	„ 29.7 „	+ 1.5 „	− 9.4 „	„ 27.1 „	− 1.9 „	0	„ 18.6 „		+ 6.7 „	„ 10.8 „
21 0	− 12.0 „	„ 12.2 „	+ 1.1 „	− 14.8 „	„ 13.9 „	− 1.2 „	− 10.6 „	„ 14.9 „		0	„ 11.8 „
30	− 23.5 „	„ 17.5 „	0	− 24.6 „	„ 18.1 „	0	− 14.6 „	„ 22.3 „		0	„ 12.8 „
22 0	− 19.0 „	„ 9.9 „	0	− 19.7 „	„ 13.2 „	0	− 18.1 „	„ 13.4 „		− 7.1 „	„ 8.8 „
30	− 18.0 „	„ 5.3 „	0	− 19.2 „	„ 6.9 „	0	− 18.5 „	„ 6.7 „		− 8.6 „	„ 4.9 „

TABLE XXX (continued).

Gr. M. T.	Zi-ka-wei			Batavia		Christchurch			Ekaterinburg		
	P_h	P_d	P_v	P_h	P_d	P_h	P_d	P_v	P_h	P_d	P_v
h m											
9 15	− 19.2 γ	W 5.0 γ	No remarkable desturbance.	− 22.1 γ	0	0	W 32.0 γ	− 1.5 γ	?	?	?
36	− 19.2 „	0		− 18.1 „	0	+ 5.9 γ	„ 3.7 „	0	?	?	?
10 0	− 26.4 „	0		− 13.2 „	0	+ 7.6 „	„ 17.1 „	0	?	?	?
14 45	+ 1.2 „	E 8.0 „		− 3.2 „	W 7.2 γ	+ 19.2 „	E 20.8 „	− 2.8 „	?	?	?
16 10	− 15.6 „	0		− 13.5 „	E 4.8 „	+ 8.7 „	W 11.9 „	0	?	?	?
17 0	+ 8.4 „	0		+ 4.6 „	0	+ 3.3 „	E 8.9 „	0	?	?	?
30	+ 3.6 „	0		+ 4.2 „	0	+ 3.1 „	„ 8.1 „	+ 0.6 „	+ 5.7 γ	E 33.1 γ	+ 3.7 γ
18 0	0	„ 1.0 „		0	„ 2.4 „	+ 6.7 „	W 1.5 „	+ 1.2 „	+ 4.5 „	„ 32.2 „	0
19 0	− 10.8 „	„ 4.0 „		− 9.6 „	„ 2.4 „	+ 13.4 „	„ 3.7 „	+ 1.5 „	− 21.0 „	„ 44.5 „	+ 1.3 „
15	0	„ 1.0 „		− 3.9 „	„ 8.4 „	− 8.5 „	0	+ 0.9 „	− 18.7 „	„ 44.5 „	− 3.2 „
30	+ 6.0 „	W 3.0 „		+ 9.9 „	„ 13.2 „	− 17.8 „	E 8.9 „	0	0	„ 38.2 „	− 9.5 „
45	+ 6.0 „	„ 5.0 „		+ 9.2 „	„ 6.0 „	− 24.1 „	„ 7.4 „	0	+ 20.0 „	„ 28.5 „	− 15.0 „
20 0	+ 7.2 „	„ 5.0 „		+ 8.9 „	„ 1.2 „	− 25.4 „	„ 8.9 „	0	+ 26.0 „	„ 24.9 „	− 17.4 „
15	+ 8.4 „	„ 2.0 „		+ 10.7 „	„ 1.2 „	− 14.3 „	„ 9.6 „	0	+ 22.0 „	„ 26.8 „	− 15.7 „
30	+ 2.4 „	E 2.0 „		+ 4.6 „	„ 3.6 „	− 4.9 „	„ 9.6 „	+ 0.9 „	+ 14.0 „	„ 35.0 „	− 13.9 „
21 0	0	„ 5.0 „		− 1.0 „	„ 3.6 „	− 4.9 „	„ 3.7 „	+ 2.8 „	+ 4.2 „	„ 44.8 „	− 11.2 „
30	0	„ 5.0 „		0	„ 3.6 „	− 5.3 „	„ 11.1 „	+ 3.7 „	− 1.3 „	„ 42.0 „	− 10.0 „
22 0	− 6.2 „	„ 3.0 „		− 5.3 „	„ 2.4 „	0	„ 14.1 „	+ 4.3 „	− 5.2 „	„ 37.7 „	− 8.7 „
30	− 7.2 „	„ 3.0 „		− 7.8 „	„ 2.4 „	?	?	?	− 7.7 „	„ 25.2 „	− 5.0 „

TABLE XXXI.

Partial Perturbing Forces on the 8th February, 1903.

Gr. M. T.	Honolulu		Baldwin		Toronto		Cheltenham	
	P'_h	P'_d	P'_h	P'_d	P'_h	P'_d	P'_h	P'_d
h m								
19 15	− 5.9 γ	0	− 12.2 γ	0	− 7.2 γ	W 3.0 γ	− 12.1 γ	0
30	− 11.5 „	0	− 30.0 „	0	− 30.2 „	„ 5.4 „	− 31.5 „	0
45	− 11.8 „	0	− 27.2 „	0	− 23.0 „	0	− 28.0 „	E 7.1 γ
20 0	− 11.8 „	0	− 33.0 „	E 12.1 γ	− 27.8 „	E 27.1 „	− 29.7 „	„ 28.5 „
15	− 8.9 „	W 5.0 γ	− 28.2 „	„ 5.7 „	− 19.4 „	„ 16.8 „	− 23.8 „	„ 14.3 „
30	− 4.1 „	„ 2.5 „	− 18.6 „	0	− 9.9 „	0	− 13.9 „	„ 4.8 „
21 0	0	E 2.5 „	− 7.5 „	W 1.9 „	0	0	− 3.1 „	0
30	− 1.0 „	„ 5.0 „	− 12.2 „	„ 6.3 „	− 8.1 „	W 12.0 „	− 8.3 „	W 5.9 „
22 0	+ 1.0 „	„ 5.8 „	− 11.1 „	„ 6.3 „	− 8.1 „	„ 7.2 „	− 5.0 „	„ 2.9 „
30	0	„ 6.6 „	− 5.4 „	„ 5.1 „	− 2.2 „	„ 7.2 „	0	„ 2.9 „

TABLE XXXI (continued).

Gr. M. T.	Pawlowsk		Stonyhurst		Wilhelmshaven		Kew	
	P'_h	P'_d	P'_h	P'_d	P'_h	P'_d	P'_h	P'_d
h m								
19 15	+ 39.2 γ	E 39.1 γ	− 9.7 γ	E 45.1 γ	+ 17.7 γ	E 62.3 γ	− 9.7 γ	E 30.9 γ
30	+ 48.2 „	„ 10.6 „	+ 41.2 „	„ 49.7 „	+ 80.3 „	„ 66.0 „	+ 34.2 „	„ 52.9 „
45	+ 44.2 „	0	+ 28.5 „	„ 39.4 „	+ 47.5 „	„ 39.1 „	+ 32.6 „	„ 41.1 „
20 0	+ 13.6 „	„ 16.6 „	+ 11.7 „	„ 8.6 „	+ 3.7 „	„ 8.5 „	+ 14.2 „	„ 15.9 „
15	+ 28.2 „	„ 32.6 „	0	„ 47.5 „	+ 24.7 „	„ 57.5 „	+ 3.0 „	„ 44.9 „
30	+ 16.6 „	„ 30.8 „	− 7.1 „	„ 31.4 „	+ 2.3 „	„ 37.3 „	− 3.0 „	„ 33.6 „
21 0	− 10.6 „	„ 15.2 „	− 6.6 „	„ 9.1 „	− 6.5 „	„ 10.4 „	− 3.5 „	„ 14.5 „
30	− 6.5 „	„ 34.5 „	− 16.3 „	„ 11.4 „	− 14.0 „	„ 19.6 „	− 14.8 „	„ 14.5 „
22 0	− 7.0 „	„ 19.8 „	− 8.1 „	„ 8.6 „	− 7.0 „	„ 12.2 „	− 9.2 „	„ 9.3 „
30	− 3.5 „	„ 10.6 „	− 6.6 „	„ 2.8 „	− 2.8 „	„ 8.6 „	− 6.1 „	„ 5.1 „

TABLE XXXI (continued).

Gr. M. T.	Potsdam		Val Joyeux		Munich		Pola	
	P'_h	P'_d	P'_h	P'_d	P'_h	P'_d	P'_h	P'_d
h m								
19 15	+ 23.0 γ	E 45.7 γ	0	E 20.8 γ	+ 6.5 γ	E 16.8 γ	+ 6.7 γ	E 29.2 γ
30	+ 61.2 „	„ 39.6 „	+ 28.7 γ	„ 62.0 „	+ 39.5 „	„ 41.2 „	+ 39.3 „	„ 29.9 „
45	+ 48.8 „	„ 29.5 „	+ 36.8 „	„ 43.5 „	+ 40.5 „	„ 28.2 „	+ 40.7 „	„ 22.9 „
20 0	+ 14.5 „	„ 13.2 „	+ 29.5 „	„ 20.8 „	+ 30.5 „	„ 7.6 „	+ 27.3 „	„ 22.9 „
15	+ 27.4 „	„ 46.6 „	+ 10.4 „	„ 50.2 „	+ 20.5 „	„ 37.4 „	+ 23.7 „	„ 35.4 „
30	+ 7.6 „	„ 30.5 „	0	„ 35.2 „	+ 10.0 „	„ 29.7 „	+ 11.2 „	„ 27.1 „
21 0	0	„ 11.7 „	0	„ 19.2 „	+ 2.5 „	„ 12.2 „	+ 4.0 „	„ 13.9 „
30	− 11.7 „	„ 20.3 „	− 12.8 „	„ 20.9 „	− 8.5 „	„ 17.5 „	− 6.7 „	„ 18.1 „
22 0	− 5.3 „	„ 12.2 „	− 8.0 „	„ 15.1 „	− 6.5 „	„ 9.9 „	− 4.4 „	„ 13.2 „
30	− 5.3 „	„ 8.1 „	− 8.0 „	„ 10.9 „	− 6.5 „	„ 5.3 „	− 4.4 „	„ 6.9 „

TABLE XXXI (continued).

Gr. M. T.	San Fernando		Tiflis		Dehra Dun		Zi-ka-wei		Batavia	
	P'_h	P'	P'_h	P'_d	P'_h	I'_d	P'_h	P'_d	P'_h	P'_d
h m										
19 15	0	E 13.1 γ	+ 7.2 γ	E 10.8 γ	+ 13.8 γ	W 7.9 γ	+ 14.4 γ	W 5.0 γ	+ 8.5 γ	E 8.4 γ
30	+ 25.8 γ	" 27.8 "	+ 44.1 "	" 10.8 "	+ 41.0 "	" 14.8 "	+ 19.2 "	" 12.0 "	+ 22.8 "	" 13.2 "
45	+ 35.5 "	" 24.6 "	+ 42.8 "	" 5.2 "	+ 32.7 "	" 9.8 "	+ 21.6 "	" 15.0 "	+ 22.8 "	" 6.0 "
20 0	+ 22.2 "	" 13.9 "	+ 32.1 "	" 6.3 "	+ 31.5 "	" 9.8 "	+ 22.8 "	" 15.0 "	+ 22.8 "	" 1.2 "
15	+ 10.4 "	" 18.8 "	+ 34.1 "	" 17.8 "	+ 31.5 "	" 4.9 "	+ 22.8 "	" 10.0 "	+ 24.2 "	" 1.2 "
30	0	" 4.1 "	+ 20.7 "	" 18.6 "	+ 19.3 "	0	+ 15.6 "	" 3.0 "	+ 18.8 "	" 3.6 "
21 0	− 2.2 "	0	+ 7.5 "	" 14.9 "	+ 11.8 "	0	+ 15.6 "	0	+ 14.6 "	" 3.6 "
30	− 12.6 "	0	+ 3.1 "	" 22.3 "	+ 7.9 "	0	+ 10.8 "	0	+ 11.4 "	" 3.6 "
22 0	− 10.4 "	0	− 0.8 "	" 13.4 "	+ 3.1 "	0	+ 10.8 "	0	+ 5.7 "	" 2.4 "
30	− 8.1 "	0	− 2.2 "	" 6.7 "	0	0	+ 3.6 "	0	+ 4.6 "	" 2.4 "

Current-Arrows for the 8th February, 1903; Chart I at $9^h 15^m$.

Fig. 87.

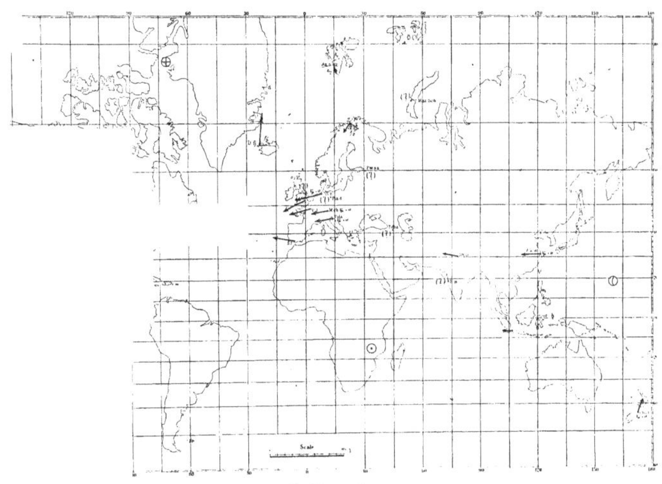

Fig. 88.

ora Polaris Expedition. 1902—1903.

ws for the 8th February, 1903; Chart IV at 14h 45m, and Chart V at 16

Current-Arrows for the 8th February, 1903; Chart VI at $18^h\ 0^m$, and Chart VII at $19^h\ 0^m$.

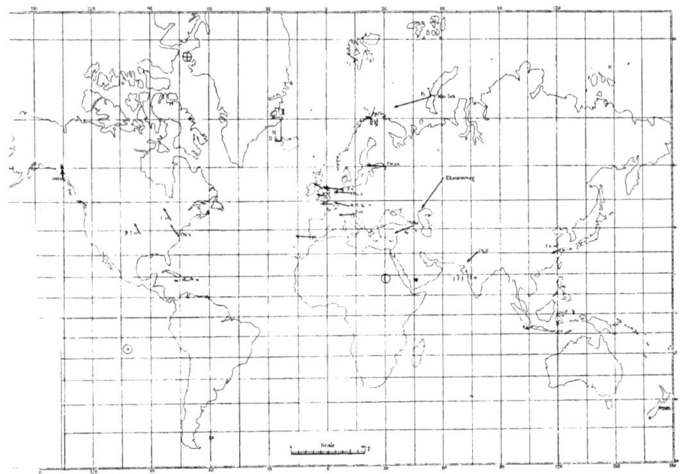

Current Arrows for the 8th February, 1903; Chart VIII at $19^h\ 15^m$, and Chart IX at 19^h 30

Current-Arrows for 8th February, 1903; Chart X at $19^h\ 45^m$. and Chart XI at $20^h\ 0^m$.

Fig. 92.

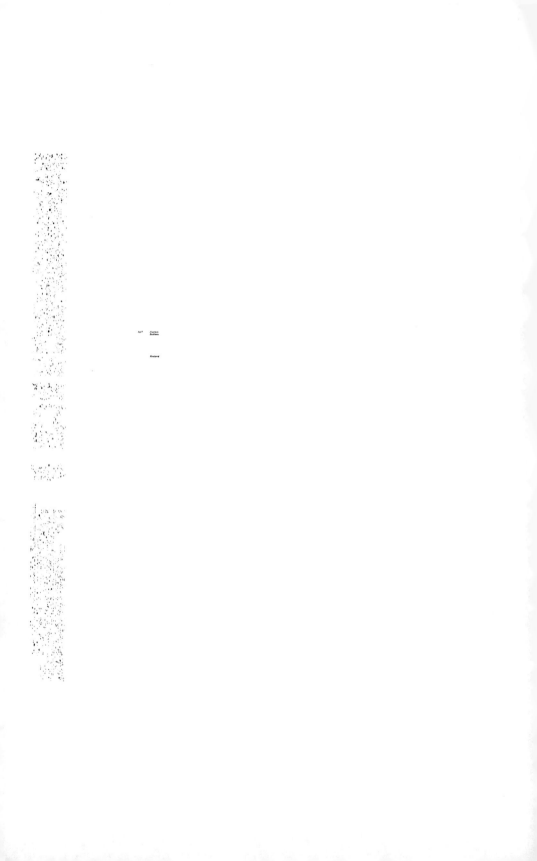

urrent-Arrows for the 8th February, 1903; Chart XIV at $21^h\ 0^m$, and Chart XV at $21^h\ 30^m$,

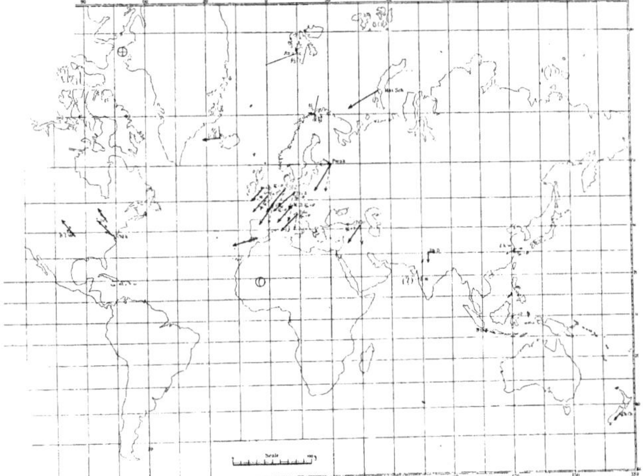

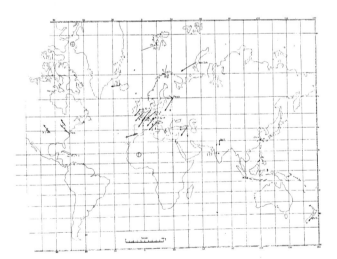

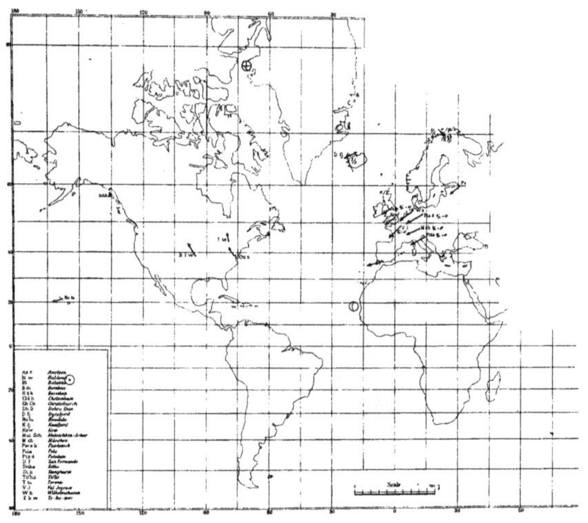
Scale

THE PERTURBATIONS OF THE 27th & 28th OCTOBER, 1902.

(Pl. IV).

61. Throughout the first half of October, there was calm as far as our arctic stations were concerned. About the 24th, however, a violent storm takes place, lasting from about 5 hours before midnight Gr. M. T. until 4 hours after. During the succeeding days, perturbations of more or less strength occur, beginning late in the evening and attaining their highest development at about midnight. As day advances, there is once more calm, but the storm returns again before midnight. This condition of things continues, and culminates in the violent storms about the 31st. From some of the stations there is included a characteristic equatorial perturbation, occurring on the 29th and 30th. This perturbation is already described Art. 54.

The time occupied by the perturbations of the 27th and 28th October is from 14^h on the 27th until about 1^h on the 28th, the curve for this period being shown on Plate IV.

At the arctic stations, the character of the conditions is that of two separate storms, one of which occurs early in the afternoon, with its maximum about 16^h. This is fairly powerful at Axeløen, while at the other Norwegian stations it is comparatively less so. The other storm is at its height at about 22^h to 23^h, and is a well-defined, fairly powerful perturbation, lasting about three hours.

In southern latitudes, the direct impression of the conditions of this perturbation is to some extent quite different. We will take, for instance, the condition at Tiflis, a station that occupies an intermediate position between the polar and the equatorial regions, and where we are therefore likely to find conditions that are characteristic of both. Here the perturbations last much longer. Even earlier than noon, there are perturbations indicating the presence of a perturbing force directed northwards. At about 13^h the force turns round, the perturbation appearing also distinctly in declination, where it is directed eastwards. With the exception of one intermediate storm, this state of affairs lasts until $20^h\ 24^m$. The interruption lasts from $15^h\ 24^m$ to $16^h\ 54^m$, and thus coincides in time with the already-mentioned perturbation in the north. The same thing is found at Dehra Dun and Batavia, but there the perturbation is chiefly in H.

Finally, from $21^h\ 40^m$ until about midnight there is a perturbation that occurs simultaneously, and is in connection with the perturbation round the Norwegian stations. It is most powerful at our Norwegian stations, but in southern latitudes it is much less than the perturbation that occurred earlier. In this way, the treatment of the perturbation falls naturally into two sections, the first from 14^h to $20^h\ 30^m$, and the second from $21^h\ 40^m$ until about midnight.

THE DISTRIBUTION OF FORCE.

62. The *first section.* $14^h - 20^h\ 30^m$.

The perturbation during this period is especially worthy of remark from its being particularly powerful at the equator, in the regions about Dehra Dun and Batavia.

While these comparatively powerful perturbations are taking place at the equator, there are also storms round the auroral zone. We see, on the other hand, that the effect in America increases towards Sitka, where there are two distinct maxima during this period. One of them coincides with the already-mentioned intermediate storm and occurs between $15^h\ 30^m$ and $17^h\ 15^m$. This is preceded by a powerful perturbation lasting from 13^h to $14^h\ 45^m$.

From this it would appear that this part of the perturbation shows, to some extent at any rate, the effect of polar systems, which this time seem to keep, in some measure, fairly near the regions to the north of Sitka.

There is much resemblance between this first section of the perturbation and that of the whole on the 15th February, which is worthy of notice, and is immediately apparent on looking at the curves. They also both occur at about the same time of day.

At Sitka the two perturbations resemble one another also in detail. On both days the conditions are those of two separate perturbations, each of about the same duration and following the same course, and each with a well-defined maximum. The chief difference is that the perturbation of the 15th February occurs on the whole about 40 minutes later in the day. The resemblance extends still farther, for about three hours before this perturbation, there are on both days two fairly powerful and well-defined, but brief perturbations; but the perturbation occurring at about midnight on the 27th October has no parallel on the 15th February.

The resemblance is not, however confined to Sitka. Both in Europe and India, the conditions exhibit surprising points of similarity. If we look, for instance, at the curves for Tiflis, we find on both days a long perturbation answering to a perturbing force towards the south and south-east. This is interrupted by another perturbation of short duration, which represents a perturbing force directed towards the north-east; and in both cases this occurs simultaneously with the latter of the two almost separate storms at Sitka.

The curves for the Norwegian stations also exhibit some similarity. At Axeloen there is the distinct effect of the system that forms the first perturbation at Sitka from 13^h to $14^h\,45^m$; this however possibly does not appear so distinctly from the copied curves, as these first begin at that time when the perturbation has reached its maximum. After this perturbation the intermediate storm commences with a strength, which relative to the preceding storm and to the storms on the other Norwegian stations, forms a good analogy to that taking place on the 15th February, 1903.

The perturbation of the 15th February has already been described at length, and most of the remarks there made with regard to the theory of the perturbation may be applied to the present case: On the whole also we find a good correspondence with the conditions for the 8th February but the details that day are here not quite so striking resemblant as on the 15th.

As on the 15th February, the distribution of force before and after the intermediate storm is about the same. This section of the perturbation therefore divides into two parts,

(1) the long storm, and

(2) the brief, intermediate storm.

The field during the long storm is shown on *Charts I, II and III* at 14^h, 15^h and $15^h\,30^m$ and after the intermediate storm on *Chart VII* at 17^h. Here too, it shows as a whole the very same conditions as the field on the 15th February.

The current-arrow at Kaafjord and at Matotchkin Schar is directed eastwards on the whole, while that at Axeloen is directed westwards. Also the same conditions which we have found (see p. 191) with the previously described storms which appear at this time of day. Farther south in Europe, the current-arrows also point in a westward direction. There is also the remarkable circumstance that the force increases southwards from Stonyhurst and Kew. At Pawlowsk, the force before the intermediate storm is almost insensible, whereas in the district between Tiflis and Batavia it is very strong, and strongest of all at Dehra Dun. In the United States the direction of the current-arrow is NNW. At Sitka the current-arrow has a typical direction, north-west. At Honolulu the conditions are very quiet during the whole twenty-four hours.

It thus appears that the strong effect found in the south of Asia is not limited to those regions only, but does not extend round the equator. We see that as on the 15th February, North America and Europe constitute an area of divergence of the perturbing force. The neutral point should be situated in a region not far from Stonyhurst. Whether there is an area of convergence on the other side of the world, we cannot say, as there is no material from those regions.

The *intermediate* storm, like the corresponding one on the 15th February, is particulary powerful at Axeløen and Matotchkin Schar, and probable less so at Kaafjord as far as we can see from the curve, which at this time has disappeared from the magnetogram-paper. The current that conditions the perturbation seems therefore now be near our north-eastern stations. The duration of this storm is also about the same. In Central Europe and southwards to Batavia, its commencement and termination are well characterised. It occurs between 15^h 30^m and 16^h 45^m. The corresponding storm on the 15th February lasted from 16^h 15^m to 17^h 45^m.

In the eastern hemisphere a decomposition has been undertaken, the result being shown on *Charts IV, V, and VI*, at 16^h, 16^h 20^m, and 16^h 30^m respectively.

Throughout the western hemisphere, with the exception of Sitka, the perturbation is somewhat less powerful than in the eastern. The effect in the United States is principally noticeable in H, showing that the current-arrow for the intermediate storm would be directed westwards. As these however are very small, we have not marked them on the charts, but only drawn the current-arrows corresponding to the total force. The eastern field in this storm is of about the same form and proportional strength as that of the 15th February. The current-arrow in Europe points south-east, and turns off towards the east through southern Asia. As Zi-ka-wei it even goes a little north, so that there is a good indication that the current-lines here form an en entire circle, as they return in the regions round the Norwegian stations, where the arrows are directed westwards along the auroral zone. On the western hemisphere, on the other hand, there is certainly an area of divergence, with, it appears a weaker perturbing force. The field in the intermediate storm is thus of the same character as that found in the polar elementary storms. This also applies to the northern stations.

At Matotchkin Schar and Axeløen there is a powerful perturbation with current-arrows directed westwards. The vertical intensity at Matotchkin Schar is very great, and is directed upwards; at Axeløen the balance moves up and down about its mean position. At first P_v is directed downwards, but in less than a quarter of an hour it has changed, and is directed upwards, after which it changes once more. There is the same variableness in P_v on the 15th February, but on that occasion it begins by being directed upwards. At Kaafjord, both now and on the 15th February, the conditions are more in accordance with those in southern latitudes, the arrow being directed towards the south-east. The circumstance of the current-arrow at Kaafjord having almost the opposite direction to those at the two north-eastern stations, is also found on the 15th February, and its probable explanation we assumed, in the description of that perturbation, that there was a precipitation on the day side.

For this storm there are unfortunately no registerings from Dyrafjord; they would have been of very great significance.

The *second section*. 21^h 40^m — about midnight.

The polar storm from 21^h 40^m to about midnight is very powerful round the Norwegian stations. Its beginning and end are fairly distinct; it is well defined and simple in its course. This time, too, the changes in the perturbation are most rapid at Axeløen, where the conditions on the whole are more disturbed. This storm manifests itself by simultaneously-occurring perturbations, that are observable all over the northern hemisphere. The table below gives the time at which the storm begins, reaches its maximum, and ends, as also the maximum value of P_1.

TABLE XXXII.

Observatory	Begins in H.	Begins in D.	Reaches Max.	P_1 Max.	Ends in H.	Ends in D.
	h m	h m	h m		h m	h m
Axeloen	22 45[1]	21 45[1]	ca. 23 0	265.0 γ	ca. 0 10	ca. 0 20
Matotchkin Schar .	21 40[1]	21 40[1]	ca. 22 20	240.0 »	23 48	ca. 23 50
Kaafjord	21 40[1]	21 40[1]	22 18	225.0 »	23 50	23 55
Stonyhurst	21 38	21 36	22 40	30.0 »	23 28	0 8
Kew	21 32	21 44	22 46	29.0 »	23 20	ca. 0 20
Wilhelmshaven . .	21 40	21 44	22 50	29.0 »	23 20	0 12
Potsdam	21 40	21 40	22 47	24.0 »	23 20	ca. 0 15
Val Joyeux	21 48	21 48	22 54	23.0 »	23 26	ca. 0
San Fernando . .	ca. 21 36	21 44	22 50	22.5 »	ca. 23 30	ca. 0 10
Munich	21 42	21 45	22 50	21.0 »	ca. 23 25	ca. 0 15
Toronto	21 40	22 10	22 58	16.0 »	ca. 23 40	ca. 23 20
Pawlowsk	21 42	21 40	23 0	14.5 »	23 20	ca. 0
Sitka	21 25	indeterm.	ca. 23	14.0 »	ca. 0	indeterm.
Baldwin	21 45	»	ca. 23	13.0 »	23 55	»
Tiflis	21 40	21 40	23 0	11.0 »	ca. 0	ca. 23 20
Honolulu	ca. 21 30	indeterm.	ca. 23	10.5 »	ca. 23	ca. 0
Batavia	ca. 21	0	ca. 22 30	4.3 »	23 20	0
Dehra Dun	ca. 21	0	23 0	4.0 »	ca. 0	0

(1) The beginning of this special storm.

This storm, as the table and the curves show, appears to be a system that occurs simultaneously at all the places at which it is in any degree observable, and has more or less the same course. The effect of the force diminishes on the whole, with increasing distance from the district surrounding the Norwegian stations. This storm must therefore be classed with the polar elementary storms, and as one of the very simplest.

The properties of the field may be briefly characterised by saying that its form is typical of the polar elementary storms that have their storm-centre about the Norwegian stations. It commences also at the usual time of day. In this way we find again the following typical properties:

(1) An area of convergence situated in the regions about Europe and western Asia.

(2) The point of convergence moves eastwards.

(3) An area of divergence in North America.

On the charts VIII and IX the hours 22^h and $22^h\ 20^m$, the point of convergence is in the regions north of Pawlowsk. P_1 is comparatively small, and P_2 is directed upwards. In the later charts, the forces show that the point of convergence has moved towards the east, the arrow having turned with the hands of a clock. The current-arrows at the Norwegian stations are directed westwards along the auroral zone. At Kaafjord and Matotchkin Schar, P_2 is directed upwards, and at Axeløen downwards, showing that the horizontal portion of the current passes to the north of the two former stations, but to the south of the latter.

TABLE XXXIII.

The Perturbing Forces on the 27th October, 1902.

Gr. M. T.	Honolulu		Sitka		Baldwin		Toronto	
	P_h	P_d	P_h	P_d	P_h	P_d	P_h	P_d
h m								
14 0	+ 5.6 γ	W 2.5 γ	− 29.0 γ	W 42.0 γ	− 7.8 γ	W 5.7 γ	− 6.7 γ	W 3.0 γ
15 0	0	" 2.5 "	− 9.7 "	0	− 4.0 "	" 14.0 "	0	" 10.8 "
30	0	" 7.5 "	− 10.1 "	W 5.4 "	− 4.0 "	" 11.4 "	0	" 9.0 "
16 0	+ 1.3 "	" 5.8 "	− 20.9 "	" 10.8 "	− 6.1 "	" 17.8 "	− 5.4 "	" 15.0 "
20	+ 3.5 "	" 9.1 "	− 22.1 "	" 44.6 "	− 7.8 "	" 24.2 "	− 5.8 "	" 16.8 "
30	+ 7.5 "	" 2.3 "	− 24.6 "	" 42.0 "	− 5.1 "	" 27.3 "	− 5.8 "	" 21.0 "
17 0	+ 10.8 "	0	− 11.0 "	" 26.2 "	0	" 21.6 "	0	" 19.8 "
22 0	− 7.0 "	0	− 7.8 "	0	− 5.8 "	" 1.9 "	− 8.1 "	0
20	− 8.9 "	" 2.5 "	− 10.6 "	E 0.9 "	− 8.5 "	0	− 13.5 "	E 8.4 "
40	− 10.3 "	" 4.2 "	− 8.3 "	W 0.9 "	?	0	− 9.9 "	" 7.8 "
23 0	− 10.8 "	" 4.2 "	− 13.8 "	" 2.7 "	− 10.8 "	E 2.5 "	− 13.5 "	" 8.4 "
20	− 9.8 "	" 4.2 "	− 10.6 "	" 3.6 "	− 6.1 "	0	− 6.8 "	" 3.6 "

TABLE XXXIII (continued).

Gr. M. T.	Axeløen			Matotchkin Schar		
	P_h	P_d	P_v	P_h	P_d	P_v
h m						
14 0	− 60.8 γ	W 28.3 γ	− 110.0 γ	+ 43.4 γ	W 6.2 γ	+ 17.5 γ
15 0	− 52.5 "	" 15.0 "	− 61.5 "	+ ca. 78.0 "	" 42.3 "	− 35.1 "
30	− 108.0 "	" 43.8 "	− 93.5 "	+ ca. 22.0 "	" 61.6 "	− ca. 168.0 "
16 0	− ca. 345.0 "	" 64.7 "	+ 61.5 "	− ca. 92.0 "	E 75.8 "	− > 168.0 "
20	− 290.0 "	" 62.5 "	− 46.7 "	− 79.0 "	" 61.5 "	− 152.0 "
30	− 19.8 "	" 49.5 "	+ 56.5 "	− 97.2 "	" 67.8 "	− 119.0 "
17 0	− 99.0 "	" 52.8 "	− 12.3 "	+ 12.1 "	" 20.0 "	− 47.0 "
22 0	− 51.5 "	E 30.2 "	+ 194.0 "	− 195.0 "	" 78.0 "	− 112.0 "
20	− 28.0 "	" 58.0 "	+ 222.0 "	− ca. 214.0 "	" 112.0 "	− 89.1 "
40	− 69.0 "	" 63.6 "	+ 266.0 "	− 194.0 "	" 63.3 "	− 70.2 "
23 0	− 253.0 "	" 81.6 "	+ 110.0 "	− 108.0 "	" 9.8 "	− 56.2 "
20	− 177.0 "	" 81.4 "	+ 295.0 "	− 119.0 "	" 48.2 "	− 70.2 "

TABLE XXXIII (continued).

Gr. M. T.	Kaafjord			Pawlowsk			Stonyhurst	
	P_h	P_d	P_v	P_h	P_d	P_v	P_h	P_v
h m								
14 0	+ 16.5 γ	W 15.1 γ	+ 29.6 γ	0	E 6.0 γ	0	− 3.5 γ	0
15 0	+ 35.2 "	" 25.9 "	+ 26.3 "	− 0.5 γ	W 5.5 "	0	− 4.6 "	W 1.1 γ
30	+ > 35.2 "	" 37.8 "	+ 35.7 "	− 1.0 "	E 4.6 "	+ 0.7 γ	0	" 5.7 "
16 0	+ > 35.2 "	E 46.2 "	+ 36.2 "	+ 12.5 "	" 42.3 "	+ 3.0 "	− 20.4 "	E 20.0 "
20	+ 23.6 "	" 33.3 "	+ 5.2 "	+ 25.1 "	" 36.8 "	0	− 3.1 "	" 29.7 "
30	+ 26.5 "	W 2.6 "	+ 17.4 "	0	" 20.7 "	0	0	" 12.6 "
17 0	+ > 35.2 "	" 33.3 "	+ 45.1 "	− 15.1 "	" 10.6 "	+ 1.5 "	− 8.3 "	W 6.6 "
22 0	− 135.0 "	E 39.2 "	− 75.2 "	+ 12.6 "	W 2.3 "	− 3.0 "	+ 14.8 "	E 11.4 "
20	− 198.0 "	" 74.0 "	− 101.0 "	+ 6.0 "	" 1.3 "	− 6.0 "	+ 10.7 "	" 10.3 "
40	− 144.0 "	" 27.7 "	− 127.0 "	+ 10.6 "	" 5.5 "	− 8.2 "	+ 12.2 "	" 26.8 "
23 0	− 100.0 "	" 53.3 "	− 127.0 "	+ 14.1 "	0	− 12.0 "	+ 8.3 "	" 24.0 "
20	− 74.9 "	" 72.2 "	− 117.0 "	− 4.5 "	E 13.8 "	− 12.0 "	− 1.0 "	" 10.8 "

TABLE XXXIII (continued).

Gr. M. T.	Kew		Val Joyeux			Wilhelmshaven		
	P_h	P_d	P_h	P_d	P_v	P_h	P_d	P_v
h m								
14 0	− 4.0 γ	0	− 4.0 γ	0		− 4.6 γ	W 2.4 γ	0
15 0	− 7.7 „	E 3.7 γ	− 5.6 „	0		− 7.0 „	0	0
30	− 3.1 „	0	− 9.6 „	W 3.3 γ		− 7.0 „	„ 6.1 „	− 2.0 γ
16 0	−23.0 „	„ 17.3 „	−16.0 „	E 15.9 „	No noticeable deflection.	−13.0 „	E 33.6 „	+ 9.0 „
20	−11.2 „	„ 35.0 „	−18.4 „	„ 29.3 „		+ 7.9 „	„ 42.8 „	+ 6.0 „
30	− 7.7 „	„ 22.5 „	− 4.0 „	„ 15.9 „		− 2.3 „	„ 16.5 „	+ 4.0 „
17 0	−15.3 „	0	−13.6 „	W 5.0 „		−20.5 „	W 10.4 „	− 3.0 „
22 0	+15.3 „	E 9.7 „	+11.2 „	E 8.4 „		+17.7 „	E 9.2 „	
20	+17.8 „	„ 4.7 „	+13.6 „	„ 3.3 „		+10.7 „	„ 3.1 „	A small pos. deflection at 22h
40	+15.3 „	„ 18.3 „	+12.0 „	„ 13.4 „		+20.0 „	„ 15.9 „	
23 0	+10.2 „	„ 18.7 „	+16.0 „	„ 16.7 „		+17.7 „	„ 17.1 „	
20	0	„ 9.7 „	+ 3.2 „	„ 11.7 „		0	„ 12.2 „	

TABLE XXXIII (continued).

Gr. M. T.	Potsdam		San Fernando		Munich	
	P_h	P_d	P_h	P_d	P_h	P_d
h m						
14 0	− 6.3 γ	0	− 4.5 γ	0	− 4.5 γ	E 1.5 γ
15 0	− 9.8 „	W 1.5 γ	− 8.3 „	0	−10.0 „	„ 3.8 „
30	− 9.8 „	„ 4.4 „	− 3.8 „	W 4.2 γ	− 8.5 „	0
16 0	−12.0 „	E 28.0 „	−13.4 „	0	−13.0 „	„ 22.8 „
20	+ 4.4 „	„ 30.5 „	− 6.4 „	E 16.4 „	− 1.0 „	„ 32.7 „
30	− 6.3 „	„ 10.7 „	− 6.4 „	„ 9.8 „	− 3.0 „	„ 21.3 „
17 0	−18.7 „	W 6.2 „	− 4.5 „	0	−12.5 „	0
22 0	+15.4 „	E 3.1 „	+14.1 „	„ 9.8 „	+12.5 „	„ 4.6 „
20	+12.6 „	0	+16.9 „	„ 8.2 „	+12.5 „	0
40	+20.9 „	„ 7.6 „	+13.1 „	„ 14.4 „	+15.0 „	„ 8.4 „
23 0	+17.8 „	„ 10.7 „	+14.1 „	„ 17.2 „	+15.0 „	„ 11.3 „
20	0	„ 10.2 „	+ 4.5 „	„ 11.5 „	+ 4.5 „	„ 12.2 „

TABLE XXXIII (continued).

Gr. M. T.	Pola			Tiflis			Dehra Dun	
	P_h	P_d	P_v	P_h	F_d	P_v	P_h	P_d
h m								
14 0	− 6.2 γ	E 2.8 γ	0	−11.3 γ	E 4.1 γ	0	−11.0 γ	E 8.8 γ
15 0	−11.2 „	0	− 0.4 γ	−16.9 „	0	+ 1.3 γ	−21.7 „	„ 3.9 „
30	−11.6 „	0	0	−14.3 „	„ 2.2 „	0	−17.3 „	„ 3.9 „
16 0	−13.9 „	E 18.7 „	+ 5.5 „	− 3.2 „	„ 20.4 „	− 1.8 „	+ 4.3 „	„ 5.9 „
20	+ 0.9 „	„ 25.0 „	+ 3.2 „	0	„ 14.8 „	+ 2.6 „	+ 2.6 „	„ 6.8 „
30	− 3.1 „	„ 14.6 „	0	− 7.9 „	„ 14.8 „	+ 1.8 „	− 7.1 „	„ 8.8 „
17 0	− 9.9 „	0	0	−21.4 „	„ 9.2 „	+ 2.8 „	−18.9 „	„ 8.8 „
22 0	?	?	?	+ 5.8 „	W 1.9 „	− 1.0 „	+ 1.6 „	
20	?	?	?	+ 6.4 „	„ 2.2 „	− 0.5 „	+ 1.6 „	
40	?	?	?	+ 8.6 „	„ 1.9 „	− 1.3 „	+ 2.4 „	Very small westerly deflections
23 0	?	?	?	+10.5 „	„ 1.1 „	− 1.3 „	+ 3.9 „	
20	?	?	?	+ 4.7 „	E 5.2 „	− 0.3 „	+ 3.1 „	

TABLE XXXIII (continued).

Gr. M. T.	Zi-ka-wei			Batavia		Christchurch	
	P_h	P_d	P_v	P_h	P_d	P_h	P_d
h m							
14 0	− 4.9 γ	E 7.2 γ	No deflection.	0	0	+14.7 γ	E 3.0 γ
15 0	−12.3 „	„ 4.1 „		−13.1 γ	0	0	0
30	− 6.2 „	„ 5.2 „		−12.8 „	W 1.2 γ	+ 2.3 „	0
16 0	+16.0 „	„ 3.1 „		+11.3 „	„ 4.8 „	− 8.3 „	„ 17.6 „
20	+ 8.6 „	„ 4.1 „		− 4.3 „	„ 2.4 „	+ 6.4 „	„ 14.9 „
30	+ 6.2 „	„ 7.2 „		0	„ 2.4 „	+11.0 „	„ 8.9 „
17 0	0	„ 10.3 „		− 7.7 „	0	+ 9.2 „	„ 3.7 „
22 0	No measurable deflection.			− 4.3 „	0	− ca. 2.3 „	0
20				− 4.3 „	0	− ca. 4.1 „	0
40				− 2.1 „	„ 1.2 „	− ca. 3.7 „	ca. „ 3.7 „
23 0				− 1.1 „	„ 3.0 „	− ca. 4.1 „	„ „ 4.4 „
20				0	„ 1.8 „	− ca. 1.8 „	„ „ 2.9 „

TABLE XXXIV.

Partiel Perturbing Forces on the 27th October, 1902.

	16h 0m		16h 20m		16h 30m	
	P'_h	P'_d	P'_h	P'_d	P'_h	P'_d
Honolulu	The intermediate storm not well defined; the effect seems to be a perturbing force directed southwards.					
Sitka						
Baldwin						
Toronto						
Axeloen	−276.0 γ	0	−216.0 γ	E 95.0 γ	−117.0 γ	E 12.8 γ
Matotchkin-Schar .	−179.0 „	E 94.0 γ	−162.0 „	„ 35.6 „	−172.0 „	„ 83.0 „
Kaafjord	?	„ 85.0 „	?	„ 74.0 „	?	„ 37.0 „
Pawlowsk	+ 20.1 „	„ 41.3 „	+ 29.6 „	„ 25.3 „	+ 12.6 „	„ 11.5 „
Stonyhurst	− 17.8 „	„ 22.8 „	+ 3.1 „	„ 34.3 „	+ 7.7 „	„ 18.3 „
Kew	− 12.2 „	„ 17.8 „	+ 2.6 „	„ 35.0 „	+ 8.2 „	„ 23.4 „
Val Joyeux	− 9.6 „	„ 19.2 „	− 12.0 „	„ 36.8 „	+ 3.2 „	„ 20.9 „
Wilhelmshaven . .	− 1.9 „	„ 40.3 „	+ 23.3 „	„ 53.1 „	+ 15.0 „	„ 26.9 „
Potsdam	+ 1.3 „	„ 35.5 „	+ 20.6 „	„ 37.0 „	+ 9.8 „	„ 16.7 „
San Fernando . . .	− 6.4 „	„ 4.1 „	− 3.2 „	„ 16.4 „	0	„ 12.3 „
München	− 3.0 „	„ 21.3 „	+ 10.0 „	„ 32.7 „	+ 9.5 „	„ 22.8 „
Pola	− 1.3 „	„ 20.8 „	+ 9.0 „	„ 27.8 „	+ 9.0 „	„ 17.3 „
Tiflis	+ 16.0 „	„ 16.3 „	+ 21.4 „	„ 11.1 „	+ 12.6 „	„ 7.8 „
Dehra Dun	+ 26.0 „	0	+ 21.2 „	0	+ 12.6 „	0
Zi-ka-wei.	+ 25.8 „	W 6.2 „	+ 20.9 „	W 5.1 „	+ 17.2 „	W 2.6 „
Batavia	+ 25.0 „	„ 4.8 „	+ 21.4 „	„ 2.4 „	+ 14.6 „	„ 2.4 „
Christchurch	− 18.4 „	E 17.1 „	− 6.0 „	E 14.9 „	0	E 9.7 „

Current-Arrows for the 27th October, 1902; Chart I at 14^h, and Chart II at 15^h.

Current-Arrows for the 27th October, 1902; Chart III at $15^h\ 30^m$, and Chart IV at 16^h.

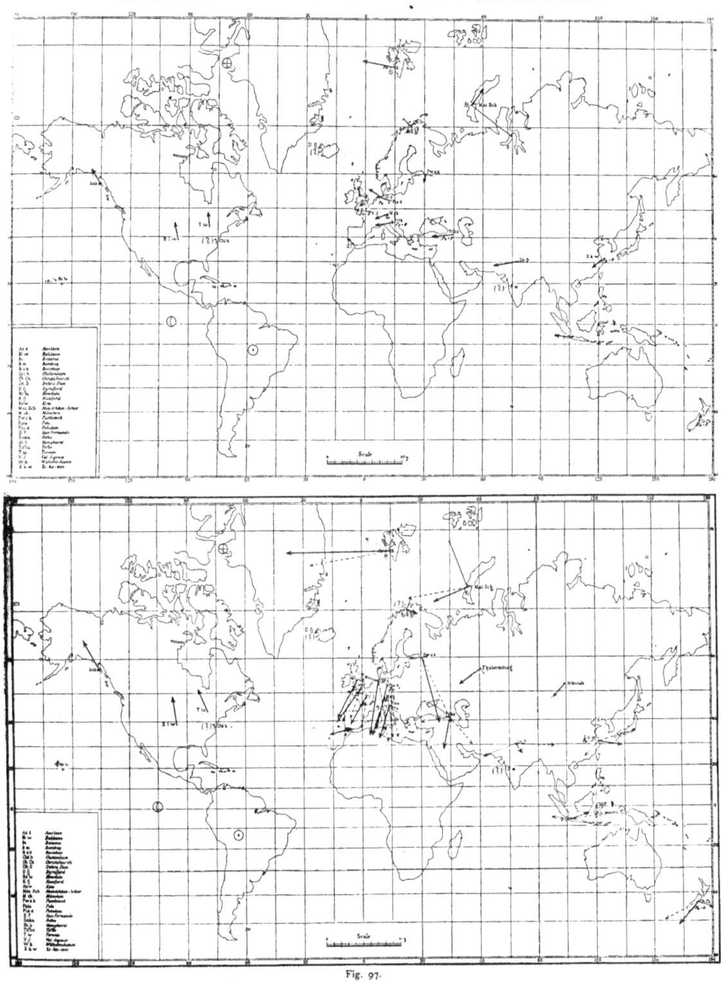

Fig. 97.

Current-Arrows for the 27th October, 1902; Chart VII at 17^h, and Chart VIII at 22^h.

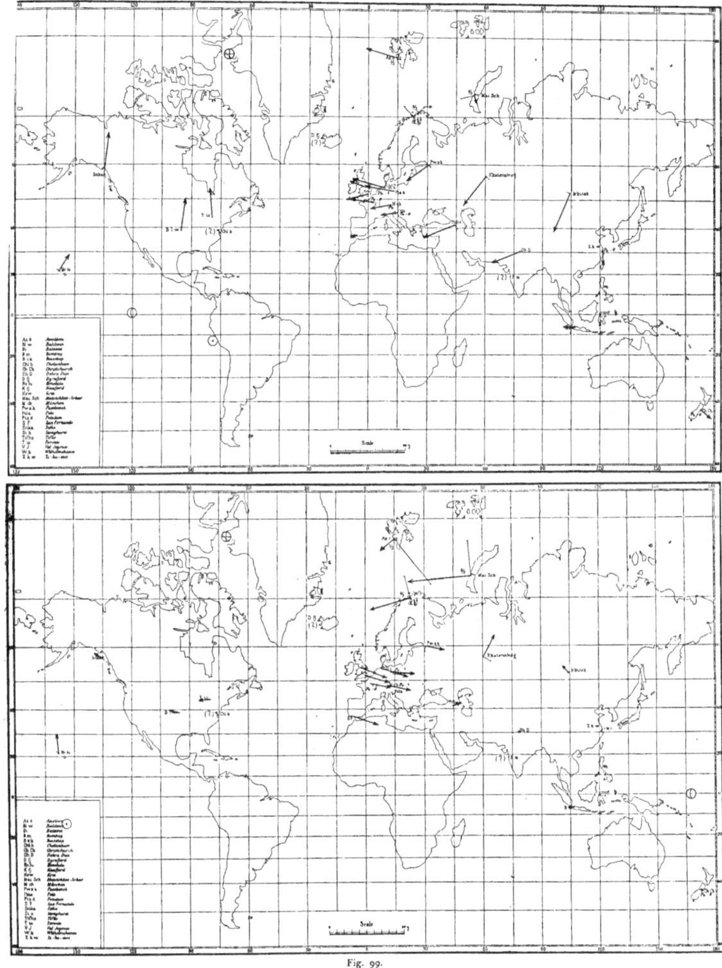

Fig. 99.

Current-Arrows for the 27th October, 1902; Chart IX at 22^h 20^m, and Chart X at 22^h 40^m.

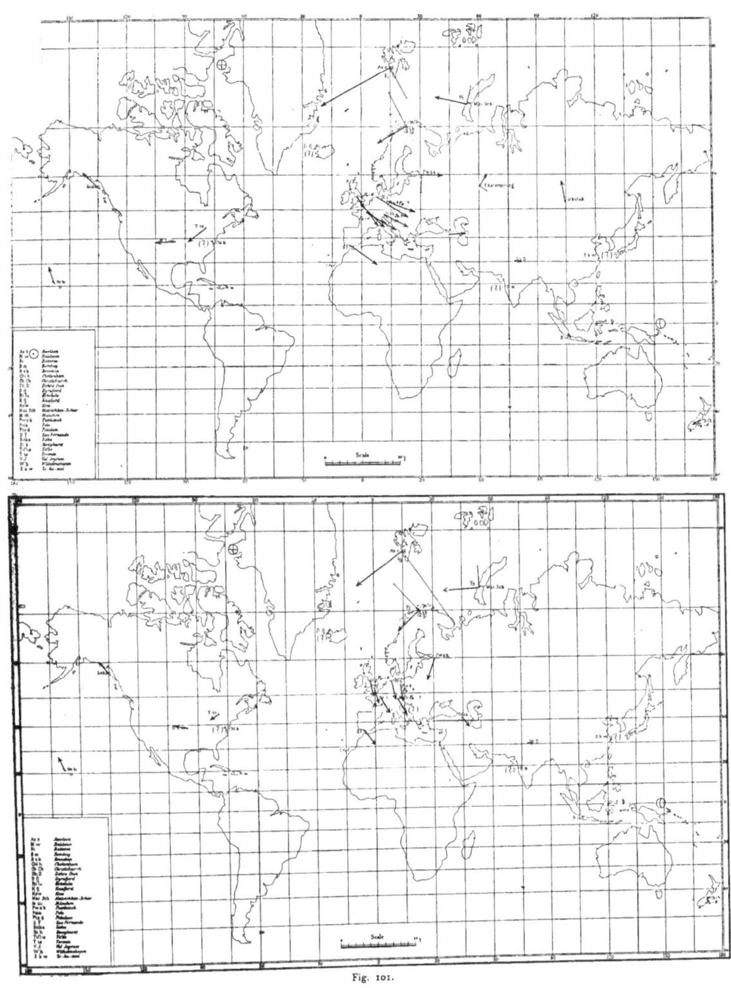

Fig. 101.

THE PERTURBATION OF THE 28th & 29th OCTOBER, 1902.

63. After the last polar elementary storm that occurred before midnight on the 27th October, the conditions once more become comparatively calm, and continue so until about 18^h the following day, when another perturbation of considerable power occurs. Sitka is the only place that forms an exception to this, as there a perturbation of a rather considerable strength occurs about midnight, local time; but its sphere of action is rather limited, as it is not noticed either at the Norwegian stations or at the other stations in North America.

The perturbation-conditions during this twenty-four hours closely resemble those of the preceding day and night. On both days, the conditions at the Norwegian stations are characterised by two separate storms; but on the 28th, these two storms are closer together, the first storm on that day being about two hours and a half later than the first on the 27th, and the second on the 28th perhaps half an hour earlier than that on the 27th.

When we come to lower latitudes, we find the conditions during the time from 14^h to 20^h rather different on the two days. There is no trace on the 28th of the long storm that occurred on the 27th, and was especially powerful at the equator; it is the intermediate storm that answers to the first storm on the 28th. On the other hand, there is an astonishing resemblance between the conditions of the two days in the last storms both at our Norwegian stations and in lower latitudes. We thus notice that the deflection in H at Kaafjord are in the same direction on both days, and the D-curve has an undulating form while the deflection in V is uniform in direction and very great. Farther south we find that the H-curve on both days is of an undulating character; there are two intermediate more or less marked maxima separated by a minimum.

It appears from the curves that the distribution of strength in the northern hemisphere is about the same on the two days. It is thus evident that on this occasion also there are two separate polar

TABLE XXXV.

Observatory	Perturb. I						Perturb. II					
	Begins in H	Begins in D	Reach. max.	P_1 max.	Ends in H.	Ends in D.	Begins in H	Begins in D	Reach. max.	P_1 max.	Ends in H	Ends in D
	h m	h m	h m		h m	h m	h m	h m	h m		h m	h m
Axeløen	18 9	18 9	18 33	248.0 γ	19 10(1)	19 15(1)	21 45	21 30	22 15	266.0 γ	ca. 24	ca. 23 20(1)
Matotchkin Schar	18 3(1)	ca. 18 15	18 50	138.0 „	ca. 19 30	19 8	ca. 20 35	21 33	21 57	209.0 „	„ 23 40	„ 23
Kaafjord	18 8	„ 18	18 45	78.0 „	„ 19 45	19 15(1)	20 40	21 33(1)	22 20	175.0 „	„ 22 50	„ 23
Pawlowsk	18 5	18 15	18 45	14.5 „	19 20	indeterm.	21 30	21 30	21 50	27.5 „	23 25	23
Wilhelmshaven	18 12	18 10	18 45	25.5 „	19 26	19 10	21 34	21 56	22 10	25.5 „	23 30	23 30
Val Joyeux	18 15	18 15	18 45	17.0 „	19 30	19 20	21 40	ca. 22	22 10	24.0 „	22 45	23 35
Stonyhurst	18 8	18 8	18 45	16.5 „	19 24	19 16	21 27	21 40	22	21.0 „	22 32	23 20
Potsdam	18 8	18 6	18 45	23.5 „	19 24	19 3	21 30	21 18	22 10	19.0 „	23 30	23 27
Munich	18 12	18 10	18 45	15.0 „	19 45	19 5	21 35	21 40	22 10	17.0 „	ca. 23 30	ca. 23 45
Kew	18 7	18 7	18 45	16.0 „	19 24	19 20	21 30	20 50	22 8	16.5 „	22 40	23 24
Pola							21 30	ca. 21 30	22 10	16.0 „	23 30	23 45
Toronto	18 15	indeterm.	18 45	3.0 „	19 45	indeterm.	21 30	21 40	21 50	15.0 „	22 40	23
San Fernando	18 10	18 12	18 45	21.0 „	19 45	ca. 19 30	31 32	21 40	22	13.5 „	23 5	23 20
Tiflis	18 5	ca. 18 15	18 45	14.0 „	20 16	indeterm.	21 35	21 35	22 10	13.5 „	ca. 24	22 30
Baldwin	18 15	no pert.	18 45	3.5 „	ca. 19	no pert.	indeterm.	21 40	22 10	13.0 „	indeterm.	ca. 23 20
Dehra Dun	18 15	indeterm.	19	11.0 „	20 15	indeterm.	21 20	21 40	22 20	8.2 „	23 30	„ 23
Sitka	17 57	indeterm.	18 45	10.6 „	indeterm.	indeterm.	ca. 21 40	indeterm.	22 20	7.5 „	22 45	indeterm.
Batavia	18 15	no pert.	19	7.0 „	20 15	no pert.	22	no pert.	22 20	2.5 „	23 15	no pert.

(1) The commencement of these special storms.

elementary storms, both with fairly simple course. The table above gives the time at which the two perturbations begin, attain a maximum, and end, and the value of P_i at its maximum. We find here a distinct confirmation of the statement that the effect of the force diminishes from the poles to the equator.

The table shows that the two perturbations differ essentially as regards distribution of strength. Although the first storm is less powerful at the Norwegian stations, and rather less powerful in Central Europe, it is nevertheless somewhat more powerful than the second when we come nearer to equator. There is a still greater difference with regard to the conditions in America, the first storm being almost imperceptible there.

We thus receive a decided impression that the current-system that conditions the field—however this may be constituted in the second storm—is situated, on the whole, farther west, a circumstance that may to some extent explain the different distribution of strength in the two storms.

THE FIELD OF FORCE.

64. The field during the *first storm* is in the main of the same form and relative strength as in the intermediate storm on the 27th, but less powerful. The current-arrows in the north are directed westwards along the auroral zone, and the effect is strongest at Axeløen and Matotchkin Schar. P_v at Kaafjord and Matotchkin Schar is directed downwards, at Axeløen upwards. There is an area of convergence with a fairly strong force in the eastern hemisphere, but an area of divergence with comparatively little force in the western. The point of convergence is situated in the regions round the north-east of Russia. The field, at those places from which we have observations, is almost stationary. At Pawlowsk, P_v is directed upwards.

The field during the *second storm* is almost exactly the same as that during the second storm on the previous day. All that has been said of the field on the 27th may be directly applied to this perturbation.

As on the previous day, there is a movement of the system towards the east. This is evident, both from the clockwise turning of the arrows in the south of Europe, and from the conditions at the Norwegian stations. If we look at the current-arrows for Axeløen and Kaafjord, we see that they are at first convergent, showing that the storm-centre is to the west of those stations. When the storm is almost at its height, they become parallel, and end by being divergent, thus indicating the eastward position of the storm-centre.

These two storms, as we see, are the very ones to afford favorable conditions for a determination of the strength of the horizontal portion of the current, and such a calculation will therefore be made.

The very interesting systems of current-arrows are shown on the Charts I to VII.

TABLE XXXVI.

The Perturbing Forces on the 28th October, 1902.

Gr. M. T.		Sitka		Baldwin		Toronto		Axeløen		
		P_h	P_d	P_h	P_d	P_h	P_d	P_h	P_d	P_v
h	m									
18	15	− 4.2 γ	0	− 0.7 γ	0	0	0	− 44.6 γ	0	+ 103.0 γ
	30	− 6.7 „	0	− 3.0 „	0	− 0.9 γ	0	− 12.8 „	W 7.6 γ	+ 96.0 „
	45	− 10.3 „	E 2.3 γ	− 3.7 „	0	− 2.7 „	0	− 89.7 „	„ 38.1 „	+ 258.0 „
19	0	− 7.8 „	„ 4.1 „	− 1.7 „	0	0	0	− 153.0 „	„ ca. 22.3 „	+ 88.5 „
21	40	− 1.2 „	„ 1.4 „	− 12.2 „	E 6.4 γ	− 6.3 „	E 3.0 γ	− 13.8 „	E 40.8 „	+ 183.0 „
22	0	− 6.6 „	„ 1.4 „	− 13.5 „	„ 3.2 „	− 9.0 „	„ 12.6 „	− 89.7 „	„ 95.2 „	+ 349.0 „
	20	− 7.1 „	0	− 8.5 „	„ 2.5 „	− 6.8 „	„ 7.2 „	− 166.0 „	„ 112.0 „	+ 352.0 „
	40	− 1.8 „	0	− ca. 7.4 „	„ 1.9 „	− 1.4 „	„ 3.6 „	− 116.0 „	„ 25.8 „	+ 246.0 „
23	0	− 0.4 „	0	− 6.8 „	0	0	0	− 172.0 „	„ 69.5 „	+ 231.0 „

TABLE XXXVI (continued).

Gr. M. T.		Matotchkin Schar			Kaafjord			Pawlowsk		
		P_h	P_d	P_v	P_h	P_d	P_v	P_h	P_d	P_v
h	m									
18	15	− 67.7 γ	0	− 37.9 γ	− 26.6 γ	E 12.6 γ	− 16.4 γ	+ 13.1 γ	W 1.8 γ	0
	30	−161.0 „	E 46.8 γ	− 49.1 „	− 49.6 „	„ 12.9 „	− 60.1 „	+ 8.1 „	„ 4.6 „	− 3.0 γ
	45	−147.0 „	„ 29.0 „	− 54.8 „	− 76.1 „	„ 18.5 „	− 68.1 „	+ 14.6 „	„ 2.8 „	− 4.5 „
19	0	−127.0 „	„ 36.7 „	− 39.3 „	− 39.5 „	„ 20.7 „	− 48.4 „	+ 7.0 „	0	− 5.2 „
21	40	−143.0 „	„ 50.8 „	− 46.3 „	− 91.5 „	W 5.5 „	− 84.6 „	+ 3.0 „	„ 15.6 „	0
22	0	−191.0 „	„ 39.7 „	− 52.6 „	−151.0 „	E 29.2 „	−147.0 „	+ 11.6 „	„ 18.4 „	− 4.5 „
	20	−175.0 „	„ 26.8 „	− 39.3 „	−147.0 „	„ 94.7 „	−132.0 „	+ 12.6 „	„ 1.4 „	−11.2 „
	40	−108.0 „	„ 22.3 „	− 35.1 „	− 57.8 „	„ 54.0 „	−119.0 „	+ 3.0 „	E 4.1 „	−11.2 „
23	0	− 63.4 „	W 4.5 „	− 35.1 „	− 17.7 „	„ 15.9 „	−103.0 „	+ 10.1 „	„ 1.8 „	− 8.2 „

TABLE XXXVI (continued).

Gr. M. T.		Stonyhurst		Kew		Val Joyeux			Wilhelmshaven		
		P_h	P_d	P_h	P_d	P_h	P_d	P_v	P_h	P_d	P_v
h	m										
18	15	+ 3.5 γ	E 14.8 γ	+ 5.1 γ	E 8.0 γ	+ 2.4 γ	E 3.3 γ	No measurable deflection.	+14.0 γ	E 12.2 γ	Slight deflections.
	30	+ 7.7 „	„ 9.7 „	+ 8.2 „	„ 5.1 „	+ 8.8 „	„ 10.0 „		+15.9 „	„ 3.7 „	
	45	+ 8.2 „	„ 14.3 „	+10.2 „	„ 12.2 „	+11.2 „	„ 12.5 „		+21.0 „	„ 14,6 „	
19	0	+ 3.5 „	„ 5.7 „	+ 5.6 „	„ 7.0 „	+12.0 „	„ 10.9 „		+10.3 „	„ 3.1 „	
21	40	+17.3 „	„ 4.0 „	+11.0 „	0	+ 3.2 „	0		+ 9.8 „	W 1.2 „	
22	0	+13.3 „	„ 16.0 „	+13.3 „	E 9.4 „	+19.2 „	„ 3.3 „		+18.7 „	E 6.1 „	
	20	+ 5.1 „	„ 16.6 „	+ 5.6 „	„ 14.5 „	+12.0 „	„ 15.9 „		+12.6 „	„ 17.1 „	
	40	− 2.5 „	„ 9.7 „	0	„ 6.1 „	+ 3.2 „	„ 8.4 „		− 2.3 „	„ 6.7 „	
23	0	0	„ 14.8 „	0	„ 12.6 „	+ 2.4 „	„ 11.7 „		+ 4.2 „	„ 16.5 „	

TABLE XXXVI (continued).

Gr. M. T.		Potsdam		San Fernando		Munich		Pola		
		P_h	P	P	P_d	P_h	P_d	P_h	P_d	P_v
h	m									
18	15	+16.8 γ	E 7.6 γ	+13.1 γ	E 8.2 γ	+ 7.0 γ	E 5.3 γ	?	?	?
	30	+13.5 „	„ 2.5 „	+16.9 „	„ 9.0 „	+ 8.5 „	„ 4.6 „	?	?	?
	45	+21.5 „	„ 9.2 „	+16.9 „	„ 12.3 „	+12.0 „	„ 8.4 „	?	?	?
19	0	+11.4 „	„ 2.5 „	+16.6 „	„ 9.0 „	+ 9.0 „	„ 4.6 „	?	?	?
21	40	+13.6 „	W 4.0 „	+ 9.0 „	„ 4.1 „	+ 7.5 „	0	+12.1 γ	W 2.8 γ	Slight deflection.
22	0	+21.2 „	0	+16.9 „	„ 8.2 „	+16.0 „	0	+13.4 „	E 6.9 „	
	20	+13.5 „	E 10.2 „	+ 9.0 „	„ 9.0 „	+12.5 „	„ 9.9 „	+ 9.0 „	„ 8.3 „	
	40	+ 1.9 „	„ 4.6 „	+ 6.4 „	„ 1.6 „	+ 3.5 „	„ 7.6 „	+ 4.0 „	„ 7.6 „	
23	0	+ 7.9 „	„ 9.2 „	+ 3.2 „	„ 5.7 „	+ 4.5 „	„ 10.6 „	+ 4.9 „	„ 9.0 „	

TABLE XXXVI (continued).

Gr. M. T.	Tiflis			Dehra Dun		Batavia	
	P_h	P_d	P_v	P_h	P_d	P_h	P_d
h m							
18 15	+ 7.1 γ	W 0.4 γ	− 1.3 γ	+ 5.9 γ	W 1.9 γ	+ 1.1 γ	No deflection.
30	+ 9.2 „	„ 1.5 „	− 0.8 „	+ 5.9 „	„ 3.0 „	+ 1.8 „	
45	+ 13.7 „	„ 1.5 „	− 1.8 „	+ 9.8 „	„ 4.9 „	+ 1.8 „	
19 0	+ 11.6 „	„ 0.7 „	− 1.0 „	+ 9.0 „	„ 3.0 „	+ 7.1 „	
21 40	+ 2.6 „	„ 5.6 „	− 1.0 „	− 1.6 „	„ 3.0 „	0	
22 0	+ 9.2 „	„ 9.7 „	− 1.8 „	+ 3.1 „	„ 6.9 „	0	
20	+ 11.3 „	„ 2.2 „	− 2.0 „	+ 6.3 „	„ 4.9 „	+ 2.8 „	
40	+ 5.4 „	0	− 0.8 „	+ 3.5 „	„ 1.0 „	+ 1.8 „	
23 0	+ 6.4 „	0	− 2.0 „	+ 5.5 „	0	+ 1.8 „	

Current-Arrows for the 28th October, 1902; Chart I at $18^h 15^m$.

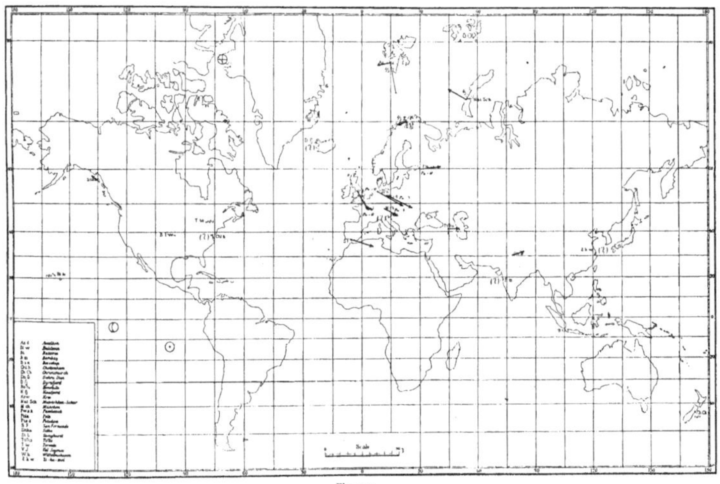

Fig. 102.

Current-Arrows for the 28th October, 1902; Chart II at $18^h\ 30^m$, and Chart III at $18^h\ 45^m$.

Current-Arrows for the 28th October, 1902; Chart IV at 19^h, and Chart V at 21^h 40^m.

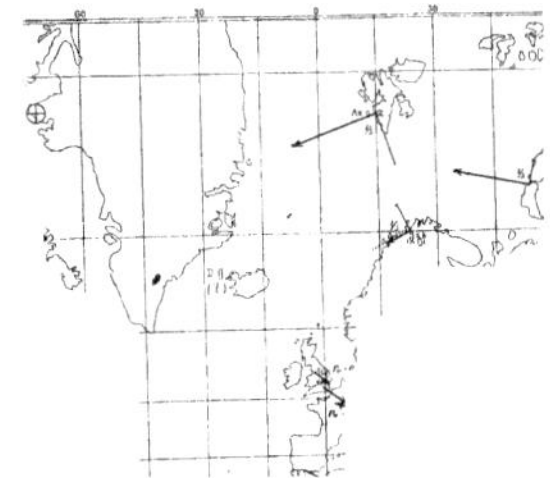

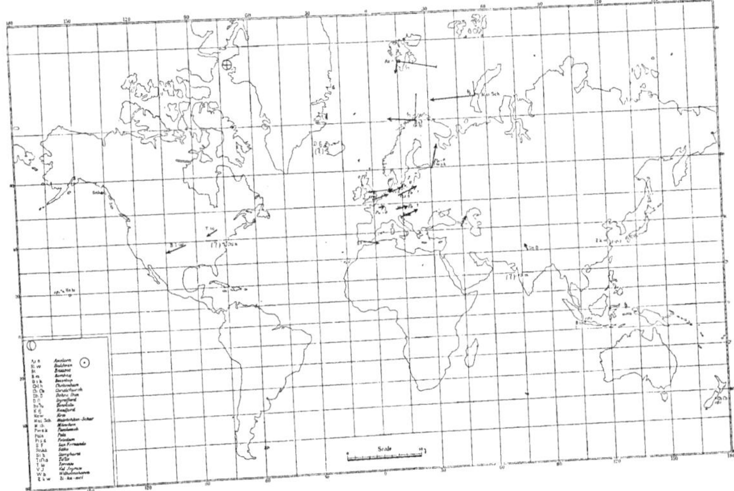

Current-Arrows for the 28th October, 1902; Chart VI at $22^h\ 0^m$, and Chart VII at $22^h\ 20^m$.

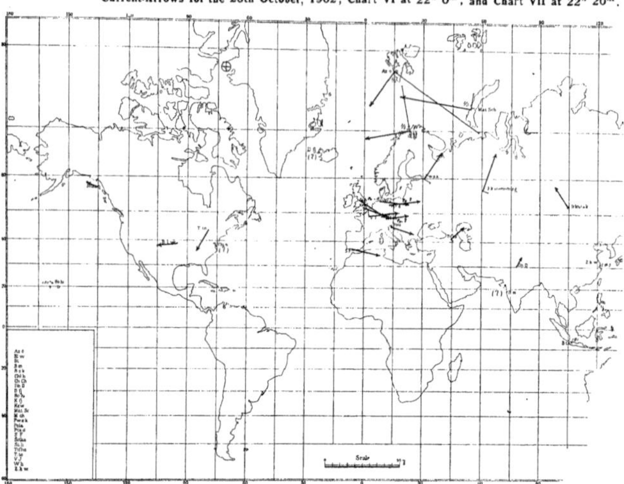

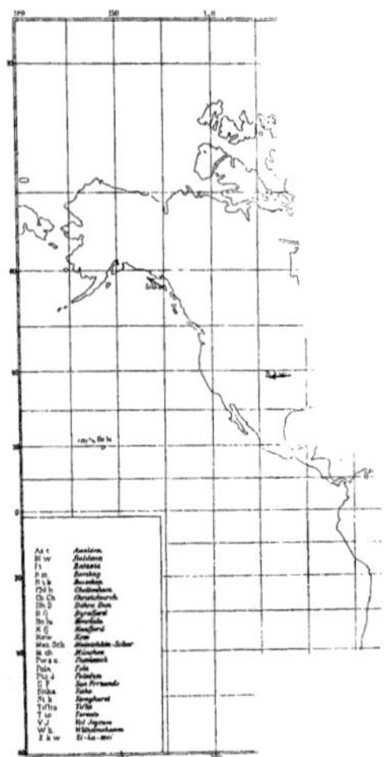

Current-Arrows for the 28th October, 1902; Chart VIII at $22^h\ 40^m$, and Chart IX at $23^h\ 0^m$.

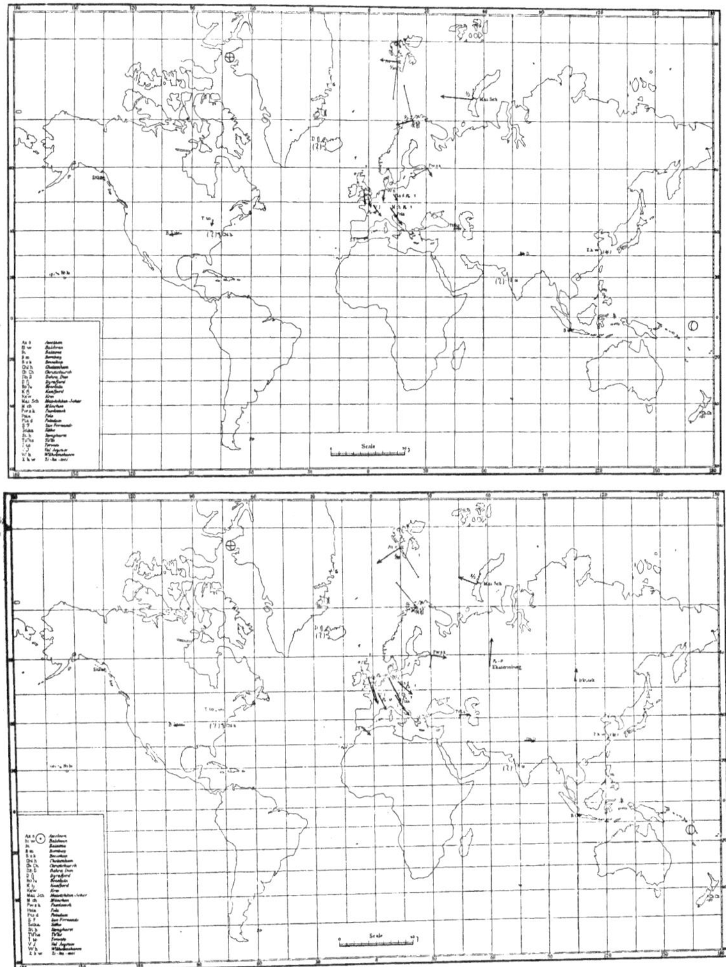

Fig. 106.

THE PERTURBATIONS OF THE 31st OCTOBER & 1st NOVEMBER, 1902.

(Pl. VII).

65. After the last storm on the 28th October, quiet conditions once more prevail; but at about 18^h on the following day, the storm bursts out again, and continues until midnight, and it seems, that the two polar perturbations, that occured rather destinctly on the 28th now come so near one another, that they form a single one (cf. Pl. VI).

On the next day again, this is repeated. At Axeløen in particular, there are powerful perturbations, but they commence at about 16^h. In southern latitudes, this twenty-four hours is fairly quiet; but during the morning of the 31st, a storm begins, which lasts uninterruptedly for nearly twenty-four hours. It appears at the poles with tremendous violence, although perhaps its strength is even more unusual at the equatorial stations. Considering its long duration and its universal distribution, we may say that it is the greatest storm that has been observed by us.

A circumstance which adds still more to the interest of this storm is that it occurs at the new moon, and what is more, there was even an eclipse of the sun during the perturbation. This eclipse began at 5^h 58.5^m on the 31st October, and ended at 10^h 2.3^m. It was only partial, and the greatest phase (0.699) occurred at 8^h 0.4^m, in longitude $100°$ $56'$ East, and latitude $70°$ $53'$ North. The eclipse cannot in itself be considered as affecting this perturbation in any essential degree. Whatever direct effect there may possibly be of the eclipse itself this must at any rate be very small as compared with the total amount of the perturbation, as no special change is observable in the curves, coinciding with the time of the eclipse. We know that powerful storms often occur at the same time as an eclipse, without being directly due to it; but it has been stated "that an observable magnetic variation makes itself felt during the time of a solar eclipse, and that this variation is analogous in its nature to the solar diurnal variation, differing from it only in degree." (1) In this case it is difficult for us to study this direct influence, as we have no material from the places at which the eclipse was greatest.

If the moon can be supposed to exert any influence on the perturbation, it must be owing to the fact that it is a new moon. We will not here, however, enter more particularly into these questions but only describe the perturbation, and find out its actual distribution and course.

It exhibits great variableness round the Norwegian stations. The curves have a very serrated appearance, resulting from great vibration in the field of perturbation. Notwithstanding this, however, the conditions of the perturbation as a whole, run a fairly simple course, which may be characterised as follows.

During the time that the perturbation lasts, namely from about 9^h on the 31st October to 3^h on the 1st November, most of the curves for the magnetic elements form a single undulation with crest and sinus. This wave differs, however, in phase at the three stations. At Kaafjord the deflections changes sign in all three elements between 18^h and 18^h 30^m. At Matotchkin Schar it changes in H at about 16^h, in D at 16^h 45^m, and in V at 19^h 15^m, thus taking place on the whole earlier than at the former station. At Axeløen, the undulating form is very marked in the declination, the change not taking place until about 22^h. The smaller variations must be regarded as ripples upon this principal undulation. Two of these shorter variations in particular are considerable and worthy of notice. One of them appears at about 14^h, the other at about midnight, with maximum about 23^h 45^m. At Axeløen, where the main undulation was somewhat less marked in H, these two intermediate storms are very prominent.

(1) L. A. Bauer: Terrestial Magnetism Vol. 7, p. 192.
W. van Bemmelen: Contribution to the Knowledge of the Influence of Solar Eclipses on Terrestrial Magnetism.
C. Nordmann, Bulletin Astronomique, Mars 1907.

At Sitka too, this storm occurs with a violence that approaches what we find at the Norwegian stations, this being greatest between $13^h\ 15^m$ and 14^h, at which time the H variometer-needle is deflected out of the field. This storm occurs at the same time as the first great intermediate storm at Axeløen. Great storms also occur at the other stations in the western hemisphere; and even at Honolulu the perturbation on that day is fairly powerful. In the United States the character of the perturbations varies more or less with time and place. Unfortunately we here only have registerings for the first part of the perturbation.

In Central and Southern Europe the perturbation is rather considerable though relative to that in the equatorial stations comparatively slight, especially the first part. Up to $17^h\ 45^m$ the conditions remain fairly uniform—a deflection in H, indicating a decrease in the horizontal intensity, and a westward deflection in D. At about $17^h\ 45^m$, the D-curve goes over to the opposite side of the mean line, while the deflection in H is increased. The D-curve of San Fernando forms an exception to this; as the change in direction here does not take place till about 2 hours later. The course somewhat resembles that at Kaafjord, as the change in D takes place at about the same time as the above-mentioned change in the amplitude. Between 23^h and $0^h\ 35^m$ there is a rather strong impulse in D, this being simultaneous with the second powerful storm at Axeløen.

In the region of Dehra Dun, Batavia and Christchurch, the storm is very powerful, the first part of it being even more powerful than in England, France and Germany. At $12^h\ 30^m$, the perturbing force at Dehra Dun attains a value of $80\,\gamma$.

The conditions on the whole are fairly simple. At Dehra Dun for instance until $13^h\ 15^m$ the perturbation is noticed principally in H and then there also is a deflection in the declination towards the east. Similar conditions we also find at the other stations. The deflection in H is uniform in direction throughout, as H is decreasing all the time. The character of the curve is quiet on the whole, without any great, sudden changes; and only at about $13^h\ 30^m$ is there such a change in the deflection.

It appears from the coincidence of the previously-mentioned powerful storm at Sitka with that on Axeløen, that these deflections are connected with one another. The perturbation on this date resembles in many respects the preceding perturbations of the 15th and 8th February and that of the 27th October. We may thus make a comparison with the perturbation of the 27th October for instance. On this day we also found a storm of long duration, that was especially powerful and of similar effect in the south Asiatic districts. During that perturbation there was an intermediate storm that was also powerful in the districts of Dehra Dun and Batavia, and was almost the reverse of the long storm.

A little before midnight there was another short storm, the effect of which was very slight at Dehra Dun, but powerful in Europe. The chief difference is that the long storm of the 31st October is much more powerful and of much longer duration, so that both the short storms come within its limits. The first intermediate storm, moreover, occurs a little earlier in the day, and the second a little later, than those on the 27th October.

Analogous with what we have done in the case of the last described storms this perturbation is divided into three principal phenomena, the long storm and two intermediate storms. There are indeed more interruptions than these two during the long storm, that might well be studied, for there are innumerable small interruptions; but as far as we can tell from our material, it is only these two that have a universal and powerful effect, and between them and the other irregularities there is a wide gulf that cannot be crossed without leading to so great a multiplicity, that the main lines would be lost, and the study of the phenomena rendered nearly impossible.

THE FIELD OF FORCE.

66. (1) *Charts I to VIII* represent the conditions during the time between 9^h and $12^h\ 30^m$.

During this comparatively long time, the form of the field in the eastern hemisphere remains almost constant. It may be briefly charaterised in the following manner:

At the equator there are powerful perturbing forces directed southwards. In Central and Southern Europe, the force is only about half as great as at Dehra Dun and Batavia, and throughout is south-west in direction. At Kaafjord and Matotchkin Schar, the current-arrow is directed all the time eastwards along the auroral zone, a circumstance that seems to have some connection with the fact that during this time these stations are situated on the day-side. At Axeløen the force is almost in the opposite direction. The current-arrow is at first directed southwards, but in the course of the above-mentioned period turns clockwise until at $12^h\ 30^m$ its direction is WSW.

In medium and northern latitudes in the western hemisphere the conditions are more variable, whereas at Honolulu there is a powerful perturbation that remains almost constant all the time. The conditions there are very similar to those at Dehra Dun; the current-arrow at both places is directed westwards, but is a little smaller at Honolulu.

The conditions in North America are very interesting, and require a fuller description.

At Sitka, as already mentioned, the perturbation is extremely violent; and the curve presents the same very serrated appearance that is so characteristic of the powerful storms about the auroral zone. On looking at the charts, we see that the perturbing force remains more or less constant in direction. The current-arrow is directed principally westwards, sometimes a little WSW. The strength too, varies but not much on the whole.

During the polar elementary storms that occur about midnight, and have their centre in the regions round the Norwegian stations, we have always found that there is only little difference between the conditions at Sitka and those at Toronto and Baldwin; but on this occasion there is a very great difference between them, and even considerable difference between Toronto and Baldwin. In the case of the last-named two stations, moreover, there is great variableness from time to time, which makes these perturbations very distinct from those in the eastern hemisphere with their more constant conditions. This circumstance is to be explained by the fact that the perturbation in the north of North America is due in a great measure to the occurrence of more or less independent storms that are confined to those regions.

In order to obtain a clear idea of the field that is produced by these storms in the north of North America, we should examine it at those times when the force is greatest, as we may then most safely disregard the other forces that are acting through other systems. Let us look then at Charts IV to VIII. We see that the arrow at Sitka remains almost constant. The arrows at Toronto and Baldwin show that there is an area of convergence there, with very great convergence of the perturbing force. We cannot help noticing that this field exhibits the same properties that characterised the field in the previously-discussed polar elementary storms with their centre at the Norwegian stations. At Sitka there is a comparatively powerful perturbation with constant direction of the perturbing force, corresponding to the conditions at the Norwegian stations; and in both cases the current-arrow is directed towards the west. The area of convergence in North America on this day corresponds with the area of convergence in the European district under the above mentioned elementary storms.

The correspondence appears still greater when we notice that the centre of these storms has about the same position in relation to the sun as the previously-mentioned polar elementary storms at the Norwegian stations, the storm-centre in these cases being in the district that has midnight at the time of the storm, or often on the morning side. In the case of the perturbation here described we also find the same. The chart for $9^h\ 30^m$ forms an exception to this. In the first place it must be remarked that the arrows are small; and as we have only taken out total forces, we cannot know how much is due to local storms. The circumstances are explained quite naturally, however, by assuming that the storm-centre now lies farther east. As the perturbing forces at Toronto and Baldwin are very small, we must then make the assumption that the point of convergence of the system is now situated in the vicinity of these stations, a little to the east of them; but as the conditions here, if minutely entered into, are rather complicated, we must not investigate the matter more closely.

In this connection we may refer to the previously-described perturbation of the 28th December, where we also met with an area of convergence in North America. On that day, however, the storm-centre seems to lie at a greater distance from Sitka, the curves having a far less disturbed character than now. There we also found that the field of precipitation was at first situated farther to the east, and then moved westwards.

(2) *Charts IX, X and XI* represent the conditions as they appear during the first powerful intermediate storm. The perturbing force at Sitka has about the same direction as before, but is much greater. This perturbation, moreover, is particularly powerful at Axeløen, with a perturbing force that is directed SSE all the time.

We have endeavoured to separate the effect of the intermediate storm from the rest, the total force being decomposed. Owing to the manner in which the decomposition has been carried out, one of the systems of arrows gives a field with almost the same form as the one already described.

With regard to the field in the intermediate storm, we first notice how rapidly the force diminishes, both in the neighbourhood of Sitka and in that of Axeløen, at any rate in the districts from which we have observations.

In the district of Zi-ka-wei, Dehra Dun, and Batavia, the direction of the intermediate perturbing force on the whole is almost the reverse of what it had been earlier, and the magnitude is very considerable. This circumstance also occurred during the intermediate storms of the 27th October, 1902, and the 8th and 15th February, 1903.

In Europe there is a peculiarity in the conditions, namely, that the effect of the intermediate storm is very small. The perturbing forces throughout are smaller than in the Asiatic district, and exhibit considerable variableness, although the current-arrows all through are directed south-west.

At Baldwin and Toronto the effect is great, but the conditions are somewhat different, as the perturbing force has rather a different direction.

(3) The remaining *charts, XII to XIX,* embrace the period from $17^h\ 45^m$ to 1^h on the 1st November.

We have no observations of this period from America and Honolulu. In the eastern hemisphere the perturbation-conditions change very slowly. During the day-period the current-arrows at the Norwegian stations Kaafjord and Matotchkin Schar are directed eastwards; at the beginning of the night-period they begin to turn. In the case of Matotchkin Schar, this has already taken place at $17^h\ 45^m$ (Chart XII). At $18^h\ 30^m$, the current-arrow for Kaafjord has its usual direction westwards along the auroral zone. Throughout this last period, Axeløen has a comparatively small horizontal component, which sometimes varies greatly in direction. The vertical component, on the other hand, is very considerable, and is directed downwards, thus indicating that it is perhaps an effect of the current that causes the powerful perturbations in H at Kaafjord and Matotchkin Schar. The vertical components at these stations indicate that the main bulk of the current is passing right over, or a little to the south of, Matotchkin Schar, and south of Kaafjord. Simultaneously with this reversal of the force, we notice a great change with regard to the force in the rest of Europe, this, on the chart for $18^h\ 30^m$, being about as powerful as at Dehra Dun; but on the other hand the force has now diminished considerably at Zi-ka-wei. The current-arrows in Central Europe on the whole at this point of time are south-west in direction.

As we come southwards towards San Fernando, we find the arrow turning more to the west. We receive the impression that the perturbation-conditions have moved westwards with the sun. This movement seems to be continued, as the magnitude of the force in Central Europe, as compared with that at Dehra Dun, is increasing, while the direction of the arrows becomes more southerly, that is to say, the turning is counter-clockwise.

On Chart XVIII, for $23^h\ 45^m$, the force is decomposed, as we have endeavoured to take out the force for the other powerful storm at Axeløen. This storm, which commences at about midnight, and is powerful at the Norwegian stations, has also, as far as may be judged from our material, the outward field that is characteristic of these storms. There is an area of convergence in the north-east of Europe and the north-west of Asia.

The last chart—at 1^h—shows the perturbations in Europe, including the Norwegian stations, to be greatly diminished, while at Dehra Dun the perturbation still continues fairly powerful for a long time. Throughout the next twenty-four hours, H has a value that is about 10γ below that of the preceding calm days, notwithstanding that the curve on the following day is of a quiet character. As the mean line has been drawn in relation to the calm days, this low value of H will affect the perturbing force, and serve to increase its total amount.

HOW THESE PERTURBATIONS MAY BE EXPLAINED.

67. In the above description we have pointed out the most important properties of this perturbation. These we will now briefly recapitulate.

(1) The perturbation is very violent at the Norwegian stations. The character of their curves is very disturbed. The curve for Sitka for that day is of the same character.

(2) The perturbation, in the eastern hemisphere especially, may be divided into one long, principal storm, whose field, in its main forms, varies only very slowly, and two intermediate, powerful, but briefer storms, that differ considerably from the first-named in the fields of force that they produce.

We will first take the conditions during *the long and more constant storm*, beginning with that part of it for which we also have material from the American stations and Honolulu.

On account of the violent nature of the storms round the Norwegian stations, we must assume that the systems come close to these places. There are thus great precipitations on the day-side, and the current-arrow during the period is directed eastwards along the auroral zone.

The effect in lower latitudes undoubtedly seems to some extent to be due to the direct influence of these polar precipitations. The fact that the perturbations in this period are all more powerful in the district of Dehra Dun and Batavia than in Europe, might make it natural to suppose that in addition to the polar systems there are also systems that have their greatest effect in the equatorial regions. This kind of storm we have already mentioned, and have referred them to the so-called negative equatorial storms (p. 83). In this perturbation we have a typical example of such a storm.

In North America the perturbation-conditions varied in a manner that was without parallel in the eastern hemisphere. This, together with the great changes in the perturbation-conditions from place to place, points to the conclusion that the perturbations here are due to systems that are relatively independent as compared with that which occurs farther east; and on a closer investigation, it also appears that the field is of the same form as that during the polar elementary storms that occur on the night-side of the earth. From the great strength of the perturbation at Sitka as compared with Toronto and Baldwin, we may conclude that the first-named station must be situated in the neighbourhood of the field of precipitation. The current-arrow also remains constant, pointing westwards along the auroral zone. It would appear that on this occasion these polar storms occur rather far south. If we were thus to assume, as we might with reason do, that these polar storms in North America, and perhaps also farther west, surround themselves with a field whose properties resemble those during the series of polar elementary storms already described, with centres near the Norwegian stations, it will be impossible to explain the strength and direction of the force at Honolulu as a direct effect of correspondent polar systems with centres in North America. The perturbation at Honolulu must mainly be conditioned by the equatorial system.

During the second part of the long storm, the Norwegian stations begin to enter the evening and night side, and we see that the current-arrows turn round. This takes place earlier at Matotchkin Schar than at Kaafjord, showing that the cause producing this change in direction moves westwards with the sun. At the Norwegian stations the perturbations have a very local character, but the conditions on the whole are almost alike at Kaafjord and Matotchkin Schar, that is to say the direction of the current-arrows; but at Axeløen they are very different. There there is a great vertical component, but a small horizontal component (e. g. Chart XVI). A possible explanation of this is, perhaps, that as the current on this occasion lies rather far south, Axeløen comes near to the neutral area.

In lower latitudes also, we see that the district of the most powerful field has moved westwards or in other words, this perturbation is of such a kind that the greater part of it follows the sun.

We have already mentioned that at the stations Dehra Dun, Bombay and Batavia, a long diminution in the horizontal intensity ensues, continuing throughout the day and night following.

At the Norwegian stations the polar storms cease, and comparatively quiet conditions supervene as early as 3^h on the 1st November.

In this manner we see that the perturbations that have appeared at the equator make themselves independent of the polar storms, and outlast them. It might indeed be argued that the perturbation is due to an after-effect of the long storm, in other words, that after the polar storms have ceased, it is not real current-systems with which we have to do, but only an induced and slowly-vanishing temporary magnetism in the magnetisable masses of the earth. This would be in accordance with the quite character of the curve on the following day.

In reality we here have before us a question of a fundamental nature, the answering of which would be of the greatest importance to our comprehension of terrestrial magnetism itself, but would require an acquaintance with these magnetisable masses such as we do not possess.

It is certainly not impossible that a storm such as this, which has been powerful and lasted long, may have after-effects. But the after-effect cannot explain it entirely; for at 5^h on the 1st November, at a time when the storm in the north has ceased, H at Bombay still amounts to $33\,\gamma$. It is true the force at Bombay has passed a value of $89\,\gamma$, and during several hours maintains a value of about $70\,\gamma$; but nevertheless an after-effect of half this amount seems improbable.

If such an after-effect at the equator were due to a temporary remnant-magnetism in the earth, and if we suppose the magnetisable masses to be arranged symmetrically with reference to the magnetic equator, we should also expect to find the direction of this effect the reverse of that of the exterior magnetising force.

In treating of the first part of the perturbation, by considering the conditions at Honolulu, we arrived at the conclusion that we must here assume the existence of a negative equatorial system (see Art. 32), as the perturbations at Honolulu did not harmonise either in direction or strength with the conditions farther north, and took no part in the great variations undergone by the perturbations in North America. According to this, we may conclude that this time there is the effect of a current-system which acts most powerfully in the regions round the equator. We are naturally led to connect this perturbation with a circular stream of electric corpuscles flowing round the earth, resembling the luminous ring round the terrella in the experiment represented in fig. 37. On account of the universal distribution of the effect, the current cannot lie near the earth, but should be at a distance of at least the same magnitude as the earth's radius. If this were the case, we should expect to find similar disturbances in the vertical intensity near the poles, and, still more, an increase in this force in the north. It is at once apparent that the form of the vertical curve for Axeløen has some resemblance to that of the *H*-curves at Dehra Dun and Batavia; and the quiet character of the curve may perhaps indicate that here we have not principally direct effects of the polar storms. The deflection really answers to an increase in *V*, and remains powerful and so constant that the probability of its being caused only by the powerful storms about the auroral zone is not very great.

A calculation of the magnetic effect produced at various places by a circular current round the earth at a considerable distance from it, may here be of some interest.

Let us first assume that such a corpuscular circular current has the same magnetic effect as a galvanic linear current. This circular current we will suppose to be situated almost in the plane of the magnetic equator, its centre coinciding with that of the earth, and its radius equal to $2R$, R being the radius of the earth.

The effect of such a current upon a magnetic mass 1 cm.$^{3/2}$ gr.$^{1/2}$ sec.$^{-1}$, situated in the plane of the current, we find to be

$$F = \int_0^{2\pi} \frac{i}{10} \frac{(a - l\cos\varphi)\, d\varphi}{(a^2 + l^2 - 2al\cos\varphi)^{3/2}},$$

where a is the radius of the current-circle, l the distance of the magnetic pole from the centre of this circle, i the current in amperes, and F the force expressed in *C. G. S.* units.

This integral may easily be transformed into elliptical integrals of the normal types.

We have here calculated it numerically for the values $a = 2R$, $l = R$, and we find that

$$F_1 = 1.23 \cdot \frac{i\pi}{10R}$$

In the centre of the current-circle we have

$$F_2 = \frac{i\pi}{10R}$$

It will be seen that the force is somewhat less at the centre of the earth than in the equatorial districts; but the difference is not very great.

We will now consider the earth as a homogeneous magnetisable sphere, situated in a uniformly magnetic field of a strength

$$F = \frac{i\pi}{10R}$$

The magnetisation produced in the sphere will give rise to the forces

$$F_p = 2KF, \quad F_e = -KF,$$

respectively at the pole and at the equator, where

$$K = \frac{\mu - 1}{\mu + 2},$$

μ being the permeability of the sphere. (See Mascart: L'Électricité et le Magnétisme. Paris, 1896; p. 417.)

The value of μ, that may be used for the earth, is very difficult to determine. F. Pockels (Wiedemanns Analen 63, p. 199, 1897) gives values of about 1.1 for basalt for the smallest field-intensities. For other minerals, however, we find values of even a hundred times greater, e. g. magnetite, pyrrhotite, hæmatite, limonite, etc.

If we take 2 as an average value of μ, we obtain

$$F + F_p = 2(F + F_e).$$

In this way we should expect to find values of P_v at the magnetic poles about double the value of P_h observed near the equator. For greater values of μ the proportion $P_v : P_h$ will increase, and vice versa.

From about 16^h to 18^h we really find conditions that seem to favour our assumptions, when we compare the values of P_v at Axeløen with the value of P_h at Dehra Dun and Batavia. Later on, however, we find that P_v increases greatly, while P_h, at the equatorial stations, is slowly diminishing and that before this period P_v is much less and even sometimes directed to the opposite side.

We cannot, of course, draw any further conclusions from this, as it is impossible to determine how great a part of P_v at Axeløen is due to polar precipitations. There is all the greater need of caution in drawing conclusions, from the fact that the conditions at Christchurch—which is in a comparatively high southern latitude—show that at that place there is only a very slight perturbation in the vertical intensity, and from about $13^h\ 30^m$ onwards, the corresponding P_v is directed downwards, not upwards as we should expect when only the equatorial perturbation is acting. We there find, moreover, comparatively powerful perturbing forces in the horizontal components, and it would thus appear that there were precipitations of a more polar character in the southern hemisphere also.

If, with the assumed value of μ, we make the force P_h at the equator equal to $75\,\gamma$, we find that

$$F + F_e = \frac{3}{4}\frac{\pi i}{10R} = 75 \cdot 10^{-5},$$

and i must then be equal to about $2 \cdot 10^6$ amperes, a value of the same order as that which we shall find in the calculation of the current-strength in the polar perturbations (see Chap. IV).

The first intermediate storm, with maximum about $13^h\ 42^m$ occurs during the same time and with great violence, at Sitka and at Axeløen. Its local character at these places shows that the current-systems are comparatively near to both stations.

It is plain from the simultaneous appearance of the intermediate storms at Sitka and at Axeløen, that these two storms must be closely connected with one another; but whether they are the effect of a single system, or of separate and more limited systems of precipitation in the vicinity of the two stations, it is impossible to decide with any certainty.

We have seen in Art. 52 (cf. fig. 68) how well the assumption of separate fields of simultaneous precipitation agrees with our theory; and circumstances are actually found here that seem to favour such a view. The maximum occurs, indeed, at about the same time, namely at $13^h\ 42^m$, but the storm begins at Sitka about a quarter of an hour before that at Axeløen, and perhaps does not end until a quarter of an hour after the latter has ceased. If we look at the declination at Baldwin, where the intermediate storm is well defined, it appears that the storm there begins at $13^h\ 8^m$, and concludes at $14^h\ 34^m$.

If we look at the H-curve for Kew or Wilhelmshaven, we notice that during this perturbation the course of the curves is as follows: first at $13^h\ 12^m$, there is a deflection answering to a diminution of H; at $13^h\ 24^m$, H has an intermediate minimum, then increases until $13^h\ 42^m$, then decreases until $14^h\ 5^m$, when it again increases, and at $14^h\ 30^m$ the effect of the impulse has ceased. The D-curve has a similar course. It may perhaps therefore be natural to interpret the conditions in Europe in the following manner.

Between $13^h\ 12^m$ and $14^h\ 30^m$ there is a perturbation of uniform direction, occurring simultaneously with the perturbation in America. P_h and P_d are directed respectively south and west, answering to a current-arrow pointing north-west or west-north-west. This is interrupted by another perturbation, which lasts from $13^h\ 24^m$ to $14^h\ 5^m$, and acts in almost exactly the opposite direction; and at the moment when this latter storm reaches its maximum at Kew, it causes the effect of the former perturbation to cease as far as Kew is concerned. Here, on account of the intermediate storms, the perturbing force thus becomes 0 at the moment when the storm is at its height. During the brief storm, the current-arrows are directed ESE, and these should be connected with the brief, powerful storm at Axeløen, just as the latter is naturally connected with the powerful impulse in the southern Asiatic district.

The assumption that a distant system such as that in the vicinity of Sitka would have so great an effect in Europe as we find here, may, however, present some difficulty; and yet more doubtful does such an explanation become when we look at the conditions at Kaafjord, where, all through, a system is acting which produces current-arrows with an easterly direction. Simultaneously with the intermediate storm at Axeløen, there appears to be an intermediate storm here, which, as far as H is concerned, begins, reaches its maximum, and ends, almost at the same times as the storm at Axeløen. The deflections, however, are the reverse of those at Axeløen, as in this case we find positive values of P_h, and the strength is considerably less. In the declination, on the other hand, there is a rather brief impulse in an easterly direction, with maximum at about $13^h\ 30^m$, being therefore almost exactly simultaneous with the maximum of the first deflection at Wilhelmshaven. The curves at Matotchkin Schar show in some respects a resemblance to the conditions at Axeløen, and in others to those at Kaafjord. In H the maximal negative deflection occurs earlier than at Axeløen, and about simultaneously with that in the declination at Kaafjord, i. e. at about $13^h\ 30^m$, while at the same time there is also a fairly powerful easterly deflection in the declination. As regards the intermediate storm, the conditions at Matotchkin Schar might seem to form a connection between the conditions at Axeløen and those at Sitka, thus indicating that we had before us a connected intermediate system with current-arrows on the night-side of the earth directed westwards. If we accept the first explanation of the conditions, we should thus have to ignore completely the effects of the system in the neighbourhood of Kaafjord, a system which seems, indeed, to be comparatively weaker, and in that respect will have a more limited sphere of action, but on the other hand is so close to the Central European stations, that its effect there will in all probability be very apparent.

It should be remarked that the effect in Central Europe of this system in the neighbourhood of Kaafjord is similar to that of the assumed system at Sitka, as they will both produce current-arrows directed westwards.

Finnally, as the conditions at Matotchkin Schar appear to indicate that the system at Sitka is continued westwards to Axeløen—a circumstance that we have previously continually met with—there is every probability that the westward-directed intermediate current-arrows are the effect of the system observed at Kaafjord. Farther west we should without doubt have found this system more fully developed; and observations from Dyrafjord would therefore have been of great importance here.

We must suppose then that the effect of the southern system near Kaafjord might first predominate, then the stronger but more distant system near Axeløen at the time when the latter is at its height, and finally the southern system once more. The fact that the conditions in the Asiatic districts are more analogous to those at Axeløen also finds a natural explanation here, the southern system at Kaafjord being of far less strength than that at Axeløen, and therefore having a correspondingly smaller area of action. We are confirmed in these assumptions by the course of the broken-lined arrows in Charts IX and X. Thus on Chart IX we find an indication of a small area of divergence on the day-side, and a larger area of convergence on the night-side; while on Chart X this area of convergence extends farther west to the western stations of Central Europe.

The storm at Axeløen is an afternoon storm, and ought therefore to be compared throughout with such storms, e. g. those of the 15th and 8th February, 1903, and the 27th October, 1902, where we also found two rather different systems acting at Axeløen and at Kaafjord.

The last great intermediate storm, from $11^h\ 12^m$ to $0^h\ 42^m$, has on the whole been already characterised, as we have previously proved that it has the same field of force as the ordinary polar elementary storms that occur about midnight, and have their centre about the Norwegian stations.

TABLE XXXVII.

The Perturbing Forces on the 31st October, and 1st November, 1902.

Gr. M. T.	Honolulu		Sitka		Baldwin		Toronto	
	P_h	P_d	P_h	P_d	P_h	P_d	P_h	P_d
h m								
9 0	−34.7 γ	E 5.8 γ	− 75.3 γ	E 117.0 γ	+ 14.9 γ	0	+ 17.1 γ	W 25.8 γ
30	−34.7 »	» 5.8 »	− 66.6 »	» 72.5 »	+ 3.4 »	E 17.1 γ	+ 5.8 »	E 6.0 »
10 0	−33.2 »	» 10.8 »	− 91.0 »	» 57.2 »	− 2.0 »	W 10.2 »	− 2.7 »	W 23.5 »
15	−35.6 »	» 11.6 »	−135.0 »	» 78.8 »	+ 19.7 »	» 34.2 »	0	» 75.8 »
30	−35.6 »	» 9.9 »	−101.0 »	» 142.0 »	+ 42.3 »	0	+ 52.2 »	» 52.3 »
11 0	−32.3 »	» 7.4 »	−111.0 »	» 105.0 »	+ 26.1 »	E 8.9 »	+ 30.2 »	» 36.6 »
30	−32.3 »	» 12.4 »	− 98.9 »	» 86.4 »	+ 5.4 »	W 14.6 »	+ 5.8 »	» 47.5 »
12 30	−31.9 »	» 7.4 »	− 97.3 »	» 100.0 »	+ 58.3 »	» 27.3 »	+ 12.1 »	» 65.0 »
13 30	−25.3 »	» 11.6 »	–> 212.0 »	» 8.1 »	− 31.2 »	» 59.0 »	− 48.1 »	» 56.5 »
42	−27.7 »	» 16.6 »	–> 212.0 »	» 84.8 »	− 10.1 »	» 76.2 »	− 36.0 »	» 71.0 »
14 0	−25.3 »	» 7.4 »	−203.0 »	» 40.5 »	+ 2.3 »	» 64.8 »	− 21.2 »	» 87.8 »
17 45	?	?		?	?	?	?	?
18 30	?	?		?	?	?	?	?
19 15	?	?		?	?	?	?	?
20 30	?	?		?	?	?	?	?
21 45	?	?		?	?	?	?	?
22 0	?	?		?	?	?	?	?
23 45	?	?		?	?	?	?	?
1 0	?	?		?	?	?	?	?

TABLE XXXVII (continued).

Gr. M. T.	Axeløen			Matotchkin Schar			Kaafjord		
	P_h	P_d	P_v	P_h	P_d	P_v	P_h	P_d	P_v
h m									
9 0	− 11.1 γ	E 39.4 γ	− 4.9 γ	+ 94.0 γ	W 22.2 γ	− 7.6 γ	+ 15.3 γ	W 13.9 γ	+ 18.8 γ
30	− 19.3 »	» 51.4 »	− 9.7 »	+111.0 »	» 53.5 »	− 15.3 »	+ 38.0 »	» 12.1 »	+ 81.5 »
10 0	− 15.2 »	» 43.0 »	− 29.2 »	+128.0 »	» 31.0 »	− 18.6 »	+ 68.2 »	0	+ 92.5 »
15	− 46.5 »	» 43.0 »	− 31.6 »	+152.0 »	» 36.3 »	− 27.2 »	+ 91.0 »	» 9.9 »	+ 89.3 »
30	− 43.2 »	» 40.5 »	− 31.6 »	+194.0 »	» 31.8 »	− 30.7 »	+138.0 »	» 15.7 »	+ 71.3 »
11 0	− 40.4 »	» 44.6 »	− 17.3 »	+201.0 »	» 52.0 »	− 66.4 »	+125.0 »	» 30.7 »	+ 39.2 »
30	− 59.0 »	» 26.3 »	− 17.3 »	+169.0 »	» 18.6 »	− 86.0 »	+100.0 »	» 24.8 »	+ 39.2 »
12 30	− 86.0 »	» 9.2 »	− 17.3 »	+142.0 »	» 56.3 »	–> 102.0 »	+219.0 »	» 50.2 »	− 40.8 »
13 30	−394.0 »	» 24.2 »	+131.0 »	−142.0 »	E 46.5 »	–> 102.0 »	+189.0 »	E 117.0 »	−317.0 »
42	−547.0 »	» 51.6 »	+383.0 »	+ 21.5 »	W 130.0 »	–> 102.0 »	+257.0 »	W 133.0 »	–> 512.0 »
14 0	−234.0 »	» 11.9 »	+ 85.0 »	+214.0 »	» 181.0 »	–> 102.0 »	+316.0 »	» 53.8 »	−269.0 »
17 45	− 42.2 »	W 80.0 »	+119.0 »	−177.0 »	0	–> 102.0 »	+ 22.1 »	E 10.2 »	− 35.2 »
18 30	− 64.0 »	» 00.0 »	+195.0 »	–> 240.0 »	E 52.2 »	–> 102.0 »	−165.0 »	» 17.6 »	+125.0 »
19 15	− 36.7 »	» 92.3 »	+217.0 »	–> 240.0 »	» 237.0 »	0	−212.0 »	» 65.2 »	+172.0 »
20 30	−124.0 »	» 32.8 »	+397.0 »	–> 240.0 »	» 257.0 »	+ 29.0 »	−282.0 »	» 159.0 »	+247.0 »
21 45	− 35.8 »	0	+421.0 »	–> 240.0 »	» 337.0 »	+147.0 »	−294.0 »	» 193.0 »	+188.0 »
22 0	− 29.4 »	» 30.2 »	+367.0 »	–> 240.0 »	» 203.0 »	+154.0 »	−263.0 »	» 95.3 »	+161.0 »
23 45	−108.0 »	E 82.9 »	+343.0 »	–> 240.0 »	» 163.0 »	+ 0.8 »	−351.0 »	» 88.3 »	− 20.3 »
1 0	+ 18.4 »	» 39.6 »	+214.0 »	− 65.0 »	» 68.2 »	−108.0 »	−113.0 »	» 75.0 »	− 7.0 »

TABLE XXXVII (continued).

Gr. M. T.	Stonyhurst		Wilhelmshaven			Kew		Potsdam	
	P_h	P_d	P_h	P_d	P_v	P_h	P_d	P_h	P_d
h m									
9 0	− 11.2 γ	W 14.3 γ	− 16.8 γ	W 18.9 γ	0	− 12.7 γ	W 12.2 γ	− 8.9 γ	W 15.7 γ
30	− 12.2 »	» 20.0 »	− 16.8 »	» 21.4 »	0	− 11.2 »	» 20.2 »	− 6.0 »	» 20.3 »
10 0	− 14.3 »	» 13.1 »	− 19.1 »	» 8.5 »	0	− 16.8 »	» 10.3 »	− 7.0 »	» 14.2 »
15	− 15.3 »	» 13.1 »	− 21.5 »	» 7.9 »	0	− 19.4 »	» 10.3 »	− 13.9 »	» 17.2 »
30	− 12.7 »	» 18.8 »	− 20.5 »	» 14.7 »	0	− 15.8 »	» 17.3 »	− 13.9 »	» 22.8 »
11 0	− 19.4 »	» 18.8 »	− 20.5 »	» 14.7 »	0	− 15.8 »	» 24.8 »	− 17.4 »	» 17.8 »
30	− 16.3 »	» 12.0 »	− 12.1 »	» 9.7 »	0	− 11.7 »	» 14.5 »	− 14.3 »	» 8.6 »
12 30	− 11.2 »	» 24.5 »	− 18.2 »	» 34.2 »	0	− 4.0 »	» 23.8 »	?	?
13 30	− 18.3 »	» 5.7 »	− 38.6 »	0	0	− 26.4 »	» 6.5 »	?	?
42	− 7.1 »	» 22.2 »	− 6.1 »	» 7.9 »	+ 14.0 γ	− 9.2 »	» 11.7 »	?	?
14 0	− 22.4 »	» 16.6 »	− 41.0 »	» 12.2 »	+ 13.0 »	− 22.8 »	» 14.9 »	?	?
17 45	− 41.8 »	» 25.7 »	− 61.5 »	» 20.2 »	+ 11.0 »	− 27.0 »	» 26.2 »	?	?
18 30	− 47.4 »	E 22.3 »	− 48.0 »	E 29.3 »	+ 22.0 »	− 46.9 »	E 12.2 »	?	?
19 15	− 57.5 »	» 31.4 »	− 54.5 »	» 45.8 »	+ 16.0 »	− 52.0 »	» 22.9 »	?	?
20 30	− 49.4 »	» 31.4 »	− 44.3 »	» 41.0 »	+ 16.0 »	− 41.3 »	» 28.6 »	?	?
21 45	− 38.2 »	» 28.0 »	− 34.5 »	» 31.8 »	+ 7.0 »	− 28.5 »	» 25.2 »	?	?
22 0	− 37.2 »	» 42.8 »	− 31.7 »	» 51.4 »	+ 7.0 »	− 31.6 »	» 40.0 »	?	?
23 45	− 13.2 »	» 57.7 »	− 14.0 »	» 57.5 »	+ 9.0 »	− 4.6 »	» 51.5 »	?	?
1 0	− 15.8 »	» 48.0 »	− 16.8 »	» 31.2 »	0	− 14.8 »	» 16.8 »	?	?

TABLE XXXVII (continued).

Gr. M. T.	Pola			San Fernando		Dehra Dun	
	P_h	P_d	P_v	P_h	P_d	P_h	P_d
h m							
9 0	− 7.1 γ	W 15.3 γ	− 7.4 γ	− 7.6 γ	W 9.0 γ	− 45.6 γ	E 6.9 γ
30	− 7.6 »	» 18.1 »	− 7.6 »	− 5.1 »	» 9.8 »	− 50.4 »	» 1.9 »
10 0	− 13.0 »	» 17.3 »	− 6.7 »	− 13.4 »	» 4.9 »	− 56.3 »	» 3.0 »
15	− 17.5 »	» 18.7 »	− 4.8 »	− 17.9 »	» 8.2 »	− 60.2 »	» 3.9 »
30	− 16.6 »	» 23.6 »	− 4.4 »	− 12.8 »	» 15.6 »	− 66.5 »	0
11 0	− 20.6 »	» 26.4 »	+ 0.4 »	− 14.7 »	» 18.8 »	− 71.0 »	0
30	− 18.8 »	» 18.7 »	+ 8.2 »	− 14.7 »	» 18.8 »	− 64.5 »	» 1.9 »
12 30	− 20.2 »	» 27.8 »	+ 9.5 »	− 19.8 »	» 18.8 »	− 81.0 »	» 3.0 »
13 30	− 46.2 »	0	+ > 20.2 »	− 49.9 »	» 14.8 »	− 36.2 »	» 21.6 »
42	− 29.6 »	» 4.2 »	+ > 20.2 »	− 40.8 »	» 13.9 »	− 48.5 »	» 12.3 »
14 0	− 42.5 »	» 4.1 »	+ > 20.2 »	− 44.8 »	» 8.2 »	− 59.0 »	» 19.7 »
17 45	− 32.6 »	» 11.1 »	+ > 21.2 »	− 38.3 »	» 13.1 »	− 59.0 »	» 15.8 »
18 30	− 44.8 »	E 11.1 »	+ > 21.2 »	− 56.2 »	» 6.5 »	− 54.0 »	» 19.7 »
19 15	− 42.6 »	» 21.6 »	+ > 21.2 »	− 60.8 »	E 4.1 »	− 47.5 »	» 17.8 »
20 30	− 30.8 »	» 26.4 »	+ > 21.2 »	− 46.6 »	» 9.0 »	− 40.2 »	» 15.8 »
21 45	− 12.5 »	» 22.8 »	+ 18.6 »	− 26.8 »	» 9.8 »	− 33.5 »	» 7.8 »
22 0	− 14.3 »	» 29.2 »	+ 18.6 »	− 33.2 »	» 13.1 »	− 33.5 »	» 8.8 »
23 45	+ 13.0 »	» 26.4 »	+ 9.9 »	− 13.4 »	» 27.0 »	− 31.8 »	0
1 0	− 17.9 »	» 7.0 »	+ 2.9 »	− 14.7 »	» 10.6 »	− 43.3 »	0

TABLE XXXVII (continued).

Gr. M. T.	Zi-ka-wei		Batavia		Christchurch		
	P_h	P_d	P_h	P_d	P_h	P_d	P_v
h m							
9 0	− 45.5 γ	E 2.0 γ	− 57.0 γ	o	− 33.2 γ	E 8.2 γ	− 4.6 γ
30	− 52.8 »	W 4.0 »	− 60.5 »	o	− 41.4 »	W 11.1 »	− 4.6 »
10 0	− 54.0 »	o	− 60.5 »	o	− 35.4 »	» 17.1 »	− 3.7 »
15	− 63.5 »	» 1.0 »	− 60.5 »	o	− 39.1 »	» 22.3 »	− 2.2 »
30	− 73.1 »	» 4.0 »	− 67.5 »	o	− 51.5 »	» 27.5 »	− 2.2 »
11 0	− 68.4 »	» 3.0 »	− 74.5 »	o	− 58.3 »	» 27.5 »	− 4.3 »
30	− 58.8 »	» 1.0 »	− 64.5 »	o	− 45.5 »	» 21.6 »	− 3.7 »
12 30	− 73.1 »	» 3.0 »	− 81.5 »	o	− 58.3 »	» 53.5 »	o
13 30	− 6.0 »	E 6.0 »	− 18.5 »	W 15.6 γ	− 4.1 »	» 25.3 »	o
42	− 28.8 »	» 5.0 »	− 33.0 »	» 15.6 »	− 13.8 »	» 44.6 »	+ 4.9 »
14 0	− 30.0 »	» 6.0 »	− 45.5 »	» 6.0 »	− 13.3 »	» 54.2 »	+ 4.9 »
17 45	− 24.0 »	» 3.0 »	− 54.3 »	» 13.2 »	− 12.4 »	E 8.9 »	+ 3.7 »
18 30	− 13.2 »	» 7.0 »	− 46.8 »	» 15.6 »	− 8.7 »	» 37.8 »	+ 2.5 »
19 15	− 13.2 »	» 4.0 »	− 41.2 »	» 15.6 »	− 18.4 »	» 43.9 »	+ 4.3 »
20 30	− 14.4 »	o	− 34.2 »	» 9.6 »	− 36.3 »	» 53.5 »	+ 4.3 »
21 45	− 7.2 »	W 8.0 »	− 38.3 »	o	− 52.5 »	» 38.6 »	+ 3.7 »
22 0	− 4.8 »	» 8.0 »	− 35.8 »	o	− 54.8 »	» 33.5 »	+ 3.7 »
23 45	− 16.8 »	» 5.0 »	− 49.0 »	?	− 60.2 »	?	?
1 0	− 9.6 »	» 10.0 »	?	?	?	?	?

TABLE XXXVIII.

Partial Perturbing Forces during the Perturbation of the 31st October, 1902.

Observatory	$13^h\ 30^m$		$13^h\ 42^m$		$14^h\ 0^m$	
	P'_h	P'_d	P'_h	P'_d	P'_h	P'_d
Honolulu	− 11.7 γ	E 14.5 γ	− 16.4 γ	E 15.8 γ	− 20.2 γ	E 14.1 γ
Sitka	−> 118.0 »	W 63.0 »	−> 123.0 »	» 22.5 »	−> 131.0 »	W 4.5 »
Baldwin	− 31.2 »	» 29.2 »	− 10.2 »	W 44.0 »	+ 2.3 »	» 35.0 »
Toronto	− 43.2 »	» 8.4 »	− 32.0 »	» 21.1 »	− 17.6 »	» 39.7 »
Axeloen	− 307.0 »	E 35.8 »	− 457.0 »	E 87.5 »	− 160.0 »	E 25.8 »
Matotchkin Schar	− 367.0 »	» 166.0 »	− 232.0 »	W 73.5 »	o	W 13.3 »
Kaafjord	o	» 163.0 »	+ 81.5 »	» 86.1 »	+ 14.8 »	» 7.3 »
Stonyhurst	− 19.9 »	» 9.1 »	o	» 4.5 »	− 14.8 »	o
Wilhelmshaven	− 24.2 »	» 18.3 »	+ 8.4 »	E 11.6 »	− 23.8 »	E 5.5 »
Kew	− 27.0 »	» 3.7 »	− 5.6 »	o	− 23.9 »	W 3.7 »
Pola	− 24.2 »	» 15.2 »	− 8.5 »	» 11.0 »	− 20.7 »	E 10.3 »
San Fernando	− 26.8 »	W 6.5 »	− 17.2 »	W 5.7 »	− 21.0 »	W 6.5 »
Dehra Dun	+ 37.0 »	E 16.7 »	+ 19.3 »	E 4.9 »	+ 8.3 »	E 6.9 »
Zi-ka-wei	+ 58.8 »	» 8.0 »	+ 34.8 »	» 3.0 »	+ 22.8 »	» 8.0 »
Batavia	+ 50.5 »	W 12.0 »	+ 34.8 »	W 9.6 »	+ 18.8 »	W 4.8 »
Christchurch	+ 38.6 »	E 11.2 »	+ 26.7 »	» 11.2 »	+ 21.6 »	» 19.3 »

Current-Arrows for the 31st October, 1902; Chart III at $10^h\ 0^m$, and Chart IV at $10^h\ 15^m$.

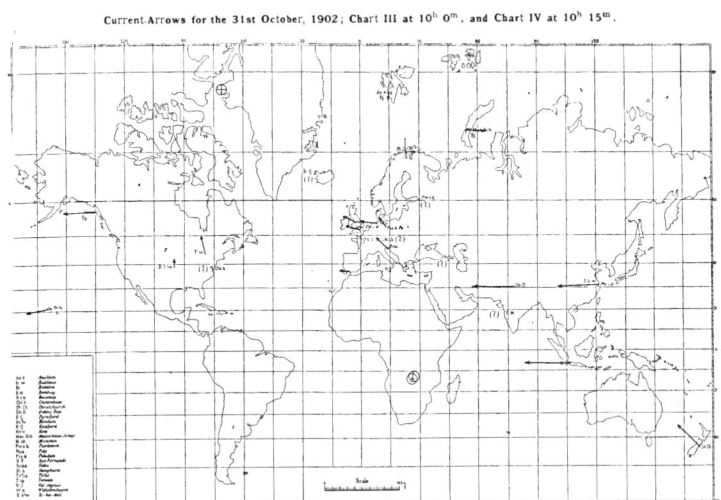

Current-Arrows for the 31st October, 1902; Chart V at 10^h 30^m, and Chart VI at 11^h 0^m.

Current-Arrows for the 31st October, 1902; Chart VII at $11^h\ 30^m$, and Chart VIII at $12^h\ 30^m$.

urrent-Arrows for the 31st October, 1902; Chart XI at $14^h\ 0^m$, and Chart XII at $17^h\ 45^m$.

Current-Arrows for the 31st October, 1902; Chart XIII at $18^h\ 30^m$, and Chart XIV at 19^h 15

Current-Arrows for the 31st October, 1902; Chart XV at $20^h \cdot 30^m$, and Chart XVI at 21^h 45^m.[1]

Fig. 114.

[1] The vertical intensity arrow on Chart XVI at Matotchkin Schar by mistake made too short; should be 10 times the given length.

Current-Arrows for the 31st October, 1902; Chart XVII at $22^h\ 0^m$(1), and Chart XVIII at $23^h\ 45^m$.

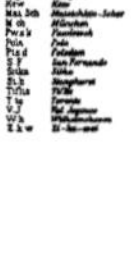

de too short; should be 10 times the given length.

Current-Arrows for the 1st November, 1902; Chart XIX at $1^h\ 0^m$.

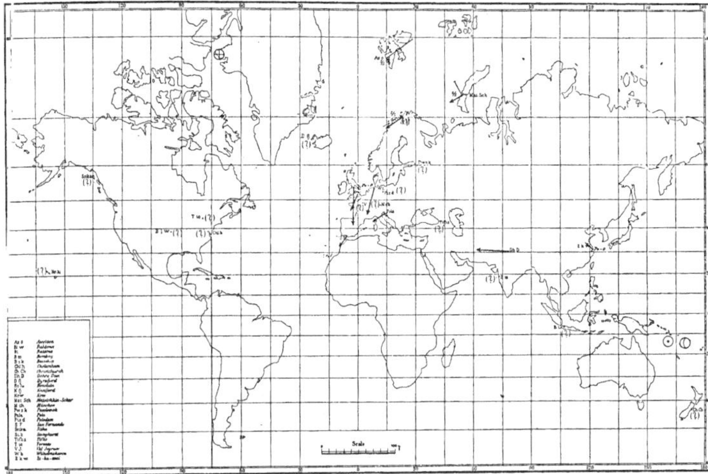

Fig. 116.

THE PERTURBATIONS OF THE 11th & 12th OCTOBER, 1902.

(Pl. II).

68. From 11^h on the 11th October, to about $0^h\ 30^m$ on the 12th, perturbations that are sometimes violent are noted at all the stations from which we have observations. They are unusually violent in the equatorial regions, where the conditions become rather complicated, as there are undoubtedly often several current-systems, sometimes even occurring simultaneously.

The perturbations seem to fall naturally into three principal sections,

The first from 11^h to $17^h.20^m$ on the 11th October,

The second from $17^h\ 20^m$ to $18^h\ 30^m$ on the 11th October, and

The third from $18^h\ 30^m$ on the 11th October, to $0^h\ 30^m$ on the 12th.

In the *first section,* it is especially in the horizontal intensity that the perturbation occurs. We see that the perturbing force allmost everywhere is directed northwards along the magnetic meridian. The way in which the force is generally distributed during this period is shown on *Chart II,* for $17^h\ 0^m$.

It appears from the copies of the curves, that this part of the perturbation is especially well developed in the equatorial regions. This, together with the serrated character of the horizontal intensity curve, and the direction of the force, would make it appear that this is mainly a positive equatorial perturbation of the well-known type (cf. e. g. Art. 27).

During this first section, polar storms also occur at our Norwegian stations; but they are not very considerable, although of sufficient strength to explain the partial loss of the typical character of the equatorial perturbation, especially as regards the northern stations.

Between $12^h\ 25^m$ and $13^h\ 15^m$, however, a considerable polar perturbation sets in.

It should be especially noticed that, as the curves show, during this interval of time there is a perturbation at Sitka that, for that place, is rather violent. The direction of the perturbing force is very nearly west all the time, and its greatest value is reached at about $12^h\ 50^m$. It is also noticed at the Norwegian stations, most distinctly at Matotchkin Schar; at Axeløen it is less, but still noticeable, and at Kaafjord it is almost imperceptible. If we look at the curve for Matotchkin Schar, we see that the force there is uniform in direction along the magnetic meridian; and we notice particularly that the maximum does not occur until $13^h\ 18^m$ almost half an hour later than at Sitka. This must either be explained by a movement of the current-system, or we must assume that the perturbation at Matotchkin Schar is due to a relatively different system.

The farther we go from the polar regions, the less perceptible does this brief polar perturbation become. It is distinctly noticeable at Baldwin and Cheltenham, but not at Honolulu. At the European stations, it is only just perceptible. At Zi-ka-wei and Dehra Dun it is distinctly noticed, at Batavia it is almost imperceptible. At Christchurch on the other hand, there is a rather violent perturbation in relation to the place, only noticeable in the *H*-curve. The perturbing force is here directed northwards along the magnetic meredian, corresponding to a current-arrow from west to east. The effect at Christchurch cannot have been produced by the same system as that which acts in the northern hemisphere; for the effect of the latter is imperceptible even at Honolulu and Batavia.

The explanation of this seems to be that simultaneously with the descent in the north, a similar phenomenon appears near the south pole, and it is the effects of the latter that we observe at Christchurch.

On Chart I, for $12^h\ 50^m$, only the current-arrows corresponding to the polar storms are shown, as we have endeavoured to separate their effect from that of the equatorial system by a decomposition of the total perturbing force.

The *second section* includes the interval from $17^h\ 20^m$ to about $18^h\ 30^m$, and it commences with the appearance of violent storms in the arctic districts. The effect is especially strong at Matotchkin Schar, but less so at Axeløen. At Sitka, on the other hand, it is very marked.

Chart III at $18^h\ 0^m$. The distribution of force seems on the whole to be conditioned by this polar storm. Judging from the serrated character of the curves, however, it seems that the effect of the equatorial storm is still perceptible.

Of arctic stations, Matotchkin Schar is the one at which the force is strongest; and its direction is there south-east. At Axeløen it is less, and is directed south-west.

If we look at the European stations from Pawlowsk to San Fernando, we find that at all of them, with the exception of Pawlowsk, the forces are rather small. Even at Stonyhurst it is less than at Tiflis and Dehra Dun. The direction of the current-arrow at Pawlowsk is about south, in the district Potsdam to Wilhelmshaven and Munich, south-west, and at Stonyhurst and Kew, almost west. If we go right across to North America, we find the direction at Cheltenham NNW, at Toronto still more northerly, and at Baldwin almost north. They form, as we see, a harmonious continuation of the directions in Europe, becoming more and more northerly as we pass from the European stations across the Atlantic to North America. Thus the current-arrows should indicate the existence of current-vortices with a clockwise motion in the North Atlantic. In reality there is something like a divergence of the horizontal component of the perturbing force out from a point in these districts. Somewhere or other, possibly near Iceland, there should be a point of divergence for P_h. At Val Joyeux and Pawlowsk, there is a distinct vertical component directed downwards.

It may further be stated that the current-arrows at Sitka, Baldwin, Toronto, and Cheltenham converge towards a point in the vicinity of Prince Albert Land.

Eastwards from Europe, the arrows turn off, but now towards the east. The directions of the arrows, in connection with that at Sitka, indicate that somewhere in the north-east of Siberia, there is a point of convergence for the perturbing force.

The *third section*, from about $18^h 30^m$ to $0^h 30^m$, is characterised by a long polar storm. The field of force of this storm is shown on the *Charts V, VI, XI, XII, XIII and XIV*, respectively for $19^h 30^m$, $20^h 30^m$, $21^h 15^m$, $21^h 30^m$, 22^h, and 23^h.

We see that the distribution of force is about the same in all of them, the strength of the field alone showing any variation.

At the arctic stations, the direction of the force is generally SSE and SE.

There is a departure from this condition at $19^h 30^m$, when the force at Axeløen and Kaafjord is almost westerly in direction. At $20^h 30^m$ the force at Kaafjord is SSE in direction, but it is still west at Axeløen.

In the rest of Europe and in Asia, the direction of the force is ESE. At San Fernando it turns a little more south, and in America the direction is south-west. This shows that in the North Atlantic there must be a point of divergence of P_h similar to that described at 18^h. At Sitka, the direction of the perturbing force is WNW.

During this period, however, there are several departures from these conditions, and it is evident from the copies of the curves, that of these there are three principal ones, the first occurring at about $18^h 34^m$ (see Chart IV), the second between $20^h 45^m$ and $21^h 20^m$ (see Charts VII to X), and the third between about $23^h 10^m$ and $0^h 25^m$ (see Chart XV).

The fact that after these short interruptions the field once more assumes its original form, makes it probable that the interruption is due to comparatively independent, brief current-systems, that occur simultaneously with the long polar storm. The correctness of this view of the matter is also confirmed by the fact that the differences do not occasion the same relative increase in strength at the various stations. If we look at the curves, we shall see that these differences occur simultaneously all over the world, even as far off as Christchurch. At the Norwegian stations also, sudden powerful perturbations are observed, some of which have a different direction from that of the long storm. The three short perturbations are thus polar storms, which intrude themselves upon the long storm. The latter we will designate as the principal storm, and the three others as intermediate storms.

The circumstances, as we see, are such as justify a decomposition of the perturbing forces into two components, each of which is the effect of a separate current-system. This decomposition has been effected in the case of the last two intermediate storms, but not of the first, as that storm commences at the time of transition from the second to the third section.

This is apparent in the curves, e. g. for Tiflis and the south-east Asiatic stations, where the *H*-curve, at about $18^h 30^m$, drops suddenly, showing that P_h, from being positive, has become negative. This circumstance makes it impossible to draw any exact normal line for the taking out of the partial forces.

We will now describe in detail the three intermediate storms.

The *first intermediate storm*, at about $18^h 34^m$.

This perturbation appears in the curves as a great, but brief, deflection at about $18^h 34^m$. At the southern stations, such as Tiflis, Dehra Dun, etc., it appears to be the long perturbation only that is at all powerful during this period. It is evident, however, from the conditions at the arctic stations, especially Kaafjord and Matotchkin Schar, that it cannot be regarded only as an increase of the principal storm, for the horizontal component of the perturbing force during this period turns round in the opposite direction.

Chart IV shows the current-arrows answering to the total perturbing force at $18^h\ 34^m$.

The current-system on the whole bears a fairly close resemblance to that of the principal storm, which has already been described.

The chief difference between them is that at Kaafjord and Matotchkin Schar the direction of the force is the reverse of that which we find during the succeeding part of this section, as the current-arrow is now directed along the auroral zone from west to east. The magnitude of the total perturbing force at Matotchkin Schar now gives a false impression of the forces that are in operation, as the total force there seems to be about equal to the difference of the forces actually present. At Kaafjord, however, the long principal storm, with current-arrows directed westwards, does not seem to have any noticeable influence until about $19^h\ 30^m$.

As regards Matotchkin Schar, we find that the current-arrows again point in the direction they had in the first section.

Unlike the distribution of force during the principal perturbation, the current-arrows in Europe are now directed westwards, and at the most northerly stations even a little north. These last, during the principal perturbation of the third section, had a more southerly direction.

In America the conditions are essentially the same as those during the long perturbation, the only exception being that the arrow at Sitka is comparatively longer and more eastward in direction.

We thus see that this time also, the perturbing forces approximately diverge from a point in the North Atlantic. The strength with which the perturbation appears in the regions round Batavia, Dehra Dun and Zi-ka-wei is especially worthy of notice.

The arrows at Irkutsk, Honolulu and Sitka indicate the formation of negative vortices corresponding to a convergence of the perturbing forces. In this case, the area of convergence would be situated in the regions surrounding the Behring Sea.

The ***second intermediate storm***, from $20^h\ 45^m$ to $21^h\ 20^m$.

In the decomposition of the total perturbing force in this storm, we have attempted to distinguish between its effect and that of the principal storm, at all the southern stations where the conditions before and after are constant.

At the arctic stations the curve shows distinctly that a particularly strong impulse occurs during this period, especially noticeable at Axeløen, where the surrounding conditions are fairly normal.

We have therefore not thought it advisable to undertake any decomposition there. The normal line for the taking out of the partial part, should be the curve as it would be drawn on paper if the principal storm only had been acting; but owing to the rapid change in the principal perturbation, this line cannot be determined with sufficient certainty.

The result of the decompositions is shown on *Charts VII—X*. The resulting arrows are here drawn entire. The arrows representing the principal storm are drawn with a dotted line, those representing the intermediate storm with a broken line.

The field in the principal storm is of course the same as that previously described.

In the field of force and its variations, this intermediate storm shows a great resemblance to the ordinary polar elementary storms, such as those of the 15th December, 1902, the 10th February, 1903, etc.

On Chart VIII — for $20^h\ 52.5^m$ — the partial current-arrows in the district Pawlowsk to San Fernando are directed south-east, and at Tiflis east, while farther east they turn more north. This indicates a convergence of the perturbing force in the north-west of Asia or the north-east of Europe.

The conditions in North America at this point of time are peculiar. At the three stations in the east of that continent, the direction of the current-arrow is east, and at Sitka south-west, or on the whole rather different from that which might be expected from its resemblance to the above-mentioned polar elementary storms. This lasts, however, only for about 10 minutes during the first part of the perturbation, whereupon P_h decreases, and for a moment is about zero; and in the two succeeding charts the directions of the arrows are the same as, for instance, on the 15th December, 1902, and the 22nd February, 1903.

The resemblance to these storms is still further increased by the circumstance that in Europe there is a corresponding positive turning of the perturbing force.

The *third intermediate storm*, from about 23^h 10^m to 0^h 25^m.

As regards the arctic regions, this polar storm is powerful at Axeløen, rather less so at Matotchkin Schar, and at Kaafjord, strange to say, it is almost imperceptible in H and D, while in the vertical intensity it is quite distinct.

At the same time there is a distinct difference in the perturbation-conditions in southern latitudes, these being particularly powerful and distinct in Europe, and noticeable also in the East and in the United States, while at Sitka the perturbation is almost imperceptible. The oscillations are on the whole uniform in direction, indicating that the forces remain in one direction all the time. We have therefore considered it sufficient to show the distribution of force at one moment during the time when the perturbation is at its height. This is represented on

Chart XV; time 23^h 45^m.

This storm, on the whole, has a great resemblance to the previously-described elementary night-storms, e. g. to that of the 23rd March, 1903. They commence at about the same time of day, i. e. a little before midnight. In both of them, the distribution of force remains constant throughout the perturbation, and is in the main similar.

The perturbing forces of southern latitudes, as the chart shows, seem to indicate that we have a point of convergence situated, in this case, very near Kaafjord, the effect of this system being there almost exclusively in a vertical direction. The horizontal arrow drawn for Kaafjord would appear, to judge from the curve, to be due mainly to the effect of the principal storm, which is still in activity. At one place in the north of Canada, perhaps near Hudson's Bay, there is a point of divergence of the horizontal component of the perturbing force.

Notwithstanding the long duration of the perturbation, and its somewhat varied character, we believe that we have succeeded, by means of the foregoing analysis, in elucidating the main features of the perturbation-conditions, and taking out the elementary phenomena that together form the present storm in all its diversity. In the course of the period of time considered, the following principal phenomena have been shown:

A positive equatorial perturbation from about 11^h—18^h, and the six following polar storms:

(1) The polar storm from 12^h 25^m to 13^h 15^m,
(2) The polar storm from about 17^h 20^m to 18^h 30^m,
(3) The main polar storm from about 18^h 30^m to 0^h 30^m,
(4) The first intermediate polar storm, maximum at 18^h 34^m,
(5) The second intermediate polar storm from 20^h 45^m to 21^h 20^m,
(6) The third intermediate polar storm from 23^h 10^m to 0^h 25^m.

TABLE XXXIX.

The Perturbing Forces on the 11th October, 1902.

Gr. M. T.	Honolulu		Sitka		Baldwin		Toronto		Cheltenham	
	P_h	P_d	P_h	P_d	P_h	P_d	P_h	P_d	P_h	P_v
h m										
12 50	+ 5.0 γ	0	−22.4 γ	W 51.7 γ	− 0.7 γ	0	?	W 5.6 γ	0	0
17 0	+ 7.6 „	0	+ 8.6 „	E 18.4 „	+ 8.1 „	W 6.4 γ	+15.7 γ	„ 6.0 „	+20.6 γ	W 8.9 γ
18 0	+ 6.5 „	W 5.8 γ	−20.8 „	W 9.0 „	− 3.7 „	„ 12.7 „	− 4.0 „	„ 18.6 „	− 5.0 „	„ 11.3 „
30	0	„ 22.5 „	− 4.6 „	„ 86.4 „	−25.0 „	„ 13.4 „	−29.7 „	„ 28.6 „	−43.9 „	„ 12.5 „
34	0 (?)	„ 12.5 „	0	„ 81.0 „	−25.4 „	„ 8.9 „	−41.0 „	„ 25.0 „	−40.0 „	„ 9.5 „
19 0	− 3.8 „	„ 12.4 „	− 8.6 „	„ 20.7 „	− 7.8 „	„ 6.4 „	− 4.5 „	„ 15.0 „	−14.7 „	„ 6.5 „
30	− 4.2 „	„ 13.3 „	− 6.9 „	„ 18.5 „	− 5.1 „	„ 2.5 „	+ 2.3 „	„ 10.3 „	− 9.4 „	„ 7.7 „
20 0	− 8.1 „	„ 16.6 „	−11.3 „	„ 20.7 „	−16.9 „	0	−15.7 „	„ 9.6 „	−25.0 „	„ 7.3 „
30	− 9.1 „	„ 19.1 „	−16.5 „	„ 29.3 „	−15.2 „	„ 4.8 „	− 9.4 „	„ 16.9 „	−22.1 „	„ 10.1 „
45	− 6.5 „	„ 19.1 „	−17.9 „	„ 29.3 „	− 8.8 „	„ 6.4 „	− 7.2 „	„ 17.2 „	−17.7 „	„ 13.7 „
52.5	+ 3.9 „	0	−20.0 „	„ 45.0 „	− 4.9 „	„ 7.6 „	0	„ 18.1 „	−10.5 „	„ 14.0 „
21 0	− 8.4 „	„ 20.8 „	−27.5 „	„ 67.6 „	−31.1 „	„ 9.5 „	−28.4 „	„ 23.5 „	−42.1 „	„ 18.4 „
7.5	− 2.4 „	„ 4.2 „	−21.0 „	„ 23.0 „	−23.6 „	„ 3.8 „	−37.0 „	„ 21.0 „	−29.4 „	„ 20.0 „
15	− 7.8 „	„ 20.0 „	−11.5 „	„ 33.4 „	−15.2 „	„ 9.2 „	−16.6 „	„ 26.6 „	−23.5 „	„ 21.4 „
30	− 6.5 „	„ 16.6 „	− 7.1 „	„ 27.0 „	−11.5 „	„ 12.1 „	−15.3 „	„ 27.7 „	−20.6 „	„ 20.8 „
22 0	− 3.9 „	„ 15.8 „	− 4.3 „	„ 23.4 „	−11.8 „	„ 12.1 „	−13.5 „	„ 22.3 „	−17.4 „	„ 16.1 „
30	− 3.9 „	„ 10.0 „	0	„ 13.0 „	−11.2 „	„ 3.2 „	−13.5 „	„ 6.6 „	−14.1 „	„ 3.6 „
23 0	− 3.9 „	„ 6.6 „	+ 1.2 „	„ 7.2 „	−11.8 „	E 2.9 „	−12.2 „	E 1.8 „	−11.8 „	0
45	0	0	0	„ 11.3 „	−12.8 „	„ 2.5 „	−15.0 „	0	−15.0 „	0
24 0	− 2.6 „	0	− 2.2 „	„ 10.4 „	− 9.1 „	W 2.6 „	−10.8 „	W 1.8 „	−11.2 „	0

TABLE XXXIX (continued).

Gr. M. T.	Axeløen			Matotchkin Schar			Kaafjord		
	P_h	P_d	P_v	P_h	P_d	P_v	P_h	P_d	P_v
h m									
12 50	+ 22.9 γ	W 26.0 γ	− 24.6 γ	+ 43.5 γ	W 11.1 γ		0	W 8.0 γ	+ 12.0 γ
17 0	+ 3.2 „	„ 34.0 „	− 81.0 „	+ 66.0 „	„ 22.0 „		?	?	?
18 0	− 92.0 „	„ 73.5 „	− 56.5 „	− 93.0 „	E 254.0 „		?	?	?
30	− 57.3 „	„ 61.2 „	+ 17.2 „	+ 30.0 „	„ 13.4 „	No values can be taken out, as the position of the normal line seems to have become a permanent change during the perturbation. (See Pl. II).	+113.0 γ	„ 29.4 „	−107.0 „
34	− 46.0 „	„ 16.2 „	+ 42.0 „	+ 88.0 „	W 7.4 „		+126.0 „	„ 80.0 „	−120.0 „
19 0	− 20.5 „	„ 42.2 „	0	− 48.8 „	„ 53.5 „		+ 35.6 „	„ 34.8 „	0
30	0	„ 35.9 „	0	− 96.0 „	„ 69.0 „		− 59.3 „	„ 73.4 „	− 13.3 „
20 0	− 20.5 „	„ 39.4 „	+228.0 „	− > 180.0 „	„ 89.0 „		−153.0 „	E 86.2 „	− 42.6 „
30	− 2.3 „	„ 42.2 „	+130.0 „	− > 180.0 „	„ 187.0 „		−296.0 „	„ 174.0 „	+ 5.6 „
45	−179.0 „	E 139.0 „	+442.0 „	− > 180.0 „	„ 414.0 „		−225.0 „	„ 176.0 „	+ 58.6 „
52.5	−137.0 „	„ 111.0 „	+290.0 „	− > 180.0 „	„ 346.0 „		−346.0 „	„ 259.0 „	+242.0 „
21 0	−238.0 „	„ 76.2 „	+ 12.3 „	− > 180.0 „	„ 348.0 „		−296.0 „	„ 238.0 „	+237.0 „
7.5	−110.0 „	„ 94.5 „	+492.0 „	− > 180.0 „	„ 190.0 „		−161.0 „	„ 116.0 „	− 32.0 „
15	−130.0 „	„ 53.0 „	+327.0 „	− > 180.0 „	„ 178.0 „		−182.0 „	„ 58.8 „	−115.0 „
30	− 41.3 „	„ 10.4 „	+287.0 „	− > 180.0 „	„ 85.0 „		−152.0 „	„ 53.3 „	− 94.0 „
22 0	− 10.1 „	„ 11.0 „	+216.0 „	− 168.0 „	„ 129.0 „		−130.0 „	„ 84.4 „	− 98.0 „
30	+ 5.0 „	W 10.7 „	+110.0 „	− 136.0 „	„ 125.0 „		−115.0 „	„ 67.8 „	− 96.0 „
23 0	+ 13.7 „	E 9.4 „	+ 86.0 „	− 91.3 „	„ 94.0 „		−118.0 „	„ 82.5 „	− 70.3 „
45	− 69.0 „	„ 65.9 „	+393.0 „	− 91.0 „	„ 53.5 „		− 44.4 „	„ 18.4 „	−122.0 „
24 0	− 43.5 „	„ 43.5 „	+182.0 „	− 51.0 „	„ 38.0 „		− 49.8 „	„ 9.2 „	−119.0 „

TABLE XXXIX (continued).

Gr. M. T.	Pawlowsk			Stonyhurst		Kew		Val Joyeux		
	P_h	P_d	P_v	P_h	P_d	P_h	P_d	P_h	P_d	P_v
h m										
12 50	−10.2 γ	W 5.0 γ	0	0	W 4.0 γ	0	W 7.0 γ	+ 4.4 γ	W 4.2 γ	0
17 0	+15.1 „	„ 12.4 „	+ 0.7 γ	+15.3 γ	„ 11.4 „	+ 18.3 γ	„ 11.7 „	+ 22.4 „	„ 7.5 „	0
18 0	− 5.0 „	E 44.2 „	+ 4.9 „	−12.2 „	0	− 11.7 „	0	− 4.8 „	E 6.7 „	+ 6.0 γ
30	−35.2 „	„ 18.8 „	+11.2 „	−35.6 „	„ 14.3 „	− 29.5 „	„ 11.7 „	− 29.6 „	W 10.5 „	+ 10.0 „
34	−44.0 „	„ 10.0 „	+12.0 „	−28.5 „	„ 17.0 „	− 25.5 „	„ 15.0 „	− 33.6 „	„ 12.0 „	+ 10.0 „
19 0	−12.6 „	„ 7.4 „	+11.2 „	− 6.6 „	„ 4.0 „	− 8.3 „	„ 4.7 „	− 4.0 „	0	+ 6.0 „
30	− 8.1 „	„ 11.5 „	+ 7.5 „	−11.2 „	0	− 10.2 „	0	− 6.4 „	E 0.8 „	+ 7.5 „
20 0	+ 7.6 „	„ 42.2 „	0	−21.4 „	E 21.7 „	− 20.8 „	E 19.6 „	− 16.0 „	„ 26.7 „	+ 9.6 „
30	−15.6 „	„ 50.6 „	− 5.6 „	−22.4 „	„ 26.3 „	− 27.0 „	„ 24.4 „	− 20.0 „	„ 26.7 „	+ 12.0 „
45	− 1.6 „	„ 55.5 „	− 5.6 „	−21.9 „	„ 54.5 „	− 22.8 „	„ 42.1 „	− 16.0 „	„ 47.7 „	+ 10.0 „
52.5	+ 4.5 „	„ 62.0 „	−10.0 „	−12.2 „	„ 54.0 „	− 12.7 „	„ 53.0 „	− 8.0 „	„ 52.5 „	+ 8.0 „
21 0	+ 5.0 „	„ 72.6 „	−15.0 „	−29.6 „	„ 58.2 „	− 30.5 „	„ 52.4 „	− 15.2 „	„ 54.2 „	+ 13.2 „
7.5	−10.6 „	„ 78.0 „	−15.0 „	−51.0 „	„ 49.0 „	− 50.0 „	„ 59.0 „	− 40.8 „	„ 54.0 „	+ 13.0 „
15	− 7.1 „	„ 52.7 „	−12.4 „	−30.0 „	„ 50.9 „	− 32.1 „	„ 46.0 „	− 24.4 „	„ 14.3 „	+ 10.0 „
30	−10.1 „	„ 30.3 „	− 7.5 „	−15.3 „	„ 32.8 „	− 15.8 „	„ 32.3 „	− 11.6 „	„ 29.2 „	+ 8.0 „
22 0	− 7.6 „	„ 27.6 „	− 7.5 „	−15.3 „	„ 28.6 „	− 13.8 „	„ 25.8 „	− 10.0 „	„ 25.5 „	+ 6.4 „
30	− 9.6 „	„ 19.8 „	− 6.0 „	−16.3 „	„ 20.0 „	− 15.3 „	„ 18.7 „	− 11.2 „	„ 14.2 „	+ 5.0 „
23 0	− 6.5 „	„ 13.3 „	− 5.2 „	−11.7 „	„ 14.3 „	− 9.7 „	„ 12.6 „	− 8.0 „	„ 11.7 „	+ 4.0 „
45	+12.8 „	0	− 2.5 „	+64.0 „	„ 28.0 „	+ 7.6 „	„ 23.0 „	+ 8.0 „	„ 21.0 „	+ 1.0 „
24 0	+ 7.6 „	0	− 5.6 „	+ 3.8 „	„ 20.0 „	+ 3.6 „	„ 20.0 „	+ 5.6 „	„ 17.6 „	0

TABLE XXXIX (continued).

Gr. M. T.	Wilhelmshaven			Potsdam		San Fernando		Munich	
	P_h	P_d	P_v	P_h	P_d	P_h	P_d	P_h	P_d
h m									
12 50	− 2.3 γ	W 11.6 γ	0	− 5.7 γ	W 2.5 γ	?	?	+ 5.0 γ	0
17 0	+ 23.3 „	„ 18.3 „	+ 2.0 γ	+20.6 „	„ 11.7 „	+20.8 γ	0	+ 14.0 „	W 9.1 γ
18 0	− 7.0 „	E 7.9 „	+ 4.0 „	− 9.5 „	E 10.2 „	− 4.6 „	0	− 10.0 „	E 11.4 „
30	− 37.3 „	W 18.9 „	+ 5.0 „	−39.5 „	W 9.1 „	−26.2 „	W 15.6 γ	− 35.0 „	W 3.0 „
34	− 46.7 „	„ 26.8 „	0	−39.0 „	„ 15.3 „	−25.0 „	„ 16.4 „	− 38.5 „	„ 7.5 „
19 0	− 4.7 „	„ 4.3 „	+ 6.0 „	− 7.6 „	„ 1.5 „	− 8.0 „	0	− 14.0 „	E 1.5 „
30	− 7.5 „	E 3.0 „	+ 7.0 „	− 7.9 „	E 3.0 „	− 9.0 „	0	− 14.0 „	„ 3.8 „
20 0	− 7.9 „	„ 33.0 „	+ 8.0 „	− 7.6 „	„ 30.0 „	−18.6 „	E 9.8 „	− 15.0 „	„ 25.1 „
30	− 19.6 „	„ 33.0 „	+ 9.0 „	−19.0 „	„ 33.5 „	−24.6 „	„ 10.6 „	− 22.5 „	„ 34.3 „
45	− 7.9 „	„ 51.3 „	+ 8.0 „	− 4.1 „	„ 48.8 „	−19.2 „	„ 27.8 „	− 14.5 „	„ 40.3 „
52.5	+ 4.7 „	„ 57.3 „	+ 8.0 „	+ 3.2 „	„ 50.0 „	−10.0 „	„ 37.0 „	− 3.0 „	„ 48.0 „
21 0	− 7.9 „	„ 64.2 „	+ 6.0 „	−20.6 „	„ 55.8 „	−38.4 „	„ 20.5 „	− 9.0 „	„ 59.3 „
7.5	− 38.7 „	„ 70.0 „	+ 3.0 „	−31.7 „	„ 58.8 „	−41.0 „	„ 22.0 „	− 32.5 „	„ 57.0 „
15	− 20.5 „	„ 51.8 „	+ 3.0 „	−18.1 „	„ 49.2 „	−28.8 „	„ 26.2 „	− 25.0 „	„ 48.7 „
30	− 13.0 „	„ 27.4 „	+ 1.0 „	−10.8 „	„ 27.5 „	−19.2 „	„ 20.8 „	− 13.5 „	„ 30.5 „
22 0	− 12.1 „	„ 21.3 „	0	−10.5 „	„ 21.3 „	−12.8 „	„ 17.2 „	− 12.5 „	„ 24.3 „
30	− 14.4 „	„ 12.8 „	0	−11.4 „	„ 14.2 „	−12.8 „	„ 9.8 „	− 12.0 „	„ 17.5 „
23 0	− 14.1 „	„ 7.3 „	0	− 9.5 „	„ 9.7 „	− 9.6 „	„ 10.3 „	− 10.5 „	„ 12.2 „
45	+ 13.5 „	„ 17.7 „	0	+13.0 „	„ 15.3 „	0	„ 17.0 „	+ 7.5 „	„ 16.8 „
24 0	+ 9.3 „	„ 13.4 „	0	+ 8.5 „	„ 8.9 „	0	„ 12.7 „	− 7.5 „	„ 12.2 „

TABLE XXXIX (continued).

Gr. M. T.	Tiflis			Dehra Dun		Bombay		
	P_h	P_d	P_v	P_h	P_d	P_h	P_d	P_v
h m								
12 50	0	W 4.8 γ	0	0	E 4.5 γ	0	0	0
17 0	+ 15.0 γ	„ 3.7 „	− 2.6 γ	+ 15.8 γ	„ 3.0 „	+ 11.2 γ	0	0
18 0	+ 8.4 „	E 24.1 „	0	+ 20.0 „	„ 13.8 „	+ 15.8 „	E 8.4 γ	0
30	− 25.7 „	„ 18.5 „	+ 9.4 „	− 25.6 „	„ 24.6 „	− 14.3 „	„ 9.6 „	+ 8.0 γ
34	− 36.0 „	„ 20.5 „	+ 12.2 „	− 39.0 „	„ 22.5 „	− 25.7 „	„ 8.5 „	+ 8.0 „
19 0	− 14.6 „	„ 11.1 „	+ 2.6 „	− 13.8 „	„ 11.8 „	− 13.8 „	„ 7.8 „	0
30	− 14.8 „	„ 12.6 „	+ 2.6 „	− 16.5 „	„ 9.9 „	− 14.3 „	„ 6.1 „	0
20 0	− 5.6 „	„ 17.4 „	0	− 7.8 „	„ 6.9 „	− 9.2 „	„ 4.8 „	0
30	− 10.7 „	„ 28.5 „	+ 1.3 „	− 9.1 „	„ 12.8 „	− 9.2 „	„ 8.4 „	0
45	0	„ 26.0 „	− 1.3 „	+ 2.4 „	„ 4.9 „	− 4.1 „	„ 1.2 „	− 4.8 „
52.5	+ 11.0 „	„ 25.0 „	− 5.1 „	+ 12.6 „	„ 1.0 „	+ 3.6 „	0	− 8.0 „
21 0	+ 7.7 „	„ 37.8 „	− 2.3 „	+ 2.4 „	„ 9.9 „	+ 8.7 „	„ 9.6 „	+ 6.4 „
7.5	− 13.5 „	„ 43.5 „	+ 2.8 „	+ 9.5 „	„ 13.0 „	− 9.4 „	„ 7.4 „	+ 2.4 „
15	− 10.9 „	„ 31.5 „	+ 2.8 „	− 8.3 „	„ 11.8 „	− 8.2 „	„ 8.4 „	0
30	− 7.3 „	„ 13.6 „	0	− 8.7 „	„ 8.9 „	− 6.4 „	„ 6.6 „	0
22 0	− 7.5 „	„ 17.1 „	0	− 9.5 „	„ 9.9 „	− 5.6 „	„ 6.1 „	0
30	− 8.8 „	„ 13.0 „	0	− 9.1 „	„ 6.9 „	− 7.7 „	„ 3.6 „	0
23 0	− 8.6 „	„ 8.5 „	0	− 8.3 „	„ 5.9 „	− 7.9 „	„ 3.0 „	0
45	+ 4.9 „	„ 1.8 „	− 2.3 „	0	0	+ 2.5 „	0	0
24 0	+ 2.4 „	„ 1.1 „	− 1.6 „	+ 2.8 „	0	+ 2.6 „	0	0

TABLE XXXIX (continued).

Gr. M. T.	Zi-ka-wei		Batavia		Christchurch		
	P_h	P_d	P_h	P_d	P_h	P_d	P_v
h m							
12 50	+ 6.4 γ	E 8.9 γ	+ 7.1 γ	W 6.0 γ	+ 23.0 γ	0	+ 1.5 γ
17 0	+ 15.5 „	„ 10.9 „	+ 12.1 „	E 2.4 „	+ 9.2 „	W 10.4 γ	+ 1.5 „
18 0	+ 33.1 „	„ 10.9 „	+ 30.3 „	W 4.8 „	+ 4.6 „	E 7.4 „	0
30	− 14.0 „	„ 13.4 „	− 13.5 „	E 6.0 „	+ 6.9 „	„ 16.3 „	0
34	− 19.1 „	„ 7.4 „	− 21.4 „	„ 6.0 „	+ 3.7 „	„ 13.4 „	0
19 0	− 14.1 „	„ 5.9 „	− 10.7 „	„ 2.4 „	− 6.4 „	W 7.4 „	+ 2.5 „
30	− 16.5 „	„ 3.5 „	− 12.8 „	0	− 12.0 „	„ 11.1 „	+ 1.9 „
20 0	− 12.8 „	0	− 12.1 „	W 2.4 „	− 18.4 „	„ 3.5 „	+ 1.3 „
30	− 7.6 „	„ 1.0 „	− 9.6 „	„ 6.0 „	− 19.3 „	„ 2.2 „	+ 1.5 „
45	0	„ 2.0 „	− 1.8 „	„ 9.6 „	− 24.2 „	„ 6.7 „	+ 3.6 „
52.5	+ 7.6 „	W 7.4 „	+ 7.8 „	„ 14.4 „	− 23.0 „	„ 5.4 „	+ 4.0 „
21 0	0	„ 3.0 „	0	„ 8.4 „	− 18.8 „	E 4.4 „	− 1.0 „
7.5	− 2.5 „	E 3.5 „	− 8.5 „	„ 7.2 „	− 17.5 „	„ 10.4 „	+ 2.5 „
15	− 5.1 „	„ 1.0 „	− 7.5 „	„ 6.0 „	− 18.8 „	„ 7.4 „	+ 3.6 „
30	− 6.4 „	„ 0.6 „	− 8.9 „	„ 4.8 „	− 23.3 „	„ 0.7 „	+ 3.3 „
22 0	− 7.0 „	0	− 9.3 „	„ 3.6 „	− 18.4 „	„ 0.7 „	+ 2.0 „
30	− 7.6 „	0	− 12.1 „	E 1.2 „	?	?	+ 0.7 „
23 0	− 7.0 „	0	− 13.2 „	„ 3.6 „	?	?	?
45	0	0	− 5.0 „	W 2.4 „	?	?	?
24 0	0	0	− 3.2 „	„ 1.2 „	?	?	?

TABLE XXXIX (continued).

Gr. M. T.	Ekaterinburg			Irkutsk		
	P_h	P_d	P_v	P_h	P_d	P_v
h m						
17 0	+ 5.0 γ	0	+ 1.0 γ	+ 17.0 γ	E 5.0 γ	+ 2.0 γ
18 0	+ 1.0 „	E 6.0 γ	+ 1.0 „	+ 15.0 „	„ 20.0 „	− 4.0 „
30	− 7.5 „	„ 12.0 „	+ 2.0 „			
34	− 9.5 „	„ 13.5 „	+ 2.0 „	+ 13.5 „	„ 19.0 „	− 5.0 „
19 0	− 15.0 „	„ 20.0 „	+ 3.0 „	+ 12.0 „	„ 14.5 „	− 4.0 „
30	− 16.0 „	„ 25.0 „	+ 3.0 „			
20 0	− 12.0 „	„ 28.5 „	+ 2.0 „	+ 7.0 „	W 4.5 „	− 1.0 „
30	− 2.5 „	„ 36.0 „	+ 7.5 „			
45	+ 1.5 „	„ 41.0 „	− 14.5 „			
52.5	+ 2.7 „	„ 43.5 „	− 16.0 „			
21 0	+ 3.0 „	„ 43.5 „	− 16.0 „	+ 3.0 „	„ 8.0 „	− 4.0 „
7.5	+ 2.7 „	„ 43.0 „	− 15.5 „			
15	+ 1.0 „	„ 41.0 „	− 14.5 „			
30	− 1.5 „	„ 32.0 „	− 12.5 „			
22 0	− 5.0 „	„ 14.0 „	− 9.0 „	+ 2.0 „	E 5.0 „	− 4.0 „
30	− 6.5 „	„ 9.0 „	− 6.0 „			
23 0	− 6.0 „	„ 4.0 „	− 5.0 „	− 9.0 „	W 4.5 „	− 1.0 „
45	− 0.5 „	„ 5.6 „	− 5.0 „			
24 0	0	„ 7.3 „	− 5.0 „	− 8.0 „	„ 3.0 „	+ 2.0 „

TABLE XXXX.

Partial Perturbing Forces on the 11th October, 1902.

Gr. M. T.	Honolulu		Sitka		Baldwin		Toronto		Axeloen	
	P'_h	P'_d	P'_h	P'_d	P'_h	P'_d	P'_h	P'_d	P'_h	P'_d
h m										
12 50	?	0	− 25.5 γ	W 63.0 γ	− 4.2 γ	0	?	0	+ 22.9 γ	W 26.0 γ
20 45	+ 1.6 γ	0	0	E 4.5 „	+ 4.4 „	0	+ 4.5 γ	E 3.6 γ	− 179.0 „	E 173.0 „
52.5	+ 3.9 „	0	0	„ 13.5 „	+ 9.2 „	E 3.2 γ	+ 14.4 „	„ 3.6 „	− 137.0 „	„ 145.0 „
21 0	0	W 4.1 γ	− 10.6 „	0	− 17.0 „	„ 2.5 „	− 13.9 „	0	− 238.0 „	„ 109.0 „
7.5	− 2.3 „	„ 4.1 „	− 6.2 „	W 27.0 „	− 8.8 „	W 5.1 „	− 20.2 „	„ 4.2 „	− 110.0 „	„ 97.5 „
15	0	E 2.5 „	0	„ 6.8 „	− 1.7 „	0	− 0.5 „	„ 1.2 „	0	0
23 45	0	0	0	„ 6.8 „	− 6.0 „	E 2.5 „	− 5.4 „	W 1.0 „	− 78.0 „	„ 53.0 „

TABLE XXXX (continued).

Gr. M. T.	Cheltenham		Pawlowsk		Stonyhurst		Kew		Val Joyeux	
	P'_h	P'_d	P'_h	P'_d	P'_h	P'_d	P'_h	P'_d	P'_h	P'_d
h m										
12 50	− 5.3 γ	0	− 10.1 γ	0	− 9.2 γ	0	− 8.7 γ	0	− 12.0 γ	0
20 45	+ 4.4 „	0	+ 8.5 „	E 8.3 γ	0	E 20.5 γ	+ 6.1 „	E 11.7 γ	+ 4.4 „	E 15.1 γ
52.5	+ 11.5 „	0	+ 14.1 „	„ 16.5 „	+ 10.2 „	„ 20.0 „	+ 13.8 „	„ 15.0 „	+ 12.4 „	„ 18.4 „
21 0	0	W 3.0 γ	+ 15.1 „	„ 32.2 „	− 10.2 „	„ 22.8 „	− 5.1 „	„ 16.8 „	+ 4.0 „	„ 20.1 „
7.5	− 13.5 „	„ 1.8 „	0	„ 36.8 „	− 30.6 „	„ 23.4 „	− 21.3 „	„ 17.3 „	− 22.0 „	„ 20.9 „
15	0	0	0	„ 17.4 „	− 10.2 „	„ 13.1 „	− 9.7 „	„ 11.7 „	− 8.0 „	„ 15.1 „
23 45	− 5.9 „	0	+ 13.5 „	0	+ 10.7 „	„ 20.0 „	+ 13.2 „	„ 15.0 „	+ 12.0 „	„ 13.8 „

TABLE XXXX (continued).

Gr. M. T.	Potsdam		Wilhelmshaven		San Fernando		Munich	
	P'_h	F'_d	P'_h	P'_d	P'_h	P'	P'_h	P'_d
h m								
12 50	− 9.5 γ	0	− 11.2 γ	0	?	?	− 8.5 γ	0
20 45	+ 12.3 "	E 15.2 γ	+ 13.0 "	E 12.8 γ	+ 5.7 γ	E 15.6 γ	+ 7.0 "	E 9.1 γ
52.5	+ 19.0 "	" 19.3 "	+ 20.0 "	" 21.3 "	+ 11.2 "	" 17.6 "	+ 13.0 "	" 16.4 "
21 0	0	" 25.8 "	+ 10.3 "	" 29.3 "	− 17.2 "	" 3.3 "	+ 10.5 "	" 19.0 "
7.5	− 19.0 "	" 31.4 "	− 22.3 "	" 35.3 "	− 19.2 "	0	− 18.8 "	" 25.1 "
15	− 4.1 "	" 22.8 "	− 6.5 "	" 21.6 "	− 10.8 "	0	− 9.5 "	" 19.0 "
23 45	+ 19.6 "	" 11.4 "	+ 21.8 "	" 17.7 "	+ 6.7 "	" 10.6 "	+ 15.0 "	" 7.5 "

TABLE XXXX (continued).

Gr. M. T.	Tiflis			Dehra Dun		Bombay		
	P'_h	P'_d	P'_v	P'_h	P'_d	P'_h	P'_d	P'_v
h m								
12 50	− 4.6 γ	E 4.8 γ	0	− 5.9 γ	E 4.5 γ	− 5.1 γ	0	0
20 45	+ 10.0 "	" 2.6 "	− 0.5 γ	+ 7.1 "	W 7.8 "	+ 6.6 "	W 8.6 γ	− 4.8 γ
52.5	+ 21.4 "	" 4.5 "	− 2.6 "	+ 15.7 "	" 12.8 "	+ 15.3 "	" 7.3 "	− 8.0 "
21 0	+ 17.1 "	" 11.1 "	+ 2.6 "	+ 15.7 "	" 4.9 "	+ 10.2 "	0	+ 6.4 "
7.5	− 2.1 "	" 14.8 "	+ 4.1 "	− 2.4 "	0	− 1.0 "	E 1.2 "	+ 2.4 "
15	− 2.1 "	" 5.6 "	+ 1.3 "	− 0.8 "	0	0	0	+ 1.6 "
23 45	+ 10.8 "	W 2.0 "	− 3.0 "	+ 7.5 "	" 3.0 "	+ 8.4 "	0	0

TABLE XXXX (continued).

Gr. M. T.	Zi-ka-wei		Batavia		Christchurch		
	P'_h	P'_d	P'_h	P'_d	P'_h	P'_d	P'_v
h m							
12 50	+ 3.8 γ	E 8.9 γ	+ 3.6 γ	W 6.0 γ	+ 23.0 γ	0	− 0.9 γ
20 45	+ 4.5 "	W 2.5 "	+ 6.0 "	" 5.4 "	− 4.4 "	W 7.1 γ	0
52.5	+ 17.5 "	" 7.4 "	+ 16.0 "	" 10.8 "	− 1.8 "	" 3.7 "	0
21 0	+ 12.8 "	" 4.0 "	+ 8.0 "	0	+ 3.2 "	E 3.7 "	− 5.8 "
7.5	+ 3.8 "	E 3.5 "	− 0.7 "	0	+ 6.0 "	" 10.8 "	− 0.5 "
15	+ 1.0 "	" 1.0 "	− 0.7 "	0	+ 3.2 "	" 5.6 "	+ 1.6 "
23 45	+ 8.7 "	0	?	" 3.6 "	?	?	0

For $12^h\ 50^m$ we have at

Kaafjord: Very slight and indistinct partial deflections.

Matotchkin Schar: $P'_h = + 43.5\ \gamma$, $P'_d = 0$, $P'_v = + 17.5\ \gamma$.

Axeløen: $P_v = - 25,0\ \gamma$.

Current-Arrows for the 11th October, 1902; Chart I — Partial values — at $12^h\ 50^m$, and Chart II at $17^h\ 0^m$.

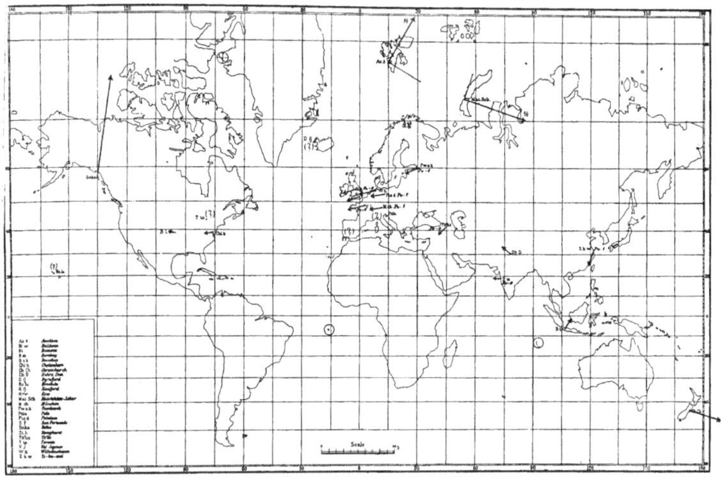

Fig. 117.

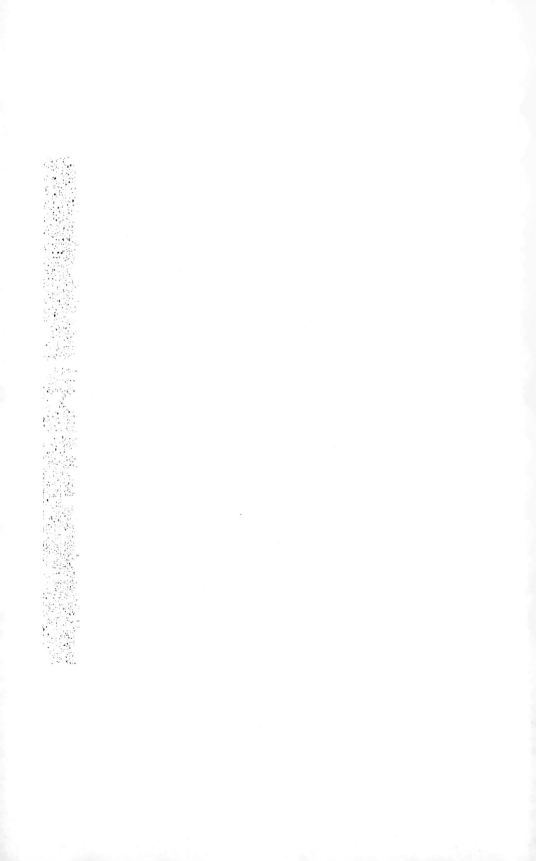

ent-Arrows for the 11th October, 1902; Chart V at 19^h 30^m, and Chart VI at 20^h 30^m.

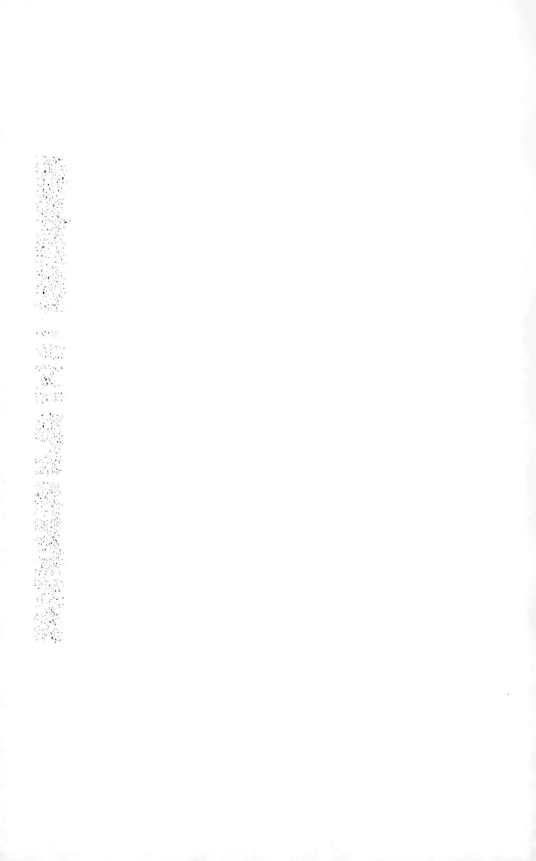

Current-Arrows for the 11th October, 1902; Chart IX at 21^h 0^m, and Chart X at 21^h 7.5^m.

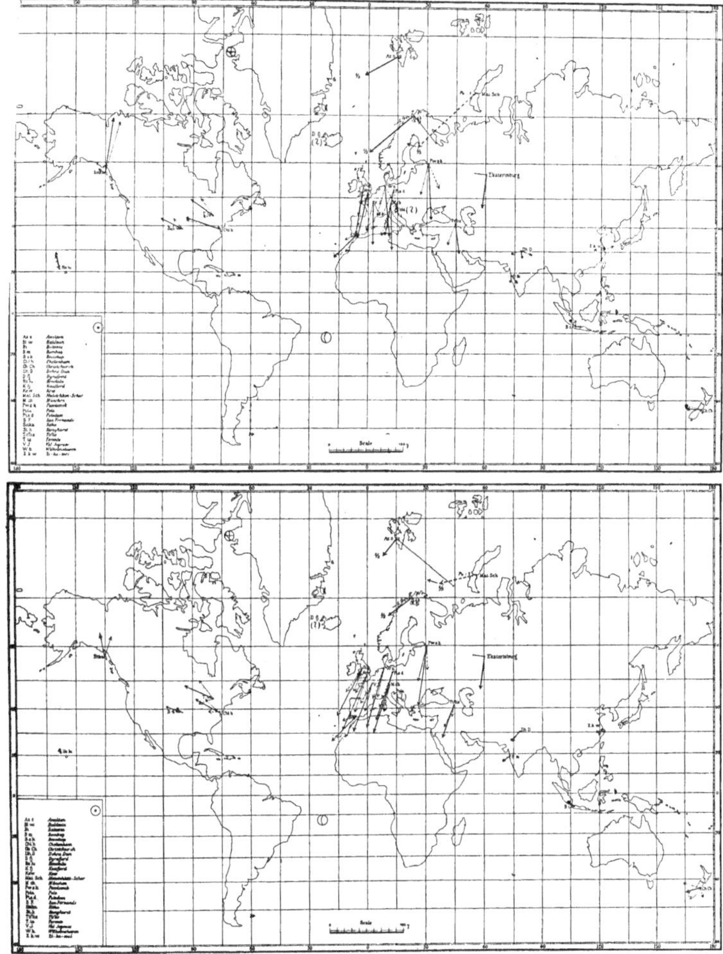

Fig. 121.

Current-Arrows for the 11th October, 1902; Chart XI at $21^h\ 15^m$, and Chart XII at $21^h\ 30^m$.

Fig. 122.

Current-Arrows for the 11th October, 1902; Chart XIII at $22^h\ 0^m$, and Chart XIV at $23^h\ 0^m$.

Fig. 123.

Current-Arrows for the 11th October, 1902; Chart XV at $23^h 45^m$.

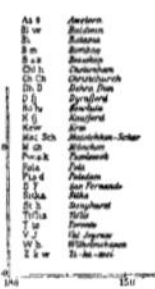

Fig. 124.

CONCERNING THE CAUSE OF THE PERTURBATIONS. POSITIVE AND NEGATIVE POLAR STORMS.

69. In describing the preceding perturbations, we have discussed more or less fully the various systems that might be supposed to be the cause of the various fields of perturbation. The results of these reflections, as regards the polar storms, may be summarised as follows: that on the night-side, and to some extent also, in very high latitudes, on the day-side (Axeløen), powerful perturbations will as a rule be formed, with current-arrows directed westwards in the area of precipitation; and that on the day-side, only a few degrees farther south, fields of precipitation will often be formed, with eastward-pointing current-arrows. There is a continual recurrence of conditions such as these, but they are often indistinct, a fact which may probably be accounted for by the small number of polar stations from which we have received registerings.

We have already touched upon the question as to how these systems may be supposed to be formed; and we will therefore here only refer the reader to Article 36, especially pp. 105 and 106, and fig. 50 a & b. From the experiment represented in fig. 38 b, there is every reason to suppose that not only the rays that descend on one side of the screen in low latitudes, but also some, at any rate, of those that descend in the polar zone of the terrella, are rays that curve round somewhat in the manner shown in fig. 39, in the equatorial plane, for rays answering to values of γ between —0.5 and —0.9, and in fig. 50 b. In the experiment shown in fig. 47 b, there is a precipitatation at the top and at the bottom of the screen, which undoubtedly turns off in a manner resembling that shown in fig. 50 a.

The two systems will now produce, in southern latitudes, their respective areas of convergence and divergence; it is these areas that are represented on our charts, and which justify us in also drawing conclusions respecting those parts of the auroral zone in which we have no stations.

These two types of perturbations thus seem to be those which characterise the polar storms; and as we are constantly meeting with them, we will give them different names. It will perhaps be practical to employ the same terms as in the equatorial storms. The characteristic difference in the polar regions between the two types, which instantly strikes the eye, is the direction there shown by P_h. We will then designate those storms which produce in their field of precipitation negative values of P_h, *negative polar storms*, and those that produce positive values of P_h, *positive polar storms*. These names are not chosen with any regard to the actual rays which we imagine will produce these fields, but only on account of the effect we find on the earth. On the other hand, however, we also see the agreement between, for instance, the positive polar and equatorial storms by comparing the figures and experiments just mentioned (figs. 39 for $0 \gtreqless \gamma \gtreqless -0.9$, and 50 b, 38 b and 68 [1, 4, 7]). In these cases the rays pass the earth in a westerly direction. A similar agreement exists between the negative polar and equatorial storms, as will be easily seen from the corresponding figures and terrella-photographs (figs. 39 for $\gamma < -1$ and 50 a, 37 & 47 b). In these last, according to our assumption, the corpuscular current passes the earth in an easterly direction, in a manner already frequently indicated.

With this circumstance before us, we shall also find that during the present perturbations all the fields formed can be explained comparatively easily. They will, of course, not be polar systems alone that act. At the outset it is more or less probable that rays will also descend in lower latitudes, and thus have an effect, that will possibly sometimes obliterate the effects of the polar systems.

As the probable cause of the first-occurring positive equatorial perturbation has been already sufficiently discussed, we need here only refer the reader to our previous remarks in Article 31.

We will first look then at the *first polar storm*, represented on *Chart I*. The time is $12^h\ 50^m$, not long, that is to say, after noon Greenwich; and we do actually find on the day-side what appears to be an area of divergence. We have here endeavoured to distinguish the effects of the polar storm from those of the equatorial, and the arrow-directions shown on the chart answer only to the former. The certainty with which the perturbing forces are determined is therefore somewhat diminished. In the next place there are no observations from Dyrafjord; and they would have been of the greatest importance here, as that station would probably have been situated not far from the storm-centre, the effects of which seem traceable in the district to the south of it. The current-arrows at Matotchkin Schar and Axeløen seem to indicate that this is the effect of a positive polar storm. The very small perturbing force at Kaafjord may possibly indicate that that station was situated in the vicinity of the point of divergence; and the positive P_v that we find is in accordance with this. It is impossible to say with any certainty what precipitation there might be on the night-side of the earth. The only northern station in this district from which we have observations, is Sitka; and there the conditions of the horizontal intensity also indicate that we are near the field of precipitation of a negative polar storm, as we find negative values of P_h. There is moreover a comparatively wide deflection in the declination, so that the current-arrow is not directed north-west along the auroral zone, but almost due north. This circumstance perhaps indicates that the storm-centre was situated a little to the west of the place. There is no distinctly-marked area of convergence in southern latitudes, and as the system can only be comparatively weak this is natural enough, as we are very badly off for stations in that part of the world.

The *second polar storm* — *Chart III* — exhibits fields, the form and nature of which are of the greatest interest. A glance at the chart shows us two distinct characteristic areas, an area of convergence in the east of Europe, Asia and the west of North America, and an area of divergence in the district from Western Europe to the east of North America. The storm-centre of the negative polar storm seems to be situated in the north-east of Asia. The arrows at Matotchkin Schar and Axeløen indicate a continuation of this system. Unfortunately we have no observations for this point of time from either Dyrafjord or Kaafjord, as the curves in this periode of time, in the case of the latter station, have disappeared, the points of light from the magnetometers having been too faint to act on the photographic paper. It is however probable that there have been positive deflections here in the horizontal intensity curve, judging partly from the course of the curve immediately after, when it is drawn once more, and partly from the conditions we have previously met with, where the fields have shown themselves on the whole almost exactly similar. In any case, circumstances such as these would agree exactly with the area of divergence found in the district Europe to America, as has already been pointed out in the preceding description. If we imagine a positive polar system in the district extending from the regions west of Greenland, across Dyrafjord, towards Kaafjord, we here recognise the form of field with which we are continnally meeting during the storms that occur at that time of day, namely in the afternoon, Gr. M. T., only that the positive system sometimes extends a little farther to the east. In this connection we need only refer the reader to the storms on the 9th December, 1902, the 15th and 8th February, 1903 (see especially p. 191), and the 27th and 31st October, 1902.

In this manner a close agreement with the first polar storm is arrived at. As may be seen, we have only to assume that the old systems have moved a little westwards and have altered, the positive storm having become less, and the negative greater, so that the latter is now the more powerful and greater in extent.

The third or *main polar storm* is shown on *Charts V, VI, XI, XII, XIII and XIV.* The form of the various fields is here the same in all essentials, and bears no small resemblance to the field during the preceding storm. We still seem to have a similar area of divergence in the same district as before. On looking at the northern stations, we find that the arrow at Kaafjord has taken a westerly direction, which would indicate that the positive polar system that is supposed to produce this area of divergence does not now extend so far east as before, a circumstance which recalls conditions found during the preceding perturbation of the 31st October and 1st November, 1902. We then found that the reversal of the direction of P_h occurred earlier at the eastern stations than at the western, as if the cause of this reversal were in some way or other moving westwards with the sun.

It now seems as though the negative polar system extends as far as Kaafjord; but if we investigate matters in lower latitudes, we find no distinctly-defined area of convergence. We do indeed find current-arrows in Europe directed southwards as we should expect, and they are of considerable strength, a fact which may possibly indicate that the two systems are here acting more or less in the same direction. At Honolulu and Sitka, we also find current-arrows such as we should expect to find on the east side of the area of convergence; but in the intermediate district we find no eastward-directed current-arrows forming a transition between these two areas. The current-arrows in the south of Asia, on the other hand, have a westward direction.

It should here be remarked, however, that if the system in the north is not very powerful, the effect in the extreme south of Asia will be comparatively slight; and if, at the same time, there occur systems whose greatest effect is at the equator, they will there easily gain the ascendancy and obliterate the effects of the polar storm. We should therefore, in order to explain the conditions during this period in such a manner, have to assume that simultaneously with the negative polar storm there occurred a storm of a kind similar to the negative equatorial storms that caused the current-arrows in the south of Asia to point westwards instead of eastwards; and there are actually circumstances that indicate that this would be the case. In the first place, the character of the horizontal intensity curve at these Asiatic stations is fairly quiet, with the exception of the districts surrounding the intermediate storms, a peculiarity which we found to be characteristic of this kind of equatorial storm. In the second place, the conditions in P_v also give a similar indication. A negative equatorial storm in the northern hemisphere will produce vertical arrows directed downwards, while the system that should form the area of convergence would produce vertical arrows directed upwards.

At first, it is true, positive values of P_v are found at Pawlowsk, Ekaterinburg and Tiflis, when the polar storm is still comparatively slight (see Chart V); but when the latter has developed considerable power, we must imagine that the greatest effect of the polar system is in the north. We now find all the time, moreover, negative values at Pawlowsk and Ekaterinburg (see Charts VI and XI—XIV); while on Chart VI P_v is still positive at Tiflis. This subsequently diminishes at Tiflis too, becoming for the most part zero (Charts XII—XIV), and sometimes turning a little round to the opposite side (Chart XI). At those stations of Western Europe from which we have observations of the vertical intensity, we find throughout positive values of P_v, though sometimes zero. We may imagine this circumstance to be partly caused by the positive polar system of precipitation, which produces positive values of P_v in the area of divergence, but also partly by the assumed negative equatorial storm, which will here operate in the same direction. One might perhaps be tempted to believe that this last polar system might possibly produce the positive values of P_v at the more eastern stations; but this is not possible if the systems are at all of the constitution we have supposed. If, for instance, on Chart V, the vertical arrow at Tiflis were solely due to this positive polar system, the horizontal arrow produced by this ought at least to be as large as the one really found there. It seems impossible to explain this circumstance by comparison with the size of the current-arrows in Europe and America; and as regards Chart VI it is still more difficult to imagine that this system, which, in all probability, should be considered as comparatively weaker than the more easterly one, should have a greater effect at Tiflis than the last-named storm, which is moreover nearer to that station.

There thus seems to be sufficient reason for supposing that this is really a storm that acts most powerfully at the equator, and is of the nature of the so-called negative equatorial storms.

We hereby also get a comparatively simple explanation of these fields as only the result of a simple coöperation between the already-described elementary phenomena.

We will in conclusion refer to the remarks that have been made concerning the positive value of P_v at Tiflis, which, in several of the storms described, has occurred in similar areas of convergence, e. g. in the perturbation of the 26th December, 1902 (Charts I and II, and especially the description on pp. 137 and 138), and that of the 15th February, 1903 (Chartes V and VI, with description on p. 178). In these earlier cases, we could not come to any definite decision regarding the systems which produced this apparent abnormal value; and we only suggested the possibility that these storms resembled the cyclo-median perturbations. Here, however, it seems more probable that the type resembles the negative equatorial storms.

The fourth polar storm, or *first intermediate storm*, is shown on *Chart IV*. The field here does not differ essentially from that described under the third polar storm. We can only imagine the alteration to be produced by the fact that the positive polar system, which we supposed existed there, now undergoes a sudden increase in power and extent, so that it reaches beyond Matotchkin Schar. The arrow at Irkutsk, moreover, in connection with those at Honolulu and Sitka, indicates, though faintly, an area of convergence in that district; and the arrow at Axeløen ought probably to be interpreted as a continuation of this more easterly system. We must here, however, be careful not to draw too certain conclusions from the conditions at Irkutsk, for we have only hourly observations to go upon.

The fact that these two systems of precipitation work into one another, is one that we have often observed before, especially in the case of Matotchkin Schar, e.g. in the intermediate storms of the 15th

February, and the 27th and 31st October (see the corresponding Plates), where the change, however, was of an opposite kind, a more easterly negative storm seeming to encroach upon the westerly positive storm for a time. On the 9th December, 1902 (Pl. IX), there is an example of still more typical conditions. At Dyrafjord and Kaafjord the arrows have strongly-marked easterly directions. The pronounced westerly directions at Axeløen are, we are inclined from the above to think, a continuation of a more easterly-situated negative polar storm. At Matotchkin Schar, on the other hand, we find that now one storm, now the other, seems to be the stronger, so that the directions of the arrows are always swinging round from west to east, or from east to west. These conditions, however, can be better studied in the material from 1882—83, where we have at our disposal observations from a larger number of polar stations.

This sudden change may be illustrated by imagining the two systems like those in fig. 50 a & b, moving together until they are lying close to each other, and imagining the rays to the east deflected as in fig. 50 a, and those to the west as in fig. 50 b. If we imagine a system such as this displaced, we shall obtain conditions at those places through which the boundary between the two kinds of polar storms passes, similar to those found at Matotchkin Schar.

The fifth polar storm, or *second intermediate storm*, shown in *Charts VII—X*, also exhibits in its main features the same peculiarities as the long storm. The explanation of the change we here see should apparently be sought in a suddenly strengthened impulse in the polar system, whereby the latter, in southern latitudes, acquires a greater effect. This causes the area of convergence here too, to appear more distinct, the effect of the polar system being for a time greater than that of the equatorial storm; and we obtain current-arrows pointing eastwards (see Chart VIII). The area of divergence also becomes stronger, and it thus appears that in this system too, there should be an impulse at the same time.

Finally, with regard to the sixth polar, or *third intermediate storm* (*Chart XV*), the conditions are quite analogous. There is an increased impulse in the polar systems, especially in the negative, an increase which is only slight, although relatively strong, the perturbing forces now being very small. The equatorial storm still seems to have an effect which acts in the very opposite direction in the south of Asia, but in America in the same direction as the polar systems.

In this way we have succeeded in explaining all the above phenomena in a manner that is exactly analogous to that employed in the preceding perturbations, and based only upon our previously-discovered simple elementary phenomena.

THE PERTURBATIONS OF THE 23rd & 24th NOVEMBER, 1902.
(Pl. VIII).

70. After the powerful storms at the end of October and the beginning of November have ceased, conditions are fairly quiet, at any rate at the Norwegian stations; and the few perturbations that do occur are of comparatively small strength. On the 19th November, however, quite a powerful perturbation appears rather suddenly. This forms the introduction to a series of powerful perturbations which develope daily for rather more than a week, the last powerful storm being on the 26th. These storms reach their maximum of strength between the 23rd and the 25th. The conditions recall those in October, when there was a similar period of powerful storms.

We remarked then that the position of the moon must have exercised an influence upon the behaviour of the perturbations, as the maximum occurred just about the time of the new moon. On this occasion too, we are in a period not far from the new moon; but the maximum does not coincide with it in time. The most powerful storms occurred, as we have said, between the 23rd and the 25th November;

whereas the new moon was on the 3oth, or at a time when the powerful storms had just ceased. Although it seems probable that the proximity of the new moon has something to do with the strength of the storms, other circumstances here seem to be of greater importance. We will not enter more fully into this question, however, but merely suggest that the time between the two maxima of about twenty-five days corresponds very nearly to the sun's period of rotation in low heliographic latitudes, a circumstance that may possibly help to explain this condition. In the case of this series of perturbations we find, moreover, a very striking harmony with the observations of the occurrence of sun-spots during the same period.

To represent this series of perturbations, we have selected those occurring during the period from the afternoon of the 23rd to the morning of the 24th, having copied the magnetograms from 15^h on the 23rd to 7^h on the 24th (see Pl. VIII).

We have observations for this day from all the stations. Unfortunately, however, the horizontal intensity curve for Matotchkin Schar has not been drawn, so that we have registerings only of the other two elements. At Dyrafjord, moreover, the registerings are somewhat defective, as they were some of the first that were made there, and can therefore only be regarded as trial registerings. The determination of the mean line is therefore a little uncertain; but as the conditions at about 17^h, or a little earlier, judging by the other stations, are more or less normal, the uncertainty is not so great after all; and as the deflections, at any rate during the greater part of the period in question, are considerable, the uncertainty will not seriously affect the current-arrows.

THE DISTRIBUTION OF FORCE.

71. The storms that occur here, as a close examination will show, may be referred to those types of perturbations with which we have become acquainted in the preceding perturbations. In order to distinguish them in some measure from one another, we will here, too, divide the perturbations into three sections,

the 1st section from 15^h 20^m to about 16^h,
the 2nd section from 16^h to about 22^h, and
the 3rd section from 22^h to 7^h on the day following.

The *first section* comprises a slight, brief perturbation that is perceived simultaneously at almost all the stations from which we have received observations. The effect is strongest at the equatorial stations in the south of Asia. In low latitudes there are deflections only in H, and P_h is positive everywhere. At the Central European and arctic stations, on the other hand, there are also deflections of varied extent in the declination curve. This then is a typical positive equatorial storm, as *Chart I* for the hour 15^h 48^m distinctly shows.

There are a few peculiarities in this equatorial perturbation that are worth noticing. The first of these is the shortness of its duration. Judging from the conditions at the stations in the south of Asia, it ends at about 16^h, and thus lasts only a little more than half an hour. If, on the other hand, we look at the district Tiflis to Stonyhurst, the storm appears to be going on for another hour and a half, the perturbing forces there having the peculiarities that characterise these storms; but the conditions, at any rate, are not so unmixed as to allow of its being on the whole characterised as such.

In the second place, the conditions in America are somewhat peculiar. There is no sudden rise of the horizontal intensity curve at about 15^h 30^m as at the other stations. It is not until somewhat later that the curve ascends, and its rise is comparatively slow. We may therefore reasonably assume that here too, other perturbing forces come into play, perhaps polar precipitation of some kind or other, acting with comparative strength. We have also previously found similar abnormal conditions during the positive equatorial perturbations in these districts, and we then suggested, that it would probably be due to polar precipitation in the north of North America (cf. pp. 67 & 128). Here, however, the abnormal condition is far more marked than in these two earlier storms.

Upon the conclusion of this equatorial perturbation, we enter upon

the *second section*, from 16^h to about 22^h.

The perturbing forces appearing here are generally small; but from about $17^h\ 30^m$ to about $18^h\ 20^m$ they are comparatively large, especially in southern latitudes.

The conditions at $17^h\ 40^m$ are shown on *Chart II.* If we look at the curves for the Norwegian stations during this period, we find, as regards the horizontal intensity, that there is a perturbing force at Axeløen directed southwards, and at Dyrafjord and Kaafjord there are perturbing forces directed northwards all the time. The declination-curve oscillates at all the stations above and below the mean line. We have unfortunately no registerings of H from Matotchkin Schar for this perturbation; but from the other three stations there is sufficient material to enable us to conclude that the field during this period is the typical one for a post-meridian storm. There are distinct effects of a positive polar storm at Dyrafjord and Kaafjord, and at Axeløen the effect of a negative storm, which, after what has been said, we are inclined to suppose extends eastwards on the night-side of the globe. This comes out clearly on Chart II. In Europe and Asia there is a distinct area of convergence; and in America and the districts east of it, there seems undoubtedly to be an area of divergence. These conditions agree well with the results we have already arrived at, regarding the appearance and formation of the systems at various times of day. As the forces, however, for the later part of this period are small, we have contented ourselves with this one chart as representative of the period.

The *third section* from about 22^h on the 23rd November, to 7^h on the 24th.

At about 22^h, the conditions begin to alter considerably. The Norwegian stations have now entered the night-side of the earth, and accordingly the deflections in H for Kaafjord and Dyrafjord swing round so that we now get the westward-directed current-arrows that are characteristic of the night-storms. The change in direction does not take place, however, until about $21^h\ 30^m$ at Dyrafjord, and an hour later — at about $22^h\ 30^m$ — at Kaafjord. This may seem to be at variance with what we have previously found to be the case, as for instance in the perturbations of the 31st October and 1st November, 1902, when we found that the cause of the change appeared to move westwards with the sun. Here, however, we find the opposite, as the change takes place earlier at the more westerly-situated Dyrafjord than at Kaafjord.

There are, however, several things to notice in this connection that may aid in a comprehension of these conditions.

In the first place, on the 31st October, we were considering the stations Matotchkin Schar and Kaafjord, both of which are situated to the south of the auroral zone; whereas here we have one station — Dyrafjord — to the north, and one — Kaafjord — to the south of the zone. It is by no means improbable that this circumstance is of some importance. It would be natural, indeed, to imagine that owing to the more northerly situation of Dyrafjord in relation to the magnetic axis, it would be easier for the system acting at Axeløen to have an influence here than at Kaafjord, which in this respect has a more southerly situation; and that on this account the positive storm of the preceding section would be able to act longer at Kaafjord than at Dyrafjord.

In the next place it should be observed that the times considered in the two cases differ very considerably from one another, a fact which is undoubtedly very important; for if we assume that the position of the sun in relation to the magnetic axis of the earth is of great importance in deciding the position of the systems of precipitation, we must also assume that the relative motion of the earth and the sun will govern the displacement of the systems from time to time.

There are two circumstances in connection with this relative motion, that must here be considered. This is easily seen by looking at the conditions at the point of intersection of the magnetic axis with, for instance, the northern hemisphere. In the first place, the sun's azimuth will increase, in the course of the day, more or less evenly by 360° in a westerly direction; and in the second place, the height of the sun above the astronomical horizon of this place during the same period, will vary periodically with an amplitude of about 23° 20′.

If we now look at these two components of the motion separately, we must in the first place assume, as regards the change of azimuth, that this by itself will cause the systems to move right round the earth in a westerly direction in the course of the twenty-four hours.

The alteration of altitude will cause a displacement of the systems in a manner characteristic of this condition; and it is quite conceivable, that this displacement may sometimes be the reverse of that due to the variation in azimuth. It is therefore probable at the outset that the displacement of the systems would be somewhat different at different times of day. When the sun is near the meridian of the magnetic axis, and the variation in altitude is therefore very slight, it might be supposed that the westward movement of the systems, caused by the variation in azimuth, will most frequently predominate.

At times when the alteration in altitude is comparatively great however, we might possibly expect to find comparatively greater effects from this second component of the motion; and it would then be natural that the conditions became rather more complicated. Nor does it appear to be impossible for the displacement due to the alteration in altitude to be sometimes greater than that due to the variation in azimuth.

We now find, when we look at these two perturbations, that the time at which we considered the conditions in this respect on the 31st October, was just about that at which the sun passed the above-mentioned meridian. There, too, we found a displacement of the systems westwards with the sun; whereas in this perturbation we are just at a time when the alteration in altitude is very great; and we find that the conditions are actually now developing somewhat differently.

It might not be out of place here, as an analogy to these conditions, to compare them with those found by Störmer's calculations. This cannot, of course, be regarded as anything more than an analogy, at any rate here; for a number of circumstances have been set aside in the calculations, which would certainly exert no small influence. In this connection we need only look at fig. 76, p. 160, to obtain a general idea of the conditions.

To every altitude, ψ, of the sun above the magnetic equator, there are one or more corresponding fields of precipitation, whose positions are determined by the corresponding value of Φ. If we imagine the sun to sink, for instance, from $\psi = 10°$ to $\psi = -10°$, we should find a field of precipitation for the negative rays that would move during this period from about $\Phi = -37°$ to $-53°$, or eastwards on the post-meridian side. The next system, which appears on the evening and night-side, will have a westward motion from about $\Phi = -157°$ to $-121°$ and thus changes place with almost double the rapidity. The third system again, will undergo an eastward displacement, from about $\Phi = -218°$ to $-259°$, that is to say with a rapidity even greater than that of the preceding one. We thus see that the displacement, on account of the alteration in the sun's altitude, of the systems of precipitation, considered from the place mentioned above, is sometimes in one direction, sometimes in the other.

In this case, that is when the sun is sinking as indicated, in the first and third systems of precipitation the two components of the motion will move the systems to opposite sides, and they will thus counteract one another. The alteration of altitude will moreover have the greatest significance for the system on the night-side. In the case of the second system, the two components will move the system in the same direction.

We see further from the figure, that near those places at which $\frac{d\psi}{d\Phi} = 0$, even a very small change of altitude will produce comparatively great displacement of the systems. It would perhaps be interesting to examine a little more closely the velocities of the displacement corresponding to the two components of the motion; but this would carry us too far. We are only considering these conditions for the purpose of finding analogies. and not in the hope of finding perfect correspondence in the details.

In conclusion we must also remark that the system with the eastward-directed arrows on the 31st October, was of far greater strength than the corresponding system in the present perturbation.

When all these circumstances are taken into consideration — and there might be many others that also exerted an influence — there is no necessity whatever for supposing that they contradict the results previously found. Nor is this in reality anything new or unknown; it is only a negative night-system, which, at the Norwegian stations, appears to move eastwards along the auroral zone, a condition that we have continually found in earlier perturbations. The storms that occur in this section prove also to be of the form that is typical of these night-storms with centre at the Norwegian stations.

As in the earlier perturbations, we might also here separate several intermediate storms from one long main storm; but as in this case in southern latitudes they do not stand out so distinctly from one another as in the previous perturbations, we have thought it better not to attempt any such decomposition, as its uncertainty would be too great. At our Norwegian stations we find, almost all the time, deflections in the horizontal intensity curve, indicating a diminution in H. Two or three times there is a slight, brief deflection to the opposite side, e. g. at Kaafjord at about $23^h\ 30^m$, and at Axeløen from 2^h to about $2^h\ 20^m$. Both the declination and the vertical intensity curve for Dyrafjord oscillate above and below the normal line all the time, while at the other three stations the deflections are nearly uniform in direction, with only a few short interruptions where the curve goes over to the other side. The direction of this long deflection is easterly at all three stations. In V the perturbing force is directed upwards at Matotchkin Schar and at Kaafjord, and downwards at Axeløen, indicating that the horizontal part of the current is situated to the north of the first two places, and to the south of Axeløen, or in a manner exactly similar to that of the preceding storms. Between 23^h and 24^h, we find a brief deflection to the west in the declination-curve for Kaafjord, corresponding to the above-mentioned brief reversal in the H-curve, but a little earlier. We also find a similar reversal of direction in the vertical intensity curve for Axeløen, the perturbing force at that time being directed upwards for a short time.

With regard to the other European stations, we find that the greatest deflections, at any rate during the greater part of the perturbation, are in the declination-curve. These deflections are in the same direction at all the stations, namely east, indicating that the current-arrows have a southerly direction. Between 2^h and 4^h however, P_h sometimes prevails over P_d. At the same time we notice at our northern stations a powerful intermediate storm, which, however, has the same direction as the main storm.

The horizontal intensity curve is very sinnous in form at all the stations, and the deflections are now positive, now negative. At Pawlowsk, however, they are positive throughout, with the exception of two or three short, slight deflections to the opposite side. In southern Asia also, comparatively powerful disturbances are distinctly observable, occurring both in H and in D. The deflections here are not in one direction all the time, but in different directions at different times. On comparing the curves with the registerings at the Norwegian stations, we find that the stronger impulses at the latter are also accompanied by similar impulses at the stations of southern Asia, a circumstance which clearly indicates that the two are closely connected.

At Christchurch there are also powerful storms at this time, both in H and in D, lasting far longer than the period we are now considering.

Finally, in America there are also powerful storms, during which the deflections in H are negative all the time, whereas in D, while sometimes very powerful, they are more variable as regards the direction of the perturbing force.

On *Charts III—VIII* are shown the various fields that appear during the various phases of the perturbations in this section.

We have already remarked that the perturbation-conditions as a whole are to be understood as a long, more or less constant, perturbation, going on all the time, accompanied by several intermediate, short, but powerful storms. The latter will now form fields, which, as a rule will differ to some extent from those produced by the long storm. The form of the field answering to the long storm will thus be more or less obliterated during these intermediate storms. In the earlier perturbations, similar long storms, interrupted by short, intermediate storms, have continnally been found, and their conditions have, as a rule, been comparatively so simple, that it has been possible to separate the two phenomena. Here, however, the conditions during the long storm are so disturbed, that it has not been possible to take out the intermediate perturbing forces, although conclusions as to their behaviour may be drawn from the form of the curves.

The conditions which we have been led to consider as the typical ones, are, as we have already said, a combination of negative and positive polar storms, the former occurring principally on the night-side, while the latter are characteristic of the day-side, and in latitudes that as a rule are a little more south than those in which the negative storms attain their greatest strength (see Art. 69). The position of these systems may of course vary somewhat, according as the conditions under which the perturbations are formed alter. In addition to these polar precipitations, there have also been, as we have often seen before, simultaneously-acting storms of types that should be due to stiffer rays, which acted most powerfully in low latitudes. Rays of this kind do not appear to have had any specially noticeable influence during this perturbation. We shall find, however, that the conditions as a whole may be referred to two polar systems of the two types mentioned above; and we shall thus receive fresh confirmation of the correctness of our former assumptions.

The resemblance between the fields is quite striking, even on a casual glance at the various charts. The typical form of the field is most clearly seen in the charts in which Ekaterinburg and Irkutsk are also shown. These charts are only marked for the full hours 23^h, 24^h and 2^h, as has generally been done when the conditions varied considerably from time to time. They distinctly show an area of convergence of most characteristic form in the district Europe and Asia, but displaced a little on the various charts in a direction east and west. We find the same conditions at the other hours in the case of most of the stations. At the stations of Southern Asia, on the other hand, the conditions are often rather peculiar, and the perturbing forces sometimes directed the opposite way to that one would expect to find as the effect of the long polar night-system. The current-arrows, however, are as a rule very small, and therefore the accuracy with which the directions are determined is considerably less. Uncertainty in the position of the normal line will exert a considerable influence. Sometimes, however, the deflections are so great that it cannot be put down to inaccuracy alone; and we are then obliged to assume that there are other forces asserting themselves. This, for instance, is the case on Chart VI for $0^h\ 50^m$ on the 24th. In order to explain these, it might be well to see whether here, too, there were not an equatorial storm such as we have often found before. Although it is not impossible that a storm such as this may be acting here, there is nothing that decidedly points in that direction. On the contrary it seems more probable that these deflections are produced by a more or less intermediate positive polar storm, such that would act in these districts. In the first place, the stations in the south of Asia have begun to move into the day-side; and we have repeatedly seen that these systems are more readily formed there. In the second place, the current-arrows in the east of North America differ a good deal in direction from the general one. Their main direction is south-east, and they thus appear to be instrumental in forming the most easterly part of the area of divergence, which we should therefore expect to find on the day-side of the globe.

Finally, in the third place, the course of the horizontal intensity curve during this period, indicates quite distinctly at Kaafjord the effects of an intermediate positive polar storm, which, however, are a little weaker than those of the long negative storm acting simultaneously in that district. A similar effect seems to be traceable at Sitka, as also at Dyrafjord. It is therefore not improbable that this is also a similar effect.

In this case, as so often before, Honolulu occupies rather a peculiar position as regards the perturbing forces. If, however, we assume that the centre of the positive storm lies comparatively far south, the conditions at Honolulu might be explained, if it were imagined to be in proximity to the point of divergence. The more northerly negative storm might then also produce current-arrows directed eastwards. It may also, and perhaps with more probalility, be imagined that purely local conditions might exert no little influence.

In addition to the great area of convergence that we have found throughout this section, the current-arrows in Western Europe and the east of North America indicate an area of divergence in that district until 2^h on the 24th. In accordance with this, we here also find positive values of P_v.

Thus the conditions do not seem to differ essentially from those we find in the second section of these storms. The systems acting appear to be on the whole the same as before, only altered as regards their strength and displaced a little. The area of divergence, which at first appeared on the day-side of the earth, has thus, during this storm, remained for a considerable time, continuing indeed on to the evening and night side. Charts VI and VII, for the hours $1^h\ 20^m$ and $2^h\ 40^m$, clearly show, however, how this area of divergence now rapidly moves westwards, until at $2^h\ 40^m$ it is in the district of North America and the east of Asia. In accordance with this, the positive vertical arrows in Europe disappear, some becoming zero, as at Val Joyeux, some turning round to the opposite side, as at Pola.

The last chart for this period, Chart VIII, shows the conditions as they appear at $6^h\ 30^m$ shortly before the termination of the storm. At Axeløen and Dyrafjord we find about this time increased strength in the deflections, and simultaneously in southern latitudes corresponding deflections in the magnetic elements. The forces on the whole are small, and from several stations we have received no observations; nevertheless there seems to be an area of convergence in the district extending from Europe to the east of North America, with a point of convergence a little south of Iceland and Greenland. The arrows, moreover, in the east of North America, together with Honolulu and Zi-ka-wei, possibly indicate an area of divergence in those districts; but as we have so few stations there, we can draw no certain conclusions in the matter.

According to this, we again appear to have the effects of the two polar storms as before, only that the storms have moved considerably westwards.

We have thus, by going through this perturbation in its various phases, succeeded in explaining all the fields that occur, from the previously-mentioned simple points of view. The conditions here have been simpler, in so far as there appear to be no particularly marked effects of equatorial systems, but on the whole only of polar systems. Although we have not, as before, thought it expedient to attempt a decomposition of the forces that appear, into the separate elementary phenomena, we have been able, by observation of the fields, to make such a separation. We thus obtain, through the study of this perturbation, a further support to our theory of the simple elementary laws that govern the apparently complicated conditions found in the great compound storms.

TABLE XLI.

The Perturbing Forces on the 23rd & 24th November, 1902.

Gr. M. T.	Honolulu		Sitka		Baldwin		Toronto		Cheltenham	
	P_h	P_d	P_h	P_d	P_h	P_d	P_h	P_d	P_h	P_d
h m										
15 32	+ 4.0 γ		+ 2.1 γ	W 8.6 γ	0	E 2.5 γ	0	E 9.1 γ	+ 3.7 γ	E 3.0 γ
48	+ 3.0 „		+ 1.7 „	E 1.4 „	+ 6.9 γ	0	+ 6.7 γ	„ 1.8 „	+ 10.3 „	0
16 30	0		− 2.8 „	0	+ 2.8 „	W 6.4 „	?	?	+ 4.7 „	0
17 40	− 3.0 „		− 15.7 „	W 9.0 „	− 6.0 „	„ 3.5 „	− 7.8 „	W 2.4 „	− 4.8 „	0
18 20	0		− 10.6 „	„ 18.0 „	0	„ 3.2 „	0	„ 4.8 „	0	0
22 0	+ 18.2 „		− 14.9 „	0	+ 4.1 „	„ 6.4 „	+ 7.6 „	„ 1.2 „	+ 6.5 „	„ 2.4 „
30	+ 12.0 „	No noticeable deflections.	− 48.5 „	E 5.0 „	− 18.0 „	„ 17.8 „	− 18.0 „	„ 25.3 „	− 19.1 „	W 14.8 „
23 0	+ 9.9 „		− 56.4 „	W 6.8 „	− 18.1 „	„ 9.5 „	0	„ 12.4 „	− 9.4 „	„ 17.2 „
20	+ 7.9 „		− 43.8 „	„ 6.3 „	− 15.5 „	„ 21.5 „	− 5.8 „	„ 39.8 „	− 18.2 „	„ 23.8 „
24 0	+ 5.2 „		− 21.1 „	„ 7.7 „	− 18.6 „	„ 12.7 „	− 17.5 „	„ 21.0 „	− 25.0 „	„ 11.6 „
0 50	+ 2.0 „		− 14.2 „	E 9.0 „	− 14.5 „	E 3.2 „	− 3.4 „	E 18.1 „	− 4.4 „	E 8.6 „
1 20	0		− 45.0 „	W 18.2 „	− 32.8 „	W 38.0 „	− 36.0 „	W 31.4 „	− 48.5 „	W 20.8 „
2 0	0		− 1.3 „	„ 18.0 „	− 21.0 „	„ 9.5 „	− 20.5 „	„ 2.4 „	− 28.0 „	0
40	− 11.8 „		− 46.0 „	„ 23.5 „	− 41.5 „	E 63.6 „	− 26.0 „	E 76.0 „	− 28.8 „	E 60.0 „
4 30	− 11.4 „		− 11.6 „	„ 35.5 „	− 23.0 „	„ 12.7 „	− 20.0 „	„ 19.9 „	− 28.5 „	„ 14.8 „
5 30	− 10.4 „		+ 4.8 „	„ 25.0 „	− 18.3 „	„ 10.8 „	− 13.5 „	„ 24.0 „	− 22.0 „	„ 14.8 „
6 30	− 10.4 „		− 28.1 „	E 28.0 „	− 2.1 „	„ 12.7 „	− 11.2 „	„ 13.0 „	− 12.0 „	„ 8.9 „

TABLE XLI (continued).

Gr. M. T.	Dyrafjord			Axeløen			Matotchkin-Schar		
	P_h	P_d	P_v [1]	P_h	P_d	P_v	P_h	P_d	P_v
h m									
15 32	?	?	?	− 10.1 γ	E 15.3 γ	+ 13.5 γ		W 15.0 γ	+ 13.0 γ
48	?	?	?	− 8.3 „	„ 8.8 „	+ 28.3 „		„ 8.0 „	− 17.0 „
16 30	?	?	?	− 12.0 „	W 51.0 „	− 54.0 „		„ 3.0 „	− 20.0 „
17 40	+ 23.0 γ	0	− 79.0 γ	− 72.0 „	E 26.0 „	+ 57.0 „		E 8.0 „	− 60.0 „
18 20	+ 90.0 „	0	− 102.0 „	− 110.0 „	„ 77.0 „	+ 71.0 „		0	0
22 0	− ca. 310.0 „	W 14.0 γ	− 288.0 „	− 225.0 „	„ 146.0 „	0	The *H*-curve is not drawn on the magnetogram.	W 39.0 „	− 168.0 „
30	− 170.0 „	„ 76.0 „	− 88.0 „	− 288.0 „	„ > 163.0 „	+ 309.0 „		E 87.0 „	− 252.0 „
23 0	− 253.0 „	E 145.0 „	− 16.0 „	− 149.0 „	„ 36.0 „	+ 420.0 „		„ 480.0 „	− > 343.0 „
20	− 272.0 „	W 23.0 „	+ 8.0 „	− 450.0 „	?	− 37.0 „		„ 172.0 „	− 186.0 „
24 0	− 112.0 „	„ 36.0 „	+ > 85.0 „	− 117.0 „	„ 126.0 „	+ 540.0 „		„ 81.0 „	− 222.0 „
0 50	− 288.0 „	„ 46.0 „	+ 60.0 „	− 67.0 „	„ 62.0 „	+ 466.0 „		„ 59.0 „	− 227.0 „
1 20	− 225.0 „	„ 30.0 „	+ 55.0 „	− 240.0 „	„ 156.0 „	+ 415.0 „		„ 92.0 „	− 278.0 „
2 0	− 181.0 „	E 53.0 „	+ > 85.0 „	− 43.0 „	„ 93.0 „	+ 360.0 „		„ 63.0 „	− 247.0 „
40	− > 800.0 „	„ 148.0 „	+ > 85.0 „	− 608.0 „	„ 150.0 „	+ 540.0 „		0	− 336.0 „
4 30	− 240.0 „	„ 84.0 „	+ 50.0 „	− 212.0 „	„ 46.0 „	+ 154.0 „		„ 4.0 „	− 173.0 „
5 30	− 80.0 „	„ 55.0 „	− 92.0 „	− 61.0 „	„ 49.0 „	+ 154.0 „		W 22.0 „	− 101.0 „
6 30	− 244.0 „	„ {> 3.0 „ ; < 43.0 „}	+ 18.0 „	− 97.0 „	„ 61.0 „	+ 98.0 „		„ 27.0 „	− 101.0 „

[1] The Value of ε_v here someWhat uncertain.

TABLE XLI (continued).

Gr. M. T.	Kaafjord			Pawlowsk			Stonyhurst		Kew	
	P_h	P_d	P_v	P_h	P_d	P_v	P_h	P_d	P_h	P_d
h m										
15 32	+ 47.0 γ	E 11.0 γ	0	+ 15.1 γ	E 3.7 γ	0	+ 10.2 γ	0	+ 10.2 γ	0
48	+ 21.0 „	W 2.0 „	0	+ 12.6 „	0	0	+ 7.6 „	0	+ 8.9 „	W 3.3 γ
16 30	+ 5.0 „	„ 8.0 „	0	+ 7.5 „	W 2.3 „	0	+ 8.2 „	0	+ 10.2 „	„ 1.9 „
17 40	+ 23.0 „	E 66.0 „	0	+ 12.1 „	E 34.0 „	− 2.2 γ	− 2.0 „	E 11.2 γ	0	E 7.0 „
18 20	+ 18.0 „	„ 29.0 „	0	− 6.0 „	„ 9.2 „	0	0	W 1.7 „	0	W 1.9 „
22 0	+ 33.0 „	„ 48.0 „	− 6.0 γ	+ 1.0 „	„ 15.6 „	0	+ 1.5 „	E 11.4 „	0	E 9.4 „
30	− 25.0 „	„ 138.0 „	− 50.0 „	+ 49.3 „	„ 59.0 „	− 14.0 „	− 17.1 „	„ 70.8 „	− 17.4 „	„ 54.0 „
23 0	−455.0 „	„ 442.0 „	+ 6.0 „	+ 5.0 „	„ 96.0 „	− 22.4 „	− 28.0 „	„ 59.5 „	− 30.5 „	„ 59.0 „
20	−172.0 „	W ca. 63.0 „	− 76.0 „	+ 5.8 „	„ 55.5 „	− 18.7 „	− 13.8 „	„ 58.0 „	− 17.8 „	„ 47.2 „
24 0	−272.0 „	E 172.0 „	−101.0 „	+ 24.4 „	„ 14.2 „	− 28.4 „	+ 8.2 „	„ 71.0 „	+ 5.1 „	„ 62.8 „
0 50	−168.0 „	„ 165.0 „	−104.0 „	+ 25.1 „	„ 22.0 „	− 37.0 „	− 15.3 „	„ 32.1 „	− 11.4 „	„ 29.0 „
1 20	−224.0 „	„ 132.0 „	−100.0 „	+ 30.9 „	„ 39.0 „	− 40.7 „	− 12.2 „	„ 68.6 „	− 11.7 „	„ 65.0 „
2 0	−197.0 „	„ 163.0 „	− 88.0 „	+ 5.0 „	„ 20.7 „	− 30.0 „	− 13.8 „	„ 25.7 „	− 17.8 „	„ 31.8 „
40	−329.0 „	„ 154.0 „	−102.0 „	+ 46.8 „	W 17.9 „	− 43.3 „	+ 23.2 „	„ 70.8 „	+ 23.1 „	„ 57.0 „
4 30	− 33.0 „	„ 6.0 „	− 81.0 „	+ 9.1 „	„ 6.9 „	− 33.5 „	0	„ 25.1 „	− 2.5 „	„ 25.7 „
5 30	+ 18.0 „	W 11.0 „	− 62.0 „	− 2.5 „	„ 13.8 „	− 23.1 „	− 4.6 „	„ 17.2 „	− 8.7 „	„ 11.7 „
6 30	− 30.0 „	E 22.0 „	− 50.0 „	?	?	?	+ 4.6 „	W 10.3 „	+ 1.5 „	W 8.9 „

TABLE XLI (continued).

Gr. M. T.	Val Joyeux			Wilhelmshaven			Potsdam		
	P_h	P_d	P_v	P_h	P_d	P_v	P_h	P_d	P_v
h m									
15 32	+ 8.8 γ	E 3.4 γ	0	+13.6 γ	E 8.5 γ	0	+13.6 γ	E 8.6 γ	0
48	+ 9.6 „	0	0	+11.6 „	„ 3.0 „	0	+12.0 „	„ 1.5 „	0
16 30	+ 8.0 „	0	0	+11.6 „	„ 3.7 „	0	+11.4 „	„ 2.5 „	0
17 40	0	„ 12.1 „	0	+ 4.7 „	„ 21.5 „	0	+ 5.7 „	„ 17.8 „	0
18 20	0	0	0	− 3.3 „	0	− 6.0 γ	0	0	0
22 0	+ 2.4 „	„ 4.2 „	0	− 3.7 „	„ 6.1 „	− 5.0 „	0	„ 7.6 „	+ 2.1 γ
30	− 3.6 „	„ 46.0 „	+ 6.3 γ	+15.4 „	„ 85.8 „	+ 4.0 „	+20.6 „	„ 66.0 „	+ 2.7 „
23 0	−24.8 „	„ 53.4 „	+10.8 „	−24.2 „	„ 73.5 „	0	−15.8 „	„ 66.0 „	+ 6.3 „
20	−12.8 „	„ 41.4 „	+ 5.4 „	−11.6 „	„ 51.3 „	− 4.0 „	0	„ 47.2 „	+ 2.1 „
24 0	+12.0 „	„ 50.0 „	+ 4.5 „	+21.0 „	„ 51.3 „	− 6.0 „	+23.7 „	„ 42.6 „	− 7.2 „
0 50	−11.6 „	„ 25.0 „	+ 4.0 „	−13.0 „	„ 18.4 „	−15.0 „	− 6.3 „	„ 21.0 „	− 7.2 „
1 20	− 6.4 „	„ 58.5 „	+ 5.0 „	+18.6 „	„ 73.5 „	− 5.5 „	+17.1 „	„ 56.0 „	−12.6 „
2 0	− 8.4 „	„ 30.0 „	+ 8.5 „	−11.6 „	„ 21.5 „	−10.0 „	−13.6 „	„ 21.5 „	− 8.7 „
40	+31.2 „	„ 50.0 „	0	+63.0 „	„ 52.3 „	−10.0 „	+55.8 „	„ 34.2 „	−23.8 „
4 30	+ 3.6 „	„ 23.0 „	0	+14.4 „	„ 22.0 „	−14.0 „	+ 9.2 „	„ 12.2 „	−18.1 „
5 30	− 4.0 „	„ 14.2 „	0	− 2.1 „	„ 9.5 „	−10.0 „	− 6.3 „	0	−13.0 „
6 30	+ 1.6 „	W 7.5 „	0	− 1.9 „	W 20.8 „	− 7.0 „	− 6.3 „	W 17.8 „	− 9.7 „

TABLE XLI (continued).

Gr. M. T.	San Fernando		Munich		Pola			Tiflis		
	P_h	P_d	P_h	P_d	P_h	P_d	P_v	P_h	P_d	P_v
h m										
15 32	+ 9.6 γ	0	+ 9.8 γ	E 3.8 γ	+10.3 γ	E 2.1 γ	0	+11.8 γ	E 1.8 γ	− 3.8 γ
48	+13.4 „	0	+12.0 „	0	+11.2 „	0	− 1.1 γ	+13.2 „	0	− 3.8 „
16 30	+ 9.6 „	0	+10.0 „	„ 2.3 „	+ 7.2 „	0	0	+ 6.3 „	„ 5.2 „	− 1.4 „
17 40	+ 4.2 „	E 7.5 γ	+ 6.0 „	„ 14.1 „	+ 2.7 „	„ 15.2 „	+ 2.1 „	+15.0 „	„ 18.2 „	− 4.1 „
18 20	+ 4.8 „	„ 1.6 „	+ 1.5 „	„ 3.0 „	0	„ 2.8 „	− 0.5 „	+ 2.1 „	„ 12.6 „	0
22 0	+ 2.9 „	„ 4.9 „	0	„ 3.0 „	− 4.7 „	„ 3.8 „	+ 2.8 „	+ 2.1 „	„ 16.5 „	0
30	− 7.6 „	„ 19.7 „	0	„ 48.7 „	+ 2.2 „	„ 41.0 „	−11.0 „	+21.4 „	„ 32.0 „	− 4.1 „
23 0	−26.2 „	„ 25.4 „	−18.0 „	„ 55.1 „	−15.6 „	„ 53.0 „	+ 9.4 „	+10.0 „	„ 45.6 „	− 1.3 „
20	−16.0 „	„ 25.4 „	−11.0 „	„ 37.3 „	−13.4 „	„ 38.0 „	+ 6.4 „	+ 3.0 „	„ 33.4 „	− 1.5 „
24 0	+ 1.6 „	„ 39.3 „	+11.5 „	„ 34.2 „	+ 9.0 „	„ 34.6 „	+ 4.1 „	+13.4 „	„ 9.3 „	− 3.1 „
0 50	− 7.6 „	„ 13.1 „	− 7.5 „	„ 16.0 „	− 9.4 „	„ 23.5 „	− 1.0 „	0	„ 11.5 „	− 3.6 „
1 20	−11.8 „	„ 32.8 „	+ 2.5 „	„ 46.4 „	− 1.1 „	„ 46.5 „	+ 4.2 „	+12.8 „	„ 20.2 „	− 6.9 „
2 0	−16.3 „	„ 13.1 „	−16.0 „	„ 16.8 „	−19.5 „	„ 25.0 „	0	−11.5 „	„ 11.0 „	− 1.5 „
40	+28.1 „	„ 36.9 „	+37.5 „	„ 30.5 „	+28.6 „	„ 26.4 „	+ 0.8 „	+31.0 „	W 13.0 „	− 8.9 „
4 30	0	„ 15.6 „	0	„ 9.5 „	0	„ 11.1 „	− 3.2 „	+ 2.3 „	„ 9.3 „	− 5.3 „
5 30	− 5.1 „	„ 7.4 „	−12.5 „	„ 3.8 „	− 8.7 „	„ 6.6 „	− 1.3 „	− 5.1 „	„ 10.6 „	− 3.1 „
6 30	+ 6.1 „	0	− 9.0 „	W 14.4 „	− 3.6 „	W 3.5 „	− 1.3 „	?	„ 18.6 „	− 1.5 „

TABLE XLI (continued).

Gr. M. T.	Dehra Dun		Bombay		Zi-ka-wei			Batavia		Christchurch		
	P_h	P_d	P_h	P_d	P_h	P_d	P_v	P_h	P_d	P_h	P_d	P_v
h m				No curve received.			No noticeable deflection.					
15 32	+15.4 γ	0	+13.0 γ		+13.1 γ	0		+11.0 γ	0	+ 5.9 γ	0	0
48	+13.4 „	0	+10.8 „		+10.1 „	E 2.0 γ		+ 9.3 „	E 5.4 γ	+ 4.6 „	W 1.5 γ	+ 0.8 γ
16 30	0	E 3.4 γ	+ 1.3 „		0	„ 5.9 „		0	„ 6.6 „	+ 1.8 „	0	0
17 40	+17.4 „	„ 10.8 „	+11.8 „		+11.3 „	„ 5.4 „		+11.6 „	„ 9.0 „	− 4.6 „	E 11.2 „	0
18 20	+ 3.1 „	„ 10.8 „	+ 1.5 „		+ 7.1 „	„ 11.9 „		+ 5.0 „	„ 6.6 „	+ 2.3 „	„ 12.0 „	0
22 0	+ 1.2 „	„ 11.8 „	+ 1.5 „		0	„ 3.0 „		− 1.8 „	„ 15.6 „	− 8.7 „	„ 11.2 „	+ 3.4 „
30	+14.2 „	„ 13.8 „	+11.2 „		0	W 1.0 „		+ 1.4 „	„ 12.0 „	− 9.2 „	„ 12.0 „	+ 1.5 „
23 0	+10.8 „	„ 11.3 „	+ 7.2 „		+ 6.5 „	„ 7.9 „		+ 6.2 „	„ 12.0 „	−18.3 „	„ 19.5 „	+ 1.5 „
20	− 1.8 „	„ 11.3 „	0		− 1.2 „	„ 9.9 „		− 0.7 „	„ 10.8 „	−11.9 „	„ 21.7 „	+ 1.8 „
24 0	− 1.0 „	0	+ 3.6 „		− 3.0 „	„ 12.9 „		0	0	− 2.7 „	„ 28.5 „	+ 1.2 „
0 50	−12.6 „	0	− 5.1 „		−11.9 „	„ 6.9 „		?	?	+14.0 „	„ 18.0 „	+ 1.4 „
1 20	+ 5.7 „	0	+ 4.1 „		0	„ 1.0 „		?	?	+11.4 „	„ 19.5 „	+ 1.5 „
2 0	−11.8 „	W 7.4 „	−10.8 „		− 5.9 „	„ 3.0 „		?	?	+17.8 „	„ 12.0 „	+ 0.9 „
40	+ 9.8 „	„ 29.5 „	+ 4.1 „		− 7.1 „	„ 17.8 „		?	?	−16.0 „	„ 8.2 „	+ 0.6 „
4 30	− 2.8 „	„ 4.9 „	+ 3.8 „		− 2.4 „	E 6.4 „		?	?	− 8.2 „	„ 3.8 „	0
5 30	?	0	?		0	„ 9.4 „		?	?	−17.8 „	0	0
6 30	?	0	?		− 9.5 „	„ 1.5 „		?	?	−18.3 „	W 9.0 „	0

TABLE XLI (continued).

Gr. M. T.	Ekaterinburg			Irkutsk		
	P_h	P_d	P_v	P_h	P_d	P_v
h m						
22 0	+29.0 γ	W 5.0 γ	− 8.0 γ	+ 4.0 γ	E 5.8 γ	− 2.0 γ
23 0	+32.0 „	„ 15.0 „	−13.0 „	+23.0 „	W 3.5 „	− 6.0 „
24 0	+36.0 „	„ 25.5 „	−19.0 „	+21.0 „	„ 20.3 „	− 6.0 „
2 0	+11.0 „	„ 8.9 „	−15.0 „	− 5.0 „	„ 9.9 „	− 7.0 „

Current-Arrows for the 23rd November, 1902; Chart III at $22^h\ 30^m$, and Chart IV at 23^h.

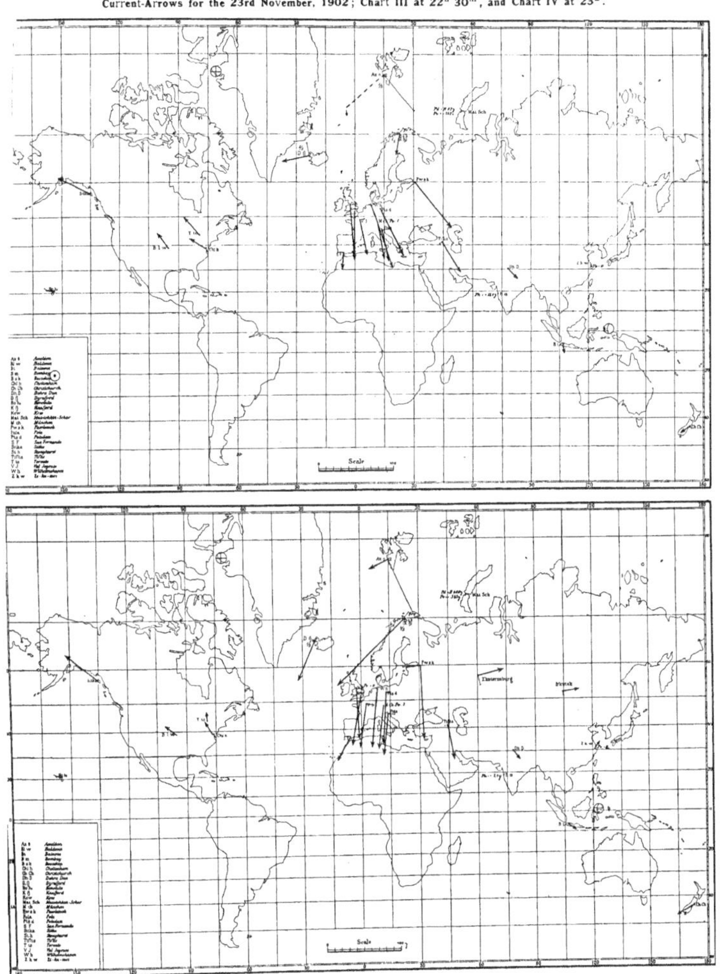

Fig. 126.

Current-Arrows for the 23rd and 24th November, 1902; Chart V at 23^h 20^m and 24^h on the 23rd, and Chart VI at 0^h 50^m and 1^h 20^m on the 24th.

Current-Arrows for the 24th November, 1902; Chart VII at 2h and 2h 40m, and Chart VIII at 6h 30m.

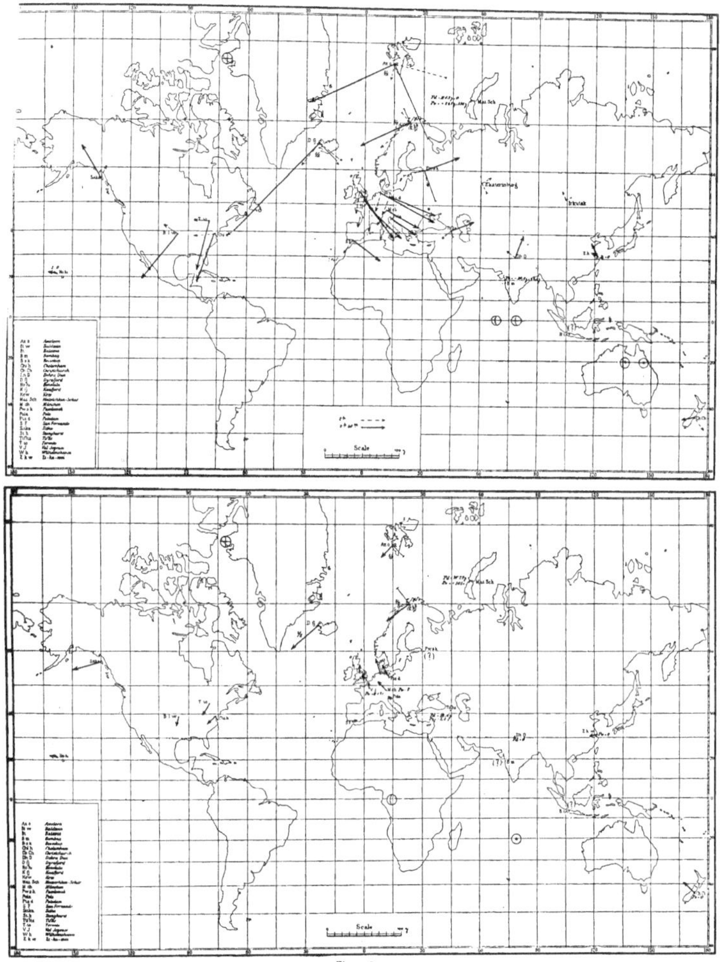

Fig. 128.

THE PERTURBATIONS OF THE 26th & 27th JANUARY, 1903.

(Pl. XV).

72. After the conclusion of the characteristic equatorial perturbation at $14^h\ 20^m$ on the 26th January (Art. 27), the conditions are comparatively quiet until about 18^h. At that hour they begin to be disturbed, especially in the north; and at about 19^h they assume the character of a powerful storm. From now on, powerful storms alternate with calmer periods, the most powerful being at about 23^h; and it is not until late in the morning of the 27th that comparative calm once more ensues.

While this is going on, there are powerful storms in low latitudes, both in the eastern and in the western hemisphere. We may at once mention, as a characteristic circumstance, that the deflections in the curves both in the western hemisphere and in Europe remain fairly uniform in direction throughout, notwithstanding the length of the storm. The strength of the perturbation diminishes greatly on the whole towards the equator. When we come as far south as Christchurch, it is very slight during the period up to 22^h on the 26th January. It subsequently becomes somewhat more powerful, though not more so than, for instance, at Dehra Dun.

(a) *Concerning the Occurrence of the Storm at the Norwegian Stations.*

The curves for Dyrafjord are indistinct, to some extent, indeed, altogether invisible. There is, however, sufficient to show that the storms have been violent. The declinometer especially has oscillated violently. From the vertical intensity curve, which is reproduced the best, we obtain an impression of two storms. The first of these commences at $18^h\ 35^m$, and lasts until about $21^h\ 0^m$. P_v is powerful here, and directed upwards. The second storm, which is of much longer duration and greater strength, reaches its maximum at about midnight. During this storm P_v is directed downwards.

From Kaafjord we have registerings only for the first part, up to $23^h\ 0^m$. Here too, a relatively independent perturbation is observable, which is particularly powerful in V, where a maximum is reached at $19^h\ 45^m$. Subsequently the storm increases, and is very powerful at about $22^h\ 30^m$, after which time it once more diminishes.

At Axeløen, very disturbed conditions commence at about $16^h\ 35^m$, and from that time storms continue until far on in the morning of the day following. The two storms already mentioned are very distinct here, and very powerful. The first is particularly powerful in H, where it begins and ends very suddenly at $19^h\ 10^m$ and $20^h\ 32^m$ respectively. This is followed by an interval of comparatively quiet conditions. The second powerful storm, which is so powerful in H that the curve runs off the paper —a thing which at this station very rarely happens—commences very suddenly at $22^h\ 24^m$. In D it begins earlier and more gradually. It is very violent between $22^h\ 30^m$ and $23^h\ 30^m$. The storm decreases until midnight, when another powerful storm commences, reaching a maximum at about $0^h\ 35^m$ on the 27th.

The first storm, as we see from the curves, occurs almost simultaneously at the above three stations. As regards the second storm there is a remarkable circumstance, in that it appears earlier at Kaafjord than at Axeløen. At 22^h its strength at Kaafjord is considerable, while at Axeløen, at the same hour, it is comparatively slight. There is a movement of the storm from Kaafjord to Axeløen; and from this too we may conclude that the cause of the storm must come comparatively near to the earth in that region.

The first part of the perturbation at Matotchkin Schar—up to $19^h\ 45^m$—is wanting. Even by that time it is exceedingly violent. It then diminishes for some time, and reaches a distinct minimum at $21^h\ 6^m$, whereupon it once more suddenly increases, and maintains a considerable strength until 2^h. It is particularly violent in the horizontal intensity. The light from the principal reflector passes, as is usual in the greater storms, out of the field, and at the same time that from the other reflector enters; but the latter also passes out of the field at $21^h\ 39^m$, and does not return until $23^h\ 25^m$. The storm is then losing strength, and at $23^h\ 54^m$ reaches a distinct minimum, after which it once more increases, and the light from the second reflector again passes repeatedly out of the field of observation. At $0^h\ 46^m$ it returns finally, and from that time the storm abates rapidly.

This perturbation, as we see, developes into one long storm, though with indications of the three maxima that were so conspicuous at Axeløen.

(b) *A General Characterisation of the Conditions in Southern Latitudes.*

As in most of the preceding compound storms there here appears to be a long perturbation in Europe, lasting from about $18^h\ 0^m$ on the 26th January to $7^h\ 0^m$ on the 27th. During this long storm, there occur some powerful intermediate storms, with a distribution of force differing from that produced by the long storm. We have here three of these sharply-defined intermediate storms; and they coincide on the whole in time with the three previously-described powerful storms at Axeløen.

The conditions at Pawlowsk are to some extent different. The *H*-curve there on the whole shows very little disturbance, there being powerful, well-defined perturbations only during the three intermediate storms. In *D*, on the other hand, there are powerful perturbations from $18^h\ 15^m$ until the morning of the day following. The conditions in the vertical intensity are especially interesting. The curve shows a deflection of long duration and uniform direction, answering to a perturbing force directed upwards.

Tiflis forms the transition from the conditions in Europe to those in the south and east of Asia, and these in their turn to the conditions at Batavia.

There is on the one hand a great resemblance between Tiflis and the district Kew to Pola; there is the same maximum, and the course of the perturbation is on the whole the same, the only difference being that the field is turned so that the conditions in the declination most resemble the *H*-curve at Tiflis. But on the other hand, the *H*-curve at Tiflis shows so great a resemblance to that at Dehra Dun, for instance, that it might almost be supposed that they were taken at the same place with apparatuses that differed a little in sensibility.

At Dehra Dun, Bombay and to some extent Tiflis, the horizontal intensity has on the whole a value that is below the normal. On the morning of the 27th, the normal line runs for a long distance almost parallel with the curve, and does not join it until about noon on that day.

The two last maxima are fairly distinct as far south as Christchurch, one at about $23^h\ 0^m$, the other at $0^h\ 38^m$. These maxima, however, are not nearly so pronounced as they are farther north; the perturbation-conditions remain more constant.

The perturbations in the western hemisphere are on the whole weaker than in the eastern, especially during the first part. The first maximum, which at Axeløen assumed the character of a brief, powerful, well-defined storm, is distinctly noticeable though not very powerful, at Sitka; while at the other stations it is almost imperceptible.

From $22^h\ 15^m$ on the 26th, right on to 8^h on the 27th, there is unrest. We here have the same two maxima as in the eastern hemisphere, namely, at about $22^h\ 55^m$ and at $0^h\ 30^m$.

There thus occurs in southern latitudes a long perturbation in *H*, with a perturbing force directed southwards; and to some extent the deviations in the curves are occurring simultaneously with those at the polar stations.

On glancing at the curves, we notice a no slight resemblance between those for Sitka and those for Christchurch. It is true that the perturbations at Sitka are much more powerful, but the course has nevertheless a great resemblance, especially noticeable in the last maximum, at about $0^h\ 35^m$. This is a resemblance not infrequently observed.

At Honolulu the conditions resemble those at Dehra Dun, the horizontal intensity remaining below the normal until far into the morning of the 27th. We cannot say when it became normal, as we have no magnetogram for the 24-hours following. It appears that the position of the curve at the conclusion of the magnetogram received is a little too low, and the normal line is therefore here put a trifle low.

THE FIELD OF FORCE.

73. The perturbation-conditions, as already mentioned, appear to some extent to be those of a long storm interrupted by powerful intermediate storms.

The decomposition of these phenomena, however, is somewhat difficult of accomplishment; and we have therefore, as in the case of the preceding perturbation, calculated only the total perturbing force. We then obtain at each place only the aggregate effect of all the simultaneously-acting forces; and it is therefore probable that the characteristic peculiarities of the polar fields will be most apparent at the times when the polar storms are most powerful, unless the other systems, equatorial or otherwise, that might be supposed to be acting, were at the same time correspondingly increased.

If we look at the various fields that occur, we find an exact resemblance to the fields in those perturbations that occurred about midnight Gr. M. T. All the systems exhibit the peculiar fields that characterise the polar storms, namely an area of convergence and an area of divergence. The first of these comes out clearly on all the charts. Its position varies indeed, but only slightly; and it remains, throughout the series of charts, in the district Europe and Asia. This indicates that the negative system of precipitation extends very far in a direction east and west along the auroral zone on the night-side of the globe, a circumstance that we have frequently met with in previous storms.

The area of divergence is often very faint and indistinct, for instance in the first three charts, in which the current-arrows in America are very small. In Europe, however, at these hours, there is a more or less distinct indication of its existence. In Chart II, for instance, the current-arrows in the west of Europe seem to be turning westwards, while those at the eastern stations turn in the opposite direction. In the subsequent charts, the perturbing forces in America attain to considerable dimensions, and the area of divergence also comes out distinctly there.

The arrow at Sitka, which throughout is directed westwards along the auroral zone, seems to indicate that the influence of the polar precipitation which produces the negative polar storm in Europe and Asia, also has some effect at that place. It might indeed be imagined that the positive storm also would predominate at Sitka, so that the current-arrow there would belong to the area of divergence; but this does not seem very probable, as in that case the positive field of precipitation would need to have a disproportionately high northerly position.

With regard to the vertical intensity we find that there are exceedingly distinct negative values of P_v in the area of convergence, especially at Pawlowsk and Ekaterinburg, near which the point of convergence, or rather the neutral district, appears to lie. This district, according to the charts, seems to be situated in the north-east of Europe or the north-west of Asia. Here the vertical arrows are comparatively powerful all the time, while the horizontal component of the perturbing force is often exceedingly small, a condition of affairs that we should expect to find in the vicinity of the point of convergence. As, therefore, this is very clearly shown by the vertical intensity curve for Ekaterinburg, we have placed on the charts current-arrows for the hours 22^h and 23^h, as well as for intermediate times, although the values interpolated between the entire hours will often be very uncertain, especially when the perturbing force is small. A similar course has been followed with respect to Irkutsk; for the field, as already mentioned, does not appear to vary much as time passes, and the uncertainty of the interpolated values is therefore smaller.

In the area of divergence, at the time when it is rather well developed in Europe, there are also positive values of P_v at the western stations. This appears on Chart II both at Potsdam, Pola and Tiflis. It may however be a little doubtful whether it is the positive polar storm that produces these values at the last-named station; it is perhaps more probable that they are brought about by a storm that was caused by perturbations of a more equatorial nature. That this was the case seems probable, moreover, from the conditions at the other stations of Southern Asia, which also appear to run a slightly abnormal course. There, however, the perturbing forces are so small that nothing certain can be said. At Pola, the positive deflections in the vertical intensity curve continue until nearly 23^h, when they go over to the opposite side.

On Chart IX, the conditions at Dehra Dun and Bombay seem once more to be a little abnormal; and a study of the curves for the succeeding period will show that the perturbing forces there continue to act far on into the 27th. These forces, as we have said, occur principally in H, which they serve to diminish. We have also already remarked that before the end of the period we find at Honolulu an abnormally low horizontal intensity curve, which thus seems to agree with the conditions at the stations in Southern Asia. The character of the curve is comparatively quiet, and it is therefore possible that this is the effect of a storm of a more equatorial nature, perhaps a negative equatorial storm.

If we now in conclusion compare the perturbation-fields that have appeared during this perturbation with those that we have found in the preceding storms, we at once notice the great resemblance. The storms here described occurred, as we have seen, about Greenwich midnight; and we found the characteristic large area of convergence on the night-side in Europe and Asia. There also appeared more or less certain indications of an area of divergence upon the day-side. And these are the very conditions that we have continually met with before.

We therefore feel justified, after having gone through this long series of perturbations, in concluding that the phenomena that we have previously described as elementary, viz. the positive and negative polar, the positive and negative equatorial, and the cyclo-median perturbations, generally are sufficient to explain the fields that will be formed during the most varied magnetic storms. All the fields that we have met with thereby receive a very simple explanation, and no serious disagreement has presented itself, although, of course, the material has very often been insufficient to allow of certain conclusions being drawn.

TABLE XLII.

The Perturbing Forces on the 26th & 27th January, 1903.

Gr. M. T.	Honolulu		Sitka		Baldwin		Toronto	
	P_h	P_d	P_h	P_d	P_h	P_d	P_h	P_d
h m								
19 30	0	0	−14.7 γ	W 7.2 γ	− 2.5 γ	W 6.4 γ	− 4.5 γ	E 2.4 γ
20 0	+ 2.1 γ	0	−17.7 "	" 17.6 "	− 7.1 "	" 11.4 "	− 6.7 "	W 1.8 "
30	+ 5.2 "	0	− 8.9 "	" 24.3 "	− 2.1 "	" 14.0 "	− 5.4 "	" 6.0 "
22 0	− 3.6 "	0	−11.5 "	0	0	" 8.3 "	−10.4 "	E 5.4 "
30	− 9.1 "	W 5.0 γ	−33.8 "	" 1.8 "	−24.1 "	" 6.4 "	−30.6 "	W 6.0 "
23 0	−17.9 "	" 5.8 "	−64.1 "	" 11.3 "	?	?	−58.4 "	E 18.1 "
30	−13.8 "	" 5.0 "	−46.0 "	" 18.0 "	−21.2 "	" 3.8 "	−35.0 "	" 6.0 "
24 0	−12.0 "	0	−26.6 "	" 4.5 "	−12.0 "	E 3.2 "	−28.0 "	" 3.0 "
0 22.5	−11.2 "	0	−28.8 "	" 4.5 "	−15.9 "	0	−36.9 "	W 3.0 "
30	−19.2 "	" 1.7 "	−35.4 "	" 18.0 "	−35.4 "	W 15.9 "	−52.5 "	" 16.9 "
45	−24.7 "	0	−26.6 "	" 9.9 "	−36.1 "	" 14.0 "	−46.4 "	" 16.9 "
1 30	−25.2 "	0	?	?	−24.8 "	0	−28.0 "	E 20.4 "

TABLE XLII (continued).

Gr. M. T.	Dyrafjord			Axeløen			Matotchkin Schar		
	P_h	P_d	P_v	P_h	P_d	P_v	P_h	P_d	P_v
h m									
19 30	+25.4 γ	E 41.6 γ	−178.0 γ	−253.0 γ	E 46.3 γ	+324.0 γ	?	?	?
20 0	?	W 18.1 „	−110.0 „	−198.0 „	„ 46.2 „	+344.0 „	− 323.0 γ	E 220.0 γ	−173.0 γ
30	+58.8 „	0	− 71.0 „	− 22.5 „	W 6.8 „	+324.0 „	− 264.0 „	„ 145.0 „	−162.0 „
22 0	−141.0 „	„ 18.0 „	0	+ 25.3 „	E 22.3 „	+300.0 „	−>408.0 „	„ 100.0 „	+ 47.6 „
30	Curve almost invisible. The points that can be seen indicate a negative deflection of considerable extent.	Curve difficult to distinguish from the V-curve. There are however principally easterly deflections somewhat greater than the preceding ones.	Rather large positive deflections.	ca. −281.0 „	„ 102.0 „	+643.0 „	−>408.0 „	„ 205.0 „	A violent positive deflection of about 440 γ, after which the curve disappears.
23 0				ca. −519.0 „	„ > 166.0 „	+648.0 „	−>408.0 „	„ 199.0 „	
30				−347.0 „	„ ca. 122.0 „	+700.0 „	− 402.0 „	„ 357.0 „	
24 0				− 91.5 „	„ 138.0 „	+635.0 „	−>408.0 „	„ 279.0 „	
0 22.5				−232.0 „	„ ca. 111.0 „	+755.0 „	−>408.0 „	„ 315.0 „	
30				−366.0 „	„ > 166.0 „	+702.0 „	− 355.0 „	„ 182.0 „	
45				−315.0 „	„ 129.0 „	+598.0 „	−>408.0 „	„ 208.0 „	
1 30				−187.0 „	„ 163.0 „	+547.0 „	− 105.0 „	„ 93.0 „	

TABLE XLII (continued).

Gr. M. T.	Kaafjord			Pawlowsk			Stonyhurst		Kew	
	P_h	P_d	P_v	P_h	P_d	P_v	P_h	P_d	P_h	P_d
h m										
19 30	− 72.3 γ	E 14.7 γ	−202.0 γ	+ 23.1 γ	E 16.8 γ	− 3.7 γ	+ 5.1 γ	E 41.7 γ	+ 4.1 γ	E 37.4 γ
20 0	− 95.0 „	„ 72.0 „	− 13.7 „	?	?	?	− 9.2 „	„ 34.9 „	− 11.2 „	„ 33.7 „
30	− 69.9 „	„ 63.8 „	−129.0 „	− 10.6 „	„ 35.0 „	− 7.5 „	− 15.8 „	„ 14.3 „	− 17.8 „	„ 16.4 „
22 0	−368.0 „	„ 41.5 „	− 4.7 „	0	„ 27.6 „	− 16.5 „	− 12.2 „	„ 46.3 „	− 15.3 „	„ 38.3 „
30	−577.0 „	„ 66.0 „	−253.0 „	+ 52.8 „	„ 22.1 „	− 36.7 „	− 17.8 „	„ 107.0 „	+ 10.2 „	„ 82.7 „
23 0	−276.0 „	„ 104.0 „	?	+ 28.7 „	„ 27.6 „	− 46.4 „	+ 14.8 „	„ 100.0 „	+ 15.3 „	„ 98.2 „
30	Curves disappeared.			+ 7.5 „	„ 57.5 „	− 48 6 „	− 27.5 „	„ 81.1 „	− 19.4 „	„ 72.5 „
24 0				0	„ 36.8 „	− 44.8 „	− 18.8 „	„ 57.1 „	− 13.2 „	„ 56.2 „
0 22.5				+ 20.1 „	„ 10.6 „	− 42.6 „	+ 20.4 „	„ 79.9 „	+ 15.3 „	„ 60.9 „
30				+ 24.1 „	„ 27.6 „	− 48.6 „	+ 11.2 „	„ 97.1 „	+ 10.2 „	„ 84.8 „
45				− 7.5 „	„ 33.1 „	− 50.1 „	− 24.0 „	„ 70.8 „	− 20.4 „	„ 73.0 „
1 30				+ 7.8 „	„ 22.5 „	− 40.3 „	− 10.7 „	„ 29.7 „	− 17.8 „	„ 56.2 „

TABLE XLII (continued).

Gr. M. T.	Val Joyeux			Wilhelmshaven			Potsdam		
	P_h	P_d	P_v	P_h	P_d	P_v	P_h	P_d	P_v
h m									
19 30	0	E 37.6 γ	From 10h to 1h there appears to be a negative deflection with maximum at about 12¼h of −11 γ; but it is not easy to determine from the magnetogram whether the curve has too great a value before, or too small a value after.	+ 23.3 γ	E 47.0 γ	+ 7.0 γ	+ 20.5 γ	E 30.5 γ	− 1.2 γ
20 0	− 6.4 γ	„ 34.3 „		− 5.1 „	„ 38.5 „	0	− 9.2 „	„ 30.5 „	− 15.8 „
30	− 18.0 „	„ 20.9 „		− 22.8 „	„ 18.3 „	0	− 20.5 „	„ 17.2 „	+ 2.7 „
22 0	− 16.0 „	„ 41.0 „		− 4.2 „	„ 42.2 „	0	− 7.3 „	„ 31.0 „	− 1.5 „
30	+ 13.6 „	„ 87.8 „		+ 42.0 „	„ 90.5 „	0	+ 46.1 „	„ 65.0 „	− 10.8 „
23 0	+ 20.0 „	„ 86.2 „		+ 39.7 „	„ 80.2 „	− 7.0 „	+ 34.1 „	„ 59.4 „	− 13.8 „
30	− 12.8 „	„ 71.0 „		− 13.1 „	„ 71.0 „	− 19.0 „	− 9.5 „	„ 56.4 „	− 12.0 „
24 0	− 12.0 „	„ 58.4 „		− 15.4 „	„ 43.4 „	− 15.0 „	− 11.7 „	„ 34.5 „	− 13.2 „
0 22.5	+ 16.8 „	„ 51.0 „		+ 28.0 „	„ 43.4 „	− 20.0 „	+ 29.0 „	„ 25.4 „	− 21.0 „
30	+ 18.4 „	„ 72.8 „		+ 33.8 „	„ 69.1 „	− 18.0 „	+ 28.4 „	„ 49.7 „	− 22.5 „
45	− 16.8 „	„ 62.7 „		− 12.1 „	„ 52.0 „	− 20.0 „	− 17.4 „	„ 35.5 „	− 15.0 „
1 30	− 4.8 „	„ 49.3 „		− 3.3 „	„ 45.2 „	− 19.0 „	− 5.4 „	„ 33.5 „	− 15.0 „

TABLE XLII (continued).

Gr. M. T.	San Fernando		Munich			Pola		
	P_h	P_d	P_h	P_d	P_v	P_h	P_d	P_v
h m								
19 30	+ 3.0 γ	E 24.6 γ	+ 8.0 γ	E 29.7 γ	0	+ 7.1 γ	E 24.3 γ	+ 4.2 γ
20 0	− 8.9 „	„ 14.8 „	− 5.0 „	„ 30.5 „	− 0.9 γ	− 2.7 „	„ 29.8 „	+ 3.0 „
30	− 18.5 „	„ 4.1 „	− 17.0 „	„ 19.8 „	0	− 14.3 „	„ 13.9 „	+ 2.1 „
22 0	− 16.3 „	„ 15.6 „	− 8.5 „	„ 29.7 „	0	− 9.4 „	„ 27.7 „	+ 4.0 „
30	0	„ 61.5 „	+ 22.5 „	„ 51.7 „	− 1.1 „	+ 17.5 „	„ 49.9 „	+ 7.0 „
23 0	+ 7.4 „	„ 59.5 „	+ 31.0 „	„ 67.0 „	− 4.2 „	+ 29.1 „	„ 52.7 „	− 2.1 „
30	− 18.5 „	„ 34.4 „	+ 1.5 „	„ 50.1 „	− 4.5 „	+ 1.8 „	„ 47.2 „	− 2.1 „
24 0	− 16.3 „	„ 29.1 „	− 6.0 „	„ 43.7 „	− 4.5 „	0	„ 34.0 „	− 5.5 „
0 22.5	+ 6.7 „	„ 49.2 „	+ 25.5 „	„ 29.7 „	− 4.7 „	+ 22.8 „	„ 23.6 „	− 4.2 „
30	− 3.7 „	„ 54.0 „	+ 24.5 „	„ 45.7 „	− 4.9 „	+ 22.0 „	„ 40.2 „	− 1.7 „
45	− 29.6 „	„ 30.3 „	− 3.5 „	„ 49.5 „	− 6.4 „	− 7.1 „	„ 40.9 „	− 6.1 „
1 30	− 17.8 „	„ 31.5 „	− 3.0 „	„ 35.8 „	− 4.5 „	− 2.0 „	„ 31.2 „	− 5.3 „

TABLE XLII (continued).

Gr. M. T.	Tiflis			Dehra Dun		Bombay		
	P_h	P_d	P_v	P_h	P_d	P_h	P_d	P_v
h m								
19 30	+ 9.3 γ	E 5.6 γ	− 2.8 γ	+ 5.9 γ	W 8.9 γ	+ 2.6 γ	W 1.8 γ	0
20 0	+ 4.2 „	„ 20.4 „	− 1.2 „	+ 7.1 „	0	+ 5.6 „	0	0
80	− 8.8 „	„ 18.6 „	+ 1.3 „	− 1.6 „	E 4.9 „	− 4.6 „	E 4.9 „	0
22 0	− 4.2 „	„ 11.1 „	0	− 5.9 „	W 4.9 „	− 10.2 „	W 6.2 „	0
30	+ 22.1 „	0	− 7.7 „	+ 5.9 „	„ 19.7 „	+ 2.6 „	„ 14.8 „	− 1.6 γ
23 0	+ 33.2 „	„ 5.6 „	− 6.4 „	+ 27.5 „	„ 19.7 „	+ 20.5 „	„ 18.4 „	− 1.6 „
30	+ 11.5 „	„ 18.6 „	− 1.8 „	+ 13.0 „	„ 4.9 „	+ 10.8 „	„ 12.3 „	0
24 0	+ 7.7 „	„ 8.2 „	− 2.6 „	+ 5.9 „	„ 7.9 „	+ 5.6 „	„ 14.8 „	− 2.0 „
0 22.5	+ 19.9 „	W 9.3 „	− 5.1 „	+ 12.2 „	„ 18.7 „	+ 10.2 „	„ 18.4 „	− 8.0 „
30	+ 19.4 „	0	− 3.3 „	+ 13.2 „	„ 14.9 „	+ 10.2 „	„ 18.4 „	− 6.4 „
45	− 5.5 „	E 13.0 „	+ 1.3 „	− 8.7 „	„ 4.9 „	− 8.2 „	„ 12.3 „	0
1 30	− 2.2 „	„ 8.9 „	− 2.6 „	− 5.9 „	„ 4.9 „	− 10.2 „	„ 20.8 „	0

TABLE XLII (continued).

Gr. M. T.	Zi-ka-wei [1]			Batavia		Christchurch		
	P_h	P_d	P_v	P_h	P_d	P_h	P_d	P_v
h m								
19 30	0	W 6.0 γ	No noticeable deflection.	+ 3.3 γ	0	+ 1.8 γ	W 3.7 γ	The V-curve seems to be a trifle too high, answering to a positive P_v until 3h, afterwhich it is a little too low, answering to negative P_v.
20 0	+ 12.6 γ	„ 9.0 „		+ 12.4 „	W 3.6 γ	+ 4.6 „	0	
30	+ 6.0 „	E 4.0 „		+ 5.3 „	0	+ 6.9 „	E 5.9 „	
22 0	− 6.0 „	W 10.0 „		− 3.5 „	0	− 11.0 „	W 13.4 „	
30	− 12.0 „	„ 21.0 „		− 1.6 „	„ 3.6 „	− 22.1 „	„ 11.9 „	
23 0	+ 13.2 „	„ 35.0 „		+ 17.1 „	„ 1.8 „	− 23.0 „	„ 3.0 „	
30	+ 14.4 „	„ 19.0 „		+ 8.9 „	0	− 16.1 „	„ 3.7 „	
24 0	+ 12.6 „	„ 16.8 „		+ 3.3 „	„ 6.0 „	− 10.1 „	„ 9.7 „	
0 22.5	+ 7.2 „	„ 14.0 „		+ 6.7 „	„ 8.4 „	− 7.8 „	„ 5.9 „	
30	+ 12.0 „	„ 16.0 „		+ 4.6 „	?	− 23.9 „	„ 3.7 „	
45	− 2.4 „	„ 15.0 „		?	?	− 20.7 „	E 2.2 „	
1 30	− 7.2 „	„ 14.4 „		?	?	− 11.5 „	0	

[1] The determination of time is here somewhat uncertain, as only midnight is marked upon the copy received, which, moreover, is reduced to half the linear size of the original magnetogram.

TABLE XLII (continued).

Gr. M. T.	Ekaterinburg			Irkutsk		
	P_h	P_d	P_v	P_h	P_d	P_v
h m						
19 30	+ 1.9 γ	E 31.0 γ	− 6.5 γ	+ 14.5 γ	0	− 1.5 γ
20 0	+ 9.5 "	" 34.0 "	− 10.0 "	+ 19.0 "	W 2.9 γ	− 2.0 "
30	+ 17.2 "	" 28.0 "	− 11.0 "	+ 11.5 "	" 1.8 "	− 3.6 "
22 0	+ 27.5 "	" 3.4 "	− 18.0 "	− 1.0 "	" 15.0 "	− 5.0 "
30	+ 30.8 "	W 5.6 "	− 22.0 "	+ 14.5 "	" 46.3 "	− 5.5 "
23 0	+ 32.5 "	" 10.3 "	− 24.0 "	+ 33.0 "	" 56.2 "	− 6.0 "
30	+ 27.0 "	" 9.0 "	− 22.5 "	+ 24.0 "	" 46.3 "	− 8.8 "
24 0	+ 18.5 "	" 7.8 "	− 20.0 "	+ 9.0 "	" 31.3 "	− 10.0 "
0 22.5	+ 11.7 "	" 7.3 "	− 18.4 "	0	" 27.5 "	− 9.5 "
30	+ 9.5 "	" 6.7 "	− 18.0 "	− 3.0 "	" 26.4 "	− 9.0 "
45	+ 6.5 "	" 6.2 "	− 17.0 "	− 7.5 "	" 24.0 "	− 8.0 "
1 30	+ 6.0 "	0	− 13.5 "	− 8.8 "	" 12.8 "	− 6.5 "

Current-Arrows for the 26th January, 1903; Chart I at $19^h\ 30^m$ (1).

Fig. 129.

(1) By an unfortunate mistake, the arrow for P_v at Axeloen in this and the eight following charts, has been given a direction the should be.

Current-Arrows for the 26th January, 1903; Chart II at 20^h 30^m, and Chart III at 22^h 0^m (1).

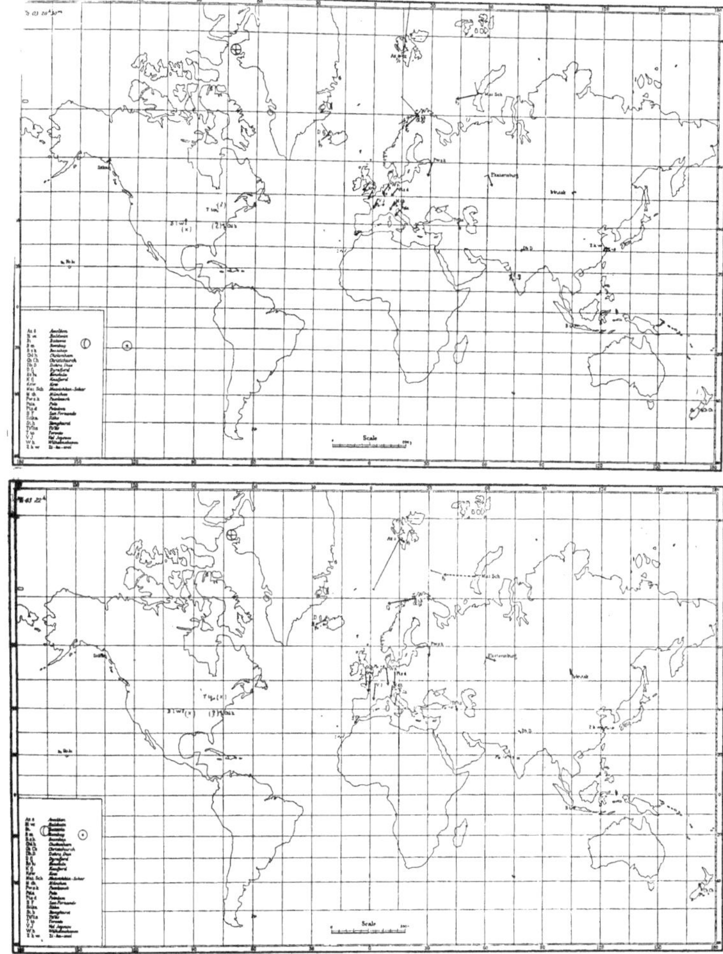

Fig. 130.

(1) Arrow for P_φ at Axeløen reversed. See note, p. 292.

Current-Arrows for the 26th January, 1903; Chart IV at 22^h 30^m, and Chart V at 23^h 0^m [1].

Current-Arrows for the 26th & 27th January, 1903; Chart VI at $23^h\ 30^m$, and Chart VII at $0^h\ 22.5^m$ (1).

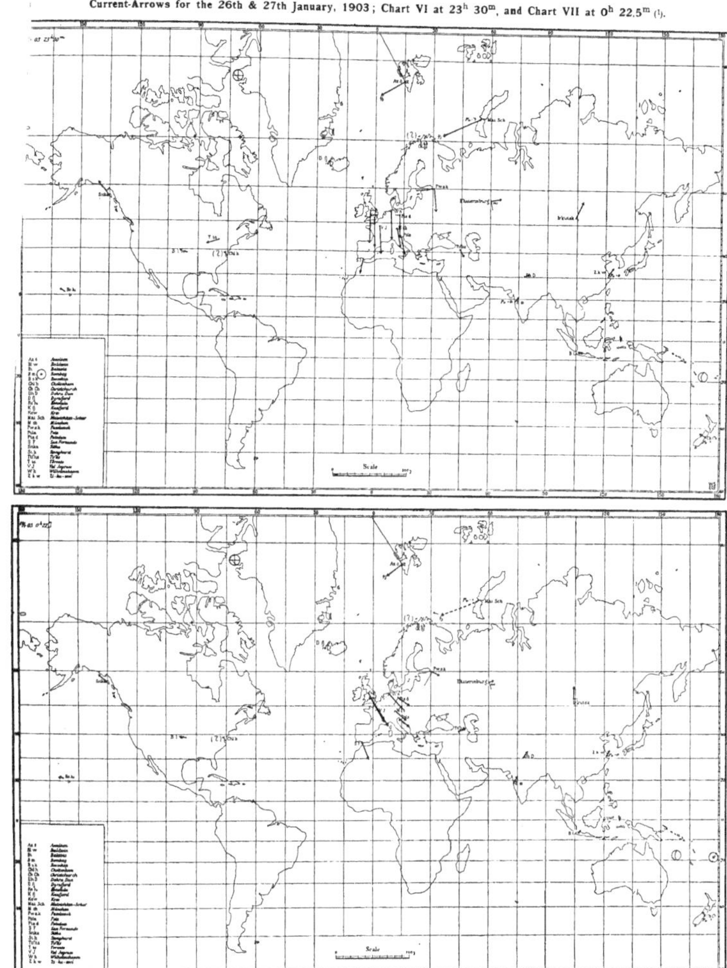

Fig. 132.

(1) Arrow for P_F at Axeløen reversed. See note, p. 292.

FURTHER COMPARISON WITH THE TERRELLA-EXPERIMENTS.

74. In order to obtain a clear idea of the way in which the various light-phenomena around our terrella appear under conditions answering to the earth's positions at the various seasons, I have made three series of experiments representing an equinox, and the summer and winter solstices.

For each of these seasons, 12 photographs have been taken in four groups of three. The position of the magnetic north pole in the four groups answers respectively to noon, 6 p. m., midnight and 6 a. m.

In order to obtain a position answering to the summer or winter solstice, the discharge-tube was inclined so that its axis was at an angle of $23^1/_2°$ below or above the horizontal position answering to the equinox. The terrella was suspended by a universal joint in such a manner that it always maintained the desired position in relation to the cathode rays during a rotation of the terrella answering to the diurnal revolution of the earth.

Thirty-six photographs have thus been taken, with the highest possible magnetisation of the terrella with a magnetising current of 33 amperes, corresponding to a magnetic moment of about 10 000 cm.$^{5/2}$ gr.$^{1/2}$ sec.$^{-1}$ (see fig. 70, p. 155).

I have also taken 36 photographs of the terrella in exactly the same positions as the above, but with a magnetising current of only 15 amperes, corresponding to a magnetic moment of about 6 200. These 72 photographs, with descriptions, will be found farther on in this work.

It will be interesting, however, to describe here some few examples of these with their photographs, because of the great significance of the light-phenomena observed, in the explanations of magnetic storms given in the preceding pages.

In the eight photographs following, the terrella has a position answering to the winter solstice and 6 a. m. at the earth's magnetic north pole.

The experiment represented is almost the same, but the photographs are taken from eight different points of view.

Fig. 134.

The pressure employed in the discharge-tube was about 0.02 mm., except in the case shown in photograph 7, where it was 0.013 mm. The current-strength was 8 milliamperes with a voltage of 3300; and lastly a magnetising current of 33 amperes was employed upon the terrella.

In taking photographs 1, 2, 3 and 4, the axes of the cameras were directed towards the centre of the terrella, and were lying in a plane that passed through the axis of the discharge-tube. This plane formed an angle of $66^1/_2{}^\circ$ with the vertical line, and thus formed the same angle with the horizontal plane as the axis of the discharge-tube.

When the angular distances to the axes of the cameras were measured from the axis of the tube in the direction of the cathode, and in the above-mentioned inclined plane with the centre of the terrella as the vertex, measuring contrary to the hands of a clock seen from above, the angles in the four positions were respectively 90°, 180°, 270° and 315°. Photograph 6 was taken with the axis of the camera horizontal in the vertical plane through the axis of the tube, and directed towards the centre of the terrella, and towards its night-side.

Photographs 5, 7 and 8 were taken in positions that may be described as follows: in three vertical planes through the centre of the terrella at angular distances of 45°, 270° and 315° respectively from the vertical plane through the axis of the tube, the axes of the cameras pointing towards the centre of the terrella, and forming an angle of 20° with a horizontal plane.

There are two different phenomena that come out very clearly in these photographs, or rather in the experiments which the photographs reproduce.

In *the first* of these, we have the luminous spirals, almost closed rings, that are formed round, and at a certain distance from, the magnetic poles of the terrella. These spirals vary in position with the rotation of the terrella; and I consider them as answering to the auroral zones on the earth. These principal spirals of light form in my opinion the most remarkable phenomenon that I have discovered in my terrella-experiments. The more highly the terrella is magnetised, the narrower does the band of light become, keeping, however, its intensity. The bands of light are here almost coherent; but different degrees of luminosity in the precipitation are easily seen, answering to the various districts of precipitation shown by the experiments given in fig. 47 a & b.

It will be seen from photographs Nos. 1 and 5, fig. 134, that the spirals begin above as a broad luminous band, indicating a great descent of rays upon the terrella. At the top, to the left of the band in No. 1, there is a slight illumination in space outside the terrella, as also in No. 7. These two illuminations are the beginning and end of the greatest precipitation of rays in the band of light. The principal bands of light can be easily followed in photographs 2 and 6, then in 3 and 7, and 4 and 8, right round the terrella, until they disappear. In No. 8 especially, we see both beginning and end of this long spiral of light round the south pole of the terrella-magnet, which answers to the terrestrial-magnetic north pole.

These continuous bands of light recall a most remarkable and ingenious hypothesis made by A. E. Nordenskiöld (¹). He assumes that the usual arc of polar aurora seen in Bering Strait was part of a ring of light situated in a plane perpendicular to the radius of the earth, which terminates in a point near the magnetic pole (lat. 81° N., long. 80° W. Gr.). He concludes that the plane which contains the auroral arc, and which is perpendicular to this radius, cuts it at a distance of 125 kilometres below the surface of the earth. In this plane the lower edge of the ring of auroral light would be about 200 kilometres above the surface of the earth.

The *second phenomenon*, which is clearly visible in the experiments shown in fig. 134, is the presence of portions of luminous rings, also almost circular, which lie considerably nearer to both poles

(¹) A. E. Nordenskiöld: Vega-Expeditionens Vetenskaplige Iakttagelser. Första Bandet, p. 417, Stockholm, 1882.

of the terrella's axis of rotation, than the previously described luminous spiral. These portions of luminous rings, with a very much smaller radius than the first rings had, have already been shown, e. g. in photographs 3, 6 and 9 in fig. 68. It will be easily seen that these small luminous half-rings are comparatively independent of the large luminous spirals round the poles, when the magnetising current for the terrella is reduced to, for instance, 15 amperes. There then appear the peculiar, triangular patches also covering the equatorial regions, that are seen in fig. 68, in place of the large polar rings; while the small rings continue almost unchanged up at the poles. On looking more closely into the phenomenon, we see that these small ring-portions are formed round a luminous point upon the terrella, this point being the apex of a cone of light that may often be seen in space outside the terrella. I have selected three photographs in which this cone of light comes out well, and reproduced them, with the contrasts brought out as clearly as possible (see fig. 135). The apex of these cones falls upon the terrella near either pole, and strange to say does not greatly change its position during the rotation of the terrella. It remains on the post-meridian side near the noon meridian through the centre of the cathode, and moves a little backwards and forwards, principally east or west, during the rotation.

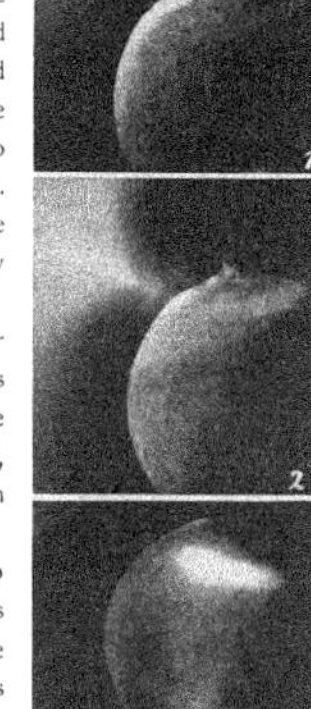

Fig. 135.

It should be remarked that the cones of light seen in the figure appear to withdraw from the terrella when the magnetisation is increased, whereas the little ring of light still strikes the terrella. To the east of the apex of the cone of light, the ring of light is seen in the air (see photograph 2, fig. 135), while to the west it is thrown upon the phosphorescent terrella in the form of a semicircle (see photographs 3, 4, 7, and 8, fig. 134).

These cones of light are extremely interesting. They are similar to those that I first described in connection with the drawing-in of cathode rays towards a magnetic pole, in the same paper (1) in which I expressed for the first time my belief that the northern lights are formed by corpuscular rays drawn in from space, and coming from the sun.

On looking closely at fig. 135., we see that the drawn-in cone really consists of several envelopes; in the original photographs, as many as three cones, with very different apical angles, are distinguishable.

This is a very interesting phenomenon, which is also demonstrated in another way in the paper just mentioned. I found by studying a series of successive inversions of a shadow-cross at the bottom of a Crookes' tube standing before a strong magnet, that the cathode rays must intersect one another several times before they reached the bottom of the tube.

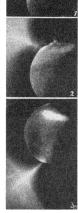

Poincaré (2) has made this drawing-in phenomenon the subject of mathematical investigation, and has demonstrated that the cathode rays move like geodetic lines upon certain cones with a common generatrix, so that each ray has its conjugate cone.

Wiedemann and *Wehnelt* (3) thought they could prove that this repeated crossing of rays in the discharge-tube was produced by the frequent intersection of the same cathode rays in the tube, and that the phenomenon recalled the circumstances connected with a vibrating cord.

(1) Archives des Sciences Physiques et Naturelles, Geneva, 4th period, vol. I, 1896.
(2) Comptes Rendus, 123, p. 930, 1896.
(3) Wiedemanns Annalen, Vol. LXIV, No. 3, 1898.

In investigations made at the same time, but not published until some months later[1] I had shown, however, that the phenomena were not so simple; it is certain, indeed, that no theoretically clear understanding has yet been arrived at with regard to the formation of the cones of light shown in fig. 135. In the above-named paper, I have shown how the theory can explain a number of discontinuously occurring luminous rings in the discharge-tube, even if we suppose the cathode to emit a whole sheaf of rays, and not only separate bundles with definite angles of emanation for the rays. It may possibly also be shown that the above-mentioned cones of light in space are formed by a maximal agglomeration of rays about certain surfaces, thus making the density of the rays there so great that the rarefied air in the tube becomes more luminous near these surfaces.

I have here touched upon this matter because these cones of light and their attendant phenomena will be found to play an important part in our theory of terrestrial-magnetic and auroral phenomena.

My special reason for here reproducing the above photographs in fig. 134 and mentioning the experiments, is my desire to indicate phenomena that may possibly afford a full explanation of a peculiar circumstance that has frequently been pointed out in the preceding pages. We have seen that during the so-called positive polar storms on the post-meridian side of the earth, the current-arrows at Dyrafjord, Kaafjord and Matotchkin Schar have often been directed eastwards, more or less along the auroral zone, while at the same time the arrow at Axeløen pointed in the opposite direction, westwards along the auroral zone (cf. the perturbations of the 11th, 27th and 31st October, 23rd November, 9th December, and 8th and 15th February).

The great spiral of light round the magnetic south pole of the terrella represents, in my opinion, the precipitation of rays on the night-side of the globe during long magnetic storms. It represents the "horizontal part" of the current generally passing between Kaafjord and Axeløen at about midnight, the breadth of which I have estimated to be not more than 500 km. While discussing the long magnetic storms, we have frequently pointed out that in the afternoon the negative storm at Axeløen seems to be closely connected with storms farther east on the night-side of the earth; while at the same time a positive storm is observed at Dyrafjord and Kaafjord.

Our photographs in fig. 134 answering to 6 a. m. at the magnetic south pole, clearly show that the spiral of light begins in a very high latitude on the post-meridian side, whence it passes round the terrella in its descent to lower latitudes. When, for instance, the terrella is turned so that it is noon at the pole the beginning of the spiral also moves down towards lower latitudes, its longitude, however, changing only slightly, measured from the cathode.

In this connection I will mention that during the observations of aurora at the Haldde observatory in mid-winter, 1899—1900, the following phenomena were observed day after day. Early in the afternoon, generally at about 5 or 6 p. m., local time, an arc would appear far to the north and close down on the horizon, and would remain through the evening, moving farther and farther south, and higher and higher in the sky. As it came nearer, it would sometimes divide into several separate arcs. At about 9^h or 10^h it would disappear, generally rather suddenly. During these auroral displays, our magnetometers were generally disturbed; but the most powerful magnetic storms almost always occurred after midnight, when there was generally no aurora to be seen. This seems to agree well with the conditions on the terrella, where the first great precipitation begins on the post-meridian side far up near the pole, and descends to lower latitudes before it ceases or becomes a faint band of light, which continues round the terrella. This greatest precipitation consists of rays that descend almost perpendicularly upon the terrella; while the slighter precipitation on the night-side must be produced by rays that rather glance past the terrella. Corresponding rays that glanced past the earth on the night-side would generally produce magnetic storms.

---

[1] Archives des Sciences Physiques et Naturelles, Geneva, 4th period, vol. IV, 1898.

It is with a view to a careful study of the conditions connected with the positive polar storms that I have endeavoured to bring out in my terrella-experiments the directions in which the rays descend tangentially to the terrella's surface at various times of day in the polar regions, by the aid of narrow phosphorescent screens.

Owing to an accident to my discharge-tube, the final results of these investigations will not appear until the next section of this work; but I nevertheless have so many photographs of experiments that I have made, that I seem already to have a tolerably clear idea of the phenomena. We will first look again at some of the experiments already described, namely those shown in figures 38, 46 and 47

These experiments show indeed perfectly clearly that there are bundles of rays that graze the terrella from east to west along the auroral zone, corresponding, in my opinion, to the conditions on the earth during positive polar storms, and also bundles of rays that graze the terrella from west to east, corresponding to negative polar storms.

Fig. 38 *b* shows a tongue of light on the screen, down towards the "auroral zone" of the terrella, which is not found on the other side of the screen in the position observed. We will call the first side of the screen the *a*-side, and the other the *b*-side. The tongue of light does not appear upon the screen in the position shown in fig. 38 *a*, but it is found on the *a*-side of the screen in fig. 38 *c*, where, however, it does not extend so far in towards the terrella; and on the other hand we also see already on the *b*-side, on the opposite part of the screen, a considerable amount of precipitation. In the position shown in fig. 46 *a*, which forms a direct continuation of the experiment in 38 *c*, the precipitation does not even extend so far on the *a*-side, while on the *b*-side it has become very marked, and goes right down to the terrella, indicating rays that glance past the terrella from west to east, though without doubt single rays curve in towards the terrella, and form narrow loops before they go out again, very much as shown in the diagram, fig. 50 *a*.

In fig. 47 *b*, we see a powerful precipitation on the *b*-side of the screen, produced by the same kind of rays.

The precipitation on the *a*-side of the screen in fig. 38 distinctly shows that a wedge-shaped tongue of rays is thrust in towards the terrella, reaching farthest on the afternoon and evening side; the rays turn back as shown in fig. 50 *b*, and in my opinion correspond to the rays that occasion positive polar storms on the earth.

These conditions are confirmed and rendered still clearer by the experiments represented in the 8 photographs in fig. 136.

The first five of these refer to an experiment in which the position answers to that of the earth in the winter solstice, and to about noon at the earth's magnetic north pole, and the last three to another experiment in which the position represents an equinox, and midnight at the same magnetic pole. From the north pole of the terrella issue three narrow, phosphorescent screens, 3 millimetres in height and about 3 centimetres long, by the aid of which it was intended to determine the direction of the rays in the various instances of precipitation in the polar regions.

The five positions of the camera, from which photographs 1 to 5 of the first experiment were taken, may be determined as follows:

The axes of the cameras pointed towards the centre of the terrella, and were situated in vertical planes, at angular distances of 45°, 90°, 180°, 270° and 315° from the vertical plane through the axis of the discharge-tube. In each case the axes of the cameras were at an angle of 20° with the horizon. In the three positions from which photographs 6, 7 and 8 were taken, the axes of the cameras were situated in three vertical planes, at angular distances of 45°, 90° and 135° from the above-mentioned vertical plane, the axes being pointed towards the centre of the terrella, and forming the same angles with the horizon as before. It will easily be understood from these last three photographs, that the object

Fig. 136.

in taking them thus was to investigate the conditions on both sides of one of the above-mentioned three screens, the one whose position answered to about 6 p. m.

Photographs 1, 2, 3 and 5 show very distinct precipitation on the *b*-side of the screen, and shadows on the *a*-side, indicating that the precipitation on the terrella has a tangential motion from west to east in the "auroral zone". Photographs 3, 4 and 5 also show, however, a slighter precipitation at the very bottom of the *a*-side of the screen.

We obtain a clearer understanding of this twofold phenomenon from photographs 6, 7 and 8. We here see quite distinctly, although not nearly so distinctly as in the experiment itself, that the broad band of light consists of two bands, one more northerly that moves from west to east, and one more southerly that moves from east to west. The northern band of light breaks off just to the east of the screen, while the southern band breaks off just to the west of the screen, in both cases because of the shadow cast by the screen.

These circumstances seem to give us the key to the apparent enigma of the simultaneous occurrence of a negative polar storm in Spitsbergen, and a positive polar storm at Kaafjord and Matotchkin Schar.

CHAPTER IV.

CONCERNING THE INTENSITY OF THE CORPUSCULAR PRECIPITATON IN THE POLAR REGIONS OF THE EARTH.

75. While discussing the magnetic storms, we have pointed out a number of such storms, affecting the whole earth, which are evidently brought about by electric currents of some kind or other, acting in the region of the auroral zone. The current-system that might explain these storms is often of a very complicated nature, as the magnetic effect round the auroral zone frequently inclines us to believe that there are precipitations of electrically-charged corpuscles over several districts simultaneously all round the auroral zone.

When the conditions are so complicated, it will be inadvisable to try to obtain a practical result by comparing the magnetic effect of the corpuscles upon the earth with the effect of galvanic currents; for generally speaking at present a direct calculation of the magnetic effect of the electric corpuscles in different parts of the earth is too difficult of accomplishment. Up to the present, the possible paths oı the electric particles have been found by numerical quadrature; but the actual distribution and density of the rays round the earth have not been found by calculation. The solution of this problem would oı course be of the very greatest importance, if by its means a calculation might be made, from the magnetic effect upon the earth, of the number of corpuscles emitted by the sun per second. It will be easily understood that the greatest interest will attach to the establishment of the relation between the energy emitted by the sun in the form of corpuscular currents, and the energy sent out in the form of heat and light, more especially for the purpose of deciding whether the amount of the latter energy might possibly have been produced by a disintegration of the sun corresponding to the calculated quantity of corpuscles. At the present standpoint of the theory, however, we must be content with rough calculations and estimates such as those we shall make in the next few articles.

In certain simple cases, especially during the perturbations that we have called elementary storms, it may, however, be useful to compare the magnetic effect of the corpuscular currents with galvanic currents of so simple a nature that a calculation of the magnetic forces is easy. It may now be regarded as an undoubted fact that in the regions round the auroral zone we sometimes have currents which, at any rate for short distances, have the magnetic effect that a more or less horizontal current above the earth's surface would have, and which is comparatively small in section.

This is especially shown in the elementary storms that we have considered, where we very often have currents that pass over the earth between Axeløen and Kaafjord. The main intensity of these currents is probably compressed into a comparatively small section, judging from the fact that the vertical components of the perturbing force at the two stations generally have contrary direction, and are of about the same magnitude as the horizontal components. In this case we could compare the magnetic effect of the corpuscular current with the effect of a galvanic current, and endeavour to determine the strength of the current, or rather, obtain some idea of its magnitude by assuming, as a first, most simple approximation, that the magnetic effect outwards might be satisfied by a linear galvanic current of a certain strength, situated at a certain height.

In describing the separate elementary storms, we were able to show that the main features of the distribution of force in those perturbations were explained by the assumption of two vertical electric currents with opposite directions, connected by a horizontal portion of current. In Art. 36 we investigated the effect of a current-system such as this, and found a very close agreement with the actual circumstances during the polar elementary storms. The results there arrived at might now be employed for the purpose of estimating the operating strength of the currents. If, however, we look at the stations that are situated at all near the auroral zone, we can there simplify the problem considerably. It is immediately manifest that observations from points on the transverse axis of the system and near the storm-centre, must be favorable for a determination of the strength of the current. The field in the immediate vicinity of a linear conductor is somewhat similar to the field about an infinitely long, rectilinear current along the tangent at the nearest point on the conductor. When both stations have the same point on the conductor as their nearest, the field for both of them, at the place under consideration, will be determined by one infinitely long, rectilinear current; and as this is horizontal, it will simplify the reckoning considerably, and at the same time furnish a calculation of the degree of proximity of the current to the earth.

Fortunately for the solution of our problem, Axeløen and Kaafjord, in a number of perturbations, occupy this very position; and we shall only take those cases in which the current passes between the two stations, as we shall thus obtain a more certain determination of the altitude.

The question now is whether it is possible to decide when the current-system is thus situated in relation to the two stations. This must be decided separately in each case. We will only mention, as a necessary condition, that the current-arrows for Axeløen and Kaafjord must point in the same direction, and their vertical components be in opposite directions.

A calculation, similar to that given below, of the currents that cause polar storms, was made by me some years ago, for the stations Bossekop and Jan Mayen, with the aid of material from the expeditions of 1882 and 1883 [1].

We shall now proceed to calculate the current-strength and altitude of an infinitely long, rectilinear, horizontal current above the surface of the earth, when we know its effect in magnitude and direction at two points on the earth's surface.

Since we cannot on the whole lay claim to accuracy, we will here assume that the surface of the earth in the district in question is a plane surface.

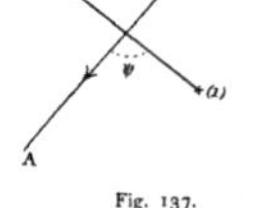

Fig. 137.

AB is the horizontal projection of the current; (*1*) and (*2*) represent respectively Kaafjord and Axeløen.

According to the above, the connecting line between the points (*1*) and (*2*) should be perpendicular to AB. This would be an ideal case, which will only approximately be attained. We will therefore assume that the lines form an angle, ψ, with one another. In cases in which the calculation will be employed, this angle will be nearly 90°.

We will further imagine the system projected upon a plane perpendicular to the line of the current. This line and the two points (*1*) and (*2*) on the earth's surface are then projected as three points, C, S_1 and S_2.

[1] Expédition Norvégienne de 1899—1900, p. 27.

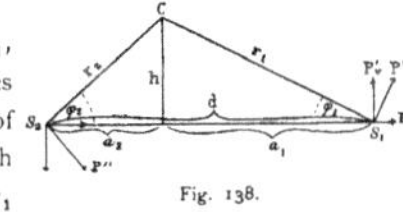
Fig. 138.

We will use the following signs:

The distance from point S_1 to the current is designated r_1, the distance from point S_2 to the current, r_2. The angles these lines make with the ground-line we will call φ_1 and φ_2, and the height of the current above the earth's surface, h. The portions into which the height-line divides the ground-line of this triangle, we will call a_1 and a_2. The distance between Kaafjord (*1*) and Axeløen (*2*) is designated D; hence the projection of this distance on the above plane is $d = D \sin \psi$. For the perturbing forces we will use the signs P', P_1' and P_v' respectively for the total, the horizontal and the vertical forces at Kaafjord, and correspondingly P'', P_1'' and P_v'' for those at Axeløen.

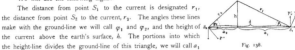

If the magnitude and direction of the perturbing forces are given, the problem will be not only determined, but over-determined, so that it affords a test of the correctness of our assumption.

The direction of the forces, for instance, is sufficient to determine the situation of the current. The strength of the current can then be determined by that of the perturbing force at the one station. The strength of the perturbing force at the other station may then serve as a check.

The calculation can be made according to the following formulæ:

$$\tan \varphi_1 = \frac{P_1'}{P_v'}, \quad \tan \varphi_2 = \frac{P_1''}{P_v''}$$

$$r_1 = \frac{\sin \varphi_2}{\sin(\varphi_1 + \varphi_2)} d$$

$$r_2 = \frac{\sin \varphi_1}{\sin(\varphi_1 + \varphi_2)} d$$

$$h = \frac{\sin \varphi_1 \sin \varphi_2}{\sin(\varphi_1 + \varphi_2)} d$$

Two values will be obtained for the strength of the current, according as the force at Axeløen or that at Kaafjord is employed:

$$i_1 = \frac{5 P' \sin \varphi_2 \, d}{\sin(\varphi_1 + \varphi_2)}$$

$$i_2 = \frac{5 P'' \sin \varphi_1 \, d}{\sin(\varphi_1 + \varphi_2)}$$

In these and the succeeding formulæ, P_1 and P_v are always to be regarded only as the numerical values of the respective perturbing forces.

As it occasionally happens that one of the vertical components is wanting, we shall also solve the droblem under that assumption. If the other vertical component is there, it may be used as a check.

If we introduce:

$$\frac{a_2}{a_1} = \delta,$$

$$\frac{P_1'}{P_v'} = p,$$

and
$$\frac{P_1'}{P_1'' \cos^2 \varphi_1} = q^2,$$

we obtain the following equations:

$$a_1 + a_2 = a_1 (1 + \delta) = D \sin \psi = d, \tag{1}$$

$$\frac{h}{a_1} = \tan \varphi_1 = p, \tag{2}$$

$$P_1' = \frac{i}{5} \frac{h}{r_1^2} = \frac{i}{5 a_1} p \cos^2 \varphi_1, \tag{3}$$

$$P_1'' = \frac{i}{5}\frac{h}{r_2^2} = \frac{i}{5}\frac{p}{a_1(\delta^2 + p^2)}\,. \tag{4}$$

If we divide (3) by (4), we find that

$$\delta = \pm\sqrt{q^2 - p^2}\,. \tag{5}$$

In the cases we shall come to examine here, however, the current is always between Kaafjord and Axeløen, so that δ will always be positive. In equation (5) p and q are known quantities; hence δ can be calculated, and from (1) and (2) we obtain

$$h = \frac{Dp}{1 + \delta}\sin\psi\,. \tag{6}$$

From (5) and (3) i can then be calculated, and we obtain

$$i = \frac{5 P_1' h}{\sin^2 \varphi_1} \tag{7}$$

With regard to the determination of the angle ψ, we should remark that if the current-arrows for Kaafjord and Axeløen make the same angle with the great circle between these two stations, ψ will simply equal that angle. The angles will generally be somewhat different. Calling them respectively ψ_1 and ψ_2, we put

$$\psi = \frac{\psi_1 + \psi_2}{2}\,.$$

NUMERICAL VALUES FOR HEIGHT AND STRENGTH OF CURRENT.

(1) *The Perturbation of the 15th December, 1902.*

76. During this perturbation the balance at Kaafjord stuck fast, so the direction can only just be distinguished. As the sensibility of the balance at Axeløen was not determined until after the return of the expedition, and may thus be not altogether free from error, we will see what can be concluded regarding height and strength of current, when we suppose that we know only the horizontal components, and the direction, but not the strength, of the vertical components.

Between $1^h\ 45^m$ and $2^h\ 0^m$, the horizontal components at Axeløen and Kaafjord are almost alike in direction; and the outer field shows that the storm-centre during this time must be somewhere near Spitsbergen. We will therefore take $1^h\ 52.5^m$ as the most favorable moment for determining the strength and altitude of the current. At this point of time, the values are as follows:

$$P_1'' = 186\,\gamma$$
$$P_1' = 30\ \text{»}$$
$$D = 896\ \text{km.}$$
$$\psi = 70^\circ.$$

We further introduce here a quantity $\varkappa$ which is thus defined,

$$\varkappa = \frac{P_1''}{P_1'} = 6.2\,.$$

If we divide the equation (4) by (3), we find

$$p = \sqrt{\frac{1 - \varkappa\delta^2}{\varkappa - 1}}\,;$$

and by employing equation (6) we obtain

$$h = \frac{D\sin\psi}{1 + \delta}\sqrt{\frac{1 - \varkappa\delta^2}{\varkappa - 1}}\,. \tag{8}$$

By inserting this value for h in (7) we obtain

$$i = \frac{5P_1' D \sin\psi\, \varkappa (1 - \delta)}{\sqrt{1 - \varkappa\delta^2}\,\sqrt{\varkappa - 1}}, \tag{9}$$

and

$$\tan\varphi_2 = \frac{h}{a_2} = \frac{1}{\delta}\sqrt{\frac{1 - \varkappa\delta^2}{\varkappa - 1}},$$

By the above equations, h and i are determined as functions of δ. On account of the direction of the vertical components, we have

$$\delta > 0 .$$

If our assumptions are correct, we must have real quantities, and the strength of the current must be finite. We then obtain

$$\delta < \sqrt{\frac{1}{\varkappa}}, \quad \text{where} \quad \sqrt{\frac{1}{\varkappa}} = 0.402 .$$

It is easily ascertained that the function for h in this interval has neither maximum nor minimum. As the function in the interval considered is continuous and finite, we may conclude that it has its extreme values at the limits of the interval, and especially in such a way that we get the greatest height when $\delta = 0$.

In the case of i we find that the function has a minimum for the value $\delta = \frac{1}{\varkappa}$, that is to say for a value within the interval considered.

Still narrower limits may be set to the interval, however, if we now make use of our knowledge of the vertical intensity at Axeløen.

The sensibility of the balance was determined, after the return of the expedition, as 24.6. If, therefore, we employ a value of 35, there is no doubt that it is too high. We then obtain

$$\tan\varphi_2 = \frac{P_1''}{P_e'} > 0.885 ,$$

or

$$\delta < 0.312 .$$

0 and 0.312 can thus be employed as the limits for δ.

In the following table, the height and strength of the current are calculated for 4 values of δ, namely $\delta = 0, \frac{1}{\varkappa}$, 0.263 and 0.312. The value $\delta = 0.263$ anwers to a sensibility of the balance of 24.6, and therefore the values we obtain there should be the nearest to the true values.

TABLE XLIII.

	$\delta = 0$	$\delta = \frac{1}{\varkappa}$	$\delta = 0.263$	$\delta = 0.312$	
h	368	286	220	177	km.
i	342,000	314,000	334,000	374,000	amperes

We see that even if we pay no attention at all to the vertical intensity for Axeløen, we may still conclude that the current cannot lie higher than 368 km., answering to a current-strength of 342,000 amperes, and also that the current-strength cannot be less than 314,000 amperes, provided our assumptions in other respects hold good.

Considering that P_e for Axeløen is known with very fair accuracy, the true values should lie near those that answer to $\delta = 0.263$.

The values found for h and i are, as we shall presently see, comparatively small in this perturbation, indicating that the perturbation is comparatively slight, and of rather a local character in the north.

(2) *The Perturbation of the 10th February, 1903.*

77. The current-arrows for Axeløen and Kaafjord remain in one direction for a considerable time, and are almost perpendicular to the arc of the great circle between the two stations. It also appears from the outer field that the storm-centre of the current-system is in the neighbourhood of Axeløen and Kaafjord. We have therefore calculated the strength and altitude of the current at several hours at which the conditions are approximately those mentioned in the introduction.

In this case we employ P_2 for Kaafjord. The vertical component for Axeløen we shall use as a check. This quantity, if our assumption is correct, will be determined by the formula

$$P_2'' = \frac{\delta}{p} P_1''.$$

In the following table, the calculation has been made for four different hours.

It may here be remarked that both in this and the succeeding Tables, the units of length and current-strength are respectively a kilometre and an ampere.

TABLE XLIV.

Time	P_1'	P_2'	P_1''	d	p	q	δ	h	i	P_2'' cal.	P_2'' obs.
23^h 22.5^m	235	229	200	890	1.026	1.553	1.166	422	966,000	227	150
37.5	200	253	308	890	0.790	1.026	0.653	426	1,109,000	254	164
45	187	258	353	890	0.725	0.889	0.532	421	1,143,000	259	158
24^h 0	120	168	146	860	0.714	1.114	0.855	328	582,000	175	172

The table shows that at the first three of the hours mentioned the current would be at the same height — about 420 km. —; and this is the more strange as the separate quantities in the formulæ differ considerably.

The values for δ seem to indicate that up to $23^h.45^m$ the current is moving towards Axeløen. While moving thus, the current, on an average, would keep at about the same height above the surface of the earth.

A comparison between the calculated and the observed values for P_2'', will show that the calculated vertical components on the whole are too large; the observed values are only about two thirds of the calculated. A result such as this is just what might be expected. Our calculations presuppose that the transverse section of the current is very small in proportion to the distance between Kaafjord and Axeløen; but considering the cosmic constitution of the current, this is not very probable.

We could make the calculation here also, assuming both the total forces to be given. The result will be found in the following table.

TABLE XLV.

Time	P'	P''	h	i_1	i_2	Mean of i_1 & i_2
23^h 22.5^m	328	250	516	1,182,000	806,000	994,000
37.5	323	349	495	1,289,000	974,000	1,131,500
45	319	387	487	1,324,000	1,033,000	1,178,500
24 0	206	225	345	612,000	600,000	606,000

From this it appears that the two calculated current-strengths are not quite alike, but the difference is not greater than would be expected. The mean gives values that agree very closely with those previously found. The height found is somewhat greater in the last case. It will easily be perceived that if the current is spread over a larger section, we shall find the height somewhat too great.

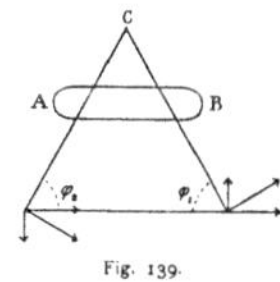

Fig. 139.

Our calculated current will lie, for instance, at C (fig. 139), whereas in reality the current may be gathered at a lower level $A\,B$.

In this way the height of the current will be rather an indefinite conception; but we believe the values found will at any rate give an approximate determination of the heights at which the greatest density of the current in each separate case must be looked for.

We will now, in conclusion, see how far the conditions at Dyrafjord and Matotchkin Schar agree with the values found. Assuming the strength of the current to be the same, we will calculate the height at which a horizontal current must pass in order to produce the magnetic disturbances that occur at the two stations. If we call the distance from the station to the nearest point in the current r, we obtain

$$r = \frac{i}{5P},$$

where P is the total perturbing force.

If we assume the current to be horizontal, we obtain

$$h = r \sin \varphi ,$$

where

$$\tan \varphi = \frac{P_1}{P_2}$$

TABLE XLVI.

Time	Dyrafjord				Matotchkin Schar		
	i	P	r	h	P	r	h
$23^h\ 22.5^m$	966,000	388	498	482	419	461	410
37.5	1,109,000	132	723	1176	332	668	614
45	1,143,000	193	1504	754	216	1058	973
24 0	609,000	124	731	947	74	1645	1574

These calculations show that if the current were horizontal, it would lie especially high above the two stations, Dyrafjord and Matotchkin Schar, particularly during the latter part of the perturbation. Our assumptions for these calculations can only, as we have already said, be regarded as a first approximation; but it is most probable that the error will be in the same direction in all the calculations, so that the relative proportions will be fairly correct. If the current were to continue with the same average strength, it could not do so at the same height as between Kaafjord and Axeløen, but would curve upwards.

This harmonises well with our view of the current-system, which maintains that the system would curve upwards. The actual circumstances at Dyrafjord and Matotchkin Schar could also be explained, however, if we assume that the current there is spread over a large section. Moreover the assumption that the average strength in the advancing current would preserve its value unchanged, owing to the undoubtedly cosmic nature of the current, can by no means be regarded as safe, as the paths of the separate electric corpuscles will be very numerous. The constancy of the average current-strength can therefore only be regarded as a very rough assumption.

A comparison of the current-strengths found for this perturbation, with those for the perturbation of the 15th December shows that the former are about three times as great as the latter. At the same time the effect of the force at corresponding places in the field outside the arctic regions is much smaller on the 15th December — only about one third.

This shows that the strength of the universal disturbances that accompany the storms in the north, stand in about the same relation to one another as the strength of the currents which we assume to be the direct cause of those storms. This accords well with our assumption; for if it be assumed that during these polar storms the form of the current-system is more or less the same, the force at corresponding places in the outer field would be proportional to the strength of the current.

In connection with these calculations of current-strengths, I will here refer to the current-strengths that I found for the stations Bossekop and Jan Mayen, given in my former report, "Expédition Norvégienne de 1899—1900" etc., pp. 27 & 28. They agree well with those now found, as they vary between 317,000 and 983,000 amperes.

(3) *The Perturbation of the 22nd March, 1903.*

78. In this perturbation, as already mentioned, the storm-centre is in the neighbourhood of Axeløen and Bossekop, whither the station at Kaafjord had now been moved (see p. 10). The current-arrows for these stations are similar in direction, and are almost at right angles to the great circle between them. The vertical components are very large and in contrary directions. It would thus appear that the conditions are such as to justify a more elaborate calculation of the strength of the current, according to the methods previously given.

In the calculation on this occasion, we shall consider the total force for Axeløen as known, as also P_1 for Bossekop. At the latter station the patch of light for the balance has moved off the paper, so that there, during the time at which the storm is most powerful, we only know the lower limit of this quantity.

In the table below are given the most important values that enter into the formulæ, as also the values found for P_v', h and i.

TABLE XLVII.

Time	ψ	d	P_1'	P_v' obs.	P_1''	P_v''	δ	h	i	P_v' cal.
$22^h\ 0^m$	71°	847	207	205	315	408	0.738	278	1,170,000	363
15	73	856	211	$>$ 205	372	492	0.674	259	1,324,000	414
30	73	856	220	$>$ 205	157	484	1.209	156	1,282,000	561
45	81	885	134	$>$ 205	148	396	0.946	156	945,000	379

The height on this occasion is not great. The strength of the current, on the other hand, is fairly great, amounting to $1^1/_3$ million amperes. If we compare the calculated values of P_v' with those observed, we also on this occasion, at $22^h\ 0^m$, find that the calculated value is too high. As regards the subsequent hours we can say nothing decided; probably they also are too high. For the explanation of these conditions, the reader is referred to the perturbation of the 10th February.

(4) *The Perturbations of the 27th & 28th October, 1902.*

79. In the storms that occur just before midnight on these two days, there are, as we said when discussing them, circumstances which justify a calculation of the strength and altitude of the horizontal portion of the current. The results of this calculation are given in the table below.

TABLE XLVIII.

Time	ψ	P_1'	P_v'	P_1''	P_v'' obs.	P_v'' cal.	δ	h	i
Oct. 27, $23^h\ 0^m$	78°	113	127	266	110		imaginary		
20	67°	104	117	195	295	91.5	0.413	522	614,000
" 28, 22 20	68°	175	132	200	352	120	0.767	608	835,000

On the 27th, at 23h, ϑ, as we see, is imaginary. This shows that the perturbation-conditions at the two stations at that moment do not satisfy the assumptions made. The reason of this is possibly to be sought in the cross-section that the actual current must have, or perhaps in the fact that the perturbation-conditions could in no way be ascribed to the effect of a more or less aggregate system. We might have several simultaneously-acting systems of to some extent more local character. We very frequently see at these stations in the north, that disturbances occur at one station that are not noticed at another. We shall never be without these local disturbances; but the thing is that they shall be slight in comparison with the total effect.

A great local disturbance seems really to occur just about 23h. There is a sharp deflection of rather long duration, which tends to increase P_1''. From the fact that there is no corresponding change at Kaafjord, we may conclude that this deflection cannot be ascribed entirely to a movement of the main system.

We also, by looking at the curve for P_v, obtain the impression of a new system, which would lie to the north of Axeløen, as the deflection is in the opposite direction.

At the second hour, 23h 20m, the great local disturbance at Axeløen is over, or at any rate fainter, and we now obtain a real solution. The calculated vertical component for the time, however, is somewhat smaller than the observed. This is also the case on the day following.

THE ENERGY OF THE CORPUSCULAR PRECIPITATION. THE SOURCE OF THE SUN'S HEAT.

80. We consider it to be beyond doubt that the powerful storms in the northern regions, both those of long duration, and the short, well-defined storms that we have called elementary, are due to the action of electric currents above the surface of the earth near the auroral zone.

These currents, as far as the elementary storms are concerned at any rate, act, in the districts in which the perturbation is most powerful, as almost linear currents, that for a considerable distance are approximately horizontal. In the preceding articles, we have attempted, in some of the magnetic storms described, to calculate the strength of horizontal currents such as might be the cause of the storms, supposing that they acted magnetically as galvanic currents. The values found, which cannot certainly lay claim to any great accuracy, will yet give an approximate idea of the strength of these currents.

In the case of the greater storms, we found current-strengths that varied between 500,000 and 1,000,000 amperes, or even considerably more.

It might be interesting to know the amount of energy per second of this current. According to my hypothesis, the currents would not, in reality, be galvanic, but be formed of cathode rays, or more generally of rays of electric corpuscles. We will make this hypothesis, then, the basis of our estimates.

By energy we in the mean time understand the kinetic energy of the corpuscular current that passes per second through a cross-section of the horizontal part of the current, and where the corpuscles are assumed to flow in the path of the before-mentioned galvanic current. In Article 36, fig. 50 *a* & *b*, we gave a diagram of the manner in which we in reality approximately imagine the corpuscles to move. With the method of calculation here employed, we obtain only a small lower limit of the energy of the corpuscular current.

If we call the number of corpuscles that pass the cross section in the time-unit n, the apparent mass of a particle μ, and the velocity v, we obtain as the energy W.

$$W = \tfrac{1}{2}\, n\mu v^2.$$

If each particle carries a charge of ε electrostatic units, we have

$$i = \frac{\varepsilon \cdot n}{3 \times 10^9} \text{ amperes}$$

and thus

$$W = \frac{3}{2} \cdot 10^9 \, i \, \frac{\mu}{\varepsilon} \cdot v^2.$$

If the C.G.S. system be employed, we obtain W expressed in ergs per second.

The energy of the current will chiefly depend upon the kind of rays that form the current. It is evident, however, from the magnetic storms previously described, that the corpuscular rays here referred to must be very "stiff" magnetically.

Leaving the question of the particular nature of these rays for the present undecided, we will make the calculation for two types.

(1) For cathode rays, whose velocity is small in proportion to the velocity of light, we have, when ε is calculated in electrostatic units,

$$\frac{\varepsilon}{\mu} = 510 \times 10^{15} \text{ cm.}^{\frac{3}{2}} \text{ gr.}^{-\frac{1}{2}} \text{ sec.}^{-1} \text{ (}^1\text{)}.$$

For rays where $v = 0.7 \times 10^{10}$, we thus find that

$$W = \frac{3}{2} \times 10^9 \, i \cdot \frac{0.49 \times 10^{20}}{510 \times 10^{15}} = 1.44 \times 10^{11} \, i \text{ ergs per second},$$

or

$$W = 19.6 \, i \text{ h.-p.}$$

(2) For β rays with velocities of

$$v = 2.59 \times 10^{10} \text{ cm. sec.}^{-1},$$

we have, by *Kaufmann's* determinations,

$$\frac{\varepsilon}{\mu} = 255 \times 10^{15} \text{ cm.}^{\frac{3}{2}} \text{ gr.}^{-\frac{1}{2}} \text{ sec.}^{-1} \text{ (}^1\text{)}.$$

Corresponding to this,

$$W = 3.94 \times 10^{12} \, i \text{ ergs per second},$$

or

$$W = 535 \, i \text{ h.-p.}$$

The energy in each separate case can then be calculated according to these expressions.

For $i = 1{,}000{,}000$ amperes, we obtain in the first case

$$W = 19.6 \times 10^6 \text{ h.-p.},$$

in the second case

$$W = 535 \times 10^6 \text{ h.-p.},$$

or 100 times more than the maximal amount of force that all the waterfalls of Norway together could deliver by a perfect regulation of all water-courses.

There is much that seems to favour the idea that the rays that come to the earth are very "stiff", and may possibly have considerably more energy than the here assumed β rays. We recollect that the apparent mass increases comparatively quickly when the velocity of the corpuscles approaches that of light. We know of β rays whose velocity is only 5 per cent. less than that of light, and whose apparent mass is 50 per cent. greater than that of the β rays assumed above.

We have moreover, in the preceding pages, during powerful magnetic storms, calculated current-strengths greater than a million amperes, which is the amount here taken as the basis.

---

(1) The values are calculated from those given in Sir J. J. Thomson's "Corpuscular Theory of Matter"; London, 1907.

We may thus take it for granted that a kinetic energy answering to 10^9 horse-power during powerful storms, will not be too high for the corpuscular current.

This is calculated, however, on the supposition that the corpuscular current moved parallel with the surface of the earth in the auroral zone.

The matter, however, as we have shown at the conclusion of Article 36, is not so simple. In order to know what kinetic energy should be ascribed to a corpuscular current that had the observed magnetic effect upon the earth, we should need to have a complete mathematical solution of the manner in which the rays from the sun would distribute themselves round the earth. Up to the present, indeed, *Størmer* has found the *possible* paths of the rays by numerical quadrature, and he may perhaps in time succeed in finding a more complete solution, from which the above-mentioned magnetic effect might be calculated. We will even now, however, make an attempt, by an estimate, to find out whether it is possible that the corpuscular current which the sun emits from a sun-spot is so large as to indicate a disintegration on the sun, which might account for the solar radiation of heat and light.

Let me say at the outset that in making certain, for the time being, purely computational assumptions, which yet may subsequently, at any rate in their aggregate effect on the result, prove to be more or less correct, we come directly upon a value of the development of heat by disintegration per square centimetre of the sun's surface, that is very near that which is deduced from the solar constant.

These assumptions, or estimates, are as follows.

In the first place it is assumed that the corpuscles issue at right angles to the sun's surface, and that their density decreases inversely as the square of their distance from the sun.

In the second place it is assumed that as the corpuscles do not move parallel with the earth's surface, but come in towards the earth more or less as shown in fig. 50 *a* & *b*, their kinetic energy is much greater than calculated for the district between Kaafjord and Axeløen during the storms under consideration; we assume 100 times greater.

In the third place we take it for granted that the quantity of rays that are drawn in towards the polar regions of the earth, is not nearly so great as the quantity of rays that would have come into contact with the earth if the latter had been non-magnetic. This we conclude from our terrella-experiments. We there see distinctly that the more strongly the terrella is magnetised, the narrower does the zone become, where the rays come in towards the terrella. And we see by the illumination that fewer and fewer come in.

If our terrella were to be magnetised so powerfully that the conditions corresponded with those on the earth, it would have to be immeasurably more magnetic than it is possible to make it (see "Expédition Norvégienne de 1899—1900", etc. p. 40).

We now assume that 100 times as many rays would fall upon the earth if it were non-magnetic, as actually do so in the auroral zone.

By these assumptions we thus arrive at the fact that a corpuscular current, of which the energy amounts to 10^{13} horse-power, would have come into contact with the earth, if the latter had been non-magnetic.

The last factor is perhaps rather large. On the other hand we have disregarded the fact that only a portion of the rays that are eventually formed by the disintegration in the sun, succeed in forcing their way into space; most of them will be absorbed into the solar atmosphere. Only the most penetrating, most inflexible rays escape into space and reach the earth. If this were also taken into consideration, it would perhaps compensate in the result for the possibly too high estimate of the above-mentioned factor 100.

We found, then, that we could put the energy of the rays that would come into contact with the earth, if the latter were non-magnetic, at 10^{13} horse-power.

We will now imagine this amount of energy radiating from a sun-spot, and that the bundle of rays is so large that the conditions, so far as the earth is concerned, are the same as if corpuscles were being steadily emitted from the entire surface of the sun. We may mention that farther on, when explaining other terrestrial-magnetic phenomena, we shall assume that corpuscles do continually radiate from the whole of the sun's surface; but they must be assumed to possess properties somewhat different to those of the corpuscles that radiate from the sun-spots.

In our calculation we shall employ the same value for the earth's radius as in Article 36, namely 6366 kilometres, the radius of the earth's orbit is taken as 23,440 times that of the earth, and the radius of the sun as 109 times that of the earth. The amount of energy that is emitted per square centimetre from the sun's surface in the form of rays will then be

$$\frac{10^{13}}{\pi \cdot 6366^2 \cdot 10^{10}} \cdot \frac{23440^2}{109^2} \cdot 7.36 \times 10^9 = 2.7 \times 10^9 \text{ ergs per second.}$$

If we keep the same designations as before, we thus obtain

$$\tfrac{1}{2} N \mu v^2 = 2.7 \times 10^9,$$

in which, employing the same β rays as before, we have the following values:

$$\mu = 1.2 \times 10^{-27} \text{ (}^1\text{)}$$

$$v = 2.59 \times 10^{10}.$$

Hence we find that

$$N = 6.7 \times 10^{15},$$

which is the number of β particles that each square centimetre of the surface of the sun-spot would emit per second.

Now 1 gramme of radium emits 7.3×10^{10} β particles per second, and at the same time gives off $\frac{100}{3600}$ gramme-calories (²).

We then obtain

$$\frac{6.7 \times 10^{15}}{7.3 \times 10^{10}} \cdot \frac{100}{3600} \text{ gr. calories, answering to about 14 h.-p.,}$$

which is thus the amount of energy that is set free by a disintegration of the sun's matter, which would answer to the quantity of rays emitted from it in the form of these corpuscular rays.

This amount corresponds, as already stated, to the amount of energy which the sun sends out in the form of light and heat. If the solar constant equals 3, we find a radiation from every square centimetre of the sun's surface of about 13 horse-power.

A disintegration such as this in the sun does not necessarily presuppose the presence there of great quantities of radium, uranium, or thorium.

Rutherford, in his work entitled "Radio-Activity" (³), says:

"There seems to be every reason to suppose that the atomic energy of all the elements is of a similar high order of magnitude. With the exception of their high atomic weights the radio-elements do not possess any special chemical characteristics which differentiate them from the inactive elements. The existence of a latent store of energy in the atoms is a necessary consequence of the modern view

(¹) See Sir J. J. Thomson's "Corpuscular Theory of Matter", pp. 16 & 33 London, 1907.
(²) See E. Rutherford's "Radio-Activity", 2nd edition, pp. 436 & 474 Cambridge, 1905.
(³) l. c., p. 475.

developed by *J. J. Thomson* (1), *Larmor* and *Lorentz*, of regarding the atom as a complicated structure consisting of charged parts in rapid oscillatory or orbital motion in regard to one another".

Under the temperature-conditions prevailing in the sun, it is possible that ordinary matter may be so radio-active, that it is not necessary to assume the presence in great quantities of the radio-elements known in ordinary temperatures.

It was pointed out by *Rutherford* and *Soddy* (2), that the maintenance of the sun's heat for long intervals of time did not present any fundamental difficulty, if a process of disintegration such as occurs in the radio-elements were supposed to be taking place in the sun.

We may perhaps succeed, in the way here indicated, in obtaining a distinct idea of the amount of heat that can be developed in the sun by disintegration; and thus an important contribution will be made to the solution of the old, and to natural philosophy so important, question of the origin of the sun's heat.

---

(1) I see with great satisfaction that Sir J. J. Thomson, in his classic research on the nature of the cathode rays (Phil. Mag. Number CCLXIX, October 1897), in which we find the first definite experimental evidence towards proving that the chemical atom is not the smallest unit of matter, has taken as his starting-point my discovery that the magnetic deviation of cathode rays depends only upon the tension between cathode and anode, if the magnetic force is constant. (See Birkeland, Compt. Rend., Sept. 28, 1896.) This theorem has been verified by Sir J. J. Thomson, l. c., and W. Kaufmann, Wied. Ann. Vol. LXI. No. 7, 1897.

(2) Phil. Mag., May, 1903.

Pl. I

The Perturbation of the 6th October, 1902

Registerings from 13h 30m to 15h 30m, Gr. M. T.

THE PERTURBATION OF THE 6th OCTOBER, 1902.

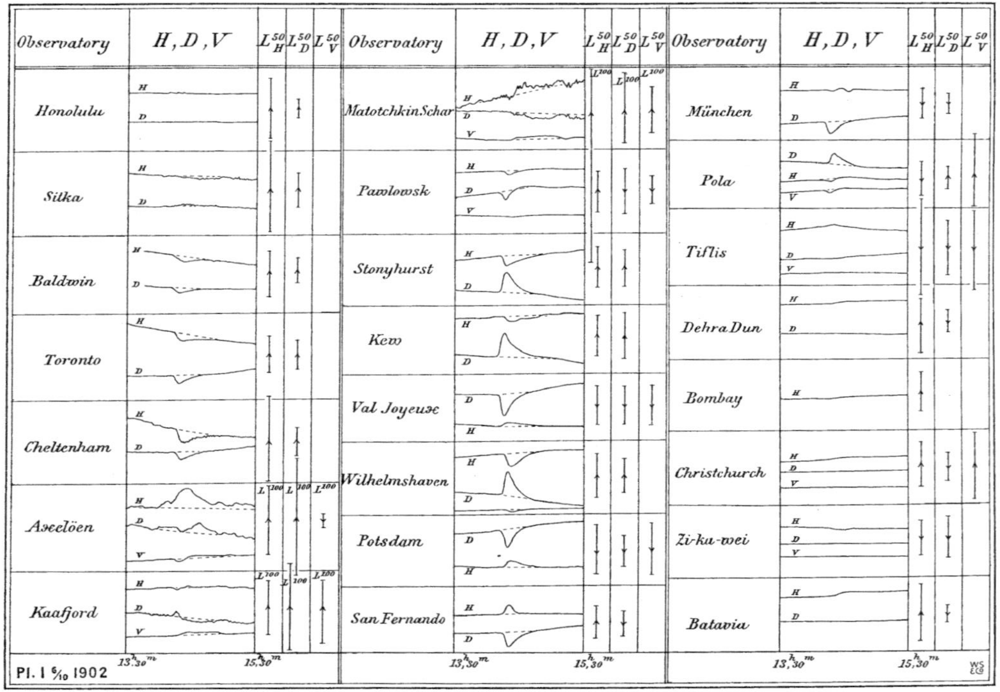

Observatory	H, D, V	L^{50}_H	L^{50}_D	L^{50}_V	Observatory	H, D, V	L^{50}_H	L^{50}_D	L^{50}_V	Observatory	H, D, V	L^{50}_H	L^{50}_D	L^{50}_V
Honolulu	H, D				Matotchkin Schar	H, D, V	L^{100}	L^{100}	L^{100}	München	H, D			
Sitka	H, D				Pawlowsk	H, D, V				Pola	D, H, V			
Baldwin	H, D				Stonyhurst	H, D				Tiflis	H, D, V			
Toronto	H, D				Kew	H, D				Dehra Dun	H, D			
Cheltenham	H, D				Val Joyeuse	D, H				Bombay	H			
Axelöen	H, D, V	L^{100}	L^{100}	L^{100}	Wilhelmshaven	H, D				Christchurch	H, D, V			
Kaafjord	H, D, V	L^{100}	L^{100}	L^{100}	Potsdam	D, H				Zi-ka-wei	H, D, V			
					San Fernando	H, D				Batavia	H, D			
Pl. I 6/10 1902	13^h30^m – 15^h30^m					13^h30^m – 15^h30^m					13^h30^m – 15^h30^m			WS&C

Pl. II

The Perturbations of the 11th October, 1902

Registerings from 12h on the 11th to 2h on the 12th, Gr. M. T.

THE PERTURBATIONS OF THE 11th OCTOBER, 1902.

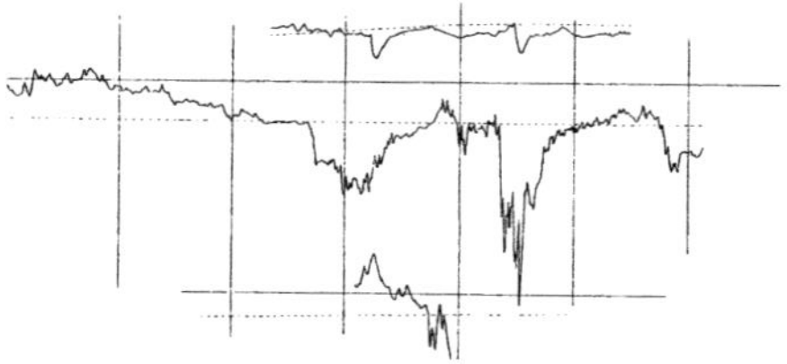

Pl. III

The Perturbation of the 23rd October, 1902

Registerings from 17^h on the 23rd to 5^h on the 24th, Gr. M. T.

THE PERTURBATION OF THE 23rd OCTOBER, 1902.

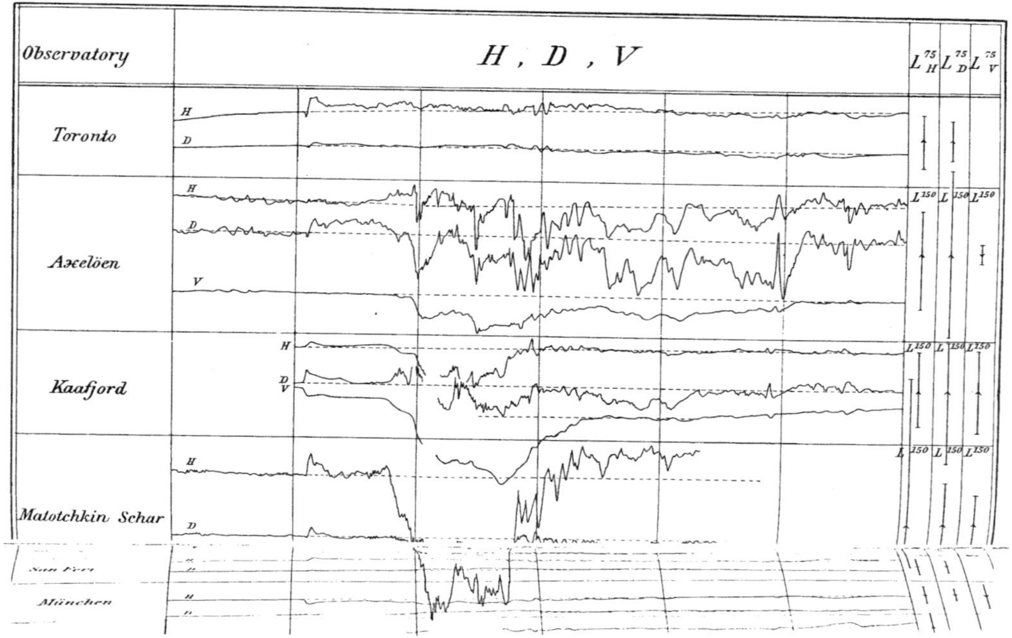

Observatory

H, D, V

L^{75}_{H} L^{75}_{D} L^{75}_{V}

Toronto

H

D

Axeløen

H

D

V

L^{150} L^{150} L^{150}

Kaafjord

H

D

V

L^{150} L^{150} L^{150}

Matotchkin Schar

H

D

L^{150} L^{150} L^{150}

München

H

D

San Fernando

München

Bombay

Dehra Dun

Christchurch

Pola

D

H

Pl. IV

The Perturbations of the 27th & 28th October, 1902

Registerings from 14h on the 27th to 1h on the 28th, Gr. M. T.

THE PERTURBATIONS OF THE 27th & 28th OCTOBER, 1902.

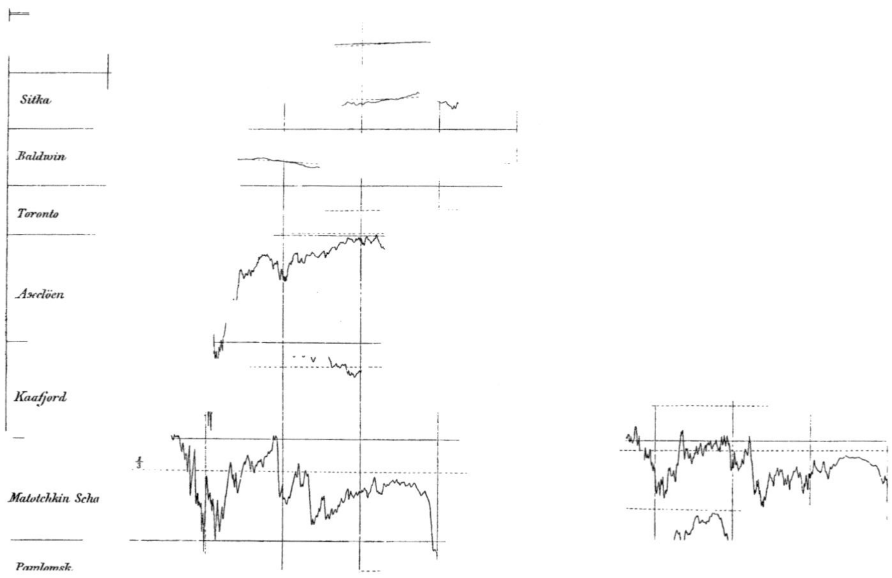

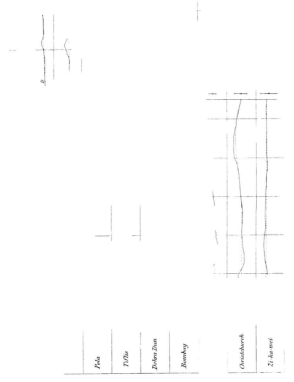

D
Pola
Tiflis
Dehra Dun
Bombay
Christchurch
Zi-ka-wei

Pl. V

The Perturbations of the 28th & 29th October, 1902

Registerings from 14^h on the 28th to 1^h on the 29th, Gr. M. T.

THE PERTURBATIONS OF THE 28th & 29th OCTOBER, 1902.

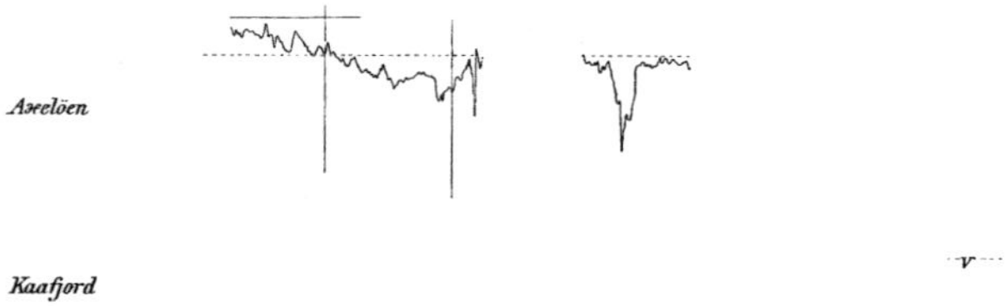

Axelöen

Kaafjord

München

Pola

Tiflis

Dehra Dun

Bombay

na

Christchurch

Pl. VI

The Perturbations of the 29th and 30th October, 1902

Registerings from 16h on the 29th to 4h on the 30th, Gr. M. T.

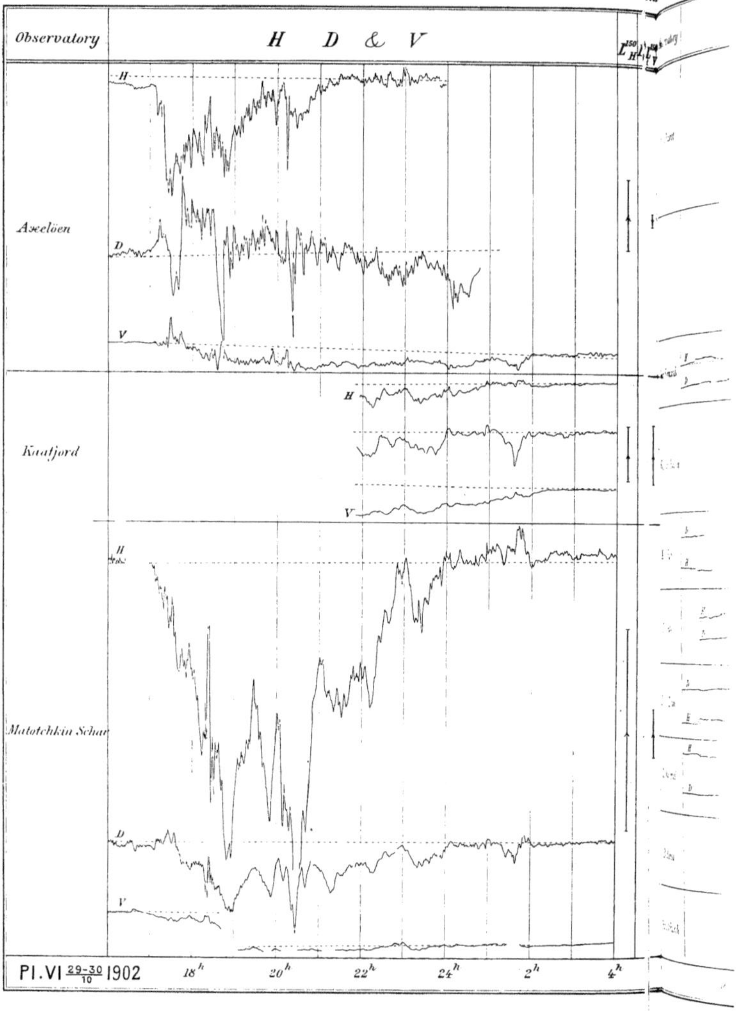

Observatory	H D & V	$L^{150}_{H\,D\,V}$
Axelöen		
Kaafjord		
Matotchkin Schar		

Pl. VI $\frac{29-30}{10}$ 1902 18^h 20^h 22^h 24^h 2^h 4^h

Pl. VII

Perturbations of the 31st October and 1st November, 1902

Registerings from 6h on the 31st to 2h on the 1st, Gr. M. T.

The Perturbations of the … 1st November, 1902.

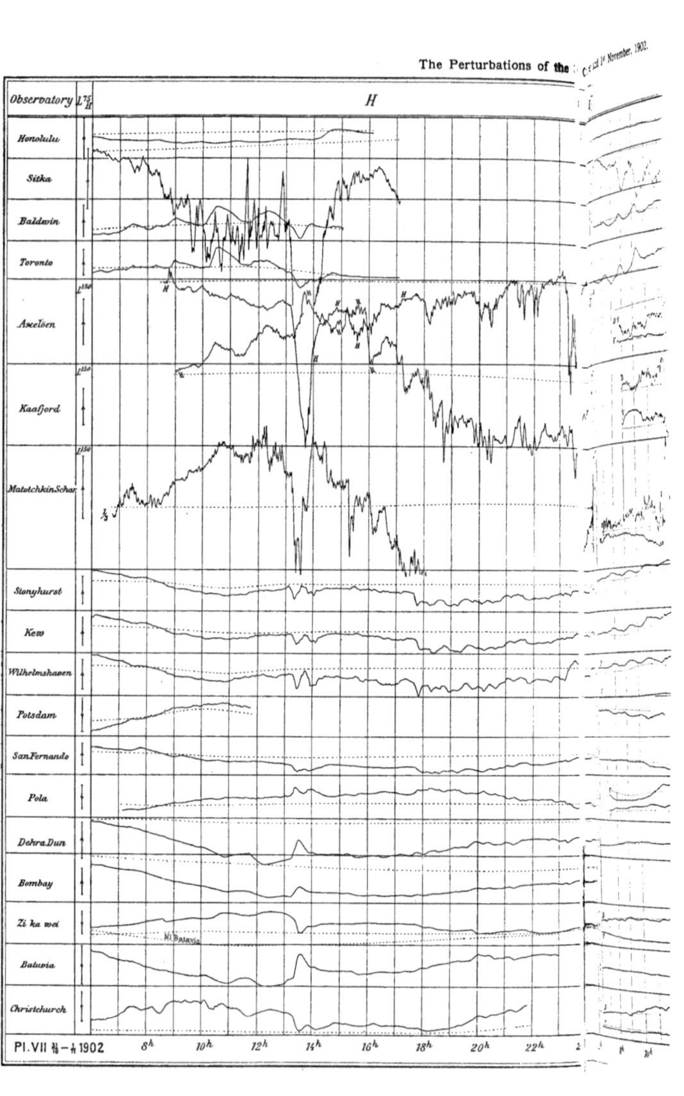

Pl. VII 31/10 – 1/11 1902

The Perturbation ober and 1st November, 1902.

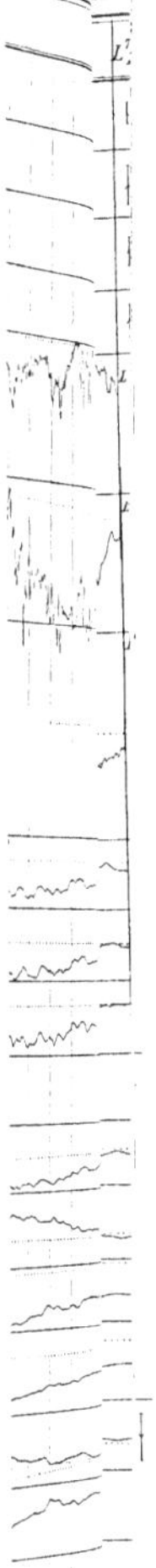

Pl. VIII

e Perturbations of the 23rd and 24th November, 1902

Registerings from 15h on the 23rd to 7h on the 24th, Gr. M. T.

The Perturbations of the 23rd and 24th November, 1902.

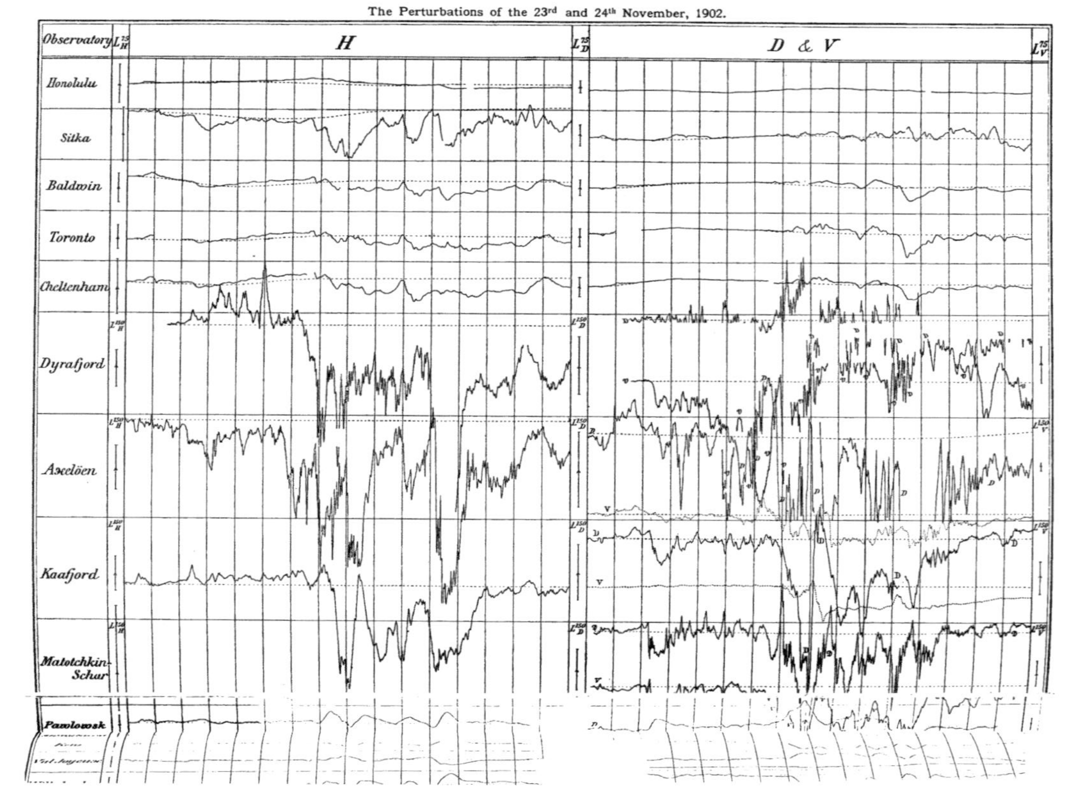

Pl. IX

The Perturbations of the 9th December, 1902

Registerings from 5h to 18h, Gr. M. T.

THE PERTURBATIONS OF THE 9th DECEMBER, 1902.

Dyrafjord

Axelöen

Kaafjord

Pl. X

The Perturbation of the 15th December, 1902

Registerings from 23h on the 14th to 5h on the 15th, Gr. M. T.

THE PERTURBATION OF THE 15th DECEMBER, 1902.

Dyrafjord

Matotchkin Schar

Axeløen

Pl. XI

The Perturbation of the 25th December, 1902

Registerings from 23^h on the 24th to 5^h on the 25th, Gr. M. T.

[illegible]	H	P
[illegible]		
Baldoon		
Toronto		
Cheltenham		
[illegible]		
[illegible]		
[illegible]		
[illegible]		
[illegible]		

Pl. XII

The Perturbations of the 26th December, 1902

Registerings from 18h on the 26th to 2h on the 27th, Gr. M. T.

THE PERTURBATIONS OF THE 20th DECEMBER 1902.

II

Observatory

Honolulu

Sitka

Baldwin

Toronto

Cheltenham

Dyrafjord

Axeløen

Kaafjord

Matotchkin Schar

Pawlowsk

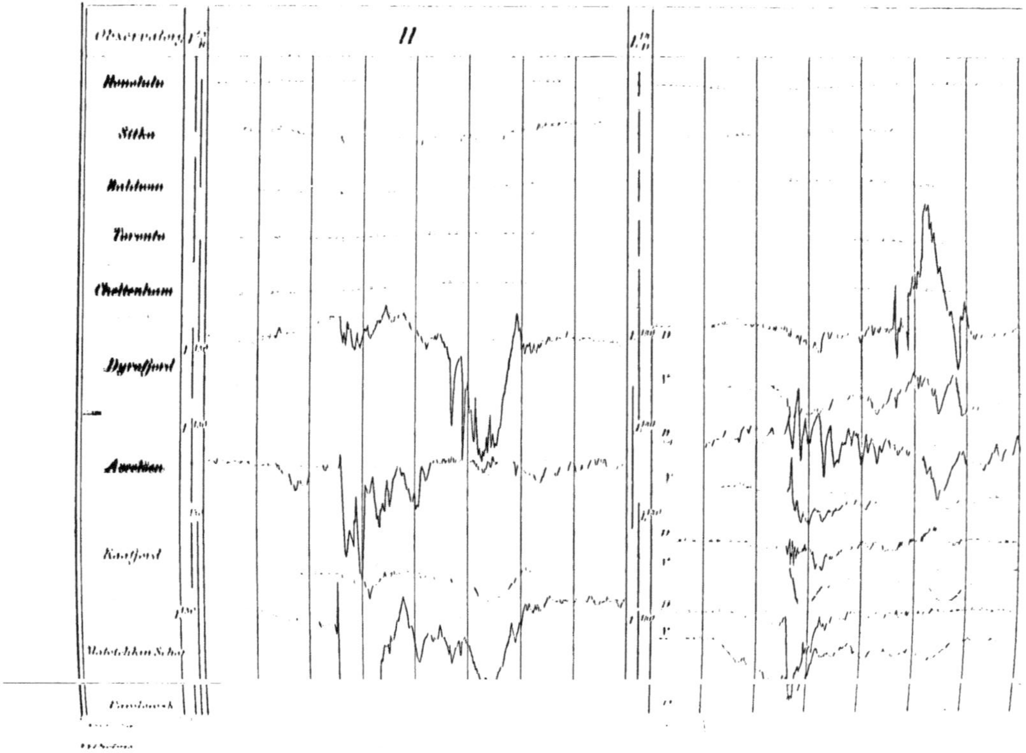

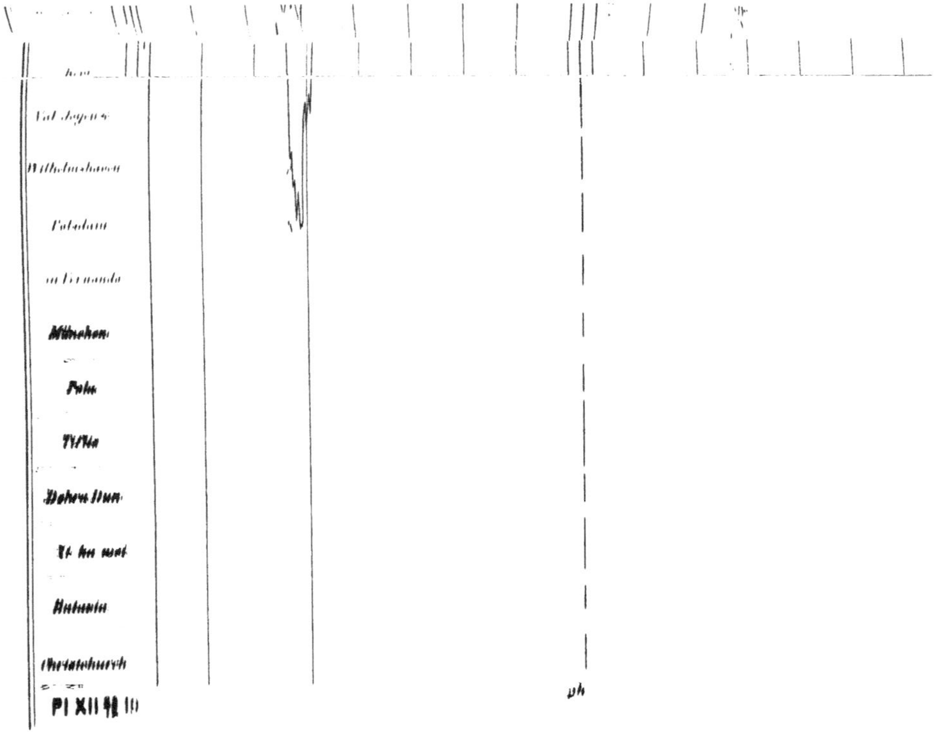

Kew

Val Joyeux

Wilhelmshaven

Potsdam

S. Fernando

München

Pola

Tiflis

Dehra Dun

Zi-ka-wei

Batavia

Christchurch

ph

Pl. XII Fig. 10

Pl. XIII

The Perturbation of the 28th December, 1902

Registerings from 3^h to 8^h, Gr. M. T.

THE PERTURBATION OF THE 28th DECEMBER, 1902.

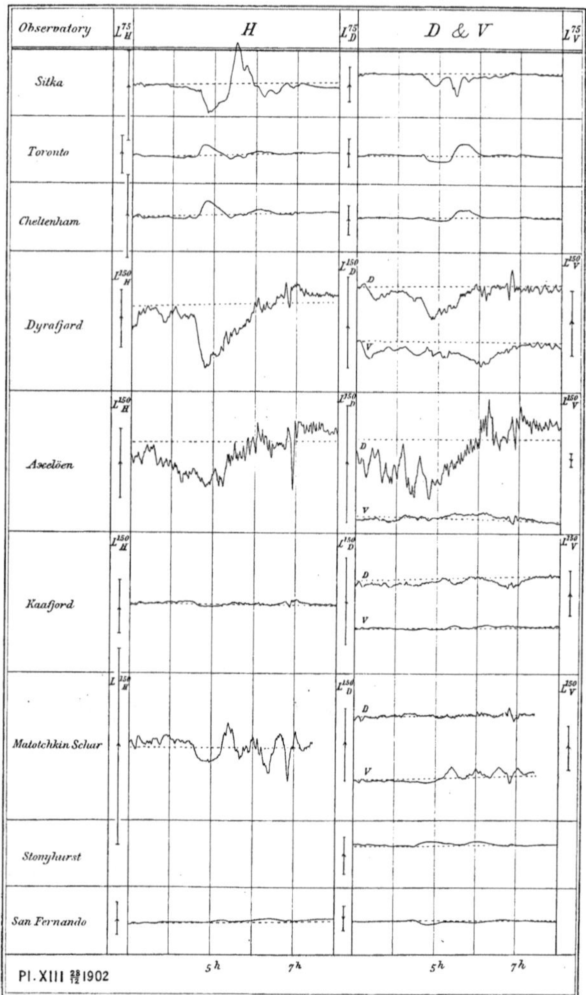

Pl. XIII $\frac{28}{12}$ 1902

Pl. XIV

The Perturbation of the 26th January, 1903

Registerings from 7^h to 15^h, Gr. M. T.

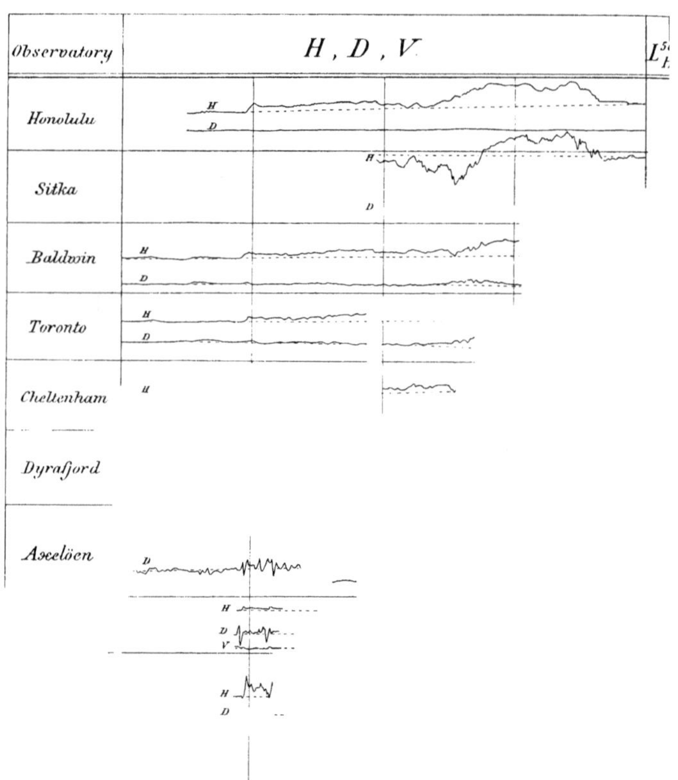
Observatory
H , D , V
Honolulu
Sitka
Baldwin
Toronto
Cheltenham
Dyrafjord
Axelöen
H
D
V

Y, 1903.

H, D, V

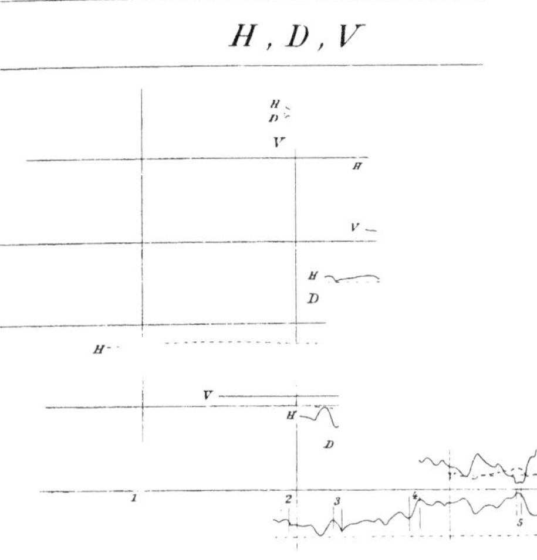

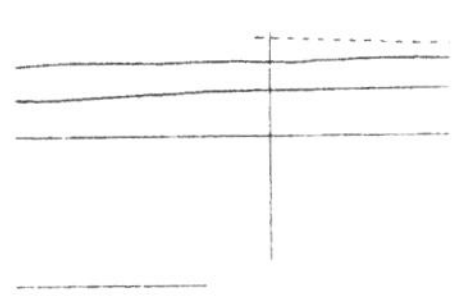

H

D

Pl. XV

The Perturbations of the 26th & 27th January, 1903

Registerings from 18h on the 26th to 7h on the 27th, Gr. M. T.

THE PERTURBATIONS OF THE 26th & 27th JANUARY, 1903.

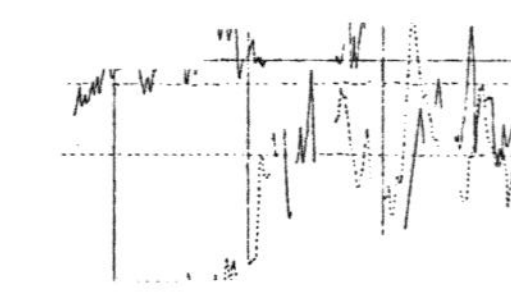

Pl. XVI

The Perturbation of the 8th February, 1903

Registerings from 8h to 12h, Gr. M. T.

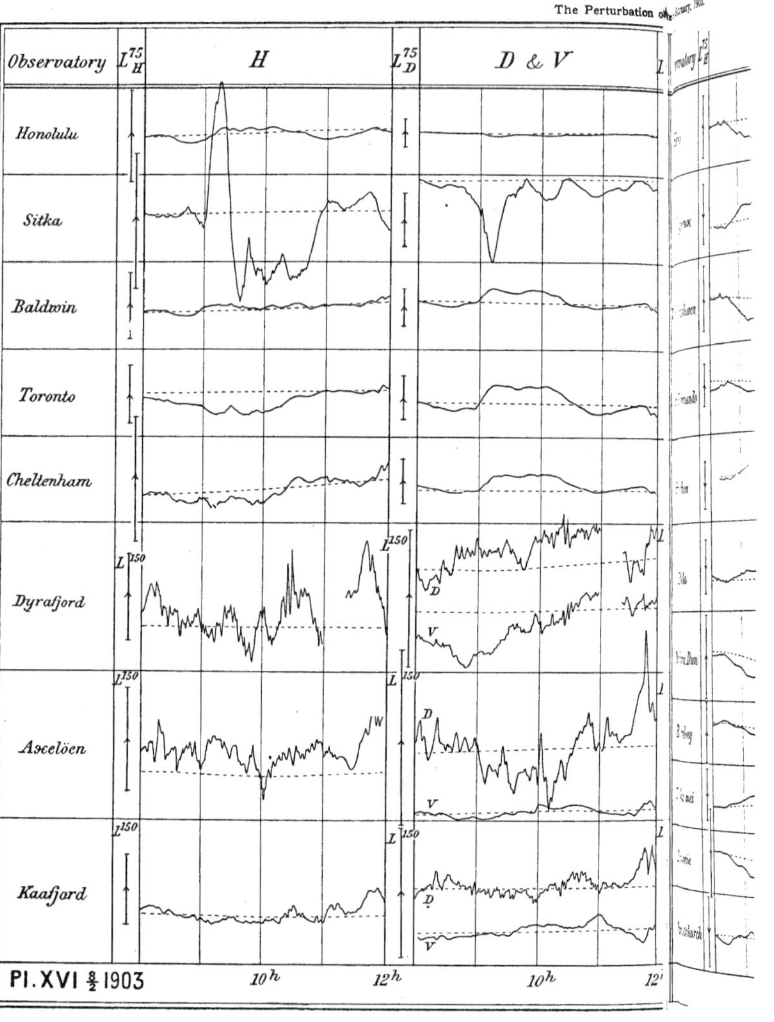
Observatory
L^{75}_{H}
H
L^{75}_{D}
D & V
Honolulu
Sitka
Baldwin
Toronto
Cheltenham
Dyrafjord
Axelöen
Kaafjord
L^{150}
D
V
W
10h
12h
10h
12
Pl. XVI 8/2 1903

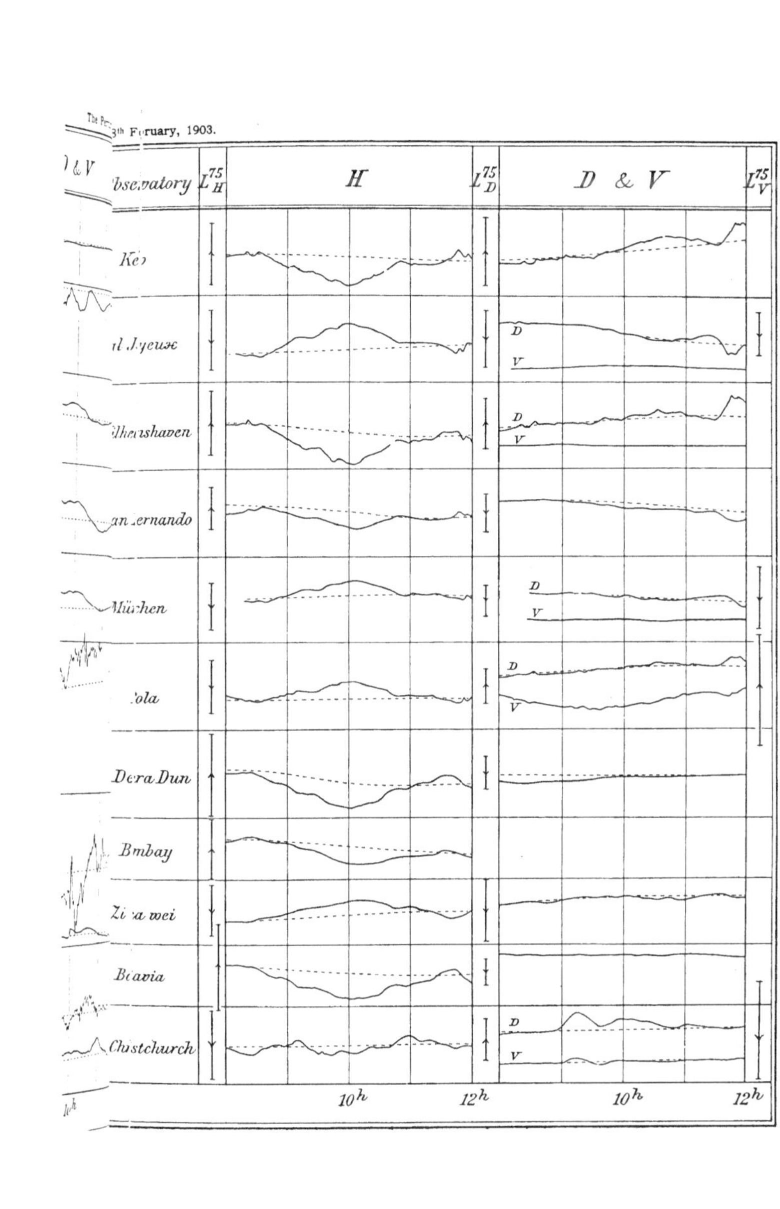
3th F[illegible]ruary, 1903.
Observatory
L^{75}_{H}
H
L^{75}_{D}
D & V
L^{75}_{V}
Kew
Val Joyeux
Wilhelmshaven
San Fernando
München
Kola
Dera Dun
Bombay
Zi-ka-wei
Batavia
Christchurch
D
V
10h
12h

Pl. XVII

The Perturbations of the 8th February, 1903

Registerings from 13h to 24h, Gr. M. T.

H

Honolulu

Sitka

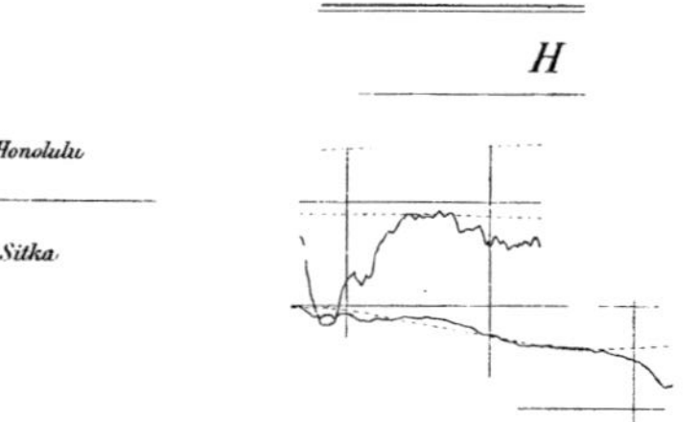

Dyrafjord

Axeløen

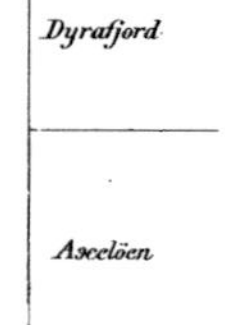

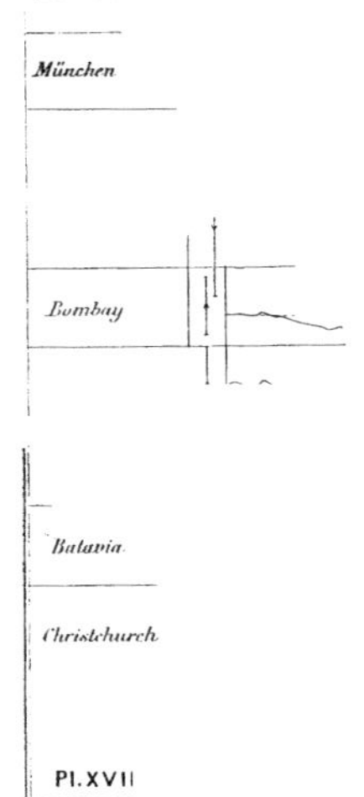
Potsdam
München
Bombay
Batavia
Christchurch
Pl. XVII

Pl. XVIII

The Perturbation of the 10th February, 1903

Registerings from 20h on the 10th to 3h on the 11th, Gr. M. T.

THE PERTURBATION OF THE 10th FEBRUARY, 1903.

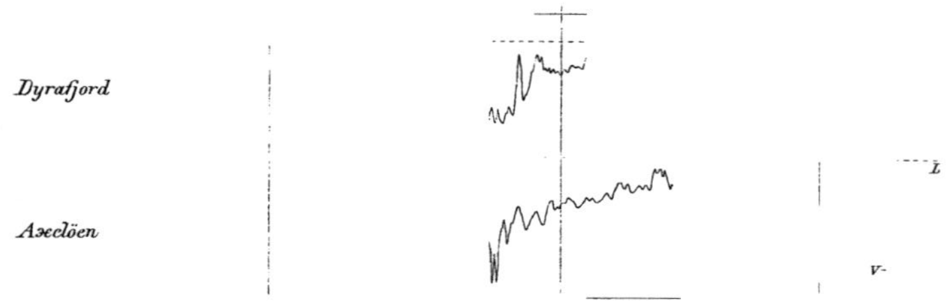

Dyrafjord

Axelöen

Pl. XIX

The Perturbation of the 15th February, 1903

Registerings from 13^h to 20^h, Gr. M. T.

The Perturbation of the 15th February, 1903.

Kaafjord

Dyrafjord

Axeløen

Pl. XX

The Perturbations of the 22nd March, 1903

Registerings from 12h on the 22nd to 1h on the 23rd, Gr. M. T.

THE PERTURBATIONS OF THE 22nd MARCH, 1903.

Dyrafjord

Pl. XXI

The Perturbations of the 31st March, 1903

Registerings from 19h on the 30th to 3h on the 31st, Gr. M. T.

THE PERTURBATIONS OF THE 31st MARCH, 1903.

Dyrafjord

Axeløen

H

D

V

Bossekop

Christchurch

THE NORWEGIAN AURORA POLARIS EXPEDITION 1902–1903

VOLUME I

ON THE CAUSE OF MAGNETIC STORMS AND THE ORIGIN OF TERRESTRIAL MAGNETISM

BY

KR. BIRKELAND

SECOND SECTION

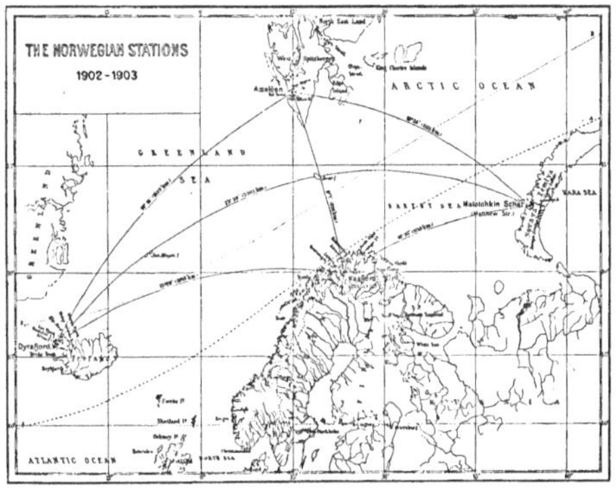

CHRISTIANIA
H. ASCHEHOUG & CO.

LEIPZIG
JOHANN AMBROSIUS BARTH

LONDON, NEW YORK
LONGMANS, GREEN & CO.

PARIS
C. KLINCKSIECK

CHRISTIANIA. A. W. BRØGGERS PRINTING OFFICE. 1913.

PREFACE.

Five years have gone by since the first Section of the present work, Volume I, was published. In spite of uninterrupted and persevering labour, we have only now succeeded in making Section II ready for publication.

The observations that formed our material were however exceedingly numerous, and the questions that in the course of our work presented themselves for solution were of a somewhat multifarious nature. The limits that were originally designed for Vol. I have therefore been overstepped, and the volume has been expanded to about double the compass at first intended.

The present section begins with the discussion of magnetic observations from 15 stations of the well-known polar investigations of 1882—1883, by which my earlier results from observations from 25 stations in medium latitudes in 1902—1903, have received a most valuable complement.

As regards the conditions during the positive and negative polar storms, and particularly the diurnal motion of the respective magnetic storm-centres, we have arrived at results that seem to us so valuable, that they have fully rewarded us for the exertions and personal sacrifices that the work has cost.

In order further to make it clear whether our results from the working-up of the above-mentioned observations from the most varied parts of the world could be brought into theoretic harmony with my previous assumptions, I have carried out a long series of experimental investigations with a magnetic globe in a large vacuum-box intended for electric discharges. I have hereby been enabled to obtain a representation of the way in which cathode-rays move singly, and group themselves in crowds about a magnetic globe such as this. Special study has been made of those crowds of rays that produce magnetic effects analogous to those observed upon the earth during positive and negative polar storms.

Those who will go through the whole labyrinth that this concatenation of experiments forms, cannot but be attracted by their scientific beauty; and in the end they will see that great difficulties have resolved themselves into a surprising clearness.

I hold that I have demonstrated that the magnetic storms on the earth — the positive and negative polar storms, and the positive and negative equatorial storms — may be assumed to have as their primary cause the precipitation towards the earth of helio-cathode rays, of which the magnetic stiffness is so great that the product $H.\varrho$ for them is most usually about 3×10^6 C. G. S. units.

On account of the magnetic condition of the earth, these new solar beams which I have discovered, will especially make their way towards the earth in the polar regions in the two auroral zones, where they also certainly produce other effects which play an important part in various meteorological phenomena.

SCHUSTER, in a later work, considers that from energy and from electrostatic considerations alike, he can prove that even originally well-defined pencils of cathode-rays from the sun cannot reach the earth. The existence of such pencils of rays was clearly presupposed to be necessary to the theory as already formulated by me in 1899; and this assumption is now said to be untenable.

From the results which are here produced, however, it will undoubtedly appear that there must be a flaw somewhere or other in the reasoning of the distinguished natural philosopher; for one is inclined to regard the descent of the above-mentioned pencils of rays to the earth as an experimental fact.

I have also endeavoured, in Chapter VI, directly to demonstrate the points in which Schuster's assumptions in no way admit of being applied to our case. I will here, moreover, with regard to the electrostatic repulsion between our helio-cathode rays, refer to formulæ by OLIVER HEAVISIDE. In his Electrical Papers, Vol. II, Part III, p. 495, mathematical investigations are to be found of electrically charged corpuscles in translatory motion, and from these it appears, on a discussion of the formulæ, that when the velocity of the corpuscles equals that of light, the electrostatic repulsion between the rays maintains the balance with the electro-dynamic attraction. And as regards our helio-cathode rays, their velocity, according to the theory, differs no more than a hundred metres from that of light.

We find, with regard to these rays that the acceleration with which an electron is repelled from the pencil of rays will not be what Schuster gives, but from the very first moment 3.3 million times less. Subsequently this acceleration decreases with very great rapidity, in so far as the longitudinal mass of the repulsed electron comes into play.

In a paper he has just published, HALE communicates some preliminary results on the general magnetism of the sun, at which he has arrived by the aid of instruments and experimental methods that are altogether admirable. He considers that the entire sun must be magnetic, with polarity like that of the earth, and with a vertical intensity at the poles of about 50 gausses.

These results seem at first sight to be quite irreconcilable with those in this work. If the sun were perceptibly magnetic in the same manner as the earth, but with an intensity 70 times as great, it is perfectly certain that no helio-cathode ray of the kind in question could ever reach the earth.

Hale, however, is of opinion that the magnetism of the sun differs radically from that of the earth.

It seems to me that the phenomena observed by Hale might be explained as the effects produced by invisible spots, or by the pores, considered as electric vortices, nothwithstanding all the reasons that Hale adduces against such an assumption.

In a note to the Comptes Rendus de l'Académie des Sciences, Paris, Aug. 25, 1913, I have given the reasons that favour my view.

The experimental investigations which at first were designed to procure analogies capable of explaining phenomena on the earth, such as aurora and magnetic disturbances, were subsequently extended, as was only natural, with the object of procuring information as to the conditions under which the emission of the assumed helio-cathode rays from the sun might be supposed to take place.

The magnetic globe was then made the cathode in the vacuum-box, and experiments were carried on under these conditions for many years.

It was in this way that there gradually appeared experimental analogies to various cosmic phenomena, such as zodiacal light, Saturn's rings, sun-spots and spiral nebulæ.

The consequence was that attempts were made to knit together all these new discoveries and hypotheses into one cosmogonic theory, in which solar systems and the formation of galactic systems are discussed perhaps rather more from electromagnetic points of view than from the theory of gravitation.

One of the most peculiar features of this cosmogony is that space beyond the heavenly bodies is assumed to be filled with flying atoms and corpuscles of all kinds in such density that the aggregate mass of the heavenly bodies within a limited, very large space would be only a very small fraction of the aggregate mass of the flying atoms there.

And we imagine that an average equilibrium exists in infinite space, between disintegration of the heavenly bodies on the one hand, and gathering and condensation of flying corpusles on the other.

I cannot conclude this great work without expressing my warmest thanks to my numerous assistants for their most able collaboration. If I mention them according to the number of years in which they have faithfully helped me, I must begin with my good old friend, now dead, schoolmaster DIETRICHSON, who for ten years continued to work every day at my side. In the next place there are some young, energetic men, a few of whom have already begun independent work — Mr. KROGNESS, now manager of the Haldde Observatory, Mr. VEGARD, now a tutor at our university, Mr. SKOLEM, a very skilful mathematician, and Mr. DEVIK, a capital experimenter. Further Captain BULL, of the Norwegian Navy, and Mr. NORBY, have done a large amount of calculation, and Mr. NATRUD and Mr. B. TOLSTAD, assistants at the Norwegian Geographical Survey, have made many drawings. The translation of also the whole of this volume has been done very satisfactorily by Miss JESSIE MUIR.

Christiania; September, 1913.

Kr. Birkeland.

CONTENTS.

PART II.

POLAR MAGNETIC PHENOMENA AND TERRELLA EXPERIMENTS.

CHAPTER I.

POLAR MAGNETIC STORMS 1882—1883.

			Page
Art.	81.	The Treatment of the Observations from the Polar Expedition of 1882 & 1883	319
„	82, 83.	The Perturbation of the 15th January, 1883	323
„	84.	The Perturbations of the 2nd January, 1883	339
„	85.	The Perturbations of the 1st November, 1882	350
„	86.	The Perturbations of the 14th and 15th February, 1883	361
„	87.	The Perturbations of the 15th July, 1883	371
„	88.	The Perturbations of the 1st February, 1883	386
„	89.	The Perturbations of the 15th December, 1882	397
„	90.	The Perturbations of the 15th October, 1882	412

CHAPTER II.

MATHEMATICAL INVESTIGATIONS. PRELIMINARY RÉSUMÉ.

„	91.	The Calculation of the Field of Force for the assumed Polar Current-System	423
„	92.	Résumé	439
„	93.	A Possible Connection between Magnetic and Meteorologic Phenomena	449

CHAPTER III.

STATISTICAL TREATMENT OF MAGNETIC DISTURBANCES OBSERVED AT THE NORWEGIAN STATIONS 1902—1903.

„	94.	Introductory	451
„	95.	The Total Storminess as a Function of Time and its Relation to Solar Activity	517
„	96.	On the Possible Influence of the Moon upon Magnetic Storms	519
„	97.	The Seat of the Radiant Source	521
„	98.	Sun-Spots and Storminess	524
„	99.	Annual Variation of Storminess	526
„	100.	On the Diurnal Distribution of Storminess	536
„	101.	Positive and Negative Storminess	536
„	102.	P and N Storminess	537
„	103.	Properties of the «Average Polar Storm»	538
„	104.	Comparison of Storminess at the four Stations	541
„	105.	Separation of Great and Small Disturbances	546
„	106.	The Distribution of Storminess and the Solar Origin of Polar Storms	547
„	107.	Application to Theory	551

CHAPTER IV.

EXPERIMENTS MADE WITH THE TERRELLA WITH THE SPECIAL PURPOSE OF FINDING AN EXPLANATION OF THE ORIGIN OF THE POSITIVE AND NEGATIVE POLAR STORMS.

Page

Art. 108. Introductory 553

STUDY OF RAYS OF GROUP A.

„ 109. Experiment in which the Terrella had only a Vertical Screen 560
„ 110. Experiments in which the Terrella is surrounded by a Horizontal Screen 566
„ 111. Equatorial Rings of Light 569

STUDY OF RAYS OF GROUP B.

„ 112. The Course of the Rays in the Polar Regions over the Terrella 572
„ 113. Experiments for determining the Tangential Component of the Polar Precipitation in Relation to the Surface of the Terrella 580
„ 114. On an Intimate Connection between Rays of the two Groups A and B 583
„ 115. On the Size of the Polar Ring of Precipitation 591
„ 116. The Value of H . ϱ for the Helio-Cathode Rays 598
„ 117. Experiments for the Determination of the Situation of the Polar Zone of Precipitation in Various Positions of the Magnetic Axis 600
„ 118. Investigations Regarding the Angle formed by the Precipitated Rays with the Magnetic Lines of Force. Application to the Polar Aurora 603

CHAPTER V.

IS IT POSSIBLE TO EXPLAIN ZODIACAL LIGHT, COMETS' TAILS, ÁND SATURN'S RING BY MEANS OF CORPUSCULAR RAYS?

„ 119, 120, 121. Zodiacal Light 611
„ 122. Appendix. Expedition to Assouan and Omdurman 624
„ 123. Magnetic Registerings, the 9th April, 1911 629
„ 124. Comets' Tails 631
„ 125. Halley's Comet, May, 1910 639
„ 126. Meteorological Observations about the Time of the Transit of Halley's Comet, 1910 . . 647
„ 127. The Saturnian Ring 654

CHAPTER VI.

ON POSSIBLE ELECTRIC PHENOMENA IN SOLAR SYSTEMS AND NEBULAE.

„ 128. The Sun 661
„ 129. Experiments showing Analogies to Solar Phenomena 662
„ 130. Application of the Analogies to further Study of Celestial Phenomena 670
„ 131. The Worlds in the Universe 677
„ 132. Investigations of the Motion of Electric Corpuscles in the Field of an Elementary Magnet especially to find the Conditions for the Approach to Boundary-Circles 678
„ 133. Study of the Approach to Boundary-Circles, when there is a Resistance in the Medium 686
„ 134. Study of the Approach to Boundary-Circles, when the Charge of the Particles is variable 693
„ 135. Study of the Approach to Boundary-Circles outside the Magnetic Equatorial Plane . . . 697
„ 136. Comparison of Boundary-Circles approached by Different Sorts of Corpuscles 706
„ 137. Experiments made with the largest Vacuum-box with a Capacity of 1000 Litres . . . 709
„ 138. On the Charge of Metallic Particles ejected from a Cathode 716
„ 139. On the Possible Density of flying Corpuscles in Space 720

PART III.

EARTH CURRENTS AND EARTH MAGNETISM.

CHAPTER I.

EARTH CURRENTS AND THEIR RELATION TO CERTAIN TERRESTRIAL MAGNETIC PHENOMENA.

		Page
Art. 140.	Introduction	725
" 141.	Strength and Distribution of Earth-Currents	728
" 142.	Diurnal Variation of Earth-Currents	729
" 143.	Earth-Currents and Magnetic Disturbances	730
" 144.	Earth-Current Registerings at Kaafjord and Bossekop, 1902—1903	731
" 145.	Constants for the Experimental Arrangements	734
" 146.	The Magnetic Effect of Earth-Currents	736
" 147.	On the Connection between Polar Storms and Earth-Currents	741
" 148.	Earth-Currents and Positive Equatorial Perturbations	746
" 149.	On the Simultaneity of Earth-Currents and Magnetic Disturbances	746
" 150.	Earth-Currents at Bossekop	748
" 151.	The Influence of the Earth-Current upon the Vertical Intensity	749
" 152.	Observations of Earth-Currents at Kaafjord, May, 1910	751
" 153.	Theoretical Investigation of the Currents that are Induced in a Sphere by Variation of External Current-Systems	757
" 154.	Numerical Calculation of the Currents	768
" 155.	Currents that are Induced by Rotation or Removal of the Systems	779
" 156.	Earth-currents in Lower Latitudes	784
" 157.	Earth-currents in Germany	784
" 158.	Earth-currents in France	788
" 159.	Earth-currents in England	791
" 160.	Earth-currents at Pawlowsk	792
" 161.	Comparison of Simultaneous Earth-Current Observations	793
" 162.	Consideration of the Conditions during Positive Equatorial Storms	794
" 163.	The Diurnal Variation of the Earth-Currents	796

PLATES.

Pl. XXII. The Perturbations of the 15th October, 1882.
Pl. XXIII. The Perturbations of the 1st November, 1882.
Pl. XXIV. The Perturbations of the 15th December, 1882.
Pl. XXV. The Perturbations of the 2nd January, 1883.
Pl. XXVI. The Perturbations of the 15th January, 1883.
Pl. XXVII. The Perturbations of the 1st February, 1883.
Pl. XXVIII. The Perturbations of the 14th and 15th February, 1883.
Pl. XXIX. The Perturbations of the 15th July, 1883.
Pl. XXX. Earth-currents and magnetic elements. Series I. Kaafjord.
Pl. XXXI. Earth-currents and magnetic elements. Series II. Kaafjord.
Pl. XXXII. Earth-currents and magnetic elements. Series II continued. Kaafjord and Bossekop.
Pl. XXXIII. Earth-currents and magnetic elements. Series II continued. Bossekop.
Pl. XXXIV. Earth-currents and magnetic elements. Series III. Kaafjord.
Pl. XXXV. Earth-currents and magnetic elements. Series III continued. Bossekop.
Pl. XXXVI. Earth-currents and magnetic elements from Germany (for Nov. 5—6 also France and England).
Pl. XXXVII. Earth-currents and magnetic elements from Greenwich.
Pl. XXXVIII. Earth-currents and magnetic elements from Parc St. Maur and Greenwich.
Pl. XXXIX. Earth-currents and magnetic elements from Parc St. Maur.
Pl. XL. Earth-currents and magnetic elements from Parc St. Maur.
Pl. XLI. Earth-currents and magnetic elements from Parc St. Maur and Greenwich.
Pl. XLII. Earth-currents and magnetic elements from Greenwich and Parc St. Maur.

ERRATA.

Page 616, line 9 from below: For $0 > \theta - \frac{\pi}{2}$, read $0 > \theta > -\frac{\pi}{2}$
— 670, line 17 from below: For the Articlenumber 129, read 130.

PART II.

POLAR MAGNETIC PHENOMENA AND TERRELLA EXPERIMENTS.

CHAPTER I.

POLAR MAGNETIC STORMS 1882—1883.

81. The Treatment of the Observations from the Polar Expeditions of 1882 & 1883. In the discussion of the magnetic storms in Part I, it was frequently pointed out that we obtained only an imperfect knowledge of the conditions round the auroral zone, owing to the fact that, with the exception of our four arctic stations, all the stations from which we had observations were in southern latitudes. We have frequently drawn conclusions as to how the phenomena up there might naturally be assumed to have developed, if the perturbation-areas that appeared in southern latitudes could be explained by the previously-mentioned simple points of view.

We will therefore, in this part of our work, subject these conditions to a closer study, and will then be able to see whether the actual conditions round the auroral zone prove to be in accordance with our previous conclusions.

It is the polar storms in particular that will make an interesting subject of study; and it will then be especially necessary to investigate the movement and formation of the various systems of precipitation in the course of the twenty-four hours.

It will be remembered that in the compound storms of 1902—03, we arrived at a very simple interpretation of the occurrence of the polar storms, and of the changes in their main features. This interpretation we now have the opportunity of verifying, and even supplementing on various points. We will here recall to the reader's mind the more important of the main features.

In the first place, we divided the polar storms into two kinds, namely, the *negative* polar storms, during which we found negative values of P_h, in the district of precipitation round the auroral zone, and the *positive* polar storms, during which we found positive values of P_h in the district of precipitation (see Art. 69. Part I).

The negative storms had, as a rule, an extensive area of precipitation on the night-side of the earth, and also on the day-side in high latitudes (Axeløen). The positive storms had a more limited district of precipitation, and as a rule appeared on the afternoon-side of the earth.

It further appeared that during the great magnetic storms, these areas of precipitation seemed to move, the movement to some extent following the sun in its apparent daily course round the earth, and being dependent upon the sun's change of altitude above the magnetic equator (see Art. 71, Part I).

In the material we are now going to study, these conditions can be investigated far more thoroughly. In the reports of the international polar expeditions of 1882 and 1883, we have a material carefully worked up, that will prove to be of the greatest interest to us in this study. It is the term-days observations that are of special importance for our purpose. We have at our disposal observations from ten stations scattered round the auroral zone, namely, Godthaab, Kingua Fjord, Fort Rae, Uglaamie (Point Barrow), Ssagastyr, Little Karmakul, Sodankylä, Bossekop, Cape Thordsen and Jan Mayen, and also from Fort Conger, a station situated in the vicinity of the magnetic axis of the earth, and from some more southerly stations, four of which have been employed, namely, Christiania, Pawlowsk, Göttingen and Kasan.

The method employed in the working-up, is exactly analogous to that used with the observations from 1902 and 1903, except that here, instead of registerings we have readings of the magnetic elements for every fifth minute.

The variation in these elements, in the case of a number of stations, is represented graphically in the respective publications, and for stations where this is not already done, we have ourselves drawn the magnetic variation curves. The same hour-length is employed throughout, namely, 15 mm. per hour, whereas the scale for the deviations varies somewhat from place to place, according to the amplitude of the oscillations.

These curves are placed under one another in plates, thereby affording a clear view of the course of the perturbations from station to station. These plates in reduced size will be found at the end of this section. Further, the perturbing forces are calculated for a series of points of time, these being represented on a polar chart by current-arrows in precisely the same manner as before. In this connection, however, it should be remarked that in calculating P_d, it is the value of H existing at the moment, that has to be employed but during the powerful storms this value may vary so considerably that the same value of H cannot be used for the whole perturbation, and a correction must be introduced. This correction is always evident during the powerful storms that take place in the regions here studied. L_D^{100} is given in the plates for the value of H, which answers to the normal line. During powerful storms in H, therefore, direct use cannot be made of this, if fairly great accuracy is desired; but as a rule the error will not be very great. For this reason, the values of P_d that we find at Fort Conger during powerful storms will be somewhat uncertain, as we there have only absolute observations of H to go upon. This is of little significance, however, in our studies.

The scale of the arrows on the charts is given, and, as will be seen, is generally about five times that employed in the previous observations. By this means the current-arrows at the polar stations are of a suitable size, while at the southern stations another scale must be used. This is indicated by there adding $\frac{5}{1}$, $\frac{5}{2}$, and so on, which means that the scale employed is $\frac{5}{1}$, $\frac{5}{2}$, etc. of the general given one. This is a reversal of our former plan of introducing the factors $\frac{1}{5}$, $\frac{2}{5}$, etc. at the polar stations in order to indicate the local scale there in each case.

On most of the charts here, moreover, there are several sets of current-arrows for one series of generally as many as three different points of time. Instead of vertical arrows, which are found upon the charts on which only one point of time is marked, a little table is here placed beside the station, giving the corresponding values of P_v. A similar table is given for P_d at Fort Conger, where only term-days observations of the declination were carried out. Further, the magnetic meridian of that place is indicated, and an idea is thus obtained of the direction, and to some extent of the magnitude of the perturbing force at the various times. A powerful westerly-directed perturbing force in D thus corresponds to a current-arrow directed westwards, more or less NW or SW, according as the perturbing force in H might be more or less powerful, positive or negative. It will be seen that the magnetic meridian and the geographical meridian at this place are nearly at right angles to one another, so that a westerly-directed perturbing force as regards the magnetic meridian, answers to a perturbing force directed southwards. If, on the other hand, P_d is only small, there is either, if P_h is also small, only a small current-arrow, or, if P_h is fairly large, a current-arrow directed northwards or southwards. In this way it is possible to make use of these observations.

For the calculation of the perturbing forces, it is necessary to have a more or less accurate knowledge of the diurnal variation. By the diurnal variation must be understood the variation that there would have been in the magnetic elements, if there had been no perturbations, in other words, if the day had been a 'quiet day'. The diurnal variation, however, in the case of certain stations, has been calculated as the mean of all the observations in a certain space of time. The results found therefrom, however, are useless in this connection, just because the perturbing forces themselves then come into the diurnal variation, and it was these we wanted to eliminate. By taking a sufficient number of observations, it might be thought that the perturbing forces would be eliminated, as the oscillations would possibly be as frequent to one side as to the other; but this is not the case. The oscillations, when they occur, generally have a definite direction for every distinct time of day. Perturbations, for instance, that occur about midnight, local time, at most of the polar stations, in horizontal intensity, will almost exclusively show negative values of P_h. By the addition of all the values, too low a mean value of H will therefore be found here. If we would use such a determination, perturbing forces would often be found, for instance, at times when the conditions were quite normal.

At several stations, however, the diurnal variation has been calculated exclusively from observations on quiet days. If the days used in these calculations really were 'quiet' we might apply these determinations. A quiet day in the Polar regions is, however, a very rare occurrence, and in most cases, on the majority of the 'quiet' days made use of, deviations having the character of minor perturbations occur. When these perturbations are not eliminated, the result would not always be applicable to our purpose.

For us, in the calculation of the perturbing forces, the best means of obtaining an approximately accurate determination of the diurnal variation on the day in question is, by means of the hourly observations made daily at the various stations, to draw the magnetic curves for the nearest quiet days before and after the fixed day; by comparing these we can draw a normal line, that is in correspondence with only the quiet parts of the curves, from which consequently the perturbations are eliminated. This is the method that has mainly been followed.

In Kingua Fjord not a single really quiet day is to be had, especially in the afternoon, Greenwich time; the conditions are always more or less disturbed. In the forenoon, however, the conditions are very often fairly undisturbed. From the most quiet days found in the material, it seems, however, to become clear, that the diurnal variation in the afternoon is but small, and that consequently the disturbed conditions here must be regarded as perturbations. As normal line, we have therefore here drawn a fairly straight line, and as the variations as a rule are somewhat considerable, the error in the position of the normal line will be of less importance.

This circumstance, that magnetic perturbations occur much more frequently at this station than at the other polar stations, is a fact of very great importance for our theory, and we will return to this later on.

At the stations where the hourly observations have not been taken, namely, Christiania, Göttingen, and Kasan, the determination of the diurnal variation becomes considerably more difficult and to some extent rather uncertain. We here have only the more or less quiet term-days to go upon, in addition to the comparisons we can draw with observations of recent years and adjacent stations. The determination of the normal line at these two stations may therefore sometimes be somewhat arbitrary, especially in the case of the vertical intensity of Göttingen, where it has occasionally been impossible to make any determination.

On the whole we may remark, that the diurnal variations that we have used must of course not be regarded as entirely correct, when the oscillations attain a certain amplitude, however, the uncertainty in the normal line is of smaller significance.

With regard to the vertical intensity, the observations are often a little unreliable, and it may perhaps be doubtful on the whole whether any conclusions at all may be drawn from these observations, especially in the case of those stations at which the method employed was that of induction in bars of soft iron.

We have thus made no use of the vertical intensity observations from Ssagastyr, as the perturbing forces constantly appearing there are of an altogether different order of magnitude to that which we find at the other polar stations, whereas the agreement in the horizontal elements is very close. In Sodankylä too, the perturbing forces taken from the observations of vertical intensity, are often apparently abnormal compared with the conditions at the stations situated in the vicinity of that station.

As stated in Art. 23 the map-projection employed for our polar chart is not orthomorphic. The deformation is not great, however, but yet sufficiently so, especially at the southern stations, to be taken into consideration in the placing of the current-arrows. These are thus not placed at the angle which they form on the earth with for instance the geographical east and west, but at a rather smaller angle. The amount by which this angle (v) is reduced has been calculated for two or three latitudes, the result being given in the following table:

TABLE XLIX.

v	0°	15°	30°	45°	60°	75°	90°
Göttingen	0	1° 3′	1° 51′	2° 13′	1° 56′	1° 8′	0
60° N. Lat.	0	39′	1° 8′	1° 19′	1° 10′	40′	0
70° N. Lat.	0	17′	30′	35′	31′	17′	0

In these charts also we have indicated the position of the sun and the moon. Their signs (☉ and ☾) are placed in the margin of the chart, that for the sun on the noon meridian, that for the moon on the meridian that it is crossing at the moment under consideration. The point in which the magnetic axis intersects the surface of the earth, has been calculated for the beginning of 1883 as situated in latitude 78° 20′ N., and longitude 68° 49′ W([1]).

In the preceding observations, *Greenwich mean time* has been employed throughout, and it will also be used now in order to facilitate comparison.

In the observations of which we make use, everything relating to the fixed days is given according to Göttingen mean time, and we have therefore effected the necessary reduction all through. The difference in time between these two places amounts to $0^h\ 39^m, 8$, or in round numbers to $0^h\ 40^m$, the latter being the figure we have employed. Lastly, the hours, as before, are counted from 0 to 24, 12 answering to Greenwich mean noon.

With regard to the arrangement of the perturbations, we have used the same method as previously; first treating of the days on which the simplest and most perspicuous conditions of perturbation prevail — those on which the typical phenomena are most prominent. The more complicated phenomena are dealt with later.

Amongst the disturbances we find here, is also an equatorial one, but, as it is the polar disturbances that interest us most, this perturbation is noticed amongst the last.

The plates of the curves are, on the other hand, arranged in chronologic order.

In conclusion, we give a table of the perturbations in the order in which they are described.

([1]) V. Carlheim Gyllensköld, "Note sur le Potentiel Magnétique de la Terre exprimé en Fonction du Temps". Arkiv för matematik, astronomi och fysik. Vol. 3, No. 7. Upsala 1906.

TABLE L.

No. of Perturbation	Date	No. of Plate
21	January 15, 1883	XXVI
22	January 2, 1883	XXV
23	November 1, 1882	XXIII
24	February 14/15, 1883	XXVIII
25	July 15, 1883	XXIX
26	February 1, 1883	XXVII
27	December 15, 1882	XXIV
28	October 15, 1882	XXII

THE PERTURBATION OF THE 15th JANUARY, 1883.
(Pl. XXVI).

82. The part of this day, which we now intend to study, is, as the Plate shows, that between 10^h and $23^h\ 20^m$ Gr. M. T., the latter hour corresponding with 24^h Göttingen mean time, at which point of time the observations on this term day cease.

The first glance at the Plate shows us that during this period a number of characteristic, well-defined and more or less powerful storms occur at the various stations.

A closer examination shows that these storms would naturally be divided into several groups.

In the first place we find in the period from 10^h to about 14^h a fairly well defined group of tolerably powerful perturbations. Before and after it, the conditions are more or less quiet at all the stations. The curves moreover indicate that for this period the perturbation-area can be divided into two parts, (1) the regions of Kingua Fjord, Fort Rae and Uglaamie, (2) Little Karmakul and Ssagastyr.

In (1), Kingua Fjord, Fort Rae and Uglaamie, it is evident that there is a negative polar storm with its centre in the neighbourhood of Fort Rae, where the deflections on the whole are greatest.

In (2), Little Karmakul and Ssagastyr, we distinctly see the effects of a positive polar storm. The forces are considerably more powerful at Ssagastyr than at Little Karmakul (note the values of ε_h at the two stations), and the storm-centre of this positive storm must thus be assumed to lie nearer the former station than the latter.

It is difficult to prove with certainty the existence of any distinct movement in these systems during this period, at any rate by only a direct consideration of the curves. The perturbation begins a little earlier at Fort Rae that in Kingua Fjord and at Uglaamie. If we look at the close of the perturbation, we find that the deflection in H lasts a little longer at Fort Rae than at the other two stations; whereas in D the deflections last longest in Kingua Fjord. It is difficult, however, to draw any conclusion from this.

Nor it is easy to find any distinct movement in the other system of precipitation. The deflections begin more or less simultaneously at the two stations, and then increase fairly evenly. To a certain extent we may speak of two maxima, the second of these being considerably greater at Little Karmakul than the first, a circumstance which may possibly indicate that a removal of the storm-centre actually takes place westwards towards this station. At Ssagastyr, however, the storm lasts a little longer than at Little Karmakul; but no conclusion can be drawn from this fact, as the conditions at Cape Thordsen are rather peculiar, and will probably exert an influence at Little Karmakul.

If we look at the conditions at Cape Thordsen during this period, we see that the curve for the horizontal intensity is very peculiar, first of all showing positive values of P_h, then negative values, and finally positive values once more. It seems evident that we here have before us the effects of a negative storm, which during the interval from 12^h to 14^h, encroaches upon a positive storm of longer duration, and that from $12^h\ 50^m$ to $13^h\ 50^m$ the effect of this negative storm is the strongest, so that negative values of P_h are found. This view seems to receive support from the conditions in declination, where, from 12^h to 14^h, there occurs a clearly defined deflection.

If we continue to follow the series of polar stations, we find during this period practically no perturbation at Bossekop and Sodankylä, nor is there any deflection in Jan Mayen until the end of the period under consideration, when a new positive storm begins there, with a very well-formed and clearly-defined deflection, during the period from 13^h to $16^h\ 40^m$. The defining of the first section, which we have previously undertaken, is thus not suitable for this station.

At Fort Conger also, there occurs a deflection which bears no small resemblance to the deflection in the horizontal-intensity curve in Kingua Fjord; only in this case the perturbing forces are small. At the southern stations there are no perturbing forces of any strength during this period.

It may be as well here, in connection with these conditions, which are read directly from the copies of the curves, to consider at once the area of perturbation, as represented in *Chart I and II* for the hours *$11^h\ 20^m$, $11^h\ 50^m$, $12^h\ 20^m$, $12^h\ 50^m$ and $13^h\ 20^m$*, Gr. M. T.

The two characteristic areas of precipitation described above, the negative in the north of America, and the positive in the north of Asia and to some extent also in Europe, are here very distinctly seen. At first it is only the negative system that has a marked effect, and its storm-centre appears to be situated in the vicinity of Fort Rae. At Uglaamie, during this first part of the time, the current-arrow has an easterly direction, the reverse of that which we find subsequently. It is as though we had before us the effects of a positive polar storm, and this may possibly be the case; but if so, it is very ill-defined, and this makes it impossible to decide the question with any certainty. At the succeeding hours moreover the current-arrow at this station swings round anti-clockwise, and remains directed westwards during the remainder of this first section which we are now considering. We may perhaps be justified in taking these conditions as a proof of a movement of the systems of precipitation in a westerly direction.

At the other stations situated in the vicinity of the areas of precipitation, the current-arrows increase more or less evenly, so that at the last of the hours of observation they attain their greatest strength, and the areas undergo no great changes. A quite distinct impression of a westerly movement in the positive precipitation area will be obtained by comparing the Chart II for $12^h\ 30^m$ with the two last times on Chart I. On Chart I $12^h\ 20^m$ it is only at Ssagastyr that the positive storm occurs with considerable violence, in Little Karmakul the perturbating force is still comparatively insignificant. At $12^h\ 50^m$, on Chart II, also in Little Karmakul, a somewhat powerful perturbating force occurs. The strength is, however, as yet greatest in Ssagastyr, but at $13^h\ 20^m$, as we see from Chart I, the strength of the perturbing forces is about equal at these two stations. The centre of the storm seems thus constantly to move westwards.

At Cape Thordsen only do we see the current-arrow turning clockwise in accordance with the peculiar conditions that we have just described.

According to what we have seen in Part I, the positive polar storm will now, in lower latitudes, produce an area of divergence.

With regard to the conditions in lower latitudes, we find only small perturbing forces at the first three hours of observation; but at $13^h\ 20^m$, the forces have increased to no small extent; and the shape of the western portion of an area of divergence is now actually recognised.

We will finally also draw attention to the agreement that we find between this and our previous results, namely, that the negative area of precipitation is formed upon the night and morning side, while the positive system is formed upon the afternoon and evening side.

In conclusion, we must also consider the values that we find of P_v. These, as we have said, will sometimes be rather uncertain, *inter alia* on account of the construction of the measuring apparatus; and we must therefore be careful not to think we can draw definite conclusions, especially where there are only slight deflections.

At Fort Rae, as we see, there are all this time positive values of P_v, which would thus imply that the main body of the current-system was situated slightly to the south of this station. At Uglaamie, on the other hand, negative forces first appear in the vertical intensity. When the horizontal current-arrow has assumed the more constant westerly direction, the vertical curve goes over to the opposite side, and the positive deflections then last for the remainder of the period under consideration.

Also on looking at P_v, it seems thus, as though at first there were perturbing forces of a more local character at Uglaamie.

At Little Karmakul, the positive values of P_v indicate that the positive system of precipitation must lie a little to the north of the place.

We will now pass on to consider the conditions that develop after the conclusion of this first period.

It would be quite possible, in the succeeding part of the term day also, to mark off several divisions; but such a marking-off would scarcely be advisable, as the perturbation-conditions, as a whole, are all the time undergoing a more or less continuous change.

Here, as in the preceding section, the perturbations admit of being arranged in two groups. On the one side we have a negative polar storm, on the other side a positive.

We will first consider the negative storm. This occurs, as will be seen from the plates, in the district about Kingua Fjord, Fort Rae, Uglaamie and Ssagastyr, and furthermore at Cape Thordsen and Fort Conger. In the preceding section, however, the storm-centre was in the vicinity of Fort Rae; and now the perturbing forces there are considerable weaker than at the other stations.

The storm-centre thus seems to have moved. In the first part of this last section, the most powerful perturbing forces seem to be concentrated upon the districts about Uglaamie and Ssagastyr; but this condition is not very apparent, as the forces round the auroral zone at these stations rarely vary much in magnitude.

Later however — at about 20^h or 21^h — there is a distinctly defined storm-centre at Cape Thordsen. At the other stations, where the negative storm occurred before, the perturbations at this hour are practically over.

It thus seems as if we here had a distinct westward movement of the negative storm-areas.

There next occurs, as already mentioned, a positive polar storm, but in a much more limited area than the negative, judging at any rate by the stations from which we have observations.

We stated in the preceding section, that at the conclusion in Jan Mayen, a positive polar storm began. In the present section, this positive storm developes greatly, and forms a system of precipitation, which at first extends from Godthaab eastwards to the regions near Little Karmakul, but is afterwards concentrated more upon the regions about Bossekop.

These are conditions which are immediately apparent from the curves. Judging from the deflections in the horizontal-intensity curves for Jan Mayen and Bossekop, it would appear that the storm-centre during this period, after lying in the vicinity of Jan Mayen while the storm is comparatively less powerful, has subsequently moved eastwards to Bossekop, the storm, at the same time, attaining its greatest strength. Whether the conditions do actually develope in this way, it is impossible to determine merely by the aid of the observations from these two stations, seeing that magnetically considered, Jan Mayen lies considerably farther north than Bossekop. Observations from the southern border of the auroral zone would here have been of great importance.

The great difference in the effects of the force at Bossekop and Sodankylä is characteristic. At the latter station the forces are on an average only about one quarter of those at Bossekop. As these stations are situated very near to one another, it may be concluded that the acting systems come fairly close to the last-named.

The conditions at Little Karmakul during this period are particularly interesting and peculiar. This station is situated, as will be seen, upon the boundary between the two districts of precipitation; to the east and north we come upon the negative polar storm, to the west there is the positive. It would therefore be natural to suppose that at this boundary-station, both these systems would act; and this proves to be actually the case.

In both the areas of precipitation, the positive as well as the negative, the deflections in horizontal intensity continue to be in one direction as long as the storm lasts. At Little Karmakul, on the other hand, the conditions are different; at one time there are wide deflections in the positive direction, at another wide deflections in the negative direction, and again smaller deflections up and down about the normal line. It thus appears from a direct consideration of the curves, that we now have a direct effect of the positive system, and then of the negative, and now again the two systems neutralise one another's effect.

Altogether analogous, although less marked, conditions are found in Jan Mayen, where at first the positive system acts almost exclusively, then mainly the negative, but only in a series of brief impulses, after which the horizontal-intensity curve returns once more to its normal height. As regards declination the conditions are somewhat similar; but there it is not possible to determine so directly which system it is that is acting at the various times.

At about 23^h, the perturbations are ended at almost all the stations, and after that time it is only at two or three places that perturbing forces of any special magnitude appear, and these should probably be regarded as more local.

Six charts have been drawn up for this period, representing in all 17 epochs, by means of which the course of the perturbations may be followed from hour to hour.

Similar fields in the main are found upon the various charts, only displaced to some extent from time to time.

Chart III; time 14^h 20^m, 15^h 20^m and 16^h 20^m.

At the first-named hour there are more or less powerful forces only in the district about Jan Mayen, Cape Thordsen and Ssagastyr; and the current-arrows there are directed eastwards. It is impossible to decide from the charts whether this is a connected system or not. The curves seem to indicate, however, that it can scarcely be an entirely connected system.

Nor has the perturbation developed any special power at 15^h 20^m; and at Ssagastyr, and Cape Thordsen, the earlier perturbing forces have almost entirely disappeared. In Jan Mayen only is there still the effect of the positive system.

It may even now be worth while to notice the conditions at Godthaab and Kingua Fjord. At these two stations we now have arrows that point in exactly the opposite direction; at one place a positive storm is evidently acting, at the other a negative, and it would thus seem as if the boundary between two such tracts just chanced to be between the two stations. This is a condition with which we shall subsequently frequently meet, and which we therefore at once point out.

We thus again meet a peculiarity in the state of things in Kingua Fjord, and further on we will have an opportunity of also coming in contact with other cases diverging somewhat from what we find at the other stations. It might therefore be well to examine here at once what might be supposed to be the natural reason.

When we have hitherto considered the polar storms, the conditions of the horizontal intensity have always been of the greatest importance, as the direction of the current-arrows was either pointing eastwards or westwards.

This is, however, not always the case as regards Kingua Fjord; on the contrary, it is in the declinations that the strongest forces frequently are shown, and the direction of the current-arrows is very frequently pointed pretty nearly due south.

These somewhat peculiar conditions are surely connected with the northerly situation, as regards magnetic conditions, of this station compared to the others with the exception of Fort Conger.

We will here refer to the terrella experiments, which will be more fully dealt with in a subsequent chapter. In order to elucidate the subject, we will however here give a copy of a photograph, Fig. 140.

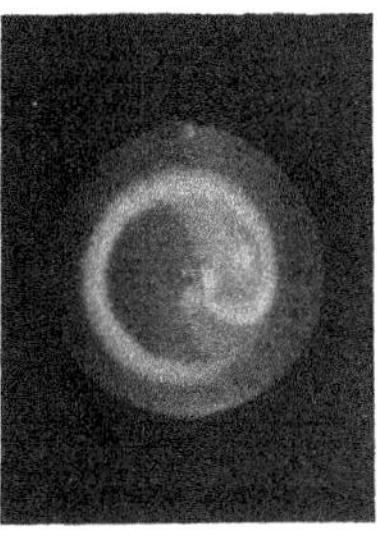
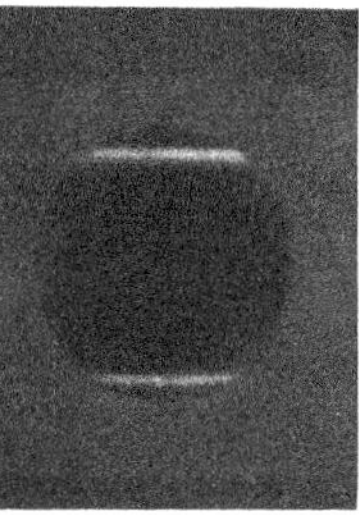
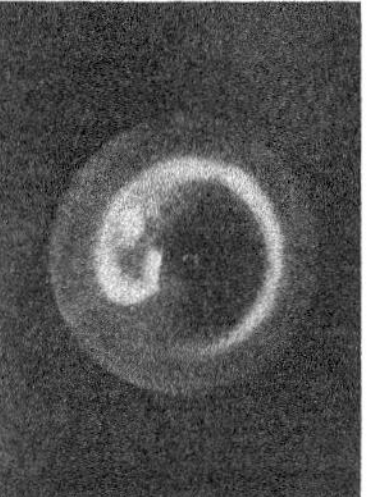

Fig. 140.

In most of the illustrations hitherto given, the terrella has been suspended on an axis, the position of which has corresponded with that of the earth, thus forming an angle with the terrellas magnetic axis of about 20°.

As this however gave a less easily seen representation of the entire polar area of precipitation, the terrella is here suspended on an axis in the magnetic equatorial plane. The position of the electrode can be thus altered as desired by turning the terrella on the axis on which it hangs and thus produce some positions which should correspond to various positions of the earth in relation to the sun.

In the experiment corresponding with the three above given photographs, the cathode is placed in the magnetic equator of the terrella and thus answers to the times when the direction from the earth to the sun is perpendicular to the magnetic axis of the earth.

On the first figure, the camera is pointed directly on the south pole of the terrella magnet, the position of which on the plate is marked with a cross. The figures of light we here see represented, should therefore correspond to the areas of precipitation which we would expect to find round the earth magnetic north pole, or, more accurately expressed, about the intersecting point of the magnetic axis with the northern hemisphere. The other picture is taken with the axis of the camera parallel to the cathode-rays' direction of issue, so that the conditions should represent the areas of precipitation we find on the night side of the earth. The third picture is meant to show the conditions around the earth magnetic south pole, the photograph being taken directly towards the terrella magnet's north pole. The position of this is also marked on the plate.

As will be seen from the picture, the areas of precipitation form a distinctly spirally shaped belt, winding upwards towards the magnetic pole.

The upper part of this spiral belt always appears sharply and clearly defined, sometimes as a more isolated patch, sometimes, as in this instance, this patch appears in connection with an elongated adherent polar belt. The patch comes out very plainly in the first and third plates, as an oval shaped figure of light within the long spiral belt. This patch does not alter its place much for different positions of the terrella in relation to the cathode, and it exists under all degrees of stiffnesses of the cathode rays. The remainder of the polar belt is, on the other hand, more variable in its formation. According as the magnetic and electric conditions are altered, this belt undergoes severe changes. At times the whole is continuous, as on the plate here, at other times several well defined figures of light can be found, and at times the whole can almost disappear. As regards further details, we must, however, here confine ourselves to referring to a subsequent chapter, in which the terrella experiments are described and in which the tangential direction of the rays nearest the earth in various parts of the area of precipitation are examined. As will be found there, we have also further succeeded in showing that the cathode rays, close to the terrella, are bent in a manner which in the main features exhibits the most complete analogy to the characteristic systems of precipitation on the earth which we constantly meet. By fixing screens at suitable places, it has likewise been possible to show that the rays which precipitate themselves in the luminous polar belts on that side which corresponds with the afternoon side in the vicinity of the terrella will be bent off towards the west—and thus corresponding rays will have magnetic actions on the earth as a current towards the east—while the other rays, especially on the night side, will be bent in the opposite direction, i. e. towards the east; to the north and south we must then imagine the direction respectively to the south and north poles of the terrella magnet. We thus find a clearly evident analogy between the actual conditions and the experiments.

The analogous system of corpuscular rays, which we imagine around the earth, will thus, by the rotation of the earth, in the course of a day be moved round, at the same time its shape will be somewhat changed owing to the sun's altered height above the magnetic equator. The only part which never disappears is the marked patch near the axis.

If we now assume that Kingua Fjord is situated just at that part of the earth where the system of precipitation corresponding to this patch is passing, we seem to get a natural explanation of the peculiar phenomena we observe here.

(1) In the afternoon, Greenwich time, which would be noon and afternoon local time, strong variations in the magnetic elements constantly occur; this corresponds with the light patch always being visible, and thus every day the corresponding system will pass the spot.

(2) That the direction of the current-arrows is frequently pointing southwards, agrees with the luminous belt in the innermost portion nearest the pole swinging strongly northwards or southwards.

(3) During a later perturbation, 15th December 1882, we find at Kingua Fjord for a prolonged period polar precipitations, while none such made themselves distinctly noticeable at the other stations. This accords with the system corresponding with the luminous patch also occurring simultaneously with the equatorial ring — compare fig. 37 Part I.

At the third hour given on Chart III, $16^h\ 20^m$, perturbations of no inconsiderable magnitude have developed at all the stations.

At Ssagastyr and Cape Thordsen, a negative polar storm is now distinctly acting, a storm that is also continued round the geographical pole to Fort Conger, Kingua Fjord, Fort Rae and Uglaamie.

On the afternoon side, moreover, south of this negative system, we have the effects of a positive system in the district embracing Godthaab, Jan Mayen and Bossekop. Little Karmakul is situated, as we see, upon the boundary between these two regions, and at the hour in question has a current-arrow directed southwards, which may be interpreted as a resultant of the effects of these two systems.

The sun has now almost reached the meridian of the magnetic axis.

We will now consider the further course of the perturbation upon the succeeding charts.

Chart IV also represents three epochs, namely, $16^h\ 50^m$, $17^h\ 20^m$ and $17^h\ 40^m$.

The fields on this chart have, in the main, exactly the same appearance, the only difference being that the strength of the perturbing forces at the various stations has undergone certain alterations.

The positive storm now appears at first only at Bossekop, and then in the district about Bossekop and Little Karmakul. The perturbing forces there are now very considerable, and at the same time the forces arrange themselves at the southern stations in a manner that accords very well with what, from our previous investigations in Part I, we should expect to find. This, at any rate, is the case at the nearest stations, Sodankylä, Pawlowsk and Christiania.

Between Bossekop and Sodankylä the forces diminish greatly, in accordance with the fact that the point of divergence is being approached. At Pawlowsk this point has been passed, and the direction of the current-arrow is the reverse of that at the two stations just mentioned. The forces at Christiania are also what they would be if there were an area of divergence in that region; and at Göttingen also, the accordance is in a measure satisfactory.

We have seen that the perturbing forces during this period first appeared with considerable power at Bossekop, and then at Little Karmakul. Whether this is a displacement of the positive system, or only owing to an increase in the size of the area of precipitation, is a question about which there may be some doubt. If we look, however, at the area at the stations situated a little farther south, the probability seems to be in favour of the first alternative. Unfortunately we have no observations from the district in, or south of, the auroral zone west of Norway; where there would undoubtedly have been marked effects of the positive system of precipitation, which would have been of some assistance in studying it. We must thus, in employing the more southerly stations, once more make use of the same method of procedure as in Part I. In the present case, however, we have a station, of which the situation in this connection is of no small interest, and which was wanting in the former observations, namely Christiania. This station, in connection with Pawlowsk, will be, as we shall see, of much service in finding a kind of limit for the positive area of precipitation.

At $16^h\ 50^m$ the arrow at Pawlowsk shows that this station is now in the eastern part of the area of divergence, while Christiania at that time is probably not far off the transverse axis of the system.

At $17^h\ 40^m$ Pawlowsk is in the vicinity of the transverse axis, while Christiania is then evidently in the western portion of the area.

These circumstances thus appear to indicate that this is rather a movement of the system, than an increase in the size of the precipitation-area of a system which does not change its position much.

The conditions at Göttingen also to some extent agree fairly well with the above, although the direction of the arrows there seems perhaps to be a little too southerly.

The conditions in Jan Mayen are rather interesting too. They show that the positive system there must lie to the south of the station. The inconsiderable forces occurring in a horizontal direction may be naturally explained by assuming that the negative system to the north, and the positive system to the south, neutralise one another's effect in a horizontal direction, but on the other hand act together in a vertical direction, so that the aggregate effects are all the greater.

As regards the vertical intensities at the other stations in the positive polar area, the conditions at Bossekop show that the area of precipitation must be looked for somewhat to the north of that station. This has also been the case in most of the previous instances of similar storms in Part I (see perturbations of 11th, 23rd and 31st October, and 9th December, 1902, and 8th and 15th February, 1903). At

Sodankylä, on the other hand, we find negative precipitation in the vertical intensity, that is to say a direction the very reverse of that which one would have expected. The easiest explanation of the circumstance — but hardly a permissible one—is, that an error has found its way in, either as a consequence of a fault in the apparatus, an error in observation, or an error in calculation; for there seems to be no local current-system at work here. Earth-currents might possibly be supposed to exert a considerable influence, but scarcely as much as in the present instance.

Conditions, however, are found at this place which may be capable of accounting for these discrepancies; we have just recently ascertained that in the regions round Sodankylä, there are enormous ironfields, the ore of which possesses magnetic properties of extraordinary strength.

This could affect the perturbing forces in vertical intensity especially, if we imagine the magnetic masses distributed in a horizontal layer. It would be easy to imagine a distribution of magnetic masses which, by means of induction, might be supposed to occasion anomalies such as these which we find here.

If we, for instance, imagine the station to be situated immediately above the one end of a horizontal magnetic shaft, then the horizontal forces in the neighborhood could be expected to induce free magnetism at the ends of this shaft, and that again would be able to produce strong effects in vertical intensity in a station situated directly above.

At Godthaab we now have no particularly noticeable effect of the positive system. The perturbing forces are of inconsiderable magnitude.

At the other stations, as will be clearly seen, negative storms are acting, which, during the three epochs here represented, remain more or less unchanged both in form and strength. Fort Conger evidently follows closely upon this series of stations, there being a westerly-directed current-arrow there of a strength similar to that at the other stations.

From the values of P_v to be found at the various stations, a few details may be concluded as to the situation of the current-system. At Fort Rae and Uglaamie, we see that the negative precipitation takes place north of the former place and south of the latter, and thus, probably more or less in the auroral zone, which just comes between these two stations.

In connection with this, we should remember the meaning of the two curves drawn, which show the position of the belt of Northern light. The more southerly, shows the places where aurora is most frequently observed. The more northerly, connects points where aurora is seen as frequently in the south as in the north.

At Cape Thordsen, we also have small negative values of P_v. We must not however, conclude directly from this, that the negative precipitation takes place north of that place, as to the south of it there is the positive polar system, which will here just produce negative values of P_v. It would therefore be a fairly probable assumption that the negative precipitation occurred a little to the south of, or possibly more or less directly over, the place. If the area of precipitation were to the north of the station, the perturbing forces in the vertical intensity would probably be greater than we here find them to be, as the two systems would then cause vertical forces directed in the same direction. In all probability, this is the case on Jan Mayen; and we also find powerful perturbations in the vertical intensity.

Chart V, $18^h\ 25^m$, $19^h\ 5^m$, $19^h\ 25^m$. The sun is now in the vicinity of the meridian of the magnetic pole, which it crosses in this period.

Here, too, we find the same areas of perturbation as before. The negative storm has now concentrated itself more upon the night-side of the globe. In the district Cape Thordsen, Jan Mayen and Kingua Fjord, however, there are quite distinct effects of a negative system which is acting there. The area of perturbation here, however, is not so well defined as before. The positive system is distinctly noticed at Bossekop, and at $19^h\ 5^m$ at Little Karmakul too. This chart also shows with extreme clearness at this station, how the two systems encroach upon one another. At $18^h\ 25^m$ they almost entirely neutralise one another's effect, at $19^h\ 5^m$ there is a strong effect of the positive system, and at $19^h\ 25^m$ a strong effect of the negative system.

The current-arrows at Pawlowsk and Christiania now seem to indicate, that this positive system does not extend so far westwards.

It is interesting to follow the movement of the arrow at Pawlowsk from $19^h\ 5^m$ to $19^h\ 25^m$, that is to say, at the time the negative system is extending its area of precipitation westwards to Little Karmakul. The arrow at Pawlowsk moves with it. Thus, at $19^h\ 5^m$ the current-arrow indicates that the station is more or less in the middle of an area of divergence somewhat to the west of the transverse axis, so that we then have principally the effects of the positive system. At $19^h\ 25^m$, on the other hand, the current-arrow shows that the station is either in the east part of an area of divergence, or in the west part of an area of convergence. This, then, indicates, that we here have either the effects of the westerly positive system that we find in the neighbourhood of Bossekop, or those of the negative system extending eastwards from Little Karmakul. It is probable, however, that both of these will exert an influence, and that the current-arrow must be regarded as the result of their united action.

The conditions here, are thus evidently governed by the polar systems, just as we supposed in Part I.

The direction of the deflections in the vertical intensity, are now, on the whole, the same as in the preceding chart. We still find the same disagreement between Bossekop and Sodankylä; and at Pawlowsk $P_z = 0$, just as in the preceding chart. There is, howewer, a slight deviation in the curve, corresponding to positive values of P_z, which are too small to allow of being taken out.

On *Chart VI and VII,* the conditions develope farther in the same direction, inasmuch as the areas of precipitation are now concentrated more on the night-side of the earth, if we may judge by the observations at our disposal. At the other polar stations, however, there are still, on the whole, more or less distinct, westerly-directed current-arrows.

It is very possible, however, that a little farther south there may be areas of precipitation that cannot be observed here. The rather abnormal current-arrows at Fort Rae, which is situated south of the auroral zone, might, perhaps, indicate something of the sort. On Chart VI too, Göttingen and Christiania seem to be situated in the eastern part of an area of divergence, and thus indicate the existence of a positive system of precipitation.

We notice such a system at Bossekop and Sodankylä, and we should therefore have to suppose that this system extended westwards along the auroral zone, and probably south of it, or into its southernmost part, so that its effect at the stations from which we have observations, and which are situated to the north of it, are not affected in any great degree by it.

On Chart VII, the negative polar system in the north of Europe seems to have got the upper hand and to be also governing the conditions in the stations in the south of Europe. As regards Christiania and Göttingen, however, a positive polar system such as that we assumed to exist on Chart VI, will also act in more or less the same direction. At Bossekop, up to $21^h\ 5^m$ on Chart VII, there are marked effects of a system such as this, although at the last hour shown, $21^h\ 20^m$, this storm is over there.

There is little to be said regarding the vertical intensities. At Fort Rae only, it may be remarked, that there is now and again a deflection in a positive direction. This is in a kind of accordance with the fact that the conditions of the current-arrows are also slightly different from those at the other neighbouring polar stations, which thus also seems to indicate that other perturbing forces are at work here.

On *Chart VIII,* for the hours $21^h\ 40^m$ and $22^h\ 40^m$, the powerful storms at the stations here under consideration, are over, although at several places there are sometimes quite powerful perturbing forces; but there is now no distinct impression of a coherent current-system.

TABLE LI.

The Perturbation on the 15th of January 1883.

Gr. M. T.	Uglaamie			Fort Rae			Kingua Fjord		Godthaab	
	P_h	P_d	P_v	P_h	P_d	P_v	P_h	P_d	P_h	P_d
h m										
11 20	+ 56 γ	W 13 γ	− 51 γ	−100 γ	E 28.5 γ	+ 35 γ	− 7 γ	W 9.5 γ	− 6 γ	W 3 γ
50	− 24 „	„ 90 „	− 82 „	−146 „	„ 50.5 „	+ 75 „	− 60 „	E 3 „	− 25 „	„ 24 „
12 20	−145 „	E 26 „	− 20 „	−218 „	„ 81 „	+ 85 „	− 62 „	W 11.5 „	− 23 „	„ 22 „
13 20	−252 „	„ 415 „	+ 41 „	−268 „	„ 155 „	+ 85 „	− 78 „	„ 31.5 „	− 40 „	„ 17 „
14 20	+ 11 „	„ 13 „	+ 31 „	− 38 „	„ 54 „	+ 15 „	− 19 „	E 22.5 „	+ 35 „	E 20 „
15 20	− 20 „	„ 37 „	+ 31 „	− 27 „	„ 32 „	− 5 „	− 35 „	W 67 „	+ 27 „	„ 20 „
16 20	− 41.5 „	„ 101 „	+ 31 „	− 53 „	„ 44 „	− 5 „	− 87 „	„ 84 „	+ 31 „	„ 18 „
50	− 80 „	„ 109 „	+ 92 „	−100 „	„ 83 „	− 25 „	− 60 „	„ 85 „	0	W 28 „
17 20	−154 „	„ 13 „	+102 „	−108 „	„ 82 „	− 55 „	− 57.5 „	„ 95 „	+ 7 „	„ 5.5 „
40	−168 „	„ 89 „	+ 61 „	− 99 „	„ 52 „	− 65 „	− 45 „	„ 91 „	+ 28 „	„ 3 „
18 25	−114 „	„ 115 „	+ 10 „	+ 3 „	„ 18 „	− 65 „	− 25 „	„ 109 „	+ 55 „	„ 10 „
19 5	−107 „	„ 149 „	0	+ 33 „	„ 1 „	− 35 „	− 12 „	„ 65 „	− 4 „	„ 34 „
25	− 71.5 „	„ 79 „	− 31 „	0	W 18 „	− 25 „	0	„ 44.5 „	+ 19 „	„ 22.5 „
40	− 61 „	„ 71 „	− 51 „	− 7 „	„ 22 „	− 15 „	+ 4 „	„ 48.5 „	+ 20 „	0
20 0	− 30 „	„ 3 „	− 61 „	− 26 „	„ 25 „	− 25 „	0	„ 57.5 „	+ 23 „	„ 3
20	− 24.5 „	W 3 „	− 82 „	+ 6 „	„ 20 „	+ 15 „	+ 9 „	„ 41.5 „	+ 8 „	„ 22.5 „
40	− 33.5 „	E 3 „	−112 „	− 38 „	„ 23.5 „	− 5 „	− 4 „	„ 48.5 „	− 8 „	„ 39 „
21 5	− 55 „	„ 8 „	−133 „	− 26 „	„ 10 „	+ 15 „	+ 4 „	„ 59 „	− 13 „	„ 53.5 „
20	− 19 „	„ 8 „	−112 „	− 22 „	„ 26 „	+ 5 „	0	„ 44.5 „	− 10 „	„ 45 „
40	− 27.5 „	W 24 „	−112 „	− 36 „	„ 32.5 „	− 5 „	− 12 „	„ 37	− 21 „	„ 39.5 „
22 20	+ 2 „	„ 40 „	− 71 „	+ 18 „	„ 24.5 „	+ 25 „	0	„ 5.5 „	+ 8 „	„ 5 „
23 15	− 5 „	„ 3 „	− 41 „	− 5 „	E 2.5 „	+ 5 „	0	„ 1 „	0	E 3 „

TABLE LI (continued).

Gr. M. T.	Jan Mayen			Bossekop			Sodankylä		
	P_h	P_d	P_v	P_h	P_d	P_v	P_h	P_d	P_v
h m									
11 20	− 1 γ	W 5.5 γ	− 5 γ	0	0	− 6 γ	0	0	− 4 γ
50	− 1 „	„ 14 „	0	0	E 3.5 γ	− 3 „	+ 3 γ	0	− 20 „
12 20	− 4 „	E 5.5 „	+ 7 „	+ 2 γ	„ 3 „	+ 2 „	+ 6 „	0	0
13 20	+ 24 „	„ 2 „	+ 4 „	+ 13 „	W 19.5 „	+ 15 „	+ 9 „	W 18.5 γ	− 10 „
14 20	+ 95 „	W 7.5 „	− 4 „	+ 16 „	0	+ 27 „	+ 8 „	„ 2 „	− 17 „
15 20	+ 64 „	„ 1.5 „	− 8 „	+ 10 „	E 3 „	+ 22 „	+ 3 „	0	− 14 „
16 20	+ 59 „	„ 18 „	− 83 „	+ 65 „	W 3.5 „	+ 72 „	+ 10 „	E 8.5 „	− 65 „
50	+ 18 „	„ 50 „	−110 „	+ 90 „	E 7 „	+118 „	+ 20 „	„ 23.5 „	− 74 „
17 20	+ 17 „	„ 3.5 „	−126 „	+171 „	„ 16.5 „	+164 „	+ 27 „	0	− 60 „
40	− 6 „	„ 3 „	−127 „	+185 „	W 32 „	+192 „	+ 24 „	W 4.5 „	− 92 „
18 25	+ 28 „	„ 31.5 „	−153 „	+126 „	„ 42.5 „	+118 „	+ 22 „	„ 15.5 „	− 72 „
19 5	−110 „	„ 17.5 „	−133 „	+157 „	„ 96 „	+110 „	+ 33 „	„ 33.5 „	− 10 „
25	− 14 „	„ 89 „	−150 „	+ 65 „	„ 32 „	+ 88 „	+ 28 „	0	− 86 „
40	− 11 „	E 7.5 „	−143 „	+128 „	„ 35.5 „	+125 „	+ 52 „	E 4.5 „	− 70 „
20 0	− 52 „	W 74 „	−167 „	+ 76 „	„ 46 „	+ 49 „	+ 25 „	W 16.5 „	− 38 „
20	+ 5 „	„ 68 „	−156 „	+ 65 „	„ 12.5 „	+ 54 „	+ 30 „	E 6 „	− 45 „
40	− 13 „	„ 14 „	−160 „	+105 „	„ 9 „	+ 70 „	+ 26 „	W 13.5 „	+ 11 „
21 5	− 78 „	E 75 „	−154 „	+ 92 „	„ 22 „	+ 44 „	+ 28 „	E 2.5 „	− 13 „
20	− 38 „	W 31 „	−164 „	− 5 „	E 24.5 „	− 82 „	+ 21 „	W 4 „	+ 22 „
40	− 46 „	„ 156 „	−156 „	− 43 „	W 33.5 „	−148 „	+ 3 „	„ 7 „	+ 34 „
22 20	+ 47 „	„ 54 „	−106 „	0	„ 3.5 „	− 66 „	+ 12 „	„ 6.5 „	+ 18 „
23 15	+ 2 „	„ 9 „	− 51 „	0	„ 7 „	− 34 „	− 2 „	„ 8 „	− 7 „

TABLE LI (continued).

Gr. M. T.	Cape Thordsen			Little Karmakul			Ssagastyr	
	P_h	P_d	P_v	P_h	P_d	P_v	P_h	P_d
h m								
11 20	+ 19 γ	W 1.5 γ	+ 11 γ	+ 17 γ	E 4 γ	0	+ 3 γ	E 15 γ
50	+ 59 „	E 18 „	− 9 „	+ 14 „	0	+ 17 γ	+ 58 „	„ 9 „
12 20	+ 3 „	„ 28.5 „	− 7 „	+ 60 „	W 9 „	+ 35 „	+107 „	W 16 „
13 20	− 52 „	„ 25.5 „	− 38 „	+143 „	„ 63 „	+ 84 „	+138 „	„ 54 „
14 20	+ 54 „	W 4 „	− 57 „	− 4 „	E 5.5 „	+ 41 „	+ 75 „	„ 12 „
15 20	+ 16 „	„ 6 „	− 31 „	+ 16 „	„ 2 „	+ 26 „	− 13 „	0
16 20	− 66 „	„ 15.5 „	− 47 „	+ 17 „	„ 68 „	+ 12 „	−288 „	„ 135 „
50	− 70 „	„ 20 „	− 25 „	0	„ 64 „	− 10 „	−203 „	„ 45 „
17 20	−129 „	„ 24.5 „	− 22 „	+261 „	W 41 „	+ 11 „	−113 „	0
40	−107 „	„ 28.5 „	− 20 „	+211 „	„ 50 „	+ 32 „	−107 „	E 6 „
18 25	− 56 „	„ 36.5 „	− 35 „	+ 16 „	„ 14.5 „	+ 31 „	−182 „	„ 29 „
19 5	− 26 „	„ 53.5 „	− 22 „	−202 „	E 34.5 „	− 6 „	−218 „	„ 41 „
25	− 22 „	„ 47.5 „	− 13 „	+135 „	W 11.5 „	+ 20 „	− 40 „	„ 46 „
40	− 95 „	„ 65 „	− 9 „	0	„ 2 „	− 35 „	− 56 „	„ 29 „
20 0	−161 „	„ 73 „	− 71 „	− 8 „	E 12.2 „	− 27 „	− 29 „	„ 32 „
20	−108 „	„ 23.5 „	− 39 „	− 27 „	„ 39.5 „	− 23 „	− 43 „	„ 44 „
40	−254 „	E 91 „	+ 51 „	− 28 „	W 35.5 „	− 30 „	− 51 „	„ 37 „
21 5	−257 „	W 108 „	−217 „	−279 „	„ 21 „	− 90 „	− 63 „	„ 49 „
20	−135 „	E 10 „	+ 22 „	−213 „	E 50 „	− 59 „	− 37 „	„ 32 „
40	−100 „	W 64.5 „	− 25 „	− 7 „	„ 12.5 „	− 45 „	+ 96 „	„ 41 „
22 20	0	„ 38.5 „	− 5 „	+ 74 „	W 11 „	− 40 „	+ 21 „	„ 39 „
23 15	+ 14 „	E 6 „	+ 27 „	+ 38 „	„ 1.5 „	+ 7 „	− 10 „	0

TABBLE LI (continued).

Gr. M. T.	Christiania		Pawlowsk		Göttingen			Fort Conger
	P_h	P_d	P_h	P_d	P_h	P_d	P_v	P_d
h m								
11 20	− 2 γ	W 3 γ	0	0	− 1 γ	E 4 γ	+ 4 γ	E 4.5 γ
50	− 3 „	„ 3.5 „	0	0	− 2 „	0	+ 3 „	W 7.5 „
12 20	0	„ 3 „	0	E 2.5 γ	− 3 „	„ 3 „	+ 4 „	„ 9 „
13 20	+ 2.5 „	„ 14.5 „	− 3 γ	W 9.5 „	− 1 „	W 8 „	+ 7 „	„ 45.5 „
14 20	+ 1.5 „	„ 4.5 „	0	0	+ 6 „	E 9 „	+ 11 „	„ 2 „
15 20	− 2 „	0	− 3 „	0	0	„ 10.5 „	+ 8 „	„ 10.5 „
16 20	− 7 „	„ 3 „	− 8 „	E 10.5 „	− 6 „	„ 8 „	+ 5 „	„ 58 „
50	− 11 „	0	− 10 „	„ 19 „	− 11.5 „	„ 12 „	+ 4.5 „	„ 73 „
17 20	− 9 „	„ 3 „	− 7 „	„ 3.5 „	− 10 „	„ 6.5 „	+ 7.5 „	„ 98 „
40	− 8 „	„ 12 „	− 10 „	0	− 8 „	W 2 „	+ 6 „	„ 66.5 „
18 25	0	„ 10 „	− 1 „	W 7 „	0	„ 2.5 „	+ 2 „	„ 57 „
19 5	+ 9.5 „	„ 1.5 „	− 7 „	„ 5 „	+ 2.5 „	E 8 „	0	„ 57 „
25	− 4.5 „	„ 7.5 „	− 5 „	E 9 „	− 4.5 „	0	+ 7.5 „	„ 38 „
40	− 1 „	E 5 „	+ 5 „	„ 9.5 „	− 5.5 „	„ 8.5 „	+ 5 „	„ 50 „
20 0	− 3 „	„ 4.5 „	+ 5 „	0	− 2 „	„ 14 „	+ 3 „	„ 54 „
20	− 3 „	„ 6 „	+ 2 „	„ 12 „	− 7 „	„ 12 „	+ 1.5 „	„ 58 „
40	− 4.5 „	„ 23.5 „	+ 5 „	„ 11 „	− 11.5 „	„ 28 „	+ 1 „	„ 51.5 „
21 5	+ 19 „	„ 28 „	+ 16 „	„ 5.5 „	0	„ 31.5 „	− 1.5 „	„ 56 „
20	+ 13 „	„ 30.5 „	+ 14 „	„ 13.5 „	0	„ 30.5 „	− 3.5 „	„ 30 „
40	+ 11 „	„ 14.5 „	+ 7 „	„ 1.5 „	+ 5.5 „	„ 18.5 „	0	„ 31.5 „
22 20	+ 4 „	„ 9.5 „	+ 11 „	0	+ 3 „	„ 8 „	− 3 „	E 11.5 „
23 15	− 7 „	W 3.5 „	− 6 „	W 3 „	− 4 „	„ 2.5 „	− 8 „	„ 42.5 „

Current-Arrows for the 15th January 1883.

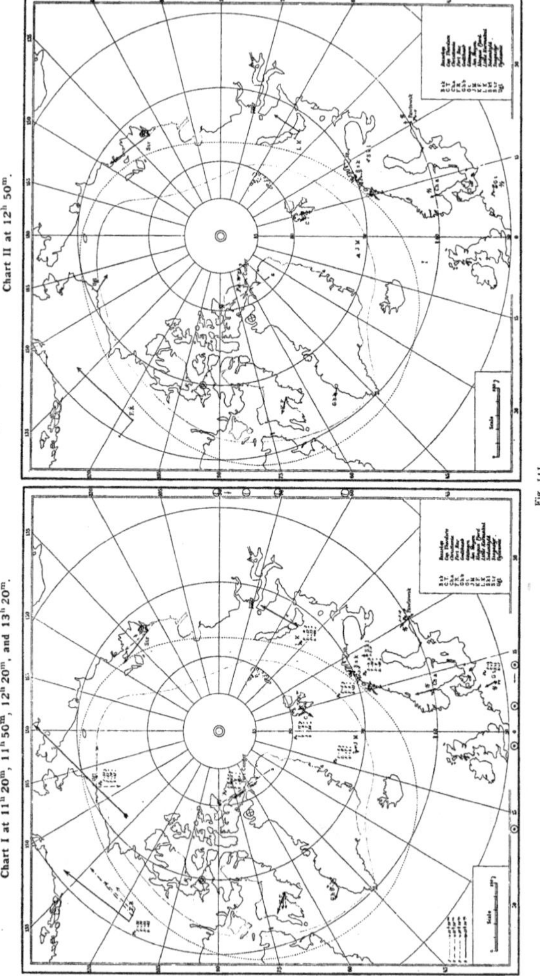

Chart I at $11^h 20^m$, $11^h 50^m$, $12^h 20^m$, and $13^h 20^m$.

Chart II at $12^h 50^m$.

Fig. 141.

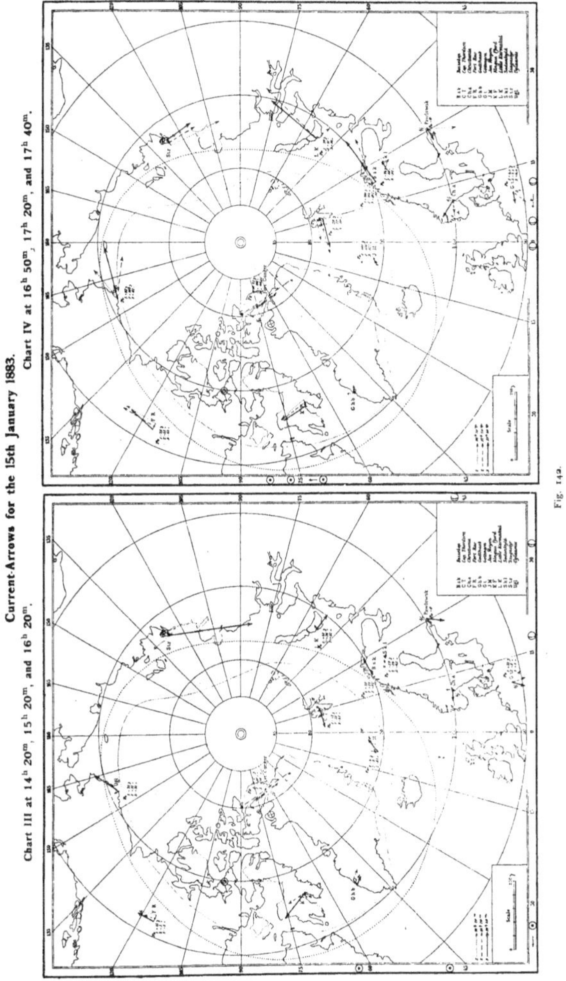

Current-Arrows for the 15th January 1883.

Chart III at $14^h\ 20^m$, $15^h\ 20^m$, and $16^h\ 20^m$.

Chart IV at $16^h\ 50^m$, $17^h\ 20^m$, and $17^h\ 40^m$.

Fig. 142.

Current-Arrows for the 15th January 1883.

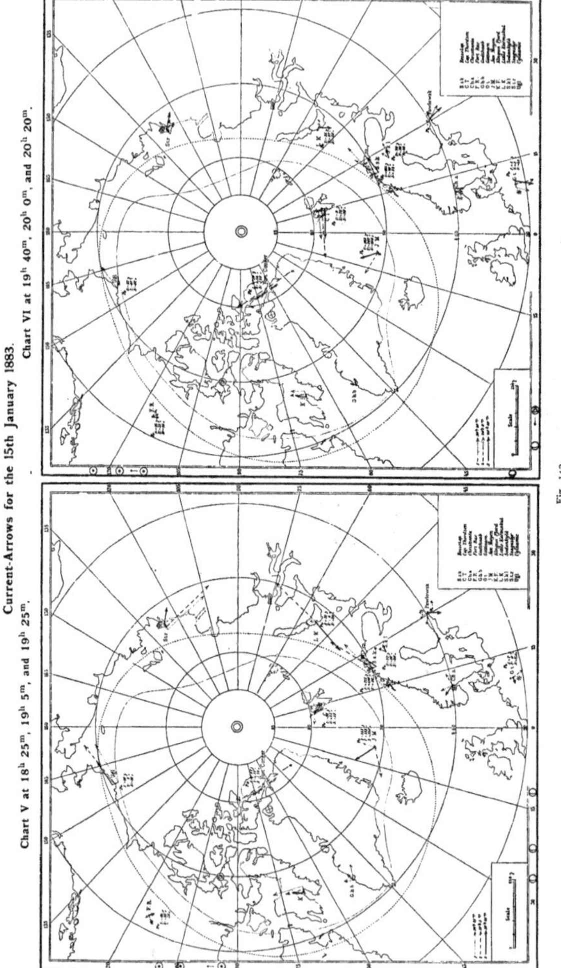

Chart V at $18^h\ 25^m$, $19^h\ 5^m$, and $19^h\ 25^m$.

Chart VI at $19^h\ 40^m$, $20^h\ 0^m$, and $20^h\ 20^m$.

Fig. 143.

Current-Arrows for the 15th January 1883.

Chart VII at $20^h\ 40^m$, $21^h\ 5^m$, and $21^h\ 20^m$.

Chart VIII at $21^h\ 40^m$, and $22^h\ 20^m$.

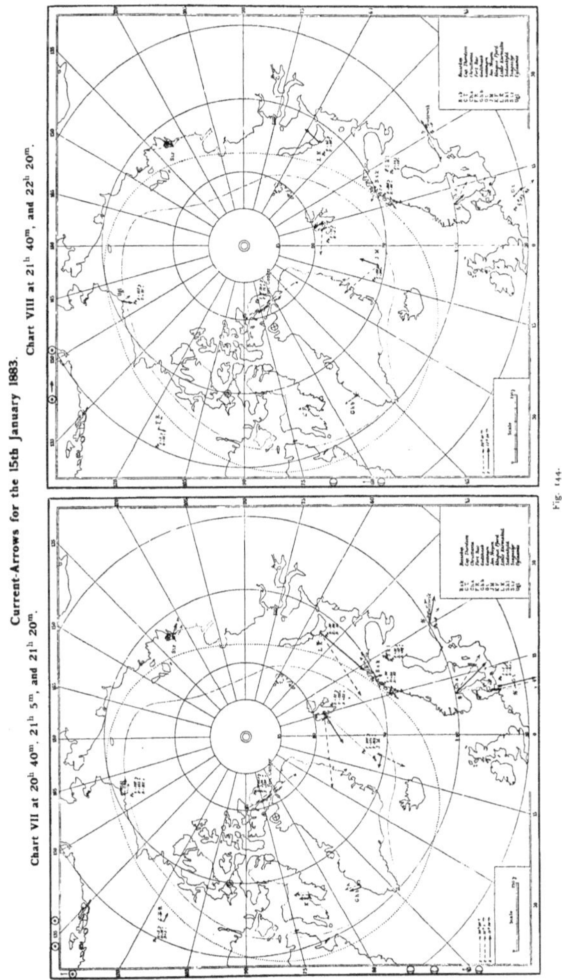

Fig. 144.

83. We may here draw a comparison with the areas of precipitation that may be calculated according to fig. 76, Part I, p. 160.

If θ is the angle that the sun's declination-circle makes with the meridian of the magnetic axis, δ the sun's declination, and $(90 - \varphi)$ the angle made by the magnetic axis with the earth's axis, i. e. $\varphi = 78^0\,20'$, we have

$$\sin\psi = \cos(\varphi - \delta) - 2\sin^2\frac{\theta}{2}\cos\delta\cos\varphi\,,$$

where ψ has the same meaning as in Art. 53 in Part I.

We will reckon the angle θ positive towards the west like Φ, θ thus standing for the time that has passed since the sun crossed the meridian of the magnetic axis.

The longitude of $\oplus$, as already stated, is $68^0\,49'$ W, so that the period during the perturbations under consideration here, namely 10^h—$23^h\,20^m$ G. M. T. corresponds to values of θ lying between about

$$-100^0 < \theta < +100^0$$

which answers to about

$$-22.5^0 < \psi < -9.5^0.$$

Thus ψ first increases from -22.5° to -9.5°, and then decreases from -9.5° to -22.5°.

We will now see from these calculations what areas of precipitation we should expect to find.

In making such comparison, we do not mean that the areas of precipitation we find by calculation should exactly correspond with the various storm centres which occur during the perturbations. The areas of precipitation found by calculation, are those in which the rays fall perpendiculary on the surface of the Earth, what are actually calculated are rays which go to the origin, where the assumed elementary magnet is situated. The regions that just correspond with these, must, in my opinion, best be compared with the places where aurora occurs, but these do not always correspond with the storm-centres of the magnetic disturbances. But we might, however, expect to find analogies and we will therefore proceed here briefly to make such comparison.

We will first consider the negative rays. For $\psi = -22.5^\circ$ we find, as fig. 76 shows, no precipitation, but as soon as ever ψ has increased a couple of degrees, an area of precipitation appears on the afternoon-side, at first spreading with considerable rapidity east and west, and subsequently dividing more into two systems, one of which moves towards the morning-side and the other towards the evening-side, as the sun approaches the meridian of the magnetic axis.

Shortly after the formation of the first area of precipitation, a new one is formed upon the morning-side, which also, as the sun rises higher, divides into two parts, one of which moves towards the night-side of the earth, the other towards the morning-side. There will moreover be areas of precipitation answering to rays that have passed round the earth before their descent, and correspond to values of $|\Phi|$ that are greater than 360°. These are not taken into consideration here.

For positive rays we find more or less the same values of Φ for the first two areas of precipitation.

After the sun has crossed the meridian of the magnetic axis, it might be supposed that the phenomena would be repeated in the reverse order, but with the whole area moved westwards. We will now see whether analogies to these conditions are actually found.

At first, then, we should expect to find two areas of precipitation, one on the afternoon-side, and one on the morning-side.

This agrees exceedingly well with what we found in the first section, where we pointed out the two areas in which the storm was concentrated. One of these, the negative, appeared on the morning and night side from Kingua Fjord to Fort Rae and Uglaamie, beginning slightly earlier at Fort Rae than at the other two stations. The other, the positive, occurred on the afternoon and evening side, from Little Karmakul to Ssagastyr. Here then there appear to be distinct analogies.

Afterwards, four areas of precipitation should be found, distributed over the polar regions. Owing to the scarcity of stations, it is of course difficult, if not impossible, to prove any agreement in detail. We will only point out that on Chart III, the perturbing forces are distributed more or less evenly about the auroral zone.

At the conclusion the negative storms are concentrated upon the night and morning side, perhaps moved a little more towards the night-side than one would expect. On the afternoon-side there are no particularly powerful areas of precipitation, but we have no observations either, from the regions south of the auroral zone.

While speaking of the repetition of the phenomena in reverse order after the sun has crossed the meridian of the magnetic axis, we will draw attention to the two deflections in the horizontal-intensity curve at Uglaamie, which seem distinctly to be almost a repetition of the same phenomenon. The second phenomenon does not, it is true, occur when the sun is exactly as far west of the meridian of the magnetic axis as it was east in the first, but only approximately so.

If this phenomenon is to be explained in this manner, it must be assumed that as, at the first deflection, the station lay to the west of the storm-centre, and as the strength of the deflections is more or less the same, at the second deflection the station must be almost equally far to the east of the storm-centre; and it is very probable that this is the case.

Similar remarks may also be made with reference to Fort Rae.

THE PERTURBATIONS OF THE 2nd JANUARY, 1883.

(Pl. XXV.)

84. The perturbation-conditions on the above day exhibit in many respects a great resemblance to the conditions during the preceding perturbation of the 15th January, 1883. This is at once evident on comparing the plates for these two days.

The period of this day which we shall discuss is from 11^h to the conclusion of the day, $23^h\ 20^m$, Gr. M. T.

During this period there occur, as on the 15th January, a series of powerful, well-defined storms, while for some time previously, it had been more or less calm.

On this occasion also, the perturbations occurring may be divided fairly distinctly into two sections, namely, a first section from 11^h to 16^h, and a second section from 16^h to $23^h\ 20^m$.

The first section is mainly characterised by the powerful negative storms that appear in North America.

At Fort Rae, there is a considerable and well-defined deflection in the horizontal-intensity curve, with a corresponding deflection in the declination curve. The deflections increase at first fairly evenly from $11^h\ 30^m$. We find the most powerful perturbing forces at about 14^h; after which the forces decrease, until about $15^h\ 30^m$, when the conditions are again more or less normal.

At Uglaamie, the conditions are somewhat more complicated. At a little before 12^h, wide deflections suddenly occur in the magnetic curves. In the horizontal intensity, they are in a negative direction, and the curve has a very jagged appearance. At about 13^h, however, they decrease, and for a time the curve oscillates over and under the normal line. In the declination, on the other hand, the deflections at this hour are very considerable, showing the presence of powerful perturbing forces, which are evidently acting in the neighbourhood of this station.

Later on there are again considerable negative deflections in the horizontal-intensity curve, these deflections now being very well-defined without any sharply projecting points. They continue to the end of the first section, the conditions at about $15^h\ 45^m$ being once more normal.

A third station, which ought to be mentioned in connection with these two, is Kingua Fjord; for these three stations together form a more or less distinct group, as a negative polar storm is now acting in this district. We have considered the effect of this storm at the two preceding stations, and we found that at the conclusion of this first section, the storm there was over. This is not the case, however, in Kingua Fjord, where the storm continues without cessation into the next section, although for a short time about $16^h\ 10^m$, the perturbing forces are very small.

At the time when the curves at Fort Rae and Uglaamie have their maximal deflection, a distinct maximum is also to be found in Kingua Fjord; but the perturbing forces there are considerably weaker.

It appears, upon the whole, as if the storm-centre must be situated in the district Fort Rae—Uglaamie, at first probably nearest to the former; at the conclusion however we find the strongest effects at Uglaamie.

It is not impossible, therefore, that we have before us a displacement, in a westerly direction, of the area of precipiation; but the conditions are probably more complicated.

In these districts then, a negative system of precipitation is acting.

If we now examine the other curves in this first period, we find at Little Karmakul and Ssagastyr quite distinct, although comparatively slight, effects of a positive system of precipitation. At Cape Thordsen there are also positive deflections in the horizontal-intensity curve at first; but at the time when the negative storm at the American stations is at its height, the curves seem to show that here too there is a negative polar storm which counteracts the effect of the positive, and makes the curve oscillate to the opposite side. The conditions in the declination and vertical intensity also indicate something similar; for at the time when the negative storm here should begin, we find distinct deflections in these two elements, lasting about as long as the negative storm seems to be acting.

In the district Godthaab to Jan Mayen, there is also a positive storm which continues into the next section, and there attains considerably greater strength.

We thus find in this perturbation also, the characteristic systems of precipitation, a negative and a positive, of which the first is fairly powerful and very pronounced, while the second is comparatively slight.

We may now at once look at the first four charts, which represent the perturbation-conditions during this first section.

Chart I shows the conditions at $13^h\ 20^m$, that is to say at a time when the negative storm at Fort Rae has about reached its height. For the time before this, in which, as already mentioned, there are fairly powerful forces at Uglaamie, while those at the other stations were comparatively small, no charts have been drawn, as the condition is clearly apparent from the curves.

The current-arrow at Uglaamie is now directed NNE, and thus indicates that the conditions are somewhat different from those that are usual in the auroral zone during the polar storms in which the current-arrow is directed either westwards or eastwards. In order to explain this condition, it might be assumed, as has previously been done, that there was here a co-operation between a positive and a negative polar storm.

In the district Kingua Fjord and Fort Rae, there are distinct effects of a negative polar storm, while at the other stations the perturbing forces are very small.

On the next charts, *Charts II—IV*, for the hours $14^h\ 5^m$, $14^h\ 20^m$, 15^h and $15^h\ 20^m$, the conditions are but little changed in the main. Now too we find a distinct negative polar system in the north of America; and in the district Godthaab eastwards to Ssagastyr, there occur more or less distinct traces of a positive system. This is most cleary apparent on Chart III, for $14^h\ 20^m$ and on Chart IV at 15^h. At the latter hour we notice especially strong effects of this system at Ssagastyr. At Cape Thordsen, on the other hand, we find at $14^h\ 5^m$ a distinct westward-pointing current-arrow, which should indicate that we had before us the effect of a negative system of precipitation, which is, indeed, in accordance with what we have already noticed when considering the condition of the curves.

After this first section, there supervenes, at most of the stations, a brief period of fairly quiet conditions. The only exceptions to this are the stations Kingua Fjord, Godthaab and Jan Mayen, where there are now quite distinct oscillations. At Cape Thordsen too, there are distinct oscillations in the declination, but the perturbing forces are very small.

This intermediate period of time, commences at about 16^h, that is to say at about the time when the sun crosses the meridian of the magnetic axis.

Fairly powerful storms, however, soon develope at all the stations from which we have observations, some of them appearing as negative polar storms, and some as positive.

The perturbations in this last section also exhibit in the main exactly the same conditions as the preceding perturbation of the 15th January.

Exclusively negative storms appear, as we see, at the stations Kingua Fjord, Fort Rae, Uglaamie and Cape Thordsen. At Godthaab, Bossekop and Sodankylä there are almost exclusively positive storms; but these have not so distinctly the character of a positive storm, as the course of the curve is fairly quiet, and the perturbing forces are comparatively small. In the declination, moreover, there are perturbing forces that exceed in magnitude the values of P_h.

Little Karmakul is now, as also in the preceding storm, situated just on the boundary between the two areas of precipitation. On the east and north of the station are the negative storms, on the west the positive. In consequence of this, the conditions here become rather peculiar, as sometimes the negative system, sometimes the positive, exerts the strongest influence, and the horizontal-intensity curve accordingly oscillates now to the one side, and now to the other.

This condition comes out very characteristically here in this period.

In Jan Mayen also, we find similar conditions. There we evidently have a negative storm, which, during the period from 17^h to 19^h, breaks in upon a positive storm. The latter is of much longer duration than the former, but of comparatively smaller strength; and therefore, when the negative storm breaks in, it will gain the ascendancy and cause the deflections in the horizontal-intensity curve to go to the negative side. In the declination also, at about the same time, there is a corresponding change in the direction of the deflections.

From about $18^h\ 30^m$ to 20^h, there are once more positive deflections, but then the curve changes again, and from the last-named hour until the close of the period, we find once more negative values of P_h. It is not easy to say, merely from a direct consideration of the curves, whether, at the close of the period, positive storms are also exerting an influence here.

At Bossekop and Sodankylä the positive deflections are only slight, and the character of the curves is fairly quiet. It might therefore possibly be assumed that the deflections were the effect of the negative system, whose area of convergence was situated to the north of these stations. Such an assumption, however, cannot at any rate be applied to the conditions in Jan Mayen, at Little Karmakul or at Godthaab, as the positive deflections there are far too considerable in amplitude.

If we endeavour to fix the position of the centres of these storms from the intensity of the deflections, we find as regards the negative storms that the greatest forces on the night-side are at Ssagastyr and Cape Thordsen at about 18^h, when the storms are at their height.

At Uglaamie, the deflections in this section are of exactly the same character as those in the preceding section, and of very nearly the same strength.

At Fort Rae, on the other hand, there is a deflection which is very distinct, but far slighter than that in the preceding section, and also considerably slighter than the deflections at Uglaamie.

In the first section we found the most powerful perturbing forces at Fort Rae, indicating the proximity to that station of a storm-centre.

This storm-centre was then situated to the east of Uglaamie. Now, in this last section, it is situated to the west of it; and the conditions at that station during these two perturbations, are in the main exactly similar.

If we look at the time of the appearance of the two perturbations, we find that the first takes place just about as long before the passage of the sun through the meridian of the magnetic axis, at about $16^h 30^m$, as the second perturbation does after it. In the description of the preceding perturbations, we also pointed out a similar circumstance; but it was not arranged quite so symmetrically with regard to the time for the sun's passage through the meridian in question, as on the present occasion. As regards the positive storm, the position of its centre cannot be so directly determined, as no district can be pointed to, about which the forces evidently concentrate themselves.

As regards the southern stations; we find there too, simultaneously with the powerful polar storms at about 18^h, a distinctly-defined perturbation, which, at Christiania and Göttingen, is particularly strong in the declination; while at Pawlowsk the deflections in horizontal intensity and declination are about equal.

In the vertical-intensity curve for Jan Mayen, we notice a particularly characteristic, well-marked deflection in a negative direction. It increases at first fairly evenly, but comparatively quickly, reaches a maximum at about $18^h 30^m$, and then once more decreases rather more slowly until about 22^h, when the conditions are almost normal.

Almost exactly the same thing is found at Little Karmakul.

At the intermediate stations, Bossekop, Sodankylä and Cape Thordsen, on the other hand, the conditions are somewhat different. At the first-named station, the forces are of comparatively smaller strength, and the deflections there are first positive, then change and become negative, after which, for the remainder of the period, the curve oscillates over and under the normal line. At Sodankylä the order is reversed, negative deflections coming first, then positive, and then small deflections, now in a positive, now in a negative direction.

At Cape Thordsen, the course of the vertical-intensity curve is peculiar. We there find, at the time when the storm is at its height, very strong but brief impulses, now to one side, now to the other, but more often in a positive direction. Later on, when the storm has diminished in strength, we find first a negative deflection, then for a time fairly normal conditions, and then finally, at the end of the period, positive deflections.

In what way these conditions in the vertical intensity are to be interpreted will best be learnt by looking at the charts, which show the perturbation-conditions for this section.

The last four *charts, V to VIII,* for the hours $17^h 20^m$, $17^h 40^m$, $18^h 20^m$ and $19^h 20^m$, represent the conditions as they develope during this period.

On Chart V, the most powerful storms have not yet begun. We see the negative system of precipitation, which extends in a ring round the north pole.

We now find the strongest perturbing forces at Ssagastyr and Kingua Fjord. The conditions at Cape Thordsen, Fort Rae and Uglaamie, seem, however, to indicate that there can hardly be several sharply-divided systems of precipitation in the negative storm, but that the whole must be regarded as a more or less coherent phenomenon. The succeding charts show this even more distinctly.

A positive system of precipitation also appears quite distinctly at Godthaab. At Bossekop, Sodankylä and Little Karmakul, at which, together with Jan Mayen, we have also seen effects of the positive polar storm, the direction of the arrows is easterly, but the arrows are small.

At the three southern stations, the current-arrows have a south-easterly direction, at the two western of them a little more south, and at Pawlowsk a little more east. These conditions indicate that the stations in question are in the western part of an area of convergence; and it therefore seems as if the influence exerted by the negative system were also predominant in these southern latitudes.

The forces here are of smaller strength, but in Charts VI and VII we see this condition developed to a very much greater degree. The form of the field has undergone no special change, but the perturbing forces have now increased considerably in strength at the great majority of the stations. This is especially the case on the night-side of the globe. At Cape Thordsen the forces have greatly increased, the most powerful being now found there, although at Ssagastyr the perturbing forces are almost of the same magnitude. We now evidently have a powerful negative system of precipitation on the night side of the globe, which also has a distinct effect in Jan Mayen. At both Little Karmakul and Godthaab, on the contrary, there is, as Chart VI shows, a positive storm at $17^h 40^m$; while Chart VII, for $18^h 20^m$, shows a distinct negative polar storm at those stations. The effects of the positive storm, however, do not come out distinctly on these two charts, as the negative storm, owing to its strength, seems to dominate the whole area; but as we have no observations from the districts south of the auroral zone on the afternoon-side of the earth, it is not possible to determine with any certainty the manner in which the conditions actually develope. We have already seen from the curves that this is in all probability a positive storm, and probably also the one that asserts itself to some extent at Bossekop and Sodankyla, and is the cause of the current-arrow having so marked an easterly direction. Finally, if we look at the conditions in the north of Europe and Asia on Chart VII, the discontinuity apparent on a comparison of the conditions at Bossekop and Sodankylä with those at the other stations, would be difficult to explain, if we do not assume that a system of precipitation actually exists there, which counteracts the strong negative system, of which the effects are so apparent everywhere else.

Lastly, there is another circumstance which should be taken into consideration, namely, the conditions in the vertical intensity. If we look at Chart VI, we see that at Bossekop there is a perturbing force, of which the vertical component is directed downwards. A circumstance such as this cannot be explained if we only assume the negative system, of which the area of precipitation falls north of the place; for this would here act in the opposite direction. On the other hand, a positive storm north of the place will actually produce positive values of P_v, and as already remarked in the account of the preceding perturbation, the positive systems will as a rule have their area of precipitation somewhat to the north of this place.

The vertical intensity at Sodankylä, however, exhibits just the opposite condition. We have already pointed out once or twice the abnormal condition appearing in the direction of the deflections in the vertical intensity at Sodankylä, and we will therefore merely refer here to what has been previously mentioned respecting the probable cause of this.

At the three southern stations, the conditions appear to be mainly affected by the negative storm, as the current-arrows indicate that this district is in the western part of an area of convergence; but it is not, of course, on this account impossible that there may be positive precipitation in the district along the southern part of the auroral zone from Norway westwards.

If we assume that such a system exists, then Christiania and Göttingen would be situated in the eastern portion of its area of divergence; here, however, the current-arrows are directed southwards. Whether there is a negative storm-centre in the district east of this, or a positive storm-centre to the west, the direction of the current-arrows at these stations will be very much the same. It may therefore be very reasonably supposed that these two systems actually existed simultaneously; the conditions at the more southerly stations would also be very much satisfactorily explained on the basis of such an assumption.

On Chart VIII, for $19^h 20^m$, the powerful storms are over, at any rate at those stations from which we have observations. Simultaneously with the decrease in the strength of the negative storm from the dominant magnitude that it had in the two preceding charts, the positive area of precipitation once more shows up distinctly, extending from Godthaab, across Jan Mayen, to Bossekop.

The shape of the negative system of precipitation is the same as before, but the forces throughout are considerably weaker, the strength being more or less uniform at all the stations of the group in which the storm is acting. The strongest perturbing force is at Uglaamie, but this is comparatively little greater than those at Ssagastyr, Cape Thordsen and Kingua Fjord.

With regard to the conditions in the vertical intensity, we notice all the time in Jan Mayen the strong negative forces. This may be explained as the effect of the negative system to the north of the place, or of the positive system, which must be situated to the south of the place, or best of all, of course, as a co-operation of these two factors.

The probability of the correctness of the last assumption is manifest. Whether the one or the other of the two systems has the greater influence in a horizontal direction, and causes the current-arrow to point to one side or the other, as these systems here counteract one another, the conditions in the vertical intensity do not change the direction of their deflections, as the two systems act in the same direction, the strength alone varying so that when the storms are at their height, the vertical arrow is also greatest.

After $19^h\ 20^m$ the magnetic elements are a little disturbed before the close of the period, but the disturbances are of little strength, and therefore do not give rise to perturbation-areas of sufficient power and coherence to make them worthy of being studied in detail. For one reason, our observations are too few, and for another these storms will have a more local character, so that the connection will not come out so clearly.

In conclusion we will point to a circumstance, which one cannot help noticing in going through this perturbation, namely that the positive storms always occurred on the afternoon-side. The negative storms formed as a rule a more or less circular or spiral area of precipitation round the geographical pole, or the pole of the magnetic axis; but when there were strongly-marked storm-centres, these were formed, as a rule, upon the night-side of the globe.

Thus far then, this perturbation also furnishes a support to the view of the behaviour and course of the perturbations, which we have previously put forward.

Unfortunately we have no observations of this day from Fort Conger, as the 1st January had been taken there as the term-day, instead of the 2nd January.

TABLE LII.

The Perturbation of the 2nd January 1883.

Gr. M. T.	Uglaamie			Fort Rae			Kingua Fjord		Godthaab	
	P_h	P_d	P_v	P_h	P_d	P_v	P_h	P_d	P_h	P_d
h m										
12 20	− 47 γ	W 105 γ	− 57 γ	− 65 γ	E 36 γ	+ 60 γ	− 27 γ	W 15 γ	+ 4 γ	W 7.5 γ
13 20	− 27 "	" 160 "	− 18 "	−215 "	" 57 "	0	− 62 "	" 38 "	− 23 "	" 3 "
14 5	−220 "	E 70 "	+ 8 "	−275 "	" 110 "	− 73 "	−123 "	" 18.5 "	+ 23 "	E 48 "
20	−158 "	" 110 "	+ 55 "	−175 "	" 45 "	− 60 "	− 65 "	" 6.5 "	+ 12 "	" 42 "
15 0	− 92 "	W 7 "	+ 52 "	−113 "	" 34 "	− 50 "	− 54 "	" 46.5 "	0	" 22 "
20	−117 "	E 78 "	+ 57 "	− 55 "	" 27.5 "	− 30 "	− 61 "	" 57 "	0	" 11 "
16 20	+ 11 "	" 0 "	+ 20 "	− 5 "	W 12.5 "	− 10 "	− 28 "	" 18.5 "	+ 9 "	" 8.5 "
17 20	− 68 "	" 72 "	+ 20 "	− 55 "	E 36 "	0	− 77 "	" 148 "	+ 74 "	" 31 "
40	− 94 "	" 93 "	+ 41 "	− 74 "	" 45.5 "	− 15 "	− 87 "	" 102 "	+ 32 "	" 5.5 "
18 20	−165 "	" 72 "	+ 61 "	− 57 "	" 33 "	− 10 "	− 65 "	" 109 "	0	W 20 "
19 20	−205 "	" 48 "	− 26 "	− 32 "	0	− 10 "	− 33 "	" 82 "	+ 42 "	E 45 "
20 20	−114 "	" 90 "	− 43 "	− 40 "	W 10 "	0	− 13 "	" 64 "	+ 8 "	W 40 "

TABLE LII (continued).

Gr. M. T.	Jan Mayen			Bossekop			Sodankylä		
	P_h	P_d	P_v	P_h	P_d	P_v	P_h	P_d	P_v
h m									
12 20	− 7 γ	E 3 γ	+ 12 γ	0	W 8 γ	0	0	0	+ 7 γ
13 20	− 5 "	W 3 "	+ 12 "	0	0	− 5 γ	− 3 γ	0	+ 7 "
14 5	+ 22 "	0	+ 18 "	+ 13 γ	" 8 "	+ 10 "	+ 9 "	W 5.5 γ	0
20	+ 24 "	0	+ 18 "	+ 5 "	" 20 "	0	+ 3 "	" 15 "	0
15 0	+ 35 "	E 10 "	+ 35 "	0	E 6.5 "	+ 19 "	− 7 "	E 8 "	− 15 "
20	+ 12 "	" 6 "	+ 22 "	0	0	+ 10 "	− 4 "	0	− 9 "
16 20	+ 13 "	0	+ 34 "	+ 5 "	0	+ 10 "	0	0	0
17 20	+ 5 "	W 31 "	− 17 "	+ 23 "	" 15 "	+ 20 "	+ 10 "	" 13.5 "	− 17 "
40	− 30 "	" 58 "	− 30 "	+ 60 "	" 78 "	+ 57 "	+ 37 "	" 53 "	− 43 "
18 20	− 55 "	E 16 "	− 47 "	+ 15 "	" 28 "	− 28 "	+ 14 "	" 29.5 "	0
19 20	+ 50 "	" 17 "	− 60 "	+ 47 "	" 9 "	+ 28 "	+ 12 "	" 13 "	− 9 "
20 20			− 65 "	− 5 "	" 21 "	− 10 "	− 5 "	" 22.5 "	− 8 "

TABLE LII (continued).

Gr. M. T.	Cape Thordsen			Little Karmakul			Ssagastyr	
	P_h	P_d	P_v	P_h	P_d	P_v	P_h	P_d
h m								
12 20	+ 42 γ	E 11 γ	− 10 γ	− 6 γ	W 2 γ	0	+ 37 γ	E 19 γ
13 20	+ 20 "	" 11 "	+ 41 "	+ 5 "	E 8 "	0	+ 12 "	" 37 "
14 5	− 67 "	" 11.5 "	− 22 "	+ 43 "	W 16.5 "	+ 12 γ	+ 24 "	" 9 "
20	− 12 "	" 8 "	0	+ 66 "	" 30 "	+ 35 "	+ 64 "	W 4 "
15 0	− 47 "	" 26 "	− 25 "	+ 15 "	E 19 "	+ 30 "	+ 225 "	" 63.5 "
20	− 28 "	" 29 "	− 19 "	+ 17 "	" 8.5 "	+ 28 "	+ 69 "	E 3 "
16 20	+ 3 "	" 12 "	− 33 "	− 16 "	" 1.5 "	+ 9 "	− 19 "	" 4 "
17 20	− 75 "	W 17.5 "	− 13 "	+ 41 "	" 23.5 "	− 23 "	− 235 "	W 16 "
40	− 30 "	E 385 "	+ 422 "	+ 93 "	W 30 "	− 79 "	− 544 "	0
18 20	− 78 "	" 68.5 "	− 302 "	− 82 "	E 70.5 "	− 174 "	− 339 "	E 26 "
19 20	− 10 "	" 17 "	− 169 "	0	" 24.5 "	− 71 "	− 93 "	" 28.5 "
20 20	− 57 "	" 5 "	+ 14 "	− 79 "	" 76 "	− 76 "	− 72 "	" 41 "

TABLE LII (continued).

Gr. M. T.	Christiania		Pawlowsk			Göttingen		
	P_h	P_d	P_h	P_d	P_v	P_h	P_d	P_v
h m								
12 20	+ 2 γ	0	0	E 2.5 γ	Perhaps small deviations, but nothing can be taken out.	+ 1 γ	E 6 γ	− 1 γ
13 20	− 1.5 "	0	0	" 3.5 "		− 1 "	0	+ 0.5 "
14 5	+ 7 "	0	+ 8 γ	0		0	" 2 "	+ 14 "
20	+ 3.5 "	W 7 γ	+ 3 "	W 5 "		− 1 "	0	+ 13.5 "
15 0	− 7 "	0	− 3 "	E 4.5 "		− 8.5 "	" 8 "	+ 6.5 "
20	− 4 "	0	0	0		− 6 "	" 8 "	+ 4 "
16 20	0	0	0	0		+ 1 "	" 8.5 "	− 1 "
17 20	+ 4.5 "	E 7 "	+ 8 "	" 6.5 "		+ 5 "	" 10 "	− 7.5 "
40	− 1.5 "	" 21 "	+ 16 "	" 20 "		+ 1 "	" 22 "	− 1 "
18 20	+ 15 "	" 37.5 "	+ 26 "	" 15.5 "		+ 12.5 "	" 36 "	− 1.5 "
19 20	− 4 "	" 7 "	0	" 5 "		− 1 "	" 2.5 "	− 0.5 "
20 20	− 11 "	" 9.5 "	0	" 13.5 "		+ 1 "	" 6 "	+ 3 "

Current-Arrows for the 2nd January 1883.

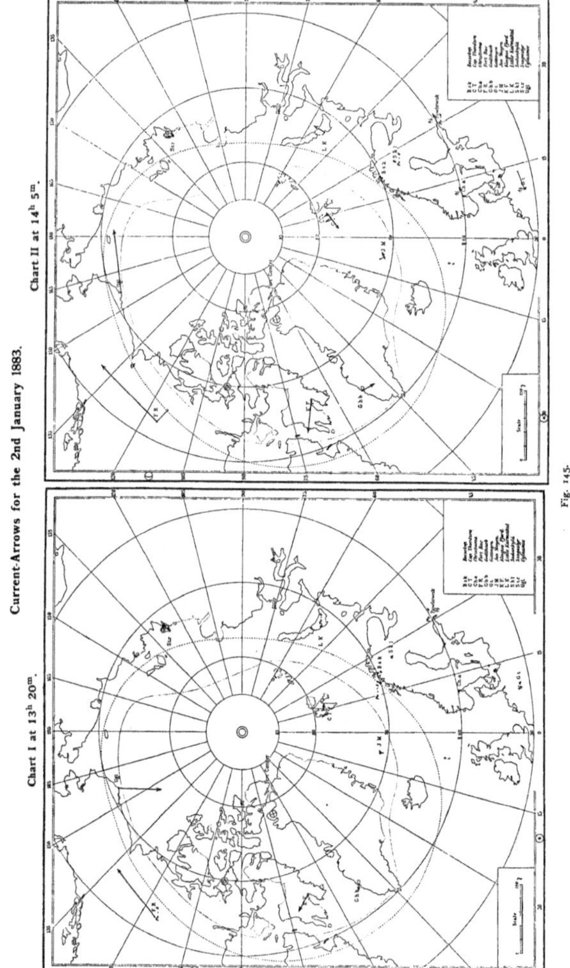

Fig. 145.

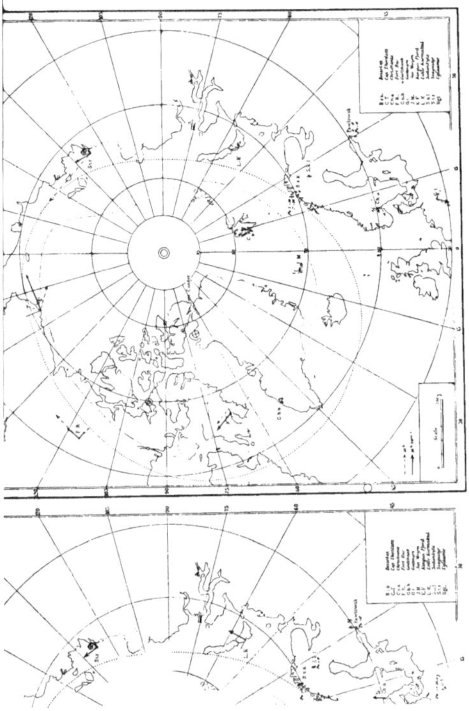

Fig. 146.

Current-Arrows for the 2nd January 1883.

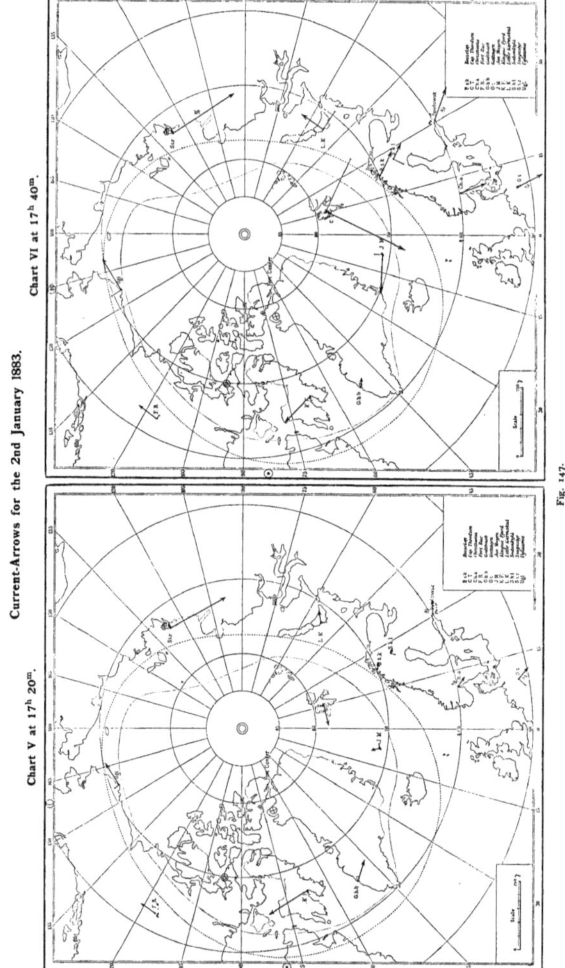

Chart V at $17^h\ 20^m$.

Chart VI at $17^h\ 40^m$.

Fig. 147.

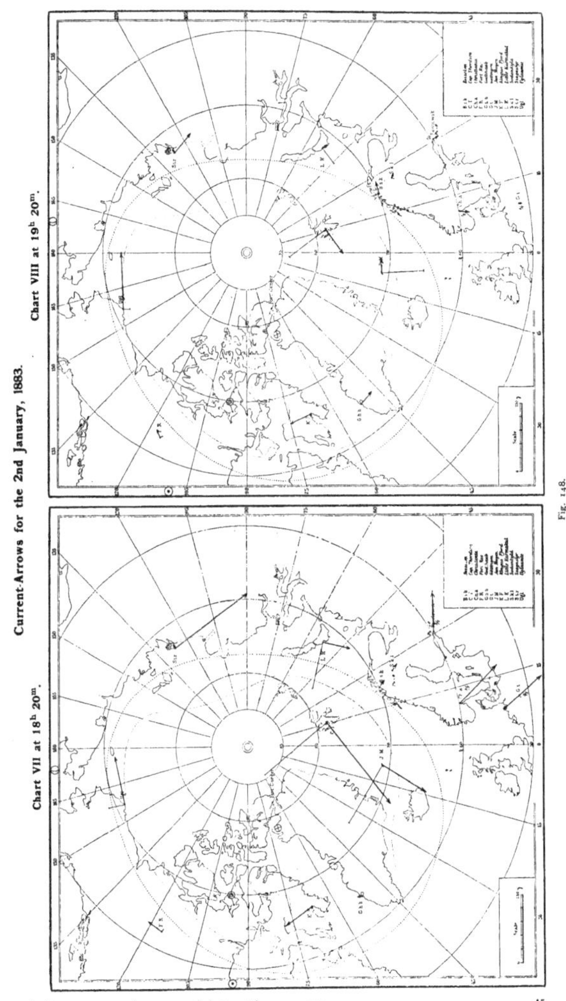

Current-Arrows for the 2nd January, 1883.

Chart VII at 18ʰ 20ᵐ.

Chart VIII at 19ʰ 20ᵐ.

Fig. 148.

THE PERTURBATIONS OF THE 1st NOVEMBER, 1882.

(Pl. XXIII).

85. The striking resemblance that these perturbations bear to the two preceding storms, is apparent on a first glance at the copies of the curves. All the storms occur at the same time of day; they are on the whole very characteristic and well-defined; the direction of their deflections is the same; they are of more or less the same strength; and they are preceded by a comparatively quiet period.

In this case, too, it will be best to divide the period into two sections, the first being from 10^h to about $16^h\ 30^m$, the second from about $16^h\ 30^m$ to $23^h\ 20^m$.

This division, however, does not, as in the case of the preceding perturbations, suit all stations equally well. The conditions at Jan Mayen and Godthaab in particular, do not admit of a natural division such as this.

The principal phenomenon in the first section is the powerful negative storm that we find in North America.

This storm is exceedingly characteristic and well-defined, and the perturbing forces, during the time when the storm is at its height, are of very considerable strength. Thus at Uglaamie, the oscillations are so great that the needle for the horizontal intensity between 14^h and 15^h is deflected beyond the field of observation, and only re-enters it now and then, namely, at $14^h\ 5^m$, $14^h\ 10^m$ and $14^h\ 20^m$, so that there are once more definite readings for these hours. The strongest perturbing forces, it will be seen, appear at Uglaamie, and we must therefore look for the storm-centre of this negative system of precipitation in the neighbourhood of that station.

The storm-centre on this occasion is a little more easterly in position than in the storms in the first section of the two preceding perturbations. At the same, the conditions at Ssagastyr are also somewhat different. We there have now distinct effects of the negative system of precipitation. The forces are not so strong as at Uglaamie, but the curve has a very jagged character. At first the perturbing forces in the declination are directed eastwards, and in magnitude considerably exceed those in the horizontal intensity. Subsequently, at $14^h\ 15^m$, the deflections are reversed, and after $14^h\ 20^m$ there are only small values of P_d, which is now east, now west; and from that hour the perturbing forces in the horizontal intensity are the predominating. This station is thus evidently situated to the west of the centre of the negative storm, although probably actually in the field of precipitation. In the first section of the two preceding storms, we did not find at Ssagastyr any special effect of the negative system of precipitation, which was also found during these two storms in North America.

We found, on the contrary, more or less distinct effects of a positive system of precipitation. At Uglaamie, on the other hand, the conditions during these two preceding storms were exactly analogous to the conditions we now find at Ssagastyr. In these regions, during the first section of the perturbations, there appears a negative system, which, in its behaviour and character, exactly corresponds with those we found during the two preceding storms; but the position of the system on this occasion has moved a little westwards, so that the conditions at Uglaamie during the preceding storms, answer to those at Ssagastyr during the present storm.

It will be well to carry the comparison still further, and see how far the conditions at the other stations are analogous to those we have formerly found. Before doing so, however, we will remind the reader of what we said in the two preceding perturbations regarding the conditions at Cape Thordsen during the first section. It appeared from the curves that simultaneously with the powerful negative storm in North America, a negative storm also occurred at Cape Thordsen, counteracting the positive storm which prevailed during the period before and after, and causing the deflections to some extent to alter, so that we found negative values of P_h at the hours at which the storm in America had its maximum. During the present perturbation, in the interval before the powerful negative storms, there is no pronounced positive storm at Cape Thordsen; and we now find simultaneously with the storms in America, very strongly marked effects of a negative polar storm of very considerable strength (compare Plates XXVI, XXV and XXIII). If we go on farther, to Fort Conger, we find there, too, quite distinct effects of a negative storm as the declination-curve there, just at the period under consideration, in which the negative storm occurs, exhibits a very distinct, well defined, westerly deflection of the declination curve of very considerable amplitude. As previously remarked, current-arrows directed westwards answer to a westerly deflection such as this.

It would appear, therefore, that this is an effect of more or less the same system as that acting at Cape Thordsen. At Kingua Fjord also, there seems to be a negative storm, judging from the deflection of the horizontal-intensity curve; but it is difficult to decide so directly here, as the absolute value of the declination in this case is fairly great, thus giving the deflections in the declination curve greater importance than at those stations at which the declination-value is only small. It seems, however as if this too were principally the effects of a negative storm, and if so, one of longer duration than at the other stations; but these conditions will be better studied by the aid of the charts.

In addition to this, or these, negative area or areas of precipitation, we find in the region about Godthaab, Jan Mayen and Bossekop, a distinctly positive system of precipitation. The effects of this system are most clearly apparent in Jan Mayen, where the positive deflections in the horizontal-intensity curve are of considerable amplitude and very well defined. The deflections, however, as already remarked, do not terminate at the conclusion of the first section, but continue, without great alteration in strength, directly into the next section. This is at any rate the case as regards Jan Mayen and Godthaab, where the storm is most powerful. At Bossekop the perturbing forces are only small, and here we find a distinct strengthening of the positive deflections, just at the time when the negative storms are at their height. Here too, however, the absolute value of the forces is not particularly great.

A positive area of precipitation such as this, was also one of the peculiarities of the first section of the two preceding storms. The position here, however, is a little different from what it was earlier; but the only way in which it differs from that of the other storms is that the area of precipitation does not extend so far eastwards as before.

At Little Karmakul, there are no perturbing forces, in this first section, of sufficient magnitude to warrant the supposition that they are due to the effect of systems of precipitation in the vicinity of the place. In declination, however, we find at about 15^h, that is to say, just at the time when the negative polar storm has its maximum, a very well defined deflection, though of comparatively little strength. In the horizontal intensity, on the other hand, the conditions during this deflection are more or less normal, and it is not until a little later that we find perturbing forces here too, and these in a negative direction.

In the vertical intensity the conditions here are interesting. Simultaneously with the deflection in declination, there is a corresponding negative deflection here. Immediately before this, there is a deflection in the opposite direction. As these deflections are very well defined, it is possible to attribute some importance to them, notwithstanding their comparatively small strength. It seems reasonable to suppose, both on account of the quiet character of the curves, and the small strength, that the conditions are due to the effect of a system that is not in the immediate vicinity of the place. The direction of the current-arrows that we find here is northerly, and will thus answer to conditions in the eastern part of an area of convergence. The vertical arrow, in accordance with this, is directed upwards. It must thus be either the negative system with district of precipitation in the neighbourhood of Cape Thordsen, which produces these characteristic perturbation-conditions at Little Karmakul or the southern positive system, which has its area of convergence to the north of the main axis, or perhaps both these two in cooperation.

If we look at the conditions at Bossekop, we find, as already mentioned, a peculiar strengthening in the positive deflections in the horizontal-intensity curve, just at the time when the negative storms are at their height. This, as we have said, may be explained directly as an effect of the positive storm; but we will here draw attention to the fact that it is also possible to explain the conditions as effects of the negative system lying to the north, if we assume the point of convergence of the system to be situated to the north of Bossekop. Lastly, it is possible that these two factors act simultaneously, and this might perhaps be the most probable explanation.

At the southern stations, the conditions seem mainly to be ruled by the positive polar storm. We here find a distinct, well-defined deflection in the horizontal-intensity curve in a negative direction; whereas in declination we find only deflections of small amplitude. These are first directed eastwards, and then, at about $15^h 20^m$, turn round. The current-arrow in these regions turns distinctly clockwise for a certain angular distance. This, it must be assumed, would indicate that as the point of divergence of the positive system is situated to the north of these stations, as P_h is negative, the system of precipitation now would be moving, although only slightly, eastwards. As we have learnt in Part I, it is just such a deviation of the current arrow that marks a movement of the system of precipitation. As, however, we have so few stations in the positive area of precipitation, it is scarcely possible to prove with any great degree of certainty the existence of such movement by the aid of our observations from the arctic regions.

If we look, lastly, at the perturbing forces in the vertical intensity, we find that at Pawlowsk they are in accordance with the fact that that place is situated in an area of divergence, as P_v there is positive. At Göttingen also, we find evidently positive deflections in the vertical-intensity curve at the time the perturbation is in progress. This is apparent on a direct consideration of the curve. We have not taken out any perturbing forces, however, as the position of the mean line is rather difficult to determine from the data at our disposal. Its determination would therefore be too uncertain, and the values obtained might possibly give misleading ideas of the actual conditions. In this first section, however, there seems to be no doubt as to the direction of the deflections, although they cannot easily be given decided values.

At Bossekop we find a well-defined positive deflection in the vertical curve. This should indicate that the positive system of precipitation exerted a distinct influence here, and was situated to the north of the place, for the negative system that is found still farther north, would here occasion deflections to the opposite side. If the actual perturbation-conditions at Bossekop are in accordance with the observation taken, it must necessarily be supposed that the effect of the positive system extends thither. This is moreover natural, to judge from the conditions at Pawlowsk, where there are strikingly clear proofs of the effect of the positive system. While there are thus positive deflections in the vertical-intensity curve at Bossekop and Pawlowsk, at Sodankylä the deflections are as usual in exactly the opposite direction. The probable explanation of this has already been mentioned.

On *Charts I and II*, for the hours *13^h 20^m*, *14^h 5^m*, *14^h 20^m*, *14^h 40^m*, *15^h 15^m* and *16^h 20^m*, all these conditions come out very distinctly. On the night side, from Fort Rae, through Uglaamie, to Ssagastyr, extends the great negative system of precipitation.

A kind of continuation of this is found at Cape Thordsen and Fort Conger, or it might be supposed that a more or less independent system is at work there.

At Kingua Fjord the direction of the arrow is distinctly southerly, but swings round from east at $13^h 20^m$—at which hour the storm thus really seems to belong to the positive system of precipitation—to a fairly decided west at the close of the period, which would indicate that a negative polar storm was then acting. The transition from the more positive to the more negative character of the storm does not, however, take place so discontinuously as we are accustomed to find at Little Karmakul, for instance where we very frequently find such reversals. On account of the fairly constant direction of the current-arrow, one might be tempted to believe that these were really systems of precipitation in which the direction of the principal axis is not so decidedly east and west, but more north and south. It is easy to imagine a connection established between such a system in Kingua Fjord, and the negative system of precipitation at Cape Thordsen and Fort Conger. Such a condition is not only conceivable, but, as previously observed, we find by the experiments, phenomena which clearly demonstrate that we should expect to find, just in these tracts, areas of precipitation the main axis of which were directed tolerably nearly due N—S; compare p. 327, fig. 140, art. 82.

The conditions at Godthaab and Jan Mayen in connection with the southern stations, show us distinctly a positive system of precipitation with accompanying area of divergence. At Pawlowsk, as we see, there are also positive vertical arrows; and we have already seen that at Göttingen, during this period, a positive deflection appeared in the vertical-intensity curve. We thus have every indication of the existence of this positive system of precipitation.

These are in the main the most characteristic conditions during the first section of the perturbation. It is difficult to prove with certainty any movement of the systems.

At several stations there now ensues a longer or shorter period of more normal conditions, after which the new perturbations belonging to the second section commence. At other stations there is no such distinct division, but the deflections continue without ceasing on into the next period.

The perturbation-conditions here prove to be rather more complicated than in the preceding section.

We will here make Ssagastyr our starting-point. The perturbing forces appear here chiefly in the horizontal intensity. The amplitude of the deflection is now about the same as during the preceding section; but its duration is here a little longer. No exact statement of the time of the appearance and termination of the perturbation can be given, but roughly speaking, the perturbation occupies the period from $16^h\ 30^m$ to 20^h. Simultaneously with this, the conditions at Uglaamie and Fort Rae are very interesting, as we there find simultaneous deflections in the curves, especially in the horizontal-intensity curve, in a negative direction; but the forces are now comparatively very weak.

At the stations west of Ssagastyr, however, there are fairly powerful perturbing forces. As before, we can follow the negative storm over Cape Thordsen and Fort Conger; and at the first of these stations, the perturbing forces are of considerable strength.

The conditions at Little Karmakul and Bossekop are now of special interest. At the first-named station we again meet with a condition of which we have so often before had instances, namely, the simultaneous action of positive and negative perturbing forces. We there find now positive, now negative deflections in the horizontal intensity, until about $18^h\ 30^m$, from which time the deflections are negative and remain so for the rest of the period. From this hour then, the effects of the negative storm predominate, and the perturbing forces are exceedingly powerful, thus indicating the proximity of a storm-centre.

At Ssagastyr, we found, it will be remembered, exclusively negative deflections in the horizontal-intensity curve, beginning at the very beginning of the period.

At Little Karmakul, it is not until considerably later that the negative storm gains the ascendancy; and this would therefore seem to indicate that the negative storm-centre, or district of precipitation, is moving westwards.

This last view of the conditions is also confirmed by a comparison with those at Bossekop. At first there is evidently a positive polar storm acting, and we cannot perceive any special trace of a negative storm. At about $19^h\ 30^m$, however, the curve for the horizontal intensity goes to the opposite side, and for the rest of the time we find fairly powerful effects of a negative polar storm, although the perturbing forces here are not so great as those we find at Little Karmakul. If we look at the time after which the negative storm acts exclusively, at the last two stations, we find here too a considerable difference in time between them, namely, of almost exactly one hour.

Thus the negative storm appears considerably later at the more westerly stations than to the east, in this district; and we feel justified in taking these circumstances as a proof that the negative storm-centre in this section of the perturbations, is moving westwards, and thus in some way or other is following the sun in its apparent diurnal motion.

It would not be right, however, to draw conclusions respecting the details of this movement from these facts, for it cannot, of course, be taken for granted that the district of precipitation moves exactly along the auroral zone as the perturbations run their course. This is all the more inadmissible from the fact that at Cape Thordsen and Fort Conger, there are distinct proofs that also polar areas of precipitation exist farther north, and that therefore in detail the conditions may be a little more complicated. It would at any rate be natural to expect that the conditions would not be so simple if, instead of comparing stations that were all situated south of the auroral zone—as was the case with the three stations just considered—we were to compare the conditions at stations lying some to the north and some to the south of that zone. This proves to be the case, when we go farther west to Jan Mayen, and compare the conditions there with those, for instance, at Bossekop. There too, it is true, there is first a positive storm, which is very powerful and pronounced, and later on the direction of the deflections in the horizontal-intensity curve change, indicating that now, instead of a positive polar storm, the effects are those of a negative storm; but the change takes place earlier than at the more easterly situated Bossekop. The cause of this may therefore naturally be looked for in the circumstance that Jan Mayen is situated to the north, and Bossekop to the south, of the auroral zone, and that therefore the northern, or north-western, branch of the negative district of precipitation—if it may so be called—might be supposed to reach Jan Mayen earlier than its eastern, or more southern part reaches Bossekop. The explanation of the conditions in Jan Mayen might thus be that it was the effect of the negative system of precipitation at Cape Thordsen, extending, as the perturbation proceeded, westwards to Jan Mayen, or possibly moving in that direction. This view is further supported by the fact that the change in Jan Mayen occurs just at the time when there is a sudden, very considerable increase in the negative deflection in the horizontal-intensity curve for Cape Thordsen. When we finally come to consider the conditions of the vertical intensity, we shall return to this subject with other circumstances that favour our view.

The negative deflections in the horizontal-intensity curve for Jan Mayen are comparatively small. In the declination, on the other hand, there is a uniformly-directed, westerly deflection, which, as a rule, exceeds those in the horizontal intensity in strength. About the time when the change in the horizontal intensity takes place, there is no special change to be observed in the deflections in the vertical intensity or the declination.

It is possible, perhaps probable, that here too, after the change has taken place, there are still effects of the positive system. The comparatively small forces in the horizontal intensity, and the comparatively powerful forces in the declination, seem to indicate something of the kind; but it is difficult, indeed impossible, to settle the point with certainty.

The other station where there were distinct effects of the positive system of precipitation was Godthaab. Here the system acts a trifle longer than in Jan Mayen; but there is no negative storm afterwards, the conditions being fairly normal.

With regard to the southern stations, we see that the conditions in the horizontal intensity, during the first part of the section, are rather variable. At those lying more to the west, such as Christiania and Göttingen, however, there are throughout perturbing forces that act in a negative direction, and are of sufficient magnitude to indicate, more or less certainly, an area of divergence which should answer to the positive system of precipitation that we find in the district Godthaab, Jan Mayen and Bossekop. At the more easterly station Pawlowsk, on the other hand, the curve for the horizontal intensity oscillates more about the normal line, without exhibiting any marked direction. It appears therefore as if the effect of the positive system of precipitation were weaker here, which is quite natural, seeing that we are approaching the negative storm-centre.

Later on, it is the deflections in the declination — which are easterly all the time — that predominate. This, as we have often seen before, is a circumstance that has to do with the moving into these southern districts of the negative system's area of convergence. We should also find the same direction of current in the eastern part of the area of divergence, which is connected with the positive system of precipitation. Of these two systems, which of course may be imagined to co-operate, the first will here have the greatest effect.

The course of the vertical-intensity curve at Pawlowsk also seems to indicate — although one cannot here venture to draw very certain conclusions — that at first it is in an area of divergence, where P_v is positive, and afterwards in an area of convergence, at the time when we find negative values of P_v there. The course of the vertical-intensity curve at Göttingen exhibits similar conditions, but there they are still more uncertain, as the normal line is very difficult to determine. It would not therefore be advisable to draw any conclusions from this.

With regard to the vertical intensity in other respects, it may be noticed that in Jan Mayen there are negative deflections all through the section, with the exception of the last few hours of the period. This is what we have found previously, and indicates that there is a negative precipitation to the north of the place, or a positive precipitation to the south, or both simultaneously. At Bossekop we first have positive deflections, as long as the positive storm is acting; and this should indicate that the positive system is situated to the north of the place. Simultaneously with the alteration in the horizontal intensity curve, there is also an alteration in the curve for the vertical intensity; and from the moment when the negative storm gains the ascendancy, we find negative values of P_v for the rest of the period. It would seem, from the above, natural enough that the conditions should actually be in accordance with this. At Sodankylä, on the other hand, we find the exact opposite; and we thus again meet with that peculiar phenomenon, to which we have several times drawn attention.

If the vertical-intensity observations at Cape Thordsen are to be relied upon, the negative system acting there should at first lie to the north of the place, but in the last part of the period to the south. This agrees very well with the conditions at Bossekop, as the supposed passage of the system over the station at Cape Thordsen, at the time when P_v there goes over from a negative value to a positive, takes place just when the negative storm gains the ascendancy at Bossekop. Thus at the time when the vertical intensity at Cape Thordsen indicates that the negative system of precipitation is approaching Bossekop, we really find there marked effects of a negative polar system.

This gives us a clear hint of the way in which the movement of the systems of precipitation up there are to be understood, and seems to confirm our previous assumptions in the matter. We found, it will be remembered, a removal of the system of precipitation towards the west, when we looked at the three stations Ssagastyr, Little Karmakul and Bossekop, which were all situated south of the auroral zone. No similar movement, however, could be traced to Jan Mayen, and we adduced, as a possible cause of this, the circumstance that magnetically considered, that island had a comparatively much more northerly situation. We further indicated that the conditions in Jan Mayen might possibly be explained by assuming that the system at Cape Thordsen was moving westwards. We see now, however, that at these hours there are also indications that the system at Cape Thordsen has a southerly movement, or at any rate that its movement will have a component in a southerly direction; and it therefore seems fairly probable that the change will take place a little earlier in Jan Mayen than at the more southerly situated Bossekop.

The simplest conception of the matter might be, that this was a to some extent connected negative system of precipitation, whose eastern part extended more or less along the auroral zone, but whose western part curved more northwards; and that the whole of this district of precipitation moved westwards with the sun.

Such an assumption also agrees with what we find by experiment. We may here, for instance, refer to fig. 140, pag. 327, where we clearly see such a deviation of the area of precipitation towards the N., and particularly to the subsequent chapter in which the terrella experiments are specially treated of.

Having discussed the conditions of perturbation so thoroughly, we need now only briefly touch upon the perturbation-areas that we find represented on the charts for this section.

On *Charts III, IV* and *V*, we find the direction of the current-arrows for the period in question shown for nine epochs.

In its main features, the movement of the negative system of precipitation that we found and have described above, can be distinctly followed.

If we considered the three polar stations mentioned above, which are situated to the south of the auroral zone, we see, on Chart III, distinct effects of the system only at the most easterly of these, namely, Ssagastyr. At Little Karmakul, the negative storm does not gain the ascendancy until Chart IV; on Chart III the current-arrow swings backwards and forwards.

Lastly, at Bossekop it appears that it is not until the last epoch represented on Chart IV that the negative storm is predominant. Before that, there are only more or less distinct effects of the positive system. We further see on Chart IV that the negative storm appears earlier in Jan Mayen than at Bossekop. As regards the negative storm in other respects, we see all the time at Cape Thordsen strong westerly-directed current-arrows. East of Ssagastyr, the strength of the current-arrows diminishes considerably, so that the boundary of the area of precipitation is probably between Ssagastyr and Uglaamie.

At the close of the section, we find the negative storm-centre in the north of Europe or the north-west of Asia.

The positive system asserts itself distinctly only on Chart III, at Godthaab, Jan Mayen, Bossekop and Little Karmakul.

With regard to the conditions in southern latitudes, we see only slight, though sometimes fairly distinct, indications that the stations are in an area of divergence. Nor is this unlikely; for, judging from the observations from the northern regions, we should expect to find the area of divergence farther west.

On the other hand, we find on Charts IV and V, quite certain indications of an area of convergence.

There is one circumstance, however, which to some extent seems to point in the opposite direction, namely, the conditions in the vertical intensity at Pawlowsk. We have already noticed that first positive, and then negative values of P_z are found here; but now we see that the positive forces also appear to last longer than the period in which the positive storm predominates, being even apparent at times when there are fairly distinct indications in a horizontal direction that we are in the area of convergence of the negative system of precipitation. It is not impossible that the conditions are actually like this; but on the other hand it should be remarked that the position of the normal line during this period, might very possibly be a little different from what it is here; and one must therefore not conclude too much from this circumstance. There is, moreover, a great possibility that in southern latitudes perturbing forces might be operating that are imperceptible here, but which may yet exert a disturbing influence upon the perturbation-conditions that we are now considering.

At Göttingen, as we have said, the vertical intensity also exhibited conditions similar to those at Pawlowsk. Here, however, they were more easy of explanation, as the station lies so much farther west, that one might well imagine the positive system to be acting as long as the positive deflections appear to continue.

The last phase of the perturbation, as will immediately be seen, is just what we have previously designated as a negative polar elementary storm, with the storm-centre in the north of Europe, a storm such as we have again and again met with in Part I. In these storms, we have learnt to understand how they are a link in a long chain of perturbations, which, it appears, steadily develope in the course of the day, in more or less the same manner. In the succeeding pages, we shall see how confirmation of this will actually be obtained.

TABLE LIII.

The Perturbations of the 1st November, 1882.

Gr. M. T.	Uglaamie			Fort Rae			Kingua Fjord	
	P_h	P_d	P_v	P_h	P_d	P_v	P_h	P_d
h m								
12 20	0	E 18.5 γ	0	− 52 γ	E 22.5 γ	+ 25 γ	0	E 17 γ
13 20	0	„ 32 „	+ 14 γ	− 44 „	„ 22 „	+ 65 „	0	„ 68 „
14 5	− 257 γ	0	+ 112 „	− 236 „	„ 103 „	+ 255 „	− 53 γ	„ 43 „
20	− 152 „	„ 23 „	+ 118 „	− 231 „	„ 102 „	+ 255 „	− 45 „	„ 26.5 „
40	− > 258 „	„ 217 „	+ 89 „	− 337 „	„ 160 „	+ 55 „	− 76 „	0
15 15	− 199 „	„ 162 „	+ 80 „	− 285 „	„ 102 „	− 10 „	− 55 „	W 57 „
16 20	− 5 „	W 37 „	+ 28 „	0	„ 38.5 „	− 35 „	− 50 „	„ 9.5 „
17 20	− 15.5 „	„ 5 „	− 5 „	− 20 „	W 1 „	− 50 „	− 12 „	„ 9.5 „
50	− 15.5 „	„ 16 „	− 33 „	0	E 4 „	− 10 „	+ 7 „	„ 47 „
18 20	− 69.5 „	E 26 „	− 56 „	− 20 „	0	− 20 „	− 7 „	„ 5 „
19 0	− 51 „	W 21 „	− 56 „	− 22 „	W 2.5 „	+ 20 „	+ 25 „	„ 32.5 „
20	− 51 „	E 16 „	− 56 „	− 27 „	„ 23 „	+ 5 „	+ 9 „	„ 29 „
20 0	+ 11.5 „	W 40 „	− 37.5 „	− 20 „	„ 28 „	− 5 „	− 4 „	„ 17 „
20	+ 4 „	„ 66 „	− 33 „	− 14 „	„ 15.5 „	0	0	„ 26 „
21 10	0	„ 64 „	− 23.5 „	0	„ 9 „	+ 25 „	+ 3 „	„ 31 „
22 20	0	„ 72 „	+ 9.5 „	− 19 „	E 1.5 „	+ 20 „	− 11 „	„ 30 „

TABLE LIII (continued).

Gr. M. T.	Godthaab		Jan Mayen			Bossekop		
	P_h	P_d	P_h	P_d	P_v	P_h	P_d	P_v
h m								
12 20	+ 5 γ	0	+ 5 γ	W 3 γ	0	0	0	0
13 20	0	E 25 γ	+ 37 „	„ 8.5 „	+ 6 γ	+ 5 γ	0	− 27 γ
14 5	+ 6 „	„ 50 „	+ 133 „	E 11.5 „	− 12 „	+ 16 „	W 12.5 γ	+ 39 „
20	+ 19 „	„ 56 „	+ 128 „	W 8 „	− 34 „	+ 37 „	0	+ 78 „
40	+ 20 „	„ 65 „	+ 127 „	0	− 40 „	+ 15 „	E 6.5 „	+ 29 „
15 15	+ 32 „	„ 42 „	+ 122 „	E 17 „	− 44 „	+ 62 „	„ 14 „	+ 90 „
16 20	+ 30 „	„ 36.5 „	+ 71 „	W 6.5 „	0	+ 14 „	W 18 „	+ 10 „
17 20	+ 15 „	„ 34 „	+ 97 „	„ 5.5 „	− 22 „	+ 7 „	„ 6 „	+ 13 „
50	+ 32 „	„ 28 „	+ 78 „	„ 12 „	− 36 „	+ 15 „	„ 7 „	+ 12 „
18 20	+ 50 „	„ 53.5 „	+ 6 „	„ 45.5 „	− 44 „	+ 50 „	„ 7 „	+ 63 „
19 0	+ 15 „	W 11 „	− 53 „	E 11.5 „	− 36 „	+ 66 „	„ 33 „	+ 18 „
20	− 5 „	„ 17 „	− 17 „	W 63 „	− 80 „	− 21 „	„ 45 „	− 21 „
20 0	− 11 „	0	− 64 „	„ 89 „	− 85 „	− 168 „	0	− 27 „
20	− 5 „	„ 11 „	− 51 „	„ 64 „	− 60 „	− 83 „	E 18 „	− 82 „
21 10	− 7 „	„ 17 „	− 28 „	„ 48.5 „	+ 23 „	− 123 „	„ 67 „	− 177 „
22 20	− 20 „	„ 25 „	− 77 „	„ 24 „	0	− 86 „	„ 25 „	− 177 „

TABLE LIII (continued).

Gr. M. T.	Sodankylä			Cape Thordsen			Little Karmakul		
	P_h	P_d	P_v	P_h	P_d	P_v	P_h	P_d	P_v
h m									
12 20	+ 7 γ	E 4 γ	+ 4 γ	+ 4 γ	E 7 γ	+ 12 γ	— 19 γ	0	+ 15 γ
13 20	+ 3 „	„ 4 „	— 2 „	+ 10 „	0	— 35 „	0	0	— 5 „
14 5	— 10 „	0	— 23 „	— 82 „	W 8.5 „	— 60 „	+ 5 „	W 6.5 γ	+ 45 „
20	— 7 „	„ 8 „	— 24 „	— 76 „	„ 10 „	— 60 „	— 4 „	E 4 „	+ 12 „
40	— 9 „	„ 11.5 „	— 25 „	—100 „	E 2.5 „	— 55 „	+ 6 „	W 53.5 „	— 10 „
15 15	— 3 „	„ 6 „	— 26 „	—147 „	„ 107 „	+ 60 „	— 5 „	„ 21.5 „	— 34 „
16 20	— 6 „	W 8 „	— 15 „	— 30 „	W 8 „	— 90 „	— 38 „	„ 3 „	— 10 „
17 20	— 11 „	E 2.5 „	— 18 „	—104 „	„ 12 „	— 52 „	+ 75 „	„ 8.5 „	+ 12 „
50	0	0	— 19 „	—110 „	E 20.5 „	— 10 „	—100 „	„ 19 „	— 50 „
18 20	0	„ 7 „	— 49 „	—119 „	„ 7.5 „	— 8 „	+ 65 „	„ 85 „	— 8 „
19 0	+ 5 „	W 11.5 „	— 25 „	—162 „	W 50.5 „	— 18 „	—316 „	E 50 „	— 59 „
20	0	„ 6 „	— 39 „	— 67 „	„ 49 „	+ 30 „	—221 „	„ 69 „	— 17 „
20 0	— 40 „	„ 4 „	+ 38 „	— 13 „	„ 10.5 „	+ 15 „	—121 „	„ 81 „	0
20	— 18 „	E 9 „	+ 7 „	— 29 „	„ 3 „	+ 10 „	— 88 „	„ 48 „	0
21 10	— 7 „	„ 45 „	+ 31 „	— 14 „	„ 42 „	+ 25 „	—438 „	„ 131 „	—170 „
22 20	— 12 „	„ 7.5 „	+ 41 „	— 58 „	E 16.5 „	+ 93 „	+ 5 „	W 12 „	—130 „

TABLE LIII (continued).

Gr. M. T.	Sagastyr		Christiania		Pawlowsk			Göttingen		Fort Conger
	P_h	P_d	P_h	P_d	P_h	P_d	P_v	P_h	P_d	F_d
h m										
12 20	0	W 14.5 γ	— 1.5 γ	0	0	0	0	0	0	0
13 20	+ 0 γ	E 25 „	— 1 „	0	0	0	0	— 5.5 γ	0	W 34.5 γ
14 5	— 7 „	„ 218 „	— 14 „	W 2.5 γ	— 20 γ	E 3.5 γ	+ 5 γ	— 22 „	W 2.5 γ	„ 121 „
20	—187 „	0	— 16.5 „	0	— 20 „	„ 10.5 „	+ 7 „	— 26.5 „	E 1.5 „	„ 113 „
40	—235 „	„ 71 „	— 18.5 „	E 9 „	— 22 „	„ 14 „	+ 7 „	— 30.5 „	„ 9 „	„ 119 „
15 15	—197 „	„ 90 „	— 18.5 „	„ 5.5 „	— 18 „	„ 9.5 „	+ 9 „	— 26.5 „	„ 3 „	„ 154 „
16 20	— 50 „	0	0	0	0	W 5 „	+ 5 „	— 7.5 „	W 2.5 „	„ 15.5 „
17 20	—141 „	„ 18.5 „	— 2 „	0	0	0	?	— 2.5 „	„ 4.5 „	„ 31 „
50	—192 „	„ 24.5 „	+ 3.5 „	„ 3.5 „	+ 7 „	0	?	+ 3 „	E 1 „	„ 68 „
18 20	—215 „	„ 36.5 „	— 4.5 „	W 4 „	0	E 3 „	+ 7 „	— 3 „	W 3 „	„ 93 „
19 0	—255 „	0	+ 5 „	E 8 „	+ 8 „	„ 4 „	+ 8 „	0	E 7.5 „	„ 80 „
20	—200 „	„ 43 „	— 4 „	„ 4.5 „	— 5 „	„ 8.5 „	+ 9 „	— 7.5 „	„ 1.5 „	„ 76 „
20 0	— 37 „	„ 50 „	+ 1 „	„ 20.5 „	+ 5 „	„ 5 „	+ 4 „	— 2.5 „	„ 18.5 „	E 5 „
20	— 52 „	„ 48 „	— 10.5 „	„ 5.5 „	— 5 „	„ 11 „	0	— 4.5 „	„ 5 „	W 18 „
21 10	—111 „	0	0	„ 40.5 „	+ 19 „	„ 28.5 „	— 3 „	0	„ 35 „	„ 8.5 „
22 20	— 10 „	0	+ 1.5 „	„ 21 „	+ 3 „	„ 7.5 „	— 5 „	+ 4 „	„ 18 „	E 10 „

Current-Arrows for the 1st November,

Chart I at 13^h 20^m, 14^h 5^m, and 14^h 20^m.

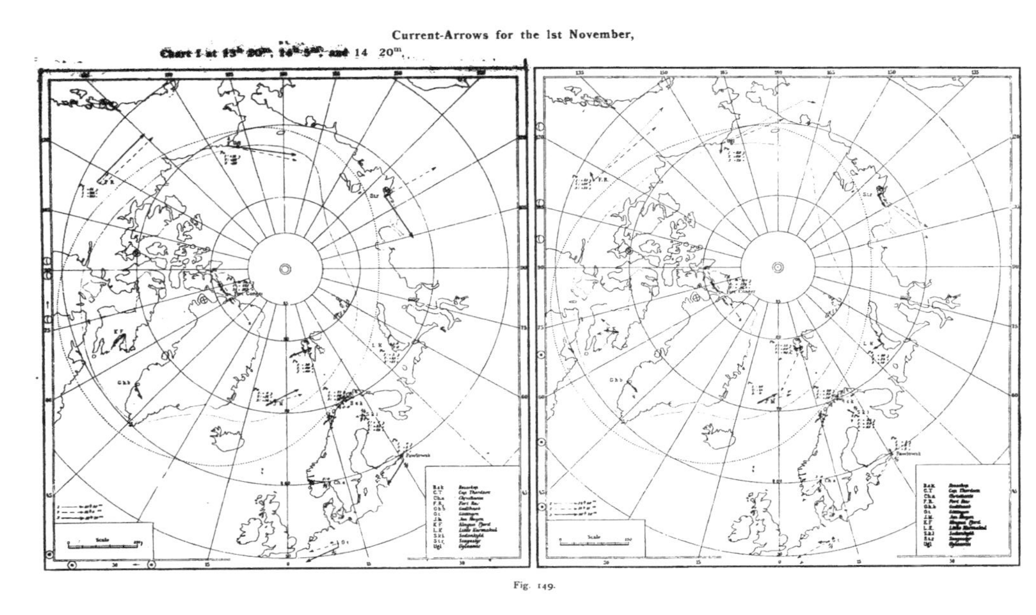

Fig. 149.

Current-Arrows for the 1st November, 1882.

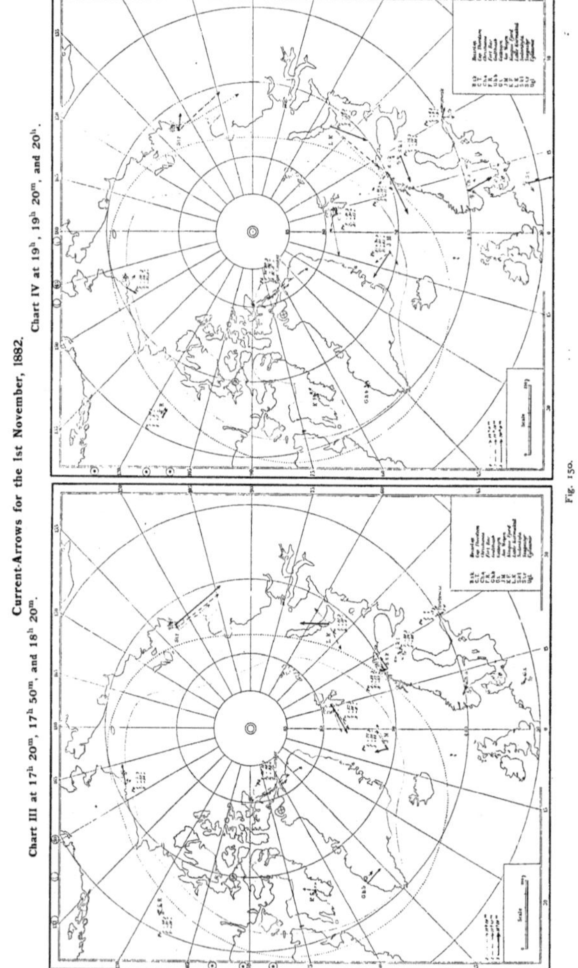

Chart III at 17^h 20^m, 17^h 50^m, and 18^h 20^m.

Chart IV at 19^h, 19^h 20^m, and 20^h.

Fig. 150.

Current-Arrows for the 1st November 1882.
Chart V at $20^h\ 20^m$, $21^h\ 10^m$, and $22^h\ 20^m$.

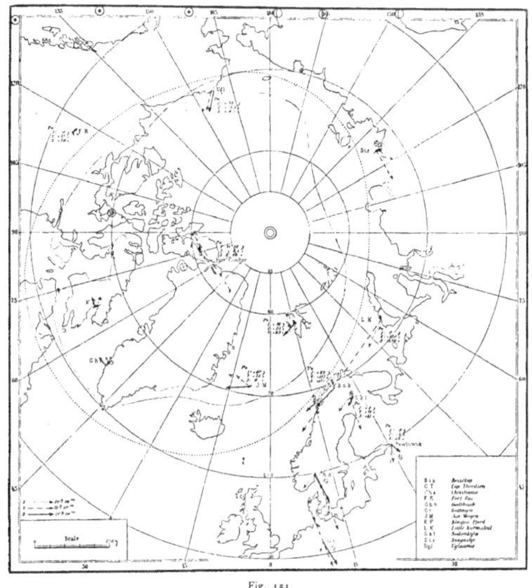

Fig. 151.

THE PERTURBATION OF THE 14th and 15th FEBRUARY, 1883.

(Pl. XXVIII).

86. The three preceding perturbations have exhibited a very great resemblance to one another in their manner of occurrence and course.

It will be remembered that in the last-described of these three perturbations, we found at the close a strong negative area of precipitation in the north of Europe, while at the other stations there were only small perturbing forces.

This last perturbation, with its rather limited area of precipitation, was of the same type as those we so often met with in Part I. It was this type of perturbation that exhibited the simplest conditions, and that we found was the usual one about Greenwich midnight. At the beginning of the present term day, we find, as the curves show, an exactly similar negative polar storm, whose district of precipitation is also restricted to the very same region. The perturbation is here exceedingly characteristic and well-defined, and the subsequent conditions are very normal, so that the day, on this account, at several places where there are no daily hourly-observations has been of great importance in the determination of the diurnal variation. At the beginning of the period, the storm, in several places, has almost reached its maximum.

It is at the four stations, Little Karmakul, Cape Thordsen, Bossekop, and Jan Mayen, that the storm developes to its greatest strength.

If we look at the curves, we see that there are several peculiarities in this perturbation that are worthy of notice.

In the first place, the maximum does not occur exactly simultaneously at these stations.

At Little Karmakul and Jan Mayen it occurs almost simultaneously at $23^h 25^m$—30^m, at any rate if we consider the conditions in the horizontal intensity, where the deflections are most characteristic. At the two intermediate stations, on the other hand, the maximum does not occur until a little later, at $23^h 40^m$—45^m. This circumstance is evidently to be ascribed to a movement in, or of, the system of precipitation. In the next place, the negative deflections in the horizontal intensity do not cease simultaneously either. At Little Karmakul the deflections decrease rather rapidly, and even go over to the other side at $0^h 15^m$, so that after that time we find almost exclusively positive values of P_h until about $2^h 30^m$, after which, for the rest of the period considered, the curve oscillates about the normal line, but with very small deflections.

Here then, the negative storm appears to be superseded by a positive storm at about $0^h 15^m$.

At the three other stations, however, there is no indication of any positive storm.

At Cape Thordsen, the conditions in the horizontal-intensity curve have once more become normal at about $0^h 50^m$; at Bossekop and Jan Mayen, on the other hand, this does not take place until about $1^h 20^m$—30^m.

It will be difficult to demonstrate any single movement of the system of precipitation, by the difference in time between the various maxima of the negative deflections; but at the conclusion of the storm, the conditions seem to be simpler. We see that the storm lasts longer at the more westerly stations than at those farther east.

By east and west, here, must not be understood geographical east and west, but rather the directions parallel with the auroral zone, and by north and south the directions perpendicular to it. If we use the geographical east and west, Cape Thordsen is of course situated to the west of Bossekop; whereas magnetically, it must be considered as lying to the east of that station. We saw too, that the storm terminated earlier at Cape Thordsen than at Bossekop.

This last fact also seems to indicate that the system of precipitation is moving westwards, more or less parallel with, or along, the auroral zone.

In the declination too, there are quite considerable perturbing forces; but the curves here have sometimes rather a disturbed character, in contrast to those of the horizontal intensity.

It is, as we have said, principally at the four stations mentioned above, that the perturbation especially asserts itself; although distinct effects of the system of precipitation are found also at Kingua Fjord and Godthaab. The conditions at the last-named station are moreover of peculiar interest, as at about 1^h there is a strong, well-defined deflection there in the horizontal intensity curve. At that hour we do not find deflections at any of the other stations, which might indicate any special connection with this deflection, and thus this storm appears to be very local.

As regards the American stations, we find at Fort Rae distinct signs of a positive polar storm. The greatest deflections are at about 3^h, at which hour there is also a distinct deflection in the other elements.

At Kingua Fjord and Uglaamie, there are also deflections at the same hour, which might be the effects of a positive system of precipitation, but they are quite small.

We have then, on this day, once more two systems of precipitation, a negative and a positive. Of these the first is the stronger, and it appears on the night-side of the globe. The positive system appears to be considerably weaker, judging from the observations we have at our disposal, and it appears upon the afternoon-side of the earth.

At Little Karmakul there seems moreover to be a positive system of precipitation. But it is especially interesting here to find the positive system of precipitation in the vicinity of Fort Rae, as this is the only station in this district situated to the south of the auroral zone, and where therefore one would expect to find effects of a positive system, if such a system actually existed in those regions. This is the first instance we have of a storm, which appears at Fort Rae at this time of day, and it thus proves to have the character of a positive polar storm. This instance is of peculiar interest, as it shows that the occurrence of positive afternoon storms, which we have so often demonstrated at the European stations, as also at Ssagastyr during the storms just described, is also found in these regions. The reason why opportunities of observing this phenomenon here are comparatively rare is probably principally that this is the only American station in a suitable position a little south of the auroral zone.

The perturbation-conditions at Sodankylä are also interesting. The horizontal forces are comparatively very small, indicating that this station is not far off the point of convergence of the negative system, a circumstance which is immediately evident on looking at the charts.

If we consider the vertical perturbing forces, we see in the negative area of precipitation, that at the two polar stations, Cape Thordsen and Jan Mayen, which are to the north of the auroral zone, there are perturbing forces directed downwards; while at the two polar stations, Little Karmakul and Bossekop, which are to the south of the auroral zone, the forces are directed upwards. This seems clearly to prove that the precipitation takes place more or less exactly in the auroral zone.

With regard to the vertical forces at Sodankylä, the conditions are just as abnormal as in the previous perturbations. The forces are positive and fairly powerful. Concerning them, we will only refer the reader to the remarks previously made about this condition. At the southern stations there are well defined perturbations in the various elements, simultaneously with the negative storm in the north.

Seven charts have been drawn for this perturbation. On the first three, we instantly recognise the principal phenomenon that was the characteristic one in this storm, namely, the strong negative area of precipitation on the night-side of the globe in the regions around Northern Europe. South of the area of precipitation, a very distinct area of convergence is formed, with all its characteristic peculiarities. The vertical intensity at Sodankylä is the only exception. In order to obtain a better impression of this area of convergence, we have also drawn a current-arrow on Charts II and III for Kasan. From this station, we have five-minutely observations in declination, but in horizontal intensity only readings at an average interval of two hours. At about $23^h\ 50^m$, Gr. M. T., we find a reading, which, when compared with the other readings, shows with tolerable certainty that at that time there is a perturbing force P_h of about $+15\gamma$. As we had drawn no chart for this hour, we have employed this value together with the two values of P_d, which can be determined directly for the two points of time. The two current-arrows are thus only to be regarded as an approximately correct expression for the respective perturbing forces; and they have only been included here in order to bring out more distinctly the form of the area of convergence.

On the other side of the principal axis in the system of precipitation, one would expect to find an area of divergence; but during the preceding storms, the conditions in these high latitudes have been so perturbed that it has been impossible to prove the existence of anything of the kind. This time, however, the area of precipitation is so local that we might perhaps expect to find it.

We do moreover actually find perturbing forces at Kingua Fjord and Godthaab, which, in strength and direction, are very much what we should expect to find in that part of the area of divergence, which comes into these districts.

At Fort Conger, there are only small perturbing forces in the declination. If the point of divergence of the system were between this station and Cape Thordsen, the direction of the current here should be northerly. As we have no observations of horizontal intensity here, we are unable to verify this; but it may perhaps be worth while to point out an interesting harmony with the conditions in the area of convergence. We see from the chart that Fort Conger and Pawlowsk are situated more or less symmetrically one on each side of the principal axis of the system. The tangents to the magnetic meridian at the two places are moreover more or less parallel. (The declination at Pawlowsk is very near 0, and we see that the line magnetic N—S, which is drawn on the chart through Fort Conger, is very nearly parallel with the meridian of 30° east longitude.)

As the forces on the two sides of the principal axis would probably be more or less symmetrical in arrangement, we might perhaps expect to find a certain amount of symmetry in the declination-deflections at the two places. When the declination-curve at Pawlowsk swings out to the west, the curve at Fort Conger should swing out to the east, and vice versa. This will be immediately apparent if we imagine the polar elementary field (fig. 40, p. 86, Part I) placed with its principal axis along the auroral zone in the north of Europe. If we here imagine the storm-centre to move from time to time, and as a consequence the current-arrow at Pawlowsk to turn clockwise, the current-arrow at Fort Conger will turn through a corresponding angle counter-clockwise, and vice versa.

It will be seen from Charts I—IV, that we now have before us considerable oscillations of the current-arrow at Pawlowsk, and it would therefore be another reason for now being able to find a corresponding movement at Fort Conger. If we compare the declination-curves at about 0^h, we do actually find a similarity in form, which at first glance may seem unimportant, but which nevertheless is quite characteristic. It is at this time, too, that the negative storm is most strong and the area of precipitation so far concentrated, that one might expect to find similar conditions as mentioned.

The reason why the normal line is situated differently at the two stations, may only be that the situation of the stations in respectively the areas of convergence and divergence, is a little different. It is the form of the curve that gives the change in the force's strength and direction from time to time, and the normal line that gives the absolute values of the force. In comparing the curves, it must of course be remembered that the scale at Fort Conger is considerably larger than that at Pawlowsk, so that the variations in the perturbing forces at work are somewhat similar in magnitude.

In the interval between Chart II and Chart IV, the current-arrow at Pawlowsk, as we see, makes a considerable turn clockwise. During the same period, P_d at Fort Conger changes from east to west, which means that the current-arrow, if assumed to have a component in a northerly direction, turns a certain angle counter-clockwise. In the interval from Chart I to Chart II, in which the movement at Pawlowsk is certainly distinct, but slight, nothing can be decided, as we do not know P_h at the other station, and there is little variation in P_d.

We must, of course, be careful not to attach too much importance to this circumstance, and the apparent harmony between the actual perturbation-conditions and theory; but on the other hand, this has a special interest, as it is one of the very few cases in which we seem able to trace the areas of both convergence and divergence of the same polar elementary storm.

This movement of the current-arrows, which we see, at any rate, distinctly in the area of convergence, should therefore indicate that the storm-centre was moving eastwards during the perturbation. The conditions at Little Karmakul, however, do not seem to indicate any such movement; on the contrary, the perturbing force diminishes here rather rapidly, and then, from Chart IV, changes. The field in the first three charts does not, however, present any difficulties, as we only need to assume that the district of precipitation to the east of the European stations is rather more northerly in situation than it is in these regions. This is not at all at variance with what we have seen before, for even in Part I we have drawn attention to the fact that the negative areas of precipitation on the day-side would be situated a little farther north than those on the night-side.

The direction of the current-arrow at Little Karmakul on Chart IV might be explained by the circumstance that the station was situated in the area of convergence of the negative system of precipitation, and south of the point of convergence; but a consideration of the course of the curve seems to make such an assumption at any rate very improbable, as the forces are much too strong, and the character of the curve too disturbed. These conditions seem to indicate more or less certainly that we have before us the effects of a positive precipitation.

The fact that it is difficult to follow the movement of the system in the polar regions, may to some extent be due to our lack of observations for the time about the beginning of the perturbations.

If we assume that the negative district of precipitation continues also in the districts to the eastward of Europe as indicated above, we have a good explanation of the perturbation-area that appears on Chart IV. If, on the other hand, we assume that it terminates somewhat to the west of Little Karmakul, it will be much more difficult to find a simple explanation of that, supposing the storm to be more or less purely polar. Altogether it is difficult to say anything more definite about the conditions here, as the observations supply only very imperfect information regarding the perturbation-conditions.

On Chart V we see however that the positive system in Little Karmakul, which hitherto have not been very prominent and which on the whole would appear to have been of mainly local character, begins to assert itself more strongly. Simultaneously with this, the traces of converging area, which we up to Chart IV find at the southerly stations, disappear.

On Chart VI the negative system in the north of Europe has disappeared, but on the other hand we now find the previously mentioned system at Godthaab very well developed. At Fort Rae the positive polar storm also begins to develope, although the forces there are still very weak.

Lastly, on Chart VII, for 2^h 15^m, the positive system at Fort Rae has attained a more or less considerable magnitude. We find moreover a negative storm that is only slight, though very distinct; and on each side of the principal axis; the two characteristic areas of convergence and divergence seem to be formed here too.

Subsequently the positive storm at Fort Rae developes further, and attains its greatest strength at about 3^h. As, however, at this hour, there are no perturbations of any great strength at the other stations, we have drawn no chart.

TABLE LIV.

The Perturbation of the 14th & 15th February, 1883.

Gr. M. T.	Uglaamie			Fort Rae			Kingua Fjord	
	P_h (1)	P_d	P_v	P_h	P_d	P_v	P_h	P_d
h m								
23 25	− 32.5 γ	W 42.5 γ		+ 11 γ	W 15.5 γ	− 10 γ	+ 11 γ	W 45 γ
40	− 21 „	„ 40 „	No deflection sufficiently well defined to allow of anything being deduced. The temperature, has also varied greatly.	+ 11 „	„ 20 „	+ 10 „	+ 7 „	„ 48.5 „
45	− 27 „	„ 34.5 „		+ 6 „	„ 22 „	0	+ 12 „	„ 50 „
55	− 35 „	„ 45 „		+ 17 „	„ 20 „	− 30 „	+ 17 „	„ 52.5 „
0 0	− 45.5 „	„ 32 „		0	„ 11 „	0	+ 13 „	„ 51.5 „
10	− 27.5 „	„ 53 „		+ 9 „	„ 9 „	+ 10 „	+ 3 „	„ 42 „
20	− 15 „	„ 8 „		+ 15 „	„ 6.5 „	0	+ 4 „	„ 27 „
50	− 18 „	0		+ 31 „	„ 4.5 „	0	− 7 „	„ 39.5 „
1 0	− 6 „	E 10.5 „		+ 21 „	E 4.5 „	+ 10 „	− 5 „	„ 43 „
10	+ 13 „	„ 2.5 „		+ 29 „	W 2 „	+ 10 „	− 5 „	„ 18 „
20	+ 20 „	W 5.5 „		+ 23 „	„ 2 „	0	+ 3 „	„ 31 „
40	+ 12.5 „	„ 8 „		+ 46 „	„ 13 „	+ 10 „	+ 10 „	„ 7 „
2 15	+ 14 „	„ 24 „		+ 70 „	„ 17.5 „	0	+ 21 „	„ 12 „
55	+ 27 „	„ 8 „		+ 70	„ 22 „	− 100 „	+ 21 „	E 3.5 „

(1) Great Variation in temperature, which has a great influence on the form of the normal line.

TABLE LIV (continued).

Gr. M. T.	Godthaab		Jan Mayen			Bossekop		
	P_h	P_d	P_h	P_d	P_v	P_h	P_d	P_v
h m								
23 25	+ 15 γ	W 54 γ	−276 γ	E 35.5 γ	+103 γ	− 95 γ	W 24.5 γ	−213 γ
40	− 10 „	„ 54 „	−279 „	„ 32 „	+ 25 „	−166 „	E 2.5 „	−280 „
45	− 8 „	„ 54 „	−274 „	W 43.5 „	+ 62 „	−153 „	„ 23 „	−246 „
55	− 8 „	„ 60.5 „	−204 „	„ 42.5 „	+ 57 „	−135 „	„ 54 „	−220 „
0 0	0	„ 60 „	−196 „	E 32.5 „	+ 47 „	−117 „	„ 62 „	−212 „
10	− 4 „	„ 47 „	−177 „	0	+ 51 „	− 83 „	„ 46.5 „	−176 „
20	+ 3 „	„ 33 „	−105 „	W 4 „	+ 15 „	− 61 „	„ 43 „	−145 „
50	− 92 „	„ 34 „	− 25 „	„ 63 „	+ 35 „	− 38 „	„ 13.5 „	− 88 „
1 0	− 54 „	„ 84.5 „	− 59 „	„ 8 „	+ 56 „	− 18 „	„ 16 „	− 51 „
10	−130 „	„ 42 „	− 37 „	„ 2 „	+ 46 „	− 6 „	„ 124 „	− 33 „
20	− 49 „	„ 56 „	− 30 „	E 18 „	+ 55 „	+ 5 „	„ 9 „	− 6 „
40	− 47 „	„ 8.5 „	+ 2 „	„ 7.5 „	+ 36 „	− 8 „	„ 15 „	− 22 „
2 15	+ 12 „	„ 14.5 „	− 38 „	„ 22.5 „	+ 38 „	− 7 „	„ 23 „	− 19 „
55	+ 14 „	„ 8.5 „	− 23 „	„ 10 „	+ 58 „	− 15 „	„ 31.5 „	− 36 „

TABLE LIV (continued).

Gr. M. T.	Sodankylä			Cape Thordsen			Little Karmakul		
	P_h	P_d	P_v	P_h	P_d	P_v	F_h	P_d	P_v
h m									
23 25	+ 3 γ	W 4 γ	+ 70 γ	− 76 γ	E 59.5 γ	+185 γ	−174 γ	E 44 γ	−139 γ
40	− 18 „	„ 4.5 „	+ 88 „	−183 „	„ 3 „	+ 96 „	−146 „	„ 35.5 „	−129 „
45	− 16 „	E 6.5 „	+ 76 „	−214 „	„ 37 „	+116 „	− 56 „	W 42.5 „	−113 „
55	− 21 „	„ 25.5 „	+ 77 „	−203 „	„ 73 „	+142 „	−108 „	E 42 „	−124 „
0 0	− 17 „	„ 32 „	+ 58 „	−168 „	„ 57 „	+133 „	− 59 „	„ 44.5 „	−111 „
10	− 11 „	„ 25 „	+ 39 „	−136 „	„ 77 „	+140 „	− 11 „	„ 42.5 „	− 97 „
20	− 12 „	„ 24 5 „	+ 62 „	− 77 „	„ 43 „	+ 88[1] „	+ 32 „	0	− 74 „
50	− 13 „	„ 7.5 „	+ 30 „	+ 15 „	W 11.5 „	+ 34 „	+ 91 „	W 4 „	− 23 „
1 0	− 10 „	„ 8.5 „	+ 24 „	− 32 „	E 64 „	+ 18 „	+ 97 „	„ 12 „	− 10 „
10	− 6 „	„ 6 „	+ 15 „	− 6 „	„ 40 „	− 27 „	+ 60 „	„ 8 „	− 13 „
20	− 12 „	„ 7.5 „	+ 5 „	− 1 „	„ 31 „	− 38 „	+ 47 „	0	− 8 „
40	− 13 „	„ 9 „	− 6 „	+ 2 „	0	+ 23 „	+ 24 „	E 3.5 „	− 17 „
2 15	+ 3 „	„ 13 „	− 7 „	− 61 „	„ 80 „	+ 89 „	+ 39 „	W 2.5 „	− 13 „
55	− 1 „	„ 16 „	+ 8 „	− 7 „	„ 48.5 „	+105 „	+ 26 „	„ 3 „	− 32 „

[1] For this hour there was no observation, and the value given is interpolated between $0^h\ 15^m$ and $0^h\ 25^m$.

TABLE LIV (continued).

Gr. M. T.	Ssagastyr	Christiania		Pawlowsk			Göttingen			Fort Conger
		P_h	P_d	P_h	P_d	P_z	P_h	P_d	P_z	P_d
h m 23 25	The perturbation can only be discerned by small oscillations in the curves, and there are no distinct deflections; nothing therefore can be deduced.	+ 32 γ	E 41.5 γ	+ 18 γ	E 8.5 γ	− 5 γ	+ 13.5 γ	E 37 γ	− 16.5 γ	E 21 γ
40		+ 29.5 „	„ 25 „	+ 15 „	W 2 „	− 8 „	+ 24 „	„ 27.5 „	− 3.5 „	„ 16.5 „
45		+ 34 „	„ 25.5 „	+ 15 „	E 3 „	− 8 „	+ 22.5 „	„ 27.5 „	− 2 „	„ 4.5 „
55		+ 13.5 „	„ 31.5 „	+ 13 „	„ 13.5 „	− 12 „	+ 19.5 „	„ 29.5 „	− 1 „	W 18.5 „
0 0		+ 8 „	„ 32 „	+ 8 „	„ 17 „	− 12 „	+ 21.5 „	„ 28.5 „	− 0.5 „	„ 19 „
10		+ 2 „	„ 23.5 „	0	„ 14 „	− 10 „	+ 7.5 „	„ 17 „	0	„ 12 „
20		+ 1 „	„ 17.5 „	− 5 „	„ 13 „	?	+ 5.5 „	„ 11.5 „	0	„ 14.5 „
50		− 6 „	W 3 „	− 10 „	„ 2 „	?	+ 1 „	W 6.5 „	+ 1.5 „	„ 16.5 „
1 0		− 3.5 „	0	− 7 „	„ 5 „	?	− 2 „	0	+ 0.5 „	E 13 „
10		0	0	− 4 „	„ 1.5 „	?	0	E 2.5 „	0	„ 16 „
20		+ 1 „	„ 0.5 „	− 3 „	„ 1.5 „	+ 2 „	− 2 „	„ 2 „	− 1 „	„ 4.5 „
40		+ 1 „	„ 1.5 „	0	„ 1.5 „	+ 3 „	+ 3.5 „	0	0	„ 6.5 „
2 15		+ 4 „	E 4 „	− 5 „	„ 5 „	+ 4 „	+ 3.5 „	„ 7.5 „	0	„ 23 „
55		− 1 „	„ 2 „	+ 2 „	„ 5 „	0	+ 1 „	0	− 3 „	„ 41.5 „

Current-Arrows for the 14th February 1883.
Chart I at 23h 25m.

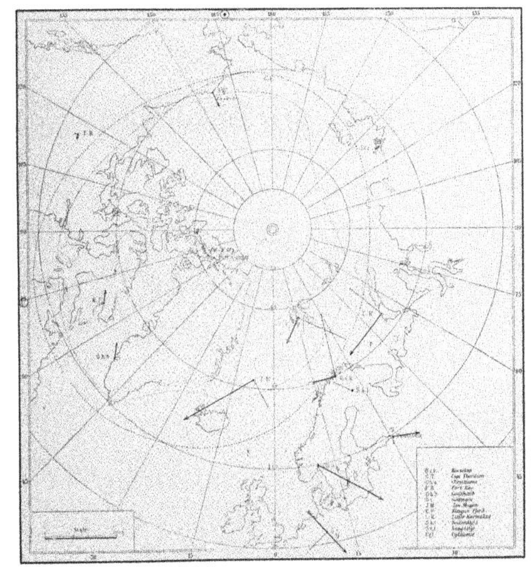

Fig. 152.

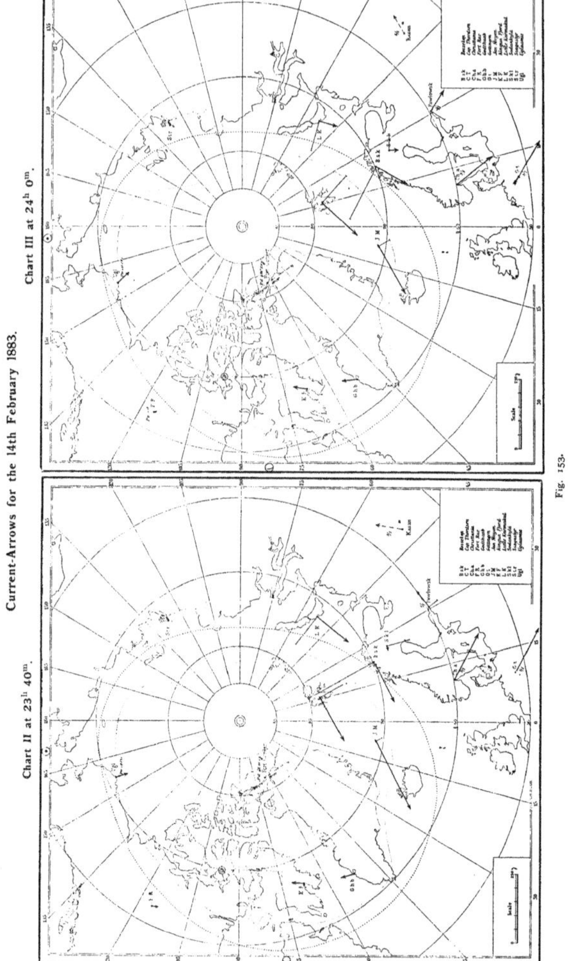

Current-Arrows for the 14th February 1883.

Chart II at $23^h\ 40^m$.

Chart III at $24^h\ 0^m$.

Fig. 153.

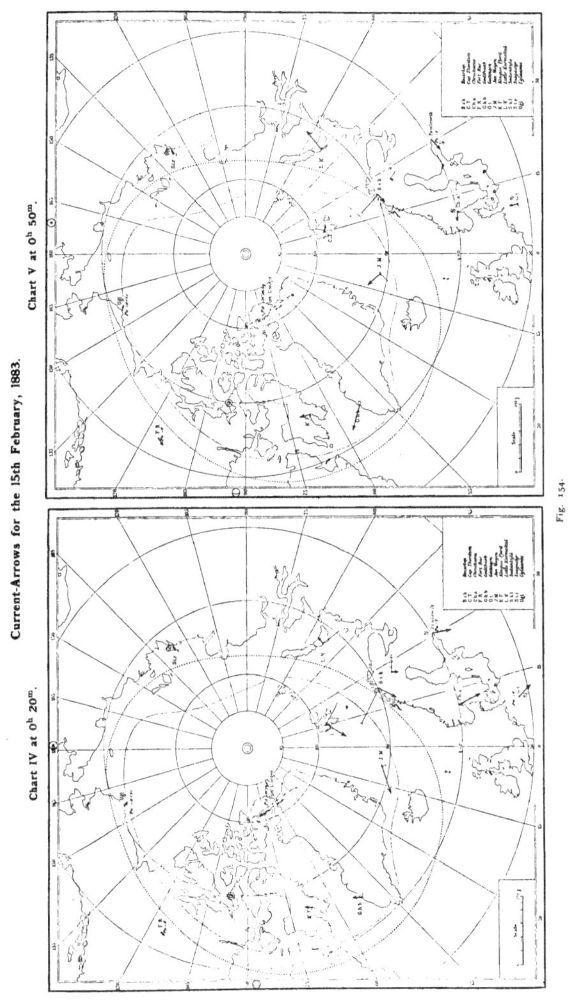

Current-Arrows for the 15th February, 1883.

Chart IV at $0^h\ 20^m$. Chart V at $0^h\ 50^m$.

Fig. 154.

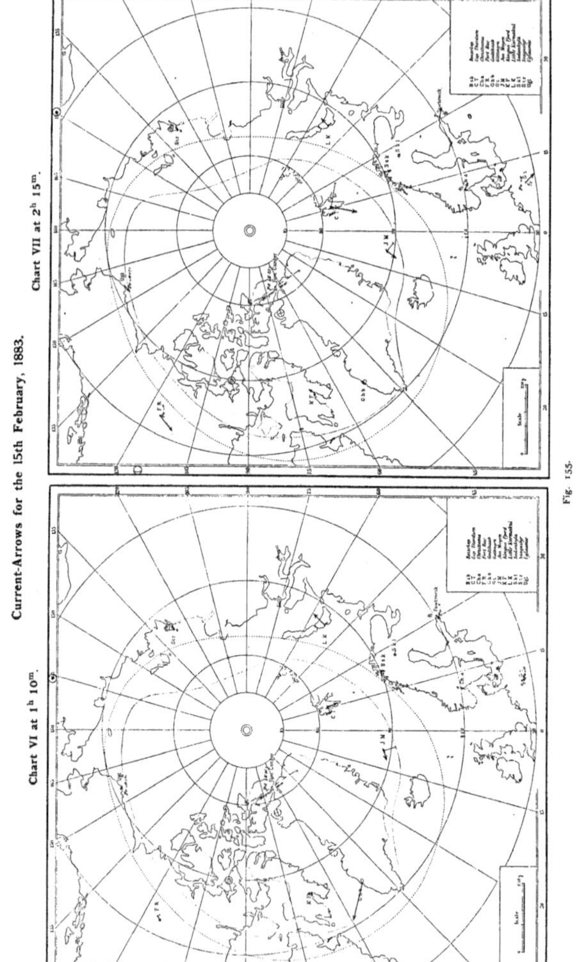
Current-Arrows for the 15th February, 1883.
Chart VI at $1^h\ 10^m$.
Chart VII at $2^h\ 15^m$.

Fig. 155.

THE PERTURBATIONS OF THE 15th JULY, 1883.
(Pl. XXIX).

87. As the curves show, the storms occurring on the above date, especially those in the polar regions, are exceedingly characteristic and well defined, and of considerable power.

We have previously described principally magnetic storms that occurred in the winter, and two or three perturbations about the spring equinox. Special interest will therefore attach to a case of a magnetic storm occurring near the summer solstice, and the storm now to be described is a good example of just such a storm.

It may at the outset seem very unlikely that the main features in the occurrence and course of the perturbations should change character; indeed one would rather expect to find the same principal features, while the details might possibly exhibit more peculiar conditions.

We will now go through the various phases, and see how well these assumptions are confirmed.

We may consider the interval from 6^h to 10^h as a first section, for during that time there occur at several places, as the curves show, perturbations that are all comparatively slight, but sometimes very well defined. The most powerful forces occur at Fort Rae, where the perturbation is a series of brief impulses taking place at about $7^h\ 30^m$, $8^h\ 20^m$, and from 9^h to $9^h\ 20^m$.

The deflections in the district Fort Conger to Cape Thordsen are particularly characteristic, and the time of their commencement there is a little earlier than in the perturbation at Fort Rae.

At Bossekop and the southern European stations, disturbances are only sometimes noticeable, and the deflections are as a rule too small to be taken out.

It may here be worth while pointing out one circumstance connected with this first perturbation, namely, that there is at the same time a deflection in one of the earth-current components at Pawlowsk, which exhibits a remarkable resemblance to the deflections in the magnetic curves to the north. Whether this is accidental, or whether a close connection between these phenomena exists, we will not attempt to decide here. In this connection we will refer to a later chapter where the earth-currents are described.

As the systems acting here are rather weak, the drawing of the corresponding current-arrows on the charts will not give a much clearer idea of the perturbation-conditions than we obtain by the direct consideration of the curves. We have not therefore drawn any chart for this period of the storm. Its field of operations appears to be rather limited, and its occurrence more or less local in the north.

At Fort Rae, where it is about midnight at this time, the storm is of the nature of a negative polar storm; but nothing decided can be said as to what it may be at the other stations.

After this slight, comparatively brief perturbation, a long period supervenes during which the conditions are normal.

At about 14^h, however, powerful perturbations begin to develope all round the polar stations. In the district Fort Rae, Uglaamie and Ssagastyr, an exceedingly characteristic, powerful negative polar storm developes, which also seems to act with considerable strength at Kingua Fjord, judging from the deflections in the horizontal intensity. At the last-named place, the system appears to be a little earlier in its occurrence than at Fort Rae. We must not, however conclude too much from the conditions in the horizontal intensity alone, as the deflections in declination have a greater significance at Kingua Fjord than at the other stations.

A perturbing force in the horizontal intensity will thus here produce current-arrows directed more or less north and south, while at the other stations the variations in the horizontal intensity will answer to current-arrows pointing east and west. It is therefore best here to keep principally to the charts for a general idea of the conditions.

In the district Jan Mayen, Bossekop and Little Karmakul, on the other hand, a fairly powerful positive polar storm developes, its effects also being at first apparent as far north as Cape Thordsen, and at Godthaab.

On Cape Thordsen and Jan Mayen, that is to say at the two stations situated to the north of the auroral zone, the conditions are a little more complicated, from the fact that later on, at about 16^h, a negative polar storm appears to break in upon the positive, which, in Cape Thordsen, it considerably exceeds in strength, causing in consequence strong negative deflections in the horizontal-intensity curve.

The negative storm that asserts itself here, also acts, and very powerfully too, at Fort Conger, where the deflections are strongly marked.

With regard to Jan Mayen, the effects of the negative storm are not so apparent, partly because the effects of the positive storm are very strongly marked, and partly because perhaps the area of precipitation of the negative storm is not so much in the immediate vicinity of this station as of Fort Conger and Cape Thordsen. The negative storm, when at its height — that is to say at about 17^h or 18^h — only succeeds in almost neutralising the effect of the positive storm as far as the horizontal intensity is concerned. In declination and vertical intensity, on the other hand, especially in the latter component, there are very marked deflections at the above-mentioned time. P_z is in one direction all the time, and negative. This is what might be expected, as both the negative system to the north, and the positive system to the south, will cause deflections in a negative direction. The character of the declination curve is more disturbed, and several powerful, brief impulses occur, now in one direction and now in another.

The perturbations are evolving, when thus looked at as a whole, exactly in the same manner as in the most typical of the cases we have already considered.

It is moreover easy here to study the movements of the systems, which stand out with peculiar distinctness in the case of the negative system of precipitation.

At Kingua Fjord, the wide deflections in the horizontal-intensity curve begin rather suddenly at $14^h\ 10^m$. At Fort Rae, on the other hand, the deflections at first increase more slowly, so that no definite time for their commencement can be given. On looking at the horizontal-intensity curve, however, we find a considerable difference in time, by comparing the beginning and the time of the maximum deflection. It is a little doubtful how great this difference is, but we may put it roughly at one hour.

We cannot, however, take it for granted that the effects observable at these two stations are those of one and the same system; but we obtain a better general idea from the charts.

At Uglaamie we also have a very characteristic deflection in H, which both begins and ends rather abruptly. It is therefore easy here to determine a difference in corresponding hours. Compared with Kingua Fjord, there is a difference of about $1^1/_2$ hours in the time of its commencement, while it ends only about three quarters of an hour later than at Kingua Fjord. Between Uglaamie and Fort Rae there is a distinct difference of about half an hour, observable both at the beginning and the end; and there seems to be no doubt that this is the effect of one and the same system. The deflections in H are here so powerful that the needle is outside the field of observation from $16^h\ 55^m$ to $19^h\ 35^m$, except at $18^h\ 25^m$ and $18^h\ 30^m$, when readings have been taken.

The next station at which the negative storm acts is Ssagastyr, where the deflections in H begin about half an hour later than at Uglaamie, and are very sharp and distinct. A comparison, as regards the time of the maximum, with Fort Rae, shows a similar condition, the difference being about one hour.

The deflections in H do not decrease regularly until the conditions have once more become normal; but for two hours after about 19^h there is a more or less constant perturbing force of about $-150\,\gamma$. The character of the curve seems to indicate that there has been some defect in the instruments, and that the needle in some way or other has become fixed; but as there are at the same time perturbing forces in the declination, it is impossible to be sure of this.

We can thus, in this district from Kingua Fjord, or at any rate from Fort Rae, through North America to Ssagastyr, trace a distinct westward movement of the system of precipitation. A powerful but not extensive system first developes in the vicinity of Kingua Fjord, and apparently spreads towards the west and forms the great, connected system of precipitation in the north of America, presumably simultaneously with the westward movement of the entire system with the sun.

No specially pronounced movement is descernible, on the other hand, in the positive system. It might appear, indeed on a cursory glance at the deflections in Jan Mayen as compared with those at Bossekop and Sodankylä, as if there were a distinct eastward movement of the system; for at about 14^h 20^m the positive deflections at the first-named station attain a considerable strength, and remain more or less constant until 16^h, when they once more diminish rapidly. At Bossekop and Sodankylä, the positive deflections begin at about the same time as those in Jan Mayen; but they increase slowly, and the most powerful forces are not found until between 16^h 30^m and 17^h 30^m, the time at which the conditions in H in Jan Mayen are fairly normal. It might thus appear as if the positive system had here moved eastwards; but we have already explained the way in which this phenomenon is to be understood, and how the negative system to the north breaks in upon the positive system first acting in Jan Mayen. This, however, does not preclude a possibly eastward movement of the system of precipitation. It is also probable that the positive storm-centre will be moved; but the observations we possess do not afford sufficient evidence of this.

Little Karmakul is now also upon the border between the two systems of precipitation; and its curves have consequently the disturbed, jagged character so often observed before. At one time the positive system is the stronger, at another the negative, although at first the positive system predominates, while from about 17^h 30^m onwards, the effect of the negative system is the more apparent.

The negative storm at Cape Thordsen and Fort Conger must on the whole be regarded as a continuation of the negative storm in North America and the north-east of Asia, although it is very possible that it forms a more independent system.

At the southern stations it is sometimes rather difficult to determine the normal line, as the diurnal variation at this season of the year is considerable, and the data from which the determination is made are as a rule few. It is therefore possible that some error will attach to the values found; but at the times when the perturbing forces are powerful, this will have no great signifiance.

At about 20^h, this perturbation is practically over. This is clearly apparent from the curves of the horizontal intensity. It is not yet quiet everywhere, however, as, in the declination especially, there are sometimes fairly powerful perturbing forces.

In the district Fort Conger to Kingua Fjord, the effects of a fairly powerful system of precipitation are still distinctly apparent, and are noticeable at Godthaab and to some extent in Jan Mayen. The perturbation is especially powerful at Kingua Fjord. At about 23^h, however, new storms begin to develope, evolving in the usual manner of the polar storms at about midnight, Greenwich time. A powerful negative storm on the night-side, from Little Karmakul, across Bossekop to Jan Mayen, forms the main system, its effect also extending westwards across Ssagastyr to Uglaamie. We find moreover distinct traces of a positive system on the afternoon-side, especially at Fort Rae; but the horizontal-intensity curve for Godthaab and possibly Kingua Fjord indicates that these stations are also affected by this positive system. Here, however, the conditions seem to be rather more complicated, perhaps because the effects of the above-mentioned system occurring in these regions are still apparent.

Special attention should be paid to the positive system of precipitation on the afternoon-side in North America. It occurs principally at Fort Rae, that is to say it is most marked at the station situated to the south of the auroral zone.

It is, as stated in the description of the preceding perturbation, comparatively seldom that the effects of positive systems of precipitation can be observed in these polar regions. This, however, is the most characteristic example of such effects, and therefore goes far towards confirming our previous assumptions. Unfortunately, only the first half of the perturbation can be studied, as the period of observation ends while the deflections are greatest.

We have now briefly reviewed the development of the perturbation by considering the curves, and have found that in the main the same conditions are repeated, and the development takes place in exactly the same manner, as in the earlier storms.

We will now pass on to consider the charts in which we have represented the various fields of perturbation. These fields are here slightly more complete, as we have also made use of observations from Kasan, from which place we have entire series of observations of the two horizontal components for the last term days from the 15th May onwards.

For this day we have drawn 14 charts representing 15 epochs in all.

As *Chart I* shows, it is the positive storm that first developes. It is especially noticeable that the positive system of precipitation appears to be situated comparatively far south, judging from the conditions at the southern stations; for if it is principally only this positive system that is acting, the stations that we have included here must lie to the north of the point of divergence of this system. There is of course also a possibility that in addition there is precipitation of stiffer rays in rather lower latitudes, these being here those with the greatest effect.

The positive system has developed most fully in the district Godthaab to Jan Mayen, while its effects farther east are comparatively slight.

There is perhaps rather more uncertainty as to the manner in which the conditions at Kingua Fjord are to be understood. The direction of the current-arrows there is almost due south. Judging from the chart, it would seem likely that the conditions might be considered as a continuation of the positive system of precipitation. When we considered the curves and compared them with those at Godthaab, we found, it will be remembered, that the character of the deflections at the two stations was sufficiently different to justify the assumption that they were not very closely connected with one another, but that on the contrary a system was acting at Kingua Fjord that was scarcely noticeable at Godthaab. This assumption also seems to be the most probable on looking more carefully at the charts. At first, however, this system at Kingua Fjord is comparatively inconspicuous and rather limited in its effects; and the positive system that has formed to the east of it sometimes seems to encroach upon it and get the upper hand. This is the case at the time of *Chart II*, when there clearly seems to be a positive system of precipitation right from Kingua Fjord eastward past Little Karmakul, possibly as far as Ssagastyr. No effects of a negative system of precipitation are noticeable. The strong current-arrows at the southern stations also seem to indicate now that in addition to the great precipitation in or about the auroral zone, there may be smaller amounts of precipitation farther south. Without such an assumption it would be difficult to find a simple explanation of these current-arrows. The jagged, disturbed character of the curves, especially the horizontal-intensity curves, is moreover a circumstance that supports this view, this fact indicating that the systems in operation cannot be very far from the station itself. At the same time, the oscillations at the polar stations to the north—Jan Mayen, Bossekop and Sodankylä—as also at Cape Thordsen, are comparatively gentle, without any sudden, violent changes backwards and forwards. At Little Karmakul, however, the curve is rather jagged.

The negative system of precipitation does not appear distinctly until $15^h\ 30^m$, *(Chart III)*, either at Kingua Fjord, where it is strongest, or at Fort Rae. The positive system is also well developed here; but at Cape Thordsen the perturbing forces in the horizontal components are rather small, this being due to the fact that the negative system, which there developes subsequently to such considerable strenght, is already encroaching upon the positive. In other respects there is little alteration in the appearance of the field, and the forces at work are only sometimes weaker than before.

The current-systems continue to develope upon the succeeding charts. On *Chart IV*, for the period $15^h\ 40^m$ to $15^h\ 50^m$, the conditions are not very different from those on Chart III, except that the forces at Fort Rae are a little more powerful.

On *Chart V* the development of the negative system can be followed. At the first hour shown, $16^h\ 15^m$, Uglaamie is in its district of precipitation, but the latter does not extend as far west as Ssagastyr. At $16^h\ 40^m$, however, the great negative system has developed all round at the various stations. This now forms a more or less continuous circuit, which can be traced from Godthaab to Kingua Fjord, across Fort Rae, Uglaamie and Ssagastyr to Cape Thordsen and Fort Conger.

The northerly position of this system on the afternoon-side is worthy of notice, as also its comparatively southerly position on the morning-side, as, judging from the vertical intensity, it should lie in the first case to the north of Cape Thordsen, and in the second to the south of Fort Rae.

We must, however, once more urge the necessity of caution in drawing conclusions from the conditions in the vertical intensity, and need only point to the vertical arrows at Sodankylä during these storms, which here too exhibit rather abnormal conditions as regards direction.

The positive area of precipitation seems now to be considerably reduced, and distinct effects are found only at Bossekop, Sodankylä and Little Karmakul. In reality, however, it may possibly extend farther west, but then farther south than the regions from which we have observations.

On Jan Mayen the current-arrow is comparatively very small, while the vertical arrow is of considerable length and is directed upwards. This is in accordance with a circumstance that we have also drawn attention to previously, namely, that the station is situated between a northern negative and a southern positive system of precipitation.

We find no special change in the form of the field in *Charts VI* and *VII*, but the forces increase considerably everywhere. The high value of P_d at Fort Conger should be especially noticed, it being about 864 γ at $17^h\ 20^m$ (Chart VII), or considerably more than any of the other perturbing forces observed.

P_h cannot be measured at Uglaamie, as the needle has swung out of the field of observation; so it may possibly have been as great or even greater here. It is interesting, however, to find that there is also powerful precipitation close to the magnetic axis.

As *Charts VIII* and *IX* show, the negative system encroaches farther upon the positive, and causes a reversal of the current-arrow at Little Karmakul; while at the same time the current-arrows at Pawlowsk and Kasan become more southerly in direction. On Chart IX, the effects of the positive system are slight at the stations under consideration.

At $18^h\ 20^m$, on *Chart X*, we once more find a fairly powerful polar positive system of precipitation from Kingua Fjord eastwards to Little Karmakul. This time, however, the system appears to be a little farther north, at any rate in Europe; as Pawlowsk, Kasan and Göttingen are now distinctly in the southern part of the area of divergence of the system. As this only lasts for a short time, it should rather be regarded as a brief impulse. The effects of the negative system still continue, however, although the forces are to some extent less powerful than before.

Chart XI, for $18^h\ 55^m$, represents the perturbation-conditions as they appear shortly before the great systems disappear. We still find distinct traces of the great negative current-circle, while on the other hand, the effects of the positive system are less distinct, although it seems to exist, judging from the conditions in Jan Mayen and the southern stations; but this cannot be decided with certainty.

When these storms have ended, there is an interval of more or less normal conditions at most places, although it is by no means quiet everywhere; but what perturbations there are, are of a more local character, and the existence of large connected systems can hardly be proved with certainty. This is clearly evident from *Chart XII*, for the hour 19^h 50^m.

Two *Charts, XIII* and *XIV*, have been drawn for the last section of the perturbations of the day under consideration, from about 22^h to the end of the period. The conditions are comparatively simple and clear. On the night-side there is a powerful negative system of precipitation, which extends from Ssagastyr westwards through the north of Europe to Godthaab and Kingua Fjord. At the two last-named stations the direction of the current-arrows is a little peculiar. The principal axis of the system seems to turn off towards the north rather abruptly. This seems to be analogous to the circumstance we have so often observed before, namely, that the negative system turns up, on the afternoon-side, into higher latitudes to the north of a positive system in the vicinity of the auroral zone (fig. 140 p. 327).

At Fort Rae too, there is certainly a positive system, while the storm-centre of the negative system is in the north of Europe.

The current-arrow at Fort Rae, which should give the direction of the positive system of precipitation, has, it is true, a rather marked southerly direction; but this is so nearly the opposite of what we find during the ordinary negative storms here, that there seems no doubt that this is a positive area of precipitation.

TABLE LV.

The Perturbations of the 15th July, 1883.

Gr. M. T.	Uglaamie			Fort Rae			Kingua Fjord		
	P_h	P_d	P_v	P_h	P_d	P_v	P_h	P_d	P_v
h m									
6 50	+ 47 γ	W 26.5 γ	+ 30 γ	+ 7 γ	W 11 γ	− 10 γ	+ 28 γ	0	0
7 30	+ 46 „	0	+ 10 „	−202 „	E 102 „	0	+ 8 „	E 22.5 γ	0
8 20	+ 27 „	E 21 „	0	−119 „	„ 15 „	−100 „	+ 5 „	„ 22.5 „	+ 22 γ
9 5	+ 4 „	„ 5.5 „	0	− 68 „	W 82 „	+ 20 „	− 23 „	„ 9.5 „	0
10 20	+ 9.5 „	0	0	0	0	0	− 15 „	0	0
11 50	+ 3.5 „	„ 2.5 „	0	0	0	0	− 52 „	0	0
13 20	− 2 „	„ 8 „	0	+ 17 „	0	0	− 55 „	W 15 „	0
14 35	+ 3.5 „	W 18.5 „	+ 25 „	− 24 „	E 13.5 „	+ 10 „	−222 „	E 55.5 „	− 27 „
55	+ 8 „	„ 90 „	+ 35 „	− 47 „	„ 11 „	+ 20 „	− 83 „	„ 184 „	− 22 „
15 30	− 17.5 „	E 2.5 „	+ 23 „	−142 „	„ 70 „	+ 45 „	−323 „	W 115 „	− 52 „
40	− 53 „	„ 2.5 „	+ 80 „	−181 „	„ 83 „	+ 65 „	−300 „	0	− 72 „
50	− 91 „	„ 53 „	+ 80 „	−202 „	„ 152 „	+ 28 „	−205 „	„ 140 „	−165 „
16 15	− 187.5 „	„ 26.5 „	+137 „	−325 „	„ 209 „	+220 „	−285 „	„ 233.5 „	− 85 „
40	− 290 „	„ 172 „	+180 „	−522 „	„ 280 „	+100 „	−343 „	„ 120 „	− 71 „
17 0	− >300 „	„ 20 „	+205 „	−613 „	„ 298 „	+140 „	−240 „	„ 62.5 „	−125 „
20	− >300 „	W 108 „	+230 „	−606 „	„ 412 „	+120 „	−364 „	„ 86 „	−126 „
35	− >300 „	E 246 „	+180 „	−545 „	„ 325 „	+ 40 „	−255 „	„ 62.5 „	−136 „
55	− >300 „	„ 609 „	+135 „	−338 „	„ 128 „	−200 „	−216 „	„ 144 „	−140 „
18 20	− >300 „	„ 218 „	+ 80 „	−134 „	W 115 „	− 55 „	− 26 „	E 174 „	− 91 „
55	− 239 „	„ 205 „	+ 57 „	+ 30 „	E 47 „	− 20 „	+ 24 „	W 270 „	−101 „
19 50	+ 30 „	W 55 „	+ 10 „	+ 11 „	W 20 „	− 20 „	− 7 „	„ 124 „	− 63 „
22 55	− 48 „	„ 8 „	− 15 „	+221 „	E 46 „	− 40 „	+182 „	„ 191 „	− 21 „
23 15	0	„ 77 „	+ 33 „	+314 „	„ 116 „	− 60 „	+197 „	„ 210 „	+ 84 „

TABLE LV (continued).

Gr. M. T.	Godthaab		Jan Mayen			Bossekop		
	P_h	P_d	P_h	P_d	P_v	P_h	P_d	P_v
h m								
6 50	+ 14 γ	W 7.5 γ	− 15 γ	E 11.5 γ	0	0	E 7 γ	0
7 30	0	E 4 "	0	0	− 10 γ	0	0	0
8 20	0	" 14 "	0	W 5.5 "	0	0	0	0
9 5	0	0	− 9 "	" 3 "	0	0	0	0
10 20	0	0	0	0	0	0	0	0
11 50	− 5 "	W 8.5 "	0	" 8.5 "	0	0	0	0
13 20	+ 10 "	0	+ 12 "	0	− 7 "	0	0	0
14 35	+ 9 "	E 83.5 "	+ 143 "	" 4 "	− 24 "	+ 73 γ	0	+ 47 γ
55	− 30 "	" 153 "	+ 249 "	E 15.5 "	− 37 "	+ 142 "	W 27 "	+ 113 "
15 30	+ 109 "	" 127 "	+ 220 "	" 33 "	− 20 "	+ 113 "	E 7 "	+ 130 "
40	+ 24 "	" 188.5 "	+ 235 "	" 15.3 "	− 35 "	+ 140 "	W 12 "	+ 118 "
50	+ 103 "	" 137 "	+ 260 "	W 19 "	− 61 "	+ 120 "	0	+ 170 "
16 15	0	W 11 "	+ 132.5 "	" 5.5 "	− 160 "	+ 200 "	" 21.5 "	+ 250 "
40	− 161 "	" 59.5 "	+ 35 "	" 52.5 "	− 183 "	+ 260 "	E 50 "	+ 208 "
17 0	− 107 "	" 61.5 "	+ 13 "	" 108 "	− 204 "	+ 255 "	W 81 "	+ 272 "
20	− 132 "	" 89.5 "	− 59 "	" 54 "	− 157 "	+ 288 "	" 16 "	+ 200 "
35	− 105 "	E 67 "	− 37 "	" 52.5 "	− 200 "	+ 247 "	0	+ 110 "
55	− 80 "	0	− 5 "	" 106 "	− 220 "	+ 76 "	" 8.5 "	− 38 "
18 20	+ 29 "	" 157 "	+ 277 "	" 52 "	− 283 "	+ 95 "	E 25.5 "	+ 100 "
55	+ 15 "	W 37 "	+ 81 "	" 53 "	− 132 "	0	W 7 "	− 13 "
19 50	+ 27 "	" 38 "	0	" 34 "	− 97 "	+ 55 "	0	+ 78 "
22 55	− 330 "	" 235 "	− 500 "	E 138 "	+ 225 "	− 577 "	" 100 "	− 207 "
23 15	+ 54 "	" 286 "	− 532 "	W 19 "	+ 223 "	− 606 "	" 23.5 "	− 213 "

TABLE LV (continued).

Gr. M. T.	Sodankylä			Cape Thordsen			Little Karmakul		
	P_h	P_d	P_v	P_h	P_d	P_v	P_h	P_d	P_v
h m									
6 50	0	4	0	+ 14 γ	E 44 γ	+ 26 γ	+ 42 γ	0	+ 35 γ
7 30	0	0	0	+ 27 "	" 5 "	0	+ 11 "	0	0
8 20	0	0	0	+ 15 "	W 4 "	0	− 13 "	0	0
9 5	0	0	0	0	" 5 "	0	− 28 "	0	0
10 20	0	0	0	+ 5 "	0	− 10 "	+ 25 "	0	− 13 "
11 50	0	0	0	0	0	0	+ 5 "	0	0
13 20	0	0	0	0	0	0	+ 10 "	0	− 10 "
14 35	+ 64 γ	W 4 "	0	+ 68 "	" 53 "	− 42 "	+ 83 "	W 41 γ	− 5 "
55	+ 112 "	" 17 "	+ 90 γ	+ 132 "	" 110 "	− 74 "	+ 237 "	" 100 "	− 5 "
15 30	+ 88 "	E 6 "	− 27 "	+ 30 "	" 57.5 "	− 150 "	+ 76 "	" 44 "	+ 2 "
40	+ 96 "	0	+ 78 "	− 20 "	" 60 "	− 200 "	+ 143 "	" 66 "	+ 2 "
50	+ 80 "	" 6 "	− 60 "	− 22 "	" 57 "	− 180 "	+ 133 "	" 57 "	+ 27 "
16 15	+ 105 "	W 4 "	− 90 "	− 105 "	" 48.5 "	− 225 "	+ 180 "	" 152 "	− 26 "
40	+ 177 "	" 25 "	− 60 "	− 165 "	" 33.5 "	− 248 "	+ 55 "	" 134 "	− 123 "
17 0	+ 142 "	" 41 "	+ 10 "	− 156 "	" 103.5 "	− 296 "	+ 82 "	" 61 "	− 30 "
20	+ 253 "	0	0	− 456 "	" 28 "	− 430 "	+ 150 "	" 108 "	− 60 "
35	+ 234 "	E 6 "	+ 81 "	− 287 "	" 178 "	− 476 "	− 212 "	" 15 "	− 272 "
55	+ 105 "	" 4 "	+ 82 "	− 178 "	" 133 "	− 237 "	− 307 "	E 128 "	− 178 "
18 20	+ 60 "	" 19 "	− 81 "	− 35 "	" 108 "	− 125 "	+ 123 "	W 90 "	− 120 "
55	+ 12 "	0	+ 32 "	− 86 "	" 115 "	− 150 "	− 273 "	E 9 "	− 30 "
19 50	+ 10 "	" 6 "	+ 10 "	− 33 "	" 39 "	− 62 "	+ 160 "	W 96 "	+ 67 "
22 55	− 460 "	" 75 "	+ 150 "	+ 60 "	E 129 "	+ 192 "	− 775 "	E 167 "	− 33 "
23 15	− 520 "	0	− 100 "	− 37 "	" 55.5 "	+ 330 "	− 658 "	" 173 "	− 70 "

TABLE LV (continued).

Gr. M. T.	Ssagastyr		Christiania		Pawlowsk		
	P_h	P_d	P_h	P_d	P_h	P_d	P_v
h m							
6 50	+ 20 γ	0	+ 2 γ	0	0	0	+ 14[1] γ
7 30	+ 10 „	W 6.5 γ	0	0	0	0	+ 14[1] „
8 20	0	„ 12.5 „	− 1 „	0	0	0	+ 8[1] „
9 5	0	„ 6.5 „	+ 1 „	0	0	0	0
10 20	0	0	0	0	0	0	0
11 50	0	0	0	0	0	0	0
13 20	− 10 „	0	+ 15.5 „	E 9.5 γ	+ 20 γ	0	0
14 35	+ 28 „	„ 10.5 „	+ 48.5 „	0	+ 52 „	W 14 γ	0
55	+ 50 „	„ 23 „	+ 86 „	W 9.5 „	+ 91 „	„ 26 „	0
15 30	+ 40 „	„ 27 „	+ 48 „	„ 2.5 „	+ 52 „	„ 13.5 „	+ 5 „
40	− 26 „	„ 73.5 „	+ 62 „	„ 4.5 „	+ 36 „	„ 12 „	+ 5 „
50	+ 36 „	„ 37 „	+ 38 „	0	+ 35 „	„ 13.5 „	+ 6 „
16 15	+ 5 „	„ 189 „	+ 42 „	„ 9.5 „	+ 33 „	„ 7 „	+ 8 „
40	− 277 „	„ 56.5 „	+ 83 „	„ 14 „	+ 62 „	„ 16.5 „	+ 15 „
17 0	− 137 „	„ 66 „	+ 81 „	„ 23.5 „	+ 52 „	„ 14.5 „	+ 18 „
20	− 224 „	„ 296 „	+ 68.5 „	0	+ 58 „	E 19 „	+ 19 „
35	− 378 „	E 91 „	+ 83 „	E 19 „	+ 75 „	„ 26 „	+ 20 „
55	− 285 „	W 150 „	+ 47 „	„ 4.5 „	+ 26 „	„ 28.5 „	+ 19 „
18 20	− 389 „	E 140 „	− 13.5 „	„ 4.5 „	− 37 „	„ 16 „	+ 10 „
55	− 111 „	W 14.5 „	+ 24.5 „	W 17.5 „	+ 18 „	W 12 „	+ 8 „
19 50	− 152 „	E 39 „	+ 3 „	„ 19 „	0	„ 9.5 „	+ 2 „
22 55	− 212 „	„ 36.5 „	− 9 „	„ 19 „	− 2 „	„ 24 „	− 34 „
23 15	− 235 „	„ 65 „	− 6 „	„ 12 „	− 14 „	„ 26 „	− 47 „

[1] uncertain values.

TABLE LV (continued).

Gr. M. T.	Göttingen			Kasan		Fort Conger
	P_h	P_d	P_v	P_h	P_d	P_d
h m						
6 50	0	W 2 γ	− 5 γ	0	0	E 51 γ
7 30	− 1 γ	„ 7 „	− 3 „	0	0	„ 40.5 „
8 20	− 2 „	0	+ 1 „	0	0	W 15 „
9 5	+ 2 „	0	0	0	0	„ 6 „
10 20	0	0	+ 4 „	0	0	E 18 „
11 50	0	„ 1.5 „	− 2 „	0	0	0
13 20	+ 3 „	E 2.5 „	+ 9.5 „	+ 11 γ	W 3.5 γ	0
14 35	+ 22 „	W 8.5 „	+ 7.5 „	+ 31 „	„ 16 „	W 25.5 „
55	+ 50 „	„ 8 „	0	+ 49 „	„ 19 „	„ 40 „
15 30	+ 14 „	„ 8 „	+ 13 „	+ 26 „	„ 16 „	E 5 „
40	+ 21 „	„ 7 „	+ 21.5 „	+ 32 „	„ 16 „	W 6 „
50	+ 3 „	„ 5.5 „	+ 26 „	+ 14 „	0	„ 46 „
16 15	+ 3 „	„ 8 „	+ 21 „	+ 11 „	„ 8.5 „	„ 99 „
40	+ 24 „	„ 11.5 „	+ 24 „	+ 28 „	„ 7 „	„ 225 „
17 0	+ 19 „	„ 13 „	+ 35 „	+ 17 „	„ 8 „	„ 549 „
20	+ 10 „	0	+ 44 „	+ 41 „	E 24 „	„ 864 „
35	+ 4 „	E 29 „	+ 51.5 „	+ 34 „	„ 25 „	„ 684 „
55	0	„ 10 „	+ 47 „	+ 19 „	„ 28 „	„ 687 „
18 20	− 53 „	„ 7.5 „	+ 49 „	− 38 „	„ 21.5 „	„ 198 „
55	+ 11 „	W 22 „	+ 25.5 „	0	W 8.5 „	„ 289.5 „
19 50	0	„ 17 „	+ 38.5 „	− 8 „	„ 3 „	0
22 55	+ 36 „	„ 4 „	− 7.5 „	− 6 „	„ 31 „	E 37.5 „
23 15	+ 33 „	E 4.5 „	− 21.5 „	− 8 „	„ 27.5 „	„ 82.5 „

Current-Arrows for the 15th July, 1883.

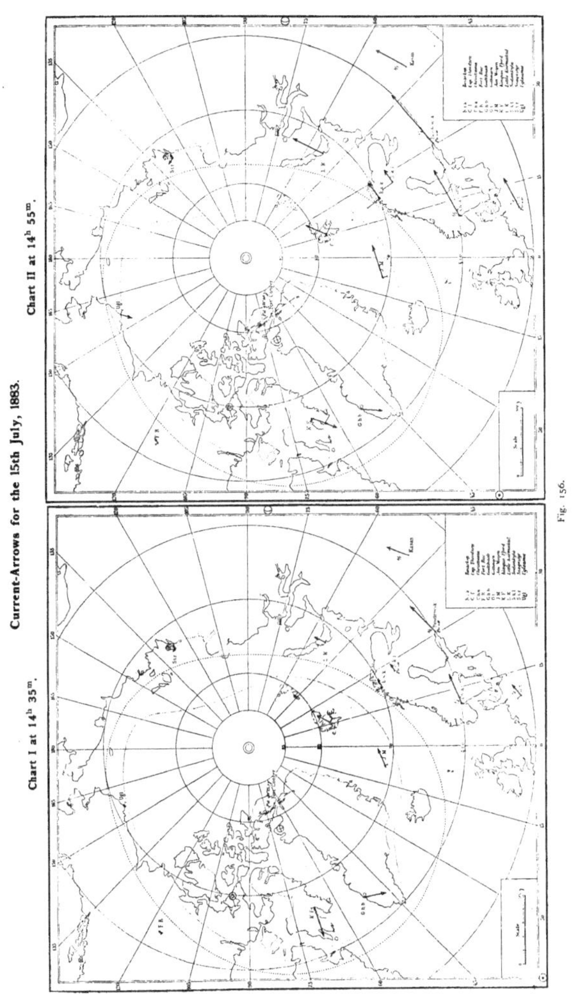

Fig. 156.

Current-Arrows for the 15th July, 1883.

Chart III at $15^h 30^m$. Chart IV at $15^h 40^m$ and $15^h 50^m$.

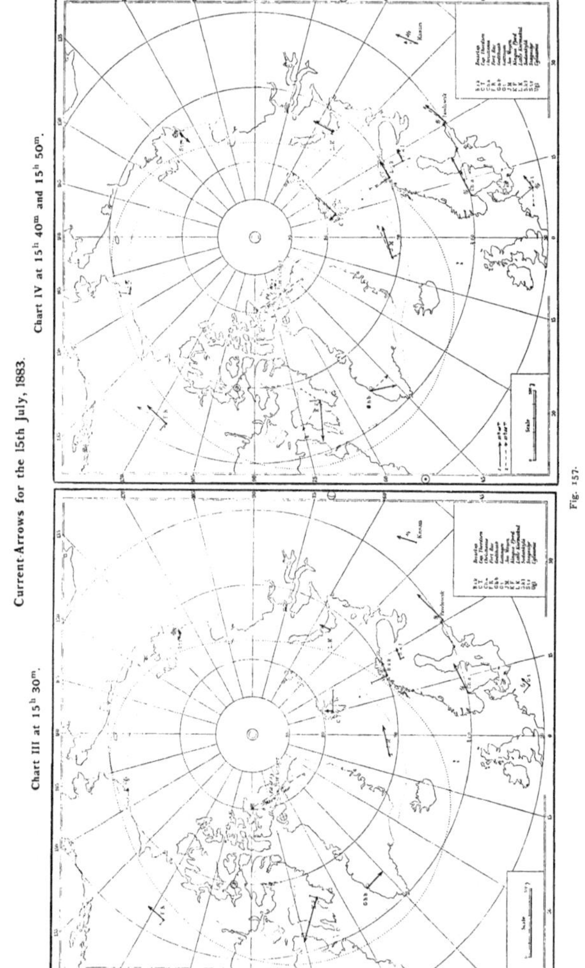

Fig. 157.

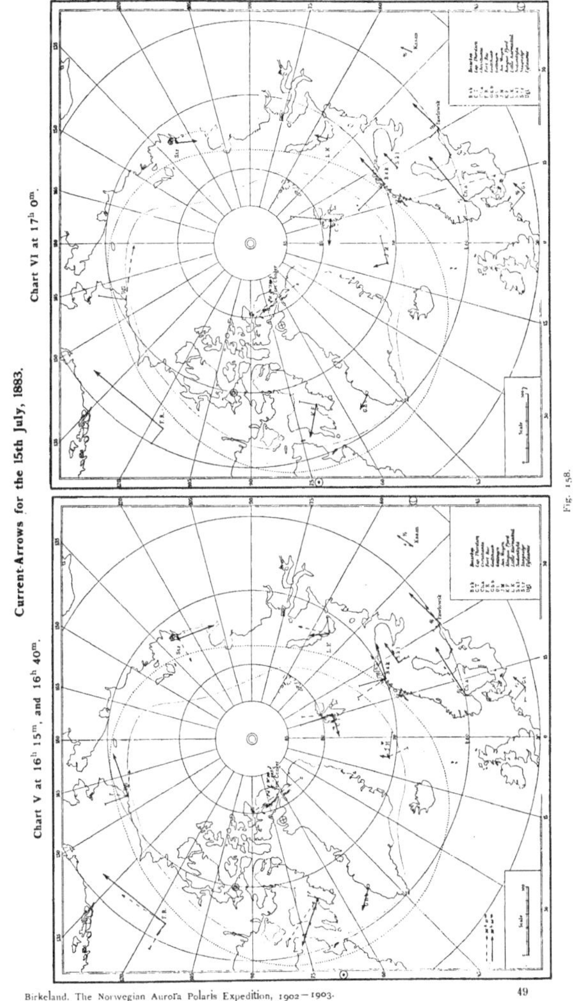
Current-Arrows for the 15th July, 1883.
Chart V at 16h 15m, and 16h 40m.
Chart VI at 17h 0m.

Fig. 158.

Current-Arrows for the 15th July, 1883.

Chart VII at $17^{h}\ 20^{m}$. Chart VIII at $17^{h}\ 35^{m}$.

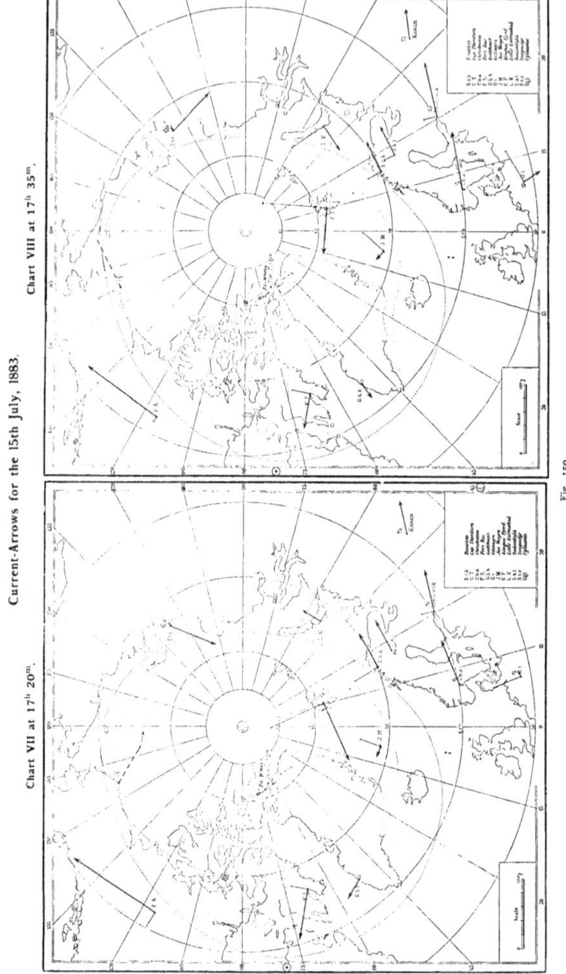

Fig. 159.

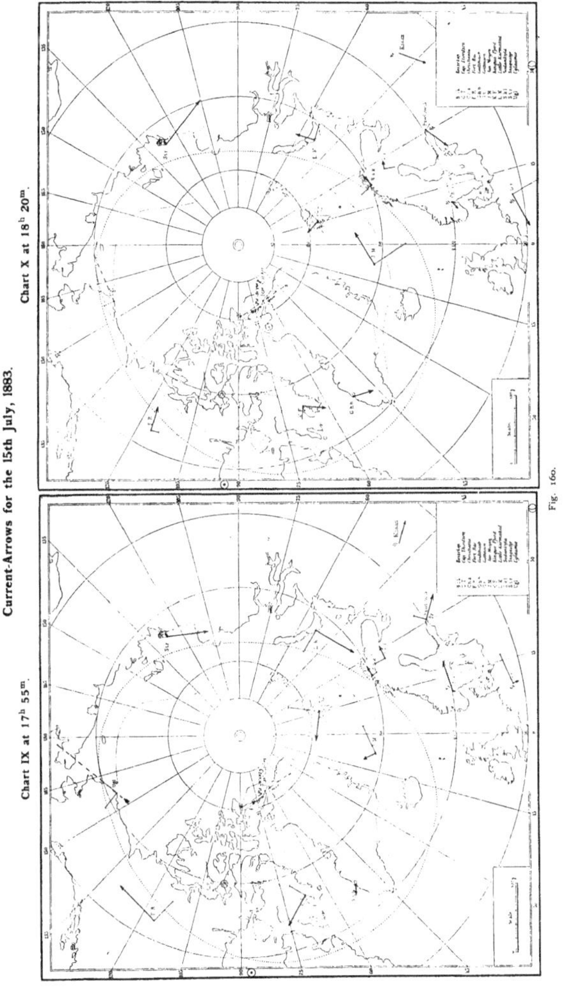

Current-Arrows for the 15th July, 1883.

Fig. 160.

Current-Arrows for the 15th July, 1883.

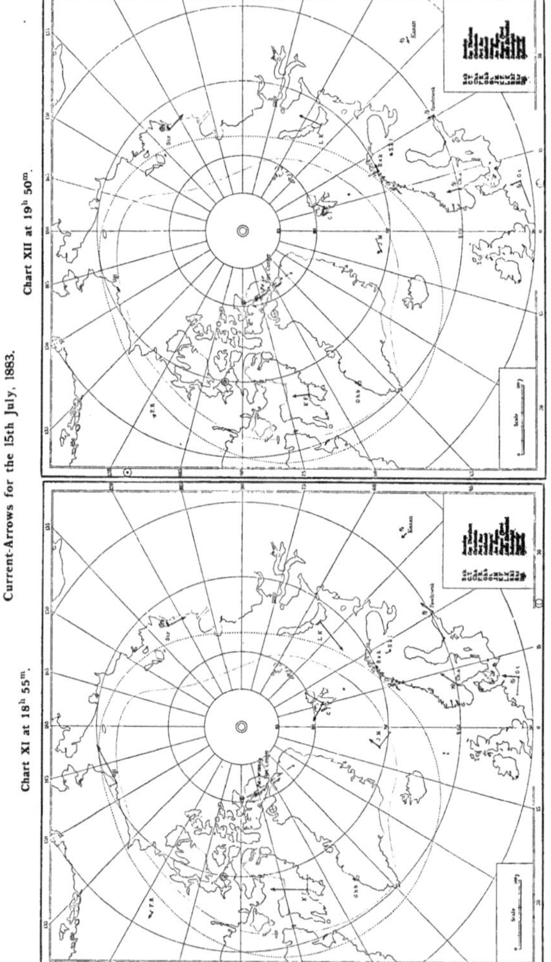

Chart XI at $18^h\ 55^m$.

Chart XII at $19^h\ 50^m$.

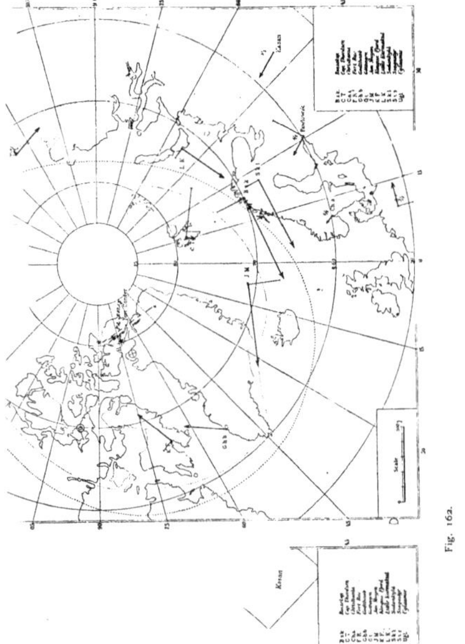

Fig. 162.

THE PERTURBATIONS OF THE 1st FEBRUARY, 1883.

(Pl. XXVII).

88. Here, as on the preceding term day, we can separate a first comparatively slight perturbation, appearing, more or less isolated, at about 11^h, from the subsequent powerful storms. Only at two stations, Fort Rae and Godthaab, do we find, during the first period, rather powerful perturbing forces. We find, moreover, distinct deflections at several other stations, but these in the first place are considerably weaker, and in the next, of longer duration, than at the two stations just mentioned; and the character of the deflections does not seem to indicate that they have any very close connection with one another.

This first perturbation is particulary distinct at Fort Rae. At 10^h 55^m the deflections suddenly increase to a maximum, and then again decrease rather rapidly. The perturbing forces are negative in the horizontal intensity, and directed eastwards in the declination; and the current-arrow, as will be seen from *Chart I*, is directed westwards along the auroral zone at the time when the deflections are strong. There is thus, certainly, negative polar precipitation in the neighbourhood of this station; and at Godthaab too, a negative polar storm seems to be acting.

If we look more carefully at the chart, there appear to be signs of positive forces at Ssagastyr, and possibly a positive system has formed on the afternoon-side, but if so, it is not very clearly developed. In this respect, however, we have not sufficient data to go upon.

After this precursor of the subsequent powerful storms, there follows an interval in which no very great forces appear. Soon, however, new storms begin, which rapidly develope until they attain considerable strength, and form the principal systems of that day.

The storms in this period will naturally be divided into two sections,

(1) those that occur between 14^h 30^m and 19^h 45^m, and

(2) the storms from 19^h 45^m until the end of the period.

Such a division of the phenomena will of course be imperfect, and may appear somewhat artificial, since we have constantly found, that one system developes from another; but it is done for practical reasons, in order, if possible to obtain a clearer general view of the conditions.

At about 14^h 30^m, some more or less powerful deflections begin at Kingua Fjord in the horizontal intensity and declination simultaneously, their direction indicating the presence of a negative polar storm. This can apparently be traced farther, over Fort Conger, where there is at the same time a distinct deflection in the declination-curve; and judging from the conditions of this curve at Cape Thordsen, this system is also at work there. There appears to be a weaker positive storm in the vicinity of Jan Mayen.

This perturbation, however, is of brief duration, and its field of operations is comparatively restricted. In the course of about an hour, it is practically over. At about 16^h, on the other hand, powerful storms begin to develope at all the stations round.

The deflections at Kingua Fjord increase most rapidly to a considerable amplitude, and attain their highest value as early as 17^h, after which they remain more or less powerful in declination, while P_h decreases fairly evenly, reaching its normal condition again at about 20^h. This negative system of precipitation apparent at Kingua Fjord, now extends as a great system westwards. It is felt at all the arctic stations, more strongly, indeed, than anything else at the time when the deflections are greatest; for here too, there occurs simultaneously a positive system of precipitation, which to some extent counteracts the negative.

The distribution of force round the auroral zone is here, too, exactly similar to that found during the earlier storms. At Kingua Fjord, Fort Rae, Uglaamie, Ssagastyr, Cape Thordsen, and possibly Fort Conger, it is almost exclusively the negative system of precipitation that acts; at the other polar stations, the positive system also asserts itself more or less strongly.

As regards Ssagastyr, however, there is one thing to be noticed. From $17^h\ 40^m$ until $18^h\ 30^m$, the deflections in the horizontal intensity are too great to allow of being observed. The direction in which the needle moved is not given, nor is the character of the curve such as to enable the direction of the deflection to be determined with certainty. Judging from previous experience, however, there would seem to be no doubt that the deflection has been in a negative direction.

In the first place, we have never met with positive perturbations here that have been powerful enough to make the needle move out of the field of observation. Further, this station lies just between Uglaamie and Little Karmakul, at both of which, it may be seen, the negative storm is very powerful. This is also the case at Cape Thordsen.

As the negative storm is powerful at all the stations surrounding Ssagastyr, it would be very improbable, judging by all that we have seen previously, that a strong positive system could act at that one station; and moreover, the part of the curve for the time immediately after this interval, indicates, although faintly, that there has been a negative, not a positive, deflection. The current-arrows we have marked, indicate, therefore, that the needle has moved out in a negative direction; but, in order to indicate the slight uncertainty, we have placed an asterisk by the arrows in question.

The perturbing forces everywhere are exceedingly powerful; and the storm-centre of the negative storm is in the district from Uglaamie to Little Karmakul, probably about Ssagastyr.

We think, however, that we can prove a distinct movement of the system. This is developed earliest round Kingua Fjord, where the forces even at $16^h\ 10^m$, have attained considerable power. The deflections here increase rather rapidly to a maximum. At Fort Rae and Uglaamie, on the other hand, the deflections at first increase more slowly; but, at both these stations the perturbing forces are of considerable magnitude as early as 17^h.

At Ssagastyr, the negative deflections do not begin until $17^h\ 40^m$; but they are then suddenly so strong, that the needle passes out of the field of observation.

The negative system thus seems to begin in the neighbourhood of Kingua Fjord, developes there with considerable rapidity, and, simultaneously with the extension of the area of precipitation and the increase of the perturbing forces, the storm-centre moves westwards. If we endeavour to trace a similar movement onwards to Little Karmakul and Cape Thordsen, it appears that the same observation may be made with regard to the first of these two stations; but consideration must be paid to the fact, that this is within the positive system's sphere of operations, and, before the negative storm gains the ascendancy, there are distinct positive forces. This is also the case afterwards. When the powerful, but brief, negative precipitation is over, positive forces appear once more, this time more powerful than before. The powerful negative forces appear a little later than at Ssagastyr, but we must beware of drawing conclusions from this condition respecting the movement of the system, the more so as there was powerful negative precipitation north of the auroral zone even earlier, as the conditions at Cape Thordsen show. The deflections in the horizontal intensity at the last-named station, resemble, in many respects, the corresponding deflections at Uglaamie. At both places we find, at about 17^h or $17^h\ 30^m$, a secondary maximum, and at about $18^h\ 30^m$ the true maximum. There is a slight time-displacement, however, especially in the first secondary maximum, so that the deflections at Cape Thordsen come a little later than those at Uglaamie. The similarity of these curves is strikingly evident at the very first glance; but if we look at the declination-curve, we find no particular resemblance, and the deflections in this component will have a greater significance at Cape Thordsen than at Uglaamie. What we will here draw special attention to, however, is that the negative deflections at Cape Thordsen begin rather early, and thus develope more or less simultaneously with those at Uglaamie, possibly a trifle later; and there are thus considerable forces at Cape Thordsen before they appear at Ssagastyr. The explanation of this must be, either, that simultaneously with the extention of the negative system of precipitation westwards through North America from

Kingua Fjord, it also spreads eastwards towards Cape Thordsen and perhaps farther, and then unites with the western branch; or the system in America will most rapidly spread on the north of the auroral zone, and will not extend farther south to Ssagastyr until later, or there may be two rather distinct areas of precipitation.

It is possible, too, that the circle which the negative system of precipiatation appears to form round the pole of the earth, is formed more or less at once, and that the displacement that we find in the deflections is occasioned by the movement and deformation of the entire circle.

The most probable cause of these phenomena, however, seems to be, that several of them separately exert influence.

A more or less circular, negative system of precipitation will be formed somewhat rapidly, in which there may be one or several districts in which the strength of the precipitation is greatest. By imagining these maximal zones to be moved from time to time, the differences in corresponding hours that appear can be simply explained.

In addition to the negative system, there is also, as already mentioned, a positive system in the district from Godthaab eastwards along the auroral zone to Little Karmakul. At these two stations, especially the former, this system is comparatively weaker to begin with; but, on the other hand, at Godthaab, at the end of the section under consideration, we find practically no effects of it, while at Little Karmakul, at the end, it is quite distinct and powerful.

The effects are strongest at the intermediate stations, Jan Mayen, Bossekop and Sodankylä; and there we find the characteristic condition that we have so frequently met with.

At first the positive system is at work, being then broken in upon by the stronger negative system, which causes a partial reversal of the direction of the deflection at the time when the storms are at their height. Finally, simultaneously with the decrease in the negative precipitation, the positive forces once more gain the ascendency, and the conditions are again such as would be found in the neighbourhood of a positive district of precipitation. It is interesting to observe the conditions at these stations, and see how they alter the farther magnetically north we go. At the three polar stations, Jan Mayen, Bossekop and Sodankylä, the perturbation-conditions are, on the whole, exactly analogous; but we can trace a continuous variation in them from Jan Mayen, through Bossekop to Sodankylä.

At the first of these three stations, the negative storm is the strongest, although the positive deflections are at first quite strong. At Bossekop, the precipitation is, on the whole, less, but the positive deflections are more numerous than the negative. Lastly, at Sodankylä, the effect of the negative storm is comparatively slight, and the positive deflections predominate. We can thus trace a continuous change; farther north the negative storm acts the more strongly, farther south the positive. If we look still farther south, at Christiania, the positive storm seems to be acting alone. At the time when the negative storm is at its height, there is a strong deflection there in a positive direction; and the curves are sufficiently jagged to make it probable that this station is not far from the district of precipitation of the positive system. The positive system therefore seems to be somewhat far south in its position.

If, on the other hand, we go still farther south to Pawlowsk and Göttingen, we seem to have passed the point of divergence, for the forces there, in the horizontal intensity, are in a negative direction, and we thus have a change. It must be principally the positive storm which also acts here, if there are not, at the same time, systems of which the greatest effect is exerted in lower latitudes.

If we look for some movement of the positive system, we find, at first, that the forces are strongest in the west, but at the close the storm is most fully developed farther east. The positive forces in the horizontal intensity also appear very much earlier at the western stations Jan Mayen and Christiania than farther east. At Godthaab, the effect is of short duration, and the storm is not very clearly developed. This might indicate a movement eastwards, such as we have frequently met with at this hour of day; but it should also be remembered that there possibly exists another movement of the systems. We might, for instance, imagine the positive system to be moved southwards, which would cause the occurrence of phenomena such as those we now have before us; for the western stations, Godthaab and Jan Mayen, are in the north of the auroral zone, while the eastern stations are in its southern part.

The order of the systems is thus exactly such as we are accustomed to find at this time of day during the most typical of the storms already described.

With regard to the movement of the systems from time to time, we find apparent traces of a westerly movement of the negative system in America, and possibly a less pronounced easterly movement of the positive storm-centre.

On the three *charts* following, *II, III,* and *IV*, the development of the perturbations in this section can be distinctly followed. That of the negative storm is the more marked. Between $15^h\ 20^m$ and $16^h\ 20^m$, it is distinctly developed only at Kingua Fjord; but at $16^h\ 50^m$ there are also distinct, strong current-arrows at Fort Rae and Uglaamie on the one side, and Cape Thordsen on the other. The most powerful forces, however, are still at Kingua Fjord.

At $17^h\ 20^m$ the great current-circle has already formed, and we find the most abundant precipitation in the district Kingua Fjord to Fort Rae, showing that the storm-centre has moved a little westwards. At both Godthaab and Jan Mayen, where previously the positive storm was the strongest, there are now powerful negative forces.

The storm is at its height from $18^h\ 15^m$ to $18^h\ 30^m$, and we find very strong perturbing forces, especially on the night-side. The most powerful are apparently at Ssagastyr; but as the deflections at both Uglaamie and Little Karmakul are too wide to be measured, it is possible that they may be just as powerful there as at Ssagastyr. The negative storm then decreases once more on Chart IV, while at the same time the storm-centre moves back to the regions about Fort Rae and Uglaamie. It may be noticed that this contrary movement of the system of precipitation, takes place after the sun has crossed the meridian of the magnetic axis.

The positive system can be followed in a similar manner. At first it extends from Godthaab eastwards as far as Little Karmakul, as shown on Chart II. On Chart III the negative system breaks in upon it, causing, in some cases, distinct reversals of the direction of the current, as, for instance, at Little Karmakul and Jan Mayen; while in others the current-arrow only swings backwards and forwards as at Bossekop and Sodankylä. At Christiania, however, the effects are still chiefly those of the positive system.

At the end, we find again stronger effects of the positive system, the force at Little Karmakul, at $18^h\ 50^m$, for instance, being of remarkable magnitude. Its effect are also distinct at the more westerly stations. At Jan Mayen, where, not long before, the effects of the negative system had been so distinct, the current-arrow has once more begun to oscillate, and at $19^h\ 20^m$ is more indicative of the positive system, although there are evident signs of the action of both systems simultaneously.

The negative values of P_v constantly found at Jan Mayen should also be noted. They show that although at one time the positive system is the stronger, and at another the negative, this variableness has no special significance as regards the vertical forces. As we have so often pointed out, the explanation of the phenomenon is, that the negative system, whose area of precipitation must be assumed to be chiefly to the north of Jan Mayen, and the positive system, whose storm-centre is certainly situated to the south of that station, will both act in the same direction, namely a negative direction.

The conditions at Little Karmakul are also somewhat variable, and on Chart IV we find both distinct negative and distinct positive forces.

The last current-arrow for $19^h\ 50^m$ comes more properly in the next section of the perturbations, which we shall now proceed to examine.

The second main section of the powerful perturbations on this date, is, as we have said, from $19^h\ 45^m$ to the end of the period.

At about $19^h\ 45^m$, it appears that comparatively quiet conditions have once more supervened at almost all the stations. At a few of them the conditions are almost normal for a short time; at others we find a more or less marked minimum in the deflections; while at others again there appears to be a transition, as the storm, which was previously positive, now changes to negative.

That which characterises this second section of the perturbations, is the powerful negative polar storms which we find at all the stations. These are certainly only to be considered as a further development of the earlier negative systems of precipitation observed. In the deflections at Kingua Fjord at this time, there is a minimum of no great distinctness. The declination-deflections, which have previously continued to be quite strong, have shown a slight indication of a minimum at about $19^h\ 45^m$; while the horizontal-intensity curve, which, since about 17^h has been more or less evenly approaching the normal line, has now reached it. The horizontal intensity then remains almost normal for a couple of hours, only oscillating slightly about the normal line.

The conditions at Uglaamie form a suitable starting-point for our reflections upon the perturbations in this section. There are, as will be seen, two strong deflections separated by an interval in which the deflections have a brief, but very marked, minimum just before 21^h. These two deflections are so strong that in both cases the needle passes out of the field of observation.

To the first of them, there are corresponding deflections at Ssagastyr and Fort Rae, as also at a number of other stations, although, the resemblance at some of them, is less marked. At Fort Conger the resemblance is quite striking. At Little Karmakul, there are also two maxima, which show some resemblance to those at Uglaamie; but the resemblance between the first pair of them is not so great. It has more the appearance of a brief but powerful impulse, a precursor of the subsequent strong deflection.

The storm thus appears as a negative polar storm; with its centre in the vicinity of Uglaamie.

On *Charts IV* and *V* there are two hours which represent the conditions during this first phase of the second section. There are fairly powerful perturbing forces at several stations.

The different systems that we here see are, of course, connected in some way or other with each other; but it seems as if the system in the neighbourhood of Uglaamie was more or less independent. It is therefore very likely that there is a large, more or less connected, negative system of precipitation, in which there are two storm-centres, one in the vicinity of Uglaamie, and the other in the region eastwards from Kingua Fjord.

The hour $20^h\ 30^m$ on Chart V, also belongs to this first phase of the perturbations. We here see the conditions at the time of the strong deflection at Little Karmakul.

The negative system of precipitation now also forms a circle round the geographical north pole, and the forces seem to be concentrated about several storm-centres. There still seems to be one at Uglaamie, one at Little Karmakul, and one less powerful one at Kingua Fjord; but whether they are in reality so clearly separated as they appear to be, it is difficult to say.

We find here no distinct traces of positive systems, although it is possible that such do actually exist, and from what we have seen, are to be looked for to the south, or in the southern part, of the auroral zone, from Europe westwards; but we have no stations there.

A distinct, though rather faint indication of such a system is to be found indeed in Jan Mayen at about 20^h, and the rapid transition from Little Karmakul to Bossekop, found on Chart V, for the hour $20^h\ 30^m$, is possibly due to the existence of positive polar precipitation to the west. The direction of the current-arrow at Gøttingen, which is a little more westerly than might be expected if the negative systems only were acting, may also possibly indicate the existence of a positive system of precipitation such as this.

The most powerful storm does not develope, however, until this first phase is past.

The second phase of the storms in this second section may be considered as coming in the interval between $20^h\ 40^m$ and the close of the period. The deflections, which at first, at any rate, correspond to the effects of a negative polar storm, are very powerful everywhere; and at Uglaamie and Ssagastyr the needle passes out of the field of observation. The various deflections, however, are not so well-defined as to make it easy to find any distinct movement of the systems.

What we will, however, draw particular attention to, is the perturbation-conditions at Fort Rae. At the close of the period, a distinct change takes place there in the direction of the perturbing force. We previously found only negative deflections in the horizontal intensity, indicating that negative systems of precipitation were at work; but now a positive system appears here. That this station is on the afternoon-side of the globe, and further that it is to the south of the auroral zone, are circumstances that agree closely with what we should have expected to find; and the positive system, the existence of which, during the last storms, we were unable to prove, and could only suggest the possibility of, appears once more just at a time when we might expect to find its effects at the stations we are considering.

At the southern stations the forces are unusually powerful.

The fields of force for this last phase of the storms, will be found represented on the last three charts, from $20^h\ 50^m$ to $23^h\ 15^m$.

We now find, as so often before during the powerful storms, a negative current-circle round the geographical pole.

The greatest forces are found upon the night-side, and they are of unusual magnitude. The storms are negative everywhere, except at $22^h\ 20^m$ and $23^h\ 15^m$ in America, where we meet with the effects of the already-mentioned positive storm. In Europe, the negative area of precipitation has moved farther south, if we may judge by the conditions in the vertical intensity; for both in Jan Mayen and at Cape Thordsen there are now positive values of P_v, whereas previously they were negative only. The precipitation seems therefore, now to take place to the south of these stations, whereas, previously it was chiefly to the north. This is in agreement with the fact that the negative area of precipitation comes farther south on the night-side than on the day-side.

In Europe, the direction of the current-arrows is rather south, even as far north as Bossekop. In Central Europe this is the normal condition during similar storms; but the forces there are now so powerful, that to a certain extent we have used the same scale as at the polar stations.

On *Chart VII*, the powerful negative storm is almost over, and only at a few places we now find perturbing forces, indicating that it is still in existence. At Fort Rae, on the other hand, we find powerful effects of the positive storm that has been mentioned as occurring there.

TABLE LVI.

The Perturbations of the 1st February, 1883.

Gr. M. T.	Uglaamie			Fort Rae			Kingua Fjord		Godthaab	
	P_h	P_d	P_v	P_h	P_d	P_v	P_h	P_d	P	P
h m										
11 20	0	W 4 γ	+ 20.5 γ	− 110 γ	E 40 γ	+ 120.5 γ	− 20.5 γ	E 7.5 γ	− 97.5 γ	W 11.5 γ
12 20	− 61 γ	„ 4 „	0	− 30 „	„ 41.5 „	+ 25 „	− 14.5 „	W 12 „	− 37.5 „	„ 10 „
13 20	0 „	0	0	+ 10 „	„ 7 „	0	− 5.5 „	E 20.5 „	− 10 „	E 3 „
14 20	+ 20 „	„ 2 „	− 10 „	+ 20 „	0	0	+ 20 „	„ 19.5 „	− 10 „	W 5.5 „
15 20	− 33.5 „	E 1 „	+ 20.5 „	− 32.5 „	„ 26.5 „	− 31.5 „	− 172 „	W 79.5 „	+ 15 „	E 31.5 „
16 20	− 23.5 „	0	0	− 25 „	„ 30.5 „	0	− 105 „	„ 75 „	+ 34 „	„ 36.5 „
50	− 101.5 „	„ 7 „	+ 20.5 „	− 90 „	„ 84 „	− 57.5 „	− 147 „	„ 92 „	+ 29 „	„ 55.5 „
17 20	− 117 „	„ 10.5 „	+ 81.5 „	− 155 „	„ 122 „	− 82.5 „	− 162.5 „	„ 87.5 „	− 132 „	W 115 „
18 15	−>304.5 „	„ 66.5 „	+ 214 „	− 255 „	„ 180 „	+ 335 „	− 92.5 „	„ 129 „	− 126.5 „	„ 118.5 „
30	−>304.5 „	„ 32 „	+ 254 „	− 450 „	„ 123.5 „	+ 268.5 „	− 91 „	„ 94 „	− 139.5 „	„ 88.5 „
50	− 276.5 „	„ 13 „	+ 197 „	− 355 „	„ 73.5 „	− 90 „	− 121.5 „	„ 94 „	− 131 „	„ 58.5 „
19 20	− 44 „	W 8 „	+ 122.5 „	− 170 „	„ 23.5 „	− 37.5 „	− 55 „	„ 94 „	− 61 „	„ 34 „
50	− 237.5 „	E 27 „	+ 41 „	− 28.5 „	„ 26.5 „	− 80 „	0	„ 132 „	+ 37.5 „	„ 135 „
20 30	− 294.5 „	„ 6 „	+ 103 „	− 64 „	„ 56.5 „	− 40 „	0	„ 100.5 „	− 15.5 „	„ 103 „
50	− 134 „	„ 9.5 „	+ 184 „	− 85 „	„ 86.5 „	0	+ 34.5 „	„ 115 „	− 6.5 „	„ 106 „
21 15	−>308 „	„ 22.5 „	+ 276 „	− 240 „	„ 142.5 „	− 140 „	− 0.5 „	„ 182 „	− 48.5 „	„ 179 „
30	−>308 „	W 6.5 „	+ 299 „	− 268.5 „	„ 72 „	− 131.5 „	+ 11 „	„ 194 „	− 54 „	„ 198.5 „
40	−>308 „	„ 47.5 „	+ 206 „	− 280 „	„ 113 „	− 258.5 „	+ 7 „	„ 190 „	− 50 „	„ 207 „
22 20	−>308 „	E 56 „	− 71.5 „	− 24 „	W 25 „	− 30 „	+ 46 „	„ 169 „	− 4 „	„ 214 „
23 15	+ 50.5 „	W 14 „	− 74.5 „	+ 170 „	„ 82 „	− 20 „	+ 60 „	„ 67.5 „	+ 50 „	„ 91.5 „

TABLE LVI (continued).

Gr. M. T.	Jan Mayen			Bossekop			Sodankylä		
	P_h	P_d	P_v	P_h	P_d	P_v	P_h	P_d	P_v
h m									
11 20	− 4 γ	0	+ 14.5 γ	0	W 7 γ	− 0.5 γ	0	W 17.5 γ	0
12 20	+ 1.5 „	0	− 5.5 „	− 1 „	0	− 1 „	0	„ 9.5 „	0
13 20	+ 22.5 „	W 3 γ	+ 6 „	+ 1 „	„ 1.5 „	0	+ 3.5 γ	„ 12 „	0
14 20	+ 3 „	0	0	0	E 1.5 „	− 1.5 „	0	„ 7 „	+ 2.5 γ
15 20	+ 70 „	„ 26.5 „	− 10 „	+ 4 „	W 16.5 „	+ 6 „	+ 3 „	„ 19.5 „	+ 9 „
16 20	+ 40 „	„ 36 „	− 10 „	+ 4 „	„ 22.5 „	+ 1 „	+ 7.5 „	„ 18 „	+ 9 „
50	+ 135 „	„ 55 „	− 50 „	+ 11 „	„ 41.5 „	+ 10 „	+ 7.5 „	„ 36.5 „	+ 5.5 „
17 20	− 9 „	„ 82 „	− 80 „	+ 26 „	„ 80 „	+ 15 „	+ 35 „	„ 58 „	− 14.5 „
18 15	− 223.5 „	„ 150 „	− 20 „	− 49 „	„ 63.5 „	− 45.5 „	− 60 „	„ 17 „	+ 103 „
30	− 240 „	„ 110 „	− 20 „	− 24 „	„ 71 „	− 60 „	+ 35.5 „	„ 88.5 „	+ 160 „
50	− 123 „	„ 84.5 „	− 54 „	+ 40.5 „	„ 109 „	− 8 „	+ 122 „	„ 132.5 „	+ 112.5 „
19 20	+ 21 „	„ 60.5 „	− 145 „	+ 25 „	„ 25 „	+ 11 „	+ 61.5 „	„ 4.5 „	0
50	+ 66 „	„ 77.5 „	− 130 „	+ 6 „	„ 35.5 „	+ 6.5 „	+ 10 „	E 8.5 „	− 40 „
20 30	+ 6 „	„ 104.5 „	− 43 „	− 5.5 „	„ 6.5 „	− 1 „	− 55 „	„ 59 „	− 39 „
50	− 51 „	„ 115 „	− 17.5 „	− 77.5 „	E 155 „	− 16 „	− 75 „	„ 58 „	+ 15 „
21 15	− 145 „	„ 166 „	+ 40 „	− 84 „	„ 222 „	− 25 „	− 104 „	„ 274 „	+ 42 „
30	− 565 „	„ 81 „	+ 317.5 „	− 115.5 „	„ 94 „	− 47 „	− 167.5 „	„ 228 „	+ 34 „
40	− 350 „	„ 80 „	+ 205 „	− 103 „	„ 173 „	− 59.5 „	− 150 „	„ 192 „	+ 106 „
22 20	− 55 „	„ 89 „	+ 115 „	− 56 „	„ 130 „	− 5 „	− 152.5 „	„ 141 „	− 55 „
23 15	+ 65 „	„ 47.5 „	+ 68 „	− 22 „	„ 85.5 „	+ 11 „	− 100 „	„ 86 „	− 46 „

TABLE LVI (continued).

Gr. M. T.	Cape Thordsen			Little Karmakul			Ssagastyr	
	P_h	P_d	P_v	P_h	P_d	P_v	P_h	P_d
h m								
11 20	+ 8 γ	W 8 γ	− 40 γ	+ 13 γ	W 7.5 γ	− 13 γ	+ 57 γ	E 2.5 γ
12 20	+ 7 „	„ 6 „	0	0	E 10 „	− 2 „	+ 28 „	0
13 20	+ 10 „	„ 10.5 „	+ 1 „	+ 8 „	0	− 14 „	+ 24 „	0
14 20	− 5 „	„ 7 „	− 4 „	− 12 „	„ 15 „	− 4 „	+ 2 „	0
15 20	+ 15.5 „	„ 45 „	− 25 „	0	W 10 „	+ 3 „	− 5 „	0
16 20	− 1.5 „	„ 30 „	− 22 „	+ 6 „	„ 10 „	+ 2 „	+ 28 „	0
50	− 89.5 „	„ 64.5 „	− 56 „	+ 64 „	„ 45 „	+ 4 „	− 14 „	W 21.5 „
17 20	− 70 „	„ 70 „	− 61 „	− 26 „	„ 40 „	− 67 „	− 95 „	„ 47 „
18 15	−177.5 „	„ 135 „	− 37 „	− >760 „	E 510 „	− 5 „	− >635 „ (1)	„ 236.5 „
30	−195.5 „	„ 116.5 „	− 20 „	− 59 „	„ 65 „	− 40 „	− >655 „ (1)	0
50	−156 „	„ 102 „	− 50 „	+ 273 „	W 148 „	−125 „	− 72 „	E 125 „
19 20	− 50 „	„ 54.5 „	− 21 „	+ 82 „	„ 76 „	− 36 „	− 42 „	„ 68 „
50	− 7.5 „	„ 65 „	− 25 „	− 50 „	E 19 „	+ 10 „	− 72 „	„ 150 „
20 30	− 34 „	„ 70 „	− 10 „	− 647 „	„ 325 „	+138 „	− 60 „	W 2.5 „
50	− 12 „	„ 125 „	− 27 „	− 632 „	„ 660 „	+137 „	− >655 „ (1)	„ 12.5 „
21 15	−170 „	„ 95 „	+172 „	− 636 „	„ 340 „	−140 „	− >655 „ (1)	„ 15 „
30	−174 „	„ 84 „	+219 „	− 633 „	„ 640 „	− 26 „	− >655 „ (1)	E 296 „
40	−205 „	„ 49.5 „	+270 „	− 630 „	„ 375 „	+ 33 „	− >655 „ (1)	„ 520 „
22 20	−112 „	E 66.5 „	+188 „	− 145 „	„ 214 „	+ 24 „	− 350 „	„ 277 „
23 15	0	W 10.4 „		− 153 „	„ 193 „	+ 68 „	+ 44 „	„ 73.5 „

(1) See description p. 387.

TABLE LVI (continued).

Gr. M. T.	Christiania		Pawlowsk			Göttingen			Fort Conger
	P_h	P_d	P_h	P_d	P_v	P_h	P_d	P_v	P_d
h m									
11 20	0	0	+ 5 γ	W 15.5 γ	0	− 3.5 γ	W 15 γ	0	E 8.5 γ
12 20	0	E 3.5 γ	+ 3.5 „	„ 7.5 „	+ 5.5 γ	− 1.5 „	„ 4 „	+ 0.5 γ	W 6.5 „
13 20	+ 5 γ	0	+ 6 „	„ 6.5 „	+ 7 „	+ 1 „	„ 6 „	+ 1 „	E 6 „
14 20	+ 9 „	„ 3 „	+ 5 „	0	+ 1.5 „	+ 6 „	„ 0.5 „	+ 7 „	W 1.5 „
15 20	+ 17 „	W 9.5 „	+ 9.5 „	„ 16 „	+ 0.5 „	+ 18.5 „	„ 11 „	+ 6 „	„ 55.5 „
16 20	+ 24 „	„ 8 „	+ 15	„ 14.5 „	0	+ 22 „	„ 7.5 „	+ 0.5 „	„ 76.5 „
50	+ 23.5 „	„ 25.5 „	+ 6 „	„ 24 „	0	+ 23.5 „	„ 19.5 „	− 2 „	„ 120 „
17 20	+ 27 „	„ 48 „	+ 1.5 „	„ 36 „	+ 5.5 „	+ 23.5 „	„ 37 „	− 6 „	„ 124.5 „
18 15	+ 45.5 „	E 2.5 „	− 45 „	E 52.5 „	+ 42.5 „	− 65 „	E 14 „	+ 24 „	„ 337.5 „
30	+ 55.5 „	0	− 29.5 „	„ 16 „	+ 80 „	− 66 „	„ 5.5 „	+ 31.5 „	„ 289.5 „
50	+ 27 „	W 58 „	− 64 „	„ 15 „	+ 95 „	− 54.5 „	W 73 „	+ 30.5 „	„ 121.5 „
19 20	− 26.5 „	E 8 „	− 27.5 „	„ 34.5 „	+ 57 „	− 42 „	0	+ 28.5 „	E 26.5 „
50	− 6 „	„ 6.5 „	− 14 „	„ 28.5 „	+ 32.5 „	− 10.5 „	E 1 „	+ 20 „	W 133.5 „
20 30	+ 8.5 „	„ 37 „	− 20 „	„ 67 „	+ 21.5 „	− 30 „	„ 39 „	+ 24.5 „	„ 114 „
50	+ 3.5 „	W 9.5 „	− 25 „	„ 71.5 „	+ 16 „	− 23 „	„ 25 „	+ 22 „	„ 93 „
21 15	− 31.5 „	E 142.5 „	− 14 „	„ 181.5 „	+ 4.5 „	− 84.5 „	„ 139.5 „	+ 45 „	„ 219 „
30	− 13.5 „	„ 109 „	− 37.5 „	„ 138.5 „	0	− 55.5 „	„ 99 „	+ 39 „	„ 214.5 „
40	− 4.5 „	„ 98 „	− 40 „	„ 125 „	− 0.5 „	− 58.5 „	„ 83 „	+ 36 „	„ 206.5 „
22 20	− 43.5 „	„ 72.5 „	− 50 „	„ 80 „	− 16.5 „	− 37.5 „	„ 64 „	+ 18 „	„ 75 „
23 15	− 19 „	„ 30.5 „	− 32 „	„ 43 „	− 11 „	− 11 „	„ 31.5 „	+ 4 „	E 0.5 „

Current-Arrows for the 1st February, 1883.

Chart I at $11^h\ 20^m$, $12^h\ 20^m$, and $14^h\ 20^m$. Chart II at $15^h\ 20^m$, $16^h\ 20^m$, and $16^h\ 50^m$.

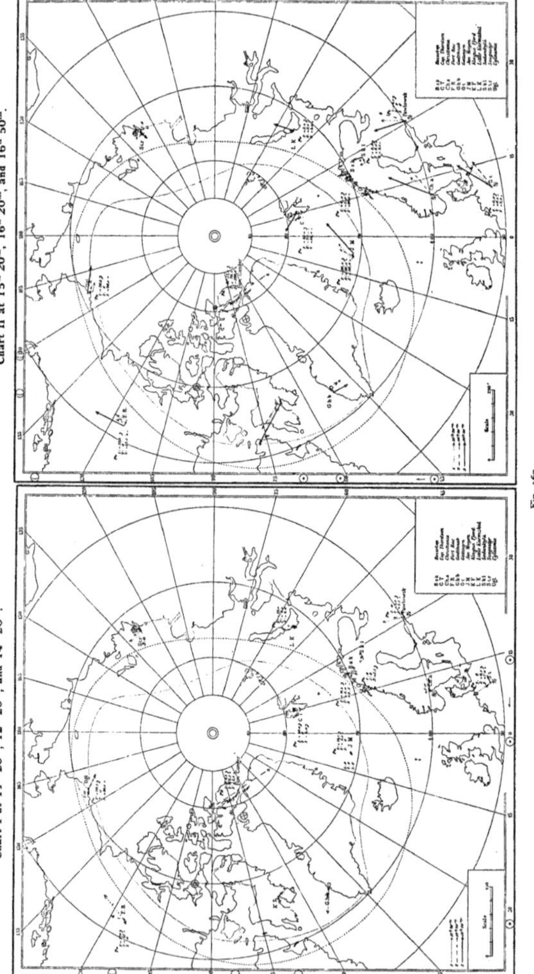

Fig. 163.

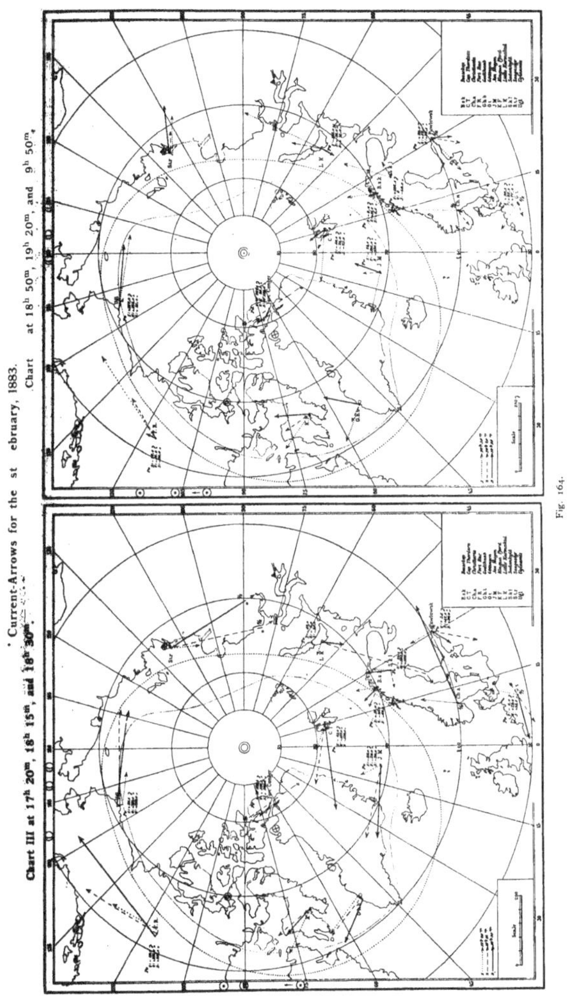
Chart III at $17^h\ 20^m$, $18^h\ 15^m$, and $18^h\ 30^m$.
Current-Arrows for the st ebruary, 1883.
Chart at $18^h\ 50^m$, $19^h\ 20^m$, and $9^h\ 50^m$.

Fig. 164.

Current-Arrows for the 1st February, 1883.

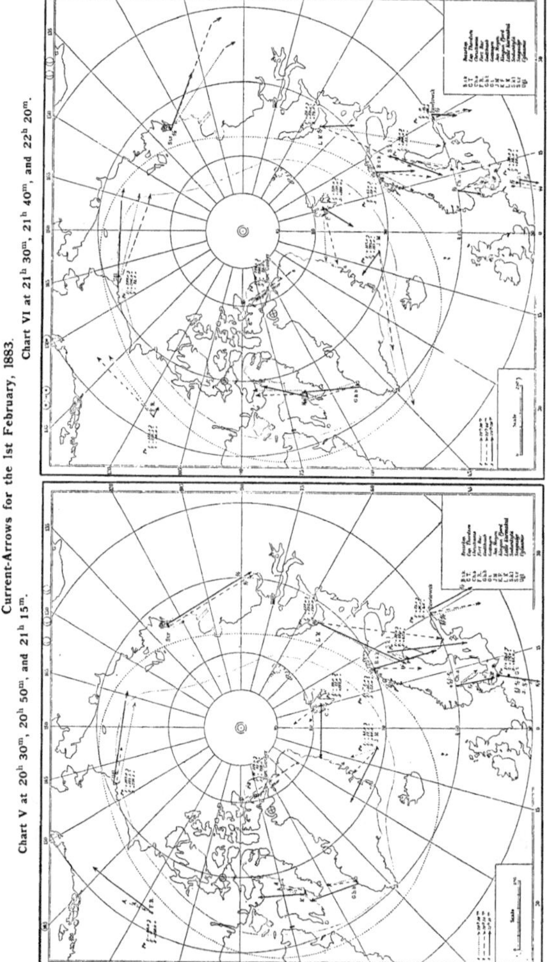

Chart V at $20^h\ 30^m$, $20^h\ 50^m$, and $21^h\ 15^m$.

Chart VI at $21^h\ 30^m$, $21^h\ 40^m$, and $22^h\ 20^m$.

Current-Arrows for the 1st February, 1883.
Chart VII at $23^h\ 15^m$.

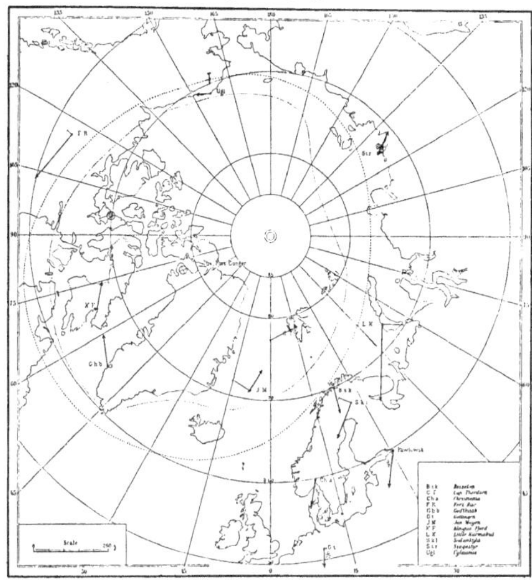

Fig. 166.

THE PERTURBATIONS OF THE 15th December, 1882.
(Pl. XXIV.)

89. The interest that attaches to the perturbations occurring on the above date, consists in the fact that we at first have a clearly developed positive equatorial storm. In the storms previously described, it was principally, at any rate, polar precipitation that showed itself, and the effects of which we studied. On this occassion, therefore, a special opportunity is afforded of studying perturbation-conditions in the polar regions about the auroral zone during an equatorial perturbation.

It may seem difficult to prove that it is really an equatorial perturbation with which we are concerned, seeing that our observations are chiefly from polar stations. It appears, however, that the more southern European stations are quite sufficient to determine this; for the perturbation-conditions that we have learnt to consider as characteristic of positive equatorial storms always come out very distinctly there.

If we compare Christiania, Pawlowsk and Göttingen, we find the conditions during the period previous to $10^h\ 15^m$ fairly normal; but then, at all three stations, there suddenly appears a perturbing force, which, at $10^h\ 20^m$ in the horizontal intensity, has a negative direction, but is then once more rapidly reversed; and from $10^h\ 25^m$ there are continuous positive perturbing forces, until the equatorial storm is interrupted by the polar storm that occurs during the last part of the period of observation. The course of the horizontal-intensity curve is the same at all three stations mentioned above, the similarity being most marked between Christiania and Göttingen. The curves for these places are drawn on the same scale, but in that for Pawlowsk the same serrations are found, notwithstanding that the scale employed is only one fifth that of the other two stations. The positive forces continue moreover all the time, as the storm remains more or less constant in strength. Further, on the most southern station, Göttingen, we find, it is true, at first from $10^h\ 15^m$ to about $10^h\ 25^m$ some small but very characteristic deflections in D; from that time, however, the declination-curve coincides fairly well with the normal line, until the polar storm sets in at the end of the period. We thus find here the well-known characteristic features always to be found in positive equatorial storms, and there is therefore no doubt that this is one of that class.

The determination of the normal line for this date is somewhat difficult, on account of the smallness of the perturbing forces and the length of the perturbation. The uncertainty thus arising is most apparent at the close of the equatorial perturbation. At its commencement, on the other hand, the uncertainty is not great, so that the forces taken out then differ very little, at any rate, from the actual values; and in the subsequent polar storm, the perturbing forces are of sufficient magnitude to make any uncertainty in the position of the normal line less important.

From Part I it will be remembered that the direction of the current-arrows in the north of Europe was not so nearly due east as at the more southern stations, but was as a rule a little more northerly, as there were also perturbing forces in declination. This is also the case now. At Pawlowsk there is a considerable deflection in the declination curve, whereas at Christiania and Göttingen this deflection in the declination is not so marked. It is possible, however, that there too there are some more powerful forces than those indicated on the plate, as the normal line is very difficult to determine, on account of the absence of daily hourly-observations.

At nearly all the polar stations, we find, at $10^h\ 20^m$, a rather sudden deflection in the curves, which indicates that the effect of the equatorial storm begins suddenly and simultaneously everywhere.

At three polar stations in the eastern hemisphere, which are situated to the south of the auroral zone, namely, Bossekop, Sodankylä, and Ssagastyr, there are positive deflections of fairly constant strength in the horizontal intensity, from the beginning of the positive storm until about $16^h\ 30^m$. Similarly we find in the declination at the first two of the above stations continual westerly deflections of fairly constant amplitude, while at Ssagastyr the deflections in this component amount to almost nothing.

At Cape Thordsen, the course of the declination-curve shows conditions very similar to those in the south, and is thus evidently due to an equatorial current-system; but polar precipitation makes its influence more felt here than at the stations just considered. This seems to be especially the case at first, when the horizontal-intensity curve has a rather more disturbed character.

The equatorial character of the perturbation disappears, however, as we go westwards. We also find the first impulse again at the other stations, and it is therefore evident that the perturbations we find here are connected with the equatorial storm, while it is equally certain that other effects seem to be present. If we look, for instance, at the conditions in Jan Mayen, we find at first only very small oscillations to either side of the normal line. In the horizontal-intensity curve these are principally above the line, thus answering to positive values of P_h, but farther on, at Godthaab, we find deflections which, though inconsiderable as regards strength, are mainly in the opposite direction, representing negative values of P_h.

Continuing westwards, we come to the station that is the most important in this instance, namely, Kingua Fjord. The disturbances there are evidently of a distinctly polar character; and the uneven nature of the curves seems to indicate that the polar system of precipitation is at no great distance from the station. This is especially evident in the declination. At first the deflections are mainly directed eastwards, but subsequently change, and from about $13^h 30^m$ until the close of the period, are directed westwards. The strength of the deflections is considerable, and as early as 15^h they have attained a magnitude of the same order that we are accustomed to find during the polar storms in these regions.

At the other polar stations there are no specially marked effects of polar storms until about $16^h 30^m$, so that up to that time the storm is concentrated about the districts surrounding Kingua Fjord. Even at Godthaab there are no distinct effects of the storm.

Continuing still westwards from Kingua Fjord through North America, we come to Fort Rae and Uglaamie. Here the effect of the equatorial storm seems once more to be more evident. In the horizontal intensity we find, at about $10^h 20^m$, an impulse exactly similar to that at the other stations at which the equatorial storm occurs; and after this we find, during the time that the equatorial storm is going on, mainly positive deflections of more or less constant amplitude, and as regards strength very much what one would expect to find them. There are, however, quite distinct effects of other systems. In two or three places, for instance, we find in the horizontal-intensity curve, deflections to the opposite side; and there are also sometimes impulses that are in all probability too powerful to be the direct effect of the equatorial current-system. This circumstance is most clearly apparent in the declination-curve. It is most natural here to assume that there is polar precipitation in addition to the equatorial system.

The most interesting feature here is, as we have said, the pronounced polar storm at Kingua Fjord. It is fairly powerful, but of very limited area, and recalls in a striking manner circumstances that we have previously found in our experiments.

We see, for instance, in this connection, in looking at fig. 37 on page 80, Part I, that in addition to the equatorial ring that is formed, there is a very distinct patch of light in the polar region, and some fainter, less distinct polar precipitation more on the noon or morning side of the terrella. This clear, sharply-defined patch answers to rays that descend towards the earth and leave it again in paths that lie comparatively close together. A system of precipitation of this form, in the immediate vicinity of the patch, will probably exert a considerable magnetic influence; but this will rapidly decrease with increasing distance from the patch. It is just an effect such as this that we appear to have at Kingua Fjord. There are, as we have said, powerful perturbing forces, which indicate comparatively abundant polar precipitation, while the effect of this precipitation at a station no farther off than Godthaab, is scarcely traceable. Lastly, if we look at the position of the patch in the figure, in relation to the magnetic pole of the terrella and the direction to the cathode, and imagine where this patch would fall if the earth and its magnetic axis were to take the place of the terrella and its magnetic axis, and the direction to the sun that to the cathode, it will easily be seen that the patch would fall more or less in the region round Kingua Fjord. It thus seems very probable that this is an in-drawing of rays such as we find by experiment. As we have said, there are also certain effects of polar precipitation at Fort Rae and Uglaamie, which may be connected with the slighter polar precipitation seen in the figure to the left of the distinct polar patch. The latter, however, may possibly be a more or less accidental resemblance; but the subsequent experiments may perhaps give fuller information regarding this circumstance.

With regard to the occurrence of the comparatively powerful polar storm at Kingua Fjord simultaneously with the equatorial storm, we may remind the reader of the various more or less abnormal conditions that we have come across at the American stations during the equatorial storms described in Part I. Of these we will mention the storms of the 23rd and 24th November, 1902, described on pages 273 and 274 in which these abnormal conditions were very greatly developed, and also the storms of the 26th January, 1903—page 67—and the 22nd March, 1903—page 128.

At the time we believed that these more or less abnormal conditions must be due to polar precipitation of some kind, concerning which we were then unable to express an opinion. Here, however, we have distinct proofs of the existence of such polar systems during equatorial storms also. It was especially in America that this precipitation occurred then, and now we find the same thing occurring here. The fact that it is in America that it occurs, is without doubt connected with the appearance of these storms at more or less the same time of day; and the situation of the magnetic pole in those regions is a circumstance of no little significance.

The equatorial storm is represented on the first four charts.

Chart I is drawn for a number of hours, to show the characteristic oscillation of the perturbing forces, which we have previously always observed simultaneously with the commencement of the effect of the equatorial storm. As the curves we have to go by are not continuous, but only readings for every fifth minute, the variation cannot be followed as it might have been if we had had photograms. It will be seen that the current-arrows at the stations in the south of Europe turn right round through an angle of about 180°, but it is not possible to determine whether the movement is clockwise or anti-clockwise. Later on, when the movement is less pronounced, it can be followed.

At the three southern stations the current-arrow moves in a direction contrary to that taken by the hands of a clock from $10^h\ 25^m$ until $10^h\ 30^m$, from which time until $10^h\ 35^m$ or $10^h\ 40^m$ it reverses its direction.

At Göttingen, where the first two current-arrows are not in quite such opposite directions as at the two other stations, it appears from the chart that the movement from $10^h\ 20^m$ to $10^h\ 25^m$ has been in the same direction as from $10^h\ 25^m$ to $10^h\ 30^m$, namely anti-clockwise, and that the principal phenomenon at the beginning of the equatorial perturbation would therefore be first a turn through 180° in a direction contrary to that of the hands of a clock, and then a smaller, slower turn back, after which the direction of the current-arrow remains constant as long as the effects of the equatorial system predominate.

At Ssagastyr, on the other hand, the current-arrow moves through a smaller angle, clockwise, and apparently more or less regularly, from $10^h\ 20^m$ to $10^h\ 30^m$, and then remains more constant for the remainder of the period represented on Chart I.

At Uglaamie too, the movement of the current-arrow is similiar to that at Ssagastyr; but its direction, unlike that at most of the other stations, is northerly.

At Fort Rae, the equatorial character of the perturbation is once more clearly apparent. The direction of the current-arrow also undergoes a great change as the perturbation begins, exactly similar to that which takes place at the southern European stations.

At the other stations too, there are great deflections, at Godthaab, for instance, as much as 180°. At Cape Thordsen the movement is less, and anti-clockwise; while south of that station it is generally clockwise, at any rate after $10^h\ 25^m$.

While there are considerable perturbing forces from $10^h\ 30^m$ to $10^h\ 40^m$ at the stations round Jan Mayen, those at Jan Mayen itself have now almost disappeared. There is evidently some connection between this circumstance and the fact that the current-arrows at Cape Thordsen and Godthaab are now almost in opposite directions. At the first of these stations, the equatorial system appears to exert a considerable influence, while in the region round Godthaab and Kingua Fjord, there seem to be other influences at work, probably polar precipitation, which, as we have seen, subsequently developes to a considerable strength in this very region.

On the other charts which represent the conditions during the equatorial storm, the current-arrow at the stations to the south of the auroral zone undergo, as a rule, little change in direction or size; and the form of the field remains fairly constant.

At the stations in and to the north of the auroral zone, on the other hand, the conditions are somewhat more variable. This is especially the case at Kingua Fjord. This is quite evident on looking at *Chart II—IV.* At $11^h\ 20^m$ the current-arrow seems in a great measure to be due to the equatorial system, although even now polar precipitation is also certainly asserting itself. At $14^h\ 20^m$, however, the polar system predominates, and at $15^h\ 20^m$ is still more evident.

With the exception of the polar storm at Kingua Fjord, there are none of any magnitude before $16^h\ 30^m$; but from that hour polar storms begin to be more and more apparent at other stations. At the same time the equatorial storm still continues to act for some time.

Between $16^h\ 20^m$ and 17^h, we find in the horizontal-intensity curve at Göttingen and Christiania a very characteristic wave; and at Pawlowsk exactly the same thing is found, although, as the scale is smaller, it is less distinct. Similar deflections are also found at the same time in the horizontal-intensity curve at Kingua Fjord, in declination and horizontal intensity at Godthaab and in declination and vertical intensity at Cape Thordsen, in Jan Mayen, and at Fort Conger, all of which exhibit so great a resemblance to one another, that there must undoubtedly be some connection between them.

In cases such as this, in which there are effects of both polar and equatorial systems simultaneously, the fact of finding conditions which seem to indicate that the two systems at the same time undergo similar changes, is in perfect accordance with what theory would lead us to expect. According to this, all the perturbing systems that appear simultaneously are due to one system of corpuscular rays, which become deformed by terrestrial magnetism, and, in their effects upon the earth, are apparently more or less separate phenomena. This however, it should be remarked, is only apparent. Theoretically there must always exist a genetic connection between simultaneous perturbations of the most varied kinds, both polar and equatorial, south polar and north polar, etc., etc. A connection such as this is often shown during equatorial storms in which, simultaneously with the serrations in the horizontal-intensity curve, there are found in the polar regions of the earth similar serrations or deflections that are too great to be ascribed to changes in the equatorial system, and which are certainly effects of polar precipitation. Another very typical example of this is to be found on this date, at about 14^h, on comparing, for instance, Christiania and Göttingen on the one hand, with Kingua Fjord or Fort Conger on the other.

We have often before pointed out simultaneous changes in positive and negative polar storms, which of course are also only indicative of the above-mentioned connection between the phenomena.

The polar storms that occur at the close of the period are both positive and negative.

The order of these polar storms on this date is the same as that so often found to be characteristic of afternoon storms, referred to Greenwich time.

At the more southern of the arctic stations in Europe, Jan Mayen and Bossekop, we find the effects of the positive storm. The storm occurs a little earlier at Jan Mayen than at Bossekop, as the horizontal-intensity curve at the former station begins, at about 15^h, to increase more or less regularly to its greatest height, which it attains at about $16^h\ 40^m$. At Bossekop the greater positive deflections do not occur until a little after 16^h; but the curve there rises somewhat more rapidly, and attains its greatest value at about $17^h\ 30^m$. After this first maximum has been reached, the positive deflection remains more or less constant in amplitude for some time, until the negative storm breaks in upon it.

At about 19^h the positive deflections in the horizontal-intensity curve for Jan Mayen begin to decrease, but at the same time the deflection in declination increases, thus forming the transition to the last portion of the observation-period, in which, as we see, the negative polar storms predominate. At Bossekop the transition from the positive to the negative storm is considerably sharper, and occurs at about 20^h.

At the other stations, wherever perturbations of any magnitude occur, we find only negative storms.

At Ssagastyr there is a short, comparatively small, but well-defined negative storm at about 17^h; and at Uglaamie there is also a negative storm. A maximum is found here a little before 19^h.

At Cape Thordsen we find negative deflections that very much resemble the positive deflections at Bossekop and Jan Mayen, for at all three stations we find deflections of fairly constant strength for a period of some length. The similarity is immediately seen on looking at the curves.

Lastly we also find considerable perturbing forces at Kingua Fjord; but, as already remarked, they will be more easily studied by looking at the charts. The most powerful storms, however, are between 20^h and 22^h. At almost all the polar stations here mentioned, there are negative deflections, as a rule very well defined. Only at Sodankylä do we find a considerable positive deflection, this being at about 20^h 30^m.

There is a certain amount of time-displacement here. The great negative deflections, for instance, begin a little earlier at Ssagastyr than at the European polar stations; but as we unfortunately have no observations from Little Karmakul, this circumstance cannot be closely studied. Moreover there are other phenomena which encroach upon it: for in all probability there will be positive storms occurring simultaneously in districts from which we have no observations. Now and then too, we find positive deflections, which may be interpreted as effects of such a system, e. g. the one just mentioned at Sodankylä, a small positive deflection in Jan Mayen at about 22^h 30^m, and two or three distinct positive deflections at Fort Rae, in the interval between 22^h and the close of the period of observation.

On looking at the declination-curve for Kingua Fjord, we are at once aware of a peculiar circumstance. This is the jagged, disturbed character of the curve before 20^h, and the wide, but regular deflection after that hour. We have seen that as a rule the curves in the polar regions during equatorial storms are of an exceedingly jagged, disturbed character, whereas the curve during well-defined polar stoms may frequently exhibit a fairly quiet course, even if the deflections are large. It may well be, therefore, that this transition to a more quiet course is an indication that the equatorial system is disappearing.

At the southern stations we find the most powerful forces in declination, and the deflection here begins at the time that the more powerful negative forces appear in the northern regions.

We now pass on to consider the last eight charts, on which the perturbation-conditions for the last part of the period are shown.

On *Chart V*, for 17^h 20^m we see evident traces of the positive polar storm, its district of precipitation extending from Godthaab across Jan Mayen to Bossekop. At all the other polar stations there are distinct effects of negative storms; while at the southern stations the equatorial system is still evidently at work.

On *Chart VI*, for 20^h 20^m, the effects of the equatorial storm have disappeared, and those of the positive polar storm are found only at Bossekop and Sodankylä. Everywhere else in the polar regions, we find more or less pronounced effects of negative precipitation, these being especially marked in the district Uglaamie to Ssagastyr. At Fort Conger there is also a more or less westerly-directed current-arrow, which in strength considerably surpasses those at Kingua Fjord, Godthaab and Cape Thordsen, This should probably be regarded as a continuation of the system of which traces were found in Jan Mayen.

The divergence of the current-arrows for Christiania, Göttingen and Pawlowsk, especially the westerly direction that they have at the first two stations, seems clearly to indicate the existence of a system of positive precipitation in the regions westward from Bossekop along the auroral zone; for these two arrows appear to be enclosed in an area of divergence corresponding to such a system. The positive vertical arrow for Göttingen is also in accordance with this.

The arrow for Pawlowsk, on the other hand, seems to be in the eastern area of convergence, answering to the negative storm in the north of Asia, but may also be considered as belonging to the area of divergence of the positive system of precipitation. The effects of the two systems will presumably be combined here, as they will each produce an arrow with a southerly direction.

On the next four *charts, VII to X,—from 20h 50m to 21h 25m*,—we find the perturbation-conditions represented as they appear at the time when there are powerful negative storms round the auroral zone. The form of the field of perturbation undergoes no particular change, but from time to time there is some variation in the strength.

The current-arrows at the European stations are directed southwards, as they usually are in the polar night-storms. This circumstance is certainly in some measure due to the more western positive system; for in the district in which it appears, this storm will diminish the effect of the adjacent negative precipitation, so that that system will in a manner be interrupted at the place where the positive precipitation appears. In this way, however, the constitution of the current-system will be such that the characteristic areas of convergence and divergence would be prominent, and this is just what these current-arrows indicate. The positive perturbing force in the vertical intensity at Göttingen also indicates the existence of such a system. Without the assumption of a system such as this, it would perhaps be rather difficult to find a simple explanation of these southern-pointing arrows, as the negative storms seem to be fairly evenly distributed about the auroral zone. We should then have to assume a more complicated constitution of the perturbing current-system, for instance, that rays came comparatively near to the earth as far south as this, and that their direct effect was of the greatest importance, or something similar. According to what we have said above, however, assumptions such as these are not necessary, our simple assumptions being apparently sufficient to explain the principal phenomena.

On the last two *charts, XI and XII*, for the hours *22h 15m, and 23h 5m*, the strength of the negative storm has considerably decreased, and we once more find traces of the positive storm, at 22h 15m in Jan Mayen, and at 23h 5m at Fort Rae. The negative storm now appears to be concentrated about the region from Ssagastyr to Cape Thordsen, i. e. on the night-side. At the southern stations there are no great changes to be discovered. We found that the vertical arrows at Göttingen must be due to the positive system; but the deflection in the curve is in striking harmony with the negative storm in the north, as the deflections begin to increase simultaneously. This may therefore only be indicative of the connection existing between the positive and the negative precipitation.

TABLE LVII.

The Perturbations of the 15th December, 1882.

Gr. M. T.	Uglaamie			Fort Rae			Kingua Fjord	
	P_h	P_d	P_v	P_h	P_d	P_v	P_h	P_d
h m								
10 20	− 13 γ	W 16 γ	0	− 19 γ	E 0.5 γ	+ 10 γ	− 30 γ	W 4.5 γ
25	− 31 „	„ 22 „	+ 10 γ	+ 15 „	W 6.5 „	+ 10 „	+ 8 „	E 32 „
30	− 13 „	„ 28 „	+ 10 „	+ 29 „	„ 9.5 „	+ 10 „	+ 9 „	„ 31 „
35	+ 3.5 „	„ 33 „	0	+ 28 „	E 14.5 „	+ 10 „	− 5 „	„ 11 „
40	+ 6 „	„ 40 „	0	+ 20 „	W 2 „	+ 10 „	− 27 „	W 2 „
11 20	+ 25 „	„ 45 „	− 20.5 „	+ 22 „	E 2.5 „	+ 10 „	− 17 „	E 24 „
12 10	− 11 „	„ 26.5 „	− 30.5 „	+ 2 „	W 4.5 „	+ 20 „	0	„ 56 „
13 0	+ 8.5 „	E 16 „	− 20.5 „	+ 18 „	„ 9 „	0	0	„ 38 „
14 20	0	„ 42.5 „	− 10 „	+ 16 „	0	0	− 33 „	W 23 „
15 20	+ 20 „	„ 21 „	− 20.5 „	+ 17 „	E 2.5 „	− 10 „	− 32 „	„ 62 „
16 20	0	„ 32 „	− 41 „	+ 9 „	„ 5.5 „	0	− 55 „	„ 95 „
17 20	− 24 „	„ 50 „	− 30.5 „	− 1 „	„ 18 „	0	− 76 „	„ 148 „
18 20	− 61 „	„ 40 „	− 91.5 „	− 6 „	„ 1 „	− 20 „	− 49 „	„ 124 „
19 20	− 77 „	„ 66 „	− 132 „	− 7 „	„ 15.5 „	− 35 „	− 52 „	„ 157 „
20 20	− 109 „	„ 87 „	− 223 „	+ 12 „	„ 28.5 „	0	0	„ 56.5 „
50	− 200 „	„ 46 „	− 145 „	− 19 „	„ 18 „	− 10 „	− 35 „	„ 120 „
21 5	− 298 „	„ 114 „	− 275 „	− 45 „	„ 16 „	0	− 35 „	„ 148 „
15	− 257 „	„ 64 „	− 315 „	− 80 „	W 2 „	− 20 „	− 43 „	„ 149 „
25	− 174 „	„ 67 „	− 275 „	− 76 „	„ 17 „	− 20 „	− 53 „	„ 132 „
22 15	− 56 „	W 2.5 „	− 295 „	+ 7 „	„ 26.5 „	0	− 15 „	„ 93 „
23 5	− 16.5 „	„ 13 „	− 295 „	+ 121 „	„ 56.5 „	+ 30 „	+ 26 „	„ 82 „

TABLE LVII (continued).

Gr. M. T.	Godthaab		Jan Mayen			Bossekop		
	P_h	P_d	P_h	P_d	P_v	P_h	P_d	P_v
h m								
10 20	+ 70 γ	E 14 γ	0	E 15 γ	+ 3 γ	− 14 γ	E 21 γ	0
25	− 27 „	„ 14 „	+ 8 γ	„ 21.5 „	+ 7 „	+ 11 „	W 20 „	0
30	− 33 „	„ 1 „	+ 1 „	0	+ 7 „	+ 13 „	„ 16 „	+ 8 γ
35	− 13 „	W 1 „	0	„ 3.5 „	+ 5 „	+ 12 „	„ 11 „	+ 2 „
40	− 7 „	E 2 „	0	„ 1.5 „	0	+ 9 „	„ 11 „	0
11 20	− 2 „	„ 15 „	+ 2 „	W 3 „	0	+ 2 „	„ 10.5 „	0
12 10	− 28 „	„ 15.5 „	+ 9 „	„ 2 „	0	+ 4 „	„ 16 „	+ 4 „
13 0	0	„ 10.5 „	+ 5 „	„ 2.5 „	− 5 „	+ 6 „	„ 14 „	0
14 20	− 7 „	„ 3.5 „	+ 10 „	„ 8.5 „	− 5 „	+ 7 „	„ 17.5 „	0
15 20	− 5 „	„ 5 „	+ 23 „	„ 13 „	− 4 „	+ 7 „	„ 14 „	− 8 „
16 20	− 9 „	„ 5.5 „	+ 32 „	„ 18.5 „	0	+ 4 „	„ 16 „	0
17 20	+ 18 „	„ 25 „	+ 45 „	„ 15.5 „	− 40 „	+ 50 „	„ 21 „	+ 64 „
18 20	− 5 „	„ 6 „	+ 56 „	„ 8.5 „	− 65 „	+ 50 „	„ 17.5 „	+ 74 „
19 20	0	W 10 „	+ 37.5 „	„ 22 „	− 61 „	+ 57 „	„ 21 „	+ 59 „
20 20	− 29 „	„ 41 „	− 65 „	„ 116.5 „	+ 10 „	+ 50 „	E 60 „	− 195 „
50	− 103 „	„ 124 „	− 230 „	„ 121 „	− 30 „	− 172 „	„ 21 „	− 245 „
21 5	− 119 „	„ 192 „	− 150 „	„ 31.5 „	− 39 „	− 100 „	„ 49.5 „	− 193 „
15	− 114 „	„ 160 „	− 205 „	„ 71 „	− 31 „	− 110 „	„ 56.5 „	− 178 „
25	− 107 „	„ 155 „	− 75 „	„ 80 „	− 44 „	− 95 „	„ 53.5 „	− 157 „
22 15	+ 5 „	„ 42 „	+ 65 „	„ 74 „	− 82 „	− 26 „	„ 28 „	− 325 „
23 5	0	„ 62 „	− 45 „	„ 27.5 „	− 75 „	− 47 „	„ 32 „	− 78 „

TABLE LVII (continued).

Gr. M. T.	Sodankylä			Cape Thordsen			Ssagastyr	
	P_h	P_d	P_v	P_h	P_d	P_v	P_h	P_d
h m								
10 20	− 5 γ	0	+ 84 γ	+ 65 γ	E 26 γ	− 9 γ	+ 59 γ	W 18.2 γ
25	+ 6 „	W 9.5 γ	+ 26 „	+ 40 „	W 18 „	− 40 „	+ 18 „	E 2.9 „
30	+ 12 „	„ 11.5 „	+ 24 „	+ 16 „	„ 12 „	− 28 „	+ 13 „	„ 7.5 „
35	+ 12 „	„ 11 „	+ 28 „	+ 4 „	„ 5 „	− 34 „	+ 12 „	„ 9 „
40	+ 12 „	„ 10 „	+ 25 „	+ 7 „	„ 9.5 „	− 9 „	+ 10 „	„ 10 „
11 20	+ 6 „	„ 7.5 „	+ 46 „	+ 17 „	„ 8 „	0	+ 11 „	W 7.5 „
12 10	+ 10 „	„ 8 „	+ 25 „	+ 37 „	„ 12 „	0	+ 26 „	„ 7.5 „
13 0	+ 12 „	„ 7 „	+ 30 „	+ 15 „	„ 7 „	0	− 22 „	„ 6.5 „
14 20	+ 8 „	„ 11 „	+ 31 „	+ 10 „	„ 10.5 „	0	+ 36 „	0
15 20	+ 7 „	„ 11.5 „	+ 24 „	+ 10 „	„ 15 „	0	+ 45 „	„ 2 „
16 20	+ 6 „	„ 9 „	+ 6 „	+ 15 „	„ 11.5 „	0	+ 26 „	0
17 20	+ 13 „	„ 15.5 „	+ 16 „	− 26 „	„ 23.5 „	0	− 95 „	E 10.5 „
18 20	+ 2 „	„ 8 „	+ 10 „	− 50 „	„ 20 „	0	+ 2 „	„ 33 „
19 20	+ 4 „	„ 4.5 „	+ 14 „	− 35 „	„ 30.5 „	0	− 48 „	„ 17 „
20 20	+ 85 „	E 36 „	− 19 „	− 5 „	„ 49.5 „	+ 43 „	− 237 „	„ 19 „
50	− 35 „	„ 40 „	− 23 „	− 249 „	„ 51.5 „	+ 110 „	ca. − 370 „	„ 20.5 „
21 5	− 11 „	„ 58 „	− 38 „	− 275 „	„ 85 „	− 100 „	− 339 „	W 10 „
15	− 32 „	„ 52.5 „	− 35 „	− 308 „	E 41 „	+ 123 „	− 301 „	„ 16 „
25	− 17 „	„ 59.5 „	+ 43 „	− 208 „	„ 69 „	+ 196 „	− 236 „	„ 3.5 „
22 15	− 4 „	„ 35 „	+ 38 „	− 123 „	„ 37 „	+ 83 „	− 102 „	E 84 „
23 5	− 17 „	„ 29 „	+ 54 „	− 153 „	„ 18 „	+ 140 „	− 44 „	„ 26.5 „

TABLE LVII (continued).

Gr. M. T.	Christiania		Pawlowsk		Göttingen			Fort Conger
	P_h	P_d	P_h	P_d	P_h	P_d	P_v	P_d
h m								
10 20	− 12 γ	E 9.5 γ	− 7.5 γ	E 4.5 γ	− 8.7 γ	E 12.5 γ	− 12.5 γ	E 16 γ
25	+ 8.5 „	W 3.5 „	+ 12 „	W 6 „	+ 11.5 „	W 3.5 „	+ 1 „	W 33.5 „
30	+ 12 „	„ 6.5 „	+ 14 „	„ 14 „	+ 18.5 „	„ 13 „	− 3 „	„ 33 „
35	+ 12.5 „	„ 4.5 „	+ 18 „	„ 8.5 „	+ 17 „	„ 7 „	− 7.5 „	„ 25 „
40	+ 10.5 „	„ 4 „	+ 17 „	„ 7.5 „	+ 15 „	„ 4.5 „	− 7.5 „	„ 14 „
11 20	+ 9.5 „	0	+ 12 „	„ 8 „	+ 11 „	0	− 5 „	E 8 „
12 10	+ 11 „	0	+ 11 „	„ 9 „	+ 12 „	„ 5 „	− 7 „	W 6 „
13 0	+ 13.5 „	0	+ 11 „	„ 5.5 „	+ 13.5 „	0	(1)	„ 8.5 „
14 20	+ 12 „	0	+ 8 „	„ 8 „	+ 11 „	0	(1)	E 2 „
15 20	+ 11.5 „	0	+ 8 „	„ 5 „	+ 11 „	0	(1)	W 8 „
16 20	+ 9 „	0	+ 4 „	„ 5.5 „	+ 9.5 „	0	(1)	„ 7 „
17 20	+ 13 „	0	+ 10 „	„ 8 „	+ 14.5 „	0	(1)	„ 28.5 „
18 20	+ 4 „	0	0	„ 5 „	+ 7 „	0	0	„ 26.5 „
19 20	+ 2.5 „	0	0	0	+ 5.5 „	0	0	„ 40.5 „
20 20	− 13 „	E 21 „	+ 5 „	E 32 „	− 13 „	E 13.5 „	+ 9 „	„ 107 „
50	+ 2 „	„ 63.5 „	+ 7 „	„ 39.5 „	− 10 „	„ 54 „	+ 16.5 „	„ 136 „
21 5	− 1 „	„ 59 „	+ 3 „	„ 44 „	− 9.5 „	„ 53 „	+ 16 „	„ 148 „
15	− 5 „	„ 59 „	+ 2 „	„ 43.5 „	− 8 „	„ 52 „	+ 18 „	„ 155 „
25	− 6 „	„ 56.5 „	0	„ 45 „	− 8 „	„ 50 „	+ 16.5 „	„ 180 „
22 15	− 17 „	„ 30.5 „	− 13 „	„ 30.5 „	− 13 „	„ 21.5 „	+ 12 „	„ 128 „
23 5	− 8 „	„ 26 „	− 7 „	„ 20.5 „	− 4 „	„ 18 „	+ 12 „	„ 71 „

(1) Small oscillations; probably negative deflections.

Current-Arrows for the 15th December, 1882.

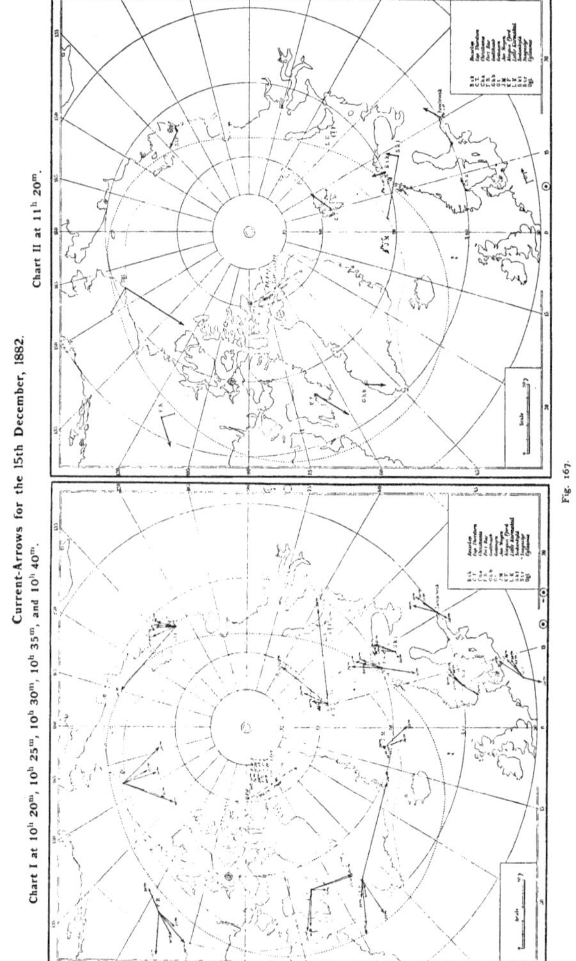

Chart I at $10^h\ 20^m$, $10^h\ 25^m$, $10^h\ 30^m$, $10^h\ 35^m$, and $10^h\ 40^m$.

Chart II at $11^h\ 20^m$.

Fig. 167.

Current-Arrows for the 15th December, 1882.

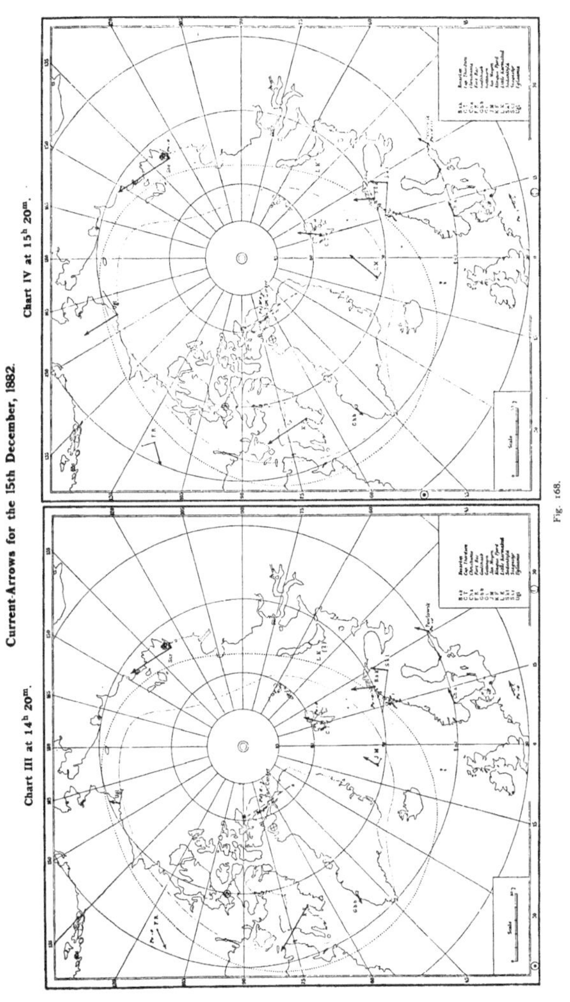

Chart III at $14^h\ 20^m$.

Chart IV at $15^h\ 20^m$.

Fig. 168.

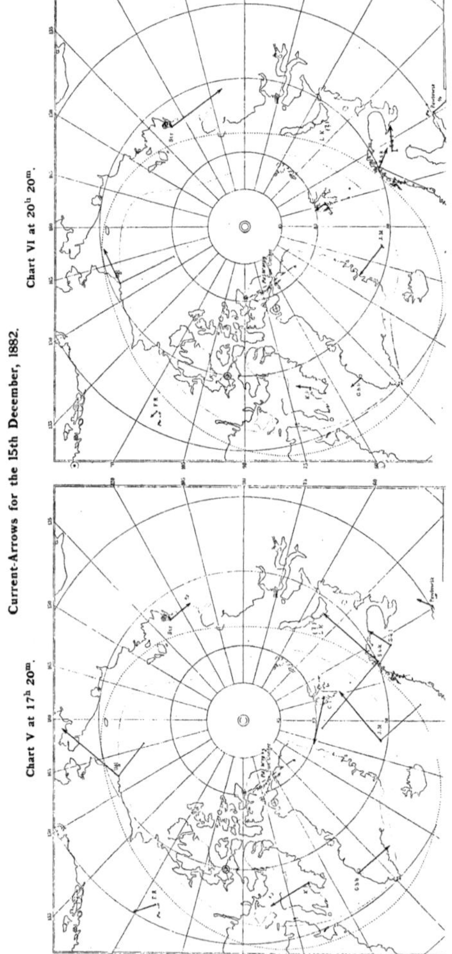

Current-Arrows for the 15th December, 1882.

Chart V at $17^h\ 20^m$.

Chart VI at $20^h\ 20^m$.

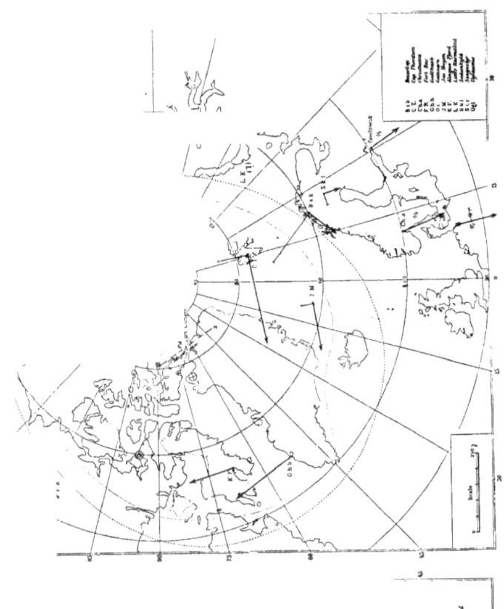

Fig. 170.

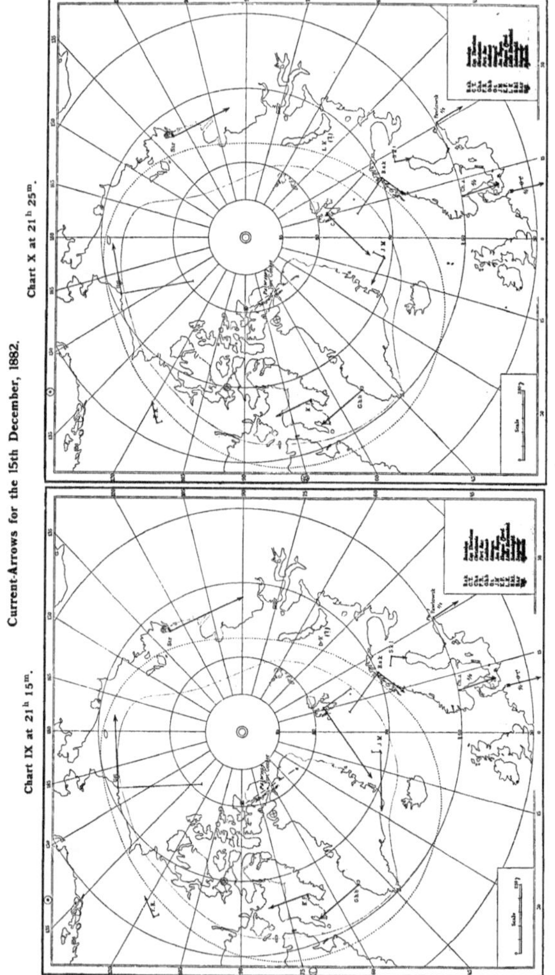

Current-Arrows for the 15th December, 1882.

Chart IX at $21^h\ 15^m$.

Chart X at $21^h\ 25^m$.

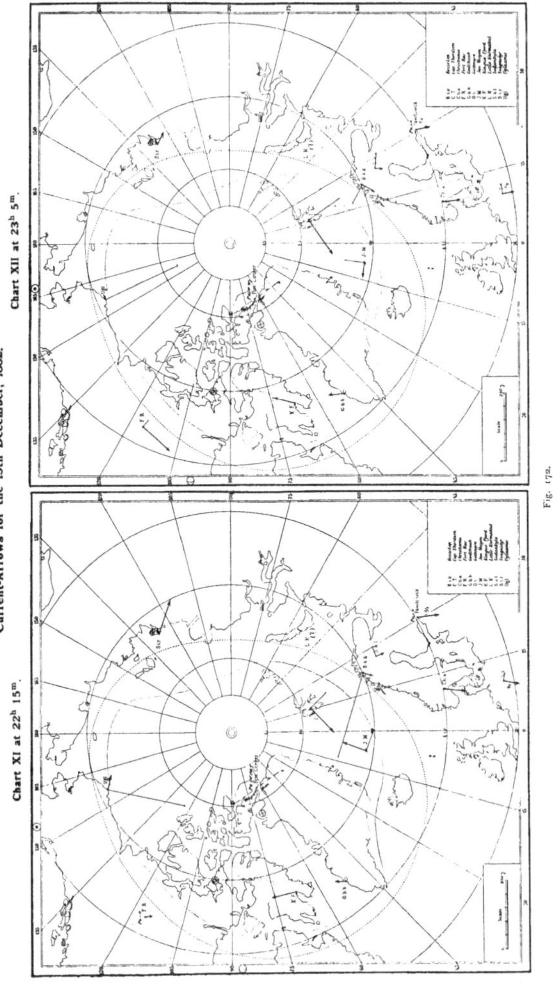

Current-Arrows for the 15th December, 1882.

Chart XI at 22^h 15^m.

Chart XII at 23^h 5^m.

Fig. 172.

THE PERTURBATIONS OF THE 15th OCTOBER, 1882.

(Pl. XXII).

90. This observation-period differs from those already described, in the occurrence of fairly powerful perturbations almost throughout the day, only the last part of the period being a little quieter. The position of the normal line is therefore to some extent difficult to determine, especially at Christiania and Göttingen, where there is less to go by. In the case of the last-named station, therefore, no such line has been drawn for the vertical intensity. It will nevertheless be possible, from the course of the curve, to determine the direction of the deflections when these are greatest. We have employed this quieter district as the starting-point for the placing of the normal line, assuming the conditions there to be more or less normal.

At the beginning of the period, a well-developed negative polar storm of considerable strength is found in the district from Jan Mayen to Bossekop. The most powerful forces appear in Jan Mayen. At the same time, we find at the stations to the south of this district, deflections which evidently appear to be governed by the same forces that produce the storm in the north. At Christiania and Göttingen we find serrations similar to those that are especially distinct in Jan Mayen. On the other hand we also find positive polar precipitation developed in America, especially at Fort Rae.

At Kingua Fjord too, there seems to be the effect of a similar system, but, as we have said, the conditions there will be better studied by the aid of the charts; for a mere consideration of the curves may possibly be misleading.

The first part of the observation-period is at a time when it is night in Europe and afternoon in North America. These storms are thus of exactly the same kind as those which we are accustomed to find at this time of day.

Chart I represents the perturbation-conditions at the above-mentioned time. The district of precipitation of the negative storm is distinctly visible in Jan Mayen and Bossekop, and the effects of the positive system at Fort Rae.

It will further be seen that round the district of negative precipitation, the current-arrows are grouped in the manner generally, if not always, found in the polar storms. The current-arrows to the south fit very well into the system of convergence, which corresponds to a negative system of precipitation. At 1^h 20^m Christiania appears to be in the immediate vicinity of the point of convergence of the system, which, at the last hour given, 2^h 20^m, seems to have moved towards Pawlowsk. At the same time the powerful forces in Jan Mayen are considerably reduced, and thus the storm-centre seems to have moved a little eastwards.

Another circumstance that may possess some interest is the direction of the current-arrows at Godthaab, Kingua Fjord and Cape Thordsen, where the forces at certain times are rather small, and there thus appears to be no particular local precipitation. It would therefore seem probable that we should here find effects of the powerful negative system acting in the neighbourhood of Jan Mayen. As effects of this there should be an area of divergence in these regions, and the arrows do indeed admit of being arranged in such a system; for if we follow a current-line in this district from Bossekop westwards across Jan Mayen, to Godthaab and Kingua Fjord, we see that it turns off here to the right and runs northwards. Fort Conger, unfortunately, cannot give satisfactory information concerning the further course of the current-line, the conditions indicating only that the direction is somewhat easterly. This too is in accordance with what we should expect.

The direction at Cape Thordsen indicates that the current-line turns southwards, back to the regions about Bossekop. Thus the course of the current-lines seems to be similar to that which we should expect to find in the system's area of divergence.

It is probable, however, that there will also be other forces in operation, and that the conditions are not so simple as here described. One circumstance, for instance, that has not been touched upon is the connection that seems to exist between the deflections at the southern stations and the system in North America.

At the hours here observed there do not, it is true. appear to be any conditions that point distinctly in this direction; but at about 3^h the deflections, especially in *H*, at Fort Rae on the one hand and Christiania and Göttingen on the other, exhibit so great a resemblance to one another that it would seem probable that a more or less close connection existed. Simultaneously with these deflections at about 3^h, we also find similar changes at several other stations, e. g. at Kingua Fjord, where there is a characteristic and well-defined deflection in declination towards the east, the forces here having previously had a westward direction. This, as will appear from Chart II, seems to indicate the intrution of a positive storm. At Fort Rae, on the other hand, it is evidently a negative storm; and at Jan Mayen we find at the same time a corresponding change in the deflections, perhaps here, too, the effects of a positive storm asserting itself, as the negative deflections diminish considerably, although none go over to the other side.

In the field of perturbation at 2^h 50^m, represented on *Chart II*, the negative system of precipitation comes out very distinctly in Jan Mayen, Bossekop and Fort Rae, while at Kingua Fjord there are signs of a positive polar storm.

The current-arrows at the southern stations, on the other hand, exhibit conditions that appear more peculiar. If they are due entirely to the negative system of precipitation to the north, even Göttingen must be situated to the north of the point of convergence of this system, or perhaps more strictly speaking to the north of the neutral area of the system.

There will, however, be some difficulties in the way of an assumption such as this, and moreover the course of the curves appears to indicate that the cause should be sought in a system that is closely connected with that which appears most distinctly at Fort Rae and Kingua Fjord, and which in all probability also causes the great diminution in the negative deflections in the horizontal intensity in Jan Mayen just at this time. We have frequently observed a similar resemblance between the conditions in Central Europe and those in North America; and in discussing our experiments in a later chapter, we shall find conditions that are apparently similar to these.

The next phenomenon that strikes one on looking at the plate is a perturbation that is especially characteristic and well defined at Cape Thordsen, more particularly in the horizontal intensity, where it appears as a negative storm. Its effects are also distinctly apparent in Jan Mayen, where the perturbing forces even exceed those at Cape Thordsen in strength. Of the arctic stations, it is only at these two that this storm is distinct; even at Bossekop there is no distinct effect of the system.

On looking at Little Karmakul, however, and comparing its horizontal-intensity curve with that of Cape Thordsen, we find, on closer examination, quite a remarkable resemblance. The deflections in *H* at Little Karmakul, from about 3^h until about 15^h, are positive nearly all the time. On the other hand there are no perturbing forces of any magnitude at the same time as the negative storm at Cape Thordsen; but the commencement of the decrease in the positive deflections is exactly simultaneous with that of the increase in the negative deflections at Cape Thordsen, and the maximum of the negative storm at Cape Thordsen with the lowest position of the horizontal-intensity curve at Little Karmakul, and we then have distinctly negative perturbing forces there. Lastly, the curves again increase at both stations simultaneously, while the deflection at Cape Thordsen decreases, and the positive deflections at Little Karmakul increase. It would thus seem reasonable to suppose that at the latter station we have before us the effects of two simultaneous storms, the positive storm continuing all the time, and the negative intruding upon it, and partly compensating the positive deflections, partly effecting their reversal. The similarity between the curves is in fact so striking that this assumption seems very probable.

At the southern stations the conditions are evidently regulated by the negative polar storm. This storm appears particularly clear, if we compare the horizontal-intensity curves.

Except at Little Karmakul, no very distinct traces of positive polar storms are to be found during this period, Uglaamie being the only station at which there appear to be any more or less evident effects of such a storm. It will be seen that this station, at the time, is on the afternoon side of the globe.

The fields of perturbation at 3^h 20^m and at 4^h 20^m are represented on Chart II.

At 3^h 20^m we find the negative storm in the polar regions developed, to any extent, only in Jan Mayen. The three southern stations indicate simultaneously by their current-arrows that they are in the eastern part of the area of convergence.

At 4^h 20^m, however, the storm-centre has moved or perhaps rather expanded eastwards, thus bringing Cape Thordsen into the district of negative precipitation. At the same time the current-arrows for the three southern stations turn clockwise through a considerable angle, just as our previous assumptions would lead us to expect.

At Pawlowsk, at this time, we find negative values of P_v, indicating the existence there of an area of convergence. At Göttingen too, the direction of the perturbing force in the vertical intensity seems to be the same; there is a distinct wave in the curve just at the time of the negative deflections at the two arctic stations.

At 5^h the positive storms in America are over, and negative storms begin everywhere, developing subsequently to a considerable strength.

At Godthaab the negative storm began to develope earlier. The negative perturbing forces here must be regarded as continuations of the powerful eastern system.

At Fort Rae too, the negative deflections become stronger, and at 6^h a fairly powerful negative storm begins to develope, and continues until about 17^h. There are two maxima here, separated by a period during which the negative forces are considerably weaker, although the direction of the deflections remains unchanged.

At Uglaamie the stronger negative forces appear somewhat later than at Fort Rae; but a little before 8^h they begin to increase rapidly until they attain considerable strength. The negative deflections then continue more or less constant in strength until about 17^h, after which they are small.

At Kingua Fjord too, negative storms appear to be at work; but we will reserve our description of the conditions there until we come to the charts.

We note that this transition from positive to negative storms in America takes place at the time when these districts enter the night-side of the earth. At the same time the districts in Asia and Europe move on to the day-side of the globe, and at the polar stations here, Cape Thordsen excepted, we also find a transition, but from negative to positive systems, and thus the reverse of that in America.

The change takes place earliest in the most easterly districts. At Little Karmakul, for instance, there seem to be positive storms as early as 3^h. At a little before 6^h, however, they begin to be more distinct, the positive deflections becoming larger and larger, until about 14^h there is a maximum for the positive deflections.

At Bossekop and Sodankylä the positive storm developes very characteristically; but the positive deflections begin a little later. At about 5^h 20^m the negative storm at Bossekop is over, and from that time until about 10^h, there are small deflections now to one side and now to the other. At 10^h the positive deflections begin to increase with comparative rapidity, and reach their maximum at about 15^h 20^m, when they decrease rapidly. The development of the storm at Sodankylä is very similar.

If we go on to Jan Mayen, we still find, at the beginning of the period, effects of negative deflections. After 10^h, the positive storm there developes powerfully. Thus while the effects of the positive storm appear more or less simultaneously at Bossekop and in Jan Mayen, the previous negative storm lasts considerably longer at the Jan Mayen station than at Bossekop. It seems evident from this that the district of positive precipitation is moving westwards.

This movement, which has so often been mentioned, and which has undoubtedly some connection with the earth's rotation, is here very distinct, as the perturbations concerned are of longer duration than usual, and perhaps also because they are at a time not very distant from the equinox.

In Jan Mayen, however, forces soon appear which seem to counteract the effects of the positive storm; a negative system seems to encroach upon the positive for a short time, and once or twice cause a reversal of the values of P_h.

This negative storm is evidently the same that appears at Cape Thordsen, but here it is far more powerful. The effects of the positive storm are slight. Before $11^h\ 20^m$ the horizontal-intensity curve at the latter station oscillates about the normal line, perhaps the result of the action of alternate slight positive and negative precipitation. After $11^h\ 20^m$, however, a very well defined negative polar storm appears, which developes and reaches its maximum simultaneously with the positive storm in the south. Simultaneous serrations are also frequently to be found, a circumstance which indicates the connection which evidently exists between these cases of precipitation.

A comparison of the horizontal-intensity curves for Cape Thordsen and Jan Mayen will give a distinct impression that it is the negative storm that breaks in upon the positive, and produces the peculiar phenomena found in Jan Mayen. That the positive storm is going on all the time seems to be clearly evident, however, from the fact that simultaneously with the disappearance of the negative storm at Cape Thordsen, the positive forces once more assert themselves, and the positive deflections then diminish just as at Bossekop. It is also characteristic that at Bossekop too, the negative storm intrudes and produces the peculiar curve that we find at about 16^h.

The horizontal-intensity curve at Little Karmakul also shows clearly a condition exactly similar to that in Jan Mayen, namely a long positive storm, upon which the somewhat shorter negative storm intrudes. For a time too, the latter is the stronger, just before it reaches its greatest height. At 16^h, however, positive forces once more appear, evidently the same strengthening of the positive system as at Sodankylä. After that hour the curve oscillates about the normal line, thus indicating the supremacy of the positive and negative forces alternately. In declination, however, the direction of the deflections is nearly always the same, namely westward; but here too, the curve is exceedingly jagged and disturbed in character.

At Fort Conger, the last of the polar stations, it will be seen that the declination-curve very much resembles that at Cape Thordsen, and we may therefore assume that the system continues westwards through that station.

At the southern stations, the deflections are evidently governed by the precipitation in the arctic regions; and we sometimes find a very distinct resemblance between the various serrations. The deflections in the horizontal-intensity curves for Christiania and Pawlowsk are not constant in any part of the period, but are sometimes in one direction and sometimes in another, although the negative deflections predominate. Farther south, on the other hand, e. g. at Göttingen, we find negative deflections all the time.

In declination we find the deflections for the most part directed westwards at Christiania and Göttingen, whereas at Pawlowsk there are no very considerable forces in that component. In the vertical-intensity curve at Pawlowsk a very distinct positive deflection appears.

The conditions at Göttingen are exactly similar. The rise in the vertical-intensity curve at about 7^h, and the fall at about 11^h, are undoubtedly connected with the diurnal variation, while the last rise with a maximum at about 16^h seems to be connected with the perturbations.

These conditions, the distribution of the districts of positive and negative precipitation over the various regions of the earth, and their intermingling, are thus in perfect accordance with our previous experience. The polar areas of perturbation are always manifested in the main in the same manner, and every part of the day has, so to speak, its characteristic area of perturbation, which will always approximately form when there are any perturbations.

At the close of the period, the conditions are, as we have said, almost normal everywhere, with the exception of Kingua Fjord, where there are still some powerful forces.

We will now look at the charts for this last section of the perturbation.

Chart III represents the conditions from 5^h 20^m to 7^h 20^m. The storms are chiefly negative. There is the powerful system in America, especially noticeable at Fort Rae, and one less powerful in Jan Mayen, a westward continuation of which is indicated by the conditions at Godthaab and Kingua Fjord.

Of the positive storms there is little observable here. At 6^h 20^m there is an indication of one at Little Karmakul, but the force is not great.

The current-arrows at the southern stations at 7^h 20^m are rather more difficult to include in a polar field of perturbation answering to the systems of precipitation appearing here. The observations we have are too few for us to determine the nature of the perturbing forces at work; we will only draw attention to the simultaneous deflections appearing in the horizontal-intensity curves for Fort Rae and Kingua Fjord on the one hand, and Christiania and Göttingen on the other: The maxima occur simultaneously, and there are also several coherent serrations. This is apparent chiefly until 11^h, after which hour the polar systems in the north of Europe also appear much more powerful, so that the phenomena in Central Europe are mainly governed by this precipitation.

This is certainly to some extent a phenomenon similar to that with which we meet at about 3^h on this day.

On *Chart IV* the positive storm appears more distinct, but has not yet extended farther than to Little Karmakul. At Godthaab and Kingua Fjord, the same negative system is at work as in Chart III; but it has now moved westwards, so that Jan Mayen is no longer in the district of precipitation.

The current-arrows in Central Europe may either belong to the area of divergence of the eastern positive system, or to the area of convergence of the western negative system. It is rather doubtful whether the system at Godthaab, and still more that at Kingua Fjord, can be regarded as a negative system of precipitation. P_h, it is true, is negative everywhere, so it therefore might be called so; but the direction of the principal axis is more north and south than usual, a circumstance that is more conspicuous later on.

On Chart IV, the arrows seem to be principally connected with the American system, while on *Chart V* they form a transition between the negative system on the west and the positive system on the east, or, as we might say, between the system at Cape Thordsen and the more southern system at Jan Mayen and Bossekop.

It is an unfortunate circumstance that on Chart IV there are no observations of horizontal intensity for Fort Conger. If there had been a strong current-arrow there, directed southwards, it would seem likely that a current-circuit had been formed from Fort Rae, through Uglaamie, Fort Conger, and Godthaab, and probably back to Fort Rae. When the system has moved a little, we find a circuit similar to this, as there is negative precipitation at Cape Thordsen; but this circuit does not appear at all distinctly until *Chart VI*. If this could have been demonstrated as early as Chart IV, a very much better survey of the perturbation-conditions would have been obtained, and a fact to hold to when seeking, by experiments, for points of similarity. A fact such as this would have brought about some modifications in our reasoning, but no essential simplification.

As the observations, that we have at our disposal, seem to show, the negative system of precipitation developes by a more or less continual extension of its area westwards.

In the second case we should have to imagine that a more or less momentary current-circuit was formed, which increased somewhat during the course of the perturbation, while at the same time, owing to the rotation of the earth, its position was changed.

Both these assumptions are possible, but it is not easy to say which is the more correct one. It will thus be a matter for future research to procure a clear understanding of this point; the present observations are too few.

The positive system, with its area of divergence, comes out very distinctly on *Chart V*, with all the characteristics of such a storm. The point of divergence of the system is evidently in the vicinity of Pawlowsk. P_v is here positive in direction, and the horizontal forces are sometimes very small.

In addition to this, the field is characterised by the negative storm, which now, as already mentioned, seems to have moved towards the west, while at Cape Thordsen we also now find negative perturbing forces.

There is nothing very new to be seen on *Chart VI*. Judging from the current-arrows in Central Europe, we should be inclined to suppose that the positive system of precipitation has extended farther westwards; but at the same time the more northerly negative storm has also increased in strength, so that the two counteract each other's effect in a horizontal direction in Jan Mayen. In vertical intensity, however, both systems at that station act in the same direction, and we therefore find powerful negative perturbing forces there.

As we have said, the negative circuit is now more distinct.

At Little Karmakul, sometimes the positive, sometimes the negative system is the more powerful.

On the last chart, *Chart VII*, the powerful systems have disappeared, and we find only faint indications of the former powerful storms.

At the first, and to some extent the second hour, there are still forces of some considerable magnitude; but at the last hour it is for the most part only at Kingua Fjord that storms are still going on.

TABLE LVIII.

The Perturbations of the 15th October, 1882.

Gr. M. T.	Uglaamic			Fort Rae			Kingua Fjord	
	P_h	P_d	P_v	P_h	P_d	P_v	P_h	P_d
h m								
0 20	+ 3 γ	W 2.5 γ	+ 43.5 γ	+ 83 γ	W 18.5 γ	0	+ 59 γ	W 56 γ
1 20	+ 30 „	0	+ 44.5 „	+ 70 „	„ 13.5 „	− 40 γ	+ 51 „	„ 28.5 „
2 20	+ 34 „	„ 12 „	+ 25.5 „	+ 26 „	E 4.5 „	− 80 „	+ 30 „	„ 13.5 „
50	+ 45 „	„ 26.5 „	+ 17 „	− 90 „	„ 135 „	− 90 „	+ 20 „	E 60.5 „
3 20	+ 35 „	„ 26.5 „	0	− 24 „	„ 19 „	−110 „	+ 46 „	„ 17 „
4 20	+ 35 „	E 5.5 „	− 4 „	− 3 „	„ 3 „	− 80 „	+ 35 „	W 0.5 „
5 20	− 0.5 „	„ 46 „	0	− 19 „	„ 32 „	− 90 „	− 25 „	E 19.5 „
6 20	− 22.5 „	W 38.5 „	− 19 „	−146 „	W 2 „	+ 90 „	− 1.5 „	„ 21 „
7 20	+ 87.5 „	E 188 „	− 66 „	−300 „	E 109.5 „	−100 „	− 66.5 „	„ 24 „
8 20	−104 „	„ 56 „	− 32.5 „	−400 „	„ 116.5 „	+170 „	− 50 „	„ 59 „
9 20	−132 „	W 106 „	− 46.5 „	− 93 „	„ 12 „	+170 „	− 59 „	„ 54.5 „
10 20	− 96.5 „	„ 26 „	+ 47 „	−182 „	„ 62.5 „	+ 10 „	− 67 „	„ 44.5 „
11 20	−142.5 „	E 19.5 „	+ 47 „	−260 „	„ 55 „	+180 „	− 56 „	„ 31.5 „
12 20	−151 „	„ 103.5 „	+ 4.5 „	−223 „	„ 120.5 „	+280 „	−120 „	„ 55 „
13 20	− 85 „	„ 49.5 „	+ 94 „	−258.5 „	„ 5 „	+390 „	−125 „	„ 27.5 „
14 20	− 57 „	„ 222.5 „	+ 98.5 „	−331 „	„ 243.5 „	+230 „	−130 „	W 21.5 „
15 20	− 81 „	„ 35 „	+118 „	−375 „	„ 159 „	+ 20 „	−113.5 „	E 5.5 „
16 20	−148 „	„ 62 „	+ 98.5 „	−237.5 „	„ 121.5 „	60 „	−100 „	W 77 „
50	− 51.5 „	„ 16 „	+ 75 „	− 89 „	„ 40 „	− 20 „	− 14 „	„ 10 „
17 20	− 45 „	W 26 „	+ 56.5 „	− 30 „	„ 14 „	+ 10 „	− 24 „	„ 70 „
18 20	− 20 „	E 31.5 „	+ 0.5 „	0	„ 12.5 „	− 20 „	− 25 „	„ 78.5 „

TABLE LVIII (continued).

Gr. M. T.	Godthaab		Jan Mayen			Bossekop		
	P_h	P_d	P_h	P_d	P_v	P_h	P_d	P_v
h m								
0 20	+ 32 γ	W 53.5 γ	−395 γ	E 27.5 γ	+155 γ	−165 γ	E 39 γ	−225 γ
1 20	− 8 „	„ 36 „	−357.5 „	„ 40 „	+227.5 „	−177.5 „	„ 54 „	−302.5 „
2 20	+ 12 „	„ 7 „	−127.5 „	„ 53.5 „	+171.5 „	− 72.5 „	„ 53 „	−163.5 „
50	− 75 „	E 50.5 „	−112.5 „	„ 62 „	+144 „	−121 „	„ 63 „	−180 „
3 20	− 20 „	W 8.5 „	−196.5 „	„ 87.5 „	+128 „	− 45 „	0	− 36 „
4 20	− 60 „	„ 13 „	−290 „	„ 65 „	+ 60 „	− 35 „	W 17.5 „	− 60 „
5 20	− 37 „	E 14 „	− 42.5 „	„ 4.5 „	− 4 „	0	E 11.5 „	0
6 20	− 50 „	„ 13 „	− 77 „	„ 17 „	− 8 „	− 4.5 „	W 15 „	0
7 20	− 73 „	„ 42.5 „	−115 „	„ 28 „	− 9 „	− 23.5 „	„ 3 „	+ 11 „
8 20	−235 „	W 11 „	− 35 „	0	− 55 „	+ 6 „	„ 20 „	+ 83.5 „
9 20	−197 „	„ 40 „	− 10 „	0	− 16 „	+ 11.5 „	„ 17.5 „	+ 50 „
10 20	−115 „	E 0.5 „	+ 49 „	W 9 „	+ 5.5 „	+ 21.5 „	„ 14 „	+ 57.5 „
11 20	− 68 „	„ 33.5 „	+ 78.5 „	E 3 „	+ 1.5 „	+ 61 „	„ 27 „	+ 90 „
12 20	− 34 „	„ 79.5 „	+144 „	„ 26.5 „	− 4 „	+101 „	„ 37 „	+147.5 „
13 20	− 86 „	„ 45.5 „	+105 „	„ 23 „	−107.5 „	+140 „	„ 38.5 „	+180 „
14 20	−106 „	W 18 „	+ 10 „	W 17 „	−190 „	+130 „	„ 3 „	+161 „
15 20	− 33 „	E 55 „	− 4.5 „	„ 58.5 „	−100 „	+185 „	„ 30.5 „	+155 „
16 20	− 85 „	W 32 „	− 21 „	„ 30 „	−115 „	+160 „	„ 28 „	+115 „
50	− 22 „	E 22.5 „	+ 84 „	„ 113 „	−110 „	+ 61.5 „	„ 48.5 „	+ 71.5 „
17 20	0	„ 6 „	+ 67.5 „	„ 34.5 „	− 82 „	+ 44 „	„ 10 „	+ 75 „
18 20	− 18 „	„ 14.5 „	− 12.5 „	0	− 42.5 „	+ 1 „	„ 5.5 „	+ 26 „

TABLE LVIII (continued).

Gr. M. T.	Sodankylä			Cape Thordsen			Little Karmakul		
	P_h	P_d	P_v	P_h	P_d	P_v	P_h	P_d	P_v
h m									
0 20	− 22 γ	E 9 γ	+ 72 γ	− 13 γ	E 18 γ	+170 γ	− 42 γ	E 26 γ	− 85 γ
1 20	− 55 „	„ 4 „	+ 90 „	− 5 „	„ 26.5 „	+130 „	− 46 „	„ 2.5 „	− 95 „
2 20	− 25 „	„ 11.5 „	+ 69 „	+ 10 „	„ 60.5 „	+176 „	+ 70 „	W 20 „	− 70 „
50	− 40 „	„ 23 „	+ 40 „	+ 10 „	„ 61.5 „	+160 „	0	E 7 „	− 45 „
3 20	− 25 „	W 8 „	+ 48.5 „	− 6.5 „	„ 75.5 „	+120 „	+138 „	W 44 „	+ 25 „
4 20	− 4 „	„ 12.5 „	+ 2.5 „	−155 „	„ 85.5 „	+100 „	− 32 „	„ 14 „	− 42 „
5 20	+ 10 „	E 15.5 „	+ 19.5 „	− 8 „	„ 4.5 „	− 27 „	+ 38 „	E 9.5 „	− 42 „
6 20	+ 4.5 „	W 7.5 „	+ 16 „	− 6.5 „	„ 33 „	− 18 „	+ 55 „	W 21.5 „	+ 16 „
7 20	− 9 „	0	+ 11.5 „	+ 27 „	„ 43.5 „	− 9 „	+ 23 „	E 1 „	+ 18 „
8 20	− 6 „	„ 10 „	0	+ 5 „	„ 28 „	− 98 „	+ 88 „	W 37 „	+ 56 „
9 20	0	„ 7.5 „	+ 21.5 „	− 15 „	„ 22 „	−100 „	+110 „	„ 21.5 „	+ 38 „
10 20	+ 17.5 „	„ 11.5 „	− 9 „	+ 21 „	W 5 „	−110 „	+102 „	„ 58.5 „	+ 43 „
11 20	+ 46 „	„ 13.5 „	− 10 „	− 24.5 „	E 7.5 „	−135 „	+154 „	„ 59 „	+ 59 „
12 20	+ 46 „	„ 23.5 „	− 30 „	− 67 „	W 47 „	−170 „	+ 46 „	„ 54 „	+ 55 „
13 20	+ 68.5 „	„ 23.5 „	− 92 „	− 51.5 „	„ 20.5 „	−130 „	+323 „	„ 58 „	+ 79 „
14 20	+ 62.5 „	E 2 „	− 90 „	− 40 „	„ 11.5 „	−294 „	+360 „	„ 81.5 „	− 63 „
15 20	+132.5 „	W 23.5 „	−125 „	− 77 „	„ 49.5 „	−148 „	−106 „	„ 47 „	−273 „
16 20	+ 61 „	„ 15.5 „	− 55 „	−115 „	„ 61 „	?	0	„ 78.5 „	−135 „
50	+ 26 „	„ 14.5 „	− 27 „	− 70 „	„ 60 „	− 62 „	− 72 „	„ 38 „	+ 41 „
17 20	+ 3 „	0	− 24.5 „	− 23 „	„ 27 „	− 60 „	− 81 „	E 3 „	+ 9 „
18 20	− 6 „	0	+ 5 „	− 18 „	„ 13 „	− 25 „	+ 8 „	W 8 „	+ 8 „

TABLE LVIII (continued).

Gr. M. T.	Pawlowsk			Christiania		Göttingen		Fort Conger
	P_h	P_d	P_v	P_h	P_d	P_h	P_d	P_d
h m								
0 20	+14 γ	W 5.5 γ	−10 γ	+12 γ	E 12 γ	+21 γ	E 23.5 γ	E 9 γ
1 20	0	" 14.5 "	−10 "	+ 1 "	W 4 "	+16.5 "	" 5.5 "	" 13.5 "
2 20	+ 1 "	0	−13 "	− 8.5 "	E 7 "	+ 4 "	" 10.5 "	" 9.5 "
50	−11.5 "	" 5 "	−15 "	−25 "	W 3.5 "	− 8 "	W 6 "	" 49 "
3 20	− 4 "	" 19 "	− 6 "	− 9 "	" 17.5 "	+ 5 "	" 11 "	" 43 "
4 20	+ 9 "	" 18 "	−10 "	+15 "	" 8.5 "	+14 "	E 4 "	" 30 "
5 20	0	E 2 "	− 8.5 "	− 5.5 "	0	− 9 "	" 5.5 "	W 11.5 "
6 20	− 1.5 "	W 9.5 "	0	+ 5 "	" 4.5 "	− 1 "	" 3.5 "	E 31 "
7 20	−16 "	" 9.5 "	0	−18 "	" 5.5 "	−19 "	" 4.5 "	" 30.5 "
8 20	−10 "	" 19 "	+ 2.5 "	−23 "	" 19 "	− 20 "	W 9 "	" 33 "
9 20	+ 4 "	" 10.5 "	+ 2.5 "	−15 "	" 14 "	−11.5 "	" 11.5 "	" 9.5 "
10 20	+12.5 "	" 5 "	+ 5 "	−11 "	" 19 "	− 9 "	" 26 "	W 21.5 "
11 20	+10 "	E 3 "	+ 7.5 "	0	" 14 "	−10.5 "	" 23.5 "	" 41.5 "
12 20	− 2.5 "	0	+15 "	+ 3 "	" 21 "	−12 "	" 18 "	" 44.5 "
13 20	− 8.5 "	0	+21 "	+ 0.5 "	" 23.5 "	−11 "	" 15 "	" 38 "
14 20	−11.5 "	0	+22.5 "	− 8 "	" 9.5 "	−21.5 "	" 8 "	" 56.5 "
15 20	−11.5 "	W 5 "	+25 "	0	" 4.5 "	−26 "	" 6 "	" 66.5 "
16 20	−15 "	" 3.5 "	?	− 8.5 "	" 4 "	−19.5 "	" 6.5 "	" 57 "
50	− 6 "	" 5 "	?	− 6 "	0	−13 "	" 4 "	" 27.5 "
17 20	− 2 "	E 3 "	+10 "	− 5 "	0	− 8 "	" 2 "	" 22.5 "
18 20	0	0	+ 7 "	− 1 "	0	− 2 "	" 1.5 "	0

Current-Arrows for the 15th October, 1882.

Chart I at $0^h 20^m$, $1^h 20^m$, and $2^h 20^m$.

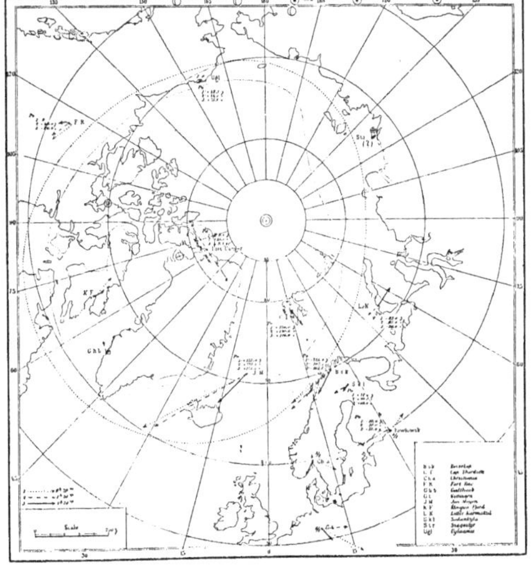

Fig. 173.

Current-Arrows for the 15th October, 1882.

Chart II at $2^h\ 50^m$, $3^h\ 20^m$, and $4^h\ 20^m$.

Chart III at $5^h\ 20^m$, $6^h\ 20^m$, and $7^h\ 20^m$.

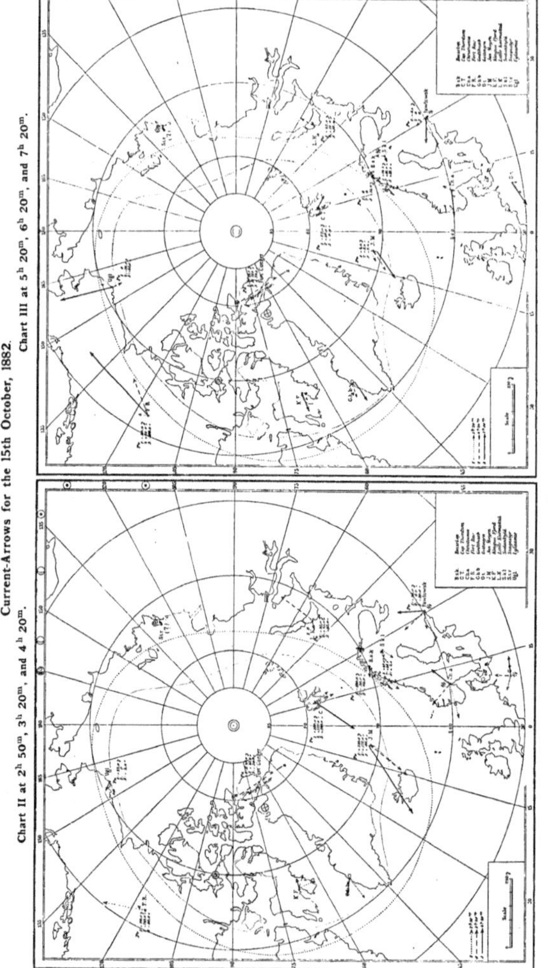

Fig. 174.

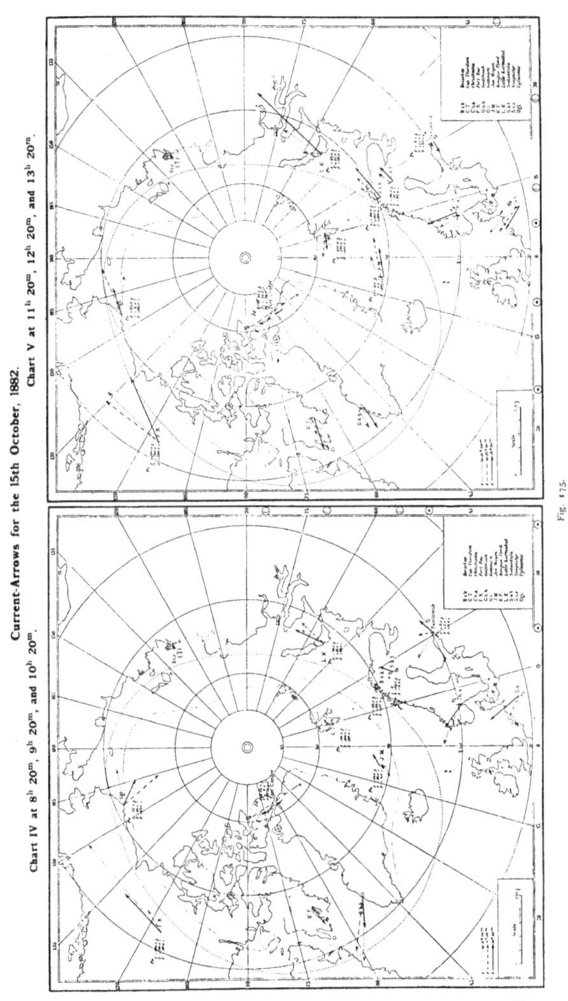

Current-Arrows for the 15th October, 1882.

Chart IV at 8^h 20^m, 9^h 20^m, and 10^h 20^m.

Chart V at 11^h 20^m, 12^h 20^m, and 13^h 20^m.

Fig. 175.

Current-Arrows for the 15th October, 1882.

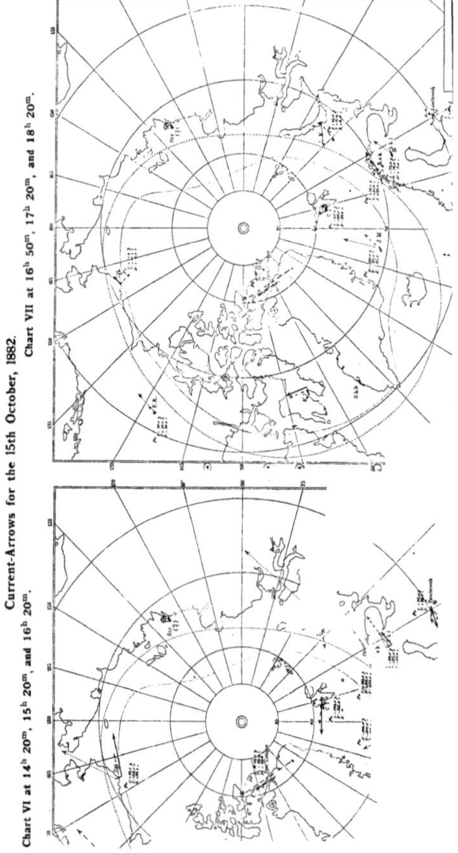

Chart VI at 14h 20m, 15h 20m, and 16h 20m.

Chart VII at 16h 50m, 17h 20m, and 18h 20m.

CHAPTER II.

MATHEMATICAL INVESTIGATIONS. PRELIMINARY RESUMÉ.

91. The calculation of the Field of Force for the assumed polar current-system. While studying polar perturbations of the most varied character, we have constantly met with what we called the typical field for an elementary polar storm. We have also indicated the kinds of current-systems that might be naturally supposed to give rise to such fields. In Art. 36 we moreover worked out a little calculation in order to obtain some idea of the distribution of intensity in this field of force. We there selected the simplest possible form of current-system, namely a linear current consisting of two vertical portions, which were connected with a third portion that was parallel with the tangent to the principal axis in the storm-centre of the current-system.

Our only aim in the earlier calculation was to prove the reversal in the direction of the force which took place in the point of convergence, or that of divergence, when one moved from the storm-centre out along the transverse axis of the system, and to obtain some idea of the proportion between the magnitudes of the forces in the storm-centre and at great distances.

A more complete calculation of the field of force for such a system might, however, be of some importance, and we will therefore make one here.

During great perturbations, the area of precipitation, as we have frequently pointed out, will extend over large parts of the auroral zone, thus causing the principal axis, or those districts in which the most powerful forces occur, to assume approximately the form of parts of a small circle. Very often, indeed, we find conditions which indicate the existence of an entire current-circle. Instead, therefore, of the current-system previously employed, it would be better to use one in which the rectilinear horizontal portion of the current is replaced by a curved portion. The actual calculation will thereby be made a little more complicated; but, as we shall see, a considerable advantage will be gained in another way.

We will consider, then, the effect upon the earth of a current-system consisting of two vertical rectilinear pieces of current, in one of which the current, from infinity, will approach the earth as far as a height h, and in the other continue, from the height h, out into infinity, the two pieces being connected by a curved piece of current lying at a constant height h above one particular small circle, whose spherical radius is ζ.

We do not, of course, mean that the separate active corpuscular rays, which we assume to be the cause of the storms, move in accordance with a diagrammatic arrangement such as this; the whole thing is only an endeavour to find out how near we can get to the true perturbation-conditions, if we assume that the integral effect of all the rays in a system of precipitation is replaced by a linear current-system of this form.

We will first look at the effects of the vertical currents.

As our system of coordinates, we will employ a rectangular Cartesian system, with its origo in the centre of the earth. We will further take the axis Z perpendicular to the plane of the current-arc.

As polar coordinates we will employ the signs ϱ, θ and ω, ϱ being the distance from the origo, θ the angle formed by the radius vector and the positive axis Z, and ω the angle between the plane XZ and the plane through the axis Z and the radius vector.

We will further, in the case of the positive directions, employ the system of coordinates used by HERTZ in his Inaugural Dissertation, "Ueber die Induktion in rotierenden Kugeln"(¹), as in a subsequent chapter we shall go into the subject of induction currents, and shall then have occasion to use the developments we here work out, and it is therefore best to introduce these signs at once. The positive directions of X, Y, Z, θ, and ω, are shown by arrows in the figure.

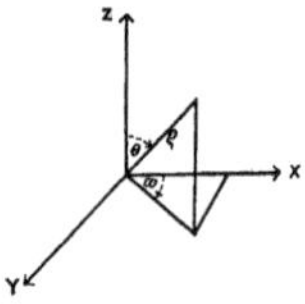

Fig. 177.

We will, then, determine the force-components along the radius vector, the meridian and the parallel circle in a fixed point upon a sphere with an arbitrary radius ϱ, (ϱ supposed $< L$). One of the vertical pieces of current produced will intersect the surface of this sphere in a point ϱ, ζ, μ.

The total effect due to a piece of current such as this (see p. 101, Part I) is

$$P = i\,\frac{1}{\varrho \sin\beta}\left|\frac{s}{\sqrt{\varrho^2 \sin^2\beta + s^2}}\right|_{L-\varrho\cos\beta}^{\infty} = i\,\frac{1}{\varrho\sin\beta}\left[1 - \frac{L-\varrho\cos\beta}{\sqrt{\varrho^2 - 2\varrho L\cos\beta + L^2}}\right] \qquad 1)$$

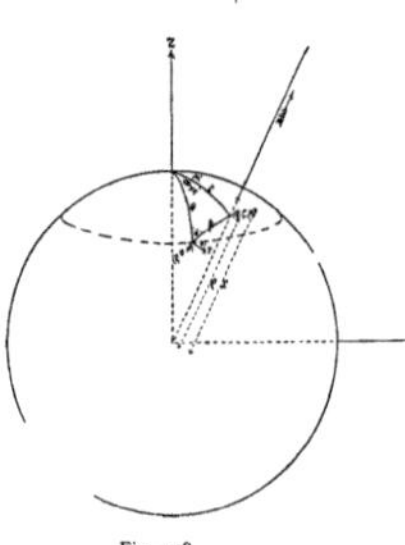

Fig. 178.

where we have put $R + h = L$, and β is the arc of the great circle between the place under consideration and the point of intersection of the produced path of the current with the surface of the sphere. We shall, moreover, when not otherwise stated, always make use of the C. G. S. system, and the electro-magnetic system of measurement.

The three components are thus

$$P_\varrho = 0,\; P_\theta = P\sin v,\; P_\omega = -P\cos v, \qquad (2)$$

where v is the angle between the direction of the magnetic force and the parallel circle, reckoned positive, as shown in the figure. In the case in which the positive current is flowing away from the sphere, i. e. in the direction of increasing ϱ, we will call the direction of the current positive.

What we have to do is to find an expression for β and v. This is given directly by the spheric triangle drawn in the figure

$$\cos\beta = \cos(\omega - \mu)\sin\zeta\sin\theta + \cos\zeta\cos\theta, \qquad (3)$$

$$\sin v = \frac{\sin\zeta\sin(\omega-\mu)}{\sin\beta}, \qquad (4)$$

and

$$\cos v = \frac{\cos\zeta - \cos\theta\cos\beta}{\sin\beta\sin\theta}.$$

By simple combination, the effect of the vertical portions of the current may be found by these formulæ.

We shall then consider the magnetic effect of the curved portion of the current.

We will call the direction of the current positive when it coincides with the direction of increasing ω.

The coordinates of the current-elements we will call L, ζ and μ, μ thus answering to ω. What we have to do, then, is to determine the effect of this element in a point ϱ, θ, ω, on the sphere.

According to Biot & Savart's law, we then have

$$dP = i\,\frac{L\sin\zeta d\mu}{d^2}\sin\alpha, \qquad (5)$$

(¹) H. HERTZ, "Gesammelte Werke", Band I.

P is the magnetic force produced by the current-element at the place, d the distance from the the element, and α the angle between this distance and the direction of the current-element.

e now have to determine the force-components. The decomposition will be effected along the ector, the meridian, and the parallel circle.

looking at the figure we obtain

$$\left.\begin{aligned} x &= \varrho \sin\theta \cos\omega \\ y &= \varrho \sin\theta \sin\omega \\ z &= \varrho \cos\theta \end{aligned}\right\} \tag{6}$$

us the direction-cosines for the radius vector are

$$\sin\theta\cos\omega, \quad \sin\theta\sin\omega, \quad \text{and } \cos\theta.$$

e direction-cosines of the tangent to the parallel circle are

$$-\sin\omega, \quad \cos\omega, \quad \text{and } 0,$$

again we obtain the direction-cosines of the meridian,

$$\cos\theta\cos\omega, \quad \cos\theta\sin\omega, \quad \text{and } -\sin\theta,$$

immediately apparent on looking at the figure.

r the distance d, we find

$$d^2 = L^2 + \varrho^2 - 2L\varrho\,[\cos\zeta\cos\theta + \sin\zeta\sin\theta\cos(\omega-\mu)].$$

e direction-cosines for this distance d are

$$\frac{L\sin\zeta\cos\mu - \varrho\sin\theta\cos\omega}{d}, \quad \frac{L\sin\zeta\sin\mu - \varrho\sin\theta\sin\omega}{d}, \quad \text{and } \frac{L\cos\zeta - \varrho\cos\theta}{d}$$

e direction of the force is now perpendicular to the current-element, of which the direction- are

$$-\sin\mu, \quad \cos\mu, \quad 0,$$

the direction towards the current-element. From this we find the cosines for the direction of netic force,

$$\frac{\cos\zeta - \varrho\cos\theta}{\triangle}\cos\mu, \quad \frac{L\cos\zeta - \varrho\cos\theta}{\triangle}\sin\mu, \quad \text{and} - \frac{L\sin\zeta - \varrho\sin\theta\cos(\omega-\mu)}{\triangle},$$

$$\triangle = \sqrt{(L\cos\zeta - \varrho\cos\theta)^2 + [L\sin\zeta - \varrho\sin\theta\cos(\omega-\mu)]^2}$$

r α we find the following expression:

$$\sin\alpha = \frac{1}{d}\sqrt{(L\cos\zeta - \varrho\cos\theta)^2 + [L\sin\zeta - \varrho\sin\theta\cos(\omega-\mu)]^2} = \frac{\triangle}{d}$$

nce we find

$$dP = i\,\frac{L\sin\zeta d\mu}{d^3}\cdot\triangle \tag{7}$$

the components

$$dP_\varrho = iL\sin\zeta\left[L\sin\theta\cos\zeta\,\frac{\cos(\omega-\mu)\,d\mu}{d^3} - L\sin\zeta\cos\theta\,\frac{d\mu}{d^3}\right],$$

$$dP_\theta = iL\sin\zeta\left[(L\cos\zeta\cos\theta - \varrho)\,\frac{\cos(\omega-\mu)\,d\mu}{d^3} + L\sin\zeta\sin\theta\,\frac{d\mu}{d^3}\right], \text{ and}$$

$$dP_\omega = -\,iL\sin\zeta\,(L\cos\zeta - \varrho\cos\theta)\,\frac{\sin(\omega-\mu)\,d\mu}{d^3}.$$

we put

$$I_1 = \int_{\mu_0}^{\mu}\frac{\cos(\omega-\mu)\,d\mu}{d^3}, \quad I_2 = \int_{\mu_0}^{\mu}\frac{d\mu}{d^3}, \tag{8}$$

h the lower limit may be chosen at pleasure, we obtain

$$P_\varrho = iL \sin\zeta \left[L \sin\theta \cos\zeta \,.\, \mathrm{I}_1 - L \sin\zeta \cos\theta \,.\, \mathrm{I}_2\right]_{\mu=\mu_1}^{\mu=\mu_2}$$

$$P_\theta = iL \sin\zeta \left[(L \cos\zeta \cos\theta - \varrho) \,.\, \mathrm{I}_1 + L \sin\zeta \sin\theta \,.\, \mathrm{I}_2\right]_{\mu=\mu_1}^{\mu=\mu_2} \tag{10}$$

$$P_\omega = -iL \sin\zeta \,.\, \frac{L \cos\zeta - \varrho \cos\theta}{\varrho L \sin\zeta \sin\theta} \left[\frac{1}{\sqrt{L^2 + \varrho^2 - 2\varrho L (\cos\zeta \cos\theta + \sin\zeta \sin\theta \cos(\omega - \mu))}}\right]_{\mu=\mu_1}^{\mu=\mu_2}, \tag{11}$$

where μ_1 and μ_2 represent the values of μ at the ends of the arc.

We may, then, say

$$\left.\begin{aligned} P_\varrho &= P_\varrho{}^0(\omega - \mu_2) - P_\varrho{}^0(\omega - \mu_1) \\ P_\theta &= P_\theta{}^0(\omega - \mu_2) - P_\theta{}^0(\omega - \mu_1) \\ P_\omega &= P_\omega{}^0(\omega - \mu_2) - P_\omega{}^0(\omega - \mu_1) \end{aligned}\right\} \qquad 2)$$

If, therefore, we calculate the quantities $P^0(\omega - \mu)$ [i. e. $P_\varrho^0(\omega - \mu)$, $P_\theta^0(\omega - \mu)$, and $P_\omega^0(\omega - \mu)$] for all values of θ and μ, we can afterwards determine the length of the piece of current.

These formulæ cannot, however, be employed for $\theta = 0$ and $\theta = 180$, as in these cases P^0 becomes, as will easily be seen, infinitely great. Here, therefore, special formulæ must be developed for the forces. The following formulæ are found for these special cases.

$$P_\varrho = \mp iL \sin\zeta \frac{L \sin\zeta}{(L^2 + \varrho^2 \mp 2L\varrho \cos\zeta)^{3/2}} \triangle\mu$$

$$P_\theta = \quad iL \sin\zeta \frac{2(\pm L \cos\zeta - \varrho)}{(L^2 + \varrho^2 \mp 2L\varrho \cos\zeta)^{3/2}} \sin\frac{\triangle\mu}{2} \cos\left(\omega - \frac{\mu_2 + \mu_1}{2}\right)$$

$$P_\omega = -iL \sin\zeta \frac{2(L \cos\zeta \mp \varrho)}{(L^2 + \varrho^2 \mp 2L\varrho \cos\zeta)^{3/2}} \sin\frac{\triangle\mu}{2} \sin\left(\omega - \frac{\mu_2 + \mu_1}{2}\right),$$

where $\triangle\mu = \mu_2 - \mu_1$, and where the upper signs will be employed for $\theta = 0$, and the lower for $\theta = 180°$.

While P_ω is expressed in algebraic form, the other two components, as we may easily convince ourselves, are expressed as elliptic integrals.

We have, then, to get these put into a practical form for the numerical calculation. This may be accomplished by using Legendre's normal forms, by means of which we can make a direct use of his tables of elliptic integrals.

We put

$$\omega - \mu = \pi - 2\tau, \text{ i. e., } \cos(\omega - \mu) = -\cos 2\tau = -1 + 2\sin^2\tau \qquad 3)$$

$$L^2 + \varrho^2 - 2L\varrho(\cos\zeta \cos\theta - \sin\zeta \sin\theta) = L^2 + \varrho^2 - 2L\varrho \cos(\zeta + \theta) = k_2^2 \qquad 4)$$

$$\frac{4L\varrho \sin\zeta \sin\theta}{k_2^2} = k_1^2 \qquad 5)$$

Hence the expression for d becomes

$$d = k_2 \sqrt{1 - k_1^2 \sin^2\tau} \qquad 16)$$

If we introduce this, we have

$$\mathrm{I}_1 = -2\int_{\tau_0}^{\tau} \frac{1 - 2\sin^2\tau}{k_2^3 (1 - k_1^2 \sin^2\tau)^{3/2}} d\tau = -\frac{4}{k_2^3 k_1^2} \int_{\tau_0}^{\tau} \frac{1 - k_1^2 \sin^2\tau - 1 + \frac{k_1^2}{2}}{(1 - k_1^2 \sin^2\tau)^{3/2}} d\tau,$$

or, if we assume μ_0 so that $\tau_0 = 0$,

$$\mathrm{I}_1 = -\frac{4}{k_2^3 k_1^2} \int_0^{\tau} \frac{d\tau}{\sqrt{1 - k_1^2 \sin^2\tau}} + \frac{2(2 - k_1^2)}{k_2^3 k_1^2} \int_0^{\tau} \frac{d\tau}{(1 - k_1^2 \sin^2\tau)^{3/2}}$$

we employ Legendre's signs,

$$E(k_1, \tau) = \int_0^\tau \sqrt{1 - k_1^2 \sin^2\tau}\, d\tau \tag{17}$$

$$F(k_1, \tau) = \int_0^\tau \frac{d\tau}{\sqrt{1 - k_1^2 \sin^2\tau}}, \tag{18}$$

tance we have, as can easily be proved (see Legendre's 'Fonctions Elliptiques', Vol. I, p. 70),

$$\ldots = \frac{1}{1 - k_1^2} E(k_1, \tau) - \frac{k_1^2}{1 - k_1^2} \frac{\sin 2\tau}{2\sqrt{1 - k_1^2 \sin^2\tau}} = \frac{1}{\cos^2\nu} E(k_1, \tau) - \tan^2\nu \frac{\sin 2\tau}{2\sqrt{1 - k_1^2 \sin^2\tau}} \tag{19}$$

can at once put

$$\sin \nu = k_1. \tag{20}$$

n angle such as this must in any case be determined, if Legendre's tables are to be used.

e have, then

$$I_1 = -\frac{4}{k_2^3 k_1^2} \cdot F(k_1, \tau) + \frac{8(2 - k_1^2)}{k_2^3 \sin^2 2\nu} \cdot E(k_1, \tau) - \frac{(2 - k_1^2)}{k_2^3 \cos^2\nu} \frac{\sin 2\tau}{\sqrt{1 - k_1^2 \sin^2\tau}} \tag{21}$$

ther,

$$I_2 = \frac{2}{k_2^3} \int_0^\tau \frac{d\tau}{(1 - k_1^2 \sin^2\tau)^{3/2}} = \frac{2}{k_2^3 \cos^2\nu} \cdot E(k_1, \tau) - \frac{\tan^2\nu}{k_2^3} \cdot \frac{\sin 2\tau}{\sqrt{1 - k_1^2 \sin^2\tau}},$$

referred,

$$I_2 = \frac{8 k_1^2}{k_2^3 \sin^2 2\nu} \cdot E(k_1, \tau) - \frac{k_1^2}{k_2^3 \cos^2\nu} \frac{\sin 2\tau}{\sqrt{1 - k_1^2 \sin^2\tau}}, \tag{22}$$

y the coefficients of corresponding terms in I_1 and I_2 have a common denominator.

n this way we have determined all the quantities that we shall require to use.

n the tables below we have given the force-components of the rectilinear portion of the current, and ues of the quantities P°, calculated for various values of θ and $\omega - \mu$. The special calculation required for values of $\omega - \mu$ between 0° and 180°, answering to values of τ between 0° and 90°.

or

$$\tau = m\pi \pm \tau_1,$$

m is a whole number, and τ_1 an arc $< \frac{\pi}{2}$, we have, for E and F,

$$E(\tau) = 2mE\left(\frac{\pi}{2}\right) \pm E(\tau_1)$$

$$F(\tau) = 2mF\left(\frac{\pi}{2}\right) \pm F(\tau_1)$$

egendre, l. c., Vol. I, p. 14). For the third term we also have exactly the same relation,

$$\frac{\sin 2\tau}{\sqrt{1 - k_1^2 \sin^2\tau}} = \pm \frac{\sin 2\tau_1}{\sqrt{1 - k_1^2 \sin^2\tau_1}},$$

ly difference being that the value of the expression, for $\tau = \frac{\pi}{2}$ is equal to zero. We therefore e relation,

$$P^\circ(\tau) = 2m \cdot P^\circ\left(\frac{\pi}{2}\right) \pm P^\circ(\tau_1).$$

inally we will also give the formula for the magnetic potential of the current. This can very simply uced from the formula for the components of the magnetic force.

As it is well known the expression for this quantity involve an additive constant, that may chosen at pleasure.

We may therefore, for instance choose such a constant, that the value of the potential at the cent of the sphere will be zero. Under this supposition we may write the potential, V, as

$$V = -\int_{o}^{\varrho} P_\varrho \, . \, d\varrho$$

as the term on the right is an expression for the work done against the field when a positive magne pole of unit strength passes from the centre of the sphere to a certain point on its surface.

P_ϱ is only due to the curved portion of the current. We find by equations (3), (8) and (9)

$$P_\varrho = iL^2 \sin\zeta \int_{\mu_1}^{\mu_2} \frac{\cos(\omega-\mu)\sin\theta\cos\zeta - \sin\zeta\cos\theta}{(L^2+\varrho^2-2L\varrho\cos\beta)^{3/2}}\, d\mu.$$

We further have, as will be easily seen

$$\cos(\omega-\mu)\sin\theta\cos\zeta - \sin\zeta\cos\theta = \frac{\cos\zeta\cos\beta - \cos\theta}{\sin\zeta}$$

By introducing this expression and by integration with respect to ϱ, we find, pag. 101, Part I

$$V = i\int_{\mu_1}^{\mu_2} \frac{(\cos\theta - \cos\zeta\cos\beta)(\varrho - L\cos\beta)}{\sin^2\beta\sqrt{\varrho^2+L^2-2\varrho L\cos\beta}}\, d\mu + i\int_{\mu_1}^{\mu_2} \frac{(\cos\theta - \cos\zeta\cos\beta)\cos\beta}{\sin^2\beta}\, d\mu.$$

or if preferred

$$V = i\int_{\mu_1}^{\mu_2} \frac{(\cos\theta - \cos\zeta\cos\beta)\,[\varrho - (L-d)\cos\beta]}{\sin^2\beta\, .\, d}\, d\mu. \qquad ($$

where d stands for the square root.

As will be seen, V may also be expressed as elliptic integrals.

For numerical calculations I think however that the above form is the most practical one.

By derivation of this expression we find the force-components of the whole current-system, we have done to control the correctnes of our calculations.

In our calculations we have imagined the current to lie at a height of about 400 kilometr (L = 1,063 R), the average height of currents, as we found by our calculations in Chapter IV of Part

In the tables, we have employed γ as the unit for forces; and $i = 10^5$ [i. e. 10^6 amperes].

TABLE LIX.

Values of $P_\varrho{}^\circ$ for the horizontal portion of the current.

θ	$\omega-\mu=0^\circ$	15°	30°	45°	60°	75°	90°	105°	120°	135°	150°	165°
0	− 135,72	− 124,41	− 113,10	− 101,79	− 90,48	− 79,17	− 67,86	− 56,55	− 45,24	− 33,93	− 22,62	− 11,3
10	− 161,11	− 122,92	− 92,80	− 71,45	− 56,21	− 44,97	− 35,98	− 28,42	− 21,91	− 15,93	− 10,35	− 5,1
20	− 62,04	− 54,26	− 43,07	− 34,91	− 28,67	− 23,62	− 19,31	− 15,51	− 12,06	− 8,86	− 5,82	− 2,8
40	+ 13,61	+ 3,13	− 3,73	− 6,96	− 7,94	− 7,79	− 7,07	− 6,07	− 4,93	− 3,72	− 2,48	− 1,2
60	+ 4,22	+ 1,44	− 0,82	− 2,32	− 3,09	− 3,40	− 3,31	− 2,99	− 2,51	− 1,94	− 1,32	− 0,6
90	+ 1,69	+ 0,81	+ 0,03	− 0,58	− 1,00	− 1,23	− 1,29	− 1,24	− 1,09	− 0,87	− 0,60	− 0,3
140	+ 0,88	+ 0,65	+ 0,42	+ 0,22	+ 0,06	− 0,06	− 0,14	− 0,18	− 0,18	− 0,16	− 0,12	− 0,0
180	+ 0,78	+ 0,71	+ 0,65	+ 0,58	+ 0,52	+ 0,45	+ 0,39	+ 0,32	+ 0,26	+ 0,19	+ 0,13	+ 0,0

TABLE LIX (continued).

Values of P_θ° for the horizontal portion of the current.

θ	$\omega-\mu=0°$	15°	30°	45°	60°	75°	90°	105°	120°	135°	150°	165°	180°
5	+ 12.77	+ 10.90	+ 9.15	+ 7.60	+ 6.27	+ 5.13	+ 4.15	+ 3.30	+ 2.55	+ 1.86	+ 1.21	+ 0.60	0
0	+ 34.59	+ 24.97	+ 17.65	+ 12.77	+ 9.55	+ 7.36	+ 5.72	+ 4.43	+ 3.37	+ 2.42	+ 1.55	+ 0.78	0
5	+ 100.55	+ 46.43	+ 23.96	+ 14.94	+ 10.56	+ 7.95	+ 6.04	+ 4.64	+ 3.52	+ 2.49	+ 1.64	+ 0.84	0
0		68	+ 23.25	+ 14.07	+ 9.92	+ 7.47	+ 5.78	+ 4.51	+ 3.40	+ 2.45	+ 1.61	+ 0.78	0
0		97	+ 12.59	+ 9.53	+ 7.54	+ 6.06	+ 4.89	+ 3.89	+ 3.00	+ 2.20	+ 1.42	+ 0.70	0
0	+ 7.42	+ 7.38	+ 7.01	+ 6.37	+ 5.61	+ 4.82	+ 4.06	+ 3.32	+ 2.62	+ 1.94	+ 1.28	+ 0.64	0
0	+ 2.00	+ 2.74	+ 3.28	+ 3.52	+ 3.49	+ 3.29	+ 2.94	+ 2.52	+ 2.05	+ 1.55	+ 1.04	+ 0.52	0
0	+ 0.67	+ 1.22	+ 1.70	+ 2.03	+ 2.20	+ 2.21	+ 2.09	+ 1.87	+ 1.57	+ 1.22	+ 0.83	+ 0.42	0
0	+ 0.17	+ 0.56	+ 0.92	+ 1.21	+ 1.41	+ 1.52	+ 1.52	+ 1.42	+ 1.24	+ 0.99	+ 0.69	+ 0.35	0

Values of P_ω° for the horizontal portion of the current.

θ	$\omega-\mu=0°$	15°	30°	45°	60°	75°	90°	105°	120°	135°	150°	165°	180°
10	− 6.69	− 6.32	− 5.52	− 4.71	− 4.04	− 3.53	− 3.16	− 2.88	− 2.68	− 2.54	− 2.44	− 2.39	− 2.37
15	− 18.21	− 14.71	− 10.36	− 7.72	− 6.14	− 5.14	− 4.47	− 4.01	− 3.69	− 3.47	− 3.32	− 3.24	− 3.21
20	− 43.17	− 24.36	− 14.08	− 9.81	− 7.59	− 6.27	− 5.41	− 4.83	− 4.43	− 4.15	− 3.97	− 3.87	− 3.84
30	− 21.91	− 18.92	− 14.32	− 11.05	− 8.94	− 7.55	− 6.60	− 5.94	− 5.47	− 5.15	− 4.94	− 4.82	− 4.78
40	− 15.65	− 14.79	− 12.89	− 10.97	− 9.41	− 8.23	− 7.35	− 6.70	− 6.23	− 5.90	− 5.68	− 5.55	− 5.51
60	− 12.78	− 12.52	− 11.82	− 10.93	− 10.02	− 9.20	− 8.51	− 7.95	− 7.53	− 7.21	− 6.99	− 6.86	− 6.82
90	− 13.25	− 13.13	− 12.81	− 12.34	− 11.81	− 11.26	− 10.75	− 10.31	− 9.94	− 9.65	− 9.45	− 9.32	− 9.28
40	− 24.14	− 24.08	− 23.90	− 23.74	− 23.30	− 22.93	− 22.51	− 22.18	− 21.86	− 21.59	− 21.39	− 21.27	− 21.23

TABLE LX.

Values of P_θ for one of the vertical portion of the current.

θ	$\omega-\mu=0°$	15°	30°	45°	60°	75°	90°	105°	120°	135°	150°	165°	180°
0	0	+ 7.86	+ 15.18	+ 21.46	+ 26.29	+ 29.32	+ 30.35	+ 29.32	+ 26.29	+ 21.46	+ 15.18	+ 7.86	0
5	0	+ 13.11	+ 23.49	+ 29.78	+ 32.11	+ 31.41	+ 28.69	+ 24.78	+ 20.21	+ 15.31	+ 10.25	+ 5.13	0
10	0	+ 24.55	+ 36.78	+ 38.32	+ 34.87	+ 29.85	+ 24.65	+ 19.75	+ 15.25	+ 11.11	+ 7.25	+ 3.58	0
15	0	+ 49.70	+ 51.20	+ 41.63	+ 32.69	+ 25.55	+ 19.91	+ 15.36	+ 11.56	+ 8.27	+ 5.34	+ 2.62	0
20	0	+ 69.83	+ 51.20	+ 36.61	+ 27.08	+ 20.49	+ 15.66	+ 11.90	+ 8.89	+ 6.33	+ 4.07	+ 2.06	0
30	0	+ 20.69	+ 23.47	+ 19.87	+ 15.80	+ 12.38	+ 9.64	+ 7.41	+ 5.56	+ 3.97	+ 2.56	+ 1.25	0
40	0	+ 7.01	+ 10.24	+ 10.37	+ 9.18	+ 7.67	+ 6.21	+ 4.90	+ 3.73	+ 2.69	+ 1.75	+ 0.86	0
60	0	+ 1.87	+ 3.19	+ 3.79	+ 3.82	+ 3.51	+ 3.05	+ 2.52	+ 1.99	+ 1.47	+ 0.97	+ 0.48	0
90	0	+ 0.60	+ 1.10	+ 1.43	+ 1.57	+ 1.57	+ 1.46	+ 1.27	+ 1.05	+ 0.80	+ 0.54	+ 0.27	0
40	0	+ 0.23	+ 0.44	+ 0.60	+ 0.71	+ 0.76	+ 0.76	+ 0.71	+ 0.61	+ 0.48	+ 0.33	+ 0.17	0
80	0	+ 0.17	+ 0.33	+ 0.46	+ 0.57	+ 0.63	+ 0.66	+ 0.63	+ 0.57	+ 0.46	+ 0.33	+ 0.17	0

TABLE LX (continued).

Values of P_{ω} for one of the vertical portion of the current.

θ	$\omega-\mu=0°$	15°	30°	45°	60°	75°	90°	105°	120°	135°	150°	165°	180°
0	+ 30.35	+ 29.32	+ 26.29	+ 21.46	+ 15.18	+ 7.86	0	− 7.86	− 15.18	− 21.46	− 26.29	− 29.32	− 30.35
10	+ 53.33	+ 44.91	+ 27.65	+ 11.89	+ 0.62	− 6.86	− 11.76	− 14.96	− 17.07	− 18.44	− 19.29	− 19.75	− 19.90
15	+ 70.59	+ 42.62	+ 12.85	− 1.65	− 8.61	− 12.20	− 14.16	− 15.28	− 15.93	− 16.31	− 16.53	− 16.64	− 16.67
20	0	− 8.65	− 12.88	− 14.25	− 14.69	− 14.77	− 14.71	− 14.60	− 14.47	− 14.35	− 14.26	− 14.21	− 14.19
30	− 53.32	− 42.95	− 29.28	− 21.39	− 17.16	− 14.74	− 13.24	− 12.26	− 11.60	− 11.15	− 10.86	− 10.70	− 10.64
40	− 30.35	− 27.78	− 22.59	− 17.95	− 14.65	− 12.45	− 10.96	− 9.95	− 9.26	− 8.79	− 8.48	− 8.30	− 8.24
60	− 14.19	− 13.68	− 12.40	− 10.65	− 9.39	− 8.19	− 7.26	− 6.56	− 6.04	− 5.68	− 5.43	− 5.30	− 5.26
90	− 6.54	− 6.41	− 6.05	− 5.55	− 5.00	− 4.47	− 4.01	− 3.63	− 3.32	− 3.10	− 2.94	− 2.85	− 2.82
140	− 2.27	− 2.24	− 2.13	− 1.97	− 1.77	− 1.56	− 1.34	− 1.14	− 0.97	− 0.84	− 0.74	− 0.68	− 0.66
180	− 0.66	− 0.63	− 0.57	− 0.46	− 0.33	− 0.17	0	+ 0.17	+ 0.33	+ 0.64	+ 0.57	+ 0.63	+ 0.66

From these quantities we can determine, by a simple combination, the distribution of force in systems with a horizontal piece of current of arbitrary length.

In the following tables we have put together the force-components of three such systems, the length of the arc in the first being 75°, in the second 180°, and in the third 270°.

ω is here always reckoned from the transversal axis.

TABLE LXI.

Values of P_{ϱ} for a current-system corresponding to $\triangle\mu = 75°$.

θ	$\omega = 7,5°$	22,5°	37,5°	52,5°	67,5°	82,5°	97,5°	112,5°	127,5°	142,5°	157,5°	172,5°
0	− 56.55	− 56.55	− 56.55	− 56.55	− 56.55	− 56.55	− 56.55	− 56.55	− 56.55	− 56.55	− 56.55	− 56.55
10	− 157.97	− 143.10	− 116.14	− 86.94	− 64.39	− 49.55	− 40.28	− 34.62	− 30.82	− 28.42	− 27.07	− 26.28
20	− 46.11	− 41.15	− 38.43	− 34.95	− 27.56	− 22 85	− 19.82	− 17.80	− 16.42	− 15.51	− 14.95	− 14.68
40	+ 37.91	+ 32.04	+ 21.40	+ 10.20	+ 2.33	− 2.03	− 4.22	− 5.30	− 5.82	− 6.07	− 6.17	− 6.22
60	+ 11.58	+ 10.09	+ 7.62	+ 4.75	+ 2.17	+ 0.18	− 1.15	− 2.08	− 2.64	− 2.89	− 3.17	− 3.27
90	+ 3.93	+ 3.57	+ 2.91	+ 2.10	+ 1.27	+ 0.50	− 0.13	− 0.63	− 0.99	− 1.24	− 1.39	− 1.47
140	+ 1.12	+ 1.06	+ 0.94	+ 0.78	+ 0.60	+ 0.41	+ 0 23	+ 0.06	− 0.07	− 0.18	− 0.25	− 0.28
180	+ 0.32	+ 0.32	+ 0.32	+ 0.32	+ 0.32	+ 0.32	+ 0.32	+ 0.32	+ 0.32	+ 0.32	+ 0.32	+ 0.32

Values of P_{θ} for a current-system corresponding to $\triangle\mu = 75°$.

θ	$\omega = 7,5°$	22,5°	37,5°	52,5°	67,5°	82,5°	97,5°	112,5°	127,5°	142,5°	157,5°	172,5°
0	− 36.80	− 34.29	− 29.45	− 22.59	− 14.20	− 4.84	+ 4.84	+ 14.20	+ 22.59	+ 29.45	+ 34.29	+ 36.80
5	− 44.47	− 36.83	− 23.76	− 8.84	+ 4.56	+ 14.61	+ 21.21	+ 25 07	+ 27.11	+ 28.08	+ 28.49	+ 28.63
10	− 36.35	− 24.71	− 2.62	+ 19.10	+ 30.25	+ 32.48	+ 30.88	+ 28.40	+ 26.01	+ 24.18	+ 22.98	+ 22.34
15	+ 69.36	+ 61.71	+ 67.05	+ 70.18	+ 55.17	+ 41.50	+ 32.48	+ 26.51	+ 22.50	+ 20.00	+ 18.53	+ 17.75
20	+ 415.77	+ 377.38	+ 242.48	+ 105.07	+ 58.03	+ 43.10	+ 28.23	+ 22.29	+ 18.60	+ 16.42	+ 15.13	+ 14.45
30	− 11.48	− 8.01	+ 8.54	+ 24.14	+ 24.76	+ 20.84	+ 17.17	+ 14.45	+ 12.57	+ 11.30	+ 10.52	+ 10.16
40	− 19.16	− 14.33	− 5.07	+ 4.12	+ 9.04	+ 10.38	+ 10.16	+ 9.47	+ 8.77	+ 8.22	+ 7.84	+ 7.66
60	− 9.78	− 7.92	− 4.81	− 1.38	+ 1.42	+ 3.27	+ 4.29	+ 4.80	+ 4.99	+ 5.04	+ 5.04	+ 5.03
90	− 4.92	− 4.72	− 3.11	− 1.72	− 0.34	+ 0.84	+ 1.76	+ 2.42	+ 2.86	+ 3.14	+ 3.31	+ 3.38
140	− 2.84	− 2.59	− 2.11	− 1.48	− 0.76	− 0.03	+ 0.65	+ 1.26	+ 1.75	+ 2.12	+ 2.37	+ 2.50
180	− 2.43	− 2.27	− 1.95	− 1.49	− 0.94	− 0.32	+ 0.32	+ 0.94	+ 1.49	+ 1.95	+ 2.27	+ 2.43

TABLE LXI (continued).

Values of P_ω for a current-system corresponding to $\triangle\mu = 75°$.

θ	ω = 7,5°	22,5°	37,5°	52,5°	67,5°	82,5°	97,5°	112,5°	127,5°	142,5°	157,5°	172,5°
0	+ 4.84	+ 14.20	+ 22.59	+ 29.45	+ 34.29	+ 36.80	+ 36.80	+ 34.29	+ 29.45	+ 22.59	+ 14.20	+ 4.84
10	+ 14.95	+ 42.01	+ 57.04	+ 53.51	+ 39.98	+ 26.93	+ 17.56	+ 11.33	+ 7.22	+ 4 42	+ 2.39	+ 0.75
15	+ 11.86	+ 42.66	+ 69.63	+ 46.54	+ 21.78	+ 10.25	+ 5.03	+ 2.51	+ 1.24	+ 0.59	+ 0.25	+ 0.07
20	− 2.91	− 10.73	− 22.12	− 12.89	− 7.54	− 5.17	− 3.78	− 2.80	− 2.04	− 1.40	− 0.81	− 0.27
30	− 11.16	− 35.77	− 52.95	− 42.03	− 25.40	− 15.37	− 9.80	− 6.49	− 4.33	− 2.78	− 1.56	− 0.50
40	− 6.56	− 18.50	− 25.33	− 24.25	− 18.82	− 13.43	− 9.38	− 6.51	− 4.46	− 2.91	− 1.64	− 0.53
60	− 2.45	− 6.79	− 9.58	− 10.43	− 9.71	− 8.20	− 6.52	− 4.96	− 3.60	− 2.43	− 1.40	− 0.46
90	− 0.97	− 2.74	− 4.05	− 4.78	− 4.93	− 4.63	− 4.06	− 3.35	− 2.59	− 1.83	− 1.09	− 0.36
140	− 0.43	− 1.24	− 1.93	− 2 43	− 2.71	− 2.77	− 2.64	− 2.35	− 1.94	− 1.44	− 0.88	− 0.30
180	− 0.32	− 0.94	− 1.49	− 1.95	− 2.27	− 2.43	− 2.43	− 2.27	− 1.95	− 1.49	− 0.94	− 0.32

TABLE LXII.

Values for P_ϱ for a current-system corresponding to $\triangle\mu = 180°$.

θ	ω = 0°	15°	30°	45°	60°	75°	90°	105°	120°	135°	150°	165°	180°
0	− 135.72	− 135.72	− 135.72	− 135.72	− 135.72	− 135.72	− 135.72	− 135.72	− 135.72	− 135.72	− 135.72	− 135.72	− 135.72
0	− 250.27	− 248.84	− 244.11	− 234.84	− 219.07	− 194.15	− 161.11	− 128.08	− 103.15	− 87.38	− 78.12	− 73.39	− 71.95
0	− 85.47	− 84.96	− 83.35	− 80.32	− 75.20	− 66.94	− 62.04	− 57.14	− 48.89	− 43.77	− 40.74	− 39.13	− 38.62
40	+ 41.35	+ 41.08	+ 40.09	+ 37.90	+ 33.44	+ 25.34	+ 13.61	+ 1.88	− 6.23	− 10.68	− 12.88	− 13.86	− 14.14
60	+ 15.06	+ 14.82	+ 14.03	+ 12.70	+ 10.57	+ 7.66	+ 4.22	+ 0.77	− 2.14	− 4.27	− 5.60	− 6.39	− 6.62
90	+ 5.97	+ 5.84	+ 5.46	+ 4.83	+ 3.95	+ 2.88	+ 1.69	+ 0.50	− 0.57	− 1.45	− 2.08	− 2.46	− 2.59
40	+ 2.04	+ 2.00	+ 1.89	+ 1.71	+ 1.47	+ 1.19	+ 0.88	+ 0.58	+ 0.30	+ 0.06	− 0.12	− 0.23	− 0.27
80	+ 0.78	+ 0.78	+ 0.78	+ 0.78	+ 0.78	+ 0.78	+ 0.78	+ 0.78	+ 0.78	+ 0.78	+ 0.78	+ 0.78	+ 0.78

Values of P_θ for a current-system corresponding to $\triangle\mu = 180°$.

θ	ω = 0°	15°	30°	45°	60°	75°	90°	105°	120°	135°	150°	165°	180°
0	− 60.97	− 58.89	− 52.80	− 43.11	− 30.48	− 15.78	0	+ 15.78	+ 30.48	+ 43.11	+ 52.80	+ 58.89	+ 60.97
10	+ 8.44	+ 7.79	+ 6.15	+ 4.55	+ 5.95	+ 15.85	+ 3459	+ 53 83	+ 63.23	+ 64.63	+ 63.03	+ 61.39	+ 60.74
20	+ 498.02	+ 496.51	+ 491.61	+ 481.44	+ 460.77	+ 411.54	+ 270.45	+ 129.35	+ 80.13	+ 59.45	+ 49.29	+ 44.39	+ 42.88
40	− 5.69	− 5.87	− 6.30	− 6.53	− 5.44	− 1.04	+ 742	+ 15.88	+ 20.28	+ 21.37	+ 21.13	+ 20.71	+ 20.53
60	− 7.99	− 7.85	− 7.35	− 6.34	− 4.48	− 1.61	+ 200	+ 5.61	+ 8.47	+ 10.33	+ 11.34	+ 11.84	+ 11.98
90	− 5.76	− 5.59	− 5.06	− 4.14	− 2.83	− 1.18	+ 067	+ 2.52	+ 4.16	+ 5.47	+ 6.39	+ 6.92	+ 7.10
40	− 4.21	− 4.07	− 3.64	− 2.95	− 2.05	− 0.98	+ 017	+ 1.32	+ 2.38	+ 3.29	+ 3.98	+ 4.41	+ 4.55
80	− 4.03	− 3.90	− 3.49	− 2.85	− 2.02	− 1.04	0	+ 1.04	+ 2.02	+ 2.85	+ 3.49	+ 3.90	+ 4.03

Values of P_ω for a current-system corresponding to $\triangle\mu = 180°$.

θ	ω = 0°	15°	30°	45°	60°	75°	90°	105°	120°	135°	150°	165°	180°
0	0	+ 15.78	+ 30.48	+ 43.11	+ 52.80	+ 58.89	+ 60.97	+ 58.89	+ 52.80	+ 43.11	+ 30.48	+ 15.78	0
10	0	+ 7.45	+ 16.33	+ 28.16	+ 43.86	+ 60.73	+ 68.91	+ 60.73	+ 43.86	+ 28.16	+ 16.33	+ 7.45	0
20	0	− 1.62	− 3.39	− 5.55	− 8.73	− 14.93	− 25.14	− 14.93	− 8.37	− 5.55	− 3.39	− 1.62	0
40	0	− 4.01	− 8.57	− 14.23	− 21.32	− 28.71	− 32.25	− 28.71	− 21.32	− 14.23	− 8.57	− 4.01	0
60	0	− 2.87	− 5.84	− 8.88	− 11.79	− 14.03	− 14.89	− 14.03	− 11.79	− 8.88	− 5.84	− 2.87	0
90	0	− 1.80	− 3.55	− 5.15	− 6.47	− 7.37	− 7.68	− 7.37	− 6.47	− 5.15	− 3.55	− 1.80	0
40	0	− 1.16	− 2.24	− 3.18	− 3 91	− 4.37	− 4.53	− 4.37	− 3.91	− 3.18	− 2.24	− 1.16	0
80	0	− 1.04	− 2.02	− 2.85	− 3.49	− 3.90	− 4.03	− 3.90	− 3.49	− 2.85	− 2.02	− 1.04	0

TABLE LXIII.

Values of P_ϱ for a current-system corresponding to $\triangle\mu = 270°$.

θ	$\omega = 0°$	15°	30°	45°	60°	75°	90°	105°	120°	135°	150°	165°	180°
0	− 203,58	− 203.58	− 203.58	− 203.58	− 203.58	− 203.58	− 203.58	− 203.58	− 203.58	− 203.58	− 203.58	− 203.58	− 203.58
10	− 290.36	− 289.97	− 288 64	− 286.25	− 282.42	− 276.36	− 266.70	− 251.32	− 227.72	− 197.09	− 167.89	− 149.01	− 142.91
20	− 106.37	− 106.21	− 105.69	− 104.78	− 103.36	− 101.23	− 98.03	− 93.08	− 85.34	− 81.35	− 77.88	− 71.74	− 69.83
40	+ 34.67	+ 34.64	+ 34 53	+ 34.29	+ 33.76	+ 32.67	+ 30.45	+ 26.02	+ 18.02	+ 6.54	− 4.67	− 11.68	− 13.91
60	+ 12.32	+ 12.26	+ 12.09	+ 11.75	+ 11.17	+ 10.21	+ 8.82	+ 6.75	+ 4.01	+ 0.91	− 1.96	− 3.91	− 4.65
90	+ 5.11	+ 5.06	+ 4.92	+ 4.67	+ 4.30	+ 3.77	+ 3.09	+ 2.26	+ 1.33	+ 0.39	− 0.42	− 0.97	− 1.17
140	+ 2.09	+ 2.07	+ 2.01	+ 1.91	+ 1.76	+ 1.58	+ 1.38	+ 1.16	+ 0.95	+ 0.75	+ 0.59	+ 0.49	+ 0.45
180	+ 1.17	+ 1.17	+ 1.17	+ 1.17	+ 1.17	+ 1.17	+ 1.17	+ 1.17	+ 1.17	+ 1.17	+ 1.17	+ 1.17	+ 1.17

Values of P_θ for a current-system corresponding to $\triangle\mu = 270$.

θ	$\omega = 0°$	15°	30°	45°	60°	75°	90°	105°	120°	135°	150°	165°	180°
0	− 43.11	− 41.64	− 37.33	− 30.48	− 21.56	− 11.16	0	+ 11.16	+ 21.56	+ 30.48	+ 37.33	+ 41.64	+ 43.11
10	+ 42.12	+ 41.76	+ 40.65	+ 38.81	+ 36.33	+ 33.57	+ 31.61	+ 33.37	+ 43.88	+ 64.96	+ 86.69	+ 98.84	+ 102.21
20	+ 523.35	+ 522.94	+ 521.63	+ 519.46	+ 515.77	+ 509.57	+ 498.99	+ 478.73	+ 430.81	+ 291.89	+ 154.48	+ 111.46	+ 101.35
40	+ 5.58	+ 5.47	+ 5.13	+ 4.57	+ 3.84	+ 3.08	+ 2.73	+ 3.93	+ 8.67	+ 17.68	+ 26.88	+ 32.05	+ 33.48
60	− 2.05	− 2.05	− 2.05	− 2.00	− 1.81	− 1.31	− 0.30	+ 1.57	+ 4.43	+ 7.99	+ 11.41	+ 13.77	+ 14.62
90	− 2.70	− 2.65	− 2.50	− 2.21	− 1.76	− 1.07	− 0.11	+ 1.15	+ 2.65	+ 4.21	+ 5.61	+ 6.57	+ 6.91
140	− 2.61	− 2.54	− 2.31	− 1.94	− 1.42	− 0.77	− 0.01	+ 0.82	+ 1.67	+ 2.44	+ 3.08	+ 3.49	+ 3.63
180	− 2.85	− 2.76	− 2.47	− 2.02	− 1.43	− 0.74	0	+ 0.74	+ 1.43	+ 2.02	+ 2.47	+ 2.76	+ 2.85

Values of P_ω for a current-system corresponding to $\triangle\mu = 270°$.

θ	$\omega = 0°$	15°	30°	45°	60°	75°	90°	105°	120°	135°	150°	165°	180°
0	0	− 11.16	− 21.56	− 30.48	− 37.33	− 41.64	− 43.11	− 41.64	− 37.33	− 30.48	− 21.56	− 11.16	0
10	0	− 1.98	− 4.29	− 7.35	− 11.74	− 18.31	− 28.16	− 41.88	− 56.44	− 61.56	− 48.99	− 25.55	0
20	0	+ 0.66	+ 1.34	+ 2.09	+ 2.96	+ 4.04	+ 5.55	+ 8.07	+ 13.59	+ 23.04	+ 11.97	+ 4.68	0
40	0	+ 1.34	+ 2.80	+ 456	+ 6.81	+ 9.91	+ 14.23	+ 19.98	+ 25.91	+ 27.69	+ 21.89	+ 11.48	0
60	0	+ 1.14	+ 2.34	+ 3.69	+ 5.22	+ 6.98	+ 8.88	+ 10.65	+ 11.68	+ 11.20	+ 8.81	+ 4.80	0
90	0	+ 0.87	+ 1.76	+ 2.66	+ 3.56	+ 4.42	+ 5.15	+ 5.60	+ 5.61	+ 5.03	+ 3.81	+ 2.05	0
140	0	+ 0.70	+ 1.38	+ 2.00	+ 2.54	+ 2.94	+ 3.18	+ 3.20	+ 2.99	+ 2.52	+ 1.83	+ 0.96	0
180	0	+ 0.74	+ 1.43	+ 2.02	+ 2.47	+ 2.76	+ 2.85	+ 2.76	+ 2.47	+ 2.02	+ 1.43	+ 0.74	0

We have moreover shown these three fields of force on charts, one for each field separately, and one giving the field for two simultaneous systems of the first kind, one in the north and the other in the south.

The fields of force in these charts are not represented by current-arrows as in the perturbations, but by current-lines (equipotential lines, see p. 85) and by lines of force for the horizontal components. Lines have moreover been drawn on another chart for constant values of P_ϱ.

In order to construct the former of these easily, when the force components at various places on the earth have been calculated, the following mode of procedure has been adopted.

The relation $\frac{P_\theta}{P_\omega}$ was determined for the various points at which the force-components were calculated, this relation being a measure for the angle that the horizontal force-component forms with the circles $\theta = \text{const.}$ or $\omega = \text{const.}$ These we may call, for the sake of brevity, parallel circles and meridians. We next drew curves for the various meridians and parallels, showing how this condition varied them. A number of points could then be determined by interpolation, upon the various sets of where this relation had a constant value. It thereby became possible to draw upon a chart in which this relation was constant. Along these "isogonic" lines[1], the lines of force or the current-orm equal angles with, for instance, the meridian. The tangent directions were now drawn in a of short, parallel strokes, which intersected the various isogonic lines; and by employing a suffi-umber of these, the chart could be as thickly covered with these small tangent directions as be desired. Lines of force and equipotential lines could then at once be drawn.

Vith regard to the equipotential lines, care must be taken that those drawn are equidistant.

Ve may here use the formula (23), or as we know that the potential along the parallel circles and ıns varies respectively as

$$\triangle_\omega V = -\int_{\omega_0}^{\omega} \varrho \sin\theta P_\omega \, d\omega \qquad \text{and} \qquad \triangle_\theta V = -\int_{\theta_0}^{\theta} \varrho P_\theta \, d\theta$$

y either by calculation or by graphic or by numerical integration easily find out the different data, ary for this purpose.

\s regards the lines of force, it will be seen that they all point in towards the two characteristic the points of convergence and divergence, so that here, in drawing, we have two fixed points so a distribution of tangent directions to hold to.

Ve must finally not omit to remark that while we have drawn equipotential lines in such a way ıe magnetic intensity in a horizontal direction is in inverse ratio to the distance of the equipotential the distance between the lines of force gives no indication of the intensity. The reason of this is ıe lines of force give only the lines for the horizontal components, and not the total magnetic force.

Field of force for a polar current-system of the assumed form.

$\zeta = 20^0$, $\triangle u = 75^0$. $\triangle V = 0{,}218\, i$.

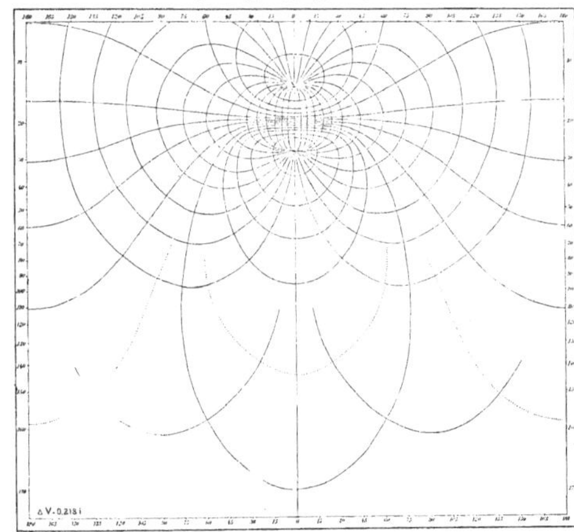

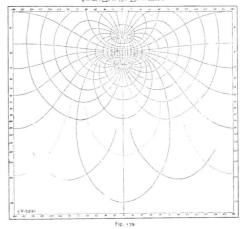

Fig. 179.

Cf. J. W. SANDSTRÖM: Über die Bewegung der Flüssigkeiten, Annalen der Hydrographie und maritimen Meteorologie, 1909, p. 242.

Fields of force for polar current-systems of the assumed form.

$\zeta = 20^0$, $\triangle u = 180^0$. $\triangle V = 0{,}349\ i$.

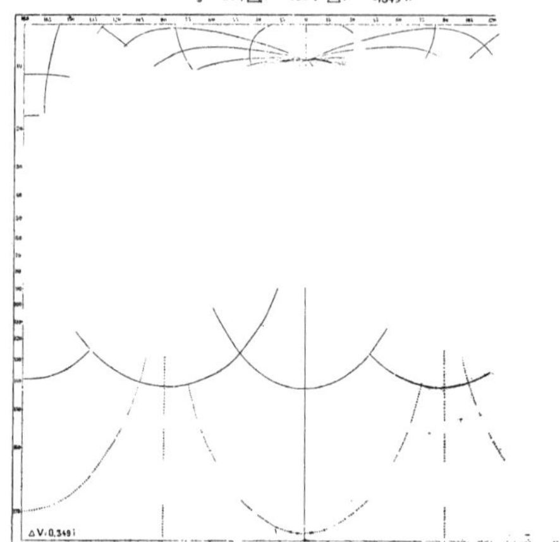

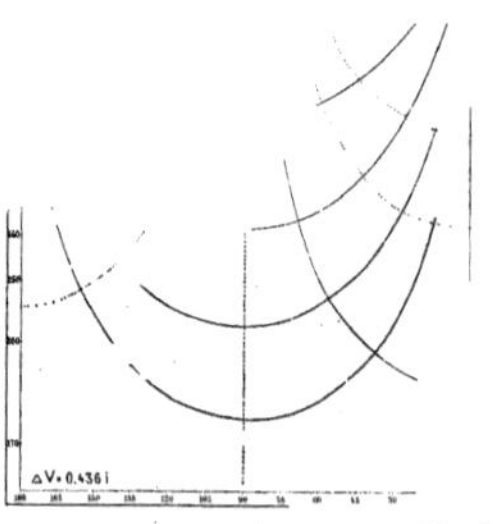

Fig. 181.

f force for polar current-systems of the assumed form.

$\zeta_1 = 20^0, \zeta_2 = 160^0, \triangle u = 75^0. \quad \triangle V = 0{,}218\, i.$

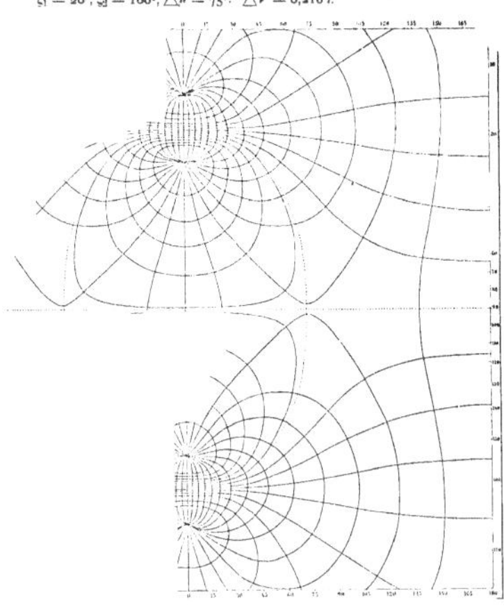

Fig. 182.

Curves for constant value of P_ϱ.

$\xi = 20^0, \triangle u = 75^0.$

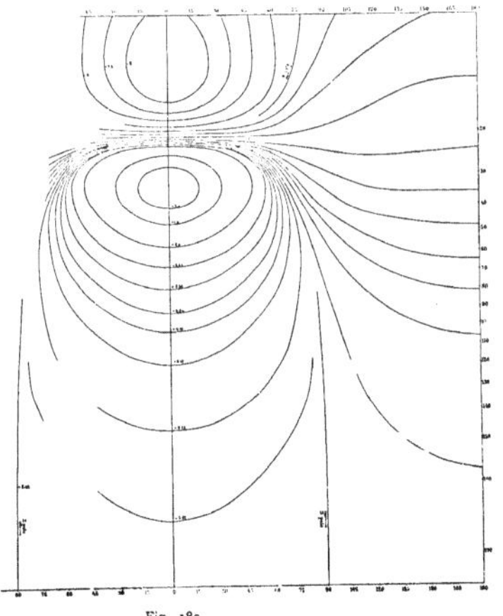

Fig. 183.

Curves for constant values of P_{ϱ}.

$\zeta = 20, \triangle u = 180^0$.

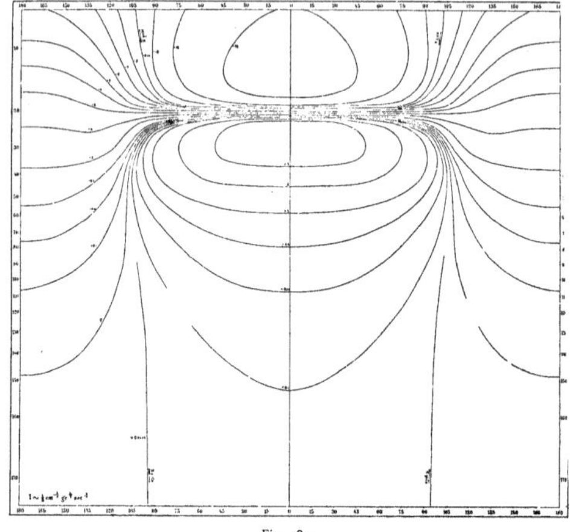

Fig. 184.

$\zeta = 20^0, \triangle u = 270^0$.

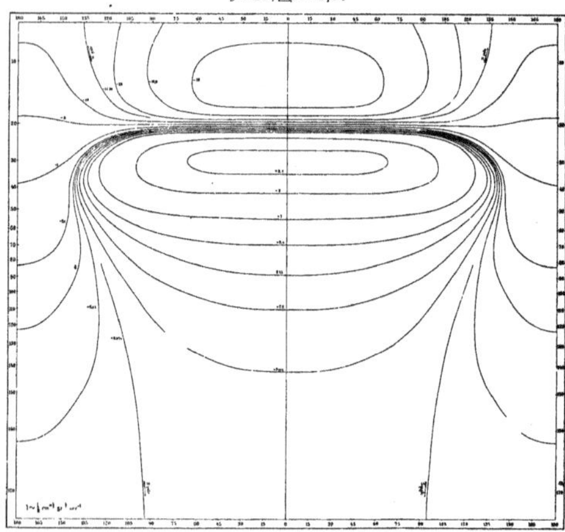

Fig. 185.

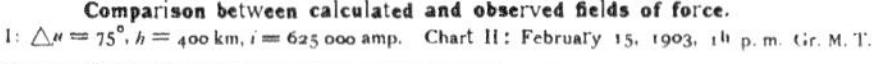

Comparison between calculated and observed fields of force.

I: $\triangle H = 75°$, $h = 400$ km, $i = 625\,000$ amp. Chart II: February 15, 1903, 1h p. m. Gr. M. T.

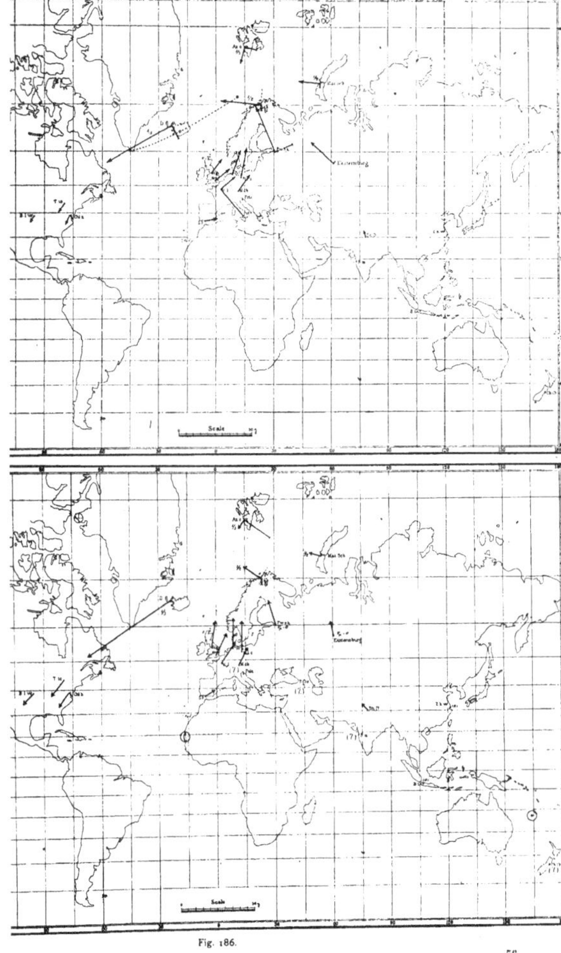

Fig. 186.

If we look at these charts, the great accordance with the observed areas of perturbation is at once apparent.

We have finally made a direct comparison with one of the observed elementary storms (see pag. 437). We have here placed our current-system with its storm-centre in the neighbourhood of Dyrafjord, $\theta = 0$ in the point of intersection of the earth's magnetic axis with the surface of the earth, and we have employed the system with the shortest horizontal piece of current, $\triangle \mu = 75°$. With this arrangement this will come very nearly along the auroral zone. The projection of the assumed current-system is indicated on the chart by a dotted line.

For this system the magnetic force-components are then calculated for the stations from which we have observations. The agreement, as will be seen, is striking as regards the horizontal current-arrows, except that the current-system employed seems to be a trifle too large. If we had taken $\triangle \mu$ a little smaller, or if the storm-centre had been chosen somewhat more westerly, the agreement would unquestionably have been still closer. In the vertical forces the arrow observed at Val Joyeux is considerably smaller than might be expected from the calculations. The direction is the same, however, in both cases. The cause of this is to be looked for partly in the fact that the constitution of the actual current-system must only with a very rough approximation be assumed to be capable of being replaced by such a system, and that the actual current-system might not possess such a strongly-marked horizontal component as is here assumed. Perhaps the agreement would have been better also in the vertical intensity if we had used a form of the current-system analogous to that given diagramatically in fig. 187. We believe, moreover, that much of the disagreement may be due to earth-currents, which would have the effect of increasing the horizontal magnetic force-components, while reducing the vertical. It is possible that these currents played the most important part. We must further draw attention to the uncertainty that may be connected with the observed values of P_v. We see this with special distinctness in Charts III and VII—X for the 15th February, in which there seem to be powerful perturbing forces in the vertical intensity at Uccle, while at the surrounding stations —Val Joyeux, Wilhelmshaven and Munich—no particularly noticeable effect is found. The uncertainty in the determination of the normal line is, as will be understood, rather great.

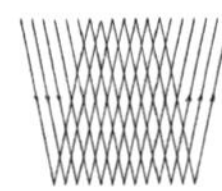

Fig. 187.

At Axeløen the observed vertical arrow is considerably greater than the calculated. This may only be due to the great uncertainty which attaches to the statement of the scale value for V at this place.

RÉSUMÉ.

92. By means of the long series of perturbations that we have now gone through, we have succeeded in obtaining a more or less clear idea of the magnetic storms, and have classified them according to their appearance and course. As, however, the material employed was large, it may be advisable to go once more briefly over the principal results at which we have arrived in the preceding pages.

The perturbing forces are calculated from the deviations from the normal daily course followed by the magnetic elements on calm days, as represented in Article 14. On the charts, the horizontal components are shown by current-arrows, of which the length is proportional to the size of horizontal component of the perturbing force, and whose direction gives the direction of an electric current over the place, which would produce a magnetic force similarly directed. These current-arrows, however, are only a geometrical representation of the perturbing forces, and indicate nothing whatever as to the existence of such currents.

In a number of places, moreover, the vertical component of the perturbing force is given by a line at right angles to the current-arrow, on the left of it—left of a person, standing on the earth and facing the direction of the current-arrow—if P_v is positive, that is to say if the force is directed towards the earth, and on the right of it if the force is directed upwards.

The storms that are first described are those which exhibited the simplest conditions, while later on, the more complicated perturbations are taken.

We succeeded, in this way, in first separating the so-called equatorial perturbations from the polar. Each of these types of perturbation have their characteristic area of perturbation, which is clearly apparent from the charts, as also from the comparison of curves which we made for each perturbation studied.

We have considered that the perturbations should be divided in all into five different types,

1. The positive equatorial storms,
2. The negative equatorial storms,
3. The positive polar storms,
4. The negative polar storms, and
5. The cyclo-median storms.

Of these it is especially the positive and negative polar storms, and the positive equatorial storms, that are most frequently met with.

The chief peculiarities of the positive equatorial storm are as follows:

Everywhere in low and medium latitudes, positive perturbing forces are met with in the horizontal intensity, while at the same time in declination no deflections, or only very small ones, are found. In the vertical intensity, only small perturbing forces occur.

If we consider the conditions in rather lower latitudes, we find the strongest perturbing forces in the equatorial regions, while the perturbing forces decrease in strength with increasing distance from the magnetic equator.

The deflections in horizontal intensity always increase at the beginning of the storm rather rapidly and to a certain height, after which the perturbing forces remain more or less constant in strength for a long period.

In the horizontal-intensity curve, there are always a number of very characteristic serrations, which are found again at all the stations situated in low and medium latitudes, and these serrations appear at any rate very nearly simultaneously all over the globe. This is also the case with the time of the occurrence of the perturbation. We have made some determinations for the purpose of finding out whether any differences in time could be proved in these at various stations. We have also found differences of some minutes; but as, in many cases, the accuracy with which the time can be determined is not as great as could be desired, we will not venture to express any certain opinion upon this foundation.

If, on the other hand, we approach the auroral zone, the perturbation-conditions alter to some extent. We also find in declination deflections like those in horizontal intensity. A peculiar impulse at the beginning of the perturbation, which was less noticeable in lower latitudes, now comes out distinctly, this being that the deflections in horizontal intensity are not first in a positive direction, but in a negative; and the current-arrow, or, if preferred, the perturbing force, oscillates here, at first quite distinctly, through a more or less considerable angle. This condition is most distinct in the immediate vicinity of the auroral zone. Here too, we find again serrations to some extent similar to those at southern stations, but often considerably larger.

At one station in polar regions, Kingua Fjord, in the only instance of such a perturbation found in the material from 1882 and 83, we came upon a storm in which the perturbing forces, which were of considerably greater strength, seemed to be distinct from the perturbations at the other stations.

Very frequently, perhaps as a rule, the positive equatorial storm is interrupted by the breaking in upon it of a polar storm.

The two best examples we have of perturbations of this type are the storms of the 26th January, 1903, and the 15th December, 1882. Plates XIV and XXIV show very clearly the above-described characteristics of this type of perturbation.

Fig. 31, on p. 69 of Part I, gives an excellent idea of the perturbation-area of such a storm.

Figs. 167 & 168, pp. 406 & 407, show the area of perturbation about the auroral zone during a positive equatorial storm. The characteristic turning of the perturbing force at the beginning of the perturbation is seen on Chart I. The same peculiar condition is also shown in fig. 57, Chart I, p. 133, at the two stations, Dyrafjord and Axeløen, where the movement is especially distinct.

Other instances of positive equatorial storms are found on the 9th December and 23rd October, 1902, the 22nd and 30th March, 1903, the 29th—30th and the 11th—12th October, 1902, and the 23rd —24th November, 1902.

We have sought for the cause of these positive equatorial storms in corpuscular rays, which we imagine issuing from the sun, their main mass being gathered in the magnetic equatorial plane of the earth. In fig. 37 we see cathode rays, under certain circumstances, may concentrate themselves in such a manner. In this case, the rays go from west to east round the earth, in such a manner that corresponding current-arrows would have to be directed as in the negative equatorial storms. It is probable, however, that the rays in the innermost parts swing round once or oftener, so that those nearest the earth pass it from east to west. In fig. 38 b, we have shown how the rays can bend round before the earth in this manner, and the nearest part will therefore produce on the earth a magnetic force-effect directed northwards, which thus answers to a positive perturbing force in the horizontal intensity. It is in rays of this kind, which turn round and pass nearest to the earth in a direction from east to west (if they are rays with negative particles), that in our opinion the cause of these positive equatorial storms must be sought. Fig. 39 shows a number of rays of this kind, lying in the magnetic equatorial plane, which STØRMER has found by calculation. The rays answering to values of γ between 0.3 and 0.9, are specially noticeable.

In reality, the constitution of the current-system which produces the magnetic storms of this type, is rather complicated, as there are at the same time perturbations in the north, which cannot be explained merely by an equatorial current-system. This is in perfect accordance with the conditions of which the experiments give a hint. Fig. 38, a and b, gives, for instance, quite distinct information of the existence of a connection between the rays which operate in the equatorial regions at a distance from the earth, and those which come in a wedge close in to the earth in the polar regions, the latter, in our opinion, being the cause of aurora polaris and the polar magnetic storms.

The patch of light in the polar regions, seen in fig. 37, is, we believe, connected with the powerful and strictly local storm in Kingua Fjord, which we found during the perturbation of the 15th December, 1882.

Similar polar precipitation, of which the existence cannot be so directly proved, should, we believe, be regarded as the cause of a number of apparently abnormal conditions that we found, for instance, in the perturbations of the 26th January, 1903 (P. 67), the 22nd March (P. 128), and the 24th November, 1902 (PP. 273 & 274).

The serrations that we find most strongly marked in the polar regions, must similarly be ascribed to polar precipitation. Simultaniously with the change in the equatorial current-system which produces the various serrations in the curve, slight polar precipitation will occur at places in the polar regions, acting locally with comparative power, but its effect decreasing rapidly outwards.

These occurrances of slight polar precipitation will always accompany a positive equatorial storm. For this reason, the character of the curves in the polar regions is very irregular in comparison with those farther south. This may be seen, for instance, by comparing Axeløen with Bombay or Batavia on Pl. XIV. We must thus, during the positive equatorial storms, imagine a constantly acting, more equatorial current-system, and a number of slight occurrence of polar precipitation in the north and south. As these two systems are undoubtedly, as we have said, connected with onea nother, a change in the one will always or at any rate as a rule be accompanied by a corresponding change in the other. We shall demonstrate this more clearly later on in the experiments, which show that the rays may run for a time more or less in the magnetic equator, but then intersect that plane at continually increasing angles, after which they finally descend in polar regions.

We next have the polar magnetic storms. In these, the most powerful forces are found in the polar regions, while the forces decrease very rapidly in strength with descent to lower latitudes.

In these storms, it will be possible, as a rule, to demonstrate in the polar regions one or more more or less distinctly defined areas, within which the most powerful perturbing forces are gathered. It appears that as a rule the character of the storm is mainly dependent on whether, in this area, there are positive or negative deflections in horizontal intensity. When the former occur, we disignate the storm as a positive polar storm, when the latter, a negative.

We will first look at the negative polar storm. It occurs very frequently, and often attains a very considerable strength. In order to obtain an insight into the nature of the storm, we looked out the very simplest of those contained in our material, and these were first discussed.

Among the chief peculiarities of the negative polar storm, it may be pointed out, in addition to that already mentioned, that the character of the curve in the polar regions is generally much serrated and irregular, which indicates that the acting forces, or current-systems, as we prefer to call them, must approach comparatively near to the place under consideration, coming nearest to the earth, now in one place, now in another. In lower latitudes, on the other hand, this disturbed character disappears, and the course of the curves is fairly even and quiet, although considerable deflections occur.

Whereas during this type of perturbation, negative values of the perturbing forces in the horizontal intensity were found in the polar regions, in lower latitudes at the same time positive values of this component were found. Here, then, we have a reversal of the component. The deflections in declination, on the other hand, may at one time be easterly, at another westerly, according to circumstances.

We find the typical perturbation-area of a negative polar storm in its most perfect and distinct form, shown in figs. 41 and 42. Areas such as these are constantly met during the negative polar storms. In the auroral zone there are very strong current-arrows directed westwards, while south of it the current arrows point in the opposite direction. In fig. 40, p. 86, we have endeavoured to give a diagrammatic representation of an ideal form for the typical perturbation-area that appears during the negative polar perturbations. The large vertical arrow, *A*, is supposed to coincide with the direction of the current arrows found in the most perturbed area, the so-called 'storm-centre'. The entire lines are the lines of force for the horizontal magnetic forces that occur upon earth; while at right angles to them are the dotted potential lines, or, as we have called them, current-lines. It is the right half of this figure that should correspond with the field represented, for instance, in figs. 41 & 42. For the sake of the general idea, we called the line that coincides with the arrow *A* in fig. 40, or along the current-arrow in the storm-centre, the principal axis of the system, and the line at right angles to it the transverse axis. It will be seen that we have supposed the area of perturbation upon the two sides of the principal axis to be exactly symmetrical, but in reality this will never altogether be the case, but only approximately. As a rule there are occurrences all over the polar regions of strong or slight polar precipitation, which easely effaces the traces that we might expect to find of such a condition. When the storms are particularly simple and well-defined, however, indications may to a certain extent be found of a perturbation-area on the other side of the principal axis, that is more or less symmetrical with the first. We believe we have found a condition such as this in the perturbation of the 14th & 15th February, 1883, where there is a simple, well-defined negative polar storm with storm-centre in the north of Europe, of which the principal axis lies more or less along the auroral zone in this district, there being no storms of any marked strength at the same time at other places round the polar zone. We here have some stations more or less symmetrically situated on both sides of the principal axis; and in the description on pp. 363 & 364, some conditions are pointed out that, although possibly only slight, would seem to confirm this assumption. In figs. 152 & 153, the current-arrows at Kingua Fjord and Godthaab indicate that such an area actually exists in the regions to the north of the principal axis.

On the transverse axis there are two characteristic points that are enclosed by the current-lines. The horizontal components of the perturbing forces in the regions round one of these points, are directed straight in towards the point, while in the other all the horizontal forces point straight out from it. In the points themselves, the horizontal force is zero.

The first of these points we have called the system's point of convergence, the second its point of divergence.

The storm-centre during a magnetic storm does not remain in the same place all the time. As a rule, a more or less distinct movement of the various storm-centres can be traced. In the polar regions this can best be seen from the horizontal-intensity curves, where a more or less distinct difference in the time of the beginning, maximum, and conclusion of the deflections at the various stations is found. We may also refer here to the perturbation of the 15th December, 1902, where this condition comes out with unusual clearness when we look at the horizontal-intensity curves for Dyrafjord and Axeløen, on Pl. X. There seems no doubt that the storm-centre here was at first situated in the vicinity of Dyrafjord, and afterwards moved eastwards along the zone, so that at the end of the perturbations, it was situated nearest to Axeløen. This is also apparent on looking at corresponding charts. At first the current-arrow at Dyrafjord is the strongest; but it then decreases, while the current-arrow at Axeløen increases. While in the polar regions, such movement of the storm-centre can be demonstrated, the current-arrows in southern latitudes will turn a certain angle, clockwise or anti-clockwise, and always in such a manner as to make it seem likely that it is produced by a movement of the whole perturbation-area in the same direction as that in which the storm-centre in the polar regions moves. Thus, simultaneously with the movement of the storm-centre the whole pertubation-area in lower latitudes will move in the same direction. As it appears from the character of the curves that the acting current-systems must come near to the polar stations, while they must be comparatively distant from the stations in lower latitudes, and also on account of the evident connection existing between the pertubations in high and low latitudes, we have considered ourselves justified in drawing the following conclusion:

During the negative polar storms, a current-system of some kind or other will be formed in the polar regions, the magnetic effects of which will be the primary cause of the perturbation-area formed.

According to this, the magnetic pertubing forces in low latitudes must be considered for the most part as very distant effects of this polar system of precipitation. The direct magnetic effect of this system then, we believe would be the primary. There might moreover be imagined a number of secondary effects, such as, in the first place, induced earth-currents, in the next, electric currents in the atmosphere occasioned secondarily by the ionisation which, especially in the upper strata of the atmosphere, must be thereby occasioned simultaneously.

The question which next comes up is: How must this polar current-system be supposed to be constituted? Here too, we believe the cause should be sought in corpuscular rays coming from the sun. These rays, when they come under the influence of the magnetic field of the earth, will be drawn in in zones round the magnetic axis. A single ray, considered by itself, will, if not under the influence of other corpuscles, move in a spiral path in towards the earth, then turn, and leave the place in a similar manner. We must thus imagine the corpuscular current as a whole, descending towards the earth in paths that are more or less vertical, then turning when near the earth, and once more leaving it, unless they are absorbed in the earth's atmosphere. How the rays, as a whole, will behave in the vicinity of the earth, is a question that cannot be decided in advance. It is a problem that requires special treatment. We have succeded in throwing much light upon the question by placing screens of various sizes and shapes upon our terrella. These, when the terrella is irradiated with cathode rays, will cast shadows, and from these shadows the course that the rays take near the earth can be directly measured. The experiments will be described in Chapter IV of the present Part. The simplest assumption we can make on the whole is that the rays in the vicinity of the earth turn round in an easterly or westerly direction. The conditions round the auroral zone also show that during the negative polar storms, there are effects like those of a horizontal electric current situated at a certain height above the earth.

Upon this basis, we have tried to find out how near we are to the actual circumstances when we assume that the polar current-system that is formed during a negative polar storm, can be replaced by a current-system consisting of two vertical infinite branches, which are connected by a horizontal piece of current. In Article 36 (pp. 102 & 103), we have made an estimate of how the horizontal forces vary when we move from the storm-centre outwards along the transverse axis of the system. This showed that as regards the horizontal forces, in the principal features even a quantitative agreement could be reached between the observed forces and those calculated as the effect of this ideal system. In the preceding Article, we have also made a minute calculation of the magnetic effects of such current-systems. A direct comparison of these areas with the observed pertubation-areas of the negative polar storms, show the close agreement that exists here. In fig. 186, for the sake of distinctness, we have placed a pertubation-area observed in one of the most characteristic elementary negative polar-storms in our material, by the side of that of such a linear current-system. In the horizontal forces, the resemblance, as we see, is striking. In the vertical intensity, the direction is also the same, but the observed forces are at a certain place, Val Joyeux, considerably smaller than the calculated. The cause of this should, we believe, be sought for principally in two circumstances. The first of these is that in our current-system the horizontal portion of current may be more conspicuous than in reality it is. If a rather different form of this had been chosen, e. g. if it had been assumed that the rays were more as if they ran in towards the earth in a point, as shown diagrammaticaly in the figure 187, the agreement would probably have been closer also as regards the vertical intensity. A system of this form would also probably be more in accordance with the actual current-system. As, however, it is a question of a very rough estimate, and the calculation of the first is considerably easier, we have employed this form.

In the second place, a no inconsiderable part of the perturbing forces observed will certainly be due to earth-currents. These, as every one is aware, will, when they have the effect of increasing the horizontal forces due to an external current-system, have the effect of decreasing the vertical component. The magnitude of the vertical intensity has been employed, ever since Gauss's time, for the purpose of determining how great a part of the magnetic effects observed must be ascribed to external forces, and how great a part to internal. In the chapter on earth-currents we shall look more closely into this proportion as regards the magnetic storms. We must, however, expressly draw attention to the great uncertainty that attaches to the determination of the perturbing force in the vertical intensity. We may, for instance, refer to Chart III for the storm of the 15th February, 1903. Here, while at Uccle there is a comparatively powerful vertical arrow, at the surrounding places, Val Joyeux, Wilhelmshaven and Munich they are too small to be measured. The values of P_v, therefore, when small, must be considered as only approximately correct.

There is another circumstance that we may point out. In the negative polar storms about midnight, Greenwich time, the horizontal portion of current will as a rule fall between Axeløen and Kaafjord, that is to say north of the latter station. In the most powerful storm we have studied, however, namely that of the 31st October and 1st November, 1902, the current seems to have moved to the south of this station, as there are now positive deflections in vertical intensity. In lower latitudes, at Wilhelmshaven and Pola, we also find at the same time positive perturbing forces in the vertical intensity, which however, we think should be considered as the effects of the negative equatorial storm, of which there are also distinct effects.

The third of the principal forms of magnetic storms, is the positive polar storm, of which the following are the chief peculiarities:

The form of the perturbation-area is on the whole the same as that of the negative polar storms; but all the forces in that area, both horizontal and vertical, are in the opposite direction. In the polar regions, in this type of perturbation, there are positive deflections in the horizontal intensity. We have here employed the same terms as in the negative polar storms — storm-centre, principal axis, transverse axis, and points of convergence and divergence. Whereas in the negative polar storms we found the system's area of convergence to the south of the storm-centre when considering the conditions in the northern hemisphere, in the positive storms we find the system's area of divergence in that region. As a rule, the perturbing forces in the positive polar storms diminish just as rapidly in strength as those in the negative polar storms; and we find here too a reversal in direction of the horizontal component of the perturbing force at about the same distance from the storm-centre as in the negative polar storms. Not infrequently, however, we meet with cases in which there are positive deflections comparatively far south.

In the matter of strength, the positive polar storms are as a rule somewhat weaker than the negative, and the character of the curves in the polar regions is not quite so disturbed in the former type of perturbation as in the latter. The field of a positive polar storm appears most distinctly in fig. 34, Chart IV for the 9th December. In fig. 83—charts for the 15th February, 1903 – the form of the field of force also comes out quite distinctly. At Uccle, for instance, is seen the powerful positive vertical component that is characteristic of the area of divergence.

It will be seen that the positive polar storms may be explained as effects of a current-system of the same form as that which we assumed as the cause of the negative polar storms, if we assume that the current flows in the opposite direction. In fig. 50, p. 105, we have given a diagrammatic representation of a system of rays, of which the effect in the main will be equivalent to the current-system that we have employed, and which possibly, on the whole, will be more like the actual positive polar current-system.

These two principal systems, the negative and the positive polar perturbation systems, rarely occur quite alone. As a rule they occur simultaneously, but in different districts. It appears that they always, on the whole, are grouped in the same manner in relation to the sun, and in the following manner:

On the morning and night sides of the globe, there is always a powerful, negative polar system of precipitation, generally fairly extensive, in which the principal axis of the system falls, as a rule along the auroral zone. This negative system continues westwards on to the afternoon side, but here the principal axis of the system turns northwards to the districts north of the auroral zone, and it looks as if the system also as a rule would be continued westwards until it joined the negative system on the morning side. What the form as a whole, of the system of precipitation would be, cannot, however, be determined; but it is conceivable that it is more or less analogous to the spiral luminous figures that are reproduced in fig. 140 on p. 327. The positive polar system developes along the auroral zone, most strongly in the southern part of the zone. It may sometimes be of very considerable extent, but as a rule is much smaller than the negative system. In this way there will be a boundary-station in the auroral zone, as a rule upon the evening side, which will be situated between the positive and negative systems. Thus, while at the stations on the afternoon side in the auroral zone, the positive storm is the principal phenomenon, and on the night side the negative, and the perturbations here occur with great distinctness and with well-defined deflections in a positive or negative direction, as the case may be, at this boundary-station now one system, now the other, will prevail, causing the deflections in horizontal intensity to be at one time positive, at another negative.

We have a very clear example of this circumstance in the perturbation of the 15th January, 1883 (Chart V, p. 336).

While in the district to the west of Little Karmakul, i. e. at Bossekop, etc., effects of a positive polar storm are apparent all the time, and to the east, at Ssagastyr and Uglaamie the effects are exclusively those of a negative polar storm, the current-arrow here oscillates backwards and forwards, is at first, $18^h\,25^m$, very small, but increases rapidly with direction easterly, $19^h\,5^m$, then turns, and at the last point of time, $19^h\,25^m$, is a powerful westward-pointing current-arrow.

At a station situated on the afternoon side a little north of the auroral zone, the northern negative system and the southern positive system will counteract one another horizontally, but co-operate in vertical intensity. Powerful perturbing forces, therefore, are very often found in vertical intensity. The current-arrows for the horizontal perturbing forces there now point in one direction, now in another, and are sometimes exceedingly small. In Jan Mayen, we constantly find this condition very marked (see Charts V—VII for the 15th January, 1883, pp. 336 & 337; Charts V—X for the 15th July, 1883, pp. 381—383; Charts V—VII for the 15th October, 1882, pp. 421 & 422).

This division of the negative and positive systems of precipitation will always appear in a more or less complete form whenever polar storms occur.

This area of perturbation will thus, as a whole, be moved westwards in the course of the perturbation, in a manner such as would be found if the systems of precipitation formed systems closely connected with the sun.

In detail, however, we shall be able to find the perturbation-conditions somewhat different from those we have here described as the typical. The forces will always, as already stated, in the extended negative area of precipitation, concentrate themselves about one or several storm-centres. At the same time, the negative systems that occur at the other places will more or less disappear. Frequently there is a single, comparatively very limited, negative system of precipitation, while the rest of the negative current-circuit has practically disappeared. This has very often proved to be the case at about Greenwich midnight. At about this hour, we very frequently find a powerful, well-defined and comparatively very limited, negative system in the north of Europe, while at other places round the arctic zone, no negative systems of precipitation are apparent, as far as we can see from our observation-material. For this reason the storms that occur at this time exhibit particularly simple areas of perturbation. It is the simplest of these that we have taken first, and thereby found the elementary type of negative polar storm.

With regard to the movement of the systems, it should be observed that it is only in its main features that this takes place as stated above, differences being very frequently found in the details. We have, for instance, just mentioned an example of the movement of an elementary negative polar system eastwards along the auroral zone, simultaneously with the development of the storm. The cause of this is, we believe, in a great measure to be found in the fact that the height of the sun above the magnetic equator varies. In fig. 76 we have shown a curve that, according to Störmer's calculations, gives the connection between the height of the centre of emanation above the magnetic equatorial plane, and the deflection undergone by the ray that goes to the origin, when we consider an elementary magnet situated in that point, with its axis along the *Z*-axis. We have thought that a similar connection must exist between the height of the sun above the earth's magnetic equator and the position of the various storm-centres, and that when this height of the sun alters, the various perturbation-centres will be moved similarly to these "distinguished" rays in the calculations. These rays, however, will move now towards the east, now towards the west, according as the height of the sun changes (see fig. 76 & Article 71). We should therefore also expect to find similar conditions at the storm-centres. The finding of deviations from the regular moving of the perturbation-systems towards the west, is thus only a conceivable consequence of our theory. In the first storm in Part II (Article 83), we have made a comparison between the positions of the storm-centres observed and the calculated areas of precipitation at the various times, and their movement from time to time. We think, too, that we have found in some cases very distinct analogies, although of course there will be no question of any exact agreement.

In Chapter I of Part II, we have principally studied the occurence and development of the various polar systems. In all the perturbations, we have not only again and again found the characteristic conditions that are touched upon here, but a number of details have also appeared that are constantly found in storms of most varied character. The manner in which the polar systems break in upon one another is always exceedingly characteristic. We recall, for instance, the relation between the effects of the positive and negative storms at Cape Thordsen in the afternoon. If we compare the afternoon storms at this station on the 15th and 2nd January, 1883, and the first November 1882 (see Pl. XXVI, XXV & XXIII), we find, as proved in detail in our previous description of the storm of the 15th January, a negative storm from 12^h to 14^h, breaking in upon a positive storm of long duration. On January 2nd there is a similar phenomenon from 14^h to 16^h, but the positive storm is much less pronounced. On the 1st November also, there is a corresponding phenomenon from $13^h\ 30^m$ to 16^h, but here the effects of the positive storm have almost entirely disappeared. A slight indication of a similar circumstance is also met with in the storm of the 1st February, 1883. It would take too long, however, to go more minutely into these matters here, and we will therefore only refer the reader to the description of two storms in which the characteristic conditions are especially conspicuous. These are the perturbations of the 15th January and the 15th July, 1883, in which the perturbation-conditions are perhaps most easily surveyed.

A positive polar storm in the auroral zone will, as we have said, always, or at any rate generally, be accompanied by a negative polar storm in rather higher latitudes. Some such idea as the following might then seem probable as the explanation of this circumstance:

We know, according to the theory, that corpuscular rays that move in the earth's magnetic field will approach the polar regions in paths that twist spirally about the magnetic lines of force. If the rays possess great magnetic stiffness, the radius of these spirals will be comparatively great. If we assume that such ray-spirals exist on the afternoon side of the earth, and that they lie close together somewhat in the manner shown in fig. 188, the connection with the southern positive system and the northern negative system, as regards the polar regions, can be explained quite simply. For rays of a stiffness answering to $H\varrho = 7 \times 10^6$, we find in the polar regions, where $H =$ about 0,5, $\varrho = 1,4 \times 10^7$ cm. The diameter of these spirals must then be 280 km. The principal features of the field in southern latitudes can probably also be brought out as effects of a spiral system such as this. Judging from our experiments, of which the results are clear enough, an explanation such as this, is not entirely satisfactory, and would at any rate have to be considerably modified. According to those experiments, the rays in the positive and negative storm-centres seem inclined to behave in a manner similar to that shown diagrammatically in fig. 50. The above explanation cannot, at any rate, be applied to the night storms, in which there is only a negative storm in the south.

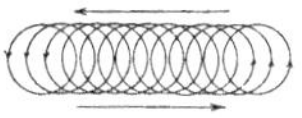

Fig. 188.

We thought of showing two more types of perturbations, namely, the negative equatorial storm and the cyclo-median storm. We have, however, only a few examples of these among our observations.

The negative equatorial storms are most powerful in the region of the equator, where the perturbing forces in horizontal intensity are negative. Their area of perturbation may be explained as the effect of a current-system of which the greater part is situated more or less in the magnetic equator, as the experiment shown in fig. 37 shows, and where the rays have a movement similar to those that are calculated in the magnetic equatorial plane for an elementary magnet answering to values of γ that are $\gtreqless -1$ (see fig. 39, p. 82). In order that these rays shall come comparatively near to the earth, their stiffness must be comparatively great. More flexible rays would be deflected like the rays in the equatorial plane for $\gamma > -1$, and thus glance by the other in the opposite direction. We believe we have effects of rays such as these in the positive equatorial storms. The rays that we believe should produce the negative equatorial storms must therefore be assumed to go round the earth and to be magnetically more inflexible than those that produce the positive equatorial storms. In accordance with this, it is only during very powerful storms that we have found these negative perturbation-areas. The fact that the active rays during specially powerful perturbations have a greater magnetic stiffness than those in the less powerful storms, is also indicated by the circumstance that has just been touched upon, namely, that the storm-centres during the latter seem to move southwards. In particularly violent magnetic storms, it is well known that the auroral zone moves southwards, so that polar aurora can be observed even in very low latitudes. The simultaneously-occurring magnetic storms have also, in lower latitudes, a completely polar character, which indicates that the acting current-systems come in to the immediate vicinity of the place. But if the corpuscular rays come in towards the earth in such low latitudes, their stiffness must be considerable. These circumstances will be explained fully in Chapter IV. The forces that occur in the negative equatorial storms are also considerable greater than those found in the positive. Among our observations, we have found only examples of negative equatorial storms, which occur simultaneously with polar storms, and it is perhaps doubtful whether this type of perturbation on the whole can occur alone. We have not sufficient material, however, for the formation of any well-founded opinion on the matter.

We find the perturbation-areas in which the negative equatorial storm is most distinctly apparent, during the perturbation of the 31st October, 1902, and the 8th February, 1903 (see figs. 107—116, with description in Art. 66 & 67; and figs. 87 & 88, with description on p. 189). In the latter case, however, we have suggested, at the foot of p. 189, another possible interpretation of the field.

The last cyclo-median type of perturbation, we have supposed would answer to effects of rays of a degree of stiffness answering to the experiments shown in figs. 66 & 68, 1—6. How the rays in these triangular figures move, is indicated in the lowest of the three figures 71, and in fig. 72. These should be rays that came comparatively near the earth in lower latitudes, and which formed fields similar in form to these figures, that is to say spirals in which the direction of the current-arrows was anti-clockwise. We find similar spiral fields in the areas of convergence in the negative polar storms. In the cyclo-median storms, however, the forces in low latitudes must be more powerful in comparison with the forces in the polar regions, than in the negative polar storms. We have only a few instances of such perturbation-fields that can be characterised as rather well defined. We believe the perturbation of the 6th October is a storm of which the field of force should be explained as the effect of such a cyclo-median system. In our discussion of the compound perturbations, we have also several times come across fields that would naturally be due to cyclo-median systems, but in which nothing certain could be decided, owing to the complicated character of the storm. Fields of this kind are to be found in figs. 78 and 79, for the 25th December, 1902, and figs. 87 and 88, for the 8th February, 1903.

These five types of perturbation, however, as the above shows, must not be considered as completely separate phenomena. There will be a genetic connection between them, and this frequently finds expression in the fact that when there are simultaneous effects of several systems, a change in one system will be accompanied by a change in the other. This is especially distinct in simultaneous positive and negative polar storms, but is also very prominent in simultaneous positive equatorial and polar storms.

For all the perturbation-areas we have studied, a natural and simple explanation of the main features has been found by the aid of these five types of perturbation. In addition to the direct magnetic effect of these corpuscular systems, there will also be effects of simultaneously-occurring earth-currents, and possible atmospheric ionic currents and secondary cathode rays. There seems to be no doubt that the first of these exert a considerable influence, and we shall study them more closely in a later chapter; but what effect the atmospheric currents might have is a rather more doubtful question.

In Chapter IV of Part I, an estimate is made of the intensity of the corpuscular currents that appear in the polar storms, and the amount of energy they carry. The making of such an estimate has been made possible by the fact that we have two stations, Axeløen and Kaafjord situated one on each side of the auroral zone, and that, as already mentioned, the current-systems form in the auroral zone, that is, between the two stations. We have assumed that in the simplest perturbations the conditions up there can be regarded approximately as effects of an infintely long, horizontal rectilinear current, situated between these two stations. We can then, by the aid of the observations at the two stations, determine both the strength of such a current, and its height. The question will indeed be over-determined, and we can thus obtain a kind of idea of the approximation with which an assumption such as this can be employed. In the simple cases that we have studied, the approximation, as a rule, must be considered as quite satisfactory. We found the average strength of the current in the storms we investigated to be about 10^6 amperes, and the average height about 400 kilometres. If we also used Dyrafjord and Matotchkin Schar, we sometimes arrived at greater heights, up to more than 1500 kilometres; but we believe these are probably due to the fact that our assumption in this case does not hold good.

We have further examined into the amount of energy which these current-systems must represent, and have come, by estimating, to figures such as about 2×10^7 h. p., if we assume that the systems are formed of ordinary cathode rays, about 5×10^8, if we assume β rays with a velocity of 2.59×10^{10} cm.sec.$^{-1}$. It is, however, reasonable to suppose—as we shall show in describing the terella experiments in a subsequent chapter—that the rays in this case are considerably stiffer than these. If we assume a stiffness 10 times as great as the stiffest α rays, or answering to $H\varrho = 7 \times 10^6$, we obtain an amount of energy of about 10^{13} h. p.. From this we have inferred backwards and proved that we come, by assumptions which are still indeed rather arbitrary, but not unreasonable, to values for the amount of energy emitted from the surface of the sun in the form of corpuscular rays, that are as great as those of the energy emitted in the form of light and heat. It does not therefore seem improbable that the disintegration of the sun's matter which is undoubtedly taking place, and which must be assumed to be the cause of the corpuscular rays observed, would be great enough to account for the emission of light and heat from the sun.

A POSSIBLE CONNECTION BETWEEN MAGNETIC AND METEOROLOGIC PHENOMENA.

93. If the view we have maintained is correct, namely, that the magnetic storms are due to corpuscular rays that are drawn in in zones round the magnetic poles, where they pass directly down into the athmosphere of the earth, it is clear that these rays, especially in the upper strata of the atmosphere, must be assumed to produce a strong ionisation in the air. In our expedition of 1902 & 3, atmospheric-electrical measurements were made, which will be gone into later on; but it may be remarked here, that the result of these measurements showed that the "Zerstreuung" of the air at those stations averaged about twice as much as in Christiania, indicating that the air up there is considerably more ionised than in lower latitudes. In an expedition which I made, in company with my assistant, Mr. Krogness, to Kaafjord at the time when Halley's comet crossed the sun's disc in May 1910, I had an opportunity of studying this matter more closely.

Instead of, as before, making the measurements at places that are at no great height above sea-level, I on this occasion investigated it at my old aurora-observatory on the top of Haldde Mountain, about 910 metres above the sea. Here there proved to be sometimes tremendous variations. On the 20th May, for instance, values were found that went up to about 500 times the normal. Unfortunately the attempt was interrupted in the middle of these measurements; but I had an opportunity of making insulation-tests twice at that time, which proved there was no perceptible leakage. If we can demonstrate this circumstance with certainty, we presumably have before us a phenomenon that is closely connected with the peculiar light-phenomena that Lemström discovered in 1882 & 3 on a mountain-top at Sodankylä.

There is no doubt that such strong ionisations will have a very great influence upon atmospheric conditions, especially upon the formation of clouds, and must thus be assumed to be a meteorological factor of no small importance, especially for the districts in the vicinity of the auroral zone. I am of the opinion that this is a very important connecting link between terrestrial-magnetic and meteorological phenomena. I have therefore recently submitted to the Norwegian State authorities, a suggestion that a permanent up-to-date-magnetic-meteorological observatory be established upon the top of Haldde, for the purpose, if possible, of throwing light upon these interesting and meteorologically important matters.

There was another phenomenon, striking examples of which we had the opportunity of seeing on this expedition in May, 1910, namely, the formation of what may be called auroral clouds. In addition to the usual polar bands, which in a clear sky, could very often be observed in the form of several evenly luminous arcs, of which, however *one* was especially conspicuous, exactly similar to parallel auroral arcs, we very frequently found formations of cirrus clouds, which exhibited the most perfect argreement with various auroral formations. Several times we had capital examples of the manner in which such clouds are formed, how drapery-formations appeared in a short time, exactly in the same manner as an auroral drapery. The first observer, who has called attention to this very interesting fact seems to be

ADAM POULSON [1]. As far as I know, no one has, however, studied this phenomenon in connectio simultaneous magnetic registrations at the same place. This we had the opportunity of doing the very interesting fact came out, that the formation of these clouds was always accompani simultaneous magnetic storms and earth-currents; and there thus appears to be no doubt that the direct cloud-forming effects of the same rays that occur in the auroral phenomena. From this it that these cirrus-clouds are directly formed by the corpuscular rays which we suppose to be the of magnetic storms and aurora. The first hypothesis that one naturally might form as to this pheno is, that the clouds are due to water-vapour brought to condensation by the ions formed by the im negative rays. It is however also a probability that some of the observed »auroral clouds« are n clouds, but merely a very strong concentration of corpuscular rays, which in the case of darkness appear luminous; in the daytime the concentration of corpuscles should have the effect of maki places where they occur less transparent, and able to diffuse light, and thus become visible. In way also possibly certain faint polar bands observed in the polar regions might be explained. Acc to circumstances these concentrations may disappear or give rise perhaps to real clouds.

---

[1] Met. Zeitschrift 12, 161 (1895).

CHAPTER III.

STATISTICAL TREATMENT OF MAGNETIC DISTURBANCES OBSERVED AT THE NORWEGIAN STATIONS 1902—1903.

INTRODUCTORY.

94. In the previous treatment of the perturbations given in the first part of this work, each disturbance has been examined individually. This investigation led us to divide the perturbations into groups, each of which possessed certain characteristic properties, especially with regard to the distribution of the perturbation in space relative to the earth.

In the following we shall proceed to study the variation in the time of the appearance at our four arctic stations of the magnetic storms occurring during the period of our observations.

In order to solve this problem, it is necessary first to fix the unit by which we are to measure the »quantity« of perturbation that has occurred during a certain interval of time. One way would be to count the perturbations, e. g. those which exceeded a certain magnitude. Such a mode of procedure is often employed to obtain a quantitative measure of phenomena of this kind; but the method is not very exact, as perturbations count equally, even when their magnitude varies within wide limits. Further we are met with the difficulty, or rather impossibility, of defining what is meant by one disturbance.

We have therefore decided to follow a more exact method, which can always be applied without ambiguity. In this method the »quantity« of perturbation is measured by what we shall call *storminess*, which is defined as follows:

We assume the perturbing force in any of the magnetic elements H, D or V in the time interval $o < t < T$ to be found as a function of time. The determination of this function from experiments only requires the possibility of finding the perturbing force at any moment, which can be done in the way described in Part I of this work.

By the *absolute* storminess in one of the components—say the horizontal component—we understand the quantity:

$$|S_H| = \frac{1}{T}\int_0^T |P_h|\,dt = S_H^a$$

It is equal to the average perturbing force P_h if the latter is always taken to be positive.

It will also be of interest to consider separately disturbances in the positive and negative direction, and for this reason we define the *positive* and *negative* storminess

$$S_H^p = \frac{1}{T}\int_0^T P_h^p\,dt$$

$$S_H^n = \frac{1}{T}\int_0^T P_h^n\,dt$$

where P_h^p is any positive value of P_h in the interval, and P_h^n any negative value. It follows:

$$|S_H| = S_H^p + S_H^n$$

Further we shall introduce a quantity representing the difference between the positive negative storminess

$$S_H^d = S_H^p - S_H^n = \frac{1}{T}\int_0^T P_h\, dt$$

For Declination and Vertical Intensity similar expressions are defined.

Finally, by the *Total* storminess in the same interval of time we are to understand the qu

$$S^T = \sqrt{|S_H|^2 + |S_D|^2 + |S_V|^2}$$

In accordance with the definition of storminess here given the positive and negative st corresponds respectively to a positive and negative direction of the perturbing force.

There are two problems which will be dealt with in the following pages and form the mai of our investigation, viz:

(1) The total storminess as a function of time.

(2) The distribution of disturbances in magnitude and direction at the different hours of the the possible diurnal variation of the storminess.

For the practical carrying out of the calculation, the following mode of procedure has been a The storminess was calculated for each period of two hours, for all three components, and the and negative storminess were taken out separately. The numbers for one day were placed in tl horizontal line as shown in the first series of tables.

For each five-day period, the mean was taken of all numbers in the vertical columns corres to the same hour-interval. This gave a horizontal line containing the distribution of the storminess various hours for a period of five days. Taking the mean of the positive and negative stormi this horizontal line, we obtain the positive and negative storminess S^p and S^n for a period of fiv and their sum $(S^p + S^n)$ gives the absolute storminess for the same period.

We think it of considerable importance that a continuous record should be given of the occ of magnetic perturbations during the whole period of observation, believing that such a record w an idea of how far we have succeeded in the first part of this work in treating the most important perturbations.

For our present purpose, however, it is the average values, that mostly concern us; and w therefore decided on publishing the following separate tables:

FIRST SERIES.

Tables for the continuous two-hourly records. These tables will be divided into groups of fiv corresponding to each five-day period. The numbers will be expressed in arbitrary units, whi differ for the three components, but will be the same for all four stations. The factors of transfo into absolute units will be given for each component.

SECOND SERIES.

Tables giving the distribution of storminess at the different hours of the day for each five-day The periods will be divided into groups of 6. The mean value for each vertical column is taken f group and placed in a horizontal line, thus giving the distribution of storminess for 30 days, whi be taken as one month.

Finally, we find the mean distribution for the whole period of our observations. For each there will be three tables, one for each component; and the numbers will be expressed *in magnetic units.*

THIRD SERIES.

Tables giving the record of storminess for each five-day period. There will be one table for each station, containing the positive, negative and absolute storminess for each component, and one column containing the *total* storminess. The numbers will be expressed in absolute units.

The method of calculating the storminess is very much the same as that employed for calculating the perturbing force. The "normal line" is drawn on the magnetogram in the way described. During the perturbations a number of areas are formed by the registred curve and the normal line. The areas on both sides of the latter are taken out for each interval of two hours, and from them, knowing the scale-value and the length of the interval, we can find the positive and negative storminess. The relative values given in the first series of tables are simply these areas given in centimetres and reduced to the same sensitiveness for all four stations.

In taking out average values, it is necessary, as we know, to have a value for every two-hour interval throughout the period. It will unavoidably happen that in some records short intervals of time may be missing, but the blank interval due to the change of paper on the cylinder will generally be so short that it will practically introduce no error; for the intermediate values can be found by connecting harmoniously the two ends of the curve. If during a perturbation, the curve is invisible for a short interval of time, we have employed the same method of completing the curve by harmoniously connecting the two parts.

In the curves it has occasionally happened that records were wanting for several hours. If considerable disturbances were occurring at the other stations during these intervals, we should have to omit the whole five-day period; but as a rule we have been able to estimate thestorminess for the blank intervals.

Values which are not found directly from the curves, and consequently cannot claim great accuracy, will be put in brackets. These values may be found in various ways e. g. by completing the curve for a short blank interval or by estimating the value from the curves of the other components at the same place or from the curves of the neighbouring stations.

During such investigations it became clear that the great storms did not show the same properties as the small ones with respect to distribution in space and time. It was therefore of interest to find the average properties of the great storms separately.

The classification of the storms into great and small is of course to a certain extent quite arbitrary. We have decided on the following procedure: To find the storminess of great storms, we take that of every two-hour period for which the positive or negative storminess is greater than 15γ in any of the components. If the condition for a great storm is fulfilled for a certain two-hour period in one component, the corresponding storminess is counted in the case of the other components even when it is less than 15γ.

FIRST SERIES.

RECORDS OF STORMINESS FOR EACH TWO-HOUR PERIOD.

The storminess in absolute units is found in the following way:

$S_H = 14.9\, F_H$ (in γ). $S_D = 9.17\, F_D$ (in γ). $S_V = 17.5\, F_V$ (in γ).

F_H is the number given in the table for the storminess in the horizontal intensity.
F_D ——»—— declination.
F_V ——»—— vertical intensity.
The columns with the heading + contain positive storminess.
—»— — —»— negative —»—

Matotchkin Schar.

TABLE LXIV.

Disturbances in Horizontal Force (F_H).

Gr. M.-T.	0—2		2—4		4—6		6—8		8—10		10—12		12—14		14—16		16—18		18—20		20—2	
Date	+	—	+	—	+	—	+	—	+	—	+	—	+	—	+	—	+	—	+	—	+	—
October 3	0	0.1	0	0.1	(0.1)	(0.1)	0.1	0	0.1	0	0.3	0	0	0	0	0.3	0.1	0	0	0.2	0	0.
4	0	0.6	0	0.2	0	0.4	0	0.2	0.1	0.1	0.7	0	0.6	0	0.2	0	0	0	0	0	0	0.
5	0	0.2	0.2	0.1	0.5	0.1	0.1	0	0	0.1	0.2	0	0	0	0	0	0	0	0.1	0	0	0.
6	0	0	0	0	0	0.1	0.1	0	0	0	0	0.1	0.2	0	0.2	0	0	0	0	0	0	0.
7	0	0	0	0	0	0	0	0	0	0	0	0	0	0	0	0	0	0	0	0	0	0
8	0	0	0	0.1	0	0	0	0	0	0	0.2	0	0.1	0	0	0.1	0.1	0	0.2	0.3	0	0.
9	0	0.1	0	0	0	0.1	0.1	0.1	0	0.2	0	0.4	0.1	0.1	1.1	0	0.1	0.2	0	0.2	0	0
10	0	0	0	0	0	0	0	0	0	0	0	0	0	0	0	0	0	0	0	0	0	0
11	0	0	0	0	0	0.1	0.1	0	0	0.2	0.1	0.3	2.1	0	0.4	0	2.3	2.0	0.5	6.5	0	17.0
12	1.4	0.3	0.7	0	0	0	0	0	0.1	0.1	0	0.1	0.2	0.1	0.1	0.1	0	0.1	0	0	0	0
13	0	0	0	0	0	0.1	0	0	0.1	0	0.4	0	0	0	0.1	0.3	0.6	0.3	0	8.5	0	1.5
14	0	0	0	0	0	0	0	0	0	0	0	0	0	0	0	0.1	0	0	0.1	0.1	0	0
15	0	0	0	0	0	0.1	0	0.1	0.1	0	0	0.1	0	0	0	0	0	0	0	0.4	0.2	0.1
16	0	0.1	0	0	0	0.2	0	0	0	0	0	0	0	0	0	0	0	0	0	0	0	0.1
17	0	0	0	0	0	0	0	0	0	0	0	0	0	0.1	0	0.2	0	0	0.1	0	0.1	0
18	0	0.1	0	0.1	0.2	0	0	0.1	0.2	0	0.1	0	0	0	0	0.1	0.2	0	0	3.4	0	0.8
19	0	1.0	0	0.7	0	0.1	0.1	0	0	0	0	0	0	0	0	0	0	0	0.1	0	0	0
20	0	0	0	0	0	0	0	0	0.1	0	0.2	0.1	0	0.2	0	0.1	0	0	0	0	0	0.1
21	0	0	0	0	0	0	0	0	0	0	0.1	0	0	0	0.1	0	0.8	0	0.1	0.1	0	0.8
22	0.1	0.2	0	0.2	0.1	0	0	0.1	0	0	0	0	0	0	0	0.1	0	0	0	0	0	0.1
23	0	0	0	0	0	0.1	0	0	0	0	0	0.1	0.1	0	0.1	0	0	0	0.1	0	0	4.3
24	0.9	0	(0.2)	(0.4)	(0.1)	(0.2)	0	0	0	0	0	0.1	0	0	0.3	0	1.8	0	0.7	1.4	0	14.5
25	0	8.3	0.3	1.5	0	0.4	0	0.2	0.4	0	1.3	0	2.0	0	0.1	0.3	0	0.6	0	0.7	0	1.9
26	0	1.0	0	0.1	0	0	0	0	0	0.1	0	0.1	0	0.2	0.8	0	0.4	0.1	0	0.7	0	0.5
27	0	0.8	0.4	0.2	0.3	0.1	0	0.2	0.5	0.1	0.1	0.3	0.9	0.1	3.2	1.2	1.0	2.7	0	5.3	0	4.0
28	0.2	0.1	0	0.3	0.1	0.1	0	0.4	0	0.2	0	0.6	0	0.3	0.4	0.1	0.1	0.5	0	6.1	0	5.5
29	0	1.0	0	0.1	0.2	0.1	0	0.2	0.2	0	0	0.1	0.1	0.1	0	0.1	0.2	1.3	0	9.8	0	8.9
30	0.4	0	0.2	0	0.1	0	0	0.4	3.3	0	2.0	0.4	0	1.0	0	0.3	0	0.4	0	0.5	0	2.3
31	0	1.2	0.1	0.5	0.1	1.3	1.4	0.2	5.2	0	12.5	0	7.0	1.3	5.2	0.4	0.3	6.7	0	16.5	0	16.0
November 1	0	7.0	0.4	0.9	1.5	0	0.9	0.1	1.4	0	0.2	0.1	0.1	0	0	0	0.1	0.1	0	0	0	0.6
2	0	0	0	0	0	0.1	0	0.1	0	0.2	0	0.1	0	0.2	0.1	0.2	0.4	1.9	0	9.5	0	3.4
3	0	0.3	0	0	0	0	0	0	0	0.1	0	0	0	0	0.1	0	0.3	0	0.1	0	0	0.6
4	(0)	(0.1)	(0)	(0)	(0)	(0)	(0)	(0)	(0)	(0.1)	(0)	(0)	(0)	(0)	(0)	(0)	(0.3)	(0.6)	(0)	(4.0)	(0)	(2.0
5	(0)	(0.1)	(0)	(0)	(0)	(0)	(0)	(0)	0	0	0	0	0	0	0	0	0	0	0.1	0	0	0
6	0	0	0	0	0	0	0	0	0	0	0	0	0	0	0	0	0.5	0.5	0	6.3	0	3.8

IV (continued). F_H Matotchkin Schar.

		2−4		4−6		6−8		8−10		10−12		12−14		14−16		16−18		18−20		20−22		22−24	
+	−	+	−	+	−	+	−	+	−	+	−	+	−	+	−	+	−	+	−	+	−	+	−
0.4	0.1	0.5	0	(0)	(0)	0	0	0	0	0.1	0	0	0	0	0	0	0	0.1	0	0	0	0	0
0	0.3	0	0	0	0	0	0	0.2	0	0	0	0	0	0	0.1	0.2	0	0	0.1	0	0.3	0	0
(0.1)	(0.1)	(0.1)	(0)	(0)	(0)	0	0	0	0	0	0.1	0	0	0	0	0	0.1	0	0	0	0	0	0.3
0	0	0	0	0	0	0	0	0	0	0	0	0	0.1	0	0.1	0	0.1	0	0	0	1.3	0	1.9
0.1	0	0	0	0	0	0	0	0	0	0	0	0	0	0	0	0	0	0	0	0	0	0	0
0	0	0	0	0	0	0	0	0	0	0	0	0	0	0	0	0.1	0	0.1	0.1	0	0.3	0	0.8
(0.1)	(0.3)	(0)	(0)	(0.1)	(0)	(0)	(0)	(0)	(0.1)	0.1	0.1	2.8	0	0.2	0	0	0.2	0	0.3	0	0.1	0	0
0	0	0	0	0.2	0	0.1	0.1	0	0.3	0.2	0.2	0.7	0	2.7	0	0.9	0.1	0.1	0.1	0.4	0.9	0	0.6
0.3	1.1	0.1	0	0.2	0	0	0	0	0	0	0.2	0	0.1	0.5	0	0.7	0	0.7	0	0.1	3.7	0.3	0.7
0.2	0	0.1	0	0.1	0	0	0	0	0	0	0.1	0	0.1	0	0	0	0	0	0	0	0	0	0.5
0	0.9	0.5	0	0.2	0	0	0	0	0.1	0	0	0	0.1	0	0	(0)	(0.5)	(0)	(0.5)	(0)	(0)	(0)	(0)
(0)	(0)	(0)	(0)	(0)	(0)	0	0	0	0	0	0.1	0	0	0	0	0	0.1	0	0.3	0	3.4	0	1.6
0	0	0	0.1	0	0	0.1	0	0.1	0	0.1	0	0	0.1	0	0	0	0	0	0	(0)	(0.6)	(0)	(0.5)
(0)	(0)	(0)	(0)	(0)	(0)	0	0	0	0	0	0	0	0	0	0	0.1	0.6	0.1	1.1	0	0.5	0	0.1
0	0	0	0	0	0	0	0	0	0	0.1	0	0.1	0	0.3	0	2.3	0	0	5.2	0	19.5	0.1	2.8
—	—	—	—	—	—	—	—	—	—	—	—	—	—	—	—	—	—	—	—	—	—	—	—
—	—	—	—	—	—	—	—	—	—	—	—	—	—	—	—	—	—	—	—	—	—	—	5.0
(0)	(5.0)	(0)	(0.5)	(0)	(0.2)	(0)	(0.1)	3.3	0	4.7	0	5.8	0.1	5.0	0.6	0	11.0	0	10.0	0	15.0	0	15.0
0.3	6.0	0.1	0.7	0	0.5	(0)	(0)	0.7	0	3.7	0	4.0	0	6.3	0.5	0.9	6.0	0	14.1	0	12.9	0	5.6
0	6.5	0	0.4	0.2	0	0	0	0.4	0	0.1	0.1	0	0	2.3	0.8	0	4.6	0.7	0.9	0.6	0.1	0	0.4
0	0.1	0	0.1	0.1	0.1	0.2	0	0.2	0	0	0	0.1	0	0	0.1	0.2	0	0	0	0	0.1	0	0.3
0	0.2	0	0	0	0.1	0	0	0	0	0.1	0	0	0	0	0	0.3	0	0	0	0	0.3	0	0.2
0	0.1	0.2	0	0	0	0	0	0	0	0	0	0	0	0	0	0	0	0	0	0	0	0	0
0	0	0	0	0	0	0	0.1	0.1	0	0.1	0	0.3	0	0.5	0	0.6	0.1	0	0.6	0	0.2	0	0.1
0	0.2	0	0	0	0	0	0	0	0	0.1	0	0.1	0.1	1.3	0	0.6	0.2	0	1.8	0	1.9	0	0
0	0.2	0	0.2	0	0.1	0	0	0	0.1	0.3	0	0.1	0.1	0.5	0	0.2	1.3	0.1	0.3	0	0	0	0
0	0	0.3	0	0.2	0	0	0	0	0	0	0	0	0	0.1	0	0.1	0	0	0.3	0	0.3	0	0
0	0	0	0	0	0	0	0	0	0	0	0.1	0	0.1	0	0	0.1	0	0	1.6	0.1	0.2	0	0.1
0	0	0	0	0	0	0	0	0	0	0	0	0	0	0	0.1	0	0	0	0	0	0	0	1.3
0.2	0.6	0	0	0	0	0	0	0	0	0	0	0	0.1	0	0	0.3	0	0	0	0	0	0	0
0	0	0	0	0	0	0	0	0	0	0	0	0	0	0	0	0	0	0	0.4	0	2.0	0.1	0.4
0	0.1	0	0	0	0	0	0	0	0	0	0	0	0	0	0	0	0	0	0.3	0.1	0.3	0	0.2
0	0	0	0	0	0	0.4	0	0.2	0	0.1	0	0.1	0	0.3	0	1.2	0.9	0.8	0	0.3	0.1	0	0.1
0	0.1	0.1	0.3	0.3	0.1	0	0	0	0.2	0.6	0.1	0.6	0	0	0	0	0.1	0.1	0.7	0.1	0.7	0	0.4
0	0	0	0.1	0	0	0	0	0	0	0	0.1	0.1	0	0.1	0	1.4	0	0.4	0.3	0.1	1.9	0	2.3
0	0.1	0.1	0	0	0	0	0	0.1	0	0	0	0	0	0.1	0	0.5	0	(0.1)	0.5	(0)	(1.6)	(0)	(2.0)
0	0.3	0	0.1	0	0	0	0	0	0	0	0.2	0	0	0	0	0.2	0.1	0.2	0	0	1.7	0	0.5
0.1	0	0	0.2	0	0	0	0	0	0	0	0	0	0	0	0	0.2	0	0.1	0	0	0	0	0.1
0	0.8	0.1	0.1	0	0	0	0	0	0	0	0	0	0	0	0	0	0	0.2	1.9	0.2	0.2	0	0
0	0	0	0	0	0	0	0	0	0	0	0.1	(0)	(0)	(0)	(0)	(0.2)	(0)	(0.1)	(0)	(0)	(1.5)	(0)	(0.5)
(0)	(0)	(0)	(0)	(0)	(0)	(0)	(0)	0	0.1	0	0	0	0	0	0	0	0	(0)	(0.3)	0	0.4	0	0
0	0	0	0	0	0	0	0	0	0	0	0	0	0	0	0	0	0	0	0	0	0	0	0.4
0	0.2	0	0	0	0.1	0	0	0	0	0	0.2	0	0.4	0	0.1	0.3	0.1	0	1.1	0	1.9	0	0.1
0	0	0	0	0	0	0	0	0	0	0	0	0	0	0.2	0.1	0	0.1	0	0	0	0	0	0
0	0	0	0	0	0	0	0	0	0	0	0	0	0	0	0	0.2	0	0.2	0	0.1	0	0	0
0	0	0	0	0	0	0	0	0	0	0.1	0.2	0	0.3	0	0.1	0.3	0	0.2	0.2	0	4.7	0	9.0
0	6.5	0.5	2.5	0.3	0.5	0.2	0.2	3.5	0	2.6	0	7.3	0	5.3	0	1.1	2.3	0	13.3	0	8.1	0.3	0.1
0.1	0.2	0.1	0.3	0.1	0	(0)	(0)	(0)	(0)	(0.1)	(0.1)	(0)	(0.1)	(0.2)	(0)	(0.4)	(0.1)	(0.1)	1.5	(0)	(2.5)	(0)	(0.5)
(0)	(0.2)	(0)	(0.2)	(0)	(0)	0	0	0	0.1	0	0.4	0	0.2	0	0.1	0.1	0.1	0	2.1	0	0.4	0	0.9
0.1	0.1	0	0.1	0	0	0	0	0	0	0.2	0.1	0.2	0	0.1	0.1	0	0.1	0	0.7	0	3.8	0	2.3

TABLE LXIV (continued). F_{II} Matotchkin Schar.

Cr. M.-T.	0—2		2—4		4—6		6—8		8—10		10—12		12—14		14—16		16—18		18—20		20—22		22—24	
Date	+	−	+	−	+	−	+	−	+	−	+	−	+	−	+	−	+	−	+	−	+	−	+	−
December 27	0	0.1	0	0.2	0	0	0	0.1	0	0.1	0	0	0.1	0	0	0	0.2	0	0.1	0.1	0	2.0	0	4.2
28	0	1.4	0.2	0	0.2	0.2	0.1	0.2	0	0.6	0.1	0.2	0.2	0.1	0.3	0	0.3	0.1	0	0.1	0	0.7	0	0.1
29	0	0	0.4	0	0.4	0	0	0.1	0	0	0	0	0	0	0	0	0.2	0.1	0.1	0.2	0	0.3	0	0
30	0	0	0.2	0	0	0	0	0	0	0.1	0	0.2	0	0	0	0	0.5	0	0.2	0	0.2	0	0	0.1
31	0	0	0	0	0	0	0	0	0	0	0	0	0	0	0	0	0.2	0	0.1	0	0	0	0	0.1
January 1	0	0	0	0	0	0	0	0	0	0	0	0	0	0.1	0	0	0	0.2	0	0	0	0	0	0.4
2	0	0.2	0	0	0	0	0	0	0	0	0.2	0	0.2	0	0	0	0	0	0	0	0	0	0	0.2
3	0	0	0	0	0	0	0	0	0	0	0	0	0	0	0	0	0.2	0	0.4	0.2	0.3	0.2	0.2	0
4	0	0.1	0	0.3	0	0.1	0.1	0.2	0.3	0	0.1	0	0	0	0	0	0.1	0.1	1.9	0	0	0.4	0	0.8
5	0	0.2	0.4	0	0.4	0	0	0.2	0.2	0.2	0.1	0.2	0.2	0.1	1.2	0	0.2	4.8	0	1.2	0	1.6	0	0.5
6	0	0	0	0.2	0	0	0	0	0	0	0	0	0	0.2	0	0.3	0.2	0	0.3	0.2	0	1.8	0	0.2
7	0	0	0	0	0	0	0	0.1	0	0	0	0.2	0	0.4	0	0.1	0.1	0.1	0.1	0.3	0	0	0	0
8	0	0	0	0	0	0.1	0	0	0.1	0	0.3	0	0.1	0	0.5	0	1.5	0	0	0.5	0	2.1	0	0.2
9	0	0	0	0	0	0.1	0	0.2	0	0	0	0	0.1	0	0.1	0	1.2	0	0.1	0.2	0	0.6	0	2.1
10	0.1	0.5	0.1	0	(0)	(0)	(0)	(0.1)	0	(0)	0	0	0.1	0.1	2.8	0	1.4	0	0.1	0.1	0	0.9	0	1.7
11	0	0.4	0	0.3	0.1	0	0	0	0	0	0	0	0.1	0	0.3	0	0.3	0	0.6	1.1	0.1	1.0	0	0.7
12	0	0.3	0.1	0	0	0	0	0	0	0	0	0.1	0.1	0	0.4	0	0.3	0	0	1.2	0	0	0	0.6
13	0.1	0.3	0.2	0	0	0	0.1	0	0	0	0	0	0	0	0	0	0.8	0	0.7	1.3	0.1	0.6	0	0.3
14	0	0	0	0	0	0	0.1	0	0	0	0	0	0	0	0	0	0	0	0	0	0	0.1	0	1.2
15	0	0.1	0	0	0	0	0	0.1	0	0.1	0	0	0	0.1	0.1	0	0.1	0	0	0.2	0	0.1	0	0
16	(0)	(0.1)	(0)	(0.1)	(0)	(0)	0	0	0	0.3	0.1	0	2.5	0	0.3	0.1	1.5	0	0.3	0.3	0	1.0	0	1.0
17	0	0.3	0	0.1	0	0	(0)	(0)	(0)	(0.1)	0	0.1	0	0.2	0	0.2	0	0	0	0.2	0	0.2	0	0.1
18	0	0	0.1	0	(0)	(0)	(0)	(0)	(0)	(0)	0	0	0	0.1	2.0	0.1	2.2	0	1.2	0	0.5	0	0	0.1
19	0	0.1	0	0.2	0	0	0	0	0	0.1	0	0.4	0.4	0	0.6	0.1	1.4	0.1	0.1	0.1	0	1.0	(0)	(0.3)
20	(0)	(0.1)	(0)	(0.1)	(0)	(0)	0	0.1	0.1	0	0	0.1	0	0.1	0	0.2	0	0	0.1	0	0.8	0.1	0	0
21	0	0	0	0	0	0.2	0	0.1	0	0	0	0	0	0.4	0.1	0.1	0	0	(0.1)	(0.2)	0	0.8	0	2.2
22	0	1.0	0.1	0	0	0	0	0	0	0	0	0	0	0.1	0	0.1	0	0.1	0.1	0.2	0	0	0	0.1
23	0	0	0	0	0	0	0	0	0	0	0.3	0	0.1	0	0.5	0	1.2	0.2	0.3	3.8	0	5.3	0.1	0.2
24	0	0	0.1	0.2	0.2	0	0	0.2	0	0.2	0	0	0	0.1	0	0.2	0.6	0	0.1	0.6	0	0.6	0	0.7
25	(0)	(0)	(0)	(0)	(0)	(0)	(0)	(0.1)	0	0.1	0	0.2	0	0.1	0	0.2	0.2	0.1	0	0	0	0	0	0
26	0	0	0	0	0	0	0	0	0	0	0.1	0	0.5	0	0.1	0	0	(0.5)	0	(26.5)	0	26.6	0	40.0
27	0	22.1	0.1	2.4	0.9	1.1	0.7	0.2	0.4	0.1	0.1	0.2	0	0.1	0.1	0	0.3	0	0.1	0.6	0	1.5	0	1.6
28	0	0.2	0	0.2	0	0	0	0	0	0	0.1	0	0	0.2	0	0	1.0	0	0.6	0.2	0	0.8	0	0
29	0	0.1	0	0.1	0	0	0	0	0	0	0.1	0	0.2	0	0.1	0.1	0	0	0	0	0	0	0	0
30	0.1	0	0.1	0	0.1	0	0	0.1	0.1	0.2	0.9	0	2.9	0	3.4	0	5.0	0	2.1	0	0.3	0.1	0	0
31	0	0.3	0.1	0	0.1	0	0	0	0	0	0	0	0	0	1.1	0	1.7	0	0.6	0	0	0	0	0
February 1	0	0	0	0	0	0	0	0	0	0	0	0	0	0	0	0	0.6	0	1.0	0	0.1	0	0	0
2	0	0.1	0	0	0	0	0	0	0	0	0	0.1	0	0.1	0	0	0	0	0.2	0	0.2	0	0	0
3	0	0	0	0	0.2	0	0	0	0	0	0.1	0	0.2	0	0	0	0	0	0	0	0	0	0	0
4	0	0	(0)	(0)	(0.1)	(0)	(0)	(0)	(0)	(0)	(0)	(0)	(0)	(0)	(0)	(0)	(0.5)	(0)	(0.5)	(0)	(0.1)	(0)	(0)	(0)
5	0	0	0.1	0.1	0	0	0	0	0	0	0.1	0	0.2	0	0.4	0	1.2	0	0.6	0	0	0.3	0	0.1
6	0	0	0	0	0.1	0.1	0.1	0	0	0	0	0	0	0.2	(0.2)	(0)	(0.5)	(0)	(0.4)	(0)	0	0	0	0
7	0	0	0	0.1	0	0	0	0	0	0	0	0	0.2	0	0.1	0	0.7	0	0.1	0	0	4.7	0	9.4
8	0.1	0.4	0.1	0.6	0.3	0.5	0.2	0.5	0.5	0	0	0	0.2	0	4.0	0	4.8	0	0.3	11.3	0	14.8	0.2	0.8
9	0	0.8	0.1	0.5	0.1	0.4	0.2	0.1	0.2	0	0.1	0	0.1	0	0	0	0.2	0	0.2	0.5	0	4.0	0	4.2
10	0.1	0.4	0.1	0	0	0	0	0.1	0	0.1	0	0	0.1	0	(0.2)	(0)	0	0	0	0.1	0	1.3	0	0.0
11	0.3	0.5	0.2	0	0.1	0	0	0.1	0	0.1	0.2	0	0.1	0.1	0.1	0.1	1.0	0	0.6	0.7	0	1.5	0	3.0
12	0	0.6	0	0.1	0	0	0	0	0	0	0	0.1	0	0	0	0	1.1	0	0.5	1.2	0	3.3	0	1.0
13	0	1.9	0.2	0.1	0.1	0.1	0	0.2	0.2	0	0.4	0	0.2	0	0.3	0	0	0.1	0	3.7	0	0.5	0	1.1
14	0.1	0.4	0.2	0.1	0	0	0	0	0	0	0	0.1	0	0.1	0.4	0	0	0.1	0.3	0.1	0	1.5	0	2.0

LXIV (continued). F_H Matotchkin Schar.

T.	0—2		2—4		4—6		6—8		8—10		10—12		12—14		14—16		16—18		18—20		20—22		22—24	
	+	−	+	−	+	−	+	−	+	−	+	−	+	−	+	−	+	−	+	−	+	−	+	−
15	0	0.3	0	0.1	0	0	0	0.1	0.1	0	0.1	0.3	0.6		3.7	0	0.8	3.6	0.9	0	0	0.2	0	0
16	0	0	0	0	0	0	0	0	0	0.1	0.1	0	0.1		0.1	0	0.2	0.4	0.1	0.5	0	0.3	0	0.1
17	0.1	0.2	0	0	0	0.1	0	0	0	0	0.1	0	0.1	1	0.5	0	0.2	0	0	0	0	0.2	0	0.1
18	0	0.4	0	0	0	0	0	0	0	0	0	0	0		0	0	0	0	0	0.1	0	0.1	0	0
19	0	0	0	0	0	0	0	0	0	0.2	0	0	0.1	8.	0	0.3	0	.3	0	0.1	0	0	0	0.1
20	(0)	(0.1)	(0)	(0.2)	(0)	(0.2)	(0)	(0.2)	0	0	0.1	0	0	0.1	0	0.1	0	0	0	0	0	0	0	0
21	0	0	(0)	(0)	(0)	(0)	(0)	(0)	0	0.1	0	0	0.4	0	0.2	0.1	0.5	0	0	0	0	0	0	0
22	0	0	0	1.4	0	1.8	0	2.0	1.0	0.2	2.2	0	0.6	0.5	0.1	0.1	0.3	0	0.1	0.1	0	0.4	0	0.4
23	0	0.1	0	0	0	0	0	0	0	0	0	0	0	0.1	0	0.1	0	0.1	0.1	0.3	0	0.8	0	0.3
24	0	0.1	0	0	0	0	0	0	0	0	0	0	0	0	0	0	0.8	0	0.2	0	0	0	0	0
25	0	0	0	0.3	0	0.8	0.1	0.6	0.6	0.1	1.1	0	0.1	0.2	0	0.3	0	0	0	0	0	0	0	0
26	0	0	0	0	0	0	0	0	0	0	0	0	0	0	0	0.1	0	0	0	0	0	1.1	0	0.5
27	(0)	(0)	(0)	(0.1)	(0)	(0.2)	0	0	0	0	0	0	0	0	0	0	0	0	0	0	0	0	0	0
28	0	0	0	0	0	0	0	0	0	0	0	0	0	0	0	0	0	0	0	0	0	0	0	0
1	0	0	0	0	0	0	0	0.1	0.1	0	0	0	0.1	0	0.3	0	2.0	0.1	1.5	0	0.4	0.1	0	2.3

TABLE LXV.

Disturbances in Declination (F_D).

T.	0—2		2—4		4—6		6—8		8—10		10—12		12—14		14—16		16—18		18—20		20—22		22—24	
	+	−	+	−	+	−	+	−	+	−	+	−	+	−	+	−	+	−	+	−	+	−	+	−
r 3	0	0	0	0	0	0	0	0	0	0.3	0.1	0.2	0	0.2	0	0.1	0	0.3	0.1	1.1	0	0.4	0	0
4	0.2	0	0	0	0.5	0	0.2	0.1	0	0.5	0.4	0.1	0.1	0.1	0.3	0.1	0.2	0	0.1	0	0.1	0.1	0.1	0.2
5	0.2	0.1	0.3	0.2	0.4	0.4	0	0	0.2	0.1	0.1	0	0	0	0	0	0	0	0.1	0	0.1	0.1	0	0
6	0	0	0	0	0.1	0.2	0	0	0.2	0	0	0	0	0	0	0.4	0	0	0	0	0.1	0	0.1	0
7	0.1	0	0.2	0	0	0	0	0	0	0	0	0	0	0	0	0	0	0	0	0	0	0	0.1	0.1
8	0.2	0	0.1	0.1	0	0	0	0	0	0	0.2	0	0	0	0	0	0	0	0	0.9	0.1	0.1	0.1	0.1
9	0.1	0	0	0	0	0	0.1	0	0	0	0.1	0.1	0.1	0	0	0.5	0.1	0.3	0.1	0.1	0	0	0	0
10	0	0	0	0	0	0	0	0	0	0	0	0	0	0	0	0	0	0	0	0	0	0	0.1	0.1
11	0	0	0.5	0	0.1	0.1	0	0.3	0.2	0	0.1	0.1	1.3	0	0.3	0	1.7	8.0	0.1	2.2	0	17.8	0	6.0
12	0.6	0.2	0.1	0.3	0	0	0	0	0.1	0.1	0.1	0	0	0	0.1	0.2	0.1	0	0	0	0	0	0	0
13	0	0	0	0	0	0	0	0	0	0	0.1	0.5	0	.8	0.1	.2	0.2	0.5	0	1.9	0	0.5	0	0
14	0	0	0	0	0	0	0	0	0	0	0	0	0		0.1		0	0	0.1	0.2	0.1	0	0.2	0.1
15	0	0	0	0	0	0	0.1	0	0.7	0	0.2	0	0		0		0	0.2	0.3	0.5	0.3	0.2	0.2	0
16	0	0	0	0	0	0	0	0	0	0	0	0	0		0		0	0	0	0	0	0	0	0.1
17	0	0	0	0	0	0	0	0	0	0	0	0.1	0	8	0	8	0	0	0.1	0	0	0	0	0.1
18	0	0.2	0.2	0.1	0.4	0.1	0	0	0.2	0	0	0	0	0	0	.1	0.1	0	0.1	4.3	0	1.5	0	0.4
19	0	0.5	0.7	0	0.2	0	0	0.1	0	0	0	0	0	0	0		0.1	0	0	0.1	0	0	0	0
20	0	0	0	0	0	0	0	0	0.1	0	0.3	0.1	0	0.1	0		0	0	0	0	0	0.3	0.1	0
21	0	0	0	0	0	0.1	0	0	0	0	0	0	0	0	0.2		0.9	0.2	0.1	0.4	0	0.4	0	0.3
22	0	0.1	0	0.1	0	0	0	0.1	0	0	0	0	0	0	0	8.1	0	0	0	0	0	0	0	0
23	0	0	0	0	0	0	0	0	0	0	0	0		0	0	0	0	0	0.1	0	0	1.5	0.1	1.5
24	0	1.3	0.3	0	0.1	0	0	0	0.1	0	0.1	0		0	0.3	0	1.3	0	0.4	3.4	0	16.0	0	13.0
25	0	8.3	0	1.6	0	0.2	0.2	0	0	0.2	0.7	0.1	8.5	0.1	0.1	0.3	0.1	0.8	0	0.2	0	0.2	0	0.2
26	0	0.4	0.2	0	0	0	0	0	0	0.1	0	0.4		0.1	0.3	0.3	0.6	0.3	0	1.8	0.1	0.1	0	0.5
27	0	2.4	0.4	0.4	0.2	0	0.5	0	0.8	0.1	0	0.4	8.2	0.3	1.6	0.5	0.7	2.5	0.1	6.3	0	3.3	0	6.1
28	0	0.6	0	0.5	0.1	0.3	0	.2	0.3	0	0.1	0	0	0.3	0.2	0.3	0.1	0.5	0	3.0	0	2.3	0	1.2
29	0.1	0	0	0	0	0	0.1		0.2	0.1	0.3	0	0	0.1	0	0.1	0.1	1.5	0	9.4	0	7.9	0	3.6
30	0.1	1.0	0	0.3	0	0.1	0.7	.1	3.6	0	2.9	0	0	0	0.4	0	0.2	0.5	0.1	0.6	0	1.7	0.2	0.1
31	0.1	0.5	3.4	0	3.	0	1.7		3.0	0	4.7	0	9.7	2.5	12.3	0	1.5	3.0	0	13.3	0	26.0	0	21.3
r 1	0	9.0	0	1.8	0.3	0	0.2	8.3	0.2	0	0.1	0.4	0	0	0	0	0	0	0	0	0.1	0.2	0	0

TABLE LXV (continued). F_D Matot[...]

Gr. M.-T.	0–2		2–4		4–6		6–8		8–10		10–12		12–14		14–16		16–18		18–20		20–	
Date	+	−	+	−	+	−	+	−	+	−	+	−	+	−	+	−	+	−	+	−	+	
November 2	0	0	0	0	0	0	0.1	0	0	0.1	0	0.4	0	0.1	0.1	0	0.6	0.7	0	6.6	0	
3	0	0.4	0	0	0	0	0.2	0	0	0	0	0	0	0	0	0	0.1	0.1	0	0.1	0.2	
4	(0)	(0.1)	(0)	(0)	(0)	(0)	(0.1)	(0)	(0)	(0)	(0)	(0.1)	(0)	(0)	(0)	(0)	(0)	(0)	(0)	(0)	(0)	(
5	(0)	(0.1)	(0)	(0)	(0)	(0)	(0)	0	0	0	0	0	0	0	0	0	0	0	0	0.1	0	
6	0	0	0	0	0	0	0	0	0	0	0	0	0	0	0	0	0.7	0.2	1.1	4.0	0	
7	0	0.4	0	0	0	0	0.1	0	0	0	0	0	0	0	0	0	0	0	0	0	0	
8	0	0.7	0	0.2	0	0	0.1	0	0	0	0	0	0	0	0	0.3	0.1	0	0.1	0.1	0.2	
9	0	0	(0)	(0)	(0)	(0)	0	0	0	0	0	0	0	0	0.2	0	0	0	0	0	0	
10	0	0.1	0	0	0	0	0.1	0	0.2	0	0	0	0	0.2	0	0.1	0.1	0	0.1	0	0.3	
11	0	0	0.1	0	0	0	0	0	0	0	0	0	0	0	0	0	0	0	0	0	0	
12	0	0	0	0	0	0	0	0	0	0	0	0	0	0	0	0.1	0	0	0.1	0	0	
13	(0)	(0.5)	(0)	(0.2)	(0.1)	(0)	(0)	(0)	(0)	(0)	0.3	0	0.9	0.1	0.1	0.1	0	0.6	0.1	0.4	0	
14	0	0	0	0	0.6	0.1	0.2	0	0.2	0	0.4	0	1.0	0	1.4	0	0.7	0.4	0.1	0.7	0.2	
15	0	1.2	0	0.7	0	0.1	0	0.1	0	0	0.1	0.3	0.3	0	0.7	0	0.2	0.3	0.3	0.3	0.1	
16	0	0.7	0	0.3	0	0.1	0	0	0	0	0	0	0	0.2	0	0	0	0	0	0	0	
17	0	0.8	0.1	0.3	0.4	0.1	0.1	0	0	0	0	0	0	0	0	0	0	0.5	0	0.4	0	0
18	0	0	0	0	0	0	0	0	0	0	0	0	0	0	0	0	0	0.3	0	0.7	0	1
19	0	0.6	0	0	0	0	0.2	0	0	0.1	0.3	0.1	0.1	0.1	0.2	0	0.3	0	0.2	0.1	0	1
20	0	0	0	0	0	0	0	0	0	0	0	0	0	0	0	0.1	0.2	1.4	0	2.6	0	1
21	0	0	0	0	0	0	0	0	0	0	0	0.1	0	0.3	0.2	0.2	0.2	0.5	0.1	3.3	0	12
22	(0)	(1.0)	(0)	(0.5)	(0.2)	(0)	(0)	(0)	(0.2)	(0)	(1.0)	(0)	(1.0)	(0.1)	(2.0)	(0.2)	(1.5)	(5.0)	(0)	(5.5)	(0)	(8.
23	(0)	(2.0)	(0)	(1.0)	(0.5)	(0.5)	(1.0)	(0.2)	0.3	0.1	2.0	0.1	0.2	0.2	0.1	0.3	0.5	2.4	1.1	0.9	1.3	1.
24	0	11.2	0.3	7.4	1.2	0.1	1.7	0.1	1.7	0	2.0	0	4.0	0.2	10.0	0.1	0.7	11.0	0	8.5	0	22.
25	0.2	5.2	0	1.2	0.2	0.5	0.6	0.1	1.3	0	3.0	0	3.0	0	2.3	0.1	3.7	2.0	0.1	11.5	0	13.
26	0	4.2	0	0.5	0	0.6	0	0.4	0.3	0	0	0.2	0	0	0.8	0.4	0.1	3.5	0.1	1.2	0.1	0.
27	0.1	0.1	0.1	0	0.1	0.2	0.1	0.2	0.1	0	0	0	0.1	0	0	0.1	0	0.2	0	0.1	0	0
28	0	0	0	0.1	0	0.1	(0.1)	(0)	0	0.1	0.1	0.1	0	0.3	0	0.3	0	0.9	0.1	0.1	0	0.
29	0	0	0	0.1	0	0.2	0	0	0	0	0.4	0	0.1	0.1	0	0	0	0	0	0	0	0
30	0	0	0	0	0	0.1	0.2	0	0	0.1	0.1	0.1	0	0.3	0.2	0.1	0	1.3	0	1.7	0	1.
December 1	0	0	0	0	0	0	0	0	0	0	0	0.2	0	0.3	0.5	0.1	0	3.0	0	1.7	0	2.
2	0	0	0	0	0.1	0	0	0	0.1	0	0.8	0	0.1	0.2	0.2	0.1	(0)	(0.7)	(0)	(2.5)	0	0
3	0.1	0	0	0.1	0	0	0.3	0	0	0	0	0	0	0	0.1	0.1	0	0	0.2	0.1	0	0.
4	0	0	0	0	0	0	0	0	0	0	0	0	0	0.1	0	0	0	0.7	0	3.1	0	0.
5	0	0	0	0	0	0	0	0	0	0.3	0.1	0	0	0	0	0	0	0	0	0.1	0	0
6	0.3	0.2	0	0	0	0	0	0	0	0	0	0	0	0	0	0.1	0	0.3	0	0	0	0
7	0	0	0	0	0	0	0	0	0	0	0	0	0	0	0	0	0	0	(0)	(0.5)	0	1.
8	0	0	0	0.1	0	0	0.1	0	0	0.1	0	0	0	0	0	0	0	0	0	0.2	0	1.
9	0	0	0	0	0	0	0.4	0	0.4	0	0.5	0	0.1	0	0.2	0	1.0	0.7	0.2	1.0	0	0.
10	0.1	0	0.2	0.1	0.4	0.3	0	0	0.1	0.2	0.1	0.3	0.5	0.2	0	0.1	0	0	0.2	1.3	0	1.
11	0	0.1	0	0	0	0	0.1	0	0.1	0	0	0.2	0.2	0	0.3	0	0.4	1.5	0.4	1.0	0.8	1.
12	0.1	0.2	0	0.2	0	0	0.1	0	0	0	0	0	0	0	0.2	0	(1.0)	(1.0)	(0.2)	(1.0)	(0)	(1.
13	0.1	0.1	0	0	0	0	0	0	0	0.1	0	0.4	0	0.2	0	0.1	0.4	4.0	0.3	0.2	0.1	1.
14	0	0.1	0	0	0	0	0	0	0	0	0	0	0	0	0	0	0.1	0.1	0.1	0	0	0
15	0.1	0.1	0	0.2	0	0	0	0	0	0	0	0	0	0	0	0	0	0	0	0.8	0	0.
16	0	0.1	0	0	0	0	0.4	0	0.1	0	(0)	(0.2)	(0)	(0.1)	(0.1)	(0)	(0.4)	(1.5)	(0.2)	(0.5)	(0)	(1.
17	(0)	(0)	(0)	(0)	(0.1)	(0)	(0)	(0)	0	0	0	0	0	0	0	0	0	0	0.1	0.1	0.1	0
18	0	0	0	0	0	0	0	0	0	0	0	0	0.1	0	0	0	0	0	0	0	0	0
19	0	0	0	0	0.1	0	0	0.1	0.1	0	0	0.1	0	0.1	0	0.2	0.2	0	0.1	0.7	0.1	1.
20	0	0	0	0.1	0.2	0.1	0.1	0	0	0	0	0	0	0	0.2	0.3	0	0	0	0	0	0
21	0	0	0	0	0	0	0	0	0	0	0	0	0	0	0.2	0	0.6	0	0.2	0	0	0

TABLE LXV (continued). F_D Matotchkin Schar.

Gr. M.-T.	0—2		2—4		4—6		6—8		8—10		10—12		12—14		14—16		16—18		18—20		20—22		22—24	
Date	+	−	+	−	+	−	+	−	+	−	+	−	+	−	+	−	+	−	+	−	+	−	+	−
December 22	0	0	0	0	0	0	0	0	0	0	0.1	0	0.2	0.1	0.2	0.1	0	0	0	1.1	0	3.3	0	5.5
	0.1	4.3	1.1	0.1	0.7	0	0.1	0.4	1.1	0.3	2.2	0	1.3	0.9	1.7	0.2	0.7	2.9	0	9.5	0	9.1	0.3	0.1
	0.3	0.1	0.4	0	0.1	0.1	(0)	(0.1)	(0.2)	(0.1)	(0.3)	(0.1)	(0.4)	(0.4)	(0.5)	(0.1)	(0.5)	(0.5)	(0.2)	(1.0)	(0)	(2.0)	(0.2)	(0.1)
	(0.1)	(0.5)	(0.4)	(0)	(0.2)	0.1	0	0	0.4	0	0.3	0.1	0.1	0	0	0	0	0.3	0	1.1	0.1	0	0.1	0
	0	0	0.1	0.1	0.1	0.1	0	0.1	0.1	0	0.1	0.2	0.1	0.5	0.2	0.2	0	1.0	0.2	0.1	0.1	3.7	0.3	0.1
	0.2	0.1	0	0.1	0	0.1	0	0	0.1	0	0	0.2	0	0	0	0	0.2	0	0.3	0.3	0	0.6	0.1	2.1
	0.4	0.2	0.3	0.1	0.2	0.1	0.3	0.1	0.2	0.3	0.1	0.2	0.1	0.2	0	0	0	1.0	0	0.2	0.2	0.2	0	0.1
	0	0	0.1	0	0	0	0	0	0	0	0	0	0	0	0	0	0.2	0.9	0.2	0.2	0.1	0	0	0
	0	0	0	0	0	0	0	0	0	0.1	0	0	0	0	0	0	0.2	0.2	0.3	0	0.1	0	0.1	0
	0.1	0	0	0	0	0	0	0	0	0	0	0	0	0	0.1	0	0.1	0.1	0.2	0	0	0	0	0.1
January	0	0	0	0	0	0	0	0	0	0	0	0	0	0	0	0	0.2	0.1	0	0.1	0	0	0.4	0
	0	0.3	0	0	0	0	0.1	0.1	0	0	0.2	0	0	0	0	0	0	0	0	0	0.1	0	0	0
	0.1	0	0.1	0	0	0	0	0	0	0	0.1	0	0	0	0	0	0.1	0	0.2	0.5	0.4	0.1	0	0
	0	0.2	0.1	0.	0.2	0	0.5	0.2	0.4	0	0.3	0	0.1	0	0	0	0.1	0	0.4	0.7	0.3	0.1	0	0.2
	0	0.6	0.2	0.	0.2	0.	0	1.2	0.2	0.5	0.1	0.2	0	0.6	0.2	0.4	0.5	2.5	0	1.5	(0.2)	(0.1)	0	0
	0.1	0	0.6	0	0.1	0	0	0	0	0	0	0	0	0	0	0.4	0	0.5	0	1.5	0.2	0.2	0.1	0
	0	0	0	0	0	0	0	0	0	0	0	0	0	0	0	0	0	0.3	0.1	0.3	0	0	0	0
	0	0	0	0	0	0	0	0	0	0.1	0.1	0.1	0	0	0	0	0.1	1.0	0	1.3	0	1.7	0	0.1
	0	0	0	0	0	0	0.1	0	0	0.1	0	0.1	0	0	0.1	0	0	1.3	0	1.9	0	1.4	0	2.1
	0	0.2	0	0	(0)	(0)	(0)	(0)	(0)	(0)	0	0	0.1	0.1	0.4	0.1	0.1	0.4	0	0	0.1	0.1	0.1	0.3
	0.6	0	0	0	0	0.1	0	0	0	0.1	0	0.1	0	0.1	0	0.8	0.1	0.4	0.4	2.4	0.6	0.8	0.5	0
	0.2	0	0.1	0	0.1	0	0	0	0.1	0	0	0	0	0.3	0.1	0.2	0	0.3	0.3	0.9	0.1	0	0.2	0.1
	0.1	0.1	0	0.	0	0	0	0	0	0	0	0	0.1	0	0	0.1	0.2	1.3	0.2	1.3	0	1.1	0.1	0
	0.1	0	0	0	0	0	0	0	0	0	0	0	0.1	0	0	0	0	0	0	0	0	0	0.2	0.1
	0.5	0	0	0	0	0	0	0	0	0.1	0	0.1	0.2	0.1	0.2	0.2	0	0.4	0.2	0.2	0	0.2	0	0
	(0.2)	(0)	(0.1	(0)	(0)	(0)	(0)	(0)	0	0.3	0	0.1	0.3	0.6	0	0.7	0.1	0.6	0.1	0.4	0.1	0.1	0.1	0.4
	0.2	0	0.1	0	0	0	(0)	(0)	(0.1)	(0.1)	0	0	0	0.1	0	0	0.1	0.1	0	1.0	0.2	0	0	0
	0.1	0	0	0	(0)	(0)	(0)	(0)	(0.1)	(0.1)	0	0	0	0	0.4	0.6	0.8	0.2	0.8	0.3	1.2	0	0.1	0
	0.2	0	0.1	0	0	0.1	0	0	0	0	0.2	0.3	0.1	0.1	0.9	0	0.6	0.9	0.3	0	0	0.2	(0)	(0.1)
	(0.1)	(0)	(0)	(0)	(0)	(0)	0	0	0.3	0	0.2	0	0	0	0	0	0	0	0	0.5	0.1	0.7	0	0
	0	0	0	0	0	0.1	0	0.2	0	0.1	0.1	0	0.1	0.4	0.2	1.0	0.4	0.1	(0)	(0.5)	0	0.6	0.6	0
	0.3	0.2	0	0	0	0	0	0	0	0	0	0	0	0	0	0.1	0.1	0.1	0	0.9	0	0.9	0.1	0
	0	0	0	0	(0)	(0.1)	(0)	(0)	0.1	0	0.1	0.2	0.1	0.3	0.1	0.4	0.2	2.3	0.1	3.5	0	5.4	0.3	0.3
	0.2	0	0.1	0.	0	0.2	0	0	0	0.1	0	0	0	0.1	0	0.2	0	1.3	0.1	1.1	0.1	0.3	0.1	0.1
	(0.1)	(0)	(0)	(0)	(0)	(0.1)	(0)	(0.1)	0	0	0	0	0	0	0	0.4	0	0.9	0	0.2	0	0	0	0
	0	0	0	0	0	0	0	0	0.1	0	0.3	0	0.6	0.2	0.2	0	(0.2)	(1.0)	(0)	15.0	0	18.1	0	28.0
	0	17.5	0	4.	0.3	1.4	0.7	0.1	0.1	0.2	0	0.1	0	0.3	0	0.2	0	1.3	0.1	0.6	0	0.7	0	0.5
	0.1	0	0	0	0	0	0	0.1	0	0.1	0.1	0	0	0.1	0.1	0	0.4	1.6	0.1	0.4	0.1	0	0.8	0
	0.2	0	0.1	0	0	0	0	0	0	0	0	0	0.1	0	0.1	0	0	0	0	0	0	0	0	0
	0.1	0.1	0	0	0	0	0	0.2	0.5	0	3.1	0	3.9	0	3.2	0	4.4	0	0.6	0.2	0	0.2	0	0.1
	0.1	0.2	0	0	0	0	0	0	0	0	0	0	0	0	0.4	0	1.4	0.1	0.3	0	0.1	0	0	0
February 1	0	0	0	0	0	0	0	0.1	0	0	0	0	0	0	0.2	0	0.6	0.1	0.2	0.2	0.1	0	0	0
	0.2	0	0	0	0	0	0	0	0	0	0.1	0	0.2	0	0.1	0	0	0.1	0	0	0	0	0	0
	0	0	0	0	0	0	0	0.1	0.5	0	0.1	0.1	0.2	0	0	0	0	0	0	0	0	0	0	0
	0	0	(0)	(0)	(0)	(0)	(0)	(0)	(0)	(0)	(0.1)	(0)	(0.1)	(0)	(0.2)	(0)	(0.5)	(0.1)	(0.1)	(0)	(0)	(0)	(0)	(0)
	0	0	0.1	0.1	0.1	0	0.1	0	0	0	0.3	0	0	0	0.1	0.1	0.4	1.3	0.1	0.5	0.3	0	0.3	0
	0	0	0.1	0	0.2	0.4	0.1	0.2	0.2	0	0	0	0	0.4	(0.1)	(0)	(0.2)	(0.6)	0.1	0	0	0	0	0
	0	0	0.1	0	0	0	0	0	0	0	0.1	0	0.1	0	0	0	0.2	0.5	0	0	0	3.2	0	7.6
	0	1	0.1	0.9	0.9	0	1.1	0	0.3	0	(0.1)	(0)	0.7	0	3.5	0	1.9	1.1	0.7	6.5	0.2	15.3	0	2.5
	0	0	0.2	0.3	0.2	0.5	0.1	0.2	0	0	0	0	0	0	0	0	0.1	0.1	0.1	1.6	0	2.4	0	2.7

:TABLE LXV (continued). F_D Matotcl

Gr. M.-T.	0—2		2—4		4—6		6—8		8—10		10—12		12—14		14—16		16—18		18—20		20—:	
Date	+	—	+	—	+	—	+	—	+	—	+	—	+	—	+	—	+	—	+	—	+	·
February 10	0	0.6	0.1	0.1	0.2	0	0.2	0.1	0	0	0	0	0	0	0	0	0	0	0	0.1	0.1	0
11	0	1.2	0.2	0.1	0.4	0	0.1	0.2	0	0.1	0.3	0	0.1	0	0.3	0	0.7	0	0.4	0.8	0	
12	0.1	0.3	0	0	0	0	0	0.1	0	0	0	0	0	0	0	0	0	2.0	0.2	2.4	0	
13	0.2	0.6	0.4	0.1	0.1	0.3	0	0.5	0.2	0	0.3	0	0.1	0.2	0	0	0	0.1	0	1.6	0.1	
14	0.2	0.2	0.1	0.2	0	0	0.1	0	0	0	0.1	0.2	0.1	0.1	0.1	1.1	0	0.3	0.1	0.1	0.2	
15	0	0	0	0	0	0	0	0	0.1	0	0.4	0	0.3	0	0.8	2.7	0.3	6.3	0	1.0	0	0
16	0	0	0	0	0	0	0	0	0	0	0	0	0	0	0.1	0	0	0.7	0	0.3	0.1	0
17	0.2	0	0	0	0	0.2	0	0	0	0.1	0.1	0	0.1	0	0	2.0	(0.1)	(0.3)	0	0	0.2	0
18	0.2	0	0	0	0	0	0	0	0	0	0	0	0	0	0.1	0	0.1	0	0	0	0	0
19	0	0	0	0	0	0	0	0	0	0.2	0	0	0	0	0	0.1	0	0.2	0.1	0	0	0
20	(0)	(0)	(0.1)	(0.1)	(0.3)	(0)	(0.3)	(0)	0	0	0.1	0	0	0	0	0	0	0	0	0	0	0
21	0	0	(0)	(0)	(0)	(0)	(0)	(0)	0	0	0.1	0	0.5	0	0.5	0.2	0.1	0.2	0	0	0	0
22	0	0.1	0.3	0.3	1.0	0.1	1.3	0.2	1.3	0	2.5	0	1.1	0	0.1	0.1	0.1	0.2	0.1	0.3	0	0
23	0	0	0	0	0	0	0	0	0	0	0	0	0.3	0	0	0	0.1	0.1	0.3	0.1	0	0
24	0	0	0	0	0	0	0	0	0	0	0	0	0	0	0	0	0.2	0.1	0.4	0	0	0
25	0	0	0.2	0	1.3	0	2.3	0	1.8	0	1.7	0	0.7	0.1	0.1	0.3	0	0.1	0	0.1	0	0
26	0	0	0	0	0	0	0	0	0	0	0	0	0	0	0	0	0	0	0	0.2	0	1
27	(0)	(0)	(0)	(0)	(0.3)	(0)	0	0	0	0	0	0	0	0	0	0	0	0	0	0	0	0
28	0	0	0	0	0	0	0	0	0	0	0	0	0	0	0	0	0	0	0.	0	0	0
March 1	0	0	0	0	0	0	0	0	0	0	0	0	0.3	0	0.3	0	1.0	0	0.7	0.6	0	0

TABLE L

Disturbances in Verti

Gr. M.-T.	0—2		2—4		4—6		6—8		8—10		10—12		12—14		14—16		16—18		18—20		20	
Date	+	—	+	—	+	—	+	—	+	—	+	—	+	—	+	—	+	—	+	—	+	
October 3	0	0	0	0	0	0	0	0	0	0	0	0	0	0	0	0	0	0	0	0	0	0
4	0	0	0	0	0	0	0	0	0.1	0.2	0	0.3	0.7	0	0.1	0	0	0	0	0	0	0
5	0	0	0.1	0	0.5	0	0	0	0.1	0	0	0	0	0	0	0	0	0	0	0	0	0
6	0	0	0	0	0	0	0	0	0	0	0	0	0	0.1	0	0.1	0	0.3	0	0	0	0
7	0	0	0	0	0	0	0	0	0	0	0	0	0	0	0	0	0	0	0	0	0	0
8	0	0	0	0	0	0	0	0	0	0	0	0	0.1	0	0	0	0	0	0.1	0.3	0	0
9	0	0	0	0	0	0	0	0	0	0	0	0	0	0	1.0	0	0.1	0.1	0	0.1	0	0
10	0	0	0	0	0	0	0	0	0	0	0	0	0	0	0.1	0	0	0	0	0	0	0
11	0	0	0	0	0	0	0	0	0	0	0	0	0.5	0	0	0	0.1	3.3	0	9.3	0	5
12	0	0	0	0	0	0	0	0	0	0	0	0	0	0	0.1	0	0	0	0	0	0	0
13	0	0	0	0	0	0	0	0	0	0	0	0	0	0	0.1	0	0.2	0.4	0	1.2	0	0
14	0	0	0	0	0	0	0	0	0	0	0	0	0	0	0	0	0	0	0	0.3	0	0
15	0	0	0	0	0	0	0	0	0	0	0.1	0	0	0	0	0	0.1	0	0	1.1	0	0
16	0	0	0	0	0	0	0	0	0	0	0	0	0	0	0	0	0	0	0	0	0	0
17	0	0	0	0	0	0	0	0	0	0	0	0	0	0	0	0	0	0	0	0	0	0
18	0	0	0	0.1	0	0	0	0	0	0	0	0	0	0	0	0	0	0	0	4.0	0	0
19	0	0.2	0	0.4	0	0.2	0	0	0	0	0	0	0	0	0	0	0	0	0	0	0	0
20	0	0	0	0	0	0	0	0	0	0	0	0	0	0	0	0	0	0	0	0	0	0
21	0	0	0	0	0	0	0	0	0	0	0	0	0	0	0	0	0.3	0	0	0	0.1	0
22	0	0.2	0	0	0	0	0	0	0	0	0	0	0	0	0	0	0	0	0	0	0	0
23	0	0	0	0	0	0	0	0	0	0	0	0	0	0	0	0	0	0	0	0	0	1
24	0	0	0	0	0	0	0	0	0	0	0	0	0	0	0	0	0.1	0	0	0.9	0	2
25	0	3.9	0.2	1.4	0.2	0	0.1	0	0.8	0	1.5	0	2.4	0	0.9	0	0.4	0	0.1	0	0	0
26	0	0.2	0	0.1	0	0	0	0	0	0	0.2	0	0.5	0	1.4	0	0.8	0	0.1	0.8	0	0
27	0.1	0.4	0	1.1	0.1	0.4	0	0.1	0.2	0	0.4	0	0.6	0	0.4	4.5	0	5.5	0.1	1.8	0.1	1

VI (continued). F_V Matotchkin Schar.

0—2		2—4		4—6		6—8		8—10		10—12		12—14		14—16		16—18		18—20		20—22		22—24	
+	—	+	—	+	—	+	—	+	—	+	—	+	—	+	—	+	—	+	—	+	—	+	—
0	0	0.1	0.2	0	0	0	0	0	0	0	0	0	0	0	0	0	0.2	0	0.9	0	0.7	0	0.6
0	0	0	0	0	0	0	0	0	0	0	0	0	0	0	0	0	0.6	0	4.5	0	4.4	0	1.9
0	0	0	0	0	0	0.1	0	1.3	0	0.6	0	0	0	0	0	0	0	0	0.1	0	0.8	0	0.3
0	0.2	0	0	0.1	0	0.9	0	1.6	0	0.4	0.6	0	2.0	0	2.0	0	2.0	3.0	0.6	9.4	0	4.9	0
0.4	0.3	0	0.7	0	0.4	0.1	0.1	0.1	0	0	0	(0)	(0)	(0)	(0)	(0)	(0.3)	(0)	(0.5)	(0)	(0.5)	(0)	(0)
0	0	0	0	0	0	0	0	0	0	0	0	0	0	0.3	0	0.3	1.0	1.7	0.9	0.1	1.5	0.1	0
0	0	0	0	0	0	0	0	0	0	0	0	0	0	0	0	0.3	0.1	0.1	0	0	0.8	0	0.1
(0)	(0)	(0)	(0)	(0)	(0)	(0)	(0)	(0)	(0)	(0)	(0)	(0)	(0)	(0.1)	(0)	(0.1)	(0.1)	(0)	(0.1)	(0)	(0.2)	(0)	(0)
(0)	(0)	(0)	(0)	(0)	(0)	(0)	0	0	0	0	0	0	0	0	0	0	0	0	0.1	0	0.2	0	0
0	0	0	0	0	0	0	0	0	0	0	0	0	0	0	0	0.1	0.8	0	9.5	0	3.6	0.1	0.1
0	0.3	0	0.1	0	0	0	0	0	0	0	0	0	0	0	0	0	0	0	0	0	0	0	0
0	0.1	0	0.4	0	0	0	0	0	0	0	0	0	0	0.1	0	0	0.3	0	0.1	0	0.4	0	0
0	0	0	0	0	0	0	0	0	0	0	0	0	0	0	0	0	0	0	0	0	0	0	0.1
0.1	0	0	0	0	0	0	0	0	0	0	0	0.2	0	0.1	0	0	0	0	0	0	1.5	0	2.2
0	0	0	0	0	0	0	0	0	0	0	0	0	0	0	0	0	0	0	0	0	0	0	0
0	0	0	0	0	0	0	0	0	0	0	0	0	0	0	0	0	0	0	0	0	0.3	0	1.6
(0)	(0.2)	(0)	(0)	(0)	(0)	(0)	(0)	(0)	(0)	0.4	0	2.4	0	0.1	0	0.2	0	0.1	0.3	0	0.2	0	0
0	0	0	0	0	0	0	0	0	0	0.1	0	0.3	0.1	0.3	0.2	0	0.7	0.2	0.1	0.1	1.2	0.1	0.4
0	0.6	0	0	0	0	0	0	0	0	0	0	0	0	0.3	0	0.3	0	0	0.2	0	1.7	0	1.2
0	0.2	0	0	0	0	0	0	0	0	0	0	0.8	0	0	0	0	0	0	0	0	0	0	0.3
0	0.6	0	0.1	0	0	0	0	0	0	0	0	0	0	0	0	0.2	0	0	0	0	0	0	0
0	0	0	0	0	0	0	0	0	0	0	0	0	0	0	0	0.2	0	0	0.2	0	3.1	0	2.5
0	0	0	0	0	0	0	0	0	0	0	0	0.5	0	0.1	0	0	0	0	0.2	0	1.6	0	1.2
0	0	0	0	0	0	0	0	0	0	0	0	0	0	0.1	0	0.5	1.0	0.2	1.6	0	0.7	0	0
0	0	0	0	0	0	0	0	0	0	0.1	0.1	0.1	0	0.3	0	0.9	1.6	0.9	0.8	1.8	3.6	0	1.7
(0)	(0)	(0)	(0)	(0)	(0)	(0)	(0)	(0)	(0)	(0)	(0)	(0)	(0)	(0)	(0)	(0.8)	(7.0)	(2.0)	(2.0)	(2.0)	(2.0)	(0)	(4.0)
(0)	(1.0)	(0)	(1.0)	(0)	(0)	(0)	(0)	0.1	0.1	3.0	0	0	1.0	0	0.2	0.1	1.7	0	1.7	0.2	1.8	0	15.0
0	15.5	0	11.5	0	7.6	0	5.0	0.2	0.6	0	2.4	0	11.0	0	15.5	3.0	10.0	5.8	0.2	12.1	0	7.3	0
1.7	0	1.5	0	(0)	(0.5)	(0)	(0.5)	0.2	0	0.6	0	0.4	0.1	0.1	3.8	0	13.0	6.5	2.5	0.2	1.2	0	(1.7)
0	(3.0)	0	0.1	0.1	0	0	0	0.1	0	0.2	0	0	0	0.3	3.3	0.1	4.0	0.3	0.4	0.5	0	0	0
0	0	0	0.2	0	0	0.1	0.1	0	0	0	0	0	0	0	0	0.2	0	0	0.1	0	0	0	0
0	0	0.2	0	0	0	0	0	0	0	0.1	0	0	0	0	0	0.2	0.5	0	0.2	0	0	0	0
0	0	0	0	0	0	0	0	0	0	0	0	0	0	0	0	0	0	0	0	0	0	0	0
0	0	0	0	0	0	0	0	0	0	0	0	0	0	0	0	0.1	0.5	0.5	0.2	0	0.1	0	0
0	0	0	0	0	0	0	0	0	0	0	0	1.4	0	0.9	0	0.1	1.0	0	1.9	0	1.5	0	0
0	0	0	0	0	0	0	0	0	0	0.2	0	0	0	0.2	0	0	2.7	0	0.5	0	0	0	0
0	0	0	0	0	0	0	0	0	0	0	0	0	0	0.5	0	0.1	0	0	0.7	0	0.3	0	0.1
0	0	0	0	0	0	0	0	0	0	0	0	0	0	0	0	0.2	0	0	1.9	0	0.5	0	0
0	0	0	0	0	0	0	0	0	0	0	0	0	0	0	0	0	0	0	0	0	0	0	0.7
0	0.4	0	0	0	0	0	0	0	0	0	0	0	0	0	0	0	0.3	0	0.2	0	0	0	0
0	0	0	0	0	0	0	0	0	0	0	0	0	0	(0)	(0)	(0)	(0)	0	(1.0)	0	2.5	0.2	0.1
0	0	0	0	0	0	0	0	0	0	0	0	0	0	0	0	0	0	0	0.5	0	1.4	0	0.1
0	0	0	0	0	0	0	0.2	0	0.1	0.5	0	0	0	0.3	0	0.6	2.2	0.1	0.6	0	0.3	0	0.1
0	0	0.1	0	0.1	0.1	0	0.1	0.1	0.6	2.1	0.2	2.1	0.2	0.1	0.2	0.1	0.3	0	2.2	0	3.2	0	0.8
0	0.1	0	0	0	0	0	0	0	0	0	0	0	0.1	0	0.9	0.4	0	0	1.3	0	4.7	0	0.9
0.3	0	0	0.1	0	0	0	0	0	0	0	0	0	0	0.1	0	0.6	0	(0)	(1.0)	(0)	(4.5)	(0)	(1.0)
0.1	0	0	0	0	0	0	0	0	0	0.4	0	0.2	0	0	0	0.2	2.4	0	1.6	0	2.9	0	0.7
0	0	0	0	0	0	0	0	0	0	0	0	0	0	0	0	0.1	0	0	0	0	0	0	0
0	0.4	0	1.2	0	0.2	0	0	0	0	0	0	0	0	0	0	0	0	0.1	1.2	0.1	0	0	0
0	0	0	0	0	0	0	0	0	0	0	0	(0)	(0)	(0)	0	(0.2)	(0.6)	(0)	(1.0)	(0)	(2.0)	(0)	(0.5)

TABLE LXVI (continued). F_γ Matotch

Gr. M.-T.	0–2		2–4		4–6		6–8		8–10		10–12		12–14		14–16		16–18		18–2			
Date	+	–	+	–	+	–	+	–	+	–	+	–	+	–	+	–	+	–	+	–	+	
December 17	(0)	(0)	(0)	(0)	(0)	(0)	(0)	(0)	0	0	0	0	0	0	0	0	0	0	0.1	0.1	0	
18	0	0	0	0	0	0	0	0	0	0	0	0	0	0	0	0	0	0	0	0	0	
19	0	0	0	0	0	0	0	0	0	0	0	0	0.3	0	0.8	0	0.3	0	0	1.1	0	
20	0	0	0	0	0	0	0	0	0	0	0	0	0	0	0.6	0	0	0	0	0	0	
21	0	0	0	0	0	0	0	0	0	0	0	0	0	0	0.2	0	0.5	0	0.2	0	0	
22	0	0	0	0	0	0	0	0	0	0	0.1	0.1	0	0	0.2	0	0	0	0	1.0	0	
23	0	3.4	0	2.7	0	0.5	0.4	0.1	1.3	0.1	1.0	0	0.5	1.3	0	1.0	0	13.5	0	19.2	0	
24	0	0	0	0	0	0	(0)	(0)	(0)	(0)	(0.2)	(0)	(0.4)	(0)	(0.1)	(0)	(0.5)	(0)	(0)	(1.0)	(0	
25	(0)	(0)	(0)	(0)	(0)	(0)	0	0	0	0	0.1	0.1	0	0	0	0	0.1	0	0	1.2	0	
26	0	0	0	0	0	0	0	0	0.1	0	0.2	0.1	1.1	0	1.5	0	0.4	0	0	0.4	0	
27	0.1	0	0	0	0	0	0	0	0	0	0	0	0	0	0	0	0.6	0	0.5	0	0	
28	0	2.4	0	0.1	0.5	0.2	0.4	0	0.2	0	0.5	0	1.5	0	0.2	0	0.6	0	0	0.3	0	
29	0	0	0	0	0	0	0	0	0	0	0	0	0	0	0	0	0.5	0.3	0	0.2	0	
30	0	0	0	0	0	0	0	0	0	0	0	0	0	0	0	0	0.1	0.2	0	0.3	0	
31	0	0.1	0	0	0	0	0	0	0	0	0	0	0	0	0	0	0.3	0	0.1	0.2	0 1	
January 1	0	0	0	0	0	0	0	0	0	0	0	0	0	0	0	0	0.4	0	0.1	0	0	
2	0	0.6	0	0.1	0	0	0	0	0	0	0	0	0	0	0	0	0	0	0	0	0	
3	0	0	0	0	0	0	0	0	0	0	0	0	0	0	0	0	0	0	0	1.7	0	
4	0	0	0.1	0.1	0	0.2	0.1	0.2	0.1	0	0	0	0	0	0	0	0.2	0	0.2	0.6	0	
5	0	0	0	1.0	0	1.3	0	0.2	1.6	0	0.9	0	0.9	0	1.7	0	0	4.5	0	2.2	0	
6	0	0	0	0	0.1	0	0	0	0	0	0	0	0	0	0.3	0	0.3	0	0	1.5	0	
7	0	0	0	0	0	0	0	0	0	0	0	0	0	0	0	0	0.2	0	0	1.0	0	
8	0	0	0	0	0	0	0	0	0	0.1	0	0.1	0	0	0.3	0	0.4	0.6	0	0.6	0	0.
9	0	0	0	0	0	0	0	0	0.2	0	0.2	0	0.2	0	0.5	0	1.2	0.2	0.5	0.2	0	0.
10	0.5	0.1	0.3	0	(0)	(0)	(0)	(0)	(0.1)	(0)	0	0	0.4	0	3.5	0	1.3	0	0.1	0	0	
11	0.1	0	0	0	0	0	0	0	0	0	0	0	0	0	0.5	0	0.5	0	0.3	3.2	0.1	1.
12	0.1	0	0	0	0	0	0	0	0	0	0	0	0	0	0.6	0	0.7	0.1	0	3.2	0	0
13	0	0	0	0	0	0	0	0	0	0	0	0	0	0	0.2	0	0.8	0.1	0.1	4.4	0	3.
14	0	0	0	0	0	0	0	0	0	0	0	0	0	0	0	0	0	0	0	0	0	0
15	0	0	0	0	0	0	0	0	0	0	0	0	0.2	0	0.8	0	0.4	0	0	0.5	0	
16	(0)	(0)	(0)	(0)	(0)	(0)	0	0	0	0	0.4	0	4.5	0	1.4	0	2.3	0	0.2	0.1	0	0.
17	0	0	0	0	0	0	(0)	(0)	(0)	(0)	0	0	0.2	0	0	0	0.1	0	0.1	0	0	0.
18	0	0	0	0	(0)	(0)	(0)	(0)	(0)	(0)	0	0	0	0	1.6	0.3	1.6	0.1	0	2.2	0	0.
19	0	0	0	0	0	0	0	0	0	0	0.2	0	0.5	0	0.3	0.1	0.1	1.9	0	0.1	0.1	0.
20	(0)	(0)	(0)	(0)	(0)	(0)	0	0	0	0	0	0	0	0	0	0	0	0	0	0.4	0	
21	0	0	0	0	0	0	0	0	0	0	0	0	0.5	0	0.8	0.1	0.1	0	(0)	(0.7)	0	1.
22	0	1.2	0	0.2	0	0	0	0	0	0	0	0	0	0	0	0	0	0.5	0	2.3	0	
23	0	0	0	0	0	0	(0)	(0)	0	0	0	0.1	0	0.2	1.1	0	0.1	1.5	0.3	2.9	0	5.
24	0.1	0	0	0	0	0	0	0	0	0	0	0	0.1	0	0	0	0.1	0.7	0	3.4	0	
25	(0)	(0)	(0)	(0)	(0)	(0)	(0)	(0)	0	0	0	0	0	0	0	0	0.2	0.8	0	0.1	0	
26	0	0	0	0	0	0	0	0	0	0	0	0	0.2	0	0	0.1	0	(3.0)	0.6	5.8	(0.5)	(5.
27	(0)	(1.0)	(0)	(1.0)	(0)	(0.5)	(0)	(0)	0	0	0	0.4	0	0	0.4	0	0	1.1	0	1.4	0	
28	0	0	0	0	0	0	0	0	0	0	0	0	0	0	0	0	0.6	0.8	0.3	0.3	0	
29	0	0	0	0	0	0	0	0	0	0	0	0	0	0	0	0	0	0	0	0	0	0
30	0	0	0	0	0	0	0	0	0	0	2.0	0	3.2	0	2.5	0	0.4	2.2	0	1.6	0	
31	0	0.1	0	0	0	0	0	0	0	0	0	0	0	0	1.7	0	0.8	0.2	0.5	0	0	
February 1	0	0	0	0	0	0	0	0	0	0	0	0	0	0	0	0	0.2	0.1	0.4	0.1	0	0
2	0	0	0	0	0	0	0	0	0	0	0	0	0	0	0	0	0.1	0	0	0	0	0
3	0	0	0	0	0	0	0	0	0	0.2	0	0	0	0	0	0	0	0	0	0	0	0
4	0	0	(0)	(0)	(0)	(0)	(0)	(0)	(0)	(0)	(0)	(0)	(0)	(0)	(0.4)	(0)	(0.3)	(0.1)	(0.2)	(0)	(0)	(0)

LXVI (continued). F_V Matotchkin Schar.

.	0—2		2—4		4—6		6—8		8—10		10—12		12—14		14—16		16—18		18—20		20—22		22—24	
	+	—	+	—	+	—	+	—	+	—	+	—	+	—	+	—	+	—	+	—	+	—	+	—
5	(0)	(0.1)	0	0.2	0	0	0	0	0	0	0	0	0.1	0	0.9	0	0.8	0.4	0.1	0.9	0.1	0.6	0	0
6	0	0	0	0	0	0.1	0	0	0	0	0	0	0.3	0	0.2	0	(0.4)	(0.2)	0.2	0	0.1	0	0	0
7	0	0	0	0	0	0	0	0	0	0	0	0	0	0	0	0.2	0.3	0.1	0.1	0	0	3.9	0	8.4
8	0	0.5	0	0.9	0	1.6	0.8	0.3	0.5	0	(0)	(0)	2.0	0	0.8	0.7	0.7	2.7	0	5.9	0.5	2.4	0.4	0.5
9	0	0.1	0.1	0.2	0.3	0.2	0	0.2	0	0.1	0	0	0	0	0	0	0	0.2	0	3.9	0	4.2	0	3.5
10	0.1	0.1	0.1	0	0	0	0	0	0	0	0.2	0	0.3	0	0	0	0	0	0	0	0.3	1.2	0	4.4
11	0.1	0.2	0.1	0	0	0	0.6	0	0	0.1	0.3	0	0.1	0	0	0	0	0.8	0.2	1.9	0.2	0.5	0	1.4
12	0.1	0.1	0	0	0	0	0	0	0	0	0	0	0	0	0	0	3.0	0	0	1.9	0.3	0.5	0.1	0.1
13	0.1	0.7	0	0	0	0.1	0.1	0	0	0	0.5	0	0.3	0	0	0	0	0	0.2	0.9	0.1	0.7	0	0.2
14	0.2	0.1	0.1	0	0	0	0	0	0	0	0.4	0	0	0	1.0	0	0	0.2	0	0.6	0	2.2	0	2.8
15	0	0	0	0	0	0	0	0	0	0	0	0	0.4	0	1.2	1.9	0	17.0	0	0.3	0	0.4	0	0.2
16	0	0	0	0	0	0	0	0	0	0	0	0	0	0	0	0	0.3	0.7	0	0.7	0	0.1	0	0
17	0	0.1	0	0	0	0	0	0	0	0	0	0	0.2	0	1.0	0.1	(0)	(0)	0	0	0	0.4	0	0
18	0	0	0	0	0	0	0	0	0	0	0	0	0	0	0	0	0	0	0	0	0	0	0	0
19	0	0	0	0	0	0	0	0	0	0	0	0	0	0	0	0	0	0	0	0	0	0	0	0
20	(0)	(0)	(0)	(0.3)	(0)	(0.4)	(0)	(0.3)	0	0	0	0	0	0	0	0	0	0	0	0	0	0	0	0
21	0	0	(0)	(0)	(0)	(0)	(0)	(0)	0	0	0	0	1.0	0	1.3	0	0.7	0	0	0	0	0	0	0
22	0	0	0	1.2	0	1.5	0.2	1.1	1.6	0	2.6	0	0.8	0	0	0	0.2	0	0.1	0	0	0	0	0
23	0	0	0	0	0	0	0	0	0	0	0	0	0.2	0	0	0	0.5	0	0.2	0.1	0	0.4	0	0
24	0	0	0	0	0	0	0	0	0	0	0	0	0	0	0	0	0.5	0	0.2	0	0	0	0	0
25	0	0	0.2	0	0	0	1.7	0	3.0	0	3.5	0	2.7	0	0.7	0	0.1	0	0	0.1	0	0	0	0
26	0	0	0	0	0	0	0	0	0	0	0	0	0	0	0	0	0	0	0	0	0	1.3	0	0.1
27	(0)	(0)	(0.1)	(0)	(0)	(0)	0	0	0	0	0	0	0	0	0	0	0	0	0	0	0	0	0	0
28	0	0	0	0	0	0	0	0	0	0	0	0	0	0	0	0	0	0	0	0	0	0	0	0
1	0	0	0	0	0	0	0	0	0	0	0	0	0.9	0	1.8	0	0.6	0.8	(0.2)	(0.2)	0	0.1	0	2.1

Kaafjord.

TABLE LXVII.

Disturbances in Horizontal Force (F_H).

T.	0—2		2—4		4—6		6—8		8—10		10—12		12—14		14—16		16—18		18—20		20—22		22—24	
	+	—	+	—	+	—	+	—	+	—	+	—	+	—	+	—	+	—	+	—	+	—	+	—
r 3	0	0.1	0.3	0	0.1	0	0	0.1	0	0	0	0	0	0	0	0.1	0	0	(0.1)	(0)	0.1	0	0	0
4	0	0	0	0	0.2	0	0.1	0.1	0.1	0.1	0.1	0.5	0.1	0	0.8	0	0.6	0	0	0	0	0	0.1	0
5	0	0	0	0	0	0	0	0	0	0	0	0	0.1	0.1	0.1	0	0	0	0	0	0	0	0.1	0
6	0	0	0	0	0	0.1	0	0	0	0	0	0.2	0	0.1	0	0.2	0	0.2	0.5	0	0	0	0	0.2
7	0.1	0.5	0	0.1	0	0	0	0.1	0	0	0	0	0.1	0	0	0	0	0	0	0	0	0	0	0
8	0	0	0	0	0	0	0	0	0	0	0	0	0	0	0	0	0	0	0	0	0	0	0	0
9	0	0	0	0	0	0	0	0	0	0	0	0	0	0	0	0	0	0	0	0	0	0	0	0
10	0	0	0	0	0	0	0	0	0	0	0.1	0	0.2	0.1	0	0	0	0	0	0	0	0	0	0
11	0	0	0	0	0	0	0	0	0.1	0	0.1	0	0.2	0	0	0	0	0	0	0.1	0	0.1	0	0.3
12	0.1	0.1	0	0	0.1	0	0.1	0	0	0	0.1	0.3	0.8	0	0.7	0	5.6	0	1.7	0.4	0	3.4	0	9.2
13	0.2	0	0	0.3	0	0.2	0	0.1	0	0.1	0	0	0	0	0	0	0	0	0	0	0	0	0	0
14	0	0	0	0	0	0	0	0	0	0	0	0	0	0	0	0	0	0	0	0	0	0	0	0
15	0	0	0	0	0	0	0	0	0	0	0	0.1	0.1	0	0	0.1	0.1	0	0	0	0	0.1	0	0.7
16	0	0	0	0	0	0	0	0	0	0.1	0	0.1	0.1	0	0	0	0	0	0	0	0	0	0	0
17	0	0	0	0	0	0.1	0	0	0	0	0.1	0	0.1	0.1	0.6	0	0.4	0	0	0	0	0	0	0
18	0	0	0.1	0	0.1	0	0	0	0.1	0	0.6	0	0.3	0	0.1	0	0	0.1	0	0.4	0.2	4.2	0	0.2
19	0	0.6	0.1	0	0	0	0	0.1	0	0.2	1.1	0	0.9	0	1.9	0	1.7	0	0.6	0.2	0	13.0	0	7.5
20	0	2.2	0	0.8	0.1	0.4	0.1	0.2	0.9	0.1	2.5	0	0.5	0	0.2	0.2	0.2	0	0.9	0.7	0	8.5	0	2.9
21	0	0.1	0	0	0	0	0	0	0.1	0	0	0	0.1	0.1	0	0.5	0.2	0.1	0	0.1	0	0	0	0
22	0	0	0	0	0	0	0	0	0	0	0	0.1	0	0	0.1	0	0.1	0.1	0.1	0.4	0	0.7	0	5.1

TABLE LXVII (continued). F_H

Gr. M.-T.	0–2		2–4		4–6		6–8		8–10		10–12		12–14		14–16		16–18		18–20		20
Date	+	−	+	−	+	−	+	−	+	−	+	−	+	−	+	−	+	−	+	−	+
September 23	0	3.1	0	2.1	0.1	0	0.1	0	0	0	0	0	0.1	0	0	0	0	0	0	1.9	0
24	0	0	0	0	0	0	0	0	0	0	0	0	0	0	0	0	0	0	0	0	0
25	0	0	0	0	0	0	0	0	0	0	0	0.1	0	0.2	0.2	0	0	0	0	0	0
26	0	0.2	0	0	0	0	0	0.1	0	0	0	0	0	0	0	0	0	0	0	0.1	0
27	0	0.2	0	0	0	0	0	0	0	0	0	0	0	0.3	0	0.3	0	0.2	0.4	0	0
28	0	0.5	0	0	0	0	0	0	0	0	0	0	0	0.1	0	0	0	0.1	0	0	0
29	0	0.1	0	0	0	0	0	0	0	0	0	0	0.1	0	0	0.1	0	0.1	0	0	0.2
30	0	0.2	0	0	0	0	0	0	0	0	0	0	0	0	0.1	0.1	0.2	0.2	0.1	1.4	0
October 1	0	5.4	0	1.8	0	0.1	0	0	0	0	0	0	0.1	0	8	8	8	8	8	8	8
2	0	0	0	0	0	0	0	0	0	0	0	0.1	0.2	0	.1			.1	.1		
3	0.1	0	0	0	0	0	0	0	0	0	0	0	0.1	0	0.2	0	0.3	0	0.1	0.1	0
4	0	0.3	0	0	0.1	0	0	0	0	0.1	0.1	0	0	0.3	0	0	0	0	0	0	0
5	0	0.1	0	0.2	0.1	0.1	0	0	0	0	0	0	0	0	0	0	0	0	0.1	0	0
6	0	0	0	0	0	0	0	0	0	0	0	0	0	0	0	0	0	0	0	0	0
7	0	0	0	0	0	0	0	0	0	0	0	0	0	0	0	0	0	0	0	0	0
8	0	0	0	0	0	0	0	0	0	0	0.1	0	0	0	0	8	8.1	8	8.1	8	8
9	0	0	0	0	0	0	0	0	0	0	0	0	0	0	0						
10	0	0	0	0	0	0	0	0	0	0	0	0	0	0	0						
11	0	0	0	0	0	0	0	0	0	0	0	0	0.3	0.1	0		(	( )	.0	.0	
12	0	0.6	0	0	0	0	0	0	0	0	0	0	0.1	0	0.1						
13	0	0	0	0	0	0	0	0	0	0	0.1	0.1	0	0	0.1	0.1	0	0	0	0.1	0
14	0	0	0	0	0	0	0	0	0	0	0	0	0	0	0	0	0	0	0	0	0
15	0	0	0	0	0	0	0	0	0	0	0	0.1	0	0	0	0	0	0	0	0	0.1
16	0	0	0	0	0	0	0	0	0	0	0	0	0	0	0	0	0	0	0	0	0
17	0	0	0	0	0	0	0	0	0	0	0	0	0	0	0	0.	0	0	0.1	0	0
18	0	0	0.1	0	0.1	0	8	8	8	8	8	8	8	0	0	0	0.1	0	0.1	0.2	0
19	0	0.8	0	0.9	0	0								0	0	0	0	0	0	0	0
20	0	0	0	0	0	0								0	0	0	0	0	0	0	0
21	0	0	0	0	0	0								0	0	0	0	0	0	0	0
22	0	0	0	0	0	0								0	0	0	0	0	0	0	0
23	0	0	0	0	0	0	0	0	0	0	0	0	8	8	8	8	8	8	0	0	0
24	0	0.3	0.1	0	0	0	0	0	0	0	0	0				.1	.6	.1	0.7	0	0.1
25	0	9.4	0.1	1.5	0	0	0	0	0	0	0	0		.4		.		.2	0	0	0.2
26	0	0.5	0	0.1	0	0	0	0	0	0	0	0				.	.1		0	0.1	0
27	0	1.4	0	0.4	0.2	0.1	0	0.1	0.1	0	0.1	0	.2		.0		.8		2.3	0.2	0
28	8.1	8.1	8	0.3	0.1	0	0	0	0.1	0	0	8	8.1	8	8	8.1	8	8	8	1.5	0
29		.		0	0.2	0	0	0	0	0	0.1						( )	(		0	0.3
30		.		0.1	0	0	0	0.7	0.4	0.6	1.3	.2		.1	0	0.2	0.1			0	0
31		.		0.3	0	1.2	0.1	0.3	1.2	0.2	.3		1 .7		15.3	8	3.7	.8		12.7	0
November 1		.	.2	1.1	0.8	0	0	0	0.1	0	.1	.2								0	0
2	0	0	0	0	0	0	0	0	0	0	0	0	0.1	0	0	0.1	0.7	0	0.6	3.7	0.1
3	0	0	0	0	0	0	0	0	0	0	0	0	0	0	0	0	0	0	0	0	0.1
4	0	0	0	0	0	0	0	0	0	0	0	0	0	0	0	0	0	0	0	0	0
5	0	0	0	0	0	0	0	0	0	0	0	0	0	0	0	0	0	0	0	0	0
6	0	0	0	0	0	0	0	0	0	0	(0)	(0)	(0)	(0)	(0)	0	0	0	0.1	0.6	0
7	0.1	0.3	0.1	0	0	0	0	0	0	0	0	0	0	0	0	0	0	0	0	0	0
8	0	0.1	0	0	0	0	0	0	(0)	(0)	0	0	0	0	0	0	0	0	0	0	0
9	0	0	0	0	0	0	0	0	0	0	0	0	0	0	0	0	0	0	0	0	0
10	0	0	0	0	0	0	0	0	0	0	0	0	0	0	0	0	0	0	0	0	0.2
11	0	0	0	0	0	0	0	0	0	0	0	0	0	0	0	0	0	0	0	0	0.2

(continued). F_H Kaafjord.

	2–4				6–8		8–10		10–12		12–14		14–16		16–18		18–20		20–22		22–24	
–	+	–	+	–	+	–	+	–	+	–	+	–	+	–	+	–	+	–	+	–	+	–
0	0	0	0	0	0	0	0	0	0	0	0	0	0	0	0	0	0	0	0.1	0	0.1	0.2
0	0.2	0.7	0.5	0	0	0	0	0.1	0	0.5	0	0.4	0	0.1	0	0.2	0.1	0.1	0	0	0	0
0	0	0	0	0.1	0	0	(0)	(0)	(0)	(0.1)	0	0	0.1	0.1	0	0	(0)	(0)	0.3	0.1	0	0.2
0.6	0	0	0	0	0	0	0	0	0	0	0	0	0.1	0	0	0	0	0	0	1.2	0.1	0.3
0	0	0	0	0	0	0	0	0	0	0	0	0.1	0	0.1	0	0	0	0	0	0	0	0.1
0.6	0	0	0	0	0	0	0	0.1	0	0	0	0	0	0	0	0	0	0	0	0.1	0	0
0	0	0	0	0	0	0	0	0	0	0	0	0	0	0	0	0	0	0	0	0.6	0	0.3
0	0	0	0	0	0	0	0	0	0	0	0	0	0	0	0	0	0	0	0	0.2	0	0
0	0	0	0	0	0	0	0	0	0	0	0	0	0	0	0.2	0.1	0.1	0.1	0.1	0.1	0	0
0	0	0	0	0	0	0	0	0	0	0	0	0	0	0	0.6	0	0.8	0.2	0	10.0	0	2.1
0	0	0	0	0	0	0	0	0	0	0.1	0.1	0.1	0.2	0	1.6	0.8	1.2	0.3	0	4.4	0	2.8
0	0.3	0	0.1	0	0.1	0.2	0.2	0	0.5	0.1	0.4	0.1	0.4	0	0.9	0.1	2.1	0	2.7	0	1.2	11.1
13.4	0	16.3	0.3	1.2	0.3	0.9	1.3	0	2.8	0	8.0	0	10.0	0	1.3	5.3	0.3	2.4	0	14.0	0	8.7
4.4	0	1.4	0.3	0.2	0.5	0.1	0	0.5	0.7	0	1.9	0	3.6	0	4.2	0.2	0	14.5	0	7.5	0	3.3
6.1	0.1	0.1	0.1	0.2	0	0	0.2	0	0	0.1	0	0	3.8	0	5.2	0	2.0	0	0.4	0	0	0.1
0	0	0	0	0.1	0	0.1	0	0.1	0	0.1	0	0	0	0.1	0	0	0	0	0.1	0	0	0
0	0.1	0	0	0	0	0	0	0	0	0	0	0	0	0	0	0	0	0	0	0	0	0
0	0	0	0	0	0	0	0	0	0	0	0	0	0	0	0	0	0	0	0	0	0	0
0	0	0	0	0	0	0.1	0	0	0	0	0	0.2	0	0.1	0	0	0	0.1	0	0	0	0.1
0.1	0	0	0	0	0	0	0	0	0.1	0	0	0.4	0	0.1	0.3	0	0	0.2	0	1.1	0	0
0.1	0	0	0.3	0	0.1	0	0	0	0	0	0	0	0.2	0.1	0.5	0	0	0	0	0	0	0
0.1	0	0	0	0	0	0	0	0	0	0	0	0	0	0	0	0	0	0	0	0	0	0
0	0	0	0	0	0	0	0	0	0	0	0	0	0	0	0	0.1	0.2	0.1	0.2	0	0	0
0	0	0	0	0	0	0	0	0	0	0	0	0	0	0	0	0	0	0	0	0	0	0.7
0.2	0	0	0	0	0	0	0	0	0	0	0	0	0	0	0	0	0	0	0	0	0	0
0	0	0	0	0	0		0	0	0	0	0	0	0	0	0	0	0.2	0	0.1	0	0	0
0	0	0	0	0	0		0	0	0	0	0	0	0	0	0	0	0	0	0.1	0	0	0
0	0	0	0.1	0	0.1		0	0	0	0	0	0	0	0	2.7	0	0.9	0	0.2	0.1	0	0.1
0	0.3	0.1	0.5	0	0		0	0	0.1	0.2	0	0.2	0	0.1	0	0	0.2	0.1	0.2	0.1	0	0.2
0	0	0.2	0	0	0	8	0	0	0	0.1	0.1	0	0.1	0	0.5	0	0.7	0	2.6	0	0	1.1
0.3	0	0.1	0	0	0	0	0	0	0	0	0	0	0.1	0	0.7	0	1.5	0	0.3	0	0	0
0	0	0	0.1	0	0.1	0	0	0	0.1	0.1	0.1	0	0	0	0.1	0.4	0.1	0.1	0.1	0.9	0.1	0.2
0.1	0	0	0	0	0	0	0	0	0	0	0	0	0	0	0	0	0	0	0	0	0	0
1.4	0.1	0.1	0	0	0	0	0	0	0	0	0	0	0	0	0	0	0.1	0	0	0	(0)	(0.1)
0	0	0	0	0	0	0	0	0	0	0.2	0.1	0.1	0	0	0.5	0	0	0	0	0	0	0
0	0	0	0	0	0	0	0	0	0	0	0	0	0	0	0	0	0	0	0	0	0	0
0	0	0	0	0	0	0	0	0	0	0	0	0	0	0	0	0	0	0	(0)	(0.1)	0	0
0	0	0	0	0	0	0	0	0	0	0	0.1	0.1	0.1	0	0.1	0	0.3	0	0	0.4	0	0
0	0	0	0	0	0	0	0	0	0	0	0	0	0	0.1	0	0	0	0	0	0	0	0
0	0	0	0	0	0	0	0	0	0	0	0	0	0	0	0	0	0.2	0	0	0	0	0
0	0	0	0	0	0	0	0.4	0	0.5	0	0	0.1	0	0.2	0.1	0	0.1	0	0	2.0	0	5.8
10.1	0	1.5	0	1.2	0	1.1	0.3	0.1	0.9	0.1	3.6	0	0.3	0.1	3.4	0	0.3	2.3	0	4.6	0	0.1
0.5	0	0.5	0.2	0	0	0	0.2	0.1	0.1	0.1	0.2	0.1	0.6	0	0.2	0	0.6	0.1	0	0.4	0	0.6
0.2	0	0.2	0.3	0	0	0	(0.2)	(0)	(0.4)	(0)	(1.0)	(0.1)	(0.2)	(0.1)	(0.9)	(0)	(0.3)	(0.6)	(0)	(2.0)	(0)	(2.0)
(1.0)	(0)	(0.5)	(0.1)	(0.2)	(0)	(0)	0	0	0	0	0	0.1	0	0.1	0	0	0	0	0	0.7	0	2.0
0	0	0	0	0	0	0	0		0	0	0	0	0	0	0	0	0	0.2	0	0.6	0	2.8
1.6	0	0.1	0.1	0.1	0.2	0	0		0.2	0	0.1	0.1	0.2	0	0	0.4	0	0	0.3	0	0	0
0	0	0	0	0	0	0	0		0	0	0	0	0	0	0.1	0	0.1	0	0	0	0	0
0	0	0	0	0	0	0	0		0	0	0	0	0	0	0	0	0	0	0	0	0	0
0	0	0	0	0	0	0	0	8	0	0	0	0	0	0	0	0	0	0	0	0	0	0

TABLE LXVII (continued). F_R

Gr. M.-T.	0—2		2—4		4—6		6—8		8—10		10										
Date	+	—	+	—	+	—	+	—	+	—	+	—	+	—	+	—	+	—	+	—	+
January 1	0	0	0	0	0	0	0	0	0	0	0	0	0	0	0	0	0	0	0	0	0
2	0	0.1	0	0	0	0	0	0	0	0	0.1	0	0.1	0	0	0	0	0	0	0	0
3	0	0.1	0	0.1	0	0	0	0	0.1	0.1	0	0	0	0	0	0	0	0	0.1	0	0.2
4	0	0.1	0	0.3	0	0.1	0.1	0.2	0	0	0	0	0	0	0	0	0	0	0.1	0	0.1
5	0	0.1	0.5	0.1	1.1	0.5	0	0	0	0.1	0.2	0.1	0.2	0	0	0	0.9	0.1	0	0	0.1
6	0	0.1	0	0.2	0	0	0	0	0	0	0	0	0	0	0	0	0	0	0	0.1	0.1
7	0	0	0	0	0	0	0	0	0	0	0	0	0	0	0	0	0	0	0.1	0.1	0
8	0	0	0	0	0.1	0	0.1	0	0	0	0	0	0	0	0	0	0	0	0	0.1	0
9	0	0	0	0	0.1	0	0	0	0	0	0	0	0.1	0	0.1	0	0	0	0	0	0.1
10	0	0.2	0	0.4	0	0	0	0	0	0	0	0.1	0	0.1	0	0.3	0.1	0.1	0	0	0.1
11	0	0.1	0	0.2	0	0	0	0	0	0	0	0	0	0	0	0.1	0.1	0	0.6	0	0.6
12	0	0.1	0	0.1	0.1	0	0	0	0	0	0	0.1	0	0	0	0	0	0	0.3	0	0
13	0	0.1	0	0	0	0	0	0	0	0	0	0	0	0	0	0	0	0	0	0.5	0
14	0	0	0	0	0	0	0	0	0	0	0	0	0	0	0	0	0	0	0	0	0
15	0	0	0	0	0	0	0	0	0	0	0	0	0	0	0	0	0	0	0.3	0.2	0
16	0	0	0	0.1	0	0	0	0	0	0	0	0	0	0.3	0	0.2	0	0	0	0	0
17	(0)	(0.1)	(0)	(0.1)	(0)	(0)	(0)	(0)	0	0	0	0	0	0	0	0	0	0	0	0	0.1
18	0	0	0	0	0	0	0	0	0	0	0	0	0	0.1	0.4	0.1	0.2	0	1.2	0	0
19	(0)	(0.1)	(0)	(0.1)	(0)	(0)	(0)	(0)	0	0.1	0	0.2	0.1	0	0.3	0	0.6	0	0.1	0	0.1
20	0	0	0	0.3	0	0	0	0	0	0	0	0	0	0	0	0.1	0	0	0.1	0	0.1
21	0	0	0	0	0	0.1	0	0	0	0	0	0	0	0.2	0	0.4	0	0	0	0	0.3
22	0	0.2	0	0	0	0	0	0	0	0	0	0	0	0	0	0	0.1	0	0.1	0	0
23	0	0	0	0	0	0.1	0	0	0	0	0.1	0	0.1	0	0	0.2	0.6	0	0.4	0.1	0.1
24	0	0	0	0.2	0	0	0	0.2	0	0	0	0	0.1	0	0.1	0	0.2	0	0.3	0.1	0.1
25	0	0	0	0	0	0.1	0	0	0	0	0	0	0	0	0	0	0	0	0	0	0
26	0	0	0	0	0	0	0	0	0.1	0	0.2	0	0.3	0	0.2	0	0	0.1	0	3.3	0
27	0	5.0	0	0.2	0	0	0	0	0	0	0	0.1	0.1	0.1	0	0	0.1	0.1	0	0.1	0
28	0	0	0	0	0	0	0	0	0	0	0	0	0	0	0	0	0.1	0	0	0	0.2
29	0	0	0	0	0	0	0	0	0	0	0	0	0	0	0	0.1	0	0	0	0	0
30	0	0	0	0	0	0	0	0	0	0.1	0	0	0.1	0.1	0.3	0.1	1.7	0	3.7	0	0
31	0	0.1	0	0	0	0	0	0	0	0	0	0	0	0	0	0.1	0.2	0	0	0	0
February 1	0	0	0	0	0	0	0	0	0	0	0	0	0	0	0	0	0.1	0	0.1	0	0
2	0	0	0	0	0	0	0	0	0	0	0	0	0	0	0	0	0	0	0	0	0
3	0	0	0	0	0	0	0	0	0	0	0	0.1	0.1	0	0	0	0	0	0	0	0
4	0	0	0	0	0	0	0.1	0	0	0	0	0	0	0	0	0	0	0	0	0	0.1
5	0	0	0	0	0	0	0	0	0	0	0	0	0	0	(0)	(0)	(0)	(0)	(0)	(0)	(0)
6	0	0	0	0	(0)	(0)	(0)	(0)	(0)	(0)	(0)	(0)	(0)	(0)	(0)	(0)	0	0	0	0	0
7	0	0	0	0	0	0	0	0	0	0	0	0	0	0	0	0	0	0	0	0	0
8	0	0.3	0.1	1.3	0	0.2	0.1	0.1	0.1	0.3	1.0	0.1	0.2	0.4	3.6	0	4.3	0	1.3	5.0	0
9	0.1	0.3	0	1.0	0	0.2	0	0	0	0	0	0	0	0	0	0	0	0	0.1	0.1	0
10	0	0.5	0	0	0	0	0	0	0	0	0	0	0	0	0	0	0	0	0	0	0.1
11	0.1	0.3	0.1	0.1	0.2	0	0	0.1	0	0	0	0.1	0	0.1	0	0.1	0	0	0.2	0.1	0
12	0	0.2	0	0	0	0	0	0	0	0	0	0	0	0	0.2	0	0.1	0	1.7	0	0.3
13	0	2.7	0.6	0	0.4	0	0.1	0	0.1	0.1	0	0.3	0	0.2	0.1	0	0	0	1.1	0.1	0.1
14	0	0.7	0.1	0.1	0	0	0	0	0	0	0.1	0.1	0.2	0	0.3	0.2	0.1	0	0.2	0	0.2
15	0	0	0	0	0	0	0	0	0	0	0	0.6	0	0.3	0.8	0	2.0	0.1	0	0	0
16	0	0	0	0	0	0	0	0	0	0	0	0	0	0	(0)	(0)	(0)	(0)	(0)	(0)	(0)
17	(0)	(0.3)	(0)	(0)	(0.2)	(0)	(0.1)	(0.1)	(0)	(0)	(0)	(0)	(0)	(0)	(0)	(0)	0	0	0	0	0.1
18	0	0.2	0	0	0	0	0	0	0	0	0	0	0	0	0	0	0	0	0	0	0
19	0	0	0	0	0	0	0	0	0	0	0	0	0	0	0	0	0	0	0	0	0

XVII (continued). F_H Kaafjord.

0—2		2—4				6—8		8—10		10—12		12—14		14—16		16—18		18—20		20—22		22—24	
+	—	+	—	+	—	+	—	+	—	+	—	+	—	+	—	+	—	+	—	+	—	+	—
—	—	—	—	—	—	—	—	—	—	—	—	—	—	—	—	—	—	—	—	—	—	—	—
—	—	—	—	—	—	—	—	—	—	—	—	—	—	—	—	—	—	—	—	—	—	—	—
—	—	—	—	—	—	—	—	—	—	—	—	—	—	—	—	—	—	—	—	—	—	—	—
—	—	—	—	—	—	—	—	—	—	—	—	—	—	—	—	—	—	—	—	—	—	—	—
—	—	—	—	—	—	—	—	—	—	—	—	—	—	—	—	—	—	—	—	—	—	—	—
0	0	0	0.1	0.1	0.2	0	0.1	0	0.2	0.1	0.1	0	0.2	0	0	0	0	0	0	0	0	0	0
0	0	0	0	0	0	0	0	0	0	0	0	0	0	0	0	0	0	0	0	0	0.6	0	0.5
0	0	0	0	0	0	0	0	0	0	0	0	0	0	0	0	0	0	0	0	0	0	0	0
0	0	0	0	0	0	0	0	0.1	0	0	0	0	0	0	0	0	0	0	0	0	0	0	0
0	0	0	0	0	0	0	0	0	0	0	0	0	0	0.1	0.1	2.3	0	0.8	0	0	0.3	0	1.2
0	0.8	0.1	0	0	0.1	0	0	0	0	0	0.1	0	0	(0)	(0)	(0)	(0)	1.4	0	2.1	0	1.1	0
0	0.4	0	0	0	0	0	0	(0)	(0)	(0)	(0)	(0)	(0)	(0)	(0)	(0)	(0)	(0)	(0)	(0.2)	(0)	(0.2)	(0.2)
(0)	(0.1)	(0)	(0)	(0)	(0)	(0)	(0)	0	0	0	0	0	0	0	0	0	0	0	0	0.2	0	0	0.5
0	0.4	0	1.6	0.1	0.2	0	0	0.1	0	0.3	0	0	0	(1.0)	(0)	1.2	0.1	0.1	0.1	0	0	0	0
0	0	0	0	0	0	0	0	0	0	0	0	0.1	0	0.1	0	0.1	0	0	0	0.3	0.3	0.2	0.4
0.4	0	0.2	0	0.4	0	0.3	0.4	0	1.2	0.1	0.1	0	0.2	0.8	0	1.3	0	0.7	0.1	0	0	0	0.4
0	1.7	0.8	0.1	0.8	0	0.2	0	0	0	0	0	0	1.0	0.4	0.1	2.7	0	0.6	0.6	0.2	6.4	0	10.0
0.6	0.1	0	1.3	0.3	0	0.5	0	0	0.1	0.1	0.1	0	0	0	0	0.1	0	0.3	0	0	0.4	0	0.2
0	0	0	0.1	0	0	0	0	0	0	0	0	0.1	0	0.2	0	0.8	0	0.6	0	0.1	0.1	0	0.2
0.1	0	0	0	0	0	0	0	0	0	0	0	0.3	0	0.1	0.1	0.1	0	0.4	0.1	0.7	0	0	0

		2—4		4—6		6—8		8—10		10—12		12—14		14—16		16—18		18—20		20—22		22—24	
+	—	+	—	+	—	+	—	+	—	+	—	+	—	+	—	+	—	+	—	+	—	+	—
0	0.2	0.2	0.1	0.1	0	0	0.1	0.2	0	0.1	0	0.2	0	0.1	0	0.6	0	(0)	(0.1)	0.3	0	0	0.1
0	0.4	0	0.2	0	0.3	1.0	0.1	0.8	0	0.1	0.2	0.5	0	0	0.4	0	0.4	0	0.1	0.1	0.1	0.1	0
0	0.1	0.1	0.1	0.1	0.1	0	0.1	0	0.1	0.1	0.1	0	0.2	0	0.1	0	0	0	0	0.1	0	0	0.3
0	0.5	0	0.2	0	0.3	0	0.1	0	0.1	0.4	0	0.2	0.1	0	0.1	0	0	0.1	0.3	0	0.1	0.1	0
0	1.4	0	0.6	0.1	0.1	0	0.1	0	0.1	0	0	0.1	0	0	0.1	0	0	0	0	0	0	0	0
0	0	0	0	0	0	0	0	0	0	0	0	0	0	0	0	0.1	0	0.1	0	0	0.3	0.3	0.1
0.4	0	0.1	0.1	0	0.1	0.2	0	0	0	0	0	0	0	0	0.1	0	0	0	0.1	0	0.7	0.1	0
0	0	0	0	0	0.1	0	0.1	0.1	0.1	0.4	0.1	0.4	0	0	0.6	0	0.2	0	0.1	0	0.5	0	0.1
0	0	0	0.1	0	0.1	0.1	0.1	0.2	0	0.3	0	0.1	0.1	0	0.6	0.1	0	0.5	0.3	0	0.7	0	0.9
0	0.9	0	0.3	0	0.2	0	0.3	0	0.1	0.3	0	1.5	0	1.6	0	4.5	0	2.1	0	0	2.3	0	4.8
0	0.4	0	0.4	0.1	0.5	0	0.8	0	0.1	0	0	0.1	0	0	0.1	0	0	0	0	0.1	0	0	0.1
0	0	0	0	0	0	0	0	0	0	0	0	0	0	0	0	0	0	0	0	0	0	0	0
0	0.1	0	0.1	0.1	0.1	0.1	0	0	0	0.2	0	0.1	0.1	0	0.1	0	0	0.1	0	0.1	0.8	0	1.6
0.1	0.2	0	0.1	0	0	0	0	0.3	0	0.3	0	0	0	0	0.3	0	0	0	0	0	0.4	0.1	0
0.1	0.1	0	0.1	0	0.5	0.1	0.1	0.1	0	0.2	0.1	0.2	0.1	0.1	0.1	0.6	0	0.6	0	0	0.2	0	0.1
0	0.1	0	0.1	0	0	0	0	0.1	0.1	0.4	0.1	0	0.7	0	0.3	0	0.2	0.1	2.1	0	8.8	0	1.7
0	1.7	0	0.6	0	0.5	0.1	0.5	0.6	0	0.8	0.2	0.4	0	1.4	0	1.3	0	0.6	0.4	0	7.3	0	6.4
0	2.0	0	2.4	0.4	0.9	0.5	0.4	0.1	0.4	1.3	0.1	1.2	0	0.1	0.2	0.2	0	0.2	3.4	0.1	2.2	0.1	0.4
0	0.2	0.1	0	0.1	0.2	0.2	0.2	0	0.1	0.1	0.1	1.2	0	0.1	0.4	0.1	0.5	0	0	0	0	0	0
0	0	0	0	0	0	0	0	0.1	0	0.1	0.1	0.1	0.1	0.1	0	0	0.9	0.1	1.9	0	1.3	0	2.1
0	4.2	0	1.2	0	0.3	0.4	0.1	0.1	0.1	0.1	0.1	0.2	0	0	0.1	0	0.6	0.1	0.6	0	2.0	0	0.1
0	0.1	0	0	0	0	0	0.1	0.1	0.1	0	0.2	0.2	0	0	0	0	0	0	0	0	0	0	0
0.1	0	0	0.1	0	0	0	0	0	0.1	0	0.2	0.2	0	0	0.7	0	0	0.1	0.1	0	0	0	0
0.3	0.1	0	0.1	0	0	0	0	0.1	0	0	0.1	0	0	0	0	0	0	0.2	0.5	0	0.5	0	0
0.2	0.1	0	0.1	0	0.1	0	0.1	0	0.1	0	0.1	0.1	0.4	0	0.3	0	0	0.1	0.2	0.1	0.1	0.2	0

TABLE LXVIII (continued). F_D

Gr. M.-T.	0—2		2—4		4—6		6—8		8—10		10—12		12—14		14—16		16—18		18—20		2
Date	+	—	+	—	+	—	+	—	+	—	+	—	+	—	+	—	+	—	+	—	+
September 28	0.4	0.1	0	0.3	0	0	0	0	0	0	0	0	0.2	0.1	0	0	0.1	0	0.1	0.1	0.1
29	0.2	0.2	0.1	0.1	0.1	0.1	0	0	0	0	0.1	0	0.2	0	0	0.1	0.2	0	0.3	0	0.4
30	0	0.7	0	0	0	0	0	0.1	0.1	0.1	0	0	0	0.1	0.6	0	1.7	0	0.2	1.2	0
October 1	0	5.7	0	2.9	0	0.1	0.2	0	0.1	0	0.1	0	0	0	0	0.1	0.1	0	0.2	0.6	0.1
2	0	0	0.1	0	0	0	0	0.1	0.6	0	0.1	0.1	0	0	0.2	0	0.2	0.2	0.2	0	0.2
3	0.1	0	0.1	0.1	0	0	0	0.1	0.1	0.1	0.2	0.2	0.1	0	0	0.1	0	0	0	0.5	0.1
4	0.3	0.1	0	0	0.1	0.2	0.2	0	0	0	0.9	0	0	0.1	0	0.1	0	0	0	0.1	0.1
5	0.3	0	0.2	0.1	0.2	0.3	0	0.2	0	0.1	0.1	0	0	0	0	0	0.1	0	0	0	0.2
6	0	0	0	0	0	0.1	0	0	0	0	0	0.1	0	0	0	0.2	0	0	0	0	0.1
7	0	0	0	0	0	0	0	0	0	0	0	0	0	0	0	0	0	0	0	0	0.1
8	0	0	0	0	0	0	0	0	0	0.1	0.4	0.1	0.1	0	0	0.1	0	0	0	0.8	0
9	0.1	0	0	0	0	0	0	0.1	0.1	0.1	0	0	0.1	0	0	0.3	0	0.1	0	0.1	0
10	0	0	0	0	0	0	0	0	0.1	0	0.1	0	0	0	0	0	0	0	0.1	0	0
11	0	0.1	0.2	0.1	0	0.3	0.2	0.1	0.1	0	0.2	0.1	0.5	0	0.1	0	0.1	(0)	2.2	1.3	0.2
12	0	0.1	0	0.4	0	0	0.1	0	0.1	0	0.1	0	0.1	0	0.1	0	0.2	0	0	0	0.1
13	0	0	0	0	0	0	0	0	0.2	0	0.7	0.1	0	0	0.1	0	0.1	0	0	0.6	0
14	0	0	0	0	0	0	0	0	0.1	0	0	0	0	0	0	0	0.5	0	0.7	0	0.1
15	0.1	0	0	0	0	0	0.1	0.1	0.1	0	0.1	0.1	0	0	0	0	0	0	0.1	0.2	0
16	0	0.1	0	0	0	0	0	0	0	0	0	0	0	0	0	0	0	0	0	0	0
17	0	0	0	0	0	0	0	0	0	0.1	0	0.1	0	0	0	0	0	0	0	0	0
18	0.1	0	0.2	0.1	0.1	0.2	0.1	0	0	0	0.1	0	0	0.1	0	0.1	0.1	0	0.1	0.9	0
19	0.1	0.8	0.1	0.2	0	0.5	0	0.2	0	0	0	0	0	0	0	0	0	0	0	0.1	0
20	0	0	0	0	0	0	0	0	0	0.1	0.2	0.2	0	0	0	0	0	0	0	0	0
21	0	0.1	0.1	0.1	0	0	0	0	0.1	0	0	0	0	0.1	0.1	0	0	0.6	0	0.1	0.1
22	0	0	0	0.2	0	0.2	0	0.1	0	0	0	0	0	0	0	0	0	0	0	0	0.1
23	0	0	0	0	0	0	0	0	0	0.1	0	0	0	0	0	0	0	0	0.1	0	0.6
24	0	1.1	0.4	0.1	0.3	0	0	0.1	0.1	0	0.4	0	0	0.1	0.1	0.1	0.8	0.1	0.5	0.4	0
25	0	11.8	0	2.6	0	0.5	0.1	0.1	0	0.4	0.1	0.5	0.4	0.2	0	0.3	0	0.1	0	0.1	0.2
26	0	0	0	0	0	0	0	0	0	0	0	0.4	0	0.3	0.3	0.2	0.9	0	0	0.5	0.1
27	0	1.9	0	1.5	0.2	0.4	0.5	0.2	0.1	0.3	0	1.3	0.1	0.5	1.9	0.3	1.6	0.8	0.6	0.3	0
28		0.3	0.1	0.3	0.1	0.1	0	0	0.1	0	0	0.4	0	0.4	0.1	0.1	0.1	0	0.2	1.1	0.1
29		0.2	0	0.2	0.2	0.1	0.1	0.1	0	0.1	0.1	0	0.1	0.1	0	0	(0.1)	(0.2)	0	0.2	0
30	.1	0.9	0	0.2	0.4	0	0.4	0.1	1.0	0.1	1.1	0.1	0	0	0	0.3	0	0.6	0	0.5	0
31		3.6	0.4	0.5	2.7	0	1.8	0	0.6	0.1	1.9	0.1	3.3	1.3	8.5	0	15.7	0	1.7	3.2	0
November 1	8.1	8.4	0	2.3	0.2	0.2	0.1	0.2	0.4	0	0.2	0	0	0	0	0	0	0	0	0	0
2	0	0	0	0	0	0	0.1	0	0.1	0	0.1	0	0	0	0.3	0	1.9	0.1	0.1	1.8	0
3	0	0.3	0	0.2	0	0	0.1	0.1	0	0	0	0	0	0	0	0	0	0	0	0	0
4	0	0.1	0	0	0	0	0	0	0	0	0	0	0	0	0	0	0	0	0	0	0
5	0	0	0	0	0	0	0	0	0	0	0	0	0	0	0	0	0	0	0	0	0
6	0	0	0	0	0	0	0	0	0	0	(0)	(0)	(0)	(0)	(0)	(0)	(0)	(0)	0	0.9	0
7	0	0.6	0	0.1	0	0.1	0	0	0	0	0	0	0	0	0	0	0	0	0	0	0
8	0.1	0.1	0	0.6	0	0	0	0	(0)	(0)	0	0	0	0	0	0.1	0.2	0	0	0	0.1
9	0	0	0	0	0	0	0	0	0	0	0	0.1	0	0	0	0	0	0	0	0	0
10	0	0.2	0	0	0	0	0.1	0	0	0	0	0	0.1	0.1	0	0	0	0	0	0	0.2
11	0	0.2	0	0	0	0	0	0	0	0	0	0	0	0	0	0	0	0	0	0	0
12	0	0	0	0	0	0	0	0	0	0	0	0	0	0	0	0.1	0	0.1	0	0.1	0
13	0	0.6	0.3	0.5	0.2	0.2	0.3	0.4	0.1	0.1	0.1	0	0.6	0.2	0	0.1	0.1	0.2	0	0.3	0
14	0	0	0	0	0.1	0	0.1	0.2	0	0	(0.1)	(0)	0	0	0	0.3	0	0.1	(0)	(0.2)	(0)
15	(0)	(0.3)	(0.1)	(0.1)	(0.1)	(0)	(0.1)	(0.1)	0.2	0	0.3	0	0.2	0.1	0.1	0.1	0	0.1	0	0.3	0.2
16	0	0.6	0	0	0	0	0	0	0	0	0.1	0	0	0.3	0	0.4	0	0	0	0	0

TABLE LXVIII (continued). F_D Kaafjord.

Gr. M.-T.	0—2		2—4		4—6		6—8		8—10		10—12		12—14		14—16		16—18		18—20		20—22		22—24	
Date	+	—	+	—	+	—	+	—	+	—	+	—	+	—	+	—	+	—	+	—	+	—	+	—
November 17	0	0.6	0	0.3	0.1	0	0	0.1	0	0.2	0	0	0	0	0	0	0.1	0.1	0	0	0.1	0	0	0
	0	0	0	0	0	0	0	0	0	0	0	0.1	0	0	0	0	0	0.1	0.1	0.1	0	1.7	0	1.4
	0	0.4	0.1	0	0.1	0	0	0.1	0.1	0	0.5	0	0.1	0	0.1	0	0.1	0.1	0	0.3	0	1.1	0	1.1
	0.1	0.1	0	0	0	0	0	0.1	0	0	0	0	0	0	0	0	0.2	0.4	0.1	0.8	0	1.0	0	0.1
	0.2	0	0	0	0	0	0	0	0	0.1	0.2	0	0.3	0	0.5	0	0.2	0.7	0.2	0	0	5.0	0	2.5
	0	1.0	0	0.4	0	0	0	0	0	0.1	0	0	0.1	0	0.1	0.3	1.3	0.7	0	3.9	0	4.3	0	3.3
	0	0.5	0	0.2	0	0.1	0.2	0.2	0.5	0.4	1.5	0	0.3	0.1	0.1	0.2	1.0	1.6	0.2	1.0	0.3	1.1	0.6	14.0
	0	6.2	0	9.3	0.5	1.0	1.8	0.1	1.5	0.1	3.1	0	4.4	0	5.6	0.1	0.9	2.9	1.4	1.6	0.4	10.2	0	9.0
	0	3.7	0	1.8	0.1	0.9	0.5	0	0.1	0.1	1.2	0	2.4	0	2.7	0	2.0	0.5	0.5	5.9	0	6.4	0	3.4
	0	2.3	0	0.2	0	0.3	0	0.2	0.2	0.1	0	0.2	0.1	0	0.4	1.1	2.7	0	1.2	0	0.1	0.1	0	0.4
	0	0.2	0	0.2	0	0.1	0.1	0.1	0	0	0.3	0	0	0.1	0	0.2	0.1	0	0	0	0	0	0.1	0
	0	0	0	0	0	0	0	0	0	0.1	0	0.2	0	0.1	0	0	0	0	0	0	0.2	0.1	0	0.1
	0	0	0	0	0	0	0	0	0	0.2	0	0	0	0	0	0	0	0	0	0	0	0	0	0
	0	0	0	0	0	0.1	0	0.1	0	0	0	0	0	0.3	0.1	0.1	0.2	0.3	0.7	0	0	0.1	0.1	0
December 1	0.	0	0	0	0	0	0	0	0	0	0.2	0	0.6	0	0.1	0	0.3	0.1	0.2	0	0.1	1.0	0.1	0
	0.	0	0	0	0	0.1	0.3	0	0	0.3	0.6	0	0.3	0	0.2	0	0	0.2	0	0.4	0	0	0	0.2
	0.	0.2	0	0.5	0	0.1	0.1	0	0	0.1	0.1	0	0	0.1	0.1	0	0	0	0	0	0	0	0	0
	0	0	0	0	0	0	0	0	0	0	0	0	0	0	0	0	0.2	0.2	0	1.2	0.2	0.6	0.1	0.1
	0	0	0	0	0	0	0.1	0	0	0.1	0.2	0	0	0	0	0	0	0	0	0	0	0	0	0.4
	0	0.7	0	0	0	0	0	0.1	0	0	0	0	0	0	0	0	0	0	0	0	0	0	0	0
	0	0	0	0	0	0	0	0	0	0	0.2	0	0	0	0	0	0	0	0	0.3	0.1	0.7	0	0.2
	0	0	0.1	0	0	0	0	0	0	0.1	0.1	0	0	0	0	0	0	0	0	0.1	0	0.7	0	0.1
	0	0	0	0	0	0	0.2	0	0.2	0.1	0.1	0	0.2	0	0.2	0	1.0	0	0.1	0.7	0.2	0.4	0.2	0
	0.1	0	0.2	0.1	0.4	0.1	0.1	0.1	0	0.3	0	0.3	0.1	0.2	0	0.1	0	0	0.1	1.5	0	1.6	0	0.5
	0	0.1	0.1	0.1	0	0	0.2	0.1	0	0	0	0.1	0.1	0.1	0	0.2	0	0.8	0.1	0.6	0	2.8	0	2.0
	0	0.5	0	0.2	0	0	0	0.1	0	0.1	0	0	0	0	0.1	0	0	2.2	0	1.8	0	0.5	0	0.4
	0.1	0.2	0.1	0.1	0.1	0	0.1	0	0	0.2	0	0.4	0	0.2	0.1	0	0	3.0	0	0.2	0.3	0.3	0	0.8
	0.1	0.1	0	0	0	0	0	0	0	0	0	0	0.1	0	0	0	0	0.1	0	0	0	0	0.1	0
	1.0	0.3	0	0.7	0	0	0	0	0	0	0	0	0	0	0	0	0	0	0.2	0.2	0.2	0.1	0	0
	0	0.1	0	0	0	0	0.1	0	0.3	0	0.6	0	0.3	0	0.6	0	0	0.8	0.3	0	0	0.6	0	0.2
	0	0.1	0.2	0	0.1	0	0	0	0	0	0.1	0	0	0	0	0	0	0	0	0.2	0.2	0	0	0
	0	0	0	0	0	0	0	0	0	0	0	0	0	0	0	0	0	0	0	0	0.2	0	0.2	0.1
	0	0	0.1	0	0.1	0	0.1	0	0	0	0	0	0	0.2	0.1	0	0.2	0	0	0.3	0.3	0.9	0	0
	0	0	0.1	0	0.1	0.1	0.1	0	0	0	0	0	0	0	0.2	0.1	0	0	0	0	0	0	0	0
	0	0	0.1	0	0	0	0	0	0	0	0	0	0	0	0	0	0.4	0	0.1	0.2	0.1	0	0	0.1
	0	0	0.1	0	0	0	0.1	0	0.1	0.2	0.4	0.1	0.1	0.1	0.2	0.1	0	0.1	0	1.0	0	3.2	0	7.1
	0	5.2	0	1.6	2.2	0	1.1	0.3	0.2	0.3	1.2	0	1.9	0.3	1.8	0	1.3	0.1	0.5	1.0	0	5.4	0	0.5
	0.4	0	0.4	0	0.2	0.1	0.2	0	0.1	0.4	0	0.4	0.2	0.1	0.5	0	0.2	0.8	0.1	1.9	0	1.7	0.1	0.4
	0.1	0.4	0.1	0.7	0.2	0	0	0	0.1	0.2	0.1	0.3	0	0.1	0	0	0.1	0	0.2	0.3	0	0.3	0.7	0
	0.2	0	0.1	0	0.1	0	0	0	0	0	0.1	0.2	0.3	0.1	0.9	0	0.1	0.2	0	0.3	0	1.6	0.2	0.4
	0	0.3	0	0.3	0	0	0.1	0	0	0.1	0	0.2	0.1	0.1	0	0	0	0	0	0.2	0	1.7	0	3.1
	0	3.3	0	0.5	0	0.5	0.2	0.2	0	0.4	0.1	0.3	0	0.2	0.1	0	0.4	0.1	0.3	0.1	0.1	0.5	0	0
	0	0	0.1	0	0	0	0	0.1	0	0	0	0	0	0	0	0	0.1	0.2	0.1	0.2	0.1	0.1	0	0.1
	0	0	0	0	0	0	0	0	0	0.1	0.1	0	0	0	0	0	0	0.4	0	0.1	0	0.1	0.2	0.1
	0.1	0	0	0	0	0	0	0	0	0	0	0	0	0	0	0	0	0	0	0	0	0	0	0.1
January	0	0	0	0	0	0	0	0.1	0	0	0.1	0	0	0	0	0	0	0	0	0	0.1	0	0	0
	0.1	0.1	0	0	0	0	0	0	0	0.1	0.1	0	0	0	0	0	0.1	0	0	1.1	0	0.3	0.1	0
	0.2	0	0.2	0	0.1	0.7	0.2	0	0	0.2	0.1	0	0	0	0	0	0.1	0	0.1	0.2	0.5	0.1	0.2	0
	0	0.2	0.1	0.5	0	0.1	0.1	0	0.1	0.3	0.6	0	0.5	0	0	0.2	0.2	0.6	0	0.3	0.1	0.3	0.1	0.1
	0.1	0.1	0.4	0	0.2	0.8	0.4	0.2	0.2	0.6	0.4	0	0.1	0.1	0.2	0.2	0.4	0.7	0.1	1.6	0.7	0.8	0.7	0.2

TABLE LXVIII (continued).

F_D

Gr. M.-T.	0—2		2—4		4—6		6—8		8—10		10—12		12—14		14—16		16—18		18—20		20—22		22—24	
Date	+	—	+	—	+	—	+	—	+	—	+	—	+	—	+	—	+	—	+	—	+	—	+	—
January 6	0	0.1	0.1	0.1	0	0	0	0	0.1	0	0	0	0	0.1	0	0.1	0.1	0.2	0	0.9	0.3	0.1	0.2	0.1
7	0	0	0	0	0	0	0	0	0	0	0.1	0	0	0	0	0	0	0.1	0	0.6	0	0.1	0	0
8	0	0	0	0	0	0	0.1	0	0.1	0	0.1	0.1	0.1	0.1	0.2	0	0.1	0.4	0	0.1	0.1	0.6	0	0.3
9	0	0	0	0	0	0.1	0.2	0	0	0.1	0	0	0	0	0.7	0	0.4	0.1	0	0.3	0.1	0.2	0.1	0.8
10	0.2	0.3	0.2	0	0.6	0	0.4	0	0.2	0	0.2	0	0.3	0	0.6	0	0.5	0.1	0.2	0	0	0.9	0	0.9
11	0.1	0.3	0.5	0	0.3	0	0.2	0	0	0.1	0.1	0.1	0.1	0.1	0.2	0.1	0.4	0.2	0.3	0.7	0.6	0.4	0.1	0.2
12	0.2	0.1	0.1	0.1	0.3	0	0.1	0	0.1	0	0	0.1	0.1	0.1	0.4	0	0.1	0	0	1.0	0	0	0.1	0.3
13	0.2	0.4	0	0.2	0.1	0	0	0	0	0	0.1	0	0	0	0	0.1	0.3	0.5	0	2.8	0	1.0	0	0.2
14	0	0	0	0	0.1	0	0	0	0	0.1	0.1	0	0.5	0	0.1	0	0.1	0	0	0	0	0	0	0.5
15	0.2	0.2	0	0	0	0	0	0	0	0	0	0	0	0	0.1	0	0	0	0	0	0.1	0	0.1	0
16	0	0	0.1	0	0	0	0	0.1	0	0.1	0.2	0	0.7	0	0.3	0	0.1	0.7	0.1	0.1	0.3	0.2	0.1	0.3
17	0.2	0.1	0	0.1	0	0.1	0.1	0	0	0	0	0.1	0.1	0	0.3	0	0	0	0.1	0.1	0	0.1	0	0
18	0.2	0	0	0	0	0	0	0	0	0	0	0	0.2	0	0.1	0.7	0.3	0.6	0.5	0.5	0.3	0	0.3	0.1
19	0.1	0	0.3	0	0	0	0	0	0	0.1	0.2	0.1	0.2	0	0.4	0	0	0.6	0	0.2	0	1.2	0.2	0.3
20	0	0.1	0.1	0.1	0.1	0	0.2	0	0.5	0	0.6	0	0	0	0.1	0.1	0	0	0	0.6	0.1	0.6	0	0
21	0	0	0	0	0.1	0	0.1	0.1	0.1	0	0.1	0	0.3	0	0.1	0.2	0.1	0	0.1	0.3	0	1.3	0.2	0.3
22	0.1	0.5	0	0	0	0	0	0	0.1	0	0.2	0	0.1	0	0	0	0	0.2	0	1.1	0	0.5	0.1	0.3
23	0	0.3	0	0	0.1	0	0	0	0.1	0	0.4	0	0.2	0	0.1	0.2	0.1	0.6	0.1	1.3	0	3.4	0	0.6
24	0.1	0.1	0.3	0	0.1	0.1	0.1	0.1	0.1	0.1	0.4	0	0.3	0	0.2	0	0.1	1.1	0	1.2	0.1	0.5	0	0.2
25	0.1	0.1	0.4	0	0.2	0	0.1	0	0	0.2	0	0	0.1	0	0.1	0.1	0.1	0.3	0.1	0	0	0	0	0
26	0	0	0	0	0	0	0	0	0.2	0.1	0.5	0	0.6	0	0.1	0	0	0.3	0	3.7	0	9.0	0	11.6
27	0	1.8	0	0.3	(0)	(0)	(0)	(0)	(0.1)	(0)	0	0.2	0.1	0.1	0.1	0.1	0.1	0.2	0.2	0.1	0.2	0.4	0	0.7
28	0	0	0	0	0	0	0	0	0	0	0	0	0	0.2	0	0.2	0.1	0.5	0.1	0.3	0	0.6	0	0.2
29	0	0.1	0	0	0	0	0	0	0	0	0	0.1	0.1	0	0	0	0	0	0	0	0	0	0	0
30	0.1	0	0.1	0	0.1	0	0.1	0.1	0.2	0.2	0.5	0	1.5	0	1.3	0	1.6	0	1.0	0.2	0	0.2	0.	0
31	0.1	0.1	0	0	0	0	0.1	0	0	0.1	0	0	0	0	0.1	0.2	0.3	0.2	0.2	0	0	0	0	0
February 1	0	0	0	0	0	0	0	0	0	0.1	0.1	0	0	0	0	0	0.1	0	0.1	0.2	0	0	0	0
2	0	0	0	0	0	0	0	0	0	0	0.2	0	0	0	0	0	0	0	0	0	0	0	0	0.1
3	0	0.1	0	0	0.1	0.1	0	0.1	0.1	0	0.1	0.1	0.2	0	0	0	0	0	0	0	0	0	0	0
4	0	0	0.1	0	0.1	0	0	0	0	0	0.1	0	0.2	0	0	0	0	0	0	0	0	0.2	0	0.4
5	0	0	0	0	(0)	(0)	(0)	(0)	(0)	(0)	(0.2)	(0)	(0.1)	(0)	(0)	(0)	(0.1)	(0.1)	(0.2)	(0.2)	(0.1)	0.2)	(0)	(0.2
6	0	0	0	0	(0)	(0)	(0)	(0)	0	0	0	0	0	0.2	0	0.1	0.1	0	0	0	0	0	0	0
7	0	0	0	0.1	0	0	0	0.1	0.1	0.1	0	0.1	0	0	0	0	0.1	0.3	0	0	0.2	1.4	0	7.7
8	0.1	0.9	0.2	2.5	0	2.6	0.1	0.4	0.2	0.2	1.4	0	0.6	0.1	2.5	0.1	0.9	1.3	0.8	2.3	0.1	9.0	0	1.3
9	0.1	0.2	0.5	0.3	0.2	0.2	0	0.2	0	0.1	0.1	0.1	0	0	0	0	0.1	0.2	0	2.0	0.1	1.2	0	2.7
10	0	0.9	0.1	0.2	0.1	0	0.1	0	0	0.1	0	0.1	0	0	0	0	0	0	0	0.1	0	0.7	0.	3.3
11	0	2.1	0	0.3	0.1	0	0.1	0.1	0	0.1	0	0.1	0.1	0	0	0.1	0.1	0.1	0	1.1	0	0.6	0	1.6
12	0.3	0.3	0	0.1	0.1	0	0	0.2	0	0	0.2	0	0.1	0	0	0	0	0.7	0.5	1.2	0	1.3	0	0.6
13	0.1	0.8	0	0.3	0.1	0.1	0	0.2	0.1	0.2	0.4	0	0.1	0	0	0	0	0.2	0	0.8	0	0.9	0.	0.3
14	0.3	0.1	0.5	0.1	0	0	0.1	0.1	0.1	0	0.2	0.1	0.1	0.1	0.1	0.3	0	0.1	0	0.8	0	0.7	0	1.0
15	0	0	0.1	0.1	0	0.1	0.1	0	0.3	0	0.5	0	0.5	0	0.1	1.4	0	4.3	0	0.1	0	0.2	0.	0
16	(0.1)	(0)	0	0	0	0	0	0	0.1	0.1	0.1	0	0	0	0	0	0.1	0.3	0.1	0.5	0.1	0.1	0	0
17	0.4	0.1	0	0	0.1	0	0	0.1	0	0.2	0.1	0	0	0	0.1	1.2	0	0.2	0	0	0	0.4	0	0.2
18	0.4	0	0	0	0	0	0	0	0	0	0	0	0	0	0	0	0	0	0	0	0	0	0	0
19	0	0	0	0	0	0	0	0	0	0.1	0.1	0	0.1	0	0	0	0	0	0	0	0	0	0	0
20	0	0.1	0	0.1	0.1	0.1	0	0.1	0	0	0.1	0	0	0.1	0	0	0	0	0	0	0	0	0	0
21	0	0	0.1	0	0.2	0	0.2	0	0.1	0	0.4	0	0.9	0	0.5	0	0.1	0.1	0	0	0	0	0	0.1
22	0	0.1	0.1	0.5	1.8	0	2.6	0	1.0	0	0.7	0	0.3	0.1	0.1	0	0.2	0	0.1	0	0	0	0.	0.2
23	0.1	0.1	0	0	0.1	0.1	0	0	0	0	0	0	0.2	0.1	0	0	0.3	0	0.2	0	0	0.6	0	0.5
24	0	0.5	0	0.1	0	0.1	0	0	0	0	0.1	0	0.1	0	0	0.3	0.1	0.1	0	0	0	0	0	0

(continued). F_0 Kaafjord.

							8—10		10—12		12—14		14—16		16—18		18—20		20—22		22—24	
−	+	−	+	−	+	−	+	−	+	−	+	−	+	−	+	−	+	−	+	−	+	−
0	0.2	0.1	0.9		2	0	0.4	0	0.4	0.1	0.4	0	0	0.3	0	0.2	0	0	0	0	0	0
0.1	0	0.1	0			0	0	0	0	0	0	0	0	0	0	0	0	0	0.1	0.5	0	0.9
0.2	0	0.2	0			0	0	0	0	0	0	0	0	0	0	0	0	0	0	0	0	0
0	0	0	0			0	0.1	0	0	0.1	0	0	0	0	0.1	0	0	0	0.1	0	0	0
0	0	0	0	8	8	0	0.1	0	0.1	0	0.1	0	0.1	0	0.3	0.5	0	0.6	0	0.2	0	2.1
0.6	0	0.2	0.1	0	0	0.2	0.8	0.3	0.3	0	0	0	0	0.1	0	0.3	0.1	0.2	(0.1)	(0.5)	(0.2)	(0.5)
(0.2)	(0)	(0.2)	(0)	(0)	(0)	(0)	(0)	(0)	(0)	(0)	(0)	(0)	(0)	(0.1)	(0)	(0)	(0.2)	(0.2)	(0.1)	(0.2)	(0.1)	(0.2)
(0.2)	(0)	(0.2)	(0)	(0)	(0)	0	0	0	0	0	0.1	0	0	0.2	0	0	0	0	0.1	0	0.3	0.4
0.4	0.2	0.7	0.2	0.1	0.1	0.1	(0)	(0)	(0)	(0)	(0.1)	(0)	(0)	(0)	1.3	0	1.0	0	0.4	0	0.1	0
0.2	0	0.1	0	0	0.2	0	0	0	0.1	0	0.4	0	0	0	0.1	0	0.3	0.5	0.2	0.9	0.4	0.5
0.3	0.2	0.2	0.4		0.5	0.2	0.3	0	1.3	0	1.3	0	0	0.8	0.8	0.2	0.1	0.3	0.1	0.2	0.2	0.3
0.3	0.2	0	0.1		0.1	0	0.4	0	0.9	0	0.6	0.2	1.0	0	0.7	0.9	0.4	1.6	0.2	3.1	0	7.2
1.7	0.6	0.3	1.6		0.6	0.1	0.2	0.1	0.1	0	0	0.1	0	0	0.1	0	0.1	0.5	0.7	0.1	0.2	0
0	0.3	0	0		0.1	0.1	0.1	0	0	0	0.1	0.1	0	0.6	0.1	1.1	0	2.5	0.1	1.1	0.3	0
0	0.1	0.1	0	8	0	0	0	0	0	0	0.2	0	0	0	0.2	0.1	0.3	0.8	0.1	0.3	0.4	0

1	2—4		4—6		6—8		8—10		10—12		12—14		14—16		16—18		18—20		20—22		22—24	
−	+	−	+	−	+	−	+	−	+	−	+	−	+	−	+	−	+	−	+	−	+	−
(0.1)	(0)	(0)	(0)	(0.1)	(0)	(0)	0	0	0	0	0.1	0	0	0.2	0	0.2	(0)	(0)	0	0.3	0	0.1
0	0	0	0	0.1	0	0	0.2	0	0.1	0	0.1	0	0.1	0	0.1	0	0	0	0	0.1	0	0
0	0	0	0	0	0	0	0.1	0	0	0.1	0	0	0	0	0	0	0	0	0	0	0	0.1
0	0	0	0	0.1	0	0	0.1	0	0	0.1	0.5	0	0.4	0	0.1	0	1.0	0	0	0	0	0
2.2	0	0.9	0	0.4	0	0	0	0	0	0.1	0	0	0	0.1	0	0	0	0	0	0	0	0
0	0	0	0	0	0	0	0	0	0	0.1	0	0	0	0	0	0	0	0	0	0.2	0	0
0.1	0	0	0	0.1	0	0.1	0	0	0	0	0	0	0	0	0	0	0	0.1	0	0.4	0	0.1
0	0	0	0	0	0	0	0	0.1	0	0	1.2	0	1.7	0	1.1	0	0.1	0	0.1	0.1	0	1.0
0.1	0	0	0	0	0	0	0	0.1	0	0	0	0.1	0	0	0.2	0	0.4	0.2	0	0.4	0	1.0
0.7	0	0	0	0	0	0	0	0	0	0.3	0.2	0.1	1.3	0	1.3	0	1.9	0	1.5	0.1	1.8	4.7
0.3	0	0.2	0	0.5	0	0.1	0	0	0	0	0	0	0	0	0	0	0	0	0	0.1	0	0.1
0	0	0	0	0	0	0	0	0	0	0	0	0	0	0	0	0	0	0	0	0	0	0
0	0	0	0	0	0	0	0	0	0.1	0	0	0.1	0	0	0	0	0	0	0	0.3	0	1.7
0.1	0	0	0	0	0	0	0	0	0	0	0	0	0	0	0	0	0	0	0	0	0	0
0	0	0	0	0.1	0	0.1	0	0	0	0.1	0	0	0.2	0.1	0.9	0	0.1	0	0	0	0	0
0	0	0	0	0	0	0	0	0	0.3	0.1	1.0	0	0.2	0	0	0.1	0.1	0	0.8	5.1	0	2.1
0.6	0.7	0	0.1	0	0	0	0.3	0.1	0.6	0	0.5	0	1.3	0	2.0	0	0.1	1.3	0.1	8.2	0	5.3
2.0	0	1.5	0	0.4	0	0.1	0.9	0	2.0	0	1.2	0	0.1	0.1	0.1	0	0.1	2.0	0	4.8	0	5.4
0.4	0	0	0	0.1	0	0	0.1	0	0	0	0.2	0	0.2	0	0.2	0	0	0	0	0	0	0
0	0	0	0	0	0	0	0	0	0	0	0	0.2	0.1	0.2	0.9	0	0.1	2.4	0	2.8	0	7.1
6.0	0	2.3	0	0.4	0	0	0	0	0	0.1	1.0	0	0.1	0.2	0.9	0	0.3	1.3	0	4.3	0	1.3
0.4	0	0.1	0	0	0	0	0	0	0	0	0	0	0	0	0	0	0	0	0	0	0	0.1
0	0	0	0	0	0	0	0	0.2	0.1	0.1	0	0	0.3	0	0.1	0	0	0	0	0	0	0
0.6	0	0	0	0	0	0	0	0	0	0	0	0	0	0	0.1	0	0.5	0	0	0.6	0	0.1
0.3	0	0.1	0	0	0	0	0	0	0	0	0.2	0	0	0	0	0	0.1	0	0	0	0	0
0.8	0	0.4	0	0	0	0	0	0	0.1		0.1	0	0	0	0.1	0	0	0	0	0	0	0
0.3	0	0	0	0	0	0	0	0	0		0	0.1	0	0.1	0	0.1	0.1	0	0.1	0.8	0	11.7
2.5	0	0	0	0	0	0	0.1	0	0	1	0	0.2	0	0.3	1.8	0	1.0	0.2	0	5.6	0	8.5
8.4	0	6.0	0	2.3	0	0.5	0	0	0		0	0.1	0	0	0	0	0	0.1	0	0.2	0	0.3
0	0	0	0	0	0	0	0	0	0	8	0	0	0	0	0	0	0.1	0	0	0	0	0.1

TABLE LXIX (continued). F_v

Gr. M.-T.	0—2		2—4		4—6		6—8		8—10		10—12		12—14		14—16		16—18		18—20		
Date	+	—	+	—	+	—	+	—	+	—	+	—	+	—	+	—	+	—	+	—	+
October 3	0	0.1	0	0	0	0	0	0	0	0	0	0.1	0	0	0	0	0	0	0.2	0	0
4	0	0.4	0	0	0	0	0	0	0.1	0	0.1	0	0.5	0	0.1	0	0	0	0	0	0
5	0	0.2	0	0	0.1	0	0.1	0	0	0	0	0	0	0	0	0	0	0	0	0	0
6	0	0	0	0	0	0	0	0	0	0	0	0	0	0	0.1	0	0	0	0	0	0
7	0	0	0	0	0	0	0	0	0	0	0	0	0	0	0	0	0	0	0	0	0
8	0	0	0	0	0	0	0	0	0	0	0	0.1	0	0	0	0	0	0	0.3	0	0
9	0	0	0	0	0	0	0	0.1	0	0	0	0	0	0	0.6	0	0.2	0	0	0.1	0
10	0	0	0	0	0	0	0	0	0	0	0	0	0	0	0	0	0	0.1	0	0	0
11	0	0	0	1.2	0	0	0.1	0	0	0	0	0	1.0	0	0.3	0	(0)	(0)	0	2.1	2.5
12	0	1.9	0	0	0	0	0	0	0	0	0.1	0	0.1	0	0.1	0	0	0	0	0	0
13	0	0	0	0	0	0	0	0	0	0	0.1	0	0	0	0	0	0.1	0	0	0.7	0
14	0	0	0	0	0	0	0	0	0	0	0	0	0	0	0	0	0	0	0	0	0
15	0	0	0	0	0	0	0	0	0	0	0	0	0	0	0	0	0	0	0	0	0
16	0	0	0	0	0	0	0	0	0	0	0	0	0	0	0	0	0	0	0	0	0
17	0	0	0	0	0	0	0	0	0	0	0	0	0	0	0	0	0	0	0	0	0
18	0	0	0	0	0	0.2	0	0	0	0	0	0	0	0	0	0	0	0	0.1	0.5	0
19	0	1.0	0	0.7	0	0.1	0	0	0	0	0	0	0	0	0	0	0	0	0	0	0
20	0	0	0	0	0	0	0	0	0	0	0.1	0	0	0	0	0	0	0	0	0	0
21	0	0	0	0	0	0	0	0	0	0	0	0	0	0	0	0	0.4	0	0	0	0
22	0	0.4	0	0	0	0	0	0	0	0	0	0	0	0	0	0	0	0	0	0	0
23	0	0	0	0	0	0	0	0	0	0	0	0	0	0	0	0	0	0	0	0.1	0
24	0	1.0	0	0.5	0	0	0	0	0.1	0	0	0	0	0	0.1	0	1.2	0	1.3	0	0
25	0	7.1	0	4.8	0	1.3	0.1	0.1	0.6	0	1.7	0	2.1	0	1.2	0	0.3	0	0	0	0
26	0	0.3	0	0.3	0	0	0	0	0	0	0	0.2	0	0	0.3	0	0.5	0	0.6	0	0
27	0	1.9	0	1.5	0	0.8	0.1	0.1	0.1	0	0.2	0	0.7	0	1.5	0.1	1.9	0	0.8	0	0
28	0	1.4	0	0.1	0	0.1	0	0.1	0	0	0	0	0.2	0	0.5	0	0.4	0	0	2.1	0
29	0	2.0	0	0.6	0	0.1	0	0	0	0	0	0	0	0	0	0	(0.2)	(0.1)	0	0.7	0
30	0	2.0	0	0.1	0	0.4	0.1	0.4	1.5	0	1.3	0.1	0	0.3	0.1	0	0.1	0.1	0	0.4	0
31	0	4.8	0	4.6	0	4.8	0	1.9	1.3	0	1.6	0	0	6.3	0	8.2	0	6.8	4.3	0.5	5.9
November 1	0	3.0	0	1.9	0.4	0	0.1	0.1	0.1	0	0.1	0	0	0	0	0	0	0	0	0	0
2	0	0	0	0	0	0	0	0	0	0	0	0	0	0	0.1	0	1.4	0	0.5	0.9	0
3	0	0	0	0	0	0	0	0	0	0	0	0	0	0	0	0	0	0	0	0	0
4	0	1.2	0	0	0	0	0	0	0	0	0	0	0	0	0	0	0	0	0	0	0
5	0	0	0	0	0	0	0	0	0	0	0	0	0	0	0	0	0	0	0	0	0
6	0	0	0	0	0	0	0	0	0	0	(0)	(0)	(0)	(0)	(0)	(0)	(0)	(0)	0.1	0.2	0
7	0	0.8	0.1	0	0	0	0	0	0	0	0	0	0	0	0	0	0	0	0	0	0
8	0	0.3	0	0.7	0	0.1	0	0	0	0	0	0	0	0	0	0	0	0	0	0	0
9	0	0	0	0	0	0	0	0	0	0	0	0	0	0	0	0	0	0	0	0	0
10	0	0	0	0	0	0	0	0	0	0	0	0	0	0	0	0	0	0	0.1	0	0.2
11	0	0	0	0	0	0	0	0	0	0	0	0	0	0	0	0	0	0	0	0	0
12	0	0	0	0	0	0	0	0	0	0	0	0	0	0	0	0	0	0	0	0	0
13	0	0.7	0	1.7	0	1.5	0	1.0	0	0.1	0	0	1.3	0	0.4	0	0	0	0	0	0
14	0	0	0	0	0	0.4	0	0.2	0	0	(0)	(0)	(0)	(0)	0.2	0.4	0	0	0	0	1.7
15	0	0.9	0	0.9	(0)	(0.1)	(0)	(0)	0	0	0	0	0	0	0	0	0	0	0	0	0
16	0	0.6	0	0	0	0	0	0	0	0	0	0	0.1	0	0.2	0	0	0	0	0	0
17	0	0.5	0	0	0	0	0	0	0	0	0	0	0	0	0	0	0	0	0	0	0
18	0	0	0	0	0	0	0	0	0	0	0	0	0	0	0	0	0	0	0	0	0
19	0	0	0	0	0	0	0	0	0	0	0	0	0	0	0	0	0	0	0	0	0
20	0	0	0	0	0	0	0	0	0	0	0	0	0	0	0.3	0	0.6	0	0.7	0	0.1
21	0	0.1	0	0	0	0	0	0	0	0	0	0	0	0	0	0	0	0	0	0	0

ontinued). F_V Kaafjord.

	2	2–4		4–6		6–8		8–10		10–12		12–14		14–16		16–18		18–20		20–22		22–24	
+	−	+	−	+	−	+	−	+	−	+	−	+	−	+	−	+	−	+	−	+	−	+	−
0	0.2	0	0	0	0	0	0	0	0	0	0	0	0	0	0	0	0.3	0.2	0.1	0	0.1	0	0.8
0	0	0	0	0	0	0	0	0	0	0.2	0	0.2	0	0	0	0.4	0	0.9	0	0.8	0	0.2	2.2
0	4.6	0	3.8	0	3.0	0	1.7	0	0	0	0	0	0.2	0	0.4	0.1	0.4	0	0.3	0.1	0.2	0	0.7
0	0.7	0	0.2	0	0.1	0	0	0	0	0.1	0	0.3	0	0.4	0	0	0.1	0	0.9	0	1.7	0	1.2
0	1.5	0	0.5	0	0.1	0	0	0	0	0	0	0	0	0.2	0	0.2	0	0.1	0	0	0	0	0
0	0	0	0	0	0	0	0	0	0	0	0	0	0	0	0	0	0	0	0	0	0	0	0
0	0	0	0	0	0	0	0	0	0	0	0	0	0	0	0	0	0	0	0	0	0	0	0
0	0	0	0	0	0	0	0	0	0	0	0	0	0	0	0	0	0	0	0	0	0	0	0
0	0	0	0	0	0	0	0	0	0	0	0	0	0	0	0	0	0	0	0	0	0	0	0
0	0	0	0	0	0	0	0	0	0	0	0	0	0	0	0	0	0	0	0	0	0	0	0
0	0	0	0	0	0	0	0	0	0	0	0	0	0	0	0	0	0	0	0	0	0	0	0
0	0	0	0	0	0	0	0	0	0	0	0	0	0	0	0	0.1	0	0	0	0	0	0	0
0	0	0	0	0	0	0	0	0	0	0	0	0	0	0	0	0	0	0	0	0	0	0	0
0	0	0	0	0	0	0	0	0	0	0	0	0	0	0	0	0	0	0	0	0	0	0	0.1
0	0.1	0	0	0	0	0	0	0	0	0	0	0	0	0	0	0	0	0	0	0	0	0	0
0	0	0	0	0	0	0	0	0	0	0	0	0	0	0	0	0	0	0	0	0	0	0	0
0	0	0	0	0	0	0	0	0	0	0	0	0	0	0	0	0	0	0	0	0	0	0	0
0	0	0	0	0	0	0	0	—	—	—	—	—	—	—	—	—	—	—	—	—	—	—	—
	—	—	—	—	—	—	—	—	—	—	—	—	—	—	—	—	—	—	—	—	—	—	—
	—	—	—	—	—	—	—	—	—	—	—	—	—	—	—	—	—	—	—	—	—	—	—
	—	—	—	—	—	—	—	—	—	—	—	—	—	—	—	—	—	—	—	—	—	—	—
	—	—	—	—	—	—	—	—	—	—	—	—	—	—	—	—	—	—	—	—	—	—	—
	—	—	—	—	—	—	—	—	—	—	—	—	—	—	—	—	—	—	—	—	—	—	—
	—	—	—	—	—	—	—	—	—	—	—	—	—	—	—	—	—	—	—	—	—	—	—
	—	—	—	—	—	—	—	—	—	—	—	—	—	—	—	—	—	—	—	—	—	—	—
	—	—	—	—	—	—	—	—	—	—	—	—	—	—	—	—	—	—	—	—	—	—	—
	—	—	—	—	—	—	—	—	—	—	—	—	—	—	—	—	—	—	—	—	—	—	—
—	—	—	—	—	—	—	—	0	0	0	0	0	0	0.1	0	0.2	0	0.2	0.1	0	1.1	0	0.2
0	0	0	0	0	0	0	0	0	0	0	0	0	0	0.1	0	0	0	0	0	0	0	0	0
0	0	0	0	0	0	0	0	0	0	0	0	0	0	0	0	0	0	0.2	0	0	0	0	0
0	0	0	0	0	0	0	0	0	0	0.2	0	0.3	0	0.3	0	0	0.2	0.5	0	0	1.3	0	7.1
0	6.9	0	5.5	0	3.0	0	1.3	0.1	0	0.6	0	2.3	0	1.8	0	1.8	0	0.2	2.4	0	4.9	0	0.3
0.1	0	0	0.3	0	0	0	0	0	0	0.1	0	0.5	0	1.1	0	0.5	0	0.1	0.8	0	1.3	0	1.0
0	0.8	0	0.8	0	0.5	0	0	0	0	0.1	0	0	0	0	0	0.2	0	0.3	0.7	0	0.3	0	0.9
0.1	0.1	0	0	0	0	0	0	0	0	0	0	0.3	0	0.6	0	0.4	0	0	0	0	2.0	0	2.8
0	0.4	0	0	0	0	0	0	0	0	0	0	0	0	0	0	0	0	0.7	0	0.3	0.4	0	2.9
0	2.9	0	0.2	0.1	0.1	0.3	0	0	0	0.1	0	0.6	0	0.2	0	0.4	0	0.2	0	0.1	0.1	0	0
0	0	0	0	0	0	0	0	0	0	0	0	0	0	0	0	0.4	0	0.2	0	0	0	0	0
0	0	0	0	0	0	0	0	0	0	0	0	0	0	0	0	0.2	0	0	0	0	0	0	0.1
0	0	0	0	0	0	0	0	0	0	0	0	0	0	0	0	0.1	0	0	0	0	0	0	0
(0)	(0)	(0)	(0)	(0)	(0)	(0)	(0)	0	0	0	0	0	0	0	0	0.1	0	0	0	0	0	0	0.3
0	0.1	0	0	0	0	0	0	0	0	0	0	0	0	0	0	0	0	0	0	0	0	0	0
0	0	0	0	0	0	0	0	0	0	0	0	0	0	0	0	0	0	0.4	0	0	0.3	0	0
0	0	0	0.2	0	0	0	0.2	0	0	0	0	0	0	0	0	0	0	1.0	0	0.7	0	0.1	0
0	0	0	1.0	0.8	0	0	0.4	0	0.4	0.1	0.2	0	1.4	0	0.7	0.4	0	0	0	0	0.6	0	0.1
0	0	0	0	0	0	0	0	0	0	0	0	0	0	0	0	0	0	0.3	0	0	1.7	0	0.5
0	0	0	0	0	0	0	0	0	0	0	0	0	0	0	0	0	0	0	0	0	0	0	0
0	0	0	0	0	0	0	0	0	0	0	0	0	0	0	0.1	0.6	0	0	0.1	0	1.1	0	0.1
0	0	0	0	0	0	0	0	0	0	0	0	0	0	0	0.2	0.6	0	0.2	0	0.2	0	0	0.1
0.1	0.2	0.1	0.2	0	0	0	0	0	0	0	0	0	0	0.9	0	0.5	0	0	0.1	0.1	0	0	0.6

TABLE LXIX (continued). F_v

Gr. M.-T.	0–2		2–4		4–6		6–8		8–10		10–12		12–14		14–16		16–18		18–20		20
Date	+	–	+	–	+	–	+	–	+	–	+	–	+	–	+	–	+	–	+	–	+
January 11	0	0.1	0	0	0	0	0	0	0	0	0	0	0	0	0.5	0	0.1	0.1	1.1	0	0.
12	0	0.2	0	0.1	0.1	0	0	0	0	0	0	0	0	0	0.1	0	0	0	0	0.7	0
13	0	0.5	0	0	0	0	0	0	0	0	0	0	0	0	0.1	0	0.4	0	0.5	0.3	0.
14	0	0	0	0	0	0	0	0	0	0	0	0	0	0	0	0	0	0	0	0	0
15	0.1	0.3	0.3	0	0	0	0	0	0.1	0	0.3	0.1	0.6	0	0.7	0	0	0	0.1	0.2	0
16	0	0	0	0	0	0	0	0.1	0	0	0	0	1.6	0	0.6	0	0.1	0	0.1	0	0
17	0	0.2	0	0	0	0	0	0	0	0	0	0	0	0	0	0	0	0	0.2	0	0
18	0	0	0	0	0	0	0	0	0	0	0	0	0	0	1.7	0	2.7	0	1.6	0	0.1
19	0	0.2	0	0	0	0	0	0	0	0	0	0.1	0	0.2	0.1	0.5	1.2	0	0	0	0
20	0	0	0	0	0	0.4	0	0.1	0	0	0	0	0.1	0	0	0	0	0	0.5	0	0.4
21	0	0	0	0	0	0.2	0	0.1	0	0	0	0	0.2	0	0.5	0	0.2	0	0.7	0	0.3
22	0	1.8	0	0.1	0	0	0	0	0	0	0	0	(0.1)	(0)	(0.2)	(0)	0.1	0	0.1	0	0
23	0	0.2	0	0	0	0	0	0.1	0	0	0	0	0.1	0	0.3	0	0.9	0	0.8	0.1	0
24	0	0	0	0.1	0	0	0	0	0	0	0	0	0	0	0	0	0.5	0	0	0.4	0
25	0	0	0	0	0	0.1	0	0	0	0	0	0	0	0	0	0.1	0	0.1	0	0	0
26	0	0	0	0	0	0	0	0	0	0.1	0	0	0	0	0	0	0.2	0	0	4.5	0
27	0	1.9	0	0.1	(0)	(0)	(0)	(0)	(0)	(0)	0	0	0	0	0	0	0.2	0	0	0.2	0
28	0	0	0	0	0	0	0	0	0	0	0	0	0	0	0	0	0.9	0	0.9	0	0
29	0	0	0	0	0	0	0	0	(0)	(0)	(0)	(0)	(0)	(0)	(0)	(0)	(0.2)	(0)	(0.2)	(0.2)	(0)
30	(0)	(0)	(0)	(0)	(0)	(0)	(0)	(0)	(0)	(0)	(0)	(0)	0.5	(0)	(1.0)	(0)	(1.0)	(0)	(1.0)	(0.5)	(0.2)
31	(0)	(0)	(0)	(0)	(0)	(0)	(0)	(0)	0	0	0	0	0	0	0.3	0	1.5	0	0.2	0	0
February 1	0	0	0	0	0	0	0	0	0.9	0	0	0.4	0	0	0	0	0	0	0	0	0
2	(0)	(0)	(0)	(0)	(0)	(0)	(0)	(0)	(0)	(0)	(0)	(0)	(0)	(0)	(0)	(0)	0	0	0	0	0
3	0	0	0	0	0	0	0	0	0	0	0	0	0.1	0	0	0	0	0	0	0	0
4	0	0	0	0	0	0	0	0	0	0	0	0	0	0	0	0	0	0	0	0	0
5	0	0	0	0	0	0	0	0	0	0	0	0	0	0	0.1	0	0.8	0	0.6	0	0
6	0	0	0	0	0.1	0.2	0	0	0	0	0	0	0.1	0	0.1	0	0	0	0	0.2	0
7	0	0	0	0	0	0	0	0	0	0	0	0	0	0	0	0	0.5	0	0	0	0
8	0	3.5	0	2.0	0	3.7	0.1	0.6	0.1	0.4	0.7	0.1	0.3	0	3.0	0	3.5	0	0.8	7.4	0
9	0	0.4	0	0.6	0	0.1	0	0	0	0	0	0	0	0	0	0	0	0	0.3	0	0
10	0	1.6	0	0.4	0	0	0	0	0	0	0	0	0	0	0	0	0	0	0.2	0	0.1
11	0	3.4	0	0.5	0	0.1	0.2	0	0	0	0.2	0	0	0	0	0	0.6	0	0.9	0	0.2
12	0	0.8	0	0.1	0	0	0	0	0	0	0	0	0	0	0	0	0.4	0	2.1	0	0.3
13	0	2.1	0	0.5	0.3	0.1	0.2	0	0	0	0	0	0	0	0	0	0	0	0.1	0.5	0
14	0	1.9	0	0.1	0	0	0	0	0	0	0.1	0	0	0.2	0.3	0	0.1	0	0	0	0
15	0	0	0	0.1	0	0	0	0	0	0	0.4	0	0.6	0	3.9	0	1.1	0.4	0.2	0	0
16	0	0	0	0	0	0	0	0	0	0	0	0	0	0	0	0	0.3	0	0.1	0	0
17	0	0.2	0	0	0	0	0	0	0	0	0	0	0	0	0.9	0	0.4	0	0	0	0
18	0	0.1	0	0	0	0	0	0	0	0	0	0	0	0	0	0	0	0	0	0	0
19	0	0	0	0	0	0	0	0	0	0	0	0	0	0	0	0	0	0	0	0	0
20	0	0	0	0	0	0.1	0	0	0	0	0	0	0	0	0	0	0	0	0	0	0
21	0	0	0	0	0	0	0	0	0	0	0	0	0.2	0	0.3	0	0.4	0	0	0	0
22	0	0	0	2.1	0	3.5	0	1.8	0.1	0.5	0.4	0	0	0.1	0	0.2	0	0	0	0	0
23	0	0	0	0	0	0	0	0	0	0	0	0	0	0.1	0	0	0	0	0	0	0.2
24	0	0	0	0	0	0	0	0	0	0	0	0	0	0	0	0	0.3	0	0.1	0	0
25	0	0	0	0	0	1.4	0	1.8	0	0.1	0.5	0.1	0.6	0	0	0	0	0	0	0	0
26	0	0	0	0	0	0	0	0	0	0	0	0	0	0	0	0	0	0	0	0	0
27	0	0.5	0	0	0	0	0	0	0	0	0	0	0	0	0	0	0	0	0	0	0
28	0	0	0	0	0	0	0	0	0	0	0.1	0	0	0	0	0	0	0	0	0	0
March 1	0	0	0	0	0	0	0	0	0	0	0	0	0	0	0.4	0	3.0	0	2.5	0	0.1

ntinued). F_V Kaafjord.

2—4							8—10		10—12		12—14		14—16		16—18		18—20		20—22		22—24	
-	+	−	+	−	−	+	−	+	−	+	−	+	−	+	−	+	−	+	−	+	−	−
.8	0	0	0	0	0.1	0	0	0	0	0	0	0	0	0	1.0	0	1.5	0	0.9	0	0	3.0
.0	0	0.3	0	0	0	0	0	0	0	0	0.1	0	0	0	0	0	0.3	0	(0.2)	(0)	(0)	(0)
.2)	(0)	(0.1)	(0)	(0)	(0)	(0)	0	0	0	0	0	0	0	0	0	0	0	0	0	0.1	0	1.6
.8	0	0.7	0.1	0.3	0	0	0	0	0.2	0	(0)	(0)	4.0	0	0	0.2	0	0.3	0.1	0.1	0	0
	0	0	0	0	0	0	0	0	0	0.1	0	0	0	0.2	0	0	0.2	0	0	2.6	0	2.0
.4	0	0.6	0	1.3	0.1	0.5	0.4	0	0.3	0	1.4	0	1.0	0	1.6	0	1.4	0.1	0	1.4	0	0.8
.5	0	3.1	0	0.5	0.1	0	0.1	0.1	0.1	0.1	1.6	0	1.4	0	1.1	0.6	0	2.4	0	6.5	0.6	4.2
.1	0	2.9	0	3.5	0.6	1.3	0.1	0	0.1	0	0.1	0	0	0	0	0	0	0.3	0	0.8	0	0
	0	0	0	0	0	0	0	0	0	0	0	0	0.9	0	1.1	0	0	0.5	0	1.0	0	1.8
.3	0	0	0	0	0	0	0	0.1	0	0.1	0.1	0	0	0	0	0	0	1.6	0	2.4	0	0.4

									0—12		12—14		14—16		16—18		18—20		20—22		22—24	
−	+	−	+	−	+	−	+	−	+	−	+	−	+	−	+	−	+	−	+	−	+	−
1.0	0	6.5	0	2.2	0	0.5	0.1	0	1.0	0	3.2	0	0.8	0	0.3	0.2	0	1.1	0.2	1.2	0	0.4
0.7	0	0.6	0	0.5	0	1.5	1.2	0.3	0	0.6	1.0	0	2.6	0	0.3	1.6	0	2.1	0.5	0	0	0.5
0.3	0	0.7	0	0.1	0	0.5	0	0.4	0.5	0.1	1.4	0	0.8	0	0	0.6	0	0.4	0.2	0	0.2	0
1.2	0	0.2	0	0.4	0.1	0.1	0.4	0	1.0	0.1	4.3	0	2.1	0	0.4	0.1	0.2	3.5	0.1	0.1	0	0.4
3.6	0	6.2	0	3.3	0	0.3	0	0.1	0.1	0.1	1.2	0	0.4	0	0.1	0.1	0.1	0	0.1	0	0.2	0
0.3	0.1	0.2	0.1	0	0	0	0	0.4	0	1.6	0	1.9	0	1.2	0	0.2	0.1	0.1	0.2	0.2	0.1	0.2
1.2	0	0.5	0	0	0	0.1	0	1.2	0	1.9	0	1.2	0.7	0.1	0.8	0	0	1.5	0.1	1.1	0	0.4
1.9	0	2.0	0.1	0.6	0.1	0	0	0	4.9	0	7.3	0	4.6	0	1.0	0.6	0.2	0.9	0.6	0	0	0.9
1.7	0	2.7	0	1.8	0.5	0.2	0.6	0.3	1.3	0	0.7	0	1.7	0	0.4	0.6	0	1.9	0	0.2	0	2.1
3.4	0	1.7	0.2	0	0.1	0.1	—	—	—	—	—	—	0.6	0.9	0	4.1	0	1.4	1.1	0	0.8	0.3
0.1	0	1.7	0	1.8	0.4	0	0	0	0	0.4	0	1.0	0	0.6	0	0.3	0	0	0	0.3	0	1.2
0.1	0	0	0	0	0	0	0	0.4	0	0.1	0	0	0	0	0	0	0	0	0	0	0.2	0
0.1	0	0.3	0	0.2	0	0.3	0.1	0.1	0.1	0.2	0.1	0.3	0.1	0.3	0	0.1	0	0.3	0	2.5	0	4.0
0.7	0	0.3	0	0	0	0	0.2	0.2	0.7	0.1	0.1	0.3	0	0.1	0.1	0.1	0.1	0.3	0.1	0.1	0.1	0.1
0.4	0	1.5	0.2	1.9	0.2	0.1	0	0	0.3	0	1.2	0	2.1	0	0.1	0.4	0	0.2	0	0	0	0.1
0	0	0.4	0.1	0.2	0	0.1	0.1	0.1	3.0	0	2.9	0	0.2	0.1	0	0.3	0.1	0.3	0	10.7	0.1	0.3
1.0	0.6	0.6	0.2	0.3	0.7	0.1	3.1	0	4.7	0	2.1	0	1.1	0.9	0	0.9	0.2	3.1	0.3	3.2	0.2	1.9
1.1	0.1	4.0	0.3	1.7	2.0	0.2	5.1	0	4.3	0	3.0	0	1.2	0	0.8	0	0	10.9	0.3	0.9	0	4.0
1.1	0	1.0	0.1	0.7	0.3	0.1	0.4	0.3	1.5	0	3.5	0	1.6	0.1	0.2	1.9	0	0.1	0.1	0	0	0.2
0.2	0	0.1	0	0	0	0	0	0.3	0	0.7	0.1	0.3	0.8	0	0.1	0.9	0.1	8.3	0	6.3	0	6.6
5.3	0	5.7	0.2	0.6	0.3	0	0.6	0	2.9	0	6.2	0	1.7	0	0.9	0.8	0	3.2	0	5.5	0	3.2
3.0	0	0.8	0	0.5	0	0.8	0	0.3	0	1.5	0.1	0.2	0	0.1	0.1	0	0	0	0.1	0	0.1	0
0	0	0.1	0.2	0	0.2	0	0	0.2	0.2	0.2	2.6	0	1.4	0.7	0.2	0.1	0	1.9	0	0.3	0	0.3
2.9	0	3.4	0.1	0.1	0.2	0	0.2	0.1	0	0.2	0	0.6	0.2	0.1	0.1	0	0	0.6	0	3.5	0	1.0
0.2	0	1.5	0	0.2	0	0	0.1	0.2	0.1	0.2	1.5	0	0.1	0.2	0	0.2	0	2.1	0.1	0.9	0.1	0
1.9	0	2.9	0.1	0.5	0.2	0.2	0.1	0.1	0.5	0	0.1	0.3	0.2	0.2	0.1	1.1	0.1	0.6	0.2	1.5	1.0	0
0.3	0	0.3	0	0.2	0.1	0.1	0.1	0	0.4	0	0.2	0.1	0.2	0	0	0.4	0.2	0	0.5	0	0.4	1.6
0.4	0.1	0	0.1	0	0	0.1	0.3	0	0.1	0.1	0.1	0.1	1.2	0	0.6	2.2	0	4.5	0	4.9	0	8.0
7.0	0	5.6	0	3.3	0	1.0	(0.1)	(0.1)	0.3	0.1	0.3	0.2	0.1	2.3	0	2.1	0	3.5	0	0.7	0	1.2
0.5	0	0.9	0.1	0.2	0.1	0.1	0	0.1	0	0.2	0.1	0.3	0.8	0	0	1.0	0	1.5	0.1	0.1	0.1	0
0.1	0	0.4	0	0	0.1	0.1	0	0.2	0.4	0	0.1	0.2	0	0.4	0.2	0.1	0	1.4	0	0.8	0.1	0.6
0.7	0.9	0	0.2	0.2	0.2	0.3	0.1	0	2.1	0	1.1	0	0	0.1	0.1	0	0	0.3	0.1	0.1	0.1	0.1
0.5	0	0.9	0.3	0.2	0	0.8	0	0.4	0	0.2	0.2	0	0.2	0	0	0	0	0.1	0	0.1	0	0.4
0.1	0	0	0	0.1	0	0.1	0	0.1	0	0.6	0	0.2	0.4	0	0	0	0	0	0	1.0	0	1.2
0.3	0.1	0	0	0	0	0	0	0.1	0	0	0	0	0	0	0	0	0	0	0.1	0	0	0.2

TABLE LXX (continued). F_{H} Axeløen.

Gr. M.-T.	0–2		2–4		4–6		6–8		8–10		10–12		12–14		14–16		16–18		18–20		20–22		22–24	
Date	+	–	+	–	+	–	+	–	+	–	+	–	+	–	+	–	+	–	+	–	+	–	+	–
October 8	0	0	0	0.3	0	0.2	0.1	0	0.3	0	0.4	0	1.1	0	0.2	0.1	0.1	0.1	0	1.1	0.1	1.5	0	1.3
9	0	0.5	0	0.1	0	0.1	0	0.1	0.2	0.6	0	0.4	0.3	0	1.0	0	0.1	0.3	0	2.3	0	1.3	0	0.1
10	0	0	0	0	0	0	0	0	0	0	0.2	0	0	0.7	0	0.1	0	0	0	0.2	0.2	0	0.1	0.1
11	0	0.5	0	1.8	0.1	0.2	0.1	0.2	0.2	0.2	1.0	0	3.2	0	1.0	0.1	0	2.7	0.1	3.1	0	2.8	0.5	0.9
12	0	1.1	0	0.7	0	0.1	0.1	0.1	0.1	0.1	0.6	0	0.3	0.1	0.2	0.2	0	0.5	0	0.1	0.1	0	0	0
13	0	0	0	0	0.1	0	0.1	0	0.1	0	0.6	0	0.5	0	0	0.3	0	1.2	0	7.1	0	1.5	0	0.2
14	0	0.1	0	0	0	0	0	0	0	0	0	0	0.1	0.1	0.1	0.1	0	1.0	0	2.6	0	0.2	0	1.1
15	0	1.7	0	0.6	0	0.6	0	0.6	0	0.4	0.1	0.1	0	0.2	0.1	0.1	0	0.1	0	2.1	0	2.2	0	0.9
16	0	1.7	0	1.4	0.1	0.1	0	0	0	0	0	0	0	0	0	0	0.1	0	0.2	0	0.1	0.4	0.2	0.2
17	0.1	0	0	0.1	(0)	(0)	(0)	(0)	0	0.1	0.1	0.1	0	0.1	(0)	(0.1)	(0)	(0.1)	(0)	0.1	0	0	0	0.1
18	0	0.7	0	1.8	0.1	2.1	0.2	0.2	0.1	0	0.2	0	0	0.1	0	0.2	0	0.2	0	2.2	0	3.7	0	0.4
19	0	2.7	0	5.5	0	1.5	0.1	0	0	0.2	0	0.2	0.1	0.1	0	0	0	0	0	1.3	0	0.1	0.1	0.1
20	0	0	0	0	0.1	0.1	0	0	0.1	0	0.4	0	1.3	0	0.1	0	0	0	0	0	0	0.4	0	0.5
21	0	1.3	0	1.0	0	0.2	0	0	0.1	0	0.3	0	0.2	0	0.7	0	0	0.9	0	0.3	0.1	0.1	0	3.7
22	0	1.6	0	0.9	0.1	0.5	0.3	0.3	0	0	0	0	0.1	0	0	0	0	0.1	0	0.1	0	0	0	1.0
23	0	0.1	0	0.4	0	0.1	0	0	0	0	0	0.1	0	0	0	0	0.1	0	0	0.1	0.5	0.1	0.1	1.7
24	0	1.8	0.1	0.8	0.1	0.3	0	0	0	0	0	0.1	0.1	0.1	1.2	0	0.5	2.1	0	2.5	0	6.6	0.5	3.0
25	0	5.4	0	5.1	0	2.6	0.5	0.2	1.9	0	4.0	0	2.9	0	0.8	0.4	0	2.2	0	2.0	0	0.4	0	1.0
26	0	1.3	0	0.4	0	0	0.1	0	0.1	0.1	0.5	0.1	2.5	0	0.6	0.8	0	1.9	0	2.1	0.1	0.3	0.1	1.1
27	0	5.4	0	6.4	0.1	3.2	0.6	1.1	1.0	0.1	0.8	0	0.7	0.6	0	6.9	0	11.4	0	2.9	0	1.6	0	6.0
28	0	2.9	0.1	2.6	0.1	1.3	0.2	0.4	0.2	0	0.2	0.1	0.9	0	0.1	1.5	0	4.3	0	3.8	0	6.2	0	7.6
29	0	5.7	0	3.1	0.1	1.2	0.3	0.1	0.5	0.2	0.2	0.9	0.2	0.1	0.1	0.2	0	5.5	0	6.7	0.1	1.9	0.3	0.1
30	—	—	—	—	—	—	—	—	0.1	2.3	1.8	0.5	0.1	0.5	0.1	0.4	0	1.5	0	2.4	0	4.3	0.2	0.5
31	0	4.6	0	9.5	0	8.1	0.3	0.6	0.1	0.9	0	4.3	0	16.2	0	10.2	0	7.1	0	5.5	0	4.3	0.1	5.2
November 1	0	2.6	0.1	3.3	0.5	0.1	1.2	0	1.4	0	0.3	0	0.1	0	0	0.1	0	0.2	0	0.1	0.1	0.2	0	0
2	0	0.1	0	0	0.1	0	0.5	0	0	0.2	0.1	0.1	0.1	0.2	0.1	0.3	0	2.1	0	2.3	0	3.7	0	0.7
3	0	1.4	0	0.9	0	0.6	0	0	0	0	0	0.1	0.1	0.1	0	0.2	0	0.6	0	0.2	0	1.0	0	1.6
4	0	0.8	0.1	0	0	0	0	0	0.1	0	0	0	0.1	0	0.1	0	0.1	0	0	0	0	0	0	0
5	0	0.1	0	0.1	0	0	0	0	0	0	0	0	0	0.1	0	0.1	0	0	0	0.6	0	1.2	0	0.1
6	0	0	0	0	0	0	0	0	0	0	0	0.1	0.3	0	0	0.1	0	0.7	0.1	12.2	0.2	1.0	0.6	0
7	0.1	1.3	0.2	0.9	0.1	0.3	0	0	0	0	0.1	0.1	0	0.1	0	0	0	0.1	0.1	0	0.1	0	0	0.3
8	0	1.4	0	2.8	0	0.4	0.1	0	0.1	0	0	0.2	0	0.2	0	0.2	0	0.8	0	0.2	0	0.1	0	0.3
9	0	0.1	0	0	0	0.2	0	0	0.1	0	0.1	0	0	0.2	0.1	0.1	0	0	0	0	0	0	0	0.2
10	0	0.8	0	0.2	0.1	0	0	0.2	0.1	0.3	0	0.8	0.3	0	0.2	0	0.1	0	0	0	0	1.6	0	4.2
11	0.1	0.2	0.3	0	0	0	0	0	0	0	0	0	0	0.1	0	0.1	0	0	0	0	0	0	0.1	0
12	0.1	0	0	0	0	0	0	0	0	0.6	0	1.4	0.1	0.3	0.2	0	0.1	0	0	0.9	0	0.3	0	3.2
13	0.3	0.5	0	6.9	0	2.2	1.1	0.1	2.5	0	1.3	0	1.7	0.2	1.6	0	0	0.9	0	1.6	0.1	1.6	0.1	0.1
14	0	1.2	0	1.8	0	2.3	0	2.6	0.3	1.0	2.0	0	1.6	0	0.5	0.3	0	0.5	0	0.9	0	1.6	0	0.5
15	0	2.3	0	1.1	0	1.3	0.5	0	0.9	0	0.3	0.2	0.8	0	0.3	0.1	0	0.4	0	1.2	0.3	1.6	0.3	1.4
16	0	0.2	0	0.2	0	0.1	0	0.1	0.1	0	0	0.2	0.1	0.3	0.1	0.3	0.1	0	0	0.1	0	0.1	0	0.4
17	0	3.3	0.3	1.1	0.3	0.2	.1	0.7	0.3	1	0	0	0	0	0	0	0.1	0.3	0	0.1	0	0.1	0.3	0
18	0	0.1	0	0	0	0		0	0		0.1	0	0	0	0	0.1	0	0.1	0	0.1	0.1	3.1	0	5.0
19	0	1.1	0.1	0.7	0.1	0.3	.1	0.1	0.2		0.6	0.1	1.6	0	1.1	0	0.3	0	0	0.2	0	2.3	0	2.4
20	0	0	0	0.4	0	0.1	.1	0	0.2		0	0.3	0	0.1	0.1	0	0	2.2	0	6.3	0	3.5	0	1.3
21	0	0.4	0	0.2	0	0	8	0.1	0	8.4	0.1	0.5	0.6	0	1.2	0	0	3.2	0	3.2	0.1	5.3	0	2.9
22		1.8		1.6	0	0.3	0	0	0.2	0.1	0	0.1	0.7	0.1	0.3	2.2	0	5.7	0	8.0	0.4	0.3	0	7.3
23		1.2	0	0	2.6	0	2.7	0	3.6	0	1.2	0.3	0.5	0.2	0.1	0.4	0	2.8	0	4.0	0	8.8	0	21.0
24		5.1		20.0	0	8.6	0.1	4.2	0.1	1.5	0	1.9	0	5.2	0	10.0	0	9.0	0	3.2	0	6.8	0	5.5
25		2.1		1.8	0.1	2.3	1.2	0	2.0	0	0.5	1.1	0.4	2.7	0	5.4	0	9.4	0	8.5	0	7.6	0	4.7
26	8	6.6	8	3.7	0.4	0.1	0.2	0	0.5	0.2	0.6	0	0.2	0	0.4	3.2	0	16.0	0	4.7	0	2.8	0	4.2

TABLE LXX (continued). F_H Axeløen.

Gr. M.-T.	0−2		2−4		4−6		6−8		8−10		10−12		12−14		14−16		16−18		18−20		20−22		22−24	
Date	+	−	+	−	+	−	+	−	+	−	+	−	+	−	+	−	+	−	+	−	+	−	+	−
November 27	0.1	0.1	0.1	0.1	0.3	0.1	0.6	0.1	0.3	0.1	0.2	0.1	0.1	0.1	0	0.2	0	0.6	0	0.2	0	0	0	0
28	0	0.1	0	0	0.1	0	0.3	0	0.3	0	0.5	0.2	0.3	0	0.3	0	0	1.2	0	1.8	0	2.1	0	1.5
29	0	1.0	0.1	0.2	0.1	0	0	0	0.1	0.3	0.1	0.1	0	0.1	0.2	0	0.1	0	0	0	0	0	0	0.3
30	0	1.0	0	1.7	0	2.9	0	1.8	0	0.7	—	—	1.2	0	0	0.6	0	3.9	0	3.1	0	2.6	0	2.5
December 1	0.1	1.7	0.3	0	0	0.2	0	0.1	0.1	0	1.1	0.1	2.3	0	0.7	0	0	4.2	0	4.2	0	2.8	0	1.0
2	0	0.8	0	1.2	0.1	0.7	1.7	0	2.8	0.3	0.6	0	2.1	0	0.8	0.3	0	5.8	0	2.2	0	1.0	0	1.1
		1.6	0	3.2	0	1.4	0	0.4	0.1	0.3	0.2	0	0.7	0	0.8	0	0.1	0.1	0	0.5	0	0.6	0	0
		0.1	0	0	0	0	0	0	0	0	0	0	0.2	0	0	0	0	0.3	0	4.6	0	2.9	0	0.3
		0	0	0.1	0	0.1	0.1	0	0	0	0.1	0	0.1	0.1	0	0	0	0.1	0	0	0	0.1	0	0.3
		3.5	0.1	0	0	0.2	0	0	0	0	0.1	0.1	0.2	0	0	0.2	0	0.7	0	0.2	0	0.2	0.1	0
		0	0	0	0	0	0	0	0.1	0	0.4	0	0.2	0	0.9	0	0.5	0	0	0.7	0	3.5	0	3.0
		0.7	0	0.8	0.5	0.1	0.2	0.1	0	0.1	0.1	0	0	0	0	0	0	0	0	0.4	0	5.4	0	1.7
		0.1	0	0	0.2	0	0	0.1	0.1	0.1	0.2	0	0.8	0	0.8	0.4	0	2.5	0	3.5	0	3.1	0	2.0
		1.4	0	1.3	0.3	0.9	0.2	0.2	1.1	0.1	2.3	0	3.3	0	1.4	0	0.1	0.2	0	1.6	0	3.0	0	1.1
		1.2	0	2.7	0	0.3	0.1	0.3	0.1	0.3	0.1	0.2	0.3	0	0.4	0.1	0	3.0	0	4.0	0	8.8	0	0.8
		0.4	0.1	0.6	0	0.5	0	0.5	0	0.1	0	0.8	0	0.7	0	0.1	0	3.0	0	7.3	0.7	0.3	0.4	0
		0.4	0	0.6	0.1	0	0	0.2	0	1.2	0.1	0.1	0.2	0.1	0.2	0.3	0	5.1	0	4.4	0	3.7	0	2.0
14	0	1.6	0.6	0.1	0	0	0.1	0.1	0	0	0	0.2	0	0.1	0	0	0	0.4	0	0.1	0	0.2	0	0.2
15	0	3.0	0	3.5	0.2	0	0.2	0	0.1	0	0.1	0.3	0.1	0	0	0.1	0	0.2	0	2.7	0	2.2	0	0.7
16	0	0.2	0	0.3	0	0.2	0.3	0	0.7	0	0.2	0.1	0	0.2	0.4	0.1	0	6.5	0	2.5	0	3.5	0	2.3
17	0	0.3	0.1	0.5	0.1	0	0.1	0	0	0.1	0.2	0	0.1	0	0	0	0	0	0	0.6	0	0.7	0	0.4
18	0	0.3	0	0	0	0	0	0	0	0	0	0.4	0	0.1	0	0.3	0	0.1	0	0	0	0	0	0.2
19	0	0.9	0.1	0.4	0.2	0.5	1.1	0	0.8	0.1	0.1	0.3	0.9	0	0.5	0	0.6	0	0	1.0	0	1.6	0.1	0
20	0	0.1	0	0.4	0	0.4	0	0.1	0	0.1	0.1	0.2	0.1	0.2	0.5	0.2	0.4	0	0	0.2	0	0.1	0	0
21	0	0	0	0.1	0.2	0	0	0	0	0	0.1	0	0	0	0.1	0	0	0.7	0	2.0	0.1	0.4	0	0.1
22	0	0.3	0	0	0	0	0.3	0.2	0.7	0	1.7	0.1	0.4	0	0.7	0	0.1	0.6	0	1.2	0	0.6	0	2.8
23	0	2.9	0	14.4	0	9.7	0.5	2.7	3.0	0	0.6	0.4	0	8.4	0.5	1.6	0	5.4	0	5.7	0	9.7	0	1.5
24	0	0.4	0	1.1	0.3	0.4	0.7	0	1.8	0.2	0.2	0.6	0.6	0.1	0.7	2.2	0	6.1	0	8.4	0	1.1	0	0.8
25	0	3.7	0	5.5	0.1	1.0	1.1	0.2	0.7	0.3	0.1	0.5	0.1	0.1	0.2	0.1	0.3	0.4	0.1	0.8	0	0.4	0	0.5
26	0.1	0.5	0.4	0.2	0.2	0.1	0.1	0.1	0.4	0.1	0.4	0.4	0.9	0	1.9	0.1	0	0.7	0	0.8	0	3.7	0	0.6
27	0	0.9	0	1.3	0.6	0.2	0.5	0.2	0.3	0.1	0.1	0.6	0.2	0.3	0	0	0	0.3	0	0.8	0.1	1.8	0	2.1
28	0	5.4	0	1.8	0	3.6	1.1	0.2	0.6	0.3	1.6	0.1	1.9	0	0.4	0	0	0.5	0	0.8	0	4.2	0.1	0.2
29	0.3	0.1	0	0.6	0.8	0	0.2	0	0.4	0	0	0.1	0.1	0.1	0	0.1	0	1.2	0	2.5	0	1.0	0	0.3
30	0	0.6	0	0.2	0.1	0.1	0	0	0.1	0	0	0.3	0	0.1	0.1	0.1	0.1	0.7	0	0.4	0	0.1	0	0.6
31	0.1	0.5	0	0.1	0	0	0	0	0	0	0	0	0	0	0	0	0	0.5	0	0.3	0.1	0.1	0	0.3
1	0.2	0	0.1	0	0	0	0	0	0.2	0	0	0.4	0	0.1	0.3	0	0.2	0	0.1	0	0	0.3	0	0.6
2	0	1.4	0.8	0	0.5	0	0.3	0	0	0	0.2	0.1	0.1	0	0.1	0	0	0.1	0	0.2	0	0.2	0	0.2
3	0	0.7	0	0.8	0	0.1	0	0	0.1	0	0.3	0	0	0	0.2	0	0.1	0.1	0	2.2	0	3.5	0.1	0.2
4	0	0.6	0	2.4	0	1.9	0.7	0.7	0.8	0.2	0	0.4	0	0.4	0	0	0.1	0.3	0	2.1	0	2.5	0	1.4
5	0	2.4	0	9.3	0	5.5	0.1	1.6	1.0	0.4	0.2	0.9	0.7	0.2	0	1.0	0	5.0	0	2.7	0	0.6	0	0.6
6	0	1.0	0	2.7	0.1	0.6	0.1	0.1	0.1	0.3	0.2	0.1	1.0	0	1.2	0	0.7	0.2	0	1.9	0.2	2.8	0	1.6
7	0.2	0.7	0	0.9	0	0.3	0	0.3	0.2	0	0.3	0	0.3	0.1	0.2	0.2	0.1	0.4	0	2.1	0	1.0	0	1.1
8	0	1.2	0	1.0	0.1	0.3	0.1	0.2	0.3	0.1	0.4	0	0.8	0	0.6	0.2	0.4	0.3	0	1.4	0	2.0	0	0.1
9	0	0.4	0	0.8	0.2	0.1	0.3	0.1	0.7	0	1.5	0	1.5	0	2.0	0	0.8	0.1	0.1	1.1	0	2.3	0	0.2
10	0.3	0.3	0	4.2	0	1.3	0.6	0	0.2	0	0.7	0	1.1	0	0.4	1.1	0.3	0.4	0.1	0.7	0	1.7	0	1.6
11	0	1.8	0	0.7	0	0.6	0	0.4	0.4	0	0.3	0	0.3	0	0.6	0.7	0.1	0.6	0	4.6	0	1.9	0	2.6
12	0	1.3	0	0.8	0.3	0	0.3	0.1	0.9	0	0.2	0.1	0.6	0	1.1	0.1	0.1	0.7	0	3.7	0	0.5	0	1.0
13	0	2.7	0	0.9	0	0.1	0	0	0	0	0	0	0.1	0.1	0.3	0.1	0	1.3	0	10.9	0	3.6	0	1.0
14	0	0.3	0.1	0	0.1	0	0.1	0.1	0	0.2	0	0	0.5	0	0.2	0	0	0.3	0	0	0	0	0.1	0.6
15	0	1.2	0	1.0	0	0.8	0	0.2	0.2	0	0.5	0	1.7	0	3.1	0	0.2	1.1	0	4.0	(0)	(1.5)	0	0.6

TABLE LXX (continued). F_H

Gr. M.-T.	0—2		2—4		4—6		6—8		8—10		10—12		12—14		14—16		16—18		18—20		20
Date	+	—	+	—	+	—	+	—	+	—	+	—	+	—	+	—	+	—	+	—	+
January 16	0	0.4	0	1.1	0.2	0.1	0.3	0	0.3	0	0.2	0.2	0.8	0.7	1.1	0	0	2.0	0	2.5	0
17	0	0.5	0	1.0	0	0.5	0.1	0	0.4	0	0	0.1	0.4	0	1.1	0	0.2	0.1	0	0.2	0
18	0.1	0.1	0.1	0.4	0	0.4	0	0.1	0.1	0	0.3	0	0.2	0.1	0.2	3.6	0.3	2.1	0	5.3	0
19	0	0.5	0	0.6	0.6	0	0.4	0.1	0.9	0	0.7	0	2.8	0	0.4	0.1	0	2.7	0	0.7	0
20	0	0.3	0	1.5	0	2.8	0.2	0.9	1.2	0	0.3	1.0	0.1	0.4	0.2	0.1	0.1	0	0	0.6	0
21	0	0.4	0	0.7	0	1.2	0.2	0.1	0.2	0.1	0.3	0.1	1.2	0	0.1	1.0	0.5	0	0	0.1	0
22	0	5.4	0.1	0.4	0	0.1	0	0.1	0	0.2	0	0.8	0	0.3	0	0.1	0	0.5	0	3.2	0
23	0	0.6	0	0.2	0	0.1	0.4	0	0.3	0.1	0.7	0.5	0.6	0.1	0.6	0	0	1.6	0	2.4	0
24	0.1	0.4	0.1	0.8	0.5	0.1	0.7	0	0.8	0	0	0.4	0.5	0	0.5	0.1	0	1.6	0	4.7	0.5
25	0.2	0.1	0	0.3	0	0.3	0	0.1	0.1	0.2	0	0.1	0.2	0	0.3	0	0	0.5	0	0.1	0
26	0	0	0	0	0	0	0	0	0.1	0.1	0.3	0.1	1.7	0	1.2	0	0	2.3	0	8.5	1.2
27	0	14.0	0	7.6	0	5.3	0.1	1.0	2.3	0	0.4	0.3	0.4	0	0.4	0	0	0.8	0	1.0	0.2
28	0	0.8	0	0.3	0	0	0.1	0.2	0.2	0.2	0.1	0	0.1	0	0.3	0	0.1	1.6	0	2.5	0
29	0	0.5	0.1	0.1	0	0	0	0	0	0	0.1	0	0.2	0.1	0.2	0.1	0	0	0	0	0
30	0	0.1	0.1	0.1	0	0	0.1	0.1	0.6	0	4.4	0	6.0	0	1.2	0.1	0.1	1.6	0	0.8	0.2
31	0	0.1	0.1	0.3	0.3	0	0.1	0	0	0.1	0	0.5	0.4	0.1	0.1	0.6	0.1	2.6	0	1.6	0.1
February 1	0.2	0	0.1	0	0.2	0.2	0.6	0	0	0	0	0.5	0.1	0.3	0.5	0	0.1	0.6	0	1.0	0.2
2	0	0.1	0	0.7	0	0	0	0	0	0	0.1	0	0.1	0	0.6	0	0.3	0	0	0	0
3	0	0.8	0	1.4	0	0.5	0	0.8	0.2	0.5	0.6	0	0.5	0.3	0.1	0.1	0	0	0	0	0
4	0	0.1	0	0.3	0	0.7	0	0.2	0	0.6	0	0.6	0	0.4	0.2	0.1	0.1	0	0	0.6	0
5	0	0.2	0	0.1	0	0.1	0	0.3	0	0.5	0	0.4	0.5	0	1.8	0	0.3	1.0	0	1.8	0.5
6	0	0.6	0	0.6	0	1.7	0.1	2.4	0.5	0.2	0.3	0	1.6	0	1.8	0	1.0	0	0.4	0.2	0
7	0	0	0	0.4	0	0.3	0	0.1	0	0.3	0	0.7	0.1	0.1	0.1	0	0.1	0.8	0	0.4	0.3
8	0.4	0.2	0.4	5.0	0	5.3	0	3.2	2.1	0	2.2	0.1	2.2	0	0	4.1	0	5.3	0	15.5	0
9	0.1	0.1	0	3.0	0.4	0.3	0.4	0.2	0	0.6	0.1	0.2	0.2	0	0.3	0	0.1	0.1	0	3.5	0.3
10	0	0.6	0	1.1	0.2	0.3	0.5	0	0.4	0	0.4	0	0.5	0	0.2	0.1	0.1	0.1	0	0	0
11	0	4.4	0	4.1	0	4.4	0.8	0.1	0.4	0.1	1.4	0	1.6	0	1.9	0	0	0.7	0	4.4	0
12	0	1.8	0	0.8	0	0.1	0.3	0	0	0.2	0.1	0.2	0.3	0.1	0.8	0	0.3	1.2	0	4.0	0
13	0	2.5	0	1.9	0.2	1.2	0.6	0.1	0.5	0.1	0.7	0.1	0.3	0	0.6	0	0.1	0	0	0.7	0
14	0.1	0.9	0.1	0.9	0	0.4	0.1	0.3	0	0.5	0.3	0.2	0.2	0.1	0.8	0.1	0.3	0	0	2.0	0
15	0.1	0.1	0	1.6	0.5	0.1	0.4	0	0.7	0.1	1.1	0	1.7	0	0	3.2	0	13.0	0.5	0.3	0
16	0	0.1	0	0	0.1	0	0.2	0	0	0.1	0.1	0.1	0	0	0.3	0	0	0.7	0	4.3	0.1
17	0.2	0.1	0.1	0.4	0.4	0	0.1	0	0.2	0	0.5	0	1.7	0	0.4	1.0	0.4	0.6	0.1	0	0.3
18	0	1.1	0	0.1	0	0	0	0	0.5	0	0.1	0.3	0.1	0.3	0.2	0	0.1	0.2	0	1.0	0.1
19	0	0	0	0	0	0	0	0	0	0	0	0.7	0	0.7	0	0.3	0	0.4	0	1.3	0
20	0	0.3	0	0.3	0	1.0	0	0.4	0.1	0	0.5	0	0.2	0.1	0	0	0.1	0.1	0.1	0	0
21	0	0	0	0.1	0.1	0.1	0.1	0.1	0	0	0.3	0	2.9	0	2.4	0	0.4	0.2	0	0.1	0
22	0.1	0.1	0	1.9	0	5.2	0	6.5	0.2	0.9	1.8	0	2.9	0	1.2	0	0.6	0	0	0.4	0
23	0.1	0.9	0.1	0.1	0.2	0	0	0	0.1	0	0.9	0	1.4	0	1.2	0	0.2	0.4	0	0.3	0.3
24	0	0.7	0	0.6	0	0	0.1	0	0	0	0	0	0.2	0	0.4	0	0	0.8	0	0.3	0
25	0	0.1	0	1.1	0	7.9	0	4.5	1.2	0	3.1	0	4.2	0	0.8	0.1	0	0.6	0	0.8	0
26	0	0.1	0	0	0.2	0	0.3	0	0.1	0	0	1.0	0	0.3	0	0.1	0	0	0	0.2	0
27	0	1.4	0	1.6	0	0.6	0.2	0	0	0	0	0.1	0.4	0	0	0.1	0	0	0	0.1	0
28	0	0	0	0	0	0	0	0	0	0	0	0	0	0	0	0	0.1	0	0	0	0
March 1	0	0	0	0	0	0	0.1	0	0.6	0	0.2	0	1.6	0	1.8	0	0.1	1.7	0	4.7	0
2	0	2.2	0	1.7	0	2.8	0.1	1.6	0.1	1.0	0.2	0.3	0	0	0.1	0	0.1	0.5	0	4.8	0.1
3	0	4.4	0	2.6	0	1.5	0.1	0.2	0	0.1	0.2	0.2	0.2	0	0.1	0	0.1	0.1	0	0.1	0
4	0	0.1	0	0.3	0.4	0	0.2	0	0	0	0	0	0	0.1	0.2	0.1	0	0	0	0.1	0.1
5	0.1	0.7	0	2.9	0.1	2.7	0.1	0.3	0.5	0	2.5	0	2.8	0	1.4	4.2	0.2	3.9	0	2.3	0
6	0	0.9	0	0.1	0	0.1	0	0	0	0	0	0.1	0	0	0	0.1	0	0.4	0	1.0	0

TABLE LXX (continued). F_R Axeløen.

Gr. M.-T.	0–2		2–4		4–6		6–8		8–10		10–12		12–14		14–16		16–18		18–20		20–22		22–24	
Date	+	−	+	−	+	−	+	−	+	−	+	−	+	−	+	−	+	−	+	−	+	−	+	−
March 7	0	3.4	0	4.2	0	4.6	0.1	1.8	2.0	0	1.7	0	2.8	0	0.6	0.8	0	3.1	0.1	2.1	0.2	0.1	0	0.7
8	0	1.7	0	5.4	0	1.5	0.5	0	1.5	0	0.8	0	1.8	0.1	0.6	0.3	0	6.9	0	13.3	0	10.5	0	2.5
9	0.1	0.8	0	5.4	0	9.0	0.2	1.2	0.5	0.1	1.1	0	0.6	0	0	0.5	0	0.1	0	1.4	0.2	2.0	0.3	0.2
10	0	0.5	0	2.6	0	0.5	0	0.3	0.1	0.1	0.2	0	1.0	0	1.8	0.1	0	2.4	0	7.2	0	4.0	0	0.4
11	0	1.2	0	1.6	0	0.6	0.1	0.1	0	0.1	0.5	0.1	1.2	0	1.3	0	0	0.7	0.1	3.2	0	1.9	0	0.3
12	0	2.6	0	2.2	0	4.3	0.1	1.2	1.5	0	0.6	0.8	3.0	0	0.7	0.2	0.2	0.4	0	0.6	0.4	0.4	0.1	9.1
13	0	3.3	0	5.5	0	9.3	0.8	1.0	1.4	0.5	1.7	0.4	2.4	0	0.4	3.8	0	13.5	0	2.2	0.4	0	0	0.5
14	0	1.0	0	2.4	0.	0.5	0.4	0.1	0.8	0.1	1.0	0	2.7	0	1.0	0.2	0	0.7	0.1	1.0	0	1.7	0	0.7
15	0	0.6	0	0.5	0.	0.5	0.6	0	0.5	0	0.7	0.2	2.0	0	2.3	0	0.9	2.4	0	4.9	0.2	0	0.1	0.2
16	0.	0	0.	0.1	0	0.5	0	0.3	0	0.1	0	0.2	0.3	0	0.2	0	0	0	0	0.3	0	3.0	0	0.4
17	0	0	0	0	0	0	0	0	0	0.3	0	0	0.2	0	0.1	0	0	0.2	0	0.1	0	0.9	0	0
		0	0	0	0	0	0	0	0	0	0.1	0.1	0.1	0.1	0.4	0	0	0.2	0	0.4	0.1	0	0.1	1.2
		2.3	0	1.7	0	0.8	0	0.6	0.1	0.1	0.4	0.1	0	0.2	0.2	0.3	0	1.2	0	6.1	0	3.9	0	1.6
		1.5	0	0.5	0	0.8	0.1	0.2	0.5	0.2	1.6	0.1	0.1	0.2	0	0.6	0	0.6	0	0.2	0.3	0	0.1	0.5
		0.9	0	2.0	0	0.4	0	0.4	0.4	0.1	0.2	1.5	0.3	0.4	0.2	0.6	0	1.8	0	0.6	0.1	2.8	0	1.7
	0	0.8	0	0.8	0	0.5	0.3	0.1	0.1	0.2	0	0.3	0	0.7	0.1	0.3	0.5	0.2	0.2	0.3	0.5	1.1	0	8.9
	0.	0.4	0	1.7	0	0.5	0.2	0	0.3	0.1	0	0.9	0.1	0.1	0	0.9	0	0.7	0	0.6	0.1	0.1	0	0.5
24	0	1.2	0.	0.3	0.1	0.1	0	0.2	2.0	0	1.2	0	0.7	0.1	0	0.5	0	0.5	0	0.1	0	0	0	0
25	0	0.3	0	0.5	0	0.6	0	0.1	0.1	0	0	0	0.1	0.1	0.2	0	0	0	0	0.1	0.2	0	0.2	0
26	0	0.1	0	0.3	0	0	0	0	0	0	0	0.9	0.1	0.1	0	0	0	0.5	0	0.4	0	0.2	0	0.2
27	0	0.1	0	0.2	0	0	0	0	0	0.4	0	0.6	0.2	0.1	0.2	0	0	0	0.1	1.5	0	2.5	0	0.8
28	0	0.3	0	0	0	0	0	0	0	0	0.5	0	0.6	0	0.1	0	0	0.5	0	1.1	0	1.7	0	0.2
29	0	0.2	0	2.9	0	1.9	0	1.1	0	1.0	0	0.5	0.2	0.2	1.4	0	0	1.9	0	2.2	0	0.2	0.3	0
30	0	1.9	0	4.6	0	1.0	0.4	0.3	1.5	0	0.6	0.6	0.1	0.4	0.2	0.5	0	0.7	0	0.5	0.1	0.7	0.7	0
31	0.	6.8	0	1.1	0.2	0.1	0.1	0.4	0.7	0.2	1.5	0	1.3	0.1	1.0	0.2	0	5.1	0	1.8	0	3.2	0	1.4
April 1	0	1.4	0	2.8	0.1	0.7	0.3	0.1	0	0.1	1.2	0	2.4	0	2.9	0	1.4	0	0.1	0	0.2	0	0	11.2
2	0	4.2	0	2.3	0.9	0	0.4	0.1	0.2	0.1	0.5	0.3	1.6	0.1	0	0.8	0	0.7	0	2.3	0	3.8	0.1	1.1
3	0	2.3	0	7.3	0	0.3	0.2	0	0	0.8	0.4	0	0.4	0.1	0.4	0.4	0.9	0	0	3.3	0	2.1	0	0.6
4	0	1.3	0	3.6	0	1.7	0	0.3	0.2	0.2	0	3.1	0.3	1.0	0.2	0.1	0	0.8	0	3.2	0	0.6	0.1	0.3
5	0.1	1.2	0	3.4	0	7.1	0	3.4	0.6	0.5	0.3	1.0	2.9	0	0.1	5.1	0	2.2	0.1	0.8	0	1.5	0.5	0.2
6	0.9	0.6	2.4	2.5	1.6	8.9	0.8	0.7	3.5	0.2	0.5	7.2	0	16.5	0	3.3	0	2.6	0.4	0.5	0.1	0.1	0.2	0.1
7	0	0.1	0	0.3	0	0.7	0.1	0.2	0.8	0.2	0.1	0.7	2.3	0.1	2.6	0	0.5	1.0	0	3.6	0	4.1	0.2	1.0
8	0.2	0.3	0	0.2	0	0	0.3	0.1	0.3	0.5	0	1.3	0.1	1.1	1.9	0	0.4	0	0.1	0.3	0	0.5	0.3	3.8
	0.3	9.0	1.0	2.5	0.1	5.2	5.0	0.3	4.5	0	1.2	0.4	1.3	0	0.2	1.1	0	6.8	0	1.0	0	4.7	0.1	8.3
	0	2.5	0	2.4	0	0.5	0.1	0.8	1.7	0	0.6	0.5	1.6	0.6	1.1	1.1	0.2	1.9	0	5.6	0.1	4.7	0	2.7
11	0	1.7	0.1	0.5	0.2	0.4	0.7	0.1	1.5	0	2.9	0	0.7	1.1	0	1.7	0	0.6	0.1	0.3	0	0.7	0.2	0.1
12	0.1	0.5	0	8.2	0.3	0.7	0.1	0.1	0.1	0.7	0.1	0.7	0.1	0.3	1.7	0	0	0.9	0	4.3	0.3	0.3	0.2	2.2
13	0	4.0	0	1.7	1.0	0	1.4	0	1.7	0.2	0.1	1.3	0	1.6	0	1.3	0.3	0.1	0.2	0.1	0.1	0.6	0.1	0.2
14	0	0.3	0.3	0.2	1.0	0.1	0.7	0.1	0.7	0.5	0	0.9	2.0	0.1	1.5	0	0.5	0.1	0	1.0	0	0.7	0	0.6
15	0.1	0.3	0.5	0	0.3	0.1	0	0.2	0.8	0	5.8	0	6.7	0	0.6	1.1	0	4.2	0	3.5	0	6.6	0	1.9
16	0.5	0	0.3	0	0	0.1	0	0.1	0	0.2	(0.6)	(0.4)	(0.5)	(1.4)	(0.2)	(1.2)	(0.1)	(3.3)	(0.2)	(2.5)	0	1.3	0	0.5
17	0	0.2	0	0	0	0.1	0.2	0.1	0	0.2	0.3	0.1	0.7	2.4	0.4	2.1	0	6.7	0	4.8	0	1.7	0	0.6
18	0.2	0.4	0.2	0.2	0.1	0.2	0	0.1	0.8	0.1	2.0	0.1	2.2	0.5	0.5	1.8	0	5.9	0.8	3.3	0.2	0.9	0.2	0.1
19	0.3	1.0	0	1.6	0.3	0.2	0.2	0.6	0	0.2	0	0	0	0.3	0	0.2	0.1	0.2	0.1	0.1	0.4	0	0.7	0
20	1.0	0	0.9	0	0.4	0	0.1	0.2	0.2	0	0	1.5	0	2.2	0	0.8	0.3	0.5	0	1.6	0	0.6	0	0.5
21	0.1	0.5	0.2	0.7	0	1.4	0.3	0	0.3	0.1	0	2.9	0	2.4	0.3	0.7	0	1.3	0	1.7	0	1.0	0	0.8
22	0.1	0.4	0	0	0.1	0.1	0	0.2	0	0.3	(0)	(2.0)	0	2.8	0	1.4	0	1.3	0	1.2	0.2	0.1	0	1.8
23	0.1	0.2	0	0.7	0.2	0.2	0.6	0	0	0	0.2	0.3	3.2	0	3.2	0	1.5	0	0.1	2.0	0	0.6	0	0.3
24	0.2	0.3	0.4	0.2	0.4	0	0.7	0	0.1	0.6	0	2.2	0.1	0.6	0	0.2	0	1.0	0	3.6	0	1.6	0	0.4
25	0	1.5	0	3.1	0	0.7	0.6	0	0.5	0	3.8	0	1.1	0.3	0	1.2	0	0.5	0.1	0.1	0.1	0	0.8	0

TABLE LXX (continued). F_H

Gr. M.-T.	0–2		2–4		4–6		6–8		8–10		10–12		12–14		14–16		16–18		18–20		20–
Date	+	—	+	—	+	—	+	—	+	—	+	—	+	—	+	—	+	—	+	—	+
April 26	0.1	0.4	0.1	0.7	0.1	0.2	0.2	0.2	0.7	0.1	0	3.0	0	1.4	0.2	0.2	0.2	0.3	0	1.1	0.1
27	0	6.3	0.1	4.9	0	5.6	0	1.4	0.4	0.2	0.8	0.2	2.0	0	0	1.8	0	1.6	0	0.5	0
28	0	0.9	0	1.6	0	0.8	0.1	0.2	0.1	0	1.0	0	3.1	0	5.0	0	3.7	0	0.2	0.2	0
29	0	2.1	0.1	1.5	0.2	0	0.2	0	0.5	0	0.6	1.1	0.1	0.8	0.9	0	0	0.8	0	1.7	0
30	0	0.9	0	0.1	0	0	0	0.1	0.1	1.0	0	3.5	0	2.5	1.1	0.2	0	2.8	0	4.5	0
May 1	0	0.2	0.2	0	0	0	0.1	0	0.3	0.2	0.2	1.8	1.4	0	0.3	0.1	0	0.9	0	2.8	0
2	0	1.6	0	1.1	0	0.5	0	0.2	0	0.8	0	0.7	0	0.9	0	0.4	0.5	0	0.4	0	0.1
3	0.1	0.3	0.2	0	0.2	0	0	0.1	0.1	0	0	0.1	0.1	0.1	0.3	0	0.4	0	0.1	0.6	0.7
4	0.3	0.3	0.3	0.3	0	0.6	0.2	0.8	0.1	0.2	1.0	0.1	2.3	0	2.2	0	0.3	0.7	0.4	0.2	0.5
5	1.0	1.4	0	17.5	0	10.5	0	3.6	0.8	0.3	0.1	0.5	0.2	0.4	3.4	0	1.7	0	0.2	0.7	0.2
6	0	1.8	0	3.1	0.5	3.6	0	6.1	0.5	0.8	—	—	—	—	—	—	—	—	—	—	—
7	—	—	—	—	—	—	—	—	—	—	—	—	—	—	—	—	—	—	—	—	—
8	—	—	—	—	—	—	—	—	—	—	—	—	—	—	—	—	—	—	—	—	—
9	—	—	—	—	—	—	—	—	—	—	—	—	—	—	—	—	—	—	—	—	—
10	—	—	—	—	—	—	—	—	—	—	0.4	0.5	2.0	0	3.0	0	0.2	0.4	0	3.2	0
11	0.4	0	0.5	0	0	0.5	0.1	0.3	0	0.8	0.1	0.3	3.1	0	2.4	0	0.7	0	0	0.9	0.1
12	0.3	0.3	0.3	0.1	0.1	0.3	0.1	0.6	0	0.3	0.1	0.3	0.9	0	2.4	0	0.4	0.7	0	0.8	0
13	0.4	0	0	0.7	0	2.0	0.1	0.9	0.1	0.5	0.6	0.1	0.7	0	2.1	0	0.4	2.3	0	6.0	0.1
14	0.5	0	0.1	0	0	0.9	0	2.1	0	1.2	2.8	0	2.7	0	0.1	2.7	0.2	0.6	0.1	3.5	0
15	0	5.4	0	10.0	0	5.0	0	1.3	0	1.5	0.7	0	0.1	0.1	0.1	0.1	0	0.3	0	0.8	0
16	0.2	0.1	0.4	0.3	0	0.6	0.2	0	0.6	0	0.5	0.8	1.5	0.1	0.8	0.5	1.1	0.4	0	1.1	0.2
17	2.9	0.4	1.3	0.8	0.1	3.2	0.1	1.2	0.5	0.2	2.9	0	5.1	0	2.0	0	0.2	0.2	1.0	0.1	1.3
18	0.3	0	0.2	0.1	0	0.7	0.1	0.1	0.4	0	0.1	0.5	0.2	0.1	1.2	0	0.7	0	0.3	0.5	0.1
19	0	0	0	1.0	0	0	0	0.5	0.3	0.1	0	0.9	1.7	0.4	4.8	0	2.4	0	0.6	0	0
20	0	0	0.2	0	0	0	0	0.1	0	0.2	(0.1)	(0.5)	(1.0)	(0.2)	(1.0)	(0)	(0.5)	(0)	(0.2)	(0.2)	0.5
21	0	1.7	0	3.7	0	1.1	0	0.3	0	0.2	0	1.1	0.2	0.5	3.0	0	0.7	0.1	0.2	1.0	0.1
22	0.9	0	0.3	0.5	0	0.2	0	0.8	0.2	0.3	0.1	1.3	0.4	0.1	2.2	0	1.3	0.2	0.4	0.5	0
23	0	0.6	0	4.2	0	5.7	0.1	2.8	3.7	0	8.0	0	3.7	0	3.3	0	0.1	1.0	0.1	1.5	0
24	0.7	0.4	0.7	0.1	0.5	0	0.3	0	0	0	0.1	0.1	0.4	0	1.7	0	1.1	0	0	0.1	0.3
25	0.2	0.8	0	4.8	0	0.6	0.3	0.2	3.7	0	3.1	0	7.5	0	7.2	0	2.8	0	0.5	5.5	0.2
26	0.1	0.3	0	5.4	0.2	0.6	0.2	0.1	0.3	0	2.0	0	4.2	0	2.1	0	1.6	0	1.7	0	0.1
27	0.1	0.9	0	1.3	1.2	0	1.2	0	0.5	0.1	0.3	0.2	4.6	0	4.2	0	4.0	0	0.5	1.4	0
28	0	3.7	0	7.7	0	14.0	0	8.4	0.5	0.5	3.3	0	1.9	0	0.8	0.2	2.5	0	2.5	0	0.6
29	0.4	0.7	0	2.3	0.1	0.6	0.5	0.2	1.2	0	7.1	0	1.9	0.2	0.8	0.3	0.4	0.9	0	4.4	0.1
30	0	4.7	0	4.2	0	6.2	0.5	1.3	2.8	0	8.7	0	10.0	0	4.1	0	1.4	0	0.2	1.0	(0.2)

TABLE LXXI.

Disturbances in Declination (F_D)

Gr. M.-T.	0–2		2–4		4–6		6–8		8–10		10–12		12–14		14–16		16–18		18–20		20–22	
Date	+	—	+	—	+	—	+	—	+	—	+	—	+	—	+	—	+	—	+	—	+	—
September 3	0	7.0	0	12.8	0	3.3	0.1	1.7	0	0.2	0.6	0	1.9	0	0.7	0.1	2.3	0	3.8	0	1.9	0
4	0.1	0.3	0	0.7	0	0.9	0	0.7	0.9	0	1.0	0.1	0.5	0	2.0	0	2.9	0	3.4	0.1	1.5	0
5	0.3	0.6	0.1	1.3	0	0.2	0.3	0.2	0.7	0	0.2	0	0.3	0	0.2	0	0	0.2	0.1	0.1	0.1	0
6	0	2.3	0	1.3	0	2.8	0	0.5	0.2	0	0.4	0	1.1	0	0.8	0.1	1.2	0	2.6	0.1	1.1	0
7	0	3.8	0	3.1	0	3.2	0.2	0.1	0	0.1	0.1	0	0.3	0	0	0.2	0	0	0.1	0	0	0
8	0.1	0.1	0	0.2	0	0	0	0	—	—	—	—	0	0.2	0.1	0.1	0.2	0	0.4	0	1.3	0
9	0	0.4	0.5	0	0.5	0	0.1	0.3	—	—	—	—	—	—	0.2	0	0.1	0	0.8	0.1	0.2	0
10	0	0.6	0	1.6	0	1.3	0.1	0.2	0.2	0.1	0.6	0	3.6	0	2.2	0	3.6	0	3.7	0	1.6	0
11	0.1	0.3	0	2.9	0	2.0	0.5	0.1	0.4	0.1	0.6	0	0	0.3	0.2	0	2.0	0	3.1	0	1.0	0
12	0	2.8	0	1.3	0.1	0	0.4	0.1	—	—	—	—	—	—	—	—	—	—	—	—	1.2	0

TABLE LXXI (continued). F_D Axeløen.

Gr. M.-T.	0–2		2–4		4–6		6–8		8–10		10–12		12–14		14–16		16–18		18–20		20–22		22–24	
	+	−	+	−	+	−	+	−	+	−	+	−	+	−	+	−	+	−	+	−	+	−	+	−
September 13	0	0.5	0	2.5	0	2.4	0.1	0.5	0.3	0	0	0	0.2	0	0	0	0	0.2	0.1	0.1	0	0.5	0	0.6
	0	0	0.3	0	0	0	0	0.1	0	0.1	0	0	0	0	0	0	0	0	0	0.1	0.1	0.1	0.1	0.1
	0.1	0.2	0	0	0.1	0.3	0.2	0.2	0.1	0	0.2	0	0	0.3	0.1	0.1	0.1	0.1	0.9	0	1.2	0.1	0.3	2.2
	0.3	0.3	0.1	0.1	0	0	0.3	0	0.1	0	0.2	0.1	0	0.3	0	0.7	0.1	0.3	0.7	0	0.2	0.3	0.2	0.1
	0.3	0.3	0	2.1	0	2.3	0.1	0.2	0	0	0.3	0.1	0.5	0.1	1.2	0.3	1.2	0	0.8	0	0	0.3	0.9	0
	0.3	0	0.1	0.5	0.3	0.3	0.1	0.3	0.2	0.1	0.5	0.1	0.6	0.2	0.2	0.2	0.1	0.3	2.6	0	1.4	5.1	0	0.9
	0.2	3.0	0.6	1.1	0.1	1.0	0.4	0.4	0.5	0.7	1.8	0	0.9	0	2.9	0	1.6	0.4	3.7	1.8	2.3	0.9	0.6	1.8
	0	4.5	0	9.2	0.3	5.4	1.1	0.3	0.1	1.6	0.3	0.7	1.1	0.1	0.3	0.4	1.8	0	4.0	2.3	0.8	2.7	0	2.9
	0.2	0.7	0.1	0.4	0.1	0.6	0.2	0.3	0.1	0.1	0.3	0.2	0.9	0	0.1	0.6	0.3	1.2	0	0.3	0	0.3	0.1	0.1
	0.4	0	0.4	0	0	0.1	0	0.3	0.1	0.1	0.1	0.1	0.1	0.2	0.7	0.1	0.8	0	0.7	1.3	0.1	2.5	0.1	3.2
	0	4.7	0	6.3	0	3.5	0.6	0.5	0.4	0	1.2	0	0.9	0.2	0.3	0.4	1.6	0	1.9	0.1	0.7	0.9	0	2.8
	0	2.5	0.1	0.4	0.2	0.1	0	0.3	0.1	0	0	0	0.2	0	0.1	0.2	0	0	0	0	0	0	0.1	0.1
	0.2	0	0.2	0.1	0	0.1	0	0.5	0	0.2	0.2	0.2	0.8	0.2	1.9	0	1.5	0	0.8	0.3	0	0.5	0.4	0
	0	2.1	0.1	1.7	0.1	0.3	0.1	0.3	0.1	0.3	0	0.2	0	0.3	0.1	0	0.2	0.1	1.8	0	0.7	0.6	0	0.5
	0.1	0.6	0	2.2	0	0.9	0	0.3	0	0	0.1	0.1	0.2	0.5	0	0.4	0.4	0.1	1.8	0	0.9	0	1.1	0
	0.1	1.6	0	2.3	0	0.5	0.1	0.2	0.4	0	0.1	0.3	0	0.4	0	0.4	0.5	0.1	0.9	0	0.4	0.1	0.6	0
	0.1	1.1	0.1	0.6	0.1	0.5	0.3	0	0.1	0	0.1	0.2	0.2	0.1	0	0.4	0.5	0.1	0.3	0	1.2	0.1	0	4.5
	1.0	0.4	0.5	0	0.1	0	0.2	0.1	0.1	0	0.2	0	0	0.2	1.0	0	—	—	—	—	—	—	—	—
	—	—	—	—	—	—	—	—	—	—	0.5	0.1	0.1	0.6	0.5	0.8	0.5	0.1	0.6	0.4	0.1	1.3	0	0.9
	0	0.3	0	1.0	0.1	0.3	0.1	0.2	0	0.2	0.1	0.2	0.1	0.3	0.2	0.1	0	0.3	0.4	0.1	0.1	0.1	0.1	0.5
	0	0.6	0.1	0.2	0.1	0.1	0.1	0.3	0.2	0	0.2	0.2	0	0	0	0.2	0.1	0.3	1.0	0.1	0.4	0.3	0.2	0.2
	0.2	1.2	1.0	0	0.1	0.9	0.5	0.1	0	0.5	0.9	0	0.2	0.5	0.2	0.1	2	0	0	0.1	0.2	0.3	0.1	0.7
	0.3	0.3	0.1	1.2	0.7	0.6	0.4	0.3	0.1	0.1	0.4	0	0.1	0	0.1	0		0.1	0.1	0	0.1	0.4	0.2	0.3
	0	0.3	0	0.4	0.1	0.3	0.2	0.1	0	0	0	0.1	0.1	0	0.1	0		0	0.1	0	0.1	0.5	0	1.2
	0	0.7	0	0.1	0	0	0.1	0	0	0	0	0	0	0	0	0	8	0	0	0	0	0.1	0	1.1
	0	0.3	0.1	0.3	0	0.1	0.1	0.1	0	0.1	0.2	0.2	0.5	0	0	0.1	0.1	0.1	0.5	0.1	0.2	0.2	0	1.7
	0.1	0.8	0	0.3	0	0.3	0.2	0.2	0.1	0.2	0	0.2	0	0.2	0.1	0.6	0.8	0.1	1.0	0.1	0.2	0.3	0	0
	0	0	0.1	0.1	0.2	0	0.3	0	0.1	0	0	0.3	0	0.5	0	0.1	0.1	0	0.9	0	0.4	0	0.6	0.1
	0.1	0.2	0	1.8	0.3	0.3	0.6	0	0.4	0.1	0.2	0.1	2.3	0.1	1.1	0	4.8	0	5.5	0	2.2	3.0	0.2	2.7
	0	2.0	0.1	1.3	0	0.3	0.4	0.1	0.1	0.5	0	0.3	0.1	0.2	0.1	0.1	0.1	0.1	0.2	0	0.1	0	0	0
	0	0	0	0	0	0.1	0	0	0.1	0	0.9	0	0	0.1	0.5	0	2.2	0	2.5	0.2	0.2	0.6	0.1	0.2
	0.2	0.1	0.1	0	0	0.1	0	0.1	0	0.1	0	0	0	0.1	0.1	0	0.9	0	1.4	0	1.3	0	0	0.6
	0	1.2	0	0.4	0.1	0.2	0.8	0	—	—	—	—	—	—	—	—	0.2	0	1.4	0	0.8	0.2	0	0.9
	0	1.6	0	1.9	0	0.6	0	0.1	0.2	0	0.1	0	0	0	0	0	0	0	0.5	0	0.5	0	0.1	0.4
	0	0.3	0	0.5	0	0.6	0	0.3	0.2	0	0.1	0.1	0	0.1	0	0	0	0.1	0.6	0	1.3	0	1.2	0
	0.3	0.2	0.1	0.4	0.4	0.5	1.9	0	0.6	0.1	0.1	0.1		0.1	0	0.2	0.3	0	3.3	0	0.3	0.7	0.7	0
	0	3.5	0	8.2	0.2	1.3	0.4	0	0.1	0	0.1	0		0.1	0	0	0.1	0	0.3	0.2	0.1	0.2	0.1	0.1
	0	0.1	0.1	0.2	0.1	0.1	0.2	0	0	0.1	0.3	0.3		0.1	0	0.3	0	0.1	0.3	0	0.3	0.1	0	0.8
	0	1.1	0.1	1.1	0.1	0.2	0	0.1	0	0.1	0.1	0.1	8	0.3	0.2	0.1	0.9	0	0.4	0.1	0.7	0.6	0	2.7
	0	2.1	0	1.8	0	1.9	0.1	0.3	0.3	0	0.1	0.1	0.1	0.1	0	0.1	0	0.1	0	0	0	0.7	0	1.0
	0.1	0	0	0.2	0	0	0.1	0.1	0	0	0.1	0.1	0.1	0.1	0	0.2	0	0.1	0.4	0.1	0.4	0.6	0	1.8
	0	2.9	0.1	1.9	0.3	0.1	0.1	0	0	0.2	0	0.2	0.1	0	1.5	0	3.9	0	5.6	0	2.1	1.3	1.2	3.0
	0	6.9	0	6.7	0.1	2.4	1.2	0	0.3	0.5	0.1	0.6	0.6	0.4	0.4	0.1	0.2	0.1	0.3	0.2	0.6	0.4	0	1.3
	0	2.0	0	0.5	0.1	0.1	0.1	0.1	0	0.2	0	0.7	0.4	0	1.2	0	2.7	0	2.9	0	1.2	0	0.3	0.8
	0	4.7	0	8.5	0.1	3.6	1.6	0.2	0.7	0.4	0.2	0.6	1.3	0.2	3.1	0	4.7	0.1	4.0	0	1.0	0.2	0	3.9
	0	1.9	0.3	2.1	0.1	0.7	0.6	0	0.5	0.1	0.3	0.1	0.6	0.1	0.5	0	0.7	0.4	1.2	0.3	0.1	4.8	0	4.8
	0	4.5	0	3.3	0.2	0.7	0.4	0.2	0.2	0.2	0.5	0.1	0.3	0.1	0	0.2	1.4	0.5	2.4	0.7	0.4	0.5	0	2.4
	0	1.7	—	—	—	—	—	—	0	5.5	0.2	1.4	0.1	0.4	0.1	0.2	0.5	0	2.1	0	0.5	1.0	0.3	1.0
	0	6.2	0	13.0	0	8.0	0	4.3	0	4.1	0	4.1	0.5	1.6	3.9	0.1	11.6	0	11.1	0	6.0	0	1.2	3.0
November 1	0	5.8	0.1	5.3	0.2	0.8	1.5	0	1.1	0	0.1	0.2	0.1	0.3	0	0.2	0	0	0	0.1	0.2	0.2	0	0.1

TABLE LXXI (continued). F_D Axeløen.

Gr. M.-T.	0—2		2—4		4—6		6—8		8—10		10—12		12—14		14—16		16—18		18—20		20—22		22—24	
Date	+	−	+	−	+	−	+	−	+	−	+	−	+	−	+	−	+	−	+	−	+	−	+	−
November 2	0	0.1	0	0.1	0	0.2	0	0.4	0.1	0	0.2	0	0.5	0	1.5	0	4.2	0	5.0	0	1.3	0.5	0	0.2
3	0	0.6	0.2	0.2	0	0.5	0.2	0	0.1	0	0	0.2	0	0.4	0.1	0	0.4	0	0.4	0	0.7	0	0.1	2.0
4	0.1	0.3	0.3	0	0.1	0	0	0	0	0	0	0	0.1	0	0.1	0	0.1	0	0	0	0	0	0	0
5	0.1	0	0.1	0	0	0	0	0	0	0	0	0	0	0	0	0	0.2	0	0.3	0	0.1	0.2	0	0.3
6	0	0	0	0	0	0	0	0	0	0	0.1	0	0.1	0.1	0.1	0.1	1.8	0	2.0	5.9	0.7	0.7	0.1	0.6
7	0	2.5	0.2	1.4	0.1	0.6	0	0.1	0	0	0.1	0	0	0.1	0	0	0	0.1	0	0.1	0	0.2	0	0.2
8	0	1.8	0	3.7	0	0.5	0	0	0.1	0	0.1	0.1	0	0.2	0	0.1	0	0.3	0.3	0	0.1	0.3	0	0.4
9	0	0	0	0.1	0	0.2	0	0	0	0	0.1	0	0	0	0.2	0	0	0	0.1	0	0	0.1	0	0.8
10	0	1.2	0	0.4	0.2	0	0.1	0	0	0	0.1	0	0.1	0.1	0.1	0.2	0	0.6	0	0.2	0.2	0.6	0	2.5
11	0	0.6	0.1	0	0	0	0.1	0	0	0	0	0	0	0	0	0	0	0	0	0	0	0	0	0
12	0	0	0	0	0	0.1	0	0	0	0	0.1	0	0.1	0	0.1	0.1	0.2	0.1	0.2	0.1	0.2	0.1	0	2.5
13	0	1.9	0	11.8	0	6.0	0.2	0.5	0.7	0.1	0.7	0	1.3	0.2	0.5	0	0.4	0.2	0.4	0.4	0.3	0.3	0	0.2
14	0.1	0.1	0	1.2	0	2.3	0	0.9	2.1	0	—	—	0.6	0	0.7	0.1	0.3	0.1	0.2	0.1	0.3	0.4	0	1.2
15	0	3.5	0	2.5	0	2.4	0	0.4	0.1	0.1	0.1	0.3	0.2	0.4	0.3	0.1	0.2	0.2	0.8	0.1	0.4	0.4	0.6	0.8
16	0	0.6	0	0.3	0.1	0.2	0.1	0.1	0.1	0	0.1	0	0.4	0.3	0	0.7	0	0.3	0	0.4	0	0.3	0	1.5
17	0	4.4	0	2.7	0.1	0.5	0.1	0.8	0.1	0.1	0.1	0	0	0.1	0	0.3	0.2	0.1	0	0.1	0.1	0.2	0.2	0.3
18	0	0.1	0	0.1	0	0.1	0	0.1	0	0	0	0	0	0	0.1	0.1	0.1	0.2	0.3	0	0.2	1.9	0	3.8
19	0	1.8	0.1	1.4	0.1	0.8	0.1	0.3	0.1	0.1	0.3	0.2	0.3	0.1	0.5	0.1	0.1	0.4	0	0.5	0.2	0.6	0.3	1.2
20	0.1	0.6	0	0.5	0.1	0.4	0.1	0.2	0	0.1	0	0.1	0.1	0	0.1	0	0.4	0.1	2.0	0	0.1	1.8	0	1.6
21	0	1.4	0	1.1	0	0.5	0	0.1	0	0.2	0.2	0.2	0.9	0	1.1	0	2.8	0	4.0	0	0.6	2.3	0	1.0
22	0	2.2	0	2.1	0	0.3	0	0.4	0	0.3	0	0.5	0.4	0.1	0.5	0	2.4	1.2	1.1	0.6	0.2	1.5	0	6.7
23	0	2.9	0.1	1.7	0.3	0.3	0.4	0.4	0.6	0.5	0.3	1.4	0.4	0.3	0.1	1.0	2.0	0.7	0.5	2.1	2.3	4.2	0.2	14.4
24	0	9.0	0	15.5	0	9.5	0	6.8	0	3.0	0	2.9	2.1	0.5	2.9	0	5.5	2.0	6.5	0	3.7	2.6	0.3	2.8
25	0	3.8	0	4.0	0	4.9	0.1	1.8	0	1.1	0	2.4	1.6	0	2.0	0	3.4	0.1	3.4	1.5	0.6	3.3	0.4	1.6
26	0	4.6	0	3.7	0.5	0.3	0.2	0.1	0.3	0.1	0.2	0.2	0.1	0	2.5	0.1	4.8	0.4	2.4	0	0.4	1.1	0	0.9
27	0.1	0.5	0.2	0.3	0.5	0.3	0.2	0.1	0.2	0.1	0.1	0	0	0.1	0	0	0.1	0	0.1	0	0	0	0.3	0
28	0.1	0	0.1	0.1	0.2	0.1	0.2	0.1	0	0.2	0.2	0.3	0.1	0.2	0	0.4	0	0.5	0.7	0.2	0.9	0.2	0	0.8
29	0	0.6	0.2	0.2	0.3	0	0.2	0	0	0.1	0.2	0.1	0	0	0	0	0	0	0	0	0	0	0	0
30	0	0	0	0.2	0	1.8	0	0.9	0	0.3	0.1	0.1	0	0.7	0	0.7	0.1	0.5	0.7	0.4	0.2	1.0	0	1.5
December 1	0.1	1.4	0.1	0.1	0.1	0.1	0.4	0	0.3	0	0.1	0.1	0.4	0.1	0.1	0.4	0.9	0.3	0.9	0.6	0.3	1.5	0.1	1.1
2	0	0.8	0	1.9	0.1	2.6	0.3	0	0.4	0.1	1.0	0	1.6	0	0.7	0.2	0.9	0.5	0.5	0.4	0	0.6	0.1	0.5
3	0	0.9	0	2.9	0.2	1.1	0.5	0	0.4	0	0.9	0	0.9	0	0.6	0	0.2	0	0.3	0.1	0.1	0	0	0.2
4	0.1	0	0.1	0.1	0	0.1	0	0.2	0	0	0	0	0.1	0.1	0	0	0.6	0	0.9	0.8	0.4	0.8	0	1.4
5	0	0.1	0	0.4	0	0.5	0.1	0.1	0	0.1	0.1	0	0	0	0	0	0	0	0	0	0	0	0	1.5
6	0	4.4	0	0.5	0	0.2	0	0.1	0	0.1	0.1	0	0.3	0	0	0	0.3	0	0.2	0	0	0.2	0	0
7	0	0	0	0	0	0	0	0	0	0	0.4	0	0	0	0.1	0	0	0	0.4	0.1	0.4	0.5	0	3.2
8	0	2.3	0	1.6	0.1	0.2	0.2	0.2	0	0.3	0	0	0	0	0	0	0	0	0.3	0	0.8	1.8	0	1.7
9	0	0.1	0.1	0.1	0.1	0	0.2	0.2	0.2	0.4	0.2	0	0.5	0	0.3	0.3	1.2	0	1.1	4.6	0.4	2.3	0.3	0.7
10	0.2	0.8	0.4	1.0	0.9	0.4	0.7	0.1	0.2	0.2	1.7	0	0.7	0.2	0.2	0.4	0	0.2	0.4	0.6	0.3	1.8	0	2.4
11	0	1.8	0	2.9	0.1	0.2	0.3	0	0.1	0	0.1	0	0.2	0	0.2	0.2	0.1	0.3	0.6	1.1	0.6	3.3	0	1.6
12	0	2.9	0.1	1.0	0.1	0.4	0.2	0.1	0	0	0.1	0	0	0.1	0.4	0	1.2	0	0.1	6.0	0.2	0.6	0.1	0.7
13	0.1	1.1	0	0.5	0.3	0	0.4	0	0.1	0	0	0.3	0.1	0.4	0.4	0	1.1	0	0.4	0.5	0.8	0.9	0	0.8
14	0	2.0	0.4	0.1	0	0.1	0.1	0	0	0.1	0	0.1	0	0.2	0	0.1	0.1	0	0	0.3	0.1	0.2	0.1	0.2
15	0	3.8	0.1	3.4	0	0.1	0.1	0.1	0	0.1	0.2	0	0.3	0	0	0	0	0.1	0.7	0.3	0.4	0.5	0.2	0.2
16	0.1	0.1	0	0.2	0	0.2	0.3	0.1	0.3	0	0	0.5	0	0.6	0.2	0.2	1.4	0.6	0.3	0.3	0.2	1.5	0	1.6
17	0	0.7	0.3	0.7	0.4	0.1	0.4	0	0	0	0	0	0	0	0	0	0	0	0.1	0.1	0	0.2	0	0.5
18	0	0.4	0.1	0	0.1	0	0	0	0	0	0.1	0	0.2	0	0.1	0	0	0	0	0	0.1	0	0.1	0.5
19	0.3	0.3	0.2	0.6	0.2	0.3	0.7	0	0.8	0	0.4	0	0.4	0.1	0.5	0	0.1	0.2	0.2	0.1	0.3	0.1	0	0.1
20	0.2	0	0.3	0.1	0.3	0.1	0.5	0	0.1	0.1	0	0	0.2	0	0.3	0.3	0.1	0.1	0	0.2	0	0.4	0	0.3
21	0.1	0	0.2	0	0.6	0	0.2	0	0	0	0	0	0	0	0	0	0.3	0	0.9	0.1	0.3	0.1	0	0.2

TABLE LXXI (continued). F_D Axeløen.

Gr. M.-T.	0—2		2—4		4—6		6—8		8—10		10—12		12—14		14—16		16—18		18—20		20—22		22—24	
Date	+	−	+	−	+	−	+	−	+	−	+	−	+	−	+	−	+	−	+	−	+	−	+	−
December 22	0.1	0.3	0.1	0.1	0	0.1	0.3	0	0.2	0.2	0.4	0.5	0.2	0.2	0.7	0.1	0.4	0	1.0	0	0.7	0.1	0	2.5
	0	4.0	0	8.6	0	6.2	0	2.2	0.8	0.1	0.3	0.3	1.1	0.2	2.5	0.2	4.2	0	1.1	1.0	1.5	0.6	0	0.8
	0.2	0.3	0.2	0.6	0.8	0.2	2.0	0	1.6	0.1	0.4	0.3	0.6	0.1	1.3	0.7	0.4	2.9	5.6	0.4	0.3	0.3	0	1.4
	0	3.4	0	6.1	0	1.4	0.3	0.1	0.4	0.1	0.1	0.5	0.1	0	0.1	0	0.4	0	0.7	0.1	0.4	0.3	0	1.2
	0.1	0.7	0.3	0.3	0.4	0	0.1	0.3	0.1	0.1	0.2	0.3	0.3	0.2	0.9	0.1	0.3	0.1	0.3	0.1	1.1	0.2	0	2.1
	0	1.4	0	1.8	0.2	0.2	0.3	0.1	0.2	0.2	0	0	0.2	0	0	0	0	0.2	0.5	0	0.6	0.2	0.6	3.0
	0.3	2.4	1.3	0.4	0	4.1	1.6	0.1	0.6	0.2	0.5	0	0.3	0.1	4.5	0	0.5	0.1	0.4	0	0.7	1.2	0.5	0.1
	0.2	0.2	0.1	0.5	0.9	0.1	1.4	0	0.9	0	0	0	0	0	0.1	0	0.5	0	1.0	0.1	0.3	0.3	0.1	0.2
	0	0.3	0.2	0.1	0.1	0.1	0	0	0	0.2	0	0	0	0	0	0	0	0.2	0.3	0.6	0.2	0.1	0	0.9
	0.2	0.4	0.4	0	0.1	0	0	0	0	0	0	0	0	0	0.1	0	0.2	0	0.2	0	0.1	0.1	0	0.3
	0.1	0	0.2	0	0.2	0.1	0.1	0	0	0	0.1	0	0.2	0	0.5	0	0.2	0.2	0	0.3	0	0.1	0	1.3
	0	2.2	0	0.2	0	0	0.1	0.2	0	0	0.1	0.1	0	0	0.1	0.1	0	0	0	0.5	0	0.6	0	0.5
	0	0.5	0.1	0.5	0	0.2	0	0.1	0	0.4	0	0.1	0	0	0	0.1	0	0.1	0.1	1.4	0.6	0.4	0.1	0.7
	0.1	0.5	0	2.6	0	2.3	0.5	0.4	0.3	0.1	0.1	0.1	0.2	0	0.1	0	0.2	0.2	0.5	0	1.1	0.2	0	0.7
	0	0.7	0	5.7	0	3.4	1.0	0	0.6	0.3	1.9	0	2.3	0	1.8	0	3.7	0	1.4	0.2	0.6	0.1	0.1	0.3
	0	0.5	0	2.2	0.1	0.2	0	0	0	0.1	0	0.1	0.1	0.1	0	0.2	0.4	0.2	0.4	0	0.1	1.0	0	1.4
	0	0.6	0	0.5	0	0.4	0	0.4	0	0.3	0.1	0	0.1	0.1	0	0	0	0	0.6	0.4	0.1	0.2	0	0.1
	(0)	0.6)	(0)	1.0)	0)	0.4	0.2)	0.3	(0)	(0.3)	(0.1)	(0)	(0.1)	(0.1)	(0.5)	(0)	(0.5)	(0)	(0.3)	(0.2)	(0.3)	(0.5)	(0)	(0.8)
	0	0.7	0	1.1	0	0.6	0.1	0.3	0	0.6	0.1	0.2	0	0.1	1.3	0	0.6	0	0.1	0.3	0.4	2.0	0.2	0.9
	0.5	1.1	0	5.7	0.4	0.7	0.6	0	0	0	0.6	0	1.2	0	1.0	0	2.1	0	0.1	0	0.7	0.1	0.1	0.9
11	0.1	1.0	0.3	0.6	0.1	0.4	0.2	0.1	0.1	0	0.4	0	0.4	0	0.4	0.5	0.4	0.6	1.6	0.1	0.4	0.6	0.2	1.5
12	0	2.1	0	1.8	0.1	0.5	0.1	0.4	0.7	0	0.1	0.2	0.5	0.1	0.5	0	0.2	0	0.8	0.6	0	0.2	0	0.8
13	0	1.9	0.1	0.6	0.6	0	0.1	0	0	0	0	0.3	0	0.2	0	0.1	0.1	0.5	0.4	4.0	0.5	1.0	0	1.3
14	0	1.0	0.1	0.1	0.3	0	0	0	0.2	0	0.1	0	0.8	0	0.6	0	0.3	0	0	0.1	0.3	0	0	1.8
15	0.1	0.5	0	0.7	0	0.5	0	0	0.2	0	0.4	0	1.0	0	2.2	0	1.3	0	0.4	0	(0.3)	(0.4)	0.3	0
16	0.1	0.1	0	0.7	0	0.4	0.5	0	0.5	0	0	0.3	0.2	0.6	0.6	0	0.1	0.5	0.1	0.5	0.1	0.2	0	0.9
17	0.2	0.7	0	0.7	0	0.8	0.2	0.1	0.3	0	0.1	0	0.1	0	0.4	0	0	0	0.5	0	0.3	0.1	0	0.1
18	0.3	0	0.1	0.4	0.2	0.1	0.3	0	0.2	0	0.2	0	1.2	0	1.5	0.4	2.0	0	2.3	0.3	0.3	0.2	0	0.4
19	0	1.2	0.1	0.7	0.1	0.4	0.1	0.1	0.3	0	0.5	0	1.8	0	2.3	0.1	1.2	0.1	0.2	0.2	0.5	0.2	0	1.2
20	0	0.5	0	2.0	0	3.4	0.2	0.2	0.8	0	0.8	0	0.5	0	0.4	0	0	0	0.8	0	0.9	0.6	0.2	0
21	0.2	0	0.1	0.1	0.1	0.6	0.8	0	0.3	0.1	0	0.3	0.3	0	0.3	0.2	0.2	0.2	1.5	0	0.2	0.7	0	3.3
22	0	4.4	0.1	0.5	0	0	0	0.1	0	0	0	0.3	0.1	0	0.7	0	0.8	0	1.3	0.1	0.1	0.3	0	0.6
23	0	0.3	0.3	0	0.5	0	0.6	0	0.4	0.1	0.1	0.3	0.2	0.3	0.5	0.1	1.3	0	2.2	0	0.8	0.8	0	1.7
24	0.2	0.8	0.2	1.4	0.2	0.5	0.3	0.4	0.3	0.1	0.1	0.1	0.1	0.1	0.3	0	0.8	0.1	0.5	2.4	0.5	0.5	0	1.6
25	0	1.0	0.2	0.4	0.5	0	0.1	0	0.1	0.2	0	0.1	0	0	0.1	0.1	0.1	0.3	0.1	0.1	0	0.1	0	0.2
26	0	0	0.1	0	0	0	0	0	0.1	0	0.2	0	0.6	0.1	0.1	0	0.4	0.2	2.3	1.1	0.9	0.6	0	10.0
27	0	3.3	0	2.4	0	9.7	0	2.9	0.2	0.3	0.3	0	0.3	0.1	0.1	0.5	0.1	0.3	0.3	0.3	0	0.6	0	1.2
28	0	0.9	0	0.1	0.2	0	0.3	0.1	0.3	0	0.1	0	0.1	0.2	0.4	0	1.4	0	1.0	0	0.3	0.1	0	1.0
29	0	0.8	0.1	0.4	0.1	0.2	0	0	0	0	0	0	0.3	0	0.2	0	0	0	0	0	0	0	0	0
30	0	0	0.2	0	0.1	0.3	0.4	0.1	0.4	0	0.2	0.6	3.3	0.1	5.3	0	4.0	0	3.0	0	0	0.4	0	0.4
31	0	0.6	0.1	0.6	0.2	0.1	0.4	0	0.2	0	0	0.2	0.5	0	0.7	0	1.5	0.1	1.0	0	0.2	1.3	0	0.5
February 1	0	0.3	0.2	0	0.1	0.2	0.1	0.1	0	0.1	0	0	0.3	0	0.6	0	1.1	0	0.9	0	0.1	0.2	0.1	0.2
2	0	0.4	0.3	0.2	0.3	0	0	0	0	0	0.4	0	0.5	0	0.8	0	0.6	0	0.1	0	0	0	0.5	0.1
3	0.1	0.3	0	0.7	0.2	0.4	0.1	0.4	0.2	0	0.9	0	1.2	0	0.5	0	0	0	0	0	0	0	0.1	0
4	0.3	0	0.2	0.2	0.1	0.4	0.1	0.5	0	0.2	0.4	0.1	0.6	0	0.1	0	0.3	0	0.8	0	0.1	0.4	0	0.7
5	0	0	0.2	0.2	0.1	0.2	0	0.3	0.1	0.3	0.6	0.1	0.8	0	1.3	0	1.4	0	1.6	0	0.9	0.2	0.1	0.3
6	0	0.9	0	3.0	0.1	2.8	0.3	0.2	0.4	0	0.3	0	0.7	0	0.8	0	0.8	0	0.5	0.1	0.5	0	0	0.2
7	0	0.1	0	0.2	0.1	0.2	0	0.2	0.1	0.2	0.2	0	0.8	0	1.1	0	1.5	0	0.5	0	1.1	0	0.1	2.1
8	0	2.2	0	7.5	0	10.0	0.1	2.2	0.5	1.0	1.6	0.7	2.5	0	5.1	0	2.5	0	3.1	4.9	1.5	4.6	0.1	0.6
9	0.1	0.7	0	3.0	0	0.4	0.2	0	0	0.1	0.1	0.1	0	0	0	0	0.5	0	0	1.6	0.9	0.5	0	1.3

TABLE LXXI (continued). F_D Axeløen.

Gr. M.-T.	0—2		2—4		4—6		6—8		8—10		10—12		12—14		14—16		16—18		18—20		20—22		22—24	
Date	+	—	+	—	+	—	+	—	+	—	+	—	+	—	+	—	+	—	+	—	+	—	+	—
February 10	0	2.2	0	2.0	0.1	0.7	0.3	0.2	0.1	0.2	0.1	0.1	0.4	0	0.4	0	0.1	0	0.4	0	0.3	0.3	0.3	5.0
11	0.3	1.7	0	5.0	0	3.6	0	0.5	0.1	0.2	0.5	0	1.0	0	1.4	0	1.2	0	0.6	0	0.6	0	0	0.6
12	0	2.1	0	1.6	0	0.5	0	0.5	0	0.4	0.1	0	0.4	0	0.3	0	(1.0)	(0)	(2.0)	(0)	0.8	0.1	0.2	0.2
13	0	3.5	0	0.2	0.2	0.4	0.2	0.4	0.3	0.2	0.5	0	0.2	0	0.4	0	0.1	0.1	1.0	0	0.1	1.0	0.6	0.2
14	0	2.3	0.2	1.2	0.2	0.2	0.3	0.1	0.2	0.1	0.4	0.1	0.2	0.2	0.4	0.3	0	0.3	0.1	1.9	0.6	0.6	0.1	1.2
15	0	1.8	0	2.9	0.1	0.4	0.2	0.2	0.4	0.2	0.4	0	2.4	0	2.5	0	3.0	2.4	0.1	0.4	0.4	0.1	0.1	0
16	0.2	0.1	0	0	0.1	0.1	0	0.1	0	0.1	0	0	0	0	0.5	0	0.9	0	0.6	1.5	0.2	0.7	0.1	0.5
17	0	1.0	0	1.3	0.2	0.6	0	0.3	0	0.1	0.4	0	1.7	0	0.9	0	0.5	0.4	0	0.1	0.2	0.1	0.2	0.3
18	0.1	0.7	0.3	0.4	0.1	0.3	0.6	0	0.6	0	0.2	0.1	0.1	0	0.5	0	0.1	0.1	0.2	0	0.1	0.3	0	0.4
19	0	0	0	0.1	0	0	0	0	0	0	0	0.2	0.1	0.1	0.1	0.1	0	0.1	0.2	0.2	0	0.6	0.1	0.1
20	1	0.3	1	0.3	0.1	1.0	0	0.3	0	0.2	0.3	0	0	0.3	0	0.1	0.1	0	0.2	0.1	0.1	0.2	0.1	0
21		0	.2	0	0.5	0	0.3	0.1	0	0.3	0.9	0	2.7	0	2.9	0	1.3	0	0.3	0	0	0.1	0.2	0
22	.2	0.1		3.6	0	6.9	0	6.9	0	2.5	1.4	0.5	1.9	0	2.2	0	1.3	0	0.3	0.2	0	0.7	0	0.4
23	.1	1.2	.1	0.2	0.1	0	0	0	0	0.1	0.1	0.2	0.4	0.2	0.6	0	0.4	0	0.6	0	0.1	0	0.1	0.9
24	8	1.3	8	0.9	0.1	0.2	0	0.1	0	0.1	0	0	0.2	0	0	0	0.6	0	0.4	0.2	0	0	0.1	0.1
25	.2	0	.2	0.9	0	6.7	0	3.6	0.5	0.2	1.6	0	2.3	0	0.7	0.2	0.4	0.1	0.1	0.1	0	0.6	0	0.2
26	.1	0.2		0	0.1	0	0	0.1	0	0	0	0.4	0	0.2	0	0	0	0	0.2	0	0.9	0.8	0	1.8
27		1.6		1.9	0.1	0.8	0.2	0	0	0	0	0.2	0.1	0	0	0	0	0	0.1	0	0	0.3	0	0.4
28		0		0	0	0	0	0.1	0	0	0	0	0	0	0	0	0	0	0	0	0	0.1	0	0.2
March 1	8	0	8	0	0	0	0	0.1	0	0	0.7	0	2.0	0	2.2	0	3.7	0	0.8	0.8	0.1	0.2	0.2	2.0
2		2.8		1.8	0	2.5	0.4	0.4	0.9	0	0.2	0	0	0	0.1	0	1.3	0	2.2	0.2	1.5	0	0.2	1.6
3		4.8		2.6	0	1.9	0.2	0.1	0	0.2	0.1	0.5	0.1	0.2	0	0	0.1	0	1.1	0	0.3	0	0	0.1
4	.1	0.1		0.3	0	0.1	0	0.3	0	0.3	0	0	0.1	0	0	0	0	0	0	0	0	0.1	0.2	1.1
5	.2	1.1		3.9	0	3.5	0.2	0.2	0.1	0.1	1.7	0	3.9	0	5.1	0	6.0	0	2.7	0	0.6	0	0	0.8
6	8	1.9	8	0.3	0.2	0.1	0.2	0	0	0	0.1	0	0.4	0	0.1	0	0	0.2	1.5	0	1.2	0.2	0.5	1.8
7		3.8		6.4	0	5.2	0.1	1.9	1.1	0.5	0.6	0	2.0	0	3.0	0	3.3	0	2.8	0	0.5	0.2	0	0.8
8		3.5		7.2	0.4	1.7	1.2	0	1.1	0	1.0	0	1.6	0	3.1	0	2.7	0.6	3.6	0.8	3.3	0.3	1.2	1.9
9	.1	1.9		9.0	0	8.0	0.1	0.8	0.6	0.2	0.4	0.1	0.2	0.1	0	0.2	0.2	0.2	0.3	0.7	0.3	1.2	0	1.5
10		1.9		2.6	0	0.3	0.2	0	0.2	0.2	0	0.1	0.6	0	1.8	0	2.2	0.1	2.1	1.5	0.7	1.1	0	1.7
11	8	2.6	8	2.9	0	0.8	0.2	0.3	0	0	0.7	0	0.6	0	0.2	0	1.2	0	1.5	0.4	0.2	1.0	0	1.1
12	0	2.8		3.1	0	6.0	0.5	0.1	1.1	0	0.5	0.1	1.0	0.1	1.0	0	3.9	0	1.3	0	2.3	0	0	4.5
13	0	3.6		6.9	0	6.1	0.6	0.5	0.7	0.2	0.5	0.1	2.2	0.1	2.7	0.8	2.6	2.0	1.4	0	0.7	0.4	0.8	0.2
14	0	3.0		2.9	0.1	0.5	0.7	0	0.6	0.1	0.1	0.4	1.7	0	2.1	0	1.8	0	0.8	0.2	0.3	0.4	0.2	0
15	0.1	0.1	1	0.4	0.6	0.7	0.6	0.2	0.3	0	0.5	0.2	1.5	0	1.1	0.4	1.1	0.1	1.2	0.8	0.3	0	0.1	0.6
16	0.1	0.2	8.	0.3	0.2	0	0.3	0.1	0.3	0	(0.2)	(0.1)	0	0.1	0	0	0	0.1	0.1	0	0.1	0.1	0	0.3
17	0	0		0	0	0	0	0	0.1	0	0	0	0.1	0	0	0	0	0	0.5	0	0.4	0.1	0	0.1
18	0	0		0	0	0	0	0	0.2	0	0	0	0.1	0.1	0	0	0	0	0.2	0	0.3	0.2	0	1.7
19	0	2.6		1.7	0	0.7	0.2	0.1	0.4	0	0.3	0	0.4	0	0.9	0	2.4	0	3.5	0	0.3	0.3	0	1.2
20	0.2	1.3	4	0.2	0.2	0.3	0	0.1	0.3	0.3	1.3	0	0.6	0	0.1	0	0.2	0	0.2	0.1	0.4	0.1	0	1.3
21	0	1.1	8.	3.0	0	5.0	0.1	0.2	0.1	0.2	0.1	0.1	1.5	0	0.8	0	1.0	0	0.6	0	1.3	0.3	0	3.3
22		1.4	0	1.6	0.1	0.5	0.2	0	0	0.1	0	0.3	0.1	0	0.4	0	1.3	0	1.5	0	1.6	0	2.8	0.4
23		7.0	0	1.3	0	2.4	0.1	0.5	0	0.4	0.2	0.6	0.8	0	0.1	0	(0.2)	(0)	(0.2)	(0.1)	(0.4)	(0.1)	0	(1.0)
24	(.1)	(0.2)	(0)	(0.3)	(0.2)	(0)	(0.3)	(0.1)	(0.3)	(0)	(0.4)	0	(0.4)	(0)	(0.1)	(0.2)	(0)	(0.2)	(0.2)	0	(0)	(0)	(0)	(0)
25	(.1)	(0.2)	(0)	(0.3)	(0.2)	(0)	(0.3)	(0.1)	(0.8)	(0)	0	0	0.1	0.1	0	0	0	0	0.1	0	0.1	0.1	0.4	0.2
26	8.1	0	0	0.3	0	0	0	0.1	0	0	0.4	0	1.0	0	0.8	0	1.0	0	1.6	0	0.6	0	0	0.1
27		0.4		0.3	0	0	0	0	0.1	0.4	0	0.1	0	0	0.1	0	0.2	0	0.9	0.1	0.4	0.2	0.1	0.4
28		0.2		0.1	0	0	0	0	0	0.1	0.8	0	1.4	0	0.5	0	0.1	0.1	0.9	0	0.3	0.1	0.6	0.1
29	.3	0.1		4.0	0	2.5	0	1.6	0	0.7	0.1	0.2	0.2	0.2	2.0	0	3.5	0	4.5	0	0.7	0.1	0	1.7
30		4.5		8.3	0	2.7	0.1	0.6	0	0.5	0.2	0.1	0.4	0	0.3	0.1	0	0.4	0.2	0.2	0.8	0.1	0.7	0.1
31	8.1	8.0	8	2.9	0	2.0	0.4	1.0	0.3	0.7	0.5	0.2	1.8	0	2.4	0	2.3	0.4	2.9	0	0.8	1.0	0.5	2.2

(continued). F_D Axeløen.

	2 − 4		4 − 6		6 − 8		8 − 10		10 − 12		12 − 14		14 − 16		16 − 18		18 − 20		20 − 22		22 − 24	
−	+	−	+	−	+	−	+	−	+	−	+	−	+	−	+	−	+	−	+	−	+	−
1.5	0	3.9	0.1	1.1	0.1	0.2	0.1	0.2	0.2	0.2	1.2	0	0.4	0.3	0	0.7	0	0.2	0.5	0.1	0.1	5.1
3.7	0	3.7	0.3	0.1	0.1	0.8	0.5	0	0.5	0.1	0.9	0.2	0	1.0	1.8	0.1	1.6	0.2	2.0	0	0.2	1.1
2.4	0	8.9	0.2	0.3	0.1	0.2	0.6	0.1	0.3	0	0.5	0.1	0	0.4	0.1	0.2	0.9	0.3	1.2	0.2	0.1	0.3
1.6	0	5.5	0	2.5	0.2	0.1	0.2	0.1	0	0.3	0.8	0	0.4	0	0.7	0	3.4	0	0.6	0	0.8	0.3
2.0	0	5.2	0	10.3	0.1	3.0	0.1	0.9	0.1	0.6	2.2	0.2	4.2	0	3.6	1.0	1.0	0	0.2	0.2	0.3	1.6
2.5	0.1	10.0	0.1	8.3	0.5	0.5	0.6	0.5	0	6.6	0	8.9	1.2	2.1	1.8	0.1	1.2	0.1	0.8	0	0.7	0.1
0.1	0.3	0	0.5	0	0.6	0.1	0.1	0.6	0.1	0.8	1.3	0.2	2.2	0	1.8	0	1.4	0	2.7	0.2	0.8	1.2
1.3	0	0.8	0	0.4	0.3	0.5	0.3	0	0.2	0	0.1	0.4	1.3	0.2	0.4	0.2	2.5	0	3.2	0	1.3	2.7
5.3	0	7.0	1.2	4.3	3.9	2.1	1.3	4.0	0.9	0.3	2.0	0	2.0	0.1	5.3	0.3	5.2	0.3	3.0	0.1	0.1	4.5
1.3	0.2	2.1	0.3	0.4	0.3	0.7	0.2	0.7	0.5	0.1	1.9	0.1	0.7	1.2	2.0	0.1	4.5	0	3.0	0.8	0	0.8
1.0	0.1	0.8	0.2	0.5	0.3	0.3	0.1	0.3	0.2	0.9	0.3	0.2	0.1	0	1.0	0	1.4	0	1.0	0.2	0.4	0.2
−	−	−	−	−	−	−	−	−	−	−	−	−	−	−	−	−	−	−	−	−	−	−
−	−	−	−	−	−	−	−	−	−	−	−	−	−	−	−	−	−	−	−	−	−	−
−	−	−	−	−	−	−	−	−	−	−	−	−	−	−	−	−	−	−	−	−	−	−
−	−	−	−	−	−	−	−	−	−	−	−	−	−	−	−	−	−	−	−	−	−	−
(0.4)	(0.1)	(0.1)	(0.1)	(0)	(0.2)	(0.2)	(0.1)	(0.1)	(0.2)	(0.1)	(1.0)	(0.3)	(2.0)	(0)	(2.0)	(0)	(2.5)	(0)	0.3	0.2	0	0.2
0.1	0	0.1	0.2	0	0.2	0.2	0.1	0.2	0.4	0	2.8	0	4.5	0	4.8	0	6.5	0	0.9	0.2	0.2	0.2
0.2	0.2	0.9	0	1.0	0.3	0	0.5	0	0.2	0.3	1.8	0.8	7.5	0	6.1	0	4.1	0	4.7	0	0.8	0.5
0.8	0.1	2.8	0.1	1.0	0.6	0.6	0	0.1	0	0	0.2	0	0	0	0.1	0	0.1	0.1	0	0.2	0.2	0.2
0.1	0	0.5	0	0.7	0.3	0.3	0.8	0	0.1	0.2	0	0.4	0.1	0.4	0.2	0.1	0.8	0	0.5	0	0.3	0.2
0.1	0.2	0.8	0.1	1.8	0.1	0.3	0.2	0.3	0.1	0.2	0.2	0.1	0.9	0	0.6	0	1.1	0	0.6	0.4	0.3	0
0.3	0	0	0.3	0	0	0	0.1	0.1	(0.1)	(0.3)	0.2	0.3	0	0.3	0.4	0	1.2	0	1.0	0.3	0.3	1.4
0.6	0	1.2	0.2	0.2	0.7	0	0.5	0	0.2	0.1	1.9	0	3.1	0	3.2	0	3.6	0	1.5	0	0.4	0.2
0.5	0.1	0.5	0.1	0.5	0.1	0.3	0.1	0.1	0	0.8	0.8	0	0.6	0	1.1	0	1.2	0	1.3	0	0.4	0.6
2.5	0	6.4	0	1.8	0	0.1	0.1	0.2	0.2	0.3	0.8	0	1.0	0	0.4	0	0.3	0	0.4	0.2	0.2	0.3
0.8	0	2.0	0	0.9	0.2	0.6	0.2	0.1	0.3	0	0	0.4	0.5	0.1	1.5	0	3.9	0	4.3	0.2	1.2	0.8
7.1	0	9.0	0	6.3	0	1.1	0.4	0	0.3	0.2	0.7	0	0	0.2	0	0.4	0.2	0.1	0.3	0.1	0	1.1
2.1	0	5.2	0	3.0	0.2	0.2	0.1	0.2	0	0.8	0	1.0	0.2	0.2	1.2	0	1.4	0.1	0.8	0.2	0	1.9
4.0	0.4	2.9	0	2.0	0	0.4	0.3	0	(0.2)	(0.3)	(0.5)	(0.4)	(1.0)	(0.1)	1.7	0	3.5	0	1.2	0.1	0	1.2
1.8	0	1.5	0	1.4	0.1	0.5	0	0.1	0.2	0	1.5	0.1	3.4	0	2.0	0	1.9	0	0.3	0	0	0.3
0.4	0	1.2	0	1.3	0.2	0.4	0.4	0	0.1	0.4	0.6	0.1	0.3	0.3	0.9	0	3.6	0	3.2	0	0.6	0.1
0.2	0	0.9	0.1	0.3	0.6	0	0.6	0	0.1	0.2	0.6	0.1	0.5	0	1.1	0	2.0	0	4.2	0	3.1	0
0.4	0	0.1	0	0.1	0	0.1	0.1	0.1	0	0.1	0.1	0.1	0.2	0	0.1	0.1	0.2	0	0.5	0.1	0.2	0.2
1.1	0.1	0.7	0	1.7	0.4	0.8	0.2	0.2	0.4	0	1.3	0	1.0	0	0.2	0.2	2.3	0	2.7	0.4	1.0	0.8
3.5	0	15.0	0	16.0	0	5.5	0.3	0.5	1.9	0	0.7	0	(0.5)	(0.1)	0.8	0	2.0	0.1	1.2	0.1	0.4	0.6
1.6	0.5	1.6	0.9	2.0	0.4	3.4	0.2	1.0	0	4.1	(1.5)	(0.1)	(0.2)	(0.2)	3.5	0	4.0	0	2.4	0.5	2.5	2.9
1.6	0	10.8	0.7	2.7	0.1	1.3	0.8	0.4	1.4	0	3.5	0	(0.2)	(0.2)	(2.0)	(0)	3.0	0	0.3	0.4	0.1	0.8
1.3	0	1.9	0	0.4	0.2	0.8	0.7	0.1	0.3	0.1	0.1	0.3	0.3	0.3	1.2	0	1.5	0.2	3.4	0	0.6	0.6
5.0	0	8.0	0	3.6	0.1	0.1	0.1	0.4	0.8	0.3	0.7	0	0.5	0.1	(1.0)	(0)	(3.0)	(0)	0.8	0.2	0.7	0
1.7	0	1.3	0	0.8	0.2	0.4	0.2	0.1	0.7	0	(1.5)	(0.1)	(0.5)	(0.2)	(2.0)	(0)	(3.0)	(0)	3.3	0	0.3	0.3
0.3	0.1	0.3	0.1	0.1	0.6	0.1	1.7	0	0.7	1	0	0.6	0	0.4	1.1	0	1.5	0	1.1	0	0.1	0.9
2.4	0	1.5	0.2	0.3	0.3	0.1	0.2	0	0.4		0.5	0	0.2	0	1.4	0	2.0	0	1.1	0	0	0.2
0.3	0	0.7	0	1.2	0.5	0.1	0.7	0	0.6		0.5	0	2.3	0	3.7	0	6.2	0	2.7	0.1	0	1.9
0.6	0	0.5	0	0.8	0.2	1.2	0.2	0.1	1.0		2.0	0	(1.5)	(0.1)	2.8	0	4.7	0.2	2.7	0	0.1	1.7
5.8	0	11.8	0	5.2	0.3	0.7	0.6	0.1	0.6	0.1	0.2	0	0	0.1	0.3	0	1.5	0	1.1	0.1	0.2	1.1
3.2	0	2.0	0.1	1.5	0.1	0.4	0.2	0.4	0.2	0.1	0.1	0.6	1.1	0.4	2.2	0	2.8	0	1.1	0	0.9	3.3
0.6	1.0	2.6	0.1	2.5	0.1	0.8	0.9	0.1	0.3	0.4	0.1	0.6	0.9	0.1	1.4	0.1	1.2	0	1.6	0.2	0.6	0.6
0.3	0.1	0.5	0	0.4	0.4	0.1	0.2	0.1	0	0.4	0	0.8	0.1	0.4	0.5	0.2	2.2	0	1.0	0	0.4	0.1
0.1	0	0.8	0.4	0	0.2	0.1	(0.3)	(0.2)	0.1	0.4	0	0.3	0.9	0	1.4	0	1.5	0	0.4	0	0.1	0.2
0	0.3	0	0.4	0.1	0.1	0	0	0.2	(0.1)	(0.3)	(0)	(0.5)	(0.7)	(0.2)	(1.4)	(0.1)	(2.0)	(0)	1.5	0	1.3	0

TABLE LXXI (continued). F_D Axeløen.

Gr. M.-T.	0—2		2—4		4—6		6—8		8—10		10—12		12—14		14—16		16—18		18—20		20—22		22—24	
Date	+	−	+	−	+	−	+	−	+	−	+	−	+	−	+	−	+	−	+	−	+	−	+	−
May 21	0.2	0.8	0	4.1	0	1.2	0.2	0.3	0	0.4	0	0.2	0.5	0	3.4	0.1	0.1	0.7	2.3	0	5.0	0	0.7	1.0
22	0.2	0.4	0	1.2	0.1	0.8	0	0.8	0.2	0.2	0	0.2	0.1	0.6	0.2	0.3	1.8	0.1	2.0	0	0.4	0.3	4.0	0.1
23	0.1	0.8	0	7.7	0	13.5	0	5.9	0	2.4	0	4.0	1.0	2.1	1.5	0	0.1	1.6	0.9	0.1	0.9	0.2	0	1.5
24	0.1	2.4	0	0.9	0.1	0.4	0	0.8	0	0.4	0.1	0.2	0	0.7	0.1	0.4	0	0.5	1.4	0	2.0	0.2	1.5	0.9
25	0	2.9	0	9.8	0	3.4	0.1	1.3	0	1.7	1.5	0	4.5	0.2	1.8	0.3	2.6	0	3.8	0.1	1.0	0.1	0.8	0.3
26	0	1.1	0	8.2	0	3.5	0.2	0.1	0.1	0.3	0.4	0	0.2	0.2	0	0.3	0.5	0.3	1.8	0	0.3	0.4	0.7	1.0
27	0	3.0	0	3.9	0	0.7	0.1	0.5	1.1	0	0.1	0.2	0.6	0.2	0.4	0.7	1.3	0	1.9	0.1	0.5	0	2.0	0.3
28	0	6.2	0	12.0	0	15.0	0.1	7.3	0.5	1.0	0.6	0.1	1.2	0	0.1	0.2	0.2	0.4	0.7	0.2	0.6	0.8	0	3.0
29	0.6	3.3	0	6.9	0	4.1	0	1.8	0	1.0	0.7	1.0	0.2	0.5	1.2	0	1.0	0	2.7	0	2.0	0.2	0.4	1.0
30	0	5.9	0	11.0	0	7.0	0	2.3	0	2.3	0.1	1.8	2.6	0.2	3.2	0	4.4	0	6.0	0	(1.0)	(0)	(0.8)	(1.3)

TABLE LXXII.

Disturbances in Vertical Intensity (F_V).

Gr. M.-T.	0—2		2—4		4—6		6—8		8—10		10—12		12—14		14—16		16—18		18—20		20—22		22—24	
Date	−	+	−	+	−	+	−	+	−	+	−	+	−	+	−	+	−	+	−	+	−	+	−	+
September 3	0	2.7	1.3	2.8	4.4	0	3.0	0	(0)	(0)	0.1	0	0.5	0	0	1.6	0.1	0.7	0	3.3	0	4.7	0.2	0
4	1.5	0	1.2	0	3.0	0	0.7	0	0	0.1	0	2.8	0	0.2	0	0.2	0.7	0.8	0.4	2.4	0	2.8	0	1.8
5	0	3.2	0	2.3	0.2	0	0.1	0.2	0.1	0.1	0.1	0	0.1	0	0	0.2	1.2	0	0.5	0	0	0.1	0	1.2
6	0	2.8	0	1.0	0	0.1	0	0.3	0.3	0	0	0	0.7	0	0.4	1.5	0	1.9	0.7	0.9	0	2.8	0.1	0.1
7	0	6.3	0.2	2.5	3.2	0	0.3	0.2	0.5	0	0	0	0	0.2	0	0	0.1	0	0	0	0	0	0	0
8	0	0	0	0	0	0	0	0	0	0.1	0	0.1	0	0.1	0	0.4	0	0.3	0	1.5	0	2.2	0	0.2
9	0	0.9	0	0.1	0	0.1	0	0	0	0	0.6	0	0.1	0.1	0	0.1	0	0.1	0	0.2	0	3.4	0	0.1
10	0	0	0	0	0	0	0	0	0	0.1	0	2.8	0.3	0.2	0.5	0.6	1.0	0.1	0.4	0.5	0	0.6	0	2.1
11	0	1.5	0.1	0.5	1.0	0	0.6	0	0	0.1	0.1	0	0	0.5	0	1.5	0	0.5	0	2.1	0	1.3	0	4.7
12	0	3.0	0	0	0	0	0.1	0.1	—	—	—	—	—	—	—	—	—	—	—	—	0.4	0.6	0	6.5
13	0	0	0.1	1.5	2.4	0	1.4	0	0	0	0	0	0	0	0	0	0	0	0	0.1	0	0.7	0.1	1.3
14	0	0	0	0	0	0	0	0	0	0	0	0	0	0	0	0	0	0	0	0	0	0	0	0.1
15	0	0.5	0	0	0.1	0	0.3	0	0	0	0	0	0.1	0	0.1	0.1	0	0.9	0.5	0.3	0.4	1.1	0	5.5
16	0.1	0.1	0.4	0	0.2	0	0	0	0	0.2	0	0.3	0	0.3	0	1.5	0	0.8	0	0	0	1.7	0.1	0
17	0.6	0.1	0.3	0.2	2.4	0	0.7	0	0	0	0.1	0.1	0	1.6	0	2.3	0	0.6	0	0.1	0.1	0	0.4	0
18	0	0	0.1	0	0.1	0.2	0	0.2	0	0.4	0	1.7	0.8	0	0	0.4	0	0.2	0.1	0.2	0	9.5	0.1	4.3
19	0	3.1	0.3	0.2	0.1	0.2	0.1	0.4	0	3.5	0	1.0	0	3.0	0.2	0.8	0	2.2	0	7.4	0	23.5	0	16.2
20	0	11.6	0	13.8	0	2.6	0	1.2	0.8	2.8	7.3	0	2.7	0.2	0	2.8	0	1.0	2.6	5.7	0	9.1	0	9.8
21	0.5	0.1	0.2	0	1.1	0	0.3	0.2	0	0.2	0	0.4	1.1	0.1	0.4	0.3	2.7	0.6	0	1.2	0	0.1	0.1	0
22	0	0	0	0	0	0	0	0	0	0.1	0	0.1	0	0.2	0.2	0.9	0.1	0.3	0.5	5.2	0	7.7	0	13.0
23	0	12.8	1.6	4.5	4.6	0	1.3	0.5	0.1	0.1	0.3	0.4	0.6	0.2	0.3	0.3	0.9	1.8	0	6.8	0	9.6	0	5.3
24	0	3.3	0	0.2	0.1	0.1	0	0	0	0	0	0	0	0	0	0	0	0	0	0	0	0	0	0
25	0	0	0.1	0	0	0.1	0	0	0.1	0	0	0.3	0	0.6	0.8	0	1.8	0	0.1	1.0	0	1.6	0	0.1
26	0	3.4	0	1.3	0	0	0	0	0.1	0	0	0	0.1	0	0	0	0.3	0	0.5	0.3	0.1	4.6	0	1.0
27	0	2.3	0	0.2	0	0	0	0	0.1	0	0	0	0	0.1	0.1	0	0	0.1	0	2.5	0	2.7	0	0
28	0	3.0	0	2.2	0	0.1	0	0.8	0.5	0.2	0	0.1	0	0	0	0	1.0	0	0	0.3	0.3	0.2	0.1	0
29	0	1.1	0.1	0.4	0.1	0.1	0	0	0	0	0.1	0.1	0	0	0.6	0.1	0.3	0.3	0	0.8	0	1.2	0	14.0
30	0	1.6	0.1	0	0	0	0	0.2	0.1	0	0	0	0	0	1.0	0	8.7	0	1.3	5.0	0	15.6	0	19.6
October 1	0	19.6	0	11.0	0.5	0.2	0.9	0	0.1	0.1	0.1	0.3	0	0.2	0.1	0.1	2.8	0	0.4	0.8	0	1.3	0	1.2
2	0	0	0	0.1	0	0	0	0	0	0.2	0.1	0	0.1	0	0.2	0	0.1	0	0	1.0	0	1.8	0	0.1
3	0	0.1	0	0	0.1	0.1	0.1	0.2	0	0	0	0	0.6	0	0.8	0	0	0	0	1.2	0	1.3	0	0.7
4	0	3.8	0	0.1	0.7	0	0.6	0	0.1	0.1	0.4	0.1	0.2	0	0.2	0	0	0	0	0.1	0	1.6	0	2.7
5	0	0.8	0.1	0.2	0.1	0.4	0	0	0.1	0	0	0	0	0	0	0	0	0	0.1	0	0	0.8	0	0.8
6	0	0.1	0	0	0.1	0	0	0	0	0	0	0	0	0	0.1	0.1	0	0	0	0	0	2.8	0	1.6
7	0	0.2	0	0	0	0	0	0	0	0	0	0	0	0	0	0	0	0	0	0	0	0	0	1.8

tinued). F_V Axeløen.

0—2		2—4		4—6		6—8		8—10		10—12		12—14		14—16		16—18		18—20		20—22		22—24	
—	+	—	+	—	+	—	+	—	+	—	+	—	+	—	+	—	+	—	+	—	+	—	+
0	0	0	0	0	0	0	0	0	0	0.1	0	0	0	0	0	0	0	0.1	0.5	0	2.1	0	3.5
0	0	0	0	0	0	0	0.1	0	0	0	0	0.1	0	2.5	0	3.2	0	1.4	0	0	0.6	0	0
0	0	0	0	0	0	0	0	0	0	0	0	0	0	0	0	0.1	0	0	0	0	0	0.1	0.2
0	0.1	0	0	0	0	0	0.2	0	0	0.2	0	4.1	0	2.0	0	5.7	0	0.3	1.0	0	15.0	0	9.3
0	2.5	0	0.1	0	0	0	0.2	0.1	0	0	0	0.1	0	0.1	0.1	0	0	0	0.1	0	0	0	0
0	0	0	0	0	0	0	0	0	0	0.1	0.2	0.3	0	0.2	0	0.1	0.1	0.6	3.4	0	5.5	0	1.0
0	0	0	0	0	0	0	0	0	0	0	0	0	0	0	0	0	0	0	1.5	0	1.9	0	3.0
0	2.2	0	0	0	0	0	0	0	0	0.3	0	0.2	0	0	0	0	0.3	0.8	0.5	0.6	0.1	0.3	0.3
0	0.2	0.1	0	0.5	0	0	0	0	0	0	0	0	0	0	0	0	0	0	0	0	2.3	0	1.5
0	0	0	0	0	0	0	0	0.2	0	0	0.1	0	0	0	0	0	0	0	0	0.1	0.1	0	0
0	0.2	0.1	0.2	3.1	0	1.7	0	0	0.1	0	0.1	0	0	0	0	0.2	0	0.1	2.9	0	5.4	0	0
0	1.0	0.3	1.0	5.0	0	3.0	0	1.3	0	1.2	0	0.2	0	0	0	0	0	0	0.2	0	1.2	0	0.3
0	0	0	0	0	0	0	0	0	0.1	0	0.1	0	0.1	0	0	0.1	0	0.9	0	0.1	1.1	0	0.9
0	0.1	0.1	0.2	0	0.2	0	0	0	0	0	0	0	0	0	0	0.1	0	0.1	0.2	0	1.2	0	7.5
0.1	0.9	0	0.3	1.1	0	0.7	0	0	0	0	0	0	0	0	0	0	0	0	0	0	0.2	0.1	0.3
0.1	0	0	0	0	0	0	0	0	0	0	0	0	0	0	0	0	0	0	0	0	4.7	0	10.0
0	5.4	0	2.6	0	0	0	0	0	0	0	0	0	0	0.9	0	5.1	0	0.8	0.1	0	11.2	0	12.6
0	17.8	0	7.7	0.4	0.5	0.4	0.1	0	0.1	0.4	0.1	5.1	0	0.3	0	0	0.1	0	0.5	0	1.2	0	3.7
0	1.6	0	0.2	0	0.3	0	0.1	0	0.1	0	0	0.5	0	4.1	0	0.8	0	0	2.7	0	2.3	0	2.1
0	6.4	0.2	5.0	3.5	0.2	9.0	0	1.2	0.1	0.2	0	3.3	0	3.4	0.8	1.2	1.1	0	3.0	0	5.0	0	12.0
0	3.7	0.1	3.3	1.2	0	0.2	0.3	0	0.1	0.2	0.1	0.3	0	3.2	0	0.9	0	0.1	6.0	0	4.6	0	10.5
0	4.0	0.1	0.8	1.0	0	0.5	0.1	0.3	0.1	0.2	0.1	0.1	0	0	0	0.9	0.3	0.1	7.8	0	10.0	0	6.9
0	5.1	0.1	0.1	0.5	0.1	0.3	0.5	3.3	0.1	4.6	0	0.1	0	0	0	1.9	0	0.9	0.9	0	8.3	0	7.2
0	14.2	0	14.2	0	12.6	0.2	3.2	0.5	0	1.5	0	0.6	4.0	0	1.5	0	7.3	0	13.0	0	21.2	0	21.1
0	16.7	0	7.6	0	1.0	0.3	0	1.4	0	0.4	0	0	0	0	0	0	0	0	0	0	0.5	0	0
0	0	0	0	0	0	0	0	0.2	0	0	0	0	0	0.3	0	0.2	0.3	0	3.9	0	8.5	0	0.9
0.3	0	0.6	0	0	0	0	0	0	0	0.1	0	0	0	0	0	0.1	0	0.2	0	0	2.8	0	3.7
0	1.9	0	0	0	0	0	0	0	0.1	0	0	0	0	0	0	0	0	0	0	0	0	0	0
0	0	0	0	0	0	0	0	0	0	0	0	0	0	0	0	0	0	0	0.3	0	1.3	0	0.1
0	0	0	0	0	0	0	0	0	0	0	0	0	0	0	0	0.5	0.1	0.9	10.2	0	9.7	0	4.7
0	6.6	0	1.5	0.3	0.2	0	0	0.2	0	0.1	0	0	0	0	0	0	0	0	0	0	0	0	0
0	1.0	0	2.1	0	0	0	0	0	0	0	0	0	0	0	0	0.1	0	0.1	0.3	0	1.1	0	0.7
0	0	0	0	0	0	0	0	0	0	0	0	0	0	0	0	0	0	0	0	0	0	0	1.6
0	1.0	0.1	0.1	0	0	0	0.1	0	0	0	0	0.2	0	0.3	0	0.1	0.2	0	0.1	0	3.0	0	7.8
0	0.8	0	0	0	0	0	0	0	0	0	0	0	0	0	0	0	0	0	0	0	0	0	0
0	0	0	0	0	0	0	0	0	0	0	0	0.1	0	0.2	0	0.6	0	0.9	0	0	0.1	0	6.8
0	1.1	0	6.7	4.1	0	1.7	0	0.2	0.2	0.7	0	3.3	0	0.1	0	0.4	0	0	4.6	0	6.5	0	4.2
0.3	0.1	1.9	0	4.2	0	4.2	0	1.4	1.0	0	1.6	0.2	0	0.2	0.1	1.2	0	0.2	0	0	2.0	0	1.0
0	4.0	0.7	0	3.8	0	3.0	0	0.9	0	0	0.4	0	0.9	0.1	0.2	0	0	0	0.1	0	4.7	0	4.2
0	0.8	0.1	0	0	0.1	0	0.2	0	0.1	0	0	2.4	0	0.5	0	0.1	0.2	0.2	0	0	0.1	0	2.0
0	8.1	0.5	1.2	2.1	0	1.6	0	0.6	0	0	0	0	0	0	0	0.4	0	0.5	0	0	1.2	0	0.2
0	0	0	0	0	0	0	0	0	0	0	0	0	0	0	0	0.1	0	0	0.1	0	3.2	0	5.3
0	0.3	0.3	0	0.5	0	0.1	0.1	0	0.1	0	0.2	0.1	0.1	1.0	0	0.5	0	0.1	0	0	1.6	0.1	2.0
0	0.1	0	0	0.1	0	0.1	0	0.1	0.1	0.1	0	0	0	0.1	0	0.9	0.1	0.8	3.5	0.1	1.7	0	1.4
0	0.4	0	0	0	0	0	0	0.2	0	0	0	0.1	0.4	1.5	0	2.8	0	0.1	1.0	0	11.6	0	8.9
0.1	1.3	0.4	0	0	0	0	0.1	0	0	0.1	0	0.1	0.2	1.9	0	1.2	0.8	0.2	8.1	0	4.6	0	11.7
0	1.8	0.5	0.3	0.1	0.1	0	0.4	0.3	0.7	8.5	0	1.8	0	0	0.3	2.3	1.2	0	2.5	1.9	4.2	1.1	14.2
0	23.7	0	19.5	0	7.7	1.0	2.0	5.0	0.3	0.1	0.2	0.1	0.4	0.3	0.6	0.1	8.3	0	9.1	0	24.4	0	16.6
0	12.1	0	8.0	0	5.4	0.1	0.5	0.2	0.1	0.8	0.5	1.7	0.4	0	1.2	0	9.0	0	19.1	0	19.4	0	13.0
0	14.8	0	4.7	0.2	0.9	0	0.2	0.2	0.3	0	0	0.1	0	4.3	0.1	4.4	1.5	0	2.3	0.4	0.7	0	0.6

TABLE LXXII (continued). F_v

Gr. M.-T.	0–2		2–4		4–6		6–8		8–10		10–12		12–14		14–16				–20		
Date	–	+	–	+	–	+	–	+	–	+	–	+	–	+	–	+	–	+	–	+	–
November 27	0	0.4	0	0.2	0	0.3	0	0.2	0	0.2	0	0	0	0	0	0	0	0	0	0	0
28	0	0.3	0	0	0.2	0	0.8	0	0	0	0.1	0.8	0	1.4	0	0.4	1.2	0.1	4.6	0	(0)
29	(0)	(0.6)	(0)	(0.1)	(0.1)	(0.1)	(0.3)	(0.1)	(0)	(0.1)	0.2	0	0.3	0	0	0	0	0	0.1	0	0
30	(0)	(0.6)	(0)	(0.1)	(0.1)	(0.1)	(0.2)	(0.1)	(0)	(0.1)	(0.1)	(0.6)	0.2	0.1	0.3	0	0.2	0.1	0.1	3.4	0
December 1	0	1.0	0	0	0	0	0	0	0	0	0	1.1	1.3	0.2	1.1	0	0	2.0	0.4	2.3	0.1
2	0	0.1	0	0.3	0.1	0.2	0	1.9	0	2.3	0.7	0	0.6	0.1	0.7	0	1.5	0.5	0.3	1.1	0
3	0	2.1	0.9	0.1	3.2	0	0.5	0	0	0.1	0	0	0	0	0	0	0	0	0	0.5	0
4	0	0	0	0	0	0	0	0	0	0	0	0	0	0	0	0	0.5	0	0.1	1.1	0
5	0	0	0	0	0	0	0	0	0	0	0	0	0	0	0	0	0	0	0	0	0
6	0	2.1	0	0	0	0	0	0	0	0	0	0	0	0	0	0	0	0	0	0.3	0
7	0	0	0	0	0	0	0	0	0	0	0	0	0	0	0	0	0	0	0	0.9	0
8	0	2.0	0	0.2	0	0	0	0	0	0	0	0	0	0	0	0	0	0	0	0	0
9	0	0	0	0	0	0.1	0.1	0.1	0	0	0	0	0	0	0	0	3.1	0	2.7	0	2.5
10	0	1.4	0	1.3	0.8	0.4	0.1	0.2	0	0.8	0	0.1	0.5	0.1	0.6	0	0	0	0.2	0.9	0
11	0	1.7	0	1.6	1.3	0	0	0	0	0	0	0.3	0.2	0	0	0	0.8	0.2	0.3	1.4	1.0
12	0	1.5	0	0.1	0	0	0	0	0	0	(0.2)	(0)	(0)	(0)	0.3	0	0	1.2	0	4.7	0
13	0	1.4	0	1.1	0	0	0	0	(0)	(0)	0.5	0	0	0	(0.1)	(0)	(0)	(0.5)	0.1	0	(0)
14	0	(2.0)	(0)	(1.3)	(0)	(0)	(0)	(0)	(0)	(0.1)	0.5	0	0	0	0	0	0	0	0	0	0
15	0	4.9	0	4.0	0	0	0	0	0	0.1	0	0	0	0	0	0	0	0	0	1.3	0
16	0.2	0	0	0	0	0	0	0	0	0.2	0	0	0	0	0.3	0.1	2.0	0.9	0	0.6	0.1
17	0	0.3	0	0	0	0	0	0	0	0	0	0	0	0	0	0	0	0	0	0	0
18	0	0	(0)	(0)	(0.2)	(0)	(0)	(0)	(0)	(0)	0.4	0	0	0	0	0	0	0	0	0	0
19	0	0.5	0	0	0.5	0	0.1	0	0.1	0	0	0	0.7	0	1.7	0	0.6	0	0	0.9	0
20	0	0.3	0	0	0.2	0	0	0	0	0	0	0	1.0	0	3.5	0	1.8	0	1.8	0	0.3
21	0	0	0	0	0	0	0	0	0	0	0	0	0	0	0	0	0.7	0	0.3	0.8	0
22	0.1	0	0	0	0	0	0.1	0	0	0.6	0.2	0.3	1.4	0	4.3	0	4.0	0	3.3	0	0.2
23	0	4.6	2.6	2.5	6.2	0	7.2	0	11.0	0	10.5	0	9.0	0	10.0	0	2.2	0.3	0.1	3.5	0
24	0	2.6	0.1	1.5	0.6	0	0.3	0	0	0.1	0.5	0	0.3	0.3	2.5	0.1	2.2	0.8	3.5	0.7	0
25	0	1.7	0	3.6	0.6	0	0	0.2	0.3	0	0	0.1	0	0	0	0	0	0	0	2.1	0
26	0	1.2	0	0.2	0	0.1	0	0.1	0	0.1	0	0	1.4	0	3.1	0	2.0	0	1.9	0	0.2
27	0	0.4	0	0.3	0	0.1	0	0	0	0	0	0	0	0	0	0	0	0	0	0	0
28	0	6.2	0	1.5	1.8	0.2	3.0	0	0.4	0.1	1.3	0	1.8	0	0.5	0	0	0	0	0	0.2
29	0	0.7	0	0	0	0	0	0	0	0	0	0	0	0	0	0	0	0.1	0	0.6	0
30	0	0	0	0	0	0	0	0	0	0	0	0	0	0	0	0	0.2	0.1	0.4	0.1	0
31	0	0	0	0	0	0	0	0	0	0	0	0	0	0	0	0	0	0	0	0	0.1
January 1	0	0	0	0	0	0	0	0	0	0	0	0	0	0	0	0	0.1	0	0	0	0
2	0	0.8	0	0	0	0	0	0.1	0	0	0	0	0	0	0	0	0	0	0.1	0	0
3	0	0	0.2	0	0	0	0	0	0	0	0	0	0	0	0	0	0	0	0	1.1	0.1
4	0	0.1	0	0.8	1.5	0	0.9	0.1	0.7	0	0	0.6	1.7	0	0.5	0	0	0	0	0.1	0
5	0	0.1	0.2	1.9	3.5	0	0.5	0.1	1.0	0.7	0.8	0	1.1	0	3.9	0	2.0	0.1	0.5	0.3	0
6	0.1	0.2	0.4	0	0	0	0	0	0	0	0	0	0	0	0	0	0.2	0	0	0.7	0
7	0	0	0	0	0	0	0	0	0	0	0	0	0	0	0	0	0	0	0	1.0	0
8	0	0	0	0	0	0	0	0	0	0	0	0	0	0.1	0.1	0	0	0	0	0	0
9	0	0	0	0	0	0	0.2	0	0	0	0	0.7	0	0.1	0.5	0	0.5	0	0	0.6	0.7
10	0	2.9	0	3.1	0	0	0	0	0	0	0	0	1.2	0	3.8	0	4.0	0	0.3	0	0
11	0	1.1	0	0.5	0	0	0	0	0	0	0	0	0	0	0	0	0	0	0	2.7	0
12	0	3.7	0	2.5	0	0.1	0	0	0	0	0	0	0	0.1	0.1	0.1	0	0	0.8	0.6	0
13	0	2.1	0	0.1	0	0	0	0	0	0	0	0	0	0	0	0	0	0	1.4	0.6	0.2
14	0	0.1	0	0	0	0	0	0	0	0	0	0	0	0	0	0	0	0	0	0	0
15	0	1.0	0	0.3	0	0	0	0	0	0	0	0	0	0.6	0	0	3.5	0	2.0	0	(0)

(continued). F_r Axeløen.

2	2–4		4–6		6–8		8–10		10–12		12–14		14–16		16–18		18–20		20–22		22–24	
+	−	+	−	+	−	+	−	+	−	+	−	+	−	+	−	+	−	+	−	+	−	+
0.1	0	0.3	0.9	0	1.1	0	0.1	0	0.4	0	4.0	0	1.0	0	0	0.6	0.3	0.1	0	0.5	0	1.2
0.5	0	0	0	0	0	0	0	0	0	0	0	0	0.1	0	0	0	0	0.5	0	2.5	0	0.3
0	0	0	0.7	0	0.3	0	0	0	0	0.1	0.3	0	0.6	0	4.8	0.1	0.2	1.0	0.4	0.4	0.8	0
0.4	0	0.2	0.4	0	0.1	0	0	0.4	0.5	0	2.0	0	2.0	0	1.8	0.2	0.5	0	0.5	0.6	0	0.6
0.4	0	2.2	0.6	0.4	0.3	0.7	0	2.6	0	0	0	0	0	1.1	0	0.9	0	0.6	0.2	1.3	0.2	0
0	0.1	0	1.6	0	0.8	0	0	0.1	2.0	0	7.7	0	11.8	0	3.0	0	(0.2)	(0.4)	(0.2)	(1.0)	0	(1.5)
(0.8)	(0)	(0.5)	(0.6)	(0)	(0.4)	(0.1)	(0.1)	(0.4)	0	0	(0)	(0)	(0)	(0)	(0)	(0)	0.1	1.1	0	0.6	0	0
0	0	0	0	0	0	0	0	0.7	0	0.8	0	0.8	1.7	0	0.4	0.1	0.1	0.6	1.1	2.8	0	1.1
0.8	0	1.2	0	0	0	0.7	0.1	0.4	0.3	0	0	0.3	0.1	0.1	0	0.6	0.4	3.5	0	3.8	0	4.9
1.6	0	0	(0)	(0)	0.2	0	0.2	0.3	0.4	0	0	0	0	0.3	0	0.2	0	0.1	0	0	0	0
0	0	0	0	0	0	0	0	0	0.5	0	0.1	0.3	0	1.6	0.8	0.4	0	11.0	0	13.5	0	34.5
30.8	0	15.3	0	6.6	0	0.2	0	1.5	0.2	0	0.2	0	0.1	0	0	0.1	0	0.3	0	1.0	0	3.3
0.4	0	0	0	0	0	0	0	0	0	0	0	0	0	0	0.2	0	0.6	0	0	1.3	0	1.2
0	0	0	0	0	0	0	0	0	0	0	0	0	0	0	0	0	0	0	0	0	0	0
0	0	0	0	0	0	0	0	0	1.3	0	5.0	0	4.5	0	1.5	0	0.4	0	0.1	0	0	0
0.5	0	0.4	0	0	0	0	0	0	0	0	0	0	1.1	0	2.7	0	0.7	0	1.0	0	0.1	0
0	0	0	0	0	0	0	0	0	0	0	0	0	0.1	0	0.1	0	0.2	0	0	0	0	0
0	0	0	0	0	0	0	0	0	0	0.9	0	0.6	0	0	0	0	0	0	0	0	0.2	0
0	0	0	0	0	0	0	0	0.1	0	0.2	0	0	0	0	0	0	0	0	0	0	0	0
0	0	0	0	0	0	0	0	0	0	0	0	0	0	0	0	0	0	0	0	0.3	0	2.2
0	0	0	0	0	0	0	0	0	0.2	0	0	0	0.8	0	1.2	0	0	0.3	0	0.8	0	0
0	0	1.0	0.4	0	0.2	0	0	0	0	0	0.2	0	0.7	0	0.8	0	0.6	0	0	0	0	0
0	0.1	0	0.1	0	0	0	0	0	0	0	0	0	0.1	0	1.5	0	0.8	0	0	1.7	0	12.5
(6.0)	0	(4.8)	0	5.3	0	2.5	0.3	0.5	2.7	0	0.5	0	2.7	0	2.0	0.5	0	6.2	0	16.2	0	3.0
0.6	0	1.0	0.3	0.1	0	0.1	0	0	0	0	0	0	0	0	0	0.1	0	1.8	0	2.7	0	6.0
2.3	0	0.3	0	0	0	0.2	0	0.2	0	0	0.1	0	0	0	0	0	0.1	0	0	0	1.0	4.2
0.6	0.1	0.7	1.2	0	0.3	0	0	0.1	0	0.5	0	0.8	0	0	0	0	0.5	0.5	0	0.7	0	1.3
2.0	0	0	0	0	0	0	0	0	0	0	0	0	0.2	0	0.8	0	0	0.5	0	2.0	0	0
2.7	0	0.9	0	0	0	0	0	0	0	0	0	0	0	0	0	0	0	0	0	0	0	1.5
3.6	0	0.8	0	0	0	0	0	0	0	0	0	0	1.4	0	(0.2)	(0)	0.3	0.4	0	2.6	0	5.7
3.1	0	1.5	0	0	0	0	0	0	0	0	0	0	0.8	0	1.9	1.2	0.3	0.2	0	0.1	0	0
0	0	0	0	0	0	0	0	0	0	0	0	0	0	0	0.1	0	1.2	0.4	0.1	0	0	0
0.4	0	0	0	0	0	0	0	0	0	0	0.4	0	2.1	0	2.4	0	0.6	0	0	0.8	0	1.4
1.3	0	0	0	0	0	0	0	0	0	0	0	0	0.2	0	0	0	0	0	0	0	0	0
0	0	0	0	0	0	0	0	0	0	0	0	0.1	0	0.1	0	0	0	1.2	0	1.0	0	0.1
0	0	0	2.1	0	2.0	0	0.3	0	0	0	0	0	0	0	0	0	0	0	0	0.1	0	0
0	0	0	0	0	0	0.1	0	0	0	0.1	1.2	0	2.1	0	3.1	0	0.2	0	0	0	0	0
0.4	0	7.1	0	10.6	0	7.7	1.1	1.1	3.8	0	0.7	0.1	0	0.1	0.6	0	0.4	0	0.1	0	0	0
0	0	0	0	0	0	0	0	0	0	0.2	0	0.5	0.2	0	1.0	0	0.5	0	0	0.5	0	1.8
1.4	0	0.3	0	0	0	0	0	0	0	0	0	0	0	0	0.9	0	0.8	0	0	0	0	0
0	0	1.2	0	3.7	0	3.4	0.3	0.2	1.4	0	0.9	0.1	0	0.4	0.4	0	0	0	0	0	0	0
0	0	0	0	0	0	0	0	0	0	0	0	0	0	0	0	0	0	0	0	1.4	0	5.6
3.8	0	1.2	0	0	0	0	0	0	0	0	0	0	0	0	0	0	0	0	0	0	0	0
0	0	0	0	0	0	0	0	0	0	0	0	0	0	0	0	0	0	0	0	0	0	0
0	0	0	0	0	0	0	0	0	0	1.1	0	1.5	0.7	0	1.7	0	2.0	0	0	0.5	0	3.4
4.9	0	1.4	0.3	0	0.6	0	0.1	0.1	0.2	0	0	0	0	0	3.2	0	0.7	1.4	0	0.7	0	5.2
7.3	0	2.3	1.6	0	0	0.1	0.2	0.3	0.2	0	0	0	0	0	0.1	0	0	0	0.1	0	0.1	0
0	0	0	0	0	0	0	0.1	0	0	0	0	0	0	0	0	0	0	0	0	0.1	0	1.4
1.1	0	4.7	2.0	0	1.8	0	0.2	0.1	1.2	0	8.2	0	18.5	0	15.4	0	7.6	0	3.6	0	1.0	0
0.9	0	0	0.1	0	0.1	0	0	0	0	0	0	0	0	0	0	0	0.4	0.1	0	10.0	0	7.8

TABLE LXXII (continued). F_v

Gr. M.-T.	0–2		2–4		4–6		6–8		8–10		10–12		12–14		14–16		16–				
Date	−	+	−	+	−	+	−	+	−	+	−	+	−	+	−	+	−	+	−	+	−
March 7	0	5.7	0	3.0	2.4	0.5	6.0	0	5.0	0	2.5	0	2.0	0.1	0.7	0.3	0.7	0.9	0	7.0	0
8	0	9.0	0	10.0	0	1.1	0	0.1	0.1	0.1	0	0	3.4	0	3.0	0	0.6	1.9	2.1	5.1	0
9	0	11.1	0	13.1	0	6.8	0.5	0.6	0.2	0.2	0	0	0	0	0	0	0	0	0.2	0	0.3
10	0.5	0.1	0.2	0	0	0	0.1	0	0	0.1	0	0	0.1	0	2.6	0	3.3	0.4	0	5.0	0
11	0	3.4	0	1.4	0	0	0	0	0	0	0	0	0	0	0	0	0.1	0.2	0	3.3	0
12	0	7.8	0	4.3	0	2.4	0.3	0	0	0.1	0	0	0	0	0	0	1.5	0	0.1	0.8	0
13	0	11.5	0	10.2	0	9.7	2.1	0.1	2.0	0	0.3	0	2.7	0	3.8	0.1	7.2	1.5	0.6	1.5	0
14	0	6.3	0	3.0	0.1	0.1	0	0.1	0	0.2	0	0	0.2	0.3	3.5	0	4.5	0	3.2	0.3	0.4
15	0.2	0	0	0.2	0.2	0.4	0.1	0.2	0	0	1.0	0	1.0	0	5.7	0	5.0	0	0.9	2.4	0
16	0	0	0	0	0	0	0	0	0	0	(0)	(0)	(0)	(0)	(0)	(0)	(1.5)	(0)	(1.0)	(1.0)	(0)
17	(0)	(0)	(0)	(0)	(0)	(0)	(0)	(0)	(0)	(0)	(0)	(0)	(0)	(0)	(0)	(0)	(0)	(0)	(0)	(0)	(0)
18	(0)	(0)	(0)	(0)	(0)	(0)	(0)	(0)	(0)	(0)	0	0	0	0	0	0	0	0	0	0	0
19	0	5.3	0	1.0	0.1	0.1	0	0	0	0	0	0	1.8	0	0.5	0	0.3	0	0.4	5.0	0
20	0.5	1.5	2.2	0	0.3	0	0	0	0	0	0	0.2	0	0.1	0	0	0	0	0.1	0	0
21	0	0.4	0	0.7	2.0	0	3.3	0	1.8	0	0.5	0	0.3	0	2.1	0	1.8	0	0	0	0
22	0	0.4	0.1	0.1	0	0	0	0	0	0	0	0	0.1	0	0.2	0	0.2	0	0.2	0.4	0
23	0	2.1	0	1.4	0	0	0	0	0	0	0	0	0.1	0	0.6	0	0.1	0	0	0	0
24	0	3.3	0	0.2	0	0	0	0	0	0	0	0	0.1	0	0	0	0	0	0	0	0
25	0	0	0	0	0	0	0	0	0	0	0	0	0	0	0	0	0	0	0	0	0
26	(0)	(0)	(0)	(0)	(0)	(0)	(0)	(0)	(0)	(0)	0.5	0	0	0	0.1	0	1.1	0	0.2	0	0
27	0	0	0	0	0	0	0	0	0	0	0	0	0	0	0.1	0	0.9	0	0.5	0.8	0
28	0	0	0	0	0	0	0	0	0	0	0	0	0	0	0	0	0	0	0	0.4	0
29	0	0.1	0	2.8	0.1	0.4	0	0	0	0.5	0	0.1	0.1	0	3.0	0	5.0	0	1.0	0	0
30	0	5.2	0	4.8	0.1	0	0.1	0	0	0	0	0	0	0	0	0	0	0	0	0.1	0
31	0	4.7	0	1.4	0.1	0	0.3	0	0.1	0.1	0	0.2	0.1	0	2.0	0	3.7	0	0.2	0	0
April 1	0	1.4	0	0	0.3	0	0	0	0	0	0	0	0	0	0	0	0	0.1	0	0	0
2	0	10.7	0	2.6	0	0	4.5	0	3.0	0	0	2.7	0	0.2	0	0	0.2	0	0.9	3.0	0
3	0	6.4	0.2	8.2	0.8	0	0.2	0	2.5	0	2.5	0	0.7	0	0	1.0	0	0.1	0.1	1.3	0.5
4	0	3.3	0	3.2	1.0	0	1.8	0	0.5	0	0	0.2	0.2	0.1	0	0	0.8	0	0.2	2.1	0
5	0	2.8	0	4.8	0.5	3.6	3.0	0	2.2	0	0.2	0.2	4.1	0	10.3	0	5.0	0.2	1.6	0	0.1
6	0	6.5	0	18.6	0	16.2	0.1	.5	2.5	0	13.5	0	11.2	0	11.9	0	2.5	0.1	0	0.2	1.0
7	0.2	0	0.1	0	0.2	0	0.		0	0	0.5	0.1	3.0	0	4.5	0	5.0	0	2.3	0	1.9
8	0.2	0.1	0	0	0	0	0.	8.1	0	0.1	0	0	(3.0)	(0)	(4.5)	(0)	(3.0)	(1.2)	2.2	0	0.8
9	0	18.3	0	11.0	0.2	4.5	2.4	1.5	1.0	1.0	1.1	0	1.1	0	2.6	0	0.5	3.6	1.1	10.8	0
10	0	5.1	0	2.0	0	0.1	0.7	0.2	0.2	0.3	0	0.4	3.6	0.2	6.8	0	2.4	0	1.0	3.6	0.3
11	0	2.8	0	0.2	0	0	0.8	0.2	0.4	0	4.0	0	4.0	0	0.3	0	0.2	0	1.7	0	0
12	0	2.4	0	8.0	0.1	0.5	0.1	0	0	0	0	0	0	0	1.8	0	2.0	0	0.5	0.1	0
13	0	5.8	0	2.9	0	0	0.7	0	1.0	0	0.6	0	0.1	0	1.0	0	2.9	0	1.0	0.1	0
14	0	0	0	0	0	0	0	0	0.2	0	0	0	0.8	0	4.4	0	4.8	0	2.3	0	0
15	0	0.1	0	0	0	0	0	0	0.6	0.1	2.5	0	7.8	0	15.0	0	13.7	0	4.8	0	0.4
16	0	0.1	0	0	0	0	0	0	0	0	(0)	(0)	(0.3)	(0)	(2.5)	(0)	(3.1)	(0)	(0.9)	(0)	1.0
17	0	0	0	0	0	0	0	0	0	0	0	0	0.8	0	7.4	0	8.5	0	1.5	0	0.1
18	0.1	0	0	0.1	0.1	0.1	0	0	0	0	4.0	0	7.6	0	10.1	0	10.1	0	3.0	0.4	0
19	0	0.9	0	0.6	0	0	0.2	0.2	0	0	0	0.1	0	0	0	0	0.5	0	0.8	0	0
20	0	0	0	0	0	0	0	0.1	0	0.1	0	0	0.1	0	0	0	0.3	0	0.5	0	0.1
21	0	0	0	0	0	0.1	0	0	1.1	0	0	1.9	0	4.5	0	1.7	0	0	1.3	0	0.5
22	0.1	0.1	0.1	0	0	0	0	0	0.1	0	2.9	0	1.4	0	0	0	1.2	0	0.8	0.1	1.0
23	0	0.1	0	0.1	0	0	0	0	0	0	0	0	0.8	0	3.5	0	6.0	0	4.1	0	1.5
24	0	0.1	0.1	0	0.2	0	0	0	0	0	0	0	0	0	0.3	0	3.5	0	3.0	0	0
25	0	4.3	0	4.0	0.1	0	0	0	0	0	0	0	(0)	(0)	(0)	(0)	(0)	(0)	0	0	0.1

TABLE LXXII (continued). F_V Axeløen.

Gr. M.-T.	0–2		2–4		4–6		6–8		8–10		10–12		12–14		14–16		16–18		18–20		20–22		22–24	
Date	−	+					−	+					+	−	+	−	+	−	+	−	+	−	+	−	+
April 26	0.3	0.3	0	0.3	0	0		0.2					0	0	0	0.1	0.3	1.2	0	1.5	0	0	10.5	0	14.6
	0	26.0	0	20.7	0		0.3	0	0.1	0.1	0	0.2	1.2	0	0.6	0	0	0.2	0	0.3	0	0	0	0
	0	2.0	0	3.2	0		0.1	0	0	0.1	0	0	0	0.1	2.0	0	2.6	0	1.4	0	0	1.4	0	5.7
	0	0.9	0.4	0			0	0	0	0	0	0	(0.4)	(0)	(1.1)	(0.1)	3.6	0	0.8	0.3	0	7.6	0	7.2
		0.2	0.1	0			0.3	0	0	0	0	0	0.4	0	1.7	0	2.0	0	1.3	0.1	0	0	0	0
May 1							0	0.1	0	0	0.1	0	0	0.1	0	0	0.6	0	1.5	0	0.1	1.6	0	4.7
							0	0	0	0	0	0	0.1	0	0	0	0.7	0	1.7	0	0.1	5.0	0	4.2
							0	0	0	0	0	0	0	0	0	0	0	0.3	0	1.3	0.3	0	0.3	0
	0	0.2	0	0	0.1	0.1	0.2	0.1	0	0	0.1	0	0	0	0.1	0.1	0	0.1	0	1.0	0	11.7	0	18.5
	0	3.5	0	32.0	0.1	13.7	0.5	0.8	0.7	0.2	(5.0)	(0)	(10.0)	0	(1.5)	(0)	(0.5)	(0.3)	1.0	0.2	1.2	0	0.8	0.1
	0	1.5	0.4	0.2	0.9	2.4	1.3	0.7	4.0	0.8	6.7	0	(0.5)	(0.1)	(1.5)	(0)	0.6	0	0.1	0.1	0	6.5	0	9.5
	0	9.6	0	11.5	0.1	0.7	0.3	0.1	0.1	0.2	0.2	1.3	0.3	0	(1.5)	(0)	(3.0)	(0)	(2.5)	(0)	0.2	0	0	0.5
	0	0.3	0	0	0.1	0.1	0.2	0.3	0.1	0.5	0	0.2	0	0.1	0.7	0.1	2.4	0	3.0	0	0.1	4.0	0	4.8
			0	2.2	1.0	0.5	0.2	0	0	0.5	0.3	0.3	0	0.2	0.6	0	3.3	0	2.8	0	0	0	0.1	0
	0	0.	0	0	0.1	0	0	0.3	0	0.2	0.1	0.1	1.1	0	2.9	0	3.2	0	0.7	0.6	1.9	0.5	0	2.0
	0	0.	0	0	0	0	0.1	0.2	0	0	0	0	0.6	0	0.9	0	1.6	0	0.2	0	0	0	0.1	0
	0	0	0	0	0	0.2	0	0	0	0	0	0	0	0.2	0.4	0.4	1.7	0	0.4	0	0	0	0	0
	0	0	0	0	0	0	0.2	0	0	0	0	0	0	0	1.7	0	5.7	0	4.2	0	0	0.9	0	1.0
	0.	0	0	0	0	0.2	0	0.5	0.1	0.1	0.6	0	2.0	0	(1.5)	(0)	5.0	0	1.6	3.2	0	12.0	0	7.5
	0	8.	0	5.7	0.6	0.2	0.9	0.1	0.7	0	0	0.1	0	0	0	0	2.0	0	2.2	0	0	4.0	0	2.4
			0.1	0	0	0.1	0.1	0	0	0	0	0	0	0	0	0	0.8	0	1.8	0.4	0	1.5	0	6.4
			0	10.2	0	6.0	0.7	0.3	0	1.5	0.2	0	1.7	0	4.0	0	4.1	0	0.6	0.3	0.1	0.5	0	4.8
	0	0	0	0.1	0	0	0	0.4	0	0.3	0	0.1	0	0	1.0	0	3.3	0	1.6	0	0	0.7	0.1	0.2
	0.	0	0	0.3	0	0	0	0.5	0	0.1	0	0	0.2	0	2.8	0	0.8	0.1	0	1.0	1.1	0	0	0
	0	0	0	0	0.1	0	0	0	0	0	(0)	(0)	(0.5)	(0)	(1.0)	(0)	(1.0)	(0)	(0.5)	(0.2)	0	0	0	0
	0	1.0	0	1.2		0	0	0.1	0	0	0	0	0		4.1	0	6.1	0	3.0	0	3.3	0	0	7.8
	0	1.7	0	0		0	1.5	0	0.5	0	0	0.1	0		0.4	0	1.7	0	1.1	0	0.2	0	0.9	0.5
	0	3.1	0	7.3		9.6	0.4	1.4	0.2	0.1	4.4	0	16.2		12.4	0	6.0	0	1.5	0	0.5	0	0	4.9
	0	1.7	0	0		0.1	0	0.5	0.1	0	0.2	0	0		0	0	1.0	0	1.4	0	0	6.0	0	10.7
	0	5.7	0	5.8	8.1	0.1	0.1	0	0.1	0.4	0.2	0	4.0	8	3.7	0	0.5	0.3	0.3	3.8	0.1	1.0	0	2.7
	0	3.5	0	3.6	0.1	0	0	0	0.2	0	0	0.2	0.7	.4	0.2	0.3	3.7	0	0.6	0	0.1	0.6	0	5.9
	0	6.6	0.2	2.3	0.2	0	0	0.1	0.1	0	0.2	0.1	0	.2	0	1.6	0.7	0.1	1.5	0.3	0.1	1.0	0.1	2.4
	0	6.6	0	12.7	0	16.6	4.5	0.1	1	0	4.0	0	4.4		0.2	0.1	0	0	0.6	0.3	0.6	1.7	0	4.0
	0	5.0	0	1.9	0.1	0.3	0.2	0.2	5	0	2.6	0.4	4.5		0.4	0.3	0.4	0.2	0.5	1.7	0.3	9.3	0	8.3
	0	9.3	0	8.0	1.6	0.1	0.5	0.1	6	0.3	5.6	0	7.1	8	4.4	0	2.2	0	0.1	5.4	0	3.0	(0)	(5.0)

Dyrafjord.

TABLE LXXIII.

Disturbances in Horizontal Force (F_H).

	0–2		2–4		4–6		6–8		8–10		10–12		12–14		14–16		16–18		18–20		20–22		22–24	
Date	+	−	+	−	+	−	+	−	+	−	+	−	+	−	+	−	+	−	+	−	+	−	+	−
December 2	0	2.8	0	2.0	0.3	0.8	0.7	0	0	0.5	0.1	0.2	0.2	0.2	1.0	0.2	2.4	0	0.2	0	0	0	0.1	0.3
		1.0	0.1	0.4	0.1	0.3	0	0.1	0	0	0	0	0.1	0	0	0.2	0	0.1	0.3	0	0.2	0	0.1	0
4	0	0	0	0	0	0	0	0	0	0	0	0	0	0	0	0	0.3	0	1.9	0	1.6	0	0.4	0.1
		0	0	0	0	0	0	0	0	0	0	0	0	0	0	0	0	0	0	0	0.1	0	0.2	2.9
6	0.9	1.8	0.1	0	0	0	0	0	0	0	0	0	0	0	0	0	0	0	0	0	0	0	0	0
7	0	0	0	0	0	0	0	0	0	0	0	0	0	0	0.1	0	0	0	0.4	0	0.5	0.1	1.6	0
8	0	2.1	0	0.4	0	0	0	0	0	0.1	0	0	0	0	0	0.1	0	0	0.3	0	0.1	0.2	0.5	0.1
9	0.1	0	0	0	0.1	0	0.1	0	0.1	0	0	0.2	0	0.4	0.3	0.2	3.8	0	1.9	0	0.3	0.1	0	0.6
10	0	3.1	0.1	3.0	0	0.4	0	0.1	0	1.4	0.5	0.5	0.1	0.5	0.2	0	0.2	0.3	0.5	0	0.8	0	0.1	1.5
11	0	4.0	0	5.5	0	0.2	0	0	0	0	0	0.2	0.1	0	0.1	0.1	1.3	0	1.5	0	1.9	0	0.2	4.0

TABLE LXXIII (continued). F_H

Gr. M.-T.	0–2		2–4		4–6		6–8		8–10		10–12		12–									
Date	+	−	+	−	+	−	+	−	+	−	+	−	+	−	+	−	+	−	+	−	+	
December 12	0.1	4.4	0.1	0.3	0	0.1	0	0	0	0	0	0	0	0	0.4	0	2.6	0	3.7	0	0.1	
13	0	0.4	0	0	0.5	0	0	0	0	0.3	0.1	0.5	0	0	0.1	0.1	0.3	0	0.3	0	1.4	
14	0.7	0.3	0.1	0	0	0	0	0	0	0	0	0	0	0	0	0	0	0	0	0	0.1	
15	0	11.6	0.7	2.3	1.0	0	0	0	0	0	0	0	0	0	0	0.1	0	0	0.6	0	0.4	
16	0	0	0	0	0	0	0	0	0.1	0	0	0.2	0	0.8	0.2	0	2.4	0	0	0.1	0.2	
17	0.1	0.1	0.1	0	0	0	0	0	0	0	0	0	0	0	0	0	0	0	0	0	0.1	0
18	0	0	0	0	0	0	0	0	0	0	0	0	0	0	0	0	0	0	0	0	0	0
19	0.2	0.1	0	0.2	0.3	0.8	0.1	0	0	0	0	0	0.2	0.3	0.1	0.1	0.4	0	0.5	0	0.6	0
20	0	0	0	0.1	0	0.5	0	0.2	0	0	0	0	0	0	0	0	0.1	0	0	0.1	0	0
21	0	0	0	0.5	0	0	0	0	0	0	0	0	0	0	0	0	0.1	0	0.8	0	0.2	0
22	0	0.1	0	0	0	0.1	0.2	0.1	0.6	0.1	0.3	0.2	0.3	0.1	0.4	0	0.8	0	0.9	0	0.7	0
23	0	14.8	0	8.8	0	10.5	0.5	2.9	—	—	—	—	—	—	—	—	—	—	0.3	0.9	2.6	0.
24	0	5.2	0	5.0	0.1	0.4	0.1	0	0	0.5	0.4	0.2	0.1	0.8	0.1	0.1	0	0.2	0.7	0.1	0.9	0.
25	0	2.8	0	8.8	0	1.1	0	0	0	0.1	0	0.5	0	0.5	0.1	0.1	0	0.1	0.5	0.1	0	0.
26	0	0.9	0.1	0	0	0	0	0.1	0.2	0	0	0.1	0	0.5	0.1	0.1	0	0.1	0.2	0.5	0.9	0.
27	0.1	0.4	0	1.6	0	0.2	0	0.7	0	0.1	0	0	0.1	0.2	0	0	0.1	0.2	0.2	0	1.0	0.
28	0	6.4	0	2.6	0	5.2	1.0	0.2	0.3	0.1	0.4	0.1	0.1	0.3	0.5	0	0.2	0.4	0	0.1	0.6	0
29	0	0.7	0	0.1	0.1	0	0	0.1	0.1	0	0	0	0	0	0	0	0	0.1	0.2	0.1	0	0.
30	0	0.5	0	0.1	0	0	0	0	0	0.1	0	0	0	0	0	0	0	0.1	0.1	0.1	0	0
31	0	0.1	0	0	0	0	0	0	0	0	0	0	0	0	0.1	0	0.2	0	0	0	0	0
Jaunary 1	0	0.2	0	0	0.1	0	0	0	0	0	0	0	0	0.1	0	0	0		0		0	0
2	0.4	0.8	0.2	0	0	0	0	0	0	0	0	0	0.1	0	0	0	0		0		—	—
3	0	1.0	0	0.1	0	0	0	0	0	0	0	0	0	0	0	0.1	0	.1	0.1	.1	0	0
4	0	0.9	0.3	6.8	0.1	6.7	0.9	0.7	0.4	0	0	0.1	0	0	0.1	0	0.2	.1	0.5		0.4	0
5	0.2	4.4	0	6.1	0.1	1.5	0.6	0	0	0.6	0.1	0.2	0.4	0.2	1.1	0	2.5	8	0	8.1	0.3	0.
6	0	2.5	0	4.7	0	0.2		0	0	0		0	0	0	0.1	0.1	0.2	.1	0.1	0.1	0.1	0.
7	0	0.2	0.1	0.1	0	0		0	0	0		0	0	0	0	0	0		0.1	0	0	0
8	0	0	0	0	0	0.1	2	0.1	0.1	0.1		0.1	0	0.1	0.1	0	0.3		0.9	0	1.2	0.
9	0	0	0	0.1	0.2	0	1	0.1	0.2			0	0.1	0	0.5	0	0.4		0.1	0	1.1	0
10	0.2	1.0	0	8.2	0.3	0.1	8.	0	0	0	8	0	0.1	0	1.2	0	1.9	8.1	0	0.5	0.9	0
11	0.3	1.0	0	4.1	0.1	0.3	1	0.3	0		0	0.1	0.2	0	0.4	0	0.4		2.3	0	1.9	0.
12	0.1	1.1	0.1	1.7	0.2	0.5		1.5	0.3		0	0.4	0	0.2	0.1	0.1	0.1	.1	0.5	0	0	0
13	0.7	0.3	0.2	0.1	0	0		0			0	0	0	0	0.1	0	0		0.5	0.1	0.3	0
14	0.1	0.1	0	0	0	0		0	8		0	0	0	0	0	0	0		0.5	0	0.6	0.
15	0.1	0	0	0	0	0	8	0	0	8	0.1	0	0.2	0	0	0	0	8	0	0.1	0	0
16	0.1	0.4	0.1	0.9	0	0		0	0.3	0.2	0.1	0.3	0.9	0.2	0.3	0	0.1	.1	0	0.4	0.6	0
17	0.3	0.3	0.4	0.2	0.1	0.1		0	0.1	0	0	0	0	0	0	0	0.1	.1	0	0	0	0
18	0	0.1	0.1	0.1	0	0	.2	0.1	0.7	0	0.1	0	0.1	0.1	0.5	0.1	1.3		4.8	0	0.3	0.
19	0	1.3	0	2.6	0.1	0		0	0.1	0.1	0	0.5	0.1	0	0.8	0.3	0.6		0	0.2	0.6	0
20	0.1	0.5	0	5.8	0	2.8	8	0.7	1.2	0	0	0	0.1	0.1	0.3	0.1	0	8.3	0.2	0.1	0.7	0
21	0	0.2	0	0.4	0	1.6	.1	0.1	0	0	0	0.1	0	0.4	0.1	0.1	0.5	.1	0.4	0	1.3	0.1
22	0	5.8	0.3	0	0	0		0	0	0	0	0	0.1	0	0.1	0	0.2		0.1	0.1	0	0
23	0.1	0	0	0	0	0.4		0.1	0.1	0.3	0.4	0.1	0.2	0.4	0.4	0.2	1.9		1.5	0	2.8	0
24	0	2.6	0	3.6	0.1	0.2	.1	0.2	0.1	0.1	0	0	0.1	0.2	0.2	0.2	0.6		0.1	0.1	0.7	0.2
25	0	0	0.1	0.2	0.2	0.2	8.1	0	0.1	0.1	0	0.1	0.1	0.1	0.1	0.1	0.1	8	0	0	0	0
26	0	0	0	0	0	0		0	0.1		0.1	0.1	0.2	0.2	0	0.2	0.1	.1	2.2	0	1.6	2.7
27	0	13.2	0	14.3	0	6.0		3.2	1.3		0.7	0	0.2	0.1	0.2	0	0.7		0.7	0	1.0	0.4
28	0.1	0.2	0	0	0	0		0.1	0		0	0	0	0.2	0.1	0	1.4		0.3	0.2	0.6	0.1
29	0.1	0.2	0.1	0.1	0	0		0	0		0	0	0	0	0	0.1	0		0	0	0	0
30	0	0.4	0	0.2	0.1	0.1	8.1	0.2	0.1	8.3	0	0.6	1.4	0.1	2.2	0	6.5	8	5.8	0	0.1	0

ntinued). F_H Dyrafjord.

		2—4		4—6		6—8		8—10		10—12		12—14		14—16		16—18		18—20		20—22		22—24	
+	—	+	—	+	—	+	—	+	—	+	—	+	—	+	—	+	—	+	—	+	—	+	—
0	2.5	0.2	0.9	0.1	0.2	0.1	0.2	0.1	0.1	0	0	0	0.1	0.1	0	1.0	0	0.3	0.2	0	0.1	0	0
0	0	0	0	0	0.2	0	0.1	0	0	0	0	0.1	0.1	0.1	0	0.3	0	0.2	0	0.1	0	0.3	0
0	1.9	0	0	0	0	0	0	0	0	0	0.1	0.1	0.1	0.2	0	0.2	0	0	0	0	0	0	0.4
0	1.5	0	0.8	0.1	0.2	0.1	0.1	0.2	0	0.2	0	0.3	0.2	0.1	0.1	0	0	0	0	0	0	0	0
0	0	0	0.3	0	0.6	0	0.2	0	0.1	0.1	0.1	0.1	0	0.1	0	0.1	0	0	0.1	0.5	0	0.5	0
0	0	0	0	0	0	0	0.2	0.1	0.2	0.1	0.1	0	0.1	0.5	0	1.1	0	0.5	0	0.1	0.3	0.4	0.3
0	2.2	0	3.4	0.1	1.0	0.1	0.3	0	0	0	0	0	0.1	0.2	0	0.1	0.1	0.1	0.2	0	0	0	0
0	0	0	0.4	0	0.2	0	0.1	0	0.1	0	0.1	0.1	0	0.2	0	0.3	0.1	0.1	0.1	0.8	1.3	0.1	4.5
0	2.2	0	9.1	0	12.8	0.9	4.0	1.0	0.5	2.8	0.2	1.5	0.2	5.2	0	6.0	0	2.9	5.5	0.2	5.2	2.5	0
0.1	2.3	0	7.5	0.3	0.2	0.1	0.3	0	0.1	0.1	0.1	0.1	0.1	0.1	0	0.1	0	0.2	0.1	0.4	1.4	0.3	2.9
0	3.6	0	1.9	0.2	0.5	0.1	0.2	0.1	0	0	0	0.1	0	0	0.1	0	0.2	0.2	0.1	0.6	0	0.1	6.8
0	3.3	0	4.7	0	2.7	0.1	0.7	0.3	0.1	0	0.5	0.1	0.6	0.1	0.1	0	0	—	—	—	—	—	—
—	—	—	—	—	—	—	—	—	—	—	—	—	—	0.1	0	0.6	0	4.9	0	3.5	0	0.3	0.7
0	8.2	0	2.1	0.1	0.9	0.7	0.1	1.2	0	0.2	0	0	0.2	0.1	0.1	0	0.1	2.0	0	0.7	0	0.3	0.6
—	—	—	—	—	—	—	—	—	—	0	0.6	0	0.3	0.4	0	0.4	0	0	0.2	1.0	0	—	—
0	1.9	0	4.2	0.4	0.2	0	0	0.2	0.1	0	0.2	0.3	0	2.5	0	4.3	0	0.2	0.2	0.1	0.4	0	0.4
0	0	0	0	0	0	0	0	0	0	0	0	0	0	0.2	0	0.1	0	0.1	0	0.1	0	0	0.1
0	1.2	0	1.4	0	0.2	0	0	0	0	0	0.1	0.1	0.1	0.1	0	0	0.7	0	0.8	0.1	0.1	0	0.7
0	3.4	0.1	0.1	0	0	0	0.1	0.1	0	0	0.3	0	0	0	0	0	0	0	0	0	0	0	0
0	0	0	0	0	0	0	0	0	0.1	0	0.1	0.1	0.2	0	0.1	0.1	0.1	0	0.1	0	0.1	0	0.3
0	1.4	0	0.4	0	1.1	0	0.2	0	0	0.1	0.1	0.1	0.2	0.3	0.2	0.5	0.3	0.1	0.1	0	0.2	0	0.1
0	0.5	0	14.2	0	21.0	0	19.3	0	5.2	0.2	2.0	0	1.9	0	0.6	0.1	0.1	0.2	0.1	0	0	0	0
0	0	0	0	0	0	0.1	0.1	0	0.1	0	0.1	0.1	0.3	0.1	0.3	0.2	0	0	0	0.1	0.1	0	0.1
0.3	0.1	0	0	0	0	0	0	0	0	0	0	0.1	0.1	0.1	0.1	0.1	0.2	0.2	0.4	3.3	0	0.9	0.1
0.1	0.4	0	0.1	0	0	0.1	0	0	0	0	0	0.1	0.1	0	0	0.1	0	(0.1)	(0.1)	(0.1)	(0.1)	(0.1)	(0.2)
(0)	—	—	—	—	—	—	—	—	—	—	—	—	—	0	0.5	0.1	0.1	0.1	0.1	0.1	0.1	0.1	0
0.1	0.1	0	0	0	0	0	0	0	0	0	0	0	0	0	0	0	0	0	0	0.7	0.9	0.2	0.7
0	3.0	0	2.5	0	1.3	0.1	0.1	0	0	0	0	0	0	0	0	0	0	0.1	0	0	0	0	0
0	0	0	0	0	0	0	0	0	0	0	0	0	0	0	0	0	0	0	0	0	0	0	0
0	0	0	0	0	0	0	0	0	0	0	0	0	0	0.5	0	3.7	0	5.5	0	0.3	0.5	0	3.4
0	10.2	0	1.5	0	3.6	0.1	2.7	0.1	0.1	0	0.4	0	0.2	0.1	0	2.3	0	3.6	0	2.1	0.1	0	3.0
0	6.3	0	1.9	0	1.4	0.4	0	0	0	0	0.1	0	0.2	0.1	0	0.3	0	0	0	0	0	0	0
0	0	0	0.5	0	0	0	0	0	0	0	0	0	0	0	0	0	0	0.1	0	0.3	0	0.4	1.4
0.2	1.9	0	12.1	0.2	3.2	0.2	0	0	0	0.1	0.1	2.7	0	10.8	0	7.3	0	2.4	0	1.1	0	1.0	0.2
0.2	0.4	0.1	0	0	0	0	0	0	0	0	0	0.1	0	0.1	0	0.2	0.2	1.1	0	0.3	0.6	0.1	2.4
0.1	3.9	0	9.1	0.1	3.9	0	4.4	0.1	2.4	0.4	0.1	0.2	0.4	4.2	0.1	5.0	0	4.6	0	2.3	0	0.1	1.9
0	13.5	0	12.0	0.1	1.4	0.3	0.1	0	0.1	0	0.1	1.7	0	2.3	0	5.4	0	7.1	0.1	1.6	1.5	0	5.9
0.1	3.3	0	15.3	0	10.2	1.2	0.8	0.2	0.1	0.1	0.1	0	0.2	0.3	0	0.3	0	0	0	0.4	0.3	0.3	0.2
0.3	0.7	0.1	1.9	0.1	0	0	0	0.1	0.1	0	0	0.4	0.1	1.3	0	2.9	0	5.4	0	5.1	0	0.9	1.0
0	3.2	0.2	0.4	0.1	0	0.2	0.1	0	0	0	0	0	0	0	0	1.0	0	1.2	0	0.3	0.2	0.1	0.2
0	2.3	0.2	1.3	0	7.4	0.3	0.1	0.2	0	0.2	0	0.3	0	0.4	0	4.3	0	2.1	0	1.6	0.4	0.2	7.8
0	3.8	0	21.0	0.1	14.2	1.4	0	0.6	0.2	0	0.2	1.2	0.4	3.5	0.1	4.1	0.1	0.6	0.2	0.3	0.3	0	1.4
0	20.0	0	2.6	0.1	0.1	0	0.1	0	0.3	0	0.6	0.6	0.1	1.6	0	1.8	0	0.3	0	0	0.1	0.1	0.1
0	0	0	1.4	0	1.7	0	0.1	0	0.1	0.2	0.2	0.7	0	1.7	0	0.6	0	1.6	0	0.1	0.1	0.1	0.2
0.1	0.1	0	0.2	0	0.3	0.1	0	0.1	0.1	0	0.1	0	0	0.1	0	0	0.1	0.1	0	0.1	0.1	0	0
0	0	0	0.1	0	0	0	0	0	0	0	0	0	0	0	0	0	0.1	0.1	0.1	0	0	0	0
0	0	0	0	0	0	0	0	0	0	0	0	0.1	0.1	0	0	0.1	0	0.5	0	1.3	0	0.5	1.1
0	3.1	0	1.2	0.1	0.2	0.1	0.1	0.1	0.1	0	0.1	0	0	0.2	0.1	1.5	0	4.0	0	1.2	0	0.1	4.1
0	8.0	0.5	0	0	0.5	0	0.2	0	0.8	0.3	0.1	0.1	0.3	0.2	0	0.1	0	0.4	0	0.2	0	0.3	1.3
0	2.5	0	3.8	0	8.2	0.4	0.5	0.2	0	0	0.1	0.2	0	0	0.2	0.2	0	0.9	0	0.7	1.3	0	3.5

TABLE LXXIII (continued). F_H

Gr. M.-T.	0—2		2—4		4—6		6—8		8—10		10—12		12—							
Date	+	—	+	—	+	—	+	—	+	—	+	—	+	—	+	—	+	—	+	—
March 22	0	0.9	0	0.3	0	0.5	0	0.1	0	0	0	0	0.1	0	0.7	0	2.2	0	1.4	0
23	0	2.0	0	4.1	0.1	0.2	0	0	0	0	0.1	0	0.2	0	0	0	0.3	0	0.2	0
24	0	5.8	0.1	0.1	0	0.3	0	0.9	0.2	0	0	0.1	0.1	0.1	0.3	0	0.1	0	0.1	0
25	0	0.1	0	0.1	0	0.1	0	0	0	0	0	0	0.2	0	0	0	0	0	0	0
26	0	0.2	0	0	0	0	0	0	0	0	0	0	0.1	0	0.1	0	0.2	0.1	0.1	0
27	0	0	0	0	0	0	0	0	0	0	0.1	0	0	0	0.1	0	0.1	0	0.8	0
28	0	0	0	0	0	0	0	0	0	0	0	0	0.1	0	0.2	0	0.1	0.1	0.3	0
29	0	1.1	0	9.1	0	2.0	0	0.8	0.3	0.1	0	0.1	0.2	0	2.8	0	8.2	0	9.8	0
30	0	5.1	0	6.4	1.9	0.2	2.1	0	0.6	0	0	0.2	0.3	0	0.4	0	0.2	0.1	0.2	0.1
31	0.1	15.5	0.1	0.8	0.2	0.3	0.2	0.6	1.1	0	0.5	0	1.9	0	2.1	0	4.1	0	2.0	0
April 1	0	3.0	0	4.0	0	0	0	0	0	0	0	0	0	0	0	0	0	0	0	0
2	0	9.5	0	4.5	0.5	0	0	1.0	0	0.5	0	0	0.5	0	0	0	4.0	0	5.0	0
3	0	15.5	0	12.5	2.0	0	0	0	0	0	0	0	0.5	0	1.0	0	0.5	0	3.0	0
4	0	4.0	0	6.0	0	2.5	0	0.5	0	0	0	0	0.5	0	0.5	0	1.5	0	3.5	0
5	0	4.0	0	5.0	0	14.5	0	6.5	0	0	0	0	1.5	0	8.5	0	7.0	0	1.0	0
6	0.5	2.0	0	26.5	1.0	21.0	1.5	0.5	0.5	3.0	3.0	2.0	0.5	5.0	2.5	0	2.0	0	1.0	0.5
7	0	0	0	0	0	0	0	0	0	1.5	0.5	0	2.5	0	3.0	0	1.5	0	1.5	0
8	0.5	0.5	0	0	—	—	—	—	—	—	—	—	0.5	0	1.5	0	0	0	1.0	0
9	0	13.5	0	10.0	0	11.0	0	13.0	1.0	2.0	1.5	0	1.5	0	3.0	0	6.0	0	5.0	0
10	0	3.0	0	6.0	0	0	0	1.0	0.5	0.5	0.5	0.5	1.0	0.5	3.0	0	3.5	0	3.0	0
11	0	2.0	0	1.0	0	1.5	0	0	0.5	0.5	0	0	2.0	0	0	0	0	0	0	0
12	0	9.0	0	17.0	0	2.0	0	0	0	0	0	0	0	0	0.5	0	1.5	0	2.5	0
13	0	7.0	0	4.0	0	0.5	0	1.5	0	0.5	0	0	0	0	0	0	1.0	0	1.5	0
14	0	1.0	0	1.5	0	0	0	0	0	0	0	0	0.5	0	1.0	0	0.5	0	0	0
15	0.5	0.5	0	0.5	0	0.5	0	0	0	1.5	1.5	0	4.5	0	3.0	0	—	—	—	—

TABLE LXXIV.

Disturbances in Declination (F_D).

Gr. M.-T.	0—2		2—4		4—6		6—8		8—10		10—12		12—14		14—16		16—18		18—20	
Date	+	—	+	—	+	—	+	—	+	—	+	—	+	—	+	—	+	—	+	—
December 2	0	0.1	0	0.2	0	0.7	0.1	0.1	0	0	0.1	0	0.3	0	0.3	0	0	0.7	0.1	0
3	1.2	0.3	0.3	0.3	0	0.2	0.1	0.1	0.1	0.1	0	0	0.3	0	0	0.1	0	0	0	0.1
4	0	0	0	0	0	0	0	0.1	0	0.1	0	0	0	0	0	0	0	0	0.1	0.2
5	0.2	0	0	0	0	0	0	0.1	0.1	0	0	0	0	0	0	0	0	0	0	0
6	0.9	0.3	0	0	0	0	0	0	0	0	0	0	0	0	0	0	0	0	0	0
7	0	0	0	0	0	0	0	0	0	0	0	0	0	0	0.2	0	0	0	0	0.1
8	0.9	0.3	0.1	0.2	0.1	0.1	0.2	0.1	0.1	0.1	0	0	0	0	0	0	0	0	0	0
9	0	0.1	0	0	0.1	0	0.1	0.2	0.1	0.2	0	0	0.1	0	0	0.4	0.2	0.8	0	0.9
10	0	0.1	0.2	0.3	0.3	0.7	0.2	0.1	0	0.1	0.2	0.2	0	0.1	0	0	0.1	0.1	0.1	0.1
11	0.3	0.3	0.2	0.2	0.1	0.1	0	0	0	0	0	0.2	0.2	0	0.1	0	0.2	0.2	0.4	0.2
12	0.6	0.2	0.2	0.1	0	0	0.1	0.1	0	0	0	0	0	0	0.2	0	0.1	0.9	0.2	0.5
13	0.1	0.1	0.1	0	0	0.1	0.1	0	0	0	0.9	0	0	0	0.1	0.1	0	0.2	0.1	0
14	0.4	0	0.1	0.1	0	0	0	0	0	0	0	0	0	0	0	0	0	0	0	0
15	0.9	1.1	0	0.8	0	0	0	0	0	0.1	0	0.1	0	0	0	0	0	0	0	0.4
16	0	0	0	0	0	0	0	0	0.1	0.1	0	0	0.4	0	0	0	0.4	0.3	0	0
17	0.1	0.1	0	0.2	0	0	0	0	0	0	0.1	0.1	0	0	0	0	0	0	0	0
18	0	0.2	0	0	0	0	0	0	0	0	0	0	0	0	0	0	0	0	0	0
19	0.1	0.2	0	0	0	1.0	0	0.1	0.1	0	0.2	0	0.2	0	0.1	0	0	0.1	0	0.3
20	0	0.1	0	0	0.1	0	0	0.3	0.1	0.1	0	0	0	0	0	0.1	0	0.1	0	0.1
21	0	0	0.1	0	0	0.1	0.1	0	0	0	0	0	0	0	0	0	0	0	0.1	0

'ABLE LXXIV (continued). F_D Dyrafjord.

Gr. M.-T.	0—2		2—4		4—6		6—8		8—10		10—12		12—14		14—16		16—18		18—20		20—22		22—24	
Date	+	—	+	—	+	—	+	—	+	—	+	—	+	—	+	—	+	—	+	—	+	—	+	—
December 22	0	0	0	0	0.2	0	0.2	0.2	0.3	0.1	0.3	0.2	0.3	0	0	0.1	0	0.2	0	0	1.3	0	3.5	0
23	0	5.7	0	3.4	0	3.3	1.0	0.3	—	—	—	—	—	—	—	—	—	—	1.5	0.6	1.8	0.8	0.1	0.1
24	1.3	0.1	1.3	0.1	0.3	0.1	0.1	0.1	0.2	0.1	0.1	0.1	0.1	0	0.4	0.1	0.3	0.2	0	1.4	0	0.6	0.8	0.1
25	0.4	0.3	0.2	0.7	0.1	0.4	0	0.1	0.1	0.2	0.1	0.3	0.1	0.1	0.1	0.1	0	0	0	0.8	0	0.3	1.3	0.1
26	0.1	0.2	0.2	0.1	0.1	0.1	0	0.1	0	0.2	0.2	0.3	0	0.1	0	0	0	0	0.2	0	0	1.2	3.5	0.2
27	0	0.3	0.3	0.1	0.1	0.2	0.1	0.1	0.1	0.1	0.1	0.1	0.4	0	0.2	0	0	0	0	0	0.1	1.4	3.5	0
28	1.9	0	0.1	0.8	0	2.5	0.1	0.7	0.1	0.6	0.2	0	0.3	0.1	0.1	0.1	0	0	0	0.1	0	0.3	0.4	0.1
29	0.5	0	0.1	0.4	0.1	0.3	0.2	0.2	0	0.1	0	0	0	0	0	0	0.2	0.1	0.1	0.1	0.1	0.1	0.1	0.1
30	0.1	0	0.1	0	0.1	0	0	0.1	0.1	0	0.1	0.1	0.4	0	0	0	0	0.1	0.1	0	0	0.1	0.1	1.1
31	0	0	0	0	0	0	0	0	0	0	0	0	0	0	0	0.1	0	0.1	0	0	0	0	0.2	0.1
January 1	0.1	0.1	0	0	0.1	0.1	0	0	0.1	0.1	0	0	0.1	0	0.2	0	0.1	0.1	0	0.1	0	0.8	0.2	0.3
2	0.4	0.2	0	0.3	0	0	0.1	0.1	0	0	0	0.1	0	0	0	0	0	0	0	0	0	0	0	0
	0.3	0.3	0.1	0.1	0.1	0.1	0	0	0	0	0	0	0	0	0	0	0.1	0.1	0	0.1	0.1	0.3	0.1	0
	0.1	0.1	0.7	0.1	0	0.3	1.0	1.1	0.4	0	0.3	0.1	0.2	0	0	0	0.1	0.1	0	0.1	0.1	0.3	0	0.4
	2.3	0.1	0.1	0.9	0.1	1.1	0.8	0	2.9	0	1.7	0.1	0.8	0	0.9	0	0.2	0.6	0.1	0	0.1	0.1	0	0.1
	0.1	0	0.8	0.1	0	0.1	0	0	0	0.2	0.	0	0	0	0.1	0.1	0.3	0	0	0.1	0	2.1	0.5	0.6
	0	0.1	0	0	0	0	0.1	0	0	0.1	0.1	0	0.1	0	0.1	0	0	0	0	0.1	0	0	0	0
8	0	0.1	0	0	0.1	0	0.1	0.1	0.1	0.2	0.1	0.1	0.1	0.1	0.1	0.1	0.2	0	0	0.6	0.1	1.4	0	0.3
9	0	0	0	0.1	0.1	0.3	0.1	0.6	0.3	0	0.1	0	0	0.1	1.2	0	0.1	0.1	0	0	0.1	0.3	1.9	0.7
10	1.0	0.2	0.8	0.7	0.2	0.1	0.3	0.1	0.1	0	0.1	0	0.1	0	0.4	0	1.0	0	0	0.1	0	0.4	0.4	0.7
11	0.5	0.1	0.8	0	0.1	0.1	0.1	0.1	0	0.1	0.1	0	0.2	0	0.2	0	0.1	0.1	0	1.3	0	2.1	0	1.9
12	1.2	0.3	0.6	0.2	0.3	0	0.4	0	0.3	0	0.5	0	0.3	0	0.1	0.1	0.1	0.1	0	0.5	0	0	0.1	0.2
13	0.2	0.8	0.1	0	0	0.1	0	0	0	0	0	0	0	0	0.3	0	0	0	0	0.8	0.1	0.2	0.1	0.2
14	0.1	0.1	0.1	0.2	0	0	0	0	0	0	0	0	0	0.1	0	0.1	0	0	0	0	2.0	0.8	0.4	0.3
15	0	0.2	0	0.5	0	0.1	0.1	0.1	0.3	0	0.2	0	0.4	0	0.1	0	0	0	0	0	0	0.1	0	0.3
16	0.2	0.1	0.4	0	0	0	0.1	0	0.5	0	0	0.7	0	0.6	0	0	0	0.1	0.1	0	0	0.3	1.0	0.4
17	1.0	0.1	0.1	0.1	0	0	0.1	0	0	0	0	0	0.1	0.1	0	0	0	0	0	0	0	0	0	0
18	0.1	0	0	0.1	0	0	0.3	0.1	0.1	0.1	0	0	0.1	0	0.2	0	1.6	0	0.2	0.4	0.2	0	0	0
19	0.5	0.1	0.7	0	0.1	0	0	0.1	0	0.3	0.5	0	0.1	0.1	0.7	0.2	0.4	0.3	0.1	0	0	0.3	0.2	0.1
20	0	0.2	0.2	0.3	0.2	0.9	0.5	1.0	1.3	0	0.3	0	0.1	0.1	0.2	0	0	0	0	0	0	0.1	0	0
21	0	0	0	0	0	0.5	0.2	0.1	0.2	0	0	0.1	0.1	0	0.1	0	0.2	0	0.2	0.1	0.1	0.1	0.1	0.4
22	0	1.4	0.1	0	0	0	0	0	0	0	0	0	0	0.1	0	0	0.1	0	0	0.2	0	0	0	0.3
23	0	0.2	0	0	0.2	0	0	0.8	0.7	0.3	0.1	0.5	0.1	0.1	0.2	0.1	0.1	0.2	0.5	0.1	0.1	1.8	0	0.6
24	0.2	0.1	0.7	0	0	0.1	0	0.1	0	0.5	0	0.1	0.1	0	0	0.1	0	0.1	0	9.3	0.1	0.1	0	1.0
25	0	0	0.3	0	0.3	0	0.1	0	0.1	0	0.1	0.1	0	0.3	0	0	0	0.1	—	—	—	—	—	—
26	—	—	—	—	—	—	—	—	—	—	—	—	—	—	—	—	—	—	0.4	0.6	3.0	0.1	2.9	4.7
27	0.5	6.1	0.2	5.5	0	4.1	0.1	2.4	0.6	0.2	0.1	0.1	0	0	0	0	0	0.1	0.2	0	0	0.9	0.2	0.7
28	0.1	0.3	0.1	0	0	0	0.1	0.1	0	0	0	0	0	0	0	0	0.1	0	0.1	0.1	0	0.3	0	0.6
29	0.1	0.4	0.1	0.1	0	0.1	0	0	0.	0	0	0.1	0.1	0	0	0	0	0	0	0	0	0	0	0
30	0	0.1	0.2	0	0	0.1	0	0.2	0.5	0	0.6	0	0.5	0.1	2.1	0	2.7	0	0.4	0.4	0	0	0	0
31	0.1	0.3	0	0.3	0	0.1	0.1	0.2	0.3	0	0	0.1	0	0	0.1	0.1	0.2	0.1	0.1	0	0	0	0	0.1
y 1	0	0	0	0	0	0.1	0.1	0	0.2	0	0	0	0.1	0	0.1	0.1	0.1	0.1	0.3	0	0	0	0	0
2	0.5	0.1	0	0	0	0	0	0	0	0	0	0	0.1	0	0.1	0	0.1	0.1	0	0	0	0	0.2	0
3	0.1	0.2	0	0.2	0.1	0.2	0.2	0.2	0	0.3	0.4	0	0.1	0.2	0.1	0.1	0.1	0.1	0	0	0	0	0	0
4	0	0	0.1	0	0.2	0.1	0	0.2	0.1	0.1	0.1	0	0.1	0	0.1	0.1	0	0	0.1	0	0	0.2	0.1	0.6
5	0	0.1	0.1	0	0.1	0	0.1	0.1	0.3	0.1	0.1	0.2	0	0.1	0.2	0.1	0.2	0.3	0	0.2	0	0.6	0.2	0.2
6	0.4	0.3	0.2	0.9	0	1.3	0.2	0.1	0	0.1	0	0.1	0	0	0	0.1	0.1	0	0	0	0	0	0	0
7	0	0	0.1	0	0	0.1	0	0	0	0.2	0.1	0	0.1	0	0.1	0.1	0.2	0	0.1	0	0.3	1.0	3.8	1.1
8	0	0.6	0.2	3.3	0.1	5.3	0	4.4	1.3	0.4	2.7	0.1	2.0	0.1	1.8	0.1	0.4	0.3	6.4	0.5	6.5	0	0.8	0.4
9	0.3	0.2	0.8	0.2	0.3	0.2	0.2	0.2	0.1	0.1	0.2	0	0	0.1	0.1	0.1	0.1	0.1	0.2	0.3	0.2	0.8	2.8	0.2

TABLE LXXIV (continued). F_D

Gr. M.-T.	0—2		2—4		4—6		6—8		8—10		10—12		12—14		14—16		16—18		18—20		20	
Date	+	—	+	—	+	—	+	—	+	—	+	—	+	—	+	—	+	—	+	—	+	
February 10	0.7	0.3	0.1	0.1	0.1	0.5	0	0.1	0.1	0.1	0.1	0	0	0	0	0	0	0.1	0.1	0.1	0	
11	0.1	1.1	0	0.8	0	2.2	1.1	0.2	0.4	0.1	0.2	0.1	0.2	0.1	0	0.1	—	—	—	—	—	
12	—	—	—	—	—	—	—	—	—	—	—	—	—	—	0.5	0	0.3	0	0.5	0.3	0.1	
13	0.7	0.3	0.3	0.8	0.1	0.6	0	0.5	0	0.9	0.1	0.3	0.1	0.1	0.1	0	0	0	0.1	0.6	0	
14	—	—	—	—	—	—	—	—	—	—	0.1	0.1	0.2	0.1	0.1	0.1	0.1	0.1	0	0.1	0.1	
15	0	0.4	0.3	0.2	0	0.4	0.3	0.2	0.4	0	1.3	0	1.0	0	0.9	0	0	2.3	0	0.1	0	
16	0	0	0	0	0	0	0	0	0	0	0	0	0	0	0.3	0	0	0.1	0	0.4	0.1	
17	0.1	0.5	0.2	0.1	0.1	0.1	0	0.1	0.1	0.1	0.2	0	0.2	0	0.3	0.1	0.1	0.1	0	0	0	
18	0	1.2	0	0	0	0	0	0	0.1	0.1	0	0.2	0	0	0	0.1	0	0	0	0	0	
19	0	0.1	0.1	0	0	0	0	0	0.3	0	0.1	0	0.2	0	0	0	0.1	0.1	0.1	0	0	
20	0.1	0.1	0.1	0.1	0	0.5	0	0.1	0	0	0	0	0.1	0	0	0	0.1	0	0	0.1	0	
21	0	0	0	0	0.1	0	0.1	0.1	0	0.1	0.1	0.1	0.3	0	0.6	0	0.5	0	0	0	0.1	
22	0	0	0.8	1.5	0	8.7	0	7.1	0.1	0.8	2.1	0	1.5	0	0.1	0	0.1	0	0	0	0.1	
23	0.4	0.4	0	0	0	0	0	0	0	0.1	0	0.2	0	0.1	0.1	0	0.1	0.1	0.1	0	0.4	
24	0.5	0	0.2	0	0	0.1	0	0	0	0	0	0	0	0	0	0	—	—	—	—	—	
25	—	—	—	—	—	—	—	—	—	—	—	—	—	—	0.2	0	0	0.1	0	0	0	
26	0	0.1	0	0	0	0	0.1	0	0	0.1	0	0	0	0	0	0	0	0	0	0	0.7	0
27	0.1	0.4	0.1	0.2	0.1	0.1	0.1	0	0	0	0	0	0	0	0	0	0	0	0	0	0	0
28	0	0	0	0	0	0	0	0	0.1	0.1	0.1	0	0	0	0	0	0	0	0	0	0	0
March 1	—	—	—	—	—	—	—	—	0	0.1	0	0.2	0.1	0.1	0.1	0	0.1	0.3	0	1.6	0.1	0
2	0.1	2.1	0.1	0.2	0.2	0.5	0.2	0.8	0.5	0	0.1	0	0	0	0	0	0	0.2	0	0.3	1.0	0
3	0.2	1.4	0.1	0.9	0.1	0.4	0.5	0	0.2	0	0	0.1	0	0	0	0	0	0	0	0.2	0	
4	0	0	0.1	0	0	0	0	0	0	0	0	0	0	0	0	0	0	0	0	0	0.1	
5	0.6	0.1	0.2	1.9	0.1	1.2	0	0.3	0	0.1	0.7	0.1	1.8	0	0.9	0	0.7	0.2	0.2	0.1	0.1	
6	0.2	0.8	0.1	0	0.1	0	0	0	0	0.1	0	0.1	0	0	0	0	0	0.3	0.1	0.3	0.4	
7	1.4	0.7	0.4	2.0	0	3.1	0	3.3	2.2	0	3.3	0	2.8	0	0.8	0	1.1	0.5	0.1	0.3	0.1	
8	1.6	0.8	0	1.7	0	1.0	0.1	0.2	0.1	0.2	1.0	0	1.2	0	0.8	0	0.3	0.6	1.7	0.6	8.3	
9	1.2	0.8	0.8	2.4	0	4.9	0.5	0.5	0.3	0.2	0	0.1	0.1	0	0.1	0	0	0	0	0.3	0.1	
10	0.4	0.1	0	1.4	0	0	0	0.1	0.1	0.1	0	0	0	0	0.1	0	0.2	0.1	0.1	0.6	0.2	
11	0.8	0	0.1	0.1	0.1	0	0.1	0	0	0.1	0	0.1	0.1	0.	0	0.1	0	0.2	0.1	1.2	0.1	
12	2.0	0	0.1	0.1	0	1.7	0.1	1.1	0.5	0	0.5	0	1.0	0	0.1	0	0.6	0.3	0	0.3	3.5	
13	0.8	0.6	1.9	2.2	0.1	10.0	1.0	0.1	0.9	0.1	0.4	0.1	1.9	0	1.2	0	0.1	1.6	0.2	0.1	0.3	
14	2.8	0.3	0.1	0.4	0.1	0	0.1	0	0.1	0.4	0.1	0.1	0.4	0	0.5	0.3	0.2	0	0	0.2	0	
15	0	0	0.1	0.1	0.1	0.4	0	0.3	0	0.6	0.1	0.6	0.2	0.1	0.1	0.1	0.2	0.2	0	0.5	0.1	
16	0	0	0	0	0	0	0	0.1	0.1	0.2	0.1	0.2	0.2	0	0.1	0	0	0	0	0.1	0	
17	0	0	0.1	0	0	0	0	0	0	0	0	0	0	0	0	0	0	0	0	0	0	
18	0	0	0	0	0	0	0	0	0	0.3	0	0.2	0	0	0	0	0	0.1	0	0	0	
19	0.2	0.5	0	0.2	0	0.3	0.2	0.1	0	0.1	0.1	0.3	0	0	0	0.1	0	0.2	0	1.2	0.2	
20	2.1	0.9	0.2	0	0.1	0.2	0.3	0.1	0.6	0.1	0.6	0	0.1	0.1	0	0	0	0	0	0.1	0.1	
21	0.1	0.1	0.1	0.1	0.1	1.0	0.2	0.5	0.1	0.1	0.1	0.1	0	0.1	0	0.1	0	0.1	0	0	0.5	0.
22	0.1	0.1	0	0.1	0	0.2	0	0	0	0.1	0.1	0.1	0.1	0.1	0	0	0.1	0.2	0.1	0.1	1.6	0.
23	0	0.2	0.1	0.3	0	0.2	0	0	0	0.2	0	0.1	0.1	0	0	0	0	0.1	0	0	0.1	0.
24	0.1	1.2	0.1	0	0	0.2	0.5	0.1	0.4	0.1	1.1	0	0.2	0	0.1	0	0	0.1	0	0	0	0
25	0	0	0	0	0	0	0	0.1	0	0	0.1	0	0	0.1	0.1	0	0	0	0	0.1	0	0.
26	0	0	0	0	0	0.1	0	0	0	0.1	0	0	0	0	0	0	0	0	0.1	0.1	0.1	0
27	0	0	0	0	0	0	0.1	0	0.1	0.1	0	0.1	0	0	0	0	0	0	0	0.6	0.2	0.
28	0	0	0	0.1	0	0	0	0	0	0	0	0	0	0	0	0	0	0.1	0	0.1	0	0
29	0	0.2	0	1.5	0	2.4	0	1.0	0.6	0.3	0.1	0.2	0.2	0	0.2	0	0.1	0.4	0.1	0.8	0.7	0.
30	0.8	0.8	0	2.5	0	0.9	0.3	0.3	0.6	0.1	1.0	0	0.2	0	0	0	0	0.1	0.1	0.1	0.1	0.
31	0.1	4.4	0	0.5	0	0.5	0.9	0.3	1.1	0	0.4	0	0.2	0	0.1	0.7	0	1.5	0	0.4	0.7	1.

TABLE LXXIV (continued). F_D Dyrafjord.

Gr. M.-T.	0—2		2—4		4—6		6—8		8—10		10—12		12—14		14—16		16—18		18—20		20—22		22—24	
Date	+	—	+	—	+	—	+	—	+	—	+	—	+	—	+	—	+	—	+	—	+	—	+	—
April 1	0.5	0	0	1.0	0	0.5	0	0	0	0	0	0	0	0	0	0	0	0	0	0	0.5	0	1.0	1.0
	0	1.0	0	0	0.5	0	0	0.5	0	0	0	0	0.5	0	0	0	0	0	0	0	2.0	0.5	2.0	0.5
	1.0	1.5	0	3.0	0	0	0	0	0	0	0	0	0	0	0	0	0	0	0	0.5	0	0.5	0	0
	1.0	0	0	0	0	0	0	0	0	0	0	0	0	0.5	0	0	0	0	0	0.5	0	0	1.0	0
	0	0.5	0	0.5	0	3.0	—	—	—	—	1.0	0	0	0.5	0	2.5	0.5	1.0	1.0	0	0	0	0	0
	0.5	0.5	0	1.5	—	—	—	—	0.5	0.5	0	8.5	0	10.5	1.0	0.5	4.0	0	4.0	0	1.5	0	0.5	0
	0	0	0	0	0	0	0	0	0	0.5	0.5	0	0.5	0	0	0.5	0	0	0.5	0.5	1.0	1.5	1.0	0
	0.5	0	0	0	0	0	0	0	0	0	0	0	0	0	0	0.5	0	0	0	0	0.5	0	0.5	0
	1.0	0.5	0	6.0	0	3.0	1.0	2.5	0.5	0	0.5	0.5	0	0	0	1.0	0	0.5	1.0	1.0	3.5	0	—	0.5
	0	0	0	1.5	0	0	0.5	0	0.5	0	0.5	0	1.0	0	—	—	0.5	0	0	0	1.0	0	0	0
	0.5	0	0	0	0	0	0	0	0	0	0	0	0	0	0	0	0	0	0	0	0	0	1.5	0
	0	0	0	1.5	0	0	0	0	0	0	0	0	0	0	0	0	0	0.5	0	0.5	1.0	0.5	2.0	0
	0.5	0.5	0	0.5	0	0	0	0	0.5	0.5	0	0	0	0	0	0	0	0	0	0	0	0.5	0.5	0
	0	0	0	0.5	0	0	0	0	0	0	0	0	0	0	0	0	0	0	0	0	0	0	0.5	0
	0	0	0	0	0	0	0	0	0	0	0.5	0	0.5	0.5	0	—	—	—	—	—	—	—	—	—

TABLE LXXV.

Disturbances in Vertical Intensity (F_V).

Gr. M.-T.	0—2		2—4		4—6		6—8		8—10		10—12		12—14		14—16		16—18		18—20		20—22		22—24	
Date	+	—	+	—	+	—	+	—	+	—	+	—	+	—	+	—	+	—	+	—	+	—	+	—
December 2	0.9	0	0	0.8	0	0.3	0	0.2	0	0.1	0	0.5	0	0.1	0.4	0	0.8	0	0.3	0.1	0	0	0	0
	0	1.6	0	1.3	0	0.1	0	0.3	0	0.1	0	0	0	0	0	0	0	0	0	0	0	0	0.2	0
	0.1	0	0	0	0	0	0	0	0	0	0	0	0	0	0	0	0.1	0	0.2	0.1	0.6	0	0.1	0.1
	0	0	0	0	0	0	0	0	0	0	0	0	0	0	0	0	0	0	0	0	0	0	0.2	1.3
	0	1.8	0	0	0	0	0	0	0	0	0	0	0	0	0	0	0.1	0	0	0	0	0	0	0
	0	0	0	0	0	0	0	0	0	0	0	0	0	0	0	0	0	0	0.2	0	0	0.1	0	2.1
	0.1	0.3	0.1	0	0.5	0	0	0	0	0	0	0	0	0	0	0	0	0	0	0.1	0.2	0	0.2	0.1
	0	0	0	0	0	0	0	0	0	0	0	0	0	0	0.2	0	0.7	0.5	1.5	0	1.3	0	0.2	0
	0.7	0	0.2	1.0	0.2	0.1	0.1	0	0.2	0.1	1.5	0	1.9	0	0.2	0	0	0	0.1	0.2	0	1.5	0	3.0
	0.1	1.0	0	2.2	0	0	0	0.1	0	0.2	0	0	0	0	0.1	0	0.2	0	0	0.6	0	4.9	1.0	1.3
	1.1	0.1	0	0.1	0	0.1	0	0.2	0	0	0	0	0	0	0.2	0	1.1	0	0.3	0.1	0	0.4	0	0.2
	0	0.2	0	0	0	0	0	0	0	0.1	0.1	0.1	0	0.1	0.2	0.1	0.5	0	0	0.1	0	3.0	0	1.5
	0	1.6	0	0.1	0	0	0	0	0	0	0	0	0	0	0	0	0	0	0	0	0	0	0	0.4
	0.5	1.4	0	1.8	1.4	0	0.1	0	0	0	0	0	0	0	0	0	0	0	0.1	0.1	0.2	0	0.1	0
	0	0	0	0	0	0	0	0	0	0	0	0	0.3	0	0.3	0	1.2	0	0.5	0	0	0	0	0.1
	0	0.2	0	0.2	0	0	0	0	0	0	0	0	0	0	0	0	0	0	0.1	0	0.1	0	0	0.2
	0	0.2	0	0	0	0	0	0	0	0	0	0	0	0	0	0	0	0	0	0	0	0	0.1	0.1
	0	0.2	0	0	0	1.0	0	0.5	0	0	0	0	0.2	0	0.1	0	0	0	0.1	0.1	0	0.2	0.1	0
	0	0	0	0.1	0	0.6	0	0.4	0	0.1	0	0	0	0	0	0	0	0.1	0	0.1	0	0	0	0
	0	0	0	0.4	0	0	0	0	0	0	0	0	0	0	0	0	0.2	0	1.0	0	0.2	0	0	0.3
	0	0.1	0	0	0	0.2	0	1.3	0	0.8	0.1	0.1	0	0	0	0	0	0.1	0.1	0.1	0.3	0.5	0.7	0.4
	5.0	0.2	0	4.0	0	5.5	0	9.6	—	—	—	—	—	—	—	—	—	—	0	2.4	0	4.5	0.2	0.2
	1.3	0.3	0.5	0.6	0.2	0.1	0	0	0.1	0	0.8	0	0.6	0	0.1	0	0.2	0	0.5	0	0.5	0.2	2.5	0
	0	0.1	0.6	0.4	0	0.5	0	0	0	0.3	0	0.2	0	0	0	0.1	0	0	0.2	0.1	0.2	0	0	1.9
	0	0.7	0.1	0.1	0	0	0	0.2	0	0.3	0	0	0	0	0	0	0	0	0.2	0	0.1	1.4	0.4	0.6
	0	0.6	0	0.7	0	0.1	0	0.9	0	1.3	0	0.2	0	0.1	0	0	0	0	0.1	0	0	1.4	1.9	1.0
	0.8	0.2	0	1.4	0	2.1	0	1.4	0	1.2	0	0.3	0	0	0	0.1	0.1	0	0	0.1	0	0.2	0.1	0.2
	0.1	0	0.1	0	0	0	0	0.5	0	0.1	0	0	0	0	0	0	0.1	0	0.1	0.1	0	0	0.1	0
	0	0.2	0	0	0	0	0	0	0	0	0	0	0	0	0	0	0	0	0	0.1	0	0	0	0.8
	0	0.1	0	0	0	0	0	0	0	0	0	0	0	0	0	0	0	0	0	0	0	0	0	0.2

TABLE LXXV (continued). F_v

Gr. M.-T.	0–2		2–4		4–6		6–8		8–10		10										
Date	+	−	+	−	+	−	+	−	+	−	+	−		−	+	−	+	−	+	−	+
January 1	0	0.3	0	0	0	0	0	0	0	0	0			0	0		0	0	0		0
2	0	1.6	0	0.2	0	0	0	0	0	0	0			0	0		0	0	0		0
3	0	1.7	0	1.0	0	0.1	0	0	0	0	0			0	0		0	0	0.1		0.1
4	0.1	0	0.4	0	0	0.2	0	2.5	0.1	0	0			0.1	0.1		0.1	0	0.4		0
5	0.7	0.9	0	2.5	0	1.3	0	0.6	0	1.1	0	8.5	0+	0.2	0.3	8	1.3	0	0.2	8	0
6	0.1	0.4	0	1.3	0	0.4	0	0	0	0	0				0		0.3	0	0.1	0.1	0
7	0	0	0.1	0.1	0	0	0	0	0.2	0	0				0	.1	0	0	0.1	0	0
8	0	0	0	0	0	0	0	0.2	0	0	0				0		0.2	0	0	0.3	0.1
9	0	0	0	0.2	0	0.1	0	1.1	0	0.1	0				0.4		0.6	0	0.1	0	0.1
10	0.1	2.2	0.1	1.5	0	7.7	0	0.6	0	0.1	0	8	8.1	8	0.7	8	0.6	0	0	0	0.1
11	0.6	1.0	0.6	0	0.3	0	0.3	0	0.4	0	0	0			0			0.1	0.1	0.5	0
12	0	1.8	0	0.6	0.5	0.1	0	0.3	0	0	0	0.2			0			0	0.1	0.1	0
13	0	1.9	0.1	0	0	0	0	0	0	0	0				0			0	0.9	0	0.5
14	0	0.1	0	0	0	0	0	0	0	0	0	8			0			0	0.1	0	0
15	0	0	0	0	0	0	0	0	0	0	0	0	8.1	8	0	8	8	0	0	0	0
16	0.1	0.1	0	0.7	0.1	0.1	0	0.6	0	1.1	0	.3	0	0.2	0		0.5	0	0.6	0	0.1
17	0	1.7	0.1	0.1	0	0.1	0	0.2	0	0	0		0	0	0		0	0	0	0	0
18	0	0.1	0	0.2	0	0.2	0	0.5	0.1	0.1	0		0	0	0.3		0.4	0	0.4	0.1	0.1
19	0	1.7	0	1.2	0	0.4	0	0	0	0	0		—	—	0.3		0.1	0.2	0	0.2	0.1
20	0	0.1	0.2	0.3	0	0.9	0	1.8	0	0.2	0	8	0.1	0	0	8.1	0	0.4	0	0.3	0
21	0	0.4	0	0.4	0	1.0	0	0.2	0	0	0	0					0.1	0	0.2	0	0.1
22	0	2.1	0	0.4	0	0	0	0	0	0	0	0					0	0	0.1	0	0
23	0	0	0	0	0.1	0.2	0	0.3	0	0.7	0.1	0.1			.1		0.4	0	0.2	0.8	0
24	0.1	0.6	0	1.6	0	0.1	0.1	0	0.1	0.1	0	0					0.2	0	0.3	0	0.5
25	0	0	0	0.3	0	0.3	0	0	0	0	0	0	8	8	8	8	0	0	0	0	0
26	0	0	0	0	0	0	0	0	0	0	0	0.1			0		0	0	0	4.4	0.2
27	11.1	0	4.9	0	0.6	1.1	0	3.6	0	4.0	0	0.8			0		0.1	0.1	0	0.4	0
28	0	0.6	0	0.2	0	0	0	0.1	0	0	0	0			0		0.4	0	0.5	0	0.1
29	0.1	0.3	0	0.2	0	0	0	0	0	0	0	0			0		0	0	0	0	0
30	0	0.7	0	0.5	0	0	0	0.6	0	1.5	0	0.4	8.0	8	1.7	8	2.0	0.1	0.1	1.4	0
31	0.6	0	0	0.5	0	0	0	0.2	0	0	0		0		.2		1.6	0	1.1	0	0.1
February 1	0	0	0	0	0	0.1	0.1	0	0	0	0		0				0.1	0	0.5	0	0
2	0	1.1	0	0	0	0	0	0	0	0	0		0				0	0	0	0	0
3	0	0.5	0	1.1	0	0.8	0	0.2	0	0.1	0.2		0.2				0	0	0	0	0
4	0	0	0.1	0	0	0.1	0.1	0	0	0	0	8	0	8.2	8	8.1	0	0	0	0.1	0
5	0	0.1	0	0	0	0	0	0	0	0	0		0	0	0		0.1	0	0	0.3	
6	0	0.7	0	1.0	0	1.2	0.2	0	0	0	0	8	0		0		0	0	0	0.1	
7	0	0	0	0.2	0	0.2	0	0	0	0	0	0	0	8	0	.1	0.1	0.1	0	0	
8	0.7	0	1.9	0	1.2	0.7	0	4.1	0	2.8	0.4	0.3	0.4	0.1	0.3		0	0.7	0.1	7.2	8
9	1.3	0.3	0.5	0.3	0.3	0	0.3	0	0	0	0	0	0	0	0	8	0	0	0	0	0
10	1.7	0	0	0.8	0.1	0.2	0.2	0.2	0	0	0	0	0	0			0	0	0	0	0
11	0	0.6	0	1.7	0	2.5	0	2.6	0	1.0	0	0.3	0	0.2			—	—	—	—	—
12	—	—	—	—	—	—	—	—	—	—	—	—	—	—			0.3	0	1.2	0	0
13	3.0	0.3	0	0.7	0	0.6	0.2	0.2	0.3	0	0	0	0				0	0	0.1	0.4	0.1
14	—	—	—	—	—	—	—	—	—	—	0	0	0	8	8	8	0	0	0	0	0.1
15	0.5	0	0	1.9	0	0.4	0	0	0	0	0		.6		1.1		0.8	0.4	0.6	0	0.2
16	0	0	0	0	0	0	0	0	0	0	0				0		0	0	0.2	0	0
17	0	2.1	0	1.3	0	0.5	0	0	0	0	0				0.2		0.6	0	0	0.2	0.1
18	0.1	1.8	0	0.2	0	0	0	0	0	0	0				0		0	0	0	0	0
19	0	0	0	0	0	0	0	0	0.1	0	0.1	8	8	8	0	8	0	0	0	0	0

TABLE LXXV (continued). F_T Dyrafjord.

Gr. M.-T.	0−2		2−4		4−6		6−8		8−10		10−12		12−14		14−16		16−18		18−20		20−22		22−24	
Date	+	−	+	−	+	−	+	−	+	−	+	−	+	−	+	−	+	−	+	−	+	−	+	−
February 20	0	1.0	0.2	0.2	0.1	0.5	0	0.4	0	0	0	0	0	0	0	0	0	0	0	0	0	0	0	0
	0	0	0	0	0	0	0	0	0	0	0	0	0	0.1	0.2	0	1.0	0	0.1	0	0	0	0	0
	0	0	1.5	0.9	0	4.2	0	11.7	0	8.9	0	4.4	0.1	1.2	0	0	0	0.1	0	0	0	0	0	0.1
	0.1	0.2	0	0	0	0	0	0	0	0	0	0	0	0	0	0	0.2	0	0	0	0	1.6	0.1	1.1
	0	0.3	0	0.2	0	0	0	0	0	0	0	0	0	0	0.2	0	1.0	0	0.5	0	0	0	0	0
	0	0	0	0.7	0	4.6	0	10.7	0	4.2	0	1.6	0.1	0.1	0.1	0	0.1	0	0	0.1	0	0	0	0
	0.1	0	0	0	0	0	0	0	0	0	0	0	0	0	0	0	0	0	0	0	0	1.9	0	3.6
	0.3	0.5	0	1.5	0	0.5	0	0.1	0	0	0	0	0	0	0	0	0	0	0	0	0	0	0	0
	0	0	0	0	0	0	0	0	0	0	0	0	0	0	0	0	0	0	0	0	0	0	0.1	0
March 1	0	0	0	0	0	0	0	0	0	0	0	0	0	0	0	0	0.9	0	2.2	0	0.5	0	0.6	0.4
	0.8	0.8	0	0.9	0	1.9	0	3.7	0	0.8	0	0	0	0	0	0	0.7	0	0.3	0.2	0	2.4	0.3	4.2
	0.6	0.2	0	0.5	0	1.1	0	0.5	0	0	0	0	0	0	0	0	0	0	0.3	0	0	0	0	0
	0	0	0	0.3	0	0.1	0	0	0	0	0	0	0	0	0	0	0	0	0	0	0	0	0	1.9
	0	2.4	4.6	0.3	0	1.0	0	0.2	0.1	0	0	0	0.8	0	0.8	0	0.6	0	0.5	0	0.2	0.1	0	1.5
	0	1.3	0	0.2	0	0	0	0	0	0	0	0	0	0	0	0	0.1	0	0.1	0	0	3.3	0	3.0
	0.6	0.9	0.7	0.3	0	2.4	0	5.2	0	6.9	0	2.1	0	0.3	0	0.5	0.5	0.2	0.1	0	0	2.5	0.4	1.1
	1.8	0.3	0.7	0	0	1.4	0	0.5	0	0	0	0	1.2	0	0.9	0	0.8	0.2	0	5.1	0	10.3	0	7.2
	1.1	0.7	4.7	0	0	2.4	0	2.5	0	0.2	0.1	0	0	0	0	0	0	0	0	0	0	1.1	0	1.2
	0	0.8	0	1.5	0	0.1	0	0	0	0	0	0	0.2	0	0.4	0	0.4	0	0.2	0.4	0	1.9	0	2.7
	0	1.0	0	1.0	0	0.1	0	0	0	0	0	0	0	0	0	0	0.4	0	0.1	1.1	0	1.3	0	2.4
	0	3.4	0	1.8	0	2.6	0	1.5	0	1.5	0	0.4	0	0	0	0.8	0.3	0.2	0.1	0.9	0	3.1	0.3	3.4
	2.7	0	4.7	0.2	0	4.7	0	2.4	0	1.5	0	1.3	0.4	0	0.7	0	0.1	1.2	0.3	0	0	1.2	0.2	0.5
	3.7	0.3	0	0.9	0	0.1	0	0.3	0	0.2	0	0	0.1	0	0.4	0	0.8	0	0.2	0	0	0	0	0
	0	0	0	0.7	0	1.1	0	0.1	0	0.2	0	0	0	0.3	0.5	0	0.6	0	1.1	0	0	0	0	0.2
	0	0.1	0	0.2	0	0.2	0	0	0	0	0	0	0	0	0	0	0	0	0	0	0	0.1	0	0
	0	0.1	0	0	0	0	0	0	0	0	0	0	0	0	0	—	—	—	—	—	—	—	—	—
	—	—	—	—	—	—	—	—	—	—	—	—	—	—	—	—	—	—	0	0	0.1	0.2	0	2.0
	0	2.2	0	1.4	0	0.4	0	0.8	0	0.1	0	0	0	0	0	0	0.8	0	1.9	0	0.3	0.4	0.1	1.4
	0.1	1.6	0	0.1	0	0.1	0	0.2	0	1.0	0	0.8	0	0.5	0	0	0	0	0	0	0	0.1	0	1.6
	0	1.1	0	0.6	0	3.3	0	2.0	0	0.1	0	0	0.1	0	0.1	0	0.3	0	0	0.1	0.5	1.1	0.4	0.4
	0	0.4	0	0.4	0	0.2	0	0.2	0	0	0	0	0	0	0	0	0.6	0	0.2	0	0.1	1.7	1.0	2.4
	0.1	0.2	0.2	0.6	0	0.7	0	0.2	0	0	0	0	0	0	0	0	0.1	0	0	0	0	0.2	0.1	1.9
	0.5	0.9	0	0.4	0	0.2	0	1.3	0	0.9	0	0.2	0.2	0	0.1	0	0	0	0	0	0	0	0	0
	0	0	0	0	0	0	0	0	0	0	0	0	0	0	0	0	0	0	0	0	0	0	0	0.5
	0	0.6	0	0.1	0	0	0	0	0	0	0	0	0	0	0	0	0.1	0	0.1	0	0	0	0	0
	0	0	0	0	0	0	0	0	0	0	0	0	0	0	0	0	0	0	0.1	0.1	0	0.2	0	0.1
	0	0.1	0	0	0	0	0	0	0	0	0	0	0	0	0	0	0	0.1	0	0.1	0	0	0	0
	0.4	0	0.9	0.8	0	1.9	0	2.2	0	2.4	0	0.5	0.1	0	0.8	0	1.7	0	0	0.7	0	2.4	0.8	1.8
	1.4	0.1	2.7	0	0	0.8	0	0.3	0	0.1	0	0.1	0.2	0	0.5	0	0.4	0	0	0.1	0	0	0.7	0
	6.0	0	0.2	0	0.3	0.1	0	1.4	0	0.3	0	0	0.1	0	0.9	0	1.7	0	0.1	0	0	2.4	0.5	1.7
April 1	0.5	1.5	0	1.5	0	0.5	0	0	0	0	0	0	0	0	0.5	0	0	0	0	0	0	0.5	0	8.0
	2.5	0.5	0.5	0.5	0.5	0	0	0.5	0	1.0	0	0	0	0	0	0	2.0	0	1.0	2.5	0	5.0	0.5	2.0
	5.0	0	8.5	0.5	0	0.5	0	0	0	0	0	0	0	0	0	0	0	0	0.5	0	0	0.5	0.5	0
	0.5	1.0	1.0	0	0	0.5	0	0	0	0	0	0	0	0	0	0	1.0	0	1.0	0	0	0	1.0	0
	1.5	0	0.5	0	5.0	2.0	0	7.5	0	0.5	0	0	0.5	0	2.5	0	1.5	0	0	0	0	0	0	0
	2.0	0	11.0	0	6.0	0.5	1.5	0	1.5	0.5	3.0	1.0	0	12.0	0	6.0	0	1.5	0	0	0	0.5	0.5	0.5
	0	0	0	0	0	0	0	0	0	0.5	0	1.0	0.5	0	0	0.5	0	0	0	0.5	0	4.5	0	3.0
	0	0.5	0	0	0	0	0	0	0	0	0	0	0	0	0.5	0	0.5	0	0	0	0	0.5	5.0	0.5
	4.0	0	11.0	0	0.5	2.0	2.0	2.0	0	1.0	0	0	0	0	2.0	0	0.5	0	0	3.0	0.5	2.5	0.5	2.5
	1.0	0.5	1.5	0	0	0	0	0.5	—	0.5	0	0	0.5	0	2.0	0	0.5	0	0	0.5	2.5	1.0	0	0.5

TABLE LXXV (continued). F_V

Gr. M.-T.	0—2		2—4		4—6		6—8		8—10		10—12		12—14		14—16		16—18		18—20		20—22		22—24	
Date	+	—	+	—	+	—	+	—	+	—	+	—	+	—	+	—	+	—	+	—	+	—	+	—
April 11	0	1.0	0	0.5	0	0.5	0	0.5	0	0	0	0	0.5	0	0	0	0	0	0	0	0	0.5	0	1.5
12	4.5	0	0.5	2.5	0	2.0	0	0	0	0	0	0	0	0	0	0	1.5	0	1.0	0	0	1.0	3.5	0
13	0	0.5	0	1.5	0	0.5	0	1.0	0	1.5	0	0	0	0	0	0	0.5	0	1.0	0	0	1.5	0	1.0
14	0	0	0	0.5	0	0	0	0	0	0	0	0	0	0	0.5	0	0	0	0	0	0	0	0	0
15	0	0	0	0	0	0.5	0	0	0	1.5	0	1.0	1.0	0	0.1	0	—	—	—	—	—	—	—	—

SECOND SERIES.

DIURNAL DISTRIBUTION OF STORMINESS.

Matotchkin-Schar.

TABLE LXXVI. S_H in γ

Hour	0—2		2—4		4—6		6—8		8—10		10—12		12—14		14—16		16—18		18—20		20—22		22—24	
Period	+	—	+	—	+	—	+	—	+	—	+	—	+	—	+	—	+	—	+	—	+	—	+	—
Oct. 3—7	0	2.7	0.6	1.2	1.8	2.1	0.9	0.6	0.6	0.6	3.6	0.3	2.4	0	1.2	1.2	0.3	0	0.3	0.6	0	5.1	0	6.0
8—12	4.2	1.2	2.1	0.3	0	0.6	0.6	0.3	0.3	1.5	0.9	2.4	7.5	0.6	4.8	0.6	7.5	6.9	2.1	21.0	0	51.9	0	21.3
13—17	0	0.3	0	0	0	1.2	0	0.3	0.6	0	1.2	0.3	0	0.3	0.3	1.8	1.8	0.9	0.6	27.0	0.9	5.1	0.3	1.8
18—22	0.3	3.9	0	3.0	0.9	0.3	0.3	0.6	0.9	0	1.2	0.3	0	0.6	0.3	0.9	3.0	0	0.6	10.5	0	5.4	0	7.2
23—27	2.7	30.3	2.7	6.6	1.2	2.4	0	1.2	2.7	0.6	4.2	1.8	9.0	0.9	13.5	4.5	9.6	10.2	2.4	24.3	0	75.6	0.6	91.5
Oct. 28 Nov. 1	1.8	27.9	2.1	5.4	6.0	4.5	6.9	3.9	30.3	0.6	44.1	3.6	21.6	8.1	16.8	2.7	2.1	27.0	0	98.7	0	99.9	0.3	92.7
Mean value	**1.5**	**11.1**	**1.3**	**2.8**	**1.7**	**1.9**	**1.5**	**1.2**	**5.9**	**0.6**	**9.2**	**1.5**	**6.8**	**1.8**	**6.2**	**2.0**	**4.1**	**7.5**	**1.0**	**30.4**	**0.2**	**40.5**	**0.2**	**36.8**
Nov. 2—6	0	1.5	0	0	0	0.3	0	0.3	0	1.2	0	0.3	0	0.6	0.6	0.6	4.5	9.0	0.6	59.4	0	29.4	0.3	1.5
7—11	1.8	1.5	1.8	0	0	0	0	0	0.6	0	0.3	0.3	0	0.3	0	0.6	0.6	0.6	0.3	0.3	0	4.8	0	6.6
12—16	1.8	4.2	0.6	0	1.8	0	0.3	0.3	0	1.2	0.9	1.8	10.5	0.6	10.2	0	5.1	0.9	2.7	1.5	1.5	15.0	0.9	7.8
17—21	0	2.7	1.5	0.3	0.6	0	0.3	0	0.3	0.3	0.6	0.3	0.3	0.6	0.9	0	7.2	3.6	0.3	21.3	0	72.0	0.3	15.0
22—26	0.9	52.5	0.3	4.8	0.6	2.1	0	0.3	13.2	0	25.5	0.3	29.4	0.3	40.8	5.7	2.7	64.8	2.1	75.0	1.8	84.0	0	78.0
Nov. 27 Dec. 1	0	1.8	0.6	0.3	0.3	0.6	0.6	0.3	0.9	0	0.9	0	1.5	0.3	5.4	0.3	5.1	0.9	0	7.2	0	7.5	0	1.8
Mean value	**0.8**	**10.7**	**0.8**	**0.9**	**0.6**	**0.5**	**0.2**	**0.2**	**2.5**	**0.5**	**4.7**	**0.5**	**7.0**	**0.5**	**9.7**	**1.2**	**4.2**	**13.3**	**1.0**	**27.5**	**0.6**	**35.5**	**0.3**	**18.5**
Dec. 2—6	0.6	2.4	0.9	0.6	0.6	0.3	0	0	0	0.3	0.9	0.3	0.3	0.9	1.8	0.3	2.1	3.9	0.3	6.6	0.3	1.5	0	4.2
7—11	0	0.6	0.3	1.2	0.9	0.3	1.2	0	0.6	0.6	2.1	0.6	2.4	0	1.2	0	7.8	3.0	3.9	5.1	1.8	15.0	0.3	10.2
12—16	0.3	3.6	0.6	1.2	0	0	0	0	0.3	0	0	0.9	0	0	0.3	0	3.3	0.3	2.1	7.2	0.6	15.0	0	9.3
17—21	0	0.6	0	0	0	0.3	0	0	0	0.3	0	0.6	0	1.2	0.6	0.6	1.5	0.6	0.6	4.2	0.3	6.9	0	1.5
22—26	0.6	21.0	1.8	9.3	1.2	1.5	0.6	0.6	10.5	0.3	9.0	2.4	22.5	1.8	16.8	1.2	5.7	7.8	0.9	53.4	0	58.5	0.9	38.4
27—31	0	4.5	2.4	0.6	1.8	0.6	0.3	1.2	0	2.4	0.3	1.2	0.9	0.3	0.9	0	4.2	0.6	1.5	1.2	0.6	9.0	0	13.5
Mean value	**0.3**	**5.5**	**1.0**	**2.2**	**0.8**	**0.5**	**0.4**	**0.3**	**1.9**	**0.7**	**2.1**	**1.0**	**4.4**	**0.7**	**3.6**	**0.4**	**4.1**	**2.7**	**1.6**	**13.0**	**0.6**	**17.7**	**0.2**	**12.9**
Jan. 1—5	0	1.5	1.2	0.9	1.2	0.3	0.3	1.2	1.5	0.6	1.2	0.6	1.2	0.6	3.6	0	1.5	15.3	6.9	4.2	0.9	6.6	0.6	5.7
6—10	0.3	1.5	0.3	0.6	0	0.6	0	1.2	0.3	0	0.9	0.6	0.9	2.1	10.2	1.2	13.2	0.3	1.8	3.9	0	16.2	0	12.6
11—15	0.3	3.3	0.9	0.9	0.3	0	0.6	0.3	0	0.3	0	0.3	0.6	0.3	2.4	0	4.5	0	3.9	11.4	0.6	5.4	0	8.4
16—20	0	1.8	0.3	1.5	0	0	0	0.3	0.3	1.5	0.3	1.8	8.7	1.2	8.7	2.1	15.3	0.3	5.1	1.8	2.4	6.9	0	4.5
21—25	0	3.0	0.6	0.6	0.6	0.6	0	1.2	0	0.9	0.9	0.6	0.3	2.1	1.8	1.8	6.0	1.2	1.8	14.4	0	20.1	0.3	9.6
26—30	0.3	67.2	0.6	8.1	3.0	3.3	2.1	0.9	1.5	0.9	3.9	0.6	10.8	0.9	11.1	0.3	18.9	1.5	8.4	81.9	0.9	87.0	0	124.8
Mean value	**0.2**	**13.1**	**0.7**	**2.1**	**0.9**	**0.8**	**0.5**	**0.9**	**0.6**	**0.7**	**1.2**	**0.8**	**3.8**	**1.2**	**6.3**	**0.9**	**9.9**	**3.1**	**4.7**	**19.6**	**0.8**	**23.7**	**0.2**	**27.6**
Jan. 31 Febr. 4	0	1.2	0.3	0	1.2	0	0	0	0	0	0.3	0.3	0.6	0.3	3.3	0	8.4	0	6.9	0	1.2	0	0	0
Febr. 5—9	0.3	3.6	0.9	3.9	1.5	3.0	1.5	1.8	2.1	0	0.6	0	2.1	0.6	14.1	0	22.2	0	4.8	35.4	0	71.4	0.6	43.5
10—14	1.5	11.4	2.1	0.9	0.6	0.3	0	1.2	0.6	0.6	1.8	0.6	1.2	0.6	3.0	0.3	6.3	0.6	4.2	17.4	0	24.3	0	51.3
15—19	0.3	2.7	0	0.3	0	0.3	0	0.3	0.3	0.9	0.9	0.9	2.7	0.3	12.9	0.9	3.6	12.9	3.0	2.1	0	2.4	0	0.9
20—24	0	0.9	0	4.8	0	6.0	0	6.6	3.0	0.9	6.9	0	3.0	2.1	0.9	1.2	4.8	0.3	1.2	1.2	0	3.6	0	2.1
Feb. 25 Mar. 1	0	0	0	1.2	0	3.0	0.3	2.1	2.1	0.3	3.3	0	0.6	0.6	0.9	1.2	6.0	0.3	4.5	0	1.2	3.6	0	8.4
Mean value	**0.4**	**3.3**	**0.6**	**1.9**	**0.5**	**2.1**	**0.3**	**2.0**	**1.4**	**0.5**	**2.3**	**0.3**	**1.7**	**0.8**	**5.9**	**0.6**	**8.6**	**2.4**	**4.1**	**9.4**	**0.4**	**17.6**	**0.1**	**17.7**
Oct. 3 March 1	*0.6*	*8.7*	*0.9*	*2.0*	*0.9*	*1.2*	*0.6*	*0.9*	*2.5*	*0.6*	*4.0*	*0.8*	*4.7*	*1.0*	*6.3*	*1.0*	*6.2*	*5.8*	*2.5*	*19.9*	*0.5*	*27.0*	*0.2*	*22.7*

TABLE LXXVII. S_D in γ **Matotchkin-Schar.**

Hour	0–2		2–4		4–6		6–8		8–10		10–12		12–14		14–16		16–18		18–20		20–22		22–24	
Period		–	+	–	+	–	+	–	+	–	+	–	+	–	+	–	+	–	+	–	+	–	+	–
Oct. 3–7	0.9	0.2	0.9	0.4	1.8	1.1	0.4	0.2	0.7	1.6	1.1	0.5	0.2	0.5	0.5	1.1	0.4	0.5	0.5	2.0	0.5	1.1	0.5	0.5
8–12	1.6	0.4	1.3	0.7	0.2	0.2	0.2	0.5	0.5	0.2	0.9	0.4	2.5	0	0.7	1.3	3.4	14.9	0.4	5.8	0.2	32.2	0.4	11.2
13–17	0	0	0	0	0	0	0.2	0	1.3	0	0.5	1.1	0	1.4	0.4	0.4	0.4	1.3	0.9	4.7	0.7	1.3	0.7	0.5
18–22	0	1.4	1.6	0.4	1.1	0.4	0	0.4	0.5	0	0.5	0.2	0	0.2	0.4	0.4	2.0	0.4	0.4	8.6	0	4.0	0.2	1.3
23–27	0	22.3	1.6	3.6	0.5	0.4	1.3	0	1.6	0.7	1.4	1.6	3.1	0.9	4.1	2.0	4.9	6.5	1.1	21.1	0.2	38.0	0.2	38.4
Oct. 28 Nov. 1	0.5	20.0	6.1	4.7	9.2	0.7	4.9	1.1	13.1	0.2	14.6	0.7	17.5	5.2	23.2	0.7	3.4	9.9	0.2	47.3	0.2	68.6	0.4	47.2
Mean value	**0.5**	**7.4**	**1.9**	**1.6**	**2.1**	**0.5**	**1.2**	**0.4**	**3.0**	**0.5**	**3.2**	**0.8**	**3.9**	**1.4**	**4.9**	**1.0**	**2.4**	**5.6**	**0.6**	**14.9**	**0.3**	**24.2**	**0.4**	**16.5**
Nov. 2–6	0	1.1	0	0	0	0	0.7	0	0	0.2	0	0.9	0	0.2	0.2	0	2.5	1.8	2.0	19.4	0.4	17.1	0.2	1.1
7–11	0	2.2	0.2	0.4	0	0	0.5	0	0.4	0	0	0	0	0.4	0.4	0.7	0.4	0	0.4	0.2	0.9	0.5	0.4	1.4
12–16	0	4.3	0	2.2	1.3	0.5	0.4	0.2	0.4	0	1.4	0.5	4.0	0.5	4.0	0.4	1.6	2.3	1.1	2.5	0.5	10.1	0	6.8
17–21	0	2.5	0.2	0.5	0.7	0.2	0.5	0	0	0.2	0.5	0.4	0.2	0.7	0.7	0.5	1.3	4.9	0.5	12.8	0	31.0	0	11.7
22–26	0.4	42.5	0.5	9.1	3.8	3.1	5.9	1.4	6.8	0.2	14.4	0.5	14.8	0.9	27.4	2.0	11.7	43.0	2.3	49.7	2.5	81.5	2.3	81.5
Nov. 27 Dec. 1	0.2	0.2	0.2	0.4	0.2	1.1	0.7	0.4	0.2	0.4	1.1	0.7	0.4	1.8	1.3	1.1	0	9.7	0.2	6.5	0	6.7	0	0.5
Mean value	**0.1**	**8.8**	**0.2**	**3.8**	**1.0**	**0.8**	**1.5**	**0.3**	**1.3**	**0.2**	**2.9**	**0.5**	**3.2**	**0.8**	**5.7**	**0.8**	**2.9**	**10.3**	**1.1**	**15.2**	**0.7**	**24.5**	**0.5**	**17.2**
Dec. 2–6	0.7	0.4	0	0.2	0.2	0	0.5	0	0.2	0.5	1.6	0	0.2	0.5	0.5	0.5	0	3.1	0.4	10.4	0	0.7	0.4	0.4
7–11	0.2	0.2	0.4	0.4	0.7	0.5	1.1	0	1.1	0.5	1.1	0.9	1.4	0.4	0.9	0.2	2.5	4.0	1.4	7.2	1.4	11.0	0	7.9
12–16	0.5	1.1	0	0.7	0	0	0.9	0	0.2	0.2	0	1.1	0	0.5	0.5	0.2	3.4	11.9	1.4	4.5	0.2	6.8	0.5	1.8
17–21	0	0	0	0.2	0.7	0.2	0.2	0.2	0.2	0	0	0.2	0.2	0.2	0.7	0.9	1.4	0	0.7	1.4	0.4	2.2	0.2	0
22–26	0.9	8.8	3.6	0.4	2.0	0.5	0.2	1.1	3.2	0.7	5.4	0.7	3.8	3.4	4.7	1.1	2.2	8.5	0.7	23.0	0.4	32.6	1.6	10.4
27–31	1.3	0.5	0.7	0.4	0.4	0.4	0.5	0.2	0.5	0.7	0.2	0.7	0.2	0.4	0.2	0	1.3	4.0	1.8	1.3	0.7	1.4	0.4	4.1
Mean value	**0.6**	**1.8**	**0.8**	**0.4**	**0.7**	**0.3**	**0.6**	**0.3**	**0.9**	**0.4**	**1.4**	**0.6**	**1.0**	**0.9**	**1.3**	**0.5**	**1.8**	**5.3**	**1.1**	**8.0**	**0.5**	**9.1**	**0.5**	**4.1**
Jan. 1–5	0.2	2.0	0.7	0.5	0.7	0.5	1.1	2.7	1.1	0.9	1.3	0.4	0.2	1.1	0.4	0.7	1.6	4.7	1.1	5.0	1.8	0.5	0.7	0.4
6–10	0.2	0.4	1.1	0	0.2	0	0.2	0	0	0.4	0.2	0.4	0.2	0.2	0.9	0.9	0.4	6.3	0.2	9.0	0.5	6.1	0.4	4.5
11–15	2.7	0.2	0.2	0.2	0.2	0.2	0	0	0.2	0.2	0	0.4	0.7	0.9	0.5	2.3	0.5	4.3	2.0	8.6	1.3	3.8	1.8	0.4
16–20	1.4	0	0.5	0	0	0.2	0	0	0.9	0.9	0.7	0.7	0.7	1.4	2.3	2.3	2.9	3.2	2.2	4.0	2.9	1.8	0.4	0.9
21–25	1.1	0.4	0.2	0.2	0	0.9	0	0.5	0.2	0.4	0.4	0.4	0.4	1.4	0.5	3.8	1.3	8.5	0.4	11.2	0.2	13.0	2.0	0.7
26–30	0.7	31.7	0.2	8.5	0.5	2.5	1.3	0.7	1.3	0.5	6.3	0.2	8.3	1.1	6.5	0.4	9.0	7.2	1.4	29.2	0.2	34.2	1.4	51.5
Mean value	**1.1**	**5.8**	**0.5**	**1.6**	**0.3**	**0.7**	**0.4**	**0.7**	**0.6**	**0.6**	**1.5**	**0.4**	**1.8**	**1.0**	**1.9**	**1.7**	**2.6**	**5.7**	**1.2**	**11.2**	**1.2**	**9.9**	**1.1**	**9.7**
Jan. 31 Febr. 4	0.5	0.4	0	0	0	0	0	0.4	0.9	0	0.5	0.2	0.9	0	1.6	0	4.5	0.7	1.1	0.4	0.4	0	0	0
Febr. 5–9	0	3.2	1.1	2.3	2.5	1.6	2.5	0.7	0.9	0	0.9	0	1.4	0.7	6.7	0.2	5.0	6.5	1.8	15.5	0.9	37.6	0.5	23.0
10–14	0.9	5.2	1.4	0.9	1.3	0.5	0.7	1.6	0.4	0.2	1.3	0.4	0.5	0.5	0.7	2.0	1.3	4.3	1.3	9.0	0.7	11.0	0.5	18.0
15–19	0.7	0	0	0	0	0.4	0	0	0.2	0.5	0.9	0	0.7	0	1.8	8.6	0.9	13.5	0.2	2.3	0.5	0.4	0.2	0.4
20–24	0	0.2	0.7	0.7	2.3	0.2	2.9	0.4	2.3	0	4.9	0	3.4	0	1.1	0.5	0.9	1.1	1.4	0.7	0	0.7	0.2	0.2
Feb. 25 Mar. 1	0	0	0.4	0	2.9	0	4.1	0	3.2	0	3.1	0	1.8	0.2	0.7	0.5	1.8	0.2	1.3	1.6	0	2.2	0	4.3
Mean value	**0.4**	**1.5**	**0.6**	**0.7**	**1.5**	**0.5**	**1.7**	**0.5**	**1.3**	**0.1**	**1.9**	**0.1**	**1.5**	**0.2**	**2.1**	**2.0**	**2.4**	**4.4**	**1.2**	**4.9**	**0.4**	**8.7**	**0.2**	**7.7**
Oct. 3 March 1	*0.5*	*5.1*	*0.8*	*1.6*	*1.1*	*0.6*	*1.1*	*0.4*	*1.4*	*0.3*	*2.2*	*0.5*	*2.3*	*0.9*	*3.2*	*1.2*	*2.4*	*6.2*	*1.0*	*10.8*	*0.6*	*15.3*	*0.6*	*11.0*

TABLE LXXVIII. S_v in γ Mat

Hour	0–2		2–4		4–6		6–8		8–10		10–12		12–								
Period	+	−	+	−	+	−	+	−	+	−	+	−	+	−	+	−	+	−	+	−	+
Oct. 3–7	0	0	0.4	0	1.8	0	0	0	0.7	0.7	0	1.1	2.5	0.4	0.4	0.4	0	1.1	0	0	0
8–12	0	0	0	0	0	0	0	0	0	0	0	0	2.1	0	4.2	0	0.7	11.9	0.4	34.0	0
13–17	0	0	0	0	0	0	0	0	0	0	0.4	0	0	0	0.4	0	1.1	1.4	0	9.1	0
18–22	0	1.4	0	1.8	0	0.7	0	0	0	0	0	0	0	0	0	0	1.1	0	0	14.0	0.4
23–27	0.4	15.8	0.7	9.1	1.1	1.4	0.4	0.4	3.5	0	7.4	0	12.3	0	9.5	15.8	4.6	19.3	1.1	12.3	0.4
Oct. 28 Nov. 1	1.4	1.8	0.4	3.2	0.4	1.4	3.9	0.4	10.5	0	3.5	2.1	0	7.0	0	7.0	0	10.9	10.5	23.1	32.9
Mean value	**0.3**	**3.2**	**0.3**	**2.4**	**0.6**	**0.6**	**0.7**	**0.1**	**2.5**	**0.1**	**1.9**	**0.5**	**2.8**	**1.2**	**2.4**	**3.9**	**1.3**	**7.4**	**2.0**	**15.4**	**5.6**
Nov. 2–6	0	0	0	0	0	0	0	0	0	0	0	0	0	0	1.4	0	2.8	7.0	6.3	37.1	0.4
7–11	0.4	1.4	0	1.8	0	0	0	0	0	0	0	0	0.7	0	0.7	0	0	1.1	0	0.4	0
12–16	0	3.5	0	0	0	0	0	0	0	0	1.8	0	12.3	0.4	2.5	0.7	1.8	2.5	1.1	2.1	0.4
17–21	0	2.1	0	0.4	0	0	0	0	0	0	0.4	0.4	2.1	0	1.8	0	6.3	9.1	3.9	9.8	6.3
22–26	6.0	68.3	5.3	44.1	0.4	28.4	0	19.3	2.1	2.5	13.3	8.4	1.4	42.4	1.4	79.8	14.0	125.0	51.1	23.8	52.5
Nov. 27 Dec. 1	0	0	0.7	0.7	0	0	0.4	0.4	0	0	0.4	0	4.9	0	3.2	0	2.1	7.0	1.8	8.4	0
Mean value	**1.1**	**12.6**	**1.0**	**7.8**	**0.1**	**4.7**	**0.1**	**3.3**	**0.4**	**0.4**	**2.7**	**1.5**	**3.6**	**7.1**	**1.8**	**13.4**	**4.5**	**25.3**	**10.7**	**13.6**	**9.9**
Dec. 2–6	0	1.4	0	0	0	0	0	0	0	0	0.7	0	0	0	2.5	0	1.1	10.5	0	11.6	0
7–11	0	0.4	0.4	0	0.4	0.4	0	1.1	0.4	2.5	9.1	0.7	7.4	1.1	1.4	4.1	3.9	8.8	0.4	19.6	0
12–16	1.4	1.4	0	4.6	0	0.7	0	0	0	0	1.4	0	0.7	0	0.4	0	3.9	10.5	0.4	16.8	0.4
17–21	0	0	0	0	0	0	0	0	0	0	0	0	1.1	0	5.6	0	2.8	0	1.1	4.2	0
22–26	0	11.9	0	9.5	0	1.8	1.4	0.4	4.9	0.4	5.6	1.1	7.0	4.6	9.5	3.5	3.5	47.3	0	79.8	0
27–31	0.4	8.8	0	0.4	1.8	0.7	1.4	0	0.7	0	1.8	0	5.3	0	0.7	0	7.4	1.8	2.1	3.5	0.4
Mean value	**0.3**	**4.0**	**0.1**	**2.4**	**0.4**	**0.6**	**0.5**	**0.3**	**1.0**	**0.5**	**3.1**	**0.3**	**3.6**	**1.0**	**3.4**	**1.3**	**3.8**	**13.2**	**0.7**	**22.6**	**0.1**
Jan. 1–5	0	2.1	0.4	4.2	0	5.3	0.4	1.4	6.0	0	3.2	0	3.2	0	6.0	0	2.1	15.8	1.1	15.1	0
6–10	1.8	0.4	1.1	0	0.4	0	0	0	1.1	0.4	0.7	0.4	2.1	0	16.1	0	11.9	2.8	2.1	11.6	0
11–15	0.7	0	0	0	0	0	0	0	0	0	0	0	0.7	0	7.4	0	8.4	0.7	1.4	39.6	0.4
16–20	0	0	0	0	0	0	0	0	0	0	2.1	0	18.2	0	11.6	1.4	14.4	7.0	1.1	9.8	0.4
21–25	0.4	4.2	0	0.7	0	0	0	0	0	0	0	0.4	2.1	0.7	6.7	0.4	1.8	12.3	1.1	32.9	0
26–30	0	3.5	0	3.5	0	1.8	0	0	0	0	7.0	1.4	11.9	0	10.2	0.4	3.5	24.9	3.2	31.9	1.8
Mean value	**0.5**	**1.7**	**0.3**	**1.4**	**0.1**	**1.2**	**0.1**	**0.2**	**1.2**	**0.1**	**2.2**	**0.4**	**6.4**	**0.1**	**9.7**	**0.4**	**7.0**	**10.6**	**1.7**	**23.5**	**0.4**
Jan. 31 Feb. 4	0	0.4	0	0	0	0	0	0	0	0.7	0	0	0	0	7.4	0	4.9	1.4	3.9	0.4	0
Fbr. 5–9	0	2.5	0.4	4.6	1.1	6.7	2.8	1.8	1.8	0.4	0	0	8.4	0	6.7	3.2	7.7	12.6	1.4	37.5	2.5
10–14	2.1	4.2	1.1	0	0	0.4	2.5	0	0	0.4	4.9	0	2.5	0	3.5	0	10.5	3.5	1.4	18.6	3.2
15–19	0	0.4	0	0	0	0	0	0	0	0	0	0	2.1	0	7.7	7.0	1.1	62.0	0	3.5	0
20–24	0	0	0	5.3	0	6.7	0.7	4.9	5.6	0	9.1	0	7.0	0	4.6	0	6.7	0	1.8	0.4	0
Feb. 25 Mar. 1	0	0	1.1	0	0	0	6.0	0	10.5	0	12.3	0	12.6	0	8.8	0	2.5	2.8	0.7	1.1	0
Mean value	**0.4**	**1.3**	**0.4**	**1.7**	**0.2**	**2.3**	**2.0**	**1.1**	**3.0**	**0.3**	**4.4**	**0**	**5.4**	**0**	**6.5**	**1.7**	**5.6**	**13.7**	**1.5**	**10.3**	**1.0**
Oct. 3 March 1	*0.5*	*4.5*	*0.4*	*3.1*	*0.3*	*1.9*	*0.7*	*1.0*	*1.6*	*0.3*	*2.8*	*0.5*	*4.4*	*1.9*	*4.7*	*4.1*	*4.4*	*14.0*	*3.3*	*17.1*	*3.4*

Kaafjord.

TABLE LXXIX. S_H in γ.

	0—2		2—4		4—6		6—8		8—10		10—12		12—14		14—16		16—18		18—20		20—22		22—24	
Period	+	—	+	—	+	—	+	—	+	—	+	—	+	—	+	—	+	—	+	—	+	—	+	—
Sept. 3—7	0.3	1.8	0.9	0.3	0.9	0.3	0.3	0.9	0.3	0.3	0.3	2.1	0.9	0.6	2.7	0.9	1.8	0.6	1.8	0	0.3	0	0.6	0.6
	0.3	0.3	0	0	0.3	0	0.3	0	0.3	0	0.9	0.9	3.6	0.3	2.1	0	16.8	0	5.1	2.4	0	10.5	0	28.5
13—17	0.6	0	0	0.9	0	0.9	0	0.3	0	0.6	0.3	0.6	0.9	0.3	1.8	0.3	1.5	0	0	0	0	0.3	0	2.1
18—22	0	8.7	0.6	2.4	0.6	1.2	0.3	0.9	3.3	0.9	12.6	0.3	5.4	0.3	6.9	2.1	6.6	0.9	4.8	5.4	0.6	79.2	0	47.1
23—27	0	10.5	0	6.3	0.3	0	0.3	0.3	0	0	0	0.3	0.3	1.5	0.6	0.9	0	0.6	1.2	6.0	0	8.4	0	0.6
Sep. 28 Oct. 2	0	18.6	0	5.4	0	0.3	0	0	0	0	0	0.3	1.2	0.3	0.6	0.6	0.6	1.5	0.6	4.2	0.6	25.5	0	53.4
Mean value	**0.2**	**6.7**	**0.3**	**2.6**	**0.4**	**0.5**	**0.2**	**0.4**	**0.7**	**0.3**	**2.4**	**0.8**	**2.1**	**0.6**	**2.5**	**0.8**	**4.6**	**0.6**	**2.3**	**3.0**	**0.3**	**20.7**	**0.1**	**22.1**
Oct. 3—7	0.3	1.2	0	0.6	0.6	0.3	0	0	0	0.3	0.3	0	0.3	0.9	0.6	0	0.9	0	0.6	0.3	0	0.3	0	1.2
	0	1.8	0	0	0	0	0.3	0	0	0	0.3	0	1.2	0.3	0.3	0	0.3	0	6.3	3.0	0	41.4	0	16.5
13—17	0	0	0	0	0	0	0	0	0	0	0.3	0.6	0	0	0.3	0.3	0	0	0.3	0.3	0.3	0	0	0.3
18—22	0	2.4	0.3	2.7	0.3	0	0	0	0	0	0	0	0	0	0	0	0.3	0	0.3	0.6	0	2.1	0.3	1.2
23—27	0	34.8	0.6	6.0	0.6	0.3	0	0.3	0.3	0	0.3	0	0.6	1.2	6.0	1.5	7.5	0.9	9.0	0.9	0.9	25.5	0.3	69.6
Oct. 28 Nov. 1	0.3	37.8	0.6	5.4	3.3	3.6	0.3	3.0	5.4	2.4	26.4	1.2	38.4	0.3	45.9	0.9	11.4	2.4	0	42.6	0.9	68.1	0	78.6
Mean value	**0.1**	**13.0**	**0.3**	**2.5**	**0.8**	**0.7**	**0.1**	**0.6**	**1.0**	**0.5**	**4.6**	**0.3**	**6.8**	**0.5**	**8.9**	**0.5**	**3.4**	**0.6**	**2.8**	**8.0**	**0.4**	**22.9**	**0.1**	**27.9**
Nov. 2—6	0	0	0	0	0	0	0	0	0	0	0	0	0.3	0	0	0.3	2.1	0	2.1	12.9	0.6	6.3	0	1.5
7—11	0.3	1.2	0.3	0	0	0	0	0	0	0	0	0	0	0	0	0	9	0	0	0	1.2	1.5	0	2.7
12—16	0	1.8	0.6	2.1	1.5	0.3	0	0	0	0.3	0	1.8	0	1.5	0.6	0.9	0	0.6	0.3	0.3	1.2	3.9	0.6	2.4
17—21	0	1.8	0	0	0	0	0	0	0	0.3	0	0	0	0	0	0	2.4	0.3	2.7	0.9	0.3	33.0	0	7.2
22—2	1.2	71.7	1.2	3.4	2.4	4.8	2.7	3.6	5.1	1.5	12.0	0.9	31.2	0.6	54.0	0	39.6	19.2	16.8	51.6	9.3	77.7	3.6	78.0
Nov. 27 Dec. 1	0.3	0.3	0.3	0	0	0.3	0	0.6	0	0.3	0.3	0.3	0	1.8	0	0.9	0.9	0	0	0.9	0.3	3.3	0	0.3
Mean value	**0.3**	**12.8**	**0.4**	**9.3**	**0.7**	**0.9**	**0.5**	**0.7**	**0.9**	**0.4**	**2.1**	**0.5**	**5.3**	**0.7**	**9.1**	**0.4**	**7.5**	**3.7**	**3.7**	**11.1**	**2.2**	**21.0**	**0.7**	**15.4**
	0	1.2	0	0	0.9	0	0.3	0	0	0	0	0	0	0	0.6	0.3	1.5	0.3	0.6	0.3	0.6	0	0	2.1
7—11	0	0	0.9	0.9	1.5	0	0.3	0	0	0	0.3	0.9	0.3	0.6	0.3	0.3	9.6	0	6.0	0.3	9.6	0.6	0	4.2
12—1	0.3	5.4	0.3	0.6	0.3	0	0.3	0	0	0	0.3	0.9	0.6	0.3	0.3	0	3.9	1.2	5.1	0.3	1.2	2.7	0.3	0.9
17—2	0	0	0	0	0	0	0	0	0	0	0	0	0.3	0.3	0.3	0.3	0.3	0	1.5	0	0	1.5	0	0
22—2	0	35.4	0	8.1	1.8	4.2	0	3.3	3.3	0.6	5.7	0.6	14.4	1.2	3.3	1.5	13.8	0	3.9	9.0	0	29.1	0	31.5
Dec. 27—31	0	4.8	0	0.3	0.3	0.3	0.6	0	0	0	0.6	0	0.3	0.3	0.6	0	0.3	1.2	0.3	0.6	0.9	1.8	0	8.4
Mean value	**0.1**	**7.8**	**0.2**	**1.7**	**0.8**	**0.8**	**0.3**	**0.6**	**0.6**	**0.1**	**1.2**	**0.4**	**2.7**	**0.5**	**0.9**	**0.4**	**4.9**	**0.5**	**2.9**	**1.8**	**2.1**	**6.0**	**0.1**	**7.9**
Jan. 1—5	0	1.2	1.5	1.5	3.3	1.8	0.3	0.6	0.3	0.6	0.9	0.3	0.9	0	0	0	2.7	0.3	0.6	0	1.2	0.9	0	2.1
	0	0.9	0	1.8	0.6	0	0.3	0	0	0	0	0.3	0.3	0.3	0.3	0.9	0.3	0.3	0.3	0.9	0.9	1.2	0	4.8
11—1	0	0.9	0	0.9	0.3	0	0	0	0	0	0	0.3	0	0	0	0.3	0.3	0	3.6	2.1	1.8	1.2	0.3	3.0
16—2	0	0.6	0	1.8	0	0	0	0	0	0.3	0	0.6	0.3	1.2	2.1	1.2	2.4	0	4.2	0	0.9	0.6	0	0.9
21—2	0	0.6	0	0.6	0	0.9	0	0.6	0	0	0.3	0	0.6	0.6	0.3	1.8	2.7	0	2.4	0.6	1.5	4.5	0	1.5
	0	5.0	0	0.6	0	0	0	0	0.3	0.3	0.6	0.3	1.5	0.6	1.5	0.6	5.7	0.6	11.1	10.2	0.6	35.7	0	69.6
Mean value	**0**	**3.2**	**0.3**	**1.2**	**0.7**	**0.5**	**0.1**	**0.2**	**0.1**	**0.2**	**0.3**	**0.3**	**0.6**	**0.5**	**0.7**	**0.8**	**2.4**	**0.2**	**3.7**	**2.3**	**1.2**	**7.4**	**0.1**	**13.7**
Jan. 31 Febr. 4	0	0.3	0	0	0	0	0.3	0	0	0	0	0.3	0.3	0	0	0.3	0.9	0	0.3	0	0.3	0	0.3	0
Febr. 5—9	0	1.8	0.3	6.9	0	1.2	0.3	0.3	0.3	0.9	3.0	0.3	0.6	1.2	10.8	0	12.9	0	4.2	15.3	0	34.5	0.3	24.6
10—1	0	3.2	2.4	0.6	1.8	0	0.3	0.3	0.3	0.3	0.3	1.5	0.6	0.9	1.8	0.9	0.6	0	9.6	0.6	2.1	1.2	0.3	24.6
15—1	0	1.5	0	0	0.6	0	0.3	0.3	0	0	0	1.8	0	0.9	2.4	0	6.0	0.3	0	0	0.3	0	0	0
20—24	—	—	—	—	—	—	—	—	—	—	—	—	—	—	—	—	—	—	—	—	—	—	—	—
Feb. 25 Mar. 1	0	0	0	0.3	0.3	0.6	0	0.3	0.3	0.6	0.3	0.3	0	0.6	0.3	0.3	6.9	0	2.4	0	0	2.7	0	5.1
Mean value	**0.1**	**3.4**	**0.5**	**1.6**	**0.5**	**0.4**	**0.2**	**0.2**	**0.2**	**0.4**	**0.7**	**0.8**	**0.3**	**0.7**	**3.1**	**0.3**	**5.5**	**0.1**	**3.3**	**3.2**	**0.5**	**7.7**	**0.2**	**10.9**
March 2—6	0	5.1	0.3	4.8	0.3	0.9	0	0	0.3	0	0.9	0.3	0.3	0	3.3	0	3.9	0.3	4.5	0.3	8.4	0.9	4.5	3.3
7—11	3.	5.4	3.0	4.5	4.5	0	3.0	1.2	0	3.9	0.6	0.6	1.2	3.6	4.5	0.6	15.0	0	7.8	2.4	3.0	20.7	0	32.4
Sep. 3 March 1	*0.*	*7.8*	*0.3*	*3.1*	*0.6*	*0.6*	*0.2*	*0.4*	*0.6*	*0.3*	*1.9*	*0.5*	*2.9*	*0.6*	*4.2*	*0.5*	*4.7*	*0.9*	*3.1*	*4.9*	*1.1*	*14.3*	*0.2*	*16.3*

TABLE LXXX. S_D in γ

Hour	0−2		2−4		4−6		6−8		8−10		10−12		12−14		14−16		16−18		18−20		20
Period	+	−	+	−	+	−	+	−	+	−	+	−	+	−	+	−	+	−	+	−	+
Sept. 3−7	0	4.7	0.5	2.2	0.5	1.4	1.8	0.9	1.8	0.5	1.3	0.5	1.8	0.5	0.2	1.3	1.1	0.7	0.2	0.9	0.9
8−12	0.7	1.6	0.2	0.9	0	0.9	0.5	0.9	0.5	0.4	1.8	0.2	3.6	0.2	2.9	2.3	8.5	0.4	4.9	0.9	0
13−17	0.4	1.4	0	1.3	0.4	2.0	0.4	1.6	0.7	0.2	1.3	0.2	0.7	0.4	0.2	1.1	1.1	0	1.3	0	0.4
18−22	0	7.2	0.2	5.6	0.9	2.9	1.4	2.0	1.6	1.1	4.9	1.1	5.2	1.4	3.1	1.6	2.9	2.9	1.8	14.0	0.2
23−27	1.1	8.1	0	2.7	0	0.7	0.7	0.5	0.5	0.7	0.2	1.3	1.3	0.7	0	2.0	0	1.1	0.9	2.5	0.2
Sep. 28 Oct. 2	1.1	12.1	0.4	5.9	0.2	0.4	0.4	0.4	1.4	0.2	0.5	0.2	0.7	0.4	1.4	0.4	4.1	0.4	1.8	3.4	1.4
Mean value	**0.6**	**5.9**	**0.2**	**3.1**	**0.3**	**1.4**	**0.9**	**1.1**	**1.1**	**0.5**	**1.7**	**0.6**	**2.2**	**0.6**	**1.3**	**1.5**	**3.0**	**0.9**	**1.8**	**3.6**	**0.5**
Oct. 3−7	1.3	0.2	0.5	0.4	0.5	1.1	0.4	0.5	0.2	0.4	2.2	0.5	0.2	0.2	0	0.7	0.2	0	0	1.1	1.1
8−12	0.2	0.4	0.4	0.9	0	0.5	0.5	0.4	0.7	0.4	1.4	0.4	1.4	0	0.4	0.7	0.5	0.2	4.1	4.0	0.5
13−17	0.2	0.2	0	0	0	0	0.2	0.2	0.7	0.2	1.4	0.5	0	0	0.2	0	1.1	0	1.4	1.4	0.2
18−22	0.4	1.6	0.7	1.1	0.2	1.6	0.2	0.5	0.2	0.2	0.5	0.4	0	0.4	0.2	0.2	0.2	1.1	0.2	2.0	0.4
23−27	0	26.6	0.7	7.6	0.9	1.6	1.1	0.7	0.4	1.4	0.9	4.0	0.9	2.0	4.1	1.6	5.9	1.8	2.2	2.3	1.6
Oct. 28 Nov. 1	0.4	24.1	0.9	6.3	6.5	0.7	4.3	0.7	3.8	0.5	5.9	1.1	6.1	3.2	15.5	0.7	28.6	1.4	3.4	9.0	0.2
Mean value	**0.4**	**8.9**	**0.5**	**2.7**	**1.4**	**0.9**	**1.1**	**0.5**	**1.0**	**0.5**	**2.1**	**1.2**	**1.4**	**1.0**	**3.4**	**0.7**	**6.1**	**0.8**	**1.9**	**3.3**	**0.7**
Nov. 2−6	0	0.7	0	0.4	0	0	0.4	0.2	0.2	0.2	0.2	0	0	0	0.5	0	3.4	0.2	0.2	4.9	0
7−11	0.2	2.0	0	1.3	0	0.2	0.2	0	0	0	0	0.2	0.2	0.2	0	0.2	0.4	0	0	0	0.5
12−16	0	2.7	0.7	1.1	0.7	0.4	0.9	1.3	0.5	0.2	1.1	0	1.4	1.1	0.2	1.8	0.2	0.9	0	1.6	0.4
17−21	0.5	2.0	0.2	0.5	0.4	0	0	0.5	0.2	0.5	1.3	0.2	0.7	0	1.1	0	1.1	2.5	0.7	2.2	0.2
22−26	0	42.7	0	21.4	1.1	4.1	4.5	0.9	4.1	1.4	10.4	0.4	13.1	0.2	16.0	3.1	14.2	10.3	5.9	22.3	1.4
Nov. 27 Dec. 1	0.2	0.4	0	0.4	0	0.4	0.2	0.4	0	0.5	0.9	0.4	1.1	0.9	0.4	0.5	1.1	0.7	1.6	0	0.5
Mean value	**0.2**	**8.4**	**0.2**	**4.2**	**0.4**	**0.9**	**1.0**	**0.6**	**0.8**	**0.5**	**2.3**	**0.2**	**2.8**	**0.4**	**3.0**	**0.9**	**3.4**	**2.4**	**1.4**	**5.2**	**0.5**
Dec. 2−6	0.4	1.6	0	0.9	0.2	0.4	0.9	0.2	0	0.9	1.6	0	0.5	0.2	0.5	0	0.4	0.7	0	2.9	0.4
7−11	0.2	0.2	0.7	0.4	0.7	0.2	0.9	0.4	0.4	0.9	0.7	0.7	0.7	0.5	0.4	0.5	1.8	1.4	0.5	5.8	0.5
12−16	2.2	2.2	0.2	1.8	0.2	0	0.4	0.2	0.5	0.5	1.1	0.7	0.7	0.4	1.4	0	0	11.0	0.9	4.0	0.9
17−21	0	0.2	0.9	0	0.5	0.2	0.4	0	0	0	0.2	0	0	0.4	0.5	0.2	1.1	0	0.2	1.3	1.4
22−26	1.3	10.1	1.3	4.1	4.9	0.2	2.5	0.5	0.9	2.0	3.2	1.8	4.5	1.3	6.1	0.2	3.1	2.2	1.4	8.1	0
27−31	0.2	6.5	0.2	1.4	0	0.9	0.5	0.5	0	1.1	0.4	0.9	0.2	0.5	0.2	0	0.9	1.3	0.7	1.1	0.4
Mean value	**0.7**	**3.5**	**0.6**	**1.4**	**1.1**	**0.3**	**0.9**	**0.3**	**0.3**	**0.9**	**1.2**	**0.7**	**1.1**	**0.6**	**1.5**	**0.2**	**1.2**	**2.8**	**0.6**	**3.9**	**0.6**
Jan. 1−5	0.7	0.7	1.3	0.9	0.5	2.9	1.3	0.5	0.5	2.2	2.3	0	1.1	0.2	0.4	0.7	1.4	2.3	0.4	5.8	2.5
6−10	0.4	0.7	0.5	0.2	1.1	0.2	1.3	0	0.7	0.2	0.7	0.2	0.7	0.4	2.7	0.2	2.0	1.6	0.4	3.4	0.9
11−15	1.3	1.8	1.1	0.5	1.4	0	0.5	0	0.2	0.4	0.5	0.4	1.3	0.4	1.4	0.4	1.6	1.3	0.5	8.1	1.3
16−20	0.9	0.4	0.9	0.4	0.2	0.2	0.5	0.2	0.9	0.4	1.8	0.4	2.2	0	2.2	1.4	1.1	3.4	1.3	2.7	1.3
21−25	0.5	1.8	1.3	0	0.9	0.2	0.5	0.4	0.7	0.5	2.0	0	1.8	0	0.9	0.9	0.7	4.0	0.5	7.0	0.2
26−30	0.2	3.4	0.2	0.5	0.2	0	0.2	0.2	0.9	0.5	1.8	0.5	4.1	0.5	2.7	0.5	3.2	1.8	2.3	7.7	0.4
Mean value	**0.7**	**1.5**	**0.9**	**0.4**	**0.7**	**0.6**	**0.7**	**0.2**	**0.7**	**0.7**	**1.5**	**0.3**	**1.9**	**0.3**	**1.7**	**0.7**	**1.7**	**2.4**	**0.9**	**5.8**	**1.1**
Jan. 31 Feb. 4	0.2	0.4	0.2	0	0.4	0.2	0.2	0.2	0.2	0.4	0.9	0.2	0.7	0	0.2	0.4	0.7	0.4	0.5	0.4	0
Feb. 5−9	0.4	2.0	1.3	5.2	0.4	5.0	0.2	1.3	0.5	0.7	3.1	0.4	1.3	0.5	4.5	0.4	2.3	3.4	1.8	8.1	0.9
10−14	1.3	7.6	1.1	1.8	0.7	0.2	0.5	1.1	0.4	0.7	1.4	0.5	0.7	0.2	0.2	0.7	0.2	2.0	0.9	7.2	0
15−19	1.6	0.2	0.2	0.2	0.2	0.2	0.2	0.2	0.7	0.7	1.4	0	1.1	0	0.4	4.7	0.2	8.6	0.2	1.1	0.2
20−24	0.2	1.4	0.4	1.3	4.0	0.5	5.0	0.2	2.0	0	2.3	0	2.7	0.5	1.1	0.5	1.3	0.4	0.5	0	0
Feb. 25 March 1	0	0.5	0.4	0.7	1.6	0	4.0	0	1.1	0	0.9	0.4	0.9	0	0.2	0.5	0.7	1.3	0	1.1	0.4
Mean value	**0.6**	**2.0**	**0.6**	**1.5**	**1.2**	**1.0**	**1.7**	**0.5**	**0.8**	**0.4**	**1.7**	**0.3**	**1.2**	**0.2**	**1.1**	**1.2**	**0.9**	**2.7**	**0.7**	**3.0**	**0.3**
March 2−6	1.3	2.9	0.4	2.5	0.5	0.2	0.5	0.5	1.4	0.5	0.7	0	1.1	0	0	0.7	2.5	0.5	2.9	1.6	1.6
7−11	0.7	4.1	2.5	1.1	3.8	0	2.3	0.7	1.8	0.2	4.1	0	4.0	0.7	1.8	2.5	3.4	4.1	1.6	10.3	2.2
Sep. 3 March 1	*0.5*	*5.0*	*0.5*	*2.2*	*0.9*	*0.9*	*1.1*	*0.5*	*0.8*	*0.6*	*1.7*	*0.5*	*1.8*	*0.5*	*2.0*	*0.8*	*2.7*	*2.0*	*1.2*	*4.1*	*0.6*

ᶜXXI. S_r in γ Kaafjord.

0—2		2—4		4—6		6—8		8—10		10—12		12—14		14—16		16—18		18—20		20—22		22—24	
+	−	+	−	+	−	+	−	+	−	+	−	+	−	+	−	+	−	+	−	+	−	+	−
0	8.1	0	3.2	0	2.5	0	0	1.4	0	0.4	1.1	2.5	0	1.8	1.1	0.7	0.7	3.5	0	0	1.4	0	0.7
0	3.2	0	0	0	0.4	0	0.4	0	0.7	0	1.4	4.9	0.7	10.5	0	9.1	0	8.4	1.1	5.6	4.2	6.3	23.8
0	1.4	0	0.7	0	2.1	0	0.7	0	0	0.4	0.4	0	0.4	0.7	0.4	3.2	0	0.4	0	0	1.4	0	6.3
0	10.5	2.5	5.3	0.4	1.8	0	0.4	4.6	0.4	10.2	0.4	10.2	0.7	6.7	1.1	11.2	0.4	1.4	20.0	3.2	73.2	0	69.7
0	25.6	0	8.8	0	1.4	0	0	0	0.7	0.4	0.7	4.2	0	1.4	0.7	3.9	0	3.2	4.6	0	17.2	0	5.3
0	42.0	0	22.4	0	8.1	0	1.8	0.4	0	0.4	0.4	0.4	1.4	0	1.4	6.7	0.4	4.2	1.1	0.4	23.1	0	72.1
0	**15.1**	**0.4**	**6.7**	**0.7**	**2.7**	**0**	**0.6**	**1.1**	**0.3**	**2.0**	**0.7**	**3.7**	**0.5**	**3.5**	**0.8**	**5.8**	**0.3**	**3.5**	**4.5**	**1.5**	**20.1**	**1.1**	**29.7**
0	2.5	0	0	0.4	0	0.4	0	0.4	0	0.4	0.4	1.8	0	0.7	0	0	0	0.7	0	0	3.9	0	7.7
0	6.7	0	0.7	0	0	0.4	0.4	0	0	0.4	0.4	3.9	0	3.5	0	0.7	0.4	1.1	7.7	8.8	10.9	0	22.1
0	0	0	0	0	0	0	0	0	0	0.4	0	0	0	0	0	0.4	0	0	2.5	0	6.0	0	2.5
0	4.9	0	2.5	0	1.1	0	0	0	0	0.4	0	0	0	0	0	1.4	0	0.4	1.8	0	5.3	0	8.8
0	36.1	0	24.9	0	7.4	0.7	0.7	2.8	0	6.7	0.7	9.8	0	10.9	0.4	13.7	0	9.5	0.4	0	28.4	0.4	62.0
0	46.2	0	25.6	1.4	18.9	0.7	8.8	10.2	0	10.5	0.4	0.7	23.1	2.1	28.7	2.5	24.5	15.1	13.0	20.7	35.0	13.7	51.5
0	**16.1**	**0**	**9.0**	**0.3**	**4.6**	**0.4**	**1.7**	**2.2**	**0**	**3.1**	**0.3**	**2.7**	**3.9**	**2.9**	**4.9**	**3.1**	**4.2**	**4.5**	**4.2**	**4.9**	**14.9**	**2.4**	**25.8**
0	4.2	0	0	0	0	0	0	0	0	0	0	0	0	0.4	0	4.9	0	2.1	3.9	0	11.2	0	6.7
0	3.9	0.4	2.5	0	0.4	0	0	0	0	0	0	0	0	0	0	0	0	0.4	0	0.7	2.8	0	12.3
0	7.7	0	9.1	0	7.0	0	4.2	0	0.4	0	0	4.9	0	2.8	1.4	0	0	0	0	6.0	0	3.5	7.7
0	2.1	0	0	0	0	0	0	0	0	0	0	0	0	1.1	0	2.1	0	2.5	0	0.4	6.3	0	8.1
0	24.5	0	15.8	0	11.2	0	6.0	0	0	1.1	0	1.8	0.7	2.1	1.4	2.5	2.8	4.2	4.6	3.2	7.0	0.7	17.2
0	0	0	0	0	0	0	0	0	0	0	0	0	0	0	0	0	0	0	0	0	0	0	0
0	**7.1**	**0.1**	**4.6**	**0**	**3.1**	**0**	**1.7**	**0**	**0.1**	**0.2**	**0**	**1.1**	**0.1**	**1.1**	**0.5**	**1.6**	**0.5**	**1.5**	**1.4**	**1.7**	**4.6**	**0.7**	**8.7**
0	0.4	0		0	0		0		0	0	0		0	0		0.4	0	0	0	0	0	0	0.4
0	0	0		0	0		0		0	0	0		0	0		0	0	0	0	0	0	0	0
0	0	0	8	0	0		0		0	0	0		0	0		0	0	0	0	0	0	0	0
0	0	0	0	0	0	8	0		0	0	0	8	0	0.7		0.7	0	1.4	0.4	0	3.9	0	0.7
0.7	27.3	0	23.1	0	12.3	0	4.6	·4	0	3.5	0	11.9	0	13.3		10.2	0.7	3.9	13.7	0	34.3	0	42.4
0	11.6	0	0.7	0.4	0.4	1.1	0		0	0.4	0	2.1	0	0.7		3.9	0	3.9	0	1.4	1.8	0	10.5
0.1	**6.6**	**0**	**4.0**	**0.1**	**2.1**	**0.2**	**0.8**	**8.1**	**0**	**0.7**	**0**	**2.3**	**0**	**2.5**	**8**	**2.5**	**0.1**	**1.5**	**2.4**	**0.2**	**6.7**	**0**	**9.0**
0	0.4		4.2	2.8	0		2.1	0	1.4	0.4	0.7	0	4.9	0	2.5	1.8	0	4.9	0	2.5	3.2	0.4	1.4
0.4	0.7	8.4	0.7	0	0		0	0	0	0	0	0		3.2	1.1	6.0	0	1.8	0.7	1.1	9.8	0	4.6
0.4	3.9	1.1	0.4	0.4	0		0	0.4	0	1.1	0.4	2.1	8	4.9	0	1.8	0.4	6.0	4.2	2.1	2.8	0	13.0
0	1.4	0	0	0	1.4		0.7	0	0	0	0.4	6.0	0.7	8.4	1.8	14.0	0	8.4	0	1.8	2.8	0	5.3
0	7.0	0	0.7	0	1.1		0.7	0	0	0	0	1.4		3.5	0	6.0	0.4	5.6	1.8	1.1	8.1	0	13.0
0	6.7	0	0.4	0	0		0	0	0.4	0	0	1.8	8	3.5	0	8.8	0	7.4	18.9	0.7	28.7	0	50.1
0.1	**3.4**	**0.3**	**1.1**	**0.5**	**0.4**	**8**	**0.6**	**0.1**	**0.3**	**0.3**	**0.3**	**1.9**	**0.9**	**3.9**	**0.8**	**6.4**	**0.1**	**5.7**	**4.3**	**1.6**	**9.2**	**0.1**	**14.6**
0	0	0	0	0	0	0	0	3.2	0	0	1.4	0.4	0	1.1		5.3	0	0.7	0	0	0.7		1.8
0	13.7	0	9.1	0.4	14.0	0.4	2.1	0.4	1.4	2.5	0.4	1.4	0	11.2		16.8	0	6.0	26.6	0	56.7		58.1
0	34.3	0	5.6	1.1	0.7	1.4	0	0	0	1.1	0	0	0.7	1.1		3.9	0	11.6	1.8	2.1	6.0		44.5
0	1.1	0	0.4	0	0	0	0	0	0	1.4	0	2.1	0	16.8		6.3	1.4	1.1	0	0	1.4		1.1
0	0	0	7.4	0	12.6	0	6.3	0.4	1.8	1.4	0	0.7	0.7	1.1	7	2.5	0	0.4	0	0.7	0		0.7
0	1.8	0	0	0	4.9	0	6.3	0	0.4	2.1	0.4	2.1	0	1.4		10.5	0	8.8	0	0.4	2.8		23.8
0	**8.5**	**0**	**3.8**	**0.3**	**5.4**	**0.3**	**2.5**	**0.7**	**0.6**	**1.4**	**0.4**	**1.1**	**0.2**	**5.5**	**8.1**	**7.6**	**0.2**	**4.8**	**4.7**	**0.5**	**11.3**	**8**	**21.7**
0.4	16.8	0	3.9	0.4	1.1	0	0	0	0	0.7	0.4	0.4	0	14.0	0.7	3.5	0.7	7.0	1.1	4.2	9.8	0	23.1
0	25.6	0	23.1	0	18.6	2.8	6.3	2.1	0.7	1.8	0.4	11.2	0	11.6	0	13.3	2.1	4.9	17.2	0	42.4	2.1	25.2
0	*9.4*	*0.1*	*4.8*	*0.3*	*3.1*	*0.1*	*1.3*	*0.7*	*0.2*	*1.3*	*0.3*	*2.1*	*0.9*	*3.2*	*1.2*	*4.5*	*0.9*	*3.6*	*3.6*	*1.8*	*11.1*	*0.7*	*15.2*

Axeløen.

TABLE LXXXII.

S_{II} in γ

Hour	0—2		2—4		4—6		6—8		8—10		10—12		12—14		14—16		16—18		18—20		20—2
Period	+	—	+	—	+	—	+	—	+	—	+	—	+	—	+	—	+	—	+	—	+	-
Sept. 3—7	0.6	20.4	0	42.6	0	19.5	0.3	8.7	5.1	2.4	7.8	2.7	33.3	0	20.1	0	3.3	7.8	0.9	21.3	3.3	3
8—12	0	25.5	0.3	21.3	1.2	7.2	2.1	1.2	2.1	7.2	23.1	13.2	30.0	11.7	22.8	6.6	6.6	16.5	0.9	17.4	6.0	4
13—17	0.6	4.2	0	11.4	0.6	11.7	1.8	1.2	0.9	2.1	3.3	2.4	4.2	4.8	6.6	3.0	0.6	2.7	0.3	2.4	0.3	8
18—22	1.5	10.2	2.1	18.3	2.1	8.7	9.0	1.5	26.1	2.1	40.5	2.1	34.8	0.9	14.7	3.3	3.3	12.0	1.2	68.1	2.1	63
23—27	0	37.2	0	34.5	1.5	4.2	2.1	2.4	2.7	2.4	9.6	6.3	31.2	2.4	10.2	3.3	3.9	3.3	0	23.4	0.6	30.
Sept. 28 Oct. 2	1.8	30.3	0.3	29.1	0.9	12.6	1.2	4.5	1.8	0.9	3.9	1.2	2.4	3.0	7.5	7.5	2.1	20.4	0.9	30.3	2.4	21.
Mean value	**0.8**	**21.3**	**0.5**	**26.2**	**1.1**	**10.7**	**2.8**	**3.3**	**6.5**	**2.9**	**14.7**	**4.7**	**22.7**	**3.8**	**13.7**	**4.0**	**3.3**	**10.5**	**0.7**	**27.2**	**2.5**	**22.**
Oct. 3—7	0.9	5.1	3.0	3.9	1.5	1.5	0.9	3.9	0.3	2.4	7.5	2.4	4.2	1.2	1.8	1.5	0.9	0.3	0	5.4	0.6	6.
8—12		6.3	0	8.7	0.3	1.8	0.9	1.2	2.4	2.7	6.6	1.2	14.7	2.4	7.2	1.5	0.6	10.8	0.3	20.4	1.2	16.
13—17	.3	10.5	0	6.3	0.6	2.1	0.3	1.8	0.3	1.5	2.4	0.6	1.8	1.2	0.6	1.8	0.3	7.2	0.6	35.7	0.3	12.
18—22		18.9	0	27.6	0.9	13.2	1.8	1.5	0.9	0.6	2.7	0.6	5.1	0.6	2.4	0.6	0	3.6	0	11.7	0.3	12.
23—27		42.0	0.3	39.3	0.6	18.6	3.6	3.9	11.7	0.6	15.9	0.9	18.6	2.1	7.8	24.3	1.8	52.8	0	28.8	1.8	27.
Oct. 28 Nov. 1	8	59.1	0.6	69.0	2.7	40.2	6.0	4.2	6.9	10.2	7.5	17.4	3.9	50.4	0.9	37.2	0	55.8	0	55.5	0.6	50.
Mean value	**0.2**	**23.7**	**0.7**	**25.8**	**1.1**	**12.9**	**2.3**	**2.8**	**3.8**	**3.0**	**7.1**	**3.9**	**8.1**	**9.7**	**3.5**	**11.2**	**0.6**	**21.8**	**0.2**	**26.3**	**0.8**	**21.**
Nov. 2—6	0	7.2	0.3	3.0	0.3	1.8	1.5	0	0.3	0.6	0.3	0.9	1.8	1.2	0.6	2.1	0.3	10.2	0.3	45.9	0.6	20.
7—11	0.6	11.4	1.5	11.7	0.6	2.7	0.3	0.6	0.9	0.9	0.6	3.3	0.9	1.8	0.9	1.2	0.3	2.7	0.3	0.6	0.3	5.
12—16	1.2	12.6	0	30.0	0	17.7	4.8	8.4	11.4	4.8	10.8	5.4	12.9	2.4	8.1	2.1	0.6	5.4	0	14.1	1.2	15.
17—21	0	14.7	1.2	7.2	1.2	1.8	0.9	2.7	2.1	1.5	2.4	2.7	6.6	0.3	7.2	0.3	1.2	17.4		29.7	0.6	42.
22—26	0	50.4	2.7	81.3	9.3	33.9	12.6	12.6	19.2	5.4	6.9	13.2	5.4	24.6	2.4	63.3	0	128.7		85.2	1.2	78.
Nov. 27 Dec. 1	0.6	11.7	1.5	6.0	1.5	9.6	2.7	6.0	2.4	3.3	6.6	1.5	11.7	0.6	3.6	2.4	0.3	29.7		27.9	0	22.
Mean Value	**0.4**	**18.0**	**1.2**	**23.2**	**2.2**	**11.3**	**3.8**	**5.1**	**6.1**	**2.3**	**4.6**	**4.5**	**6.6**	**5.2**	**3.8**	**12.0**	**0.5**	**32.9**	**0.1**	**33.9**	**0.7**	**31.**
Dec. 2—6	0.6	18.0	0.3	13.5	0.3	7.2	5.4	1.2	8.7	1.8	3.0	0.3	9.9	0.3	4.8	1.5	0.3	21.0		22.5	0	14.4
7—11	0	10.2	0	14.4	3.0	3.9	1.5	2.1	4.2	1.8	9.3	0.6	13.8	0	10.5	1.5	1.8	17.1		30.6	0	71.4
12—16	0.3	16.8	2.1	15.3	0.9	2.1	1.8	2.4	2.4	3.9	1.2	4.5	0.9	3.3	1.8	1.8	0	45.6		51.0	2.1	29.7
17—21	0	4.8	0.6	4.2	1.5	2.7	3.6	0.3	2.4	0.9	1.5	2.7	3.3	0.9	3.3	1.5	3.0	2.4		11.4	0.3	8.4
22—26	0.3	23.4	1.2	63.6	1.8	33.6	8.1	9.6	19.8	1.8	9.0	6.0	6.0	25.8	12.0	12.0	1.2	39.6	3	50.7	0	46.5
27—31	1.2	22.5	0	12.0	4.5	11.7	5.4	1.2	4.2	1.2	5.1	3.3	6.6	1.5	1.5	0.6	0.3	9.6		14.4	0.6	21.6
Mean Value	**0.4**	**16.0**	**0.7**	**20.5**	**2.0**	**10.2**	**4.3**	**2.8**	**7.0**	**1.9**	**4.9**	**2.9**	**6.8**	**5.3**	**5.7**	**3.2**	**1.1**	**22.6**	**0.1**	**30.1**	**0.5**	**32.0**
Jan. 1—5	0.6	15.3	2.7	37.5	1.5	22.5	3.3	6.9	6.3	1.8	2.1	5.4	2.4	2.1	1.8	3.0	1.2	16.5	0.3	21.6	0	21.3
6—10	1.5	10.8	0	28.8	1.2	7.8	3.3	2.1	4.5	1.2	9.3	0.3	14.1	0.3	13.2	4.5	6.9	4.2	0.6	21.6	0.6	29.4
11—15	0	21.9	0.3	10.2	1.2	4.5	1.2	2.4	4.5	0.6	3.0	0.3	9.6	0.3	15.9	2.7	1.2	12.0	0	68.1	0	22.5
16—20	0.3	5.4	0.3	13.8	2.4	11.4	3.0	3.3	8.7	0	4.5	3.9	12.9	3.6	9.0	11.4	1.8	20.7	0	27.9	0	18.9
21—25	0.9	20.7	0.6	7.2	1.5	5.4	3.9	0.9	4.2	1.8	3.0	5.7	7.5	1.2	4.5	3.6	1.5	12.6	0	31.5	1.5	33.9
26—30	0	46.2	0.6	24.3	0	15.9	0.9	3.9	9.6	0.9	15.9	1.2	25.2	0.3	9.9	0.6	0.6	18.9	0	38.4	4.8	14.1
Mean Value	**0.6**	**20.1**	**0.8**	**20.3**	**1.3**	**11.3**	**2.6**	**3.3**	**6.3**	**1.1**	**6.3**	**2.8**	**12.0**	**1.3**	**9.1**	**4.3**	**2.2**	**14.2**	**0.2**	**34.9**	**1.2**	**23.4**
Jan. 31 Febr. 4	0.6	3.3	0.6	8.1	1.5	4.2	2.1	3.0	0.6	3.6	2.1	4.8	3.3	3.3	4.5	2.4	1.8	9.6	0	9.6	0.9	2.7
Febr. 5—9	1.5	3.3	1.2	27.3	1.2	23.1	1.5	18.6	7.8	4.8	10.8	4.2	13.8	0.3	12.0	12.3	4.5	21.6	1.2	64.2	3.3	46.5
10—14	0.3	30.6	0.3	26.4	1.2	19.2	6.9	1.5	3.9	2.7	8.7	1.5	8.7	0.6	12.9	0.6	2.4	6.0	0	33.3	0	19.8
15—19	0.9	4.2	0.3	6.3	3.0	0.3	2.1	0	4.2	0.6	5.4	3.3	10.5	3.0	2.7	13.5	1.5	44.7	1.8	20.7	1.5	6.0
20—24	0.6	6.0	0.6	9.0	0.9	18.9	0.6	21.0	1.2	2.7	10.5	0	22.8	0.3	15.6	0	3.9	4.5	0.3	3.3	0.9	1.5
Feb. 25 Mar. 1	0	4.8	0	8.1	0.6	25.5	1.8	13.5	5.7	0	9.9	3.3	18.6	0.9	7.8	0.9	0.6	6.9	0	17.4	0	7.2
Mean Value	**0.7**	**8.7**	**0.5**	**14.2**	**1.4**	**15.2**	**2.5**	**9.6**	**3.9**	**2.4**	**7.9**	**2.9**	**13.0**	**1.4**	**9.3**	**5.0**	**2.5**	**15.6**	**0.6**	**24.8**	**1.1**	**14.0**
March 2—6	0.3	24.9	0	22.8	1.5	21.3	1.5	6.3	1.8	3.3	8.7	1.8	9.0	0.3	5.4	13.2	1.2	14.7	0	24.9	0.6	20.1
7—11	0.3	22.8	0	57.6	0	48.6	2.7	10.2	12.3	0.9	12.9	0.3	22.2	0.3	12.9	5.1	0	39.6	0.6	81.6	1.2	55.5
12—16	0.9	22.5	0.3	32.1	0.9	45.3	5.7	7.8	12.6	2.1	12.0	4.8	31.2	0	13.8	12.6	3.3	51.0	0.3	27.0	3.0	12.9
17—21	0	14.1	0	12.6	0	16.8	0.3	3.6	3.0	2.1	6.9	5.4	2.1	2.7	2.7	4.5	0	12.0	0	22.2	1.5	22.8
22—26	0.3	8.4	0.6	10.8	0.3	5.1	1.5	1.2	7.5	0.9	3.6	6.3	3.0	3.3	0.9	5.1	1.5	5.7	0.6	4.5	2.4	4.2
27—31	0.3	27.9	0	26.4	0.6	9.0	1.5	5.4	6.6	4.8	7.8	5.1	7.2	2.4	8.7	2.1	0	24.6	0.3	21.3	0.3	24.9
Mean Value	**0.4**	**20.1**	**0.2**	**27.1**	**0.6**	**24.4**	**2.2**	**5.8**	**7.3**	**2.4**	**8.7**	**4.0**	**12.5**	**1.5**	**7.4**	**7.1**	**1.0**	**24.6**	**0.3**	**30.3**	**1.5**	**23.3**

TABLE LXXXIII (continued). S_H in γ. **Axeløen.**

Hour	0–2		2–4		4–6		6–8		8–10		10–12		12–14		14–16		16–18		18–20		20–22		22–24	
Period	+	−	+	−	+	−	+	−	+	−	+	−	+	−	+	−	+	−	+	−	+	−	+	−
April 1–5	0.3	31.2	0	59.4	3.0	29.4	2.7	11.7	3.0	5.1	7.2	13.2	22.8	3.6	10.8	19.2	6.9	11.1	0.6	28.8	0.6	24.0	2.1	40.2
6–10	4.2	37.5	10.2	23.7	5.1	45.9	18.9	6.3	32.4	2.7	7.2	30.3	15.9	54.9	17.4	16.5	3.3	36.9	1.5	33.0	0.6	42.3	2.4	47.7
11–15	0.6	20.4	2.7	31.8	8.4	3.9	8.7	1.5	14.4	4.2	26.7	8.7	28.5	9.3	11.4	12.3	2.4	17.7	0.9	27.6	1.2	26.7	1.5	15.0
16–20	6.0	4.8	4.2	5.4	2.4	1.8	1.5	3.3	3.0	2.1	8.7	6.3	10.2	20.4	3.3	18.2	1.5	49.8	3.3	36.9	1.8	13.5	2.7	5.1
21–25	1.5	8.7	1.8	14.1	2.1	7.2	6.6	0.6	2.7	3.0	12.0	22.2	13.2	18.3	10.5	10.5	4.5	12.3	0.6	25.8	0.9	9.9	2.4	9.9
26–30	0.3	31.8	0.9	26.4	0.9	19.8	1.5	5.7	5.4	3.9	7.2	23.4	15.6	14.1	21.6	6.6	11.7	16.5	0.6	24.0	0.3	32.4	1.8	32.7
Mean value	**2.2**	**22.4**	**3.3**	**26.6**	**3.7**	**18.0**	**6.7**	**4.9**	**10.2**	**3.5**	**11.5**	**17.4**	**17.7**	**20.1**	**12.5**	**13.9**	**5.1**	**24.1**	**1.3**	**29.4**	**0.9**	**24.8**	**2.2**	**25.1**
May 1–5	4.2	11.4	2.1	56.7	0.6	34.8	0.9	14.1	3.9	4.5	3.9	9.6	12.0	4.2	18.6	1.5	8.7	4.8	3.3	12.9	4.5	23.4	6.9	25.2
6–10	(3.0)	(30.0)	(3.0)	(45.0)	(7.5)	(54.0)	(0)	(45.0)	(7.5)	(9.0)	(6.0)	(7.5)	(30.0)	(3.0)	(45.0)	(3.0)	(9.0)	(6.0)	(1.5)	(30.0)	(1.5)	(36.0)	(3.0)	(15.0)
11–16	4.8	17.1	2.7	32.4	0.3	26.1	0.9	15.6	0.3	12.9	12.9	2.1	22.5	0.3	21.3	8.4	5.1	11.7	0.3	36.0	0.6	40.5	1.8	22.2
17–20	10.2	1.5	6.3	6.6	0.3	13.5	1.2	5.7	5.4	1.5	10.8	8.1	28.5	2.4	29.4	1.5	14.7	1.8	6.3	5.7	6.3	2.7	6.0	7.2
21–25	5.4	10.5	3.0	39.9	1.5	22.8	2.1	12.3	22.8	1.5	33.9	7.5	36.6	1.8	52.2	0	18.0	3.9	3.6	25.8	1.8	24.9	4.8	21.0
26–30	1.8	30.9	0	62.7	4.5	64.2	7.2	30.0	15.6	1.8	64.2	0.6	67.8	0.6	36.0	1.5	29.7	2.7	14.7	20.4	3.0	29.7	2.7	25.5
Mean value	**4.9**	**16.9**	**2.9**	**40.6**	**2.5**	**35.9**	**2.1**	**20.5**	**9.3**	**5.2**	**22.0**	**5.9**	**32.9**	**2.1**	**33.8**	**2.7**	**14.2**	**5.2**	**5.0**	**21.8**	**3.0**	**26.2**	**4.2**	**19.4**
Sep. 3 May 30	*1.2*	*18.6*	*1.2*	*24.9*	*1.7*	*16.6*	*3.2*	*6.4*	*6.7*	*2.7*	*9.7*	*5.4*	*14.7*	*5.6*	*10.9*	*7.0*	*3.4*	*19.0*	*0.9*	*28.7*	*1.3*	*24.2*	*1.5*	*20.2*

TABLE LXXXIII. S_D in γ Axeloen.

Hour	0–2		2–4		4–6		6–8		8–10		10–12		12–14		14–16		16–18		18–20		20–22		22–24	
Period	+	−	+	−	+	−	+	−	+	−	+	−	+	−	+	−	+	−	+	−	+	−	+	−
Sep. 3–7	0.7	25.2	0.2	41.8	0	18.7	1.1	5.8	3.2	0.5	4.1	0.2	7.4	0	6.7	0.7	11.5	0.4	18.0	0.5	8.3	0.5	5.9	1.3
8–12	0.4	7.6	0.9	10.8	1.1	5.9	2.0	1.3	2.7	0.9	5.4	0	9.0	2.2	6.1	0.2	13.3	0	18.0	0.2	9.5	1.1	3.2	7.7
13–17	1.3	2.3	0.7	8.5	0.2	9.0	1.3	1.8	0.9	0.2	1.3	0.4	1.3	1.3	2.3	2.0	2.5	1.1	4.5	0.4	2.7	2.3	2.7	5.4
18–22	2.0	14.8	2.2	20.2	1.4	13.3	3.2	2.9	1.8	4.7	5.4	2.0	6.5	0.9	7.6	2.3	8.3	3.4	19.8	10.3	8.3	20.7	1.4	16.0
23–27	0.5	17.8	0.7	19.3	0.5	8.8	1.3	3.4	1.1	0.9	2.7	0.9	3.8	2.2	4.3	1.8	6.7	0.4	11.3	0.7	4.1	3.6	2.9	6.1
Sep. 28 Oct. 2	2.7	7.6	1.4	8.8	0.7	4.1	1.6	1.1	1.4	0.5	1.8	1.4	0.7	2.9	3.1	3.1	3.6	1.4	5.0	1.6	4.1	3.6	1.6	13.5
Mean value	**1.3**	**12.6**	**1.0**	**18.2**	**0.7**	**10.0**	**1.8**	**2.7**	**1.9**	**1.3**	**3.5**	**0.8**	**4.8**	**1.6**	**5.0**	**1.7**	**7.7**	**1.1**	**12.8**	**2.3**	**6.2**	**5.3**	**3.0**	**8.3**
Oct. 3–7	0.9	5.6	2.2	3.4	1.8	3.4	2.3	1.4	0.5	1.1	2.7	0.5	0.7	0.9	0.7	0.5	0.5	0.7	2.2	0.4	1.4	2.9	0.9	6.3
8–12	0.4	5.9	0.5	6.8	0.9	1.8	2.9	0.7	1.3	1.6	0.7	2.0	5.2	1.8	2.3	1.6	10.6	0.5	14.6	0.4	5.6	6.3	1.4	8.1
13–17	0.4	5.8	0.2	5.0	0.2	2.9	1.4	0.9	1.1	0.2	2.2	0.2	0	0.7	1.3	0	5.9	0.2	11.5	0.4	7.4	1.4	2.5	3.8
18–22	0.5	12.6	0.5	21.1	1.4	7.2	4.7	0.7	1.8	0.5	1.3	1.1	0.2	1.3	0.4	1.3	2.3	0.4	7.7	0.5	2.5	4.1	1.4	8.3
23–27	0.2	29.7	0.2	32.0	1.1	11.2	5.6	0.7	1.8	2.3	0.7	4.0	4.5	1.3	11.2	0.5	20.7	0.5	23.8	0.5	9.5	4.5	2.7	19.4
Oct. 28 Nov. 1	0	35.3	0.9	52.2	1.1	21.6	5.4	9.0	3.2	22.7	2.0	10.6	2.9	4.5	8.1	1.3	25.6	1.6	30.2	2.0	13.0	11.7	2.7	20.3
Mean value	**0.4**	**15.8**	**0.8**	**20.1**	**1.1**	**8.0**	**3.7**	**2.2**	**1.6**	**4.7**	**1.6**	**3.1**	**2.3**	**1.8**	**4.0**	**0.9**	**10.9**	**0.7**	**15.0**	**0.7**	**6.6**	**5.2**	**1.9**	**11.0**
Nov. 2–6	0.4	1.8	1.1	0.5	0.2	1.3	0.4	0.7	0.4	0	0.5	0.4	1.3	0.9	3.2	0.2	12.1	0	13.9	10.6	5.0	2.5	0.4	5.6
7–11	0	11.0	0.5	10.1	0.5	2.3	0.4	0.2	0.2	0	0.7	0.2	0.2	0.7	0.5	0.5	0	1.8	0.7	0.5	0.5	2.2	0	7.0
12–16	0.2	11.0	0	28.4	0.2	19.8	0.5	3.4	5.4	0.4	2.7	0.7	4.7	1.6	2.9	1.8	2.0	1.6	2.9	2.0	2.2	2.7	1.1	11.2
17–21	0.2	14.9	0.2	10.4	0.5	4.1	0.5	2.7	0.4	0.9	1.1	0.9	2.3	0.4	3.2	0.9	6.5	1.4	11.3	1.1	2.2	12.2	0.9	14.2
22–26	0	40.5	0.2	48.6	1.4	27.5	1.3	17.1	1.6	9.0	0.9	13.3	8.3	1.6	14.4	2.0	32.6	7.9	25.0	7.6	13.0	22.9	1.6	47.5
Nov. 27 Dec. 1	0.5	4.5	1.1	1.6	2.0	4.1	1.8	2.0	0.9	1.3	1.3	1.1	0.9	2.0	0.2	2.7	2.0	2.3	4.3	2.2	2.5	4.9	0.7	6.1
Mean value	**0.2**	**14.0**	**0.5**	**16.6**	**0.8**	**9.9**	**0.8**	**4.4**	**1.5**	**1.9**	**1.2**	**2.8**	**3.0**	**1.2**	**4.1**	**1.4**	**9.2**	**2.5**	**9.7**	**4.0**	**4.2**	**7.9**	**0.8**	**15.3**
Dec. 2–6	0.2	11.2	0.2	10.4	0.5	8.1	1.6	0.7	1.4	0.5	3.8	0	5.2	0.2	2.3	0.4	3.6	0.9	3.4	2.3	0.9	2.9	0.2	6.5
7–11	0.4	9.0	0.9	10.1	2.2	1.4	2.5	0.9	0.9	1.6	4.3	0	2.5	0.4	1.4	1.6	2.3	0.9	5.0	11.5	4.5	17.5	0.5	17.3
12–16	0.4	17.8	1.1	9.4	0.7	1.4	2.0	0.5	0.7	0.4	0.5	1.6	0.7	2.3	1.8	0.5	6.8	1.3	2.7	13.3	3.1	6.7	0.7	6.3
17–21	1.1	2.5	2.0	2.5	2.9	0.9	3.2	0	1.6	0.2	0.9	0	1.4	0.2	1.6	0.5	0.9	0.5	2.2	0.9	1.3	1.4	0.2	2.9
22–26	0.7	15.7	1.1	28.3	2.2	14.2	4.9	4.7	5.6	1.1	2.5	3.4	4.1	1.3	9.9	2.0	10.3	5.4	15.7	2.9	7.2	2.7	0	14.4
27–31	1.3	8.5	3.6	5.0	2.3	8.1	5.9	0.4	3.1	1.1	0.9	0	0.9	0.2	8.5	0	2.2	0.9	4.3	1.3	3.4	3.4	2.2	8.1
Mean value	**0.7**	**10.8**	**1.5**	**11.0**	**1.8**	**5.7**	**3.4**	**1.2**	**2.2**	**0.8**	**2.2**	**0.8**	**2.5**	**0.8**	**4.3**	**0.8**	**4.4**	**1.7**	**5.6**	**5.4**	**3.4**	**5.8**	**0.6**	**9.3**

Hour	0—2	2—4	4—6	6—8	8—10	10—12	12—14	14—16	16—18	18—20
[illegible]	[illegible]									
Mean value	[illegible]									
[illegible]	[illegible]									
Mean value	[illegible]									
[illegible]	[illegible]									
Mean value	[illegible]									
[illegible]	[illegible]									
Mean value	[illegible]									

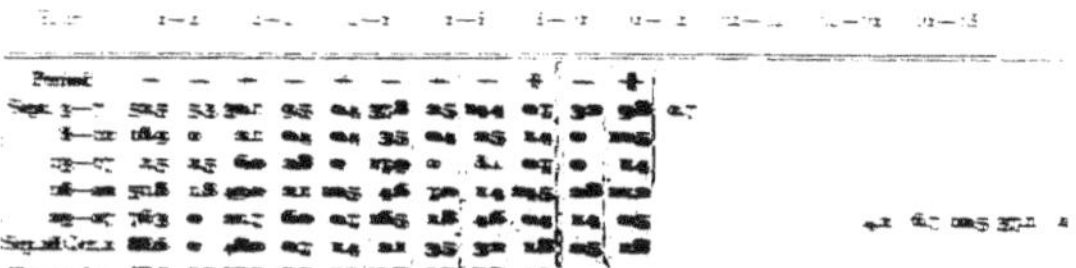

IV (continued). S_{γ} in γ Axeloen.

−2	2−4		4−6		6−8		8−10		10−12		12−14		14−16		16−18		18−20		20−22		22−24	
−	+	−	+				+	−	+	−	+	−	+	−	+	−	+	−	+	−	+	−
0	1.1	0.4	1.8	3.5	0.7	2.5	0.4	0.7	0.4	1.4	0	2.8	0.4	3.9	0	0	4.6	0.4	22.8	0	26.6	0
0	0.4	0	0	0	1.8	0	0	0.4	0	1.1	0	15.1	0.4	16.1	0	31.5	5.6	6.3	62.0	0	45.5	0.4
0	0	0.4	0	1.8	0	0	0	0.7	1.1	1.4	0	1.8	0	0.7	1.4	0.4	18.9	4.9	34.7	2.5	20.3	1.1
0.4	6.0	1.8	0.7	32.2	0	18.9	0.7	4.6	0.7	4.2	0.4	0.7	0	0	0	1.4	11.6	3.9	31.9	0.4	31.5	0.4
0.4	54.3	0.7	3.5	13.7	0.7	32.9	1.1	4.2	0.4	2.1	0	31.2	2.8	30.5	4.2	24.9	22.1	2.8	85.4	0	141.4	0
0	91.0	1.1	48.0	9.5	14.4	5.3	1.1	19.3	0.7	24.2	14.0	3.9	5.3	11.2	26.6	13.0	97.0	3.9	156.1	0	160.0	0
0.1	**25.5**	**0.7**			**2.9**	**9.9**	**0.6**	**5.0**	**0.6**	**5.7**	**2.4**	**9.3**	**1.5**	**10.4**	**5.4**	**11.9**	**26.6**	**3.7**	**65.5**	**0.5**	**70.9**	**0.3**
1.1	0	2.1			0	0	0.4	0.7	0	0.4	0	0	0	1.1	1.4	2.8	50.4	3.9	78.1	0	32.9	0
0	13.0	0.4			0.4	0	0	0.7	0	0.4	0	0.7	0	1.1	0.7	0.7	1.4	0.4	14.4	0	35.4	0
1.1	23.5	9.5	0.4	42.4	0.7	31.2	4.6	8.8	7.0	2.5	3.2	21.0	1.1	3.9	0.7	8.1	16.5	4.6	46.9	0	63.7	0
0	4.2	2.8	0	9.5	0.4	6.3	0.7	3.2	0.7	0.4	1.8	0.7	0	9.1	0.4	16.5	16.1	5.3	67.6	0.4	62.3	0.4
0	113.8	3.2	49.4	1.1	11.2	3.9	4.9	20.0	2.5	33.3	3.5	13.3	7.7	22.8	72.8	28.0	143.9	0.7	126.6	8.1	196.4	3.9
0	1.4	0	1.8	1.4	1.4	4.6	1.4	0	8.8	1.4	6.0	6.3	1.4	4.9	7.7	4.9	20.0	18.2	28.7	0.4	10.5	0.4
0.4	**26.0**	**3.0**	**8.7**	**9.3**	**2.4**	**7.7**	**2.0**	**5.8**	**3.2**	**6.4**	**2.4**	**7.0**	**1.7**	**7.2**	**14.8**	**10.2**	**41.4**	**5.5**	**70.4**	**1.5**	**66.9**	**0.8**
0	1.4	3.2	0.7	11.6	6.7	1.8	8.4	0	0	2.5	0.4	2.1	0	2.5	1.8	7.0	10.5	1.4	14.4	0	15.4	0
0	10.9	0	1.8	7.4	1.1	0.7	2.8	0	1.4	0	0.4	2.5	0	2.1	0.7	13.7	11.2	11.2	65.5	12.3	46.9	0.4
0.7	22.8				0.4	0	1.4	0	0	4.2	0	0	0.4	2.5	9.1	7.0	23.1	0.4	28.4	0.4	22.8	1.1
0	0				0	0.4	0	0.4	0	1.4	0	6.0	0	18.2	0	10.9	6.0	7.4	30.1	1.1	11.6	0.4
0.4	27.3				1.1	26.6	2.8	39.6	0.4	39.2	1.1	42.4	0.4	69.7	3.9	76.4	22.1	30.8	85.4	1.4	60.9	0
0	6.3	0	1.1	6.3	0	10.5	0.4	1.4	0	4.6	0	6.3	0	1.8	0.7	0.7	2.5	1.4	11.6	1.1	26.3	0
0.2	**11.5**	**2.1**	**0.7**	**9.1**	**1.6**	**6.7**	**2.6**	**6.9**	**0.3**	**8.7**	**0.3**	**9.9**	**0.1**	**16.1**	**2.7**	**12.6**	**12.6**	**8.8**	**30.2**	**2.7**	**30.7**	**0.3**
0	9.5	1.4	0	17.5	1.1	4.9	2.5	6.0	2.1	2.8	0	9.8	0	15.4	0.4	7.4	5.3	2.1	21.7	0.4	10.2	0
0.4	10.9	1.4	0	0	0	0.7	0	0	2.5	0	0.7	4.2	0	15.4	0	16.5	8.1	1.1	28.0	2.5	25.6	0
0		0	0.4	0	0	0	0	0	0	0	2.5	0	0.4	0.4	0	12.3	13.7	14.7	16.8	0.7	31.5	1.8
1.8		0	2.1	9.1	2.5	6.3	10.5	0.4	0.4	3.2	0	22.1	3.9	13.0	6.3	23.1	7.7	3.5	18.6	3.9	7.4	3.5
0.7		0.4	0	7.7	2.8	4.9	6.7	1.4	2.8	9.5	3.9	27.0	1.4	47.6	3.2	11.9	20.0	2.8	28.7	4.6	26.3	0
0	53.6	0	23.1	0	0.7	0	5.3	0	0	7.0	1.1	18.6	5.6	16.1	1.8	8.8	39.6	3.5	55.3	0.4	136.5	0
0.5	**16.9**	**0.5**	**4.3**	**5.7**	**1.2**	**2.8**	**4.2**	**1.3**	**1.3**	**3.8**	**1.4**	**13.6**	**1.9**	**18.8**	**2.8**	**13.3**	**15.7**	**4.6**	**28.2**	**2.1**	**30.6**	**0.9**
1.8	1.4	0	0	0	0	0	0.4	0	3.9	0	2.1	0	0	4.2	0	9.8	0	3.2	1.1	3.5	7.7	1.1
0	93.8	0.4	18.9	3.5	9.1	0.7	1.8	1.1	0	10.2	0	2.5	0	15.1	2.1	19.3	29.1	4.9	74.9	0	75.3	0
0.4	9.5	0.4	0	4.2	0.7	1.1	1.1	0	1.8	0	2.8	0.4	0	5.6	0	3.5	4.9	3.2	18.6	0	44.5	3.5
0	5.8	0	0	0	0	0	0	0	0	0	0.4	1.4	0.4	10.9	4.2	15.4	6.3	7.4	6.7	0.4	5.3	0
1.1	65.9	0	37.1	7.4	27.3	0.7	3.9	4.9	1.1	13.3	2.1	6.7	0.4	8.1	0	19.6	0	6.7	2.1	0.4	6.3	0
0	8.4	0	13.0	0	11.9	0	0.7	1.1	3.9	4.9	5.6	3.2	1.4	2.5	0	7.4	0	7.0	6.7	0	31.5	0
0.6	**12.4**	**0.1**	**11.5**	**2.5**	**8.2**	**0.4**	**1.3**	**1.2**	**1.8**	**4.7**	**2.2**	**2.4**	**0.4**	**7.7**	**1.1**	**12.5**	**6.7**	**5.4**	**18.4**	**0.7**	**28.4**	**0.8**
0	39.4	0	0	14.0	0.4	8.8	1.8	2.1	0	5.6	0	28.7	0	64.8	0	65.5	5.3	30.5	37.8	13.0	50.4	3.9
1.8	96.3	0.7	29.4	8.4	2.5	23.1	1.4	18.6	0	8.8	0.4	19.3	1.1	22.1	11.9	16.5	71.4	8.1	137.9	1.1	129.5	0
0.7	60.0	0	44.1	1.1	1.4	8.8	1.1	7.0	0	4.6	1.1	13.7	0.4	45.5	5.3	69.0	21.0	20.3	34.7	1.4	70.4	3.2
1.8	6.0	7.7	0.4	8.4	0	11.6	0	6.3	0.7	1.8	0.4	7.4	0	9.1	0	7.4	17.5	1.8	57.1	0	73.5	0
0	6.0	0.4	0	0	0	0	0	0	0	1.8	0	1.1	0	3.2	0	4.9	1.4	1.4	18.9	0	72.1	0
0	31.5	0	1.4	1.1	0	1.4	2.1	0.4	1.1	0	0	0.7	0	17.9	0	33.6	4.6	6.0	35.4	0	25.6	0
0.7	**35.2**	**1.5**	**12.8**	**5.5**	**0.7**	**9.8**	**1.1**	**5.7**	**0.3**	**3.8**	**0.3**	**11.8**	**0.3**	**27.1**	**2.9**	**32.8**	**28.2**	**11.4**	**53.6**	**2.6**	**70.3**	**1.2**
0	65.8	0.7	12.6	9.1	0	33.3	0	28.7	10.9	9.5	1.1	17.5	3.5	36.1	1.4	21.0	22.4	9.8	13.2	1.4	62.7	2.8
1.4	110.6	0.4	72.8	1.4	8.1	11.9	4.9	13.0	1.8	52.9	0.7	76.7	0	106.1	17.2	46.9	51.1	23.1	62.3	14.0	129.5	8.1
0	38.9	0	1.8	0.4	0.7	5.6	0.4	7.7	0	24.9	0	44.5	0	78.8	0	82.6	0.7	36.1	34.7	1.4	10.6	1.1
0.4	2.5	0	0.4	0.4	1.1	0.7	0.4	0	0.4	14.0	0	30.8	0	70.0	0	78.8	1.4	23.5	6.0	4.2	3.2	0.7
0.4	14.4	0.7	0.4	1.1	0	0	0	4.2	6.7	10.2	15.8	7.7	6.0	13.3	0	37.5	0.4	32.2	9.1	10.9	18.9	2.1
1.1	84.7	1.8	19.3	0.7	0.7	2.5	0.7	0.4	0.7	0	0.4	7.0	1.4	19.3	0.7	32.9	2.5	17.5	68.3	0	96.3	0
0.6	**52.8**	**0.6**	**16.7**	**2.2**	**1.8**	**9.8**	**1.1**	**9.8**	**3.4**	**18.6**	**3.8**	**30.7**	**1.8**	**53.9**	**3.2**	**58.8**	**13.1**	**23.7**	**32.9**	**5.3**	**58.5**	**2.5**

TABLE LXXXIV (continued). S_V in γ Axeløen.

Hour	0–2		2–4		4–6		6–8		8–10		10–12		12–14		14–16		16–18		18–20		20–22		22–24	
Period	+	−	+	−	+	−	+	−	+	−	+	−	+	−	+	−	+	−	+	−	+	−	+	−
May 1–5	95.2	0.7	112.0	0	48.3	1.1	3.5	2.5	0.7	2.5	0	18.2	0.4	35.4	0.4	5.6	2.5	6.3	8.8	14.7	64.1	6.0	96.3	3.9
6–10	46.9	0	48.7	1.4	13.0	7.7	4.9	7.0	7.7	14.7	6.7	25.6	1.4	6.7	0.4	25.2	0	43.8	2.5	31.9	38.5	7.7	58.8	0.4
11–15	29.1	0.4	20.0	0	2.1	2.1	2.8	4.2	0.4	2.8	0.4	2.1	0.7	9.1	1.4	15.8	0	56.0	11.2	30.1	59.2	0	38.2	0.4
16–20	40.6	0.4	37.1	0.4	21.4	0.4	4.2	2.8	6.7	0	0.4	0.7	0	8.4	0	30.8	0.4	35.0	6.7	15.8	9.5	4.2	39.9	0.4
21–25	81.2	0	50.1	0	34.3	0.4	7.0	7.0	1.8	3.2	0.4	16.8	0	70.7	0	72.1	1.1	53.6	13.3	25.6	24.5	14.4	93.1	3.2
26–30	108.5	0	99.8	0.7	50.5	7.0	1.8	18.2	1.1	20.7	2.5	43.4	2.1	58.5	8.1	18.2	1.1	24.5	27.0	11.6	54.6	3.9	89.6	0.4
Mean value	**66.9**	**0.3**	**61.3**	**0.4**	**29.8**	**3.1**	**4.0**	**7.0**	**3.1**	**7.3**	**1.7**	**17.8**	**0.8**	**31.5**	**1.7**	**28.0**	**0.9**	**36.5**	**11.6**	**21.6**	**41.7**	**6.0**	**69.3**	**1.5**
Sep. 3 May 30	*43.8*	*0.5*	*29.7*	*1.4*	*10.6*	*6.8*	*2.8*	*6.5*	*2.3*	*4.9*	*2.1*	*8.3*	*1.9*	*13.4*	*2.1*	*19.0*	*4.4*	*21.4*	*19.7*	*9.9*	*46.7*	*2.5*	*55.3*	*1.0*

Dyrafjord.

TABLE LXXXV. S_H in γ

Hour	0–2		2–4		4–6		6–8		8–10		10–12		12–14		14–16		16–18		18–20		20–22		22–24	
Period	+	−	+	−	+	−	+	−	+	−	+	−	+	−	+	−	+	−	+	−	+	−	+	−
Nov. 23–26	0	105.0	0	141.0	3.0	63.0	0	39.0	0	12.0	0	3.0	3.0	3.0	27.0	0	66.0	12.0	45.0	33.0	12.0	54.0	0	78.0
Nov. 27 Dec. 1	0.3	16.2	0	1.2	0.3	9.9	3.3	0.6	3.9	0	0.6	0.6	1.2	1.5	2.1	0.9	4.8	0.3	2.1	0.6	8.1	0.3	0.6	8.7
Dec. 2–6	4.8	16.8	0.6	7.2	1.2	3.3	2.1	0.3	0	1.5	0.3	0.6	0.9	0.6	3.0	1.2	8.1	0.3	7.2	0	5.7	0	2.4	9.9
7–11	0.3	27.6	0.3	26.7	0.3	1.8	0.3	0.3	0.3	4.5	1.5	2.7	0.6	2.7	2.1	1.2	15.9	0.9	13.8	0	10.8	1.2	7.2	18.6
12–16	2.4	50.1	2.7	7.8	4.5	0.3	0	0	0.3	0.9	0.3	2.1	0	2.4	2.1	0.6	15.9	0	13.8	0.3	6.6	2.7	4.2	2.7
17–21	0.0	0.6	0.3	2.4	0.9	3.9	0.3	0.6	0	0	0	0	0.6	0.9	0.3	0.3	1.8	0	3.9	0.3	2.7	0.3	0.3	0.6
22–26	0	71.4	0.3	67.8	0.3	36.3	2.4	9.3	3.0	2.7	2.7	3.9	1.5	7.2	2.7	1.2	3.0	1.5	7.8	4.8	15.3	6.0	1.8	62.7
27–31	0.3	24.3	0	13.2	0.3	16.2	3.0	3.0	1.2	0.9	1.2	0.3	0.6	1.5	1.8	0	1.5	2.4	1.5	0.9	4.8	0.6	2.1	19.5
Mean value	**1.5**	**31.8**	**0.7**	**20.9**	**1.3**	**10.3**	**1.4**	**2.3**	**0.8**	**1.8**	**1.0**	**1.6**	**0.7**	**2.6**	**2.0**	**0.8**	**7.7**	**0.9**	**8.0**	**1.1**	**7.7**	**1.8**	**3.0**	**19.0**
Jan. 1–5	1.8	21.9	1.5	30.0	0.9	6.6	4.5	2.1	1.2	1.8	0.3	0.9	1.5	0.9	3.6	0.3	8.1	0.6	1.8	0.6	2.1	0.9	3.6	2.4
6–10	0.6	11.1	0.3	30.3	1.5	1.2	0.9	0.6	0.0	0.3	0	0.3	0.6	0.3	5.7	0.3	8.4	0.6	3.6	1.8	9.9	1.8	2.4	5.1
11–15	3.9	7.5	0.9	17.7	0.9	2.4	0.3	5.4	0.9	0	0.3	1.5	1.2	0.6	1.8	0.3	1.5	0.3	11.4	0.6	8.4	1.8	4.8	2.4
16–20	1.5	7.8	1.8	28.8	0.6	8.7	0.6	2.4	7.2	0.9	0.6	2.4	3.6	1.2	5.7	1.5	6.3	1.5	15.0	2.1	6.6	0.9	3.6	3.3
21–25	0.3	25.8	1.2	12.6	0.9	7.2	0.9	1.2	0.9	1.5	1.2	0.9	1.5	3.3	2.7	1.8	9.9	0.3	6.3	0.6	14.4	0.9	4.8	11.7
26–30	0.6	42.0	0.3	43.8	0.3	18.3	0.3	10.5	4.5	0.9	2.4	2.1	5.4	1.8	7.5	0.9	26.1	0.3	27.0	0.6	9.9	9.6	6.0	29.1
Mean value	**1.5**	**19.4**	**1.0**	**30.2**	**0.9**	**7.4**	**1.3**	**3.7**	**2.6**	**0.9**	**0.8**	**1.4**	**2.3**	**1.4**	**4.5**	**0.9**	**10.1**	**0.6**	**10.9**	**1.1**	**8.6**	**2.7**	**4.2**	**9.0**
Jan. 31 Febr. 4	0	17.7	0.6	6.0	0.6	3.6	0.6	1.8	0.9	0.6	0.9	0.6	1.8	1.5	1.8	0.3	4.8	0	1.5	0.9	1.8	0.3	2.4	1.2
Febr. 5–9	0.3	20.1	0	61.2	1.2	42.6	3.3	14.7	3.3	2.7	9.0	1.5	5.1	1.5	18.6	0	22.8	0.6	11.4	17.7	4.5	4.6	9.9	23.1
10–14	0	60.0	0	36.0	1.2	15.0	3.0	3.0	6.0	0.6	0.9	4.5	0.9	4.5	3.0	0.9	3.0	0.9	24.0	1.2	21.0	0	3.0	30.0
15–19	0	10.5	0.3	17.1	1.2	1.2	0	0.3	0.9	0.6	0	2.1	1.5	0.9	8.4	0.3	13.5	2.4	0.9	3.3	0.9	1.8	0	4.5
20–24	1.2	7.2	0	14.1	0	66.3	0.6	58.8	0	15.9	0.9	6.6	1.2	7.8	1.5	3.6	3.0	1.8	1.8	2.1	10.2	1.2	3.0	1.5
Febr. 25 Mar. 1	0.3	9.3	0	7.5	0	7.8	0.3	0.3	0	0	0	0	0	0	1.5	1.5	11.4	0.3	17.1	0.3	3.3	4.5	0.9	12.3
Mean value	**0.3**	**22.3**	**0.2**	**28.7**	**0.7**	**22.8**	**1.3**	**13.2**	**1.9**	**3.4**	**2.0**	**2.6**	**1.8**	**2.7**	**5.8**	**1.1**	**9.8**	**1.0**	**9.5**	**4.3**	**7.0**	**5.4**	**3.2**	**12.1**
March 2–6	1.2	56.4	0.3	48.0	0.6	24.6	2.1	8.1	0.3	0.3	0.3	1.8	8.4	1.2	33.3	0	30.3	0.6	21.6	0	11.4	2.1	4.5	21.0
7–11	1.5	73.8	0.9	116.1	1.2	46.5	5.1	16.2	1.2	8.1	1.5	0.9	6.9	2.1	24.3	0.3	43.8	0	54.9	0.3	29.1	6.0	4.2	27.6
12–16	3	78.6	0.6	79.5	0.6	71.1	5.4	0.9	2.7	2.1	1.2	3.3	8.4	1.5	21.9	0.3	32.4	0.6	14.1	0.6	6.3	3.0	1.2	28.5
17–21		40.8	1.5	15.3	0.3	26.7	1.5	2.4	0.9	2.7	0.9	0.9	1.2	1.2	1.2	0.9	5.7	0.3	17.7	0.3	10.2	3.9	2.7	30.0
22–26		27.0	0.3	13.8	0.3	3.3	0	3.0	0.6	0	0.3	0.3	2.1	0.3	3.3	0	8.4	0.3	5.4	0	6.0	9.9	0	37.8
27–31		65.1	0.3	48.9	6.3	7.5	6.9	4.2	6.0	0.3	1.8	0.9	7.5	0	16.8	0	38.1	0.6	39.3	0.3	18.6	5.1	3.3	30.3
Mean value	**0.6**	**57.0**	**0.7**	**53.6**	**1.6**	**30.0**	**3.5**	**5.8**	**2.0**	**2.3**	**1.0**	**1.4**	**5.8**	**1.1**	**16.8**	**0.3**	**26.5**	**0.4**	**25.5**	**0.3**	**13.6**	**5.0**	**2.7**	**29.2**
April 1–5	0	108.0	0	96.0	7.5	51.0	0	24.0	0	1.5	0	0	9.0	0	30.0	0	39.0	0	37.5	0	15.0	0	4.5	39.0
6–10	3.0	57.0	0	127.5	3.0	96.0	4.5	43.5	6.0	21.0	16.5	7.5	18.0	16.5	39.0	0	39.0	0	34.5	1.5	16.5	19.5	4.5	102.0
11–15	1.5	58.5	0	72.0	0	13.5	0	4.5	1.5	7.5	4.5	0	21.0	0	13.5	0	12.0	0	15.0	0	13.5	6.0	0	60.0
Dec. 2 March 31	*0.9*	*32.6*	*0.6*	*33.3*	*1.1*	*17.6*	*1.9*	*6.2*	*1.8*	*2.1*	*1.2*	*1.7*	*2.6*	*1.9*	*7.3*	*0.7*	*13.5*	*0.7*	*13.5*	*1.7*	*9.2*	*3.7*	*3.3*	*17.3*

TABLE LXXXVI. S_D in γ Dyrafjord.

Hour	0–2		2–4		4–6		6–8		8–10		10–12		12–14		14–16		16–18		18–20		20–22		22–24	
Period	+	–	+	–	+	–	+	–	+	–	+	–	+	–	+	–	+	–	+	–	+	–	+	–
Nov. 23–26	3.6	9.0	0	18.0	1.8	19.8	3.6	9.0	3.6	0	0	0	0	0	3.6	0	12.6	7.2	23.4	5.4	16.2	3.6	18.0	5.4
Nov. 27 Dec. 1	0.2	2.0	0.2	0	0	2.7	0	3.2	0.9	0.5	0.2	0.2	0.9	0.4	0.7	0	0.2	0.5	0.5	0.5	0.7	2.0	1.3	3.4
Dec. 2–6	4.1	1.3	0.5	0.9	0	1.6	0.4	0.7	0.4	0.4	0.2	0	1.1	0	0.5	0.2	0	1.3	0.4	0.5	0.4	0.4	4.0	2.3
7–11	2.2	1.4	0.9	1.3	1.1	1.6	0.9	0.7	0.4	0.7	0.4	0.7	0.5	0.2	0.5	0.7	0.9	2.0	0.9	2.3	1.3	4.7	9.5	3.2
12–16	3.6	2.5	0.7	1.8	0	0.2	0.4	0.2	0.2	0.4	1.6	0.2	0.7	0	0.5	0.2	0.9	2.5	0.5	1.6	0.7	5.8	2.0	1.8
17–21	0.4	1.1	0.2	0.4	0.2	2.0	0.2	0.7	0.4	0.2	0.5	0.2	0.4	0	0.2	0.2	0	0.4	0.2	0.7	0	2.5	0	2.2
22–26	3.2	1.3	3.1	7.7	1.3	7.0	2.3	1.4	1.4	1.4	1.6	2.0	1.1	0.5	1.1	0.7	0.7	0.9	3.1	5.0	5.6	5.2	16.6	0.9
27–31	4.5	0.5	1.1	2.3	0.5	5.4	1.4	2.0	0.5	1.4	0.7	0.4	2.0	0.2	0.5	0.4	0.4	0.5	0.4	0.4	0.4	3.4	7.7	2.5
Mean value	**3.0**	**3.0**	**1.1**	**2.4**	**0.5**	**3.0**	**0.9**	**1.0**	**0.6**	**0.8**	**0.8**	**0.6**	**1.0**	**0.2**	**0.6**	**0.4**	**0.5**	**1.3**	**0.9**	**1.8**	**1.4**	**3.7**	**6.6**	**2.2**
Jan. 1–5	5.8	1.4	1.6	2.5	0.5	2.9	3.4	2.2	6.1	0.2	3.6	0.5	2.0	0	2.0	0	0.9	1.6	10.2	0.5	0.5	2.7	0.5	1.4
6–10	2.0	0.7	2.9	1.6	0.7	0.9	1.1	1.4	0.9	0.9	0.7	0.2	0.5	0.4	3.4	0.4	2.9	0.2	0	1.6	0.4	7.6	5.0	4.1
11–15	3.6	2.7	2.9	1.6	0.7	0.5	1.1	0.4	1.1	0.2	1.4	0	1.6	0.2	1.3	0.4	0.4	0.4	0	4.7	3.8	5.8	1.1	5.2
16–20	3.2	0.9	2.5	0.9	0.5	1.6	1.8	2.2	3.4	0.7	1.4	1.3	0.7	1.6	2.0	0.4	3.6	0.7	0.7	0.7	0.4	1.3	2.2	0.9
21–25	0.4	3.1	2.0	0	0.9	1.1	0.5	1.8	1.8	1.4	0.4	1.4	0.5	0.9	0.5	0.4	0.7	0.7	1.4	1.4	0.5	3.8	0.2	4.7
26–30	1.4	3.0	0.9	0.1	0	7.9	0.4	5.0	2.2	0.4	1.3	0.4	1.1	0.2	3.8	0	5.0	0.2	2.0	2.0	5.4	2.3	5.6	10.8
Mean value	**2.7**	**3.6**	**2.1**	**2.8**	**0.6**	**2.5**	**1.4**	**2.2**	**2.6**	**0.6**	**1.5**	**0.6**	**1.1**	**0.6**	**2.2**	**0.3**	**2.3**	**0.6**	**0.7**	**1.8**	**1.8**	**3.9**	**2.4**	**4.5**
Jan. 31 Febr. 4	1.3	1.1	0.2	0.9	0.5	0.9	0.7	1.1	1.1	0.7	0.9	0.2	0.7	0.4	0.9	0.7	0.9	0.7	0.9	0	0	0.7	0.5	1.3
Febr. 5–9	1.3	2.2	2.5	7.9	0.9	12.4	0.9	8.6	3.1	1.6	5.6	0.7	3.8	0.5	4.0	0.9	1.8	1.3	12.1	1.8	12.6	4.3	13.7	3.4
10–14	4.5	5.0	1.3	5.0	0.5	9.9	3.2	2.3	1.4	3.2	1.1	1.1	1.1	0.7	1.3	0.4	0.9	0.5	1.6	2.5	0.5	4.5	8.1	4.5
15–19	0.2	4.0	1.1	0.5	0.2	0.9	0.5	0.5	1.6	0.4	2.9	0.4	2.5	0	2.7	0.4	0.4	4.7	0.2	0.9	0.2	0.5	0.4	0.5
20–24	1.8	0.9	2.0	2.9	0.2	16.7	0.2	13.1	0.2	1.8	4.0	0.5	3.4	0.2	1.4	0	1.8	0.2	0.2	0.2	1.3	1.1	1.4	0.4
Feb. 25 Mar. 1	0.4	1.3	0.4	0.5	0.4	0.4	0.5	0	0.4	0.7	0.2	0.5	0.2	0.2	0.5	0	0.2	0.7	0	2.9	1.4	0.9	1.4	1.8
Mean value	**1.6**	**2.4**	**1.3**	**3.0**	**0.5**	**6.9**	**1.0**	**4.3**	**1.3**	**1.4**	**2.5**	**0.6**	**2.0**	**0.3**	**1.8**	**0.4**	**1.0**	**1.4**	**2.5**	**1.4**	**2.7**	**2.0**	**4.3**	**2.0**
March 2–6	2.0	7.9	1.1	5.4	0.9	3.8	1.3	2.0	1.3	0.4	1.4	0.5	3.2	0	1.6	0	1.3	1.3	0.5	1.6	2.9	1.4	13.3	4.0
7–11	9.7	4.3	2.3	13.7	0.2	16.2	1.3	7.4	4.9	1.1	7.7	0.4	7.6	0	3.2	0.2	2.9	2.5	3.6	5.4	15.8	5.0	14.4	3.1
12–16	0.1	1.6	4.0	5.0	0.5	21.8	2.2	2.9	2.9	2.3	2.2	1.8	6.7	0.2	3.6	0.7	2.0	3.8	0.4	2.2	7.0	1.6	6.3	7.2
17–21	4.3	2.7	0.7	0.5	0.4	2.7	1.3	1.3	1.3	1.1	1.4	1.1	0.2	0.4	0	0.4	0	0.7	0	2.3	1.4	1.8	8.6	1.4
22–26	0.4	2.7	0.4	0.7	0	1.3	0.9	0.4	0.7	0.9	2.3	0.4	0.7	0.4	0.4	0	0.2	0.7	0.4	0.5	3.2	0.9	12.1	0.9
27–31	1.6	9.7	0	8.3	0	6.8	2.3	2.9	4.3	0.9	2.7	0.5	1.1	0	0.5	1.3	0.2	3.8	0.4	3.6	3.1	2.9	8.5	2.0
Mean value	**4.7**	**4.8**	**1.4**	**5.6**	**0.3**	**8.8**	**1.6**	**2.8**	**2.6**	**1.1**	**3.0**	**0.8**	**3.3**	**0.2**	**1.6**	**0.4**	**1.1**	**2.1**	**0.9**	**2.6**	**5.6**	**2.3**	**10.5**	**3.1**
April 1–5	4.5	5.4	0	8.1	0.9	6.3	0	0.9	0	0	1.8	0	0.9	1.8	0	4.5	0.9	1.8	1.8	1.8	4.5	1.8	7.2	2.7
6–10	3.6	1.8	0	16.2	0	5.4	2.7	4.5	2.7	1.8	2.7	16.2	2.7	18.9	1.8	5.4	8.1	0.9	9.9	2.7	13.5	2.7	3.6	0.9
11–15	1.8	0.9	0	4.5	0	0	0	0	0.9	0.9	0.9	0	0.9	0.9	0	0	0	0.9	0	0.9	1.8	1.8	8.1	0
Dec. 2 March 31	*3.0*	*3.5*	*1.5*	*3.4*	*0.5*	*5.3*	*1.2*	*2.6*	*1.8*	*1.0*	*1.9*	*0.6*	*1.8*	*0.3*	*1.5*	*0.4*	*1.2*	*1.4*	*1.3*	*1.9*	*2.9*	*3.0*	*6.0*	*2.9*

TABLE LXXXVII. S_V in γ Dyrafjord.

Hour	0–2		2–4		4–6		6–8		8–10		10–12		12–14		14–16		16–18		18–20		20–22		22–24	
Period	+	–	+	–	+	–	+	–	+	–	+	–	+	–	+	–	+	–	+	–	+	–	+	–
Nov. 23–26	15.5	3.5	28.0	0	21.0	3.5	3.5	17.5	0	10.5	0	0	0	0	3.5	7.0	3.5	56.0	3.5	70.0	7.0	84.0	17.5	21.0
Nov. 27 Dec. 1	0.4	8.4	0	1.4	0	8.1	0	6.7	0.4	2.5	0	0.7	0	0.7	0	0	2.5	0.7	0.4	3.5	0.7	10.2	0	5.6
Dec. 2–6	3.5	11.9	0	7.4	0	1.4	0	1.8	0	0.7	0	1.8	0	0.4	1.4	0	3.5	0	1.8	0.7	2.1	0	1.8	4.9
7–11	3.2	4.6	1.1	11.2	2.5	0.4	0.4	0.4	0.7	1.1	5.3	0	6.7	0	1.8	0	3.2	1.8	6.3	3.2	5.3	22.8	4.9	22.8
12–16	5.6	11.6	0	7.0	4.9	0.4	0.4	0.7	0	0.4	0.4	0.4	1.1	0.4	2.5	0.4	9.8	0	3.2	1.1	0.7	11.9	0.4	7.7
17–21	0	2.1	0	2.5	0	5.6	0	3.2	0	0.4	0	0	0.7	0	0.4	0	0.7	0.4	4.2	0.7	1.1	0.7	0.7	2.1
22–26	22.1	4.9	4.2	17.9	0.7	22.1	0	38.9	0.4	7.0	4.6	1.4	2.8	0	0.4	0.4	1.1	0.7	3.5	9.1	3.9	23.1	13.3	10.9
27–31	3.8	3.9	0.4	7.4	0	7.7	0	9.8	0	9.1	0	1.8	0	0.4	0	0.4	0.7	0	0.7	1.1	0	5.6	7.4	7.7
Mean value	**6.3**	**6.5**	**1.0**	**8.9**	**1.4**	**6.3**	**0.1**	**9.1**	**0.2**	**3.1**	**1.7**	**0.9**	**1.9**	**0.2**	**1.1**	**0.2**	**3.2**	**0.5**	**3.3**	**2.7**	**2.2**	**10.7**	**4.8**	**9.4**

TABLE LXXXVII (continued). S_V in γ

Hour	0—2		2—4		4—6		6—8		8—10		10—12		12—14		14—16		16—18		18—20		20—22		22—24	
Period	+	—	+	—	+	—	+	—	+	—	+	—	+	—	+	—	+	—	+	—	+	—	+	—
Jan. 1—5	2.8	15.8	1.4	13.0	0	5.6	0	10.9	0.4	3.9	0	1.8	0	1.1	1.4	0	4.9	0	2.5	0	0.4	4.2	0	10.2
6—10	0.7	9.1	0.7	10.9	0	28.7	0	6.7	0.7	0.7	0	0	0.4	0	3.9	0.4	6.0	0	1.1	1.4	1.1	37.5	0	23.8
11—15	2.1	16.8	2.5	2.1	2.8	0.4	1.1	1.1	1.4	0	0	0.7	0.4	0	0	0	0	0.4	4.2	2.1	1.8	17.9	1.1	8.4
16—20	0.4	13.0	1.1	8.8	0.4	6.0	0	10.9	0.4	4.9	0	1.1	0.4	0.7	2.1	0.4	3.5	2.1	3.5	2.1	1.1	2.8	0	18.9
21—25	0.4	10.9	0	9.5	0.4	5.6	0.4	1.8	0.4	2.8	0.4	0.4	0	0	0.4	0	2.5	0	2.8	2.8	2.1	22.8	0.4	11.2
26—30	39.2	5.6	17.2	3.2	2.1	3.9	0	15.1	0	19.3	0	4.6	3.5	0	6.0	0	8.8	0.7	2.1	21.7	1.1	13.7	41.3	6.0
Mean value	**7.6**	**11.9**	**3.8**	**7.9**	**1.0**	**8.4**	**0.3**	**7.8**	**0.6**	**5.3**	**0.1**	**1.4**	**0.8**	**0.3**	**2.3**	**0.1**	**4.3**	**0.5**	**2.7**	**5.0**	**1.3**	**16.5**	**7.1**	**13.1**
Jan.31 Febr.4	2.1	5.6	0.4	5.6	0	3.5	0.7	1.4	0	0.4	0.7	0	0.7	0.7	0.7	0.4	6.0	0	5.6	0.4	0.4	1.1	0.4	2.5
Febr. 5—9	7.0	3.9	8.4	5.3	5.3	7.4	1.8	14.4	0	9.8	1.4	1.1	1.4	0.4	1.1	0.4	0.7	2.8	0.4	26.6	0	39.9	2.1	40.6
10—14	26.3	5.3	0	18.6	0.7	19.3	2.5	17.5	1.8	5.6	0	1.4	0	1.1	0	0	1.4	0	5.6	1.8	1.1	4.9	1.8	19.3
15—19	2.1	13.7	0	11.9	0	3.2	0	0	0.4	0	0.4	0	2.1	0	4.6	0	4.9	1.4	2.8	0.7	1.1	1.1	0.7	1.1
20—24	0.4	5.3	6.0	4.6	0.4	16.5	0	42.4	0	31.2	0	15.4	0.4	4.6	1.4	0	7.7	0.4	2.1	0	0	5.6	0.4	4.2
Feb.25 Mar.1	1.4	1.8	0	7.7	0	17.9	0	37.8	0	14.7	0	5.6	0.4	0.4	0.4	0	3.5	0	7.7	0.4	1.8	6.7	2.5	14.0
Mean value	**6.6**	**5.9**	**2.5**	**9.0**	**1.1**	**11.3**	**0.8**	**18.9**	**0.4**	**10.3**	**0.4**	**3.9**	**0.8**	**1.2**	**1.4**	**0.1**	**4.0**	**0.8**	**4.0**	**5.0**	**0.7**	**9.9**	**1.3**	**13.6**
March 2—6	4.9	16.5	16.1	7.7	0	14.4	0	15.4	0.4	2.8	0	0	2.8	0	2.8	0	4.9	0	4.2	0.7	0.7	20.3	1.1	37.1
7—11	12.3	13.0	21.4	9.8	1.8	22.4	0.4	28.7	0	24.9	0.4	7.4	4.9	1.1	4.6	1.8	7.4	1.4	1.4	23.1	0	59.9	1.4	51.1
12—16	22.4	13.3	16.5	13.3	0	30.5	0	15.1	0	11.9	0	6.0	1.8	1.1	5.6	2.8	6.3	4.9	6.0	3.2	0	15.4	1.8	14.4
17—21	0.4	17.5	0	7.4	0	13.3	0	10.5	0	4.2	0	2.8	0.4	1.8	0.4	0	3.9	0	6.7	0.4	3.2	6.3	1.8	18.9
22—26	2.1	7.4	0.7	5.3	0	3.9	0	6.0	0	3.2	0	0.7	0.7	0	0.4	0	2.8	0	1.1	0	0.4	6.7	3.9	16.8
27—31	27.3	0.7	13.2	2.8	1.1	9.8	0.7	13.7	0	9.8	0	2.1	1.4	0	7.7	0	13.3	0.4	0.7	3.5	0	17.5	7.0	12.6
Mean value	**11.6**	**11.4**	**11.3**	**7.7**	**0.5**	**15.7**	**0.2**	**14.9**	**0.1**	**9.5**	**0.1**	**3.2**	**2.0**	**0.7**	**3.6**	**0.8**	**6.4**	**1.1**	**3.4**	**5.2**	**0.7**	**21.0**	**2.8**	**25.2**
April 1—5	35.0	14.0	18.3	8.8	19.3	12.3	0	28.0	0	5.3	0	0	1.8	0	10.5	0	15.8	0	8.8	8.8	0	21.0	7.0	35.0
6—10	24.5	3.5	82.3	0	22.8	8.8	12.3	8.8	5.3	8.8	10.5	7.0	3.5	42.0	15.8	22.8	5.3	5.3	0	4.0	0.5	31.5	21.0	24.5
11—15	15.8	5.3	1.8	17.5	0	12.3	0	5.3	0	10.5	0	3.5	5.3	0	5.3	0	8.8	0	8.8	0	0	12.3	14.0	10.5
Dec.2 March31	*8.0*	*8.9*	*4.6*	*8.4*	*1.0*	*10.4*	*0.4*	*12.7*	*0.3*	*7.0*	*0.6*	*2.4*	*1.4*	*0.4*	*2.1*	*0.3*	*4.5*	*0.7*	*3.8*	*4.5*	*1.2*	*14.5*	*4.0*	*15.3*

THIRD SERIES.

THE STORMINESS AS A FUNCTION OF TIME.

Matotchkin Schar.

TABLE LXXXVIII a. (Unit γ)

Interval.	S_H^p	S_H^n	S_D^p	S_D^n	S_V^p	S_V^n	S^T
Sep. 3—7	—	—	—	—	—	—	—
8—12	—	—	—	—	—	—	—
13—17	—	—	—	—	—	—	—
18—22	—	—	—	—	—	—	—
23—27	—	—	—	—	—	—	—
Sep.28 Oct.2	—	—	—	—	—	—	—
Month.	—	—	—	—	—	—	—
Oct. 3—7	1.0	1.7	0.7	0.8	0.5	0.8	3.3
8—12	2.5	9.1	1.0	5.7	0.6	6.6	15.2
13—17	0.5	3.3	0.4	0.9	0.2	1.4	4.3
18—22	0.6	2.7	0.6	1.5	0.1	2.1	4.5
23—27	4.1	20.8	1.7	11.3	3.5	9.9	31.1
Oct.28 Nov.1	11.0	31.3	7.8	17.2	6.7	7.4	51.1
Month.	**3.3**	**11.5**	**2.0**	**6.2**	**1.9**	**4.7**	**18.2**

Kaafjord.

TABLE LXXXVIII b. (Unit γ)

Interval.	S_H^p	S_H^n	S_D^p	S_D^n	S_V^p	S_V^n	S^T
Sep. 3—7	0.9	0.7	0.9	1.2	0.9	1.6	3.6
8—12	2.5	3.6	2.0	2.3	3.7	3.0	10.0
13—17	0.4	0.5	0.6	1.2	0.4	1.2	2.5
18—22	3.5	12.5	1.9	7.9	4.2	15.3	27.6
23—27	0.2	3.0	0.4	2.1	1.1	5.4	7.7
Sep.28 Oct.2	0.3	9.2	1.2	5.2	1.0	14.5	19.8
Month.	**1.3**	**4.9**	**1.2**	**3.3**	**1.9**	**6.8**	**11.7**
Oct. 3—7	0.3	0.4	0.6	0.6	0.4	1.2	2.2
8—12	0.7	5.3	0.9	3.5	1.6	4.1	9.3
13—17	0.1	0.1	0.5	0.4	0.1	0.9	1.4
18—22	0.1	0.8	0.3	1.3	0.2	2.0	2.8
23—27	2.2	11.8	1.8	7.3	4.5	13.4	24.5
Oct.28 Nov.1	11.1	20.5	6.4	10.1	6.5	23.0	46.2
Month.	**2.4**	**6.5**	**1.7**	**3.9**	**2.2**	**7.4**	**14.4**

Matotchkin Schar.

TABLE LXXXVIII a (contin.). (Unit γ)

Interval.	S_H^p	S_H^n	S_D^p	S_D^n	S_V^p	S_V^n	S^T
Nov. 2–6	0.5	8.7	0.5	3.5	1.0	5.6	12.0
7–11	0.5	1.3	0.3	0.5	0.2	1.6	2.6
12–16	3.0	2.8	1.2	2.5	1.7	2.8	8.2
17–21	1.0	9.7	0.4	5.5	1.7	6.0	14.5
22–26	9.8	30.7	7.7	27.1	14.4	44.3	79.4
Nov.27 Dec.1	1.3	1.8	0.4	2.5	1.1	1.8	5.1
Month.	**2.7**	**9.1**	**1.8**	**6.9**	**3.4**	**10.4**	**20.3**
Dec. 2–6	0.7	1.8	0.4	1.4	0.4	2.4	4.1
7–11	1.9	3.1	1.0	2.8	2.0	7.3	11.2
12–16	0.6	3.1	0.6	2.4	0.7	6.2	8.5
17–21	0.3	1.4	0.4	0.5	0.9	0.9	2.5
22–26	5.9	6.4	2.4	7.6	2.7	19.8	33.2
27–31	1.1	2.9	0.7	1.2	1.8	3.7	7.1
Month.	**1.7**	**4.8**	**0.9**	**2.6**	**1.4**	**6.7**	**11.1**
Jan. 1–5	1.7	3.1	0.9	1.6	1.9	5.0	8.7
6–10	2.3	3.4	0.4	2.4	3.1	3.8	9.4
11–15	1.2	2.6	0.8	1.8	1.6	5.3	8.3
16–20	3.4	2.0	1.2	1.3	4.0	2.5	8.8
21–25	1.0	4.7	0.6	3.5	1.1	7.0	10.7
26–30	5.1	1.5	3.1	4.0	3.1	8.6	42.0
Month.	**2.5**	**7.9**	**1.2**	**4.1**	**2.5**	**5.4**	**14.7**
Jan.31 Feb.4	1.9	0.2	0.9	0.2	1.4	0.2	2.8
Feb. 5–9	4.2	3.6	2.0	7.6	2.9	2.6	25.5
10–14	1.8	9.1	0.9	4.5	2.7	6.4	15.2
15–19	2.0	2.1	0.5	2.2	0.9	6.4	8.8
	1.7	2.5	1.7	0.4	3.0	1.6	6.5
Feb.25 Mar.1	1.6	1.7	1.6	0.8	4.5	1.4	7.2
Month.	**2.2**	**4.9**	**1.3**	**2.6**	**2.6**	**4.8**	**11.0**
March 2–6	—	—	—	—	—	—	—
7–11	—	—	—	—	—	—	—
Oct. 3 March	*2.5*	*7.6*	*1.4*	*4.5*	*2.3*	*6.4*	*15.1*

Kaafjord.

TABLE LXXXVIII b (contin.). (Unit γ)

Interval.	S_H^p	S_H^n	S_D^p	S_D^n	S_V^p	S_V^n	S^T
Nov. 2–6	0.4	1.8	0.4	1.7	0.6	2.2	4.1
7–11	0.2	0.5	0.1	0.6	0.1	1.8	2.2
12–16	0.4	1.3	0.5	1.9	1.4	3.1	5.4
17–21	0.5	3.6	0.5	2.8	0.5	1.4	5.6
22–26	14.9	30.3	6.0	16.7	1.3	7.6	51.3
Nov.27 Dec.1	0.2	0.8	0.5	0.6	0	0	1.5
Month.	**2.8**	**6.4**	**1.4**	**4.1**	**0.7**	**2.7**	**11.7**
Dec. 2–6	0.4	0.4	0.4	0.9	0	0.1	1.5
7–11	2.4	0.7	0.7	2.3	0	0	4.2
12–16	1.1	1.0	0.7	2.2	0	0	3.6
17–21	0.2	0.2	0.5	0.4	0.2	0.4	1.1
22–26	3.9	10.4	2.6	5.6	3.7	13.2	23.5
27–31	0.3	1.5	0.3	2.1	1.2	2.1	4.4
Month.	**1.4**	**2.4**	**0.9**	**2.2**	**0.9**	**2.6**	**6.4**
Jan. 1–5	1.0	0.8	1.2	1.6	1.1	1.7	4.4
6–10	0.3	1.0	1.0	1.2	1.1	1.5	3.6
11–15	0.5	0.7	1.0	1.5	1.7	2.1	4.7
16–20	0.8	0.6	1.2	1.2	3.2	1.2	5.3
21–25	0.7	1.0	0.9	2.3	1.5	2.7	5.5
26–30	1.8	11.1	1.4	4.7	1.9	8.8	17.8
Month.	**0.8**	**2.5**	**1.1**	**2.1**	**1.7**	**3.0**	**6.9**
Jan.31 Feb.4	0.2	0.1	0.4	0.3	0.9	0.3	1.4
Feb. 5–9	2.8	7.3	1.4	5.8	3.3	15.2	22.2
10–14	1.7	3.7	0.7	3.5	1.9	7.8	11.8
15–19	0.8	0.4	0.6	1.5	2.3	0.5	3.6
20–24	0	0	1.6	0.6	0.6	2.5	(3.9)
Feb.25 Mar.1	0.9	0.9	0.9	0.9	2.1	3.4	6.0
Month.	**1.1**	**2.1**	**0.9**	**2.1**	**1.8**	**4.9**	**8.2**
March 2–6	2.2	1.3	1.2	1.3	2.6	4.8	8.5
7–11	3.8	6.3	2.5	3.8	4.2	13.5	21.3
Sep. 3 March 1	*1.6*	*4.1*	*1.2*	*2.9*	*1.5*	*4.6*	*9.9*

Axeløen.

TABLE LXXXIX a. (Unit γ)

Sept. 3–7	6.3	11.1	5.6	8.0	16.0	7.6	32.3
8–12	8.2	12.0	6.0	3.2	12.3	1.7	26.2
13–17	1.7	5.9	1.8	2.9	6.5	3.2	13.2
18–22	11.5	19.1	5.7	9.3	49.6	6.6	65.6
23–27	5.2	13.6	3.3	5.5	19.9	4.1	31.8
Sep.28 Oct.2	2.5	16.2	2.3	4.1	30.7	5.8	41.4
Month.	**5.9**	**13.0**	**4.1**	**5.5**	**22.5**	**4.8**	**35.1**
Oct. 3–7	1.9	3.4	1.4	2.3	6.4	1.3	10.0
8–12	3.0	6.8	3.9	3.1	10.4	5.9	20.3
13–17	0.7	7.4	2.8	1.8	7.1	1.3	12.6
18–22	1.2	9.1	2.1	4.9	7.6	5.7	18.2
23–27	5.4	23.2	6.8	8.9	35.4	12.0	57.5
Oct.28Nov.1	2.6	40.8	7.9	16.1	63.9	7.6	84.9
Month.	**2.4**	**15.1**	**4.2**	**6.2**	**21.8**	**5.6**	**33.9**

Dyrafjord.

TABLE LXXXIX b. (Unit γ)

Interval.	S_H^p	S_H^n	S_D^p	S_D^n	S_V^p	S_V^n	S^T
Sept. 3–7	—	—	—	—	—	—	—
8–12	—	—	—	—	—	—	—
13–17	—	—	—	—	—	—	—
18–22	—	—	—	—	—	—	—
23–27	—	—	—	—	—	—	—
Sep.28 Oct.2	—	—	—	—	—	—	—
Month.	—	—	—	—	—	—	—
Oct. 3–7	—	—	—	—	—	—	—
8–12	—	—	—	—	—	—	—
13–17	—	—	—	—	—	—	—
18–22	—	—	—	—	—	—	—
23–27	—	—	—	—	—	—	—
Oct.28Nov.1	—	—	—	—	—	—	—
Month.	—	—	—	—	—	—	—

Axeløen.

TABLE LXXXIX a (contin.). (Unit γ)

Interval.	S_H^p	S_H^n	S_D^p	S_D^n	S_V^p	S_V^n	S^T
Nov. 2—6	0.7	8.4	3.2	2.0	14.2	1.0	18.5
7—11	0.6	4.8	0.4	3.0	8.2	0.5	10.8
12—16	4.4	11.3	2.1	7.1	15.8	11.1	32.4
17—21	2.0	13.0	2.4	5.3	15.5	4.6	26.2
22—26	5.0	58.8	8.4	20.5	81.7	11.5	116.6
Nov.27Dec.1	2.6	11.4	1.5	2.9	8.3	3.5	18.9
Month.	**2.5**	**17.9**	**3.0**	**6.8**	**23.9**	**5.4**	**37.2**
Dec. 2—6	2.8	8.9	1.9	3.7	6.2	2.7	15.7
7—11	3.7	15.0	2.3	6.0	13.4	4.2	26.9
12—16	1.2	16.0	1.8	5.1	11.9	1.4	22.8
17—21	1.7	3.5	1.6	1.0	4.3	4.1	10.2
22—26	5.0	27.6	5.4	8.0	20.1	26.8	57.8
27—31	2.5	9.2	3.2	3.1	6.2	2.8	16.1
Month.	**2.8**	**13.4**	**2.7**	**4.5**	**10.4**	**7.0**	**24.9**
Jan. 1—5	1.9	13.6	2.9	4.3	4.7	5.6	19.9
6—10	4.6	10.4	2.1	4.3	7.2	3.5	18.5
11—15	3.1	13.6	2.8	4.1	8.8	2.5	21.3
16—20	3.6	10.3	3.7	2.9	6.2	7.5	20.5
21—25	2.5	11.5	2.8	3.9	9.4	9.9	24.8
26—30	5.8	19.1	4.2	8.9	36.0	4.5	49.3
Month.	**3.6**	**13.1**	**3.1**	**4.7**	**12.0**	**5.6**	**25.7**
Jan.31 Febr.4	1.7	5.0	2.7	1.4	1.5	2.0	8.6
Febr. 5—9	5.0	20.3	5.7	8.0	21.5	4.8	39.0
10—14	3.9	14.0	2.9	6.5	10.3	1.9	23.5
15—19	3.0	8.7	2.9	2.9	3.8	3.0	14.7
20—24	5.0	5.9	3.4	4.6	9.4	5.7	20.2
Feb.25 Mar.1	3.8	8.6	2.6	3.7	8.0	2.2	17.3
Month.	**3.7**	**10.4**	**3.4**	**4.5**	**9.1**	**3.3**	**20.6**
March 2—6	2.6	13.7	5.1	5.4	14.6	19.7	39.4
7—11	5.5	27.9	7.1	11.6	48.7	10.7	70.7
12—16	7.1	20.9	6.2	7.5	27.6	14.6	52.4
17—21	1.4	11.2	3.0	3.9	15.1	5.3	24.9
22—26	1.9	7.0	2.9	3.0	9.9	1.1	15.4
27—31	3.0	13.4	4.9	7.5	11.4	5.1	26.4
Month.	**3.6**	**15.7**	**4.8**	**6.5**	**21.2**	**9.4**	**38.2**
April 1—5	5.0	23.1	5.0	11.0	26.6	14.2	52.1
6—10	9.9	31.5	10.2	12.9	47.0	29.7	92.7
11—15	9.0	14.9	4.9	4.1	13.1	23.6	44.7
16—20	4.1	14.0	9.2	2.3	1.6	18.6	29.4
21—25	4.9	11.9	5.0	3.7	7.3	10.0	25.6
26—30	5.7	19.8	5.5	9.5	31.0	6.9	48.0
Month.	**6.4**	**19.2**	**6.6**	**7.2**	**21.1**	**17.2**	**48.8**
May 1—5	5.8	16.9	6.4	8.2	36.0	8.1	51.7
6—10	(9.8)	(23.6)	8.5	9.6	19.1	14.3	50.6
11—15	6.1	18.8	7.7	6.3	13.8	10.3	37.3
16—20	10.5	4.9	5.5	4.1	13.9	8.3	28.6
21—25	15.5	14.3	7.1	12.2	25.6	22.3	59.6
26—30	20.6	22.6	6.5	18.4	38.0	17.3	74.4
Month.	**11.4**	**16.8**	**6.9**	**9.8**	**24.4**	**13.4**	**50.4**
Sept.3 May30	*4.7*	*15.0*	*4.3*	*6.2*	*18.5*	*8.0*	*35.0*

Dyrafjord.

TABLE LXXXIX b (contin.). (Unit γ)

Interval.	S_H^p	S_H^n	S_D^p	S_D^n	S_V^p	S_V^n	S^T
Nov. 2—6	—	—	—	—	—	—	—
7—11	—	—	—	—	—	—	—
12—16	—	—	—	—	—	—	—
17—22	—	—	—	—	—	—	—
23—26	13.0	45.3	7.2	6.5	11.1	22.8	68.7
Nov.27Dec.1	2.3	3.7	0.5	1.3	0.4	4.0	7.6
Month.	—	—	—	—	—	—	—
Dec. 2—6	3.0	3.5	1.0	0.8	1.2	2.6	7.7
7—11	4.5	7.4	1.6	1.6	3.5	5.7	15.3
12—16	4.4	5.8	1.0	1.4	2.4	3.5	12.1
17—21	1.0	0.8	0.2	0.9	0.7	1.5	3.0
22—26	3.4	22.9	3.4	3.7	4.8	1.4	31.7
27—31	1.5	6.9	1.7	1.6	1.0	4.6	10.7
Month.	**3.0**	**7.9**	**1.5**	**1.7**	**2.3**	**4.9**	**13.4**
Jan. 1—5	2.6	6.5	3.1	1.3	1.2	5.5	12.1
6—10	2.9	5.2	1.7	1.7	1.2	9.9	14.2
11—15	3.0	3.4	1.6	1.8	1.5	4.2	9.2
16—20	4.4	5.1	1.9	1.1	1.1	6.0	12.3
21—25	3.8	5.7	0.8	1.7	0.9	5.7	11.7
26—30	7.5	13.3	2.4	4.4	10.1	7.8	28.3
Month.	**4.0**	**6.5**	**1.9**	**2.0**	**2.6**	**6.5**	**14.6**
Jan.31 Febr.4	1.5	2.9	0.7	0.7	1.5	1.8	5.7
Febr. 5—9	7.5	17.5	5.2	3.8	2.5	12.7	30.6
10—14	5.5	13.1	2.1	3.3	3.4	7.9	22.4
15—19	2.3	4.5	1.1	1.1	1.6	2.8	8.4
20—24	2.0	18.1	1.5	3.2	1.6	10.9	24.0
Feb. 25 Mar. 1	2.9	3.7	0.5	0.8	1.5	8.9	12.4
Month.	**3.6**	**10.0**	**1.9**	**2.2**	**2.0**	**7.5**	**17.2**
March 2—6	9.5	13.7	2.6	2.4	3.2	9.6	26.9
7—11	14.6	24.8	6.1	4.9	4.7	20.4	48.0
12—16	7.9	22.5	4.0	4.3	5.0	11.0	35.4
17—21	3.7	10.5	1.6	1.4	1.4	6.9	16.7
22—26	2.2	8.0	1.8	0.8	1.0	4.2	11.8
27—31	12.1	13.6	2.1	3.6	6.0	6.1	29.0
Month.	**8.3**	**15.5**	**3.0**	**2.9**	**3.6**	**9.7**	**27.9**
April 1—5	11.9	26.6	1.9	2.9	9.7	11.1	44.0
6—10	15.4	41.0	4.3	6.5	17.8	14.8	66.0
11—15	6.9	18.5	1.2	0.9	5.0	6.4	27.9
16—20	—	—	—	—	—	—	—
21—25	—	—	—	—	—	—	—
26—30	—	—	—	—	—	—	—
Month.	—	—	—	—	—	—	—
May 1—5	—	—	—	—	—	—	—
6—10	—	—	—	—	—	—	—
11—15	—	—	—	—	—	—	—
16—20	—	—	—	—	—	—	—
21—25	—	—	—	—	—	—	—
26—30	—	—	—	—	—	—	—
Month.	—	—	—	—	—	—	—
Dec.2 Mar.31	*4.7*	*10.0*	*2.1*	*2.2*	*2.6*	*7.1*	*18.3*

Variation of Total

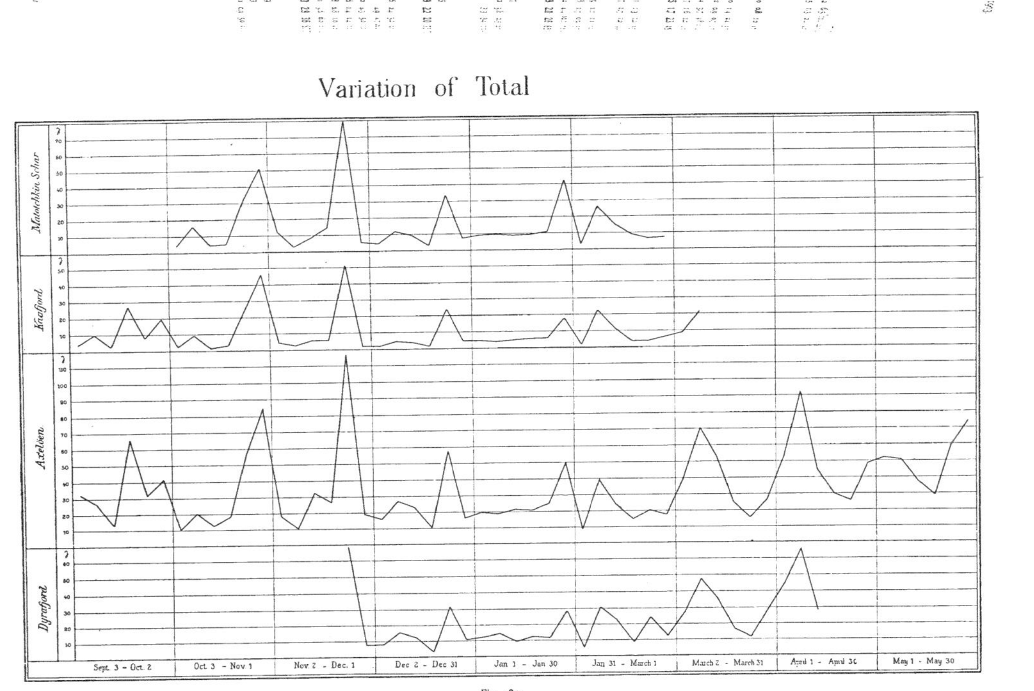

Fig. 189.

B

THE TOTAL STORMINESS AS A FUNCTION OF TIME AND ITS RELATION TO SOLAR ACTIVITY.

95. The main object of the investigation with regard to the total storminess was to find any possible regularity in the occurrence of magnetic storms, especially as regards a monthly period. The existence of such a period has been recognised by many authorities, but various opinions are held with regard to its length.

Mr. E. W. MAUNDER [1], from records made at Greenwich, deduced a period of 27.275 days. Mr. ARTHUR HARVEY [2], from a study of storms at Toronto, found independently about the same period namely, 27.246 days. Dr. AD. SCHMIDT [3], however, from observations at Potsdam, deduced a period of 29.97 days.

A period of about the same length — about 26 days [4] — is found for most magnetic elements, for atmospheric electricity and northern lights, and for a great variety of phenomena connected with meteorology. All these periods may in some way be related; but it is not my intention to overload the problem by treating such possible connections. We shall in the following pages confine our attention to the treatment of the period for the »polar storms«, as this period has actually manifested itself during the period of our observations.

The variation of total storminess at the Norwegian stations is given in tables 88 and 89 and graphically represented by the curves fig. 189. As might be expected, the curves for the four stations show almost exactly the same variation, there being merely a difference as regards absolute magnitude. If, for one station, the absolute storminess for each component were represented by curves, we should see — what seems almost a matter of course — that the storminess varies in very nearly the same way for all three components.

We notice that a period of about one month is extremely well marked at all four stations. The maxima seem to fall into two groups, the first of these having the first maximum at the end of September, and the last at the end of January, while the second group has its first maximum at the beginning of February and the last observed maximum at the beginning of May.

If we do not divide the maxima into groups, but consider the two occurring at the end of January and the beginning of February as belonging to the same maximum, we deduce a period of 30.7 days as the average of 7 periods, the first period beginning with the maximum on September 30 and the last one ending with the maximum on May 3.

Considering each group separately we get:

From the first group 30.0 days, mean of 4 periods
— » second — 28.3 — » » 3 —

Mean 29.2 days.

---

(1) E. W. MAUNDER: The „Great" Magnetic Storms, 1875—1903, and their Association with Sun-spots, as recorded at the Royal Observatory, Greenwich. Monthly Not. 64. 1904.
— » — „Magn. Disturb. 1882—1903, etc." Monthly Not. 65, 1904.
— » — „The Solar Origin of Disturb. of Terr. Magn." Astron. Nachr. 167. 1904.
— » — „Magn. Disturb. as recorded, etc." Monthly Not. 65, 538—559 and 666—681, 1905.
— » — „The Solar Origin of the Disturb. of Terr. Magn." Astroph. Journ. 21. 1905.
— » — Journ. Brit. Astr. Assoc. 16. 1905.

(2) Nature. Vol. 83, p. 354. 1910.

(3) AD. SCHMIDT: Ergebnisse magnetischer Beobachtungen in Potsdam im Jahre 1907, p. 29. Published 1910.

(4) ARRHENIUS: Lehrbuch der kosmischen Physik p. 146.

These numbers are deduced from the curves giving the storminess of each five day period. In these smothed curves there may be an error in the determination of the actual time of occurrence of the maxima.

In curve *A*, fig. 190 the variation of storminess is represented for each day during the period of observations. The curve represents the *absolute* storminess S''_H or $|S_H|$ for Axeløen. The curve for the total storminess would not be essentially different.

In the following table are given the most marked maxima, the principal maxima belonging to the two groups are marked in the third column.

TABLE XC.

Time of Max.	Size of Max. arbitrary unit.	Principal Max. Group I	Principal Period Days	Time of Max.	Size of Max. arbitrary unit.	Principal Max. Group II	Principal Period Days
Sept. 9.5	10.7			Jan. 29.5	6.5		
„ 11.5	8.5			Febr. 7.5	25.1	No. 1''	
„ 19.2	16.8			„ 10.5	11.3		
„ 22.2	15.9			„ 14.5	9.8		
„ 30.0	11.6	No. 1'		„ 21.5	9.5		
Oct. 10.5	7.8			„ 24.5	10.8		28.9
„ 21.2	12.3		30.5	March 1.5	7.0		
„ 26.6	20.7			„ 7.5	19.8	No. 2''	
„ 30.5	32.1	No. 2'		„ 12.2	10.6		
Nov. 5.5	6.4		24.0	„ 18.5	8.2		28.1
„ 13.0	9.5			„ 31.0	10.6		
„ 23.5	33.9	No. 3'		April 5.4	22.4	No. 3''	
Dec. 0.5	7.9			„ 8.6	22.1		
„ 10.5	9.5		29.2	„ 11.5	9.1		
„ 15.2	7.3			„ 14.5	13.6		29.1
„ 22.7	27.9	No. 4'		„ 26.5	11.1		
„ 27.5	9.5			May 4.5	18.0	No. 4''	
Jan. 4.5	13.4		34.3				
„ 12.5	8.8						
„ 26.0	15.4	No. 5'					
	Mean Period of Group I		29.5		Mean Period of Group II		28.7

The two groups show a characteristic difference; each of the maxima of the first group consists mainly of a single top, those of the second group consist of pairs. This fact must strengthen the assumption that the maxima within each group are closely related to one another. In the fourth column are given the intervals between successive maxima. The average period becomes 29.1 days, or about the same as found from the five-day curve. There seems no interpretation of the results possible leading to a period of less than 29.1 days.

The period found is very nearly equal to one synodic month, as the time from one opposition of the moon to the next is 29.53 days. This coincidence would naturally suggest a connection between the polar storms and the position of the moon in relation to the sun.

On the other hand we know that the polar storms are closely connected with the conditions existing on the sun, and this connection must point to the rotation of the sun about its own axis as the cause of the monthly period of polar storms. Now it has been found that different parts of the sun rotate with a different angular velocity. The least synodic period of rotation is about 26.04 days, which is the period of facula near the equator; the period, however, becomes longer as we get deeper into the sun's layers, or towards its pole. In the table below is given the synodic period for faculæ, for sun-spots, and for the photosphere.

TABLE XCI.

Heliographic Latitude	Faculæ (Stratonoff)	Sun-Spots (Carrington)	Photosphere (Dunér)
0°	26.0 days	26.8	27.4
15°	27.1 —	27.3	28.4
30°	27.8 —	28.6	29.8
45°	29.5 —	29.8	32.7

The numbers in table XCI, are taken from ARRHENIUS' Cosmical Physics [1]. They indicate that for equal heliographic latitudes the period of rotation increases towards the interior. According to PRINGSHEIM the angular velocity of faculæ, photosphere and sun-spots [2] should be the same for the same latitude. However this may be, it is commonly assumed that the angular velocity decreases from the photosphere towards the interior.

We notice that the period found for the storminess cannot be explained merely by the time of rotation of the sun-spots. The greatest number of sun-spots are found betwen 15° and 20° heliographic latitude. From this we should expect a period corresponding to that latitude, or about 27.3 days. This is about the period found by MAUNDER and HARVEY. Such a period of disturbance may well exist, but it is too small to explain the essential feature of the variation of storminess in our case. If the period of polar storminess is to be explained by the rotation of the sun, we shall either have to go to points deep down in the sun's layers, or to points near the poles, for the source of magnetic storms.

As both the moon and the sun give rise to a period such as that found for the magnetic storminess, the problem of finding out by exact methods the cause of the period becomes a rather difficult one; and it is hardly possible, by means of purely statistical methods, to decide from which of the two sources the monthly period originates. At any rate, if a statistical method could give any answer to this question, we should have records covering a long period. I think, however, we can get a step further by utilising our knowledge about the *physical conditions* which might produce the observed changes of storminess.

ON THE POSSIBLE INFLUENCE OF THE MOON UPON MAGNETIC STORMS.

96. There are two main sources of influence to consider:

(1) The moon is the seat of a magnetic field.

(2) The moon is the source of primary or secondary "electric radiation".

It is well known that the direct influence of the moon's magnetic field must at any rate be extremely small, and would cause variations of quite another type than those considered, in the magnetic storms. But there is still a possibility of an *indirect influence*, as the presence of the moon's magnetic

(1) ARRHENIUS: Lehrbuch der kosmischen Physik p. 125.

(2) E. PRINGSHEIM: Physik der Sonne p. 61.

field will produce a change in the orbits of the cathode-ray particles coming from the sun. It has been found by **Störmer** that rays which are to arrive at the earth must start in directions that lie within very narrow limits. Now the magnetic field of the moon might change the direction of the rays, and thus a number of rays may reach the earth, which otherwise would escape from it. At present, mathematical investigation has not been carried so far that the magnitude and variation of such an effect can be exactly calculated.

From a simple consideration, however, we are able to estimate the character and magnitude of the indirect effect of the moon's field, compared with the direct effect of the radiation from the sun.

The earth and the moon are put into an almost uniform field of electric radiation. Let us imagine a sphere (S) drawn with the earth as centre and with a radius equal to the distance between the earth and the moon. The radiation which must consist of very stiff rays, will enter mostly on one hemisphere and pass out of the sphere on the opposite side. On the surface of this sphere there will be a number of spots $a_1 . a_2 . . . . a_n$ where those rays enter that reach the earth.

Let us first consider the case of the moon being so far from the areas $a_1 \; a_2 \ldots a_n$ that its magnetic field at those spots is very weak. This only requires the distance from the moon to the spots to be of the order of the radius of the sphere, because we know that the direct magnetic effect of the moon upon the earth is very small.

On this assumption, the moon has no appreciable influence on the rays that come directly from the sun to the earth; but we nevertheless have to consider the effect of those rays which pass near to the moon and are so greatly deflected that their previous history, so to speak, is totally wiped out so as to leave the moon in every variety of direction. The earth will be exposed to the action of two fields of radiation, one from the sun and the other from the moon. But the rays, of which the history is effaced are scattered in all directions, so that the field of radiation from the moon must be extremely weak as compared with that from the sun. As the plane of the moon's orbit forms a comparatively small angle with the ecliptic, the directions from the moon are distributed very nearly in the same way in relation to the earth's magnetic field, as the directions from the sun; so that on an average the magnetic effects produced by the two fields must be in proportion as the intensities of the radiation.

A similar consideration will show that the effect of any secondary radiation caused by the impact of electric radiation on the moon must be very small compared with the direct effect from the sun.

If, however, the secondary electric rays are caused by radiations such as light or γ rays which do not produce magnetic effects themselves, or if the moon is the source of primary electric radiation, we are *a priori* unable to say anything definite about the order of magnitude of the effect of the moon as compared with that of the sun. We shall have to look at the observed magnetic effects for information regarding this point, and, in fact, the diurnal distribution of disturbances will give us some information in this respect.

In the case of the moon being near to the areas $a_1 \; a_2 \ldots a_n$ the effect of the rays of which the previous history is wiped out, will be of the same order as before, but now the moon may have an appreciable effect on the rays which would otherwise have reached the earth. The moon's field in this case will act as a shield for the rays, and thus be able to diminish the effect of an already existing radiation. It might be possible that the perturbations consisted in a diminution of a radiation which was constantly being given out; effects of this kind are not impossible. But we cannot suppose that the great polar storms here considered have been caused in this way. That the polar storms are due to something positively occurring is evident from the connection with aurora borealis and sun-spots, and besides great storms are found in the most varied positions of the moon.

THE SEAT OF THE RADIANT SOURCE.

97. The eruptive character of the occurrence of magnetic storms, indicates that the period might be explained by a periodic change in the intensity of the source, just as certain periods have been found for the eruption of geysers.

But such an explanation cannot be maintained; for, owing to the rotation of the sun, the radiation would have to issue from a number of sources, and it is hardly conceivable that a large number of sources would vary with the same period and be in the same phase.

The only possible explanation left, seems to be that the period of storminess is the synodic period of revolution of some layer of the sun. From this view it follows, that if the storminess is to show only one distinct monthly maximum, there must be a fairly limited region of the sun, the activity of which as a radiant source, is predominant, and we see from the curves fig. 190 that it must maintain its predominance during several revolutions of the layer to which the source is attached.

If such a source on the surface of the sun gave out electric radiation from a surface element according to the same law as for the radiation of light, the storminess due to such a source ought to vary approximately according to a sine or a cosine law, or

$$S^T = A \sin 2\pi \frac{t}{T}$$

in which T is the time of revolution of the source; and in which it is to be remembered that only positive values have a physical interpretation. We should get a number of separate waves according to this sine law, the effect of the predominant source would be felt during half the period, or 14.6 days. The effect would increase somewhat rapidly, but in the neighborhood of the maximum it would keep nearly constant for several days.

The curves of storminess, however, show a very marked difference from the sine form. The effect of the predominant source, far from being felt during half a period, is generally only felt for a few days at the time of the maximum, which occurs suddenly, and assumes a very pointed form.

How can this discrepancy be explained? There are only two possible explanations,

(1) That the suddenness is due to an eruptive character of the source, or

(2) That the radiation is greatly predominant in certain directions.

In view of the violent changes observed in the upper layers of the sun, great and sudden changes in the ray-emission will probably take place and influence the character of the phenomenon; but such changes alone are insufficient to explain the character of the variation in storminess. Above all, it can hardly account for the comparative regularity with which the maxima occur.

On an average, the source must be quite as active when it is turned away from the earth as when it is turned towards it. If the maxima were solely determined by the eruptive changes, there would be far greater changes in the length of the period of storminess than are actually observed. We see, from the curves, that the periods within each group of maxima only show comparatively small differences. It is, of course, conceivable that there might be a period of variation of the source, which could produce the observed effect; but from a physical point of view it is scarcely probable that a period of eruption would be so regular, and farther that it would coincide with another quite independent period — the period of the sun's rotation.

On the other hand, the second assumption, namely that the radiation is greatly predominant in certain directions, gives at once a simple explanation of the variation of storminess.

According to this view, the radiation would be mainly restricted to certain narrow pencils starting from the source.

When the earth comes sufficiently near to such a pencil, there will be a perturbation.

Let us suppose, that the position of the source is such that the pencils strike the earth and produce a perturbation. Let at the moment considered the heliographic longitude of the centre of the sun's disc be λ_0, and that of the source λ. To explain the observations we must assume that the angle $\lambda - \lambda_0$ cannot vary greatly for the various pencils of rays which strike the earth; for if the pencils could start from the source in the most varied directions in relation to the surface of the sun, the effect would be the broadning out of the maxima, or the causing of an enormous variation in the interval between successive maxima. As long as the maxima keep their pointed form, and occur at fairly constant intervals we are justified in assuming that the final directions of the pencils relative to the sun's surface are, roughly speaking, the same.

As the only singular direction from a plane is its normal, and as there is only one predominant direction of the pencils, I think there can be little doubt that the radiation *starts* almost perpendicular to the surface of the sun. If, after starting, the rays were not exposed to any deflecting field of force from the sun, $\lambda - \lambda_0$ would be nearly equal to zero.

It has been found by many observers, that there is a lag, or interval between the passage of a sun-spot across the central meridian, and the occurrence of the magnetic storms.

On the assumption that sun-spots act directly as a source, and the velocity of propagation of the radiation is at least as great as that of ordinary cathode rays, the lag would mean that $\lambda - \lambda_0$ had a positive value. In order, then, that the radiation, starting normally, shall reach the earth, the existence of deflecting forces is necessary. Assuming that the deflection is due to a magnetic field, and knowing, from other considerations, the stiffness of the rays, I have recently(¹) calculated the intensity of magnetisation of the sun, that would account for the observed lag.

The active area must be comparatively limited, for it is very seldom that a storm lasts for more than 24 hours. Very often several storms occur in succession at the time of the maximum, which indicates that the active area is more like a group of active spots.

The theory of the confinement of the electric radiation causing magnetic polar storms, to certain sources, which send out narrow pencils of rays, was deduced as a natural consequence of the character of the variation in storminess, and is the one that I have adopted in my previous works on these matters, as, for instance, in "Recherches sur les Taches du Soleil", read before the Christiania Videnskabsselskab on Feb. 24, 1899, where, on page 2, the view is clearly expressed in the following terms:

"Dans un mémoire inséré aux Archives des sciences phys. et. nat. Genéve, juin 1896, j'ai cherché à expliquer la relation existant entre les taches du Soleil d'une part, et les aurores boréales et les perturbations magnétiques de l'autre. Dans mon hypothèse le Soleil émet de longs faisceaux de rayons cathodiques, qui sont en partie l'objet d'un succion dans l'atmosphère terrestre de la part des pôles magnétiques, chaque fois qu'un des faisceaux cathodiques en question frôle notre planète d'assez près."

It is a matter of great interest that subsequently Mr. Maunder, from a study of perturbations observed at Greenwich, was led to the very same conclusions.

Nor does the physical side of the question present any serious difficulties. In the corona and the comets' tails, we are actually examining radiations having definite directions. The difficulty in this respect is not that we are in want of a possible explanation, but rather that we have too many of them.

In order to explain the properties of the corona and the tails of comets, it has long been supposed that the sun should possess an electric field, in which case the cathode rays might leave the sun in a direction perpendicular to its surface, just as, in a vacuum-tube, the cathode rays start perpendicularly to the cathode surface.

(¹) K. Birkeland: C. R. 1910.

Further, in the pressure of light, we have a force that would also be able to carry small dust-particles, charged or uncharged, away from the sun. This force undoubtedly plays an important part in the economy of the universe, and has been utilised by ARRHENIUS to explain aurora borealis and magnetic disturbances. Our hypothesis does not, however, require the influence of radiant pressure. If the rays are suddenly brought into being e.g. as cathode or β rays of great penetrating power, and consequently with a velocity very nearly equal to that of light, *the influence of the light-pressure on the orbits of the rays will be insignificant.*

The recent discovery by HALE of strong magnetic fields, existing in the neighborhood of sun-spots, furnishes us with a new possible explanation; for it has been found that the lines of force are nearly normal to the surface of the sun, and in order to get out, the rays would have approximately to follow the lines of force.

The most usual way of obtaining a beam of nearly parallel rays, is to let the radiation from the source pass through an aperture. Applied to the sun, it would mean that the radiation originating mostly from the interior, could only get out through an aperture in the sun's upper layers.

We are not in possession of sufficient data to tell which of these is the right explanation. It may even be that all of them may be present and play a part in the phenomenon. I think, however, that a discussion of the various possibilities will be necessary, if we shall hope to attain to a more intimate knowledge of the mechanism of the solar activity giving rise to the magnetic storms and aurora; for it is through the conclusions drawn from each hypothesis that we are able, by comparison with experiments, to test it.

The last purely mechanical explanation by means of apertures is really a very simple and a very fascinating one, which I think is deserving of attention. The advantage of the "aperture hypothesis" is that it not only explains that the radiation escapes in a certain direction, but also the fact of its being confined to narrow pencils. Through the sun-spot-hypothesis of Mr. WILSON, we have long been familiar with the idea of apertures in the sun's outer layers, and recently EMDEN, in his theory of the sun, has assumed the existence of vortices with their vortex-filaments ending on the surface of the sun, so as to form a kind of opening into the interior; and the existence of vortices has been brought to full evidence trough the spectroheliographic researches by HALE at the Mount Wilson Observatory.

The length of the period of storminess leads us to suppose, that the source, if situated near the photosphere, would have a latitude of about $\pm 30^0$. As the sun's equator forms quite a small angle with the ecliptic, and since the radiation, as we have seen, most probably issues in narrow pencils perpendicular to the sun's surface, radiation from sources in this latitude would not strike the earth at all. If it can be taken as a general rule, that the time of rotation increases towards the interior, the source, if situated nearer the equator, would be below the photosphere, which is what would be expected if the radiation were limited by apertures.

If we do not accept the assumption of apertures, the question then arises, how are the rays able to penetrate the great layers of matter above the source? The rays, which produce the magnetic storms and aurora must have a great penetrating power compared with that of other known electric radiations; but still they are unable to penetrate more matter than the earth's atmosphere. In order then, that the radiation from a source situated below the photosphere shall get out, the source must produce radiation, as a kind of secondary effect, from matter nearer the surface of the sun. One possibility is, that the source is sending out active matter of some kind, which floats above the source. We expect that important information in this respect may be obtained from the spectroheliographic observations of the sun's disc.

The distribution of calcium is especially interesting from the fact that this metal at high temperatures is found to give out a large amount of corpuscles.

SUN-SPOTS AND STORMINESS.

98. In fig. 190 the storminess is compared with the occurrence of sun-spots. The sto that of the horizontal Component at Axelöen put up for each day. The sun-spot curves *B*, *C* deduced from the "Results of Measures made at the Royal Observatory, Greenwich, of Phot the Sun taken at Greenwich, in India and in Mauritius".

The curves *D* give the total visible area of sun-spots for each day during the peri observations.

If the radiation started perpendicular to the surface of the sun, it would not be the t area of sun-spots that would be significant with regard to magnetic storms and aurora; bu spots which at the time under consideration were near to the central meridian of the sun. represents for each day the number of sun-spots for which $|\lambda_s - \lambda_0| \lesseqgtr 10^0$, where λ_s is the h longitude of the sun-spot centres as given in the Greenwich records. The dotted curves in represent the area of the umbra, the curves drawn in ful indicate the area of the whole spot.

Finally the graph *B* represents the time of passage of the central meridian of the vario of sun-spots given in the Greenwich records. At the time of the passage, an ordinate is dra length is proportional to the largest total area which the group has attained during the time i observed. Thus the graph does not give the area that the group actually had at the time of th We have even gone so far as to put up groups, which have not been visible at all at the p the central meridian. The reduction to central meridian has been done by interpolation, or if only appears on one side of the central meridian, we have extrapolated by means of the sync of the revolution of sun-spots.

A comparison between magnetic storms and sun-spots *shows that the appearence of large sun-spots does not take place so regularly as the principal maxima of storminess.* Very often lar of storminess are not accompained by any sun-spots at all.

In the following table are given a number of sun-spot groups for which there seems undoubted coincidence with magnetic storms.

TABLE XCII.

Sun-spot Group	Time of passage of Central Merid.		Time of Max. of Mag. Storminess		Lag.
4980	Sept.	19.8	Sept.	22.2 & 19.2	0.9 days
4981 & 4982	"	28.2 & 28.0	"	30	1.9 "
4983	Oct.	10.3	Oct.	10.5	0.2 "
4986	"	24.9	"	26.6	1.7 "
4987	"	29.6	"	30.5	0.9 "
4990	Nov.	20.0	Nov.	23.5	3.5 "
4999	Jan.	1.7	Jan.	4.5	2.8 "
5001	"	24.7	"	26.0	1.3 "
5002 & 5003	"	29.0 & 28.9	"	29.5	0.6 "
5013, 5014, 5016	March	27.1, 28.2, 29.1	March	31.0	1.9 "
5015	April	2	April	5.4	3.4 "
5017	"	8.7	"	8.6	—0.1 "
			Mean of the Lag: + 1.6 days		

In those cases for which a coincidence exists, the storms, as usually found, occur somewhat later The average lag 1.6 days gives $\lambda_s - \lambda_o$ equal to 21°, where λ_s and λ_o are respectively the longitudes of sun-spot and central meridian at the time of the maximum of storminess. The lag here found is only half as large as that found by RICCÒ (¹) for a number of very great storms.

As regards the principal maxima, those of September, October and November coincide with quite large and distinct groups of sun-spots. After that a marked maximum of storminess reappears quite regularly at the end of December; but the sun-spots have disappeared. Nor do the great principal maxima of February and March coincide with sun-spots. Not until April does there seem to be an apparent coincidence.

Regarding the connection between sun-spots and storminess, it seems improbable that the sun-spots can be the direct cause of the magnetic storms; for the sun-spots appear to be rather irregular in their occurrence and with a somewhat different period of revolution than the source of electric radiation. If, then, the source were formed in any way by sun-spots, we should hardly find the variation of storminess so regular as it was actually found to be during the period of our observations. The results suggest that sun-spots and magnetic storms are both of them manifestations of the same primary cause.

The storminess seems to go on whether there are sun-spots or not. But also from our point of view we shall expect to find that the passage of sun-spots is accompanied by magnetic storms; for the existence of sun-spots is to be considered as a visible sign of a great activity of the primary source. The effect will undoubtedly in a number of cases be the same as if the sun-spots themselves were sending out pencils of electric radiation. The strong magnetic fields near the sun-spots show that violent currents of electricity are actually operating in the sun-spots, and these currents may only be another effect of the same electric activity which produces the magnetic storms and aurora.

As we saw, the existence of one well-defined monthly maximum would require that there were one single complex of sources which was greatly predominant with regard to emission of electric radiation. It must, however, by no means be regarded as a matter of necessity that the same source should always maintain its predominance; but it is quite possible that the intensity of one source may diminish, and that of another increase so as to take the lead for a certain number of revolutions of the sun, until a new one is called into play to become the principal source.

In fact we saw that the results of our observations were best explained by dividing the principal maxima into two groups, and in view of the previous considerations these two groups correspond to two different complexes of sources. The first group has its last principal maximum at about January 28 and the second one its first principal maximum at about February 7, consequently the difference in heliographic longitude of the two sources should be about 120°.

This change in position of the source must be taken into account when we are dealing with the determination of the monthly period of storminess. In fact, if the period were deduced in the ordinary way from material covering a great many years, the shifting of the source would have the effect of masking the "real" period, or the period deduced from a very long interval of time might be quite different from that here found from the intervals between successive maxima. It is really no wonder then that various authorities have found a different monthly period.

(¹) Nature 82, p. 8. 1909.

ANNUAL VARIATION OF STORMINESS.

99. Observations have not been made for a sufficiently long time for an exact determination of the annual inequality. The longest period of observations, that of Axelöen, only covers a time of about nine months.

The average total storminess at Axelöen for each month during this period is given in the following table.

TABLE XCIII.

Month	S^T	L
Sept.	35.07 γ	+ 7.0°
Oct.	33.90 „	+ 5.5
Nov.	37.21 „	+ 2.5
Dec.	24.92 „	— 1.1
Jan.	25.71 „	— 4.5
Feb.	20.55 „	— 6.6
March	38.18 „	— 6.9
April	48.75 „	— 5.4
May	50.37 „	— 2.5

The numbers indicate two maxima, one in the autumn and one in the spring, or about the same type of variation as found for the annual variation of aurora borealis. Under the heading L is given the average latitude of the centre of the sun's disc for each month. We notice that the main feature of the variation of S^T follows that of the absolute value of L; but the maxima and minima of S^T seem to occur somewhat later than the corresponding ones for L.

It thus seems as if the annual inequality may be explained by assuming that the intensity of this electric radiation on an average is weaker at the sun's equator than at some distance from it; for if the radiation leaves the sun perpendicular to its surface, and if the sun's magnetic axis forms an insignificant angle with its axis of rotation, the rays which at any time shall reach the earth must start from points having about the same heliographic latitude as the centre of the sun's disc. It must however also be taken into account that the pencils consisting of diverging rays from the solar spots of radiation are probably somewhat bent towards the magnetic equator of the sun. We shall return to this question in the chapter on the results of the experimental investigations with a magnetic cathode-globe in vacuum-cases.

A résumé of the above investigations on "Storminess" at our four polar stations has already been published in a communication to the Congrès de Radiologie in Brussels, 1910. See also Arch. de Genève XXXII, August, 1911, pp. 97—116.

Since writing the above, I have seen a paper by Dr. BIDLINGMAYER, published by the Kaiserliches Observatorium at Wilhelmshaven (Berlin, 1912), in which the author has introduced the idea "terrestrial-magnetic activity", which has certain points of resemblance to that of "storminess" introduced here. Dr. Bidlingmayer has employed the idea for observations from Wilhelmshaven in the year 1911.

It is a highly interesting fact that Dr. CHREE in his most valuable "Studies in Terrestrial Magnetism", London, 1912, Chap. XVII, makes some reflections concerning sunspot relations that agree well with the results obtained by our analysis.

Our results on storminess here given were printed as early as 1910, and only the last two pages have been reprinted in 1913, two lines having been removed and a few lines added in conclusion.

Diurnal Distribution of Storminess
Matotschkin Schar

Fig. 191.

Diurnal Distribution of Storminess

Kaafjord

Diurnal Distribution of Storminess

Axelöen Pl. I

Diurnal Distribution of Storminess

Axelöen Pl. II

Fig. 194.

Diurnal Distribution of Storminess

Dyrafjord

Diurnal Distribution of Storminess
(Dec. 2 to March 1)

Fig. 196.

:tor Diagrams for „the Average Polar Storm"

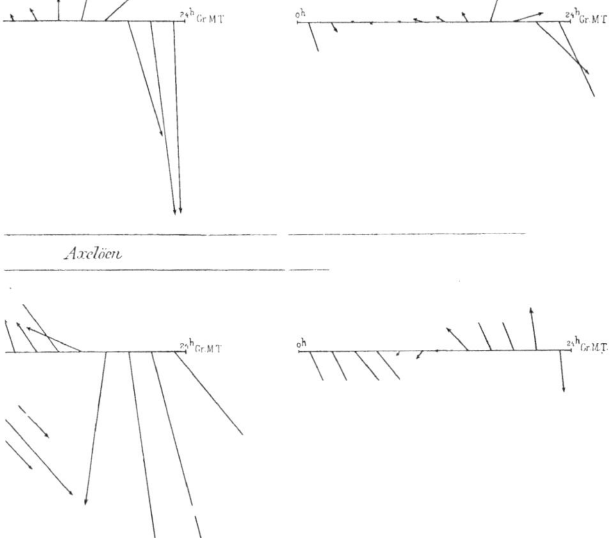

Fig. 197.

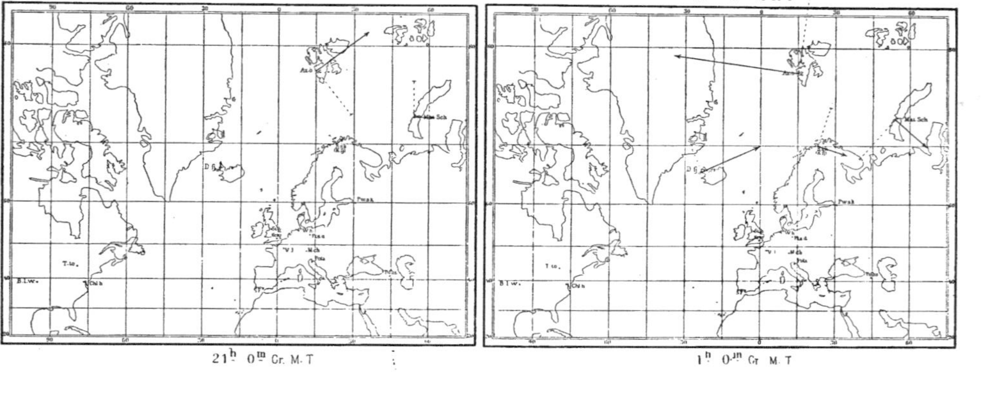

Scale
13h 0m Gr. M. T.
21h 0m Gr. M. T.
17h 0m Gr. M. T.
1h 0m Gr. M. T.
Azo.
Mat. Sch.
D.G.
Pwsk.
V.I.
Mch.
Pola
Tiflis
T.to.
Blw.
Chlh.

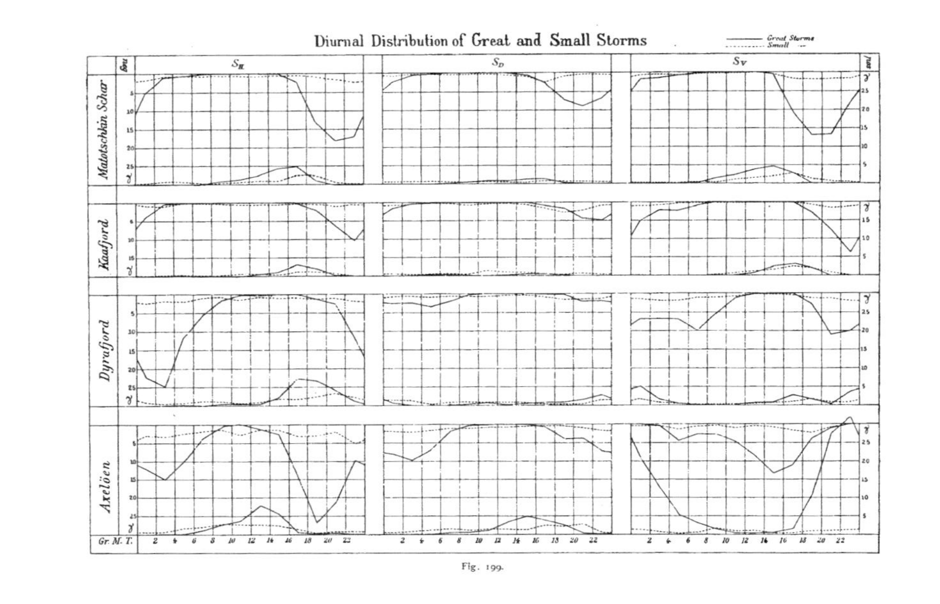
Diurnal Distribution of Great and Small Storms
Great Storms
Small
S_{II}
S_D
S_V
neg
pos
Malotschkin Schar
Kaafjord
Dyrafjord
Axelöen
Gr. M. T.
2 4 6 8 10 12 14 16 18 20 22

Fig. 199.

ON THE DIURNAL DISTRIBUTION OF STORMINESS.

100. The distribution of storminess in the various hours of the day is represented in the plates (figs. 191, 192, 193, 194 and 195).

The arrangement will be seen from the plates. Curves are given for each thirty days' period, and also one series of curves at the bottom of each plate giving the mean storminess-distribution for the whole period of observation.

For each period the following curves are given:

(1) The positive storminess S^p is represented by ordinates going upwards from the bottom line which is taken as the time-axis.

(2) The negative storminess S^n is represented by ordinates going downwards from the top line which is taken as the time axis.

(3) The values of $S^p - S^n = S^d$ are represented by ordinates taken positive upwards, and these curves are drawn in full on the diagrams, while S^p and S^n are dotted lines. The vector S^d whose scalar quantity is equal to

$$\sqrt{\left(S^d_H\right)^2 + \left(S^d_D\right)^2 + \left(S^d_V\right)^2}$$

has a very simple physical interpretation. It gives at any hour of the day the perturbing force for the average magnetic storm for the period considered To fix the ideas let us assume that all storms occurring during a certain period took place on the same day, but in such a way that the hour of the day was unaltered. We should then get a certain disturbance, the perturbing force of which at any hour would be given by the equations:

$$P_h = n\,S^d_H$$
$$P_d = n\,S^d_D$$
$$P_v = n\,S^d_V$$

where n is the number of days in the period in question.

On looking at the curves, we notice immediately that the storminess shows a very marked diurnal variation. Comparing curves for the same station and the same magnetic element we see that for different monthly periods they show very nearly the same course.

The absolute magnitude of the storminess may vary from one month to another, but the type of variation is always the same, namely that which is represented by the average curve at the bottom of each plate. This constancy of distribution of polar storminess is a matter of great interst. It shows that the amplitude and form of the average curve is by no means an accidental one, for the same type is found, and almost equally well marked, for curves representing a very short period.

POSITIVE AND NEGATIVE STORMINESS.

101. The positive and negative storminess is defined quite arbitrarily from the sign of the component of the perturbing force. There is then no necessity of any connection between, say, the positive storminess of the various components for the same station.

And further, in view of the local character of the storms near the auroral zone, the distances between the stations are fairly large, and therefore, even for the same magnetic element we may not get correspondence in positive or negative storminess at the various stations. The regularity actually shown by the elements at the various stations is rather to be considered as a strange coincidence than as a matter of necessity.

With only a few exceptions, the positive storminess shows the same type of curve for all three elements at the four stations. The negative storminess also, with the exception of a few cases, shows roughly the same type of variation at all stations and for all elements.

The cases that do not follow this rule are especially the vertical components for Axelöen and Dyrafjord. For S_V at Axelöen the diurnal period is extremely well marked, but conditions are reversed. S_V^n for Axelöen varies in a way which corresponds to that of the positive storminess; and S_V^p for Axelöen corresponds to the negative storminess found elsewhere.

For Dyrafjörd the storminess in H is greatly predominant. For S_D the amplitudes are very small, and for S_V the curve of negative storminess shows two distinct maxima, one of which corresponds to the maximum of the positive, and one to that of the negative storminess of the horizontal force.

The reason for these similarities in the variation of the two types of storminess, as well as the exceptions mentioned, will become clear through the treatment of the "average storm".

As a result of the comparison of curves, it appears that at the four stations there are mainly two types of storminess, which we shall call P and N storminess, and which, with the few exceptions mentioned, correspond respectively to the positive and negative storminess.

P AND N STORMINESS.

102. The diurnal period of the P storminess is less marked than that of the N storminess. The P storminess sets in gradually, and gradually disappears. The N storminess begins and ends more suddenly, and obtains a much greater maximum value than that of the P storminess.

While the time of the maximum of storminess is usually well defined, the exact hour of minimum is difficult to tell. In fact the minimum is more to be characterised as a calm period lasting for several hours. As a rule there will be small P storminess when the N storminess has its maximum, and a small N storminess during the interval of great P storminess; but this is merely what should be expected, and it shows that at the time of day when one type of storminess is operating, there will be little storminess of the other type. What is more remarkable, however, is the existence of an interval which *is absolutely calm*, where both N and P storminess are small.

The time of occurrence of the maxima of P and N storminess, and the interval of calmness, are given in the following table. The numbers are taken from the average curves at the bottom of each plate, which give the mean for the period of observation.

TABLE XCIV.

	Gr. M. T.								Longitude	Local Time			
	P Storminess				N Storminess					Calm Period		Maximum	
	S_H	S_D	S_V	Mean	S_H	S_D	S_V	Mean		Interval	Mean	P. st.	N. st.
Matotchkin-Schar	15	15	15	15	21	21	20	20.7	3.6^h E	$8-13^h$	10.5^h	18.6^h	0.3^h
Kaafjord	17	17	17	17	22.5	22	23	22.5	1.5 "	7–14	10.5	18.5	0.0
Axelöen	13	19	17	16.3	$\frac{19+27}{2}$	27	23	24.3	1.0 "	7–10	8.5	17.3	1.3
Dyrafjord.	18			18	2.2			2.2	1.5 W	8.5–12.5	10.5	16.5	0.7
											Mean	17.7^h	0.6^h

For Dyrafjord it is only in the horizontal force that the two types are well separated; for the other two magnetic elements the storminess is much smaller, and each group of storminess is divided between the positive and negative storminess, so that the N and P groups get mixed up. In consequence the time of maximum has been determined from the horizontal force only.

For Axelöen S^n_H has a large value during a long interval, and has in fact two maxima. The number given is the mean of the time for the two maxima.

The calm period is long and well defined for Matotchkin-Schar and Kaafjord, for Dyrafjord shorter and not so quiet, while for Axelöen it has more the character of a minimum than that of a quiet interval.

We see from the table that in spite of the rather large differences in longitude of the stations, the N storminess for the various stations has its maximum very nearly at the same local hour, about half past twelve at night. Also the P storminess has its maximum at the same time of day about six o'clock in the evening, and the calm period is always found in the forenoon. Except at Axelöen, the middle of the calm interval is at half past ten in the morning.

This result shows that the storminess near the auroral zone follows the diurnal motion of the sun.

PROPERTIES OF THE "AVERAGE POLAR STORM".

103. To obtain comparable numbers we must consider the storminess for a period common to all four stations. The storminess for the three months December, January and February, is given in table XCV and graphically represented in fig. 196. As we have already mentioned S^d can be considered as a vector representing the perturbing force of the average polar storm, and in the usual way it can be represented by a vector diagram of some sort.

TABLE XCV.

Mean Storminess for the Period December 2—March 1.

Station	Gr. M. T.	0 – 2		2 – 4		4 – 6		6 – 8		8 – 10		10 – 12	
		+	−	+	−	+	−	+	−	+	−	+	−
Matotchkin-Schar	S_H	0.25	7.27	0.73	2.03	0.72	1.13	0.38	1.05	1.28	0.60	1.85	0.68
	S_D	0.67	3.04	0.62	0.87	0.81	0.48	0.90	0.47	0.95	0.37	1.60	0.37
	S_V	0.37	2.31	0.25	1.82	0.21	1.36	0.85	0.53	1.72	0.27	3.22	0.22
Kaafjord	S_H	0.06	4.79	0.33	1.47	0.68	0.52	0.20	0.33	0.28	0.22	0.72	0.51
	S_D	0.67	2.32	0.48	1.13	1.01	0.64	1.11	0.34	0.57	0.67	1.46	0.39
	S_V	0.08	6.13	0.08	2.93	0.28	2.64	0.16	1.27	0.27	0.30	0.77	0.21
Axelöen	S_H	0.53	14.90	0.65	18 33	1.57	12.22	3.13	5.22	5.72	1.78	6.35	2.85
	S_D	0.59	9.94	0.93	11.77	1.29	8.21	2.13	2.77	1.78	1.24	2.60	0.88
	S_V	22.24	0.40	13.58	0.93	5.48	5.77	3.63	3.30	4.04	3.13	1.13	5.71
Dyrafjord	S_H	1.07	24.48	0.62	26.57	0.93	13.48	1.30	6.37	1.75	2.02	1.25	1.83
	S_D	2.44	3.02	1.49	2.71	0.55	4.11	1.10	2.46	1.48	0.93	1.58	0.59
	S_V	6.81	8.10	2 41	8.59	1.12	8.65	0.40	11.93	0.37	6.22	0.74	2.08

TABLE XCV continued.

Station	Gr. M. T.	12–14 +	12–14 −	14–16 +	14–16 −	16–18 +	16–18 −	18–20 +	18–20 −	20–22 +	20–22 −	22–24 +	22–24 −
Matotchkin-Schar . . .	S_H	3.27	0.88	5.25	0.62	7.52	2.72	3.43	13.97	0.60	19.57	0.15	19.38
	S_D	1.39	0.72	1.73	1.39	2.27	5.11	1.16	8.02	0.70	9.22	0.62	7.16
	S_V	5.13	0.36	6.49	1.11	5.45	12.48	1.29	18.79	0.50	18.42	0.22	9.74
Kaafjord	S_H	1.18	0.54	1.55	0.50	4.24	0.24	3.30	2.41	1.25	6.99	0.09	10.79
	S_D	1.40	0.33	1.45	0.68	1.26	2.62	0.72	4.21	0.65	6.49	0.50	5.83
	S_V	1.78	0.39	3.94	0.34	5.49	0.16	3.99	3.78	0.77	9.06	0.02	15.08
Axelöen	S_H	10.55	2.67	7.98	4.13	1.92	17.42	0.25	29.90	0.92	23.10	0.73	14.78
	S_D	4.83	0.58	6.34	0.59	6.38	1.27	5.81	4.31	3.23	4.59	0.68	8.79
	S_V	1.29	8.62	0.79	13.95	1.90	12.82	11.67	6.26	28.59	1.84	32.89	0.66
		1.58	1.10	4.10	0.90	9.17	0.82	9.43	2.12	7.72	3.28	3.47	7.67
		1.33	0.34	1.51	0.36	1.24	1.08	1.38	1.65	1.97	3.20	4.44	2.88
		1.16	0.57	1.58	0.15	3.83	0.59	3.34	4.22	1.39	12.35	4.40	12.02

Vector diagrams of the horizontal component of S^d are given in fig. 197 for the four stations. The vectors are drawn from points on a time-axis for every second hour.

At Matotchkin-Schar, Kaafjord and Dyrafjord, there is an interval of several hours in the forenoon with very small forces corresponding to the quiet period. At Axelöen the interval is very short. In the afternoon the perturbing force increases, and assumes at each station a nearly constant direction towards the north-west, which is maintained for several hours. Then, all of a sudden, the perturbing force turns round, takes up a direction nearly opposite to what it was in the afternoon, and assumes a comparatively large value.

If the average storm is represented by current-arrows, we should for each station get two typical current systems.

(1) One system with maximum about six o'clock in the afternoon with its current-arrows turned eastwards along the auroral zone.

(2) A second type with its current-arrows turned westwards along the auroral zone, and with its maximum about midnight.

In the following table is given the time-interval for a small perturbing force, and the times of maximum of the horizontal component of the perturbing force of the average positive and negative storms.

TABLE XCVI.

Station	Local Time			
	Small Force		Maximum	
	Interval	Mean	Pos. Storms	Neg. Storms
Matotchkin-Schar . . .	7–14	10.5	18.6	24.6
Kaafjord	6–14	10.0	18.5	24.5
Axelöen	8–10	9.0	14.0	23.2
Dyrafjord	7–12	9.5	16.0	25.5
		Mean	16.8	24.5

The positive and negative average storms have respectively their maxima at about the same time as the P and N type of storminess. The two types of storminess are merely another aspect of the existence of the two types of polar storms. It is, however, by no means a matter of course that the maxima of storminess should fall on the same hours as the maximum of perturbing force of the average storm; for from the equation $S_H^d = S_H^p - S_H^n$ it follows that S^d might be small even when both S^p and S^n are large. The coincidence regarding the occurrence of maxima of N and P storminess on the one side, and the perturbing force of the positive and negative average storm on the other, is a consequence of the fact that the occurrence of the two types of storms does not greatly overlap, *but that each type is mainly restricted to its own time of day.*

Some fields of the average storm are represented on the four charts, fig. 198.

The first chart gives the field at $13^h\ 0^m$ (Gr. M. T.) corresponding to the beginning of the positive storm. We notice that it is breaking in from the north-east. It is strongest at Axelöen and Matotchkin-Schar. The current-arrows are directed eastwards along the auroral zone. For Dyrafjord, Kaafjord and Matotchkin-Schar, the vertical component of the perturbing force is directed downwards, but upwards for Axelöen, showing that the current goes to the north of the three former stations, but to the south of the latter.

The second chart gives the field at $17^h\ 0^m$, when the negative storm is on the point of breaking in from the east. At Dyrafjord only the effect of the positive storm can be noticed. At Kaafjord the arrow is slightly turned, and at Matotchkin-Schar even more so. At Axelöen, however, it is almost completely turned to the west. It looks as if the force at Axelöen should at this hour be mostly due to current-systems different from those producing the effect at the other stations, and as we shall see later, this is also the case.

On the third chart for $21^h\ 0^m$, the negative storm dominates at Axelöen and Matotchkin-Schar and almost completely at Kaafjord; but the effect of positive storms is still most prominent at Dyrafjord.

On the last chart for $1^h\ 0^m$, the negative storm dominates at all four stations; but it is now strongest at Dyrafjord. The vertical component is directed downwards for Axelöen and upwards for the other three stations, showing that the currents on an average at this time are running above the earth's surface, and between Kaafjord and Axelöen.

Through the treatment of separate perturbations we were led to the assumption of two types of polar storms, which we called the positive and the negative polar storms. The statistical treatment of the whole material shows exactly the same two types.

The average storm in the afternoon has the properties of a typical positive polar storm; the midnight average storm has the properties of a negative polar storm; and we see that the predominant part of the storminess, at least at the three southern stations, is made up of these two types.

The cause of the singular character of S_D and S_V for Dyrafjord and S_V for Axelöen, will now become evident. The reversal of the conditions of storminess of S_V for Axelöen only means that the storm-centres of the two types of polar storms, positive as well as negative, pass between Kaafjord and Axelöen. The small amplitude in the S_D curve for Dyrafjord shows that the current on an average is nearly perpendicular to the magnetic meridian at this place. The storminess of the vertical intensity at Dyrafjord shows that the current-systems usually pass near the zenith, usually somewhat to the north of the station.

COMPARISON OF STORMINESS AT THE FOUR STATIONS.

104. In the following table is given the total storminess for the months December, January and February, and the mean of the whole three months' period. S_R is the storminess expressed in relative numbers, the total storminess of Kaafjord being put equal to one.

θ is the angular distance to the magnetic axis.

TABLE XCVII.

Station	Dec.		Jan.		Feb.		Dec. 2—March 1		θ
	S^T	S_R	S^T	S_R	S^T	S_R	S^T	S_R	
Matotchkin-Schar .	11.1 γ	1.73	14.7 γ	2.13	11.0 γ	1.36	12.2	1.72	25.3°
Kaafjord	6.4 „	1.00	6.9 „	1.00	8.1 „	1.00	7.1	1.00	24.7
Axelöen	24.9 „	3.89	25.7 „	3.73	20.6 „	2.54	23.7	3.34	16.3
Dyrafjord . . .	13.4 „	2.09	14.6 „	2.12	17.2 „	2.12	15.1	2.13	18.1

The magnitude of the storminess follows in the order Axelöen, Dyrafjord, Matotchkin-Schar, and Kaafjord. The two stations Axelöen and Dyrafjord with the smallest angular distance θ have the greatest storminess. The storminess, however, is not quite symmetrically arranged with regard to the magnetic axis, for Kaafjord, with an angular distance of 24.7° has only about half the storminess of Matotchkin-Schar with a still greater angular distance; and both stations are situated to the south of the auroral zone. Dyrafjord and Matotchkin-Schar have nearly the same storminess, although their angular distances are greatly different.

The relative storminess of Axelöen is the most remarkable. Although the great storms have their centres between Axelöen and Kaafjord, and generally quite as near to the latter station as the former, the storminess at Axelöen is more than three times as great as that at Kaafjord.

One possible explanation of the great storminess at Axelöen is, that besides the large storms with their centres between Axelöen and Kaafjord, there are a number of smaller storms which have their centres nearer to the magnetic axis. If so, we should expect the principal maxima of storminess — which are mostly due to the occurrence of large storms — compared with the *average storminess* of the stations to be smaller for stations situated near the pole and the magnetic axis.

The distinctness of the principal maxima is illustrated in the following table.

TABLE XCVIII.

Station	S^T		S^T_1		S^m		S^m/S^T		S^m/S^T_1	
	I	II	I	II	I	II	I	II	I	II
Matotchkin-Schar .	12.2 γ	15.1 γ	8.0 γ	8.8 γ	33.6 γ	46.2 γ	2.8	3.1	4.2	5.2
Kaafjord	7.1 „	9.9 „	4.3 „	5.6 „	21.2 „	28.8 „	3.0	2.9	4.9	5.1
Axelöen	23.7 „	35.0 „	18.7 „	28.6 „	48.7 „	67.0 „	2.1	1.9	2.6	2.3
Dyrafjord . . .	15.1 „	18.3 „	12.1 „	11.2 „	30.2 „	45.5 „	2.0	2.5	2.5	4.1

The columns (I) correspond to the period of three months common to all four stations. The columns (II) correspond to the whole period during which observations have been made at the various stations.

S^m is the average storminess for the principal maxima. There is one principal maximum for each thirty-day period, and for each principal maximum we have taken the storminess of the five-day period, which contains the principal maximum, and which will be the same for all stations.

S_1^T represents the total storminess left when the maximum five-day periods are taken out. We have

$$S_1^T = \frac{n\,S^T - m\,S^m}{n - m}$$

n is the total number of five-day periods in the interval, m is the number of those five-day periods which contain the principal maxima.

If all storms, large and small, had their centres distributed around the same zone, we should expect S^m/S_1^T to be about equal for all stations. We see from the table, however, that the values of S^m/S_1^T show great differences, and in such a way that the ratio is smallest for Axelöen and Dyrafjord with the smallest angular distance θ.

Consequently in between the principal maxima there are a number of storms which have their centres situated nearer the magnetic axis than those of the great storms producing the principal maxima.

Thus the very great storminess at Axelöen compared with that of the other stations is partly due to a number of storms, generally quite small, which have their centres to the north of the auroral zone.

Axelöen also takes up a singular position with respect to the diurnal variation. To show this we shall introduce a quantity, which we shall call the calmness of the station (c), and which is defined as follows:

$$c = \Sigma l \quad \text{and}$$

$$\Sigma = \frac{S^T - S_c^T}{S_c^T}$$

S^T is the total storminess
S_c^T » » » » for the calm period only
l is the length of the calm period expressed in hours.

TABLE IC.

Station	S_c^T		Σ		l	c	
	I	II	I	II		I	II
Matotchkin-Schar . .	2.7 γ	3.2 γ	3.5	3.7	5^h	17.5	18.5
Kaafjord	2.0 „	2.5 „	2.6	3.0	7	18.2	21.0
Axelöen	11.5 „	14.0 „	1.1	1.5	3	3.3	4.5
Dyrafjord.	4.2 „	5.0 „	2.6	2.7	4	10.4	10.8

The numerals I and II have the same meaning as in table XCVIII.

The calmness is about equal for Matotchkin-Schar and Kaafjord. For Dyrafjord it is about half the value of the two former stations, and for Axelöen only about $^1/_5$ of that value.

It is very remarkable that this peculiarity in the position of Axelöen—as will be seen from the curves, fig. 196 — is almost entirely restricted to the negative storminess in the horizontal force, while the storminess in the vertical direction follows the same characteristic course as that found for the southern stations, the N storminess in vertical intensity showing only *one* well-defined maximum at midnight local time.

The negative storminess in H, however, shows two instead of one maximum. There is one maximum four hours before midnight, and one four hours after midnight. The occurrence of these maxima is not accidental, but they are repeatedly found for each monthly period (see figs. 193 and 194).

It is to be expected that the effects of storms with their centres chiefly in the maximum zone of aurora would be felt during a longer time-interval of the day at a station situated nearer the pole. This will be evident from the fact that a place near the pole and magnetic axis will have about the same distance to the centre of the average storm at any hour of the day. Consequently we should expect for Axelöen to get a broadening out of the maxima. But we cannot in this way explain the occurrence of the two distinct maxima at Axelöen; for at the time they occur, we have no corresponding maxima at Kaafjord, which is just on the opposite side of the auroral zone and situated almost on the same meridian. The two maxima of S_H^n for Axelöen must therefore be caused by systems *of a very local nature*, in other words by electric currents near the station. The effect of those systems dies away so suddenly towards the south, that even at Kaafjord their effect is inappreciable.

Remembering that the auroral zone passes between Axeløen and Kaafjord and nearest to the latter station, we conclude that these local current-systems occur at a considerable distance to the north of this zone, and much farther north than the somewhat great midnight storms, which have their centres usually midway between the two stations.

We see that also through the study of the diurnal distribution of storminess we *are led to assume the existence of local storms with their centres to the north of the auroral zone.*

These centres of local storms occurring in the vicinity of the poles, show quite another diurnal distribution than the greater storms in lower latitudes. At Axelöen they are strongest and most frequent at eight o'clock in the evening and four o'clock in the morning; but small local disturbances are here frequently found also during the day-time.

Thus we come to the following conclusion: The *great storminess* and the *very small calmness* of Axelöen compared with the other stations is chiefly due to local disturbances with their centres nearer the poles, and showing another diurnal distribution than the greater and usually more universal storms, which have their centres in the auroral zone.

To judge from the direction of the horizontal component of the perturbing force, these small disturbances should belong to the type of negative storms, because the current-arrows are turned towards the west. I think, however, it will be best to restrict the class of negative polar storms to those which have their centres near the auroral zone; for it is evident that it is only at some distance from the pole that we may expect to find distinct types of positive and negative storms.

Further, when we compare the storminess in the vertical direction, the similarity between the negative storms and these northerly local storms will be difficult to maintain.

Comparing the curves of storminess for Axelöen we find for S_V no sign of maxima corresponding to the two distinct maxima of S_H^n. Thus the *local centres near this station produce practically no disturbance in the vertical direction.*

The simplest explanation of this fact is that the station is placed near a horizontal current-sheet, in other words that the currents producing the local disturbances extend over an area, which has great dimensions compared with the smallest distance between the station and the current-sheet.

The storminess S_V at Axelöen follows exactly the same diurnal distribution as that shown by the storminess at Kaafjord, which indicates that the storminess S_V for Axelöen is mainly due to the positive and negative storms passing between the two stations, and which we found to be caused by currents mostly restricted to a comparatively small cross-section.

TABLE C.

Date	Matotchkin-Schar.						Kaafjord				
	Great Storms				All	Small	Great Storms				All
	$(S^a_H)_g$	$(S^a_D)_g$	$(S^a_V)_g$	S^T_g	S^T	$S^T - S^T_g$	$(S^a_H)_g$	$(S^a_D)_g$	$(S^a_V)_g$	S^T_g	S^T
Dec. 2—6	1.12 γ	0.99 γ	1.70 γ	2.26 γ	4.11 γ	1.85 γ	0 γ	0 γ	0	0 γ	1.48 γ
7—11	3.31 „	2.03 „	7.18 „	8.16 „	11.23 „	3.07 „	1.59 „	0.89 „	0	1.83 „	4.23 „
12—16	2.34 „	1.59 „	5.48 „	6.16 „	8.45 „	2.29 „	0.82 „	1.12 „	0	1.39 „	3.59 „
17—21	0.75 „	0.29 „	0.75 „	1.10 „	2.54 „	1.44 „	0.10 „	0.14 „	0.32 γ	0.36 „	1.12 „
22—26	19.64 „	8.03 „	20.69 „	29.63 „	33.18 „	3.55 „	11.00 „	5.77 „	14.14 „	18.82 „	23.54 „
27—31	2.12 „	0.49 „	3.33 „	3.97 „	7.07 „	3.10 „	1.24 „	1.24 „	1.80 „	2.51 „	4.41 „
Jan. 1—5	2.91 „	0.93 „	5.22 „	6.04 „	8.74 „	2.70 „	0.33 „	0.05 „	1.23 „	1.27 „	4.35 „
6—10	4.02 „	1.33 „	4.81 „	6.41 „	9.40 „	2.99 „	0.37 „	0.23 „	0.84 „	0.95 „	3.56 „
11—15	1.59 „	1.01 „	4.60 „	4.97 „	8.25 „	3.28 „	0.52 „	0.58 „	1.37 „	1.57 „	4.69 „
16—20	2.80 „	0.75 „	4.85 „	5.65 „	8.82 „	3.17 „	0.60 „	0.28 „	2.57 „	2.66 „	5.25 „
21—25	3.92 „	2.18 „	5.86 „	7.38 „	10.65 „	3.27 „	0.45 „	0.64 „	1.70 „	1.87 „	5.51 „
26—30	34.27 „	15.92 „	10.33 „	39.17 „	42.04 „	2.87 „	11.97 „	4.69 „	8.87 „	15.62 „	17.79 „
Jan. 31—Feb. 4	0.70 „	0.28 „	0.74 „	1.05 „	2.76 „	1.71 „	0.05 „	0.05 „	0.44 „	0.44 „	1.42 „
Feb. 5—9	15.90 „	7.54 „	12.04 „	21.32 „	25.50 „	4.18 „	8.90 „	5.77 „	16.42 „	19.54 „	22.17 „
10—14	8.12 „	3.39 „	6.60 „	11.00 „	15.15 „	4.15 „	3.75 „	1.62 „	6.95 „	8.06 „	11.80 „
15—19	1.94 „	1.81 „	6.18 „	6.72 „	8.78 „	2.06 „	0.72 „	0.67 „	1.58 „	1.86 „	3.62 „
20—24	2.24 „	0.82 „	3.01 „	3.84 „	6.46 „	2.62 „	—	0.75 „	2.15 „	—	—
Febr. 25—March 1	2.27 „	1.77 „	4.94 „	5.71 „	7.18 „	1.47 „	1.27 „	0.84 „	4.52 „	4.77 „	6.03 „
				Mean 9.47 γ	12.24 γ	2.77 γ			Mean	4.91 γ	7.33 γ

Date	Axelöen						Dyrafjord				
	Great Storms				All	Small	Great Storms				All
	$(S^a_H)_g$	$(S^a_D)_g$	$(S^a_V)_g$	S^T_g	S^T	$S^T\ S^T_g$	$(S^a_H)_g$	$(S^a_D)_g$	$(S^a_V)_g$	S^T_g	S^T
Dec. 2—6	8.87 γ	3.10 γ	6.62 γ	11.49 γ	15.71 γ	4.22 γ	4.43 γ	0.44 γ	2.57 γ	5.14 γ	7.73 γ
7—11	15.50 „	6.24 „	15.84 „	23.02 „	26.92 „	3.90 „	9.36 „	1.22 „	7.44 „	12.02 „	15.28 „
12—16	13.98 „	4.50 „	12.13 „	19.05 „	22.80 „	3.75 „	7.55 „	0.99 „	4.24 „	8.72 „	12.06 „
17—21	1.86 „	0.44 „	6.45 „	6.72 „	10.23 „	3.51 „	0.40 „	0.17 „	0.58 „	0.72 „	3.02 „
22—26	28.96 „	10.93 „	45.15 „	54.74 „	57.81 „	3.07 „	21.49 „	4.33 „	12.72 „	25.34 „	31.65 „
27—31	7.23 „	3.08 „	7.90 „	11.14 „	16.05 „	4.91 „	5.48 „	1.77 „	3.90 „	6.96 „	10.66 „
Jan. 1—5	11.67 „	4.84 „	8.52 „	15.24 „	19.94 „	4.70 „	6.56 „	1.91 „	5.30 „	8.65 „	12.11 „
6—10	9.59 „	3.63 „	9.54 „	14.00 „	18.47 „	4.47 „	6.21 „	1.63 „	9.40 „	11.38 „	14.21 „
11—15	12.84 „	3.92 „	10.24 „	16.88 „	21.25 „	4.37 „	3.83 „	1.26 „	3.55 „	5.37 „	9.18 „
16—20	9.14 „	3.33 „	9.07 „	13.30 „	20.53 „	7.23 „	5.41 „	0.92 „	4.20 „	6.91 „	12.25 „
21—25	9.74 „	3.51 „	15.63 „	18.75 „	24.75 „	6.00 „	6.04 „	0.76 „	4.11 „	7.34 „	11.71 „
26—30	22.05 „	11.19 „	39.37 „	46.50 „	49.33 „	2.83 „	18.28 „	4.73 „	15.79 „	24.61 „	28.33 „
Jan. 31—Feb. 4	1.84 „	0.60 „	2.04 „	2.81 „	8.57 „	5.76 „	1.98 „	0.17 „	1.58 „	2.54 „	5.65 „
Feb. 5—9	20.98 „	10.34 „	24.76 „	34.06 „	38.95 „	4.89 „	21.98 „	7.47 „	13.51 „	26.86 „	30.59 „
10—14	13.51 „	5.98 „	10.53 „	18.14 „	23.53 „	5.39 „	11.52 „	1.79 „	6.00 „	13.11 „	22.40 „
15—19	7.87 „	3.40 „	6.04 „	10.49 „	14.73 „	4.24 „	4.69 „	0.85 „	2.75 „	5.51 „	8.37 „
20—24	7.40 „	5.50 „	13.71 „	16.52 „	20.24 „	3.72 „	17.51 „	3.37 „	11.25 „	21.08 „	24.02 „
Feb. 25—March 1	10.23 „	5.16 „	9.71 „	15.02 „	17.28 „	2.26 „	5.54 „	0.32 „	9.50 „	11.01 „	12.36 „
				Mean 19.33 γ	23.73 γ	4.40 γ			Mean	11.29 γ	15.09 γ

TABLE CI.

	0–2		2–4		4–6		6–8		8–10		10–12		12–14		14–16		16–18		18–20		20–22		22–24	
	+	–						–	+	–	+	–	+	–	+	–	+	–	+	–	+	–	+	–
Mat.-Schar																								
	0	5.5									1.2		2.4	0	4.3	0	4.9	2.3	1.1	13.0	0	18.1	0	17.0
" Small .	0.												0.9	0.9	0.9	0.6	2.6	0.5	2.3	1.0	0.6	1.5	0.1	2.4
S_D Great .	0	2.								0	1.0	0	0.8	0.1	1.3	0.5	1.3	2.6	0.3	7.0	0	8.8	0	6.7
" Small .	0.	0.							0.5	0.4	0.6	0.4	0.6	0.6	0.5	0.9	0.9	2.5	0.9	1.0	0.7	0.4	0.6	0.5
S_V Great .	0								1.4	0	2.2	0	3.6	0.2	4.5	0.6	2.8	10.6	0.1	16.8	0	16.8	0	8.1
" Small .	0.					0.6	0.5	0.3	0.3	0.3	1.1	0.2	1.5	0.1	2.0	0.5	2.7	1.8	1.2	1.9	0.5	1.6	0.2	1.6
Kaafjord																								
S_H Great .			0	0.5	0.2	0.3	0	0.2	0	0	0	0	0.6	0	1.1	0	3.1	0.1	2.1	2.0	0.5	6.1	0	11.0
" Small .			0.3	1.0	0.5	0.2	0.2	0.1	0.3	0.2	0.7	0.5	0.5	0.5	0.5	0.5	1.2	0.1	1.2	0.4	0.8	0.9	0.1	0.6
S_D Great .	0	1.6	0	0.5	0.4	0.3	0.5	0	0	0	0	0	0.3	0	0.6	0	0.6	1.0	0.3	1.5	0	4.3	0	4.9
" Small .	0.7	0.7	0.5	0.7	0.6	0.4	0.6	0.3	0.6	0.7	1.5	0.4	1.1	0.3	0.8	0.7	0.6	1.6	0.4	2.7	0.7	2.2	0.5	0.9
S_V Great .	0	5.1	0	2.1	0.2	2.3	0	0.9	0	0	0	0	0.7	0.3	2.4	0	3.1	0.1	2.0	3.0	0	9.5	0	13.8
" Small .	0.1	1.1	0.1	0.9	0.1	0.4	0.2	0.4	0.3	0.3	0.8	0.2	1.0	0.1	1.5	0.3	2.4	0.1	2.0	0.8	0.8	1.6	0	1.3
Axeløen																								
S_H Great .	0	11.9	0	15.0	0	9.9	1.1	3.6	2.8	0.4	3.7	0	8.0	1.4	5.5	2.6	0.6	14.1	0	26.9	0.2	21.1	0	9.7
" Small .	0.5	2.9	0.6	3.4	1.6	2.3	2.1	1.6	2.9	1.4	2.7	2.8	2.6	1.3	2.4	1.5	1.3	3.3	0.2	3.0	0.7	2.0	0.7	5.1
S_D Great .	0	8.0	0	9.7	0.1	6.7	0.5	1.8	0.6	0.3	1.1	0	3.3	0	4.9	0.1	4.0	0.8	2.6	3.9	0.3	3.7	0	7.2
" Small .	0.6	1.9	0.9	2.1	1.2	1.5	1.6	0.9	1.2	0.9	1.5	0.9	1.5	0.6	1.4	0.5	2.4	0.4	3.2	0.4	2.9	0.8	0.6	1.6
S_V Great .	20.8	0	12.8	0.5	5.2	4.4	3.1	2.7	1.5	2.7	0.2	4.7	0.3	8.2	0.5	13.3	1.6	11.1	10.7	4.0	27.6	1.1	31.7	0.2
" Small .	1.4	0.4	0.7	0.4	0.2	1.4	0.6	0.6	2.6	0.5	0.9	1.0	1.0	0.4	0.2	0.7	0.3	1.8	1.0	2.3	1.0	0.7	1.2	0.5
Dyrafjord																								
S_H Great . .	0.1	22.2	0	24.8	0.2	11.5	0	5.6	0.6	1.4	0.6	0.3	0.5	0.4	2.2	0	7.3	0	6.9	1.1	4.2	2.4	1.2	11.7
" Small . .	1.0	2.3	0.6	1.8	0.8	2.0	1.3	0.8	1.1	0.6	0.7	1.5	1.1	0.7	1.9	0.9	1.8	0.8	2.5	1.1	3.5	0.9	2.3	1.7
S_D Great . .	0.8	2.5	0.1	2.2	0	3.2	0	1.8	0.5	0.2	0.7	0	0.3	0	0.6	0	0.8	0.2	0.9	0.2	1.5	1.9	2.9	1.8
" Small . .	1.6	0.5	1.4	0.5	0.5	0.9	1.1	0.6	1.0	0.7	0.9	0.6	1.1	0.3	0.9	0.4	0.4	0.8	0.4	1.4	0.5	1.3	1.5	1.1
S_V Great . .	5.1	6.8	1.6	6.7	0.5	6.9	0	9.9	0.1	5.3	0.4	1.2	0.6	0.2	0.9	0	2.8	0.1	1.6	3.0	0.5	11.1	3.6	10.1
" Small . .	1.7	1.3	0.8	1.9	0.6	1.8	0.4	2.0	0.3	0.9	0.4	0.9	0.5	0.3	0.7	0.1	1.0	0.5	1.7	1.2	0.9	1.3	0.8	1.9

SEPARATION OF GREAT AND SMALL DISTURBANCES.

105. The separation of perturbations into great and small storms has been performed according to rules given in the introduction to this chapter, and for the period of three months common to the four stations.

In table **C** is given the storminess for great and small storms for each five-day period.

The quantity $S^T - S^T_g$ will be taken as representing the storminess of small storms.

The storminess of small storms only shows small and quite irregular variations from one five-day period to another, showing that the cause of small storms is almost constantly present. In view of our theory this would mean *that almost at any time pencils of electric rays from the sun are striking the earth, and we have to suppose a great number of sources of electric radiation spread over the surface of the sun.* On the ground of this fairly constant supply of disturbance, the great storms, from the principal sources of the sun, are superposed.

TABLE CII.

Station	$\frac{S^T_g}{S^T - S^T_g}$	$S_1^T = \frac{6 S^T - S^m}{5}$			S^m			$\frac{S^m}{S_1^T}$		
		All	Great	Small	All	Great	Small	All	Great	Small
Matotchkin-Schar . .	3.42	7.97	5.36	2.62	33.57	30.04	3.53	4.21	5.61	1.35
Kaafjord	2.05	4.56	2.29	2.26	21.17	17.99	3.18	4.64	7.85	1.40
Axelöen	4.39	18.74	14.18	4.56	48.70	45.10	3.60	2.60	3.18	0.75
Dyrafjord	2.97	12.07	8.43	3.64	30.19	25.60	4.59	2.50	3.04	1.26

The ratio of the storminess of great storms to that of the small ones, given in table **CII**, is seen to vary between 2.05 for Kaafjord and 4.39 for Axelöen, or, in other words, most of the storminess **is due** to storms belonging to the group of great storms. The ratio $\frac{S_m}{S_1^T}$ is considerably greater for the **group** of large storms than in the case when all storms are counted.

The distribution over the day of large and small storms is given in table CI, **and** graphically represented in fig. 199. Comparing these curves with those in fig. 196 we notice that ***the** characteristic diurnal period found from the treatment of all storms is even more marked for **the group of la**rge storms.*

We further notice that the greater part of the calm period of the day is due to small disturbances.

Table CIII gives the conditions for the period of four hours during which the large storms have the smallest storminess. We notice that the quantity Σ is greatly increased in the case in which the small storms are left out.

TABLE CIII.

Station	Calm Period	$(S^T_c)_g$	$(S^T_c)_s$	S^T_c	S^T_g	$\frac{S^T - S^T_c}{S^T_c} = \Sigma$		$\frac{S^T - S^T_g}{S^T}$
						All	Great	
	Gr. M. T.							
Matotchkin-Schar . . .	4 – 8h	0.81 γ	1.83 γ	2.58	9.47 γ	3.74	10.7	0.23
Kaafjord	8 – 12h	0.00 „	1.94 „	1.94	4.91 „	2.68	∞	0.33
Axelöen	8 – 12h	5.74 „	5.98 „	11.35	19.3 „	1.08	2.36	0.19
Dyrafjord	10 – 14h	1.58 „	2.69 „	4.15	11.3 „	2.64	6.13	0.25

$(S^T_c)_g$ is the total storminess of great storms for the calm period

$(S^T_c)_s$ „ „ „ small „ „ „ „ „

The result for Kaafjord of the separation of small and large storms is the most remarkable. If 33 % of small storms are removed, the quantity Σ becomes infinitely large, that is to say that out of the 67 % of storms greater than a certain value, *no one appears during the calm period.*

For the other three stations Σ has a finite value. The reason for this is partly the fact that the percentage of small storms taken out is smaller for these stations.

For Matotchkin-Schar Σ has the large value of 10.7, although the small storms removed only make up 23 % of the whole storminess. If we were to increase the upper limit for small storms so as to make the percentage a little greater, we might expect to find perfectly calmness.

Also for Dyrafjord and Axelöen Σ increases rapidly with the percentage of small storms taken out but not quite so fast as for the two former stations.

The rapid increase of Σ for great storms shows that the storminess of the calm period is almost entirely due to comparatively small storms showing a diurnal distribution somewhat different from that of large storms.

The two maxima for S_H^n at Axelöen, one four hours before, another four hours after, midnight, follow the group of large storms. Thus in spite of the fact that these storms are so limited in their sphere of action that they produce practically no effect at Kaafjord, they appear at Axelöen as fairly great storms. I think this will clearly show that the current-systems causing the disturbances carry a comparatively small amount of energy, so that they can only produce the great effect at Axelöen by passing near to the station.

THE DISTRIBUTION OF STORMINESS AND THE SOLAR ORIGIN OF POLAR STORMS.

106. The existence of a well marked diurnal period of polar storms was already shown in my previous work "Expedition Norvegienne 1899—1900". See p. 16, and Pl. II. The type of variation is found for Bossekop for the period 1899—1900, and for a number of other stations for term days during the polar year 1882—83. The diagrams are in good accordance with the present results found from a more complete statistical treatment.

From the study of the occurrence and motions of the perturbing fields, I was led to the conclusion that these fields followed the diurnal motion of the sun (See: "Exp. Norv. etc." p. 29), a result which has been brought to full evidence through the present investigation.

The diurnal distribution of perturbations shows immediately that some part of the storminess, in some way, must be connected with the sun.

One of the most interesting features of the diurnal variation is the existence of a well-marked calm period. I think this is a property of the storminess which may serve as an important test for any possible explanation, and we shall subsequently see how far it is in agreement with our own theory.

The properties of the diurnal distribution of storminess for a certain interval of time can be expressed in the following simple way: At any moment there is a region of the earth with great disturbedness. This region is not symmetrical with respect to the axis of the earth, but is mainly restricted to the night and evening side, and extends from places near the magnetic axis to places some distance to the south of the auroral zone, where the storminess rapidly diminishes. At the night end of the disturbed area we have negative, at the afternoon end positive storms, and nearest the pole we have an area of very local disturbances. On the evening side, *between the two* types, there is no calm region, but the two types will to a certain extent overlap.

From the diurnal distribution we can draw some interesting conclusions regarding the amount of storminess which is a direct effect of the sun.

Suppose part of the storminess was produced by something showing a period different from 24 hours; the storminess from such a cause would be evenly distributed over all hours of the day.

We are consequently justified in assuming that the total storminess due to causes other than the sun, cannot be greater than the storminess of the calm period. In other words, the quantity S_c^T in table CIII is an upper limit for the sum of all the storminess of the place that is not due to the sun, and $S^T - S_c^T$ is a lower limit for the sun-storminess. The quantity Σ given in table CIII will then in each case give a lower limit for the ratio of the sun-storminess to that which is due to any other cause.

We can also express the sun-storminess in percent of the whole amount. In table CIV the lower limit for the percentage of sun-storminess is given for all storms and for the great storms separately.

TABLE CIV.

Lower Limit for Sun-storminess.

Station	All Storms	Great Storms	
	$100 \frac{S^T - S_c^T}{S^T}$	$100 \frac{S_g^T - (S_c^T)_g}{S_g^T}$	$100 \frac{S_g^T - (S_c^T)_g}{S^T}$
Matotchkin-Schar . . .	79 percent	91.5 percent	70.5 percent
Kaafjord	73 "	100 "	67 "
Axelöen	52 "	70 "	57 "
Dyrafjord	72 "	86 "	63 "

The first column gives the limit for sun-storminess of all storms as a percentage of total storminess.

The second column gives the limit for sun-storminess of great storms as a percentage of the total storminess of *great storms.*

The third column gives the sun-storminess of great storms as a percentage of the total storminess of *all storms.*

The second column shows that *nearly all the storminess of great storms is caused by the sun.*

Comparing the first and third columns, we notice that the sun-storminess of great storms forms about as large a portion of the whole total storminess as that given in the first column, which is calculated from the diurnal distribution of all storms. If, then, the numbers in the first column represent the true value for the sun-storminess, it would mean that almost all the small storms were not of solar origin; but this is certainly not the case.

In calculating the values in the first column, it was assumed that the storminess of the calm period, consisting mostly of small storms, was entirely due to causes other than the sun. But we know that at least part of this storminess must be of solar origin, and through the knowledge gained about the properties actually shown by the magnetic disturbances, we are able to estimate a lower limit for the sun-storminess of the calm period.

We know from the treatment of separate storms that a polar disturbance with its centre near the auroral zone will be accompanied by small disturbances at considerable distances from the storm-centre.

To fix the idea suppose it is 8 o'clock in the morning Gr. M. T. At that time Kaafjord will be situated in the calm region. But on the opposite side of the magnetic axis polar storms are operating which are bound to produce a certain effect at Kaafjord.

We found that the elementary polar storm produced a field of a fairly regular type (see p. 86, Part I). We are justified in assuming that points situated symmetrically with respect to the axis of the eld, will have perturbing forces of the same order of magnitude. Thus if we had a storm-centre at yrafjord, we should find that this storm would produce about the same strength of field in south urope as at a place on the opposite side of the magnetic axis and with angular distance equal to that f Dyrafjord.

Table CV gives the ratio ($\varkappa$) of the strength of the field near the centre to that in southern urope (San Fernando and Munich) for a number of polar storms.

TABLE CV.

Storm	$\varkappa$	Remarks
Dec. 15	23	Calculated from the fields at 1^h and $1^h\,15^m$ p. 90
Feb. 10	12	— „ „ table XVI „ 107
March 31	17	— „ „ field at $0^h\,45^m$ „ 123
„ 22	11	— „ „ „ $22^h\,15^m$ „ 135
Dec. 26	17	— „ „ „ $23^h\,30^m$ „ 143
Feb. 15	12	— „ „ „ $16^h\,45^m$ „ 184
„ 8	13	— „ „ „ $20^h\,0^m$ „ 205
Oct. 27	12	Table XXXII „ 212
Mean	14.6	

The quantity $\varkappa$ varies between 23 and 11. The larger value corresponds to an elementary storm when he ratio is taken for the storm-centre and a point near the transverse axis of the field. If the principal xis is turned more towards the south, as on the 22nd of March, it will have the effect of making $\varkappa$ maller; for, at equal distances from the centre, the forces will be greatest along the principal axis. urther, the ratio $\varkappa$ is smaller for compound than for elementary storms, which is easily understood if ve take into account the local character of storms near the centre.

The effect produced at our stations by storms on the opposite side of the magnetic axis will be mallest for those stations which have the greatest angular distance from that axis. In view of the results xpressed in the table, we can put

For Matotchkin-Schar and Kaafjord . . . $\varkappa < 20$
„ Axelöen and Dyrafjord . . . . $\varkappa < 15$

The greatest storminess in the disturbed region for the period of our observations is known only or that part of the auroral zone extending from Dyrafjord to Matotchkin-Schar; but from the treatment of storms from the polar year 1882—1883, we have seen that storms occur with about equal strength nd frequency all round the auroral zone.

Let $(S)_m^T$ represent the diurnal maximum of total storminess. This quantity is very nearly equal to he maximum perturbing force of the average storm. For the three-month period considered we obtain

For Matotchkin-Schar . . . $(S)_m^T = 30\ \gamma$
„ Kaafjord . . . . „ $= 20$ „
„ Axelöen . . . . „ $= 43$ „
„ Dyrafjord . . . . „ $= 29$ „
Mean $30.5\ \gamma$

At the time when one of the Norwegian stations has calmness, we can assume the maximum strength of the average storm on the opposite side of the pole to be at least 30 γ.

As an upper limit for the storminess which is not of solar origin we obtain

For Matotchkin-Schar . . . . . $\sigma = S_e^T - {}^{30}/_{20} = 1.1\ \gamma$

" Kaafjord . . . . $\sigma = S_e^T - {}^{30}/_{20} = 0.4$ "

" Axelöen . . . . $\sigma = S_e^T - {}^{30}/_{15} = 9.4$ "

" Dyrafjord . . . . $\sigma = S_e^T - {}^{30}/_{15} = 2.2$ "

In table CVI is given the lower limit of sun-storminess $(S_e^T - \sigma)$ as compared with σ and also expressed in percent of the total storminess.

TABLE CVI.

Lower Limit for Sun-storminess.

Station	$\frac{S^T - \sigma}{\sigma}$	$100\ \frac{S^T - \sigma}{S^T}$
Matotchkin-Schar . . .	9.2	91 per cent.
Kaafjord	13.2	94 "
Axelöen	1.5	60 "
Dyrafjord	5.9	86 "

The quantity σ cannot be entirely due to storms which are not of solar origin. In the first place it is not impossible that every now and then we also have centres of polar storms on the morning and forenoon side of the earth. At the southern stations Matotchkin-Schar and Kaafjord, forenoon centres of any magnitude are seldom observed; but nearer the magnetic axis and the north pole, at Dyrafjord and still more frequently at Axelöen, centres of fairly small and local polar storms are also found on the forenoon side.

In the second place σ contains some storminess due to disturbances which do not belong to the polar type, but may still be of solar origin. In our opinion the equatorial storms as well as the cyclo-median perturbations are effects of the same solar agency as that producing the polar storms; but at the polar stations these storms show quite a different diurnal distribution from that of the polar storms. Thus the equatorial storms would produce about equal effects all round the auroral zone, and the cyclo-median perturbation in the instance investigated was strongest on the day side.

In view of these facts, table CVI shows that at the three southern stations *practically all the storms which occurred during the interval of our observations were caused by some agency coming from the sun.*

The fact that the lower limit of sun-storminess is smaller for Axelöen than for the other stations, must not lead us to the conclusion that a smaller part of the storminess should be of solar origin at this station. It only means that by the method used we are unable to *prove* the solar origin of a great part of the storminess at Axelöen. The efficiency of the method depends on the calmness of the station. On account of the character of the distribution of storms on the earth's surface, it is, on the northern hemisphere, merely at stations on the southern border of the auroral zone that we can expect this calm period to be well marked.

For a place in a lower latitude the storminess will be due mostly to distant systems with their centres near the auroral zone. In the night the disturbed region would be on the same side of the pole as the place considered, while in the day it would be on the opposite side of the pole; but for places far from the poles the difference in the effect of the day and night systems would be diminished. This is in accordance with the fact that the great polar storms are accompanied by disturbances in lower latitudes, which are not by any means restricted to the night side. Also if we go to regions near the poles and the magnetic axis, we shall find a rapid diminution in the calmness.

As regards Axelöen, we found the small calmness to be due, not so much to its more northerly position, as to a number of centres of local storms which showed their own characteristic diurnal distribution; but judging from the very great diurnal maxima of these storms, they must be mainly of solar origin, and moreover we must suppose that stations so near each other as Kaafjord and Axelöen have disturbances of essentially the same origin.

The result of this investigation regarding the amount of sun-storminess has an interesting bearing on the question regarding the possible influence of the moon on magnetic disturbances. If the agency—of whatever kind it was—came from the moon the effect should show a period of a lunar day, different from 24 hours. Then the moon storminess must be contained in the quantity σ, and thus be extremely small compared with that of the sun.

It ought to be remembered that the *storminess* only contains variations of a somewhat abrupt and irregular character. Then if the sun or the moon gave out a magnetically effective agency at a constant rate, the effect of such an agency would not enter into the quantity we have called storminess.

APPLICATION TO THEORY.

107. The previous results regarding the amount of storminess due to the sun are obtained without any assumption regarding the mechanism connecting cause and effect. We have merely made use of facts actually found for the distribution of storms with regard to time and space.

The next question is: How do the properties found agree with our theory?

The characteristic properties to be explained are mainly the following:

(1) The great storminess on the night and evening side of the earth, and in the region near the auroral zone.

(2) The calm region on the day side.

(3) For the somewhat great storms, with their centres passing between Axelöen and Kaafjord, we can distinguish two types, which we have called positive and negative polar storms.

(4) A number of local storms occur in the vicinity of the pole to the north of the auroral zone.

We found during the treatment of separate storms that the main features of the field of an elementary storm could be explained by a system consisting of a vertical current coming in from space, and bending round in the direction of the auroral zone at a height of some hundred kilometres above the surface of the earth, and leaving the earth as another vertical branch. Now we have seen that the average storm, which is made up of numbers of such systems, is almost entirely caused by the sun. There seems then no escape from the assumption that these current systems-coming in from space, are currents directly produced by electric radiation emanating from the sun.

We can also use another line of argument. In order that the *sun-effect* shall mainly make itself felt only within *narrow regions* on the earth's surface, and almost entirely on the side *turned away from the sun*, it is necessary that the sun-agencies descend in comparatively narrow streamers, and are deviated in some way by some field of force possessed by the earth.

Now *electric rays* from the sun, diverted by the *magnetic field* of the earth, will be just what is necessary to explain this very peculiar kind of solar action. Regarding this point I must refer to the publications of my experiments with the magnetic terrella in a vacuum-tube.

Corresponding to the disturbed region near the auroral zone, we have areas of precipitation of cathode particles on the terrella, forming bright spots or bands along a certain magnetic parallel which in a striking way corresponds to the auroral zone. The bright areas on the terrella usually show the characteristic property of being restricted to the side turned away from the cathode or the source of the radiation.

The existence of a calm period is a mere consequence of the distribution of storm-centres on the night and evening side, and the rapidity with which the magnetic effect diminishes with the distance from storm-centre. As we have seen, both these properties were consequences of our radiation theory, which will then also explain the calm period, which is especially well marked on the southern border of the auroral zone.

The existence of two types of polar storms, each restricted to its own time of day, is I think a matter of the greatest interest for the question regarding the cause of magnetic storms.

In order to find whether the P and N storminess could be explained from our theory, I have tried by means of screens placed in various positions outside the terrella, to trace the direction of the rays before they strike the terrella. In this connection it was of special interest to regard the direction of motion of the horizontal component of the corpuscles just before they struck.

It is of course difficult, not to say impossible, to reproduce in the limited space of a vacuum-tube exactly the conditions that govern the formation of magnetic storms. We have, however, been able to show from the terrella experiments the existence of two types of precipitation which I think will show the way in which the two types of storms are to be explained.

In one type most strongly developed on the night side, the horizontal components of the velocities of the corpuscles are turned towards the east; and at the same time, with a proper adjustment of magnetisation and stiffness of rays, we get precipitation on the evening side with the horizontal component of motion turned towards the west.

I think these two types correspond respectively to the negative and positive polar storms. Thus the typical distribution and direction of the two types of storms can be explained when we assume the polar disturbances to be a direct effect of electric radiation from the sun.

The local storms with their centres to the north of the auroral zone, which had a great effect at Axeløen, will, I think, be explained by our radiation theory, when we remember that the precipitations approach the magnetic axis when the rays become softer, and we should merely have to assume that the sun gives out rays of different stiffness.

We found that the large storms usually had their centres in lower latitudes than the small storms. This indicates that probably the rays given out by the very powerful sources in the sun, are stiffer than those from the many small sources producing the small storms occurring between the great maxima.

When the corpuscular currents strike the atmosphere, secondary processes may be called into play; but if these secondary processes, of whatever kind they might be, are to produce magnetic effects of the same order as the impinging rays, they must follow quickly after the primary action. For if secondary effects of the same magnitude and frequency as the primary effects were present, and these secondary effects could show up several hours after the sun agency had left the place, it would be difficult indeed to explain the existence of a calm period and the great rapidity with which the negative storms cease after midnight.

We saw that the typical field of polar storms could be explained by a current-system coming in from space corresponding to the precipitation of electric rays from the sun. This would strongly support the view that by far the greater part of the disturbance effect observed is caused *directly* by the currents of sun-radiation.

CHAPTER IV.

EXPERIMENTS MADE WITH THE TERRELLA ESPECIALLY FOR THE PURPOSE OF FINDING AN EXPLANATION OF THE ORIGIN OF THE POSITIVE AND NEGATIVE POLAR STORMS.

108. In the following pages, we shall describe a series of experiments that were made for the purpose of gaining a clear idea of the course of the rays about our magnetic terrella.

It is, of course, of great importance to calculate, as STØRMER has done, the separate *possible* paths that electric corpuscles from a distant cathode may describe about an elementary magnet, under the influence of the magnetic forces originating from the magnet; and in so doing he has thrown much light upon my earlier experiments, and, on some essential points, has supported the theory which it is my intention to work out. But as long as the mathematical problem is not entirely solved, so that the distribution of *all* paths in space is found, the utility of such calculations as an endeavour to explain, for instance, the positive and negative polar storms, is very limited. The experimental investigations with a magnetic terrella in a large discharge-tube are another matter. There it is possible, by various means, to see how the rays group together round the terrella, and even to photograph the phenomena.

It is apparent that in this way a full, clear idea of the phenomena may be obtained, so that the results, as we shall now see, may be successfully transferred to the relations between the sun and the earth, as regards the various terrestrial-magnetic and auroral phenomena that have been observed.

We can, as will be seen, guard against the liability of our discharge-tube, owing to its comparatively narrow proportions, having any injurious influence upon the range of the conclusions that can be drawn from any of the results, and those who will closely follow the entire series of elaborate experiments which have been made, will end by seeing how great difficulties resolve themselves into nearly perfect lucidity. Some of the experiments last described, were made some time after the first series; I have not, however, on that account, omitted any of the previous results, as I considered it best that the method adopted could be plainly traced.

The experiments now to be described have nearly all been made with the machine shown in fig. 67 (Section I), generating a direct current with a tension of up to 20 000 volts. The arrangemen of the sets of apparatus was also similar to that shown in the figure, but the discharge-tube now was not cylindrical but prismatic, composed of flat plates of glass, so that the photographs taken of the terrella should not be contorted by the passage of the light through the curved glass of unequal thickness. The prismatic discharge-tube, which is shown in fig. 200, was formed of plates of glass, 20 mm. in thickness, cemented together with »cementium«, and finished outside with »piceïn«. There was no great difficulty in keeping the tube air-tight, even if it were exhausted down to a pressure answering to 0.0005 mm. of mercury; but with a low pressure such as this, there was vapour in the discharge-tube, of which the pressure may well have been several times as great as that mentioned above; but in the experiments here described it had no disturbing effect upon the results, as we generally worked with greater pressure.

In order to obtain clear phenomena in the experiments, it is important that the discharge-tube shall have been exhausted for several days, and that during this time the terrella shall have been frequently magnetised, thus becoming heated and giving off gases. The discharges, moreover, must have taken place abundantly, so that superfluous gas is removed from the electrodes and the inner surfaces of the discharge-tube.

One drawback in the photographing of the various light-phenomena was the rather bright, disturbing reflections from the plate-glass sheets. They have been to some small extent removed in the retouching of the prints.

Fig. 200.

Fig. 200 is a photograph taken during an experiment with a terrella No. 5, which was 5.5 cm. in diameter, and suspended in such a manner that the magnetic axis coincided with the axis of rotation. The first series of experiments in the following pages, until stated otherwise, have been made with this terrella, of which the magnetic moment for different magnetising currents is given graphically in fig. 70 (Part I).

In the experiment shown in fig. 200, the terrella is provided with two fixed screens, one horizontal round the equator, and one vertical. On the horizontal screen moreover, there are fixed 5 short thick pieces of metal wire, coated, as are also the screens and the terrella, with tungstate of lime. The picture is interesting in that it shows how the rays from the cathode are thrown upon the walls of the discharge-tube; and it also shows how the rays are drawn in towards the terrella in the form, previously often mentioned, of two luminous horns near the poles. The dark space between these luminous horns widens greatly if the magnetisation of the terrella is increased. The rays are thrown down in abundance upon the under surface of the discharge-tube, and similarly up towards the top surface. The rays are moreover thrown forcibly against the left side surface, looking from the cathode towards the terrella; and the terrella is magnetised with the south pole uppermost. On the right side surface there is no appearance of any corresponding great precipitation of rays. It will be seen that here no perceptible rays reach the back surface of the discharge-tube.

A great number of experiments have been made with terrella No. 5, and photographs have been taken from various points simultaneously, during each separate experiment. Such photographs have been taken, for instance, of 12 different positions, with the vertical screen turned right round, 30° each time.

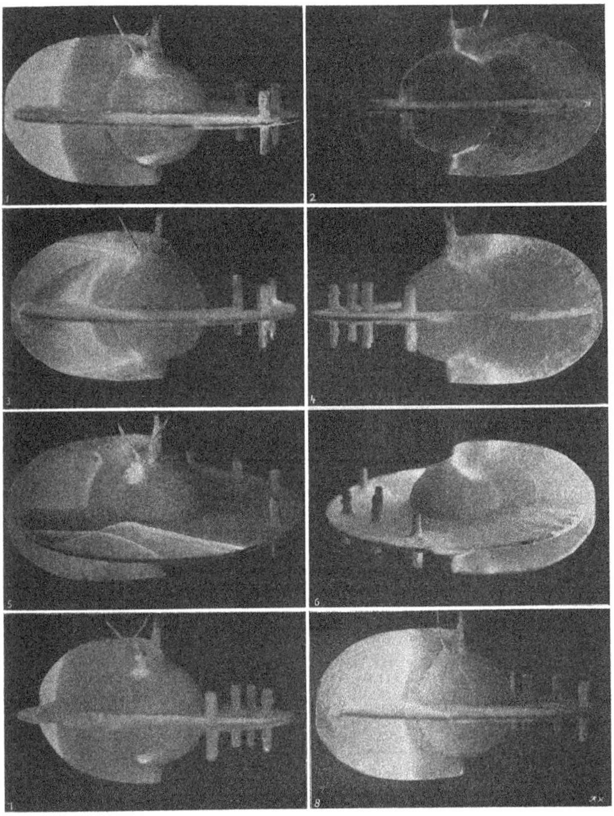

Fig. 201.

It has not been necessary, however, to reproduce more than a few of these photographs, as other experiments that have subsequently been developed from the above, more easily show clear results.

In the following pages we speak of the north and south sides, i. e. respectively the upper and under sides of the horizontal screen, and the east and west sides of the vertical. We calculate the angle between the vertical screen and the centre line between the terrella and the cathode positive eastwards from 0° to 360°.

In order to have an unmistakable manner of indicating the angles which we have occasion to mention in the following pages, we shall refer them to the axis about which the terrella can be rotated — which, in these experiments, is always vertical — and a horizontal plane through the centre of the terrella.

We employ the designations easterly hour-angle and declination to indicate the position of a place. The hour-angle is then calculated in the horizontal plane eastwards from the centre line between the centres of the cathode and of the terrella, to the projection of the place upon the horizontal plane, and the declination is an angle with its vertex in the centre of the terrella, and one side passing through the place in question, and the other through the projection of the place upon the horizontal plane. The northern declination is positive, the southern negative.

In the eight photographs reproduced in fig. 201, the experiments were made under a pressure of about 0.002 mm., with 25 milliamperes through the discharge-tube, and 30 amperes upon the terrella.

In Nos. 1 and 2, the easterly hour-angle of the vertical screen was 330°, No. 1 being photographed 90° east of the screen, and No. 2 90° west of it. The terrella, it will be noticed, is seen very little from above. In Nos. 3 and 4, the easterly hour-angle of the vertical screen was 30°, No. 3 being taken 90° east of the screen, and No. 4 90° west of it. In Nos. 5 and 6 the hour-angle of the vertical screen was 0°, the photographs being taken as before, but from a place with a declination of 25°, so that the terrella is seen from considerably above.

It should be remarked that the light-figures here seen upon the northern side of the horizontal screen are of course exactly repeated upon the southern side, since the axis of the magnet coincides with the axis of rotation; but they are not visible here.

In Nos. 7 and 8, the hour-angle of the vertical screen was respectively 300° and 60°, and the photographs were taken respectively 120° and 55° east of the screen.

With regard to the luminous precipitation upon the phosphorescent screens, that upon the vertical screen shows that a very considerable part of the cathode rays are deflected towards the left before they reach the terrella, and then, as we have seen, thrown against the left side surface of the discharge-box, looking from the cathode towards the terrella (fig. 200).

This phenomenon, in which a large proportion of the rays are carried past the terrella, and are nearest to it some way out on its afternoon side in a direction opposite to that of the earth's rotation, must be regarded as very important. We shall frequently return to it in the course of the experiments. It is deflected rays from the sun such as these, that we have previously assumed to be the principal cause of the positive equatorial storms. We shall also return to this ray-phenomenon in discussing the diurnal variation of the terrestrial magnetism and the zodiacal light.

In photograph No. 5, there are two places, A and B, in which rays descend upon the horizontal screen, and it is these two instances that we shall first consider here. We shall see that the more abundant of the two, which we will call A, and whose eastern boundary is very nearly a straight line, is due to rays which, if not arrested by the screen, would travel round the terrella in a direction from west to east, oscillating alternately above and below the plane of the magnetic equator, and most of them descending at last in "the auroral zones"; but they never seem to come into contact with the terrella to the north of the northern auroral zone or to the south of the southern auroral zone. The smaller precipitation, *B*, is due to quite another class of rays, which, unlike those of *A*, operate close to the terrella north of the northern auroral zone or south of the southern auroral zone.

The aim of the experimental investigations here described, is to obtain a clear idea of the general course of these two classes of rays. We shall first show, by numerous experiments, how the rays forming *A* curve round the terrella, rising and falling above and below the equator, when not arrested by the screens. In the next place, the rays forming *B* will be investigated by an altogether different series of experiments. We shall see, among other things, that rays of the first class will give us a natural explanation of the negative polar storms, while the rays belonging to the second group will help to explain the positive polar storms. In the following pages we shall speak of these two classes of rays as rays of group *A* and rays of group *B*.

In fig. 202, eight more photographs are given, representing various experiments. No. 1 of these answers to No. 5 of fig. 291, the only difference being that here the terrella is magnetised with 10 amperes instead of 30. Otherwise everything is the same. Only the effect of the precipitation *A* is visible, that of *B*, for reasons that will be made clear later, being no longer found on the horizontal screen. Some of the rays forming *A* now fall upon the terrella, and we obtain a figure upon the front of it that resembles the luminous figures shown in figs. 66 and 68 in Section I.

Photograph No. 2 also shows conditions similar to those of No. 5 of the preceding Plate, except that the hour-angle of the vertical screen is 180°. In the precipitation *B* there appears the shadow of one of the cylindrical pegs. In some of the experiments it sometimes happens that two shadows of the same peg are seen, one of them being cast by rays of the precipitation *A*, the other by *B*.

Photograph No. 3 was taken for the purpose of examining the sharp line of precipitation that forms the eastern limit of *A* on the horizontal screen. The experiment was made under the same pressure and with the same discharge-current as before, and the magnetising current to the terrella was of 30 amperes. The hour-angle of the vertical screen was 90°, and the photograph was taken 45° to the west of it. The experiments were made by turning the terrella in such a manner that the vertical screen came near the line of precipitation on the horizontal screen. If the vertical screen were turned ever so little more to the east than the line of precipitation on the horizontal screen, no precipitation was found upon the vertical screen. On the other hand, if it were turned less to the east than that line, precipitation appeared upon the vertical screen in the form shown in this photograph. We at once get the impression that the rays bend down towards the line of precipitation, where, if they could get through the horizontal screen, they would cross one another, so that the rays that at first were *above* the screen would go *below* it, and vice versa. It is then a natural proceeding, if we wish to study the rays in *A*, to experiment, as we have done, with a horizontal screen alone, in which there is a slit parallel with the powerful line of precipitation; and also with a vertical screen alone, containing a radial slit at the magnetic equator. This vertical screen must be so bent that the line of precipitation, right from the terrella, can be made to fall upon the slit in the screen, along the entire length of the line, so that all the rays can get through the slit simultaneously. The arrangement of the experiments is also clearly seen from their accompanying photographs, which will soon be described. A preliminary experiment was made, and this is shown in No. 4 of fig. 202. Here a slit was made in the horizontal screen, which, however, the first time, was not given the right direction along the line of precipitation. Next, a hole was made in the vertical screen, near the auroral zone. We at once discover that close to the slit in the horizontal screen, the rays leave the under side and form a second precipitation upon the north side of the terrella, while the rays from above go through the screen, and form corresponding precipitation upon the south side of the terrella. We have thus brought out the second of the remarkable instances of precipitation represented in fig. 68, Section I.

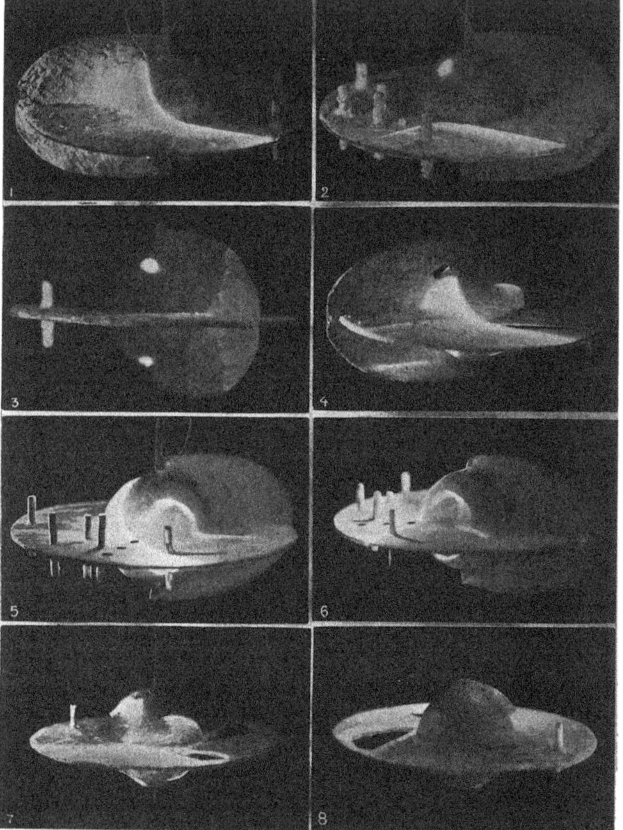

Fig. 202.

We will now pass on to mention some experiments that were made gradually, first with three round holes in a horizontal screen, next with two slits and then with three, in the horizontal screen, as it appeared that when a part of the above-mentioned marked line of precipitation belonging to *A* fell across a slit, a second line of precipitation appeared, which was also almost a right line, at an angular distance from the first of 110°. When this second line of precipitation was also placed over a second slit, the rays once more passed through the screen and formed a third line of precipitation, which was also turned about 110° in relation to the second line of precipitation.

These experiments were both troublesome and lengthy, for every time an alteration was to be made, the bottom had of course to be taken out of the discharge-box, and after the alteration had been effected, the glass box had once more to be exhausted for several days, with frequent discharges, as already mentioned, before it was again in perfect order.

Photographs 5 and 6 refer to an experiment in which 3 holes were bored in the horizontal screen, with their centres situated radially, as the figures show. The experiment was made with a pressure of about 0.002 mm., with a discharge-current of 23 milliamperes through the tube, and a magnetising current of 8 amperes upon the terrella. The eastern hour-angle of the vertical screen was 240°; and the photographs were taken respectively 120° and 60° to the west of the screen.

By this slight magnetisation, beautiful precipitation was obtained on the terrella, when the hole farthest in on the screen was brought over the first line of precipitation. It was easy to prove by a slight displacement, that it was the rays that came from above and passed through the hole, that formed the precipitation on the south part of the terrella, and vice versa.

It will be seen that the precipitation does not only fall upon the terrella, but continues in the second line of precipitation across the horizontal screen. By the employment of 11 amperes, the rays through the first hole formed the western part of this precipitation upon the terrella, while, if the rays were allowed to pass through the second hole — by a slight turn, so as to bring the first line of precipitation over the second hole — they formed the eastern part of the precipitation upon the terrella, with a continuation in the second line of precipitation on the horizontal screen. In these figures, 5 and 6, we see distinct shadows of the pegs that are fixed in the horizontal screen. Much can of course be concluded from the directions of these shadows, with regard to the course of the rays; but as the same thing comes out more distinctly in another manner in subsequent experiments, we shall here only make a few remarks. The shadow of the peg that stands on the first line of precipitation is faint, but often extends some distance, and is curved almost like an arc of a circle with its centre in the centre of the terrella. The shadows of the three pegs standing close together point outwards, and are formed of rays belonging to precipitation *B*, which, however, is not distinctly outlined in the figure.

Photographs 7 and 8 are taken with the same pressure as before, with 20 milliamperes through the discharge-tube, and with 16 amperes on the terrella.

There are now two slits in the horizontal screen, which here too have not been given quite their correct form and position in relation to the first and second lines of precipitation. The angular distance between the slits is, as will be seen from the precipitation, somewhat too small; for the slits could only be determined by successive approximations, as the second line of precipitation does not appear until the first slit is correctly cut, and the third line of precipitation until the second slit is correctly placed. In the next experiment with three slits in the horizontal screen, however, the position and shape of the slits are correct.

In this experiment, Nos. 7 and 8, both the second and the third precipitation came out distinctly upon the terrella, but not so well as in fig. 68 in Section I. The photographs are taken from places with an hour-angle of 90° and 270° respectively. Although the first line of precipitation lies on the west side of the first slit, the second line of precipitation, it will be seen, falls a little to the east of the second slit.

It should be remarked that the position of the various lines of precipitation upon the horizontal screen, depends somewhat upon the magnetising of the terrella. If, for instance, the current upon the terrella is increased from 8 amperes to 30, the first line of precipitation will move back (westwards) a little, with a parallel movement. The second and third lines of precipitation move at the same time; but the angle between the first and second, and between the second and third lines of precipitation continues more or less to be about 110°—100°. This angle, however, also diminishes somewhat under stronger magnetisation.

In this experiment, another circumstance was also investigated. A thin screen, about 3 mm. in height, formed of a strip of copper, was placed on its edge upon the terrella, and running from the latter's north pole a little way down a meridian. The screen was then divided into three branches, and was also coated with tungstate of lime. It was so placed that the rays which came through holes 1 and 2 would strike the terrella in the polar regions just where the little three-armed screen was. The intention was to determine the direction in which those rays moved which struck the terrella in its polar regions. It appeared from the experiments that the rays in the second polar precipitation, which belonged to the very northernmost part of the precipitation up in the auroral zone, come fairly perpendicularly in towards the terrella, though with some slight movement eastwards, while the rays both in the south-western and south-eastern parts of the precipitation on the terrella to the north of the horizontal screen, had a strong tangential movement, with direction from west to east.

STUDY OF THE RAYS OF GROUP A.

109. Experiment in which the Terrella had only a Vertical Screen. We shall begin by describing a series of experiments which were made with the same terrella as before, in which the magnetic poles coincide with the geographical; but the terrella now has only *one* screen.

This screen, which maintains a vertical position during the rotation of the terrella, was produced by an abrupt bending of the former vertical screen, so that the latter comes to consist principally of two plane portions, which intersect one another at an angle of about 100° in a vertical line, which is in contact with the terrella in its magnetic equator. We will call the screen the vertical screen. The photographs here reproduced give a sufficiently clear idea of its form.

Fig. 203 shows 12 pictures from experiments with this arrangement with a vertical screen provided with a horizontal slit.

Nos. 1, 2 and 3 are from experiments in which the terrella was magnetised with 8 amperes, the discharge-current was of 25 milliamperes, and the pressure answered to 0.001 mm. The outer plane part of the vertical screen formed an angle of 45° with the central line between the centre of the terrella and that of the cathode. We shall simply, in the following pages, express this by saying that the screen had an hour-angle of 45°, referring only to the outer plane part of the screen.

The photographs were taken in a horizontal plane through the centre of the terrella, from places with hour-angles of respectively 90°, 180°, and 270°.

When the screen here has an hour-angle of 45°, it does not to any great extent shut off the rays, and the light-figures on the terrella (Nos. 1, 2 and 3) are very much the same as if there had been no screen. We recognise them from fig. 68 in Section I.

Photographs 4, 5 and 6 are from experiments made under very nearly the same conditions as the preceding, except that the hour-angle of the screen is 135°. The photographs are taken from the same positions respectively.

We here obtain a capital representation of the way in which the screen acts when the slit does not fall near one of the lines of intersection of the rays, those lines which, on the horizontal screen, we called lines of precipitation.

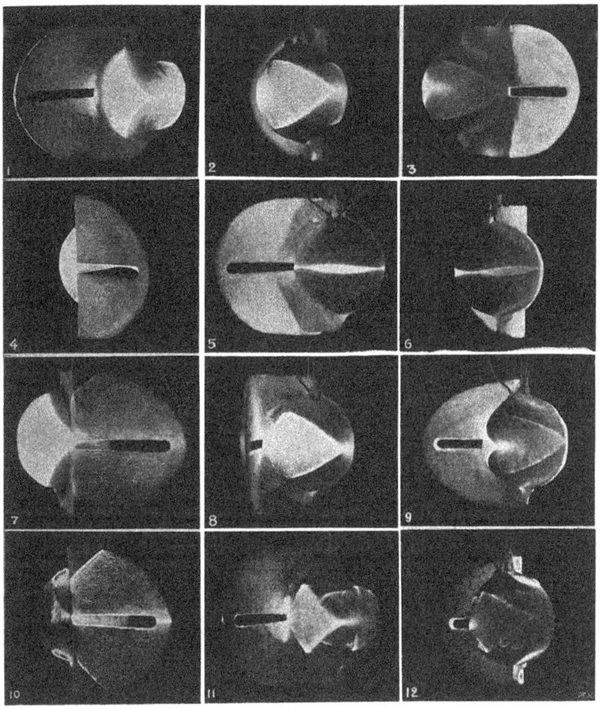

Fig. 203.

Here only a narrow pencil of rays falls through the slit in the screen. These rays continue their course round the terrella, undulating above and below the plane of the magnetic equator. There are distinct nodes and loops visible in the light-figures upon the terrella, which resemble vibrating strings. These figures become fainter according to their number, exactly like those in the preceding photographs; and by magnetising the terrella more highly, they disappear from the equatorial regions, a circumstance which is in accordance with what we found with regard to the light-figures in fig. 68. The reason of this must be that the corresponding rays, on higher magnetisation, go farther out from the terrella, while those that were nearer turn right round, some of them striking against the terrella.

Photographs 7, 8 and 9 were taken under the same physical conditions as the previous ones, only that now the hour-angle of the terrella screen is 175°, so that it nearly coincides with the first line of intersection of the rays.

We see at once that now the majority of the rays pass through the slit in such a manner that the second and third light-figures upon the terrella become very much as they were in Nos. 1, 2 and 3, where the screen did not act perceptibly. It will be seen, indeed, that something is wanting in some of the uppermost polar percipitation in the second and third light-figures here, and this probably arises from the slight precipitation of rays seen on the screen in No. 7; for it is certain that the rays that keep nearest to the magnetic equator in their journey round the terrella, have not exactly the same lines of intersection as those rays which intersect the equator at large angles. The angular distance between the consecutive lines of intersection of the former rays is greater than that of the latter. The rays that form precipitation in the auroral zone, however, are just such rays as, in their discursion above and below the magnetic equator, intersect that plane at great angles.

In these light-figures there are nearly always 110° between corresponding points in the precipitation when 8 amperes are employed as the magnetising current. The rays in the third precipitation, which are farthest up in the polar regions, intersect one another at the equator at an hour-angle that is smaller by from 15° to 20° than that of the rays belonging to the more equatorial parts of the precipitation.

When a stronger magnetising current is applied to the terrella, several instances of secondary precipitation appear on it, as we shall see; but there will always be *three* principal districts of precipitation in the polar regions, lying about 110°—100° from one another. The fact that the position of these districts is so independent of the magnetising conditions, is an exceedingly important one, as we may thus venture to transfer the results to the earth, where the magnetic moment is so enormously great. There is, in fact, on the earth, with regard to aurora, something which distinctly points to these fixed districts of precipitation in the polar regions. In the north of Norway, for instance, from about 9 to 10 p. m., and sometimes also between 4 and 5 a. m., there is a distinct culmination in the auroræ. Whether there is any aurora at about 2 in the afternoon it is impossible to say, on account of the light-conditions; but at any rate, during the darkest time of the year I have observed aurora several times at 4 in the afternoon rom the top of Haldde in 1899—1900, aurora which grew fainter and disappeared, only to return again later in the evening with increased strength. I think we are justified in concluding, from analogy with the experiments, that the rays that descend in the auroral zone are just those that come most perpendicularly down to the earth, and therefore those that make their way farthest down into the atmosphere.

Photographs 10, 11 and 12 in fig. 203 were taken during experiments in which the terrella was magnetised with 28 amperes. The pressure, indeed, according to measurement, was somewhat lower than before, namely, 0.0005 mm.; but it was subsequently proved that the statements of pressure here below 0.002 mm. are very unreliable, as there was vapour in the discharge-tube, which we had not troubled to condense, as it was of little consequence, in these experiments, whether it were there or not. A current of 23 milliamperes was sent through the discharge-tube, and the hour-angle of the screen was 155°. The photographs were taken as before.

Photograph 10 shows distinctly how the first line of intersection of the rays falls just over the slit. The continuation of the line of light is seen upon the screen in a lengthening of the slit.

Photograph 11 shows the second principal precipitation and the beginning of the third; but between them are two instances of secondary precipitation, which are especially distinct in the polar regions.

Photograph 12 shows distinctly the third principal precipitation, and in addition a number of others, fainter, and following one upon another, closer and closer, with increasing hour-angle. We have occasionally been able to count nearly 20 of them, fairly distinct.

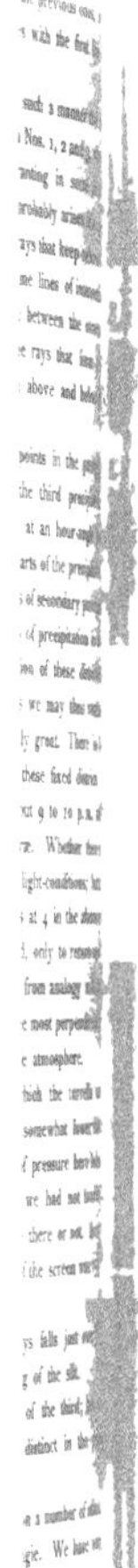

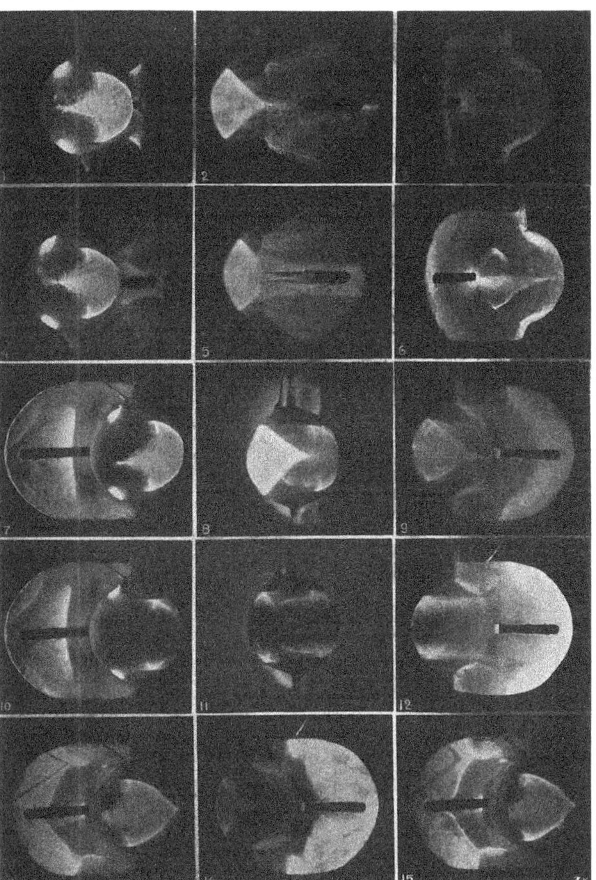

Fig. 204.

Nos. 1, 2 and 3 of fig. 204 were obtained under the same conditions as the three preceding photographs, except that the hour-angle of the screen was 265°. No. 2 shows that it is the second principal line of the rays' intersection that falls upon the screen. Two secondary precipitations are also seen upon the screen; but we shall return to these later.

The last two experiments show that the angles 155° and 265° correspond to the first and second lines of the rays' intersection with the horizontal plane (the magnetic equator), when the magnetising current is 28 amperes. By other experiments, the angles corresponding to the first three lines of intersection were found to be 155°, 265° and 365°, for the same magnetisation, and the angular distance between the second and third lines of intersection is thus only 100°.

By experiments with a magnetising current of 8 amperes on the terrella, the angles were found to be 168°, 272° and 370°

In experiments with the terrella highly magnetised, it was very interesting to watch the changes in the phenomena as the terrella became warm and gave off gas. To begin with, 8 distinct secondary precipitation-figures were once observed upon the night-side of the terrella, partly overlapping one another, and coming closer together towards the morning-side. The number of the patches of precipitation increased as the terrella grew warmer and gave off more gas, and finally there appeared continuous polar bands, answering to the north and south auroral zones.

Nos. 4, 5 and 6 were obtained under the same conditions as the preceding photographs, except that the hour-angle of the screen is now 250°, and therefore somewhat less than what would answer to the second line of intersection of the rays. It is also clearly seen in photograph 5 that the rays have not yet drawn together so that all pass through the slit. The third patch of precipitation on the terrella in No. 6 also bears evident signs of this.

The next six photographs are from a series of experiments that were made with the screen in the same position, but with pressures of 0.0009 mm., 0.0019 mm., 0.0052 mm., 0.012 mm., 0.02 mm. and 0.05 mm. The only pressures represented here are 0.0009 mm. and 0.012 mm. In the cases of the lowest pressures, vapour has certainly, as already mentioned, played an important part.

Nos. 7, 8 and 9 are from experiments with a pressure of 0.0009 mm. and the screen at an hour-angle of 40°. The photographs were taken from the same three positions as before, with respectively 90°, 180° and 270° east hour-angle.

The strength of the current through the discharge-tube varied from 18 to 22 milliamperes, and the tension between the electrodes from about 4500 to 3500 volts. The current magnetising the terrella was of 28 amperes.

No. 7 shows precipitation of returning rays upon the screen. In No. 8 the second precipitation is seen solitary, but with the third precipitation there are several secondary patches.

Nos. 10, 11 and 12 were taken under a pressure of 0.012 mm., with a current of 24 milliamperes through the discharge-tube, a tension of about 2500 volts, and a magnetising current of 28 amperes. The figures give a hint of the transition to the continuous band of light round the poles of the terrella, which appears with softer cathode rays; and it will be seen that the parts about the magnetic equator become more and more free from precipitation.

Experiments were made for the purpose of determining what tangential motion in relation to the terrella those rays had which formed the precipitation in the polar regions on the night and morning side of the terrella. The experiments were made with various pressures, and both the primary and the secondary precipitation was examined by means of the screen. It appeared in every case that the rays had a motion parallel with the auroral zone in a direction from west to east. Corresponding precipitation upon the earth would thus give rise to negative polar storms, as the various cases of secondary precipitation summed themselves up in their magnetic effects very much as shown in the case of the rays in the diagrammatic figure 50 a, in Section I.

It will be seen that almost everywhere the uppermost, polar part of the light-figures on the terrella, consists of the point of intersection of two strips of light which intersect one another at often a considerable angle, sometimes, indeed, more than 90°. It was interesting to see that the corresponding angle between the strips of light when they fell upon a vertical screen, was always much smaller than on the terrella, being quite acute. It was enlarged on the terrella by the oblique projection of the strips of light from west to east. The apex, when it touched the vertical screen, looked like a section of the wedge of light or the horn, often seen in the air about the auroral zone during experiments.

Figures 7—12, when considered as a connected group, give an indication of the reason for the appearance of all these secondary precipitations when the rays are soft and the magnetic force great. We receive the impression that some of the rays in the great bundle of rays that is working its way round the terrella from west to east, turn back once or oftener near the polar zones, describing something that resembles an epicycloidal curve. The stronger the magnetism, the more loops do the rays make, and the steeper the incline at which they intersect the magnetic horizontal plane. We shall return to these cases of secondary precipitation in the next section of the experiments.

We shall now in passing mention some experiments that are closely connected with the preceding ones, but which nevertheless originally formed the transition to the study of rays of group *B*.

When the screen had an hour-angle of about 90°, there might sometimes be noticed on its east side a remarkable shadow of the wire that conveyed the current to the terrella, this being caused by rays that have come over the polar regions of the terrella, and have then turned right round so that they come near the earth in the auroral zone with a tangential motion from east to west.

Nos. 13, 14 and 15 of fig. 204, are from experiments such as these. In both experiments the pressure was 0.01 mm., 21 milliamperes passed through the discharge-tube with a tension of 3200 volts, and 25 amperes were employed for magnetising. The positions of the screen were with hour-angles of 82° and 87°; and in both cases the photographs were taken from places with hour-angles of 130° and 310°.

Nos. 13 and 15 show distinctly how the shadow of the metal wire at a distance of 3 or 4 centimetres to the east, is thrown upon the screen in the form of two lines meeting in a point, which runs farther in towards the terrella in No. 13 than in No. 15. No. 14 gives the corresponding view of the phenomenon from the opposite side. What is particularly interesting about this last-mentioned photograph is that part of the conducting wire coated with phosphorescent matter is distinctly seen above the screen, illuminated by the rays, and thus casting a shadow back upon the east side of the screen. The rays which cause the formation of the shadow of the conducting wire come from above and strike a part of the wire that is more than 2 cm. above the north pole of the terrella. They then shoot down and bend westwards, coming in contact with the screen as the photographs show. It is a striking fact that in spite of the bending and twisting of the separate rays, the pencil of rays succeeds in throwing relatively clear shadows.

Several experiments of this nature were made without photographing them, and the particularly sharp shadow of the conducting wire, with the characteristic point directed towards the auroral zone was always noticeable. The experiments were, as we have said, an introduction to the study of what we called rays of group *B*, which give us the foundation for the explanation of the positive polar storms.

In photographs 13 and 15, the characteristic light-figures on the east side of the screen will have been noticed. These are of another kind than the precipitation upon the east side of the vertical screen, which we saw when the hour-angle of the screen was small. Upon closer investigation it appeared that a slight precipitation of returning rays also took place upon the east side of the screen when the hour-angle of the latter was about 250°. It would thus seem as if this phenomenon could be obtained in three positions of the screen, although the last, with 250°, was certainly very inconspicuous.

An endeavour was made to determine the positions of the screen in which the precipitation in these three cases was nearest to the terrella at the equator.

The first precipitation was nearest at an hour-angle of about 30°. In the second precipitation it was not easy to determine the position, as the precipitation became much fainter as the screen was turned; but it appeared to be at from 100° to 120°. The position in which the third precipitation was nearest to the equator was hardly capable of determination, but it must have been somewhere between 250° and 310°.

Finally, it was observed that when the screen had a position answering to an hour-angle of 153°, there was very marked precipitation upon its east side, nearest the corner where the screen is bent. This is also visible in No. 11 of fig. 203. It consists of returning rays. It is possible, as we have said, that they turn right round and give rise to the secondary precipitation upon the terrella. There is yet another circumstance which we will mention here. When there was comparatively much gas in the discharge-tube, there appeared, as already mentioned, continuous, luminous polar bands. These were not closed circles, but were somewhat spiral in form, as they lay at a higher latitude on the day-side than on the night and morning-side. This circumstance we have previously shown in photographs, but it is also applicable here where the magnetic axis is the axis of rotation.

110. Experiments in which the Terrella is Surrounded by a Horizontal Screen. The terrella was surrounded by a horizontal screen of aluminium after the vertical screen had been removed. The new screen, which is shown in fig. 205, had three holes or slits cut in it, so situated in relation to one another that the angle between the median lines of the first and second slits was 110°, of the second and third 110°, and of the third and first consequently 140°.

Fig. 205.

To the terrella itself were attached two almost radially projecting wires, as fig. 206 shows. They were placed there in order that conclusions might be drawn, from their shadows upon the terrella and screen, respecting the course of the rays.

Nos. 1, 2 and 3 of fig. 206, are from an experiment in which the pressure was 0.0012 mm., the discharge current 20 milliamperes, the tension 3600 volts, and the magnetising current 8 amperes. All the photographs of experiments with this screen were taken from positions in which the screen was viewed from above at an angle of 20°. The first slit is so placed that its median line forms an angle of 147° with the central line between the centres of the terrella and the cathode. For the sake of brevity, we will say that the hour-angle of the median line was 147°. The photographs were taken from positions with eastern hour-angles of 90°, 180° and 270°.

It will be seen that the first line of precipitation falls quite to the east of the slit, with the result that no second or third precipitation appears on the terrella. Nos. 4, 5 and 6 are from an experiment where the pressure was 0.0018 mm., the discharge-current 22 milliamperes, the tension 2800 volts, and the magnetising current 8 amperes. The first slit is placed so that the hour-angle of the median line is 155°, and the photographs were taken from positions of which the hour-angles were 60°, 180° and 310°. The first line of precipitation falls more or less over the first slit, so that the rays pass through it

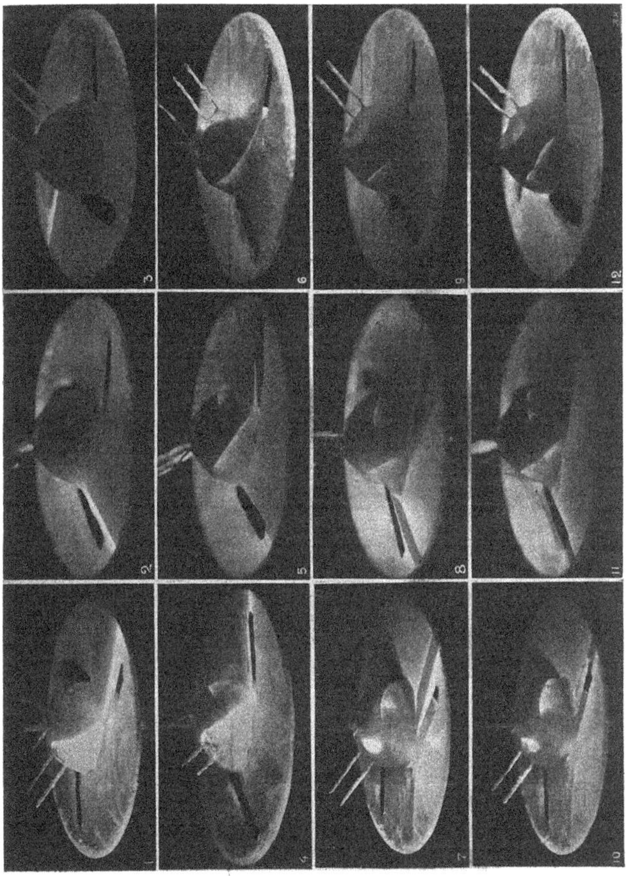

Fig. 206.

and form a new line of intersection above the second slit, through which the rays also pass. We see all three precipitations upon the terrella, and the third line of precipitation upon the horizontal screen could also be distinguished, although faintly, during the experiment. It fell on the margin of the third slit.

Some interesting experiments were made for the purpose of throwing light upon the origin of the secondary precipitation, of which frequent mention has been made in describing the experiments with a vertical screen (see figs. 203 & 204), and which was found again here under different conditions.

Nos. 7, 8 and 9 are of experiments with a pressure of 0.0014 mm. a discharge-current of 18.5 milliamperes, a tension of 3300 volts, and a magnetising current of 24 amperes. The photographs were taken from places with hour-angles of 90°, 180° and 270°.

Slit 1 was placed at 125°, just so that a small pencil of rays fell through the screen at the end nearest the terrella. This little pencil of rays which thus passed through the screen, at once gave rise to a distinct, but faint, precipitation upon the terrella. Even the third precipitation was single, without any secondary precipitation; but there is a strange precipitation upon the horizontal screen in which our attention is especially attracted by a line of precipitation almost parallel with the first line of precipitation, and only a few millimetres east of it.

That this secondary precipitation on the horizontal screen is produced by rays that have passed through slit 1 at the end nearest the terrella, is apparent from the fact that if the terrella is turned so that the first line of precipitation falls either *entirely* on the west side of the slit, or *entirely* on its east side, the secondary precipitation *completely* disappears in both cases.

The next experiment was to let the rays of the secondary precipitation through slit 1 at the same time as the first main precipitation came through it; for the distance between the two precipitations was rather less than the width of the slit.

It appeared that as soon as the secondary precipitation also passed through the slit, new precipitation made its appearance both on the horizontal screen and on the terrella, a secondary precipitation suddenly appearing in the polar regions on the night-side, similar to the primary precipitation lying immediately to the west.

Nos. 10, 11 and 12 are from an experiment which shows this. The pressure was 0.0008 mm., the discharge-current 17 milliamperes, and the magnetising current 24 amperes. The photographs were taken from the same positions as the preceding ones.

We see distinctly that the third polar precipitation consists of two consecutive precipitations. That on the east is the secondary.

The experiment was repeated several times without being photographed. Again and again it appeared that when the first secondary precipitation upon the screen passed through the slit, a new precipitation was formed nearer the second slit, the innermost part of it falling through that slit at the end nearest the terrella, thereby producing the secondary precipitation upon the night-side of the terrella, in the polar regions, farther out on the night-side than the first, which was there already.

It was distinctly seen that the second secondary precipitation upon the horizontal screen formed a much smaller angle with the first precipitation than did the second principal precipitation. It was the outer part of the first secondary precipitation which, by passing through the first slit, produced a new line of precipitation not more than 50° farther east upon the horizontal screen. Only because the end of the second slit nearest to the terrella was comparatively wide, did a pencil of rays from this precipitation pass through there, and occasion the secondary precipitation after the second polar precipitation upon the night-side.

We have previously touched upon the possibility that the connected polar precipitations upon the terrella (in the auroral zone) were composed of a whole series of close-lying secondary precipitations (see p. 552); and we assumed, after discussing our experiments, that the rays again and again looped back on themselves and described curves that more or less resembled epicycloids. Rays such as these would be able to pass at the equator and nearest to the terrella with a velocity-component from east to west, i. e. a direction the reverse of that of the primary rays.

It is interesting here to call to mind (cf. pp. 82 & 83, Section I) that the negative equatorial storms were explained by the bending round of rays in the vicinity of the magnetic equator, so that they encircled the earth from west to east, while the positive equatorial storms were explained by rays with a component motion from east to west, nearest the earth at the equator.

We are not able to see, in the photographs as reproduced here, any distinct signs of shadows cast by the two parallel wires upon the terrella, although, during the experiments, such shadows were easily discernible, though always faint.

In both precipitations, *A* and *B*, upon the horizontal screen, there occurred in certain positions of the terrella, curved but parallel shadows of these wires. These shadows have been especially useful in investigations for the purpose of coming to an understanding regarding the rays of group *B*, as we shall presently see.

111. Equatorial Rings of Light. In connection with the ray-phenomena just described, belonging to group *A*, we will discuss a phenomenon which has already been mentioned several times, and called equatorial rings. The phenomenon is described in "Expédition Norvégienne 1899—1900", p. 41; but unfortunately on that occasion the luminous rings were not photographed. There is, however, a photograph of one in Section I of the present work, p. 80, fig. 37.

The equatorial ring is formed of rays that curve round the terrella from west to east. Under specially favourable experimental conditions, the concentration of rays near the plane of the equator is so great that the rarefied gas is rendered luminous. It is not only rays that move exactly in the plane of the equator that form the ring, but more especially rays that move alternately above and below the plane of the equator in its immediate vicinity. We will here point to photographs 5 and 6, fig. 203, where ust such rays as these are made distinct by their precipitation upon the terrella about the equator. Even the rays that come nearest to the terrella in the polar regions, and which thus, in their passage through the plane of the equator, intersect it at large angles, will perhaps serve to produce the luminous ring, as they bring about a powerful concentration of rays just at the magnetic equator. We have seen indeed that the rays from one primary pencil, have numerous lines of intersection in the equator. When such rays, by a suitable proportion between the magnetism and the stiffness of the rays, are free to move a great many times round the terrella near the equator, the gas there becomes luminous, and we may have the equatorial ring. As may be expected from what has been stated, the appearance of the ring is almost a chance phenomenon; it is unstable, and many fruitless attempts may be made to induce it to show itself.

The three photographs forming fig. 207 were taken several years ago, and the experiments on that occasion were made with a powerful influence-machine. The strength of the current with these machines, however, is so small that the phenomena are not bright. The rarefied gas itself, moreover, plays a very important part if the phenomenon is to be successful. It seems as if impurities were an assistance. The experiments were made with a tension of about 6000 volts, and with the employment of about 10 amperes upon the terrella — No. 2, with a diameter of 10 cm. The magnetic moment with this current-strength was about 50,000 C. G. S. The ring is distinctly seen to be rather thin and broad, its outer margin often extending far beyond the terrella. The inner margin of the ring often comes right up to the terrella; but I have several times observed the ring standing unattached in the gas, with a dark interval between it and the terrella.

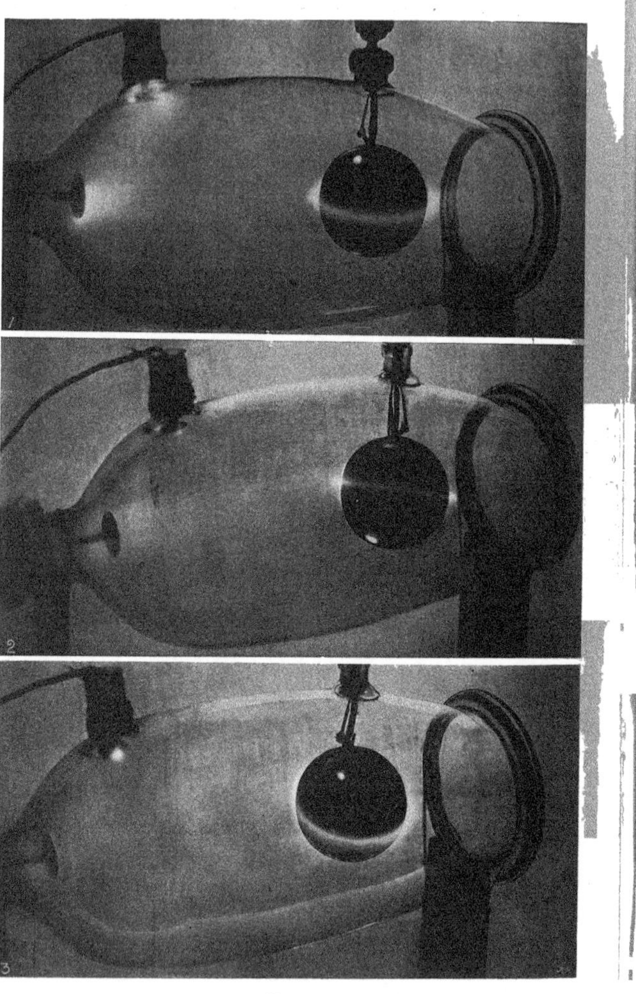

Fig. 207.

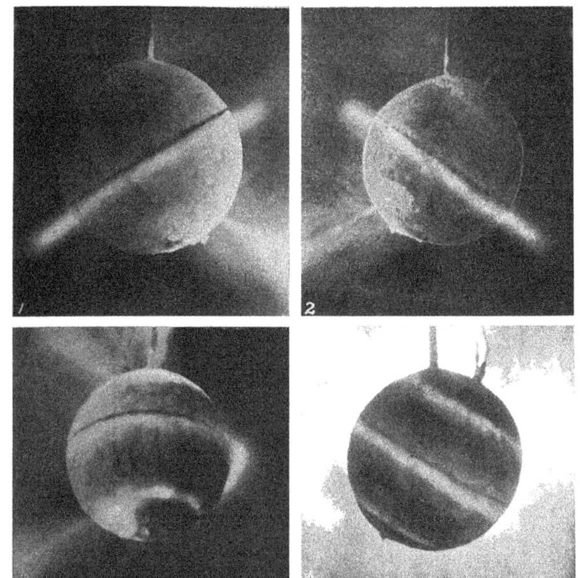

Fig. 208.

Of late, I have made these experiments partly by the employment of high-tensioned electric waves, produced by DUDELL vibrations, through the discharge-tube. It seems to be a comparatively easy way of producing them. I have, however, preferred a direct current from the previously-mentioned 20,000-volt machine (fig. 67), and have employed discharges of up to 35 milliamperes through the receptacle.

The four photographs in fig. 208 were taken from experiments such as these. The terrella No. 4 employed was 8 cm. in diameter, and a current of from 10 to 12 amperes was employed upon it (M = 28000 C. G. S.). The magnetic axis was set at an angle of about 30° with the axis of rotation, and the magnetic equator was drawn in pencil upon the terrella, as we were to see whether the ring coincided with the magnetic equator in all positions of the terrella. The first two photographs were taken while the magnetic north pole (below) had an hour-angle of 0°. They were taken from places with hour-angles of 90° and 270°.

It will be seen that although the angle between the magnetic axis and the axis of rotation is made so great, the equatorial ring lies fairly parallel with the magnetic equator. The ring here is most powerfully developed farthest from the cathode.

The next two photographs were taken during another experiment, from positions with hour-angles of 90° and 180°.

The equatorial ring has not come out particularly well here, but on the other hand the polar rings are quite distinct. We shall give better illustrations of the polar phenomena, however, later on, and will therefore not dwell upon them now. These earlier photographs were taken during experiments in which the discharge-tube was cylindrical, and not with the prismatic discharge-receptacle, which we used subsequently. In No. 3, there are indications of the equatorial ring having been brighter out from the terrella than close to it.

As will be seen later on, in the chapter on zodiacal light, I easily succeeded in producing these equatorial rings round a magnetic globe, which itself served as cathode. It was sufficient to employ a difference of a few hundred volts in the tension between the electrodes, in order to produce the discharge under these conditions.

It is therefore not impossible that these rings in every case occur owing to the magnetic globe having become negative in relation to its nearest surroundings in the discharge-tube.

In the meantime it is a fact that a considerable number of rays move round the terrella, from west to east, close to the equator; this has been demonstrated by nearly all the numerous experiments which have just been described.

STUDY OF RAYS OF GROUP B.

112. We now pass on to experiments made with terrella No. 5 provided with a vertical screen over its north pole, the plane of this screen passing through the axis of rotation, which still coincided with the magnetic axis of the terrella. The screen was placed thus in order that the course of the rays in the polar regions over the terrella could be studied. At the same time, the former horizontal screen was retained (see fig. 205), now, however, entire, without the three slits, in order to prevent the formation of polar precipitation by rays of group *A*.

The two radial wires, about 4 cm. in length, standing out from the terrella, were also retained, in order that their shadows thrown upon the screens might give information as to the course of the rays. The photographs that are reproduced here distinctly show the position of the screens and wires. At first a round hole was made in the vertical screen, and later on a slit was added.

It now appeared that when the plane of the vertical screen formed an angle of about 30° with the line of direction from the centre of the terrella to the cathode, characteristic precipitation became visible upon the screen, extending far over the screen towards the axis, when the terrella was magnetised with 8 amperes. When, on the other hand, 25 amperes were employed upon the terrella, the precipitation had moved right out to the right margin of the screen, seen from the cathode. With the employment of 14 amperes, the precipitation was so situated that its innermost edge lay farther in than the above-mentioned hole in the screen.

The nine photographs in fig 209 represent various results of the experiments made.

Nos. 1 and 2 represent experiments in which the hour-angle of the vertical screen was 30°, this angle being reckoned to the wing of the screen in which was the hole. The photographs were taken from positions with hour-angles of 300° and 120°, and looking from above at an angle of from 15° to 20° with the horizon. The pressure was 0.0014 mm., the discharge-current 24 milliamperes, and the magnetising current 14 amperes.

Nos. 3 and 4 are of a similar experiment, in which the vertical screen, with the terrella, is turned 160°, in order to obtain clearer precipitation. We say then that the hour-angle of the screen is 190° and the photographs were taken from places with hour-angles of 310° and 90°. The pressure and magnetising current were as before, but the discharge-current and the tension were respectively 20 milliamperes and 2700 volts.

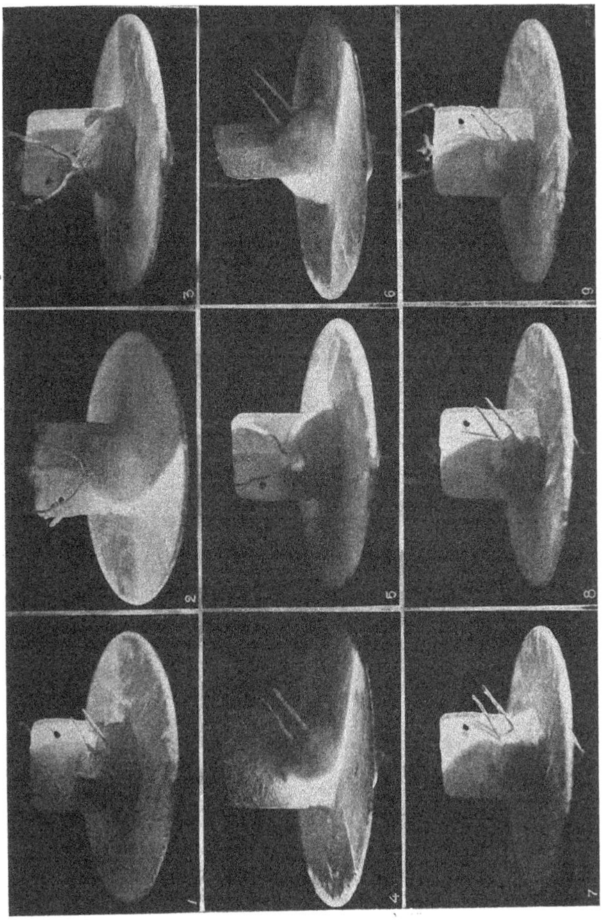

Fig. 209.

We distinctly see the shape of the precipitation upon the vertical screen, answering to a magnetisation of 14 amperes. The lowest curved edge of the precipitation upon the west side of the vertical screen lies higher than the uppermost edge of the precipitation upon the east side of the screen. For the sake of comparison we would observe that in photographs 7, 8 and 9—to which we shall return later—the precipitations are seen with very low magnetisation, namely 6.5 amperes.

In the above-mentioned experiment with 14 amperes to the terrella, it was ascertained that precipitation *B* upon the horizontal screen, had disappeared in the position given to the vertical screen of an hour-angle of 190°, but the western part of precipitation *B* appeared when the vertical screen was turned 15° or 20° either west or east. This shows that the cathode rays which produced this part of precipitation *B*, were stopped by the vertical screen in the position shown in the photograph, but that the rays slipped past and descended upon the horizontal screen as soon as the vertical screen was turned a little. It still appeared that if the vertical screen were turned eastwards to an hour-angle of about 230°, the luminous line of precipitation bounding the easternmost part of precipitation *B* upon the horizontal screen, also made its appearance.

Photographs 5 and 6 were taken from an experiment with a pressure of 0.001 mm., 14 amperes to the terrella, 3200 volts tension, and 19 milliamperes to the discharge-tube. The vertical screen has an hour-angle of 240°, and the photographs are taken from positions with hour-angles of 310° and 90°.

The rays here are fairly stiff, but the westernmost part of precipitation *B* is seen sufficiently clearly, while the easternmost has not come out distinctly in the photograph.

Experiments were made with 8 and 24 amperes to the terrella. With low magnetising — 8 amperes — the precipitation on the day-side (that turned towards the cathode) of the vertical screen was of great extent when the screen had an hour-angle of, for instance, about 200° (or 180° less, see, for example, Nos. 7—9 of fig. 209). On turning the screen eastwards, so that the angle became greater, the precipitation moved out; but there was still a little left on the uppermost right corner of the screen, looking from the cathode, right until an hour-angle of 260° had been reached (see No. 5).

With a magnetisation of 24 amperes, the precipitation was always far out on the screen, and had already disappeared with a turning of the screen to an hour-angle of 220°.

When the terrella was turned so that the vertical screen had an hour-angle of about 225°, all precipitation of light disappeared from the day-side of the screen when the magnetisation was 14 amperes, and did not return to that side until the screen had been turned about 135° farther, i. e. when the hour-angle of the screen was about 360°, and the former night-side was about to become the day-side. It was otherwise with the night-side of the screen. There was at first no light there either, when the light had disappeared from the day-side, with an hour-angle of 225°; but after turning the screen 75°, there was the maximum of a faint precipitation upon the night-side on the wing of the screen in which was the hole, and which then had an hour-angle of 300°. This precipitation is closely connected with the small, faint half-ring of light that passes through the pole (see fig. 134 and p. 298 in Section I).

Further experiments were made for the purpose of explaining precipitation *B* upon the horizontal screen, when employing 8 amperes to the terrella. It was observed that precipitation *B* originated in rays which, if the vertical screen were in a suitable position (an hour-angle of about 15°) and caught them, fell near the lowest, curved border of the precipitation of light. Precipitation *B* could be partly or entirely removed from the north side of the horizontal screen, by adjusting the vertical screen in a suitable manner. At the same time, as was to be expected, the corresponding precipitation *B* on the south side of the horizontal screen was in all cases unchanged and just as bright, as there was no vertical screen in the south polar regions.

It should be remarked that while the terrella was being turned, a distinct shadow of the right edge (looking from the cathode) of the vertical screen often appeared in precipitation *A* upon the north side of the horizontal screen. Precipitation A is in direct connection with the first precipitation upon the terrella (see fig. 66, p. 151, Section I).

Photographs 7, 8 and 9, from three experiments, were taken from *one* position with an hour-angle of 320°, and from above at an angle of 15° with the horizon. The experimental conditions were the same in all three experiments, except that the hour-angle of the screen was respectively 70°, 50° and 40°. The pressure in all three was 0.0105 mm, the magnetising current 6.5 amperes, the discharge-current 22 milliamperes and the tension 2600 volts. The photographs were taken for the purpose of studying the shadows of the two vertical wires upon the vertical screen.

No. 7, with the hour-angle of the screen 70°, shows two distinct shadows, comparatively far down in the precipitation. If the angle were made greater than 70°, the shadows sank still lower, and suddenly also made a partial appearance in precipitation B on the horizontal screen. It was, as we have said, quite clear that the lowest rays on the right of the vertical screen were rays that would have fallen upon the horizontal screen — precipitation B — if they had not been intercepted by the vertical screen.

No. 8 shows two coincident shadows of the two wires. A plane through these wires, in this position, passed approximately through the centre of the cathode. The impression given was that the rays which threw the shadow upon the vertical screen in this position, fell normally upon the screen. For the next experiments, therefore, a slit was cut in the screen in very much the same direction as that in which the shadow fell.

No. 9, which is taken with the hour-angle of the vertical screen 40°, shows that the shadows have now gone towards the left margin (looking from the cathode) of the precipitation. If, during the experiment, the angle were made less than 40°, the shadows drew up towards the edge, and became very long. The rays here evidently soon bend straight up, and they are seen to strike against the roof and floor of the discharge-box (see photograph of this during discharge, fig. 200).

In order to investigate more closely the rays that went in at right angles to the vertical screen, a slit was cut, as we have said, at the place in question. A new wing was moreover added to the screen at an angle of about 110° with the original screen, and in the manner shown in the photograph, where it appears with sufficient distinctness. The purpose of this enlargement of the screen was to catch the returning rays that had passed through the slit that had just been cut. The terrella was moreover furnished with a small movable screen, also to be seen in the photographs. This screen could be turned from outside by magnetic means, and also served in the investigation of the course of those rays which passed through the slit. The way that the rays went, however, made it difficult to observe them upon this movable screen; at any rate no photograph was obtained that could be of any use, so this small, movable screen on the whole did little service.

Nos. 1, 2 and 3 in fig. 210 were taken during experiments with a pressure of 0.0095 mm., a discharge-current of 20 milliamperes, and 6.5 amperes to the terrella. The photographs were taken from places with hour-angles of 130°, 180° and 320°.

Precipitation of returning rays that have come through the slit, is distinctly visible in No. 1. A faint continuation of the luminosity upon the screen nearest the terrella is observable; a clear wedge of light could be seen running right in towards the surface of the terrella. The position of this precipitation answered to about *3 p. m.*, and the precipitation was of such a kind that these returning rays of group B could very well have given an explanation of the positive polar storms. (Compare also the previously-described beautiful experiments shown in Nos. 13—15, fig. 204, in which 25 amperes were employed for the terrella.)

At Kaafjord, however, positive storms, with sharply-defined maximum occurred at *6 p. m.*, during the six winter months for which we have the material for judging of the conditions there. (See Chap. III, Table XCVI, p. 539).

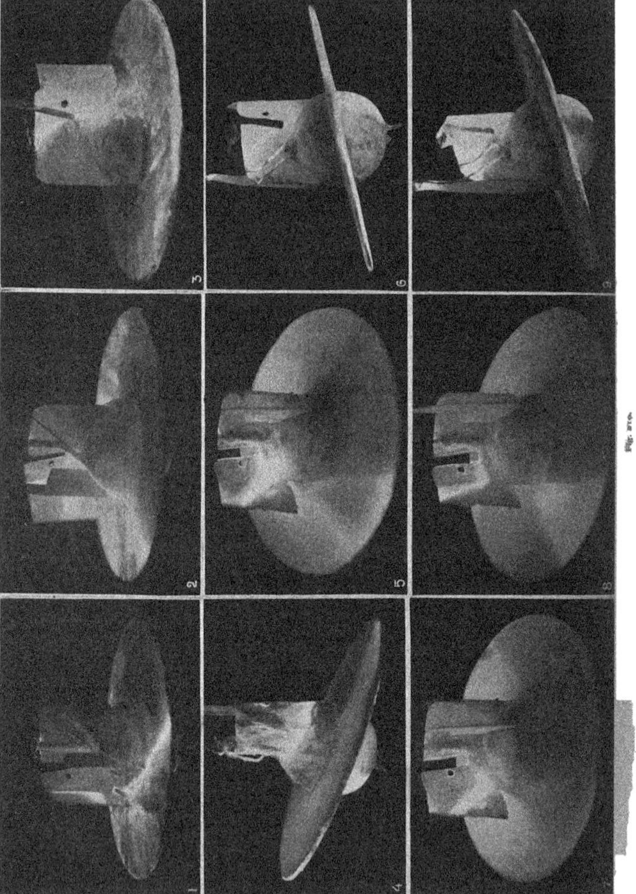

Fig. 219.

It now, however, appeared to be impossible to obtain, by means of rays through the slit, any precipitation on the terrella at a place answering to *6 p. m.*, even if the magnetisation of the terrella were altered, as long as the magnetic axis coincided with the axis of rotation.

It is possible that if the discharge-box had been much larger, returning rays of this kind might have been made, by high magnetisation, to descend upon the terrella in places answering to 6 p. m. This question will be taken up again for thorough investigation, later on. In the mean time, experiments were made in letting the magnetic axis of the terrella form an angle of about 20° with the axis of rotation, once so that the south pole turned towards the cathode, and another time so that it turned away from the cathode. This latter position must answer more or less to the condition of the magnetic axis upon the earth in winter. Experiment showed that if the magnetic south pole were turned towards the cathode, the precipitation from the rays through the slit was nearest the terrella in places answering rather to earlier hours than 3 p. m. than to later.

On the other hand, the experiments showed decidedly that when the magnetic south pole was turned away from the cathode, an abundant precipitation fell upon the terrella in places answering to 6 p. m.

Nos. 4, 5 and 6 were from an experiment in which the pressure was 0.0012 mm., the discharge-current 21 milliamperes, the tension 2100 volts, and the magnetic current to the terrella 7 amperes. They were taken from places with hour-angles of 90°, 180° and 330°.

The screen, with the slit and the hole in it, had a position answering to an hour-angle of 80°. The magnetic axis formed an angle of 20° with the axis of rotation, and the south pole was in the position of a place having an hour-angle of 180°. No. 5 shows how the rays through the slit and the hole have turned back and strike the screen.

We have seen that in all the numerous experiments mentioned here, the rays divide into two groups, which we have called *A* and *B*. The first group comprises rays whose course is about the equatorial plane, and which turn alternately up and down, above and below that plane, twisting about the terrella in a direction from west to east. The boundaries of the group upon the terrella are formed of those rays which turn so far out from the equator that they form polar precipitation. We have assumed that corresponding precipitation upon the earth forms what we have called the negative polar storms.

The second group of rays approaches the terrella in the north and south polar regions, and the rays descend in the polar belt with a velocity-component tangential to the terrella in a direction opposite to that of the rays of group *A*.

We may therefore assume that rays of this kind on the earth glance off into the auroral zone with a movement from east to west, and thus occasion what we have called positive polar storms.

That the rays about the equator must curve in the reverse way to those over the polar regions of the terrella, is a consequence of the fact that the magnetic lines of force run in opposite directions in the two places.

We will now go on to the further experiments that were made for the purpose of studying the polar rays.

Photograph 7 was taken from a place with an hour-angle of 180°, with the screen at 85°. It shows a bright precipitation of rays that have returned after passing through the slit and the hole. The pressure during the experiment was 0.0014 mm., the discharge-current 21 milliamperes, the tension 3000 volts, and the magnetising current 7 amperes.

Nos. 8 and 9 were taken under similar conditions, except that the position of the screen had an hour-angle of 90°, and the photographs were taken from places with hour-angles of 180° and 320°.

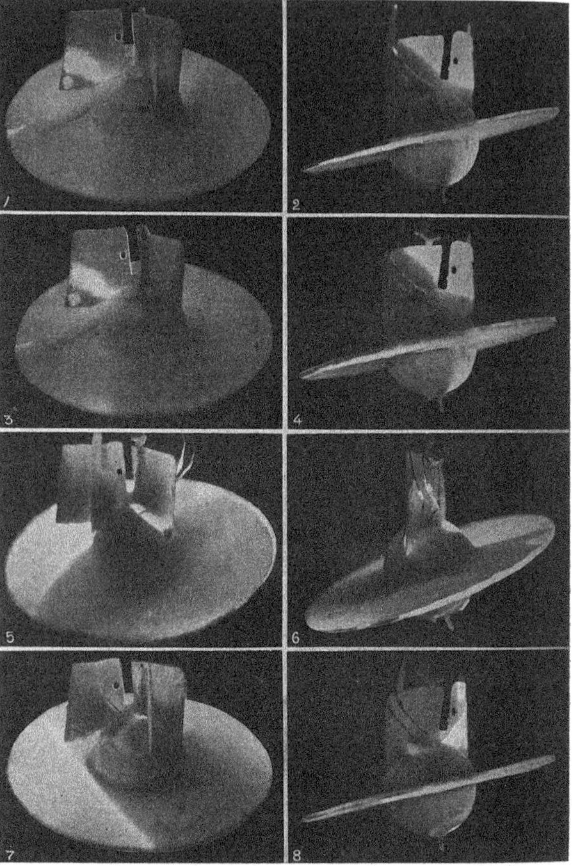

Fig. 211.

A great difference is discernible between the precipitation in Nos. 7 and 8, although the position of the screen is very little changed.

Nos. 1, 2, 3 and 4 of fig. 211 were taken in order to determine more exactly the position of the ray-precipitation now under discussion. In 1 and 2, the conditions are the same as in 8 and 9 respectively of fig. 210, except that the hour-angle of the screen is 75°; and in Nos. 3 and 4, they are also the same, except that the hour-angle of the screen is 60°. The white wedge of light on the horizontal screen (Nos. 1 & 3) is a patch belonging to precipitation *B*, formed by rays which have passed through the slit.

It will be seen from all these photographs that under these conditions the precipitation is well defined on the eastern side, and its strength is greatest on the terrella at a place answering to between 5 and 6 p. m.

Nos. 5, 6, 7 and 8 were taken from a series of experiments, made for the purpose of finding out whether rays that come in towards the polar regions of the terrella from the left side, seen from the cathode, could also form precipitation of the same kind as the rays that came through the slit on its right side, looking from the cathode.

It appeared that with the highest magnetising that the terrella could stand, a quantity of rays were drawn in towards the terrella on the left side too, descending fairly perpendicularly, so as to give the distinct impression that even the large discharge-box of sheets of plate-glass, which was employed in all the experiments described here, was not large enough, i. e. high enough above the poles, to allow of the position of the precipitation upon the terrella being accurately determined, as it might have been if the rays could have moved towards the terrella, unhindered by the sides of the discharge-box. A great many experiments were made, however, so the results described below may be considered sufficiently certain.

Photographs 5 and 6 are of experiments in which the pressure was 0.001 mm., the discharge-current 20 milliamperes, the tension 3000 volts, and the magnetising current 25 amperes. They were taken from places with hour-angles of 250° and 315°. The hour-angle of the screen was 115°. With this high magnetisation of 25 amperes, and still more with 35 amperes, which was used subsequently, the small luminous patch, described in Section I of this work at the bottom of page 298, came out. In the present case, this little ring became a rather compressed oval, a great part of it being visible upon both sides of the screen. In No. 5 we distinctly see the one part, but in No. 6 the continuation of the precipitation is no more than just visible. With 33 amperes and rather softer rays, this half of the oval was just as bright as the other part on the other side of the screen (see No. 5).

In this photograph there is also distinctly seen in the precipitation, the shadow of the conducting-wire for the current to the terrella. The shadow shows how the rays descend almost perpendicularly towards the terrella; but a twisting of the rays can also be proved resembling that of a helix.

Photographs 7 and 8 show results of experiments made with a pressure of 0.009 mm., a discharge-current of 22 milliamperes, and a magnetising current of 30 amperes. The screen has an hour-angle of 70° (it is still the wing with the hole in it from which the angle is measured), and the photographs were taken from places with hour-angles of 180° and 320°.

In No. 7 we see the continuation of the precipitation which produced the oval in No. 5. The precipitation now entirely disappears from the vertical screen where there had previously been precipitation from the returning rays that passed through the slit.

The shadow of a conducting wire is now seen in the precipitation, showing that the rays have curved round from the left side of the screen, looking from the cathode, to far back on the right side.

EXPERIMENTS FOR DETERMINING THE TANGENTIAL COMPONENT OF THE POLAR PRECIPITATION IN RELATION TO THE SURFACE OF THE TERRELLA.

113. In the preceding pages, it has frequently been stated that the polar precipitation in neighbourhood of the auroral zone was produced by rays that came in to the terrella fairly perpendicula By the previous investigations, therefore, it was not made clearly apparent how the tangential compor in the precipitation was directed at the various places on the surface of the terrella.

By the experiments illustrated in fig. 212, however, the matter has been given, by special arrar ments, all possible clearness, and it will be seen what a remarkably striking analogy comes out betw the situation and direction of the various instances of precipitation upon the terrella, on the one s and the situation of the positive and negative districts of precipitation during the polar magnetic sto on the earth, described in Chapter I of the present part, on the other.

The photographs of which these illustrations are reproductions, were unusually successful. As t were to make clear one of the most important points in the theory, they were chosen with care f a great number of more or less good ones. Any one with experience of similar experiments, will ea understand the labour that this entailed.

The experiments were made with terrella No. 4, with a diameter of 8.2 cm., *which was suspe by the magnetic equator*, so as to give the best possible opportunity of photographing the polar pre tation from the side of the discharge-box. Upon one magnetic pole—in this case the south pol a star-shaped screen was placed, consisting of 8 branches of a height of about 15 millimetres, stan on their edge.

Nos. 1, 2 and 3 were taken from an experiment in which the discharge-current was 24 milliampe the magnetising current 20 amperes, and the tension 2500—2300 volts. The pressure was 0.006 mn

The first two photographs were taken, looking towards the centre of the terrella, in a plane an easterly hour-angle of 270°, the first with a declination of + 24°, the second with —24°. third photograph was taken in the plane of the horizon from a place with an hour-angle of 240°.

Nos. 4, 5 and 6 were taken during a similar experiment, in which the discharge-current was milliamperes, the magnetising current 20 amperes, and the tension 2500 volts. The pressure was 0.009

The terrella was turned 15°, so that the line from the centre to the magnetic south pole had hour-angle of 285°. The photographic apparatuses were in the same position as before. It was inten that the conditions should answer more or less to the position of the earth in summer.

Nos. 7, 8 and 9 are of a similar experiment with a discharge-current of 24 milliamperes, a mag ising current of 20 amperes, and a tension of 2400 volts. The pressure was 0.009 mm. This t however, the terrella was turned 15° in the opposite direction, so that the hour-angle of the line to magnetic south pole was now 255°. The purpose of this was similarly to make the conditions ans to some extent to the position of the earth in winter.

At the top of all the photographs, there is a hook, which has nothing to do with the suspen of the terrella, and ought not to have been there at all, as it has nothing to do with the present ex ments. The cathode in the discharge-tube is, as will be understood, on the right of the terrella. left side will therefore answer to the night-side. For the purpose of easy reference, we will nun the eight branches that form the star-shaped screen, beginning with the middle branch on the righ the picture — the branch which, as we have said, points towards the cathode — and continuing in reverse direction to the hands of a clock.

It will at once be noticed that the principal precipitation on the three branches, 4, 5 and 6, on night-side, is found on the west side of each branch. There is no precipitation on the east side, a dark, narrow shadow is to be seen in the polar band of light on the terrella itself. In No. 2 ther

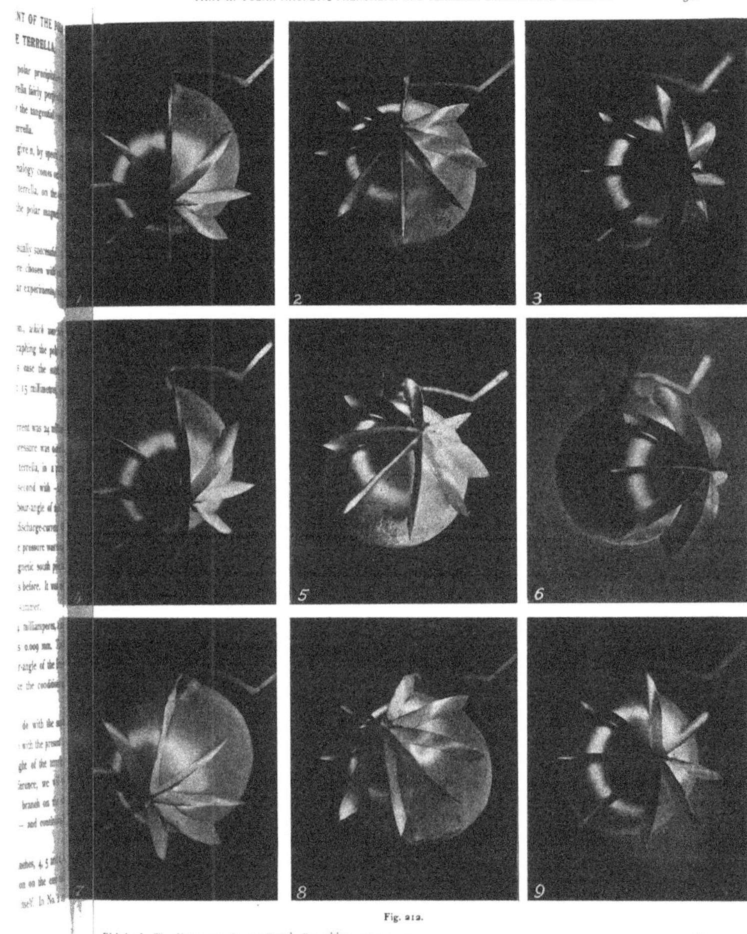

Fig. 212.

even a little light coming *under* the branches of the screen, as they do not lie close to the terrella, bu leave a millimetre here and there open between themselves and the terrella. These shadows and stripe of light tell us the average straightness with which the rays descend towards the terrella. We sha return to this, as the figures in fig. 221 are meant for such investigations.

What we here first of all substantiate is that the precipitation on the night-side of the terrella i the polar band has a tangential component eastwards. The magnetic effect of corresponding precipitatio over the earth would thus be a positive current directed westwards, just as we have always found th current-arrows directed in the *negative* polar storms in the auroral zone.

It is not only on the night-side of the terrella that we find precipitation on the west side of th eight branches, but right round the connected luminous spiral, which we shall briefly call the auror zone. Even at the beginning of the spiral nearest the pole, where, in fig. 140, p. 327, we saw a sudden curv in the luminous band, we now see precipitation in two places on the west side of screen 1, which point towards the cathode (see Nos. 2 and 8). But it is the precipitation on the night-side that is the stronges and which comes out better, even when there is no precipitation on the day-side (see fig. 204); and also has comparatively the greatest tangential component. This is thus in accordance with the fact tha the negative polar storms are generally found on the night-side of the earth.

It is also easy, however, to demonstrate in our photographs precipitation upon the screen-branche exactly analogous to the precipitation on the earth which occasions *positive* polar storms. With regar to branches 3 and 2 especially (see, for instance, photograph 3), we also find on their east side a grea precipitation of rays, which, close up to the terrella, has a strong tangential component westwards.

The magnetic effect of corresponding precipitation over the earth would thus be a positive curren directed eastwards along the auroral zone, just as we have always found the current-arrow directed i the positive polar storms. The time of day also suits these cases of precipitation exceedingly well, fo the positive polar storms occur with a maximum in the afternoon, and, as is seen, branch 3 just answer to a place on the terrella corresponding to 6 p. m.

At the extreme end of branches 5 and 4 also, there is precipitation on the east side similar t that on 3 and 2, but not going down so close to the terrella. It occurs in much lower latitudes, whil on branches 3 and 2 it has come quite up to the auroral zone.

The photographs show plainly that the precipitation on the east side of branch 4 occurs in a muc more southerly latitude than that on the west side. On branch 3, too, the precipitation on the east sid is farther south than that on the west side; but the two are considerably nearer to one another than o branch 4. On branch 2 they are still nearer to one another, looking as if they to some extent covere one another. These conditions correspond in an astonishing degree with those on the earth durin magnetic storms. We have frequently, indeed generally, seen that while there is a positive polar storn in the southern part of the auroral zone, there is at the same time a negative polar storm in th northern border of the zone. (See p. 445). These storms counteract one another in a *horizontal* direc tion, and may sometimes neutralise one another's effect in the case of stations lying between the tw precipitations; but in a *vertical* direction the two storms act together. This has often been shown i discussing the observations from Jan Mayen, for instance.

In the preceding pages, we have repeatedly put forward the opinion that this precipitation of ray with a tangential component westwards along the auroral zone, was due to rays of group *B*, that is t say, rays that are first drawn down towards the terrella in its polar regions, and then deflected an some of them thrown back. That certain rays have such a course is evident from the experiments tha are described with photographs 13, 14 and 15, fig. 204. The distinct shadows of the conducting wir that are thrown upon the screen cannot be interpreted in any other way; and the experiments describe in Art. 112 are also very conclusive.

According to the above, however, it is also conceivable that some rays of the group called A, which more especially bend round the terrella above and below the plane of the magnetic equator may also be made partially responsible for the precipitation found upon the east side of the branching screens; for we have seen that some of these rays will loop upon themselves, and it is then clear that *one* branch of the ray-trajectory will be turning back. This branch may then just occasion precipitation on the screen in more southerly latitudes of the terrella with a tangential component the reverse in direction of that to which the ray would originally have given rise.

ON AN INTIMATE CONNECTION BETWEEN RAYS OF THE TWO GROUPS A AND B.

114. In continuation of the experiments which have last been described, I have succeeded at length in obtaining complete clearness as to the relative connection between rays of group A and those of group B.

Further experiments were first made with an eight-armed star-screen with arms 3 centimetres in height instead of 1.5 centimetres as they had previously been, the purpose being to see whether the precipitation and shadows on the two star-screens corresponded.

The first eight photographs of fig. 213 show the conditions. The first four are from an experiment in which Nos. 1 and 3 were taken from directly opposite the magnetic poles, north and south, from position with hour-angles of 90° and 270° without elevation, while No. 2 was from a position with hour-angle 235° and 24° declination, and No. 4 with an hour-angle 295° and 24° declination. The discharge-current employed was 22 milliamperes with a tension of about 3000 volts and a magnetising current of 20 amperes to the terrella. The pressure sank from 0.022 mm. before the experiment, to 0.043 after it.

The next four photographs were taken from the same respective positions, with discharge-current of 5 milliamperes and tension about 3000 volts, while the magnetising current was 36 amperes. The pressure was 0.012 mm. before the experiment, and 0.066 mm. after it.

A comparison with the phenomena represented in fig. 212, in which the star-screen was about 1.5 cm. in height, shows, on the whole, a similarity. One difference that may be mentioned is that the positive precipitation does not extend so far down towards the terrella itself, as when the height of the screen was less. The negative precipitation, on the other hand, extends right in, and the polar ring on the terrella itself is now quite as well formed as with the lower screens. One especially characteristic feature is that the dark shadows in the ring of light on the terrella just behind the screening branches, are no longer now than when the screens were only 1.5 cm. high, but are, if anything, narrower. This shows that the rays do not strike so straight down towards the terrella as might be thought from the previous experiments, but that rays that come in contact with the higher parts of the screen first move a little *away* from the screen, and then turn in *towards* it again.

It will further be observed from the extremely interesting negative precipitation on the screens (from which an idea can actually be formed of the manner in which the rays approximately move from the northern polar light-ring to the southern, see Nos. 4 and 8), that the precipitation nearest the terrella is fainter and thinner than farther out. This suggested the thought that possibly one of the eight branches of the star-screen might cast a shadow upon the neighbouring branch, that again upon the next and so on. In order to determine this question, one of the branches was cut off, as shown in Nos. 9 and 10. The positions here are similar to those in Nos. 2 and 4, and the discharge-current employed in the experiment was of 23 milliamperes. It will at once be seen from the photographs that the already-mentioned narrow precipitation of light nearest the terrella upon the branching screens is not caused by the casting of the shadow by one branch upon its neighbour.

Photographs 11 and 12 were taken in two experiments, both in the same position as in No. 2. The experiments were made very much as before, but with 10 and 20 amperes to the terrella. The tension

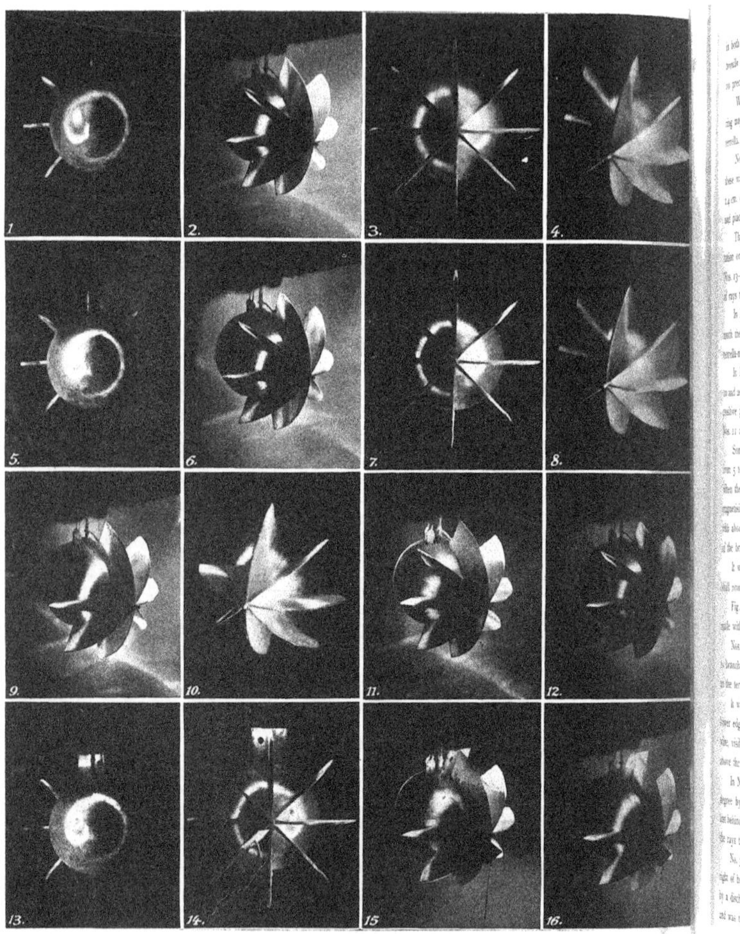

Fig. 213.

in both cases was about 2500 volts. The photographs show the important fact *that the more highly the terrella is magnetised, the farther does the positive precipitation reach towards the evening side.* There is no precipitation on branch 3 with 10 amperes' magnetisation of the terrella, but it is there with 20 amperes.

With conditions corresponding to those on the earth, where the spherical diameter of the auroral ring may be put at about 45°, the positive precipitation might reach far on into the evening side of the terrella.

Nos. 13—16 are from experiments in which two additional small screens were introduced. One of these was square, and placed at right angles to branch 3. It was pierced with a hole, and extended 1.4 cm. on each side of the branch. The other small screen was also square, was furnished with a foot, and placed radially in relation to branch 5.

The purpose of these small screens was to find out whether the rays forming the positive precipitation on the branches are only such as come by way of the poles (see the experiment in fig. 204, Nos. 13—15 and Art. 112), or whether that precipitation is due to other rays belonging to the system of rays that first intersect the magnetic equatorial plane several times.

In Nos. 13 and 14, the positions are similar to those in Nos. 1 and 3. The conditions are very much the same, with from 2800 to 3000 volts between the electrodes, and about 25 amperes to the terrella-magnet.

In Nos. 15 & 16, the position is the same as in No. 2. The magnetising current to the terrella is 10 and 20 amperes respectively, with 22 milliamperes at 3000 volts in the discharge. The absence of positive precipitation on branch 3 in No. 15 will be understood on comparing that photograph with Nos. 11 and 12.

Some experiments were made without photographing, the magnetising of the terrella being changed from 5 to 15 amperes. It then appeared that the little screen at the pole was illuminated from the right when the magnetising current was 5 amperes, the light gradually moving nearer to the pole as the magnetising was increased to 6, 8, 9, 10, &c. amperes. On branch 3, positive precipitation first appeared with about 12 amperes, and when the magnetising current was weakened, moved out from the extremity of the branch on to the left flap of the small additional screen, and finally disappeared.

It will be seen that these experiments did not throw much light upon this circumstance; but we shall now see how the facts of the case stand.

Fig. 214 shows eighteen photographs and fig. 215 sixteen photographs of a series of experiments made with this object in view.

Nos. 1, 2, and 3, fig. 214, are from experiments made with a larger screen attached at right angles to branch 3, the positions being similar to those in Nos. 1, 2 and 3 of fig. 213. The magnetising current to the terrella was about 25 amperes, and the tension in the discharge about 3000 volts.

It will be noticed, in No. 1, how the light falls upon the upper side of the new screen, with its lower edge more or less sharply defined. It should also be observed that the shadow of the suspending wire, visible in the polar light-ring shows that the rays that come into the ring seem to have passed above the new screen, that is to say at some considerable distance from the terrella's equator.

In No. 2 we first notice that the positive precipitation on branch 3 is not affected in any special degree by the new screen. On the other hand, it will be seen that part of the negative polar ring is lost behind the screen, showing that on that side the screen has been high enough to intercept some of the rays that would have helped to form the polar ring.

No. 3 shows the same shadow in the negative precipitation, and also a peculiar light-effect to the right of branch 3, and on the terrella behind the screen. This may perhaps be foreign light produced by a discharge at a point on the terrella itself, a discharge that was found out during these experiments, and was the occasion of their being broken off.

A new arrangement of screens for the terrella was now carried out, as the succeding photographs distinctly show. A vertical annular plate, coated with tungstate of lime, was soldered to the ends of the eight arms of the star-screen.

Nos. 4 and 5 were taken from positions with hour-angles of respectively 235° with 15° declination and 180° without incline. The magnetising current to the terrella was only 6 amperes, and the discharge-current was 23 milliamperes with a tension of 3000 volts.

In No. 4 are seen *continuations of the positive precipitation on branch 2, and this continuation seems to be formed from the same rays that formed the first line of precipitation on our earlier equatorial screen.* When the magnetising is increased, the precipitation spreads over the screen farther from the terrella.

No. 5 shows one of the characteristic luminous triangles that we saw in fig. 68 of Section I; but here there are also shadows of the suspending and current-conducting wires.

The position in No. 6 answers to an hour-angle of 90°. It will be observed that the polar light has been reduced, and we see two peculiar lines of precipitation on the vertical screen to the left. The magnetising current was 10 amperes, the tension 2800 volts. The shadow of the suspending wire in the polar ring of light seems to show that the rays forming the latter pass above the screen.

The conditions in No. 7 are similar to those in No. 6, except that the magnetising current is 6 amperes. In this case, with the slighter magnetisation, the peculiar lines of precipitation on the vertical screen have moved anti-clockwise, and the polar ring of light is even fainter than before. This shows, as we have already seen, that with slight magnetisation the rays go closer to the terrella at the equator.

No. 8 was taken during the same experiment as No. 7; but the hour-angle of the position is 270°. Here too we see, as in No. 4, the very remarkable continuation on the annular screen of the positive precipitation on branch No. 2.

Nos. 9—12 are all from one experiment, in which the magnetising current was 20 amperes, and the tension in the discharge 2900 volts. The hour-angles of the positions were 90°, 235° (with 15° declination), 270°, and 295° (with 20° declination). The polar ring of light on the night-side is fainter in No. 9; but the shadow of the suspending wire is very clear. No. 10 shows the positive precipitation upon branches 1, 2 and 3; but there is no distinct negative polar ring.

There is a faint negative polar ring in No. 11. In this photograph, the great peculiarity is perhaps the shadows behind branches 6 and 7.

In No. 12 there is scarcely any of the usual negative precipitation on branches 4 and 5.

Nos. 13—18 are from a very important experiment with a very small terrella of only 2.5 cm. diameter. The iron core in it was cylindrical, and measured 10 mm. in diameter, and was wound round with 240 turns of 0.4 mm. copper wire covered with silk.

This terrella was placed in the middle of a flat screen, in such a manner that the magnetic axis was at right angles to the screen. The object of the experiments made with this tiny terrella in the vacuum-box of 22 litres, was to prove that the lines of precipitation that appeared on the screen had nothing to do with the enclosing plates of the vacuum-box. It was possible that our former terrellas were too large in proportion to the vacuum-box; but it will be seen that the experiments with this little terrella show our previous results to be unaffected as far as the distribution of the rays nearest the terrella are concerned.

No. 13 shows the terrella with screen seen edge-wise. The hour-angle of the position was 180°. The luminous ring outside the terrella is only from the cathode in the backgound.

No. 14 shows discharge without magnetisation of the terrella, the hour-angle of the position being 270°. There are shadows behind the terrella. The discharge took place with 2700 volts and 23 milliamperes.

No. 15 shows the conditions with a magnetising current of 2 amperes, 3000 volts.

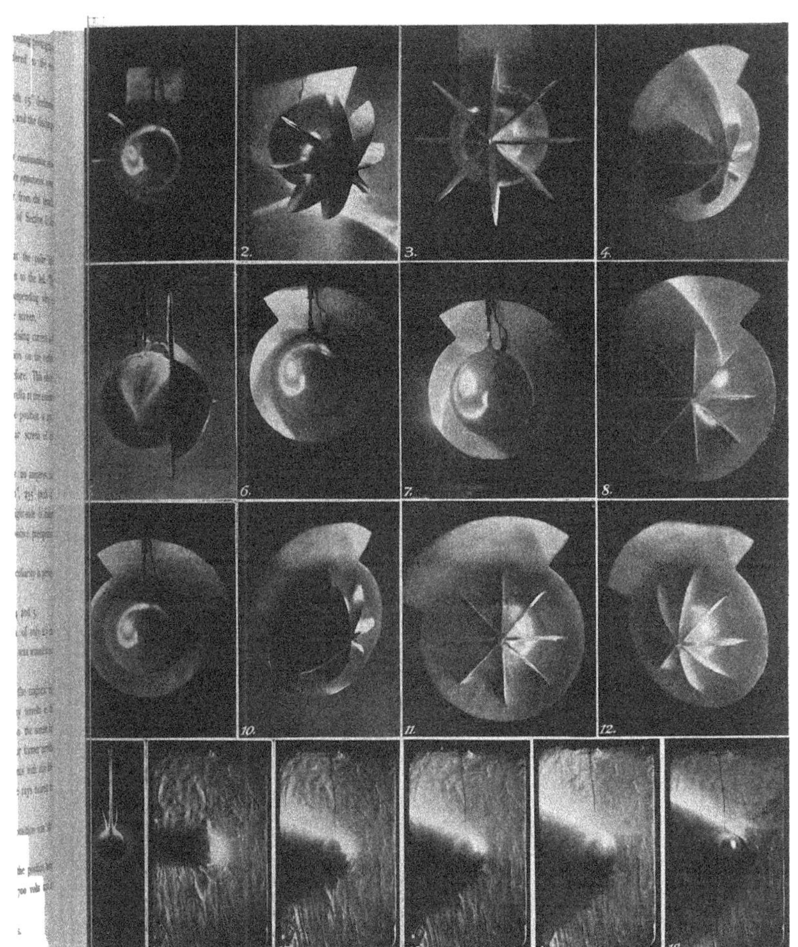

Fig. 214.

No. 16 shows the conditions with a magnetising current of 4 amperes, 3000 volts and milliamperes.

No. 17 shows the conditions with a magnetising current of 6 amperes, 3000 volts and 23 r amperes. The pressure fell from 0.014 mm. to 0.022 mm.

Lastly No. 18 shows the conditions with a magnetising current of 10 amperes and a discha current of 23 milliamperes with 3000 volts. The pressure fell from 0.017 mm. before the experin to 0.021 mm. after it.

It is a noticeable fact in all these experiments, that the remarkable occurrences of precipitation we have previously designated A and B, are also found here when the magnetisation is sufficiently sti (see No. 18). Their shape is so exactly the same as that with the larger terrellas, that we may conc that for these experiments at any rate, the vacuum-tube was large enough in our earlier experiment

In addition to these distinct, characteristic instances of precipitation on the afternoon side of ter and screen, we find upon the morning side that the pencil of rays is sharply defined, although the evidently only graze the vertical screen. In reality it is, as we shall see, the greater part of the from the cathode that are bent downwards in front of the terrella. This is immediately seen if screen is turned a little, so that the rays strike at an angle. This will be illustrated in the next p

These experiments will be of service to us, as a subsequent paragraph will show, in explainin zodiacal light.

In order to find out what became of the luminous patches upon the screen, when the plane o latter no longer passed through the centre of the cathode, the screen was turned 23° in a pos direction, and photographs were then taken.

Nos. 1—4 of fig. 215 were taken from places with hour-angles of 90° and 270°. Nos. 2 a show how the rays that turned off in front of the terrella, and only grazed the screen in its fo position, form a strong, sharply-defined precipitation in the new position. This shows that whil rays near the magnetic equator almost follow that plane, those outside the equator curve more and *away* from it. We have seen this before, having found a bright precipitation of rays respectively a and below the two magnetic poles, upon the floor and ceiling of the vacuum-box (fig. 200).

Nos. 1 & 2 are of experiments with a magnetising current of only 1.5 amperes to the ter A discharge-current of 23 milliamperes at 2800 volts. The pressure about 0.015 mm.

Nos. 3 & 4 are of a similar experiment, the only difference being that the magnetising cu was 5 amperes.

Nos. 5—10 are of important experiments in which a small screen was introduced in front o south pole of the terrella, at about right angles to the magnetic axis. The introduction of this sc was for the purpose of studying more closely the above-described precipitation of light. The large sc was turned back 23° to its original position.

The hour-angles of the several positions corresponding to these photographs were 90°, about and 270°. The first three are of experiments in which the magnetising current was 2 amperes, discharge-current 22 milliamperes with 2900 volts. The pressure was 0.017 mm.

Nos. 8—10 are of experiments like the above, with the difference that the magnetising current 10 amperes and the tension 3000 volts.

It will be noticed that the precipitation on the small screen moves outwards with increased netisation.

When we compare Nos. 7 and 10 here with Nos. 4 and 8 in fig. 214, full light will be at thrown on a hitherto somewhat obscure point. *We perceive how it is that rays of group B and ray group A, before they have reached the terrella, form a single coherent group, but that the rays which nearest to the poles of the terrella when this is sufficiently magnetised, are thrown round and ac*

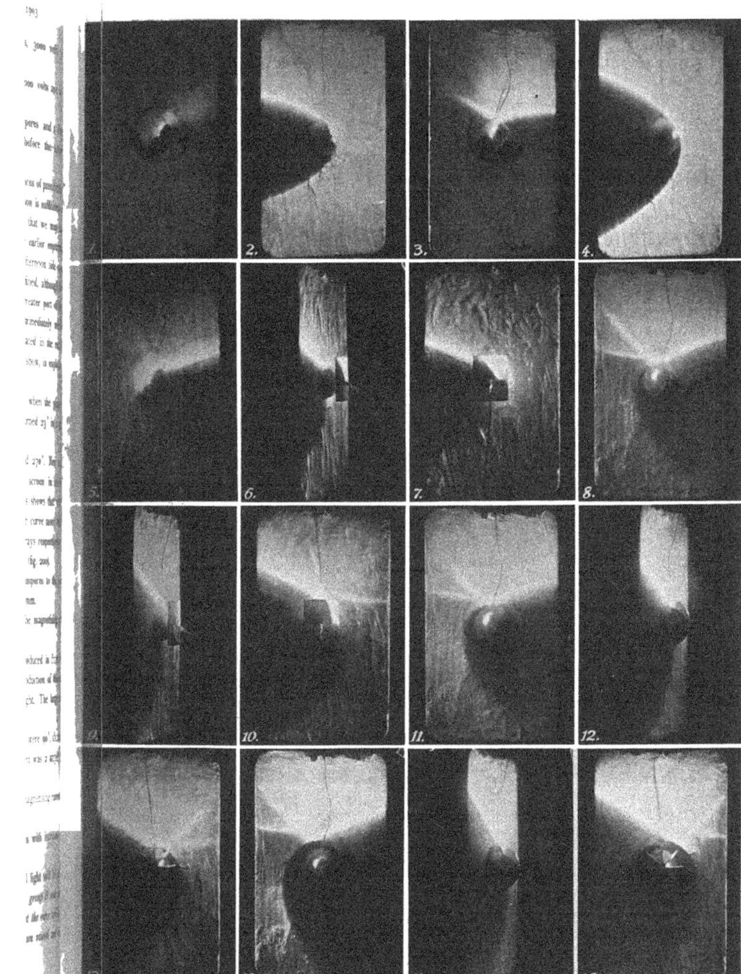

Fig. 215.

a retrograde motion. In the course of this they have an opportunity of positive precipitation on our star. screen and of the precipitation which we have called B on the equatorial screen in our earlier attempts. When we after this look back on the photographs from earlier experiments—take for instance fig. 204—we shall be able to see and understand them with much greater clearness than before. Look at the admirable pictures in the first column (Nos. 1, 4, 7, 10 and 13). We see directly how much of the spherical triangular light pictures are wanting, it is rays that have turned before they have struck out for the terrella, and we find them again in the precipitation on the eastern side of the vertical screen.

The more highly the terrella is magnetised, the greater will be the number of the rays of what we call group A that will be converted into rays of group B. We have also seen that the end of the first line of precipitation on the equatorial screen has moved away from the terrella, when this is magnetised to an exceptionally high degree, the bulk of the rays nearest the terrella in the line of precipitation, have been obliged to turn completely back.

It is interesting to observe that in Nos. 7 and 10 we have a section of the ray-masses over the poles at right angles to that shown in fig. 209, Nos. 3 and 7.

We may now conclude by analogy that it is not only rays belonging to the first triangular figure of precipitation that can be made to turn round by stronger magnetisation.

We have mentioned that such precipitation appeared three times on the eastern side of the vertical screen when the screen was turned in a positive direction through 360°. The first precipitation was strong and well defined, the second less strong, and the third slighter still. It is in this way that the bulk of the rays in the middle of the three triangular figures of light disappear from the terrella, the rasy being thrown back before they reach the terrella, when the magnetism is sufficiently strong.

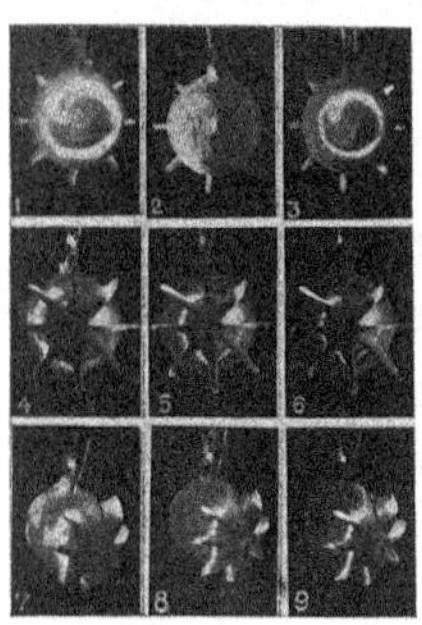
Fig. 216.

Applying this fact to the earth, we should expect that a station of medium latitude, for instance 65°, would not only have powerful positive magnetic storms attaining a maximum at 6 p. m., but would also have slighter ones about 1 a. m., and a very slight one about 8 a. m. (see p. 566). I hope to have an opportunity later on to investigate this matter.

Nos. 11—16, fig. 215, are of experiments with a small eight-armed screen, placed above the south pole of the terrella. In the first three of these, the magnetising current was 10 amperes, the discharge-current 23 milliamperes, and the tension 2400 volts. The positions have the same hour-angles as before.

In the last three photographs the magnetising current was 20 amperes. Discharge-current 22 milliamperes with a tension of 2700 volts. In the record of these experiments, the following account is given: "Experiments were also made with a current of 12 amperes to the small terrella. With this arrangement of an equatorial screen there was no trace of negative precipitation on the night-side. Great positive precipitation, on the other hand, was found on branch 2, but on none of the other branches".

Subsequent experiments have also proved that if the equatorial screen is large enough, the negative precipitation on the night-side in the polar light-ring disappears.

In Nos. 14 & 16, precipitation A and B are exceedingly distinct upon the screen, and exactly as with the large terrella (figs. 201 and 202). As we have already remarked, the circumscribing surfaces of the vacuum-box have therefore nothing to do with the shape of this precipitation.

Fig. 216 shows some results with the small terrella after the removal of the large screen through the equator.

Nos. 1, 2 & 3 were all taken from a position with an hour-angle of 90°. In No. 1 the magnetic current was 10 amperes, in No. 2 there was no magnetisation, and in No. 3 the current was 20 amperes. The full development of the polar ring of precipitation in No. 3 will be observed.

In Nos. 4—9, the arrangement was the same, except that the magnetising current was 10, 20 and 30 amperes respectively for Nos. 4 and 7, Nos. 5 and 8, and Nos. 6 and 9, and the tension 2700, 2600 and 2700 volts respectively.

Nos. 4—6 were taken from a position with an hour-angle of 270°, and Nos. 7—9 from a position with hour-angle 235° and declination 24°. The positive precipitation on branch 2 of the star-shaped screen is seen, whereas no positive precipitation appears on branch 3. Some experiments where made without photographing, for the purpose of studying this circumstance more carefully; and it then appeared that at the end of the positive side of branches 4 and 5, precipitation also occurred on our tiny terrella. When this result is also compared with that obtained when there was a large equatorial screen, it will be understood that it can hardly be only the rays that come in right across the polar regions of the terrella that produce positive precipitation.

ON THE SIZE OF THE POLAR RING OF PRECIPITATION.

115. We will now pass on to describe experiments that were made for the purpose of determining how the size of the rings of polar precipitation was dependent upon the magnetising of the terrella and the magnetic stiffness of the cathode rays employed. The intention of the experiments was to procure a basis for the judgment of the magnetic flexibility of the corpuscular rays coming from the sun and producing aurora and magnetic disturbances upon the earth in the manner we have supposed them to do.

In the experiments from which the photographs in fig. 217 were taken, the discharge-current in every case was about 25 milliamperes, and the pressure in the discharge-tube 0.046 mm. The tension difference between anode and cathode was 1800 volts in the experiments represented in the first and second rows, and it went from 1800 to 1700 volts in those in the third row. The tension remains comparatively constant here, because the pressure was so high that the amount of gas disengaged during the experiment did not alter the conditions as much as it does when the pressure is small to begin with.

The magnetising current in the three experiments was respectively 10, 20 and 30 amperes.

The position of the terrella — No. 4 — was unchanged during the three experiments, this being with the magnetic axis horizontal and at right angles to the central line to the cathode. The magnetic south pole had an easterly hour-angle of 270°, and photographs 1, 4 and 7 were taken from a place outside with the same hour-angle, photographs 2, 5 and 8 from a place with an hour-angle of 180°, and photographs 3, 6 and 9 from a place with an hour-angle of 90°.

In fig. 218 there are 9 similar photographs from 3 experiments in which the discharge-current was again about 24 or 25 milliamperes throughout, and the pressure in the discharge-tube about 0.008 mm. The tension in the three experiments was respectively 2400 volts, from 2400 to 2300, and from 2500 to 2300 volts, while the magnetising current was 10, 20 and 30 amperes. As will easily be understood, our endeavours were aimed at keeping the tension constant in each series of experiments; in the first series the tension aimed at was about 1800 volts, and in the second series about 2400 volts.

From the two series of photographs answering one to 1800 volts and the other to 2400 volts, we find in the first place that the stiffer the rays employed and the less the magnetisation of the terrella, the larger are the polar precipitation-rings. The idea originally was to magnetise the terrella so strongly that the polar precipitation-ring would acquire a spherical diameter of 45°, very much as one imagines the

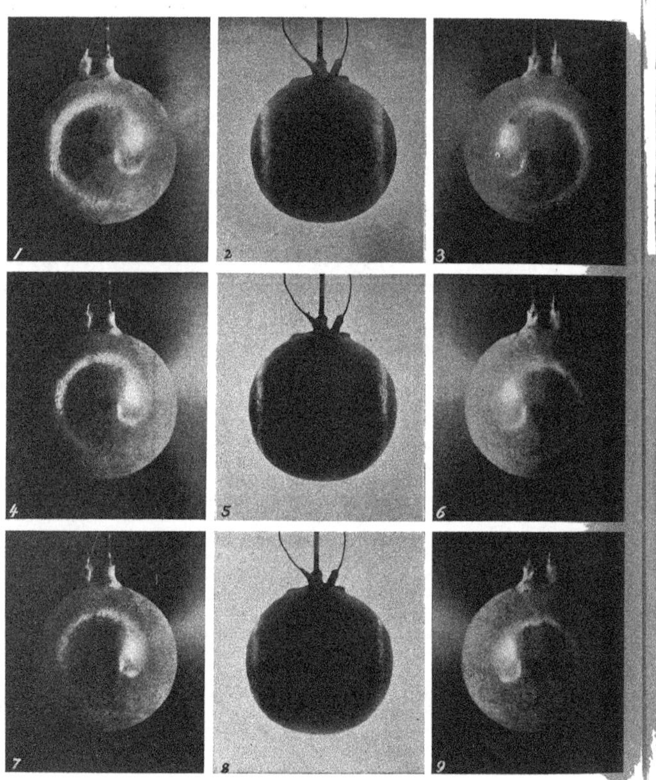

Fig. 217.

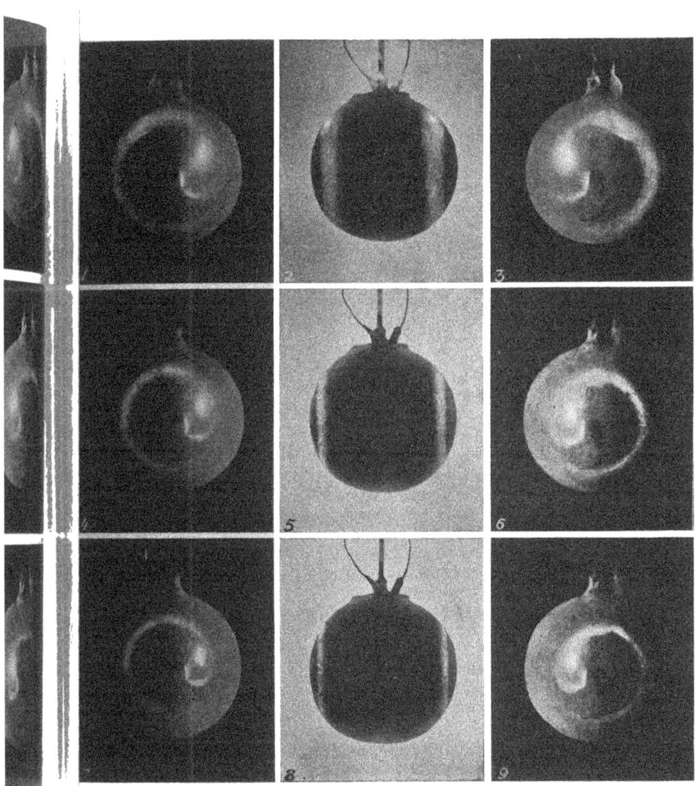

Fig. 218.

auroral zone on the earth, forming almost a circle round the point of intersection of the elementary magnetic axis with the earth's surface. The terrella employed, however, could not be magnetised sufficiently strongly, but, as we shall see, we can easily form an idea as to how much the terrella must be magnetised in order that the ring shall have its correct size. It is also no doubt possible that by selecting a somewhat stronger iron core for the terrella than the one here employed, and employing a stronger magnetising current, a precipitation-ring with a diameter of 45°, might be obtained, which would remain long enough to allow of its being photographed before the terrella became too hot. Indeed I have already, as will be seen below, realised the conditions necessary for this purpose.

There is another result which may also be directly deduced from our photographs, a result which we have moreover demonstrated many times under the most varied conditions.

It appears that *the more the terrella is magnetised, the narrower or thinner does the band of light in the ring become*, and the smaller the number of rays that are drawn in towards the terrella in the precipitation-ring. This last circumstance may be at any rate partly accounted for by the fact that the discharge-tube was not large enough for the highest magnetising of the terrella, as the rays describe large arcs before they go in towards the precipitation-ring.

With reference to photograph No. 2 in fig. 218, I would point out, as being of interest in this connection, that aurora that occurs in low latitudes on the earth, must, according to our theory, be due to stiffer rays than aurora that only occurs in the ordinary auroral zone; and the farther the northern aurora extends towards southern latitudes, the greater will be its width and we should expect that it will be seen simultaneously in the zenith over a greater area of the earth. Theory, in this case, is in harmony with experience.

In order, as we have said, to obtain an estimate of the extent to which the terrella must be magnetised to give the precipitation-ring a spherical diameter of 45°, the magnetic intensity was measured at the poles of the terrella by means of a LENARD spiral. An intensity of 1600 C. G. S. answered to a magnetising current of 10 amperes, 2400 C. G. S. to 20 amperes, and 2800 C. G. S. to 30 amperes. The relative proportions of the intensities were controlled by induction experiments with a small, flat coil, which was also placed at the pole, exactly where the LENARD spiral had been used.

The size of the precipitation-rings was then measured from the photographs, and their spherical diameter calculated in the various experiments, measuring along the middle of the band of light, the middle photographs, Nos. 2, 5 and 8, in figs. 217 and 218, being taken for this purpose. In this way the following values were obtained for the spherical diameters:

Answering to 1800 volts, 73°, 68°, and 63°;
—»— » 2400 » , 88°, 72°, » 66°,

for magnetising currents to the terrella of respectively 10, 20, and 30 amperes.

From these values we may conclude by extrapolation that with cathode rays answering to 2000 volts and a field-intensity of 4500 C. G. S. at the poles of the terrella, we should certainly obtain a small precipitation-ring with a spherical diameter of about 45°. The error in this determination is probably no greater than that in the assumption that the auroral zone upon the earth has a spherical diameter of 45°. We shall later on have an opportunity of controlling experimentally the result of this extrapolation.

We will now assume that with the above-mentioned magnetisation of the terrella, corresponding to 4500 lines of force at the poles, and with rays of 2000 volts, we obtain a comparatively correct idea of what takes place when the earth is irradiated by corpuscular rays from the sun; and upon this basis we will see what degree of stiffness these rays from the sun may then be assumed to possess. We presuppose then, that the magnetic field of force round the earth is similar in form to the field of force round our terrella, and that thus the magnetic field at great distances from the earth is not in any very essential degree affected by possible current-systems outside the earth.

We have then on the one hand a magnetic terrella with a radius of 4.1 cm., near whose poles the magnetic intensity amounts to 4500 C. G. S., and round which circle cathode rays whose velocity is $\frac{1}{13}c$, answering to 2000 volts, when c indicates the velocity of light (cf. LENARD, Ann. d. Physik, 1903, 317, p. 732).

On the other hand we have the earth, with a radius of 6.4×10^8 cm., and with a magnetic intensity in the neighbourhood of the magnetic poles that may be put at 0.68 C. G. S., and round which circle corpuscular rays with a velocity of βc.

I now believe, that when the terrella is so strongly magnetised that the polar light-rings have the same spherical diameter as the auroral zone on the earth, the cosmic ray system about the earth, which occasions aurora and magnetic storms, is similar to the cathode ray system around the terrella. Thus all details can be elaborated from our terrella-experiments and the results be applicable to the earth with a suitable proportional factor. We shall also make repeated use of this important proposition.

Now if the conditions in the one case are, so to speak, a true copy of those in the other, the radii of curvature of the corpuscular rays at all corresponding places must be as much larger than those of the cathode rays as the proportion between the radii of the earth and those of the terrella. Thus

$$\frac{\varrho}{\varrho_0} = \frac{6.4 \times 10^8}{4.1} = 1.56 \times 10^8$$

The proportion between the magnetic intensity at sets of places in the vicinity of the earth and in the vicinity of the terrella will be

$$\frac{H}{H_0} = \frac{0.68}{4500} = 1.51 \times 10^{-4}$$

Now we have, as is well known,

$$H.\varrho = \frac{m.u}{e},$$

where H is the intensity of the magnetic field, ϱ the radius of curvature of the rays, m the mass of the electric particle, e its charge, and u its velocity.

For the corpuscular rays round the earth we have therefore

$$H.\varrho = \frac{m.\beta.c}{e},$$

and for the cathode rays round the terrella

$$H_0 \varrho_0 = \frac{m_0 c}{e.13}.$$

From this we obtain the important relation,

$$H\varrho = 2.35 \times 10^4 . H_0 \varrho_0 = 3.1 \times 10^6$$

Even from this we may conclude that the rays in question must be unusually stiff magnetically. $H\varrho$ must be between 1 and 10 millions. We know only slightly penetrating positive rays which have approximately so great an inflexibility, as $H.\varrho$ for α rays from radium may have a value of 4×10^5.

W. WIEN observed that on the *negative* side also of the magnetic spectrum of kanal rays, there was a slightly deflected patch of fluorescence. These may possibly be almost inflexible negative ion-rays.

The γ rays hitherto not magnetically deflected are presumably very much like Röntgen rays in their nature. The opinion has been put forward that they are exceedingly stiff β rays (PASCHEN), or that they consist of neutral corpuscles (BRAGG). Possibly the corpuscles are not absolutely neutral either. Even rays in which $H.\varrho$ equals ten millions, there is hope of being able to deflect perceptibly by means of gigantic magnets. I shall soon have at my disposal a 30-tons magnet, with which, for a distance of from 1.5 to 2 metres, I can hold a field strength of 20,000 C. G. S.; and with this I shall try if it is possible to deflect rays such as these.

From the equations given, we obtain

$$\frac{H}{H_0} \cdot \frac{\varrho}{\varrho_0} = \frac{m}{m^0} \cdot 13\,\beta\,.$$

Further by the aid of the values for $\frac{H}{H_0}$ and $\frac{\varrho}{\varrho_0}$, we obtain $\frac{m}{m_0} \cdot \beta = 1.82 \times 10^3$

If we now assume that our corpuscular rays are formed of ordinary electrons, and that we may venture to employ LORENTZ's formulæ (¹) for the extreme case we have before us, then

$$\frac{m}{m_0} = \frac{1}{\sqrt{1-\beta^2}},$$

from which we obtain

$$\frac{m}{m_0}\beta = \frac{\beta}{\sqrt{1-\beta^2}} = 1.82 \times 10^3.$$

If we here say that $\beta = 1 - \frac{1}{x}$, we obtain approximately

$$\sqrt{\frac{x}{2}} = 1.82 \times 10^3 \text{ or } x = 6.7 \times 10^6\,.$$

We thus find that the velocity of the corpuscular rays should be $u = \beta\,.\,c = c - \frac{c}{x}$, i. e. only 45 metres less than the velocity of light. The transversal mass of the corpuscular rays, m, equals $1.82 \times 10^3\, m_0$, and is thus of an order one thousand times as great as the mass of an electron with small velocity (²).

Recently LENARD (³) has also treated this very important question, and has arrived at similar conclusions as to the stiffness of the cosmic corpuscular rays.

Although we may probably take it for granted that LORENTZ's formula in this extreme case no longer holds good, we may nevertheless conclude that the corpuscular rays from the sun, which should be capable of giving rise to such precipitation-phenomena upon the earth as are manifested in aurora and magnetic storms, must be extraordinarily penetrating and exceedingly inflexible to magnetic forces. As, on this earth, we are not acquainted with any rays possessing such properties, the above result must at first sight seem discouraging; but if we look into the matter, we soon find several observations that are in complete harmony with it.

We know, for instance, that in the polar regions aurora very frequently descends to within 50 kilometres of the earth, indeed there are good observations of its descending to within 10 kilometres and considerably lower. Auroral *rays* may sometimes be seen with a length of 30 kilometres.

It is thus clear that the rays which produce auroral phenomena, and which we assume to originate in the sun, must be capable of penetrating considerable strata of our atmosphere. They must be supposed capable of penetrating a layer of mercury more than 100 millimetres in thickness, if the rays follow the law, Equal penetrability for equal masses. This moreover agrees with the idea that these same rays, before reaching the earth, have been obliged to penetrate a certain stratum of the solar atmosphere, since they issue from the regions in the vicinity of the sun-spots.

(¹) A. H. LORENTZ, The Theory of Electrons, 1909, p. 212, equation 313.

(²) BIRKELAND, Sur la déviabilité magnétique des rayons corpusculaires provenant du Soleil. Compt. Rend. de l'Académie des Sciences, Paris, le 24 Janvier, 1910.

(³) LENARD, Ueber die Strahlen der Nordlichter, Heidelberger Akademie der Wissenschaften, Jahrgang 1910, 17. Abhandl.

At present we are acquainted with β rays, which pass through about 1 millimetre of mercury; and they are accompanied by γ rays, much more penetrating still.

LENARD (¹) has made investigations for the purpose of finding a relation between the velocity of an electron and the coefficients of absorption for corresponding rays in different substances.

He arrived at the result that the absorption increases more than a million times when we pass from β rays of radium to cathode rays with a velocity equal to a hundredth part of that of light.

It seems probable, however, that the penetrability of our rays should be much greater than that of the β rays of radium; but no simple law has yet been found that can be employed for calculating the absorption when the velocity is known.

Several physicists have found that the β rays are absorbed according to an exponential law, and that the velocity does not change when the rays pass through matter; but it would appear that these results are not certain.

We can point to yet another circumstance that indicates that the corpuscular rays coming from the sun must be extremely inflexible. After HALE'S discovery of the comparatively powerful magnetic field that is found round the sun-spots, it is an obvious conclusion that the sun on the whole is magnetic. This conclusion is also obvious for other reasons. The corona's rays in the polar regions of the sun have led several investigators to believe that the sun is magnetic, with poles near those of the axis of rotation.

It now appears that no rays can emanate from the equatorial regions of the sun out into space, if the sun is assumed to have a magnetisation that can be compared with that of the earth, and the rays are supposed to be no more inflexible than the hitherto known corpuscular rays, i. e. if Röntgen rays and γ rays are not corpuscular rays.

It is another matter altogether when we assume that the rays actually *have* the inflexibility that we have above inferred that they must have, from aurora and terrestrial magnetic phenomena on the earth. We are then even able to give a plausible explanation of a phenomenon that has been studied by RICCÒ (²), and which has to do with magnetic storms. RICCÒ has observed that there is a difference of time of from 40 to 50 hours between the passage of a large spot to the central meridian and the maximum of a magnetic perturbation that it produces on the earth. He concludes from this that the velocity with which the corresponding rays are propagated ought to be between 900 and 1000 kilometres per second.

It is easy, by quite simple calculations, to determine the path that a corpuscular ray going straight out, *with the velocity of light*, from the sun's magnetic equator will describe when the stiffness of the rays is that assumed above, and the sun is supposed to act upon the rays like an elementary magnet with a definite moment M.

I have calculated from STÖRMER'S (³) formulæ that the sun should have a magnetic moment of order 10^{28}, or about 150 times greater than that of the earth and inversely magnetic, in order to deflect our rays by an angle corresponding to this retardation of from 40 to 50 hours.

The probable existence of such corpuscle-rays from the sun as those here treated of, is even now admitted by several men of science, and it will certainly be soon acknowledged that these new solar rays, which I have thus discovered, enter deeply into many terrestrial conditions, even if they cannot compare in importance with the wondrous rays we have hitherto been acquainted with. Owing to the magnetic condition of the earth, the new solar rays, as we have seen, principally enter the polar regions.

(¹) Annalen der Physik, t. XII, 1903, p. 714.
(²) Nature, November 4, 1909.
(³) Archives des Sciences physiques et naturelles, Vol. XXIV, Chap. IV, 1907, p. 121.

116. A new glass box was constructed to contain more than 70 litres, in order that all discharge-experiments might be made with a terrella so strongly magnetised that the cathode-ray system round it would be similar to the corpuscular-ray system round the earth. The thickness of the plate-glass sheets was 22 mm.; and in order to guard against the great external pressure, the sheets forming the ends of the box were specially strengthened. The sheets that were perforated were double. The internal dimensions of the box were 36 × 36 × 55 cm. The terrella employed was 8 centimetres in diameter, and was constructed with the object of procuring more than 4500 lines of force per centimetre across the poles when the strongest magnetising current was employed. The iron core was 3 cm. in diameter, and was closely wound round with well-insulated layers of copper wire, of which the total

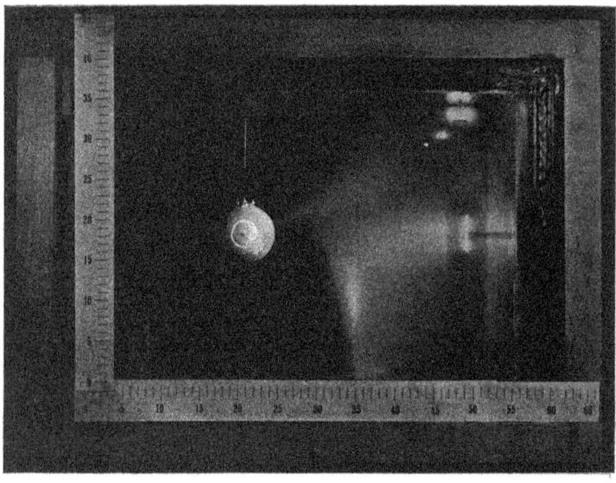

Fig. 219.

resistance was 2.6 Ω. The wire could, without injury, be charged for a few seconds with 40 amperes thereby imparting to the terrella an amount of energy equal to between 5 and 6 horse-power.

The magnetic moment M for 10 amperes was found to be 61300. At a distance of 4.5 mm. from the terrella, immediately above the pole, the number of lines of force with 10, 15 and 30 amperes magnetisation was respectively 2075, 2760 and 4200. At a distance of 7.5 mm. from the pole of the terrella the measurements were $H = 1647$, 2460 and 3280 with 10, 15 and 20 amperes respectively of magnetic current. These measurements were taken with a Leduc apparatus. To ascertain if it were correct, this apparatus was compared with a Lenard's bismuth spiral which gave the following corresponding sets of values:

H by Lenard spiral	5750	5600	4800	3550
H by Leduc apparatus	5950	5650	5040	3720

As will be seen, the respective values correspond fairly well and the records of the Leduc apparatus must thus be considered reliable. From this it must be supposed that immediately above he pole of the terrella with 20 amperes magnetisation the H has been approximately equal to 4100. f we then calculate the stiffness of the corpuscle-rays on the earth in the same manner as above, we hall see below that we are finding very nearly the same values as before.

The new terrella was first placed in our former smaller vacuum-tube (fig. 200), but it appeared that he cathode rays were thrown in such numbers against the walls of the tube, that hardly any reached the errella, and the above-described large vacuum-box was therefore, with much labour, constructed, and the errella placed in it as shown in fig. 219. A number of test experiments were also made with these ipparatuses, but unfortunately no photographs were taken except the one here reproduced, in which he magnetising current was 20 amperes and the tension about 2000 volts. This shows exceedingly well now the polar ring approximates the proper dimensions as compared with the conditions on the earth, he angular diameter of the ring being here 49°; and with a magnetising current of 30 amperes we obtained a polar ring with about the same angular diameter — judging by the eye — as the auroral zone on the earth, i. e. rather less than 45°. There is, however, no photograph of this magnetisation. A ew days later, a leak appeared in the vacuum-box, which a couple of months' work failed to stop. In case anyone should hereafter like to construct such a large vacuum-box, I would advise the use of glass sheets of 25 mm. thickness and not as here 22 mm. as the enormous pressure is liable to bend thinner plates too much.

There are two important conclusions that we can draw from the polar light-ring here photographed. Firstly, we can by this experiment control our earlier calculation of $H\varrho$ for the cosmic corpuscle-rays around the earth. If we then by a very little extrapolation calculate the stiffness of the corpuscle-rays on the earth corresponding to a circle with a diameter of 45° in the same manner as above, we find that $H\varrho = 3.1 \times 10^6$, or exactly the same value as before. The second important question we can now solve is that of the breadth of the band of precipitation on earth of the rays which occasion the polar magnetic storms. For various reasons I have hitherto assumed[1] that the width of this zone of precipitation between Kaafjord and Jan Mayen is less than 500 kilometres. The measurement of the width here on the night side of our terrella gives for these somewhat stiff rays that the breadth is 2.5° which corresponds to 280 kilometres on the earth.

The photograph reproduced shows an experiment (pressure 0.01 mm.) in which the south pole of he terrella is turned directly towards the observer. The two horns of light that are drawn in towards he polar regions of the terrella are here seen coincident with one another. In the photograph in fig. 200 the poles were above and below, and these two in-drawn horns of light were separate.

In this photograph we also see the exceedingly interesting manner in which the greater number of the rays are thrown in a direction away from the terrella on the morning side. It is this collection of rays which, in my opinion, plays an important part in occasioning the zodiacal light seen in the morning. In our photograph, on the other hand, the rays that cross one another in what we have called the first and second lines of intersection, or lines of precipitation (see figs. 201—207 and 214 & 215), are not visible. I think we should easily get the regions about these two lines of intersection — he first by preference — self-luminating in the vacuum-tube, if we so arrange it that the rays that go round the terrella on the evening side are sufficiently intense. This can be attained either by bending he cathode slightly upwards, so that several of the rays pass above the terrella, or by the equally simple method of making the cathode exceedingly large, almost as large as the vacuum-box permits. n the latter case, the conditions will be as nearly as possible like those between the earth and the sun, as the pencil of parallel rays will be the largest possible.

As will appear later on, I consider the first line of intersection (line of precipitation) of the rays on the afternoon side to be of importance in connection with the zodiacal light visible in the evening,

(1) Expédition Norvégienne de 1899—1900, p. 26, 2⁰.

while I consider the fainter collection of rays that cross one another in the second line of inte
tion to be the primary cause of the nocturnal zodiacal light-phenomenon known by the name of "Gegensch

EXPERIMENTS FOR THE DETERMINATION OF THE SITUATION OF THE POLAR ZO OF PRECIPITATION IN VARIOUS POSITIONS OF THE MAGNETIC AXIS.

117. The experiments are made with the object of obtaining more detailed material for ju whether the situation of the zone of precipitation on the terrella in the various instances can ser a guide for understanding the occurrence of the auroral draperies in the polar regions, and the situ of those polar precipitations which give rise to magnetic storms on the earth.

We shall first go through the different conditions under which the pictures 1 to 16 of fig. 220 are t

Nos. 1 and 2 are taken in the course of an experiment, in which the discharge-current w milliamperes, the tension 2500 to 2300, and the magnetising current 20 amperes. The pressure 0.008 mm. The south pole of the terrella lies in the plane of the horizon through the centre, wi hour-angle of 290°. The photographs are taken from places with hour-angles of 290° and 110°, sit on the prolongation of the magnetic axis. The pictures 3 and 4 are taken under the same experin conditions, only that the hour-angle of the south pole is 250°, and the photographs are again taken places on the prolongation of the magnetic axis with hour-angles of 250° and 70°. We see at once these 4 pictures, how the so frequently mentioned luminous patch is round and lies within the ri light when the magnetic pole turns towards the cathode (1 and 4), while the patch is drawn ou merges with the ring of light in the positions 2 and 3, in which the pole turns away from the cat

One thing in connection with this patch of light is particularly deserving of attention, that is the rays which cause it are rays that have gone the shortest way from the cathode to the terrell figs. 200 and 219 the rays which form these polar patches will be seen, showing themselves in the ra gas, like two luminous horns, as we repeatedly have mentioned.

This circumstance is of importance when we imagine the conditions transferred to the earth. sudden flare-up or eruption of corpuscle-rays take place in the sun, these would make themselve on earth *first* by a precipitation corresponding to the above-mentioned polar patch of light.

Stations on the day side of the earth which happen to be near this first precipitation, will the receive from it a first impulse announcing a coming magnetic storm.

When then, an instant later, the polar precipitation on the night side of the earth or the equ ray-formations are produced, it may appear as if there was a noticeable difference in time at the di stations on the earth for the commencement of the one and identical magnetic storm. In reality are several impulses which act in places very locally. I believe that perhaps some observ that have been made when magnetic storms were commencing, can be explained by the view here set

The pictures 5 and 6 are again taken from places on the prolongation of the magnetic axi the south pole is now given a declination of 19°, and the hour-angle is 270°. The conditions experiment are the same as before, the tension, however, being 2500 volts and the pressure 0.00

In the pictures 5, 9, 11 and 15, it will be seen that the phosphorescent coating on the t has a defect uppermost by the luminous ring. Something like a shadow appears there which has n to do with the precipitation.

The pictures 7 and 8 are taken under exactly the same conditions as 5 and 6, but with the ma poles reversed.

The pictures 9 and 10 are taken under similar conditions as before, but the tension is 2 2500 volts and the pressure in the discharge-box 0.006 mm.

The magnetic axis is turned, so that the north pole has an hour-angle of 285° and declinatio No. 9 is taken straight out from the north pole, and No. 10 out from the south pole from places prolongation of the magnetic axis.

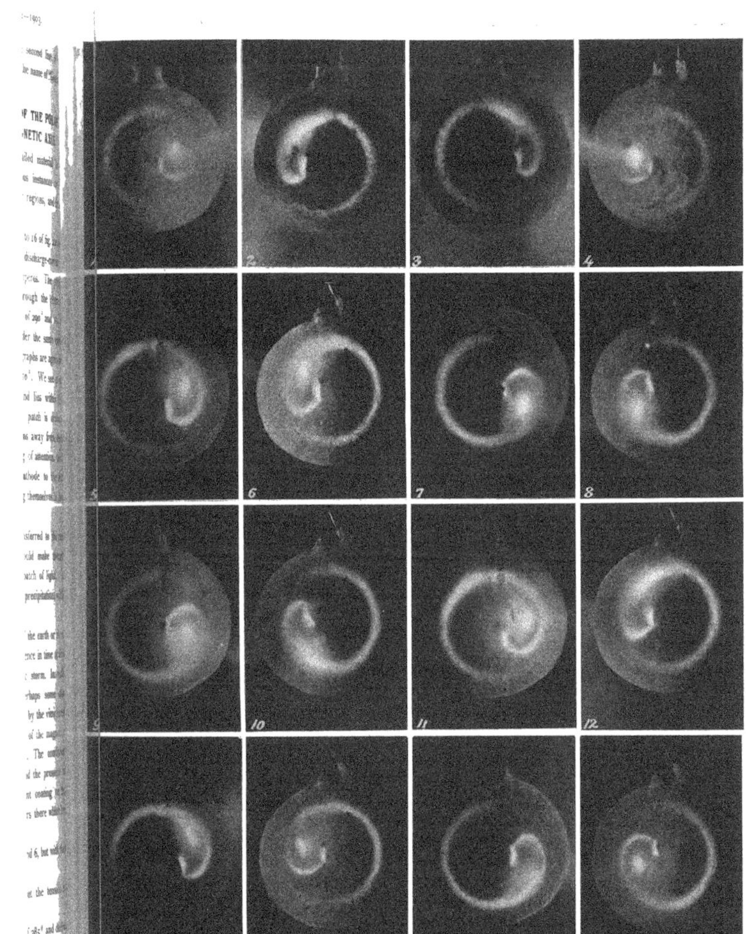

Fig. 220.

The pictures 11 and 12 are taken under similar conditions to 9 and 10 but with the magnetic poles reversed, moreover the tension is now 2900—2500, the pressure being 0.005 mm.

The pictures 13 and 14 are taken during experiments where the hour-angle of the south pole was 255° and declination 19°. The pictures are taken from places on the prolongation of the magnetic axis, the tension was 2600—2400 volts and the pressure 0.006 mm.

The pictures 15 and 16 are taken during similar conditions, but the magnetic poles are reversed (the terrella re-magnetised) and the tension was 2700—2300, under a pressure of 0.007 mm.

The magnetising current for the terrella was, as will have been understood, 20 amperes in all the experiments, and the discharge-current about 25 milliamperes.

It is also seen by the 4 last pictures how the luminous patch referred to takes different shapes in different positions and encroaches upon the luminous ring.

The most striking result of these experiments is that the polar spiral of light always forms itself, in surprisingly nearly the same manner around the magnetic poles without regard to whether the position of the magnetic axis is altered at all in relation to the central line between the terrella and cathode. The difference between the spiral round a magnetic north pole and the spiral round a magnetic south pole is easily recognised, as the spiral seen from above a north pole winds itself in the direction of the hands of a clock, while the spiral over a south pole winds the opposite way.

On the other hand, the position of the polar luminous patch is more sensitive to changes in the position of the magnetic axis, as the light patch with such alterations had changed place and shape to a certain degree.

When we apply the results described above to the earth, we would expect to find that similar spirals of precipitation to those here depicted formed around the magnetic poles or perhaps nearest around the points in which the magnetic axis of the earth intersects the earth's surface (see p. 58 of this work, Section I, and STÖRMER's Memoir in Arch. de Genève, l.c. § 17).

These spirals of precipitation must in the course of the daily rotation of the earth, swing round the true poles of the earth, while they, however, always retain their direction in relation to the line of direction to the sun, and their position in relation to the magnetic poles.

As we have seen before, the north pole spiral can, as regards the earth, with some degree of resemblance be compared with a circle of from 40 to 45° spheric diameter and with the centre in a point with latitude 78° 20′ N, longitude 71° 11′ w. (New year 1903) which was the northern point of intersection with the axis just mentioned. If the corpuscle rays from the sun happen to be specially flexible. the spherical diameter can be less than 40°.

It is obvious what ample opportunity is here afforded for testing the correctness of our theories The theoretic positions of both the precipitatons which occasion polar magnetic storms and the precipitations which occasion auroral arcs, are, as may be seen, hereby ascertained by a simple construction, after which it is merely necessary to observe the hour and place.

We get a theoretical daily and annual motion in these phenomena, by which the theory can be controlled.

Owing to the relation of the auroral spiral to the direction towards the sun, the spiral will, when compared with a fixed point of observation, appear to turn with the sun, in addition to also periodically shifting in relation to the spot in other ways.

A thorough study of these questions will be made and the results be made known in the second volume of this work. By that time other new experiments will be made as to the correct size of the polar ring of precipitation (45° angular diameter), and the situation of this at the various positions of the terrella will be determined with the utmost possible precision.

On a preliminary comparison with the investigations of the ordinary position of the auroral arc at several polar stations, it appears as if these arcs have the direction corresponding to the experiments, and the translatory motion of the arcs in a corresponding manner also makes its appearance.

INVESTIGATIONS REGARDING THE ANGLE FORMED BY THE PRECIPITATED RAYS WITH THE MAGNETIC LINES OF FORCE.

118. We will now proceed to the description of the experiments represented in fig. 221, and discuss the facts resulting from them.

The experiments were made in order to make it somewhat clear how steeply towards the terrella the rays are precipitated in the "auroral zone" under the different experimental conditions, especially when the magnetic stiffness of the rays is modified in proportion to the magnetisation. The plate is unfortunately not so good as could be desired.

These investigations are of great importance to our present theory on the auroral draperies, as we suppose that the auroral rays in the draperies are formed by those pencils of rays which come as steeply as possible towards the earth, where they are entirely absorbed by the atmosphere after having rendered the air luminous over a more or less wide expanse.

This is to some extent a modification of the opinion I have previously expressed, as I formerly supposed that the *rays* of an auroral drapery were formed by secondary beams produced in the atmosphere by the influence of the primary cosmic and corpuscular current which forms the auroral arc itself. In a certain degree, something valid will remain in this older theory; but it seems more natural to suppose that rays with such tremendous power of penetration as that dealt with here, must be the same stiff rays that we suppose to be emitted from the sun. The state of the atmosphere of the earth is hardly such as to permit the formation of such stiff rays. I have therefore been brought to take a different view of the matter, which was further confirmed by my terrella experiments, namely, that auroral rays are formed by the rushing in of distinct pencils of cosmic rays towards the earth almost exactly along the magnetic lines of force, without any turning, worth mentioning about those lines. These cosmic rays, which thus penetrate the atmosphere, are entirely absorbed, and therefore never return into space.

During the experiments about to be described, the terrella maintained an unaltered position in the discharge-tube, the line from the centre to the magnetic south pole being in a horizontal plane with an eastern hour-angle of 270°. The photographs have been taken from a place in that plane which also has an hour-angle of 270°, so that the eight branches of the screen are seen edgewise.

The discharge-current, during all the experiments, was about twenty milliamperes. The photographs 1, 2 and 3 were taken with a magnetisation current of 10 amperes, the first at a tension of 2800—2600 volts and a pressure of 0.009 mm., the second at 2200—2100 volts and a pressure of 0.017 mm., and the third at 1800—1700 volts and a pressure of 0.05 mm. The photographs 4, 5 and 6 are from experiments during which the magnetisation current was 20 amperes and the tension respectively 2500—2100 volts, 2200—2000 and 2000—1800 volts, and the pressure respectively 0.007 mm, 0.017 mm. and 0.025 mm. The photographs 7. 8 and 9 are of experiments during which the magnetisation current was 30 amperes and the tension respectively 3000—2600 volts, 2400—1800, and 1700—1500 volts, while the pressure was respectively 0.007 mm., 0.022 mm. and 0.026 mm.

In some of the photographs, for instance Nos. 4, 6, 8 and 9, on the left of the third branch of the screen, the shadow of the brass rod from which the terrella was hung in the magnetic equator will be observed. We have seen this shadow rather more clearly on a large number of the previous photographs, and it immediately gives us an idea of the steepness with which the rays here pass through the plane

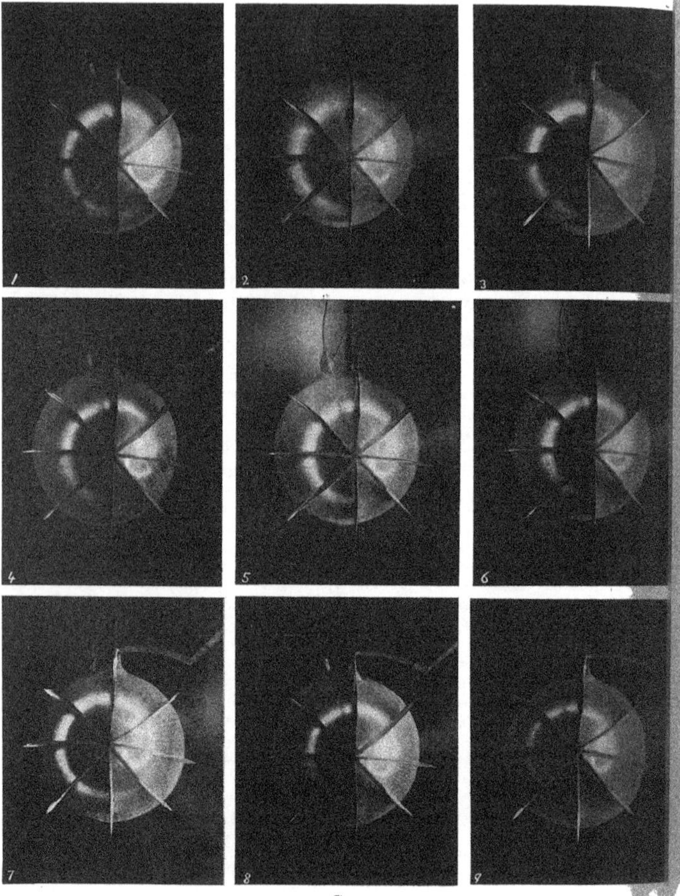

Fig. 221.

of the equator to be precipitated in the "auroral zone". It was this shadow of the suspension rod that first suggested the idea of constructing the high eight-armed screen and making these experiments which have been of such great importance to the theory of the positive and negative polar storms, as seen in Article 113. Such screens, placed on the edge above the polar regions of the terrella, have already been used previously, but the results of the experiments were not so clear (see fig. 136, page 302, Section I), because the screens were far too low.

The illustrations show us the angle at which the most perpendicular rays fall towards the terrella. The shadows behind the branches of the screen show, further, that the rays are most perpendicular in the middle of the "auroral zone". On the southern edge of the zone, the rays fall most obliquely, and on the northern edge more obliquely than in the middle, but less so than on the southern edge. It appears moreover, although not positively, from the photographs, that the rays, at about the same tension, descend somewhat more perpendicularly towards the terrella with strong than with slight magnetisation. With the same magnetisation, the rays are also somewhat more perpendicular with low than with high tension; but the difference does not appear to be so great. There are here, however, several things to be taken into consideration. It must not be forgotten, for instance, that the shadow-producing part of the screen does not remain the same in all cases, a fact of which proof is found in the form of the precipitation on the western side of the screen-branches (see fig. 212, Nos. 1, 4 and 7).

We have endeavoured in the foregoing pages, by numerous experiments, to show how the rays move round our terrella. It would have been of great interest if these experiments had been repeated with our last terrella No. 7, which was highly magnetic, in the new large discharge-box measuring 70 litres, as we might then have chosen the magnetic conditions so that the luminous polar band would have had an angular diameter of 45°. We could then at once have transferred the results to the earth, and in particular determined the perpendicularity with which, according to the theory, the auroral rays might be expected to come towards the earth. We propose to make these more extensive experiments, and the results obtained will be published in the second volume of the present work.

In a general way, it can even now be established as a fact, that rays which are finally precipitated in the "auroral zone", have first passed round the terrella, oscillating above and below the plane of the magnetic equator.

In the foregoing pages, we came, as a consequence of our experimental results, to the conclusion that the continuous luminous ring in the "auroral zone" was produced by a countless succession of secondary precipitations overlapping one another in such a manner that the luminous ring *appeared to be continuous*. We remember, for instance, having once counted, on the night side of the terrella, about 20 distinct secondary precipitations, of which those of a higher order lay to the east of those of a lower. The number of these precipitations was greatly multiplied in proportion to the increase of the magnetisation of the terrella. It is this opinion of the constitution of the luminous ring which we shall firmly maintain in endeavouring to develope a theory as to the formation of auroral draperies.

The rays which are precipitated, for instance, on the night side of the terrella, a little eastward of the place where other contiguous rays, originally from the same bundle of rays, are precipitated, will thus have travelled considerably farther than those rays which are precipitated on the west side, close by. They may, in fact, have been deflected below the level of the equator towards the south pole, and then have risen again and been precipitated in the northern "auroral zone". It will consequently be observed that the rays in the precipitation-zone are formed from separate, relatively small groups of rays which have intersected the plane of the equator several times, before they are at last precipitated. We take then first a group of rays in the northern "auroral zone", which have passed n times through the equator. The nearest companion group which had *nearly* been precipitated in this zone, has subsequently to pass through the equator once more, viz. $n + 1$ times in all, and is afterwards precipitated in the southern "auroral zone", while a corresponding bundle of rays, which had ***nearly*** been precipitated in the southern "auroral zone", passes through the equator and is precipitated beside the bundle of rays which had passed n times through the equator. The next contiguous group of rays has passed through the equator $n + 2$ times before being precipitated in the "auroral zone". As the rays now fall symmetrically above and below the magnetic equator, the corresponding process of selection will have taken place in the southern "auroral zone", so that in the northern and southern zones, auroral rays will be produced successively one after another, each one having passed through the equator *once oftener* than the nearest preceding auroral ray.

Fig. 222.
Aurora borealis observed at Bossekop on the 6th January, 1839, according to Bravais.

Although it is not our intention to deal with the auroral phenomena until we come to Volume II of the present work, where we shall see how the different forms of auroral light are to be explained, we shall, however, now show, as an illustration connected with the terrella experiment just described, how the formation of auroral draperies is to be understood. As a characteristic feature of this perhaps the most peculiar form of auroral light, we would remind the reader that the aurora borealis frequently appears as a vertically hanging curtain consisting of densely co-ordinated parallel *rays*. The curtain has most frequently its longitudinal direction in the magnetic east and west.

As further characteristics we would mention that the auroral curtain is frequently formed from east to west, or vice versa, in such a way that the rays, one after another, seem to be precipitated from the sky, and this so rapidly that the curtain can be completely formed and extend right across the heavens in a few seconds.

Another phenomenon, which is most closely related to the above, is that of the so-called ***luminous waves*** which may rush through the auroral drapery. The rays blaze up and go out, and the phenomenon is repeated successively on every ray from one end of the curtain to the other, the wave appearing to pass through the entire length of the drapery. The waves move most frequently from west to east, but also very often in the opposite direction.

The auroral curtain may have characteristic undulating folds and eddies; and from one fold luminous waves may pass along the curtain eastwards and westwards simultaneously (1).

We will now try to combine these facts with the experimental results at which we arrived through our terrella experiments.

First we must suppose that the auroral rays do not exactly follow the magnetic lines of force, but that, in what we call negative precipitation, they form a small angle towards the east with the lines of force, while in what we in analogy with the polar storms call positive precipitation, they form a small angle towards the west with the lines of force. We shall subsequently show how these angles towards the east and the west are to be understood. The angles, however, are very small, because the auroral rays are only formed by those rays from space which fall as vertically as possible along the lines of force, and they penetrate, therefore, most deeply into the atmosphere and create the auroral rays.

There are unfortunately not many observations which can be referred to with regard to this supposed inclination between auroral rays and magnetic lines of force, but in the well-known work of PAUL GAIMARD, "Voyages en Scandinavie, en Laponie etc.: Aurores boréales", page 505, we note the following remark: "We certainly are justified in stating that the rays are not always strictly parallel with the line of inclination".

In the same work BRAVAIS makes the following remarks: "We admitted one of the two following hypotheses: either the average orientation of the auroral arcs is not perpendicular to the magnetic meridian, or the average direction of the rays is not strictly parallel with the line of inclination".

We shall now see that both these hypotheses must be assumed at the same time.

CARLHEIM GYLLENSKIÖLD recapitulates, l. c., page 69, *his* result as follows:

"The disagreement in our observations is rather great. When not taking into consideration the doubtful positions, the difference of the average position is, in two cases, 22° 54′ and 20° 31′; it exceeds 10 degrees in eight others. The average difference is 6° 34′ and the probable error of the average is ± 42′.1. The members of the French expedition on board the corvette "La Recherche" have made, at Bossekop, 43 observations of the centre of the corona; the average difference is 5° and the probable error of the average is 0° 30′. The greatest difference is 15°; it exceeds 12° in two other cases. Our observations consequently agree less with each other than those of the French expedition. However, our observations are probably not in reality less exact than those made at Bossekop; we are inclined to believe that the position of the corona is subject to greater variation in a latitude of 78 degrees than in Finmark".

Mr. SIRKS OF DEVENTER (2) arrives, through 16 observations made in Europe during the great aurora borealis on February 4th, 1872, at the result that "the corona in almost all places was some degrees inferior to the magnetic inclination; the azimuth of the corona was also less than the magnetic declination".

When discussing the angle made by the auroral rays with the magnetic lines of force, the angle always meant is that between the tangents of the magnetic line of force and the axis of the auroral ray through the foot-point of its orbit.

Such an angle will generally have a projection on the plane of the magnetic meridian, through the foot-point, and on a plane through the tangent of the line of force perpendicularly on the meridian.

---

(1) See CARLHEIM GYLLENSKIÖLD: Aurores boréales. Observations faites au Cap Thordsen, Spitzberg, 1882—1883. Stockholm, 1886, Vol. II: 1, p. 136.

(2) POGGENDORFF's Annalen, Band 149.

When we mention above an angle to the east, we mean an angle whose projection on the latter plane falls to the east.

Rays of group A which have intersected the magnetic equator must, according to the theory and in conformity with the observations made by Mr. SIRKS, be supposed to form auroral coronæ situated some degrees lower than the magnetic zenith. We refer to the form of the precipitation on the west side of the screens, fig. 212, Nos. 1, 4 and 7, and fig. 213, Nos. 4 and 8.

Rays of group B, on the contrary, would be expected to create auroral coronæ situated higher than the magnetic zenith.

Thus we see that the theory gives reasons explaining that the different observations vary as to the situation of the auroral coronæ, as stated by BRAVAIS and CARLHEIM GYLLENSKIÖLD. We shall return to this important question in Volume II.

Another question which we shall soon deal with is this: Can we suppose that the cosmic rays which produce the luminous auroral rays can return to space, or are they at once *absorbed* by the atmosphere?

We will suppose that they are at once absorbed, because if the cosmic rays should return, then this must take place in and from the foot-point of the auroral ray nearest the earth. But as $H.\varrho$ for the cosmic rays is between 1 and 10 millions, the lowest value that the radius of curvature can have—namely when the ray moves perpendicularly to the lines of force above the magnetic poles of the earth—will be between 15 and 150 kilometres. The thickness of the aurora at the foot-point should then be between 30 and 300 kilometres. Now we know that even in the aurora which approaches to within a couple of kilometres(1), or very close, to the surface of the earth, the rays have a proportionally small angular diameter at the foot-point. GYLLENSKIÖLD states the value to be between 10′ and 3° (l. c., page 132).

It must consequently be considered as certain that the cosmic rays which come vertically towards the earth in such way as to form auroral rays, are entirely absorbed by the atmosphere.

Let us now see to what our experimental results will lead us, when they are applied to the auroral curtain formed by the auroral rays.

The cosmic rays approach the earth in the same manner as our cathode rays approach the terrella. We must now suppose that the auroral rays are formed by just such distinct, proportionally small groups of cosmic rays, which successively detach themselves from a larger bundle of rays after having passed through the magnetic equator, n, $(n+1)$, $(n+2)$, $(n+3)$, etc. times.

It is relatively easy, from our experiments with the terrella, to calculate, in some measure, the difference of time which in this manner should correspond to the entrance into the atmosphere of the nth and $(n+p)^{th}$ auroral rays at the moment when the auroral curtain is formed, provided that the velocity of the cosmic rays be known. This will be done later on in Volume II, but even now we may form an idea to the effect that we shall be led to results which are not in contradiction with the experience which we have now acquired.

Supposing that $H.\varrho$ is between 1 and 10 millions, and that the velocity of the cosmic rays is equal to that of light, we can conclude from the experiments that it is only a question of a fraction of a second between the formation of one auroral ray and the next one.

We proceed in the same manner as regards the so-called luminous waves which pass through an auroral curtain. If the original bundle of rays from the sun suddenly increases or decreases, this increase or decrease will be shown successively through the rays, one after another. If the rays produce precipitation corresponding to that found on the night side of our terrella, the wave will move from west to east; if the precipitation corresponds to the so-called positive precipitation, the wave should go

(1) ADAM PAULSEN: Aurores boréales observés à Godthaab 1882–1883, pages 8 and 13. Copenhagen, 1893.

from east to west. GYLLENSKIÖLD indicates about 39′ as an average of six observations of the angular velocity of the luminous wave. Supposing the thickness of the rays to be 10′, we obtain about a quarter of a second as the time which the corpuscular rays take to pass from near the southern to the northern auroral zone, and vice versa. We suppose here rays of a certain rigidity. In reality, rays of a somewhat different rigidity will, of course, occur, and the conditions will then be correspondingly more complex. We will not here enter more closely into the theoretical problems as to the explanation of the so-called folds and whirls in an auroral curtain. We will only say that we suppose that where such phenomena occur, the angle between the rays and the magnetic lines of force is nearly 0, or the angle lies in the magnetic meridian.

We have taken for granted that the auroral drapery is formed by *negative* corpuscular rays of a kind similar to β rays, and have thus assumed that α or other similar *positive* rays take no part in the formation.

There might in itself be much that would lead one to think of α rays in connection with auroral draperies, but there are decisive points that to my mind contradict such an assumption.

In the first place the auroral draperies appear, as a rule, in the time between the positive polar storms in the afternoon and the negative storms at night, i. e. just at the time when the negative corpuscular rays fall most vertically and farthest in towards the earth. During the positive storms in the afternoon, the rays are bent westwards along the auroral zone, and in the night, during the negative storms, they are bent eastwards, always supposing that our results from the terrella experiments can be transferred to the earth.

A precipitation towards the earth of α rays or other positive rays from the sun, would come in on the morning side of the earth, not on the evening side as the negative rays do; and it would be a remarkable coincidence if the positive rays were to go right round the earth and descend farthest into the asmosphere on the *evening side*, at the very place where all experience would lead us to expect the lowest precipitation of *negative* rays.

The way in which the phenomena are here compared, furnishes an explanation of an observation that is sure to be made whenever bright draperies are seen near the zenith in the neighbourhood of the auroral zone. The magnetic needles in the magnetometers then always, as far as I can learn, oscillate backwards and forwards, with alternately great positive and negative deflections.

From these points of view, it will be easily understood that the connection between the magnetic perturbations and aurora cannot be either simple or direct. Very early observers have proved that they are not the very same conditions that give rise simultaneously to the most powerful magnetic storms and the brightest aurora; but it is certain that when one of these phenomena manifests itself with great intensity, the other infallibly occurs, although there is not on that account any easily definable proportion to be found between their intensities.

During the last couple of years, attempts have been made in different ways, upon the basis of the corpuscular rays, to obtain a plausible explanation of the formation of the auroral curtains.

VILLARD [1] has tried, upon the basis of some beautiful experiments, to conceive the auroral drapery as formed by cathodic rays emanating from cirrus clouds, and afterwards drawn towards a terrestrial magnetic pole, e. g. the north pole, whence the ray returns after having penetrated far into the atmosphere and formed an *auroral ray*. He conceives then that the ray returns and goes towards the south pole, where the same ray penetrates far into the atmosphere and forms a *southern auroral ray*. The ray then returns again and goes towards the magnetic north pole, and forms there a new auroral ray by the side of the first one, and so on, times out of number.

[1] VILLARD: Les rayons cathodiques et l'aurore boréale. Paris, 1907.

On account of the absorbing power of the atmosphere, it does not appear that this theory can be maintained.

Other reasons telling against this theory are advanced by **Störmer** (¹) in his well-known essay, 'Sur les trajectoires des corpuscules électrisés dans l'espace". The only circumstance that Störmer finds in favour of Villard's theory and against mine, is that the auroral zone has a diameter of about 45°, while according to his own calculations it should be much smaller (4 to 12 degrees for cathodic rays and β-rays of radium, and for α-rays 24 to 36 degrees). In order to explain this disagreement, Störmer takes up for discussion the idea that the terrestrial magnetic field outside the earth is greatly modified by exterior currents, especially by the equatorial ring discovered through my experiments. This supposition is less natural, it appears to me, than the one advanced by me as to the rigidity of the rays, viz. that $H.\varrho$ must be between 1 and 10 millions.

Lenard also makes the same suggestion in a recent paper on this subject, as stated on p. 596.

Further, Störmer, in the same essay, paragraph 19, has advanced a very interesting theory on the creation of the auroral curtains based upon his mathematical studies on my theories.

In admitting an average value of $H.\varrho$ of 315 for cathodic rays, he finds (l. c., page 119), the theoretical dimensions of an auroral drapery. He arrives, for instance, at a length of 275 kilometres corresponding to a thickness of 72 metres.

In going through the same calculations and choosing $H.\varrho = 3.1 \times 10^6$, I find the length of the drapery almost unaltered, while the thickness has to be multiplied by 10. It will consequently be quite 700 metres. Nothing has here been added for the thickness of the auroral rays, as is done by Störmer. It cannot be conceived here, in fact, that the auroral rays can be formed as Störmer supposes, as in that case they would have a thickness of about 100 kilometres, which is contrary to all experience. It will be observed that the dimensions, calculated in the manner indicated above, do not fit so badly to a real auroral drapery; but it must be remembered that Störmer has here only calculated the space in which the rays going to the centre of the elementary magnet, approach the earth. He presupposes that the rays which in reality occur in the auroral curtains keep close to such rays through the centre. We have seen from the experiments, however, that the cosmic rays lying nearest to those which penetrate the auroral curtain, can swing entirely underneath the magnetic equator and penetrate the southern auroral zone.

From certain positions of the magnetic axis of the terrella in relation to the cathode, we observe, however, that the luminous spot which always occurs on the afternoon side to the north of the luminous ring, stretches itself into a ribbon (see fig. 220). These spots are formed by râys which are drawn directly towards the polar regions of the terrella without swinging above or below the equator, and it is perhaps these rays which are most likely to agree with the bundle of rays in Störmer's interesting calculation.

(¹) Störmer: Archives des sciences physiques et naturelles, juillet, août, sept. et oct. 1907.

CHAPTER V.

IS IT POSSIBLE TO EXPLAIN ZODIACAL LIGHT, COMETS' TAILS, AND SATURN'S RING BY MEANS OF CORPUSCULAR RAYS?

119. Zodiacal Light. In several of the experiments with a phosphorescent terella with different screens, in a large discharge-tube, we have come upon phenomena which appeared capable of serving as starting-points for an explanation of *Zodiacal light*.

Zodiacal light is the name given to a brightness which appears in the western sky after sunset, and in the eastern before sunrise, nearly following the line of the ecliptic in the heavens, and stretching upwards to various altitudes according to the season of the year.

Moreover, at certain periods of the year, what is called "Gegenschein' (discovered by BRORSEN), occurs almost directly opposite to the position of the sun.

Accurate observations have now shown that the axis of the zodiacal light diverges somewhat noticeable from the ecliptic, and recent work has assumed that it is rather a question of the sun's equator, than of the ecliptic.

The great cosmologist, CASSINI, concluded after only ten observations — the first detailed observations ever made — that the axis of the zodiacal light has a relation to the sun's equator, rising and sinking with it.

Before I proceed further with the elucidation of this question, I will here mention a peculiarity of the zodiacal light, which no attempt has ever been made to explain in anything approaching a satisfactory manner by the various theories that have been advanced. This is a pulsation in the intensity and shape of the light which has at times been noticed, a pulsation which surely testifies to an electric origin; and I am therefore of opinion that the phenomenon is akin to the pulsation which is sometimes seen in auroral lights and the oscillations in terrestrial magnetism.

HUMBOLDT writes: "I have occasionally been astonished in the tropical climates of South America, to observe the variable intensity of the zodiacal light . . . . . . When the zodiacal light had been most intense I have observed that it would be perceptibly weakened for a few minutes, until it again suddenly shone forth in full brilliancy" (Cosmos, vol. I).

Mr. BIRT, Kew Observatory, noticed in March, 1850, "One evening there was a sudden brightening of the light for an instant, and also variations in its lustre of an intermittent character. These intermissions of brightness were observed on the same evening by Mr. LOWE at Nottingham" (Am. Journ. of Sc., XV, second series, p. 121).

The Rev. GEORGE JONES, a most diligent observer of zodiacal light, relates in March, 1854: "I was surprised, one evening, at seeing the zodiacal light fade sensibly away, dimmed to almost nothing, and then gradually brighten again. This was repeated several times; but the effect, after all, was to leave me only in amazement and doubt. Subsequent nights, however, gave abundant exhibitions of this kind, of which, with the times and changes, I have made ample records with the particularity that the case required. — — —

"My records, however, will show that there is a regularity of appearance and the closing off of these pulsations, which proves that they do not belong to so uncertain a cause as atmospheric changes, but to the nebulous substance itself. They seem to intimate a great internal commotion in the nebulous matter, for they were too rapid to be occasioned by irregularities in its exterior surface.

"I noticed them again the following year, but must refer the reader to my records and charts. The changes were a swelling out, laterally and upwards, of the zodiacal light, with an increase of brightness in the light itself; then in a few minutes, a shrinking back of the boundaries, and a dimming of the light; the latter to such a degree as to appear, at times, as if it was quite dying away; and so back and forth for about three quarters of an hour; and then a change still higher upwards, to more permanent bounds". (Observations of the zodiacal light by JONES, vol. 3 of the Report on the United States Japan Expedition, 1856, page XIII).

The pulsations of the zodiacal light thus recorded cause one involuntarily to think of the regular, often almost sinusoidal magnetic pulsations and simultaneous oscillating earth-currents which so frequently occur, and markedly in the month of March. (See Part III of this Section.)

As an example, I shall quote an observation of JONES, not, it is true, from March, but from the evening of the 30th January, 1854, "The pulsations of the zodiacal light were very distinct". At the end of his series of observations we find: "7^h 54^m, its boundaries had risen to *b* again and . . . . bright: 7^h 55^m at *a* and very dim: 7^h 56^m at *b*, and bright: 7^h 57^m at *a*, and very dim: 7^h $58\frac{1}{2}^m$ at *b* and bright: 7^h $59\frac{1}{2}^m$ still at *b* and bright: it seemed now to be permanent at *b*".

Here we have plainly a period of about 2 minutes.

From another observation of JONES: "These lateral changes of the whole body of the stronger zodiacal light are very remarkable. I cannot see any room for mistake, as there might have been, had the light been more inclined to the horizon. But the horizon and ecliptic made nearly a right angle".

For comparison I shall adduce that, at the Haldde observatory, in March, 1900, I observed beautiful magnetic oscillations with a period of 128 seconds.

In May, 1910, I again registered at Kaafjord beautiful magnetic waves and simultaneous earth-current oscillations of very nearly 119 seconds, as will be seen in the subsequent part of this volume.

I quite perceive that it is easy to imagine that what are called magnetic elementary waves, which have specially been studied by ESCHENHAGEN, have their origin in oscillations of electric ray masses.

It may be worth mentioning in connection with this, that the earth in March and September is at the farthest possible distance from the nodes of the sun's equator.

It appears to me very probable, in view of the properties above described, that the zodiacal light must be primarily occasioned by electrical phenômena.

We shall now further analyse the most important attributes that the zodiacal light has been observed to possess, and see if they can be put together and explained by the supposition of an emanation of corpuscular rays from the sun. The question whether the axis of the zodiacal light is situated in the ecliptic or in the equator of the sun has been carefully considered in two important treatises by ARTHUR SEARLE. In the first of these, "The Zodiacal Light" (Proceedings of the American Academy of Arts and Sciences, 1883) as well as in the second, "The Apparent Position of the Zodiacal Light", 1885, he has made extensive researches by making special use of numerous observations from the classic and admirable volume by JONES.

In the following pages I shall endeavour to interpret all the results of observations with which I am acquainted, by starting with the supposition to which I shall subsequently come, in order to explain the diurnal variation and the origin of terrestrial magnetism, viz: that the corpnscle-rays *continually* radiate from the sun's surface (see Section I, p. 314). But *these* continuous rays must be assumed to possess properties somewhat different to those of the very stiff corpuscle-rays that radiate in short periods from the sun-spots, and which, we supposed, specially occasioned magnetic storms on the earth.

I now assume that these corpuscle-currents, which are continuously given out and probably most strongly from the neighbourhood of the sun's equator, are somewhat less stiff as regards magnetism than the rays which come in eruptions from the portions in greatest activity around the sun-spots. The constant rays are thus less penetrative through matter, and come probably from lesser depths in the atmosphere of the sun.

I have recently in a note (¹) in C. R. de l'Académie des Sciences, Paris, in explanation of certain phenomena in the magnetic storms, advanced the opinion that the sun is magnetic, with a magnetic moment of the order 10^{28} or about 150 times as great as that of the earth.

If this is the case, the corpuscles which are constantly given out will principally issue both from the regions near the magnetic poles of the sun, and moreover the rays will to a very great extent be concentrated and form a ring in the plane of the sun's *magnetic* equator, which probably only forms a small angle, or is perhaps identical, with the heliographic equator.

There is no reason, as we shall see further on, to suppose that the sun's magnetic axis should not be identical with the axis of rotation, as there can hardly be magnetisable masses with permanently fixed positions in the sun.

In the plane of the sun's magnetic equator the corpuscle-rays will doubtless, as an elementary study shows, bend comparatively sharply, near the sun; but they will keep constantly in the plane.

This question, the examination of how the corpuscle-rays move in the magnetic equator of a magnetic globe, I have investigated experimentally and have obtained very successful results. See fig. 223.

By allowing a smooth magnetic sphere (without phosphorescent coating) to be the cathode in a discharge-container, a wonderfully developed luminous ring is easily obtained around the globe.

The photographs here reproduced have been taken with a magnetic ball of 8 cm. in diameter in the smallest of the prismatic discharge-containers. It was seen that the ring expanded immensely with the stiffness of the rays and with the magnetic globe's magnetic momentum. I was unable in this instance to attain a difference in the tension between anode and cathode of more than 700 volts, when the brass ball was the negative pole; but even at this tension and a magnetising current of 21 amperes from an isolated storage battery the ring became so large that it at times reached to the glass walls of the container. I shall repeat the experiment with my largest discharge-box, when I get it repaired again, for I am convinced that I shall be able to obtain a perfectly flat ring of light of 30 cm. diameter around my strongest magnetic globe No. 7, which is also of 8 cm. diameter.

If I were in possession of sufficient quantities of pure radium-bromide, I would coat the equatorial portions of my strongest magnetic globe with that substance. It would be of interest to see if rings of β and α rays could then be made visible.

I will here observe, that when I have on previous occasions produced a luminous ring round my terrella by cathode rays from a somewhat distant cathode, it is possible that I have been mistaken. It may be that the magnetic ball has been sufficiently negative compared with the surroundings for an emanation of negative rays to take place at the same time as the ball is illuminated and surrounded by cathode rays from the real cathode. There are two reasons for this. In the first place, it was, as already mentioned, only under quite exceptional circumstances that the ring was formerly produced, while it is now produced in the way here described never wanting, and in the second, the difference in the tension need only be very small before the negative radiation from the ball occurs, so that such a difference in tension can very easily have taken place in the course of the earlier experiments.

But this condition does not, of course, affect our previous main results, in which, by the aid of various screens, we have proved how, amongst other things, rays from the cathode circulate around the terrella, bending above and beneath the plane of the magnetic equator.

(¹) C. R. 24 jan. 1910.

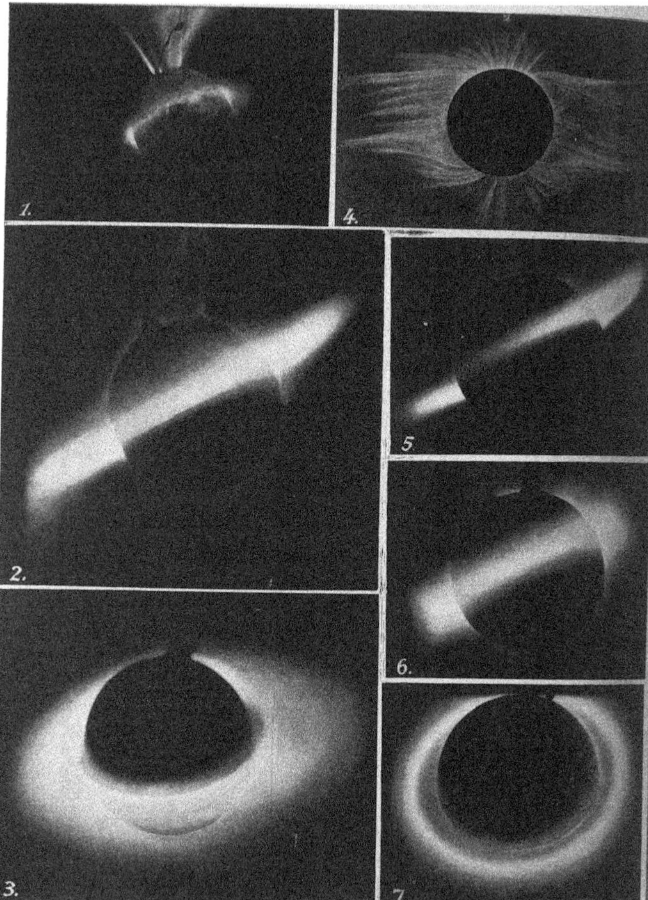

Fig. 223.

All the trials represented on the adjoining figure have been made with terrella No. 4 of a diameter of 8 cm. or perhaps 7.8 cm. when without the phosphorescent coating, and weighing 977 gr. The resistance of its magnetising coil is 1.72 Ω. The magnetic momentum at 10 amperes magnetising current was 27200.

The first experiment (see fig. 223) was made with a discharge-current of 20 milliamperes under pressure of about 0.005 mm. and with a magnetising current of 21 amperes.

In addition to the equatorial ring, discharges will be seen from the northern polar zone. This polar discharge is easily produced if there are any sharp points or unevennesses, but, on the other hand, it is difficult to obtain it when the surface is smoothly polished, as was the case in all the other experiments represented on the plate.

The picture No. 4 has been placed beside this for the sake of comparison. It is a view of the sun during an eclipse, May 17th, 1901. The picture is drawn by H. R. Morgan from the negatives. I will, later on, by the aid of points in the magnetic polar regions, both N. and S., produce a more perfect example corresponding to photograph No. 1, as this is obviously of great interest.

Photographs 2 and 3 are from an interesting experiment seen from the side and from above, in which the pressure was brought as in the first experiment, but the discharge-current was only 2 or 3 milliamperes and the magnetising current 26 amperes.

The tension was 1500 volts before the magnetisation of the spherical cathode, and the radiation from it could be seen to take place evenly from the entire surface of the sphere. After the magnetising current was put on, the tension sank immediately to 600 volts and the radiation then took place only from the equatorial regions of the spherical cathode. This could be plainly observed from the minute glowing spots from which the rays issued, near the metal ball's equator.

In the experiment represented in photograph No. 5 the pressure was as before, the magnetising current 21 amperes, and the discharge-current 3 milliamperes.

Photographs 6 and 7 are from an experiment with a magnetising current of only 2 amperes. The ring is seen from the side and from above. Pressure 0.02 mm. and the discharge-current was 5 milliamperes. It is the low magnetising current that occasions the ring to be broad and small in extent. A dark band is plainly visible between the magnetic sphere and the ring. It has happened on several occasions that the luminous ring has been divided into two concentric rings by a dark circular band.

We can find the conditions for electric radiation's getting out towards infinity from the surface of a magnetic sphere.

Suppose a magnetised sphere is giving out electric radiation of some kind. In the regions near the poles the radiation will be able to get out by passing nearly along the lines of magnetic force.

For rays in the magnetic equator, however, the magnetic force is perpendicular to the orbit of the ray-particle, and unless certain conditions are fulfilled the radiation will not be able to emerge in this place.

It will be of interest for a number of questions in cosmic physics, to find the exact conditions for rays in the place of the equator emerging into space.

Let R and φ be polar co-ordinates in the plane of the equator with the centre of the sphere as origin.

We suppose the magnetic force to be perpendicular to the plane of the co-ordinates, and outside the sphere given by the relation

$$F = \frac{M}{R^3}$$

This relation is the same as that which determines the fields of an elementary mag then apply the results of STÖRMER's mathematical analysis of the orbits of corpuscles in th elementary magnet.

According to STÖRMER, the orbits are determined by the following equation:

$$R^2 \frac{d\varphi}{ds} = 2\gamma c + \frac{c^2}{R}$$

s is the length of the orbit, γ is a constant of integration.

$$c = \sqrt{\frac{M}{H_0 \varrho_0}}$$

$H_0 \varrho_0$ is a quantity which depends on the stiffness of the rays. ϱ_0 is the radius of the corpuscular orbit, when the magnetic force perpendicular to the orbit is H_0.

Introducing the angle θ which the direction of the orbit forms with the radius vector

$$\sin\theta = \frac{2\gamma c}{R} + \frac{c^2}{R^2}$$

From the condition that $\sin\theta$ must have values between -1 and $+1$, STÖRMER fi each value of γ the orbits must be restricted to certain regions of space.

Suppose at first γ is negative and numerically greater than 1, or

$$\gamma_1 = -\gamma, \text{ where}$$
$$\gamma_1 > 1.$$

In this case we shall have an interior and an exterior region for the orbits.

The inner region is limited by the two circles,

$$R_1 = c\,(\sqrt{\gamma_1^2 + 1} - \gamma_1)$$
$$R_2 = c\,(\gamma_1 - \sqrt{\gamma_1^2 - 1})$$

The exterior region goes from infinity to the circle

$$R_3 = c\,(\gamma_1 + \sqrt{\gamma_1^2 - 1})$$

If γ_1 is less than unity, the exterior circles R_2 and R_3 cease to exist.

The rays issuing from points on the equator circle can have any direction inside th rants $0 < \theta < \frac{\pi}{2}$ and $0 > \theta - \frac{\pi}{2}$.

It is of special importance for us to examine the range of those rays which reach distance.

It will be those going out in a direction corresponding to $\sin\theta = +1$ or for these r $R = a$ when $\sin\theta = +1$, which give

$$R_1 = a = c\,(\sqrt{\gamma_1^2 + 1} - \gamma_1).$$

If the rays shall not go towards infinity

$$\gamma_1 > 1 \quad \text{or}$$
$$\frac{a}{c} < \sqrt{2} - 1.$$

If condition (3) is fulfilled, the range of the radiation will be R_2 as given by equation 2 b.

We suppose a to be constant and let c vary.

When c decreases towards $\frac{a}{\sqrt{2}-1}$, R_2 will increase and approach the value $R_2 = c$.

The greatest range which the rays can have without going towards infinity will be

$$R_2 = \frac{a}{\sqrt{2}-1} = 2.414\ a.$$

We then get the very simple result:

If electric radiation starting from the surface of a sphere in the plane of the magnetic equator, and only subject to the influence of the magnetic field of the sphere, reaches a distance from the centre greater than 2.414 times the radius to the sphere, the radiation will not be able to return to the sphere, but will pass on towards infinity. This result will hold independent of the magnetic moment of the sphere and the stiffness of the electric rays.

This result supposes that relation (1) holds good close up to the surface of the sphere. This relation actually holds good provided the sphere is uniformly magnetised or it will be more or less true for any magnetisation which makes the magnetic force in the magnetic equator a function of the distance from the centre.

If the radiation shall return to the sphere, the following condition must hold:

$$c > 2.414\ .\ a \qquad \text{or}$$

$$\sqrt{\frac{M}{H_0 \varrho_0}} > 2.414\ .\ a.$$

This result corresponds to the rays starting in the direction $\theta = \frac{\pi}{2}$.

If we consider the radiation starting normally we get

$$R_{2(\max.)} = 2\ a,$$

or if radiation starting normally reaches a distance greater than $2\ a$ from the centre, it will pass on to infinity.

If the radiation starting normally shall return to the sphere, we must have

$$c > 2\ a \qquad \text{or}$$

$$\sqrt{\frac{M}{H_0 \varrho_0}} > 2\ a.$$

Application to the sun.

In order that radiation shall emanate from the sun

$$M < 2.86 \times 10^{22}\ H_0\ \varrho_0$$

when starting in the direction $\theta = \frac{\pi}{2}$ and

$$M < 1.96 \times 10^{22}\ H_0\ \varrho_0$$

when starting normally.

For the stiffest β rays starting normally we get

$$M < 0.9 \times 10^{26}$$

and for α rays

$$M < 8 \times 10^{27}.$$

When M is of the order 10^{28} as estimated by me in *C. R.* Jan. 24, 1910, it supposes that $H_0\ \varrho_0 > 5 \times 10^5$ for normally starting rays if the rays shall be able to emerge into infinity.

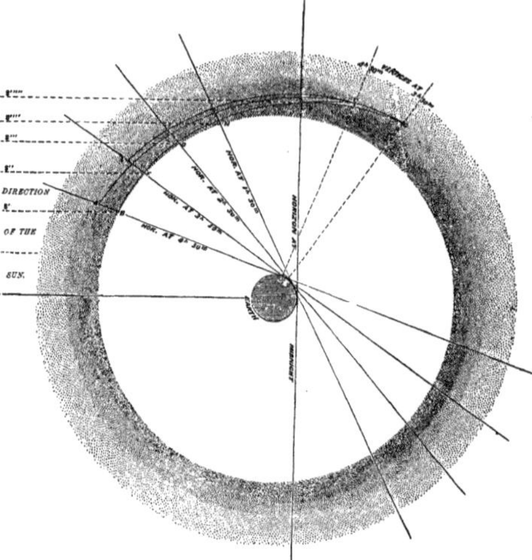

Fig. 224. Nebulous ring, with the Earth for its centre, according to Jones.

120. We now return to the radiations emanating from the sun. From my experience obtained from the experiments, I regard it as very possible from a physical point of view, that a ring of radiant matter has been formed round the magnetic equator of the sun, the dimensions of this ring being greater than those of the earth's orbit. We must recollect that in the case of the sun, it is a question of corpuscular rays of very great stiffness, as the mathematical calculations also have shown. I assume that these corpuscular rays from the sun partially consist of atoms and molecules, and not merely of electrons, thus that the radiant matter in thick layers is both slightly luminous and is capable of absorbing and scattering solar light. When treating of the formation of the tails of comets, we come back to the same idea. Possibly krypton, which seems to cause the well known auroral line in the auroral spectrum is thus emitted from the sun, and that we may be able in this manner to answer a question put by Ramsay(¹): "Is there any process which will tend to increase the relative amount of krypton in the upper regions of the atmosphere?"

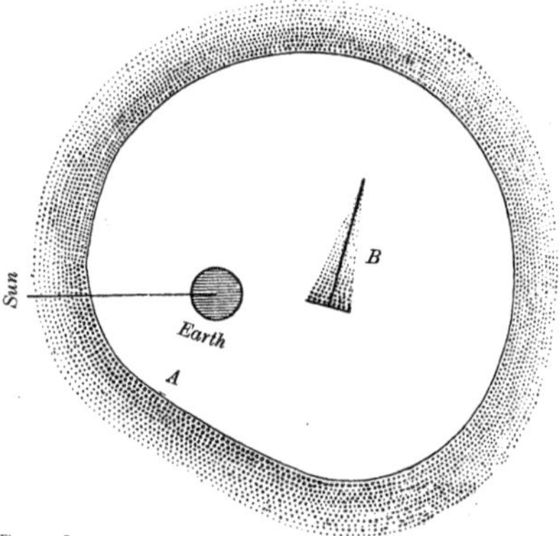

Fig. 225. Space round the Earth into which the radiant matter from the Sun does not enter.

(¹) Ramsay: The Aurora Borealis. Essays Biographical and Chemical p. 214. London 1908.

Let us now see how we can explain the characteristics that have been observed in the zodiacal light, by supposing that in the sun's equatorial plane there exists a flat ring of radiant streams of matter, consisting principally of primary rays and streams of atoms from the sun, and perhaps also of secondary rays emitted from cosmic dust moving in the same plane and which are irradiated by the primary beams from the sun.

If these corpuscle-rays and streams of atoms either themselves emit luminous rays or scatter the light of the sun, we will, as we shall soon find, be enabled more satisfactorily than ever to explain the characteristics of the zodiacal light.

It will be at once observed that my idea of this flat ring about the sun has a certain resemblance to what is called the *meteoric theory*, as it also presupposes that a ring of cosmic dust exists which encircles the sun, more particularly in the plane of the solar equator.

The idea perhaps equally resembles the theory advanced by MAIRAN in 1731, that the zodiacal light is reflected from the sun's atmosphere, stretched out into a flattened spheroid or lenticular shaped body revolving with the sun; an idea which LAPLACE has for ever set at rest by demonstrating that the sun's atmosphere "can extend no further than to the orbit of a planet whose periodical revolution is performed in the same time as the sun's rotary motion about its axis, or in twenty-five days and a half; that is only as far as $^9/_{20}$ of Mercury's distance from the sun".

We shall, however, soon see that my theory has an equally great resemblance to an entirely different view of these phenomena, namely, to the idea arrived at by JONES after discussing the results of his excellent observations: "I offer now, as a last conclusion, the hypothesis of a nebulous ring with the earth for its centre". In reality my theory combines the advantages of all earlier hypotheses, and it succeeds in explaining phenomena which none have elucidated previously, for instance the phenomenon of the counter-glow—Gegenschein—and the pulsations in the brightness and outline of zodiacal light.

From what we have learnt from our experiments we can foresee what will happen when our magnetic earth advances in the assumed ring of radiant matter that surrounds the sun.

The earth magnetism will cause there to be a cavity around the earth in which the corpuscles are, so to speak, swept away, a space around the earth from which a portion of the radiant matter has disappeared.

This cavity round the earth is doubtless not circular in such a way as JONES supposes with his nebulous ring hovering about the earth in the sun's equatorial plane, but the space has a somewhat different form which we can describe and note particulars of very closely, owing to our earlier experiments as will be seen in the following.

Diagrams showing a section of JONES's and my spaces respectively round the earth, spaces that are free from corpuscles, will be seen in figs. 224 and 225.

We shall now easily understand that we have here an explanation of "the brightness in the east before sunrise", owing to the streams of corpuscules from the sun, when they approach the earth sufficiently, becoming deflected in the same manner as the rays shown in the fig. 219, and as it is further plainly shown on the morning side of the picture No. 16, fig. 215, which is as seen taken from the south pole.

In like manner it makes it easy to explain naturally "the brightness which appears in the western sky after sunset" by referring to numerous pictures with the 1st sectional line, also called the 1st line of precipitation, of the rays; this is in a manner a boundary-line along which the cathode rays begin to travel around the terrella in curves regulated by the stiffness of the rays and by the magnetic condition of the terrella. See particularly the same picture 16, fig. 215, on the evening side.

In transferring the results of the experiments to the earth, we must recollect our above-mentioned supposition, that the rays are approaching the earth from the sun, forming a flat ring of radiant matter *travelling in the sun's magnetic equator.*

If these corpuscle-rays either emit luminous rays, or the radiant matter scatters the solar light, the brightness in the western sky will appear, because we see into the deep layers of radiant matter situated in the sun's magnetic equator, and the brightness will disappear at the boundary line where the rays spread out to travel round the earth over and under the earth's magnetic equatorial plane as mentioned in the preceding pages.

We may now in analogy with our experiments conclude that the rays round the earth, after spreading on the first sectional line, will gather again to a second sectional line (the 2nd line of precipitation), in which, however, the density of the rays will be much less than in the first sectional line, but nevertheless considerable.

In the course of our experiments we have seen that the concentration in this second sectional line is greatest by far when the magnetic axis of the terrella stood perpendicular to the direction of the rays from the cathode.

The position of this second sectional line has been somewhat varied, according to the terrella's magnetisation, but it is always approximately on the magnetic equator of the terrella, and originates and is most powerful not far from the direction opposite to the cathode.

We shall closely point out below to what a high degree the results of these last terrella experiments, transferred to the earth, serve to explain the hitherto known characteristics of the counter-glow or Gegenschein.

Jones, in his work which we have quoted, has mentioned this phenomenon, which he, however, at first did not believe to be zodiacal light. It was not till after his return from his long journey that it became clear to him that this counter-glow was a phenomenon of the zodiacal light, which was first observed by Humboldt in 1803; but *he* supposed the phenomenon to be only a reflection from the western zodiacal light, then shining with exceeding brilliancy (See Astronomische Nachrichten No. 989).

In No. 998 of the same journal is another paper on this subject by Brorsen of Serptenberg in Germany, who calls this eastern evening light by the appropriate name of "Gegenschein", and informs us that he had seen it regularly at that place during the two previous years. His paper concludes as follows: "The Gegenschein is visible, not only at the vernal, but also at the autumnal equinox; at the former time more distinctly. A faint trace of it becomes visible in January, from which time it grows stronger till March, when, and in April and the early part of May, it is quite distinct and broad.

"A much smaller and fainter Gegenschein appears in September, October and November. I have become convinced, by frequently repeated observations, that in both cases the brightest part of the Gegenschein is directly opposite the place of the sun, so that a calculation of the greatest light frequently coincides to a degree with the point of opposition to the sun.

"The observations proved that the vernal Gegenschein about the middle of April, joins the westerly zodiacal light by a stripe or belt of light, which is at first very faint, but becomes by degrees more luminous; the autumnal Gegenschein appears, in the first part of November, to be elongated along the ecliptic by a faint zone of light as far as the western horizon, which zone of light is by degrees transformed, by increasing luminosity and more distinct basis, into the well-known phenomenon of the western zodiacal light."

We shall, before we go more into the theoretical comparisons with our terrella experiments, further quote data respecting the counter-glow from a particularly important work by Arthur Searle — "The Zodiacal Light, discussed by means of the Records of Harvard College Observatory".

In this paper we find a very interesting table of collected results from observations of Gegenschein at various stations by Brorsen, Schmidt, Heis, Eylert, Busch, Groneman, Backhouse, Lewis, Barnard, and at Harvard College.

TABLE CVII.

	Jan.	Feb.	Mar.	April	May	June	July	Aug.	Sept.	Oct.	Nov.	Dec.
No. Obs. . .	1	40	45	31	7			3	29	51	17	4
Mean λ . . .	134	144	174	204	227			334	353	22	45	74
Mean $\Delta\lambda$ . .	+ 3	− 4	− 2	+ 1	− 2			+ 4	0	0	− 1	− 2
Mean β . . .	+ 4	+ 2	+ 2	+ 2	0			0	0	0	+ 1	− 2
Ext. λ	20	14	13	15	10				11	16	10	
Ext. β	8	8	10	11	7				9	12	7	6

The first line gives the total number of observations.

The other five lines give the longitude, its excess over that of the point in opposition to the sun, and the latitude, of the observed light, with its extent in longitude and latitude, so far as this can be estimated by means of the sketches or descriptions. These quantities are given only in entire degrees.

From northern stations, it appears from the table that Gegenschein has most frequently been seen in October, but the number of observations in February and March is also relatively large. According to experience at the Harvard College Observatory, the phenomenon to be observed is often difficult to distinguish, in March, from a part of the luminous band crossing the ecliptic nearly at right angles on the borders of Leo and Virgo, while in October, as in the other autumn months, it is perceptible only as a reinforcement or as an extension of the band from *Aquila* to the *Pleiades*.

During February and March the observed light has a position a few degrees preceding the point in opposition to the sun, generally north of the ecliptic.

In the autumn of 1886 the general remarks made by Barnard and those made at Harvard Observatory concur in describing Gegenschein as only a very elongated patch of light, instead of a round or elliptical spot.

Brorsen's above-quoted observations may thus in the main be said to be confirmed by all subsequent researches.

We shall now see that these results of observations of counter-glow (Gegenschein) in nature can be explained by the results of my terrella experiments.

Just about the time of the equinoxes the "second sectional line" of the corpuscle-rays round the earth, should, in analogy with the experiments, be most strongly present and as it moreover will fall in the *earth's* magnetic equatorial plane about 180 degrees from the direction to the sun, it will also fall somewhat in the plane of the ecliptic or near the sun's equatorial plane.

The experiments referred to concerning the second sectional line, are those described on pages 560 to 564. The vertical screen used in these experiments consisted principally of two plane portions which intersected one another at an angle of about 100° in a vertical line. When, therefore, it is recorded that the second sectional line, with magnetising currents 8 and 28 amperes respectively, fell on the screen when its hour-angles were 272° and 265°, it must be remembered that the plane part of the screen nearest the terrella and passing through the magnetic axis had then a length of about 92° and 185°.

The commencement of the sectional line, where by far the most of the rays cross each other, would then from the centre of the terrella be seen with lengths of 192° and 185° respectively. These angles must however be calculated along the magnetic equator of the terrella.

If it be further remembered that all three sectional lines referred to, by stronger and stronger magnetisation of the terrella, draw back somewhat, being pushed outwards almost parallelly with a quite slight reduction in length, it will be evident that with so intense a magnetisation as to correspond with the conditions on earth, the second sectional line would begin and be most strongly developed at a length but a few degrees less than 180° reckoned on the magnetic equator.

Returning now to the earth, supposed to be travelling in the ring of radiant matter round the sun.

At the equinoxes we shall see the places where the corpuscle rays are intersecting each other in the "second line of precipitation" in a line with the ring of radiant matter in the *sun's* magnetic equator, which ring must be assumed to continue also beyond the earth's orbit.

In this manner we shall be able to see through the radiant matter into a considerably thicker stratum opposite the sun, as shown diagramatically in fig. 225, and more light will be diffused, by reason of which Gegenschein may be imagined to be caused.

121. We now pass on to mention how the spectral analysis investigations which have been made of these phenomena look in view of the theory advanced here.

The spectrum of the zodiacal light has been observed for many years, but owing to its faintness the observations are very difficult to make. Among the first observers were LIAIS(1), VOGEL(2), PIAZZI-SMYTH(3), and WRIGHT(4). LIAIS at times suspected dark lines, but could not be certain of their existence. WRIGHT detected the presence of the atmospheric band at λ 5780. Other observers had thought the bright aurora line at λ 4571 a part of the zodiacal light spectrum, but the work of the last three of the above-mentioned observers seems quite conclusively to show that this belongs to the aurora alone, although it may at times appear superimposed upon the spectrum of the zodiacal light. This frequently occurs if the auroræ are at all common at the place of observation. HALL(5), observatory in Jamaica, found the spectrum continuous even when using a slit sufficiently narrow to show absorption lines in the spectrum of daylight. In other respects, all observers agree in finding the spectrum continuous with an intensity curve quite similar to that of daylight.

The fact that the intensity curve of this spectrum closely resembles that of the sun, and the existence of from 15 to 20 per cent of polarized light, as shown by the careful observations of WRIGHT(6), are in accordance with the meteoric theory(7). The above-mentioned observations were all visual.

The first successful attempt to photograph the spectrum of the zodiacal light is described by FATH, Lick Observatory Bulletin No. 165, from which the above-cited résumé is taken. The results are summed up as follows.

"Upon developing the plate a spectrum was obtained which resembles the solar spectrum exactly, in so far as can be judged from so small an object.

"Two absorption-lines could be seen with certainty. A comparison of the plate with one of the sky spectrum taken with the same slit-width showed these lines to be G and the blend of H and K of the solar spectrum.

(1) Comptes Rendus 74, 262, 1872.
(2) Astron. Nach. 79, 327, 1872.
(3) Mont. Not. 32, 277, 1872.
(4) Amer. Journ. of Sci. Ser. 3 8. 39, 1874.
(5) Observatory 13, 77, 1890; Mon. Weather Rev. 34, 126, 1906.
(6) Amer. Journ. of Sci., Ser. 3, 7, 451, 1874.
(7) Cf. SEARLE, Mem. Amer. Acad. 11, 135, 1888.
„ SEELIGER, Münch. Ber., 31, 265, 1901.
„ GEELMUYDEN, Bulletin Astron., 19, 416, 1902.

"These are the only two lines shown in the sky comparison plate within the spectrum obtained on the zodiacal light. Thus in so far as spectra of such low dispersion and resolving power can be trusted, we would seem to have good evidence to support the claim that the zodiacal light is reflected sunlight".

After these results we must ask: Is it conceivable that radiant matter can reflect sunlight as we have supposed in our theory of the zodiacal light?

Although analogies may often be misleading, there will undoubtedly be a certain value in the recollection that the atmosphere, even in quite clear weather, diffuses the daylight to so great an extent that even the most powerful stars are invisible. The light is sent back either from the air-molecules themselves, or from microscopic dust-particles that are found in the atmosphere, as by the blue of the sky.

Physical investigations of the power of electrically luminous gases to absorb and diffuse sunlight, have not, as far as I am aware, been made on any large scale; but during the last few years some very interesting results have been obtained, which will be discussed in these pages. With regard to direct experimental research into the properties of *radiant matter* in the above respects, I do not think anything has been ascertained.

In the meanwhile, I have made some observations at Kaafjord in Finmarken, which will possibly afford us some guidance in the question.

I have in broad daylight and at times in sunshine been able to observe rapidly-changing "clouds" formed like draperies with radiant structure appearing at that time of the evening in which, in winter, corresponding draperies of aurora are frequently seen.

I have thought that these must be, not real clouds, but auroral rays scattering the sunlight and therefore appearing like clouds. At all events it seems to me little likely that the condensation of moisture could take place so rapidly in the highest regions of the atmosphere, and a moment afterwards revert to vapour again (see page 450).

I have found in literature certain investigations by R. Ladenburg and R. W. Wood, of the optical conditions in electrically luminous gases and in vapour, which are of great importance to the questions we here touch upon. Ladenburg, in a treatise entitled "Ueber Absorption und Magnetorotation in leuchtendem Wasserstoff"(1), demonstrates that the number of absorbent "dispersion-electrons" is proportional to the amplitude of the transfluent current. Now the intensity of the light is also proportional to that of the current, and the number of ions at constant pressure is proportional to the strength of the current. All this should confirm the hypothesis that the bearer of the spectral hydrogen-series is the positive atomion.

Wood, after a number of interesting investigations of "Die vollständige Balmersche Serie im Spektrum des Natriums"(2), "Die selective Reflexion monochromatischen Lichtes an Quecksilberdampf"(3), and "The Ultra-violet Absorption, Fluorescence and Magnetic Rotation of Sodium Vapour"(4), is of opinion that the Balmer lines and the accompanying spectra are produced by atoms that have lost one, two, three, four, and so on, electrons.

There is now certainly very good reason for supposing that in the radiant matter which we assume to have been radiated from the sun, there is comparatively a very large number of dispersion-electrons that can take up and be in resonance with light-waves from the sun, and that *possibly here too, this number of dispersion-electrons is proportional to the enormous electric current-intensity that emanates from the sun in the manner here assumed.*

It will perhaps after this no longer be considered improbable that the mighty strata of radiant matter we have imagined we could see into when we observe zodiacal light, are capable of diffusing sufficient sunlight to occasion this slight brightness in the sky. Subsequent spectroscopic investigations may possibly prove that the zodiacal light also contains a weak light of its own, which some observers have thought to show, and thus does not merely reflect sunlight.

(1) Physikalische Zeitschrift. 10 Jahrgang. 1909; p. 497.
(2) " " " " " p. 89.
(3) " " " " " p. 425.
(4) " " " " " p. 913.

It would be natural here, under the theory of the zodiacal light, to lay great weight upon un electric evaporation of the sun's surface, which must be assumed to accompany the emission of rays in accordance with our experience of electric discharges from a cathode in high vacuum.

In the following articles on comets' tails and Saturn's ring, due consideration has been paid to these conditions. Experiments have shown that considerable quantities of matter are in this way flung out into the plane of the equator. It can be imagined that these grains of dust, moving under the influence of gravitation and electromagnetic forces, become massed together by collision into greater and greater globules.

This brings us to the assumption of a dust-ring round the sun, undergoing constant renewal from the central body; and we thus come nearer to the hypothesis most current at the present time, namely the so-called meteoric theory.

That the spectrum of the zodiacal light suggests reflected sunlight can then also be explained by the reflection of the light from these tiny particles originally produced by the radiant matter.

122. Appendix. Since the above was printed, I, together with Mr. KROGNESS, have undertaken a journey to Egypt and the Soudan, for the purpose of beginning to make personal observations of the zodiacal light.

Of the expenses of this expedition one tenth was borne by the University, one tenth by my fri Mr. SCHIBSTED, and eight tenths by myself.

For the time being, our object was to find out whether the pulsations in the light discovered by JONES were accompanied by simultaneous magnetic pulsations.

During two months, March to May, 1911, observations and attempts to photograph the light were carried on at Assouan by Mr. KROGNESS, and at Omdurman, near Khartoum, by myself.

As the then much discussed question of the simultaneity of certain abruptly-beginning magnetic disturbances seemed likely to be also of importance in connection with these observations, I published in "Nature" for March 16, 1911, (No. 2159, p. 79), a letter requesting other observatories, especially near the equator, to take "quick-run" registrations at the same hours at which we did so.

At Assouan the instruments were set up in the depths of an ancient Egyptian tomb, in which the temperature was fairly constant. Thanks mainly to Mr. KEELING, the superintendent of the Khedivial Observatory at Helouan, we enjoyed all the facilities for our work that we could desire.

Our observations of the zodiacal light were made every evening and night in favorable weather, from camps out in the desert west of Assouan and south of Omdurman, where the light from the towns in no way hampered the observations. It was a strange occpuation these observations every dark night in the Soudanese Desert, accompanied only by a chance Abyssinian servant.

The time however was not favorable, according to the general opinion of several inhabitants. The zodiacal light could often be seen much brighter there than we saw it.

As a rule the desert wind raised fine sandy dust, which caused the air to become thick, especially near the horizon. Venus, moreover, at that time was very bright, and was situated near, and sometimes in the very middle of, the cone of zodiacal light, where its presence was highly embarrassing. It was impossible, for instance, to be sure of the pulsations in the zodiacal light, although we thought now and then that we saw slight, rather sudden changes.

Only on the 30th April, the last day I was in Khartoum, just as I was about to leave it, the light was unusually strong and right up in the zenith, and I was almost sure that I could see decided changes.

I had no opportunity of noting the exact times of these changes when I unpacked my photographic apparatus; but during that last hour at the railway-station before leaving, I succeeded in getting the best photographs of the light that we took throughout the expedition.

It has subsequently appeared, after all the magnetic curves have been developed, that an unusual magnetic calm has happened to prevail during all the times at which we obtained serviceable observations of the light; there were hardly any perceptible magnetic changes at those times. Only on the 30th April did it appear that there had been a magnetic storm during the hours in which the observations of the zodiacal light were made on my departure from Khartoum.

Our attempts to obtain good photographs of the zodiacal light were at first without result. We tried altogether five or six combinations of lenses, some of the lenses being very expensive. At last we succeeded, by telegraphic order, in obtaining from Cairo and Dresden some simple cinematograph lenses, which gave fairly satisfactory results.

We then took, both at Omdurman and Assouan, at exactly the same hours, two dozen plates each evening during the last few days of our stay.

The times were photographically recorded from an electrically illuminated watch upon each plate at the beginning and end of each exposure.

There is at present nothing more to say about our results here, but it was at any rate ascertained that it was possible to obtain good photographs with our simple cinematograph lenses, by employing Hauff's "Ultra-Rapid" plates, which ought by preference to be illuminated in before the exposure according to Wood's method (1).

It is my intention as soon as possible to continue these investigations, perhaps with two stations, in the Andes in South America. By photographing the zodiacal light simultaneously from two such stations, it might be possible to obtain a parallax determination. According to Humboldt, the conditions there should be especially favorable, for in his "Cosmos", Vol. I, he remarks: "I have seen it shine with an intensity of light equal to the Milky Way in Sagittarius". Judging from our photographs, this should answer to an intensity of the zodiacal light from 5 to 10 times greater than that which we observed in Egypt and Soudan.

As we thus obtained a negative result with regard to the pulsations of the zodiacal light by our observations, we determined instead to study the magnetic curves at Greenwich for the period during which Jones had carried on his observations. This observatory is presumably the only one in which, as early as 1853, continuous magnetic registerings were made.

On going through Jones' observations, we find a considerable number of days on which he seems to have noticed pulsations of light. On two occasions he is absolutely convinced of their existence, namely, on the 30th January, and the 27th March, 1854. On the first of these we read, in italics: "There can be no doubt that there are pulsations in the zodiacal light"; on the second he remarks: "It *certainly* does pulsate".

The curves at Greenwich are drawn by instruments with great sensitiveness and comparatively long time-periods, so that possible magnetic pulsations would be more easily discovered than by the ordinary daily magnetograms. But the curves have been faint and have been gone over with ink, and have thus lost something of their character.

There are here reproduced four plates with magnetograms from Greenwich, first, two answering to the above-mentioned dates, the 30th January and the 27th March, 1854, belonging to Jones, next, two answering to the 25th February and 25th April of the same year, when in Jones' observations too, distinct pulsations are recorded. This comprises the most certain pulsations observed in the zodiacal light.

We have further chosen 5 days with light-pulsations, for which we have copies of the curves at Greenwich, which distinctly show magnetic pulsations simultaneously with those observed in the zodiacal light.

---

(1) Phys. Zeit. 1908, p. 355.

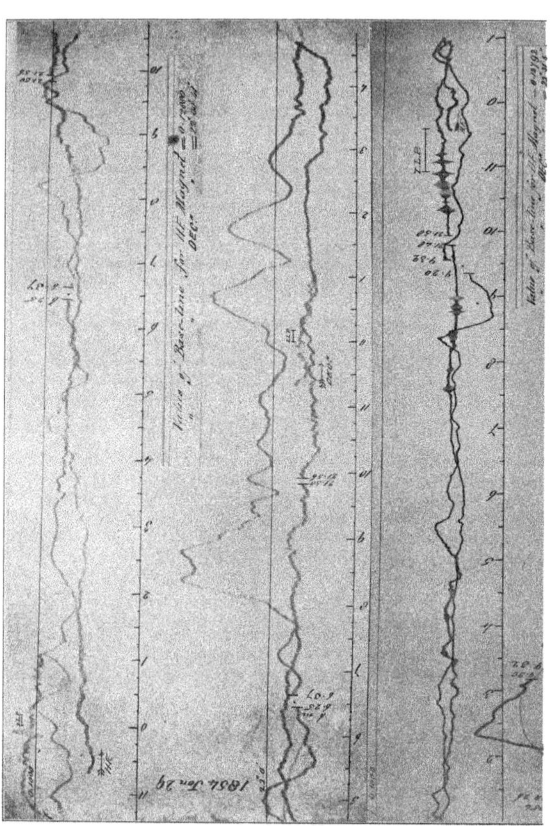

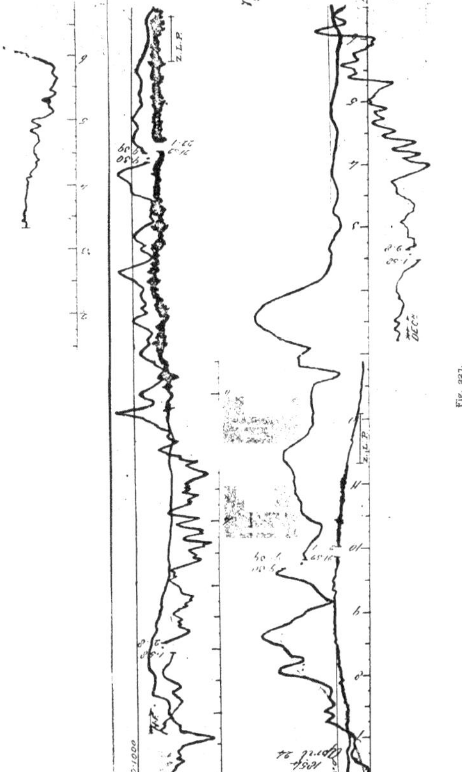

Fig. 227.

It should be stated, however, that on several occasions when JONES believes he has seen pulsations, no corresponding magnetic pulsations were traceable at Greenwich. Whether the reason of this is that the original photographic curves have been much obliterated, can scarcely be determined.

On the Plates, where Göttingen mean solar time is employed, the time is marked when JONES has observed pulsations in the zodiacal light (Z. L. P.).

It will be seen that this period—that for the 27th March—falls at the end of a series of exceedingly distinct magnetic pulsations, which are in quick-run magnetograms usually called Eschenhagen oscillations. These are especially distinct in H, but they would also certainly have been distinct in D in the original curve. Here, however, they have been fainter, and the curve has been drawn principally as a mean line, whereby these oscillations have been eliminated.

On the 30th January too, the pulsations occur at the end of a series of particularly characteristic magnetic pulsations. The latter are especially distinct in the period immediately preceding Jones' observations, but also undoubtedly seem to continue, although less powerful, during that period. The curves here, however, have been somewhat obliterated and are difficult to follow in detail.

In the magnetic curves on the other two Plates, there are rapid oscillations of comparatively long duration. These, however, are not such typical elementary waves as the preceding ones.

We finally append JONES' notes from the first two days mentioned. *a*, *b*, and *d* here indicate the special boundaries of the zodiacal light, which are given in the figures in his work.

JANUARY 30th 1854: EVENING. Lat. 26° 10′ N. Lon. 127° 42′ E.

Sun set $5^h\ 38^1/_2{}^m$.

Stronger Light $7^h\ 50^m$. &c: Diffuse, [7]$^h\ 50^m$.

Sun's Lon. 310° 20′.

There can be no doubt that there are pulsations in the Zodiacal Light. I noticed them last evening (the sky being very clear); but, it being Sunday, made no particular record of them. They were, however, distinctly to be seen; and when I called the attention of one of the quartermasters to them, he very easily made them out. His language about the Light was: "Now it seems to be dying away"; "now it is brightening again", &c. All this applied, however, only to the Stronger Light: it occurred between $7^h\ 30^m$ and 8 o'clock. This evening I was on the careful lookout for them, and, with watch in hand, made record of the changes and their times. Clouds interfered till $7^h\ 50^m$, when, this part of the sky having cleared up, I got observations. The pulsations were very distinct; observable, however, only in the Stronger Light. This, at $7^h\ 50^m$, had its boundaries as in the line *b* (see chart), and was very bright: $7^h\ 52^m$ it had sunk to the boundaries marked *a* and was very dim: $7^h\ 54^m$ had risen to *b* again, and was bright: $7^h\ 55^m$ at *a*, and very dim: $7^h\ 56^m$ at *b*, and brigth: $7^h\ 57^m$ at *a* and very dim: $7^h\ 58^1/_2{}^m$ at *b*, and bright: $7^h\ 59^1/_2{}^m$ still at *b*, and bright: it seemed now to be permanent at *b*; but clouds soon after spread over the sky, and shut out everything from sight.

These pulsations, in order to be seen, seem to require that the ecliptic should be at a high angle with the horizon; at which time the Stronger Light is very brilliant.

MARCH 27th 1854: EVENING. Lat. 35° 26′ N.: Lon. 139° 42′ E.

Sun set $6^h\ 12^1/_2{}^m$.

Stronger Light $\left\{ \begin{matrix} 7^h\ 30^m \\ 9^h\ 30^m \end{matrix} \right\}$ Diffuse at $7^h\ 30^m$, &c.

Sky remarkably clear. The following are my notes:—$7^h\ 15^m$ a whiteness running up with Zodiacal Light boundaries as far as the Pleiades, but its limits are not distinct: $7^h\ 24^m$, the light decided, but its boundaries not reliable: $7^h\ 30^m$, got boundaries of both Diffuse and Stronger Light—the latter, then, strong up to *b*, and gradually tapering, dimming off to *c*.

h. m.
At 7 35, at *a*, and dim.
7 38, do. do.
7 39, at *b*, and bright.
7 43, do. do.
7 44, at *a*, and certainly dimmed.
7 45, at *b*, and bright.
7 47, at *a*, and dim.
7 48½, do. do.
7 49, brightening.
7 50, at *b*, and bright.
7 51, at *b*, and quite bright.
7 52¼, dimming.
7 52½, at *a*, and dim.
7 53½, brightening.

h. m.
At 7 54½, at *b*, and bright.
7 55½, at *b*, and quite bright.
7 57¼, at *a*, and quite dim, as if dying away.
7 58¼ do. do. do.
7 58¾, brightening.
7 59½, at *b*, and bright.
8 0, do. and quite bright.
8 3, brighter than at any time yet, and has clearly ascended to the Milky Way by lines *d d*.
8 4½, dimmed and sunk to *b*.
8 7, brightening.
8 8, very bright, and at *d d*.
8 15, still as last, and seems to be permanent now.
9 30, boundaries to *x*.

I think I can know when it is going to be permanent, by the upper portion of the Light brightening more than at any time previously in the evening, and the strong brightness ascending higher. The first appearance of the Zodiacal Light seems to be a white light — *i. e.* when the twilight has not quite gone; afterwards it changes to a warm yellowish light. The reverse of this happens in the morning. The Diffuse Light is now very dim; in the morning it is very strong, for it.

This evening was remarkably fine for observations, and in my notes is the remark: "It *certainly* does pulsate".

123. Only one abruptly-beginning magnetic disturbance occurred in the period when we were observing with "quick-run" registrations in Assouan, namely, the 9th April.

I have, unfortunately, not received any intimation of quick-run registrations having been taken except in Samoa, where Prof. Dr. ANGENHEISTER commenced the registrations on April 10, *i. e.* one day too late. Mr. TITTMANN, superintendent of the U. S. Coast and Geodetic Survey, has been good enough to send me some copies of slow-run registerings for April 9 from Cheltenham, Porto Rico, Tucson, Sitka and Honolulu. Of these, the curves from Honolulu (158° W) and Porto Rico (65° W) are of special interest, because these stations, together with Assouan (33° E) form a particularly happy distribution of stations about the Earth.

Figure 228 shows that on this day an equatorial perturbation occurred, the character of which is very similar at the three stations. The times of commencement in H are as follows:

Honolulu	Porto Rico	Assouan
$10^h\,20^m$, 7 p. m. Gr. M. T.	20^m, 8	$20^m\,44^s$

The changes in D at the same time were very small, as might be expected would be the case with this kind of perturbation.

The first notices of time are given in a letter from the Coast and Geodetic Survey, — the last value is found by the "quick-run" magnetograms from Assouan as shown by the magnetogram.

The last time-determination is given in seconds, because of the greater accuracy that can be reckoned upon in "quick-run" registerings.

The time-marks here, which refer to the central point for the obliterated parts, are certainly correct to one second, but a greater uncertainty arises when it is a question of determining when the perturbation shall be said to have commenced. I consider we may be safe when we estimate the possible error at ± 4 seconds. But the values of the slow-run magnetogram *lie within this margin at Honolulu and Porto Rico,* where, however, the readings are naturally not so trustworthy as those of the quick-run magnetograms from Assouan.

The curves from Sitka, Tucson, and Cheltenham show that the perturbations in those places have had a somewhat different character from those at the three first-named stations, for it appears as though a magnetic polar storm interferes. The curves for D and V show the same thing.

The times we have been given from the Coast and Geodetic Survey for these stations are: — for Sitka $10^h\ 21^m$, Tucson $10^h\ 20^m$, and Cheltenham $10^h\ 21^m$, 9, and these refer to the "larger displacement" in H. This occurs shortly after the first abrupt beginning, and the times are, as may be seen, with the exception of Tucson, slightly greater than the others.

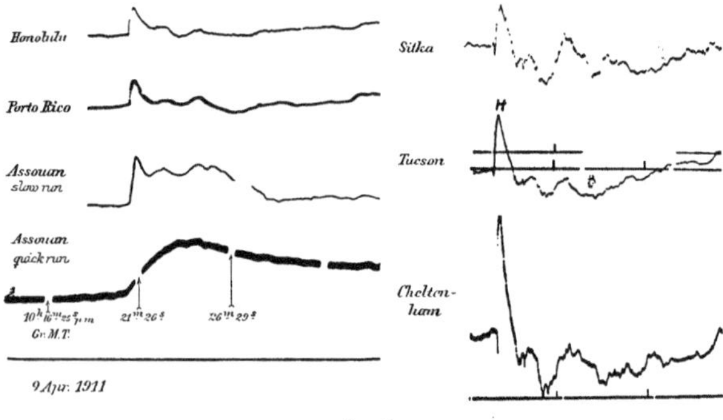

Fig. 228.

As regards Tucson, we notice that the first time-mark is considerably smaller than the later ones; for this reason, I think, this value should perhaps be taken with some reservation.

In Trondhjem, under the direction of Professor SÆLAND, "quick-run" registerings were made simultaneously with our observations at Assouan, though not between 10 and midnight, Greenwich mean time. As the above perturbation occurred just in this period, we unfortunately have only "slow-run" registerings from this station.

At this station the polar character of the storm is distinctly apparent, as might be expected from so high a latitude.

On the occasion of the magnetic storm we are here studying, the similar sudden changes occurred around the terrestrial equator *simultaneously,* within the limits of error in the observations.

When several observations of such magnetic storms around the equator obtained by quick-run registerings, are available, as I hope may soon be the case, this important question of simultaneity will be finally determined.

It may be of interest in connection with this to call to mind that in 1900, quick-run registerings were taken simultaneously in Potsdam and at my observatory at Haldde, near Bossekop. In my work "Expédition Norvégienne de 1899 - 1900 pour l'étude des aurores boréales" (Christiania, 1901), photographs of these registerings are given, which show that corresponding small sudden alterations in D were simultaneous within three seconds in Potsdam and Bossekop.

According to my theories of magnetic storms, it might be expected that sudden similar magnetic changes which occur in different parts of the earth arise rather simultaneously. When the sun suddenly sends forth a strong pencil of cathode rays towards the earth, this pencil, owing to earth-magnetism, will be broken up in such a way as to form different partial systems of magnetic impulses — polar and equatorial. The various groups of rays have to travel different way-lengths in space before reaching their nearest to the earth, and may arrive at very different regions of the earth for the different groups. But the difference in time between the various impulses affecting any particular locality on the earth can scarcely be more than a couple of seconds, while the difference in the intensity of the effects can be very considerable. We know of corresponding phenomena in the case of Aurora, which will be treated later on.

COMETS' TAILS.

124. The theory here set forth, of the emanation of electrical corpuscle-rays from the sun, might be thought to present a new point of departure in the study of the physical nature of comets, and more especially of comets' tails.

It seems evident from their spectrum that comets consist of an accumulation of cosmic dust, with various carbonaceous substances, concentrated about one or more nuclei, which are surrounded by a highly rarefied vaporous envelope in which possibly carbonaceous gases are comparatively strongly represented.

As regards more especially the particular phenomenon of the comet's tail, it has been found that it does not make its appearance until the comet aproaches the sun, and is most highly developed a little while after passing the perihelion.

If, now, this vaporous envelope surrounding the more solid part of the nucleus, be exposed to the radiation of a multitude of corpuscle-rays from the sun, it could easily be imagined that in their passage through the exceedingly rarefied gas, these rays would change their nature. The simplest assumption one is inclined to make is that some of the corpuscles that pass through the coma have acquired an appendix of gaseous atoms or molecules, which have thereby become luminous. As these rays may be supposed to continue their way in more or less the same direction as before, but with a different velocity and mass, this would be a comparatively simple explanation of the luminous tail of the comet, which is almost always directed away from the sun.

Arrhenius has also, as we know, maintained a similar theory, only that instead of electric corpuscle-rays of the kind here considered, he imagines rays of electrically-charged atoms, moving under the influence of light-pressure.

It is possible, however, that there are also other, just as natural, ways of looking at the matter. It might be imagined that after great heating by direct insolation, the comet is charged negatively by cathode-rays from the sun, and that the charging reaches so high a potential that the comet discharges itself electrically, so to speak in the direction of its own shadow. These discharges may also be imagined to be due to some extent to an emission of secondary rays from the cosmic dust of the comet.

I have been led to this thought by experimental analogies which will be described farther on. Answering to the idea that a comet is an accumulation of carbonaceous cosmic dust almost without atmosphere, I have carried out experiments in which the cathode in a vacuum-tube consisted of a carbonaceous material. The most recent investigations of the comet-spectrum seem to indicate that the radiation from a comet may be compared to that from a cathode in a Crookes' tube (Deslandres, Fowler).

TABLE CVIII.

1863 IV. Passage of Perihelion Nov. 9

	Nov. 12	Nov. 13	Nov. 14	Nov. 15	Nov. 17	Nov. 22	Nov. 25	Dec. 3
r	0,709	0,711	0,714	0,717	0,721	0,725	0,755	0,778
α	48°,80	48°,47	48°,23	48°,08	48°,01	49°,27	50°,84	56°,76
β	0°,79	2°,99	5°,19	7°,36	11°,63	21°,79	27°,42	40°,53
x	0,010	0,037	0,065	0,092	0,146	0,280	0,358	0,556
l	0,053	0,042	0,059	0,095	0,081	0,148	0,028	0,034

1862 III. Passage of Perihelion Aug. 22

	Aug. 16	Aug. 18	Aug. 21	Aug. 24	Aug. 27	Aug. 29	Aug. 31	Sept. 12
r	0,968	0,965	0,963	0,963	0,967	0,970	0,975	1,028
α	34°,10	30°,55	26°,17	22°,54	19°,92	19°,19	19°,36	32°,78
β	39°,59	36°,77	32°,99	29°,17	25°,05	22°,37	19°,78	4°,45
x	0,617	0,578	0,524	0,469	0,409	0,369	0,330	0,080
l	0,010	0,062	0,069	0,114	0,166	0,134	0,054	0,021

1861 II. Passage of Perihelion June 11

	June 30	July 2	July 4	July 6	July 8	July 10	July 12	July 14
r	0,897	0,912	0,929	0,946	0,965	0,984	1,005	1,026
α	4°,15	6°,86	9°,97	13°,12	16°,23	19°,27	22°,24	25°,10
β	7°,18	10°,27	13°,26	16°,13	18°,89	21°,56	24°,12	26°,57
x	0,112	0,163	0,213	0,263	0,312	0,362	0,411	0,459
l	0,277	0,470	0,445	0,565	0,369	0,375	0,297	0,186

1860 III. Passage of Perihelion June 16

	June 24	June 25	June 28	July 2	July 6	July 8	July 11	July 12
r	0,447	0,470	0,543	0,647	0,751	0,803	0,880	0,906
α	46°,79	42°,30	31°,87	23°,94	21°,53	21°,88	23°,47	24°,25
β	42°,76	38° 50	27°,68	16°,97	9°,17	6°,01	1°,98	0°,78
x	0,304	0,292	0,252	0,189	0,120	0,084	0,030	0,012
l	0,262	0,189	0,164		0,063	0,042	0,007	0,004

1858 V. Passage of Perihelion Sept. 30

	Sept. 16	Sept. 19	Sept. 22	Sept. 24	Sept 26	Sept 28	Sept. 30	Oct. 2	Oct. 4	Oct. 6	Oct. 8	Oct. 11	Oct. 14	Oct. 16	Oct. 19
r	0,658	0,629	0,606	0,594	0,586	0,580	0,579	0,580	0,586	0,594	0,606	0,629	0,658	0,680	0,716
α	88°,25	79°,34	69°,92	63°,44	56°,92	50°,53	44°,38	38°,74	33°,89	30°,22	28°,08	28°,02	31°,24	34°,50	40°,17
β	61°,16	62°,56	61°,86	60°,06	57,26	53°,63	49°,39	44°,73	39°,84	34°,87	29°,94	22°,77	16°,05	11°,87	6°,08
x	0,576	0,558	0,534	0,515	0,492	0,467	0,439	0,408	0,375	0,340	0,302	0,243	0,182	0,140	0,076
l	0,185	0,177	0,172	0,166	0,149	0,261	0,323	0,371	0,355	0,471	0,472	0,543	0,319	0,102	0,068

1857 V. Pass. of Perih. Sept. 30

	Sept. 12	Sept. 15	Sept. 17	Sept. 22
r	0,694	0,658	0,636	0,593
α	48°,59	55°,16	73°,74	89°,77
β	58°,82	60°,14	60°,22	56°,64
x	0,594	0,571	0,552	0,495
l	0,013	0,029	0,058	0,098

1853 III. Passage of Perihelion Sept. 1

	June 26	Aug. 3	Aug. 20	Aug. 23	Aug. 25	Aug. 26	Aug. 28	Aug. 30
r	1,566	0,838	0,470	0,410	0,374	0,358	0,331	0,313
α	105°,21	109°,68	84°,49	73°,96	65°,69	59°,95	48°,43	34°,63
β	29°,83	46°,35	59°,37	57°,12	52°,55	49°,15	40°,07	28°,30
x	0,779	0,606	0,405	0,344	0,297	0,271	0,213	0,148
l	0,004	0,011	0,035	0,117	0,170	0,174	0,181	0,203

1811 I. Pass of Perih. Sept 12

	June 10	Oct. 15	Jan. 1 1812
r	1,815	1,173	2,005
α	111°,62	67°,86	125°,30
β	·22°,52	73°,16	32°,10
x	0,695	1,123	1,066
l		0,6—0,75	

1817 I. Passage of Perihelion March 30

	March 5	March 8	March 9	March 10	March 15	March 16	March 17	March 18
r	0,900	0,823	0,797	0,770	0,629	0,599	0,569	0,537
α	57°,94	58°,48	58°,55	58°,62	58°,60	58°,39	58°,18	57°,79
β	41°,84	41°,50	41°,38	41°,24	40°,28	40°,01	39°,71	39°,36
x	0,601	0,545	0,527	0,508	0,407	0,385	0,363	0,341
l	0,004	0,008	0,011	0,013	0,028	0,054	0,060	0,064

1618 III. Passage of Perihelion Nov. 8

	Nov. 29	Dec. 1	Dec. 9	Dec. 17	Dec. 22	Dec. 24	Dec. 29	Jan. 7
r	0,669	0,708	0,864	1,017	1,111	1,148	1,240	1,400
α	19°,88	21°,36	25°,31	26°,86	27°,17	27°,16	27°,07	27°,08
β	3°,28	5°,12	10°,68	14°,36	16°,07	16°,67	17°,97	19°,81
x	0,038	0,063	0,160	0,252	0,308	0,329	0,382	0,474
l	0,28—0,35	0,205	0,43	0,37	0,072	0,315	0,45	0,125

1769. Passage of Perihelion Oct. 7

	Aug. 9	Aug. 24	Aug. 27	Aug. 30	Sept. 2	Sept. 3	Sept. 4	Sept. 5	Sept. 7	Sept. 9	Oct. 25	Nov. 1	Nov. 8	Nov. 15	Nov. 17	Nov. 21
r	1,554	1,260	1,198	1,134	1,068	1,046	1,023	1,001	0,955	0,908	0,655	0,835	1,001	1,155	1,198	1,280
α	41°,98	30°,81	28°,42	26°,42	24°,63	24°,20	23°,59	23°,09	22°,36	21°,31	108°,32	107°,86	109°,56	112°,35	113°,31	115°,40
β	5°,57	3°,04	2°,38	1°,64	0°,81	0°,51	0°,21	−0°,12	−0°,82	−1°,59	43°,03	42°,93	42°,52	42°,06	41°,93	41°,68
x	0,151	0,067	0,050	0,032	0,015	0,009	0,004	−0,002	−0,014	−0,025	0,447	0,569	0,676	0,774	0,800	0,851
l	0,199	0,061	0,225	0,335	0,480	0,524	0,533	0,621	0,618	0,566	0,070	0,279	0,107	0,063	0,102	0,114

In this connection it would be natural first to find out whether the length of a comet's tail has any special relation to the distance of the comet from the plane of the sun's equator, since we have seen, in treating of the zodiacal light, that it must be assumed that there is a ring of radiant matter round the sun in that plane.

Table CVIII gives the results of a series of calculations that have been made in order to make this matter clear. Here β is the comet's heliocentric latitude, x its distance from the plane of the

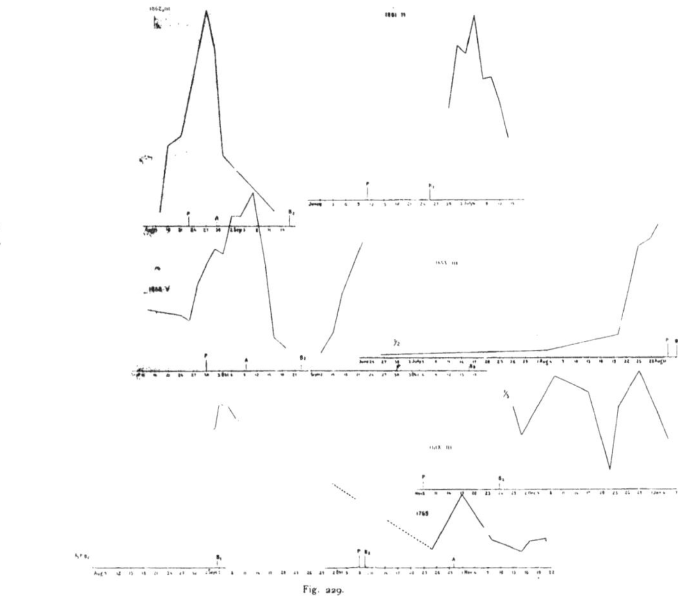

Fig. 229.

its distance from the sun, l the length of its tail, and α the angle between the radii vectores of the comet and the earth. r, x, and l are measured in radii of the earth's orbit, l being only approximate, as the tail is imagined to extend radially out from the sun, which in this connection is sufficiently accurate.

The orbit-elements employed in the calculations are taken from PH. CARL'S "Repertorium der Cometen Astronomie".

The angle between the plane of the sun's equator and the ecliptic is put at 7°, in accordance with Arrhenius' "Lehrbuch der cosmischen Physik" (p. 153); and the angle between the line of intersection of these two planes, and the line of equinox is put at 70°, also in accordance with the last-named authority.

It is impossible, however, to discover from the above Table any distinct increase in the length of the tail when the comet is in the vicinity of the plane of the sun's equator. The greatest length of tail is always found after the passage of the perihelion, and this indicates that a prominent part is played by evaporation of the constituents of the nucleus, brought about by the radiant heat of the sun.

It would appear, however, from the graphic representation (fig. 229) of the variation with time in the length of the tail, that it is not the passage of the perihelion alone that is decisive. The passage of the perihelion is marked P, the time when $\beta = 0$ is marked B_1 or B_2, answering respectively to the first and second intersections with the plane of the equator. Finally we have the point of time, A, at which the angle α has its minimum, in those cases in which this point falls within the period of time under consideration.

It may be remarked as a general characteristic, that the curves about the maximum of length of tail have very steeply ascending and descending branches. Further, this maximum sometimes occurs a comparatively short time after the passage of the perihelion – e. g. the comet 1862 III—and sometimes a comparatively long time after—e. g. the comet 1861 II. On the whole, the length of this interval varies considerably, and there does not appear to be any simple connection; the impression is rather, that the great development in the length of the tail about the maximum takes place at the time when the comet is passing certain especially favorable strata or zones. This is especially marked, for instance, in the comet 1862 III.

There are two other circumstances in particular to be considered here, namely, whether the light of the moon can obliterate the faint light of the comet's tail, and whether, during the period under consideration, the tail of the comet has moved much farther from, or much nearer to, the earth.

In only the first of the cases considered is it noted that the light of the moon has interfered, and this is shown in the curve.

With regard to the second of the above-mentioned circumstances, it is easy to estimate from [illegible] angles α and r whether the distance from the tail of the comet varies so greatly as to have any [illegible] ficance in judging of the light. In no case does it appear to exert any real influence during the [illegible] about the various maxima.

On closer inspection it appears that the great development of the tail occurs most frequent[illegible] certain distance from the sun's equator, answering to values of β of between 15° and 30°.

In this connection, one recalls how the sun-spots also occur most frequently in about 20° [illegible] centric latitude.

The comet 1618 III exhibits a peculiar circumstance, the curve for the length of its tail [illegible] distinct intermediate minimum. This might be due to the comet's having passed through two lay[illegible] pencils of rays from the sun, one immediately after the other; but it is perhaps just as likely th[illegible] peculiar condition might be due to internal causes in the comet, or to the disturbing influence of [illegible] light, or to unfavorable atmospheric conditions.

It would be natural, therefore, to compare the above-mentioned layers that were favorable to [illegible] development of comet's tails with the pencils of the strongest and magnetically stiffest corpuscle-rays which we imagine to emanate from the region surrounding the sun-spots, and which, when they sweep past our earth, produce powerful magnetic disturbances. It may be that it is these very rays, with their abundance of energy, that can charge the comet mass to a high negative tension, and thus occasion the secondary electric discharge from the comet into space.

One circumstance that speaks strongly in favour of a hypothesis such as this, is the greater development thought to have been found in years of sun-spot maxima than in years of sun-spot minima. This has been demonstrated, for instance, in Encke's comet, by Berberich and Bosler, the latter having given an exceedingly interesting graphic representation of this condition, which is reproduced here.

The agreement, as will be seen, is so striking that it seems to leave little room for doubt that we here have phenomena that must be intimately connected with one another.

For the purpose of seeing and studying how a substance containing carbon is discharged as a cathode in a vacuum-tube, I have made, as already mentioned, numerous experiments with cathodes of ordinary coal, coke, graphite, and piceïn over a metallic cathode. I have further employed an extremely

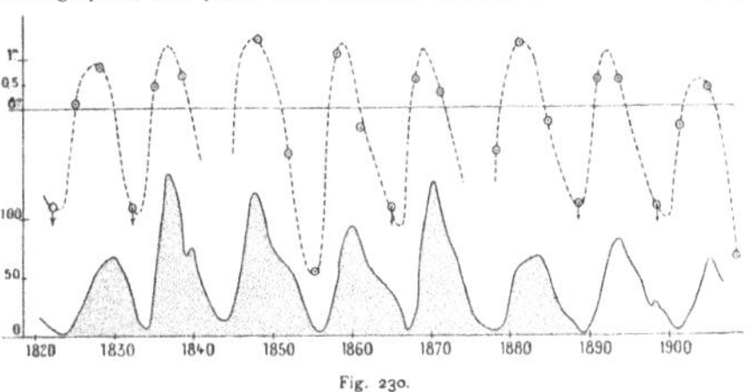

Fig. 230.

fine jet of CO_2, which was introduced through a very narrow capillary tube, and flowed out from the end of a narrow silver tube which served as cathode.

I succeeded several times in making this jet luminous, so that it had the appearance of a fine needle of light shooting out from the cathode, sometimes as much as 5 cm. in length.

A cathode of coal also sent out similar long needles of light from various points on its surface, round which the coal even became glowing.

Piceïn emitted long, thin pencils of light, often more than 10 cm. in length, one after another, as if by violent eruptions. These light-phenomena gave the impression that the electric discharge from

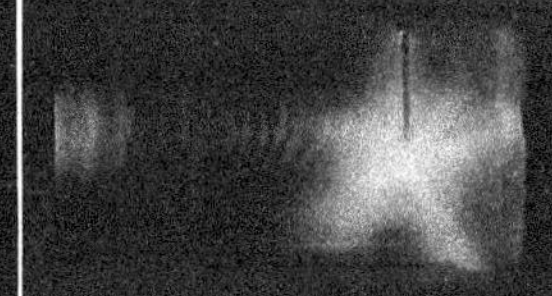

Fig. 231.

both the coal cathode and the cathode with piceïn, was accompanied by eruptive outbreaks of gaseous rays, that were made luminous in the same way as the above-mentioned carbonic acid jet. Fig. 231, 1 and 2 show discharges of this kind.

From a cathode of graphite there came long, steady pencils of light, which greatly resembled the so-called eruptions or jets in comets.

Fig. 231, 3 shows an experiment with graphite.

In these experiments with cathodes containing carbon, the rapid disintegration of the cathode was especially remarkable. In the course of two or three minutes, large dark patches appeared on the glass walls just where the long pencils of light had come in contact with them. Fig. 1 shows an instance of this in the two dark tongues side by side above on the right. To this phenomenon, which is of peculiar importance to our theory, we shall have frequent occasion to return, for instance in the article on Saturn's ring, where we assume that material particles are constantly being emitted in the plane of the ring by electric evaporation (disintegration), analogously to certain experimental observations to be described farther on.

In connection with the above-mentioned experiments with carbonaceous cathodes, experiments were also made with cathodes of platinum thinly coated with lime. This was for the purpose of finding out whether rays from a cathode such as this — which, as is known, emits exceedingly soft rays — might be repelled by electric forces, and bent right round, just as the radiation from the head of a comet appears to be by apparent repulsion from the sun.

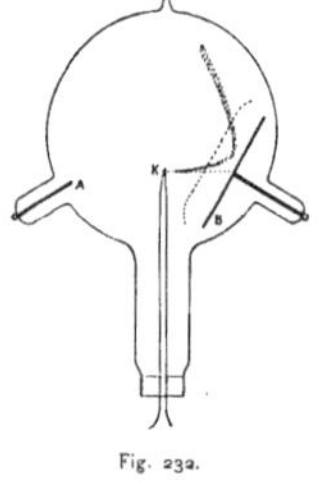

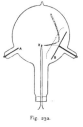

Fig. 232.

Fig. 231, 4 shows how the rays from a coated platinum cathode such as this, turn away from a large cathode-plate of brass on its right. The bending of the rays was sufficiently evident, and changed with changes in the tension employed upon the brass cathode; but there was no appearance of any backward-streaming as in the tail of a comet, as the light ceased at a short distance from the cathode. It is very possible that better results might be obtained by an arrangement somewhat different to the one here employed. In J. J. THOMSON's "Conduction of Electricity through Gases", Second Edition, p. 632, the diversion of these rays by electric force is illustrated by a drawing, reproduced here in fig. 232, which shows how the rays can be turned right back.

It will be of interest for the present question to cite, and reproduce a drawing of, an experiment described by J. STARK in "Die Elektrizität in Gasen" published in WINKELMANN's "Handbuch der Physik", B. 4, p. 582: "If a cathode-ray with a certain initial velocity enters an electric field that is at right angles to its direction, it will be deflected out of its course from points of lower to points of higher tension. If its initial velocity is very small, it soon takes exactly the direction of the electric line of force in which it lies; if, on the contrary, it is great, it will be deflected more or less in the direction of the line of force, the less so the greater its velocity, the more so the greater the strength of the field.

"Let us consider the case in which rays from *one cathode* fall upon a *second*. In figs. 233 a & b, S is the transverse section of a metal pin that can be connected with the cathode outside the tube. If, together with the wire-anode beside it, it is connected with the earth, the primary rays cast a sharp shadow of it (233 a). This immediately increases when the pin is connected with the cathode; for there is then formed about it the powerful electric field of the dark space of the cathode, and through this the approaching cathode-rays are turned aside (233 b)".

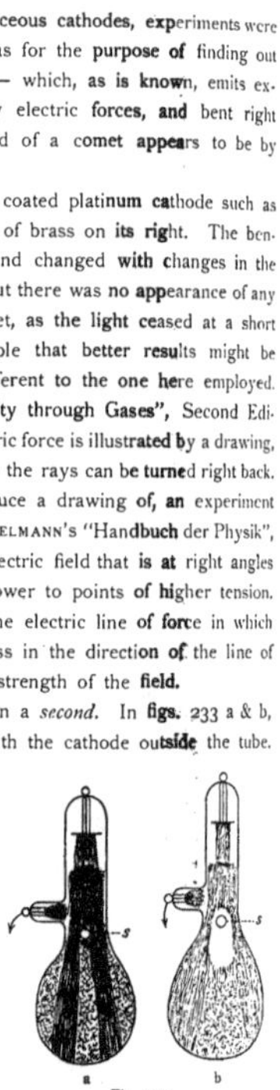

a b
Fig. 233.

According to this, it might well be imagined that luminous pencils of rays, emitted by electric discharges from a comet, are bent backwards by the electric force of cathode-rays from the sun, in such a manner that the discharges pursue their course almost in the direction of the comet's shadow, forming approximately a cone, possibly on account of the mutual repulsion of the pencils of rays emitted.

Another circumstance favorable to the assumption of the existence of such negative discharges from comets, is that of the various envelopes separated by dark interspaces so often observed in the heads of comets. Fig. 234 shows the head of Donati's comet (1858). For several weeks the coma exhibited in unrivalled perfection the development and structure of concentric envelopes. It is easy to produce, round a globe as cathode in a large vacuum-tube, several concentric luminous envelopes separated by dark spaces. These different envelopes are more distinctly seen when the globe used as cathode is magnetised. In this case the originally spherical envelopes will be flattened so as to form a ring in the magnetic equator. Fig. 235 gives a representation of such an experiment. Such envelopes, as we know, contract or expand according as the gas-pressure in the vacuum-tube becomes greater or less. The very singular phenomenon of the contraction of the comet's head with the approach of the comet towards the sun can be reasonably explained by this view. Instead of expanding, as one would naturally expect it to do under the action of solar heat, the comet's head contracts when near the sun, just because the gas pressure about the comet becomes higher there, and the electrically-formed luminous envelopes therefore contract.

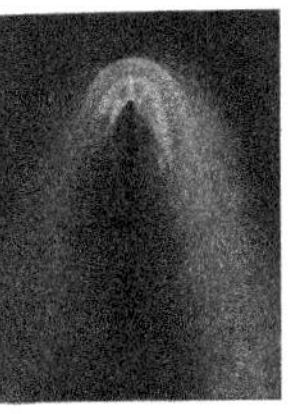

Fig. 234.

On some occasions comets have been furnished with several tails in a manner that is not quite easy to explain by the assumption that an emanation of tail-material from the comet could directly give rise to all the tails.

Figs. 236 a & b show respectively the famous Donati's comet (1858) from a drawing by BOND, and the comet of 1744 by Mlle KIRCH at the Berlin Observatory. It seemed to me it would be worth while examining whether all the luminous streaks or tails that were seen were perhaps not separate tails, but might possibly be compared with positive strata in the electric discharge from the negative comet-head such as in the discharge represented in fig. 231, 2.

I have taken two ways for determining this. First the angle α was calculated, the angle that a plane through the centre of the earth and a luminous streak in the tail, formed with the plane of the comet's orbit. The result for Donati's comet was

$$\alpha = 58.99^\circ$$

for a streak that passed over α and $\varkappa$ *Coronæ Borealis* on October 9th. The calculation is based upon a description by WINNECKE, quoted in Bond's "Account of the Great Comet of 1858" (p. 61), Annals of the Astronomical Observatory of Harvard College, Vol. III, Cambridge, 1862.

It was further found that

$$\alpha = 69.53^\circ$$

for a streak that issued from the head, and kept separate from the tail, passing over δ *Serpentis* and β *Herculis*, according to a drawing of the comet on October 9th (l. c.).

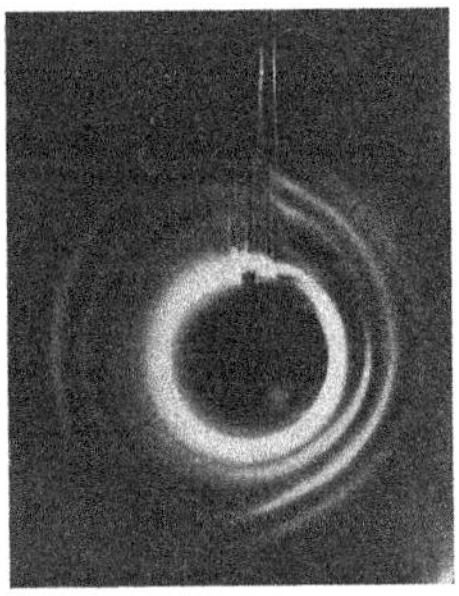

Fig. 235.

For the comet of 1744 it was found that

$$\alpha = 87.36^\circ$$

for a streak that, according to a description by LOYS DE CHÉSEAUX at Lausanne, of the appearance of the comet on the night of the 7th March (quoted in JÆGERMANN'S "Mechanischen Untersuchungen über Cometenformen", pp. 397 & 398), passed through the middle of EQUULEUS and ended in a point of which the longitude was 319° 55′, and latitude + 34° 35′.

In the second place a calculation was also made of the angle between a line from the earth to a middle-point in the streak, and a perpendicular in the plane of the comet's orbit to the streak at this point. The result was

$$\alpha = 89.08^\circ$$

For this a drawing from JÆGERMANN's above-mentioned work, Pl. VII, was employed.

In addition to this streak, which was in the middle, calculations were also made for one on each side, No. 3 on the left and No. 3 on the right, according to M. KIRCH's observation. The results were

$$\alpha = 74.29^\circ \text{ and } \alpha = 84.64^\circ.$$

The idea upon which the investigation was based was that if the streaks of light observed in the comet's tail answered to positive strata in a discharge, one would expect these layers to be at

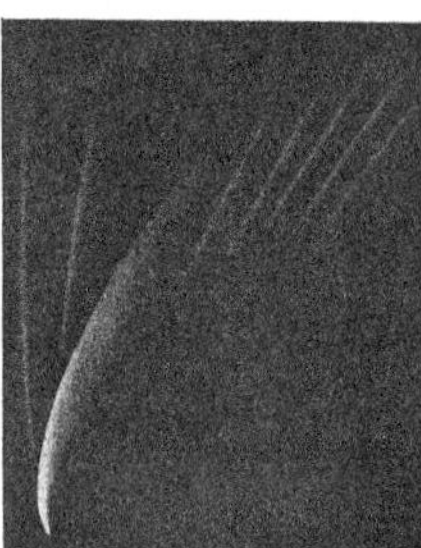

Fig. 236.

right angles to the axis of the comet's tail, which, in its turn, would be supposed to lie in the plane of the comet's orbit.

The calculations for the comet of 1744 harmonised, the angle being nearly 90°, but for Donati's comet, for which the calculations were made later, a negative result was obtained.

It is not, moreover, so entirely certain that the projections of possible positive layers that might be seen from the earth, answer to a plane at right angles to the plane of the comet's orbit. The layers are not always plane in reality (see fig. 231, 2).

The great disintegration of a cathode coated with some carbonaceous substance, by which all the carbon-particles may be thrown off from the cathode in the course of a few minutes, recalls a phenomenon observed with regard to comets, namely, that they gradually lose their ability to form tails. *Bredichin* says([1]) of the comet 1873 V, for instance, that "the emissions appear to be exhausted before the perihelion"; and of the telescopic comets he says that "as a rule it must be admitted that in the periodic comets with short period, the force that produces the emissions and the tails is relatively exhausted".

From what we have seen before, the explanation of the phenomenon of the comet 1873 V, according to our view, is that the comet had come out of the main body of cathode-rays from the sun — the active layer — before reaching its perihelion.

([1]) See Jægermann, l. c., p. 229.

It would be interesting to find out whether the pretty results obtained by *Bredichin* in his mechanical investigations of comets' tails could be made to harmonise with the theory of electric discharges through rarefied gases. The formation of several distinct tails from one comet would then possibly have causes corresponding more or less to those of the formation of the various distinct pencils of cathode-rays in an electric or magnetic field (cathode-ray spectrum).

It is now generally assumed that comets belong to our solar system, because no comet has an undoubted hyperbolic orbit. This also agrees with the fact that the spectra of comets exhibit on the whole a great similarity.

In a subsequent article we shall see how our theory of an electric radiation of matter from the sun can give a satisfactory explanation of the comet's formation, even when its orbit carries it to a distance of 1000 or 10 000 astronomical units from the sun.

125. Halley's Comet, May, 1910. An exceptionally favorable opportunity of testing the views here brought forward regarding comets' tails presented itself in May, 1910, when Halley's comet crossed the sun's disc at so comparatively short a distance from the earth that there was a possibility of the earth's passing through the comet's tail. When a magnet as great as the earth came into the comet's tail, there would surely be magnetic effects to be observed upon and from the tail, if the latter consisted of some kind of electric corpuscle-rays.

It was Herr Krogness, who, happening to read in an astronomical journal that Halley's comet would come so near to the earth, suggested that we should go up to my observatory on Haldde Mt. for the purpose of studying the possible effects of the passage. This was arranged, when I had succeeded in getting a friend of mine, Herr Schibsted, to share the expenses equally with me.

In order to secure a more widespread interest in these observations, I sent out, in March, 1910, the following circular to a number of observatories and a few periodicals (e. g. 'Nature', April 21). The figures that were reproduced in the circular are here omitted, the reader being referred to the same or better figures already printed in the present work.

"I beg to direct your attention to the following.—

"It is my intention, at Kaafjord in Finmarken (in the N. of Norway), together with my Assistant Mr. O. Krogness, to take magnetic and atmospheric observations during the period 7th May to 1st June next in connection with the transit of Halley's comet across the sun's disc on the 18th—19th May.

"The thing is, that it is conceivable that the tail of the Comet may chiefly consist of electrical corpuscular rays, and, if this be so, we would expect that these rays, owing to Earth magnetism, would be drawn in, in the Polar regions, in zones analogous with the Aurora zones, assuming the tail of the comet to be of sufficient length to reach the Earth.

"These rays will then, in such case, exercise, amongst other things, magnetic influences and electric inductionary effects, especially strong in the Polar regions, and it is particularly such effects we are desirous of tracing. The tail of the Comet, if it should consist, as above assumed, of such radiant matter, will alter its shape at a very considerable distance from the Earth, and we may expect to see similar formations of light to those which occur during my experiments with cathode rays around a magnetic terrella.

"In my work, "The Norwegian Aurora Polaris Expedition 1902—1903", descriptions will be found in several places of these phenomena, but to elucidate the subject here, I append a few new illustrations, which very plainly show the shape of these formations of light.

"Figures 1 (217) and 2 (218), show how the rays are drawn in, in belts around the magnetic poles of the terrella, correspondingly, with the Polar-light zones on the Earth. They are taken looking along and perpendicular to the magnetic axis. Fig. 1 show the spiral rings of light around a magnetic

N. pole, corresponding to the S. pole of Earth magnetism. We find these belts of light sometimes, as here, with a tolerable, even strength of light like a continuous band, and at other times we find the rays concentrated in three limited streaks, with well definable positions around the magnetic poles of the terrella.

"Figure 3 (208) also shows an equatorial ring. This phenomenon of light is magnificent, but unstable; it is difficult to produce; it may suddenly appear and suddenly vanish, as the rays which run round the terrella at the equator are difficult to get sufficiently concentrated for the rarefied gas to illuminate([1]). At the lower part of Fig. 3 and on Fig. 4 (135), a characteristic pointed tongue of light will be seen, which is drawn in, and shows the manner in which the rays here come in to the terrella. The magnetic equator is drawn on the terrella with a dark line."

(Fig. 200 & 219 give a capital picture of these pointed tongues of light. In fig. 219, the two tongues appear as *one*, the one being immediately over the other).

"It may now be imagined, that analogous formations of light might be observable, around the Earth, of the rays from the Comet's tail on the 18th—19th of May. The downward rays in the Polar regions, will, it is true, be difficult to observe in northern parts, owing to the northern declination of the sun, but in antarctic regions there could be more hope of being able to do so, and the phenomenon would then probably appear somewhat similar to the Aurora australis. At night, in low latitudes, one could conceive the possibility of a ring like the equatorial ring being observable as a sort of zodiacal light.

"About the 2nd of May, the comet will be in the vicinity of Venus (see *Bulletin de la Société Astronomique de France. Février 1910*, p. 57), and it is not impossible that indications of an alteration in those parts of the Comet's tail nearest the Planet might be noticeable.

"We may then possibly expect to find traces of the rays being drawn in towards the Polar regions of Venus, in a manner similar to that demonstrated by the experiment shown in Fig. 4 (135), or a more or less distinct bending of the Comet's tail, assuming Venus to be magnetic.

"The probability of such being visible must, however, be admitted to be small, as the central line of the tail, if it is directly away from the sun, will be at a considerable height above the Planet; but I will nevertheless call the attention of Astronomers to these conditions, as Venus, if equally as strongly magnetised as our Earth, must be expected to exercise a noticeable influence on the tail of the Comet at a distance of several million kilometres, especially if the rays in the tail are easily deviated by magnetic force.

"This phenomenon might, in case it were present, be determined by astronomical observations of the Comet's tail and Venus in the period from 1st to 3rd May and I beg therefore, dear Sir, respectfully to ask you, in the interests of science, if you would kindly have the necessary observations made, if possible, and that you would favour me with a short account of the results."

The matter awakened interest in many quarters, and from Göttingen an expedition similar to mine was sent to my former station at Dyrafjord in Iceland, under the direction of Dr. G. ANGENHEISTER. Both the Norwegian and the German stations were chosen out of regard to the fact that experience had been gained there from previous observations, especially of magnetic storms.

In addition to magnetic registerings, earth-current registerings were made and measurements taken of atmospheric electricity. Meteorological observations were also made.

Before entering upon a description of the experiments that were made, and discussing the results that may apparently be deduced from the observations at the Haldde Observatory at Kaafjord, I will attempt to give an epitome of the astronomical and meteorological observations that I have succeeded in collecting from various quarters of the globe; for it is not from observations from one place that

([1]) See Articles 111 and 119.

decisive conclusions can be drawn, but from united observations from the entire globe, and when looked upon in this way the result appears in the present case to be of a decided, positive nature.

We will first take the astronomical observations among which those of Mr. INNES, at the Transvaal Observatory, are the most fully reported — with the definite object in view of discovering whether the forms observed of some part of the comet's tail can be ascribed to the electro-magnetic influence of the earth.

By the 22nd May, Mr. INNES had already sent the following letter to 'Nature':

The Earth and Comets' Tails.

"In spite of the unreserved predictions of astronomers, the Earth did not pass through the tail of Halley's Comet on the 18th—19th May, nor subsequently. The tail as seen in the morning sky, previous to the transit of the comet across the Sun's disc, appeared like a long and straight beam of light stretching from the horizon to Aquila. It was noticed from day to day that the tail was practically fixed in position in the sky. We rather expected the tail to get nearer to Venus and Saturn as the comet approached the ecliptic, but it remained stationary. On the morning of transit, 18th—19th May, the tail was unchanged, but a second branch to the south was now noticed. It joined the northern branch to the east of the Square of Pegasus. Unfortunately this southern branch was near the zodiacal light and only distinguished from it with difficulty. Both of these tails were seen morning by morning, including this morning (22nd May, civil day), but they have diminished in brightness and were difficult to see. Further observation of these will be impossible, because of the Moon remaining above the horizon until after dawn during the next ten days. The whole eastern horizon where the tails meet, and where the zodiacal light is, was suffused with a dim and indefinite glow which was particularly noticeable on the 18th—19th and 20th—21st. This glow was not so definite in boundary as the zodiacal light. When the comet was seen on the *evening* of the 20th, we were surprised to see it had the ordinary tail pointing away from the Sun as usual. It had been noticed for several days that in the neighbourhood of the Sun the sky was not so blue as usual, but this was the case even a week before the transit and is probably merely a meteorological phenomenon. This brief summary of the facts will suffice here; the observations in detail will be published elsewhere.

"We have now to explain the reason why the Earth did not pass through the tail of the comet and why the tail broke up so that some of it was left in the morning sky, where it remains and is slowly losing its luminosity, and some (or another tail) appeared in the evening sky. It is well known that a comet under the Sun's radiant action (I do not attempt to define it more closely) expels corpuscles towards the Sun, which the Sun repels, and these luminous corpuscles form the tail. This process goes on even when (as in the case of Halley's Comet) the distance between the comet and the Sun exceeds the distance of the Earth from the Sun. If the nearer planets do not show tails it is because these corpuscles have been shed by the planets ages ago. In short, a comet and a planet under the radiant action of the Sun, and the Sun itself, all repel these corpuscles. This being so, it is impossible for the Earth to go through the tail of a comet; it simply repels the tail, and as a consequence, instead of a passage through it, a disruption near the time of passage must occur, one part being left in the (in this case) morning sky, whilst a new one is developed in the evening sky. Here I might remark that on the evening of the 20th the measured length of the new tail was 19°, on the 21st 32°, and on the 22nd it was 40°. Again, the Earth is bombarded with meteorites which are also throwing off corpuscles. These will be repelled by both Earth and Sun, so that if we look at the part of the sky opposite to the Sun we should and do see the faint tail thus formed, which is known as the Gegenschein. This simple theory explains all the facts of observation, and if it is correct, will save nervous individuals some worry when the next near approach of a comet's tail is imminent.

"P. S.—Mr. H. C. REEVE, of Lorentzville, under date of 22nd May, has sent me a letter conveying the same idea. He says: 'Whatsoever nature the stress between the Sun and the comet may be which causes the repulsion of the tail . . . . the same stress must also exist between the Earth and the comet. . . . . Under these circumstances the Earth could not possibly pass through the comet's tail'."

Dr. CHAS. F. JURITZ of the Government Analytical Laboratory, Capetown, under date of 21st May, 1910, writes:—

"The last time that I saw the nucleus previous to transit was on the morning of Tuesday, the 17th. The nucleus was then not far from the θ Arietis, and the tail stretched right away to the neighbourhood of the θ Aquilae.

"On Wednesday, the 18th, the sky was entirely overcast. The comet could therefore not be seen.

"On Thursday, the 19th, at 5 a. m., i. e. while the transit, as originally expected, was supposed, to be in progress, and the Earth in course of passage through the tail, the tail was longer and wider than ever . . . . extending right into the Milky Way, the northern edge of the tail grazing γ Pegasi. But this time the main tail was flanked by *two attendant* shorter shafts of light. The fainter of these was north of the main tail, and inclined more to the north than even the main tail did; the brighter of the two subsidiary tails stood up almost vertical from the north-eastern horizon, and seemed to extend some 8° or 9° above Venus, the planet, which was right in the middle of the beam, twinkling through it like a fixed star. Between this tail and the principal one there was a distinct circular-pointed wedge of dark sky. These two fainter tails were apparently between 15° and 20° long. The appearance of *the three beams of light* produced on me exactly the impression of the mouth of a great transparent cone into which the Earth was rushing. Imagine a stupendous glass filter funnel, down the sides of which, from stem to edge, three streaks of luminous material had been painted; *they converged towards the horizon and diverged towards the zenith.* The continued base of the three beams extended along the north-eastern horizon some 35°.

"On Thursday evening the comet was not yet visible in the west, but on Friday morning, the 19th, the main tail was still practically in its former position, although somewhat fainter. Its northern companion had disappeared, but the southern subsidiary tail was more distinct than before, and also longer, while the dark wedge separating it from the principal shaft of light was better defined than on the previous morning."

Father E. GOETZ, of the Bulawayo Observatory, writing on the 21st May, says:—

. . . . "Might it not be that the tail was more westwards than we expected, and that we passed it during the day on the 19th, and that the faint tail we saw on the 20th was a *streamer distinct from the main tail.* The slight curvature which was noticeable when the comet passed near Venus makes me think that the Earth may also have had some kind of repelling effect on the tail which would have sent it a little further west than anticipated and account for our delayed passage" . . . .

Mr. W. H. FINLAY, M. A., writes that he and Professor RUDGE, observing at Bloemfontein, saw the tail near Aquila undergo a rupture on the morning of 18th—19th May, and that he considers this was due to the tail meeting the Earth's atmosphere and being unable to penetrate it.

"It will be remembered (see Circular No. 3) that the eastern or morning tails were actually seen here on the morning of the 21st May, almost exactly three days after the transit of the comet across the sun's disk (see sketch fig. 237). At that time the north branch of the morning tail ended in 20 h. R. A., whilst the head was in 6 h. R. A. and the end of the western tail in 8 h. R. A., or roughly the angular distance from the end of one tail to the other was 240°. But there was then no connection between the comet and the eastern tails. It is highly probable that a rupture had occurred and this probably before the 18th May, as on that morning the main or northern tail got thinner as its distance from the horizon increased (see sketches p. 15 of Transvaal Observatory Circular No. 3).

"As to the actual and unbroken length of the tail, this was measured on the 17th May and found to be 107°. On the 18th the nucleus was invisible, but the tail ended 140° from the place of the head. When the whole comet was visible the greatest length seen here was thus 107°. I cannot find any authentic measure of the angular length of a comet's tail which exceeded or was even as great as this, but references to authorities are limited at this Observatory. It may be said that it would require much imagination to desire a more impressive and brilliant spectacle than that presented by Halley's Comet on the morning of the 15th, 16th, and 17th May. It was indeed a «Great Comet», such as the writer had never seen before and can hardly expect to see again.

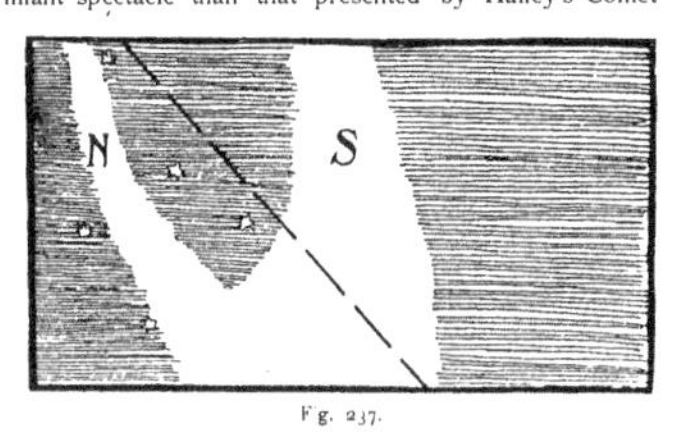

Fig. 237.

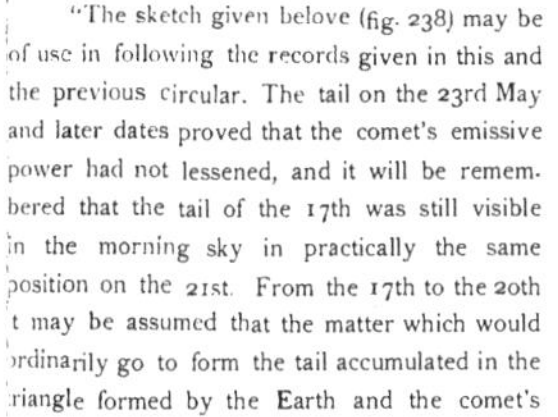

"The sketch given belove (fig. 238) may be of use in following the records given in this and the previous circular. The tail on the 23rd May and later dates proved that the comet's emissive power had not lessened, and it will be remembered that the tail of the 17th was still visible in the morning sky in practically the same position on the 21st. From the 17th to the 20th t may be assumed that the matter which would rdinarily go to form the tail accumulated in the riangle formed by the Earth and the comet's positions on the 17th and 20th May; this matter eing visible as the extensive glow involved vith the Zodiacal Light.

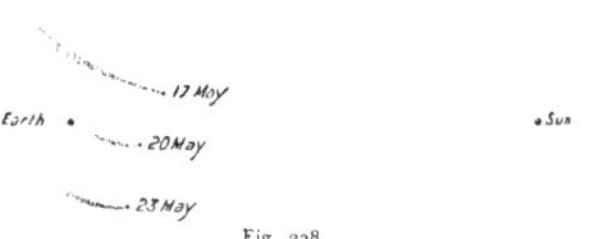

Fig. 238.

The lengths of the tails shown on the sketch are: —

1910	Units	Miles
May 17	0.30	27 000 000
» 20	0.09	9 000 000
» 23	0.16	14 000 000"

The following is an account of a peculiar observation by EGINITIS[1] on the evening of the 20th May.

"On the evening of Friday, May 20, 1910, on looking at the head of Halley's comet through our reat equatorial Gautier (0.40 m.), we found it had completely changed its appearance since the last bservation made in Athens (May 12); it was in the form of a *crescent*, resembling that of the moon a ttle before its first quarter. The *length* of the axis of the head was about 2′, almost four times less ian its *breadth*; one would have said the comet had been *truncated* or partly *occulted*. The outline of ie head towards the apex appeared very smooth and very bright, and was in the form of a parabolic rc, very luminous, not fringed externally, having its apex tangent interiorly to the *nucleus*. During the bservation, this outline became smoother and smoother, while the *tail*, of which only *a few traces*, carcely more than the beginning, were visible in the concavity of the head, showed no perceptible rolongation in the direction of the axis, unless it were a little at its margins (fig. 239).

[1] "Ciel et Terre" XXXII, March 1911, p. 94.

"The *concave* side of the crescent, which was *the first to enter the field of the telescope*, appeared to be turned towards the *west*. This peculiarity, which has struck us ever since the beginning of the observation, *has been verified by us on several occasions*, on account of its importance, to assure ourselves of it. We have concluded from it that this evening, as also this morning, the tail, which ought to be in the concavity of the head, was *apparently* directed, *in consequence of its great curvature*, towards the sun (¹).

"The same evening, however, but one or two hours later than ourselves, Dr. HARTMANN, observing the comet through a sweeper of 8 cm. on Mt. Sonnwendstein in Austria, together with Drs. WEISS and RHEDEN of Vienna, saw the same appearance of a *crescent*, but with its concavity turned in the opposite direction, namely towards the east (fig. 240) (²).

"How is this *difference* in the direction of the head of the comet to be explained? Does it arise from an *error* of observation? *Certainly not!* The observations, the one as much as the other, possess all the elements of guarantee necessary to convince us of their exactness, that of HARTMANN corroborated by the data of two other eminent observers, is indisputable; our own, that *we have verified six or seven times in succession*, by causing the comet to enter the field of the telescope, and seeing it cross it *with its concavity in front* is *as certain* as the other.

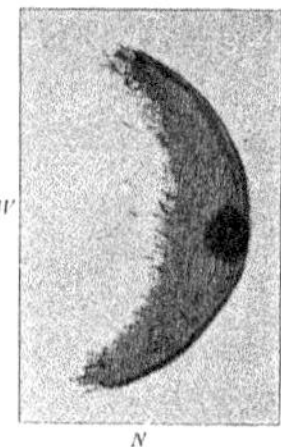

Fig. 239.

Fig. 240.

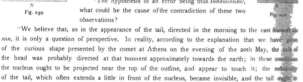

"Is it then possible to make a mistake in such a simple observation as this? It is not a question of measuring angles of position, or other slightly complicated observations, where an error might be *possible*; it is sufficient only to *see* if the crescent enters and moves in the field with its *convex* or its *concave* side in front.

"The hypothesis of an error being thus *inadmissible*, what could be the cause of the contradiction of these two observations?

"We believe that, as in the appearance of the tail, directed in the morning to the east *towards the sun*, it is only a question of perspective. In reality, according to the explanation that we have given of the curious shape presented by the comet at Athens on the evening of the 20th May, the axis of the head was probably directed at that moment approximately towards the earth; in these conditions the nucleus ought to be projected near the top of the outline, and appear to touch it; the nebulosity of the tail, which often extends a little in front of the nucleus, became invisible, and the tail ought to disappear almost completely in the telescope.

"In this hypothesis, the *difference* of the two observations might then be explained as the result of a change in the apparent direction of the convexity of the head in consequence of the rapid rotation of its axis, *relatively* to the earth; and this *relative* rotation is evidently the result, on the one hand, of the at first very rapid movement of the comet, on the other, of the contrary movement of the earth."

According to ANTONIADI (³), the observations of Eginitis must be altogether wrong, as he does not find them verified by the observations of WOOD and HARTMANN, as seen in the following sketch fig. 241 taken from Antoniadi's paper.

We do not think there is sufficient reason in this statement for disqualifying the observations of Eginitis.

(¹) Astr. Nachr., 4414 and 4421 — Comtes rendus t. CL., pp. 1408 and 1578.
(²) Drawing by Dr. HARTMANN, published in Astr. Nachr., 4431.
(³) "Ciel et Terre", December 1911, p. 425.

We will now compare the astronomical observations here quoted, of the forms of the comet's tail, with the forms that, according to our hypothesis and experiments, it would present if the tail-material consisted of electrically radiant matter.

In the first place it is obvious that here, as we maintained in our theory of the zodiacal light, we must allow that the earth's magnetism will try to keep the electric radiant matter away from the earth except in the polar regions. In the plane of the earth's equator, the negative electric corpuscle-rays that come out of space straight towards the earth, even when at a distance of millions of kilometres, are deflected westwards — as seen from the earth — and this in inverse degree to the stiffness of the rays. Compare with this the bulk of the rays in the experiment illustrated in fig. 219. The fact that the material of the comet's tail has to some extent given rise to phenomena that could not be distinguished from the phenomena of zodiacal light — as a number of accounts state — is therefore in perfect accordance with our theories. In the same way the astonishingly short tail of the 20th May as compared with that of the 17th and of the 23rd may be explained (see sketch fig. 238).

What should we have expected to see on the morning side, when the huge comet's tail was approaching the earth from May 17—21, if the tail had consisted of negative electric radiation?

We obtain clear information on this point by a comparison with fig. 219.

a
Photograph by Wood at $4^h 30^m$, G. M. T.

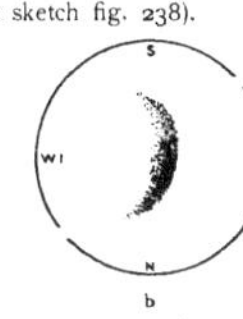

b
Eginitis at $6^h 40^m$.
Fig. 241.

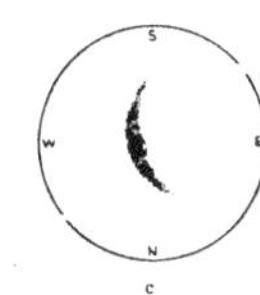

c
Hartmann at $7^h 9^m$.

The bulk of the rays must be deflected westwards. This at once explains the fact that the tail of the comet appeared morning after morning in almost the same position, although the comet had crossed the sun's disc.

We should further expect two branches from the tail, extending north and south and pointing towards the poles of the earth. This is seen on a closer inspection of the experiments shown in fig. 200, where the rays strike the floor and roof of the vacuum-box, in fig. 215 — especially Nos. 4 (see letterpress p. 588) and 14—and in fig. 219.

From the position of the earth's axis, one would have expected the in-drawing towards the north pole of the earth; and INNES' observations seem to indicate this. There appear to have been two such branches in the comet's tail, one with a north, the other with a south direction. Dr. JURITZ says in his account (see above): *"These two fainter tails were apparently between 15° and 20° long. The appearance of the three beams of light produced on me exactly the impression of the mouth of a great transparent cone into which the earth was rushing".*

We have reproduced here (fig. 237) one of the figures from Innes' account, to which we have added the position of the magnetic equator. This, it will be observed, falls just in the dark space between the two branches of the tail, which is in itself a very remarkable fact. It is doubtful, however, whether there is much to be concluded from this circumstance; but it calls to mind the phenomena illustrated in the above-mentioned figures, where there is a similar division of the cathode-rays in the magnetic equator on the morning side of the terrella.

The fact that the main direction of the comet's tail, i. e. of the rays, is oblique in relation to the earth's magnetic equator, makes the whole thing a little less clear, as a comparison with the experiments shown in figs. 215 & 219 is in this case rather imperfect.

The tail that at the same time was observed in the evening, pointing away from the sun, is also in perfect accordance with our theory and experiments.

If, in the experiment shown in fig. 219, the cathode had been bent considerably upwards — an arrangement that I have carried out several times — a correspondingly strong pencil of rays would have passed *over* the terrella, but in such a manner that the nearest rays would have curved themselves round it.

This condition answers to that of the comet having passed between the earth and the sun, when the greater part of the tail will become visible on the evening side, while the retarded or deflected tail on the morning side will become fainter and finally fade away, as the observations showed. It is not easy to see what Mr. INNES means when he says that "the angular distance from the end of one tail to the other was 240°". The distance referred to is perhaps that between the extreme points of the morning and evening tails. Regarding the observations made by EGINITIS and HARTMANN, the most natural explanation seems to be that at the time of observation there has been a narrow, fan-shaped tail, with off-shoots to north and south, which have pointed very nearly towards the earth, the one under EGINITIS' observation a little west of the earth, and that observed by HARTMANN having swung over until it pointed a little east of the earth. This fan-shaped tail with direction towards the earth, calls to mind the two in-drawn tongues of light in fig. 219, which are just off-shoots from a fan-shaped mass of light such as this. A calculation of the direction of the terrestrial-magnetic lines of force, looking from the earth towards the place in which the comet stood at the time of observation, gives a direction almost due north and south, and thus symmetric in relation to the two crescent-shaped formations observed.

Of some other remarkable observations of the comet's tail about the 20th May, the following mention may be made.

EVERSHED, in Southern India, saw the comet in the morning sky like a huge search-
not visible while passing across the sun's disc.

W. VAN BEMMELEN writes from Batavia: "I saw it before dawn on the 18th and 19th. The tail was enormous; *it rose with a high inclination to the north from the eastern horizon, like a search-light, and reached by its curvature the zenith.* I began watching it at 4.30 a. m., but saw no auroral display, nor could I detect anything of the comet's head passing the sun."

There are similar accounts from Aden, St. Thomas, and Malta.

From more northerly stations, on the other hand, there has been little to relate about the comet's tail or any luminosity that might have some connection with it. The time of year, the unfavorable position of the moon, and the atmospheric conditions, have contributed to this result.

Concerning light-phenomena seen in Norway, it may be mentioned that at Fredriksstad, at 10.30 p. m. on the 19th May, a luminous band was seen in the northern sky at a height of about 45° above the horizon, extending from east to west. It was narrowest in the west, and could not be seen quite down to the horizon, as the sky there was too light; but as far as could be seen, the radiant band pointed straight to the sun, and extended in a slight curve right across the sky to about 50° above the horizon in the north-east, where it was broader and very faint. The observer did not think that the band was an auroral band, but he was inclined to connect it with the comet's tail.

From the telegraph-office at Tana it was reported that at 3.30 a. m. on the 19th May, a light was seen, which resembled aurora, and could not have been a gleam of sunshine. A few strokes on the operator's alarm-bell were also noticed once or twice after the light had disappeared.

At Tjärstad, in Sweden, similar auroral arcs were seen at the same hour. Judging from their position, they were probably the same arcs (see STENQUIST's "The Light-Phenomena, May, 1910" (1), p. 11).

(1) Arkiv för Matematik, Astronomi och Fysik. Stockholm, 1912.

The probability is that these light-phenomena have been intense auroral arcs that were visible in spite of the bright sky; but it is not impossible that this unusual aurora had something to do with the comet.

126. We will now pass on to the meteorological observations that were made at about the time of the transit, in order to see whether, in those at our disposal, any trace could be found of effects that might reasonably be ascribed to the tail-material of the comet.

At the outset we must state that a number of observations made at some of the leading observatories, have yielded a negative result. As an instance, a series of balloon-observations, where air-samples were taken in high strata, revealed nothing of interest. On the other hand, there are other observers, who, on the days in question, noted meteorological phenomena of a peculiar and unusual nature.

In the Transvaal Observatory's Circular No. 4, of July 11, 1910, there are some particulars given by observers, of phenomena that they saw, from which it would appear that a Bishop's ring was seen. Mr. OTTO MENZELL, of Pretoria, writes:

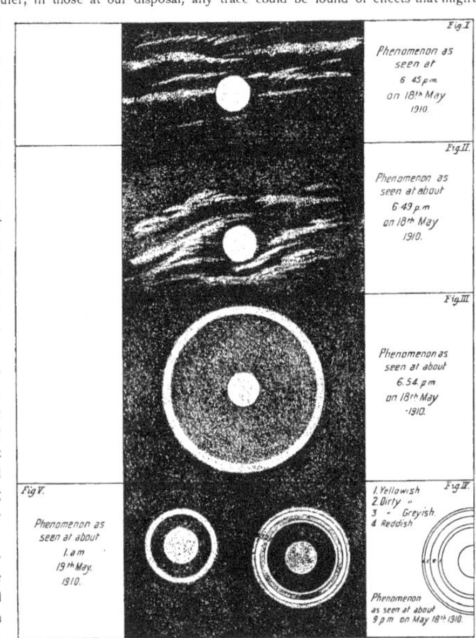

Fig. 242.

.... "At about a quarter to seven I looked at the Moon, which was then well up in the sky. I noticed a haze over it, but when looking through my glasses it shone as clear as ever. Some clouds were gathered round the Moon at that time, as shown in the first sketch fig. 242. A few minutes afterwards these clouds started moving in a peculiar circular fashion round the Moon (sketch) and continued doing so for at least five minutes until they formed a broad ring round the Moon, as shown in the third sketch. I then went home, but returned at about nine o'clock; the ring had narrowed down, as shown in the fourth sketch. The colours of the ring are described in the sketch. When looking again at the Moon at one o'clock in the morning of the 19th May, the ring had narrowed down a little more, as seen in the fifth sketch, and it seems to have remained so."

Mr. G. R. Hughes, of Pretoria, sends the following report: —

. . . . "The Moon, which was in its second quarter, was surrounded by a ring which had a metallic appearance. It was of considerable diameter and, to my recollection, the inner edge of the ring (nearest the Moon) was yellowish (the yellow of the Sun), then merging into a dirty brown. The outer edge was dull grey, like the clouds, that covered the sky. The ring appeared to have walls, if one may so distinguish from a 'flat' surface. It was exceptionally well-defined. I observed the phenomenon until nine o'clock, when I ceased to give it attention and am unable to say when it finally disappeared.

"One remarkable phase was an inner ring which manifested nothing metallic. It was faint and flat in contra-distinction to the outer and larger ring. The inner ring was dull grey in colour.

"The Moon on the night of the 19th May was again surrounded by a ring; the latter was much more clear-cut that on the previous night. There were less clouds in the sky. When I first observed the Moon between 6.30 and 7 p. m., it was clear of the clouds and had no halo. It, however, appeared to be less distinct than usual. The features on its surface were not so sharply defined; while no haze was visible to the eye, I am confident that some influence was present in the atmosphere. I tried a view with binoculars, but still the features lacked sharpness in definition. As the Moon approached the clouds, which previously were scattered, they seemed to break, and I saw the ring evolve. The area within the ring, unlike the previous night, was clear of cloud. The ring was decidedly metallic in appearance, but I did not observe so much yellow colour as on the 18th. My note reads: 'Moon surrounded by ring of dark brown material'. I observed the phenomenon for half an hour. The weather conditions at the time were restful, but later in the evening the wind arose."

Dr. Franz Linke writes in a preliminary statement in "Meteorologische Zeitschrift", June, 1910:

"The Meteorological Geographical Institute of the Physical Association at Frankfurt a. M. had erected by May 12th a temporary observatory on the Feldberg in the Taunus (880 m.), where arrangements were made for atmospheric-electric and terrestrial-magnetic registerings and observations, the results of which will be published later. At present, attention will only be drawn to the quite abnormal phenomena. Since the 12th May, we have had high-pressure weather; an evenly warm, dry current of air out of the eastern continent continued uninterruptedly, and apart from some thunderstorms, brought continuous warm, clear, summer weather.

Only on the afternoon of the 19th, a few hours, that is to say, after the passage through the comet's tail, there occurred a remarkable cirrus-overclouding with lunar halo and ring, which, if it had been observed in an ordinary way, I should have ascribed to the influence of the ions expelled into the atmosphere. I did not, it is true, even at 2 p. m., at a height of 8500 m. notice anything of these ions; on Gerdien's conductivity-instrument, a strong, but for such heights not abnormal conductivity was observed.

The same evening there first appeared the following abnormal twilight phenomena. In the southern sky a broad, reddish yellow stripe extended southwards from the sun more than 100°. In the north there was nothing similar to be observed. On Friday evening, however (May 20), a similar luminous band, about 10° in width, and of the same horizontal extent, appeared on the northern horizon. The twilight had also all the characteristics of the disturbance, such as a Bishop's ring, a reddish brown colour, unusual clearness and duration. In the course of the next few days, from the 21st to the 24th May, the clear light in the north constantly spread over the entire sky; not until the 24th was the twilight symmetrical with the sun.

If we make the cosmic dust of the comet's tail expelled into the earth's atmosphere, responsible for the twilight anomaly, we must assume that on the day of the transit, Thursday, the 19th May, principally in the equatorial regions, the cosmic dust reached the strata in which the twilight is found (a height of from 10 to 20 km.). It must however quickly disperse or fall upon the earth. Great quantities that are more slowly diffused then appear to have streamed into the polar regions, whence they are slowly distributed over lower latitudes and deeper strata.

A further observation that I believe I have made is in accordance with this, namely, that at the beginning of the twilight the northern sky is first illuminated, and it is not until later in the evening that the maximum of clearness occurs in the south. Consequently the light-deflecting and light-reflecting strata lay deeper in the north than in the south, and have thus sooner, or more rapidly, penetrated downwards, when they come from without.

I need hardly say that this would prove the electric nature of the comet's tail as a current of ions deflected by the earth's magnetism, as BIRKELAND has formulated it.

The various phases of the twilight are not so easily recognised after the commencement of the perturbation, as before. I missed in particular the first and second purple-lights.

MAX WOLF, Königstuhl-Heidelberg, has sent me a copy of his observation-notes. They are as follows:

From the night of the 17th May, 1910, a cirrus-veil developed, which, up to the afternoon of the 19th, continued to increase in fulness and form.

The veil consisted of quite peculiar forms, nothing similar having ever been seen either before or since. In addition to the complicated thick and thin, stratified and fan-shaped interpenetrating forms, there was present an all-penetrating structure of narrow, smoke-like bands, such as previously (and since) have only once been observed, namely, on the 30th June, 1908.

The colour of these exceedingly high-lying bands was entirely different from that of the tangled cirrus-covering; and this colour, combined with the apparent, quite unobstructed penetration of the two kinds of formations, produced the astonishing cloud-picture that reached its maximum on the 19th May, and roused the attention of numerous observers, all of whom were situated in the centre of the area of high pressure that at that time covered certain parts of our land. The direction of the smoke-like bands was S 20° E to N 20° W.

Late in the afternoon of the 19th May, a Bishop's ring was first observable round the sun.

There then developed, after only comparatively unimportant twilight phenomena had for some time been observed, on the evening of the 19th May, a twilight of quite unimagined intensity, extent and duration.

Three successive purple lights could be observed—distinctly purple up to $9^h\ 20^m$ local time, later for a long time red in the north-west, with all the colour-phenomena (including the wonderful turquoise-blue and ruby-red) seen earlier in the eruptions of Krakatoa and Mont Pelée, and occurring on the 1st July, 1908.

Round the moon there appeared a Bishop's ring with an intensity such as we had never seen. I determined the external radius to be 28° at the time of the culmination of the moon (at a height of 37°).

The cirrus cloud-covering then steadily decreased. But in the higher strata there still remained a very faint, tangled granulation, which made it possible to see the Bishop's ring, distinct and bright, on the 20th May, this being only visible when clearly-illuminated parts of the sky are observed through a shadow.

All the phenomena decreased very rapidly. If we call the ordinary intensity of the twilight 1, by the 17th May it had already risen to 3. I estimate the course of the intensity roughly as follows:

May 17,	I = 3
18,	6
19,	30
20,	16
21,	9
22,	6
23,	4
24,	3

The course of the intensity and extent of the discs round the sun and moon was analogous to this in this district.

The whole phenomenon was analogous, down to its *smallest* details, to that of the 30th June or 1st July, 1908, which I have described in Astr. Nachr. 4266 Bd. 178, 1908.

In a paper by D. STENQUIST (l. c., p. 14), which we only saw after the above was written, a number of interesting observations have been collected from the period from the 17th to the 21st May. The author summarises his results as follows:

"From the twilight-phenomena observed, from the abundant occurrence of cirri, in which coronæ and halos were produced, and from the existence of aurora borealis, it would seem that the earth, at the heliocentric passage, was enveloped by not inconsiderable quantities of cosmic dust (probably charged with negative electricity)".

Concerning the meteorological observations at the Haldde observatory and at Kaafjord during the transit, I will first of all emphasise the fact that many extremely characteristic polar bands were formed in a striking manner, and in more rapid succession than I ever remember to have seen before. The significance of such polar bands in connection with the theories here propounded, has been dwelt on in an article "Sur la Formation des Nuages Supérieurs", p. 75 of "Expédition Norvégienne de 1899—1900 pour l'Étude des Aurores Boréales". It is assumed that the polar bands are produced by the in-drawing, through terrestrial magnetism, of negative corpuscles from space in a manner similar to that in which the corpuscle-rays that produce auroral arcs are drawn in.

The weather on the 18th and 19th May was very unfavorable for observing, as thick mist frequently prevailed, with snow and ice-spicules in the air. Now and then, however, it was clear for some time, for instance on the evening of the 18th up on Haldde Mt., and down in Kaafjord on the morning of the 19th, beginning from midnight.

At about 8 p. m. on the 18th, I saw from the mountain at one time 4 parallel, very marked, polar bands, curving from west to east over the northern sky, with their highest point about 30° above the horizon. They changed considerably and developed rapidly, but were soon hidden by the mist. On the morning of the 19th, Krogness saw many cloud-formations of the same kind from Kaafjord, concerning which he says:

"Although the cloud-covering was not favorable for the observation of cirrus-clouds on the night of the 18th May, there were several opportunities at Kaafjord of observing very peculiar cloud-formations. There was a most unusually abundant variety of cirrus-bands. Their shape and manner of forming showed an unmistakable resemblance to those of aurora. Great drapery-like clouds would frequently appear quite suddenly, or large portions of the sky be covered with clouds in the form of a corona; and more or less bright polar bands were continually visible. The following are some of the notes made at the time (Gr. M. T. is employed):

May 18, 1910, $11^h\ 32^m$ p. m. 6 polar hands in a direction WNW—ESE passing the zenith and north and south of it.

$34—35^m$. Two or three small draperies were formed, which, however, soon disappeared. In the southernmost band numerous stripes.

39^m. A bright band suddenly makes its appearance a little north of the zenith.

40^m. A tassel with striped figures appears, and spreads eastwards in the form of a drapery along the above-mentioned polar band.

Above the mountain in the west several faint, evenly luminous, very characteristic bands. In the north brighter bands with from 2 to 3 peculiar, bright, awl-shaped, striped figures pointing downwards and westwards. These are almost due north.

45^m. In the above-mentioned band in the north, a little lower, occurs a very marked, evenly ddish light. The highest point of the band is in a direction N 35° E.

48^m. Several evenly bright bands in the form of great circles converging towards a point on the rizon in a direction about S 45° E.

To the south many fine, awl-shaped stripes massed more or less in bands.

$11^h\ 51^m$ p. m. Dark cumulo-stratus rising in the west and hiding the cirrus clouds. In the south sll many threadlike bands . . . . ."

The above extracts are sufficient for our present purpose. The observations were continued troughout the night, with interruptions occasioned by the overclouding of the sky.

The following accounts of exactly similar observations, made at the same hour—8.30 p. m.—on the 18th May, in the tn of Tönsberg, and at Blakjer, about 90 kilometres north-north-east of Tönsberg, read almost like a fairy-tale.

The account from Tönsberg was given by the magistrate of that town, his wife and son, with a statement of their readins to confirm by oath what was written. At Blakjer the phenomenon was observed and described by several trustworthy psants without any knowledge of what had been seen in Tönsberg.

The magistrate's account is as follows: "We were all three walking along the quays. The sun was near the horizon, a we saw suddenly appear round it a number—I suppose from 50 to 100, possibly more—of dark (blackish grey) circles about a arge in diameter as the moon, and these then spread out on both sides of the sun".

The people at Blakjer saw at the same hour, in the direction of the sun, bubbles the size of a child's head and smaller sdenly descending towards the earth, shining in all the colours of the rainbow.

It is not, of course, easy to say what has caused the unusual phenomena here observed, but it can only be supposed th there have been certain foreign bodies in front of the sun that have produced the various light-effects.

The magnetic registerings in Kaafjord were begun on the 7th May and continued by us until the 2l June, after which date they were carried on more or less completely by Herr L. Heitmann until th middle of July, in order that a general idea might be obtained of the course of the variations through a eriod of some length.

In addition to the ordinary slow-run registering-apparatus, we also took with us one for quick-run risterings. The latter were begun on the 18th at about 4 a. m. Gr. M. T., and continued until about 3 . m. on the 20th.

During the same period, magnetic registerings were also undertaken at Teisen near Kristiania, by Hr O. Devik.

The special interest of these observations, and our reason for mentioning them here, is the conntion that their results may be supposed to have with the passage of the earth through the cet's tail.

With regard to the magnetic curves, this period may be characterised as follows: The period from th 7th May until noon on the 18th was fairly quiet magnetically, the storms that occurred being of co paratively little strength.

At about 1 p. m. on the 18th May, an unusually powerful magnetic storm suddenly began, develog in the afternoon into a positive polar storm, then changing later, and appearing in the evening an night as a negative polar storm.

From about 4 a. m. on the 19th, the storm decreased in strength, and at about 6 a. m. the condins were once more quiet, and continued so for some days.

On the 24th, at about 9.30 a. m., an unusually powerful magnetic storm occurred once more, with a urse similar to that of the storm of the 18th—19th May, but of somewhat longer duration.

As the passage of the comet was to take place on the morning of the 19th, it seemed reasonable at rst sight to suppose the storm of the 18th—19th to be caused by the comet. Just at this time, hoever, matters were complicated by the appearance of a large group of sun-spots near the sun's centra meridian. These too, then, might be the cause of the storm, as in its main features the course of thperturbation was like that of ordinary perturbations.

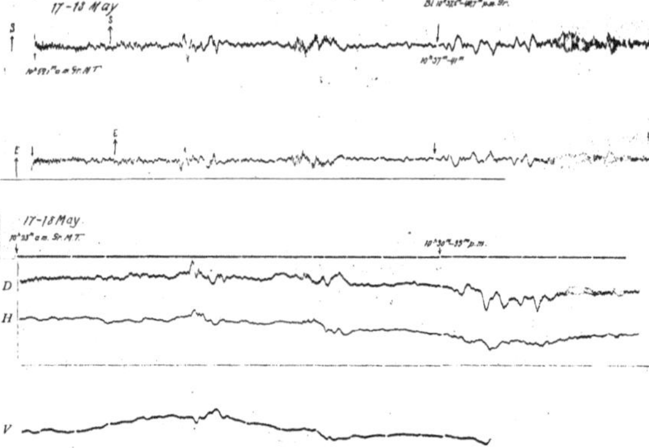

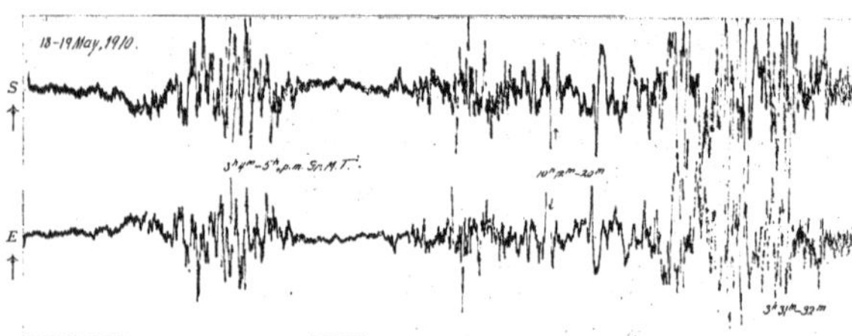

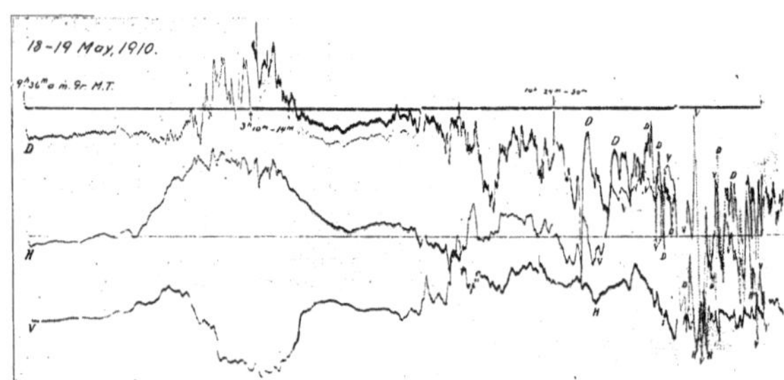

Earth currents and magnetic elements 18—19 May, 1910.

Fig. 243.

There is one circumstance, it is true, that may seem remarkable in this storm, namely, that it was ot repeated on the following day, as is generally the case with powerful magnetic storms in the polar egions. On the contrary, calm supervened very quickly, and the next day was very quiet.

At Dyrafjord, however, we find the storm repeated with diminished strength the day after as well, n the usual manner.

It should be remarked, however, that it was not until two days later that the above-mentioned sun-spots reached the central meridian of the sun. It is well known that several scientists have thought hey could show that powerful perturbations do not as a rule occur until from 40 to 50 hours after the assage of the corresponding sunspot over the central meridian. If, therefore, we apply this here, it

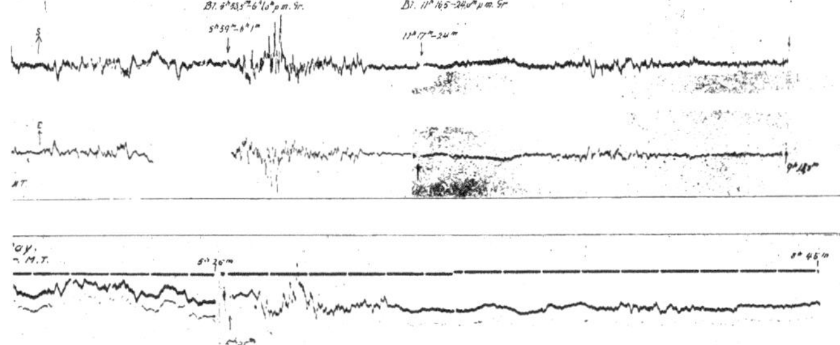

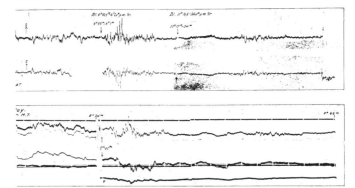

Earth-currents and magnetic elements, 19 - 20 May, 1910.

Fig. 244.

ould appear that the magnetic storm came about 4 days too early. On the 23rd May, when, according this manner of looking at the question, the influence of the group of sun-spots might be expected, was very calm magnetically. But on the morning of the 24th, as mentioned above, powerful storms ice more occurred. It thus appears to be difficult to deduce the magnetic storms from the sun-spots the usual manner; but on the other hand the above-mentioned difference in time may obviously be garded as an average value of a large number of cases, and in reality the connection between sun-ots and magnetic storms is not so simple (cf. Art. 98). It is interesting to see that Dr. ANGENHEISTER is believed he can prove a greater accordance between the appearance of the sun-faculæ and the mag-tic storminess on the earth in the month under consideration [1].

The earth-current registerings will be discussed in the next chapter. As the magnetic disturbances e exactly repeated in the earth-current curves, the reader is referred, as regards the latter, to what

(1) Cf. Angenheister's "Die Island-Expedition im Frühjahr 1910. Die erdmagnetischen Beobachtungen. Nachrichten der K. Ges. d. Wiss. zu Göttingen, Math.-phys. Kl. 1911.

has been said above. I here reproduce some of the most characteristic curves from the period between the 17th and 20th May.

The atmospheric-electric measurements included measurement of the conductivity by readings, i. e. not registerings. These readings were principally taken at Haldde Observatory.

In these measurements I received capable assistance from Herr Feyling, telegraph-director at Bossekop, who kindly accompanied me in order to assist in the observations. Several long series of 5-minute readings were taken, alternately with positive and negative potential. The observations made on the 20th May were of special interest. Their results are described in Art. 93 on p. 449.

The above is a comparison of the available observation-material concerning Halley's comet, May, 1910, with our experimental results, and may serve to strengthen the theory that the comet's tail-material consists of electric corpuscles radiated from the comet.

THE SATURNIAN RING.

127. That Saturn's rings cannot be rings of coherent matter, either solid or liquid, has long been well established by theory, which showed that the equilibrium of such an object would necessarily be unstable.

The alternative hypothesis that the rings are clouds of minute satellites, or perhaps mere particles, too small to be individually visible, but so numerous as to look, in our telescope, like a continuous mass, was investigated by Maxwell in his Adams prize essay, published in 1859. Although the stability of such a ring of particles can hardly be said up to the present to have been strictly proved, Maxwell's hypothesis has gained more and more adherents among astronomers, especially since the noteworthy addition to our knowledge of the rings of Saturn, made by Keeler, that the different parts of the rings have a rotation in conformity with Kepler's third law. The extreme thinness of the rings has been demonstrated at the times at which the plane of the rings passes through the earth. Even with the 36-inch telescope of the Lick Observatory, the rings were completely invisible in these circumstances. This shows that the entire ring must be so thin that its edge is quite invisible, even in the full light of the sun, at the distance which separates us from the planet. On the other hand, the objects composing it must be completely opaque, as is shown not only by their disappearance in the circumstances we have mentioned, but by the darkness of the shadow which they cast upon the planet when the sun illuminates them obliquely. The cloud of these very small satellites seems to be so dense that a ray of light cannot penetrate the mass.

At present Maxwell's hypothesis seems to be a strong one, although it seems almost incredible that such a ring of cosmic dust should be able to exist for ever, so to speak, without other governing forces than gravitation, when the ring is less than 21 kilometres in thickness (1), with an external radius of 135,100 kilometres.

Some astronomers, however, appear to be beginning to doubt this hypothesis.

Herman Struve, after having proved that their total mass is certainly less than $^1/_{26720}$ of that of Saturn (2), says that these rings appear to be composed solely of an "immaterial light", mere dust-films or wreaths of mist.

---

(1) Russell, Astrophys. Journ, vol. XXVII, 1908, p. 233.

(2) Publications de l'Observatoire Central Nicolas, série II, t. XI, 1898, p. 232; and Young, General Astronomy, p. 395 Boston, 1900.

Dr. Barnard, after his examination, in 1907, of the illumination of the dark side of Saturn's rings, uggests the explanation[1] that the rings are auto-luminous; but he rejects the idea by conjecturing that uch a hypothesis would not be compatible with the presumed physical composition of the rings.

I think it will be quite possible to satisfy all the results of the observations hitherto made of these ings by a hypothesis entirely different from the above-mentioned meteoric theory.

On p. 613 of the present volume, I have described some experiments that have served as a starting-oint for an explanation of the zodiacal light.

Round a highly magnetic globe, 8 cm. in diameter, in a vacuum-tube with a capacity of 70 litres, have produced a ring with a diameter of up to 34 cm., and long luminous rays in the polar regions f the globe, the whole bearing a considerable resemblance to pictures of the sun during an eclipse.

Now if the discharge-current, which in the above experiment was from 10 to 30 milliamperes, be educed to 1 milliampere or less, the polar phenomena cease, and the ring becomes exceedingly thin nd sometimes assumes an appearance almost exactly like that of Saturn's rings.

Round the magnetic equator of the globe, and touching it, a luminous zone appears, then a dark pace, which, farther from the globe, is gradually formed into a flat, dimly-luminous ring resembling the rape ring of Saturn. This dimly-luminous ring farther away increases in strength and a light-ring ppears.

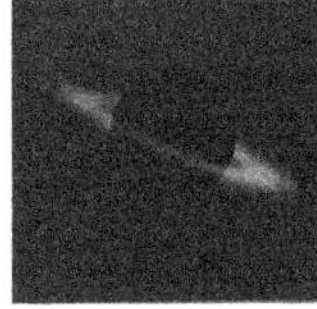

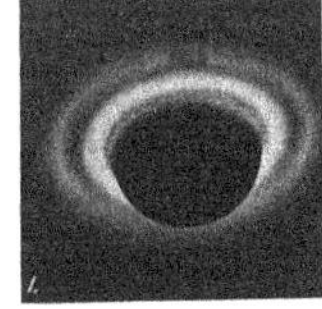

Fig. 245.

Fig. 245, 1 shows the rings from the side, and fig. 245, 2 a little from above, thus making the ark space between the globe and the ring distinctly visible. Fig. 245, 3 shows, in addition to a brightly ıminous ring, a fainter ring outside the former, and separated from it by a dark division that might nswer to Cassini's division in Saturn's ring. Fig. 235 shows that by a special arrangement it has een possible to get as many as 5 rings, one outside another, round the globe. In this case, however, e rings are not flat, as the outer ones are in the form of cylinders, which increase in height with eir distance from the globe. When the magnetisable globe *is not magnetic*, but is still a cathode, it is ften seen surrounded by several luminous spherical envelopes. It is perhaps these that, when the globe magnetised, become changed in shape and flattened.

How are the phenomena of Saturn's rings to be explained, supposing the rings to be due to imilar electric radiation from the planet, the latter being considered to be magnetic?

With regard to physical investigations of the power of an electrically luminous gas, and of radiant atter, to absorb and diffuse solar light, we have mentioned some few known facts on page 623, to hich the reader is again referred.

I think there are also here good reasons for admitting that in the radiant matter which we suppose have been radiated by Saturn, there is a comparatively very great number of electrons of dispersion,

[1] Astrophys. Journ., vol. XVII, 1908, p. 39.

which may serve as receivers and resonators of luminous waves coming from the sun, and that here too, it is quite possible that the number of electrons of dispersion is proportional to the intensity of the electric current emanating from Saturn in the manner admitted by us.

An electric radiation from Saturn such as that here assumed may certainly also be imagined to be accompanied secondarily by an ejection of tiny material particles resembling what CROOKES has called electric evaporation or volatilisation from a cathode.

A metallic cathode is so disintegrated during discharges that the material may be deposited in the form of a reflecting layer upon the neighbouring glass wall.

Different metals disintegrate in very different degrees when they form a cathode, circumstances being equal.

In arbitrary units, CROOKES[1] gives the loss of weight by disintegration in cold cathodes as follows:

Pd	Au	Ag	Pb	Sn	Pt	Cu	Cd	Ni	Ir	Fe	Al	Mg
108	100	83	75	57	45	40	32	11	10	6	0	0

For incandescent cathodes it is a different matter altogether; the disintegration is then much greater. Under the influence of magnetic forces too, there is a great difference in the amount of the disintegration. I have, for instance, shown that in such a case even a cathode of aluminium can in a short time throw off a reflecting deposit upon an adjacent glass wall (C. R., Feb. 21, 1898). The disintegration from a carbon cathode is very great. From one such, in a large exhausted vacuum-tube, I have seen half a gramme of matter thrown off in a few minutes and deposited firmly on the glass wall of the tube.

The cause of this cathodic disintegration has not yet been clearly determined. It is possibly to some extent a kind of evaporation by which the disintegration is brought about, by the high temperature that the rapid positive ions (channel rays) produce where they strike the surface of the cathode. The dependence of the disintegration on the strength of the current and of the cathode-fall is in accordance with this explanation; for *with the same duration of current, the cathode's loss of weight by disintegration is proportional to the product of the strength of the current and the potential-fall from the cathode.* But this product equals the electric work performed upon the positive ions between the negative column of light and the surface of the cathode, that is to say, proportional to the kinetic energy carried by the positive ions in the time-unit to the cathode.

This phenomenon of disintegration seems to offer a very important field for future investigation, for an accurate knowledge of these things is of fundamental importance for the theories here propounded.

HOLBORN and AUSTIN[2] have made some very interesting experiments on the amount of disintegration of cathodes of different metals under similar electrical conditions. When the tube used was filled with air, they found that y, the loss of weight in 30 minutes from circular cathodes 1 cm. in diameter could for platinum, silver (one sample), copper and nickel, be represented by the formula

$$y = 0.0016 \frac{A}{n} (V - 495) \quad (1)$$

for silver (another sample), bismuth, palladium, antimony and rhodium, the relation was

$$y = 0.0018 \frac{A}{n} (V - 495) \quad (2)$$

V is the cathode fall of potential in volts, A the atomic weight of the metal, and n its valency. Other metals such as iron, aluminium and magnesium, do not follow either of these laws. For those metals which follow the laws (1) or (2) we see that with the same current and cathode fall, the weight of

(1) See WINKELMANN, Handbuch der Physik, 2. Aufl., b. 4, p. 629.

(2) See J. J. THOMSON, Conduction of Electricity through Gases, p. 549.

ıthode disintegrated is proportional to the weight of those metals which would be deposited in voltaeters placed in series with the discharge-tube.

GOLDSTEIN (1) has discovered that when channel rays come in contact with a metal, they cause it disintegrate. If, for instance, channel rays are allowed to fall upon a gold mirror deposit on plateass, the gold disappears from the place where the most intense rays strike, so that in a short time e sheet of glass becomes transparent again. Silver and nickel also disintegrate very quickly, alumium less so.

According to GOLDSTEIN, cathode-rays also have the power to disintegrate metal plates, but in a uch smaller degree than channel rays.

The above-mentioned disintegration from a cathode is in all probability closely connected with this ghly disintegrating effect of channel rays.

It appears that all metals that undergo great disintegration when employed as cathodes, are also sintegrated in a high degree when under the influence of channel rays. That the velocity of the metal ırticles flying out from the cathode must be considerable has been shown by KAEMPF by means of tical investigations of double refraction by a metal mirror, produced by electric disintegration. According Kaempf, the particles expelled from the cathode are de-formed and brought into tension on striking ıe mirror-surface.

Up to the present, the fact of an electric or magnetic deflection of the metal particles expelled from cold cathode has not been established. It is interesting, in this connection, to know that it has been ıund that a great emission of positive ions from incandescent solid bodies is frequently accompanied by distinctly appreciable loss of weight in the emitting bodies.

I have myself of late made some experiments which give promise of throwing light upon the ıestion of the electric charging of the metal particles thrown off from a cathode.

The difficulty in these experiments is that if a vacuum-tube is introduced into a very strong magnetic ‹ld (as is here necessary), in such a manner that the direction of the discharge-current is perpendicular the lines of force, the character of the discharge is changed, the discharge-current being thrown to e side and concentrated in a narrow path.

It is a different matter altogether when the vacuum-tube is placed axially in relation to the ıgnetic field.

The character of the discharge-current is then altered, it is true, as the cathode emits the so-called ıgneto-cathode rays; but, as I have already shown, under these circumstances the disintegration of the hode is very great, and the discharge-current often seems to flow with normal density through the tire cross-section of the vacuum-tube.

I arranged my experiments in a manner very similar to that in which I first discovered these gneto-cathode rays (2).

A cylindrical vacuum-tube had a cathode in the form of a cross 18 mm. from the bottom of the e, which was a plane sheet of plate-glass cemented upon it. The cross was cut out of a thin sheet palladium, its surface being parallel with the sheet of glass. The anode was circular, and was placed nmetrically round the axis of the tube about 10 cm. behind the cathode.

The vacuum-tube was placed axially in front of a powerful cylindrical electro-magnet, in such a nner that the sheet of plate-glass at the bottom of the tube was close to the end-surface of the gnet.

(1) See E. GEHRCKE, Die Strahlen der positiven Elektrizität, p. 69. Leipzig, 1909.
(2) See Archives des Sciences Phys. et Nat. Genève, June, 1896, p. 506.

When the magnet was in operation during the discharge, a luminous column of magneto-cathode rays, cruciform in section, was sent out towards the sheet of glass, where it formed a reduced representation of the cross.

The question that interested me here, however, was how the metal corpuscles expelled from the palladium cross would be deposited.

It appeared that if the cathode were cold, the discharge-current being kept small, a light cross upon a darker ground was thrown upon the sheet of glass at the bottom of the tube. There was very little deposit upon the sheet of glass where the column of light with cruciform section had heated the glass, while beyond this there was the normal deposit of palladium.

The whole thing was different when we employed up to 30 milliamperes and more per square centimetre of the entire surface of the cathode, which thereby became highly incandescent.

After the experiment, which lasted about one minute, there was an intensely metallic cruciform deposit where the column of light had struck the glass, a cross of reduced size, opaque and shining, while outside it was a dark, semi-transparent, normal palladium-deposit.

In addition to this normal deposit upon the plane sheet of glass, there was a strongly-marked ring of evenly dark deposit upon the cylindrical surface of the vacuum-tube, nearest the cathode-cross.

There were thus distinctly two kinds of metal corpuscles ejected by the cathode, first the normal corpuscles that seem to be expelled from the cathode without being influenced to any great extent, either by electric or magnetic forces; and secondly a kind of corpuscle that accompanies the magneto-cathode rays, and these corpuscles are capable of attaching themselves to the glass wall, provided the velocity with which they reach it is sufficiently great.

This circumstance may possibly favour Righi's idea that these magneto-cathode rays consist of almost neutral "double stars" of positive and negative ions, expelled in exactly the direction of the magnetic lines of force, thus possibly a combination of a negative electron with a positive metal ion, which, under certain circumstances, can be deposited and form a metallic coating upon the glass wall of the vacuum-tube.

I then went on to find out whether I could discover any twisting of this cruciform metallic deposit, such as I have proved in the case of a cross of cathode-rays under similar circumstances (see the previously-mentioned paper in Archives des Sciences Phys. et Nat.). These crosses of cathode-rays turn clockwise when a magnetic north pole is employed and the cross looked at from the pole. We can thus find out whether the metal corpuscles in the cruciform deposit were negatively or positively charged, by noting the direction in which the cross was eventually turned.

It soon appeared that the twisting was at any rate too small to be demonstrated directly by these experiments. The experiment was therefore modified by forming the cathode as a plane, long rectangle of thin palladium, which was attached in such a manner that it stood edgewise upon the sheet of glass, with its *long side* parallel to it.

The intention with this arrangement was to obtain a sharp linear deposit where the magneto-cathode rays came in contact with the glass.

The experiment was carried out at first with a south pole in front of the sheet of glass, and then, at the same distance, a north pole, while the discharge was going on evenly all the time.

Neither was it possible, however, in this way to obtain a double turning-angle of measurable size, perhaps because the deposit-lines were not particularly sharp.

Photographs of the cruciform and linear deposits are here reproduced, from three experiments in fig. 246.

In the first and second experiments 15 milliamperes was used in the discharge and 10 amperes to the magnet. In the third the discharge-current was the same, but the current to the magnet was first, 22 amperes in one direction, during one minute, and then in the reverse-direction for one minute, so that the pole before the cathode changed from S. to N.

In the course of these experiments, however, my attention was directed more and more towards he normal metal deposit which was thrown with special abundance and evenness upon the cylindrical heet of glass, right in front of, and nearest to, the strip of palladium.

It seems as if it might be worth while to experiment with a mica shade with a slit in it, placed lmost over the palladium-sheet, to see whether any deflection of the corpuscles could be demonstrated n this way. The corpuscles moved here almost at right angles to the magnetic lines of force, therefore he chances of a deflection were very much greater than in the above-mentioned investigations. I shall eturn to these experiments in a subsequent article.

Now even if corpuscle-rays were so stiff that we could only bend them slightly with our strongest nagnets, a planet with a magnetic moment such as, for instance, that of the earth, would easily compel uch rays, emitted from equatorial regions, always to move near the plane of the planet's magnetic equator.

It has been shown that the material particles that are thrown off from a magnetised globe that is athode in a vacuum-tube, are thrown off by preference near the plane of its magnetic equator like the lectric rays. This can be seen upon a sheet of glass placed near the globe, the glass being blackened n such a manner as to make it improbable that any mere evaporation can produce the disintegration.

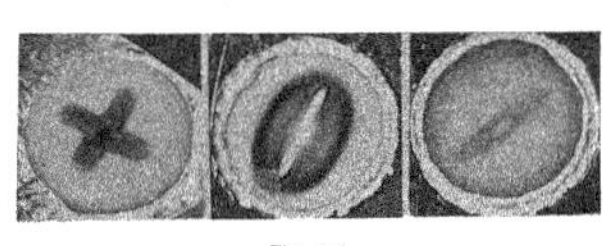

Fig. 246.

It is my opinion therefore, that in analogy with this, Saturn throws off tons of matter every day n the plane of the rings, and that it did so to a still greater extent formerly. The rings are renewed, o to speak, every moment. I have indeed gone so far in my hypothesis — as my notes to Comptes Rendus de l'Academie des Sciences show (1)—as to assume that the moons were originally formed from uch electrically ejected matter, just as the planets from matter electrically thrown off from the sun.

Whether Saturn's rings consist of radiant matter or of electrically ejected material particles, they will certainly diffuse and absorb the light of the sun, and thus give rise to light-effects and shadow-ormations similar to those now observed. Even if the ring consisted only of electrically luminescent aseous atoms, there is reason to suppose, as shown above on p. 523, that it would cast a shadow.

I would especially refer the reader to the observations at Kaafjord, where it must be assumed that he rapidly-changing cloud-formations were not real, ordinary clouds, but were electrically luminescent iry masses that had the power of reflecting and absorbing solar light, and thus had the appearance of clouds.

Auroral arcs, observed at night, have been seen after daybreak as arches of cloud; and it is ossible that this is a corresponding phenomenon.

My explanation of Saturn's rings may also be looked upon as an extension of MAXWELL'S theory, n attempt to indicate the manner in which the fine cosmic dust in the ring has formed round Saturn.

By spectroscopic examination of Saturn's ring, KEELER (2), as is known, has shown that the various arts of the rings rotate in accordance with Kepler's third law. These results can be made to agree

(1) Comptes Rendus, 7 août 1911: Les anneaux de Saturne sont-ils dus à une radiation électrique de la planète?
C. R., 21 août 1911: Le soleil et ses taches.
C. R., 4 septembre 1911: Sur la constitution électrique du soleil.
C. R., 13 novembre 1911: Phénomènes célestes et analogies expérimentales.

(2) Astrophys. Journ., vol. I, 1895, p. 416.

with our hypothesis, if the very natural assumption be allowed, that the small particles, molecules, gaseous or vaporous atoms thrown off, separated by comparatively great distances from one another, have come in the course of time to perform their mean rotation about the planet in obedience to Kepler's law. We return to this question further on.

If, after all, future investigation should confirm BARNARD's previously-mentioned suggestion that the rings are auto-luminous, it would probably in a great degree strengthen the electrical theory here brought forward of the genesis of Saturn's rings.

It appears that of late other scientists have also felt unconvinced of the correctness of MAXWELL's purely mechanical view of the nature of Saturn's rings. My colleague, M. GUILLAUME at Meudon, writes in a letter to me:

"On the subject of Saturn's ring, I draw your attention to an interesting publication by Mr. AUGUSTE SCHMIDT, the most distinguished meteorologist of Stuttgart, who, on the basis of the extreme thinness of the ring, puts forward the idea of a directing force, and he found this force, as a suggestion in a magnetic field centred on the axis of the planet, and acting on diamagnetic matter. You will understand what value there may be in a quotation of this anteriority to which, however, Mr. SCHMIDT attaches importance only as a preposterous hypothesis".

CHAPTER VI.

ON POSSIBLE ELECTRIC PHENOMENA IN SOLAR SYSTEMS AND NEBULAE.

128. The Sun. The series of experiments that I have made with a magnetic globe as cathode in a large vacuum-box, for the purpose of studying analogies to the zodiacal light and Saturn's ring, have led to discoveries that appear to be of great importance for the solar theory.

We have already several times had occasion to give various particulars regarding the manner in which these experiments were carried out. It is by powerful magnetisation of the magnetisable globe that the phenomenon answering to Saturn's rings is produced. During this process, polar radiation and disruptive discharges at the equator such as that shown in fig. 247 a (which happens to be a unipolar discharge) may also occur, if the current intensity of discharge is great. If the magnetisation of the globe

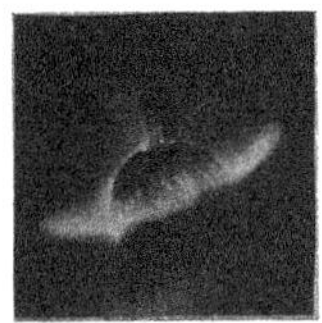

a Fig. 247. b

be reduced (or the tension of the discharge increased) gradually, the luminous ring round the globe will be reduced to a minimum size, after which another equatorial ring is developed and expands rapidly (fig. 247 b). It has been possible for the ring to develope in such a manner that it could easily be demonstrated by radiation on the most distant wall of my large vacuum-tube (see fig. 217). The corresponding ring would then have a diameter of 70 cm., while the diameter of the globe was 8 cm.

It is a corresponding primary ring of radiant matter about the sun that in my opinion can give an efficient explanation of the various zodiacal light-phenomena. In the above-mentioned experiments, it is seen how the rays from the polar regions bend down in a simple curve about the equatorial plane of the globe, to continue their course outwards from the globe in the vicinity of this plane. An aureole is thereby produced about the magnetic globe, with ray-structure at the poles, the whole thing strongly resembling pictures of the sun's corona.

Rarefied gases, rendered luminous by similar discharges from the sun, would first emit a light of their own, and then diffuse that of the sun.

It is well known that the spectrum of the corona contains above all a brilliant ray of coronium $\lambda = 5304$, and besides this there is a faint continuous spectrum, probably due to reflected solar light.

If the sun's corona is of an electric origin such as we have here assumed, we might perhaps expect to see an enormous ring of light about the sun every time the earth, during an eclipse of the

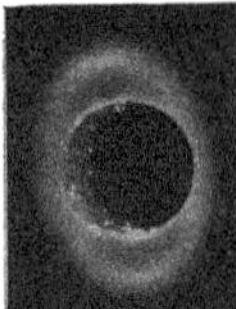
e

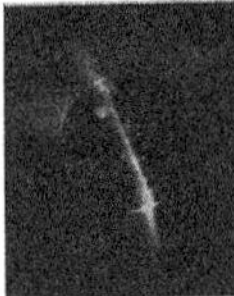
d

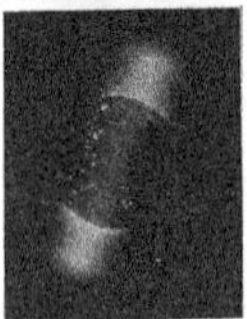
c

Fig. 248.

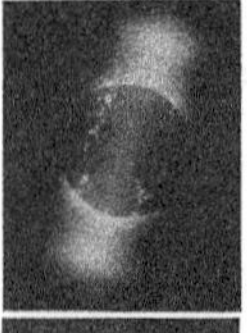
b

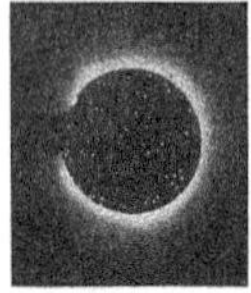
a

sun, stood very nearly in the plane of the sun's equator. This would have to be upon the assumption that in the spaces far from the sun, there is a gas that can become electrically luminescent, or, in an electric state, able to reflect sunlight.

It is possible to believe, however, that the sun's chromosphere, which is a sharply-defined envelope of hydrogen, is again surrounded by an envelope of coronium, of almost limitless extent.

Analogies from the earth's atmosphere, whose nature has been made clearer through the latest researches of HANN, HUMPHREY[1] and WEGENER, seem at any rate to indicate that the above-mentioned assumption is probable.

Wegener has recently[2] shown that there must be new fundamental layer-limits in the earth's atmosphere. Above a covering of hydrogen, which prevails from a height of 75 to 200 kilometres, a new gas is to be found, which he calls geocoronium, extending up to such heights that the steady auroral arcs, for instance, that are observed as much as 600 kilometres above the earth, would be due to electric luminescence in this gas.

129. We will now pass on to experiments that in my opinion have brought about the most important discoveries in the long chain of experimental analogies to terrestrial and cosmic phenomena that I have produced. In the experiments represented in figs. 248 a—e, there are some small white patches on the globe, which are due to a kind of discharge that, under ordinary circumstances, is disruptive, and which radiates from points on the cathode. If the globe has a smooth surface and is not magnetised, the disruptive discharges come rapidly one after another, and are distributed more or less uniformly all over the globe (see a). On the other hand, if the globe is magnetised, even very slightly, the patches from which the disruptive discharges issue, arrange themselves then in two zones parallel with the magnetic equator of the globe; and the more powerfully the globe is magnetised, the nearer do they come to the equator (see b, c, d). With a constant magnetisation, the zones of patches will be found near the equator if the discharge-tension is low, but far from the equator if the tension is high.

Fig. 248 e shows the phenomenon seen from below.

If the pressure of the gas is very small during these discharges, there issues (fig. 249, globe not magnetised) from each of the patches a narrow pencil of cathode-rays so intense that the gas is illuminated all along the pencil up to the wall of the tube. This splendid phenomenon recalls our hypothesis according to which sun-spots sometimes send out into space long pencils of cathode-rays.

SCHUSTER has recently[3] made some serious objections to the hypothesis that sun-spots emit direct, rather well-defined pencils of cathode-rays, a hypothesis which was put forward by me in 1899 and 1900, and by MOUNDER in 1904.

(1) HUMPHREY, Distribution of Gases in the Atmosphere. Bull. of the Mount Weather Observatory, II, 2.
(2) WEGENER, Zeitschrift für anorganische Chemie, B. 75, p. 107. 1912.
(3) The Origin of Magnetic Storms. Proc. of Roy. Soc., 1911.

Schuster considers that the velocity of such cathode-particles, as they sweep past the earth, is duced to about nine kilometres per second, and that the passage between the sun and the earth would ke about a year, so that the magnetic effects of such rays could not reproduce, even roughly, the aracteristic features of a magnetic disturbance.

He does, it is true, say at the conclusion of his paper:

"It is otherwise with the more refined form in which the theory has been presented by Prof. rkeland, who, qualitatively at any rate, has shown that an agreement might be reached, if we can agine the particles to be drawn in towards the earth by its magnetic forces, so that for the time being eir motion is regulated by the position of the earth's magnetic poles. Nevertheless, the argument from ergy and from electrostatic considerations alike, has now been shown to be fatal to the theory in y form".

I do not think, however, that Schuster's objections have any serious bearing on my theory, if we nsider the properties which the new sunbeams must be assumed to possess.

I have shown that cathode-rays from the sun, which are to strike down towards the earth in the rora polaris zones, must have a transversal mass about $m = 1.83 \times 10^3 \times m_0$. In other words, the ngitudinal mass of our particles is 6 milliard times greater than the mass of the particles upon which huster calculates in his energy-comments. Thus these cathode-rays will pass the earth, not with a locity of 9 kilometres, but with a velocity very little short of that of light.

In his further development, Schuster shows that ordinary cathode-rays that issued from the sun in well-defined, narrow pencil, would instantly be dispersed; for the electrostatic repulsion to which a rticle near the limits of the pencil would be subjected from the other particles in the pencil of rays, ould, according to Schuster's calculation, impart to an electron an acceleration so great that in the very st second it would fly over a distance of astronomic magnitude.

If the calculation is applied to *our* rays, this acceleration would have to be divided by 3.3 millions. t even with such an acceleration, an electron would move to a great distance in the 500 seconds that ray with the velocity of light takes in passing from the sun to the earth.

There is still, however, another point of great importance to be considered, and that is that in my eory the magnetic storms on the earth are *not* caused by a great, more or less cylindrical pencil of ys at a great distance from the earth, but generally a small, fine pencil of rays is drawn in in an arc wn to a minimum distance of from 200 to 300 km. from the earth in the aurora polaris zones. These drawn pencils of rays act partly directly over the earth, partly indirectly by the earth-currents which ey induce.

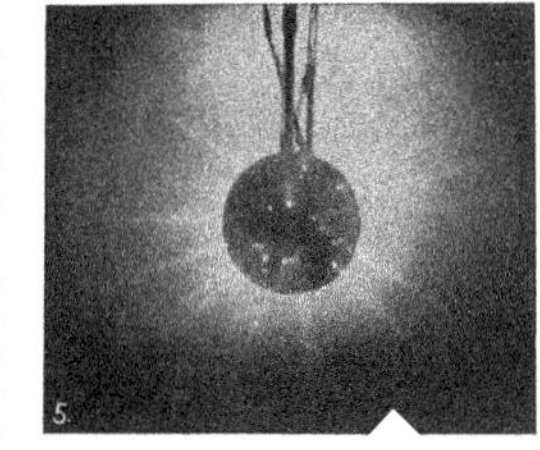

Fig. 249

Let us return to our experiments. If the globe is ghtly magnetised, the patches of eruption are seen to range themselves in zones, with long pencils issuing into ace, almost as in fig. 249; only these pencils are bent the magnetism, which is exactly analogous to what we ve assumed regarding the cathode-rays issuing from e sun.

These centres of eruption for the disruptive discharges come more marked by the addition of some Leyden jars rallel to the discharge-tube; but care must be taken not add too much capacity, as the discharge may then become oscillatory. I have generally employed about 10 to milliamperes as the discharge-current for the globe of centimetres diameter.

If the metallic globe surrounding the electro-magnet is not smooth, but has sharp points on its surface, for instance near the poles, the disruptive discharges would issue at these points, and it will be necessary to use a stronger magnetisation to make the patches arrange themselves in zones round the equator.

From the results obtained by SWABE, WOLF, CARRINGTON and SPOERER, we know that the sun-spots arrange themselves just in two zones between 5° and 40° N and S latitude, in such a manner that in the minimum-period of the spots, they begin to show themselves in high latitudes, and then descend until at their maximum-period they have reached a latitude of about 16° north and south. If we remember especially that the spots are the centres of emission of very stiff cathode-rays ($H\rho = 3 \times 10^6$ *C. G. S.*), which give rise to auroras and magnetic perturbations on our earth, it would appear as if the sun-spots were the foot-points of disruptive electric discharges from the sun. The possible depressions in the enveloping photosphere by the sun-spots, which many astronomers believe to exist, can be easily explained by reference to an experiment with discharges from a quicksilver cathode in a vacuum-tube (see fig. 201. Winkelmann's Handbuch der Physik, 4, p. 530). The pressure that the discharge here exerts upon the surface is probably proportional to the energy of the discharge, which, as we shall see, must be enormous in the case of the sun.

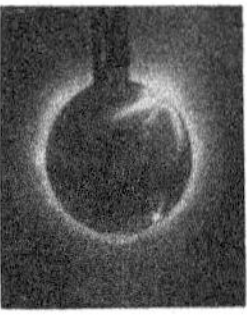

Fig. 250.

If the pressure of the gas increases, the pencils of rays no longer issue radially from the globe, as in fig. 249, but the disruptive discharges are often seen to manifest themselves in the shape of a star with four or five arms (see fig. 250), coming from an eruptive spot, and almost following the surface of the non-magnetic globe, to meet often at a point on the globe diametrically opposite.

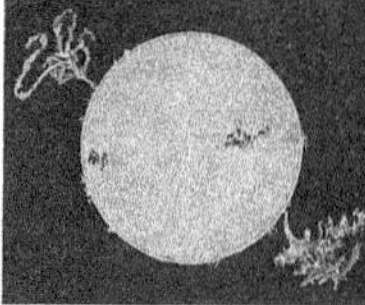

Fig. 251.

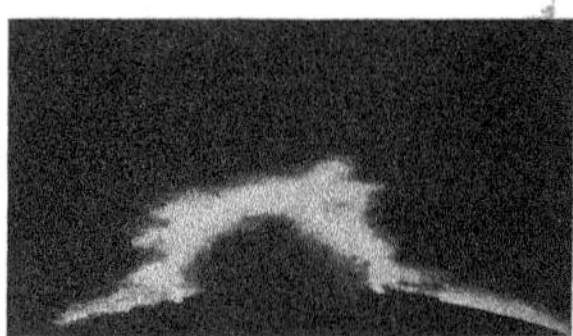

Fig. 252.

These discharges from opposite points (this is not clearly seen in fig. 250, however) brought to my mind a very strange picture of some enormous eruptions on the Sun (see fig. 251), reproduced from "Marvels of the Universe". On June 26th, 1885, M. TROUVELOT saw two huge prominences, each more than three hundred and fifty thousand miles in height, rising from the sun. Flames of such dimensions are exceedingly rare; it is therefore all the more significant that they rose exactly opposite to each other from the ends of the same diameter.

It almost always happens too, in the experiment in which the cathode-globe is magnetised, that there are two or three luminous branches turning in a spiral about the eruptive spot and near the surface of the globe. These vortices move in the opposite direction to that of the hands of a watch on the hemisphere containing the magnetic north pole, and in the same direction on the opposite hemisphere.

This corresponds exactly with the results recently obtained by HALE, ELLERMAN, and FOX relative to vortices in the hydrogen filaments and calcium vapour round a sun-spot, provided it is admitted, as I have found, that the sun and the earth are inversely magnetised (Comptes Rendus, Jan. 22, 1910).

These vortices round the spots on the magnetic globe, I have not succeeded in photographing with he present arrangements. On account of the importance of all these phenomena, however, I have constructed vacuum-vessel of 320 litres' capacity, and can employ a magnetic globe with a diameter of 24 centimetres. Vith this new apparatus, I have succeeded in obtaining good photographs, which will be mentioned below.

The discharges of the cathode-globe are partly continual discharges all over the surface, and partly lisruptive at intervals; in the latter case they issue from the eruptive spots.

Fig. 253 shows how a branch of discharge issuing from the spots sometimes follows the magnetic ines of force in the neighbourhood of the equator, giving rise to a phenomenon which greatly resembles he black filaments on the sun, studied by HALE, ELLERMAN, FOX, EVERSHED, DESLANDRES and D'AZAMBUJA.

It will be of considerable interest to compare this experiment with some photographs of quiescent rominences on the sun. Fig. 252 is a reproduction of one of Prof. HALE's earliest prominence photographs taken at KENWOOD's Observatory. I have unfortunately no data to enable me to decide whether his prominence follows more or less the lines of magnetic force on the sun.

Fig. 253.

I have sought by various methods to find value for the very singular capacity of this globe orresponding to disruptive discharges, a capacity vhich seems to vary perceptibly according to the onditions of the discharge. In the case of this lobe (8 cm. in diameter), this capacity varies bout $\frac{1}{100}$ of a microfarad, and if I assume that he sun has a corresponding capacity C in the elation of the square of the diameters, I find that $C = 3 \times 10^{18}$ microfarads.

In calculating the tension of the solar discharges according to the value $H\varrho = 3 \times 10^{6}\, C.G.S.$ (see M. Abraham, Theorie der Elektrizität, B. II, p. 183, equation (120 bis)), I find that $E = 6.4 \times 10^{8}$ olts. The energy $\frac{1}{2} E^2 C = 5.9 \times 10^{30}$ ergs, transformed into heat, will be sufficient to heat to 175° $C.$ globe of iron the size of the earth.

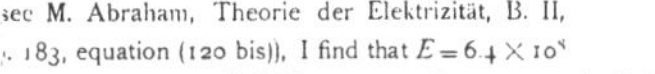

Sun-spots may be considered as the eruptive centres of similar disruptive discharges, and the uestion then immediately arises: Where shall we seek for the positive pole of these discharges, in which the spots, or that which surrounds them, represent the cathode?

There are several possible solutions to this question.

In the first place, it might be imagined that the *interior* of the sun formed the positive pole for normous electric currents, while perhaps the faculæ, in particular, round the spots, formed the negative oles. Or it might be imagined that the positive poles for the discharges were to be found *outside* the hotosphere, for instance in the sun's corona, the primary cause of the discharge being the driving away f negative ions from the outermost layers of the sun's atmosphere in some way or other—for instance, s ARRHENIUS has assumed, by light-pressure after condensation of matter round them. Finally, it might e assumed—and this, according to the experimental analogies, seems the most probable assumption— hat the sun, in relation to space, has an enormous negative electric tension of about 600 million volts.

The *first* assumption has the advantage of appearing to give a natural explanation of the movement of the sun-spots in various latitudes, provided that the sun's magnetisation is the opposite to that f the earth.

In this case the origin of the sun-spots must be that the presumptive more or less photospheric envelope was sometimes pierced by disruptive discharges, thus forming great elec That the tension necessary to pierce the photosphere would be very great would not be s this alone being sufficient to explain the very great rigidity of the cathode-rays emitted.

The temperature of the spots should, upon this hypothesis, be very high. This, it is s not seem to be well confirmed by the measurements; but the temperature of a spot cannot be by STEFAN's law, because under high degrees of dispersion the spectrum of the spots is not co it contains nothing but lines.

It may be imagined that under the action of these violent arcs the photosphere tends t more insulating (thicker?), and that after the maximum of the spots, the discharges cannot pen photosphere as easily as after a certain cooling by radiation. The discharges then begin agai latitudes as long as the necessary tension is at its maximum.

We do not know sufficiently how electric arcs move in gases, but it is at any rate not di magnetic forces, to attain a transversal velocity of 200 metres per second for an electric arc ir

In order to be able to some extent to form an estimate of the manner in which the electric arcs in the sun would move, we ought to know how the sun's magnetism is distri rather its cause. In my opinion it is the pencils of cathode rays appearing at indefinite interva outbreak and in the development of the sun-spots, that give rise to solar magnetism by creatin constant currents by induction in the conductive interior of the sun.

I have several times begun the calculations that should serve to verify my hypothesis, but not yet completed.

We know that the electric currents circulating in great spheres have a very great persist LORBERG, Crelles Journal, vol. 71, 1870, and LAMB, Phil. Trans., 1883). Lamb finds that in sphere of the size of the earth, the time necessary for a current to fall to $\frac{1}{e}$ of its initial valu million years.

The induction impulses originating in the cathode-rays emitted at intervals from the sun, be able, in the course of time, to create a perceptibly constant current.

In support of my calculations, I am making experiments with a rotating sphere made of softest magnetisable steel. The diameter of the sphere is 70 cm. The results of these investiga be included in the next volume.

If, to obtain a clearer conception, we assume a circular current round the centre of the s plane of the equator, and with a radius equal to half the solar radius, it becomes easy to calc magnetic effects in different latitudes of the photosphere. In assuming spherical currents, we c same degree of conformity with the currents circulating much nearer the solar surface.

The table gives F_β divided by $\cos\beta$ for each ten degrees of latitude comprised between 0° where F_β is the component of the magnetic force in an arbitrary unit, the length of the merid purposes of comparison, $\cos^2\beta$ is given, which, according to FAYE, should be perceptibly prc to the variation of the angular diurnal motion of the spots.

β	0°	10°	20°	30°	40°	50°
$F_\beta \sec\beta$	1.17	1.10	0.88	0.69	0.54	0.41
$\operatorname{Cos}^2\beta$	1.00	0.97	0.88	0.75	0.59	0.41

These figures have perhaps a certain interest, although, as we have said, we do not y well how electric arcs move in gases, under the action of magnetic forces.

The *second* assumption may indeed, from a physical point of view, be possible, but it is probable that any process of this nature will play a decisive part in these phenomena. It wou

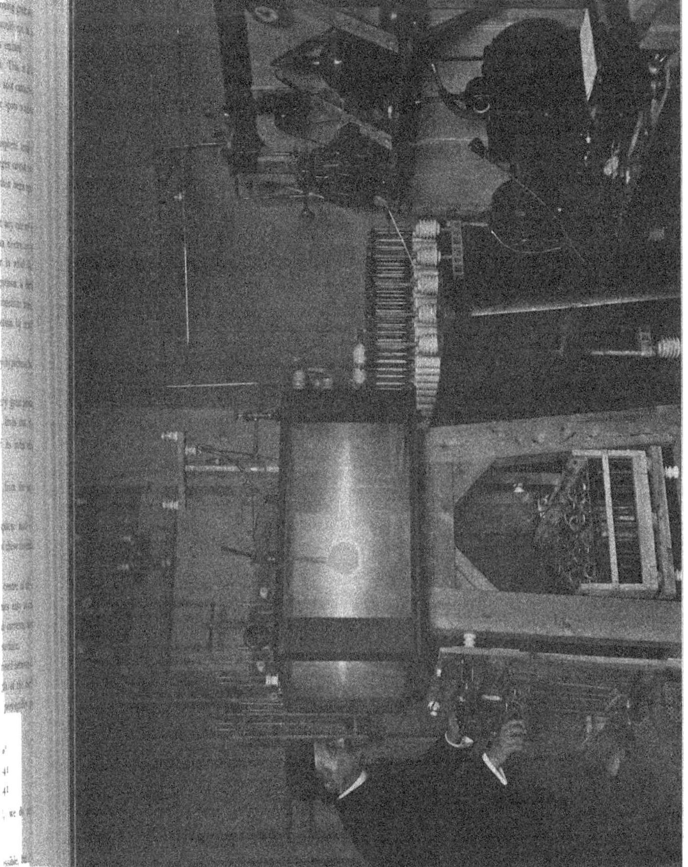

Fig. 254.

that the sun's nucleus received a positive charge, of which, it must be imagined, it would to som gradually get rid in the interval between two outbreaks of sun-spots.

There is one circumstance that is perhaps in favour of this assumption, as also of the f that is the peculiar capacity that the sun, in analogy with our magnetic globe, must have. It s if an electric condensation must take place, so that the opposed masses of electricity are found as lying close to one another.

The *third* assumption seems the most natural when the matter as a whole, is looked at f point of view of the experimental analogies. It is then a question of the manner in which this charge on the surface of the sun has been produced in interaction with space. If to the negati of electricity on the external surface of the sun, there are to some extent corresponding masses tive electricity in the interior of the sun, the first and third assumptions may be combined, whi would allow of the mysterious movements of the sun spots in the various latitudes being expl an electromagnetic action.

It must moreover be admitted, even in the third case only, that a magnetic influence on t ment of the sun-spots was to be expected, if, as has here been done, the arrangement of the sun-spo parallel rows, one on each side of the equator, is assumed to be the effect of the magnetic co The question then is whether it is possible, by an estimate, to show the probability of an exp of the actual motion of the spots—in the third case as well—only as a magnetic influence. This to be difficult. It is true that the pencils of cathode rays that radiate from sun-spots in higher l curve rapidly down towards the equator, thereby causing the component of the magnetic force angles to the current-element to be comparatively much greater in the third case than assume first; but whether this can cause the magnetic retrograde motion eventually produced to be more in the case of sun-spots in higher latitudes, than in that of spots in lower latitudes, is doubtful. T bution of the sun's magnetism may perhaps be rather different from what we assumed in the fi and thus a fairly good explanation could be given. At any rate, the rotation of the sun's body it be greater than the apparent rotation of any sun-spot, and this really agrees with the actual circu

SPOERER's discovery that groups of sun-spots are inclined to be drawn out in length in a along a parallel circle on the sun, so that the spots appearing last come to the west of those in existence, speaks most in favour of a combination of the first and third assumptions.

The same may be said of SECCHI's discovery with regard to the characteristic leaps in th rotation of a sun-spot, as the leaps usually take place in the direction of the rotation.

It is at present not easy to see how a negative tension should be continually created by in relation to space.

It is of course possible to imagine that a surplus of positive ions is always being carri from the sun or that negative ions are always being carried towards the sun, and that the tension is produced in this manner; and that the balance is maintained to some extent by dis ruptive discharges, as we have presupposed.

It seems a natural thing, however, to connect the creation of this tension with the sun's of light and heat. But as MAXWELL's electro-magnetic light theory at present stands, there is r opportunity of assuming that light-energy is carried over into electric energy, and that for tha the rays of light are absorbed into space.

It is thought by several that Maxwell's equations require a correcting term. Such a ter perhaps have influence just when there was question of a disturbance that spread into infinite

RIEMANN's discoveries in the transition from infinitely small to finite amplitudes in soun might possibly afford some information.

Fig. 255.

The idea of an unknown transformation, in space, of radiant light and heat from the sun into another form of energy, seems to have occurred recently to other scientists.

In a paper, just published, by JULIUS, on the results from the "Netherlands Eclipse Expedition, 1912",(1) the following conclusion is found:

"*Less than $\frac{1}{1000}$ of the total* (ultra-violet, visible, and infra-red) *solar radiation proceeds from those parts of the celestial body which lie outside the photospheric level.*

"This result proves that it is impossible to maintain the theory which considers the photosphere to be a layer of incandescent clouds, whose decrease of luminosity from the centre toward the limb of the solar disk would be caused by absorption and diffusion of light in an enveloping atmosphere ("the dusky veil"). For if this theory were right, then, according to the calculations made by PICKERING, WILSON, SCHUSTER, VOGEL, SEELIGER and other astrophysicists, such an atmosphere should absorb an important fraction ($\frac{2}{3}$ to $\frac{1}{3}$) of the sun's radiation. Now, as the fraction emitted appears to be smaller than $\frac{1}{1000}$, and yet the atmosphere must be in a stationary condition, one would be forced to conclude that the main part of the absorbed energy is continually being dissipated through space in some absolutely unobserved form. This necessary inference not being acceptable, we must look for another interpretation of the photosphere."

However this may be, it would be very interesting if the energy of the light and heat rays could to some extent return to the sun from space. The electric rays possibly reach as far out as the light rays, or at any rate exceedingly far, and the greater part of the energy in an electric discharge such as this may gather at the cathode, i. e. on the surface of the sun, where the electric arcs in their turn would create new heat for the radiation of more light.

In this way the age of the sun, which HELMHOLTZ and KELVIN, according to the mechanical heat theory, put at not more than 50 million years, may perhaps be put at so many hundred million years, as geologists, after researches on the earth, absolutely require. There are doubtless other sources and reservoirs of energy than those with which we are now acquainted. HELMHOLTZ was not acquainted with radium, for instance, which has of late been made use of to make the sun old enough.

129. It will be immediately apparent what far-reaching consequences are here built upon our experimental analogies. There seems to be a constantly increasing appreciation of the fruitfulness of the method established by the representation of such analogies for the study of celestial phenomena.

In 1860, HUGGINS made a laboratory, where numerous physical experiments were made for the interpretation of astronomical observations. The advantage of imitating the celestial phenomena in laboratory experiments, a method which forms exactly the base of the present studies, was thus known and appreciated half a century ago. The method has been followed by many, and has of late yielded marvellous results, HALE having discovered the existence of powerful magnetic forces in the solar vortices, and DESLANDRES having in this way made some very interesting experiments on the solar corona.

The important phenomena, which I have discovered, of disruptive discharges from points on a magnetic cathode-globe, have especially occupied my attention.

In order to investigate closely the electric analogies to the vortex-formation about the sun-spots, and to study the wonderful capacity that the globe seems to have in these disruptive discharges, I have recently resumed the whole of my experimental series with an entirely new arrangement, in which a magnetic cathode-globe of 24 cm. diameter could be employed.

I will here only give a schematic description of these experiments, of which good photographic reproductions are found below.

(1) Koninklijke Akademie van Wetenschappen te Amsterdam, May 23, 1912.

Fig. 254 shows the whole arrangement with the new vacuum-box of 320 litres. Floor and ceiling re here made of 12 mm. steel plates, the pillars between are of bronze, and the sheets of plate-glass t the sides are 30 mm. in thickness.

The experiment shows the "zodiacal-light ring". It requires little magnetising of the globe (11.3 cm. ı diameter), but a great discharge-current (up to 100 milliamperes). Similar experiments are shown in gures 255, 1 and 2. In the former the magnetic globe is only 2.5 cm. in diameter; but it was easy, specially with greatly rarefied hydrogen gas in the box, to obtain a plane of rays about the globe that ut all four glass walls in brightly phosphorescent, straight stripes from 5 to 10 millimetres wide.

It is easy to prove that the plane of rays is partly formed of rays from the upper hemisphere of ıe cathode, that are bent down towards the equator, and rays from the lower hemisphere that are bent

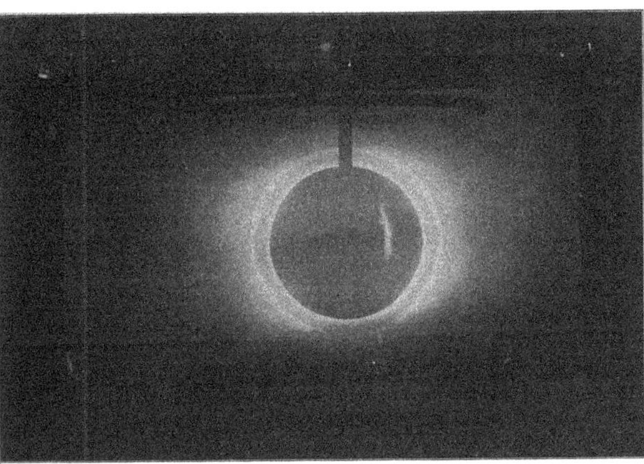

Fig. 256.

ɔwards. It will without doubt be possible to produce, with a very small cathode-globe, a ring greater proportion to the globe than is the real zodiacal-light ring in proportion to the sun, even if the latter ng be assumed to go right outside the earth's orbit. It is only by careful adjustment of the magnetit ion of the globe, however, that the ray-masses are made to coincide, so to speak, exactly in one plane.

In general, the ray-masses from above and from below intersect one another in the plane of the juator; and it is easy to form round the circle of intersection a strongly luminous ring, floating in ıace round the globe, and resembling a nimbus such as painters in olden times painted round the heads saints.

The intersecting groups of rays may often be found upon the glass walls in the form of two seırate parallel phosphorescent bands of light that can be moved to and fro by slight variation in agnetisation. I believe I have seen these groups of rays twice form circles of intersection (node-circles), when the magnetising was so arranged that the groups formed only a very small angle with one another. It would thus appear that the rays move preferably above and below the plane of the equator.

As the magnetisation is made stronger and stronger, the "node-line" in the form of a luminous circle, will approach the cathode-globe, but suddenly the balance will be disturbed, and the phenomenon will go over into a secondary ring—"Saturnian ring"—which only developes into full beauty with strong magnetisation and small discharge-current, as represented in fig. 255, 3, where the cathode-globe is 24 centimetres in diameter. Applied to the sun, our experiments would imply that we must here assume a comparatively low magnetisation, but comparatively high electric radiation.

One can imagine that among the various kinds of cathode-rays that the sun can emit, there are especially a great many that will be brought by solar magnetism to move near the plane of the sun's magnetic equator, possibly bending alternately above and below it.

Fig. 257.

Fig. 256 shows phenomena with the large 24 cm. cathode-globe — a light that resembles the sun's corona. (1)

Applied to Saturn (fig. 257), our experiments must lead us to infer that the quantity of rays emitted by the planet was comparatively small, while the magnetisation was comparatively greater than that of the sun.

Our experiments with the large cathode-globe (see fig. 255, 3) show that if it is desired to have the ring very thin, it is better to go down to about $^1/_{10}$ milliampere; but in that case the light will also be faint. The ring looks now, however, quite as thick and distinct as with $^1/_{10}$ milliampere and with one of the small cathode-globes.

Let us now simply assume that the current issuing from Saturn is as many times greater than $^1/_{10}$ milliampere, as the radius of the planet is greater than that of our globe-cathode. This gives us about 50000 amperes from Saturn. Let us assume the tension to be 100 million volts. We then find

(1) As all these figures show, the apparatus has been illuminated beforehand with ordinary light, and the experiments then made and the electric light-phenomena photographed. In this way various reflexions appear in the figures that have nothing to do with the phenomena, but they will not give rise to misunderstanding.

it the radiation from Saturn would answer to 5 milliard kilowatts. This is comparatively no great amount energy, for the lightning on our earth probably represents on an average from 4 to 5 milliard kilowatts.

This last figure I obtain in the following manner.

ARRHENIUS computes the amount af combined nitrogen falling upon the land-surface of the earth (36 million sq. km.) in the form of nitrate and nitrite of ammonia, at about 400 million tons per annum. we take for granted that a comparatively similar amount also falls upon the sea, this gives us *one* rt out of every *three million* of the nitrogen of the atmosphere as the amount that is thus combined ery year, and this, we may say with practical certainty, almost exclusively by electric discharge.

Fig. 258.

Now as we know by experiment that by the most effective electric discharges 600 kg. of nitric id is formed by the air per kilowatt-year, we can calculate that the lightning that produces nitric acid pour in the atmosphere must at least answer to an average force-supply of 4 milliard kilowatts.

We will return to our experiments with the large cathode-globe in our 320-litre vacuum-box, as as the previously-mentioned disruptive point-discharges are concerned, these, it will be remembered, ing compared with sun-spots.

It was soon evident that the quite smooth, silver-coated, large globe of 24 centimetres' diameter, is not by any means a success when it was a question of getting these negative point-discharges upon The smaller globes were much better, but it was apparent that the nature of their surface had much say in the matter. These experiments showed that with the smallest globe (2.5 cm. in diameter), it s easy to obtain, instead of the brief disruptive point-discharges, lengthy discharges from such points, ly provided there was a high vacuum, and that the current-strength of the discharge was great. These ncil-discharges would suddenly change place, and arrange themselves near the equator like the earlier spots. This possesses considerable interest, inasmuch as the sun-spots in reality represent phenomena of long duration, and not brief discharges.

With the large, silver-coated globe, it was thus very difficult to obtain point-discharges when the globe was cathode; they were pre-eminently continuous discharges from the entire surface or large portions of it.

When, however, the globe is made the anode, and the metal walls of the box, which are comparatively rough, unpolished, cast or rolled plates, are the cathode, a perfect firework-display of point-discharges takes place, in rapid succession, from the inner walls of the box. Not only were the points

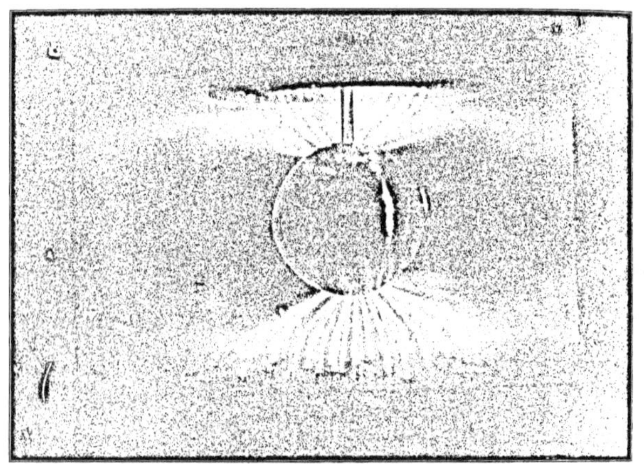

Fig. 259.

luminous, but long pencils of rays passed from the points (almost like a kind of lightning) in to the globe. Glowing metal particles were often torn from the points, especially from the steel plates, whence particles shot inwards along the path of the current.

In fig. 258, only the foot-points are visible, for when the anode-globe was non-magnetic, the flashes in towards the globe, though fairly powerful, were too brief and of too little intensity too be fixed upon the photographic plate with the camera used. When, on the other hand, the anode-globe was magnetised, the flashes became more intense (see fig. 259), and the points of discharge were congregated in the vicinity of the magnetic poles of the globe. The discharge-rays gathered in two zones about the poles of the anode-globe, as might be expected; but there also appeared a faint band of light, of which an indication may be seen, round the magnetic equator of the anode-globe.

In order to obtain point-discharges with my globe-cathode of 24 centimetres' diameter, I took two hemispherical shells of aluminium, and had them "sand-blasted" outside at a glass factory in the manner employed in the production of ground glass.

As soon as these shells were put on outside the silvered globe, I obtained point-discharges in great umbers; but they were not so intense as I had expected, not even when a large condenser was placed ı parallel with the vacuum-tube. It was only after having exhausted my discharge-box for a long time nd filled it with hydrogen, and again and again exhausted it, that these point-discharges began to be owerful.

Figs. 260 a, b, and c show three photographs of discharges under varied conditions.

The first is of an experiment with a considerable gas-pressure and very slight magnetisation of the lobe. It shows an interesting radiation from the polar regions, but the point-discharges, which, it is ue, are most numerous in the equatorial regions, have not separated into two zones as they usually id when the surface of the cathode was smooth.

The third photograph is of an experiment in hydrogen gas with a very high vacuum.

The phenomena here were powerful and sometimes of a distinct duration, that is to say not istantaneous discharges. Another interesting circumstance is that under the above-mentioned experimental

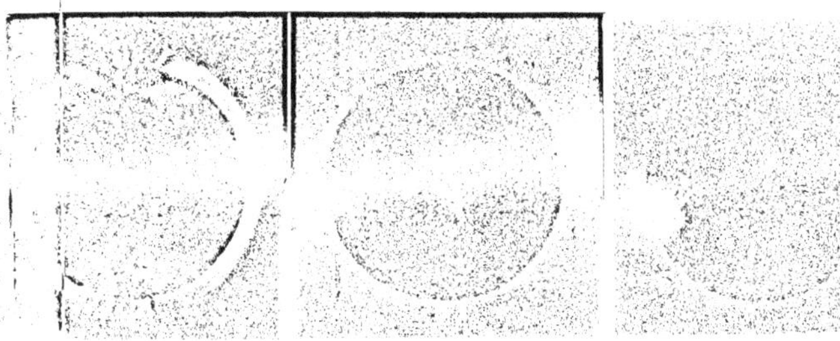

a b

Fig. 260.

nditions it was distinctly seen that the patches are not always single spots, but often consist of a group spots. For instance, on the original photograph answering to Fig. 260 c, the spot above to the right distinctly a group of 5 separate spots. We thus have here another analogy to the sun-spot conditions.

As I have frequently mentioned, I have tried in vain to photograph the vortices that sometimes evelope into great beauty round the points of light in these point-discharges on the globe-cathode. I ive said that the motion of these vortices is always counter-clockwise on the upper hemisphere, and ockwise on the lower, supposing the globe to have been magnetised so as to have a magnetic *north* *le* in the upper hemisphere. In the reverse case, the conditions are of course reversed.

A chance occurrence has now enabled me to produce these vortices with much greater brilliancy ian before. It was as follows. The vacuum-box was exhausted by a rotary mercury-pump (Gaede ımp), with a rotary oil-pump in series, both pumps being worked by small electric motors that were nnected with the electric current system of the town. Sometimes, in cases of necessity, I left the pumps orking while we were absent from the Institute.

On one occasion the tension was broken off, so that the motor stopped; and notwithstanding my lf-closing valves, the vaseline-oil from the oil-pump passed through the mercury-pump and into my rge vacuum-box.

It required considerable labour to put everything into order again, but, after renewed pumping, it was found that a little oil trickled out on to the floor of the box, thus showing that it had not all been removed.

After filling the box with hydrogen and emptying it several times, the point-discharges from the globe-cathode were much more marked than before, being peculiarly intense, even without being coupled to any external capacity. The vacuum-box too, now happened to be so air-tight, that after letting it stand untouched for a week, it was impossible to detect the entrance of any foreign gas.

The most striking feature, however, of these point-discharges—which, as I have shown, have a preference for a hydrogen atmosphere—was that the frequently-mentioned branches radiating from the point of light were so intense that they could easily be photographed by the aid of a cinematographic lens. It is evident that vapours from the vaseline-oil or decomposition gases here play a part.

When the cathode-globe was *not* magnetised, the light-tracery that appeared round the point-discharge resembled a many-armed starfish (fig. 261 a). On rare occasions it happened that the arms of light could

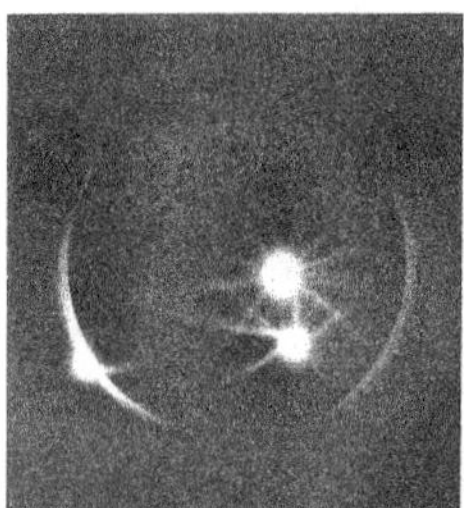

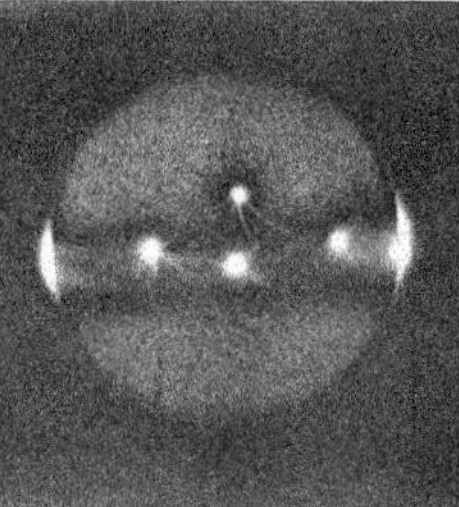

b
Fig. 261.

be followed right round the globe, where they met at a point diametrically opposite to the point of discharge. These meeting-points of the arms of light might also have the appearance of a faint point of discharge. This calls to mind TROUVELOT's drawing, which is reproduced in fig. 251.

When the cathode-globe is magnetised with the north pole uppermost, the points of discharge move near to the magnetic equator. The arms of light about these points still exist, but they have received a twist so that the vortices created have a counter-clockwise motion on the upper hemisphere (fig. 261 b), and clockwise on the lower (fig. 261 c). With a magnetised globe also, the light from a point of discharge seemed to radiate and as it were meet in a diametrically opposite point on the globe; the light runs at any rate right round the equatorial regions every time a point-discharge occurs. It is understood from the direction of the twist, that the arms of light radiating from the points of discharge, and sometimes encompassing the globe, are a *negative* radiation and thus of the same kind as that which issues almost perpendicularly from the globe (see fig. 249).

If, therefore, we take for granted that the sun and the earth are oppositely magnetised, as, for other reasons, I have previously assumed (C. R., Jan. 24, 1910), then, if the analogies are correct, *negative* electric radiation will give rise to the vortices round sun-spots, studied by HALE and ELLERMAN.

In some spectrographic researches on prominences on the solar disc, Fox makes the following tatement (1): "Examination of all the H_α (hydrogen) plates and the record of earlier observed whirls in he calcium vapours results in assigning the direction as counter-clockwise in the northern hemisphere nd clockwise in the southern. This is in agreement with the demands of FAYE's theory."

In analogy with our experiments, these whirls should not be due to *cyclones* or *whirlpools*, as Faye upposes, but to negative electric emission from certain centres of electric eruption. It should be remarked hat as this electric emission is connected with calcium vapours and with hydrogen, it is to be expected hat its velocity will not be nearly so great as that of light.

Owing to the good experimental results, which already give certain promise of the attainment of a full nderstanding of the two above-mentioned important phenomena—the vortex-formation and the apparent reat capacity of the cathode-globe,—I have begun to construct a vacuum-vessel of 1000 litres' capacity, with eiling and floor of bronze, and glass sides of 50 mm. thickness. There have proved to be disadvantages n having the floor and ceiling magnetisable (of steel) and in their not being far enough from the polar arts of the large cathode-globe. The magnetic cathode-globe is to be 40 centimetres in diameter, for ischarges of 500 milliamperes at 15 000 volts, which is the maximum delivery of my machine (see fig. 67). t will be easily understood that in addition to the purely scientific reasons for doing this, I have also a econdary object, which is to give myself the pleasure of seeing all these important experiments in the nost brilliant form that it is possible for me to give them.

131. The Worlds in the Universe. From the conceptions to which our experimental analogies ead us, it is possible to form, in a natural manner, an interesting theory of the origin of the worlds. This theory differs from all earlier theories in that it assumes the existence of a universal directing orce of electro-magnetic origin in addition to the force of gravitation, in order to explain the formation ound the sun of planets—which have almost circular orbits and are almost in the same plane—of moons nd rings about the planets, and of spiral and annular nebulæ. Even the newly-discovered, most distant noons of Jupiter and Saturn, with their retrograde revolution, do not place the theory in any doubtful ght; on the contrary, the discovery would seem to predict that if planets are still discovered round the un sufficiently far outside Neptune, they might also have a retrograde revolution.

The fundamental assumption with which we shall start will correspond with one of the three abovenentioned assumptions regarding the sun. For the sake of simplicity, we will assume, in conformity ith case 3 above, that all suns in relation to space have an enormous negative electric tension, different for the different stars, but which, as regards order, might be somewhere about a milliard volts for tars of a class similar to our sun.

In this way electric discharges will be produced, among them being disruptive discharges from comparatively small areas (spots). One might imagine that radiation from these will give rise to circular urrents in the star, parallel with the plane of the equator of the rotating central body, whereby the entral body becomes magnetic.

We can then begin, for instance, to seek for an explanation of the formation of spiral nebulæ.

POINCARÉ, at the conclusion of the preface to his book, 'Hypothèses Cosmogoniques', says:

"Un fait qui frappe tout le monde, c'est la forme spirale de certaines nébuleuses; elle se rencontre beaucoup trop souvent pour qu'on puisse penser qu'elle est due au hasard. On comprend combien est incomplète toute théorie cosmogonique qui en fait abstraction. Or aucune d'elles n'en rend compte d'une manière satisfaisante, et l'explication que j'ai donné moi-même un jour, par manière de passe-temps, ne vaux pas mieux que les autres. Nous ne pouvons donc terminer que par un point d'interrogation."

(1) Astrophys. Journ., November, 1908, p. 257.

Now we know that of the 120 000 nebulæ scattered over the sky, at least half are of a spiral form. The most remarkable thing about them is that there are very often two spirals issuing symmetrically from two diametrically opposite parts of the nebula.

We have previously seen how the continuous discharges round the magnetic cathode-globe in our experiments, could assume a shape that recalled Saturn's ring. These continuous discharges round the globe may, however, with higher gas-pressure in the almost exhausted vessel, take the form of two spirals, curved in the plane of the equator, issuing symmetrically from two diametrically opposite points on the globe.

The accompanying figure (fig. 262) represents an experiment such as this with two such spirals. The photograph was obtained by accident, and I have seen still more interesting pictures appear, several of which I shall publish at some future time.

In the above-named work of Poincaré, a number of older cosmogonic theories, almost all of which are founded upon purely mechanical conceptions, are compared. Those of LAPLACE, LIGONDÈS and ARRHENIUS are of special interest. In the last-named, the so-called light-pressure plays a conspicuous part side by side with the force of gravitation.

In Poincaré's work, all theories are in turn subjected to kindly criticism, with demonstration of the difficulties to which each one leads. It seems to be the celebrated old Laplace's nebular theory that is still considered to be the strongest.

Fig. 262.

Let us now look a little more closely at the idea here put forward, namely, that the sun each day emits by electric evaporation or disintegration considerable quantities of matter in the plane of its equator which forms the part of the electric ring already mentioned, and that in earlier ages this emission of matter has been still greater.

It is not necessary to admit at first the original nebula extended to the orbit of Neptune, as the matter is radiated by electric forces outside the system at its equator. It is very probable, moreover, that the greater part of the matter thus radiated leaves the system, and in any case takes no part in the formation of the planets.

Our analysis will show that particles from the central body may be so ejected that they afterwards move in approximately circular paths near those in which the centrifugal force due to the revolution movement counterbalances the attraction of gravitation; and one could naturally believe that it is just these globules which condense and form large spheres.

Our explanation will be applicable, not only to the planets round the sun, but also to all satellites round the planets. One can imagine Saturn's moons, and Jupiter's, down to the outermost, newly-discovered ones that move round the planet in the opposite direction to the inner, originating in a natural manner from matter, which, under the action of an electro-magnetic directing force, has been ejected from the planets in the plane of the equator.

Looked at in this way, Saturn may still be engaged in making moons by electric radiation. Mimas, almost touching the circumference of the rings, is perhaps the youngest of the satellites.

132. The equations of motion for an electrically charged particle that is in the plane of the equator of a magnetic globe (x, y plane), and is moreover influenced by the gravitation of the globe, are

$$\text{(I)} \qquad \begin{aligned} \frac{d^2x}{dt^2} &= \frac{\lambda M}{r^3}\frac{dy}{dt} + \frac{\mu}{r^3}x \\ \frac{d^2y}{dt^2} &= -\frac{\lambda M}{r^3}\frac{dx}{dt} + \frac{\mu}{r^3}y, \end{aligned}$$

where λ, μ and M (the magnetic moment of the globe) are constants.

From these equations we obtain in the first place

$$\frac{dx}{dt}\frac{d^2x}{dt^2}+\frac{dy}{dt}\frac{d^2y}{dt^2}=\frac{\mu}{r^2}\frac{dr}{dt},$$

whence

$$\left(\frac{dx}{dt}\right)^2+\left(\frac{dy}{dt}\right)^2=-\frac{2\mu}{r}+C,$$

or

$$\text{(II)} \qquad \frac{ds}{dt}=\sqrt{C-\frac{2\mu}{r}}.$$

In the second place we obtain from (I)

$$y\frac{d^2x}{dt^2}-x\frac{d^2y}{dt^2}=\frac{\lambda M}{r^2}\frac{dr}{dt},$$

whence

$$y\frac{dx}{dt}-x\frac{dy}{dt}=-\frac{\lambda M}{r}-a,$$

or in polar co-ordinates,

$$r^2\frac{d\varphi}{dt}=\frac{\lambda M}{r}+a$$

and

$$\text{(III)} \qquad \frac{d\varphi}{dt}=\frac{\lambda M}{r^3}+\frac{a}{r^2}.$$

By dividing (II) by (III), we obtain

$$\frac{ds}{d\varphi}=\frac{r^2\sqrt{Cr^2-2\mu r}}{\lambda M+ar}.$$

Now, however,

$$\left(\frac{dr}{d\varphi}\right)^2=\left(\frac{ds}{d\varphi}\right)^2-r^2,$$

and thus

$$\left(\frac{dr}{d\varphi}\right)^2=\frac{r^4(Cr^2-2\mu r)-r^2(\lambda M+ar)^2}{(\lambda M+ar)^2},$$

and hence

$$d\varphi=\frac{\lambda M+ar}{r}\cdot\frac{dr}{\sqrt{Cr^4-2\mu r^3-(ar+\lambda M)^2}}.$$

Now it is evident that the particle must move in such a manner that the square root in the last expression is always *real*. The radicand must thus be either positive or zero, and hence it follows that those values of r which cause the vanishing of the radicand, define limiting circles, which the particle in its motion can never cross.

It will be seen, moreover, from the expression for $\left(\frac{dr}{d\varphi}\right)^2$, that $\frac{dr}{d\varphi}$ is always and only then 0 (apart from the value $r=0$), when

$$\text{(IV)} \qquad Cr^4-2\mu r^3-(ar+\lambda M)^2=0,$$

that is to say when the particle is on a boundary-circle. Hence it follows that if the particle at a certain moment is retreating from the globe, it will continue to do so until it comes to a boundary-circle; but it will touch this and then turn inwards.

If we imagine a particle that is expelled from the magnetic equator of the globe, and assume that after a limited time it comes to the nearest boundary-circle, it will move back to the globe again, along a path that is symmetrical to the one by which it moved out, i. e. the outward and inward going paths lie symmetrically about the radius vector to the point on the boundary-circle in which the tangent takes place.

The correctness of this is immediately seen when it is remembered that to a given value of r there are only 2 values of $\frac{dr}{d\varphi}$, which are equally great with opposite signs.

If therefore an ejected particle is not to return to the globe, it must move in such a manner as never to reach the nearest boundary-circle. Thus it will move along a spiral with constantly increasing distance from the globe, approaching the boundary-circle asymptotically.

Let us now consider the integral

$$\varphi - \varphi_0 = \int_{r_0}^{r} \frac{\lambda M + ar}{r} \cdot \frac{dr}{\sqrt{Cr^4 - 2\mu r^3 - (ar + \lambda M)^2}},$$

where r_0 is the radius of the globe, and φ_0 the value of φ for $r = r_0$, and endeavour to condition for the existence of such a spiral curve. If $r = r_1$, indicates the smallest bound (provided there are any such, i. e. that (IV) has at least 1 positive root), then the integral infinite for $r = r_1$.

Now it will immediately be seen that if r_1 is a single root in (IV), the function under the sign may be written in the form

$$\frac{1}{\sqrt{r - r_1}} f(r),$$

where $f(r)$ remains ordinary in the vicinity of $r - r_1$, so that we may put

$$f(r) = a_0 + a_1 (r - r_1) + a_2 (r - r_1)^2 + \ldots,$$

whereby the function under the integral sign assumes the form

$$a_0 (r - r_1)^{-\frac{1}{2}} + a_1 (r - r_1)^{+\frac{1}{2}} + a_2 (r - r_1)^{+\frac{3}{2}} + \ldots.$$

If we multiply by dr and integrate indefinitely, it will at once be seen that the function resulting from the integration will not be infinite for $r = r_1$, and we therefore have no spiral the kind required.

If, on the other hand, $r = r_1$ is a double root in equation (IV), the function under the sign will have a pole of the first order for $r - r_1$, and then, as is known, the function will be mically infinite for the same value. In this case, then, we obtain a curve of the required nat also, as will be easily seen, if $r = r_1$ were a root of higher multiplicity.

The problem is thus reduced to finding the condition for equation (IV) having a dou which is $> r_0$.

If we confine ourselves to the consideration of particles that are expelled *normally* from the in its magnetic equator, with an initial velocity v_0, we obtain

$$v_0 = \sqrt{C - \frac{2\mu}{r_0}} \text{ and } \left(\frac{d\varphi}{dt}\right)_0 = \frac{a}{r_0^2} + \frac{\lambda M}{r_0^3} = 0,$$

hence

$$C = \frac{2\mu}{r_0} + v_0^2 \text{ and } a = -\frac{\lambda M}{r_0}.$$

If these values are introduced into (IV), we obtain

$$\left(\frac{2\mu}{r_0} + v_0^2\right) r^4 - 2\mu r^3 - \frac{\lambda^2 M^2}{r_0^2}(r - r_0)^2 = 0.$$

If $\frac{1}{r} = x$, $\frac{1}{r_0} = x_0$ is introduced, the equation changes to

(A) $$\lambda^2 M^2 x^2 (x - x_0)^2 + 2\mu (x - x_0) - v_0^2 = 0.$$

If x is to be a double root in this equation, the equation

(B) $$\lambda^2 M^2 x (x - x_0)^2 + \lambda^2 M^2 x^2 (x - x_0) + \mu = 0$$

must take place at the same time. The last equation may also be written

$$(2x - x_0)\, x (x - x_0) + \frac{\mu}{\lambda^2 M^2} = 0;$$

as μ is negative, it will at once be seen that the double root must be positive. But we can prove that it must also be $< x_0$; for from (A) and (B) we obtain

$$\lambda^2 M^2 x^2 (x - x_0)^2 - v_0^2 = 2\lambda^2 M^2 (2x - x_0)\, x (x - x_0)^2,$$

$$- v_0^2 = \lambda^2 M^2 x^2 (x - x_0)^2 + 2\lambda^2 M^2 x (x - x_0)^3.$$

As v_0^2 is positive, it is evident that this equation cannot take place unless $x < x_0$. We see then, that if there is any double root at all in (A), it is positive and $< x_0$, and a double root in equation (IV) is necessarily positive and $> r_0$, as it should be.

In order to find the condition for the double root and its value, we must eliminate x from (A) and (B). This is easily done in the following manner.

By multiplying by $x(x - x_0)$ on both sides in (B), we obtain, on substituting the value of $x^2(x - x_0)^2$ from (A),

$$(2x - x_0)(v_0^2 - 2\mu (x - x_0)) + \mu x (x - x_0) = 0,$$

(C) $$- 3\mu x (x - x_0) + 2\mu x_0 (x - x_0) + (2x - x_0)\, v_0^2 = 0.$$

If we multiply here by $2x - x_0$, and substitute the value of $x(x - x_0)(2x - x_0)$ from (B), we obtain

$$\frac{3\mu^2}{\lambda^2 M^2} + 2\mu x_0 (x - x_0)(2x - x_0) + (2x - x_0)^2 v_0^2 = 0,$$

(D) $$4(\mu x_0 + v_0^2)\, x (x - x_0) - (2\mu (x - x_0) - v_0^2)\, x_0^2 + \frac{3\mu^2}{\lambda^2 M^2} = 0.$$

Then when $x(x - x_0)$ is eliminated from (C) and (D), we obtain

$$4(\mu x_0 + v_0^2)(2\mu x_0 (x - x_0) + (2x - x_0)\, v_0^2) + 3\mu \left(\frac{3\mu^2}{\lambda^2 M^2} - (2\mu (x - x_0) - v_0^2)\, x_0^2\right) = 0,$$

hence we obtain

$$x = \frac{x_0 (2\mu^2 x_0^2 + 9\mu x_0 v_0^2 + 4 v_0^4) - \frac{9\mu^3}{\lambda^2 M^2}}{2\mu^2 x_0^2 + 16\mu x_0 v_0^2 + 8 v_0^4}.$$

By the substitution of this value in (C), we find the conditional equation, which, after some reductions, assumes the form

$$x_0(-2\mu^4 x_0^4 v_0^2 - \mu^3 x_0^3 v_0^4) - \frac{2\mu^3}{\lambda^2 M^2}(\mu^3 x_0^3 - 15\mu^2 x_0^2 v_0^2 - 24\mu x_0 v_0^4 - 8v_0^6) + \frac{27\mu^7}{\lambda^4 M^4} = 0.$$

If then $x_0 = \frac{1}{r_0}$ is substituted, and we multiply by $\frac{r_0^6}{\mu^5}$, we obtain

$$-\frac{2r_0 v_0^2}{\mu} - \frac{r_0^2 v_0^4}{\mu^2} - \frac{2r_0^3\mu}{\lambda^2 M^2}\left(1 - 15\frac{r_0 v_0^2}{\mu} - \frac{24 r_0^2 v_0^4}{\mu^2} - \frac{8r_0^3 v_0^6}{\mu^3}\right) + \frac{27\mu^2 r_0^6}{\lambda^4 M^4} = 0.$$

If we now put

$$-\frac{r_0 v_0^2}{\mu} = u \quad \text{and} \quad -\frac{\mu r_0^3}{\lambda^2 M^2} = v,$$

the conditional equation becomes

$$2u - u^2 + 2v(1 + 15u - 24u^2 + 8u^3) + 27v^2 = 0.$$

On the other side we may eliminate x_0 from (A) and (B), and then obtain an equation that gives the connection between the radius of the boundary-circle and v_0.

By multiplying (B) by x and substituting

$$x^2(x_0 - x)^2 = -\frac{2\mu}{\lambda^2 M^2}(x - x_0) + \frac{v_0^2}{\lambda^2 M^2},$$

we obtain

$$x^4 - x^3 x_0 + \frac{2\mu}{\lambda^2 M^2}x_0 + \frac{v_0^2}{\lambda^2 M^2} - \frac{\mu}{\lambda^2 M^2}x = 0,$$

whence

$$x_0 = \frac{-v_0^2 + \mu x - \lambda^2 M^2 x^4}{2\mu - \lambda^2 M^2 x^3}.$$

By substituting this in (B), we obtain the desired eqnation

$$-v_0^2 x^5 - \frac{\mu^2}{\lambda^2 M^2}x^3 + \frac{4\mu v_0^2}{\lambda^2 M^2}x^2 + \frac{v_0^4}{\lambda^2 M^2}x + \frac{4\mu^3}{\lambda^4 M^4} = 0,$$

or, if we again introduce r instead of x,

$$\frac{4\mu^3}{\lambda^4 M^4}r^5 + \frac{v_0^4}{\lambda^2 M^2}r^4 + \frac{4\mu v_0^2}{\lambda^2 M^2}r^3 - \frac{\mu^2}{\lambda^2 M^2}r^2 - v_0^2 = 0.$$

We shall now deal with the problem in a general way,. that is to say with an arbitrary value α_0 of the angle of expulsion or in other words the angle between the radius vector and the direction of motion at the initial moment.

We then have, as will easily be seen,

$$\frac{ar_0 + \lambda M}{r_0^3} = \frac{v_0 \sin\alpha_0}{r_0},$$

whence

$$a = r_0 v_0 \sin\alpha_0 - \frac{\lambda M}{r_0}.$$

By substituting this value in (IV), we obtain

$$\left(\frac{2\mu}{r_0} + v_0^2\right)r^4 - 2\mu r^3 - \left(rr_0 v_0 \sin\alpha_0 - \lambda M\frac{(r - r_0)}{r_0}\right)^2 = 0.$$

If we introduce, as before, $r = \frac{1}{x}$, $r_0 = \frac{1}{x_0}$, and put

$$r_0^2 v_0 \sin \alpha_0 = \lambda M (1 + k),$$

he equation takes the form

$$(1) \qquad x^2 (k x_0 + x)^2 + \frac{2\mu}{\lambda^2 M^2} (x - x_0) - \frac{v_0^2}{\lambda^2 M^2} = 0.$$

A double root in this must then also satisfy the equation

$$(2) \qquad x (k x_0 + x)(k x_0 + 2x) + \frac{\mu}{\lambda^2 M^2} = 0.$$

If we further put

$$x = \frac{x_0}{n},$$

ve obtain from (1) and (2)

$$-\frac{\mu}{\lambda^2 M^2} = \frac{(kn + 1)(kn + 2) x_0^3}{n^3},$$

$$\frac{v_0^2}{\lambda^2 M^2} = \frac{2(n - 1)(kn + 1)(kn + 2) + (kn + 1)^2}{n^4} x_0^4 .$$

In order that the whole shall have a physical significance, $\frac{-\mu}{\lambda^2 M^2}$, $\frac{v_0^2}{\lambda^2 M^2}$, and x_0 must all be positive, and x must be $< x_0$, and thus

$$(a) \qquad n > 1$$

$$(b) \qquad (kn + 1)(kn + 2) \geqq 0 ;$$

nd as

$$\sin \alpha_0 = \frac{\lambda M (1 + k)}{v_0} x_0^2 ,$$

follows that

$$\frac{\lambda^2 M^2 (1 + k)^2}{v_0^2} x_0^4 \leqq 1 .$$

The last relation may however be written

$$(1 + k)^2 n^4 - 2 k^2 n^3 + (k^2 - 6 k) n^2 + (4 k - 4) n + 3 \gtreqless 0 ,$$

$$(n - 1)^2 ((1 + k)^2 n^2 + (2 + 4 k) n + 3) \gtreqless 0 ,$$

as $n > 1$, more simply,

$$(c) \qquad (1 + k)^2 n^2 + (2 + 4 k) n + 3 \gtreqless 0 .$$

The 2 conditions (b) and (c) may be simplified by putting

$$kn = -l .$$

They thereby assume the form

$$(b') \qquad (l - 1)(l - 2) \geqq 0 ,$$

$$(c') \qquad l^2 - (2n + 4) l + n^2 + 2n + 3 \gtreqless 0 .$$

The discriminant for the function of the second order of l in (c') is

$$(n+2)^2 - n^2 - 2n - 3 = 2n + 1.$$

When, in accordance with (a), $n > 1$, the discriminant will become > 0, and conse are real values of l, which satisfy (c'). These values are determined by

$$n + 2 - \sqrt{2n+1} \leqq l \leqq n + 2 + \sqrt{2n+1}.$$

Hence it is seen that these values of l are positive. We see moreover that ther values of $l > 2$ which satisfy (c'), and then (b') is also satisfied. Hence it follows *that choice of the amount of magnetism and gravitation, initial velocity and angle of expulsion, obtain an annular formation at any desired distance from the globe.*

It is further seen that for sufficiently large values of n there are permissible values o and $> n$.

For $l < n$, we obtain

$$1 + k = 1 - \frac{l}{n} > 0,$$

that is to say

$$\alpha_0 > 0.$$

For $l > n$, we obtain

$$1 + k = 1 - \frac{l}{n} < 0,$$

that is to say

$$\alpha_0 < 0.$$

The particle can therefore approach a boundary-circle both when the direction of positive and when it is negative.

As this applies to negative particles, it of course also applies to positive particles.

It might be interesting to see, however, what direction an expelled negative particle will when the globe is so magnetised that $\lambda M > 0$.

If we assume that

$$\alpha_0 \gtreqless 0,$$

the angular velocity $\frac{d\varphi}{dt}$ is negative at the initial moment, and it will then always continu for the change from a negative to a positive revolution-direction, or vice versa, can only take

$$\frac{d\varphi}{dt} = 0,$$

that is to say when

$$ar + \lambda M = 0,$$

or

$$r = \frac{\lambda M r_0}{\lambda M - r_0^2 v_0 \sin \alpha_0};$$

but as this value of r is $\leqq r_0$, *no such reversal can take place.*

A positive direction of revolution can thus only take place when $\alpha_0 \geqq 0$.

Now we have seen that (c) cannot be satisfied with other than negative values of we have

$$r_0^2 v_0 \sin \alpha_0 < \lambda M.$$

In order that the particle shall not change from a positive to a negative direction of r is necessary that the double root r, which is the radius of the boundary-circle, shall be less

$$\frac{\lambda M r_0}{\lambda M - r_0^2 v_0 \sin \alpha_0}.$$

It will be seen, however, that this is equivalent to

$$\frac{x_0}{n} > -kx_0 .$$

As we further have, in this case,

$$-1 < k < 0 ,$$

we can suitably put

$$k = -\frac{1}{m} ,$$

and then obtain

(d) $$m > n .$$

The condition (c) assumes the form

$$(m-1)^2 n^2 + (2m^2 - 4m) n + 3m^2 \gtreqless 0 .$$

Hence it follows that

$$\frac{n^2 + 2n - n\sqrt{2n+1}}{n^2 + 2n + 3} \gtreqless m \gtreqless \frac{n^2 + 2n + n\sqrt{2n+1}}{n^2 + 2n + 3} .$$

This, in connection with (d) then gives

$$n < \frac{n^2 + 2n + n\sqrt{2n+1}}{n^2 + 2n + 3} ,$$

whence

$$n^4 + 2n^3 + 3n^2 < 0 ,$$

while at the same time $n > 1$, which is absurd.

The particle must thus change to negative direction of revolution before it approaches the boundary-circle.

Let us return for a little to the equations

$$-\frac{\mu r_0^3}{\lambda^2 M^2} = \frac{(l-1)(l-2)}{n^3} ,$$

$$\frac{v_0^2 r_0^4}{\lambda^2 M^2} = \frac{2(n-1)(l-1)(l-2) + (l-1)^2}{n^4} .$$

It follows from these that

$$\frac{r_0 v_0^2}{-\mu} = 2 - \frac{1}{n} + \frac{1}{n(l-2)} .$$

If

$$2 < l < 3 ,$$

$$\frac{r_0 v_0^2}{-\mu} > 2 \quad \text{or} \quad v_0^2 > -\frac{2\mu}{r_0} ,$$

is to say, a velocity which, *if gravitation acted alone,* would remove the particle infinitely, thus a erbolic velocity. If on the other hand, the initial velocity is hyperbolic, l cannot have other values between 2 and 3; for if $l > 3$, then

$$\frac{1}{n(l-2)} < \frac{1}{n} ,$$

d consequently

$$2 - \frac{1}{n} + \frac{1}{n(l-2)} < 2 .$$

It will further be seen that for these hyperbolic velocities, n cannot be greater than 4; for it llows from (c') that

$$l - 1 - \sqrt{2(l-1)} \leqq n \leqq l - 1 + \sqrt{2(l-1)} ;$$

and when

$$2 < l < 3,$$

then

$$1 + \sqrt{2} < l - 1 + \sqrt{2(l-1)} < 4.$$

Thus n is then always less than 4.

Values of $n > 4$ can thus only be obtained for elliptical velocities, i. e. when

$$v_0^2 < -\frac{2\mu}{r_0}.$$

Then, moreover, $l > 3$, whence it follows that

$$2 - \frac{1}{n} < \frac{r_0 v_0^2}{-\mu} < 2.$$

Hence it will be seen that great values of n can only be obtained for elliptical velo very near the parabolic, i. e. when v_0^2 is only a *little* less than $-\dfrac{2\mu}{r_0}$.

133. It might now be interesting to find out whether a negative particle could approacl circle with positive direction of revolution, if we were to assume that there was a resi medium. We have seen that if there were *no* resistance, such a motion was impossible.

When an electrically charged particle moves in the plane of the magnetic equator o globe, subject to the magnetism and gravitation from the globe, and moreover a resistance i we have the following equations of motion:

$$\frac{d^2x}{dt^2} = \frac{\lambda M}{r^3}\frac{dy}{dt} + \frac{\mu}{r^3}x - m\frac{dx}{ds},$$

$$\frac{d^2y}{dt^2} = -\frac{\lambda M}{r^3}\frac{dx}{dt} + \frac{\mu}{r^3}y - m\frac{dy}{ds},$$

where m is the resistance.

From this we obtain

$$\frac{dx}{dt}\frac{d^2x}{dt^2} + \frac{dy}{dt}\frac{d^2y}{dt^2} = \frac{\mu}{r^2}\frac{dr}{dt} - m\frac{dt}{ds}\left(\left(\frac{dx}{dt}\right)^2 + \left(\frac{dy}{dt}\right)^2\right),$$

or, if we put $\dfrac{ds}{dt} = v$,

(I) $$v\frac{dv}{dt} = \frac{\mu}{r^2}\frac{dr}{dt} - m\frac{ds}{dt}.$$

We obtain moreover

$$x\frac{d^2y}{dt^2} - y\frac{d^2x}{dt^2} = -\frac{\lambda M}{r^2}\frac{dr}{dt} - \frac{m}{v}\left(x\frac{dy}{dt} - y\frac{dx}{dt}\right),$$

or

(II) $$\frac{d}{dt}\left(r^2\frac{d\varphi}{dt}\right) = -\frac{\lambda M}{r^2}\frac{dr}{dt} - \frac{m}{v}r^2\frac{d\varphi}{dt}.$$

Now it is clear that whatever the nature of the resistance may be, it can at any ra stood as a continually positive function of r (possibly multiform, but if the particle wer retreating from the globe, it would be uniform). If the particle is able to move in such to be always retiring from the globe (and approaching a boundary-circle), $\dfrac{dr}{ds}$ is moreover positive function of r. For a path such as this then, it should be allowable to put

$$m = f(r)\frac{dr}{ds},$$

here $f(r)$ is a continually positive function of r. Let us now see what are the consequences to which iis will lead.

Equation (I) may now be written

$$v\,dv = \frac{\mu}{r^2}dr - f(r)dr\,,$$

hence we obtain

$$\frac{v^2}{2} + \frac{\mu}{r} = \frac{v_0^2}{2} + \frac{\mu}{r_0} - F(r)\,, \quad \text{when } F(r) = \int_{r_0}^{r} f(r)dr\,.$$

Further

$$r^2\frac{d\varphi}{dt} = r^2\frac{d\varphi}{ds}\frac{ds}{dt} = vr\sqrt{1-\left(\frac{dr}{ds}\right)^2}\,,$$

$$r^2\frac{d\varphi}{dt}\Bigg) = v\left(\frac{dr}{ds}r\sqrt{1-\left(\frac{dr}{ds}\right)^2} + v\frac{dr}{ds}\sqrt{1-\left(\frac{dr}{ds}\right)^2} - \frac{vr\cdot\frac{dr}{ds}\frac{d^2r}{ds^2}}{\sqrt{1-\left(\frac{dr}{ds}\right)^2}}\right),$$

ıd

$$v\frac{dv}{ds} = \frac{\mu}{r^2}\frac{dr}{ds} - f(r)\frac{dr}{ds}\,.$$

By substitution in equation (II) we then obtain

$$\sqrt{\;\dots-\left(\frac{dr}{ds}\right)^2} + v^2\frac{dr}{ds}\sqrt{1-\left(\frac{dr}{ds}\right)^2} - \frac{v^2r\frac{dr}{ds}\frac{d^2r}{ds^2}}{\sqrt{1-\left(\frac{dr}{ds}\right)^2}} = -\frac{\lambda M}{r^2}v\frac{dr}{ds} - f(r)\frac{dr}{ds}r\sqrt{1-\left(\frac{dr}{ds}\right)^2}\,,$$

by multiplying by $\dfrac{\sqrt{1-\left(\frac{dr}{ds}\right)^2}}{\frac{dr}{ds}}$,

$$\text{(III)} \qquad \left(1-\left(\frac{dr}{ds}\right)^2\right)\left(\frac{\mu}{r}+v^2\right) + \frac{v\lambda M}{r^2}\sqrt{1-\left(\frac{dr}{ds}\right)^2} - v^2r\frac{d^2r}{ds^2} = 0\,.$$

Here we will put

$$\frac{dr}{ds} = x \quad \text{and} \quad \frac{d^2r}{ds^2} = x\cdot\frac{dx}{dr}\,,$$

ıence we obtain

$$(1-x^2)\left(\frac{\mu}{r}+v^2\right) + \frac{v\lambda M}{r^2}\sqrt{1-x^2} - v^2rx\frac{dx}{dr} = 0\,.$$

We then put

$$\sqrt{1-x^2} = y\,,$$

ereby the equation assumes the form

$$\text{(IV)} \qquad \left(\frac{\mu}{r}+v^2\right)y + \frac{v\lambda M}{r^2} + v^2r\frac{dy}{dr} = 0\,.$$

By integration of this, we obtain

$$y = -e^{-\int \frac{\frac{\mu}{r} + v^2}{v^2 r} dr} \cdot \int \frac{\lambda M}{v r^3} e^{\int \frac{\frac{\mu}{r} + v^2}{v^2 r} dr} dr .$$

If we put $\int \frac{\frac{\mu}{r} + v^2}{v^2 r} dr = g(r)$, we can write

$$y = -e^{-g(r)} \left(\int_{r_0}^{r} \frac{\lambda M}{v r^3} e^{g(r)} dr + C \right) .$$

For $r = r_0$,

$$y = \sqrt{1 - \left(\frac{dr}{ds}\right)_0^2} = \sqrt{1 - \cos^2 \alpha_0} = \sin \alpha_0 = -Ce^{-g(r_0)} .$$

Thus

$$y = e^{-g(r)} \left(e^{g(r_0)} \sin \alpha_0 - \int_{r_0}^{r} \frac{\lambda M}{v r^3} e^{g(r)} dr \right) .$$

Since moreover

$$\left(\frac{ds}{dr}\right)^2 = 1 + r^2 \left(\frac{d\varphi}{dr}\right)^2 ,$$

then

$$r^2 \left(\frac{d\varphi}{dr}\right)^2 = \frac{1}{1 - y^2} - 1 = \frac{y^2}{1 - y^2} ,$$

whence

$$\frac{d\varphi}{dr} = \frac{y}{r\sqrt{1 - y^2}} ,$$

and

$$\varphi - \varphi_0 = \int_{r_0}^{r} \frac{y\,dr}{r\sqrt{1 - y^2}} .$$

If now the particle approaches a boundary-circle, then necessarily

$$\int_{r_1 - \varepsilon}^{r_1} \frac{y\,dr}{r\sqrt{1 - y^2}} = \infty$$

when r_1 is the radius of the boundary-circle. As $\lim y = 1$, and $\lim r = r_1$, then also

$$\int_{r_1 - \varepsilon}^{r_1} \frac{dr}{\sqrt{1 - y}} = \infty .$$

But as y certainly possesses a continuous 1st derivative (see (IV)), the necessary condition for the last integral being infinite is that

$$y = 1 \text{ and } \frac{dy}{dr} = 0 \quad \text{for} \quad r = r_1 .$$

We obtain then

$$\text{(V)} \qquad e^{-g(r_1)} \left(e^{g(r_0)} \sin \alpha_0 - \int_{r_0}^{r_1} \frac{\lambda M}{v r^3} e^{g(r)} dr \right) = 1 ,$$

being a positive root in the equation

$$\text{(VI)} \qquad \frac{\mu}{r} + v^2 + \frac{v\lambda M}{r^2} = 0,$$

ıich is obtained from (IV) by putting $y = 1$, and $\frac{dy}{dr} = 0$.

If (V) is to be possible for real α_0, then of necessity

$$\text{(VII)} \qquad \int_{r_0}^{r_1} \frac{\lambda M}{v r^3} e^{g(r)} dr + e^{g(r_1)} < e^{g(r_0)}.$$

Let us now look at the function

$$z = \int_{r_0}^{r} \frac{\lambda M}{v r^3} e^{g(r)} dr + e^{g(r)}.$$

We obtain

$$\frac{dz}{dr} = \frac{\lambda M}{v r^3} e^{g(r)} + e^{g(r)} \frac{\left(\frac{\mu}{r} + v^2\right)}{v^2 r} = \frac{e^{g(r)}}{v^2 r}\left(\frac{\mu}{r} + v^2 + \frac{v\lambda M}{r^2}\right).$$

If we put $u = \frac{\mu}{r} + v^2 + \frac{v\lambda M}{r^2}$, then $u = 0$ for $r = r_1$, according to (VI). But further

$$\frac{du}{dr} = \frac{\mu}{r^2} + 2v\frac{dv}{dr} + \frac{\lambda M}{r^2}\frac{dv}{dr} - \frac{2v\lambda M}{r^3} = \frac{\mu}{r^2} - 2f(r) + \frac{\lambda M}{v r^2}\left(\frac{\mu}{r^2} - f(r)\right) - \frac{2v\lambda M}{r^3} < 0,$$

membering that

$$\frac{v dv}{dr} = \frac{\mu}{r^2} - f(r).$$

Consequently $u > 0$ for $r < r_1$, and then also $\frac{dz}{dr} > 0$ for $r < r_1$, and consequently

$$z \text{ for } r = r_1 > z \text{ for } r = r_0,$$

ıich is at variance with (VII).

It is thus quite generally proved that the particle cannot *from within* approach a boundary-circle l· a positive orbit-direction.

It might now be imagined that the ejected particle first changed from out-going to in-going motion, ıd approached a boundary-circle from *without*.

We can here distinguish between two cases.

Case 1. The direction of the path of the particle is *positive* at the moment the change to in-going ɔtion takes place.

In this case y must remain positive along the entire in-going path. From the expression for $\frac{dy}{dr}$ been seen that y cannot become 0, unless $\frac{dy}{dr} < 0$. If y became negative somewhere along the ng path, it must then, owing to the continuity, as its value at the change is 1, also become 0 for r more values of r, and among these there must be a greatest value r^1. Then of necessity, how- for $r = r^1$,

$$\frac{dy}{dr} > 0 \quad \text{and} \quad y = 0,$$

is impossible.

Let us now compare the value of y for a point on the in-going orbit, with the s as for a point on the out-going.

If there had been no resistance, we should have had

$$v_{in} = v_{out};$$

but, owing to the resistance a diminution of the kinetic energy, or of the velocity, has i taken place, so that

$$v_{in} < v_{out}.$$

Consequently

$$\frac{\mu}{rv}_{in} < \frac{\mu}{rv}_{out},$$

and then also

$$\left(\frac{\mu}{rv} + v\right)_{in} < \left(\frac{\mu}{rv} + v\right)_{out}.$$

If now

$$y_{in} = y_{out},$$

then

$$\left(\left(\frac{\mu}{rv} + v\right) y + \frac{\lambda M}{r^2}\right)_{in} < \left(\left(\frac{\mu}{rv} + v\right) y + \frac{\lambda M}{r^2}\right)_{out},$$

that is to say,

$$\left(v \frac{dy}{dr}\right)_{in} > \left(v \frac{dy}{dr}\right)_{out};$$

and as

$$v_{in} < v_{out},$$

it follows that

$$\frac{dy}{dr_{in}} > \frac{dy}{dr_{out}}.$$

From this it evidently follows, that the inequality

$$y_{in} < y_{out}$$

must take place for lesser values of r than the one in question. Now for the value of r f change to in-going motion takes place, is of course

$$y_{in} = y_{out};$$

and thus the inequality

$$y_{in} < y_{out}$$

occurs for all smaller values of r. As, moreover, y_{in} always remains < 0, this means tha returns to the globe again along a steeper path than the out-going.

This proof holds good, if $\frac{dy}{dr_{out}}$ is *always* > 0. But we may prove that the result is $\frac{dy}{dr_{out}}$ were < 0 for certain values of r, at any rate if the outgoing orbit has no point of i will not here, however, go farther in the discussion of this problem.

Case 2. The direction of the path of the particle is *negative* at the moment when t in-going motion takes place.

In this case it is certain that the resistance might be of such a nature, that the di path of the particle during its in-going motion, changed to positive. Let us suppose, for for a moment it is suddenly subjected to a very great resistance at the point at which in-going motion. The path will then at first very nearly coincide with the radius vector then, as the velocity increases, the magnetism will deflect it in a positive direction.

It might therefore possibly happen that the particle in this case would approach a b the positive way.

The above-found expression for y now holds good, of course, whatever value r_0 and α_0 may have; in other words, we can quite imagine an arbitrary point in the path as the point of commencement. We may then write r_1 instead of r_0, and α_1 instead of α_0, and obtain

$$y = e^{-g(r)}\left(e^{g(r_1)} \sin \alpha_1 - \int_{r_1}^{r} \frac{\lambda M}{v r^2} e^{g(r)} dr\right).$$

If the particle then approaches, by the positive way, a boundary-circle with radius r_2, then of necessity

$$e^{-g(r_2)}\left(e^{g(r_1)} \sin \alpha_1 - \int_{r_1}^{r_2} \frac{\lambda M}{v r^2} e^{g(r)}\, dr\right) = 1,$$

r_2 being a positive root in the equation

$$u = \frac{\mu}{r} + v^2 + \frac{v\lambda M}{r^2} = 0.$$

If we put

$$z = e^{g(r)} + \int_{r_1}^{r} \frac{\lambda M}{v r^2} e^{g(r)}\, dr,$$

we obtain

$$z \text{ for } r = r_2 < z \text{ for } r = r_1,$$

no matter how little greater r_1 is than r_2; but then

(VIII) $$\frac{dz}{dr} > 0$$

for values of $r = r_2 + \varepsilon$, where ε is a positive quantity that can be chosen as small as desired.

We found further that

$$\frac{du}{dr} = \frac{\mu}{r^2} - 2f(r) + \frac{\lambda M}{v r^2}\left(\frac{\mu}{r^2} - f(r)\right) - \frac{2v\lambda M}{r^3}.$$

As $f(r)$ along the in-going path is negative, we cannot here, as before, conclude that the value of $\frac{du}{dr}$ must be negative. But if the function $f(r)$ is assumed to be such that

$$\frac{du}{dr} < 0 \quad \text{for} \quad r = r_2,$$

then, as $u = 0$ for $r = r_2$,

$$u < 0 \quad \text{for} \quad r = r_2 + \varepsilon,$$

where ε has the same signification as before. Then too, however,

$$\frac{dz}{dr} < 0 \quad \text{for} \quad r = r_2 + \varepsilon,$$

and this is at variance with (VIII).

Hence it follows that the negatively-charged particle cannot approach a boundary-circle from without by the positive way, unless

$$\frac{du}{dr} > 0 \quad \text{for} \quad r = r_2.$$

On the other hand we can prove that if, for $r = r_2$,

$$u = 0 \quad \text{and} \quad \frac{du}{dr} > 0,$$

the particle can, from without and the positive way, approach a boundary-circle with r_2 as radius.

We then obtain

$$\frac{dz}{dr} > 0 \quad \text{for} \quad r = r_2 + \varepsilon,$$

and it is thus certain that there are values of $r > r_2$, for instance r_1, such that

(IX) $$\frac{dz}{dr} > 0, \quad \text{when} \quad r_2 < r < r_1,$$

and thus also

$$z \text{ for } r = r_2 < z \text{ for } r = r_1,$$

or for an arbitrary r, we have

$$e^{g(r_2)} + \int_r^{r_2} \frac{\lambda M}{v r^3} e^{g(r)} dr < e^{g(r_1)} + \int_r^{r_1} \frac{\lambda M}{v r^3} e^{g(r)} dr.$$

If we then put $r = r_1$, we obtain

$$e^{g(r_2)} + \int_{r_1}^{r_2} \frac{\lambda M}{v r^3} e^{g(r)} dr < e^{g(r_1)}.$$

We can then, however, find an angle α_1 such that

(X) $$e^{g(r_2)} + \int_{r_1}^{r_2} \frac{\lambda M}{v r^3} e^{g(r)} dr = e^{g(r_1)} \sin \alpha_1.$$

We further put

(XI) $$v_1^2 = v_2^2 + \frac{2\mu}{r_2} - \frac{2\mu}{r_1} - 2\int_{r_2}^{r_1} f(r)\, dr.$$

Since r_1 may be chosen as little greater than r_2 as desired, it may certainly be both $\sin \alpha_1$ and v_1 can be found as positive quantities.

It is then clear that if we imagine the negative particle placed at a distance r_1 from the magnetic globe, and possessing a velocity v_1, forming an angle α_1 with the radius ve be chosen between $\frac{\pi}{2}$ and π), it will then, from without and the positive way, approach the circle with radius r_2. For since y for $r = r_1$ has the positive value $\sin \alpha_1$, we ca Case 1, that y must remain positive along the entire in-going path. The particle cannot th to out-going motion again for a value r_3 of r, unless $y = +1$ for $r = r_3$; that is to s

$$e^{g(r_3)} + \int_{r_1}^{r_3} \frac{\lambda M}{v r^3} e^{g(r)} dr = e^{g(r_1)} \sin \alpha_1;$$

but according to (X) we obtain therefrom

$$z \text{ for } r = r_3 \text{ equal to } z \text{ for } r = r_2;$$

and that, on account of (IX), cannot be, if $r_3 > r_2$.

For the value $r = r_2$, the velocity v will be determined by the fact that

$$v_1^2 = v^2 + \frac{2\mu}{r_2} - \frac{2\mu}{r_1} - 2\int_{r_2}^{r_1} f(r)\, dr;$$

but if we compare this with (XI) we obtain

$$v = v_2.$$

r_2 and v_2 were so chosen, however, that

$$\frac{\mu}{r_2} + v_2^2 + \frac{v_2 \lambda M}{r_2^2} = 0.$$

We thus have $y = 1$ and $\frac{dy}{dr} = 0$ for $r = r_2$; but this means that the particle is asymptoti-
lly approaching the circle with radius r_2.

In order to obtain the fulfilment of the condition

$$\frac{du}{dr} > 0 \quad \text{for} \quad r = r_2,$$

i is only necessary that $f(r_2)$ shall satisfy the relation

$$\frac{\mu}{r_2^2} - 2f(r_2) + \frac{\lambda M}{v_2 r_2^2}\left(\frac{\mu}{r_2^2} - f(r)\right) - \frac{2 v_2 \lambda M}{r_2^3} > 0;$$

t this, it is evident, can be done in an endless number of ways, if $f(r)$ is always to be negative.

The only remaining question is, then, whether the particle expelled from the globe can come to
ove in this manner. We have not yet succeeded in finding a complete solution of this problem; but
e have found that the resistance must be so great that the velocity must be diminished during the
going motion, in spite of gravitation.

134. We will now see whether a negatively-charged particle with positive direction of revolution
n approach a boundary-circle, if we imagine the charge decreasing to 0.

The equations of motion for an electrically-charged particle in the plane of a magnetic globe's
uator, influenced by gravitation and magnetism, are

$$\frac{d^2x}{dt^2} = \frac{\lambda M}{r^3}\frac{dy}{dt} + \frac{\mu}{r^3}x$$

$$\frac{d^2y}{dt^2} = -\frac{\lambda M}{r^3}\frac{dx}{dt} + \frac{\mu}{r^3}y.$$

We will now imagine the charge to be variable, in such a manner that it diminishes towards 0,
i the length of path increases infinitely. We can then make λ equal a function of r, but this will of
c irse be multiform if the particle should anywhere change from out-going to in-going motion or
v e versa.

We obtain, in the same way as before,

$$v^2 - v_0^2 = 2\mu\left(\frac{1}{r_0} - \frac{1}{r}\right),$$

a d

$$r^2\frac{d\varphi}{dt} = r_0 v_0 \sin \alpha_0 - M\int_{r_0}^{r}\frac{\lambda dr}{r^2}.$$

By putting

$$M\int_{r_0}^{r}\frac{\lambda dr}{r^2} = F(r),$$

w obtain therefrom

$$\left(\frac{dr}{dt}\right)^2 = v_0^2 + 2\mu\left(\frac{1}{r_0} - \frac{1}{r}\right) - \frac{1}{r^2}(r_0 v_0 \sin \alpha_0 - F(r))^2;$$

a d by dividing by

$$\left(\frac{d\varphi}{dt}\right)^2 = \frac{1}{r^4}(r_0 v_0 \sin \alpha_0 - F(r))^2,$$

w obtain

$$\left(\frac{dr}{d\varphi}\right)^2 = \frac{r^4\left(v_0^2 + 2\mu\left(\frac{1}{r_0} - \frac{1}{r}\right) - \frac{1}{r^2}(r_0 v_0 \sin \alpha_0 - F(r))^2\right)}{(r_0 v_0 \sin \alpha_0 - F(r))^2},$$

or

$$\frac{d\varphi}{dr} = \frac{r_0 v_0 \sin \alpha_0 - F(r)}{r\sqrt{\left(v_0^2 + \frac{2\mu}{r_0}\right) r^2 - 2\mu r - (r_0 v_0 \sin \alpha_0 - F(r))^2}},$$

whence

$$\varphi - \varphi_0 = \int_{r_0}^{r} \frac{(r_0 v_0 \sin \alpha_0 - F(r))\, dr}{r\sqrt{\left(v_0^2 + \frac{2\mu}{r_0}\right) r^2 - 2\mu r - (r_0 v_0 \sin \alpha_0 - F(r))^2}}.$$

If the particle is to approach a boundary-circle, then of necessity, when r_1 is the radi

(a) $$\left(v_0^2 + \frac{2\mu}{r_0}\right) r_1^2 - 2\mu r_1 - (r_0 v_0 \sin \alpha_0 - F(r_1))^2 = 0$$

(b) $$\left(v_0^2 + \frac{2\mu}{r_0}\right) r_1 - \mu + (r_0 v_0 \sin \alpha_0 - F(r_1)) \frac{\lambda_1 M}{r_1^2} = 0,$$

where λ_1 is the value λ gets for $r = r_1$. As the charge is assumed to diminish towards

$$\lambda_1 = 0,$$

whence, according to (b),

(c) $$r_1 = \frac{\mu}{v_0^2 + \frac{2\mu}{r_0}}.$$

From (a) we then obtain

$$-\mu r_1 = (r_0 v_0 \sin \alpha_0 - F(r_1))^2,$$

or

(d) $$F(r_1) = r_0 v_0 \sin \alpha_0 - \sqrt{-\mu r_1},$$

noting that $F(r_1) - r_0 v_0 \sin \alpha_0$ must be < 0, if the motion is supposed to take place direction.

From (d) we obtain

$$F(r_1) + \sqrt{-\mu r_1} < r_0 v_0,$$

and as $F(r_1)$ is certainly > 0, we obtain *a fortiori*

$$-\mu r_1 < r_0^2 v_0^2,$$

and by the aid of (c),

$$\frac{-\mu^2}{v_0^2 + \frac{2\mu}{r_0}} < r_0^2 v_0^2,$$

noting, from (c), that

$$v_0^2 + \frac{2\mu}{r_0} < 0.$$

Then

$$v_0^4 r_0^2 + 2\mu v_0^2 r_0 + \mu^2 < 0,$$

or

$$(v_0^2 r_0 + \mu)^2 < 0,$$

which is absurd.

Since the particle cannot approach a boundary-circle in a positive direction, it is does not change to a negative direction, it must either continue to travel out indefinitely, direction change from an out-going to an in-going motion.

Let us look for a little at the last case. The expression

$$r_0 v_0 \sin \alpha_0 - F(r)$$

ill then remain positive during the in-going motion; but if we compare 2 points with the same value f r, one on the out-going, and one on the in-going path, then

$$(r_0 v_0 \sin \alpha_0 - F(r))_{in} < (r_0 v_0 \sin \alpha_0 - F(r))_{out} ,$$

nd

$$\left(\left(v_0^2 + \frac{2\mu}{r_0}\right) r^2 - 2\mu r - (r_0 v_0 \sin \alpha_0 - F(r))^2\right)_{in} > \left(\left(v_0^2 + \frac{2\mu}{r_0}\right) r^2 - 2\mu r - (r_0 v_0 \sin \alpha_0 - F(r))^2\right)_{out},$$

nd consequently

$$\left|\frac{d\varphi}{dr}\right|_{in} < \left|\frac{d\varphi}{dr}\right|_{out} .$$

The particle will thus return to the globe again by a steeper path than that by which it went out om it.

Setting aside the case in which the particle recedes indefinitely, only those cases are left in which, ith negative direction of revolution, it either changes to an in-going motion again, or approaches a oundary-circle.

We will look at the former of these cases.

If we compare the value of $\frac{d\varphi}{dt}$ in 2 points with the same value of r, one on the out-going and e on the in-going path, it is evident that

$$\frac{d\varphi}{dt}_{in} < \frac{d\varphi}{dt}_{out} .$$

It might then happen that $\frac{d\varphi}{dt}$ became positive when the particle came in sufficiently near to the obe again; but then $\frac{d\varphi}{dt}_{in}$ would certainly also be positive for smaller values of r. Then as $\frac{d\varphi}{dt}_{in}$ and $\frac{d\varphi}{dt}_{out}$ would both be positive for these sufficiently small values of r, we may prove in the same way as ove that

$$\left|\frac{d\varphi}{dr}\right|_{in} < \left|\frac{d\varphi}{dr}\right|_{out}$$

it then the particle must return to the globe again. On the other hand, if it does not end in the globe, en, with negative value of $\frac{d\varphi}{dt}$, it will either turn out again, or approach a boundary-circle. It is then rtain, however, that $\frac{d\varphi}{dt}$ will continue to be negative for all time. Along an eventual out-going path, $\frac{d\varphi}{dt}$ will certainly remain negative; and if it turns in again, will also, by virtue of the relation

$$\frac{d\varphi}{dt}_{in} < \frac{d\varphi}{dt}_{out} ,$$

negative along the in-going path, and so on.

In conclusion we will see whether the particle with negative direction of revolution can approach boundary-circle from within, when $\alpha_0 > 0$. If we call the radius of the circle r_1, then

(c′)
$$r_1 = \frac{\mu}{v_0^2 + \frac{2\mu}{r_0}},$$

and

(d′)
$$F(r_1) = r_0 v_0 \sin\alpha_0 + \sqrt{-\mu r_1}\,.$$

The effect of the equation (c′) is that

$$-\mu < v_0^2 r_0 < -2\mu,$$

if r_1 is to have a positive value $> r_0$.

(d′) will certainly be satisfied if

$$\sqrt{-\mu r_1} < F(r_1) < r_0 v_0 + \sqrt{-\mu r_1} = r_0 v_0 - \frac{\mu}{\sqrt{-v_0^2 - \frac{2\mu}{r_0}}}\,.$$

Moreover r_1 must be the *smallest* positive value of r that causes the expression under the root sign in the φ-integral to vanish.

The 2nd derivative of the radicand has for $r = r_1$ the value

$$2\left(v_0^2 + \frac{2\mu}{r_0} - \frac{\lambda_1^2 M^2}{r_1^4} - (r_0 v_0 \sin\alpha_0 - F(r_1))\left(\frac{d}{dr}\frac{\lambda}{r^2}\right)_{r=r_1} M\right).$$

As however

$$\left(\frac{d}{dr}\frac{\lambda}{r^2}\right)_{r=r_1} \gtreqless 0, \qquad \text{and } r_0 v_0 \sin\alpha_0 - F(r_1) < 0,$$

the second derivative is negative. The 1st derivative then becomes positive for $r < r_1$, and consequently the radicand itself negative for values of $r < r_1$. But then the particle cannot approach any boundary-circle.

On the other hand, the orbit can certainly become a conic section at last.

Let us consider the simple case in which the particle retains its charge until it comes very near the boundary-circle, assuming that it tends towards one, but then suddenly loses its charge.

The changes from in-going to out-going motion, or vice versa, will then take place for those values of r that satisfy the equation

$$\left(\frac{2\mu}{r_0} + v_0^2\right) r^2 - 2\mu r - r_1^2 v_1^2 \sin^2\alpha_1 = 0,$$

where r_1 is very nearly the radius, r_g, of the boundary-circle, v_1 very nearly the velocity, v_g, in the boundary-circle, and α_1 very nearly $-\frac{\pi}{2}$. For the sake of simplicity we may put

$$r_1 = r_g, \quad v_1 = v_g, \quad \alpha_1 = -\frac{\pi}{2},$$

whereby we obtain

$$\left(\frac{2\mu}{r_0} + v_0^2\right) r^2 - 2\mu r - r_g^2 v_g^2 = 0\,.$$

The discriminant of this equation is

$$\mu^2 + r_g^2 v_g^2 \left(\frac{2\mu}{r_g} + v_g^2\right) = \mu^2 + 2\mu r_g v_g^2 + r_g^2 v_g^4 = (\mu + r_g v_g^2)^2 > 0,$$

and consequently the roots in the equation are always real.

If

$$\frac{2\mu}{r_0} + v_0^2 > 0,$$

ie one root is positive, and the other negative; and at the same time the velocity is *hyperbolic*, so that ie particle retires indefinitely.

If, on the contrary,

$$\frac{2\mu}{r_0} + v_0^2 < 0,$$

th the roots will be positive. Let us call the smallest r_p, and the largest r_a. Then

$$r_p = \frac{-\mu - (\mu + r_g v_g^2)}{-\left(\frac{2\mu}{r_0} + v_0^2\right)}, \qquad r_a = \frac{-\mu + (\mu + r_g v_g^2)}{-\left(\frac{2\mu}{r_0} + v_0^2\right)}.$$

Hence we obtain

$$r_p = \frac{-2\mu - r_g v_g^2}{-\frac{2\mu}{r_g} - v_g^2} = r_g, \qquad r_a = \frac{2\mu\, r_0}{2\mu + r_0 v_0^2} - r_g,$$

at is to say, the particle will move in an ellipse, of which the perihelion is just on the boundary-circle.

The eccentricity will be

$$e = \frac{r_a - r_p}{r_a + r_p} = \frac{\frac{2\mu\, r_0}{2\mu + r_0 v_0^2} - 2 r_g}{\frac{2\mu\, r_0}{2\mu + r_0 v_0^2}} = \frac{\mu r_0 - r_g(2\mu + r_0 v_0^2)}{\mu r_0}.$$

If we substitute as before

$$\frac{r_g}{r_0} = n,$$

obtain

$$e = \frac{\mu - n(2\mu + r_0 v_0^2)}{\mu} = 1 - n\left(2 + \frac{r_0 v_0^2}{\mu}\right) = 1 - 2n - n\frac{r_0 v_0^2}{\mu}.$$

We have, however (cf. p. 683),

$$\frac{r_0 v_0^2}{-\mu} = \frac{1}{n}\left(2n - 2 + \frac{kn+1}{kn+2}\right),$$

consequently

$$e = 1 - 2n + 2n - 2 + \frac{kn+1}{kn+2} = -\frac{1}{kn+2} = \frac{1}{l-2},$$

ıen we put, as before,

$$kn = -l.$$

As l may be as great as may be desired, the eccentricity may be as small as may be desired. ł the same time n must have greater values. Thus *at a great distance from the globe, the orbit will b almost circular.*

135. We have discussed above the problem of the mouvement of an electrically charged particle aout a magnetic and gravitating sphere, when the particle is ejected in the plane of the magnetic euator, and thus always remains there. We saw that there were boundary-circles towards which the particles, under certain conditions, could draw nearer and nearer, this giving rise to the planets. It still remains for us to investigate the conditions outside the plane of the equat the formation of planets is also possible there, when the particles are flung out anywhere or not. This investigation has been carried out as follows.

The equations of motion for the particle are

$$(1)\qquad \begin{aligned} \frac{d^2x}{dt^2} &= \lambda\left(Z\frac{dy}{dt} - Y\frac{dz}{dt}\right) + \mu\frac{x}{r^3} \\ \frac{d^2y}{dt^2} &= \lambda\left(X\frac{dz}{dt} - Z\frac{dx}{dt}\right) + \mu\frac{y}{r^3} \\ \frac{d^2z}{dt^2} &= \lambda\left(Y\frac{dx}{dt} - X\frac{dy}{dt}\right) + \mu\frac{z}{r^3} \end{aligned}$$

From these it is easily found that

$$(2)\qquad v = \frac{ds}{dt} = \sqrt{C - \frac{2\mu}{r}} \qquad C = \text{constant}$$

and if the magnetic field originates in a potential V,

$$(3)\qquad R^2\frac{d\varphi}{dt} = \Phi + a \qquad a = \text{constant},$$

when Φ is a certain function of $R = \sqrt{x^2 + y^2}$ and z.

If we assume that the sphere acts as an elementary magnet, i. e.

$$V = M\frac{z}{r^3},$$

then

$$(4)\qquad \Phi = \lambda M\frac{R^2}{r^3}.$$

Moreover

$$X = \frac{\partial V}{\partial x} = -\frac{3Mz}{r^5}x, \quad Y = \frac{\partial V}{\partial y} = -\frac{3Mz}{r^5}y,$$

whence

$$Y\frac{dx}{dt} - X\frac{dy}{dt} = \frac{3Mz}{r^5}\left(x\frac{dy}{dt} - y\frac{dx}{dt}\right) = \frac{3Mz}{r^5}R^2\frac{d\varphi}{dt}.$$

By substitution in the third equation of motion (1), we obtain

$$\frac{d^2z}{dt^2} = \frac{3\lambda Mz}{r^5}R^2\frac{d\varphi}{dt} + \mu\frac{z}{r^3}$$

and by substitution of the value of $R^2\frac{d\varphi}{dt}$ from (3) and (4) we obtain

$$\frac{d^2z}{dt^2} = \frac{3\lambda Mz}{r^8}(\lambda MR^2 + ar^3) + \frac{\mu z}{r^3}.$$

As moreover we have

$$ds^2 = dR^2 + R^2d\varphi^2 + dz^2,$$

we obtain by the aid of (2), (3) and (4)

$$\left(\frac{dR}{dt}\right)^2+\left(\frac{dz}{dt}\right)^2=C-\frac{2\mu}{r}-R^2\left(\frac{d\varphi}{dt}\right)^2=C-\frac{2\mu}{r}-\frac{(\lambda MR^2+ar^3)^2}{R^2r^6}.$$

We thus have to study the following system of equations:

$$\frac{d^2z}{dt^2}=\frac{3\lambda Mz}{r^8}(\lambda MR^2+ar^3)+\frac{\mu z}{r^3}$$

$$(5)\qquad \left(\frac{dR}{dt}\right)^2+\left(\frac{dz}{dt}\right)^2=C-\frac{2\mu}{r}-\frac{(\lambda MR^2+ar^3)^2}{R^2r^6}$$

$$R^2\frac{d\varphi}{dt}=\frac{\lambda MR^2+ar^3}{r^3}.$$

If, for the sake of brevity, we put

$$C-\frac{2\mu}{r}-\frac{(\lambda MR^2+ar^3)^2}{R^2r^6}=P,$$

he first equation in (5) may be written in the form

$$\frac{d^2z}{dt^2}=\frac{z}{2r}\frac{\partial P}{\partial r}.$$

From the second equation in (5),

$$\left(\frac{dR}{dt}\right)^2+\left(\frac{dz}{dt}\right)^2=P$$

e obtain by derivation as regards t,

$$2\frac{dR}{dt}\frac{d^2R}{dt^2}+2\frac{dz}{dt}\frac{d^2z}{dt^2}=\left(\frac{\partial P}{\partial R}+\frac{\partial P}{\partial r}\cdot\frac{R}{r}\right)\frac{dR}{dt}+\frac{\partial P}{\partial r}\frac{z}{r}\frac{dz}{dt}.$$

If we introduce the expression just found for $\frac{d^2z}{dt^2}$, we obtain

$$2\frac{dR}{dt}\frac{d^2R}{dt^2}=\left(\frac{\partial P}{\partial R}+\frac{\partial P}{\partial r}\frac{R}{r}\right)\frac{dR}{dt}$$

nd as often as $\frac{dR}{dt}\neq 0$, we can from this again conclude that

$$\frac{d^2R}{dt^2}=\frac{1}{2r}\left(\frac{\partial P}{\partial R}r+\frac{\partial P}{\partial r}R\right).$$

Owing to the continuity, however, this equation also retains its validity, even if $\frac{dR}{dt}=0$ for certain pecial values of t.

We can also eliminate t and find a differential equation in only z and R; this determines a surface f rotation, upon which the particle will always remain.

We have

$$\frac{dz}{dt}=\frac{dz}{dR}\cdot\frac{dR}{dt}\quad\text{and}\quad\frac{d^2z}{dt^2}=\frac{d^2z}{dR^2}\left(\frac{dR}{dt}\right)^2+\frac{dz}{dR}\frac{d^2R}{dt^2}$$

and as

$$\left(1+\left(\frac{dz}{dR}\right)^2\right)\left(\frac{dR}{dt}\right)^2 = P$$

we obtain

$$\frac{dR}{dt} = \frac{\sqrt{P}}{\sqrt{1+\left(\frac{dz}{dR}\right)^2}}, \quad \frac{dz}{dt} = \frac{\frac{dz}{dR}}{\sqrt{1+\left(\frac{dz}{dR}\right)^2}} \cdot \sqrt{P}$$

and finally by substitution of the expressions found for $\frac{d^2z}{dt^2}$ and $\frac{d^2R}{dt^2}$

$$\frac{z}{2r}\frac{\partial P}{\partial r} = \frac{d^2z}{dR^2} \cdot \frac{P}{1+\left(\frac{dz}{dR}\right)^2} + \frac{dz}{dR} \cdot \frac{\left(\frac{\partial P}{\partial R} r + \frac{\partial P}{\partial R} R\right)}{2r}$$

or

$$(6) \quad 2rP\frac{d^2z}{dR^2} + \left(\frac{\partial P}{\partial R} r + \frac{\partial P}{\partial r} R\right)\frac{dz}{dR}\left(1+\left(\frac{dz}{dR}\right)^2\right) - z\frac{\partial P}{\partial r}\left(1+\left(\frac{dz}{dR}\right)^2\right) = 0.$$

The coefficients in this equation are irrational functions of R and z. If r is introduced i
z as a dependent variable, we obtain a differential equation in r and R with rational coefficients. This differential equation is

$$2r(r^2-R^2)P\frac{d^2r}{dR^2} - 2P\left(r - R\frac{dr}{dR}\right)^2 + \left(\left(\frac{\partial P}{\partial R}r + \frac{\partial P}{\partial r}R\right)\frac{dr}{dR} - \frac{\partial P}{\partial R}R - \frac{\partial P}{\partial r}r\right)\left(r^2\left(1+\left(\frac{dr}{dR}\right)^2\right) - 2rR\frac{dr}{dR}\right) =$$

It will be seen from the second equation in (5) that the surface $P = 0$ is a boundary-surface through which the particle can never penetrate during its movement. The line of intersection of this boundary-surface with the rotation-surface upon which the particle is found, then becomes a boundary-curve, which the particle can never cross. This boundary-curve is always a circle parallel with the plane of the equator.

It will be seen that for a given point (z, R) upon the surface of rotation upon which the particle lies, there are 2 values of $\frac{dR}{dt}$ which are equally great, but have contrary signs, and similarly for $\frac{dz}{dt}$, while there is only one value of $\frac{d\varphi}{dt}$. From this it will be seen that if the particle moves in such a manner that, after a limited time, it reaches the boundary-surface, it will thence turn inwards to the sphere again along a path that is symmetrical with the outward-going path, with reference to the meridian plane through the point upon the boundary-surface at which the reversal took place.

We will next see whether the particle could approach this boundary-circle asymptotically. It is clear that no matter how the particle moves, we may put

$$\left(\frac{dz}{dt}\right)^2 = f(z)$$

and consequently

$$t = \int_{z_0}^{z} \frac{dz}{\sqrt{f(z)}}.$$

As t must increase infinitely when z approaches the value that answers to the bounda
we must have for this value of z

$$f(z) = 0 \quad \text{and} \quad f'(z) = 0$$

hat is to say

$$^{2} = 0 \quad \text{and} \quad ^{2} = 0 \quad \text{or} \quad \frac{dz}{dt}\cdot\frac{d}{dz}\left(\frac{dz}{dt}\right) = 0 \quad \text{or} \quad \frac{d^2z}{dt^2} = 0.$$

In the same way we also find that $\frac{d^2R}{dt^2} = 0$ for the **boundary-circle.** We have, then

$$\frac{z}{2r}\frac{\partial P}{\partial r} = 0$$

$$\frac{1}{2r}\left(\frac{\partial P}{\partial R}r + \frac{\partial P}{\partial r}R\right) = 0\,.$$

Disregarding the plane of the equator $z = 0$, we have

$$\frac{\partial P}{\partial r} = 0 \quad \text{and} \quad \frac{\partial P}{\partial R} = 0\,,$$

nd of course also

$$P = 0\,.$$

If we introduce the actual expressions for P, $\frac{\partial P}{\partial R}$ and $\frac{\partial P}{\partial r}$, it becomes

$$\text{(a)} \qquad C - \frac{2\mu}{r} - \frac{(\lambda M R^2 + a r^3)^2}{R^2 r^6} = 0$$

$$\text{(b)} \qquad \frac{3\lambda^2 M^2 R^2}{r^7} + \frac{3a\lambda M}{r^4} + \frac{\mu}{r^2} = 0$$

$$\text{(c)} \qquad \lambda^2 M^2 R^4 - a^2 r^6 = 0\,.$$

From (c) we obtain

$$\lambda M R^2 = \pm\, a r^3$$

nd from (b)

$$\lambda M R^2 = -\,a r^3 - \frac{\mu}{3\lambda M} r^5\,.$$

As of necessity $r \neq 0$, it must be that

$$\lambda M R^2 = +\, a r^3$$

nd thus

$$-\frac{\mu}{3\lambda M} r^5 = 2 a r^3$$

$$r^2 = \frac{6 a \lambda M}{-\mu}\,.$$

From (a) we obtain

$$\left(C - \frac{2\mu}{r}\right) R^2 = 4a^2$$

ence

$$\left(C - \frac{2\mu}{r}\right) r^3 = 4 a \lambda M\,.$$

From this we obtain by substitution of the value found for r^2

$$(Cr - 2\mu)\frac{6 a \lambda M}{-\mu} = 4 a \lambda M$$

whence

$$Cr - 2\mu = -\frac{2}{3}\mu$$

or

$$r = \frac{4\mu}{3C}.$$

If this is compared with the expressions already found for r^2, we obtain

$$\frac{16\mu^2}{9C^2} = \frac{6a\lambda M}{-\mu}$$

or

$$-16\mu^3 = 54C^2a\lambda M$$

or

$$8\mu^3 + 27C^2a\lambda M = 0.$$

Let us next consider a value-system $R + R_1$, $r + r_1$, R and r indicating the boundar
According to Taylor's development, we then have

$$P(R+R_1, r+r_1) = P(R, r) + \frac{\partial P}{\partial R}R_1 + \frac{\partial P}{\partial r}r_1 + \frac{1}{2!}\left(\frac{\partial^2 P}{\partial R^2}R_1^2 + 2\frac{\partial^2 P}{\partial R \partial r}R_1 r_1 + \frac{\partial^2 P}{\partial r^2}r_1^2\right)$$

In an infinitely small region surrounding the point (R, r), in which $P = 0$, $\frac{\partial P}{\partial R} = 0$,
obtain then

$$P(R+R_1, r+r_1) = \frac{1}{2!}\left(\frac{\partial^2 P}{\partial R^2}R_1^2 + \frac{\partial^2 P}{\partial R \partial r}R_1 r_1 + \frac{\partial^2 P}{\partial r^2}r_1^2\right).$$

For $r_1 = 0$ we obtain in particular

$$P(R+R_1, r) = \frac{\partial^2 P}{\partial R^2}\cdot\frac{R_1^2}{2}.$$

Now

$$\frac{1}{2}\frac{\partial^2 P}{\partial R^2} = -\frac{\lambda^2 M^2}{r^6} - \frac{3a^2}{R^4} < 0.$$

Hence it follows that P has negative values as near the point under consideration
desired.

If the discriminant of the quadratic form

$$\frac{\partial^2 P}{\partial R^2}R_1^2 + 2\frac{\partial^2 P}{\partial R \partial r}R_1 r_1 + \frac{\partial^2 P}{\partial r^2}r_1^2$$

is negative, it follows that P must have negative values all over the area surrounding the
consideration. Then a particle cannot move towards the circle under consideration. The
must therefore be positive, that is to say,

$$\left(\frac{\partial^2 P}{\partial R \partial r}\right)^2 - \frac{\partial^2 P}{\partial R^2}\frac{\partial^2 P}{\partial r^2} > 0$$

or

$$\left(\frac{6\lambda^2 M^2 R}{r^7}\right)^2 + \left(\frac{\lambda^2 M^2}{r^6} + \frac{3a^2}{R^4}\right)\left(-\frac{2\mu}{r^3} - \frac{12a\lambda M}{r^5} - \frac{21\lambda^2 M^2 R^2}{r^8}\right) > 0$$

or, since $\lambda M R^2 = ar^3$,

$$\frac{36\lambda^3 M^3 a}{r^{11}} + \frac{4\lambda^2 M^2}{r^6}\left(-\frac{2\mu}{r^3} - \frac{12a\lambda M}{r^5} - \frac{21a\lambda M}{r^5}\right) > 0$$

r

$$36a\lambda M + 4(-2\mu r^2 - 33a\lambda M) > 0$$

r

$$-24a\lambda M - 2\mu r^2 > 0$$

d, as $r = \frac{4\mu}{3C}$,

$$12a\lambda M + \frac{16\mu^3}{9C^2} < 0$$

$$27C^2a\lambda M + 4\mu^3 < 0 .$$

But we found above that

$$27C^2a\lambda M + 8\mu^3 = 0 .$$

Hence it followed that $\mu > 0$, which however is not the case. It is hereby proved that the particle ınnot describe a path that asymptotically approaches the boundary-circle.

The question might now arise as to whether the particle could move in such a manner that it did ɔt reach the boundary-surface, either after a finite or an infinite length of time. This would only be f the integral curve we obtain from (6) — which may be said to be the curve of projection circles r = constant in a meridian plane of the path of the particle — has an infinite length within boundary-surface. It would then be an important point to decide whether the path of the particle uld approach asymptotically a closed curve.

We have however not yet succeeded in solving this problem quite generally.

Let us now at last try to find out, whether trajectories could exist in the plane

$$z = kx .$$

The equations of motion are

$$\frac{d^2x}{dt^2} = \frac{\lambda M}{r^5}\left((r^2 - 3z^2)\frac{dy}{dt} + 3yz\frac{dz}{dt}\right) + \frac{\mu x}{r^3}$$

$$\frac{d^2y}{dt^2} = \frac{\lambda M}{r^5}\left((3z^2 - r^2)\frac{dx}{dt} - 3xz\frac{dz}{dt}\right) + \frac{\mu y}{r^3}$$

$$\frac{d^2z}{dt^2} = \frac{3\lambda Mz}{r^5}\left(x\frac{dy}{dt} - y\frac{dx}{dt}\right) + \frac{\mu z}{r^3} .$$

By substitution of $z = kx$ we obtain

$$\frac{d^2x}{dt^2} = \frac{\lambda M}{r^5}\left((r^2 - 3k^2x^2)\frac{dy}{dt} + 3k^2xy\frac{dx}{dt}\right) + \frac{\mu x}{r^3}$$

$$(1) \qquad \frac{d^2y}{dt^2} = \frac{\lambda M}{r^5}\left((3k^2x^2 - r^2)\frac{dx}{dt} - 3k^2x^2\frac{dx}{dt}\right) + \frac{\mu y}{r^3}$$

$$\frac{d^2x}{dt^2} = \frac{3\lambda Mx}{r^5}\left(x\frac{dy}{dt} - y\frac{dx}{dt}\right) + \frac{\mu x}{r^3} .$$

Here we must assume, that $k \neq 0$.

From the 1st and 3rd equation we obtain

$$(2) \qquad r^2\frac{dy}{dt} - 3(1 + k^2)x\left(x\frac{dy}{dt} - y\frac{dx}{dt}\right) = 0 .$$

The 2nd equation may be transformed to

$$(3) \qquad \frac{d^2y}{dt^2} = -\frac{\lambda M}{r^3}\frac{dx}{dt} + \frac{\mu y}{r^3}.$$

From (2) we obtain

$$x\frac{dy}{dt} - y\frac{dx}{dt} = \frac{r^2}{3(1+k^2)x}\frac{dy}{dt}$$

whence

$$(4) \qquad \frac{dx}{dt} = \frac{(2r^2 - 3y^2)\cdot x}{3(r^2 - y^2)\cdot y}\frac{dy}{dt}.$$

Further we have

$$r\frac{dr}{dt} = x\frac{dx}{dt} + y\frac{dy}{dt} + z\frac{dz}{dt} = (1+k^2)x\frac{dx}{dt} + y\frac{dy}{dt},$$

whence

$$r\frac{dr}{dt} = \frac{(2r^2 - 3y^2)}{3y}\frac{dy}{dt} + y\frac{dy}{dt} = \frac{2r^2}{3y}\frac{dy}{dt}$$

or

$$(5) \qquad \frac{1}{y}\frac{dy}{dt} = \frac{3}{2r}\frac{dr}{dt}.$$

Hence we obtain by integration

$$(6) \qquad cy = r^{\frac{3}{2}} \qquad (c = \text{constant}).$$

Hence it follows, that

$$(1+k^2)x^2 = r^2 - y^2 = r^2 - \frac{r^3}{c^2}$$

or

$$(7) \qquad x = \frac{r\sqrt{c^2 - r}}{c\sqrt{1+k^2}}.$$

The 3rd equation in (1) may be written in the form

$$\frac{d^2x}{dt^2} = \frac{\lambda M}{(1+k^2)r^3}\frac{dy}{dt} + \frac{\mu x}{r^3}.$$

From this equation in connection with (3) we obtain

$$x\frac{d^2y}{dt^2} - y\frac{d^2x}{dt^2} = -\frac{\lambda M}{r^3}\left(x\frac{dx}{dt} + \frac{y}{1+k^2}\frac{dy}{dt}\right) = -\frac{\lambda M}{(1+k^2)r^2}\cdot\frac{dr}{dt}.$$

Hence we obtain by integration

$$(8) \qquad x\frac{dy}{dt} - y\frac{dx}{dt} = \frac{\lambda M}{1+k^2}\cdot\frac{1}{r} + a.$$

Now, however, we also have

$$x\frac{dy}{dt} - y\frac{dx}{dt} = \frac{r^2}{3(1+k^2)x}\cdot\frac{3y}{2r}\frac{dr}{dt} = \frac{r}{2(1+k^2)}\frac{y}{x}\frac{dr}{dt},$$

and as

$$\frac{y}{x} = \frac{\sqrt{1+k^2}\sqrt{r}}{\sqrt{c^2 - r}}$$

ve obtain

$$(9) \qquad x\frac{dy}{dt} - y\frac{dx}{dt} = \frac{r^{\frac{3}{2}}}{2\sqrt{1+k^2}\sqrt{c^2-r}}\frac{dr}{dt}.$$

From (8) and (9) it follows, that

$$(10) \qquad \frac{r^3}{4(1+k^2)(c^2-r)}\left(\frac{dr}{dt}\right)^2 = \left(\frac{\lambda M}{1+k^2}\cdot\frac{1}{r}+a\right)^2.$$

The equation of the kinetic energy gives

$$\left(\frac{dx}{dt}\right)^2 + \left(\frac{dy}{dt}\right)^2 + \left(\frac{dz}{dt}\right)^2 = (1+k^2)\left(\frac{dx}{dt}\right)^2 + \left(\frac{dy}{dt}\right)^2 = -\frac{2\mu}{r} + \gamma \qquad (\gamma = \text{constant})$$

$$\left((1+k^2)\left(\frac{dx}{dr}\right)^2 + \left(\frac{dy}{dr}\right)^2\right)\left(\frac{dr}{dt}\right)^2 = -\frac{2\mu}{r} + \gamma.$$

According to (5) and (6) we have

$$\frac{dy}{dr} = \frac{3y}{2r} = \frac{3\sqrt{r}}{2c}$$

nd from (7) we obtain

$$\frac{dx}{dr} = \frac{2c^2-3r}{2c\sqrt{1+k^2}\sqrt{c^2-r}}$$

/hence

$$(1+k^2)\left(\frac{dx}{dr}\right)^2 = \frac{(2c^2-3r)^2}{4c^2(c^2-r)}.$$

Consequently

$$\left(\frac{(2c^2-3r)^2}{4c^2(c^2-r)} + \frac{9r}{4c^2}\right)\left(\frac{dr}{dt}\right)^2 = -\frac{2\mu}{r} + \gamma$$

$$(11) \qquad \frac{4c^2-3r}{4(c^2-r)}\left(\frac{dr}{dt}\right)^2 = -\frac{2\mu}{r} + \gamma.$$

From (10) and (11) we obtain by elimination of $\frac{dr}{dt}$:

$$\frac{r^3}{4(1+k^2)(c^2-r)}\left(-\frac{2\mu}{r}+\gamma\right) - \frac{(4c^2-3r)}{4(c^2-r)}\left(\frac{\lambda M}{1+k^2}\cdot\frac{1}{r}+a\right)^2 = 0.$$

By multiplication with $4(1+k^2)^2(c^2-r)r^2$ this equation assumes the form

$$(1+k^2)r^4(-2\mu+\gamma r) - (4c^2-3r)(\lambda M + a(1+k^2)r)^2 = 0.$$

If r is not constant, this equation must be identically satisfied. Consequently

$$\gamma = 0 \qquad \mu = 0 \qquad a = 0 \qquad c = 0.$$

But, when $c = 0$ it follows from (6), that $r = 0$.

Consequently r must be constant in all cases. Then it follows from (4) and (5) that x, y and z ust be constant, and it is seen from the original system of differential equations that x, y and z ust be 0.

Hereby it is proved, that trajectories do not exist in any plane passing through the centre of the sphere except in the equatorial plane.

Hence we may conclude, that formation of planets will hardly be possible outside the equatorial plane. If after all a multitude of trajectories could approach asymptotically a common curve outside the equatorial plane, this curve as we have shown could not lie in a plane passing through the centre of the sphere, and as further the particles certainly very soon will lose their charge, they will come to move in the most different directions. The only possibility for formation of planets must be, that the particles approached a common curve lying in a plane through the centre of the magnetic sphere, and this we have proved to be impossible.

136. Our mathematical investigations have shown as their result that if boundary-circles exist for all the velocities with which material corpuscles are expelled from the central body, the corpuscles will either return to the central body (this being what will happen in the great majority of cases), or the particles will continue to approach nearer and nearer to the boundary-circles. Possibly some velocities may also be sufficiently great to cause the particles in question to leave the system and retire indefinitely.

Concerning the charge of the particles, we may imagine three cases:

I. When the particles are not charged. They will then either retire indefinitely, or fall down again.

II. When the particles are so highly charged that the electrostatic influence dominates that of gravitation.

III. When the particles carry a charge of medium strength, so that the electrostatic influence plays an important part side by side with gravitation, which, however, is the dominating force.

If we consider negative particles in case II, we shall easily be able to prove that they can indeed approach boundary-circles, but the radius of these circles must be $< (1+\sqrt{2})\, r_0$.

The necessary and sufficient condition for the approach of a particle to a boundary-circle in this case is that the following relations shall be satisfied:

$$\text{(a)} \qquad n > 1$$

$$\text{(b)} \qquad (l-1)(l-2) < 0, \text{ or, otherwise expressed, } 1 < l < 2$$

$$\text{(c)} \qquad n - \sqrt{2n+1} \leqq l - 2 \leqq n + \sqrt{2n+1}$$

$$\text{(d)} \qquad 2 - \frac{1}{n} + \frac{1}{n(l-2)} < 0$$

$$\text{(1)} \qquad \frac{r_0 v_0^2}{-\mu} = 2 - \frac{1}{n} + \frac{1}{n(l-2)}$$

$$\text{(2)} \qquad \frac{-\mu r_0^3}{\lambda^2 M^2} = \frac{(l-1)(l-2)}{n^3}$$

From (b) and (c) we find that

$$n - \sqrt{2n+1} < 0, \text{ or } n^2 - 2n - 1 < 0,$$

that is to say, we obtain in connection with (a)

$$\text{(e)} \qquad 1 < n < 1 + \sqrt{2}.$$

Necessarily, moreover,

$$2 - \frac{1}{n} + \frac{1}{n(l-2)} < 0,$$

hence

$$\frac{1}{n(l-2)} < \frac{1-2n}{n},$$

nd consequently

$$l-2 > \frac{1}{1-2n},$$

r, if preferred,

$$l > \frac{3-4n}{1-2n}.$$

Thus on the whole $l-2$ must satisfy the following inequalities:

$$n - \sqrt{2n+1} \leqq l-2 \quad (\text{and } l-2 \leqq n+\sqrt{2n+1}, \text{ which is satisfied according to (b)})$$

$$-1 < l-2 < 0 \qquad l-2 > \frac{1}{1-2n}$$

Now if n satisfies (e), it is evident that l, in an infinite number of ways, can be so determined at these last inequalities are satisfied. Then, however, the relations (a), (b), (c) and (d) are also satisfied, d we can consequently find positive values of r_0, v_0, μ, λ^2 and M^2, which satisfy equations (1) and (2). er suitable conditions therefore, the particle can approach an arbitrary circle, of which the radius r fies the inequalities

$$r_0 < r < r_0(1+\sqrt{2}).$$

Let us, upon the assumption that gravitation dominates the effect of electric force (case III), see the radii of the boundary-circles depend upon the relation between charge and mass, or, in other s, upon the quantity λ.

The necessary and sufficient condition for the approach of a particle to a boundary-circle, is the ultaneous satisfaction of the following relations:

(a) $\quad n > 1$

(b) $\quad (l-1)(l-2) > 0$

(c) $\quad n - \sqrt{2n+1} \leqq l-2 \leqq n + \sqrt{2n+1}$

(1) $\quad \dfrac{r_0 v_0^2}{-\mu} = 2 - \dfrac{1}{n} + \dfrac{1}{n(l-2)}$

(2) $\quad \dfrac{-\mu r_0^3}{\lambda^2 M^2} = \dfrac{(l-1)(l-2)}{n^3}$

If we confine ourselves to a consideration of those values of n that are $> 1+\sqrt{2}$, (a) is satisfied, (b) is satisfied, provided (c) is. The three conditions (a), (b) and (c) may therefore be contracted the equation

$$l-2 = n + \vartheta\sqrt{2n+1}$$

ere ϑ can be given all possible values between -1 and $+1$.

If we substitute this expression for $l-2$ in (1), we obtain

$$\frac{r_0 v_0^2}{-\mu} = 2 - \frac{1}{n} + \frac{1}{n(n+\vartheta\sqrt{2n+1})}$$

and by multiplication of (1) by (2), we obtain

$$\frac{v_0^2 r_0^4}{\lambda^2 M^2} = \left(2 - \frac{1}{n} + \frac{1}{n(n + \vartheta\sqrt{2n+1})}\right)\frac{(n + \vartheta\sqrt{2n+1})(n + 1 + \vartheta\sqrt{2n+1})}{n^3}$$

The result thus attained is that the necessary and sufficient condition for the approach to a boundary-circle with radius $n r_0$, when $n > 1 + \sqrt{2}$, is that the last three equations ta a value of ϑ between -1 and $+1$.

If we were to imagine r_0, v_0 and M maintained, we can find those values of μ and λ the boundary-circles. It is at once seen that for great values of n, $\frac{r_0 v_0^2}{-\mu}$ will keep very nea values of $-\mu$ that give rise to great values of n will thus approximately be $\frac{r_0 v_0^2}{2}$. It w however, that the greater n is, the nearer to 2 will $\frac{r_0 v_0^2}{-\mu}$ be, and then $-\mu$ must be *a lit* for smaller values of n. *Under otherwise similar circumstances therefore, boundary-circles ap negative particles will be of greater radius than those approached by positive particles.* More be seen from the last equation that for great values of n, λ^2 will be approximately propor i. e. *that particles with small mass in proportion to their charge will give rise to boundary greater radii than particles with great mass in proportion to their charge.*

The particles that approach a boundary-circle may continue to move there for all time. ceivable, however, that the number of particles will gradually become so great that they will of collecting into large globules, which in their turn at last unite to form a planet, as the ele in the original particles may conceivably be supposed to have been lost.

In the case of the *sudden* loss of the charge, the mathematical investigation has sho particles will afterwards move in ellipses about the central body with perihelion in the bour and with eccentricity

$$e = \frac{1}{l - 2},$$

$$\text{where } l = -k \cdot n,\ n = \frac{r}{r_0},\ \text{and } r_0^2 v_0 \sin\alpha_0 = \lambda M(1 + k).$$

That l is great and thus the eccentricity e small, when r is great in relation to the radi central body, will be seen from the following relation, which must be satisfied:

$$n + 2 - \sqrt{2n+1} \leqq l \leqq n + 2 + \sqrt{2n+1}.$$

If the electric charge of the particles is *gradually* lost, it seems evident that the finite pletely circumscribes the boundary-circle and very nearly becomes a circle, if the boundary-ci particles has a large radius in proportion to r_0.

Let us now, on the supposition that the entire mass of particles near a boundary-circ directly about the central body (like the planets), consider the question as to how a planet, in the massing together of the globules here assumed, can acquire a *direct* rotation about its not a *retrograde*, as at first sight one would imagine.

As regards this question, I subscribe to the explanation given by POINCARÉ in his "I Cosmogoniques" (p. 51), in which he shows how planets, in the event of their having or LAPLACE's ring-formations, can acquire a direct rotation. This explanation is based upon G. H important investigations on the effect of tidal reaction between a central mass and a body revolvi

The Trowbridge explanation of the direct axial rotation may perhaps be equally applicable to our theory. He shows that if the ring be nearly of the same density throughout, the resulting planet must have a *retrograde* rotation like Uranus and Neptune. But if the particles are more closely packed near the inner edge of the ring, so that the resulting planet would be formed *much within* the middle of its width, its axial rotation must be *direct*.

Our results summarised above seem both simple and well fitted to aid in constructing a new and satisfactory cosmogonic hypothesis, based on experimental analogies (see experiments represented in fig. 255).

137. We will here take the opportunity of mentioning some more recent experiments that have been made with the largest vacuum-box with a capacity of 1000 litres. We have already referred to their commencement.

Our experiments, as might be expected, prove to be more and more interesting as we increase the scale on which they are performed.

Fig. 263 a gives a good idea of the dimensions of the vacuum-box, and the various arrangements for the experiments.

The glass walls of the box, each of which supports a pressure of about 7000 kg., are 46 mm. in thickness. No firm of makers would supply any thicker, but it was calculated that they should have been 50 mm. in order to be safe. The floor and roof of the box are constructed of brass.

The largest cathode-globe employed is 36 cm. in diameter, and the maximal discharge-current has been about 400 milliamperes. Fig. 263 b shows how the rays in the magnetic equatorial plane may be very pronounced when the magnetism is weak and the discharge-current comparatively strong (150 milliamperes). With a stronger discharge-current, a peculiar electric corona frequently occurs round the cathode, sometimes with rays out from the polar regions, the whole thing having a striking resemblance to photographs of the sun's corona during an eclipse.

If we desire to produce the phenomenon which we think may be regarded as analogous to Saturn's ring, only 1 or 2 milliamperes is required, and the magnetisation of the cathode-globe must be somewhat stronger than in the former experiment.

Fig. 263 c shows powerful and characteristic spot-discharges from the magnetic cathode-globe.

It will be necessary to give some information as to the way in which these disruptive discharges may best be brought about.

With a polished metal globe like the cathode, disruptive discharges will not easily be formed. An almost continuous discharge with electric corona is then obtained, even if, as previously mentioned, the steam of vaseline-oil be introduced into the box.

If, on the contrary, the globe is cast and not polished, such disruptive discharges will nearly always occur; but the difficulty here is that the casting of so large a globe as the one here employed never yields a homogeneous result, so that the patches keep to certain parts of the globe, even when the latter is not magnetised.

The best way in which to obtain with certainty a continuous discharge, interrupted at definite intervals by powerful disruptive discharges, seems to be the following:

The surface of the globe, after polishing, should be sand-blown and then painted over with a thin coating of vaseline-oil, which is afterwards wiped off again. This painting over, which seems to be advantageous to the phenomena, is not necessary if the farthest corners of the vacuum-box are greased with a little vaseline-oil before the box is exhausted.

When a suitable vacuum has been obtained, a short discharge of about 10 minutes with a current-strength of 200 milliamperes will completely dry up the oil on the cathode. Without discharges the globe will remain oily for many days, even in a high vacuum produced by a Gaede's molecular pump.

Fig. 263 a b c.

is probable that in this case an electric disintegration of the oil on the cathode takes place, possibly companied by a partial decomposition. This we conclude from the following experiment, which more-er is important in more respects than the one here mentioned. During our experiments, the floor d ceiling of our vacuum-box had received rather too abundant a coating of oil. In order to correct is, discharges were sent through for several hours with the floor and ceiling as cathode. As these ent on, the floor and ceiling became practically dry, whereas the glass walls received a powerful ecipitation of oil or fatty decomposition products in a zone about 3 cm. in width, the edge on one side ing somewhat diffuse, but clearly marked towards the cathode, the limit beginning on a level with

Fig. 264.

t external boundary-surface of CROOKES' dark space. A similar coating also appeared upon the in-slated tube by which the cathode-globe was suspended, after the corresponding drying of the globe; ad the same one-sided sharply-defined coating was also found on the glass vessel that contained phos-poric acid (see figure 264).

It seems from this that just about this boundary-surface all round the cathode there is formed during t discharge an atmosphere of complicated ions, while at the same time a high tension polarisation layer i working up and at last gives occasion for a disruptive discharge. This is also shown by the fact that a ertain time always elapses before the disruptive discharges begin and then attain to a stationary con-ion of frequency. The author had already put forward this assumption before the above-mentioned eperiment was made, and it will be found in a paper previously quoted here (C. R., March 17, 1913).

The experiment seems also to indicate that a great number of the corpuscles ejected in a straight li from the cathode by disintegration are stopped again at the end of the dark space.

Next some experiments were made in which the cathode-globe with a diameter of 24 cm. was sur-runded with a well amalgamated zinc shell. The vacuum-box was now cleaned from oil and fat, except f a small part of one of the glass walls, where a white coating of fat remained. This patch of fat gve rise, as we shall soon see, to an important discovery.

Concerning the experiments themselves and the light-phenomena observed, we shall only state that the radiation from the polar regions of this cathode were often particularly beautifully developed. It was further demonstrated that the quicksilver on the surface of the cathode disintegrated greatly, after which the surface of the globe gave rise to magnificently iridiscent rings of colour, of which we succeeded in taking several good colour-photographs.

We here reproduce two interesting photographs that were taken during some experiments in which the globe was no longer the cathode, but acted as a terrella, just as in the experiments described in Chapter IV.

Fig. 265 a.

Fig. 265 a shows an experiment in which an aluminium plate in one corner of the box acted as cathode, while the metallic parts of the box were the anode. It will be seen how the rays strike downwards all round the auroral zones of the terrella. A glance is sufficient to show the occurrence of phenomena that have previously with much trouble been educed from a long series of experiments.

By varying the magnetism of the globe, the radius of the light-zone was altered within wide limits, diminishing with strong magnetism.

Fig. 265 b shows something similar, but in this case the globe itself was the anode, and the phenomena were even more magnificent; for in the belt in which the rays from the cathode descended upon the terrella, "positive light" radiated from the latter, giving a remarkably beautiful effect to these light-zones about the poles.

We will now return to the above-mentioned patch of grease upon the glass wall of the vacuum-box, as it occasioned the discovery of a phenomenon that is highly worthy of attention.

After the amalgated globe had been acting as cathode for a couple of hours, it appeared that that part of the glass wall on which there had originally been a white coating of fat, gradually became grey and then very dark in colour, without any change taking place in the rest of the clear glass wall.

When the box was opened, the passing of the finger over the patch of grease produced an abun-lance of tiny drops of quicksilver. We thus see that the quicksilver corpuscles from the greatly disinte-rating cathode-globe in this large vacuum-box (1000 litres) are thrown against the glass walls, as a rule vith the result that they are reflected back again. *It was only where the surface of the glass was greasy hat the corpuscles adhered.*

This result made me think of all my former vain attemtps to make the corpuscles thrown off from palladium cathode produce a shadow upon the glass wall of the vacuum-tube, of an object standing etween the cathode and the wall. It seems natural to suppose that corpuscles that are disintegrated

Fig. 265 b.

om a cathode have not generally sufficient velocity to adhere when they strike a wall, but that at first ey generally rebound, only a few of them adhering immediately. When a coating is once formed, the her corpuscles have a better opportunity of adhering, possibly on account of electric attraction.

In order to test the correctness of this assumption, a former experiment was repeated, in which shadow-formation had been obtained. On this occasion all interior surfaces were greased with vase-ne-oil, whereas before they had been dry and clean. The result, as the accompanying reproduction of photograph (fig. 266) shows, was in astonishing conformity with the assumption.

The vaseline-oil employed soon stiffens, it is true, under the influence of the cathode-rays, but the tty substance formed "catches" the corpuscles and prevents them from being reflected back from the alls. It is thus possible in this way to demonstrate the course of the "metal rays" from the cathode, ithout having complications introduced into the phenomena by reflected rays.

Exceedingly peculiar conditions would arise in a vessel filled with flying corpuscles, if a patch on one its walls had the property of intercepting all particles that had the greatest kinetic energy, while those ith small velocity rebounded. It is assumed that the walls of the vessel, apart from the patch, throw ack all corpuscles. Might we perhaps replace the famous little Maxwell's demon with a patch of grease?

On a previous occasion we have in reality shown under somewhat different circumstances that the rays of corpuscles ejected from the cathode are reflected from the walls. In a paper in C. R., March 17, 1913, I have said that the long pencils of rays emitted from the cathode and carrying a quantity of disintegrated matter from the cathode, are reflected from a wall like rays of light from a mirror.

If, therefore, it is desired to study the course of these bundles of rays, or their ability to pass through, for instance, thin aluminium-foil, employment should be made of a layer of fat to intercept the corpuscles after their passage, as the ray-phenomena are then more easily demonstrated.

Fig. 266.

After the treatment with grease mentioned above, the cathode is well fitted for patch-experiments, and intensely powerful disruptive discharges are formed even without additional external capacity.

In a high vacuum the patches generally consist of groups, but may also consist of a single patch at each place. They may often remain in one place for a measurable length of time. They are surrounded by the previously-mentioned vortices (see fig. 261), rotating in opposite directions on the two magnetic hemispheres. Round the single patches more particularly, these vortices attain a surprising clearness and regularity.

It appears on comparison of the above with HALE's photographs of sun-spots with vortex-formations, that I have been guilty of a misunderstanding.

The experimental vortices are in the reverse direction to HALE's, supposing the magnetic north pole to be on the top of the cathode-globe. In my descriptions I have reckoned the vortex from the centre outwards, contrary to Hale, who has considered them in the more usual way.

But the consideration of the experimental whirls and the solar vortices as analogous phenomena does not seem to involve any contradiction.

In my experiment the magnetic power in the spot is determined by the magnetisation of the cathode-globe. The current-strengths carried by the discharges are too small to produce any marked local field. In a sun-spot, on the contrary, the local magnetic field predominates, and it may very well be due to the enormous conditions on the sun. In some way or other with which we are not now acquainted, vortices may arise from the discharges. The current-strengths are so great that the magnetic forces formed by them will be able to entirely reverse the original magnetic field which was due to the general magnetisation of the sun.

Here it should possibly be considered that the current-paths in the photosphere around a spot are "selected", so to speak, at the first moment, before the current-strength in the discharge has attained to any magnitude worth mentioning. Later, when it becomes perhaps millions of times greater, the current-paths retain to some extent their orientation, and produce a corresponding magnetic field.

We have repeatedly pointed out the resemblance that exists between the light-phenomena about a magnetic cathode-globe and corresponding solar phenomena, such as the corona with the radiating off-shoots in the polar-regions, and the sun-spots.

The light-phenomena about a magnetic *anode*-globe, on the other hand, are quite different, except that the radiation in the polar regions is sometimes nearly like that from a *cathode*-globe, and resembles the polar radiation of the sun.

It might at first sight appear as if this were an indication that perhaps the sun is negative in the juatorial regions and positive in the polar. If so, it would suggest the thought whether a difference in ectric tension might eventually be produced by rotation of the magnetic solar body in space.

It is easy to make an estimate here. MASCART(1) has calculated that the rotation of the magnetic irth must give rise to an aggregate electromotive force of an order 10^6 volts, acting from the poles the equator.

If we wanted to make a similar calculation with regard to the sun, we must first of all have a ilue for the amount of the magnetic force near the sun. This is still unknown. HALE(2) is at present aking attempts to measure it. In the mean time SCHUSTER(3), with certain assumptions, has calculated a recent paper that the intensity in the sun should be 440 times greater than on the earth. If we ekon with an even magnetisation of the globes, this would make the sun's magnetic moment 440×102^3 reater than that of the earth.

Now we have shown (see p. 617) that if cathode-rays from the sun with a huge moment such as is were to reach the earth, they must have a magnetic stiffness answering to

$$H \cdot \varrho > 10^{12},$$

hile from the situation of the earth's auroral zones we may infer that the helio-cathode rays which produce aurora and magnetic storms have generally a value of 3 million C. G. S. And to this last result e are inclined to attach the importance of an experimental fact.

On the other side we have calculated, from the retardation of up to 50 hours of the magnetic orms in relation to a sun-spot's passage of the central meridian, that the magnetic moment of the sun from about 100 to 150 times as great as that of the earth, or of the order 10^{28} C. G. S.

We cannot of course from this conclude anything about the magnitude of the magnetic force near te sun, for the sun is certainly no evenly magnetised globe. The general magnetisation of the sun is obably produced by electric currents in relatively thin layers round the solar equatorial regions inside outside the sun's surface. In this way the value of the magnetic force near the sun's surface may be latively great, without any overwhelming magnetic moment for the sun being assumed(4). If we start with

(1) MASCART, Traité de Magnétisme Terrestre, p. 74. Paris, 1900.

(2) Mount Wilson Solar Observatory; Annual Report, 1912, p. 179.

(3) SCHUSTER, Proc. Phys. Soc. of London, 1912, p. 127.

(4) The rays emitted from the sun will certainly to some extent serve to increase the sun's magnetism.

We have shown that if the magnetic moment of the sun is of an order 10^{28} (see p. 617), all rays of which the product $H_0 \varrho_0 < 5 \times 10^5$, will return to the sun and fall down again, or must circulate about the sun, the negative rays clockwise, the positive anti-clockwise, seen from above and assuming that the sun's magnetic north pole is uppermost.

The figures 248 b and c show how such flexible rays are moving in almost cylindrical rings about the magnetic cathode-globe. The radius of such a ring seems never to come up to 2.5 times the radius of the central sphere, as the theory predicted (p. 617).

Perhaps the "dusky veil" of the sun (see p. 670) is due to such a cylindrical ring of corpuscles moving about the sun.

Under *all* circumstances (even if the sun's magnetisation is the reverse of that here supposed), a part, and perhaps the more considerable part, of the rays emitted by the sun will thus serve to magnetise the sun; but there are perhaps also electric currents in the interior that act in the same way. We may even imagine the sun's magnetism to have originated in this manner, if we start with the assumption that the initial velocity of the negative rays is greater than that of the positive.

Suppose that the sun had originally been non-magnetic, but rotated in the same way as it now does. It is evident that the positive rays, of which the bearers may be assumed to be positive material corpuscles, will then be deflected by the rotation of the central body, even if the electric forces at first tried to eject the particles normally from the surface of the sun. Owing to gravitation (and by electric attraction if the sun were negative), the ejected particles now make their way back to the sun. The total magnetic effect of all the positive rays must then be that of a positive current circulating in the same direction as the rotation. The sun would thus be north-magnetic above, that is to say provided no other, greater forces have been acting in the reverse direction. The negative rays consisting of electrons with great velocities would probably be deflected by the rotation of the central body in the same direction as the positive rays; but the deflection would be less and the electrons would not at first return to the sun. Not until the sun was magnetic would they be reversed, and then serve, as shown above, to augment the sun's magnetism.

my assumption of a sun-moment of an order 10^{15} C.G.S. and opposite to that of the earth, an estimate will easily show that the induced electromotive force in space about the sun will not be so great as about the earth, aud its direction will be from the equator to the poles. There will then, of course, be no question of explaining the great discharges from the sun, in which the tension goes up to 600 million volts ([1]) (see p. 665), as an induction-phenomenon of this kind. The most reasonable hypothesis therefore seems to be that the sun and the stars are negative all over in relation to surrounding space.

It is otherwise if we calculate with SCHUSTER's purely hypothetical value for the intensity of 440 times the intensity on the earth. We should then most probably come to an electromotive force of about 2 milliard volts, acting from the poles to the equatorial regions, which would thus have to be regarded as the cathode in the eventually produced discharges, while the poles were anodes.

138. It is a circumstance in my planet-theory which has given me much trouble, as it looked at first as though the planets, if formed, would come to revolve the wrong way round the central body. I considered it at first most probable that the material particles expelled by disintegration from the negative-electric central body, took with them a *negative* charge.

It was soon evident, however, that if, for instance, the magnetisation of the sun—as I have had to assume—is the reverse of that of the earth, the negatively-charged particles would hardly be able to approach the boundary-circles in the same way as the planets move in their orbits. At any rate they must first change from out-going to in-going motion, and be subjected to a suitable resistance. For one thing this resistance must be such that the velocity, in spite of gravitation, would be diminished during the in-going motion, and it seems physically unreasonable to assume the existence of such a resistance. It would be far easier for the particles to approach the boundary-circles the opposite way. It therefore appeared probable from the theory, that we should be compelled to assume that the expelled particles would, partially at any rate, be positive.

This led me to think that possibly the electric disintegration from a cathode had some points of resemblance to the disintegration of radio-active substances, which emit α-particles, even if the emitting substance is charged negatively.

It occurred to me, moreover, that since it has been decided that the particles in K_1-rays emitted by a cathode, are positively charged, it might be well worth finding out whether the material particles expelled by disintegration also carried a *positive* charge.

By examining the literature on the subject, I soon saw that there were no definite results that could decide the question, although our idea of the constitution of matter presupposes that the atoms in a non-electric piece of metal are positively charged, and that there are corresponding free negative electrons between the molecules. From a theoretical point of view it might thus be conceivable that the atoms ejected from a cathode were positively charged.

It was for this reason that I commenced these investigations of the disintegration of cathodes, some of which have been described in the Article on Saturn's Ring, while others will now be described.

As mentioned on p. 659, my attention, while experimenting on the disintegration of the cathode, was increasingly drawn to the more or less normally expelled particles which formed an evenly-reflecting deposit of palladium on the cylindrical glass wall of the vacuum-tube, right round the cathode. A number of experiments were therefore made by introducing little, flat screens of mica, with or without a slit, at various distances from the cathode. The palladium cathode was in the form of a long rectangle, whose long centre line coincided with the axis of the tube. It appeared, however, that notwithstanding

([1]) In my experiments with an electric corona I use from 0.1 to 0.2 milliamperes per sq. cm. of the surface of the globe-cathode. If we suppose a similar value of the current from the sun, and the tension to be 600 million volts, this corresponds to about 100 kw. per cm. Such an energy would easily account for all heat and light radiation from the sun.

hat the particles expelled from the cathode must now be assumed to move more or less at right angles the magnetic lines of force, and that the field-strength was about 1800 lines of force per square centietre, it was not possible to prove any turning aside of the particles, first of all because the field was ot strong enough, but also because it was not possible in these preliminary experiments to obtain sharp hadows of the screen in the metallic deposit upon the glass wall.

On the other hand it appeared that the deposit came not only on the front of the screen, but also undantly on the back of it, especially if it stood near the cathode, a fact which indicated that the par-cles could acquire a retrograde motion after they had retired from the cathode.

From the appearance of the deposit upon the back of the mica screen farthest removed from the thode, I received the distinct impression that the particles had struck almost parallel with the surface the screen. I therefore made, on each side of the long side of the mica screen, a raised edge one or vo millimetres in height. Both edges turned away from the cathode. It then appeared that there was longer any deposit upon the back of the screen, although on the front and on the protecting side lges there was an abundant deposit of palladium.

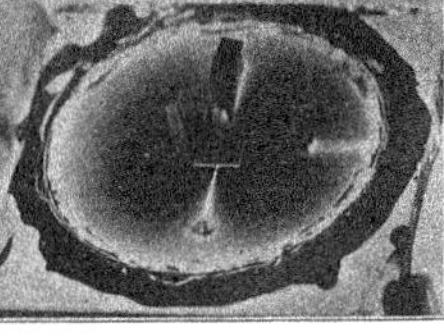

Fig. 267.

The next arrangement was as follows. A long, rectangular cathode of palladium was attached to a ick brass wire that passed through a quartz tube with walls 2 millimetres in thickness. Only the palla-um plate reached beyond the quartz tube, which was placed axially in the vacuum-tube. The anode as annular in shape, and was placed 10 cm. behind the cathode, which again was only 2 mm. above e sheet of plate-glass that was cemented to the end of the vacuum-tube. In order to prevent the acking of the sheet of plate-glass with the heat from the cathode-rays, a small square of mica was emented to the sheet of glass just under the cathode.

There were further cemented to the sheet of glass some half-cylinders at various distances, with eir convex side towards the cathode.

By these means it was shown that the palladium particles to a very great extent made their way to the concave side of the half-cylinder, if it was placed near the cathode, whereas if it was far from e cathode, the particles hardly entered it at all, although they abundantly covered the convex side with lladium. Figs. 267 a and b show how these little half-cylinders were arranged, and also that the most stant half-cylinder has cast an almost straight shadow behind it on the sheet of glass (farthest to the ht in fig. 267 b), where therefore the palladium has not been deposited. The nearest half-cylinder has rown no distinct shadow, or at any rate it is only by careful examination that there is seen to be less lladium deposited just behind it than beside it. It seemed as though some of the expelled palladium rticles had a tendency to return to the cathode again, just as if they were positively charged.

In a subsequent experiment, the cathode was a fairly thick platinum plate with a surface of a few suare millimetres, while the anode was a brass plate with a surface measuring a couple of hundred square centimetres. A high vacuum was maintained in the vacuum-tube; the tension was over 15,000 volts, and the temperature of the cathode was kept up by the current near the melting-point of platinum.

In three hours the brass anode was completely coated with a shining mirror of platinum. On the glass wall of the vacuum-case there was a fairly sharp shadow of a screen that stood between the cathode and the wall, so that in this experiment we are fully justified in speaking of "platinum rays".

In the same way, in many and varied experiments, rays of palladium and uranium were produced, with the employment of as much as from 15,000 to 20,000 volts to the cathode (the positive pole was earthed) and temperatures of from 600° to about 1800° C. The reader is here referred to the remarks in connection with the above-mentioned experiments, in which the whole of the inner surface of the vacuum-tube was greased.

The experiments seem to show that these positive rays have several of the most characteristic properties of α-rays. Both the way in which they are formed in the firm material of the c the way in which they are formed in the firm material of the cathode, and the way in which they

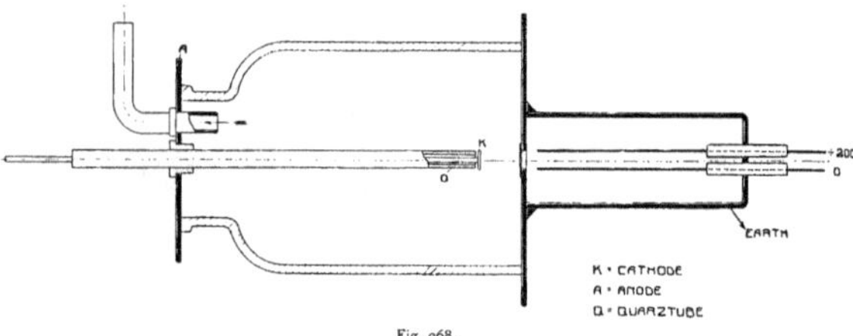

Fig. 268.

and stop in the surrounding medium indicate this. We also succeeded in sending platinum rays, and more particularly rays of metallic uranium, right through thin aluminium foil, just as can be done with α-rays.

We will describe two of these experiments more particularly. Rays from a small palladium cathode were sent through a little hole into an otherwise closed metal capsule (which was earthed) and on between two parallel brass plates at a short distance from one another, as shown in fig. 268(¹). One of the plates had — 200 volts, the other was earthed. After the experiment had been going on for 3 hours, the coating of palladium showed itself to be quite different upon the two plates. On the — 200 plate, a long, narrow, more or less well-defined pencil of rays was found, where the precipitation was very abundant. But in addition to this, there was a very thin coating of palladium *all over* this plate (which was 6.5 cm. long and 4.5 cm. wide), even on the back, especially if the plate were smaller. On the other plate, which was of the same size, there was a short, broad, fan-shaped precipitation of quite another kind than that on the —200 plate, and there was no other deposit upon the plate, either on the front

(¹) This Vacuum-tube, in several experiments, was placed between the poles of a large electro-magnet, which was just sufficiently magnetised to prevent ordinary cathode-rays from forcing their way through the hole into the otherwise closed metal capsule.

The discharge-current from the anode was thereby pressed by the force of the magnet into a thin, luminous cord along one side of the Vacuum-tube.

Every time during these experiments, after working for one or two hours, it appeared that palladium corpuscles from the cathode were driven against the electric current right up towards the anode, the glass under the luminous cord being thickly coated with a metallic band. This shows that negative palladium ions have moved up against the current. There are thus both positive and negative metal ions from the cathode.

the back. This experiment was made in varied forms more than 20 times with in the main the me result.

My explanation of the thin, diffuse deposit upon the —200 plate is that after the positive rays have st their velocity, they are drawn in electrostatically towards the plate, evenly right round it.

A reservation must be made here, however, as there is a possibility that this explanation of the *en* coating of palladium round the —200 plate is incorrect.

The cathode-rays, as might be expected, were drawn in upon the o—plate. Rays such as these ould, as we know, cause an already produced precipitation of palladium to decompose again. It may erefore be imagined that eventually it was the palladium corpuscles detached from the cathode-rays and ted for the second time, that were positive and were thus evenly drawn in towards and all round e — 200 plate.

The experiments that were made to show that metal rays went through aluminium foil, were car-d out in the following manner.

A small cassette of brass had 4 small holes, 0.5 mm. in diameter bored side by side in the lid. ut of the thinnest aluminium-foil of about one-thousandth part of a millimetre, small entire portions were arched for with a microscope, and laid in one, two, three or four layers over the four holes. Under e whole there was a sheet of glass. A little steel magnet was placed behind the cassette, for the pur-se of deflecting ordinary cathode-rays from the cathode, which was placed at a distance of 20 mm. ght in front of the holes in the cassette.

After the discharge had been going on one or two hours, the cassette was opened and the sheet glass studied. The precipitation of metal through the foil was not so considerable that it could be en without doubt with the naked eye; but by breathing on the glass a sharply-defined, well-marked ot appeared beneath the hole with *one* layer of aluminium-foil. Under the hole that was covered with o layers there also appeared a distinct spot; under that with three layers the deposit could scarcely distinguished; but under the hole that was covered with four layers, not even traces were found in y case.

Since cathode-rays as stiff as those in these discharges would easily pass through even four layers such thin aluminium-foil, and as these rays in most cases were deflected with a steel magnet, it must obably be assumed that they are metal rays that have penetrated through the foil, but in very different grees through the four holes. These experiments, however, will be continued, as also those that have en made for the determination of charge and mass of the metal corpuscles.

There is yet another point in these experiments that will be touched upon here. It has been entioned above that under certain conditions marked oscillations might occur in an oscillatory circuit nnected(¹) in parallel with the anode and cathode in the vacuum-tube as poles.

It appears that the disintegration of the cathode is much greater under these conditions, and that this case thin, luminous pencils of rays are emitted by the cathode. At the foot-points of these pencils particular, the cathode-material becomes so greatly disintegrated, that under the microscope the surface the cathode gives the impression of having been corroded with a quantity of tiny cavities. It also appears at it is not necessary to keep the temperature so high as that given above in order to obtain a powerful velopment of positive metal rays from the cathode when it is connected with such an external oscillatory circuit.

The rays that have hitherto been called α-rays, consist, as is well known, of positive helium-atoms, ected with enormous velocity from a radio-active substance, e. g. radium.

There seem, from the discoveries here mentioned, to be good grounds for extending the conception rays to include rays formed of all positive atoms that are ejected with such velocity as to give rise the properties of α-rays.

(¹) l. c., C. R., March 17, 1913.

The processes whereby such rays were formed we might call radio-activity in an extended sense, or electro-radio-activity.

We have not yet succeeded, however, in spite of continual experiments, in producing a proof that by this extended radio-activity chemical elements might be transformed into one another, or that heat was developed by the disintegration of a cathode, in the same way as when radium is transformed. The last question would acquire a fundamental importance in the problem of the heat-store and longevity of the sun and stars.

139. According to our manner of looking at the matter, every star in the universe would be the seat and field of activity of electric forces of a strength that no one could imagine.

We have no certain opinion(1) as to how the assumed enormous electric currents with enormous tension are produced, but it is certainly not in accordance with the principles we employ in technics on the earth at the present time. One may well believe, however, that a knowledge in the future of the electrotechnics of the heavens would be of great practical value to our electrical engineers.

It seems to be a natural consequence of our points of view to assume that the whole of space is filled with electrons and flying electric ions of all kinds. We have assumed that each stellar system in evolutions throws off electric corpuscles into space. It does not seem unreasonable therefore to think that the greater part of the material masses in the universe is found, not in the solar systems or nebulæ, but in "empty" space.

Let us see how thickly we should have to imagine iron atoms, for instance, distributed in space between the sun and the nearest star, *α Centauri*, if, in a sphere with the distance 4.4 light-years as radius we assumed a mass equal to that of our solar system to be evenly distributed.

The mass of our solar system may be estimated at 2×10^{33} grammes (see Young, General Astronomy, pp. 97 and 603). The distance to *α Centauri* is 4×10^{18} centimetres, and the volume of the said sphere about the sun would thus be 2.7×10^{56} cubic centimetres.

If the mass of our solar system be distributed over this sphere, there will be 7.5×10^{-24} grammes per cubic centimetre.

If the mass of an iron atom be put at 5.6×10^{-23} grammes, we find that there will fall 1 iron atom upon every 8 cubic centimetres of the sphere in question.

It seems as if no known facts can prevent us from assuming by hypothesis that the average density of these flying ions and uncharged atoms and molecules might very well be, for instance, 100 times greater than that found above.

The electron theory assumes that the ponderable atoms are surrounded by some bound electrons which oscillate about certain positions of equilibrium and with definite periods. These atoms or ions cannot then, considered optically, have properties that are very different from the optical properties in a dielectric medium.

Let us therefore imagine that we have on an average 10 iron atoms per cubic centimetre in empty space, and try to form some idea as to whether such a density would be at variance with the optical properties of space, and in the next place whether this density would be irreconcilable with the assumption that the sun sends cathode-rays down to the earth.

The latter question seems the easier to decide when we consider that there must be a row of $\frac{10^{23}}{5.6}$ cubic centimetres, one after another, to contain one gramme of iron. A column such as that would be traversed by light in 1900 years. If we assume that the stiff helio-cathode rays of which we are now

(1) See "Sur la Source de l'électricité des étoiles", C. R. Dec. 23, 1912.

ɔcaking would also in this case be absorbed in accordance with the law of traversed masses, we see once that on this point our hypothesis will scarcely meet with any difficulty.

With regard to the first question, namely how the light in the case supposed would be absorbed empty space, it is not so easy to say what influence electric atoms, dispersed through space in such ultitudes, would have upon the light that comes to us from the stars; but it is hardly credible that any ꞁeoretic investigation would show as its result that the stellar heavens would then be darkened in a anner that is at variance with reality.

Atoms with bound electrons may be imagined to absorb light and heat waves by co-oscillations of e bound electrons. The absorption conditioned in this way does not attain a noticeable value until the eriod in the entering waves agrees with that of the oscillations proper of the bound electrons. It may ɔw be imagined that the oscillations of these bound electrons may in their turn be transferred to ther waves, or that one or more electrons may separate and form cathode-rays. We know that cathode-ys can be emitted by a metal surface by irradiation with ultra-violet light, and that electrons can be t free from a metallic surface when that surface absorbs rays of light.

There is also another question which naturally presents itself for investigation: Will the assumed ensity of flying corpuscles in space bring about any appreciable resistance to the motion of the heav-ıly bodies?

Let us look at the case as regards the earth, when it was assumed that there were 10 iron atoms er cubic centimetre in space.

We will assume the least favorable case, namely that the earth intercepts all the atoms it meets. uring a revolution round the sun, the earth encompasses a volume of 1.2×10^{32} cubic centimetres. If e mass of the iron atom be put at 5.6×10^{-23} gr. and 10 atoms be assumed per cubic centimetre, ıe earth will intercept 6.7×10^{10} gr. in one year.

According to the equation $(M + \Delta M)V_1^2 = M V_0^2$, the velocity of the earth will then be dimin-hed by

$$V_0 - V_1 = \frac{1}{2} V_0 \frac{\Delta M}{M}$$

If the earth's velocity be put at 3×10^6 cm. per second, its mass at 6.06×10^{27} gr., then $V_0 - V_1$ 1.7×10^{-11} cm. per second. According to this, the earth, supposed for the sake of simplicity to be sting in its orbit, would be retarded 5.4×10^{-4} cm. per annum, or the length of the year would be creased by 1.8×10^{-10} seconds.

We see from the above that it is not impossible that future investigations will show that without oming into conflict with experience in any way here mentioned, we may reckon that there are more ıan ten thousand times greater masses gathered as flying corpuscles in "empty" space than the masses the stars and nebulæ.

And it may be imagined that an average equilibrium exists between disintegration of the heavenly ɔdies on the one side, and gathering and condensation of flying corpuscles on the other[1].

[1] In a paper, "De l'origine des mondes", Archives des Sciences, Geneva, June 15th, 1913, the author has made the views here set forth the subject of detailed consideration.

PART III.

EARTH CURRENTS AND EARTH MAGNETISM.

CHAPTER I.

EARTH-CURRENTS AND THEIR RELATION TO CERTAIN TERRESTRIAL MAGNETIC PHENOMENA.

INTRODUCTION.

140. As soon as the discovery of OERSTED, in 1819, of the effect produced by galvanic currents on magnets was made known to the world, attempts were made to explain the earth's magnetism and its variations by means of currents circulating in the earth.

As early as 1821 DAVY (1) suggested that the variation in declination might possibly be due to such currents, and some years after the same view was taken up and carried further by CHRISTIE (2) and P. BARLOW (3).

These ideas seem to have met with general acceptance, and soon became the current explanation for the pulpit.

The theory of magnetism as caused by earth-currents was merely founded on speculation, and years had passed before the question was put to an actual experimental test.

The first attempts at measuring currents in the earth's crust were made in mines in Cornwall (4). It seems, however, hardly possible to decide whether the currents measured were real earth-currents or not.

Experiments of a similar kind were made by BECQUEREL (5) in the salt-mines of Dieuze. He observed the currents called into play when various layers of the earth were connected by conducting wire.

W. H. BARLOW (6) seems to have been the first to show that currents were almost always cirenlating in the earth's crust. He used four telegraph lines starting in different directions from the same central station at Derby. About simultaneously, earth-currents were observed by BAUMGARTNER on the line between Vienna and Gratz.

It was found that earth-currents ordinarily circulating in the earth were very variable in strength. The first result of the actual test of earth-currents was that the view put forward by P. Barlow, that the earth's magnetism was directly caused by currents circulating in the earth, was not confirmed by experiment. This conclusion, as far as I know, was first positively stated by AIRY.

But there still remained for investigation the question as to whether, or to what extent earth-currents produce the magnetic variations. As the result of comparison of currents with the variation of magnetic elements, Barlow finds that simultaneous observations showed no marked similarity in the path described by the magnetic needle and the galvanometer, LLOYD, however, from the same obser-

---

(1) Sir H. DAVY: Phil. Trans. 1821 p. 7.
(2) C. H. CHRISTIE: Phil, Trans. 1827 p. 308.
(3) P. BARLOW: Phil. Trans. 1831 p. 99.
(4) R. W. FOX: Phil. Trans. 1830. R. W. FOX, HUNT, PHILLIPS: Annual Report of the Roy. Polytechnic Institution of CornWall, 1836, 1841, 1842.
(5) BECQUEREL: Comptes Rendus XIX, p. 1052.
(6) Phil. Trans. 1849, p. 61.

vations, but taking the average of several days, found curves for the diurnal variation of earth-currents, which seemed to show some similarity with corresponding magnetic variations. The similarity is not a striking one, and it is doubtful how a similarity which is not shown in simultaneous observations should be interpreted; so it was at last to be considered as rather doubtful whether the diurnal variation of the earth's magnetism was due to earth-currents.

Thus the actual test of the theory of the electric origin of the earth's magnetism had not given any trustworthy confirmatory results.

Then an event occurred which should show definitely that a connection of some kind existed between the variation of the earth's magnetism and earth-currents.

From the 29th August to the 3rd September, 1859, a great magnetic perturbation took place, accompanied by aurora borealis, and simultaneously the telegraph-lines were disturbed by currents of an extraordinary strength, which were observed at the most various parts of the world.

This event gave a great impulse to the study of earth-currents. Earth-current observations were carried on for several years on the English telegraph-lines, and were collected and worked out by C. V. WALKER[1].

About simultaneously earth-current measurements were undertaken by LAMONT[2].

In his first publication, Walker treats the earth-currents observed during magnetic disturbances. In spite of the fact that the two phenomena accompany each other, he does not venture to draw the conclusion that magnetic storms are entirely caused by earth-currents. His statements are of special interest when looked upon in the light of recent research. He says[3]: "Other influences than those exerted by electric currents upon magnets may or may not be in play; but one thing is very certain that at least a large portion of the motion presented by the magnetometers on storm days is connected with the then prevalence of earth-currents; and doubtless some portion of all the more regular and less violent disturbances may be more or less due to the same causes. At any rate, although we are considerably in the dark as to the forms of force in operation to make up the *whole* of the causes concerned in magnetic disturbances, we are yet quite certain that the current form of force is at least in *part* concerned."

In a subsequent work Walker deals with the ordinary currents found on undisturbed days. He finds that the currents observed are real earth-currents and are not due to the earth-plates or other local conditions. They are not equally frequent in all directions, but appear mainly in the two opposite quadrants N-E and S-W. This result has been confirmed by later observers.

Lamont seems to be of the opinion that magnetic storms are produced by earth currents, but he does not consider it to be proved that all variations in the earth's magnetism are due to earth-currents.

Up to this time all observations had been carried out by taking readings at intervals. In this way it was very difficult to follow the many sudden changes of earth-currents, which accompany the magnetic disturbances.

Walker has pointed out the importance of having continual photographic records of earth-currents in connection with magnetic records. The matter was taken up by Airy, Astronomer Royal, and earth-current registerings were commenced at Greenwich in 1865, and were continued for two years. The results are contained in two papers by Airy communicated to the Royal Society in 1868 and 1870; and the conclusions he has drawn from his observations have to a great extent formed the basis of later discussions.

---

(1) C. V. WALKER: Phil. Trans. 1861, p. 89, and 1862, p. 203.
(2) LAMONT: Der Erdstrom und der Zusammenhang desselben mit dem Magnetismus der Erde, 1862.
(3) loc. cit., p. 114.

His results were in short the following:

(1) He thinks that on repeatedly examining the agreements of the two systems of curves it is impossible to avoid the conclusions that the magnetic disturbances are produced by terrestrial galvanic currents below the magnets. There still remain some points to be explained before we can prove that galvanic currents, as we observe them, will account for all that we observe in magnetometer records([1]).

(2) Regarding the total magnetism he says:

"On one point we can speak with confidence; they do not explain the existence of the principal part of terrestrial magnetism"([2]).

(3) The general agreement of curves, especially in the bold inequalities, is very striking particularly in the curves relating to northerly force([3]).

(4) The small irregularities in the curves of galvanic origin are more numerous than those in the curves of magnetic origin([3]).

(5) The irregularities in the curves of galvanic origin usually precede, in time, those of magnetic origin, especially as regards westerly force([3]).

(6) The proportions of the magnitudes of rise and fall in the curves often differ sensibly, especially as regards westerly force([4]).

(7) The northerly force appears, on these days of magnetic storms, to be increased, whereas general experience leads us to expect that it would be diminished([4]).

(8) In agreement with Walker he finds that the earth-currents observed on calm days are real earth-currents, and finds that they show a well-marked diurnal period; but he says that neither in magnitude nor in law are these inequalities, consequent on galvanic currents, competent to explain the ordinary diurnal inequalities of magnetism([5]).

(9) At present we are unable to say whether the records of the galvanic currents throw any light on the origin of the diurnal variations of the magnetic elements([6]).

The next great step in earth-current research was inaugurated by the Electrical Congress at Paris 1 1881. It was decided that earth-current observations ought to be carried out simultaneously in as nany countries as possible. Partly as a result of the work of the committee, partly in connection with 1e international polar expeditions of 1882—83, a great amount of work was next done to investigate 1e laws of terrestrial currents.

In France registerings were undertaken by BLAVIER([7]), in England at the Greenwich Observatory([8]), 1 Russia by H. WILD([9]), in Finland by LEMSTRÖM([10]), in Italy by BATELLI([11]), in Bulgaria by BACHMETJEW([12]), t Kingua Fjord near the auroral zone by GIESE([13]), and in India by E. O. WALKER([14]).

([1]) loc. cit., 1868, p. 471.
([2]) " " 1868, p. 472.
([3]) " " 1870, p. 216.
([4]) " " 1870, p. 216.
([5]) " " 1870, p. 226.
([6]) " " 1868, p. 472.
([7]) E. BLAVIER: Études des Courants Telluriques, 1884.
([8]) Greenwich Magnetical and Meteorological Observations, 1882 and 1883.
([9]) H. WILD: Beobachtungen der elektrischen Ströme der Erde in kürzern Linien. Mem. Acad. Imp. Sci. St. Petersburg, 1883.
([10]) Expédition Polaire Finlandaise: 1882—83 et 1883—84.
([11]) A. BATELLI: Sulli correnti telluriche. Atti R. Acad. Lincei 1888.
([12]) BACHMETJEW: Der gegenwärtige Stand der Frage über elektrische Erdströme. Mem. Acad. Imp. Sci. St. Petersburg.
([13]) Beobachtungsergebnisse der deutschen Stationen, 1882—83, I. p. 411.
([14]) E. O. WALKER: Earth-currents in India, Journal Soc. Tel. Eng. XII, 1883; XVII, 1888; XXII, 1893.

An extensive series of registerings were undertaken in Germany on two long lines, one from Berlin to Dresden, 120 km. and the other from Berlin to Thorn, 262 km., and are treated by B. WEINSTEIN (1). Continual photographic records were kept from 1893—97 at Parc Saint Maur, and in recent years earth-current registerings have been made in Java by W. VAN BEMMELEN (2).

STRENGTH AND DISTRIBUTION OF EARTH-CURRENTS.

141. In spite of the great amount of work done on the subject, the earth-current problem is still in a somewhat unsatisfactory state. We are still very far from having attained a full comprehension of the various causes producing the galvanometer deflections.

The deflections observed are only to be considered as the algebraic sum of deflections due to a great variety of causes, some of which are due to experimental arrangements, and even the true earth-currents may be summed up from a very different origin.

Wild, Blavier and Weinstein found that on an average the electromotive foree between two points in a certain direction is proportional to their distance, and Wild estimates that on undisturbed days the electromotive force per kilometre is of the order $^{1}/_{1000}$ volt. Batelli finds 0.00068 volts per kilometre along the magnetic meridian, and 0.00081 normal to it.

During perturbations we shall find much greater values, Wild has found values up to 0.05 volts per km., and at certain moments during the disturbance of September, 1859, the electromotive force in telegraph-lines in France obtained values of about 1 volt pr. km. In 1881, PREECE found, in English telegraph-lines, 0.3 volts per km.

Some attempts at comparing simultaneous observations at various places were made by Lemström, who coördinated his own observations for Sodankylä with those of Wild from Pawlowsk.

He found that in the greater number of cases the conditions at the two stations were similar, so that great disturbances at the one station were accompanied by great disturbances at the other; but there were also cases where no similarity was found, and Lemström concludes that besides the more universal currents there are a number of quite local ones which are strong at the place but soon die off.

He also makes an interesting comparison of the absolute magnitude of earth-currents at the two places and finds as the average of 24 term-days that the amplitude at Pawlowsk is 0.0008 volt/km. and for Sodankylä 0.06 volt/km., or corresponding amplitudes are 75 times as large at Sodankylä. From this rapid increase in the earth-currents towards the arctic regions Lemström was led to the suggestion that probably there is a maximum zone of earth-currents similar to the auroral zone.

One point on which most authorities seem to agree is that the earth-currents at a certain place mostly run along a certain line of direction, either in the one direction or in the opposite. To this circumstance it is to some extent, at any rate, due that the earth-currents will run along the lines in which the earth's conductivity is greatest. In addition to this there are other reasons, e. g. an eventual marked direction of the electromotive force, which causes certain marked directions to be found in various districts.

C. V. Walker has thought he could show that this constancy of direction was not due to local causes, as he found the direction to be about the same for various places in England, viz: NE—SW.

Wild in Russia, Blavier in France, Batelli and Palmieri in Italy, and Bachmetjew in Bulgaria, also found more or less the same direction — NE—SW; but Weinstein in Germany found it NNW—SSE, and Lamont almost E—W in the neighbourhood of Munich.

---

(1) B. WEINSTEIN: Die Erdströme im Deutschen Reichstelegraphengebiet und ihr Zusammenhang mit erdmagnetischen Erscheinungen. BraunschWeig, 1900.

(2) W. VAN BEMMELEN: Koninklijke Akademie van Wetenschappen te Amsterdam, 1908.

As Lamont and Bachmetjew employed only short lines, and Palmieri made his observations on the ope of Vesuvius, their determinations in this respect must be treated with great reservation. If we ut them on one side, we may draw from the above the conclusion that the earth-currents as a whole ill be inclined to flow in a direction N—S in Europe; but local circumstances at the various places ill often cause the currents to deviate considerably from this main direction.

In the *United States* of America it has been found that during magnetic storms it is the lines inning E—W, or NE—SW which are most strongly affected.

In *India* the directions of the earth-currents, from a number of observations on telegraph-lines, as found to be N—S.

DIURNAL VARIATION OF EARTH-CURRENTS.

142. As first shown by Barlow and later by Airy, the earth-currents recorded on calm days show a ery marked diurnal period. On this point all authorities who have entered into the question seem to gree. The result is confirmed by Wild. Tromholdt, observing on telegraph-lines in Norway, found a rincipal maximum at about 7—9 p.m.

The most extensive and complete treatment of the diurnal variation is that of B. Weinstein. He und the average diurnal variation for the five years from 1884—1888, and also the variation for the ur seasons of the year, and finally the diurnal variation for each month. The type of variation is ery similar all through the year, but the amplitude is greatest in the summer and smallest in the winter ason The diurnal period, whatever may be its cause, appears to be a very definite thing, showing uite definite properties. In accordance with Walker and Airy, Weinstein says:

"After this I think that we already by looking at these curves can draw no other conclusion than at the phenomenon with which we here have to deal is a real one, and that its origin is due to a rocess of a more general character[1]".

From the comparison made with the diurnal period of terrestrial magnetism, it appears, as the sult of all the efforts made to find a connection, that no simple relation is found between the two henomena.

Weinstein finds a similarity as regards variation of earth-currents and that of the total intensity, ut such a similarity seems very difficult to interpret physically, for the effect of a surface-current tending over a large area should distinctly be felt in a similar manner in the horizontal elements, i. e. eclination and horizontal intensity. Weinstein, however, is of the opinion "that nearly all the total move-ent observed on the magnetometers generally named terrestrial magnetic variations, are only caused y variations of the earth-current, which affect the magnetometers in the same way as galvano-eters"[2]. But this result of Weinstein's does not seem very convincing while he takes for granted at "when the current-sheet has a horizontal position, there should not exist any horizontal magnetic rces worth mentioning[3]". Lately van Bemmelen[4], from records observed in Java determined e diurnal variation and found by comparison with magnetometer records "that the direction the earth-current is such that it can be regarded as causing the variations of the magnetic com-ponent and that the vibrations for them correspond"; but he finds "that the magnetic component is tarded with respect to the earth-current", and finally "that the ratio of the amplitudes of corresponding brations decreases with the duration of that vibration, so that those of the earth-current are relatively rger with a shorter duration".

---

(1) loc. cit. p. 18.
(2) loc. cit. p. 78.
(3) loc. cit. p. 69.
(4) VAN BEMMELEN: loc. cit. p. 513.

The difference in phase as well as the change in the relative magnitude of the amplitudes of the two phenomena with variation in length of period of vibration, is against the view that the diurnal variation of terrestrial magnetism is entirely due to earth-currents. In fact most authorities—Barlow, Airy, Wild, Lemström, Ellis—consider it very doubtful whether the earth-currents can explain the diurnal variation of terrestrial magnetism.

We are not at present going to discuss fully the problem of the diurnal variation of terrestrial magnetism, which will be reserved for a subsequent chapter; but I think we may say that in spite of of the most elaborate researches into the laws of terrestrial galvanic currents, no one has been able to show that these currents form the principal cause of the diurnal variation of terrestrial magnetism. Moreover recent investigation on magnetic diurnal variations, especially by A. SCHUSTER(1), von BEZOLD(2) and SCHMIDT has led to the result that the currents causing the diurnal variation must have their seat above the surface of the earth.

EARTH-CURRENTS AND MAGNETIC DISTURBANCES.

143. Most investigators in the field of earth-currents since 1880, have confirmed the result of Airy with regard to the connection between these currents and magnetic disturbances. It is in particular Blavier who has got results essentially different from those of Airy. Lemström, Wild and Bachmetjew, however, all agree with Airy, who considers the earth-currents to be the cause of magnetic disturbances. Most investigators, however, consider that there are certain exceptions yet to be explained.

Blavier, on the other hand, found that the earth-currents and magnetic disturbances are not related in such a way that the earth-currents have produced the magnetic variations; but he takes rather the opposite view that earth-currents are produced by the changes of magnetism. According to him the magnetic disturbances were mainly due to extraterrestrial currents above the place, while the earth-currents are produced by induction due to changes in the extraterrestrial currents.

This assumption is based on the fact that from his records he found the amplitudes of the accidental earth-current to be proportional to the rate of change which at the time considered is found for the corresponding magnetic elements.

Although Blavier, in a way, is certainly on the right track, I should consider it probable, in view of the results of the other investigators, that he is giving his conclusions too great generality. It might even be possible, as Blavier himself admits, that his induced currents are not altogether real earth-currents, but are partly currents induced in the cable system. Such currents, indeed, may have been present and may have influenced the results so as to give the impression that the induction-relation holds more general than it actually does. In order to find out whether the induction in the cable-system exerted any real influence, Blavier made simultaneous observations over the same areas in underground cables and in aerial lines. As the two curves thus obtained were identical, he thought himself justified in concluding that the currents observed were due to actual earth-currents.

Quite recently the question regarding the connection between earth-currents and magnetic disturbances has been treated by J. BOSLER(3), who has examined a number of disturbances recorded at Parc Saint-Maur. He finds for the cases considered that the relation is such as would be expected if the perturbing forces were directly due to the earth-currents flowing underneath the magnets.

(1) A. SCHUSTER: Phil. Trans. of the Roy. Soc. Vol. 180, p. 467, 1889.
(2) W. VON BEZOLD: Sitzungsberichte der Kgl. Akad. d. Wissenschaften zu Berlin, 1897.
(3) J. BOSLER: Comptes Rendus, p. 342, 1911.

The opinion expressed by Wild, Weinstein and others on the one side, and Blavier on the other, epresent the two extremes. We think that the right explanation will be one which unites the two xtreme cases into one theory.

In fact we think that recent investigations on terrestrial magnetism have already made it possible o look into the complexity of earth currents with a keener eye than it was possible for those who rere working some years ago.

Through the works of A. SCHUSTER[1], von BEZOLD and AD. SCHMIDT, we are already familiar with ne idea of extraterrestrial currents. The existence of such currents is a necessary consequence of the ypothesis that magnetic disturbances are the effects of electric radiation from the sun.

My previous research[2] as to the cause of various disturbances has shown that at any rate at places ear the poles, most magnetic disturbances are due to peculiar current-systems above the surface of the earth.

The view we take as regards the cause of magnetic disturbances will necessarily influence the view e take as regards their connection with earth-currents. If our hypothesis is right, we shall certainly et currents induced in the earth on account of changes in the external currents.

Recently van Bemmelen in his paper, "Registrations of the earth-currents at Batavia for the investiation of the connection between earth-current and force of earth-magnetism", treats the earth-currents rom the point of view, that they may be considered as currents induced by external currents.

He finds the ratio of the amplitudes in the earth-current registerings to the corresponding magnetoeter-records to increase as the time of oscillation diminishes and usually finds a difference in phase etween the earth-current oscillations and those of the magnetometer.

ARTH-CURRENT REGISTERINGS AT KAAFJORD AND BOSSEKOP, 1902—1903.

144. At our stations Kaafjord and Bossekop, the earth-currents were recorded in cables, 400 metres ng and resistance 1.55 Ω, one directed along the magnetic meridian and another perpendicular to it. he cable-system formed a cross with equal branches, in the centre of which the instruments were troduced.

The galvanometers employed were of the type Deprez-d'Arsonval, and were placed as a shunt on e principal line, as indicated in the accompanying figure.

The current measured in this manner on the galvanometer, will be a standard for the component the earth-current which goes in the direction of the connecting line between the two earthplates, , if preferred, for the component of the electromotive rce occurring in this district.

The earth-current conditions will be to some extent ianged when the cable is introduced. This might be ssumed to have special influence if the resistance in the ble is small in comparison with the earth-resistance. If, n the contrary, a great resistance is introduced into the rmer, it will not have any appreciable influence.

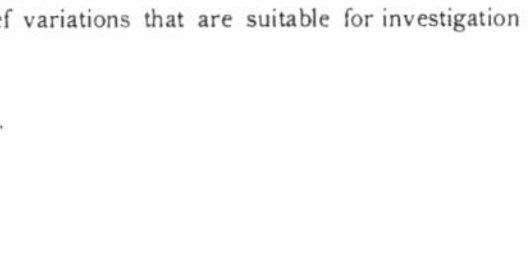

Fig. 269.

How important a part this may play it is not easy say; but in any case it will not exert any *essential* influence in the main phenomena.

The influence of the polarisation of the earth-plates will be very considerable where the lines, as this case, are short. Here, therefore, it is only the brief variations that are suitable for investigation

(1) A. SCHUSTER: Phil. Trans. of R. S. 180, p. 467, 1889.

(2) Expédition Norvégienne 1899—1900. Part I of the present work.

by this arrangement, and it is also these that are of special interest to us. They will be only slightly influenced by the plate currents, as the changes that might take place in the polarisation conditions must be assumed, as a rule, to take place comparatively slowly.

Possibly occurring thermo-electric forces will as a rule also undergo only slow, gradual changes.

Finally, we have left the effect of the direct induction in the cable-system, produced by the magnetic variations. This is made as small as possible by placing the cables on the ground.

In order to obtain an idea of its amount, we may make the following estimate. We will assume that we have a surface of flow of 400 sq. metres. Further we will assume that the component of the magnetic field at right angles to this surface varies with a velocity of 100 γ per minute. In the system there will than be induced an electromotive force with magnitude

$$10^{-8}\,\frac{0.001 \times 400 \times 10^4}{60} \text{ volts}$$
$$= 6.7 \times 10^{-7} \text{ volts.}$$

Now the earth-resistance between the plates has been measured and found to vary between 150 Ω and 1500 Ω. If we employ a mean value of 670 Ω we find the strength of the current to be

$$i = \frac{6.7 \times 10^{-7}}{670} = 10^{-9} \text{ amp.}$$

This current is divided between the galvanometer and the shunt, generally in the propo 1 : 300 Thus through the galvanometer there will pass

$$\frac{10^{-9}}{300} = 3 \times 10^{-21} \text{ amp.,}$$

and a current of this size will produce a deflection on the photographic paper of about

$$\frac{3 \times 10^{-12}}{3 \times 10^{-9}} = 0.001 \text{ mm.}$$

Thus, even for so powerful a variation in the magnetic field, there will if our assumptions hold good be only an imperceptible deflection, whereas in reality very considerable deflections are found with variations of such magnitude. The surface of flow must therefore be of an altogether different order of magnitude, if this kind of induction is to have any disturbing influence.

It appears from this estimate that what we observe must be produced by actually existing earth-currents.

As regards the nature of the soil, the following may be said.

It will be seen from the maps on p. 15 of Section I, that the observation-place in Kaafjord is situated in a region that is inclosed on all sides by high, steep mountains.

Alten Fjord, moreover, sends a narrow branch, Kaa-Fjord, up into this mountain mass; and the earth-current cables were laid upon the terraces above this branch-fjord, just at the foot of the exceedingly steep slope of Grytbotten Mountain.

These mountains are probably very rich in well-conducting veins of copper ore.

At Bossekop, the region surrounding the observation-place is flatter, and it would appear that there the local conditions play a less important part.

The reader is further referred to the description on p. 14, Part I.

Records were kept at Kaafjord from the middle of November, 1902, to the end of February, 1903, then the registerings were continued at Bossekop until April 2.

It is beyond our power to give the complete series of records of the earth-currents during this period; but we shall attempt to give, as far as possible, a true representation of the typical cases of earth-current phenomena by selecting a number of disturbed days for which we have successful records.

We are of course aware that a complete representation would have been preferable; but such a procedure in our case is excluded from the very fact that owing to difficulties with the galvanometers successful earth-current registerings are wanting during considerable intervals, and unfortunately records of earth-currents are wanting for a number of the very greatest disturbances. Being unable to give a complete representation, I think our procedure will be the best one, because very little would be gained by giving curves for intervals during which nothing of particular interest has happened.

The curves treated will be represented at the end of this volume in a series of plates giving a direct reproduction of the curves recorded photographically. In addition to the earth-current curves, the magnetometer registerings will be given for the same interval. The curves were copied partly photographically, partly by drawing on transparent paper directly from the photograms.

On each of the earth-current curves an arrow is drawn giving the direction of the galvanic current which produces a deflection in the direction of the arrow.

The plates are divided into three series.

The first series contains, in chronological order, a number of 24-hourly records representing moderate variations.

The second series contains 24-hourly records of a number of comparatively great storms, in fact the series contains all the great storms for which earth-currents have been successfully recorded.

The third series contains a number of two-hourly records.

Although we are unable to find absolute values of the earth-currents, it may still be of interest to find relative numbers for the current-changes which accompany the magnetic variations. In this way we may for instance be able to form vector diagrams for the currents, and compare them with the corresponding ones for the magnetic elements.

The determination of the somewhat rapid changes of earth-currents only lasting for a few hours can be done in a similar way as for the determination of the perturbing force, by placing on the photogram a normal line harmoniously connecting the quiet parts of the curve.

The change of current ΔI in the cable is given by the equation

$$\Delta I = \varepsilon \frac{G + s}{s} \Delta n$$

s is the shunt-resistance, G is the galvanometer-resistance, Δn is the deflection measured on the photogram, ε is the scale-value for the photogram, and gives the current through the galvanometer coil which corresponds to a deflection of 1 mm.

The corresponding electromotive force Δe between the cable terminals will be approximately

$$\Delta e = (\varrho + s) \Delta I,$$

where ϱ is the resistance of the cable, and is equal to $1.55\,\Omega$, as throughout s is small compared with G, or with sufficient accuracy

$$\Delta e = \frac{G}{s} (\varrho + s) \cdot \varepsilon \cdot \Delta n.$$

The quantity Δe is probably not equal to the electromotive force ΔE between the same points in case the cable was removed. We may put

$$\Delta E_{E-W} = q_1 \Delta e_{E-W}$$
$$\Delta E_{N-S} = q_2 \Delta e_{N-S}$$

where q_1 and q_2 are quantities which depend on the resistance of the cable and the soil and on the way in which the cables are connected with the ground. These quantities, q_1 and q_2 may easily be very large numbers.

During the stay at Kaafjord and Bossekop the resistance of the soil between the earth plates was repeatedly measured by using a Wheatstone bridge arrangement with alternating current and telephone.

The results of measurements are here given in tabular form.

TABLE CIX.

Date	Resistance		Date	Resistance	
	N–S	E–W		N–S	E–W
Nov. 15	1 000	1 200	Jan. 10	1 250	1 100
„ 19	700	750	„ 12	800	800
„ 21	700	700	„ 15	150	150
„ 26	1 150	1 000	„ 17	600	500
Dec. 4	850	750	„ 31	600	600
„ 6	400	400	Feb. 11	450	400
„ 15	1 400	2 000	„ 27	650	600
„ 17	1 500	1 500	March 7	400	350
„ 22	1 600	1 700	„ 23	430	430
„ 27	500	500	April 1	400	400
Jan. 8	1 500	1 500			

We notice that the resistance of the soil undergoes great variations, but always in such a way that the resistance is about equal in both circuits.

As the earth-connections for the two cables were made as equal as possible, we should probably at any moment be able to put

$$q_1 = q_2.$$

Thus I think when we take $\Delta \varepsilon$ as a relative measure of the earth-current, we ought to get approximately the right direction of the current. Values of $\Delta \varepsilon$ found at different times with different conditions of the soil however need not be exactly comparable.

CONSTANTS FOR THE EXPERIMENTAL ARRANGEMENTS.

143. The three galvanometers used we shall call A, B and B^1.

The sensitiveness of the galvanometers was measured by disconnecting them from the cables and exposing the instruments to a current of known strength. The deflections were indicated by marks made by the spot of light on the photogram. The scale-values for the instruments were observed at intervals and were found to keep constant within the limits of experimental error.

The scale-value and inner resistance of the galvanometers are given in the following table:

TABLE CX.

Instrument	Resistance	Scale-value
A	540 Ω	5.5×10^{-9} amp./mm.
B	735 „	2.1 „ —
B^1	57 „	2.2 „ —

The scale-values give the current in amperes corresponding to a deflection of 1 mm. on the original magnetogram.

In order to calculate Δe from the copies of curves given in the plates we must further know the shunt-resistance used in each case, and the galvanometer used in the two directions. These data are given in the following table for the various plates.

TABLE CXI.

Series	Plate Number	N—S		E—W	
		Galv.	S_2	Galv.	S_1
			ohm		ohm
I	1	A	1	B	1
	2—8	„	2	B^1	0.1
II	1	A	1	B	0.8
	2—7	„	2	„	2
	8—10	„	2	B^1	0.1
	11—13	„	2	„	0.2
	14—15	„	2	„	0.1
	16	„	2	B	1
	17	„	4	„	4
	Date				
III	Nov. 14	A	1	B	1
	„ 24	„	2	„	2
	„ 25	„	2	„	2
	March 31	„	4	„	4

The shunts will also be put up on each plate, where

S_1 is the shunt in E—W circuit

S_2 „ „ — „ N—S —

With the exception of the two-hourly records of November, the direction can be found from the following rules:

For the N—S curve a deflection upwards corresponds to a current from north to south, and for the E—W curve a deflection upwards corresponds to a current from east to west.

The perturbing forces can be calculated from the curves in the usual way by using the scale-values given in Table II, Part I, p. 50. The direction can be found from Table VIII, p. 59, or from the rule that on the plates a deflection upwards corresponds to increasing H. I., increasing westerly declination and increasing numerical value of V. I.

The sensitiveness given in the tables corresponds to the curves on the original photogram. The sensitiveness to be employed in each case can easily be found by measuring on the base-line of the plate the length (l) which corresponds to one hour. Then we obtain for the scale-value to be employed

$$\varepsilon' = \frac{2}{l}\,\varepsilon \text{ for 24-hourly records}$$

$$\varepsilon' = \frac{24}{l}\,\varepsilon \text{ „ 2- — —}$$

ε is the scale-value corresponding to the original photogram; l is to be measured in cm.

In calculating the scale-values we can with sufficient accuracy put the length of one hour on the original photogram equal to 2 cm. for all twenty-four-hourly records and equal to 24 cm. for all two-hourly records.

These values are true for the magnetograms within the limits of error of determination. For the earth-current photograms the hour-length is a little greater, for 24-hourly records it is about 2.015 cm., for 2-hourly records about 24.18 cm.

In copying the curves we have decided not to make reductions for the small differences in hour-length. The curves have been copied directly partly photographically partly on transparent paper, and then the whole plate is reduced to its proper size.

The time-marks given are as a rule first determined on the magnetic curves, as there the determination of time is easist and surest. In the next place, the time-marks are transferred to the earth-current curves, by the aid of synchronous serrations in them and in the declination-curve. This, as we shall show later, is permissible, and is the surest method when there are not simultaneous time-marks on both sets of curves.

In the rapid registerings on the contrary, we have by an electric arrangement exactly simultaneous time-marks on both sets of curves. These are marked on the plates, and the time is given below for the first and last break.

THE MAGNETIC EFFECT OF EARTH-CURRENTS.

146. Regardless of the way in which the earth-currents are produced, they must have some effect on the magnetometer, and thus in a way it may be said that magnetic disturbances are due to earth-currents.

In fact looking at the records we find, especially for the fairly moderate perturbations, that there is often an almost exact correspondence between the earth-current and the magnetometer curves, which shows that in these cases a considerable or rather the greater part of the magnetometer deflections are directly due to earth-currents.

Unfortunately this circumstance is not so distinctly shown on the copies as it appears in the original curves.

It is principally in the very small jags that the resemblance is most striking, and it has been found difficult to make an exact reproduction of these by drawing them on tracing paper.

Some of the curves have been copied photographically, these being both sets of curves for January 26 and February 10, and the earth-current curves for March 30—31.

In these it is easy to see the great similarity between earth-current and magnetism in their small, rapid oscillations.

In the curve for the 10th February especially, given as No. 13 in Series II, the characteristic oscillations at about 20^h are noticeable, these being apparently identical in the earth-currents and the horizontal magnetic elements, only shown in different scales.

There seems, therefore, in this case to be no doubt that the oscillations in the magnetic curves are to be understood in the main as the direct magnetic effect of the earth-currents.

If the time for the various jags be determined, it is also found that they are simultaneous within the limit of error to be taken into account here.

If we compare the amplitude of the deflections by these jags, we have a means of finding the effect of the earth-current. As, further, the total effect of the earth-current should be approximately proportional to the deflections measured on our galvanometers, we can, with this to aid us, eliminate the effect of the earth-current on the magnetograms.

Now forces of other origin will always be asserting themselves, but if we take into consideration only those in which the similarity is greatest, and employ a large number of jags, the mean of all these will give a more or less correct result, provided that an approximate proportion is always found to exist between the deflections in respect of amplitude.

For this purpose we have measured about 400 jags for Kaafjord, and about 100 for Bossekop.

We give here some of these determinations, as also the calculated mean values.

TABLE CXII.

Date	$\frac{P_h}{\Delta e_{EW}}$	$\frac{P_h}{\Delta e_{NS}}$	$\frac{P_d}{\Delta e_{EW}}$	$\frac{P_d}{\Delta e_{NS}}$	$\frac{\Delta e_{EW}}{\Delta e_{NS}}$	Date	$\frac{P_h}{\Delta e_{EW}}$	$\frac{P_h}{\Delta e_{NS}}$	$\frac{P_d}{\Delta e_{EW}}$	$\frac{P_d}{\Delta e_{NS}}$	$\frac{\Delta e_{EW}}{\Delta e_{NS}}$
0—3 Oct.		0.35	0.40	0.60	1.53	19—20 Dec.	0.24	0.34	0.40	0.56	1.40
			0.35	0.78	2.25		0.20	0.35	0.28	0.48	1.72
			0.19	0.32	1.68		0.24	0.37	0.46	0.70	1.51
			0.23	0.41	1.80		0.30	0.37	0.38	0.47	1.25
			0.32	0.48	1.50				0.54	0.88	1.61
			0.28	0.48	1.74		0.39	0.49	0.64	0.80	1.25
			0.42	0.66	1.56						
			0.36	0.61	1.68						
			0.45	0.66	1.50	24—25 Dec.	0.26	0.15	0.59	0.35	0.59
			0.58	0.92	1.62		0.39	0.26	0.34	0.21	0.64
			0.29	0.48	1.68		0.22	0.18	0.43	0.36	0.84
			0.50	0.96	1.92		0.46	0.34	0.39	0.29	0.73
			0.18	0.42	2.40		0.35	0.33	0.36	0.34	0.92
			0.49	1.10	2.28		0.46	0.32	0.34	0.23	0.69
			0.39	0.68	1.74		0.33	0.25	0.74	0.55	0.74
			0.38	0.62	1.68		0.54	0.36	0.55	0.36	0.67
			0.42	0.68	1.65				0.44	0.20	0.46
			0.45	0.83	1.86				0.98	0.43	0.43
			0.36	0.55	1.56						
			0.52	0.72	1.38						
			0.57	0.96	1.74	29—30 Jan.	1.39	0.78	1.67	0.93	0.57
			0.48	0.69	1.46		1.39	0.27	2.23	0.43	0.19
			0.36	0.65	1.83		2.51	0.36	2.60	0.38	0.14
		0.42	0.35	0.59	1.71		1.00	0.39	1.83	0.70	0.38
			0.31	0.45	1.50		0.95	0.35	1.50	0.56	0.38
			0.41	0.69	1.68		0.72	0.23	1.33	0.42	0.32
			0.40	0.59	1.50				0.93	0.50	0.55
			0.39	0.60	1.56		0.72	0.39	1.76	0.93	0.53
			0.26	0.43	1.68		0.50	0.15	1.80	0.53	0.29
			0.42	0.73	1.74		0.39	0.09	2.00	0.53	0.26
	0	0.20	0.45	0.72	1.62		0.61	0.21	1.67	0.57	0.34
	0	0.51	0.66	0.94	1.59		1.50	0.35	1.83	0.42	0.23
	0	0.20	0.55	0.84	1.56		1.50	0.23	2.43	0.37	0.15
			0.36	0.61	1.74		2.12	0.21	2.50	0.25	0.10
			0.36	0.54	1.50		2.23	0.26	1.33	0.15	0.12
			0.23	0.35	1.56		2.51	1.17	2.96	1.40	0.47
			0.32	0.67	2.07				2.76	0.70	0.25
			0.41	0.75	1.83		1.67	0.44	2.66	0.70	0.26
			0.51	0.82	1.62		0.39	0.13	1.67	0.55	0.33
			0.36	0.56	1.56		1.50	0.44	1.50	0.44	0.29
							2.12	0.74	2.83	0.96	0.34
							1.84	0.66	1.33	0.47	0.35
9—2 Dec.		0.20	0.48	0.64	1.30				3.83	1.32	0.34
			0.35	0.44	1.25				1.67	0.88	0.53
			0.31	0.62	2.03				2.66	1.40	0.53
			0.34	0.30	0.88				2.76	0.88	0.32
			0.28	0.70	2.44				2.03	0.60	0.29
			0.38	0.70	1.87				2.36	1.05	0.44

TABLE CXII (continued).

Date	Δe_{EW}	Δe_{NS}	Δe_{EW}		$\frac{\Delta e_{EW}}{\Delta e_{NS}}$
7—8 Febr.	1.00	0.64	1.73	1.08	0.63
	1.06	0.59	1.57	0.88	0.57
	1.45	0.63	1.73	0.76	0.44
	1.45	0.74	2.26	1.13	0.50
	0.56	0.26	1.93	0.90	0.46
	0.95	0.49	1.60	0.82	0.50
	0.72	0.42	1.76	1.08	0.61
	1.34	1.01	1.33	1.00	0.76
	1.11	0.47	2.16	0.91	0.42
			1.43	0.93	0.65
	1.34	0.67	1.37	0.70	0.50
	1.00	0.50	1.37	0.70	0.50
			2.00	0.84	0.42
			2.00	0.58	0.29
			1.33	0.62	0.46
			1.57	0.55	0.36
			1.67	0.81	0.48
			1.40	0.62	0.44
			1.53	0.76	0.48
			1.67	0.70	0.42
9—10 Febr.			2.50	1.05	0.42
			1.83	0.84	0.46
			2.86	1.05	0.37
			2.66	0.93	0.36
	0.72	0.47	1.00	0.70	0.63
	1.39	0.59	2.16	0.91	0.42
	1.39	0.39	1.83	0.60	0.32
	3.50	0.97	1.67	0.47	0.27
	0.33	0.26	0.60	0.44	0.71
		0.33		0.45	0.25
	1.39	0.36	0.83	0.22	0.26
	2.00	0.42	3.03	0.64	0.21
	1.72	1.02	2.00	1.17	0.59
	1.11	0.42	1.80	0.67	0.38
	0.61	0.22	1.53	0.53	0.36
	0.56	0.21	1.93	0.73	0.38
10—11 Febr.	0.84	0.35	1.37	0.57	0.42
	1.11	0.46	1.10	0.45	0.42
	0.95	0.67	1.03	0.70	0.69
	0.89	0.42	1.46	0.70	0.48
	0.95	0.42	1.10	0.51	0.46
	1.17	0.58	0.97	0.49	0.50
	0.73	0.39	1.13	0.61	0.52
	0.89	0.44	1.07	0.52	0.50
	1.00	0.58	1.07	0.63	0.59
	0.89	0.42	0.73	0.35	0.48
	0.72	0.27	1.46	0.54	0.38
	0.84	0.46	1.00	0.54	0.52
	0.50	0.23	1.43	0.65	0.46
	0.61	0.26	0.93	0.39	0.42
	0.78	0.34	1.60	0.70	0.44
			1.27	0.55	0.44
			1.10		

Date	Δe_{EW}			
10—11 Febr.			1.66	
			1.00	0.4
			1.10	0.5
	1.22	0.40	1.70	0.5
	1.34	0.28	1.36	0.2
	1.00	0.29	1.83	0.5
			1.33	0.7
			1.17	0.4
			1.67	0.9
			1.00	0.6
			1.27	0.7
			1.87	1.0
			1.10	0.7
			1.96	0.9
			0.73	0.4
			1.10	0.5
			1.43	0.7
			1.43	0.7
			1.33	0.7
			1.33	0.7
			1.27	0.4
			1.33	0.7
			2.13	0.6
12—13 Febr.	0.36	0.18	1.05	0.54
	0.80	0.41	1.03	0.52
	0.74	0.43	0.82	0.48
	0.62	0.49	0.59	0.47
	0.39	0.29	0.94	0.70
	0.68	0.35	0.96	0.49
	0.92	0.65	0.89	0.62
	0.71	0.53	0.84	0.62
	0.92	0.58	0.55	0.35
	0.42	0.23	0.77	0.42
	0.89	0.70	0.53	0.42
	0.30	0.20	0.53	0.35
	1.22	0.68	0.73	0.41
	1.10	0.49	0.66	0.40
	0.89	0.55	0.53	0.37
	1.07	0.58	0.64	0.35
	0.39	0.29	0.59	0.44
	0.50	0.28	1.09	0.59
	0.74		1.00	
	0.50		1.19	
	0.59	0.44	1.12	0.83
	0.56	0.37	0.98	0.65
14—15 Febr.			0.75	
			0.68	
			1.16	
			1.05	
	0.53		0.84	

TABLE CXII (continued).

Date			$\frac{P_d}{\Delta e_{EW}}$	$\frac{P_d}{\Delta e_{NS}}$	$\frac{\Delta e_{EW}}{\Delta e_{NS}}$
— ebı	0.5		0.47		
	0.50		0.30		
	0.27		0.96		
			1.30		
	0.33		1.07		
	0.47		0.78		
			1.33		
	0.44		0.78		
	0.50		1.01		
			0.78		
			0.71		
			1.25		
15—1 Febr.	2.37		0.94		
	0.62		0.80		
	0.77	0.65	1.40	1.17	0.82
			1.03	0.85	0.82
			0.93	0.80	0.86
	0.68	0.50	0.96	0.70	0.73
	0.45	0.35	1.16	0.91	0.78
	0.45	0.32	1.42	1.01	0.70
	0.56	0.39	0.86	0.61	0.70
			1.07	0.70	0.66
			0.77	0.53	0.70
	0.50	0.54	0.77	0.80	1 05
	0.74	0.58	0.80	0.63	0.78
	0.80	0.70	0.69	0.61	0.86
	0.92	0.85	0.77	0.70	0.90
			0.68	0.57	0.82
			0.87	0.80	0.92
			0.87	0.70	0.82
			1.07	0.73	0.66
			0.98	1.10	1.13
	0.27	0.15	0.89	0.53	0.59
			0.91	0.67	0.70
	0.33	0.28	1.05	0.92	0.86
	0.33	0.26	0.93	0.72	0.78
			0.94	0.76	0.78
			0.77	0.60	0.78
			1.00	0.88	0.88
			0.73	0.64	0.86
			0.52	0.41	0.78
			0.96	0.92	0.94
	0.86	0.85	0.69	0.70	0.97
			1.05		
			1.25	0.98	0.78
			0.89	0.84	0.94
			0.80	0.79	0.98
			1.00	0.78	0.78
			1.07	0.97	0.90
			0.78	0.77	0.98
			1.01	0.93	0.90
			0.80	0.79	0.97

Date	$\frac{P_h}{\Delta e_{EW}}$	$\frac{P_h}{\Delta e_{NS}}$	$\frac{P_d}{\Delta e_{EW}}$	$\frac{P_d}{\Delta e_{NS}}$	$\frac{\Delta e_{EW}}{\Delta e_{NS}}$
16—17 Febr.			0.84	0.55	0.66
			0.64	0.54	0.86
			0.77	0.64	0.86
			0.66	0.49	0.74
			1.48		
			0.84	0.82	0.98
			1.48	0.70	0.47
			1 37	0.82	0.59
			0 77	0.53	0.68
			0 53	0.52	0.98
			0 89	0.58	0.66
			0 77	0.53	0.68
			0 77	0.70	0.90
			0 77	0.60	0.78
			1 25	0.70	0.55
			1 19	0.76	0.62
			1.21	0.90	0.74
			0.77		
			1.00		
			0.66		
			0.62		
			1.16		
			0.82		
			0.98		
			0.89		
			1.48	0 88	0.59
			0.59	0 35	0.59
			2.10	1.30	0.63
			0.89	0.50	0.55
			1.26	0.81	0.62
			0.45	0 28	0.62
			1.65	1.03	0.62
			0.25	0.23	0.47
			0.73	0.52	0.70
			0.52	0.41	0.78
			0.96	0.49	0.51
			1.37	0.70	0.51
			1 00	0.58	0.59
17—18 Febr.	0.65	0.47	1.28	0.91	0 70
	0.44	0 23	1.12	0.60	0.53
	0.42	0.16	1.78	0.70	0.70
	0.59	0.29	1.42	0.70	0.49
	1.12		1.51		
	0.74		1.42		
	1.48	0.55	1.91	0.70	0.36
	0.36	0.15	1.26	0.56	0.43
	1.07	0.58	1.41	0.77	0.55
	2.24	0.15	1.96	0.62	0.62
			0.80	0.56	0.70
			1.42	0.93	0.65
			0.89	1.12	1.25
			1.28		

TABLE CXIII.

Mean Values for Kaafjord.

Date	$\frac{P_h}{\Delta e_{EW}}$	N	$\frac{P_h}{\Delta e_{NS}}$	N	$\frac{P_d}{\Delta e_{EW}}$	N	$\frac{P_d}{\Delta e_{NS}}$	N	$\frac{\Delta e_{EW}}{\Delta e_{NS}}$	S
30–31 Oct. . . .	0.21	5	0.35	5	0.38	47	0.65	45	1.69	0.
2– 3 Nov. . . .					0.49	13	0.70	13	1.39	0.
19–20 Dec. . . .	0.27	6	0.36	6	0.40	12	0.61	12	1.51	2
24–25 " . . .	0.38	8	0.28	8	0.51	10	0.33	10	0.64	2
4– 5 Jan.							0.80	7		2
29–30 "	1.16	15	0.37	15	2	22	0.71	24	0.36	0.
7– 8 Feb. . . .	1.05	12	0.52	11	1.64	27	0.80	24	0.49	0.
9–10 "	1.28	8	0.37	10	1.91	15	0.71	16	0.38	0.
10–11 "	1	22	0.38	22	1.37	51	0.59	49	0.43	0..
11–12 "	0.53	22	0.41	22	0.75	36	0.67	38	0.89	0.1
12–13 "	0.68	22	0.43	20	0.82	22	0.50	20	0.61	0.1
14–15 "	0.44	8			0.89	17				0.1
15–16 "	0.59	14	0.49	13	0.91	39	0.77	37	0.84	0.1
16–17 "					0.96	38	0.64	29	0.66	0.1
17–18 "	0.77	9	0.33	8	1.32	14	0.74	11	0.56	0.1
Weighted Mean	0.74		0.40		1.02		0.66		0.82	

The number of jags used in the calculations are indicated in the columns "N". T
$P/\Delta e$ are expressed in the units γ/microvolt; Δe corresponds to a distance of 400 metres.

In the calculation of the above, a number of the jags that agreed ill were left out of co
The table also gives the relations $P_d/\Delta e_{EW}$ and $P_h/\Delta e_{NS}$.

These quantities only acquire physical importance if we assume that the currents within which they influence the magnetometers, are of so local a character that the observed Δe_{NS} cannot be said to represent the corresponding earth current components, or in other words, i currents here flow along comparatively very sinuous current-lines. We have here include they can be employed for the purpose of eliminating the effect of the earth-current in case we have only successful records of one earth-current component.

These figures show how great accuracy we attain by this method.

Among similar synchronous oscillations may be noted those occurring in the interv 21^h and 22^h on February 10.

Here too, however, there are evidently considerable direct effects of the extra-terrestri systems; but they do not appear to have so rapidly changing a character as the variations t to the earth-currents.

An examination of the remaining curves will show similar synchronous oscillations I will here draw attention to a few of the more characteristic in Series I.

January 13, time about 16^h
— 18, » » $15^h—15^1/_2{}^h$
— 20, » » $19^h—21^1/_2{}^h$
February 13, » » $19^h—21^h$
— 17, » » $4^h—6^h$ and
$21^h—22^1/_2{}^h$
— 23, » » $16^h—24^h$,
especially about 17^h.

In Series II we have collected a number of powerful storms. Here the external perturbing forces terfere largely, so that the effects of the earth-currents only appear, as a rule, as secondary waves on e main deflections. Here too, however, they are generally very distinct.

We will indicate a few.

Dec.	26,	time about	$20^{3}/_{4}{}^{h}$
Jan.	23,	» »	$17^{2}/_{3}{}^{h}$
Feb.	7,	» »	17^{h}—18^{h}
		and	21^{h}—22^{h}
—	9,	» about	18^{h}
—	10 & 11,	» »	23^{h}— 1^{h}.

The last two sets of curves in Series II are from Bossekop. We will examine them a little more osely later on.

On looking at the reproduced curves from Kaafjord, and especially the intervals mentioned above, : notice in the first place that the two earth-circuits exactly correspond in every detail; and as the ble CXII shows, the relation between the deflections in the two components for one and the same day a very nearly constant, whereas it varies somewhat from day to day.

In the next place, the resemblance between the earth-current curves and the declination curves is troughout considerably greater than between the former and the horizontal intensity.

These facts are, I think, accounted for by the small sentiveness of the *H. I.*-magnetometer compred with that of the declinometer, and further by the fact that the direct effect of extra-terrestrial crent-systems is much more pronounced in H than in D.

Owing to the smallness of the oscillations and the difficulty of identification P_h is only found in ratively few cases.

If we could put $q_1 = q_2$, we should expect to find that the relation $P_h/\varDelta e_{EW}$ would equal $P_d/\varDelta e_{NS}$ If we compare the mean figures, we also find that such is the case; but while the relation $P/\varDelta e_{NS}$ remains nearly constant all the time, relation $P_h/\varDelta e_{EW}$ varies very considerably. As long as th shunt is kept unaltered, however, the relation is fairly constant.

Before January 12, in the *NS* line, galvanometer B was employed, after that date galvanometer E. With the change, a very distinct leap in the values of the relation $P_h/\varDelta e_{NS}$ is observable. A simila leap is observable at the change from shunt-resistance 0.2 Ω to 0.1 Ω.

In the last case the relations are reduced to very nearly half the value, which probably indicates th contact-resistances have here played a decisive part, and they must be assumed to occur in the shnt-circuit itself. An explanation may also be found for the discontinuity here found in the conditions on changing the galvanometers on January 12, merely by assuming that an influence is exerted by cotact-resistances. In such case it must be assumed to occur at the points where the cable is connected wh the galvanometer and shunt circuits. Its effect will be the same as if the resistance of the cable were increased by a corresponding amount.

It is therefore doubtful whether any great importance can be attached to the agreement between th mean figures.

ON THE CONNECTION BETWEEN POLAR STORMS AND EARTH-CURRENTS.

147. As above mentioned the second series of plates, Pl. XXI—XXIII, contains a number of simultaneous records of earth-currents and the magnetic elements during a number of comparatively grat storms. The conditions during polar storms are also given in greater detail in some of the rapid records contained in the third series of plates, e.g. Pl. XXXIV and XXXV.

The first important conclusion which we can draw from the curves is the following:

The earth currents as manifested by the galvanometer deflections during polar storms which have their centres in the vicinity of the station, cannot explain the main part of the perturbing force.

The justification of this conclusion will be immediately apparent on looking at the curves; for while the magnetometers can maintain a large deflection in a certain direction for hours, the galvanometers will change direction of deflection relative to the normal line usually a great many times during the same period. Indeed the galvanometer curves have often the appearance of oscillations round the normal line (see f. i. the curves for Nov. 2 and Febr. 12).

Now we saw in the previous article that the earth-currents produce magnetic variations according to rules given in table CXIII.

In consequence we always find that, superposed on the main wave of the magnetometer curve, which is probably due to extra-terrestrial currents, there are a number of waves and oscillations which, as regards occurrence and form, coincide with the galvanometer oscillations, a phenomenon, that is well illustrated in Series I and II, Plates XXX—XXXIII, and even better in Series III, Plates XXXIV and XXXV, giving the copies of a number of rapid-registerings. From the coincidence in form and phase I think we may safely conclude that these synchronous and similar rapid magnetic changes are direct effects of earth-currents flowing underneath the magnets. This conclusion is also confirmed by the fact that the curves of vertical intensity run more smoothly than those of the horizontal elements; for if the rapid changes are mainly due to earth-currents spread over a considerable area, such currents would produce very little effect in the vertical direction.

The rapid synchronous oscillations in the two sets of curves will always occur with greatest strength simultaneous with the strongest magnetic disturbances, and from this it is evident, that these briefer variations must be due to the same primary cause as the magnetic storms themselves, i. e. according to our assumption to an extra-terrestrial corpuscular current-system. The most natural way of explaining the connection between the outer current system and the earth-currents is that the latter are induced by variations in the former. From this, however, we cannot draw the conclusion that we always must find such a simple connection between the two sets of curves as that expressed by the rule of Blavier. On regarding the curves one would also see, that such a connection in far the most cases does not exist.

From the relations given in Table CXIII we should be able to subtract from the magnetometer-records the effect of the earth-currents. But even this corrected curve would hardly be competent to explain from the rule of Blavier the many oscillations of the earth-current curve. Looking at the curves for Kaafjord we shall often find that the corrected curve for this place will apparently run rather smoothly compared with that of the earth-currents.

This circumstance is easily explained when we consider how the perturbation-conditions develope in the polar regions.

We have become acquainted with the typical arrangement of the polar systems. On the afternoon side in latitudes that are not too high, we meet with the positive polar storms, on the night side with the negative. We have seen that the positive system answers to the effect of rays that deseend towards the earth, are deflected westwards, and again leave the earth; the negative to rays that are deflected eastwards. Both these current-systems presumably lie at a comparatively great height above the auroral zone, and their smaller and more rapid changes will therefore be less evident, and the curves in consequence are characterised by comparative smoothness. Among these systems, however, rays are met with, which descend directly earthwards to within comparatively small heights above the surface of the earth. Here the magnetic curves are exceedingly serrated. There are very rapid and comparatively strong variations, some of which are due to displacement of the districts of positive and negative precipitation, and some o the fact that every change occurring here will be felt comparatively powerfully in places where the ays come very near to the earth.

Even if, as the observations seem to show, the magnetic forces, in absolute value, are only omparatively small, or at any rate are more restricted in their effect as compared with the forces at vork in the great perturbation-systems, these rapid changes will now be assumed to generate particularly owerful induced currents, as the strength of these currents approximately is only proportional to the apidity with which the change takes place, and not to the strength of the external current.

The apparently more rapid decrease outwards in the effect of these rays than in the other systems, nay also be explained by the fact that here the rays will leave the earth in paths lying very near hose by which they came in, whereas in the other systems the contrary is the case.

It may therefore reasonably be assumed that the rapidly alternating currents observed in the earth-urrent curves, accompanied by synchronous oscillations in the magnetic curves in which it is difficult r impossible to trace the influence of external forces that might be assumed to generate these currents, re mainly created by induction of the above-mentioned systems of rays which descend towards the arth between the positive and negative polar systems of precipitation, *and far from the place of egisterings.*

In more northerly latitudes too, there are possibly local storm-centres, which will have a power-ul inductive effect. In fact the curves for Axeløen are disturbed almost at any time.

There is moreover another most important point, namely, that the relation between the magnetic ffect of an extra terrestrial system of the form we find during polar storms, and the effect of the in-luced current-system, decreases with increasing distance from the inducing current-system, *and thus he farther we get from the external current-system, the more strongly would the induced current be felt.* We shall prove this relation more fully later on. In other words the earth-currents are able to bring to ower latitudes a message of a great many distant perturbations with their centres in the vicinity of the oles in cases where the external systems are too weak to cause any appreciable *direct* effect on our nagnetometers.

In this way we may understand that in lower latitudes most observers in a great number of cases iave found the magnetometer variations to be such as would be produced by the earth-currents flowing ınderneath the magnets, and still external currents may be the primary cause of the magnetic distur-ances.

From what has been said it will be evident that we cannot usually expect to be able to trace out he cause of the earth-currents at a certain station from a comparison with magnetometer-records from he same station.

Very often the galvanometers during polar storms merely perform rapid oscillations about the normal ine (see Series II, Nos. 1, 3 & 14, Pl. XXXI and XXXII); but in some cases of somewhat small and regular olar storms with their centres in the vicinity of the stations, earth-currents were observed varying in regular way; and in accordance with the view that earth-currents are induced from changes in the rimary external systems.

The most typical instance of such a regular curve is the perturbation of the 10th of February, but he same type of correspondence between earth-current variations and polar storms is very well brought ut in a number of other storms. I will direct attention to a few of these. In Series I, January 13, about 18^h—$18^h\ 30^m$, February 13, about 19^h—20^h; in Series II, December 26, about 23^h—24^h, January 24, about 18^h, February 9, about 18^h.

In all these cases the following typical correspondence is found.

VECTOR DIAGRAMS.

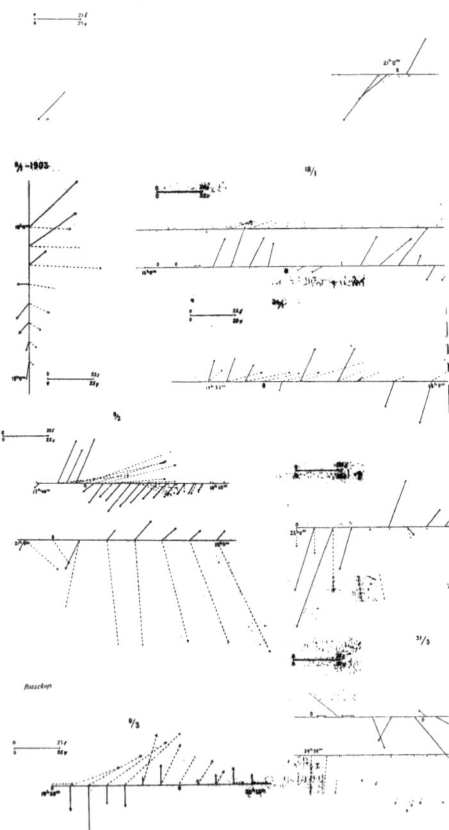

Fig. 270.

During the time of the most rapid increase of the perturbation the earth-currents obtain a maximum. When the disturbance is at its maximum the galvanometer has nearly its normal position, and when the disturbance diminishes at the greatest rate we get another maximum of galvanometer deflection, but now to the opposite side of the normal line. The storms which show this type of variation are especially those which we called polar elementary storms.

A number of elementary storms showing a correspondence of this type are graphically represented in vector diagrams (fig. 270).

Fig. 271.

If we look at these vector diagrams we notice that the current vector when passing from one direction to the opposite is not turned round quite gradually, but the vector is kept in the same line of direction. This peculiarity with regard to direction will be seen from the plates (Pl. XXX—XXXIII), and is even better illustrated by some of the rapid records, e.g. Pl. XXXIV and XXXV.

Even the rapid oscillations seem to pass along the same direction which is seen from Table CXII, which shows that the ratio between corresponding amplitudes in the two directions is about constant.

If we try to deduce the direction of the earth-currents from the variation of the magnetic force at Kaafjord by applying Lenz's law, we find a current-direction nearly opposite to that actually observed. This circumstance may seem remarkable. In order to prove that there was not some error in the determination of sensitiveness, I again, in May, 1910, made earth-current measurements with earth connections in exactly the same places as before, and found the condition confirmed.

It is to some extent doubtful where the cause of this peculiar circumstance is to be sought; but it seems reasonable to assume that the local conditions in the ground have a very essential part to play.

A consideration of the country in which the earth-current measurements were made, confirms this assumption.

We have previously described this country (p. 732), and from the description it is evident that local conditions would probably exert a great influence on the earth-current conditions, taking into account that the earth-current lines are only 400 metres long.

If we compare the current-directions found in Kaafjord with the sketch-map on p. 15 and with fig. 271, we see that the direction of the earth-current is parallel with that in which the mountain-ridge and the branch-fjord run. Now in inductions of this kind, the main direction of the earth-current should be E—W; but if we look at the shape of the fjord and of the mountain mass on the maps, we see that if on the whole the earth-currents are influenced by local conditions of this kind, it would be by no means unlikely that in the regions surrounding Kaafjord, a peculiar deflection of the current-lines such as we have here observed might take place. In order, therefore, to come to a clear understanding of this question, it would be better to observe the earth-currents with considerably longer earth-connections, and in more level country.

It was chiefly for this reason indeed that at the beginning of March we moved our station to Bossekop, where the ground is less rough.

There proved to be a considerable difference in the earth-current conditions We no longer find such a marked constancy in the current-directions. As the vector diagram for March 31 shows, the currents may here flow under various azimuths. From the same diagram it appears moreover that when the magnetic force varies in strength, the directions are throughout in accordance with those we should expect to find according to Lenz's law, especially as regards the currents with direction NW—SE.

Unfortunately, however, we have only very few successful records of typical perturbations from Bossekop.

EARTH-CURRENTS AND POSITIVE EQUATORIAL PERTURBATIONS.

148. The characteristic properties of the positive equatorial perturbations are given in the first part of this work:

Discussing the various systems which might produce these perturbations, we found it very difficult to explain their properties by supposing that earth-currents were the primary cause of these disturbances.

In lower latitudes the perturbing force is directed towards the north nearly along the magnetic meridian, and it can maintain a considerable value for a great many hours.

At Kaafjord successful earth-current records have been obtained for the E—W circuit during the most typical equatorial perturbation observed by us, namely, that of January 26. The galvanometer in the N—S line being in some way out of order no oscillations were recorded in this line.

Looking at the curves in No. 10, Pl. XXXII, we notice that the H and D curves show small, but still quite noticeable deflections lasting for several hours. In the earth-current curve there is absolutely no deflection of long duration to be noticed, but merely sudden oscillations about the normal line.

ON THE SIMULTANEITY OF EARTH-CURRENTS AND MAGNETIC DISTURBANCES.

149. The question regarding the simultaneity of the occurrence of earth-currents and magnetic storms was first discussed by Airy and since then it has been subject to considerable attention from most authorities.

We know that in a number of cases the magnetometer oscillations are direct effects from earth-currents underneath the magnets, and for these oscillations, at any rate when their beginning is abrupt and well marked, we should expect to find simultaneity within the limits of experimental errors, because the delay caused by the periods of the apparatuses can only be a question of seconds.

To be clear of this question, we must have recourse to our rapid registerings. Of these we have a great number, but as, for this purpose, the occurrence of especially characteristic serrations is required, there are not very many that are of use to us. We find a number of these reproduced in Series III. We have taken a number of the most characteristic notches on these curves, and the time-differences found between deflections in the earth-currents and the declination are given in the following table.

TABLE CXIV.

Point	24/11 4h–6h p.m. Diff. in sec.	25/11 5h–7h p.m. Diff. in sec.			25/11 7h–9h p.m. Diff. in sec.			2/4 3h 55m–5h 47m a.m. Diff. in sec.			2/4 5h 52m–7h 46m a.m. Diff. in sec.		
	A–D	*A–B*	*B–D*	*A–D*	*A–B*	*B–D*	*A–D*	*A–B*	*B–H*	*A–H*	*A–B*	*B–H*	*A–H*
1	− 0.9	0	− 10.6	− 10.6	+ 2.0	+ 2.7	+ 4.7	− 4.2	+ 5.6	+ 1.4	− 7.6	+ 1.0	− 6.7
2	+ 1.8	0	− 18.4	− 18.4	− 3.4	0	− 3.4	− 4.2	− 1.4	− 5.6	− 5.7	+ 1.9	− 3.8
3	− 2.7	0	+ 6.6	+ 6.6	+ 4.1	− 0.7	+ 3.4	0	− 1.4	− 1.4	− 3.8	+ 1.9	− 1.9
4	− 6.3	0	+ 5.9	+ 5.9	0	− 5.4	− 5.4	0	− 8.8	− 8.8	− 3.8	− 3.8	− 7.6
5	− 0.9	0	+ 0.7	+ 0.7	+ 2.7	− 2.0	+ 0.7	− 3.2	− 6.4	− 9.6	− 1.0	− 4.8	− 5.7
6	− 4.5	0	− 6.6	− 6.6	− 1.6	+ 3.9	+ 2.4	− 6.0	+ 4.4	− 1.6	0	− 4.8	− 4.8
7	− 8.1	0	− 1.4	− 1.4	− 5.5	+ 6.3	+ 0.8	− 1.4	− 4.2	− 5.6	− 1.0	− 4.8	− 5.7
8	− 16.2	0	− 10.3	− 10.3	+ 4.7	+ 3.2	+ 7.9	− 2.8	− 1.4	− 4.2	− 1.0	− 5.7	− 6.7
9	− 16.2	0	0	0	+ 7.1	− 4.5	+ 2.6	− 4.2	0	− 4.2	− 5.7	− 2.9	− 8.6
10	+ 2.1	+ 3.5	− 4.6	− 1.2	− 1.9	+ 3.9	+ 1.9	0	− 7.0	− 7.0	0	− 3.6	− 3.6
11	+ 2.1	+ 1.2	− 5.8	− 4.6	0	− 3.9	− 3.9	− 11.1	+ 5.6	− 5.6	− 3.6	− 5.1	− 8.7
12	− 1.0	+ 1.2	+ 1.2	+ 2.3	− 1.3	+ 1.9	+ 0.6	0	− 1.4	− 1.4	0	− 10.9	− 10.9
13	− 2.1	0	− 4.6	− 4.6	+ 0.7	− 0.7	0	0	0	0	− 1.5	− 6.5	− 8.0
14	− 4.3	+ 2.3	− 7.5	− 5.2	+ 3.2	0	+ 3.2	− 5.6	+ 4.2	− 1.4	+ 1.5	− 5.1	− 3.6
15	− 6.9	0	− 1.2	− 1.2	+ 5.2	+ 0.7	+ 5.8	− 2.3	− 1.2	− 3.5	—	—	—
16	− 0.8	+ 4.0	− 5.2	− 1.2	+ 3.2	− 3.9	− 0.7	− 7.0	+ 3.5	− 3.5	—	—	—
17	+ 3.5	0	− 4.0	− 4.0	+ 0.6	0	+ 0.6	0	− 3.5	− 3.5	—	—	—
18	− 10.6	0	0	0	− 1.1	+ 4.6	+ 3.4	− 5.9	− 2.3	− 8.2	—	—	—
19	+ 2.6	+ 4.5	+ 7.0	+ 11.5	− 0.6	+ 2.9	+ 2.3	− 2.3	+ 1.2	− 1.2	—	—	—
20	− 1.8	0	− 6.2	− 6.2	+ 2.9	0	+ 2.9	− 2.3	− 4.7	− 7.0	—	—	—
21	− 3.0	0	+ 0.6	+ 0.6	− 0.6	+ 1.7	+ 1.1	− 2.3	− 2.3	− 4.7	—	—	—
22	− 1.4	0	+ 1.1	+ 1.1	+ 1.1	− 5.7	− 4.5	0	− 12.8	− 12.8	—	—	—
23	+ 8.3	− 1.2	− 1.8	− 3.0	+ 1.1	− 1.7	− 0.6	− 3.5	− 11.7	− 15.2	—	—	—
24	− 2.0	+ 4.1	− 4.1	0	+ 3.4	− 5.1	− 1.7	0	− 4.7	− 4.7	—	—	—
25	+ 4.8	0	+ 1.8	+ 1.8	+ 1.7	− 8.5	− 6.8	—	—	—	—	—	—
26	0	0	− 9.6	− 9.6	+ 5.1	− 4.0	+ 1.1	—	—	—	—	—	—
27	+ 7.7	0	− 7.0	− 7.0	+ 2.8	− 6.2	− 3.4	—	—	—	—	—	—
28	+ 2.8	+ 5.1	+ 4.0	+ 9.1	− 2.8	− 4.5	− 7.3	—	—	—	—	—	—
29	− 3.5	0	− 5.7	− 5.7				—	—	—	—	—	—
30		0	− 2.4	− 2.4				—	—	—	—	—	—
Mean	− 2.05	+ 0.82	− 2.94	− 2.12	+ 1.17	− 0.89	+ 0.28	− 2.85	− 2.11	− 4.97	− 2.37	− 3.80	− 6.16

We will look at the accuracy that we can here count upon. Both the earth-current curves and the magnetic elements are registered with a rapidity of 4 mm., a minute. On both curves, at suitable intervals, exactly simultaneous time-breaks are produced by an electric contact. Now the curves can hardly be measured with greater accuracy than 0.1—0.2 mm., nor the serrations fixed more sharply than at about 0.2 mm. When therefore the time-breaks are clear, the limit of error should be 0.3—0.4 mm. or 5—6 seconds; but as we have the difference between two such measurements, the error may amount to twice that figure under otherwise favorable circumstances. Add to this the possible indistinctness of the time-break, and the difficulty of fixing the point upon the curve, and it will appear that we cannot reckon upon a greater accuracy than of about 10 sec. in the measurement of the difference. When, with this in view, we look at the figures we have obtained for the time-difference, we notice at once that of the 125 measured differences, only 10 have gone above 10 seconds, the remainder being all considerably less.

For November 24, about 4^h—6^h, only the N—S curve has been drawn, as galvanometer B for some reason would not work. Here there is therefore only *one* series of differences.

How much may we venture to conclude from these comparisons? The difference generally seems to keep below 5 seconds. The differences between the serrations in the various earth-current components are as a rule less than the difference between the latter and the magnetic elements; but a personal equation evidently plays an important part, as we can see when we compare the results of $^{25}/_{11}$ 7^h—9^h p.m. with the others, the former having been determined by one person, the remainder by another. While in the named interval there are practically no differences worth mentioning, and the difference $A-B$ between the earth-current components themselves is the greatest, the reverse is the case throughout with the others, and the negative differences, which answer to those in which the earth-current deflections come first, predominate there.

As the number of differences of more than 10 seconds is so few, and the personal equation so considerable, there seems to be little doubt that in reality the deflections are practically exactly simultanous, and that the greater time-differences that occur are only due to the chance accumulation of errors.

We thus venture to say that it is not impossible that a time-difference does exist between the variations in the earth-current and the corresponding variations in the magnetic elements; but if so, it is so small that we cannot prove it in our registerings with 4 mm. to the minute.

We learn something from this however, for we see that in our ordinary registerings ($1^h = 20$ mm.) we may consider brief variations as absolutely simultaneous on the earth-current curve and the magnetogram, so exactly, indeed, that we can quite well check the time-determination by a comparison of characteristic small serrations (5 seconds here answering to 0.028 mm.). Thus our previously-advanced assumptions (p. 736) are justified.

EARTH-CURRENTS AT BOSSEKOP.

150. It may be mentioned as characteristic of the earth-current conditions in Kaafjord, that the currents which occurred there ran backwards and forwards in the same direction in the earth, this being very nearly the direction of the adjacent coast-line.

The consequence of this is that the curves of the two earth-currents exhibit a very great resemblance in all their details. All simultaneous brief deflections are approximately proportional in the two curves.

The details in the declination, moreover, show a striking resemblance to the earth-current curves.

If, however, we look at the earth-current curves from Bossekop, we find the resemblance not nearly so great. The deflections in the two earth-current components are not always synchronous, which again indicates that the direction of the current may vary.

This, as already pointed out, is also apparent from the vector-diagram for the 31st March (fig. 270).

If we endeavour to find corresponding serrations in the earth-current curves and the magnetic ırves, we can, as regards the declination, show a number in which the correspondence is quite atisfactory.

In the curves for March 30—31, in Series II, f. i., we find a number of serrations in which the orrespondence is comparatively good.

The resemblance here is striking if D and the N—S curve are compared during the time from bout 2^{h} onwards.

The serrations in the E—W curve, on the other hand, have no distinct counterparts in the mag-tic; but unfortunately we have no observations of H at this time.

As regards D, we have compared with the earth-currents, in all, 107 serrations, which showed the eatest similarity; and in this way we found more or less constant values for the relation $P_d/\varDelta e_{SN}$ that here an elimination of the effect of the earth-currents could be made.

With regard to H, we have only succeeded, during the same period of time, in identifying 10 rrations with any certainty.

THE INFLUENCE OF THE EARTH-CURRENT UPON THE VERTICAL INTENSITY.

131. We have hitherto only considered the connection between the earth-currents and the varia-ons of the horizontal magnetic elements.

On looking at the vertical curves, however, we also frequently find very characteristic points of semblance between them and the earth-current curves.

In the case of Kaafjord, where the direction of the current is constant, it is easy to form there-om an idea as to the quarter in which the main mass of the current is to be found.

Identification is very much more difficult here than in the horizontal elements, but if we look at te curves for the 15th February in Series II, a close examination will reveal a number of small simul-neous deflections in the V-curve and the earth-current curves. An upward deflection in the V-curve swers in every case to a downward deflection in the earth-current curves, and vice versa.

As the sensitiveness for the vertical intensity is comparatively small, the resemblance in the small flections will be difficult to demonstrate, especially in the copied curves. In the original photographs te identification is easier.

In the stronger deflections the resemblance is more striking; and if we compare the course of the rtical curve at the times when the earth-current curves show considerable deflections, very characteristic ints of resemblance will as a rule be found between the two systems of curves.

In these powerful deflections, however, external current-systems will always exert a considerable ect influence, so that the phenomenon becomes less perfect. We may here point to a number of the re powerful deflections, which give a distinct impression of this resemblance.

From Series I

Jan. 13, time about 18^{h} —18^{h} 30^{m}.
— 18, » — 15^{h} 30^{m}—16^{h} 30^{m}.
Feb. 11, » — 18^{h} 20^{m}.

From Series II.

Dec. 24, time about 15^{h}45^{m} & 18^{h} 30^{m}.
Jan. 5, » — 16^{h} 30^{m}—17^{h} 30^{m}.

Jan. 23, time about $17^h 30^m—18^h 30^m$, and
$19^h 30^m—20^h 30^m$.
— 24, » — $17^h 30^m—18^h 30^m$. (Note especially the secondary deflection at about $17^h 57^m$.)
Feb. 7, » — $17^h 25^m$
— 9, » — 18^h.
— 12, » — $17^h 40^m—18^h 20^m$.

At all the places mentioned, the same condition is found as has been pointed out in the small serrations, namely, an upward deflection in the V-curve answering to a downward deflection in the A- and B-curves.

The resemblance is throughout so great that there seems no doubt that to a considerable extent the deflections are due to the direct influence of earth-currents.

We have endeavoured to determine on the original curves the relation $P_v/\Delta e$ for some small oscillations.

We have found that the numbers oscillate in such a way that the mean values of two consecutive numbers attain a satisfactory constancy. The reason of this is to be found in the fact that during the period of observation the external force changes considerably. By taking the mean this external effect will be more or less eliminated. By the aid of these numbers, we can then approximately eliminate the influence of the earth-current upon the vertical curve.

We have effected an elimination such as this for March 30—31, 1903. Unfortunately we have no earth-current registerings for the time about the commencement of the perturbation. It may perhaps seem that little that is of interest has been gained; but one fact at any rate is very apparent, namely, that the effect of the earth-currents on the vertical intensity curve is very small compared with that of the extraterrestrial currents.

In order to find out where the main body of the current is to be sought for, we may first consider one of the smaller deflections, e. g. the serrations at about $17^h 5^m$ and $17^h 13^m$ on the 15th February, Series II.

At $17^h 5^m$ we find a current that flows from SW to NE. It seems to occasion an upward deflection in the vertical curve, which answers to a magnetic force directed vertically downwards.

A horizontal current that would produce such an effect and have the the direction observed, must now be looked for in NW, i. e. in the direction of the mountain-ridge.

At $17^h 13^m$ both the earth-current and the corresponding deflection in the vertical curve have changed their direction. The current is therefore still to be looked for in the same direction, i. e. NW of the place.

This seems to agree with the assumption that the current follows the well-conducting veins of copper in Grytbotten Mountain.

At Bossekop, on the other hand, the vertical intensity is apparently more strongly affected by the earth-currents than in Kaafjord. This is easily seen by comparing the part of the vertical curve about $0^h—2^h$ for the 23rd March with the corresponding part of the N—S curve for the earth-currents; and the resemblance between the curves for March 30—31 is still more distinct.

It is principally in the N—S curve that we find agreement with the magnetic curves at Bossekop. If we here, in the same way, try to determine where the main body of the earth-current is situated, we meet at the outset with the difficulty that the current may flow under various azimuths, which may possibly indicate that the current-line in the neighbourhood of the place of observation is much curved.

It would appear from the general survey map that the geographical conditions would favour such view.

It will be seen that Bossekop is situated on a peninsula bounded partly by Alten Fjord with its o arms, Kaafjord and Rafsbotten, partly by the comparatively broad mouth of the Alten River.

The soil itself does not seem to contain any metal strata which would be more favorable to one rrent-direction than to another.

On the border-line between land and sea, however, there will always, on account of the difference the electric conductivity, be an unsymmetrical distribution of the earth-current density.

If we look at the serrations at about $2^h\ 1^m$ and $2^h\ 8^m$ on the 31st March, we see that the direction the current at the first hour mentioned is more or less from N to S. At the same time there is a rresponding force westwards in D, and in V a force vertically upwards. This last might indicate at the main body of the current was situated to the west of the place, i. e. out in the fjord; but it ight also be imagined to be produced by currents in the east that had a contrary direction, a ndition of things that would not be impossible. To decide this question, simultaneous observations ith short cable-lengths at various places is required.

If we assume that the first alternative is correct—which the greater conductivity of sea-water as mpared with soil perhaps makes probable—it might seem remarkable that in Kaafjord the main body the current is found in the land and not in the sea.

In reality, however, this is easily explained, as the upper branch of the Kaafjord, which lies near ur observing-place, is connected with the lower fjord only by a very narrow channol, while the Bossekop eninsula is surrounded by the great, wide Alten Fjord.

In the case of the second deflection at about $2^h\ 8^m$, the deflections in the magnetic curves are versed, as also in the N—S curve. We find, moreover, a distinct current-component in a direction W—E.

The direction of the current is thus now more or less SW—NE.

The same two alternatives may also be employed for the explanation of this phenomenon. It is oubtful which of the two is to be preferred; perhaps they act in concert.

OBSERVATIONS OF EARTH-CURRENTS AT KAAFJORD, MAY 1910.

152. During the expedition which I, accompanied by Mr. Krogness, made to Kaafjord at the time passage of Halley's comet across the sun's disc in May, 1910, we also, as has been stated, took servations of earth-currents with earth connections, as far as possible in exactly the same places as 1902—03.

The arrangement was the same as at that time, but for reasons already touched upon, we inserted each of the earth-connections a great resistance.

As the galvanometers previously employed had proved to be rather too sensitive, and, more partiilarly, to have no constant zero-point, we used, in their stead, two new school-instruments from DELMANN.

In the N—S line, the resistance added was 55 300 Ω, and the galvanometer employed—which I ill call a - had an internal resistance equal to 152 Ω.

In the E—W line, the resistance added was 53 000 Ω, and the galvanometer b had an internal sistance equal to 187 Ω.

The galvanometers a and b were set up at respective distances of 172 cm. and 115 cm. from the gistrator-cylinder.

The two galvanometers were introduced as a shunt upon a circuit with a resistan this arrangemement was employed the whole time.

The earth-resistance was determined from time to time, and the following values

TABLE CXV.

Earth-resistances.

Date	N–S	E–W
May 9	4700 Ω	3850 Ω
— 13	4200 „	4800 „
— 25	3700 „	6200 „
— 30	3700 „	5700 „

The sensitiveness was determined in the manner previously employed, and the follows:

TABLE CXVI.

Scale-values for one millimetre deflection; unit volt per 400 m.

Date	N–S	E–W
May 9	2.3×10^{-3}	4.2×10^{-3}
— 13	2.2	3.9
— 25	2.0	3.6
— 31	1.3	3.5

The new instruments, it appeared, maintained a very constant zero-point, but, as the of sensitiveness show, the temperature-coefficient was comparatively high.

The curves otherwise exhibit the same characteristic peculiarities as those previously

We have here, too, determined the relation between a number of synchronous s will be seen from the following table, the conditions are very constant, especially as reg three days.

The numbers $\frac{P}{\Delta e}$ are given in units $\frac{\gamma}{\text{millivolt}}$.

The numbers for June 1—2 are perhaps not so valuable as for the other days; th for the sensitivenes seems to indicate that contact-resistances have played a considerable p

TABLE CXVII.

						Date	$\frac{P_d}{\Delta e_{EW}}$	$\frac{P_d}{\Delta e_{NS}}$	$\frac{\Delta e_{EW}}{\Delta e_{NS}}$
9–10			1.74	2.12	1.22	May 16–17	1.75	2.47	1.42
			1.04	1.59	1.52		0.66	0.74	1.15
			1.04	1.46	1.41		0.81	1.16	1.42
			3.10	2.65	0.86		0.70	0.89	1.26
			2.03	2.42	1.19		0.70	0.84	1.22
			1.24	1.77	1.42		0.88	1.06	1.20
			1.25	3.05	2.43		0.55	0.74	1.35
			2.03	2.23	1.09		0.94	1.30	1.42
			1.38	1.72	1.24		0.91	1.11	1.24
			0.63	0.92	1.48		1.02	1.43	1.40
		0.33	0.70	0.83	1.17		0.62	0.84	1.38
		1.86	1.58	1.91	1.20		1.15	1.53	1.35
	1.17	1.46	1.16	1.49	1.30		0.50	0.64	1.31
			0.49	0.54	1.11		0.98	1.36	1.40
			1.19	1.45	1.22		1.40	1.97	1.42
	0.55	0.67	0.83	1.06	1.26		0.76	1.01	1.35
	0.40	0.60	0.67	1.00	1.50		0.50	0.67	1.35
	1.28	1.40	0.47	0.51	1.09		1.75	2.29	1.51
	1.72	1.93	0.90	1.03	1.15		0.87	1.16	1.35
			0.70	0.97	1.37		1.33	2.02	1.52
			0.81	1.14	1.42		0.69	0.96	1.44
		0.67	0.83	1.02	1.22		0.94	1.30	1.40
		1.53	0.95	1.16	1.22		1.09	1.50	1.38
			0.67	1.14	1.70		0.83	1.13	1.40
			0.87	1.30	1.50		0.42	0.52	1.22
			1.32	2.17	1.62		0.62	0.81	1.36
			1.12	1.62	1.44		0.56	0.71	1.26
			0.92	1.35	1.48		0.70	0.89	1.28
			0.93	1.26	1.35		0.74	0.98	1.33
			0.78	1.02	1.30		0.73	1.01	1.42
			0.82	0.91	1.09		0.59	0.81	1.40
			0.98	1.08	1.11		0.83	1.18	1.43
			0.74	0.98	1.35		0.83	1.06	1.29
			0.95	1.55	1.63	May 27–28	0.95	1.30	1.37
			0.70	0.89	1.26		0.89	1.09	1.21
			1.40	1.77	1.28		0.92	1.27	1.37
			0.60	0.86	1.45		0.89	1.30	1.45
			0.87	1.23	1.44		0.63	0.91	1.41
			0.52	0.71	1.40		1.24	1.69	1.37
			1.27	1.90	1.51		0.59	0.82	1.39
			1.05	1.26	1.22		0.79	1.09	1.39
			0.80	1.03	1.31		0.67	0.79	1.21
			0.83	1.13	1.40		0.68	0.82	1.21
			1.01	1.53	1.52		0.44	0.61	1.39
			0.77	0.96	1.22		1.01	1.27	1.25
			0.99	1.30	1.31		0.85	1.18	1.37
			0.70	0.86	1.22		0.86	1.00	1.15
			0.94	1.23	1.33		0.67	0.94	1.37

TABLE CXVII
(continued).

Date	$\frac{P_d}{\Delta e_{EW}}$	$\frac{P_d}{\Delta e_{NS}}$	$\frac{\Delta e_{EW}}{\Delta e_{NS}}$	Date	$\frac{P_d}{\Delta e_{EW}}$	$\frac{P_d}{\Delta e_{NS}}$	$\frac{\Delta e_{EW}}{\Delta e_{NS}}$
May 27—28	1.01	1.27	1.25	May 27—28	0.91	1.09	1.21
	0.76	0.97	1.29		0.54	0.79	1.39
	0.73	0.88	1.19		0.79	1.03	1.31
	0.95	1.24	1.31		0.88	1.09	1.23
	0.70	0.97	1.39		0.92	1.24	1.33
	0.56	0.76	1.39		1.15	1.57	1.35
	0.76	1.00	1.33		0.98	1.18	1.21
	0.68	1.36	1.51		0.69	0.88	1.23
	1.36	1.36	0.99		0.92	1.24	1.35
	0.80	1.00	1.25		0.35	0.45	1.29
	0.56	0.79	1.39		1.12	1.72	1.53
	0.92	1.24	1.35		0.80	1.27	1.57
	0.53	0.70	1.31		1.24	1.69	1.37
	0.57	0.79	1.37		1.35	1.84	1.37
	0.62	0.82	1.29		0.44	0.64	1.47
	0.74	1.00	1.35	June 1—2	1.43	2.68	1.86
	0.92	1.30	1.39		1.15	2.09	1.83
	1.03	1.45	1.41		1.34	2.43	1.83
	0.88	1.21	1.35		0.78	1.38	1.80
	0.94	1.33	1.43				

TABLE CXVIII.
Mean values.

Date	$\frac{P_h}{\Delta e_{EW}}$	$\frac{P_h}{\Delta e_{NS}}$	$\frac{P_d}{\Delta e_{EW}}$	$\frac{P_d}{\Delta e_{NS}}$	$\frac{\Delta e_{EW}}{\Delta e_{NS}}$
May 9—10	0.98	1.13	1.09	1.43	1.37
May 16—17			0.87	1.16	1.33
May 27—28			0.82	1.18	1.33
June 1—2			1.17	2.13	1.83

As regards absolute magnitude, however, the figures we have here determined are form those previously found. We see here that Δe has throughout values about 400 times greater than before, which indicates that q_1 and q_2 in the previous instances have really had an order of magnitude of several hundreds, as mentioned on p. 734.

In the period of about one month, during which we made observations, there is no case induction-phenomena are so conspicuous as in the storm of the 10th February, 1903; but i distinct in a number of storms.

We have previously reproduced some of the registered curves (see p. 652—653).

The induction-phenomenon appears perhaps most clearly in the vector diagrams that we have drawn and which are represented here.

We have also determined the relation between the maximal effect of the earth-currents and of the external current-system upon the magnetic elements.

We found the following figures:

May 26		May 27
Storm I; duration 1.5^h	Storm II; duration 18^m	Duration of storm 2^h,
0.18	0.09	0.15

VECTOR DIAGRAMS.

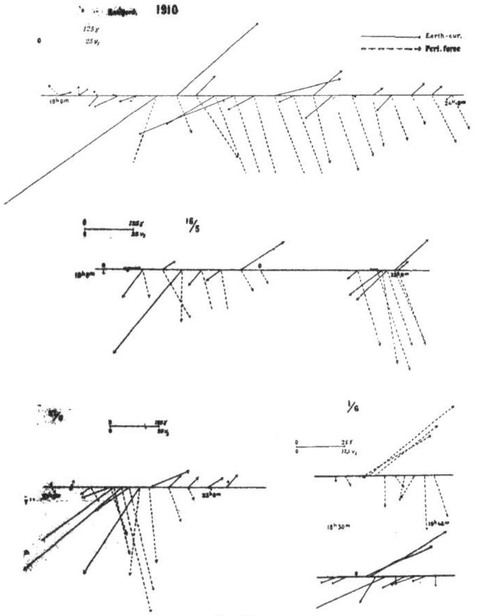

Fig. 272.

We have previously determined this relation for some of the storms from 1902—1903 and found alues varying between 0.12 and 0.52.

This last investigation therefore serves to show that the small cable-resistance that we employed reviously did not occasion any essential change in the phenomena. This should therefore justify us in conclusions that we drew from that material.

Finally, in this connection, I would touch upon a phenomenon that we observed during this ex. ition, and which may be of special interest.

In my earlier work, "Expédition Norvégienne 1899—1900" (P. 7), I drew attention to a number of very regular sinusoidal oscillations that were observed at the Haldde Observatory on March 19 and 20, 1900. I here reproduce on an enlarged scale the previously published curves showing this condition (fig. 273).

On the 18th May, 1910, we had the opportunity of observing exactly similar rapid, regular oscillations simultaneously in two sets of magnetic apparatus, which were placed at a distance of about 300 metres from one another. They proved to be accompanied by exactly similar oscillations in the earth-currents, and the two appear to be exactly synchronous, although an eventual small phase-alteration could scarcely be demonstrated. I here reproduce those curves in which these oscillations are noticeable.

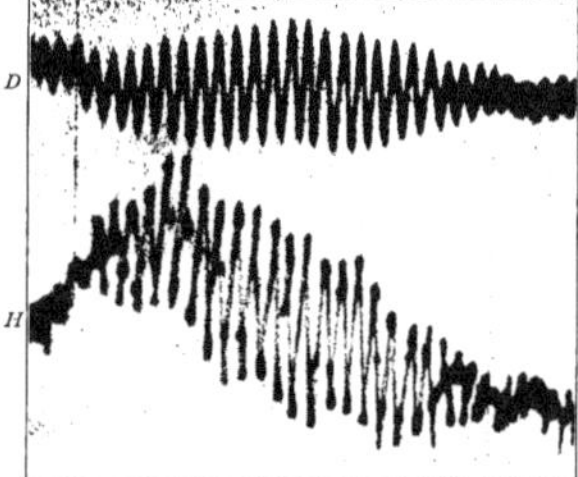

Fig. 273.

It will be seen that the oscillations occur in two epochs. At the end of these epochs there is also rapid registering with one set of magnetometers. It is here therefore that the period of oscillation can best be determined.

For this we find the following values:

119 sec.	118 sec.
122 »	113 »
128 »	121 »
124 »	109 »

Mean value: 119.3 sec.

With regard to the cause of these oscil we will only refer the reader to Art. 122. would especially call attention to here is that these oscillations occur simultaneously and probably exactly synchronously in earth-current and magnetism.

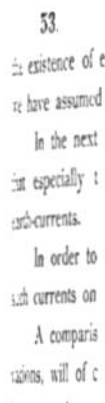

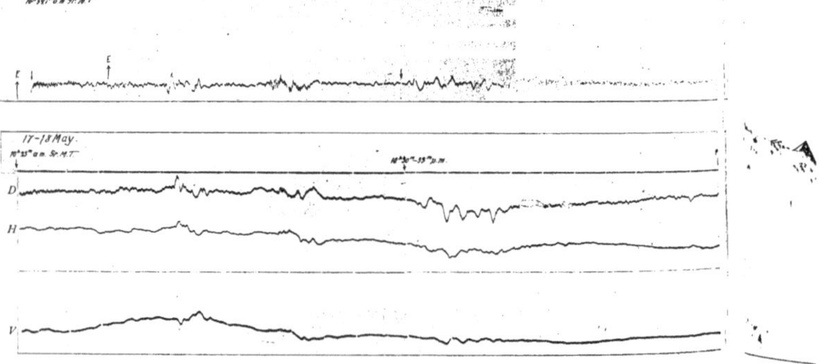

Earth currents and magnetic elements 17—18 May, 1910.

Fig. 274.

There will hereafter be more frequent opportunities of studying these phenomena, as the Norwegian ate, at my request, has conceded means to keep the Haldde Observatory in continual activity. Figure 276 ıows the observatory as it looked in 1912; but at the present time large new buildings are being added, ıd it is very well equipped with up-to-date instruments.

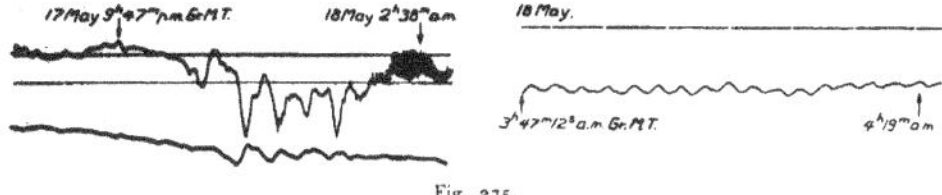

Fig. 275.

THEORETICAL INVESTIGATION OF THE CURRENTS
HAT ARE INDUCED IN A SPHERE BY VARIATION OF EXTERNAL CURRENT-SYSTEMS.

In the foregoing Article, we have had occasion to draw attention to conditions which indicate existence of earth-currents that are induced by variations in the outer polar current-system, which have assumed as the cause of the polar magnetic storms.

In the next place these currents exerted a considerable influence upon the magnetic apparatus, so at especially the smaller details in the phenomena had mainly to be regarded as the effect of the rth-currents.

In order to arrive at greater clearness, it may be interesting to make some calculations as to how ch currents on the whole will flow in the earth, and what magnetic effects they will produce.

A comparison of the results that can be obtained by the aid of the theory and the actual obser- tions, will of course only hold good of the main features of the phenomena, as in the calculations we ve to make a number of simplifying assumptions, which in reality are by no means exact.

Fig. 276. Haldde Observatory.

What we shall thus have to do is to study the currents that are induced in a sp
in an external magnetic field.

This problem has been studied by a number of scientists, some of whom have
from a general, others more from a special, point of view.

The investigations of LORBERG (1), NIVEN (2), and LAMB (3) are of great interest. If
assumption that the specific resistance of the earth is constant and equals $\varkappa$, we may d
formulæ previously developed by them.

We assume that we can write the magnetic potential of the inductive current-sy.

$$V = \Sigma_s \Sigma_n \Omega_{ns} e^{2\pi i p_s t}$$

where n may run through all whole positive values from 0 to ∞, and the summation
extends over a series of p_s, which in the special cases are to be determined.

Ω_{ns} is a solid harmonic of positive degree n, t is the time, $i = \sqrt{-1}$, and p_s is
employ LAMB's formulæ, and the same system of coördinates as before (cf. fig. 177,
then express the currents induced in the following manner. (LAMB has employed the syn

$$u = \Sigma_s \Sigma_n \frac{k^2}{4\pi} \frac{1}{(n+1)} \frac{\chi_n(k\varrho)}{\chi_{n-1}(kR)} \left(z\frac{\partial}{\partial y} - y\frac{\partial}{\partial z} \right) \Omega_{ns} e^{2\pi i p_s t}$$

$$v = \Sigma_s \Sigma_n \frac{k^2}{4\pi} \frac{1}{(n+1)} \frac{\chi_n(k\varrho)}{\eta_{n-1}(kR)} \left(x\frac{\partial}{\partial z} - z\frac{\partial}{\partial x} \right) \Omega_{ns} e^{2\pi i p_s t}$$

$$w = \Sigma_s \Sigma_n \frac{k^2}{4\pi} \frac{1}{(n+1)} \frac{\chi_n(k\varrho)}{\chi_{n-1}(kR)} \left(y\frac{\partial}{\partial x} - x\frac{\partial}{\partial y} \right) \Omega_{ns} e^{2\pi i p_s t}$$

where u, v, w, are the components of the electric current. Further,

$$k^2 = -\frac{8\pi^2 i p_s}{\varkappa}$$

and

$$\chi_n(\zeta) = 1 - \frac{\zeta^2}{2(2n+3)} + \frac{\zeta^4}{2.4(2n+3)(2n+5)} - \ldots = (-1)^n 3.5\ldots(2n+1)\left(\frac{d}{\zeta d}\right.$$

By these formulæ the induction-currents can always be determined, but the above form is
well adapted to practical calculation.

As, however,

$$xu + yv + zw = 0,$$

the currents will run in concentric spherical shells, and these may be more simply expre
of a current-function, ψ. This current-function we will define in the following manner: If,
shell with radius ϱ, we move a little way ds, and ψ, on this piece, increases from
then the component of the current at right angles to the direction of this element fr
when the observer is imagined to be standing on the spherical shell at the point i
looking in the direction of the motion, equals

$$\frac{d\psi}{ds}.$$

(1) Crelle, Vol. 71, p. 53.
(2) Phil. Trans. 1881, p 307.
(3) Phil. Trans. 1883, p. 519.

r the current-components i_θ and i_ω along respectively, meridians (ω = constant) and parallels = constant), we then obtain the following equations, changing to polar coordinates by the aid of uations (6) on p. 425:

$$i_\theta = -\frac{1}{\varrho \sin\theta}\frac{\partial\psi}{\partial\omega}, \quad i_\omega = \frac{1}{\varrho}\frac{\partial\psi}{\partial\theta}; \tag{5}$$

ow

$$i_\theta = u \cos\theta \cos\omega + v \cos\theta \sin\omega - w \sin\theta$$

$$i_\omega = - u \sin\omega + v \cos\omega$$

$$0 = u \sin\theta \cos\omega + v \sin\theta \sin\omega + w \cos\theta ,$$

ience we find

$$i_\theta = -\frac{w}{\sin\theta} = \frac{\partial}{\sin\theta\,\partial\omega}\Sigma_s \Sigma_n \left(\frac{k^2}{4\pi}\frac{\chi_n(k\varrho)}{\chi_{n-1}(kR)}\frac{\Omega_{ns}e^{2\pi i p_s t}}{n+1}\right)$$

$$i_\omega = \frac{uy - vx}{\varrho \sin\theta} = -\frac{\partial}{\partial\theta}\Sigma_s \Sigma_n \left(\frac{k^2}{4\pi}\frac{\chi_n(k\varrho)}{\chi_{n-1}(kR)}\frac{\Omega_{ns}e^{2\pi i p_s t}}{n+1}\right)$$

ie expression for the current-function will therefore be

$$\psi = -\Sigma_s \Sigma_n \frac{k^2}{4\pi}\frac{\chi_n(k\varrho)}{\chi_{n-1}(kR)}\varrho\frac{\Omega_{ns}e^{2\pi i p_s t}}{n+1} \tag{6}$$

The numerical calculation according to the above formula will be rather troublesome for an ordiry case in which the serial developments are not particularly simple, more especially if the series nverge only slowly. This will be the case with the field of the polar storms, as the acting currentstems come comparatively near to the earth.

The formulæ can, however, be simplified and put into a better form in the two extreme cases,

(1) where $|k|.R$ is very small, and

(2) where $|k|.R$ is very great.

e will especially consider these two extreme cases.

(1) $|k|.R$ *is assumed to be very small.*

We may then put, cf. (4)

$$\chi_n(k\varrho) = \chi_{n-1}(kR) = 1 \ ,$$

d we may also assume that this equation holds good for $n = 0$. We then obtain

$$\psi = -\Sigma_s \Sigma_n \frac{k^2}{4\pi}\varrho\frac{\Omega_{ns}e^{2\pi i p_s t}}{n+1} \ ;$$

t now we can write

$$\frac{\Omega_{ns}}{n+1} = \frac{1}{\varrho}\int_0^\varrho \Omega_{ns}\,d\varrho \ ,$$

d

$$\frac{k^2}{4\pi}e^{2\pi i p_s t} = -\frac{1}{\varkappa}\cdot\frac{\partial}{\partial t}e^{2\pi i p_s t} \ ;$$

d thus

$$\psi = \frac{1}{\varkappa}\,\Sigma_s\,\Sigma_n\,\frac{\partial}{\partial t}\int_0^{\varrho} \Omega_{ns}\, e^{2\pi i p_s t}\, d\varrho$$

$$= \frac{1}{\varkappa}\,\frac{\partial}{\partial t}\int_0^{\varrho} \Sigma_s\,\Sigma_n\,\Omega_{ns}\, e^{2\pi i p_s t}\, d\varrho$$

$$= \frac{1}{\varkappa}\int_0^{\varrho} \frac{\partial V}{\partial t}\, d\varrho\ .$$

We have hereby succeeded, in this case, in making all serial development superfluous.

(2) $|k|.R$ is very great.

If we look at the conditions near the surface, we find that there, too, $|k|.\varrho$ is 've from the last expression for χ_n in equation (4) it appears that for great values of the may put approximately

$$\chi_n(k\varrho) = (-1)^n\, 1 \cdot 3 \cdot 5 \ldots (2n+1)\,\frac{\sin\left(k\varrho + \frac{n\pi}{2}\right)}{(k\varrho)^{n+1}}\ .$$

From this we find, since

$$k = \pm\, 2\pi(1-i)\sqrt{\frac{p_s}{\varkappa}} \text{ when } p_s \text{ is positive, and}$$

$$k = \mp\, 2\pi(1+i)\sqrt{\frac{-p_s}{\varkappa}} \text{ when } p_s \text{ is negative,}$$

after some reduction, that

$$\psi = \Sigma_s\,\Sigma_n\,\frac{2n+1}{n+1}\sqrt{\frac{p_s}{2\varkappa}}\left(\frac{R}{\varrho}\right)^n e^{-2\pi\sqrt{\frac{p_s}{\varkappa}}(R-\varrho)}\,\Omega_{ns}\, e^{2\pi i\left(p_s t - \sqrt{\frac{p_s}{\varkappa}}(R-\varrho) +\right.}$$

when p_s is positive, and

$$\psi = \Sigma_s\,\Sigma_n\,\frac{2n+1}{n+1}\sqrt{\frac{-p_s}{2\varkappa}}\left(\frac{R}{\varrho}\right)^n e^{-2\pi\sqrt{\frac{-p_s}{\varkappa}}(R-\varrho)}\,\Omega_{ns}\, e^{2\pi i\left(p_s t + \sqrt{\frac{-p_s}{\varkappa}}(R-\varrho) -\right.}$$

when p_s is negative.

The expressions may also be written in the following form, as

$$\Sigma_n\,\frac{2n+1}{n+1}\,\Omega_{ns}(R) = 2\,\Omega_s(R) - \frac{1}{R}\int_0^R \Omega_s\, d\varrho$$

$$\psi = \Sigma_s\sqrt{\frac{\pm p_s}{2\varkappa}}\, e^{-2\pi\sqrt{\frac{\pm p_s}{\varkappa}}(R-\varrho)}\left[2\,\Omega_s(R) - \frac{1}{R}\int_0^R \Omega_s\, d\varrho\right] . \, e^{2\pi i\left(p_s t \mp \sqrt{\frac{\pm p_s}{\varkappa}}(R-\varrho) \pm\right.}$$

where the upper signs are to be employed when p_s is positive, and the lower when p_s is this case therefore, in order to find the currents at the surface, we need only make a development of the potential.

It appears from equation (9) that ψ, and with it the strength of the current, diminishes very rapidly as one moves inwards into the sphere. The currents are thus concentrated in the outermost layers of the sphere, and in this case we may imagine, as LAMB has already shown [1], that all the currents are replaced by the currents in a spherical shell with radius R. If ψ_1 stands for the current-function for the currents in this spherical shell, we shall have

$$\psi_1 = \int_{\varrho_0}^{R} \psi \, d\varrho ,$$

where ϱ_0 is a value of ϱ, where the strength of the current is insignificant.

Now

$$\int_{\varrho_0}^{R} e^{2\pi(1+i)\sqrt{\frac{p_s}{x}}(\varrho - R)} d\varrho = \frac{1}{2\pi(1+i)\sqrt{\frac{p_s}{x}}} = \frac{e^{-i\frac{\pi}{4}}}{4\pi\sqrt{\frac{p_s}{2\pi}}} ,$$

and thus

$$\psi_1 = \Sigma_s \Sigma_n \frac{2n+1}{n+1} \frac{\mathfrak{L}_{ns}(R)}{4\pi} e^{2\pi i p_s t} = \frac{1}{4\pi}\left(2V(R) - \frac{1}{R}\int_0^R V d\varrho\right) . \tag{11}$$

Thus no serial development is necessary for the determination of this current-system.

Our next important task is to determine the magnetic effect of the induction-currents. From LAMB's expression for the magnetic components in space, we can easily omit the expression for the potential of the induced currents. We find, if we call this V_i, that

$$V_i = -\Sigma_s \Sigma_n \frac{n}{n+1} \frac{\chi_n(kR) - \chi_{n-1}(kR)}{\chi_{n-1}(kR)} \frac{R^{2n+1}}{\varrho^{2n+1}} V_{ns} . \tag{12}$$

If $|k| . R$ is very small, we may write

$$\frac{\chi_n(kR) - \chi_{n-1}(kR)}{\chi_{n-1}(kR)} = \frac{k^2 R^2}{(2n+1)(2n+3)}$$

and if $|k| . R$ is very great, we may put

$$\frac{\chi_n(kR) - \chi_{n-1}(kR)}{\chi_{n-1}(kR)} = -1 .$$

Special interest attaches to the value of the potential at the surface. There, too, we can condense, so as to avoid serial developments.

In the first extreme case then, we have, for $\varrho = R$,

$$\left.\begin{aligned} V_i(R) &= -\Sigma_s \Sigma_n \frac{n}{n+1} \frac{k^2 R^2}{(2n+1)(2n+5)} V_{ns}(R) \\ &= \frac{4\pi R^2}{x} \frac{\partial}{\partial t} \Sigma_n \frac{n}{n+1} \frac{V_n(R)}{(2n+1)(2n+3)} \end{aligned}\right\} \tag{13}$$

but now

(1) loc. cit. p. 537.

$$\frac{n}{n+1} V_n = V_n - \frac{1}{\varrho}\int_0^\varrho V_n d\varrho$$

$$\frac{n}{n+1}\,\frac{V_n}{n+\frac{1}{2}} = \frac{1}{\sqrt{\varrho}}\int_0^\varrho \frac{d\varrho}{\sqrt{\varrho}}\left(V_n - \frac{1}{\varrho}\int_0^\varrho V_n d\varrho\right)$$

$$\frac{n}{n+1}\,\frac{V_n}{(n+\frac{1}{2})(n+\frac{3}{2})} = \frac{1}{\varrho\sqrt{\varrho}}\int_0^\varrho d\varrho \int_0^\varrho \frac{d\varrho}{\sqrt{\varrho}}\left(V_n - \frac{1}{\varrho}\int_0^\varrho V_n d\varrho\right) .$$

Here the summation with regard to n can be made direct, and we can therefore in this case wri

$$V_i(R) = \frac{\pi\sqrt{R}}{\varkappa}\,\frac{\partial}{\partial t}\int_0^R d\varrho \int_0^\varrho \frac{d\varrho}{\sqrt{\varrho}}\left(V - \frac{1}{\varrho}\int_0^\varrho V d\varrho\right) .$$

In the second extreme case, where $|k|.R$ is very large, $\varrho = R$ is simpler,

$$V_i(R) = \Sigma \frac{n}{n+1} V_n(R) = V(R) - \frac{1}{R}\int_0^R V d\varrho .$$

If we now look at the conditions on the earth during the magnetic storms, we can assume that the earth-current conditions as a whole exhibit a greater or less resemblance to the idealised case that we have here studied. Whether the conditions followed either of the two extreme cases, and if so, of them, would mainly depend upon the specific resistance and the length of period. If they with neither case, it might still be assumed that they will answer to something intermediate between the two.

If we assume the length of period to be 2 hours, i. e. $p = \frac{1}{7200}$, then

$$8\pi^2 p R^2 = 4.5 \times 10^{15} .$$

For sea-water we may put $\varkappa$ = about 10^{10},
for rain-water about 6×10^{13},
and for purest distilled water about 10^{15}.

The corresponding values of $|k|.R$ are

$$7 \times 10^2, \quad 9, \quad 2 .$$

The specific resistance in the outermost strata of the earth may probably now be assumed to have an order of magnitude corresponding to these figures. It should therefore be assumed that the earth-current conditions answer to something between the two extreme cases.

In order to obtain a general view of the course of the earth-currents during a polar elementary storm, we will determine the course of the induction-currents at the surface for the current-system previously employed in Art. 91, answering to the first and second extreme cases, assuming that the position of the system is fixed in relation to the earth, and that the strength of the current, i, varies. We have previously found for the potential of this system [see equation (28), p. 428],

$$V = i \int_{\mu_1}^{\mu_2} \frac{(\cos\theta - \cos\zeta\cos\beta)\,[\varrho - (L-d)\cos\beta]}{\sin^2\beta\,.\,d}\,d\mu \quad .$$

(1) $k \mid . R$ is very small.

In accordance with equation (7), we then have

$$\psi = \frac{1}{\varkappa}\frac{di}{dt}\int_0^{\varrho}\int_{\mu_1}^{\mu_2} \frac{[\cos\theta - \cos\zeta\cos\beta]\,[\varrho - (L-d)\cos\beta]}{\sin^2\beta\,.\,d}\,d\mu\,d\varrho \quad ,$$

hat is,

$$\psi = \frac{1}{\varkappa}\frac{di}{dt}\int_{\mu_1}^{\mu_2} \frac{\cos\theta - \cos\zeta\cos\beta}{\sin^2\beta}\,[\varrho\cos\beta + d - L]\,d\mu \quad . \tag{16}$$

The integration could be carried further, but the above form is the most practical for the numerical calculation.

For the current-components i_θ and i_ω, we find

$$i_\theta = -\frac{1}{\varrho\sin\theta}\frac{\partial\psi}{\partial\omega} = \frac{1}{\varrho\sin\theta}\frac{\partial\psi}{\partial\mu} \quad ,$$

hat is

$$i_\theta = \frac{1}{\varrho\sin\theta}\cdot\frac{1}{\varkappa}\frac{di}{dt}\left|\frac{\cos\theta - \cos\zeta\cos\beta}{\sin^2\beta}\,.\,[\varrho\cos\beta + d - L]\right|_{\mu=\mu_1}^{\mu=\mu_2} \quad , \tag{17}$$

nd

$$i_\omega = \frac{1}{\varrho}\frac{\partial\psi}{\partial\theta} = \frac{1}{\varkappa}\frac{di}{dt}\int_{\mu_1}^{\mu_2}\left[a_1\frac{\varrho\cos\beta + d - L}{\varrho} + a_2\frac{d-L}{d}\right]d\mu \quad , \tag{18}$$

where we have put

$$a_1 = \frac{\partial}{\partial\theta}\frac{\cos\theta - \cos\zeta\cos\beta}{\sin^2\beta} = -\frac{\sin^2\theta + (\cos\theta\cos\beta - \cos\zeta)\cos\zeta}{\sin\theta\sin^2\beta}$$

$$+ \frac{2(\cos\theta - \cos\zeta\cos\beta)(\cos\theta\cos\beta - \cos\zeta)\cos\beta}{\sin\theta\sin^4\beta} \quad , \tag{19}$$

nd

$$a_2 = \frac{\cos\theta - \cos\zeta\cos\beta}{\sin^2\beta}\cdot\frac{\partial\cos\beta}{\partial\theta} = \frac{(\cos\theta - \cos\zeta\cos\beta)(\cos\theta\cos\beta - \cos\zeta)}{\sin^2\beta\sin\theta} \quad . \tag{20}$$

For the magnetic potential V_i, we have, according to equation (14),

$$V_i(R) = \frac{\pi\sqrt{R}}{\varkappa}\frac{\partial}{\partial t}\int_0^R d\varrho\int_0^{\varrho}\frac{d\varrho}{\sqrt{\varrho}}\left(V - \frac{1}{\varrho}\int_0^{\varrho}V\,d\varrho\right) \quad .$$

We now have

$$V - \frac{1}{\varrho}\int V d\varrho = i\int_{\mu_1}^{\mu_2} \frac{\cos\theta - \cos\zeta\cos\beta}{\sin^2\beta}\left[\cos\beta + \frac{\varrho - L\cos\beta}{d} - \cos\beta - \frac{d -}{\varrho}\right.$$

$$= L i \int_{\mu_1}^{\mu_2} \frac{\cos\theta - \cos\zeta\cos\beta}{\sin^2\beta}\,\frac{\varrho\cos\beta + d - L}{\varrho\,.\,d}\,d\mu \quad .$$

Further we have

$$\int_0^\varrho\left(\int_0^\varrho \frac{\varrho\cos\beta + d - L}{\varrho\sqrt{\varrho}\,.\,d}\,d\varrho\right)d\varrho = \varrho\int_0^\varrho \frac{\varrho\cos\beta + d - L}{\varrho\sqrt{\varrho}\,.\,d}\,d\varrho - \int_0^\varrho \frac{\varrho\cos\beta + d \quad L}{\sqrt{\varrho}\,.\,d}$$

Now

$$\int \frac{d\varrho}{\varrho\sqrt{\varrho}\,d} = \frac{1}{L^2}\int \frac{\varrho\,d\varrho}{\sqrt{\varrho}\,.\,d} - \frac{2}{L^2}\frac{d}{\sqrt{\varrho}} ,$$

therefore

$$\int_0^\varrho \frac{d - L}{\varrho\sqrt{\varrho}\,d}\,d\varrho = \left[-\frac{2}{\sqrt{\varrho}} - L\int \frac{d\varrho}{\varrho\sqrt{\varrho}\,.\,d}\right]_{\varrho=0}^{\varrho=\varrho} = \frac{2(d - L)}{L\sqrt{\varrho}} - \frac{1}{L}\int_0^\varrho \frac{\varrho\,d\varrho}{\sqrt{\varrho}\,.\,d} ,$$

as

$$\lim_{\varrho=0}\frac{d - L}{\sqrt{\varrho}} = \lim_{\varrho=0}\frac{2(\varrho - L\cos\beta)\sqrt{\varrho}}{d} = 0 \quad .$$

Thus we find that

$$\int_0^\varrho d\varrho\int_0^\varrho \frac{\varrho\cos\beta + d - L}{\varrho\sqrt{\varrho}\,.\,d}\,d\varrho = \frac{2(d - 2L)}{L}\sqrt{\varrho} + (\varrho\cos\beta + L)\int_0^\varrho \frac{d\varrho}{\sqrt{\varrho}\,.\,d}$$

$$- \frac{L\cos\beta + \varrho}{L}\int_0^\varrho \frac{\varrho\,d\varrho}{\sqrt{\varrho}\,.\,d} \quad .$$

We therefore put

$$J_1 = \int_0^R \frac{d\varrho}{\sqrt{\varrho}\,d} \quad , \quad J_2 = \int_0^R \frac{\sqrt{\varrho}\,d\varrho}{d}$$

and obtain

$$V_1(R) = \frac{\pi\sqrt{R}}{\varkappa}\,\frac{d\,i}{d\,t}\int_{\mu_1}^{\mu_2} \frac{\cos\theta - \cos\zeta\cos\beta}{\sin^2\beta}\,[2(d_R - 2L)\sqrt{R} + L(R\cos\beta + L)J_1 - (L\cos\beta + R)J_2$$

where d_R is d's value for $\varrho = R$.

Here J_1 and J_2 are elliptic integrals. If, in the numerical calculation, the employment o tables is desired, they must be put into LEGENDRE's normal form. If this is done by kno we find, if φ is introduced as a new variable determined by the equation

$$[illegible] = L \frac{\cos^4\left(\varphi + \frac{\beta}{4}\right)}{\cos^4\left(\varphi - \frac{\beta}{4}\right)} \tag{25}$$

ıfter some reduction, that

$$J_1 = -\frac{F(k, \varphi_2) - F(k, \varphi_1)}{\sqrt{L}\cos^2\frac{\beta}{4}}\ , \tag{26 a}$$

$$J_2 = \frac{\sqrt{L}}{\cos^2\frac{\beta}{4}}\left[\cos^2\frac{\beta}{2}\left(F(k, \varphi_2) - F(k, \varphi_1)\right) - 4\cos^4\frac{\beta}{4}\left(E(k, \varphi_2) - E(k, \varphi_1)\right)\right.$$

$$\left. + \frac{\cos^2\frac{\beta}{4}}{\sqrt{L}} \frac{2\left[\sqrt{L}\cos\frac{\beta}{2} - \sqrt{R}\right] d_R}{R - 2\sqrt{RL}\cos\frac{\beta}{2} + L} - 2\cos^2\frac{\beta}{4}\cos\frac{\beta}{2}\right]\ , \tag{26 b}$$

vhere $F(k, \varphi)$ and $E(k, \varphi)$ are defined as before [see equations (17) and (18), p. 427].

Further,

$$\varphi_1 = \frac{\pi}{2} - \frac{\beta}{4}\ , \quad \varphi_2 = \sin^{-1}\frac{(\sqrt{L} - 1)\cot\frac{\beta}{4}}{\sqrt{(\sqrt{L}+1)^2 + (\sqrt{L}-1)^2\cot^2\frac{\beta}{4}}}\ ,$$

nd finally

$$k = \sqrt{1 - \tan^4\frac{\beta}{4}}\ .$$

m this we can deduce by derivation the corresponding expressions for the force-components elves.

By equation (24) we find directly

$$_i(R) = \frac{\pi}{\varkappa\sqrt{R}}\frac{di}{dt}\frac{\cos\theta - \cos\zeta\cos\beta}{\sin^2\beta\sin\theta}\left[2(d_R - 2L)\sqrt{R} + L(R\cos\beta + L)J_1 - (L\cos\beta + R)J_2\right]\Bigg|_{\mu=\mu_1}^{\mu=\mu_2}\ . \tag{27}$$

'or the determination of $P_{\theta i}$ we may start from equations (14) and (21), whence we easily deduce

$$P_{\theta i}(R) = -\frac{\pi}{\varkappa\sqrt{R}}\frac{\partial}{\partial t}\int_0^R d\varrho\int_0^\varrho \frac{d\varrho}{\sqrt{\varrho}}\cdot\frac{\partial}{\partial\theta}\left(V - \frac{1}{\varrho}\int_0^\varrho V d\varrho\right)\ .$$

3y employing the abbreviations we introduced into equations (19) and (20), we obtain

$$\frac{\partial}{\partial\theta}\left(\frac{\cos\theta - \cos\zeta\cos\beta}{\sin^2\beta}\cdot\frac{\varrho\cos\beta + d - L}{\varrho\cdot d}\right) = a_1\frac{\varrho\cos\beta + d - L}{\varrho\cdot d} + a_2\frac{\varrho(\varrho - L\cos\beta)}{d^3}\ .$$

Ve will then determine the integral

$$\int_0^\varrho \frac{\sqrt{\varrho}(\varrho - L\cos\beta)}{d^3}d\varrho = R\int_0^R\frac{\sqrt{\varrho}(\varrho - L\cos\beta)}{d^3}d\varrho - \int_0^R\frac{\varrho\sqrt{\varrho}(\varrho - L\cos\beta)}{d^3}d\varrho\ ,$$

but as

$$\frac{\partial}{\partial \varrho}\left(\frac{\sqrt{\varrho}}{d}\right) = -\frac{\sqrt{\varrho}\,(\varrho - L\cos\beta)}{d^3} + \frac{1}{2\sqrt{\varrho}\,.\,d}$$

and

$$\frac{\partial}{\partial \varrho}\left(\frac{\varrho\sqrt{\varrho}}{d}\right) = -\frac{\varrho\sqrt{\varrho}\,(\varrho - L\cos\beta)}{d^3} + \frac{3\sqrt{\varrho}}{2\,d} ,$$

we then obtain

$$A = \frac{1}{2} R J_1 - \frac{3}{2} J_2 - \frac{R\sqrt{R}}{d_R} + \frac{R\sqrt{R}}{d_R} = \frac{1}{2}[R J_1 - 3 J_2]$$

and

$$P_{\theta i}(R) = -\frac{\pi}{\varkappa\sqrt{R}}\frac{di}{dt}\cdot\int\limits_{\mu_1}^{\mu_2}\Big\{a_1[2(d_R - 2L)\sqrt{R} + L(R\cos\beta + L)J_1 - (L\cos\beta +$$

$$+ \frac{a_2}{2} L(R J_1 - 3 J_2)\Big\} d\mu \quad .$$

If we desire to determine $P_{\varrho t}$, we must go back to equation (12), from which it is easy t

$$P_{\varrho i} = -\Sigma_s \Sigma_n n \frac{k^2 R^2}{(2n+1)(2n+3)} \frac{R^{2n+1}}{\varrho^{2n+2}} V_{ns}$$

and for $\varrho = R$,

$$P_{\varrho i}(R) = \Sigma_n \frac{\pi R}{\varkappa} \frac{\partial}{\partial t} \frac{n}{(n+\frac{1}{2})(n+\frac{3}{2})} V_n(R)$$

$$= \frac{\pi}{\varkappa\sqrt{R}} \frac{\partial}{\partial t} \int\limits_0^R d\varrho \int\limits_0^\varrho \sqrt{\varrho}\, \frac{\partial V}{\partial \varrho}\, d\varrho = \frac{\pi}{\varkappa\sqrt{R}} \frac{\partial}{\partial t} \int\limits_0^R (R - \varrho)\sqrt{\varrho}\, \frac{\partial V}{\partial \varrho}\, d\varrho$$

$$= \frac{\pi L^2}{\varkappa\sqrt{R}} \frac{di}{dt} \int\limits_{\mu_1}^{\mu_2} (\cos\theta - \cos\zeta\cos\beta)\, d\mu \int\limits_0^R \frac{(R - \varrho)\sqrt{\varrho}}{d^3}\, d\varrho$$

or by aid of the formulæ

$$\int\limits_0^R \frac{\varrho\sqrt{\varrho}}{d^3}\, d\varrho = J_1 + 2L\cos\beta \int\limits_0^R \frac{\sqrt{\varrho}\, d\varrho}{d^3} - L^2 \int\limits_0^R \frac{d\varrho}{\sqrt{\varrho}\, d^3} ,$$

$$\int\limits_0^R \frac{d\varrho}{\sqrt{\varrho}\, d^3} = \frac{1}{L^2 \sin^2\beta}\left[-\frac{1}{2} J_1 - \frac{\cos\beta}{2L} J_2 + \frac{\sqrt{R}}{L\,.\,d_R}(R\cos\beta - L\cos 2\beta)\right] ,$$

$$\int\limits_0^R \frac{\sqrt{\varrho}}{d^3}\, d\varrho = \frac{1}{L^2\sin^2\beta}\left[\frac{L\cos\beta}{2} J_1 - \frac{1}{2} J_2 + \frac{\sqrt{R}\,(R - L\cos\beta)}{d_R}\right] ,$$

we obtain, after some reduction,

$$P_{\varrho i}(R) = \text{[illegible]} \cdot \int_{\mu_1}^{\mu_2} \frac{\cos\theta - \cos\zeta\cos\beta}{2\sin^2\beta} \left[(R\cos\beta + 3L) L J_1 \right.$$

$$\left. - (R - L\cos\beta) J_2 + 2 d_R \sqrt{R}\right] d\mu \quad . \tag{31}$$

2) $k \mid R$ is very great.

If we here put

$$V = i\Omega = i_0 \Omega \Sigma_s a_s \sin 2\pi p_s t \quad , \tag{32}$$

e can then write

$$\Omega = \int_{\mu_1}^{\mu_2} \frac{\cos\theta - \cos\zeta\cos\beta}{\sin^2\beta} \cdot \frac{\varrho - (L - d)\cos\beta}{d} d\mu \quad ,$$

nd we have

$$2\Omega(R) - \frac{1}{R}\int_0^R \Omega\, d\varrho = \int_{\mu_1}^{\mu_2} \frac{\cos\theta - \cos\zeta\cos\beta}{\sin^2\beta} \frac{R^2 - L^2 + d_R(R\cos\beta + L)}{R \,.\, d_R} d\mu \quad . \tag{33}$$

f this is inserted in equation (10) or (11), we have the current-function's expression for this extreme case,

$$\psi = T_s \,.\, \int_{\mu_1}^{\mu_2} \frac{\cos\theta - \cos\zeta\cos\beta}{\sin^2\beta} \cdot \frac{R^2 - L^2 + d_R(R\cos\beta + L)}{R \,.\, d_R} d\mu \quad , \tag{34}$$

$$\psi_1 = \frac{i}{4\pi} \int_{\mu_1}^{\mu_2} \frac{\cos\theta - \cos\zeta\cos\beta}{\sin^2\beta} \cdot \frac{R^2 - L^2 + d_R(R\cos\beta + L)}{R \,.\, d_R} d\mu \quad , \tag{35}$$

here

$$T_s = i_0 \,.\, \Sigma_s a_s \sqrt{\frac{p_s}{2\varkappa}}\, e^{-2\pi\sqrt{\frac{p_s}{\varkappa}}(R - \varrho)} \,.\, \sin 2\pi\left(p_s t - \sqrt{\frac{p_s}{\varkappa}}(R - \varrho) + \frac{1}{8}\right) \quad . \tag{36}$$

or the current-components we find

$$i_\theta = \frac{T_s}{\varrho\sin\theta} \left| \frac{\cos\theta - \cos\zeta\cos\beta}{\sin^2\beta} \cdot \frac{R^2 - L^2 + d_R(R\cos\beta + L)}{R \,.\, d_R} \right|_{\mu = \mu_1}^{\mu = \mu_2} \quad , \tag{37}$$

$$i_\omega = \frac{T_s}{\varrho} \int_{\mu_1}^{\mu_2} \left\{ a_1 \frac{R^2 - L^2 + d_R(R\cos\beta + L)}{R \,.\, d_R} + a_2 \frac{d_R^3 + L(R^2 - L^2)}{d_R^3} \right\} d\mu \quad , \tag{38}$$

here a_1 and a_2 have the same significance as before [see equations (19) and (20)].

In order to obtain the expressions for $i_{1\omega}$ and $i_{1\theta}$, we need only put $\frac{i}{4\pi}$ in the place of T_s in last two lines. The expression for the value of the potential at the surface is given by equations and (21), and we find, for $P_{\omega i}(R)$ and $P_{\theta i}(R)$,

$$P_{\omega i}(R) = i \left| \frac{\cos\theta - \cos\zeta\cos\beta}{\sin\theta\sin^2\beta} \cdot \frac{L(R\cos\beta + d_R - L)}{R^2 \cdot d_R} \right|_{\mu=\mu_1}^{\mu=\mu_2} \tag{39}$$

and

$$P'_{\theta i}(R) = -iL \int_{\mu_1}^{\mu_2} \left\{ a_1 \frac{R\cos\beta + d_R - L}{R^2 d_R} + a_2 \frac{R - L\cos\beta}{d_R^3} \right\} d\mu \quad . \tag{40}$$

Finally we obtain, according to equation (12),

$$P_{\varrho i}(R) = \Sigma_n n \frac{V_n(R)}{R} = \left(\frac{\partial V}{\partial \varrho}\right)_{\varrho = R} = -P_\varrho(R) \quad .$$

In this case then, the magnetic effect along the radius vector is equal, and in the reverse di to the direct effect of the external system, which is a fact well known from the theory of magnetic screens.

154. In accordance with the above formulæ, we have calculated the current-field at the surface for a system like the previously employed $\zeta = 20°$ and $h = 400$ km. $= 0.063$ R.

It appears from the formulæ that here too we may calculate o-functions similar to those in that is to say, functions by the aid of which we can find by subtraction the various quantities an to the external current-system with an arbitrary $\Delta\mu$.

In the tables below, a series of such quantities are given The index o has the same significance as in Art. 91. Further the value of the current-function and the current-components are calculated for the same three values of $\Delta\mu$ as before, namely, 75°, 180°, and 270°.

On the charts (figs. 277—282), the current-lines are drawn for equidistant values of ψ answering to the two extreme cases in these systems.

For the magnetic effect of the earth-currents, the components P_θ and P_ω are calculated for lati-tudes, $\theta = 20°$, 40° and 90° (see Table CXXV).

TABLE CXIX.

Values of ψ in the first extreme case.

$\Delta\mu = 75°$

θ	$\omega = 0$	7°.5	22°.5	37°.5	52°.5	67°.5	82°.5	97°.5	112°.5	127°.5	142°.5	157°.5	172°.5	180°
0	−0.315	−0.315	−0.315	−0.315	−0.315	−0.315	−0.315	−0.315	−0.315	−0.315	−0.315	−0.315	−0.315	−0.315
10	−0.282	−0.282	−0.280	−0.277	−0.274	−0.275	−0.278	−0.283	−0.288	−0.293	−0.297	−0.301	−0.303	−0.303
20	−0.040	−0.013	−0.058	−0.090	−0.127	−0.162	−0.193	−0.218	−0.240	−0.257	−0.269	−0.277	−0.282	−0.282
40	+0.252	+0.247	+0.211	+0.148	+0.072	−0.001	−0.065	−0.117	−0.158	−0.190	−0.213	−0.227	−0.235	−0.236
60	+0.250	+0.246	+0.220	+0.173	+0.114	+0.051	−0.009	−0.061	−0.105	−0.139	−0.165	−0.181	−0.190	−0.19[illegible]
90	+0.201	+0.199	+0.183	+0.153	+0.115	+0.071	+0.026	−0.015	−0.051	−0.081	−0.104	−0.119	−0.127	−0.128
140	+0.109	+0.109	+0.103	+0.092	+0.078	+0.061	+0.042	+0.023	+0.006	−0.009	−0.021	−0.029	−0.033	−0.033
180	+0.038	+0.038	+0.038	+0.038	+0.038	+0.038	+0.038	+0.038	+0.038	+0.038	+0.038	+0.038	+0.038	+0.038

TABLE CXIX (continued).

$\Delta\mu = 180°$

θ	$\omega = 0°$	15°	30°	45°	60°	75°	90°	105°	120°	135°	150°	165°	180°
0	−0.755	−0.755	−0.755	−0.755	−0.755	−0.755	−0.755	−0.755	−0.755	−0.755	−0.755	−0.755	−0.755
10	−0.664	−0.665	−0.668	−0.673	−0.678	−0.683	−0.686	−0.690	−0.694	−0.699	−0.704	−0.708	−0.709
20	−0.258	−0.263	−0.278	−0.304	−0.340	−0.387	−0.442	−0.498	−0.545	−0.581	−0.607	−0.622	−0.627
40	+0.266	+0.256	+0.226	+0.174	+0.099	+0.004	−0.105	−0.214	−0.309	−0.384	−0.436	−0.467	−0.477
60	+0.340	+0.330	+0.298	+0.245	+0.174	+0.087	−0.009	−0.104	−0.191	−0.263	−0.315	−0.347	−0.358
90	+0.317	+0.308	+0.282	+0.241	+0.186	+0.120	+0.050	−0.020	−0.085	−0.141	−0.182	−0.208	−0.217
140	+0.202	+0.198	+0.187	+0.168	+0.144	+0.115	+0.085	+0.055	+0.026	+0.002	−0.016	−0.028	−0.032
180	+0.091	+0.091	+0.091	+0.091	+0.091	+0.091	+0.091	+0.091	+0.091	+0.091	+0.091	+0.091	+0.091

$\Delta\mu = 270°$

θ	$\omega = 0°$	15°	30°	45°	60°	75°	90°	105°	120°	135°	150°	165°	180°
0	−1.132	−1.132	−1.132	−1.132	−1.132	−1.132	−1.132	−1.132	−1.132	−1.132	−1.132	−1.132	−1.132
10	−1.011	−1.012	−1.015	−1.020	−1.025	−1.032	−1.038	−1.042	−1.043	−1.042	−1.040	−1.038	−1.037
20	−0.550	−0.552	−0.560	−0.572	−0.590	−0.613	−0.641	−0.675	−0.714	−0.757	−0.794	−0.818	−0.825
40	+0.067	+0.063	+0.049	+0.027	−0.006	−0.050	−0.107	−0.177	−0.259	−0.345	−0.421	−0.471	−0.489
60	+0.205	+0.201	+0.186	+0.160	+0.124	+0.077	+0.020	−0.047	−0.119	−0.189	−0.248	−0.288	−0.302
90	+0.249	+0.244	+0.230	+0.208	+0.176	+0.137	+0.091	+0.040	−0.011	−0.059	−0.098	−0.123	−0.132
140	+0.207	+0.205	+0.198	+0.186	+0.170	+0.151	+0.130	+0.108	+0.087	+0.069	+0.054	+0.045	+0.042
180	+0.137	+0.137	+0.137	+0.137	+0.137	+0.137	+0.137	+0.137	+0.137	+0.137	+0.137	+0.137	+0.137

TABLE CXX.

Values of ψ in the second extreme case.

$\Delta\mu = 75°$

θ	$\omega = 0°$	7°.5	22°.5	37°.5	52°.5	67°.5	82°.5	97°.5	112°.5	127°.5	142°.5	157°.5	172°.5	180°
	−2.154	−2.154	−2.154	−2.154	−2.154	−2.154	−2.154	−2.154	−2.154	−2.154	−2.154	−2.154	−2.154	−2.154
1	−3.148	−3.126	−2.957	−2.664	−2.342	−2.070	−1.875	−1.742	−1.648	−1.595	−1.558	−1.533	−1.522	−1.521
2	−0.561	−0.567	−0.626	−0.801	−0.959	−1.038	−1.079	−1.104	−1.120	−1.131	−1.138	−1.142	−1.144	−1.144
4	+1.614	+1.583	+1.340	+0.919	+0.456	+0.067	−0.213	−0.404	−0.533	−0.618	−0.675	−0.709	−0.726	−0.728
6	+0.974	+0.959	+0.848	+0.655	+0.421	+0.191	−0.008	−0.168	−0.290	−0.378	−0.440	−0.479	−0.497	−0.500
9	+0.556	+0.550	+0.503	+0.418	+0.308	+0.188	+0.072	−0.032	−0.119	−0.188	−0.239	−0.272	−0.288	−0.290
14	+0.239	+0.237	+0.225	+0.201	+0.168	+0.130	+0.090	+0.050	+0.013	−0.018	−0.042	−0.059	−0.067	−0.069
18	+0.078	+0.078	+0.078	+0.078	+0.078	+0.078	+0.078	+0.078	+0.078	+0.078	+0.078	+0.078	+0.078	+0.078

$\Delta\mu = 180°$

θ	$\omega = 0°$	15°	30°	45°	60°	75°	90°	105°	120°	135°	150°	165°	180°
0	−5.169	−5.169	−5.169	−5.169	−5.169	−5.169	−5.169	−5.169	−5.169	−5.169	−5.169	−5.169	−5.169
10	−6.062	−6.040	−5.969	−5.837	−5.624	−5.313	−4.927	−4.540	−4.229	−4.016	−3.884	−3.813	−3.791
20	−2.000	−2.004	−2.017	−2.041	−2.083	−2.167	−2.360	−2.552	−2.637	−2.679	−2.703	−2.716	−2.720
40	+1.728	+1.696	+1.594	+1.405	+1.099	+0.651	+0.094	−0.462	−0.910	−1.216	−1.405	−1.508	−1.540
60	+1.292	+1.262	+1.168	+1.008	+0.779	+0.490	+0.164	−0.162	−0.451	−0.680	−0.840	−0.934	−0.964
90	+0.860	+0.839	+0.775	+0.671	+0.531	+0.364	+0.182	−0.001	−0.168	−0.308	−0.412	−0.476	−0.497
140	+0.437	+0.429	+0.404	+0.364	+0.312	+0.251	+0.186	+0.120	+0.059	+0.007	−0.033	−0.058	−0.066
180	+0.186	+0.186	+0.186	+0.186	+0.186	+0.186	+0.186	+0.186	+0.186	+0.186	+0.186	+0.186	+0.186

TABLE CXX (continued).

$\Delta u = 270^\circ$

θ	$\omega = 0°$	15°	30°	45°	60°	75°	90°	105°	120°	135°	150°	165°	180°
0	−7.753	−7.753	−7.753	−7.753	−7.753	−7.753	−7.753	−7.753	−7.753	−7.753	−7.753	−7.753	−7.753
10	−8.026	−8.019	−7.998	−7.961	−7.901	−7.809	−7.671	−7.464	−7.173	−6.825	−6.502	−6.287	−6.213
20	−3.365	−3.367	−3.371	−3.377	−3.388	−3.405	−3.430	−3.471	−3.552	−3.737	−3.896	−3.965	−3.984
40	+1.045	+1.035	+1.005	+0.949	+0.862	+0.729	+0.529	+0.233	−0.184	−0.686	−1.149	−1.457	−1.562
60	+0.916	+0.905	+0.870	+0.808	+0.716	+0.587	+0.416	+0.198	−0.056	−0.320	−0.554	−0.713	−0.768
90	+0.701	+0.691	+0.661	+0.610	+0.538	+0.444	+0.330	+0.200	+0.062	−0.069	−0.179	−0.252	−0.277
140	+0.449	+0.443	+0.428	+0.404	+0.371	+0.331	+0.286	+0.239	+0.193	+0.152	+0.120	+0.099	+0.092
180	+0.280	+0.280	+0.280	+0.280	+0.280	+0.280	+0.280	+0.280	+0.280	+0.280	+0.280	+0.280	+0.280

TABLE CXXI.

Values of $i_{\omega,o}$ in the first extreme case.

θ	$\omega = 0°$	15°	30°	45°	60°	75°	90°	105°	120°	135°	150°	165°	180°
0	0	+0.001	+0.001	+0.002	+0.002	+0.002	+0.002	+0.002	+0.002	+0.002	+0.001	+0.001	0
10	+0.781	+0.703	+0.616	+0.503	+0.388	+0.291	+0.209	+0.148	+0.101	+0.068	+0.041	+0.020	0
20	+2.109	+1.220	+0.863	+0.622	+0.451	+0.327	+0.237	+0.171	+0.120	+0.081	+0.050	+0.024	0
40	+0.436	+0.429	+0.398	+0.348	+0.292	+0.239	+0.191	+0.149	+0.113	+0.081	+0.052	+0.026	0
60	+0.177	+0.198	+0.210	+0.209	+0.197	+0.177	+0.153	+0.127	+0.101	+0.075	+0.050	+0.025	0
90	+0.071	+0.095	+0.115	+0.128	+0.133	+0.130	+0.120	+0.106	+0.088	+0.068	+0.046	+0.023	0
140	+0.019	+0.042	+0.062	+0.079	+0.090	+0.095	+0.094	+0.088	+0.076	+0.061	+0.042	+0.022	0
180	0	+0.022	+0.042	+0.059	+0.072	+0.081	+0.083	+0.081	+0.072	+0.059	+0.042	+0.022	0

Values of $i_{\theta,o}$ in the first extreme case.

θ	$\omega = 0°$	15°	30°	45°	60°	75°	90°	105°	120°	135°	150°	165°	180°
0	−0.002	−0.002	−0.002	−0.002	−0.001	−0.001	0	+0.001	+0.001	+0.002	+0.002	+0.002	+0.002
10	+4.138	+4.158	+4.196	+4.221	+4.223	+4.209	+4.185	+4.156	+4.128	+4.104	+4.086	+4.074	+4.070
20	+2.747	+2.686	+2.589	+2.496	+2.413	+2.339	+2.277	+2.224	+2.182	+2.149	+2.126	+2.112	+2.108
40	+1.836	+1.794	+1.696	+1.586	+1.487	+1.404	+1.337	+1.284	+1.242	+1.211	+1.190	+1.177	+1.173
60	+1.338	+1.321	+1.276	+1.217	+1.154	+1.095	+1.044	+1.002	+0.968	+0.942	+0.924	+0.913	+0.910
90	+1.108	+1.101	+1.079	+1.048	+1.012	+0.975	+0.940	+0.909	+0.883	+0.862	+0.848	+0.839	+0.836
140	+1.599	+1.596	+1.585	+1.568	+1.547	+1.524	+1.500	+1.477	+1.456	+1.440	+1.427	+1.419	+1.417
180	+0.083	+0.081	+0.072	+0.059	+0.042	+0.022	0	−0.022	−0.042	−0.059	−0.072	+0.081	−0.083

TABLE CXXII.

Values of $i_{\omega,o}$ in the second extreme case.

θ	$\omega = 0°$	15°	30°	45°	60°	75°	90°	105°	120°	135°	150°	165°	180°
0	0	+ 1.004	+1.939	+2.743	+3.359	+3.747	+3.879	+3.747	+3.359	+2.743	+1.939	+1.004	0
10	+ 3.623	+ 5.596	+6.313	+6.004	+5.266	+4.447	+3.657	+2.930	+2.269	+1.655	+1.080	+0.534	0
20	+32.325	+14.887	+8.616	+5.831	+4.261	+3.233	+2.492	+1.921	+1.444	+1.036	+0.673	+0.338	0
40	+ 0.509	+ 1.402	+1.799	+1.784	+1.591	+1.352	+1.116	+0.897	+0.695	+0.508	+0.333	+0.165	0
60	+ 0.077	+ 0.388	+0.612	+0.721	+0.734	+0.689	+0.610	+0.515	+0.413	+0.309	+0.206	+0.103	0
90	+ 0.014	+ 0.137	+0.241	+0.312	+0.347	+0.351	+0.331	+0.294	+0.245	+0.189	+0.128	+0.064	0
140	+ 0.002	+ 0.059	+0.111	+0.153	+0.181	+0.195	+0.196	+0.183	+0.159	+0.126	+0.088	+0.045	0
180	0	+ 0.045	+0.087	+0.123	+0.150	+0.167	+0.173	+0.167	+0.150	+0.123	+0.087	+0.045	0

TABLE CXXII (continued).

Values of $i_{\theta.o}$ in the second extreme case.

θ	$\omega = 0°$	15°	30°	45°	60°	75°	90°	105°	120°	135°	150°	165°	180°
0	− 3.879	−3.747	−3.359	−2.743	−1.939	−1.004	0	+1.004	+1.939	+2.743	+3.359	+3.747	+3.879
10	−10.087	−9.072	−7.013	−5.135	−3.788	−2.885	−2.286	−1.884	−1.614	−1.434	−1.319	−1.255	−1.235
20	+ 2.748	+1.517	+0.844	+0.567	+0.425	+0.340	+0.285	+0.249	+0.224	+0.207	+0.196	+0.190	+0.188
40	+ 4.021	+3.626	+2.822	+2.096	+1.577	+1.228	+0.995	+0.837	+0.731	+0.659	+0.613	+0.587	+0.577
60	+ 2.096	+2.014	+1.808	+1.555	+1.317	+1.118	+0.963	+0.846	+0.760	+0.699	+0.659	+0.636	+0.628
90	+ 1.411	+1.387	+1.322	+1.230	+1.128	+1.028	+0.940	+0.865	+0.806	+0.761	+0.730	+0.711	+0.705
140	+ 1.763	+1.755	+1.730	+1.692	+1.645	+1.593	+1.542	+1.493	+1.450	+1.416	+1.390	+1.375	+1.370
180	+ 0.173	+0.167	+0.150	+0.123	+0.087	+0.045	0	−0.045	−0.087	−0.123	−0.150	−0.167	−0.173

TABLE CXXIII.

Values of i_ω in the first extreme case.

$\Delta\mu = 75°$

θ	$\omega = 7°.5$	22°.5	37°.5	52°.5	67°.5	82°.5	97°.5	112°.5	127°.5	142°.5	157°.5	172°.5
0	−0.003	−0.002	−0.002	−0.002	−0.001	−0.000	+0.000	+0.001	+0.002	+0.002	+0.002	+0.003
10	+0.443	+0.470	+0.490	+0.494	+0.468	+0.401	+0.320	+0.250	+0.189	+0.148	+0.122	+0.109
20	+2.733	+2.547	+1.782	+0.983	+0.692	+0.502	+0.370	+0.277	+0.214	+0.171	+0.144	+0.131
40	+0.126	+0.151	+0.197	+0.238	+0.249	+0.235	+0.211	+0.186	+0.165	+0.149	+0.139	+0.133
60	−0.066	−0.041	−0.000	+0.045	+0.083	+0.108	+0.121	+0.127	+0.128	+0.127	+0.126	+0.125
90	−0.102	−0.087	−0.059	−0.025	+0.009	+0.040	+0.065	+0.084	+0.097	+0.106	+0.112	+0.114
140	−0.102	−0.093	−0.076	−0.052	−0.026	+0.002	+0.029	+0.053	+0.073	+0.088	+0.098	+0.103
180	−0.101	−0.094	−0.081	−0.062	−0.039	−0.013	+0.013	+0.039	+0.062	+0.081	+0.094	+0.101

$\Delta\mu = 180°$

θ	$\omega = 0°$	15°	30°	45°	60°	75°	90°	105°	120°	135°	150°	165°	180°
0	−0.004	−0.004	−0.004	−0.003	−0.002	−0.001	0	+0.001	+0.002	+0.003	+0.004	+0.004	+0.004
10	+1.143	+1.123	+1.073	+0.991	+0.904	+0.838	+0.781	+0.724	+0.658	+0.571	+0.489	+0.439	+0.419
20	+3.743	+3.720	+3.647	+3.515	+3.305	+2.974	+2.109	+1.244	+0.913	+0.703	+0.571	+0.498	+0.475
40	+0.490	+0.484	+0.467	+0.443	+0.421	+0.417	+0.436	+0.455	+0.451	+0.429	+0.405	+0.388	+0.382
60	+0.048	+0.050	+0.056	+0.069	+0.093	+0.131	+0.177	+0.223	+0.260	+0.285	+0.298	+0.304	+0.306
90	−0.099	−0.095	−0.080	−0.055	−0.020	+0.023	+0.071	+0.119	+0.161	+0.196	+0.221	+0.236	+0.241
140	−0.150	−0.144	−0.128	−0.101	−0.066	−0.025	+0.019	+0.063	+0.104	+0.140	+0.166	+0.183	+0.188
180	−0.167	−0.161	−0.144	−0.118	−0.083	−0.043	0	+0.043	+0.083	+0.118	+0.144	+0.161	+0.167

$\Delta\mu = 270°$

θ	$\omega = 0°$	15°	30°	45°	60°	75°	90°	105°	120°	135°	150°	165°	180°
0	−0.003	−0.003	−0.003	−0.002	−0.002	−0.001	0	+0.001	+0.002	+0.002	+0.003	+0.003	+0.003
10	+1.426	+1.420	+1.394	+1.353	+1.292	+1.215	+1.127	+1.047	+1.007	+0.990	+0.994	+1.005	+1.005
20	+4.056	+4.048	+4.024	+3.980	+3.914	+3.817	+3.677	+3.475	+3.168	+2.346	+1.548	+1.314	+1.243
40	+0.710	+0.706	+0.697	+0.681	+0.659	+0.632	+0.605	+0.587	+0.592	+0.627	+0.668	+0.690	+0.696
60	+0.203	+0.203	+0.202	+0.201	+0.202	+0.207	+0.220	+0.245	+0.283	+0.330	+0.375	+0.407	+0.418
90	+0.006	+0.007	+0.012	+0.021	+0.035	+0.055	+0.081	+0.114	+0.152	+0.191	+0.225	+0.248	+0.256
140	−0.084	−0.080	−0.071	−0.056	−0.035	−0.009	+0.021	+0.053	+0.084	+0.113	+0.137	+0.152	+0.157
180	−0.118	−0.114	−0.102	−0.083	−0.059	−0.031	0	+0.031	+0.059	+0.083	+0.102	+0.114	+0.118

TABLE CXXIII (continued).

Values of i_θ in the first extreme case.

$\Delta\mu = 75°$

θ	$\omega = 7°.5$	22°.5	37°.5	52°.5	67°.5	82°.5	97°.5	112°.5	127°.5	142°.5	157°.5	172°.5
0	−0.000	−0.001	−0.002	−0.002	−0.002	−0.003	−0.003	−0.002	−0.002	−0.002	−0.001	−0.000
10	−0.025	−0.065	−0.071	−0.026	+0.039	+0.092	+0.119	+0.123	+0.110	+0.086	+0.054	+0.018
20	+0.093	+0.274	+0.408	+0.409	+0.365	+0.314	+0.263	+0.213	+0.164	+0.116	+0.069	+0.023
40	+0.109	+0.306	+0.431	+0.456	+0.412	+0.344	+0.276	+0.215	+0.160	+0.111	+0.065	+0.021
60	+0.059	+0.167	+0.242	+0.277	+0.275	+0.249	+0.212	+0.172	+0.131	+0.092	+0.055	+0.018
90	+0.031	+0.089	+0.134	+0.161	+0.171	+0.166	+0.150	+0.127	+0.101	+0.073	+0.044	+0.015
140	+0.017	+0.049	+0.076	+0.096	+0.108	+0.112	+0.107	+0.097	+0.080	+0.060	+0.037	+0.013
180	+0.013	+0.039	+0.062	+0.081	+0.094	+0.101	+0.101	+0.094	+0.081	+0.062	+0.039	+0.013

$\Delta\mu = 180°$

θ	$\omega = 0°$	15°	30°	45°	60°	75°	90°	105°	120°	135°	150°	165°	180°
0	0	−0.001	−0.002	−0.003	−0.004	−0.004	−0.004	−0.004	−0.004	−0.003	−0.002	−0.001	0
10	0	+0.053	+0.095	+0.116	+0.110	+0.084	+0.068	+0.084	+0.110	+0.116	+0.095	+0.053	0
20	0	+0.115	+0.231	+0.347	+0.463	+0.574	+0.640	+0.574	+0.463	+0.347	+0.231	+0.115	0
40	0	+0.121	+0.245	+0.375	+0.506	+0.616	+0.663	+0.616	+0.506	+0.375	+0.245	+0.121	0
60	0	+0.094	+0.187	+0.275	+0.353	+0.408	+0.428	+0.408	+0.353	+0.275	+0.187	+0.094	0
90	0	+0.066	+0.129	+0.186	+0.232	+0.262	+0.273	+0.262	+0.232	+0.186	+0.129	+0.066	0
140	0	+0.047	+0.091	+0.129	+0.158	+0.176	+0.183	+0.176	+0.158	+0.129	+0.091	+0.047	0
180	0	+0.043	+0.083	+0.118	+0.144	+0.161	+0.167	+0.161	+0.144	+0.118	+0.083	+0.043	0

$\Delta\mu = 270°$

θ	$\omega = 0°$	15°	30°	45°	60°	75°	90°	105°	120°	135°	150°	165°	180°
0	0	−0.001	−0.002	−0.002	−0.003	−0.003	−0.003	−0.003	−0.003	−0.002	−0.002	−0.001	0
10	0	+0.043	+0.082	+0.114	+0.135	+0.138	+0.116	+0.067	+0.002	−0.046	−0.051	−0.028	0
20	0	+0.056	+0.112	+0.169	+0.227	+0.286	+0.347	+0.407	+0.462	+0.471	+0.347	+0.177	0
40	0	+0.052	+0.106	+0.164	+0.227	+0.297	+0.375	+0.453	+0.510	+0.499	+0.389	+0.208	0
60	0	+0.044	+0.089	+0.135	+0.182	+0.230	+0.275	+0.309	+0.319	+0.293	+0.225	+0.122	0
90	0	+0.035	+0.070	+0.104	+0.136	+0.165	+0.186	+0.197	+0.192	+0.169	+0.126	+0.067	0
140	0	+0.029	+0.058	+0.083	+0.104	+0.120	+0.129	+0.128	+0.119	+0.100	+0.072	+0.038	0
180	0	+0.031	+0.059	+0.083	+0.102	+0.114	+0.118	+0.114	+0.102	+0.083	+0.059	+0.031	0

TABLE CXXIV.

Values of i_ω in the second extreme case.

θ	$\omega = 7°.5$	22°.5	37°.5	52°.5	67°.5	82°.5	97°.5	112°.5	127°.5	142°.5	157°.5	172°.5
0	− 4.682	− 4.363	− 3.747	− 2.875	−1.807	−0.616	+0.616	+1.807	+2.875	+3.747	+4.363	+4.682
10	− 5.071	− 3.616	− 0.824	+ 1.938	+3.383	+3.735	+3.612	+3.367	+3.123	+2.930	+2.804	+2.734
20	+50.202	+45.502	+29.091	+12.395	+6.696	+4.986	+3.224	+2.560	+2.154	+1.921	+1.782	+1.709
40	− 2.565	− 1.976	− 0.843	+ 0.286	+0.902	+1.089	+1.082	+1.019	+0.951	+0.897	+0.860	+0.841
60	− 1.179	− 0.967	− 0.612	− 0.222	+0.098	+0.309	+0.424	+0.484	+0.507	+0.515	+0.515	+0.515
90	− 0.524	− 0.456	− 0.337	− 0.194	−0.053	+0.067	+0.159	+0.224	+0.267	+0.294	+0.309	+0.316
140	− 0.259	− 0.236	− 0.193	− 0.136	−0.072	−0.006	+0.055	+0.107	+0.151	+0.183	+0.204	+0.214
180	− 0.209	− 0.195	− 0.167	− 0.128	−0.081	−0.028	+0.028	+0.081	+0.129	+0.167	+0.195	+0.209

TABLE CXXIV (continued).

$\Delta\mu = 180°$

	$\omega = 0°$	15°	30°	45°	60°	75°	90°	105°	120°	135°	150°	165°	180°
0	− 7.758	− 7.494	− 6.719	− 5.486	− 3.879	− 2.008	0	+ 2.008	+3.879	+5.486	+6.719	+7.494	+7.758
0	− 0.069	− 0.131	− 0.289	− 0.412	− 0.147	+ 1.116	+ 3.623	+ 6.130	+7.393	+7.658	+7.535	+7.377	+7.315
0	+59.66	+59.50	+58.94	+57.78	+55.36	+49.42	+32.32	+15.22	+9.289	+6.866	+5.705	+5.154	+4.985
0	− 1.215	− 1.231	− 1.268	− 1.274	− 1.114	− 0.550	− 0.509	+ 1.567	+2.132	+2.292	+2.286	+2.249	+2.232
0	− 1.065	− 1.049	− 0.992	− 0.876	− 0.663	− 0.336	+ 0.077	+ 0.491	+0.818	+1.031	+1.146	+1.204	+1.220
0	− 0.634	− 0.617	− 0.564	− 0.472	− 0.340	− 0.174	+ 0.014	+ 0.202	+0.369	+0.500	+0.592	+0.645	+0.663
0	− 0.387	− 0.374	− 0.336	− 0.275	− 0.195	− 0.100	+ 0.002	+ 0.104	+0.199	+0.279	+0.340	+0.378	+0.391
0	− 0.347	− 0.335	− 0.300	− 0.245	− 0.173	− 0.090	0	+ 0.090	+0.173	+0.245	+0.300	+0.335	+0.347

$\Delta\mu = 270°$

	$\omega = 0°$	15°	30°	45°	60°	75°	90°	105°	120°	135°	150°	165°	180°
0	− 5.486	− 5.299	− 4.751	− 3.879	− 2.743	− 1.420	0	+ 1.420	+ 2.743	+ 3.879	+ 4.751	+ 5.299	+ 5.486
0	+ 3.937	+ 3.897	+ 3.781	+ 3.589	+ 3.334	+ 3.059	+ 2.897	+ 3.202	+ 4.580	+ 7.280	+10.04	+11.58	+12.01
0	+62.58	+62.53	+62.39	+62.16	+61.75	+61.06	+59.86	+57.48	+51.68	+34.82	+18.12	+12.88	+11.66
0	+ 0.001	− 0.010	− 0.044	− 0.099	− 0.170	− 0.240	− 0.258	− 0.086	+ 0.512	+ 1.625	+ 2.754	+ 3.390	+ 3.567
0	− 0.464	− 0.464	− 0.463	− 0.455	− 0.432	− 0.373	− 0.258	− 0.045	+ 0.281	+ 0.687	+ 1.077	+ 1.346	+ 1.443
0	− 0.349	− 0.344	− 0.330	− 0.303	− 0.259	− 0.191	− 0.095	+ 0.033	+ 0.185	+ 0.345	+ 0.489	+ 0.588	+ 0.624
0	− 0.248	− 0.243	− 0.223	− 0.191	− 0.146	− 0.089	− 0.020	+ 0.052	+ 0.128	+ 0.198	+ 0.255	+ 0.292	+ 0.305
0	− 0.245	− 0.237	− 0.212	− 0.173	− 0.123	− 0.063	0	+ 0.063	+ 0.123	+ 0.173	+ 0.212	+ 0.237	+ 0.245

Values of i_θ in the second extreme case.

$\Delta\mu = 75°$

θ	$\omega = 7°.5$	22°.5	37°.5	52°.5	67°.5	82°.5	97°.5	112°.5	127°.5	142°.5	157°.5	172°.5
0	−0.616	−1.807	−2.875	−3.747	−4.363	−4.682	−4.682	−4.363	−3.747	−2.875	−1.807	−0.616
10	−1.878	−5.283	−7.192	−6.786	−5.129	−3.522	−2.355	−1.566	−1.030	−0.649	−0.358	−0.115
20	+0.277	+1.093	2.409	1.232	0.596	0.344	0.218	0.144	0.096	0.061	0.034	+0.011
40	0.726	2.049	2.794	2.631	1.984	1.365	0.918	0.615	0.407	0.260	0.143	0.046
60	0.253	0.698	0.978	1.051	0.962	0.795	0.617	0.459	0.327	0.217	0.124	0.040
90	0.092	0.260	0.383	0.448	0.457	0.424	0.367	0.299	0.228	0.160	0.994	0.031
140	0.038	0.110	0.170	0.213	0.237	0.241	0.229	0.203	0.167	0.123	0.076	0.025
180	0.028	0.081	0.129	0.167	0.195	0.209	0.209	0.195	0.167	0.129	0.081	0.028

$\Delta\mu = 180°$

θ	$\omega = 0°$	15°	30°	45°	60°	75°	90°	105°	120°	135°	150°	165°	180°
0	0	−2.008	−3.879	−5.486	−6.719	−7.494	−7.758	−7.494	−6.719	−5.486	−3.879	−2.008	0
10	0	−1.001	−2.175	−3.702	−5.694	−7.816	−8.842	−7.816	−5.694	−3.702	−2.175	−1.001	0
20	0	+0.091	+0.201	+0.361	+0.649	+1.327	+2.561	+1.327	+0.619	+0.361	+0.201	+0.091	0
40	0	+0.390	+0.846	+1.437	+2.209	+3.039	+3.444	+3.039	+2.209	+1.437	+0.846	+0.390	0
60	0	+0.272	+0.556	+0.856	+1.149	+1.378	+1.468	+1.378	+1.149	+0.856	+0.557	+0.272	0
90	0	+0.163	+0.322	+0.469	+0.592	+0.676	+0.706	+0.676	+0.592	+0.469	+0.322	+0.163	0
140	0	+0.101	+0.194	+0.276	+0.339	+0.380	+0.394	+0.380	+0.339	+0.276	+0.194	+0.100	0
180	0	+0.090	+0.173	+0.245	+0.300	+0.335	+0.347	+0.335	+0.300	+0.245	+0.173	+0.090	0

TABLE CXXIV (continued).

$\Delta\mu = 270°$

θ	$\omega = 0°$	15°	30°	45°	60°	75°	90°	105°	120°	135°	150°	165°	180°
0	0	−1.420	−2.743	−3.879	−4.751	−5 299	−5.486	−5.299	−4.751	−3.879	−2.743	−1.420	0
10	0	−0.295	−0.628	−1.050	−1.630	−2.469	−3.702	−5.399	−7.188	−7.792	−6.186	−3.225	0
20	0	+0.028	+0.059	+0.098	+0.150	+0.229	+0.361	+0.621	+1.268	+2.463	+1.177	+0.420	0
40	0	+0.118	+0.250	+0.417	+0.641	+0.964	+1.437	+2.091	+2.789	+3.027	+2.398	+1.245	0
60	0	+0.101	+0.210	+0.334	+0.482	+0.658	+0.856	+1.048	+1.168	+1.133	+0.896	+0.491	0
90	0	+0.076	+0.154	+0.234	+0.317	+0.398	+0.469	+0.516	+0.522	+0.471	+0.359	+0.194	0
140	0	+0.060	+0.118	+0.172	+0.218	+0.255	+0.276	+0.279	+0.262	+0.222	+0.161	+0.085	0
180	0	+0.063	+0.123	+0.173	+0.212	+0.237	+0.245	+0.237	+0.212	+0.173	+0.123	+0.063	0

TABLE CXXV.

Values of $P_{\theta,e}$ and $P_{\omega,e}$ due to the external current-system, and of $P_{\theta,i}$ and $P_{\omega,i}$ due to the induced current.

$\Delta\mu = 75°$

θ		$\omega = 7°.5$	22°.5	37°.5	52°.5	67°.5	82°.5	97°.5	112°.5	127°.5	142°.5	157°.5	172°.5
20°	$P_{\theta,e}$	+26.47	+24.02	+15.44	+ 6.689	+ 3.694	+ 2.444	+ 1.797	+ 1.419	+ 1.184	+ 1.046	+ 0.963	+ 0.920
	$P_{\theta,i}$	+ 0.245	+ 0.225	+ 0.187	+ 0.144	+ 0.106	+ 0.076	+ 0.054	+ 0.037	+ 0.024	+ 0.015	+ 0.010	+ 0.007
	$\frac{P_{\theta,e}}{P_{\theta,i}}$	108.0	106.9	82.6	46.7	34.9	32.0	33.4	38.8	49.4	68.6	100.0	134.8
40°	$P_{\theta,e}$	− 1.220	− 0.913	− 0.323	+ 0.262	+0.576	+ 0.662	+ 0.647	+ 0.603	+ 0.558	+ 0.523	+ 0.499	+ 0.487
	$P_{\theta,i}$	+ 0.061	+ 0.061	+ 0.061	+ 0.058	+0.054	+ 0.047	+ 0.041	+ 0.035	+ 0.030	+ 0.026	+ 0.024	+ 0.023
	$\frac{P_{\theta,e}}{P_{\theta,i}}$	−19.9	−14.9	− 5.34	+ 4.48	+10.7	+14.0	+15.9	+17.4	+18.7	+20.0	+21.1	+21.7
90°	$P_{\theta,e}$	− 0.313	− 0.272	− 0.198	− 0.110	− 0 022	+ 0.053	+ 0.112	+ 0.154	+ 0.182	+ 0.200	+ 0.211	+ 0.215
	$P_{\theta,i}$	− 0.020	− 0.017	− 0.011	− 0.004	+ 0.004	+ 0.010	+ 0.016	+ 0.020	+ 0.024	+ 0.026	+ 0.027	+ 0.028
	$\frac{P_{\theta,e}}{P_{\theta,i}}$	+15.8	+16.3	+18.2	+29.3	− 5.97	+ 5.12	+ 6.99	+ 7.54	+ 7.73	+ 7.77	+ 7.78	+ 7.79
20°	$P_{\omega,e}$	− 0.185	− 0.683	− 1.408	− 0.821	− 0.480	− 0.329	−0.240	−0.178	−0.130	−0.089	−0 052	−0.017
	$P_{\omega,i}$	+ 0.018	+ 0.052	+ 0.077	+ 0.085	+ 0.084	+ 0.078	+0.068	+0.056	+0.045	+0.032	+0.019	+0.006
	$\frac{P_{\omega,e}}{P_{\omega,i}}$	−10.2	−13.1	−18.4	− 9.60	− 5.71	− 4.24	−3.54	−3.16	−2.91	−2.74	−2.70	−2.73
40°	$P_{\omega,e}$	− 0.417	− 1.178	− 1.612	− 1.544	− 1.198	− 0.855	−0.597	−0.415	−0.284	−0.185	−0.104	−0.034
	$P_{\omega,i}$	+ 0.019	+ 0.054	+ 0.078	+ 0.087	+ 0.084	+ 0.075	+0.063	+0.051	+0.039	+0.028	+0.016	+0.006
	$\frac{P_{\omega,e}}{P_{\omega,i}}$	−21.7	−21.8	−20.7	−17.9	−14.3	−11.4	−9.46	−8.10	−7.25	−6.74	−6.40	−6.07

TABLE CXXV (continued).

θ		$\omega = 7^\circ.5$	$22^\circ.5$	$37^\circ.5$	$52^\circ.5$	$67^\circ.5$	$82^\circ.5$	$97^\circ.5$	$112^\circ.5$	$127^\circ.5$	$142^\circ.5$	$157^\circ.5$	$172^\circ.5$
90°	$P_{\omega,e}$	− 0.062	− 0.174	− 0.258	− 0.304	− 0.314	− 0.295	−0.258	−0.213	−0.165	−0.116	−0.069	−0.023
	$P_{\omega,i}$	+ 0.007	+ 0.021	+ 0.032	+ 0.039	+ 0.041	+ 0.040	+0.037	+0.031	+0.025	+0.018	+0.011	+0.004
	$\dfrac{P_{\omega,e}}{P_{\omega,i}}$	− 8.39	− 8.30	− 8.11	− 7.89	− 7.61	− 7.30	−7.03	−6.78	−6.57	−6.42	−6.32	−6.25

As regards form, the current-charts exhibit, as might be expected, a great resemblance to the harts for the equipotential curves on the earth's surface and the curves for constant values of P_ϱ. It vill further be seen that the current-lines in the second extreme case draw closer together about the torm-centre than in the first extreme case. The form, however, in its main features, is very similar.)uring a polar storm, therefore, we should suppose, if the conduction-conditions in the earth were as deal as we have assumed, that a current-system would be formed, of which the form at the surface vould be something between these two extreme cases.

For large values of $\Delta\mu$, we see that the current-lines from the neighbourhood of the stormcentre ollow more closely the parallel-circles than for small values of $\Delta\mu$. In a latitude of about 40° in particular, we notice that it is often in a N—S direction that the comparatively powerful earth-urrents occur.

This may possibly have some significance in explaining the peculiar fact that in Germany, for nstance, the direction of the earth-currents is so markedly N—S. It may even be remarked that the nain direction for the earth-currents in Germany is approximately perpendicular to the auroral zone.

Another peculiarity in the occurrence of the active systems of precipitation, which also certainly ›lays a part in this respect, is the ease with which the systems of precipitation appear to form at the Norwegian stations at about midnight, Greenwich time, a circumstance which we have frequently pointed ›ut before.

If the current-strength in the outer system varies sinusoidally, there is in the first extreme case a bbase-difference of 90° between the strength of the current in the outer system and that in the inner ystem.

In the second extreme case there is a phase-displacement of 45° at the surface, and changing ery rapidly inwards. The whole current-system might approximately be imagined replaced by a system hat was concentrated in an infinitely thin globular cup, and the current-strength in this imaginary urrent-system must be assumed to oscillate in time with the current-strength in the outer system. The direction of this current will be the reverse of that in an outer current-sheet, which we may magine replacing the outer system.

If phase-displacement can be observed, there should be a means of forming a conception of the arth's conductivity. The observations in the north seem to show that the conditions follow the first ather than the second extreme case; but I think that here one ought to be very careful in drawing ny conclusions whatever concerning this circumstance, especially as $\varkappa$ must be supposed to vary within ery wide limits. In the next place, as regards the magnetic effect of the induction currents, we can specially point out how the relation $\frac{P_e}{P_i}$ varies when one retires from the current-system. In the first ase, it decreases greatly as one retires from the storm-centre; in the second extreme case, this is not so.

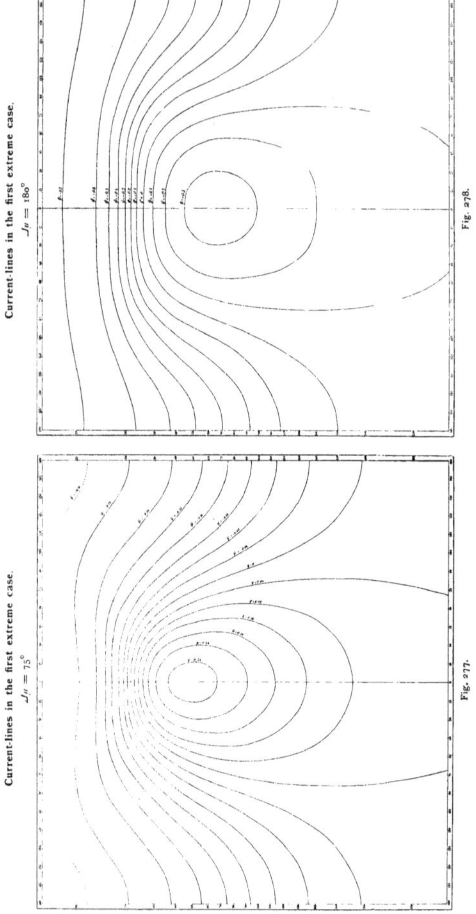

Current-lines in the first extreme case.
$\Delta u = 75^\circ$

Fig. 277.

Current-lines in the first extreme case.
$\Delta u = 180^\circ$

Fig. 278.

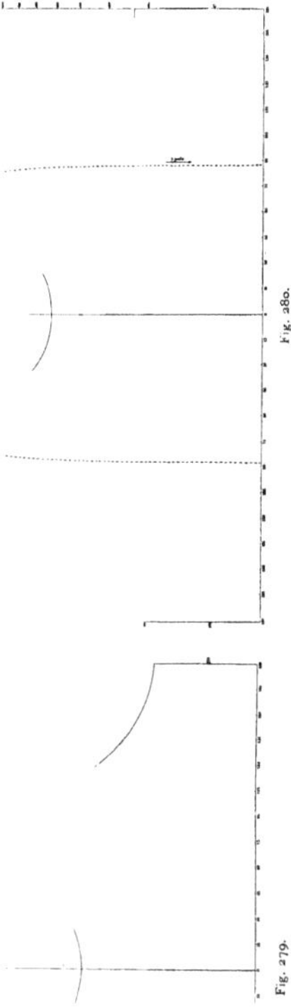

Fig. 280.

Fig. 279.

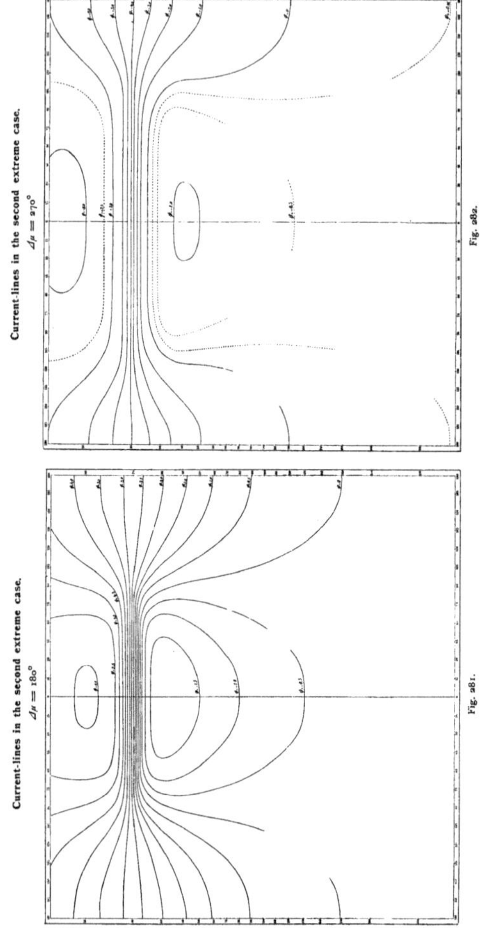

Current-lines in the second extreme case.
$\Delta\mu = 270^\circ$

Fig. 282.

Current-lines in the second extreme case.
$\Delta\mu = 180^\circ$

Fig. 281.

In the first extreme case therefore, the effect of the earth-current will be relatively greatest at some distance from the storm-centre; in the second extreme case, on the other hand, this distance will have less significance in this respect.

At Kaafjord we found that the greatest effect of the earth-currents amounted to about $^1/_6$ of the greatest effect of the outer system in the horizontal components for storms of about 2 hours' duration.

As we shall see in the next chapter, we can make a similar estimate for southern latitudes, whereby it will be possible to draw a comparison between observation and calculation, and also, by this relation, to obtain certain information concerning the earth's conductivity, although here too the uncertainty will be very great. We shall return to this later.

Finally, the condition of the vertical intensity also gives information that might afford indications of the conductivity, but here the uncertainty will be still greater.

Upon the basis of the figures that might be determined in this way, the most suitable value of $\varkappa$ might be sought. In the first place, however, such a calculation will be rather complicated, especially as one would have to include comparatively many terms of the series developments, as the systems in operation come very near to the earth. In the second place, the result of such a calculation would be *a priori* very uncertain.

It may however be mentioned in this connection that the requirement for having conditions answering to the first extreme case in the terms of higher order will be fulfilled for greater values of $|k|.R$ than in the lower terms, the requirement being that

$$\frac{|k^2|R^2}{2(2n+3)} \quad \text{or} \quad \frac{4\pi p_1 R^2}{\varkappa(2n+3)}$$

shall be a small quantity. The importance of the higher terms will thus cause the approach of the conditions to the first extreme case. *Vice versa*, in the higher terms $|k|R$ must be comparatively greater, in order to satisfy the conditions for the second extreme case, than in the lower, as we have here set aside terms of the order

$$\frac{\binom{n}{2}}{|k|.\varrho}$$

The condition of the vertical intensity might also be employed to separate inner and outer magnetic forces during the perturbations, but I think the result of such an investigation would not be nearly so certain as the method here employed of comparing synchronous serrations, as all deflections occurring in this component are very slight, and the earth's permeability certainly has a great influence here.

CURRENTS THAT ARE INDUCED BY ROTATION OR REMOVAL OF THE SYSTEMS.

155. In the preceding pages we have frequently pointed out that the systems of perturbation may be moved, especially along the auroral zone. A removal such as this will also induce currents in the earth, and it will be interesting to study the course of these currents.

As the movement takes place approximately along a small circle, the same currents will be induced as would be, supposing the system were fixed in space and the sphere rotated in relation to it about an axis perpendicular to the plane of the small circle. For this case Hertz has deduced special formulæ (1), but these are already contained in the expression given in the preceding article.

We can choose the Z-axis perpendicular to the plane of the small circle. If we then designate the angular velocity with which the system moves, or the corresponding rotation-velocity, as ω, and

(1) Cf. Hertz, Gesammelte Werke, Vol. I, p. 37.

reckon it positive if the movement of the system takes place in the direction of increasing ω, or the rotation of the sphere takes place in the direction of decreasing ω, if we imagine the outer current-system fixed in space. We need then only put

$$-\omega \frac{\partial V}{\partial \omega} \quad \text{instead of} \quad \frac{\partial V}{\partial t}$$

If we confine ourselves to the first extreme case, we therefore have, according to equation (7),

$$\psi = -\frac{\bar{\omega}}{\varkappa} \int_0^{\varrho} \frac{\partial V}{\partial \omega} d\varrho \tag{42}$$

For our polar current-system we found (cf. eq. 16)

$$\int_0^{\varrho} V d\varrho = i \int_{\mu_1}^{\mu_2} \frac{\cos \theta - \cos \zeta \cos \beta}{\sin^2 \beta} (\varrho \cos \beta + d - L) \, d\mu$$

Thus we find

$$\psi = -\frac{\bar{\omega}}{\varkappa} \frac{\partial}{\partial \omega} \int_0^{\varrho} V d\varrho$$

$$= -\frac{\bar{\omega}}{\varkappa} i \left[\frac{\cos \theta - \cos \zeta \cos \beta}{\sin^2 \beta} (\varrho \cos \beta + d - L) \right]_{\mu_1}^{\mu_2} \tag{43}$$

and according to equation (24)

$$V_i(R) = -\frac{\omega}{\varkappa} i \pi \sqrt{R} \frac{\cos \theta - \cos \zeta \cos \beta}{\sin^2 \beta} [2 (d_R - 2L) \sqrt{R} + L (R \cos \beta + L) J_1 - (L \cos \beta + R)$$

Here too we have calculated the value of the current-function for systems answering to $\Delta\mu = 75°$, 180°, 270°, and given similar σ-functions which we have previously defined, and we have also sented the induced current-system on four charts.

TABLE CXXVI.

Values of ψ_0, due to rotation.

θ	$\omega = 0°$	15°	30°	45°	60°	75°	90°	105°	120°	135°	150°	165°	180°
0	0.699	0.699	0.699	0.699	0.699	0.699	0.699	0.699	0.699	0.699	0.699	0.699	0.699
10	0.719	0.722	0.729	0.733	0.733	0.731	0.727	0.722	0.717	0.713	0.710	0.707	0.707
20	0.940	0.919	0.886	0.854	0.825	0.800	0.779	0.761	0.746	0.735	0.727	0.723	0.721
40	1.180	1.153	1.090	1.020	0.956	0.903	0.859	0.825	0.798	0.779	0.765	0.757	0.754
60	1.158	1.144	1.105	1.054	1.000	0.949	0.904	0.867	0.838	0.816	0.800	0.791	0.788
90	1.108	1.101	1.079	1.048	1.012	0.975	0.940	0.909	0.883	0.862	0.848	0.839	0.836
140	1.028	1.026	1.019	1.008	0.994	0.979	0.964	0.949	0.936	0.925	0.917	0.912	0.911
180	0.969	0.969	0.969	0.969	0.969	0.969	0.969	0.969	0.969	0.969	0.969	0.969	0.969

TABLE CXXVI (continued).

Values of ψ, due to rotation.

$\Delta\mu = 75°$

θ	$\omega = 0°$	7°.5	22°.5	37°.5	52°.5	67°.5	82°.5	97°.5	112°.5	127°.5	142°.5	157°.5	172°.5	180°
0	0	0	0	0	0	0	0	0	0	0	0	0	0	0
10	0	−0.004	−0.011	−0.012	−0.005	0.007	0.016	0.021	0.021	0.019	0.015	0.009	0.003	0
20	0	0.032	0.094	0.140	0.140	0.125	0.107	0.090	0.073	0.056	0.040	0.024	0.008	0
40	0	0.070	0.197	0.277	0.293	0.265	0.221	0.177	0.138	0.103	0.071	0.042	0.014	0
60	0	0.051	0.144	0.210	0.240	0.238	0.216	0.184	0.149	0.113	0.080	0.047	0.016	0
90	0	0.031	0.089	0.134	0.161	0.171	0.166	0.150	0.127	0.101	0.073	0.044	0.015	0
140	0	0.011	0.031	0.049	0.062	0.069	0.072	0.069	0.062	0.052	0.039	0.024	0.008	0
180	0	0	0	0	0	0	0	0	0	0	0	0	0	0

$\Delta\mu = 180°$

θ	$\omega = 0°$	15°	30°	45°	60°	75°	90°	105°	120°	135°	150°	165°	180°
0	0	0	0	0	0	0	0	0	0	0	0	0	0
10	0	0.009	0.017	0.020	0.019	0.015	0.012	0.015	0.019	0.020	0.017	0.009	0
20	0	0.039	0.079	0.119	0.158	0.196	0.219	0.196	0.158	0.119	0.079	0.039	0
40	0	0.078	0.158	0.241	0.325	0.396	0.426	0.396	0.325	0.241	0.158	0.078	0
60	0	0.081	0.162	0.238	0.305	0.353	0.371	0.353	0.305	0.238	0.162	0.081	0
90	0	0.066	0.129	0.186	0.232	0.262	0.273	0.262	0.232	0.186	0.129	0.066	0
140	0	0.030	0.058	0.083	0.101	0.113	0.117	0.113	0.101	0.083	0.058	0.030	0
180	0	0	0	0	0	0	0	0	0	0	0	0	0

$\Delta\mu = 270°$

θ	$\omega = 0°$	15°	30°	45°	60°	75°	90°	105°	120°	135°	150°	165°	180°
0	0	0	0	0	0	0	0	0	0	0	0	0	0
10	0	0.007	0.014	0.020	0.023	0.024	0.020	0.012	0.000	−0.008	−0.009	−0.005	0
20	0	0.019	0.038	0.058	0.078	0.098	0.119	0.139	0.158	0.161	0.119	0.060	0
40	0	0.034	0.068	0.105	0.146	0.191	0.241	0.291	0.328	0.320	0.250	0.134	0
60	0	0.038	0.077	0.117	0.158	0.200	0.238	0.267	0.276	0.254	0.195	0.106	0
90	0	0.035	0.070	0.104	0.136	0.165	0.186	0.197	0.192	0.169	0.126	0.067	0
140	0	0.019	0.037	0.053	0.067	0.077	0.083	0.083	0.076	0.064	0.046	0.024	0
180	0	0	0	0	0	0	0	0	0	0	0	0	0

In fig. 283 the current-system answering to a simple inductive system is given, in fig. 284 the ompound effect of two simultaneously occurring systems, situated on the same meridian, each at the ame distance from its pole, i. e. $\zeta_1 = 20°$, $\zeta_2 = 160°$.

The current-fields given on the charts will have, during the rotation, or displacements, a fixed osition in relation to the outer current-system. We may remark in particular that here in mean titudes, the direction principally found for the induced currents is N—S. These systems will pro- ably have something to say in the explanation of the diurnal variation of the earth-currents, to hich we shall return later.

The values of ψ placed upon the charts answer to a current-system of which the horizontal ortion has a direction W—E, and moves from E to W.

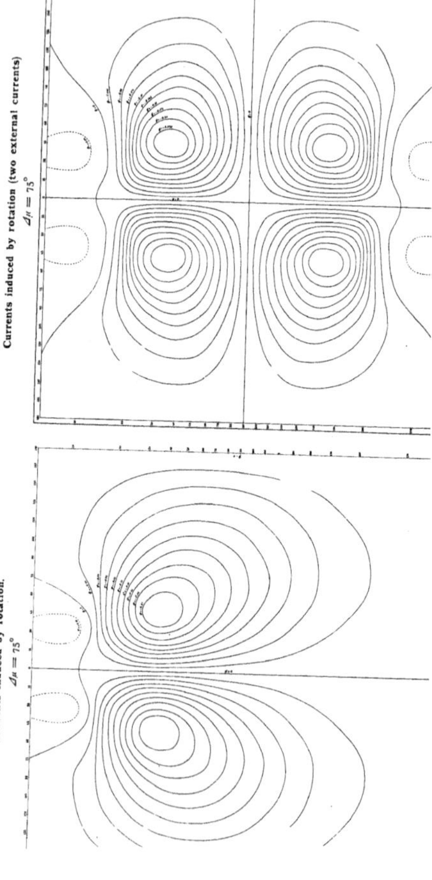

Currents induced by rotation (two external currents)
$\Delta\mu = 75^\circ$

Currents induced by rotation.
$\Delta\mu = 75^\circ$

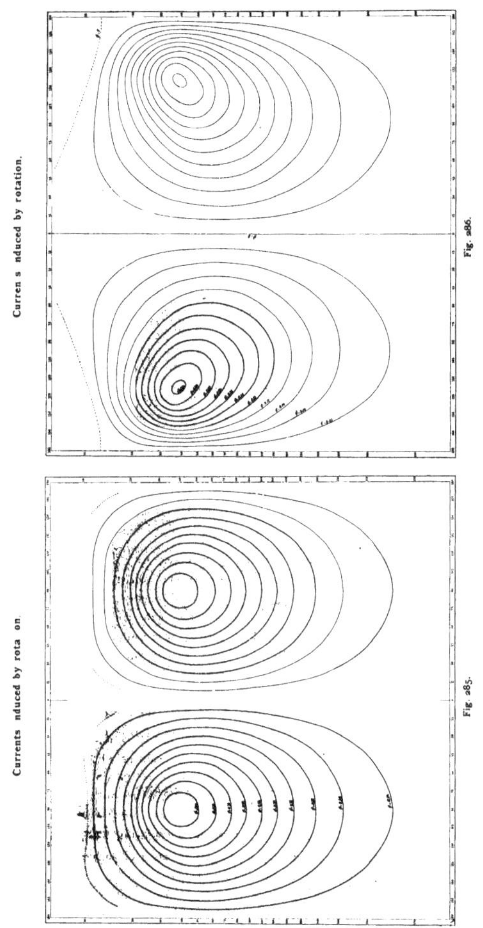

Curren s nduced by rotation.

Fig. 286.

Currents nduced by rota on.

Fig. 285.

EARTH-CURRENTS IN LOWER LATITUDES.

156. In the preceding pages, we have tried to obtain by theoretical considerations a general idea of the way in which the earth-current conditions develope in the vicinity of the auroral zone.

In order to come to a better understanding, however, it will be necessary also to consider the conditions in lower latitudes somewhat more closely.

This seems to be all the more necessary from the fact, already mentioned, that the views on the subject of earth-current phenomena in these regions, held by those scientists who have studied them, are very conflicting.

For the purpose of undertaking an investigation such as this, Mr. Krogness, with the aid of a grant from the University, went to Germany in the summer of 1910, in order to study the original curves from Professor Weinstein's material, and compare them with simultaneous magnetic curves from Wilhelmshaven or Potsdam. An investigation such as this, based upon the points of view maintained above, would be of peculiar importance, especially as Professor Weinstein himself, after similar studies, had arrived at a result that appeared to be at variance with our view of the phenomena. A great part of this material proved to be accessible, but unfortunately there were only a few days on which there were simultaneous observations of the two earth-current components. The one component, however, as we shall presently see, seems to be sufficient for our investigation.

Through the kindness of Professor B. Weinstein and Professor Ad. Schmidt in Potsdam, where this material is at present preserved, Mr. Krogness obtained the loan of a number of original curves with copies of simultaneous magnetic curves from Wilhelmshaven.

In the spring of 1911, Krogness and I, as already mentioned, made an expedition to Egypt and the Soudan for the purpose of studying the zodiacal light. On the way home, we spent a few days at Parc St. Maur and Greenwich, in order to go through some of the original earth-current registerings made at these observatories. Krogness had the opportunity of making photographic copies from a series of characteristic perturbations. The observatories further had the kindness to send us copies of a number of other selected storms.

Finally, we have had sent us from Pawlowsk a couple of photographic copies of the earth-current registerings made at that station.

As the working up of this material is inseparably connected with the investigations of the earth-current conditions in the polar regions described in the preceding articles, Krogness has kindly handed over the material he collected, so that the whole can be studied together.

In order to obtain as comprehensive a view as possible of the connection between earth-current and magnetism, we will here produce a number of copies, principally photographic. The magnetic curves from Wilhelmshaven are the only ones for which drawing on transparent paper has been employed. For the sake of the reproduction, however, we have had to darken with Indian ink those parts of the curves that were faintly reproduced; but this has been done as little as possible, and always on the photographic copy itself. We thought that in this way the curves would best preserve their character, which is here of importance, as it is often in the small details that the greatest resemblance is found.

EARTH-CURRENTS IN GERMANY.

157. We will first consider the curves from Germany.

Two earth-circuits were employed here, namely, Berlin to Thorn (E—W), and Berlin to Dresden (N—S), and the scale-value was determined daily by the interpolation of known electromotive forces, as more fully described by Weinstein in his treatise (l. c., p. 11).

If we compare the curves for the two earth-current components for November 1—2, we find throughout the most striking resemblance between the two curves. Every single jag and deflection in the one curve is accompanied by so exactly corresponding a deflection in the other curve, that by altering the sensitiveness we should be able to get all the briefer deflections to become very nearly identical. [The part of the curves just after 20^h (Gr. M. T.) answers to a time-mark, the earth-current here being interrupted for about 5 minutes in both lines (not exactly simultaneous)].

In other words we here find again the same peculiarity in the earth-currents that we found at Kaafjord.

We shall also find the same thing on looking at the perturbation on the 5th November, 1883; but in this case the curves are not photographically reproduced, but are drawn with Indian ink, as the originals were too faint and rubbed out.

If we look at the remaining curves from which we have simultaneous registerings in the two earth-current components — which we have not reproduced here — we make everywhere the same observation.

From this we may conclude, as at Kaafjord, that the earth-currents in the district here observed, follow very nearly the same direction in the earth. As a consequence, however, the one component in the brief variations — with which we are principally concerned — will be sufficient to characterise the course of the earth-currents. The want of the second component is therefore not of great importance.

We have at the end of the present volume reproduced a number of examples of various typical magnetic storms with their attendant earth-currents, from 1883.

On looking at these, several things are at once apparent. In the first place there is always a great resemblance between the course of the earth-current curve and the *D*-curve in nearly all details, which seems to indicate that the latter component is strongly influenced by earth-currents; but on the other hand we very often meet with conditions that indicate induction-currents. This is most noticeable in the simplest polar storms. The following are some examples where the conditions are especially distinct:

1883, Nov. 5, Dec. 9, March 1, Oct. 11, Nov. 28—29.

While in these cases the deflections in the *D*-curve as a whole increase comparatively evenly to a maximum, only to decrease once more to 0, a change takes place in the earth-current.

At the beginning of these perturbations the current flows in one direction, then turns, and during the last half we find the direction to be the reverse.

If, however, we examine the time of the change in the earth-current curves, we find, that it does not as a rule coincide with the time of the maximum of the deflection in *D*.

The reason of this, however, is easy to demonstrate. We need only look at the curves for the storm of the 5th November.

It is easy to prove the presence of a number of small serrations both on that day and on other days on which we see without doubt the effects of almost exclusively earth-currents. With their assistance we can now, as before, eliminate the effect of the earth-currents, leaving only the direct effect of the outer system.

In the case of the *D*-curve the agreement is so distinct that it presents no special difficulties. It is often difficult, however, to measure the small serrations in the magnetic curves, as we have only blue copies, in which, as a rule, the small details are not at all sharp. We have determined the relation between the deflections in the *D*-curve and the two earth-current curves.

In one day the figures found exhibit a (fairly) satisfactory constancy; but from one day to another the conditions vary somewhat. In this way we found the following figures:

TABLE CXXVII.

Date	23/6	10/7	30/7	16/9	28—29/11	Weighted mean	
$\frac{P_d}{\Delta e}$	1.46	1.29	2.44	0.95	2.13	1.39	2.13
Number of serrations	11	12	15	35	30	73	30
Line used	B—D	B—D	B—D	B—D	B—T	B—D	B—T

If by the aid of these numbers, we eliminate the effect of the earth-currents for instance, on the *D*-curve the 28—29th November, *we find that the change in the earth-current takes place very nearly at the time when the perturbing force attains its maximum.*

The same thing will be found in a number of other simple storms when we operate in the same manner, e. g. in those of

Dec. 8—9, Nov. 5, March 1, Oct. 11.

From this it appears that the effect of the earth-current in the declination is considerable, so considerable in fact, that the first distinct maximum in the earth-currents seems to produce the principal maximum of the *D*-curve.

In that time-interval the curve corresponding to the direct effect of the outer system varies only slightly as it approaches a maximum, whereas the variation in the earth-current curve is very marked.

As the effect of the earth-current in this district brings about an increase of the deflection, it will be easily understood that the two maxima may be very nearly simultaneous, a result at which Professor Weinstein has long since arrived; but it does not follow, as he seems inclined to suppose, that the induction-phenomenon is out of the question. It is our opinion, on the contrary, that in these storms it comes out very clearly and distinctly.

I can also here point out a peculiarity about the deflection in *D* after the distinct maximum, which, though it may seem unimportant, is yet very characteristic as regards both this storm and a number of others. I refer to the slight bulging exhibited by the descending branch of the curve.

This occurs at the times when the change takes place in the earth-current curves. Here their effect is only slight, and the reason for the somewhat altered character that the curve has here acquired is evidently that in this region the curve will mainly represent the variation in the outer current-system, while before it was also influenced to a great extent by the earth-currents. This little peculiarity we find again in most similar storms, the phenomenon being in some of them more distinct than here, in others less so. I will only refer the reader to those storms mentioned above.

A number of examples of this kind can also be shown in the material from 1902—03, as for instance, on Pl. XVIII, the course of the *H*-curve from about $23^h\ 20^m$ to 24^h at the western Central European stations in connection with the simultaneous maximum at about $23^h\ 40^m$; on Pl. XIX, the course of the *H*-curve at Tiflis just before 17^h in connection with the intermediate maximum at Matotchkin Schar, etc.

On a comparison of the variations in the horizontal intensity with the earth-currents, we may to some extent make remarks similar to those we have just made regarding the declination.

The agreement here, however, is not nearly so great; indeed, in the less powerful storms it is often impossible to demonstrate distinct synchronous serrations. In more powerful storms, the agreement is often somewhat better. Thus an elimination of the effect of the earth-current in the *H*-curve is attended with considerable difficulty, and probably cannot invariably be performed with the material at our disposal.

As the main direction of the earth-currents very nearly coincides with the direction of the mag- etic meridian, it will be easily understood that the effect of the earth-current is more distinct in D ıan in H.

We have also in the case of H attempted to determine the relation between synchronous deflec- ons in the earth-current curves and the H-curve and found the following:

TABLE CXXVIII.

Date	16/7	30/7	16/9	22–23/11	Weighted mean	
$\frac{P_h}{\Delta e}$	1.39	0.94	1.02	1.46	1.10	1.46
Number of serrations	6	3	16	20	25	20
Line used	B–D	B–D	B–D	B–T	B–D	B–T

It is doubtful, however, whether any special significance should be attached to these figures, parti- ılarly as the deflections have not always themselves the same direction.

Sometimes distinct induction-phenomena may also be found in H, e. g. on Nov. 28—29, Sept. 4, March 1.

In these storms we also find a peculiarity similar to that in the D-curve, namely a more or less ıarked bending-out of the curve simultaneous with the reversal of the earth-current curves. In such ses we can distinctly see the effect of the earth-current also in the H-curve; and the amplitude of the flection harmonises well with the figures found in Table CXXV.

Finally I may here draw attention to the fact that in cases where this bending-out is distinct, we n infer directly from the course of the curve the effect of the earth current, without at the same time ving registerings of the earth-current. This is immediately apparent from what has just been said.

In the storm of the 10th February, we found that the greatest effect of the earth-currents at Kaa- frd amounted to about 1/6 of the greatest effect of the outer current-system. By the aid of the out- ırd bends shown by the magnetic curves in southern latitudes, it is now easy, in accordance with the ove, to estimate the greatest effect of the earth-current. If we compare this with the greatest effect the outer system answering to the magnetic force at the time about the characteristic bending-out, rather perhaps at the beginning of the latter, we find

	$\frac{P_{H\,max}}{P_{Hi\,max}}$		$\frac{P_{H\,max}}{P_{Hi\,max}}$
Potsdam	0.53	Kew	0.50
Wilhelmshaven	0.52	Stonyhurst	0.54
Pawlowsk	1.12	Val Joyeux	0.44
Tiflis	0.64	Munich	0.40

If we compare this with the values we found in the theoretical argument in Art. 154, we see that f the first extreme case the relation $\frac{P_{H\,max}}{P_{Hi\,max}}$ varies when one moves away from the storm-centre to a dtance of about 20° from it, on an average from 100 to 20, or if preferred from 5 to 1, that is to s, the effect of the earth-current at the last place should be comparatively about 5 times as strong as a the first place.

In the second extreme case, however, the conditions are more or less constant.

We now find that when one moves from Kaafjord to Wilhelmshaven, the relation varies from about $^1/_6$ to $^1/_2$, that is to say, the effect of the earth-currents is relatively about 3 times as strong at the latter place as at the former. For this reason, therefore, the conditions during these storms seem to resemble extreme case No. 1 more than extreme case No. 2, which also seems to agree with the phase-displacement between earth-current and the outer inducing system, this apparently being nearly 90°.

This is most easily shown by the curves in the north; at Wilhelmshaven such a determination becomes more uncertain on account of the relatively greater importance of the earth-currents.

We may remark that the current here, in all cases, flows in such a manner that it is in harmony with the general law of induction. This should therefore be a confirmation of our view that at Kaafjord, for instance, there is really a kind of eddy in the earth-currents.

In conclusion I would point out a condition that might possibly sometimes give rise to mistakes. In Part I we have often shown that while, during a polar elementary storm, the one horizontal magnetic curve has a single bend, the other, owing to the moving of the systems of precipitation, may have a double bend.

During a simple storm of this kind, the earth-current curve will also take the form of a double undulation, owing to the induction. It may then be that these two double undulations, which of course are essentially different from one another, may yet exhibit so great a similarity that one might be tempted to assume — incorrectly — that the double undulation in the magnetic curve was an effect of the earth-currents. We appear to have such a case, for instance, on the 5th Nov., where a closer inspection shows that the double bend in H certainly *cannot* be an effect of the earth-current. In such cases therefore, one should be careful in drawing conclusions.

EARTH-CURRENTS IN FRANCE.

158. In France there are two sources in particular from which important material is obtained, namely, Blavier's work, and the earth-current registerings at Parc St. Maur. From the first of these a number of curves have been published in »Études des Courants Telluriques« (Paris, 1884); from the second a number of curves have been published in »Annales du Bureau Central Météorologique de France«. All the curves published have been reproduced from drawn copies. As this method of reproduction may easily, as we have already said, destroy a number of small details which are here of considerable interest, this may, in certain respects, perhaps be a somewhat uncertain foundation for conclusions of the kind with which we are occupied. This will especially be the case when we have to compare and determine very small, synchronous serrations, and calculate the relation between the amplitudes. It is moreover comparatively only a few days that are reproduced in these reports, and it was therefore not impossible that a number of perturbations might exist which were not reproduced, and which might be of greater interest in our investigations.

It was in order to procure the best possible basis for our study therefore, that Krogness went through the original curves, and selected a number of characteristic storms, of which we have obtained photographic copies. These copies are reproduced in Pl. XXXVIII to XLII.

The earth-wires at Parc St. Maur were in a straight line, both 14.8 kilometres in length, the one placed exactly in the direction E—W, the other exactly in the direction N—S.

By automatic disconnection there were further, except for the first couple of months of 1893, introduced exactly simultaneous time-marks on the earth-current curves and the magnetic curves. The galvanometers were shunted out, by which means the apparatus went back to its zero position, while at the same time an electric current produced oscillations in the magnetic curves.

We have here, therefore, a capital means of making exact comparisons of the points of time of the deflections in the two sets of curves.

The reader is further referred to MOUREAUX's description in "Annales du Bureau Central Météorologique de France", 1893, p. B. 25.

If we here compare the two earth-current curves, we at once discover that they do *not*, as at Wilhelmshaven, go together in every detail. As a rule, however, the resemblance is very close in the principal features, but it is frequently found, especially in the smaller details, that the character of the deflections differs a good deal in the two curves.

Nor, in accordance with this, is the relation between corresponding deflections in the two curves constant.

From this it would appear, in the first place, that the direction of the current in these regions is not so constant as at Kaafjord or in those parts of Germany in which Weinstein made his observations. The conditions, indeed, are more in accordance with those at Bossekop. The cause of the greater constancy in the direction in east Germany than in France, is probably to be found mainly in the different natural character. It may possibly be assumed that the considerably shorter length of the circuits at Parc St. Maur may play a decisive part; but such an explanation is certainly not sufficient, as in the curves published in Blavier's previously cited work, we find a similar disagreement between the circuits that make different angles with the meridian. We here too, however, in more powerful storms, find a marked principal direction for the earth-currents (cf. Bosler, Comptes Rendus, 6 février, 1911, or his Dissertation, Paris 1912, p. 67).

In the next place we find throughout a very striking resemblance between the E—W curve and the *H* curve. This condition is thus in accordance with what we found in Germany, and, as in the case of that country, we may conclude from this that the influence of the earth-current upon the horizontal intensity is comparatively great, although possibly other conditions during certain storms may act. I refer here to the changes that are caused by displacements of the systems of precipitation along the auroral zone.

Another circumstance that may also possibly cause the earth-current conditions in these two districts to be somewhat different is that — as we have often pointed out — the polar districts of precipitation frequently have quite a definite geographical position, e. g. at about midnight (Greenwich), when the storm-centre is situated, as a rule, between the four Norwegian stations.

In relation to this storm-centre or to the corresponding area of convergence, the two districts here under discussion will have a somewhat different position, and it might be imagined that this had something to do with the matter. Possibly too, the distribution of land and water has some significance, and this should then be more evident in France than in Germany.

A comparison between the *D* and N—S curves reveals throughout conditions that clearly point to induction-phenomena, for in the great majority of cases there exists, as a closer investigation shows, a more or less approximate proportion between the rate of change in the *D*-curve and the deflections in the N—S curve. The direction of the current is reversed as the *D*-curve attains its maximum or minimum; and the current-curve reaches its extremes at the time when the *D*-curve varies most. This condition is here very clearly marked. In the smallest serrations, however, we think we again find an undoubted synchronism. At the same time we may remark that if we imagine the N—S curve, for instance, moved a little to the right, its resemblance to the *D*-curve will in many cases be striking.

I need here only refer the reader to the perturbation of March 30—31, 1894, where there are specially-marked variations in both curves, or to November 24, 1894, where the perturbation-conditions are simpler. We also meet with similar examples, of which we can easily convince ourselves, in a number of the other storms given.

From this it will not be difficult to understand how a number of scientists, such as AIRY, WILD, and others, have thought they could demonstrate a difference in time between the deflections in earth-current and magnetism, the former being in advance of the latter. It is at any rate by no means impossible that without their knowing it their conclusions have really been based upon induction-phenomena similar to those here pointed out. If we measure the difference in time between various maxima in D and N—S curves of the same set — for instance, of April 30—31, 1894 — we find for the most part time-differences that vary between about 5 minutes and 20 minutes, or an average of about 12 minutes, and thus of an order of magnitude just such as the above-mentioned scientists have found.

It may perhaps be unnecessary to point to special cases of induction-phenomena, as the exhibit such a multiplicity of them; but I may mention a few of the simplest and most distinct.

October 2, 1893, about 22^h
November 2, —, — 18^h
— 3, —, — 16^h
January 5, 1894, — 4^h
March 1, —, — 23^h
May 28, —, — $22—23^h$
September 19, —, — $19—20^h$
October 16, —, — $20—21^h$
— 27, —, — $20-21^h$
November 24, —, — $19—20^h$
&c. &c.

Here too, we find that the direction of the current is what we should expect to find it according to the general law of induction. Such examples can be multiplied considerably. These same conditions are, however, not found in all perturbations. For instance, in the storm of the 28th January, 1893, $0—3^h$, the D and N—S curves appear to keep more or less together, while the change in the earth-current curves, which is here very nearly simultaneous, occurs at about the time of the maximum in the H-curve.

The induction-phenomenon is therefore here seen by comparing the earth-currents with the H-curve.

In the preceding article we pointed out that from our calculated fields of perturbation for the polar storms, we should sometimes expect to find such a condition, but that it must only be regarded as exceptional.

Unfortunately we have no time-marks on the earth-current curves, so the time cannot be determined so accurately as in the later perturbations; but by the aid of the short interruptions in the curve we have determined it as accurately as possible, and it seems to show that the conditions here indicated an exceptional case of this kind. No certain opinion can be expressed until more is known of the details of the field of perturbation. Something similar may possibly assert itself, for instance in the storms of the 6th March and 8th November, 1893, where the change in the N—S curve does not occur so exactly simultaneously with the maximum of the deflection in D.

We thus find here too the same chief peculiarities in the earth-current conditions as in Germany — conditions that agree exactly with those which, according to our theory, we should expect to find.

It is only in certain unimportant details that the conditions in Germany and France differ from one another.

Here too, we have endeavoured to determine the extent of the magnetic effect of the earth-currents; but the conditions are more difficult to deal with from the fact that the earth-currents may flow under different azimuths.

We here give the results of the comparisons of synchronous serrations.

$$\frac{P_h}{\Delta e_{EW}} : 2.59 \frac{\gamma}{\text{millivolt}} \qquad \frac{P_h}{\Delta e_{NS}} : 4.32 \frac{\gamma}{\text{millivolt}}$$

$$\frac{P_d}{\Delta e_{EW}} : 0.58 \quad — \qquad \frac{P_d}{\Delta e_{NS}} : 0.87 \quad —$$

EARTH-CURRENTS IN ENGLAND.

159. We have also received from Greenwich photographic copies of registerings of earth-currents ınd the horizontal magnetic elements for a number of selected days. They are given in Pl. XXXVI to XXXVIII, XLI and XLII.

We also give a series of examples of storms taken from curves of 1883, published in the Greenvich Observatory Reports. The curves selected are taken from various periods, but principally from he more recent years, from which there are also observations from other stations. One example in:luded is from AIRY's observations, namely, a storm on the 21st September, 1866.

In the more recent years the earth-current curves are so greatly perturbed by wandering currents, hat in the majority of cases only the night registerings are of importance to our investigations.

If we now look at these registerings from the same points of view as before, we see in the first blace that it is chiefly only one of the earth-current curves, the A-curve, that has powerful deflections.

It further appears that the deflections in the two components very nearly go together. It is, however, difficult to follow the details, partly on account of the apparently slight sensitiveness, and partly ›ecause of the strong, disturbing influence of local causes.

It therefore seems as if the direction of the current here once more remains fairly constant, and hus in accordance with the conditions at Kaafjord and in East Germany.

Here too, as at Parc St. Maur, an automatic arrangement introduces exactly synchronous time.narks into all the curves.

If we now attempt to compare the earth-current curves with the magnetic curves, we find that ıe conditions here appear to be more variable than those of the two sets of registerings previously escribed. If we first look at the storm of the 21st September, 1866, we there find the induction pheomenon extremely distinct when we compare the earth-current curve Greenwich-Croydon with the D-urve; while in the H-curve there is evidently a very marked effect of the earth-currents, the deflections ı the two curves appearing very nearly to go together.

The other earth-current component, Greenwich to Dartford, exhibits only very small deflections.

In the next storm, however, on July 21 and 22, 1889, the D and H curves seem to have changed ɔles. The deflections in earth-current and declination seem to be very nearly synchronous, while on omparing the earth-current curves with the H-curve, we find displacements which indicate inductionhenomena.

Conditions such as those in the first of these two are found strongly marked in a number of ases, e. g. the storms of

March 26—27, 1883,
October 5, 1883,
October 16, 1883,
November 1—2, 1893,
January 11—12, 1894,
November 24—25, 1894, and
November 7—8, 1893.

Examples of the other type, besides those already mentioned, are found in the storms of

February 26—27, 1893,
January 27—28, 1893, and
August 25—26, 1895.

In several of these, however, the phenomena seem to be of a very mixed character, and still more so in a number of other storms, e. g. of

November 3—4, 1889,
March 5—6, 1893, and
January 7—8, 1895.

We thus see that the earth-current conditions in these districts exhibit throughout exactly the same chief peculiarities as in Germany and France; but at the same time the cases that we have characterised as exceptional may possibly occur somewhat more frequently here.

This, however, only agrees with what, according to the above, we should consider probable.

The districts in which the observations were made here, have of course a somewhat more northerly position magnetically than the two corresponding districts in Germany and France, and a removal in a direction N—S in relation to the points of convergence of the perturbation-systems, the respective vortex-centres of the earth-current systems, must be assumed to bring about just such deviations as we have here observed.

In reality, these earth-current conditions at Greenwich may be regarded as an indication of a change from the conditions in Germany to the current-conditions in the auroral zone. In these last districts we have seen that the conditions are practically always as in the above-mentioned exceptional cases.

EARTH-CURRENTS AT PAWLOWSK.

160. Some examples of earth-current registerings and simultaneous magnetic registerings of two magnetic storms were also sent us, as already mentioned, from Pawlowsk; but as we have had no opportunity of going through the greater number of these, we have been unable to form any well-grounded opinion as to the nature of the conditions here. Two of the sets of curves that have been sent us show conditions during rather powerful storms, and the curves are of so jagged and disturbed a character that it is very difficult to follow them. Local disturbances also seem to exert a great influence. It was our intention, however, to give a reproduction of the third storm sent, namely, that of March 17 18, 1889. Here too there are great local disturbances, but nevertheless the principal course of the curves can be clearly followed. Unfortunately it appears at the last moment that the original curves are missing.

If these curves are considered from the same points of view as before, it will be easily discovered, on looking at the course of the curve as a whole, that there exists an approximate proportion between the N—S and E—W curves, and the rate of change in the D and H magnetic curves respectively. Changes in the direction of the brief deflections in the earth-current curves correspond in time with the extremes of the magnetic curves; and the most powerful deflections in the earth-current curves take place simultaneously with the greatest variations in the magnetic curves. Thus the induction-phenomenon comes out clearly on consideration of both components.

The two components of the earth-current exhibit a fairly strong resemblance; but the direction appears to be a little more variable than in Germany.

Here too, the same remark may be made as before, namely, that the resemblance between the two sets of curves becomes quite striking if the earth-current curve is moved slightly along the time-axis in the same direction as before. The direct effect of the earth-currents upon the magnetic apparatus is here more difficult to trace, but seems to be noticeable especially on comparison of the H and E—W curves, where the induction-phenomenon is not quite so distinct as in the other components.

As far as can be concluded from the observations at our disposal, it would appear that the same chief peculiarities are to be found in the earth-current conditions at Pawlowsk as at the stations previously studied.

The distinctness of the induction-phenomenon in both components here, may be partly due to the rather more northerly situation of this station, partly to the probably homogeneous nature of the soil there.

We cannot, as we have said, have any well-founded opinion as to whether the circumstances here pointed out are the usual ones, as we have so few curves to refer to.

COMPARISON OF SIMULTANEOUS EARTH-CURRENT OBSERVATIONS.

161. In selecting the storms given here, we have, in a number of instances, paid especial regard to those cases in which we have simultaneous observations from several places. In this way we can obtain some idea of the course of the earth-currents within a somewhat larger district. A number of such cases are shown in the Plates.

Of two days we have simultaneous observations from Germany, France and England, these days being November 1—2, 1883, and November 5—6, 1883. The observations from France are Blavier's and are published in his previously-mentioned work. As, however, his curves for the first of these storms are exceedingly jagged and their course in consequence not very clear, we have here given copies only of the second perturbation. It will be seen that there is a very great resemblance between the earth-current curves in Germany and the one earth-current component in France, namely, the curve for the line Paris to Dijon. We here find a very striking resemblance both in the principal course of the strong deflection that, as we have seen in Germany, indicates the effect of induction from the outer system, and in a number of details.

As regards the details, I can only point to a number of undoubtedly synchronous serrations, which are numbered on the various curves with figures from 1 to 10.

The change in the principal deflection of this current-component takes place at any rate almost simultaneously with the change in Germany.

The effect of the earth-current upon the horizontal magnetic elements cannot unfortunately, as in Germany, be eliminated, as the point of light at the time of the maximum had passed out of the field; but it seems probable, from the course of the curves, that if such elimination had been effected, the result arrived at would have been the same as in Germany, as all the characteristics of Weinstein's, curves are also found here. The change in the earth-current component takes place a little while after the D-curve has reached its maximum, and just at a place where the descending branch of the curve has an outward bend exactly similar to that found on the curve at Wilhelmshaven, and which we considered to be probably produced by the more marked direct effect from the outer systems where the effect of the earth-current was only slight.

It will be seen that in the line Paris to Dijon a shunt of $^1/_{40}$ of the galvanometer's resistance is employed, while in the others a shunt of $^1/_{20}$ is employed.

As the resistance in the lines is very nearly equal, and the distance between the earth-connections also approximately equal, we see from this that if the deflections in the various curves were to be compared, those in the line Paris to Dijon would have to be imagined increased to twice the number. We then see that it is the currents in this line that greatly predominate in stength.

Thus the earth-current moves principally in the direction given by this earth-circuit, i. e. from NW to SE and *vice versa*, that is to say, on the whole as in Germany.

The deflections that occur in the other two curves Paris—Nancy and Paris—Bar-le-Duc, and that show a somewhat different course, seem therefore to have to do only with details in the phenomenon.

As it is very difficult, if indeed possible at all, to find sufficiently distinct points of agreement between these last curves and those from Germany, it would seem probable that the variations here observed might be contingent in a comparatively greater degree upon the local geographical conditions in this country.

But in all essential phenomena we find a satisfactory agreement between the conditions in Germany and those in France.

In England too, in the various curves, we can to some extent find the same peculiarities as those here pointed out. The principal deflection in the earth-current here, however, is not nearly such a good example as in the material previously dealt with, but seems to be of a similar character. The deflections, however, are considerably smaller, and comparatively strong effects of wandering currents evidently break in and efface many of the smaller deflections with their characteristic peculiarities.

The first and most powerful deflection, simultaneous with an increase in the deflections in D, is here too, exceedingly distinct in both components, while the last, most marked bend during the time when P_d is diminishing, is extremely inconspicuous.

As regards the details, there can be seen, especially in the magnetic curves, a number of the same small, characteristic jags as in the two previously-considered regions; and they are also exceedingly typical here.

They can also be observed in the earth-current curves, but only sometimes distinctly, on account of the small degree of sensitiveness and the great local disturbances.

In all the other cases here brought together, in which we have simultaneous registerings from two of the three districts, exactly the same conclusions may be drawn as here, namely, that the earth-currents behave in all cases, in the main, uniformly throughout the district. Characteristic deflections both large and small, are followed, as a rule, synchronously, and the magnetic influence of the earth-currents upon the magnetic elements has its outcome in the exact uniformity in all the details, especially of the course of the horizontal intensity curve, at the various stations within the district under consideration.

The fact that the course of the declination-curve is so strikingly similar at the various stations should be accounted for, according to what has been said, by the almost identical effect of the combined extraterrestrial and intraterrestrial current-systems upon this magnetic element at the various places within the district. Concerning this, I need only refer the reader to the various comparisons of curves given in the plates.

162. In the preceding pages, we have principally considered the conditions during polar storms, and throughout have found our former precisely-defined view of the phenomena confirmed.

We have, however, also included a number of examples of positive equatorial storms.

The chief peculiarity of these storms consisted, it will be remembered, in the rather sudden increase in the horizontal intensity all over the earth, the deflection thus obtained remaining more or less constant for a period of varying length, until, as a rule, other forces of a more polar nature interfered.

At first, also, a deflection in the H-curve to the opposite side was very frequently found.

The currents that will be induced at the beginning of such a storm will of course, everywhere in rather lower latitudes, have a direction E—W, as will instantly be seen from the formulæ for i_ω i_θ, which, according to equations (5) and (6) on p. 759 may be written in the form

$$i_\theta = \Sigma_s \Sigma_n A_{ns} P_{\omega_{ns}} = 0$$
$$i_\omega = - \Sigma_s \Sigma_n A_{ns} P_{\theta_{ns}}$$

where A_{ns} is a certain function of n, p_s, s, ϱ and t, of which the analytical expression is easily found by equation (6). At the beginning of the perturbation, therefore, one would expect to find a deflection in the E—W curve — which is uniform in direction — that quickly increases to a maximum, and again quickly decreases towards 0.

If the E—W and N—S curves answered to the earth-current components in the magnetic E—W and N—S, the latter of these two should not exhibit a similar condition.

We have seen however, that simultaneously with the commencement of such a perturbation, one or more rather locally circumscribed polar systems of precipitation are formed.

A system such as this, however — as we remember in medium latitudes — will act throughout most strongly in the N—S component. As the polar systems of precipitation, which we have ordinarily seen to be of a briefer nature, so that the deflections in the magnetic curves increase to a maximum only to decrease again immediately afterwards, the earth-current curves answering to them will as a rule be in the form of twofold undulations.

We should expect, therefore, that at the beginning of the perturbation deflections of such a nature might sometimes be found, especially in the comparatively high latitudes here under consideration.

If we now look at the examples of such storms given in our material from the three southern stations, we see, for instance at Parc St. Maur, on the 11th January, 1894, a very characteristic example of a case of this kind.

In the E—W curve we find at first a uniformly-directed deflection, while in the N—S curve the deflection has the character of twofold waves. We see that the first earth-current impulse in this latter component must undoubtedly, at any rate to a very great extent, produce the "starting impulse" that appears so distinctly in the D-curve.

For the rest, fairly strong polar systems of precipitation are acting here all through the further course.

We thus see here that the character of the deflections at first in the E—W and N—S curves is rather different.

As it will very often be difficult or impossible to separate the deflections that are due to equatorial perturbations from those that are due to simultaneous polar systems of precipitation, when considering the magnetic curves, it will of course be so to a still greater extent if we were to try to separate the deflections in the earth-current curves that were due to the variations in these two systems.

As, however, we have seen a distinct example of the great difference between the character of the deflections in the N—S and the E—W curves, where two such systems are acting, it seems likely that this might frequently have something to say; in other words, the difference in the two earth-current curves, that we have before pointed out and ascribed chiefly to local causes, might to some extent, possibly a very great extent, be caused by the different induction-effect of simultaneously occurring polar and equatorial perturbation-systems.

If we look at the other examples that we have of equatorial storms, we find everywhere these same conditions confirmed.

From Germany and England we have two examples of such storms, on the 16th and 20th October, 1883.

In Germany there is only the one component, Berlin to Dresden; and this very nearly coincides with the direction of the magnetic meridian.

In both cases, at the beginning of the perturbation, there is a double oscillation, which should indicate the influence of polar systems.

The "starting impulse" observed in the magnetic curves at the beginning of the perturbation, southwards in H, and eastwards in D, seems to be caused by the magnetic effect produced by these induction-currents.

The direct effect of this polar precipitation must be assumed to decrease very rapidly at rather great distances, as the strength of the current in these systems of precipitation can only be comparatively small, but the changes take place with comparatively great rapidity.

If, on the other hand, we look at Greenwich for these days, we see here too an indication of a double wave; but the principal phenomenon at the beginning of the perturbation is a uniformly-directed deflection in the current-component E_1, that is to say, conditions that must have been produced by the induction-effect of the equatorial system.

The deflection in E_1 answers to a current-direction from NE to SW, and is thus fairly what we should expect, as the current-direction in the outer equatorial system is from W to E.

The deflection in the E—W curve for Parc St. Maur, on the 11th January, 1894, answers to a current-direction from E to W also in accordance with what we should expect.

We further find in all cases that the maximal deflection in the earth-currents occurs at the time when the deflection in the H-curve increases most, that is to say quite in accordance with what we should expect to find from our former experience.

These examples may easily be multiplied, but in this connection I need only refer the reader to the curves published from Parc St. Maur and Greenwich.

I will, however, draw attention to a difficulty that might possibly sometimes lead to misunderstanding. In certain cases the variation will be exceedingly strong, and both the magnetic and the earth-current curves may then be very faintly reproduced upon the photographic papers, often so faintly indeed, that it may be impossible to follow the curve in its sudden and most rapid movements. It will therefore sometimes be very easy to overlook certain small serrations.

At the beginning of the storm of the 11th January, 1894, we have a case in which the photographic curves were very faint and difficult to follow; and here, in order to indicate the uncertainty arising therefrom, we have represented these parts of the curves with dotted lines. If, therefore, in certain cases, a disagreement may be found in this respect, this uncertainty should be kept in mind.

During the positive equatorial storms also, we thus find confirmation of our previously expressed view of the connection between the magnetic perturbations and the earth-current phenomena.

THE DIURNAL VARIATION OF THE EARTH-CURRENTS.

163. In the previous articles, we have studied the connection between the earth-currents and the magnetic storms. In addition to these, however, there are certain other, more regular variations, one of which in particular, the diurnal variation, has been carefully studied. As regards earth-currents Weinstein has made a very thorough investigation of the phenomenon, based upon his observations in Germany. The principal result at which he has arrived is given in a series of curves and vector diagrams in his previously-mentioned work.

In England too, similar investigations have been carried out (see Airy, Phil. Trans. 1870, p. 215 and Pl. XXIV).

Although we will not here enter upon a detailed treatment of this phenomenon, but will reserve it for a subsequent chapter, in which the diurnal variation of magnetism at our four Norwegian stations will be discussed, it will yet be natural, in connection with what has been said, even now to point to a few circumstances regarding the diurnal variation of earth-currents, particularly as Dr. L. STEINER (1), upon the basis of Weinstein's curves, has drawn some very interesting conclusions. He finds that while the diurnal variations in the magnetic X-component — i. e. the force-component in the direction N—S — very closely follow the E—W curve, so that the variations in the former may be supposed to be the direct effect of the corresponding earth-current component, the deflections in the N—S curve are approximately proportional to the rate of change of the component Y (in the direction E—W).

It will be seen that this is in the main the result at which we have here arrived by a consideration of the earth-current conditions in Germany during the magnetic perturbations; and it would be natural to look for an explanation of the diurnal variation similar to that of magnetic storms. In these more slowly passing variations, however, other forces will exert an influence to a much greater extent than in the briefer variations. The thermo-electric forces in the earth's surface may perhaps play a very important part; and as STEINER suggests it may possibly be these currents that are mainly the cause of X so closely following the E—W curves. Of the other phenomenon, however, he gives no satisfactory explanation, but remarks that "these connections — as far as they are not due merely to chance — still await explanation". Our points of view naturally lead us to explain these conditions in the following manner.

Both our observations and our experiments have shown us that broadly speaking a purely geometric connection must always exist between the position of the sun and the situation of the systems of perturbation. In other words, what has been said seems with undoubted certainty to show that the earth will as a rule rotate in relation to an external corpuscular current-system with a more or less fixed position in space. The strength of the current, especially during magnetic storms, may vary within very wide limits; but its form has always proved to be approximately constant. I further assume, as already stated, that from the entire surface of the sun, a comparatively regular radiation of corpuscle-rays goes on, similar to the stronger and more irregular pencil-radiation of probably stiffer rays from the regions of the sun-spots.

This corpuscular field of radiation from the entire surface of the sun will now constantly surround the earth, and it is obvious that its shape will in the main be the same as that which we have found to be characteristic of the magnetic storms. When the earth now rotates in relation to this system of rays with its approximately fixed position in space, earth-currents will be induced.

The formulæ necessary for the calculation of these are given in Article 155, equations 42—44.

In this chapter we have also calculated the earth-current system that is induced by a polar system of precipitation of the previously-described form (see Table CXXVI).

Fig. 283 is a chart of the current-lines on the surface, answering to this; and we may here once more draw attention to the peculiarity already pointed out, namely, that the direction of the current-lines in medium latitudes such as those regions of Germany in which Weinstein made his observations, is practically only N—S.

It further appears very distinctly from the experiments that the rays in the equatorial regions are concentrated in such a manner that the main body of the ray-system swings round in front of the earth and passes nearest just before noon (see fig. 219). As the greater number of the rays run here, this system will in all probability also play an important part in this connection.

(1) Terr. Magn. XIII, p. 57.

In order to obtain a general idea of the course of the earth-currents induced by the rotation of the earth in relation to a system such as this, we have made a calculation of this current-system upon the assumption that the equatorial system can be replaced by an infinitely long, rectilinear current situated outside the earth in the plane of the equator.

For the potential of a current such as this, that flows at right angles to the XZ-plane, and intersects it in the points $x = x_1$, $z = z_1$, we have, as is well known, the following expression:

$$V = -i \,.\, \tan^{-1} \frac{z - z_1}{x - x_1} = -i \,.\, \tan^{-1} \frac{\varrho \cos\theta - z_1}{\varrho \sin\theta \cos\omega - x_1},$$

where the direction of the current is reckoned positive when it coincides with the direction of incr

If we say that the rotation-velocity equals $\overline{\omega}$, and further, for the sake of brevity, that

$$\left.\begin{aligned} a &= 1 - \sin^2\omega \sin^2\theta \\ b &= x_1 \sin\theta \cos\omega + z_1 \cos\theta \\ c &= \cos\theta \\ d &= \sin\omega \sin\theta \\ L^2 &= x_1^2 + z_1^2 \end{aligned}\right\}$$

we find, after some reductions, that

$$\frac{\partial V}{\partial \omega} = -i \,.\, d \frac{c\varrho^2 - z_1\varrho}{a\varrho^2 - 2b\varrho + L^2}$$

and

$$\psi = -\frac{\overline{\omega}}{\varkappa} \,.\, i \frac{c \,.\, d}{a} \left\{ \varrho + \frac{2bc - az_1}{2ac} \log \text{nat} \frac{a\varrho^2 - 2b\varrho + L^2}{L^2} \right.$$

$$\left. + \frac{2b^2c - abz_1 - acL^2}{ac\sqrt{aL^2 - b^2}} \tan^{-1} \frac{\varrho\sqrt{aL^2 - b^2}}{L^2 - \varrho b} \right\}.$$

If we here put $z_1 = 0$, we obtain the expression for the current-function answering to the torial position of the current.

We have calculated the current-system answering to an extra-terrestrial current such as result being given in the table below.

TABLE CXXVII.

Values of the current-function ψ answering to an extra-terrestrial curren situated in the plane of the equator.

x_1 is here put $= 2\varrho$. The multiplicator $-\frac{\overline{\omega}}{\varkappa}\varrho\frac{i}{10}$ is left out.

θ	0°	10°	30°	50°	70°	90°	110°	130°	150°	170°	180°
20°	0	0.052	0.145	0.211	0.241	0.237	0.207	0.158	0.099	0.033	0
40°	0	0.107	0.288	0.392	0.414	0.378	0.307	0.222	0.134	0.045	0
60°	0	0.128	0.330	0.419	0.410	0.344	0.274	0.186	0.109	0.036	0
80°	0	0.064	0.158	0.190	0.176	0.144	0.106	0.072	0.041	0.014	0

For $\theta = 90°$ we have $\psi = 0$. Further $\psi(\pi - \theta) = -\psi(\theta)$.

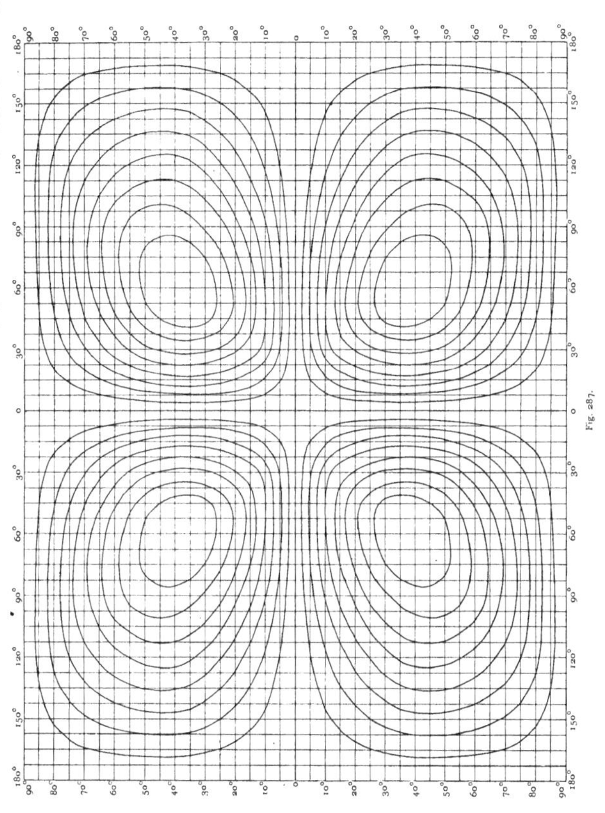

Fig. 287.

In fig. 287 we have also drawn a chart of the course of the current-lines at the surface.

There is to be noticed here a striking agreement with the main features of Schuster's chart of the potential lines for the diurnal variation (cf. his admirable memoir in Phil. Trans., 1889, p. 508).

In addition to this equatorial system, there are polar systems also at work, and we ought therefore by rights to bring such together if we want to represent the earth-current system that is induced by

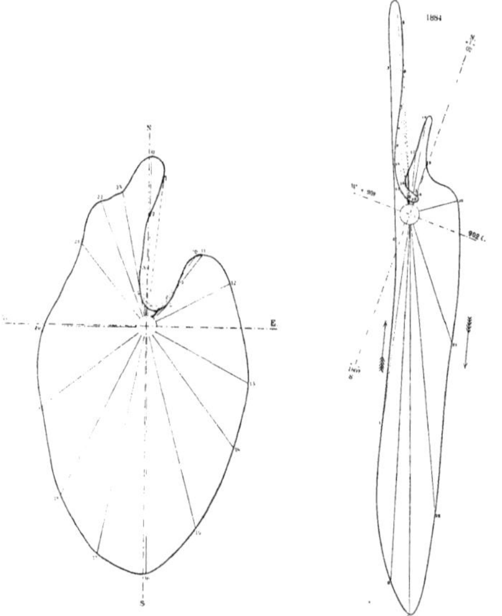

Fig. 288. Fig. 289.

rotation in relation to the entire external system. In this way it would be easy, by a suitable choice, to find current-fields that in their details too, exhibited a more perfect agreement with Schuster's chart.

A composition such as this, however, will in the first place always be rather arbitrary, and in the second place it will only answer to a part of the earth-current system that characterises the diurnal variation of these currents; while in the third place a chart of the earth-current lines and the magnetic potential lines are only comparable in certain respects.

As there is, moreover, no chart of the diurnal variation of the earth-currents all over the earth, we will not here undertake any such composition as regards the entire earth. All that exist are the determinations of the diurnal variation in Germany and England.

To enable a comparison to be made with these, we have put together, for $\theta = 40°$, which about answers to the position of Berlin, the current-components for two systems of this kind.

One of them answers to an external inducing current at the equator, where $L = 2R$, and which lies nearest to the earth at a place answering to Noon; the second answers to a rectilinear current parallel with the plane of the equator, lying at a least height, $h = 0.25\, R$, above a point in a small circle round the pole with a spherical radius of 20°, where the time is 2^h a. m., a night-system corresponding to a negative polar storm.

The strength of the current in the equatorial system is put at 20 times greater than the strength of the current in the polar system.

The vector diagram that has been drawn (fig. 288) shows us the suggestive agreement that exists in the main between this and the vector diagrams that Weinstein has calculated from observations, one of which we here reproduce (fig. 289).

We must emphasise the fact that in this first experiment we have not taken into consideration the group of rays that at about $6^h - 7^h$ p. m. must penetrate into the polar regions just where we have been led to assume that the rays which produce the *positive* polar storms descend towards the earth.

This group of rays will be included in our future calculations, as a preliminary investigation seems to show that in this way a surprisingly close agreement may be obtained between calculated and observed diagrams.

It may further be noticed that in the equatorial system in a latitude of about 50°, i. e. $\theta = 40°$, the currents are as a rule only in a direction N—S.

These currents will now approximately be proportional to $\frac{\partial \psi}{\partial \omega}$, i. e. to $\frac{\partial}{\partial \omega} \frac{\partial V}{\partial \omega}$, or, in other words, to $\frac{\partial Y}{\partial t}$, in accordance with what Dr. Steiner has found.

We can therefore, from our points of view, find a natural explanation of all the hitherto known principal features of the diurnal variation of earth-currents.

We will not at present, however, go more thoroughly into the matter of the diurnal variation of terrestrial magnetism, but will reserve it for a subsequent chapter.

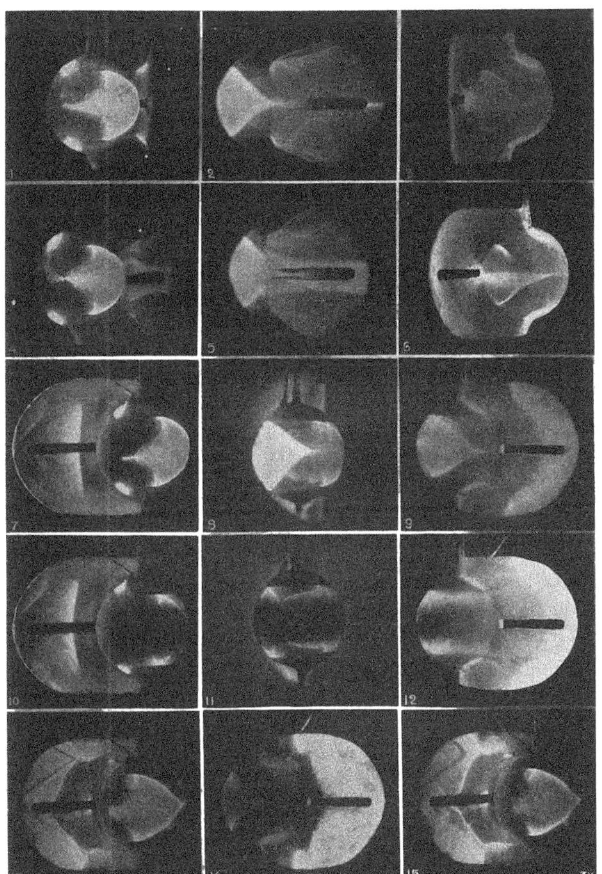

Reprint of fig. 204.

Earth-currents and magnetic elements from Pawlowsk, 17—18 March 1889.
Local Mean Time.

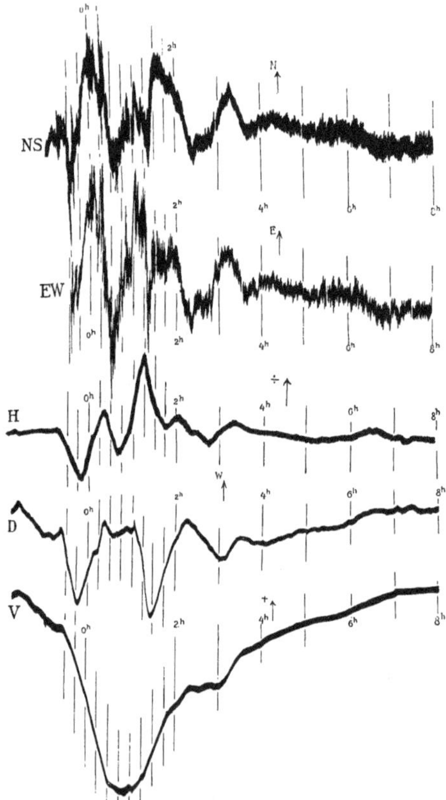

The Pawlowsk-curves, which were missing, as mentioned in Article 160 (p. 792), were found just before publication and are here reproduced.

Pl. XXII

The Perturbations of the 15th October, 1882

Term-day observations from 23h 20m on the 14th to 23h 20m on the 15th, Gr. M. T.

Pl. XXIII

The Perturbations of the 1st November, 1882

Term-day observations from 10^h to 23^h 20^m, Gr. M. T.

THE PERTURBATIONS OF THE 1st NOVEMBER, 1882.

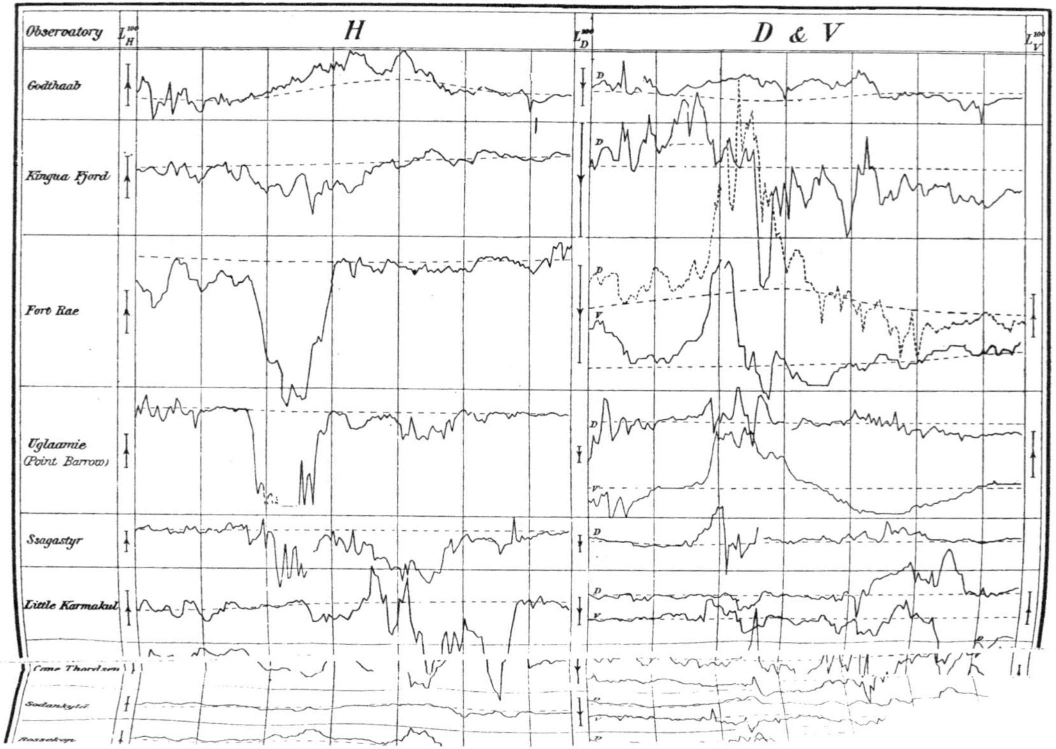

Observatory	L_H^{100}	H	L_D^{100}	D & V	L_V^{100}
Godthaab				D	
Kingua Fjord				D	
Fort Rae				D, V	
Uglaamie (Point Barrow)				D, V	
Ssagastyr				D	
Little Karmakul				D, V	

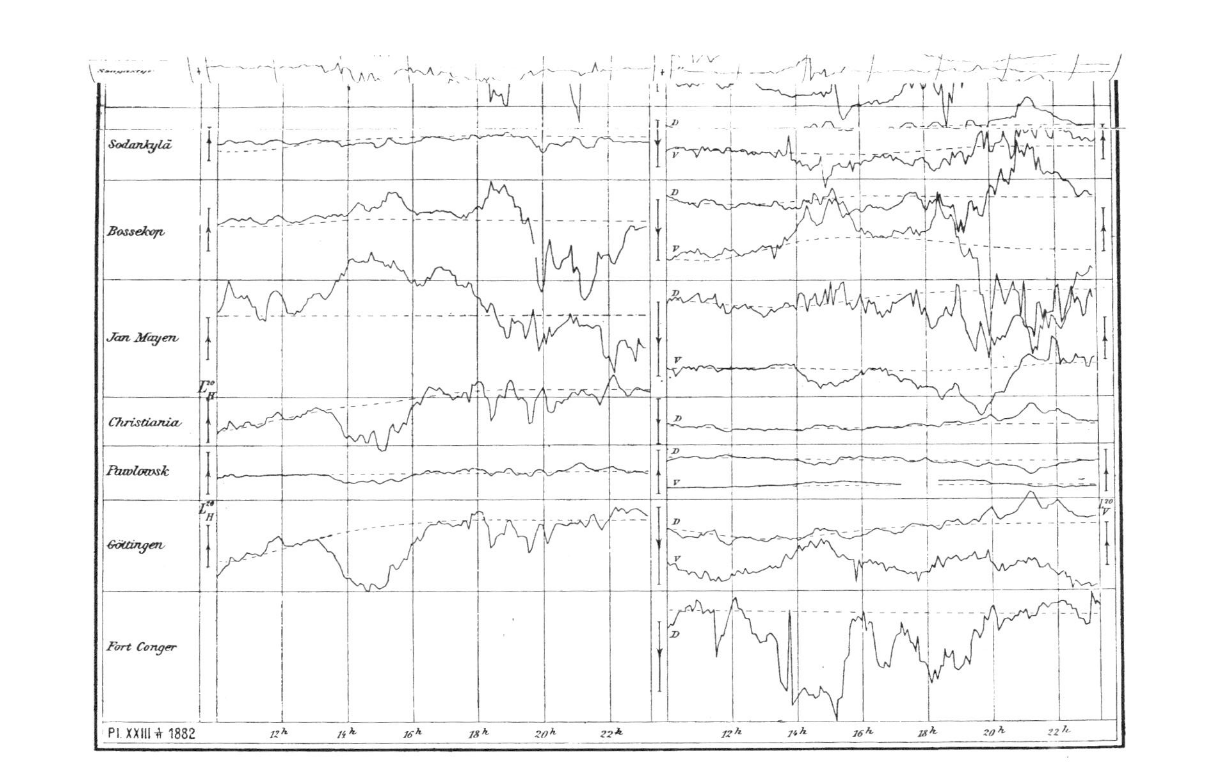

Pl. XXIII 1882

Pl. XXIV

The Perturbations of the 15th December, 1882

Term-day observations from 8^h to 23^h 20^m, Gr. M. T.

THE PERTURBATIONS OF THE 15th DECEMBER, 1882.

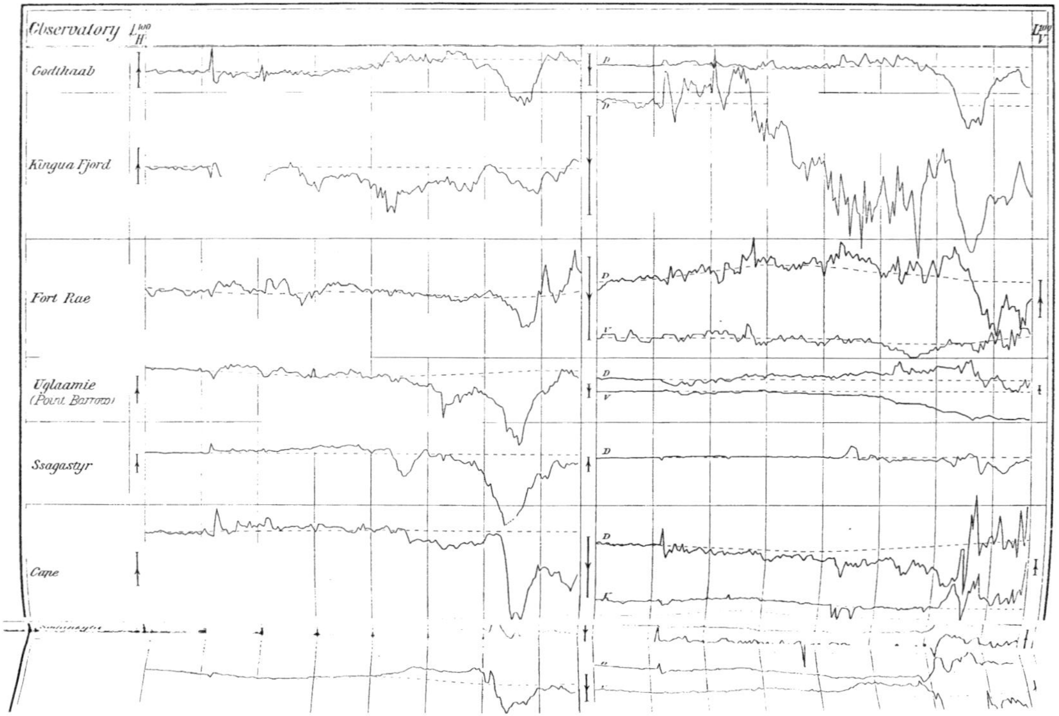

Observatory L_{H}^{100} L_{V}^{100}

Godthaab

Kingua Fjord

Fort Rae

Uglaamie
(Point Barrow)

Ssagastyr

Cape

D

V

D

V

D

D

V

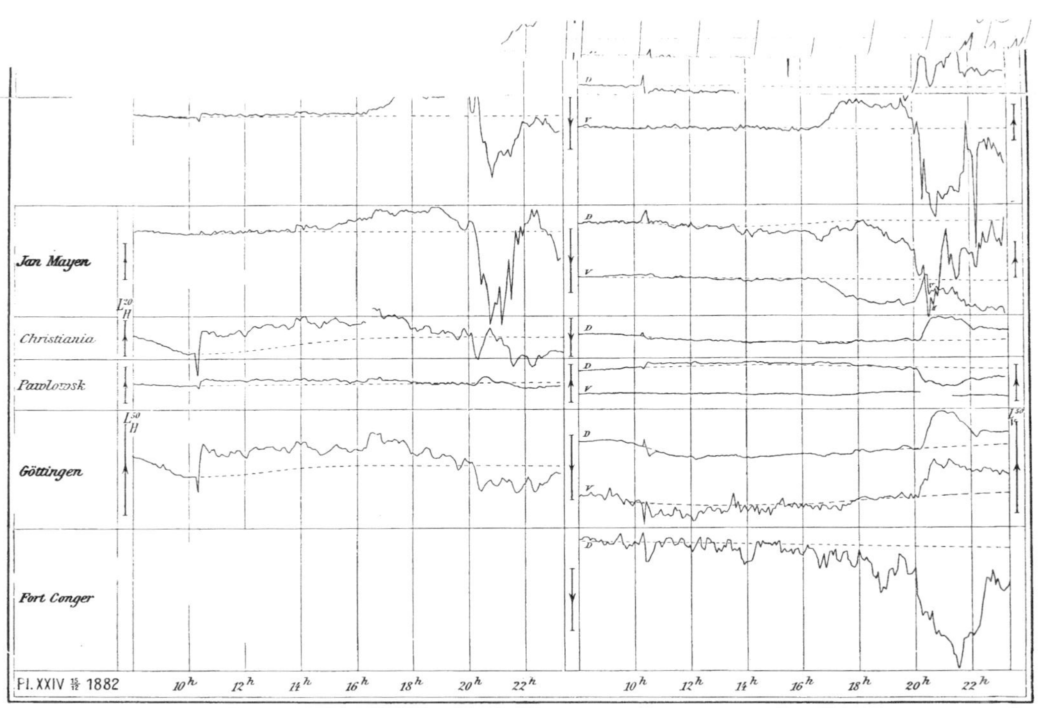

Pl. XXIV 15/12 1882

Pl. XXV

The Perturbations of the 2nd January, 1883

Term-day observations from 11^h to 23^h 20^m, Gr. M. T.

THE PERTURBATIONS OF THE 2nd JANUARY, 1883.

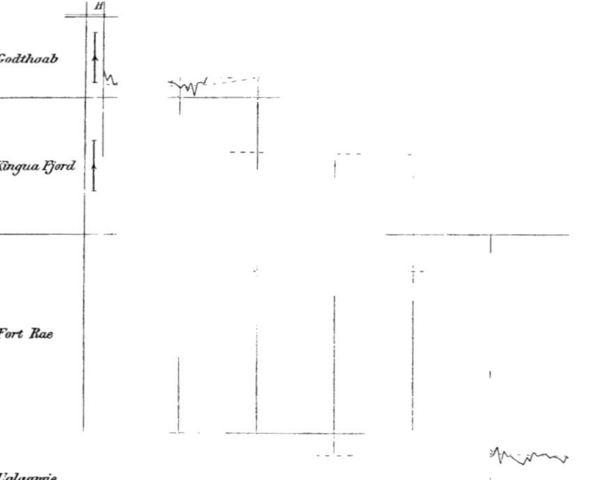

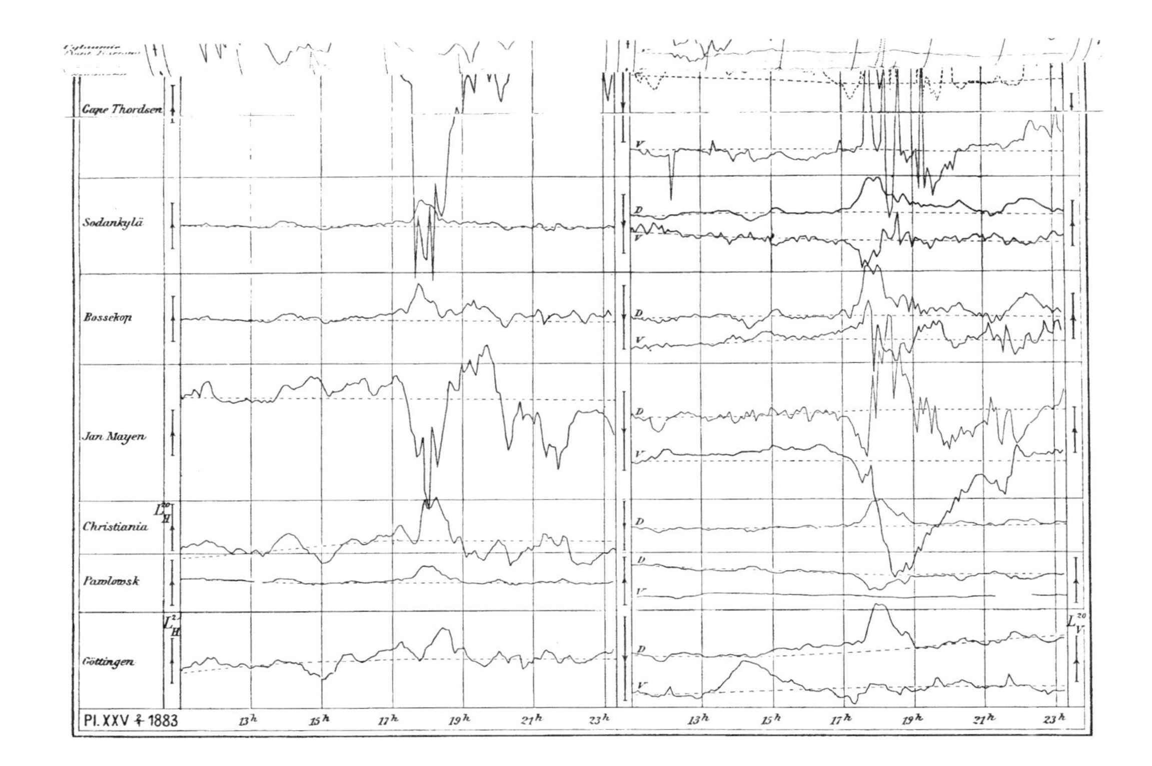

Pl. XXV 2/1 1883

Pl. XXVI

The Perturbations of the 15th January, 1883

Term-day observations from 10^h to $23^h\ 20^m$, Gr. M. T.

THE PERTURBATIONS OF THE 15th JANUARY, 1883.

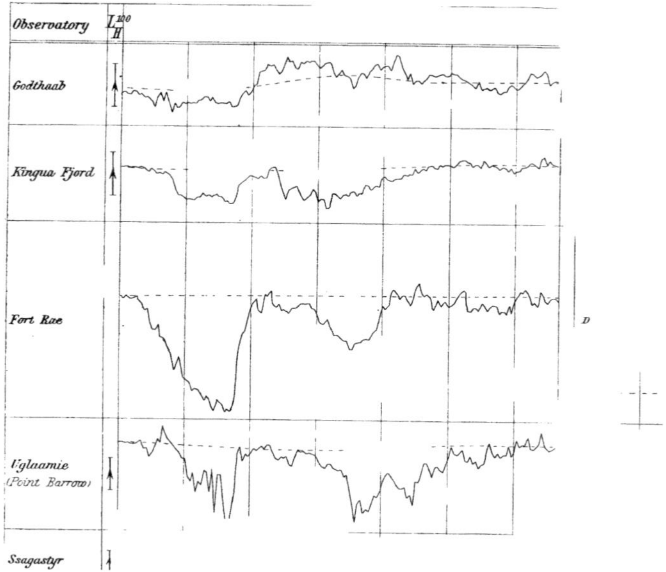

Observatory $L\frac{100}{H}$

Godthaab

Kingua Fjord

Fort Rae

D

Uglaamie
(Point Barrow)

Ssagastyr

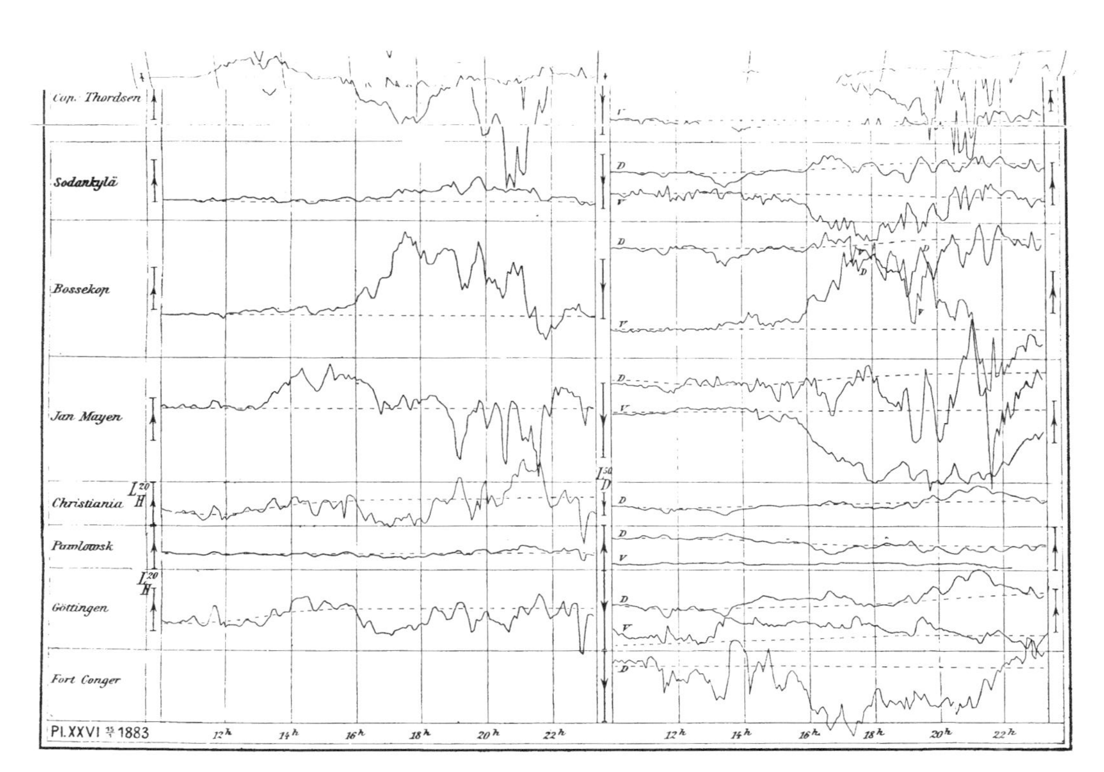

Pl. XXVI 15/1 1883

Pl. XXVII

The Perturbations of the 1st February, 1883

Term-day observations from 10^h to $23^h\ 20^m$, Gr. M. T.

THE PERTURBATIONS OF THE 1st FEBRUARY, 1883.

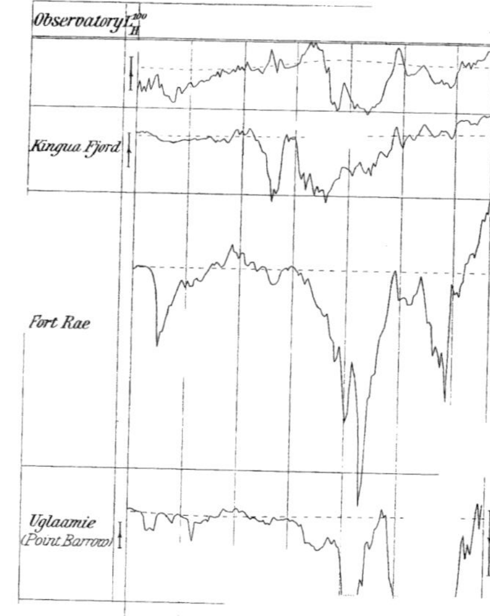

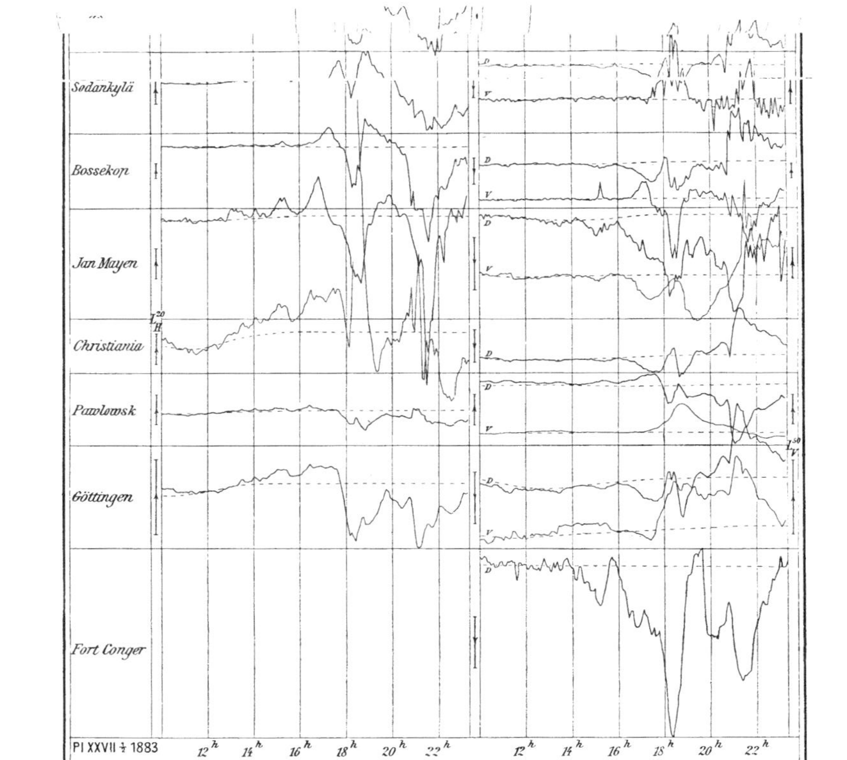
Sodankylä
Bossekop
Jan Mayen
Christiania
Pawlowsk
Göttingen
Fort Conger
D
V
12h
14h
16h
18h
20h
22h
Pl XXVII ½ 1883

Pl. XXVIII

The Perturbations of the 14th and 15th February, 1883

Term-day observations from $23^h\ 20^m$ on the 14th to $6^h\ 20^m$ on the 15th, Gr. M. T.

THE PERTURBATIONS OF THE 14th AND 15th FEBRUARY, 1883.

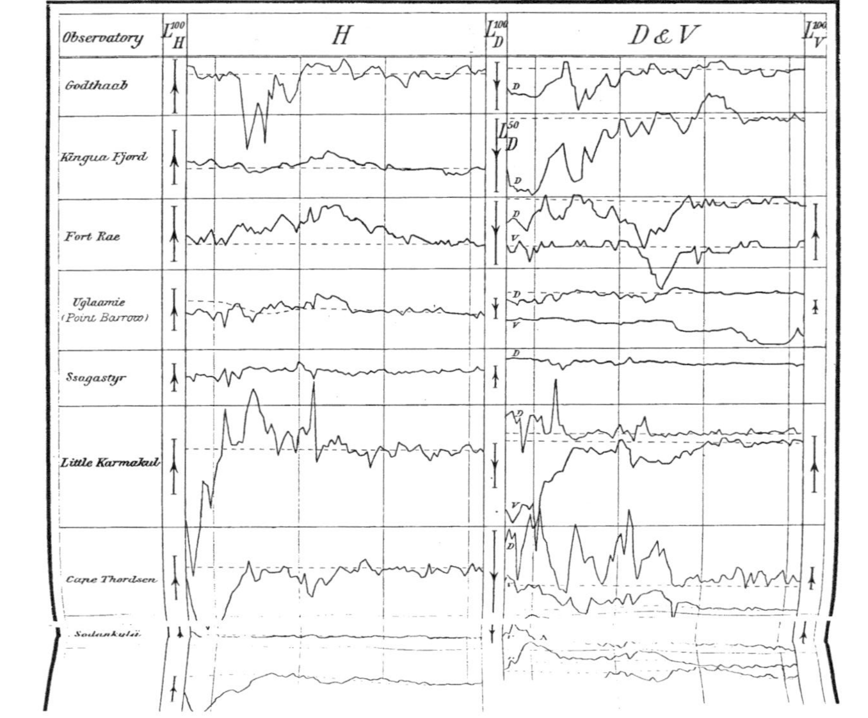

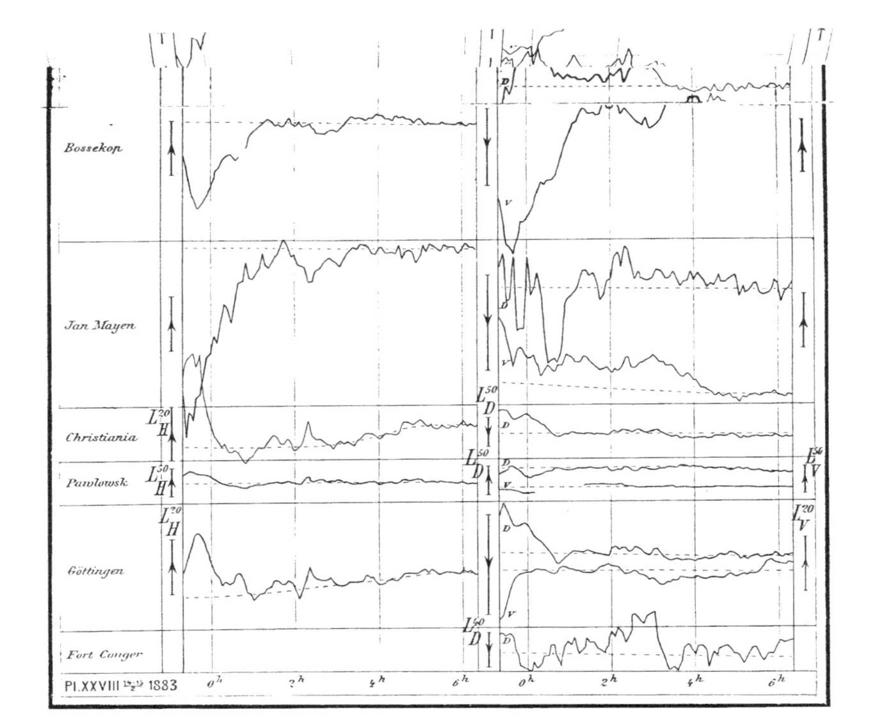

Pl. XXVIII 1883

Pl. XXIX

The Perturbations of the 15th July, 1883

Term-day observations from 6^h to 23^h 20^m, Gr. M. T.

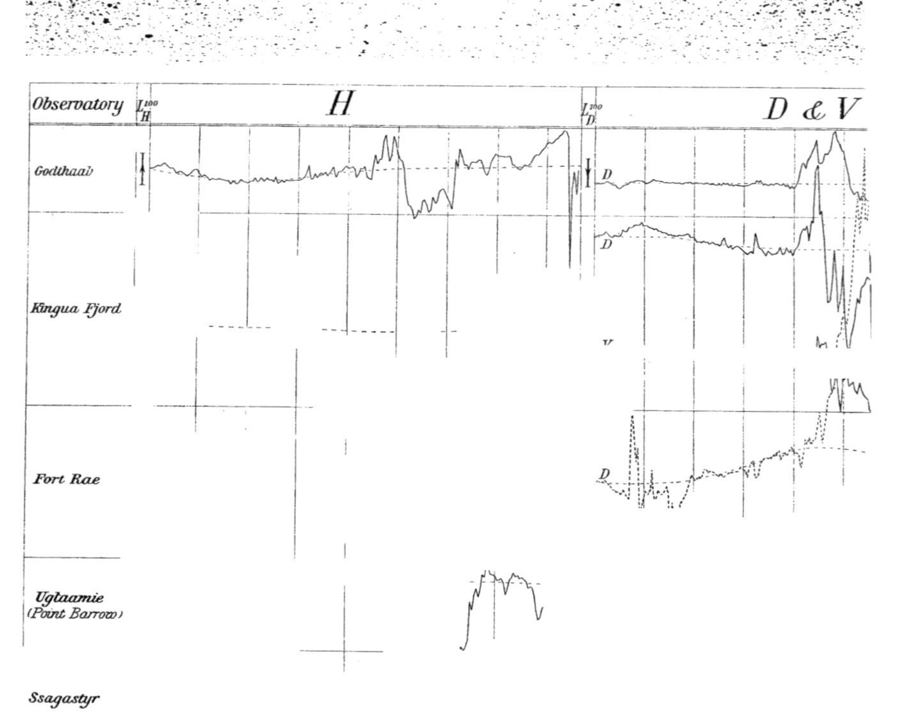

Observatory
H
D & V
Godthaab
D
D
Kingua Fjord
V
Fort Rae
D
Ugtaamie
(Point Barrow)
Ssagastyr

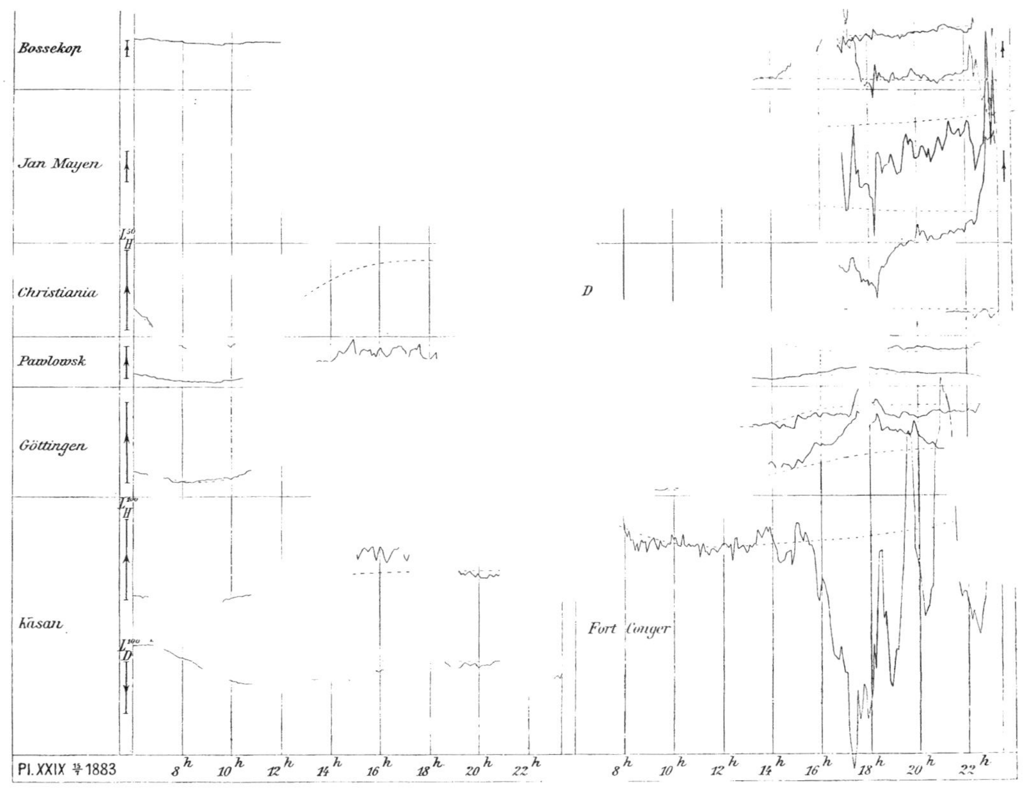

Pl. XXIX 15/7 1883

Pl. XXX

Earth currents and magnetic elements. Series I.

Kaafjord.

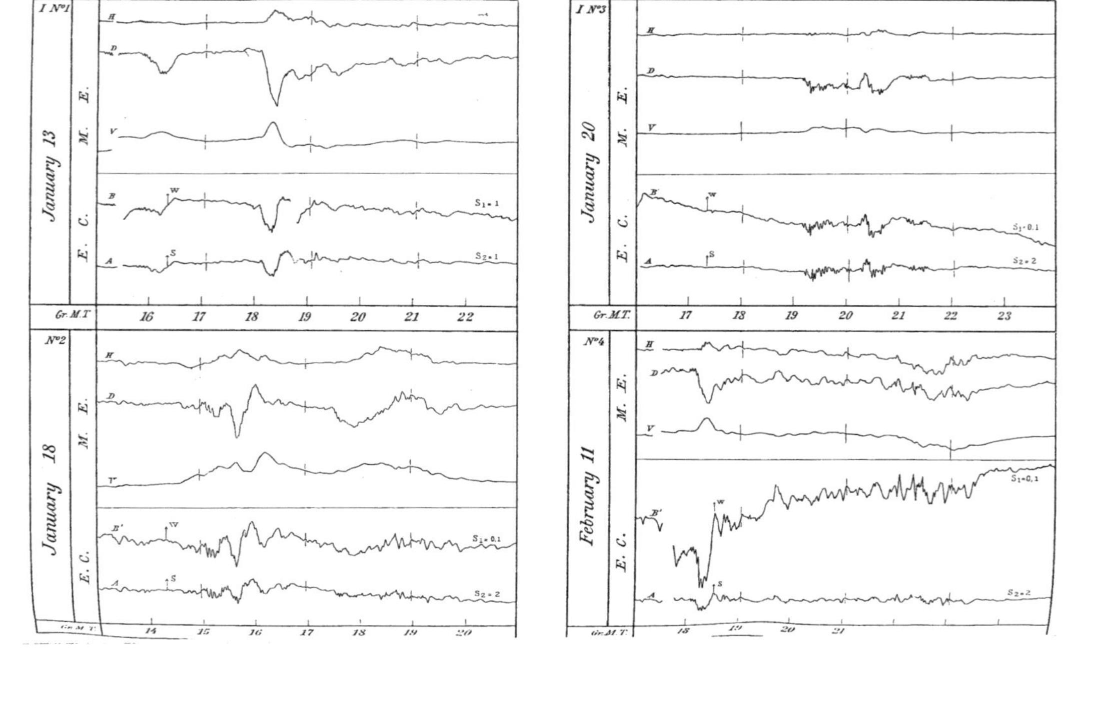

I N°1
January 13
M. E.
E. C.
H
D
V
B
A
W
S
$S_1 = 1$
$S_2 = 1$
Gr. M.T.
16 17 18 19 20 21 22
I N°3
January 20
M. E.
E. C.
H
D
V
B
A
W
S
$S_1 = 0.1$
$S_2 = 2$
Gr. M.T.
17 18 19 20 21 22 23
N°2
January 18
M. E.
E. C.
H
D
V
B'
A
W
S
$S_1 = 0.1$
$S_2 = 2$
Gr. M.T.
14 15 16 17 18 19 20
N°4
February 11
M. E.
E. C.
H
D
V
B'
A
W
S
$S_1 = 0.1$
$S_2 = 2$
Gr. M.T.
18 19 20 21

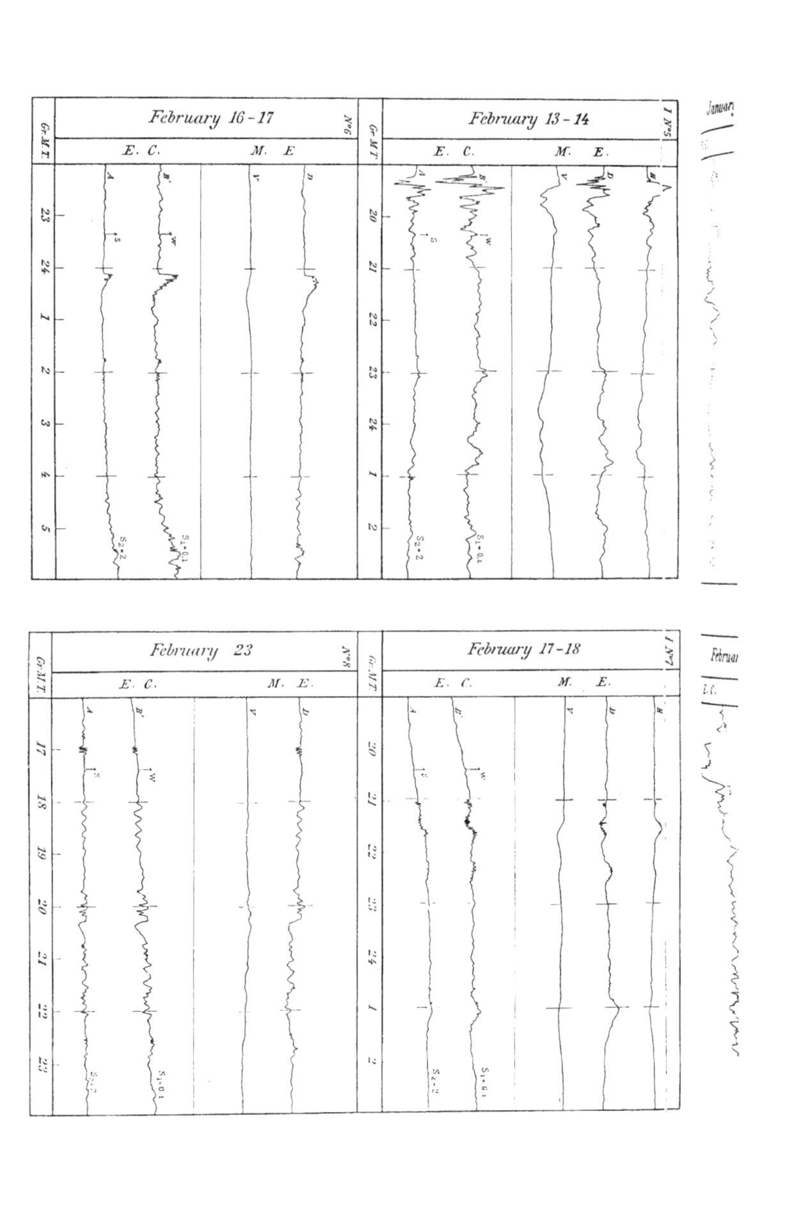

February 16-17
February 13-14
February 23
February 17-18
E. C.
M. E.
Gr.M.T.

Pl. XXXI

Earth currents and magnetic elements. Series II.

Kaafjord.

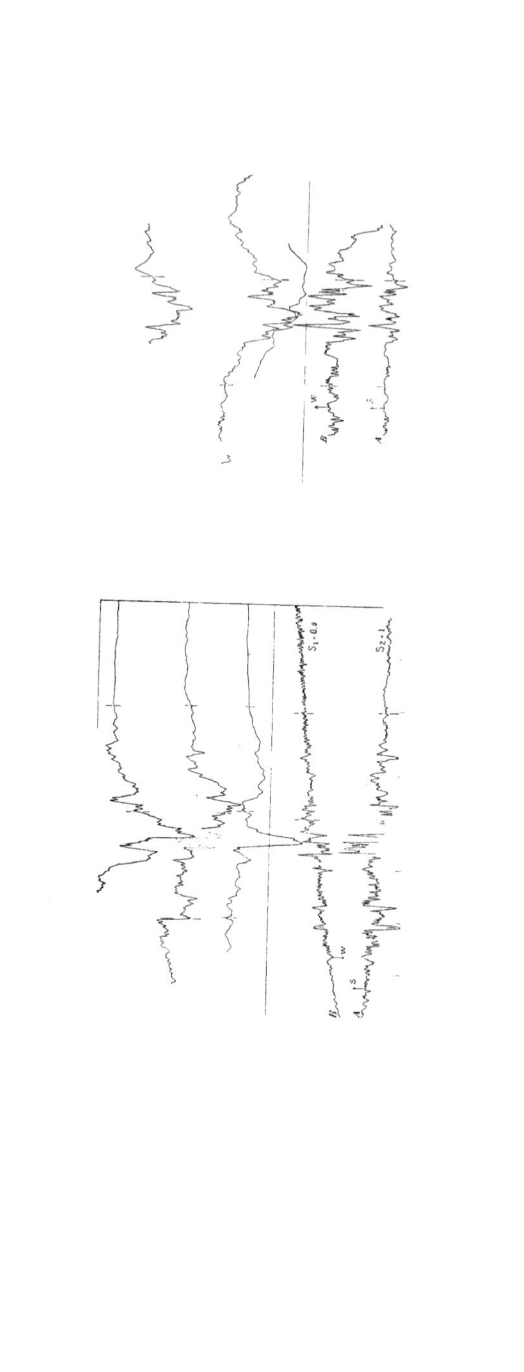

Dec.26-27

Pl. XXXII

Earth currents and magnetic elements. Series II continued.

Kaafjord and Bossekop.

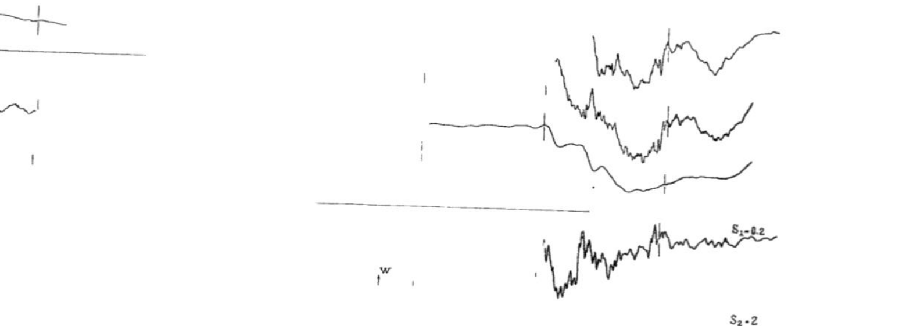
W
S
W
S₁=0.2
S₂-2

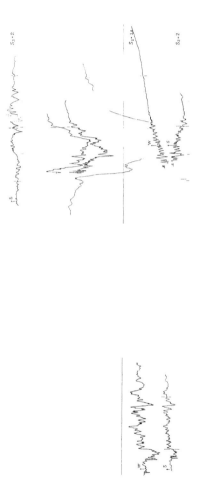

The dotted curve for Febr. 10—11 indicates the result of an elimination of the effect of the earth current upon the D-curve.

Pl. XXXIII

Earth currents and magnetic elements. Series II continued.

Bossekop.

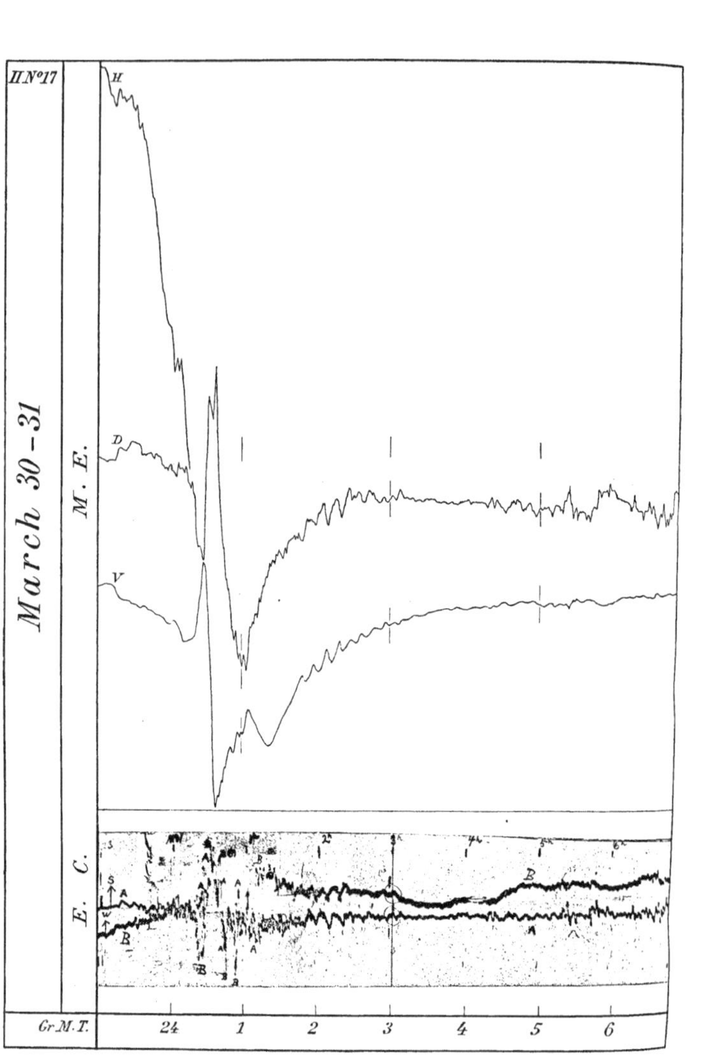
II N°17
March 30 – 31
M. E.
H
D
V
E. C.
Gr.M.T.
24
1
2
3
4
5
6

Pl. XXXIV

Earth currents and magnetic elements. Series III.

Kaafjord.

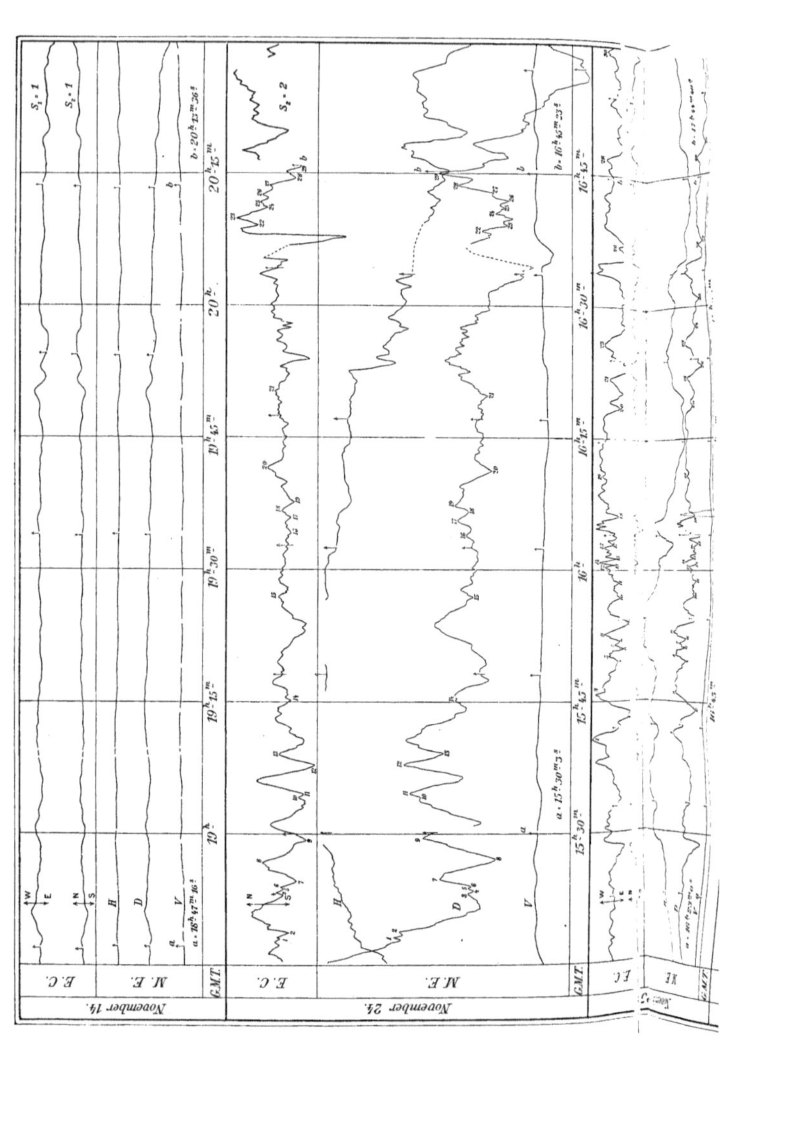

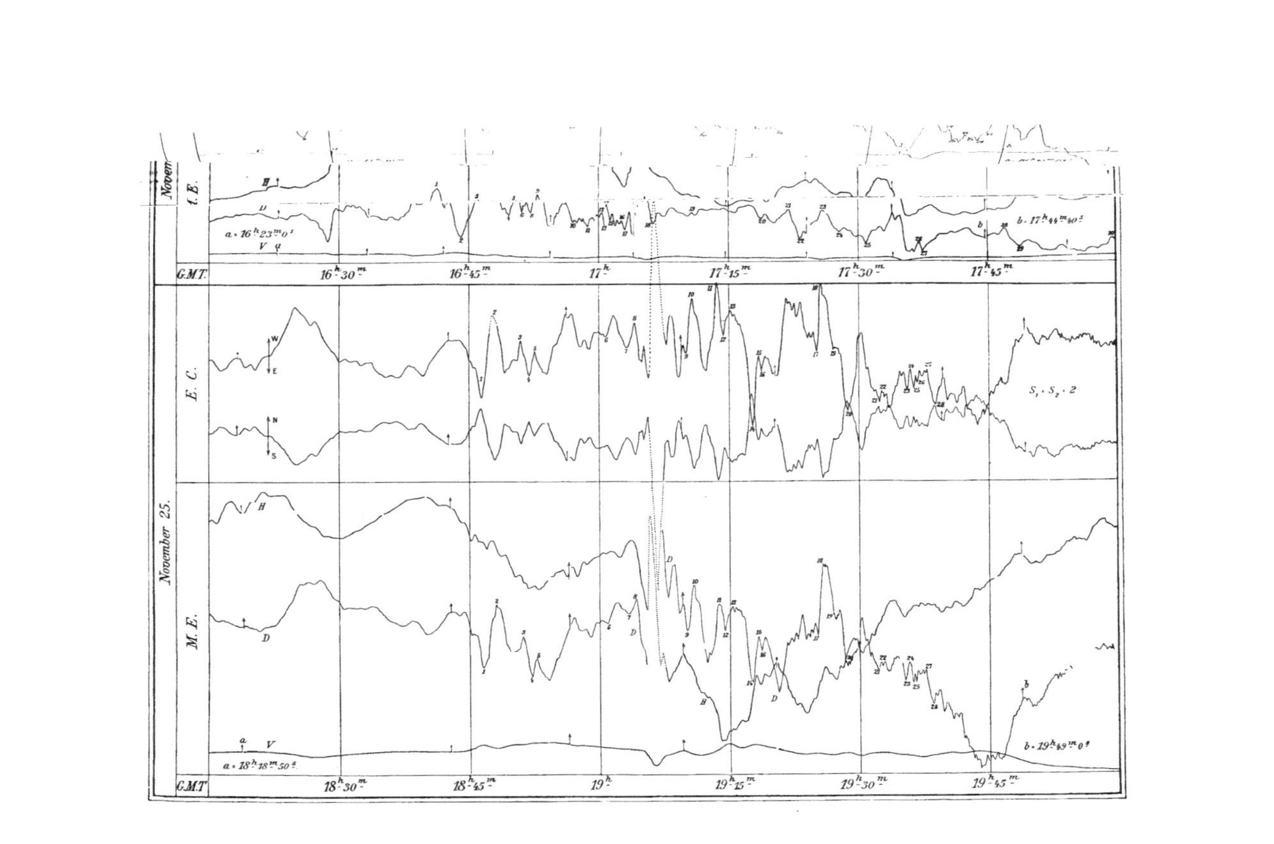

Novem
t. E.
a = 16h 23m 0s
b = 17h 44m 40s
GMT.
16h 30m
16h 45m
17h
17h 15m
17h 30m
17h 45m
E. C.
W
E
N
S
S1 : S2 = 2
November 25.
M. E.
H
D
a = 18h 18m 50s
b = 19h 49m 0s
GMT
18h 30m
18h 45m
19h
19h 15m
19h 30m
19h 45m

Pl. XXXV

Earth currents and magnetic elements. Series III continued.

Bossekop.

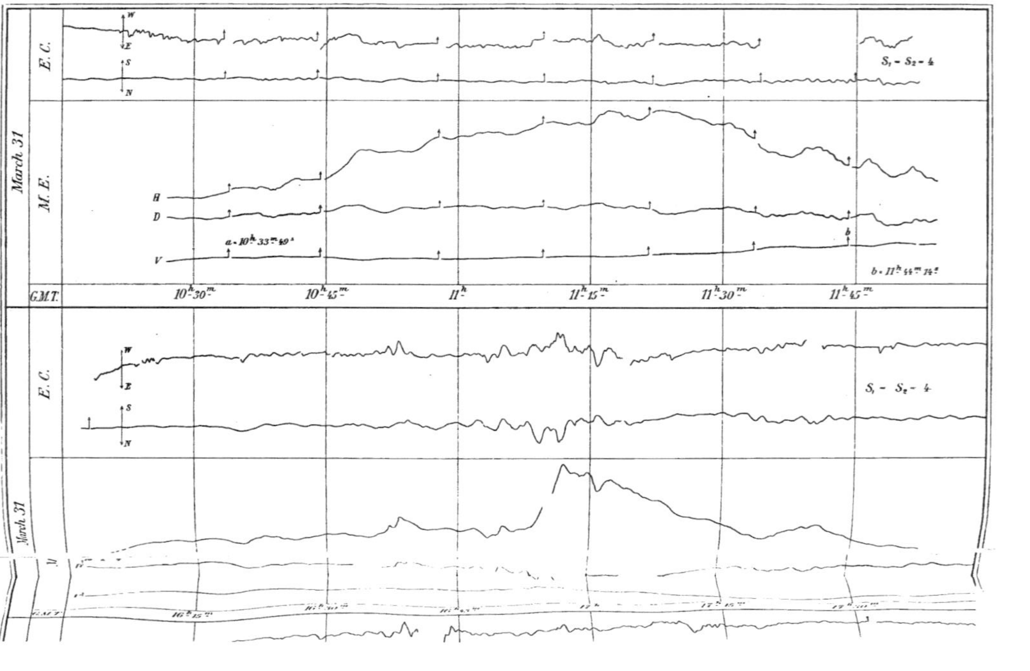
March 31
E.C.
W
E
S
N
$S_1 = S_2 = 4$
M.E.
H
D
V
$a = 10^h\,33^m\,49^s$
b
$b = 11^h\,44^m\,14^s$
G.M.T.
$10^h\,30^m$
$10^h\,45^m$
11^h
$11^h\,15^m$
$11^h\,30^m$
$11^h\,45^m$
E.C.
W
E
S
N
$S_1 = S_2 = 4$
March 31